TRIAZOLES
1,2,4

Carroll Temple, Jr.

SOUTHERN RESEARCH INSTITUTE

BIRMINGHAM, ALABAMA

Edited by

John A. Montgomery

SOUTHERN RESEARCH INSTITUTE

BIRMINGHAM, ALABAMA

AN INTERSCIENCE ® PUBLICATION

JOHN WILEY & SONS

NEW YORK . CHICHESTER . BRISBANE . TORONTO

An Interscience ® Publication

Copyright © 1981 by John Wiley & Sons, Inc.

All rights reserved. Published simultaneously in Canada.

Library of Congress Cataloging in Publication Data:

Temple, Carroll.
  Triazoles 1,2,4.

  (The Chemistry of heterocyclic compounds; v. 37
ISSN 0069–3154)
  Includes indexes.
  1. Triazoles. I. Montgomery, John A. II. Title.
QD401.T27   547′.593   80-15637
ISBN 0-471-04656-6

Printed in the United States of America

10 9 8 7 6 5 4 3 2 1

The Chemistry of Heterocyclic Compounds

The chemistry of heterocyclic compounds is one of the most complex branches of organic chemistry. It is equally interesting for its theoretical implications, for the diversity of its synthetic procedures, and for the physiological and industrial significance of heterocyclic compounds.

A field of such importance and intrinsic difficulty should be made as readily accessible as possible, and the lack of a modern detailed and comprehensive presentation of heterocyclic chemistry is therefore keenly felt. It is the intention of the present series to fill this gap by expert presentations of the various branches of heterocyclic chemistry. The subdivisions have been designed to cover the field in its entirety by monographs which reflect the importance and the interrelations of the various compounds, and accommodate the specific interests of the authors.

In order to continue to make heterocyclic chemistry as readily accessible as possible new editions are planned for those areas where the respective volumes in the first edition have become obsolete by overwhelming progress. If, however, the changes are not too great so that the first editions can be brought up-to-date by supplementary volumes, supplements to the respective volumes will be published in the first edition.

ARNOLD WEISSBERGER

Research Laboratories
Eastman Kodak Company
Rochester, New York

EDWARD C. TAYLOR

Princeton University
Princeton, New Jersey

v

Note to the Reader

The text and tables of this volume on 1,2,4-triazoles, triazolines, and triazolidines have been comprehensively compiled through the use of *Chemical Abstracts*. The text was composed from abstracts and references present in volumes 51–85 (1957–1976) of *Chemical Abstracts*, which covers the period from the reviews of K. T. Potts (1961) and J. H. Boyer (1961). The tables list the compounds indexed by *Chemical Abstracts* for the period of time from the Second Supplement of Beilstein (1929) through 1975 and also the majority of compounds for 1976. Because of the large number of compounds, only a small number of 1,2,4-triazoles, triazolines, and triazolidines are listed individually in the Subject Index, which, however, will refer to the text for a particular group of compounds. Also a particular group of compounds can be located readily by inspection of the Contents. For convenience in cross-checking, the sections of the text and the tables are denoted by the same title and number. The arrangement of the sections of text and tables has been classified according to the nature of the atom connecting the substituent to the ring, which follows the order of carbon (alkyl, aryl, carboxylic acid, acyl), nitrogen (amino, azo, nitro, etc.), oxygen (alkoxy, aroxy, etc.), sulfur (alkyl- and aryl- thio, sulfinyl, etc.), and halogen. Those compounds that contain more than one representative function are discussed and assembled, respectively, in the appropriate sections and tables. In the text, references are indicated by arabic numbers, which refer to *Chemical Abstracts* numbers listed at the end of each section. Also, the reference for a compound listed in the tables is denoted by a *Chemical Abstracts* number. The latter, listed numerically in the master reference section, will identify either the patent or journal reference. Prior to volume 66 (1967), *Chemical Abstracts* denoted the position of compounds within an abstract by a page number and a letter, either one or both of which might not correspond exactly to the *Chemical Abstracts* number that identifies the journal reference. For patents or journals that might not be readily available, *Chemical Abstracts* can be consulted. In the text, footnotes are used for those journal references that appeared prior to *Chemical Abstracts.*

Compounds described by the title of the table are listed as *parent*. *N*-alkoxy-, *N*-hydroxy-, *N*-, *P*-, and *S*-oxides, and organometallic radicals substituted on either carbon or nitrogen are not considered a representative function, and these compounds are listed with the parent triazole. All compounds are listed alphabetically. When isomeric compounds are listed, the isomer with the lowest number appears first. The tables containing the alkyl or aryl compounds include the substituted alkyl or aryl derivatives; for example, triazole-3-acetic acid is listed under alkyl rather than carboxylic acid derivatives. Also, the tables containing the amino compounds include alkyl- or arylamino, acylamino, and hydrazino type compounds. The names of compounds that are derivatives of another class have been changed and listed in the tables as triazoles. For example, *triazolyl*pyridine has been listed as *pyridyl*triazole; (triazolyl)-2-propanone as acetonyltriazole. Further, the names of imides and derivatives of guanazine have been changed, when possible, and listed as the corresponding aminotriazoles. In contrast, hydroxytriazoles, triazolol, and triazolones are listed as triazolinones; mercaptotriazoles, triazolethiol, and triazolethione as triazolinethiones. Similar considerations as described above apply to the triazolines and triazolidines. Compounds containing conjunctive names are listed alphabetically. For compounds with conjunctive names of mixed function, the chief function is determined by the grouping.

CARROLL TEMPLE, JR.

Birmingham, Alabama
June 1980

Acknowledgments

For his help and encouragement in the preparation of this book, I am indebted to Dr. John A. Montgomery. Many other colleagues at Southern Research Institute have also contributed their time to this work for which I am very grateful. In addition, my thanks go to a large group of typists for carrying out the arduous task of preparing the typescript, to the library staff for providing most of the material from which the references and tables are composed, to Chemical Abstracts Service for the availability of a retrospective substructure search on triazoles for Volumes 68–83 of Chemical Abstracts, and to Southern Research Institute for allowing the use of their facilities.

C. T.

Contents

TRIAZOLES

1,2,4

This is the Thirty-Seventh Volume in the Series

THE CHEMISTRY OF HETEROCYCLIC COMPOUNDS

Introduction

The heteroaromatic triazole ring system is composed of five atoms, two carbons, and the three nitrogens, which can be arranged in two combinations to give either 1,2,3-triazole or 1,2,4-triazole. Although two NH (**1** and **2**) and one CH$_2$ (**3**) tautomeric forms are possible for 1,2,4-triazole, this

$$\text{H}_2\text{NNH}_2 \xrightarrow[-\text{NH}_3]{\text{H}_2\text{NCHO}} \text{H}_2\text{NNHCHO} \xrightarrow[-\text{NH}_3]{\text{H}_2\text{NCHO}} \overset{\text{OCH}}{\underset{\text{NHNH}}{}}\overset{\text{CHO}}{} \xrightarrow{\text{NH}_3} \mathbf{1}$$

structure is best represented as a positively charged hydrogen associated with the resonance stabilized triazole anion.[1] In *Chemical Abstracts* 3-substituted and 3,5-disubstituted 1,2,4-triazoles are usually indexed as *s*-triazoles. The 1,2,4-1H-triazole notation is used to describe a 1*N*-substituted triazole, whereas 1,2,4-4H-triazole is used to describe a 4*N*-substituted triazole. Trivial names such as guanazole(3,5-diamino-*s*-triazole), guanazine(3,4,5-triamino-1,2,4-4H-triazole), and either bicarbamimide or urazole(1,2,4-triazolidine-3,5-dione) have been replaced by systematic names throughout this chapter. Additional information on nomenclature is described in the introduction to the tables. In addition to reviews by Potts[1] and Boyer,[2] the relationship of the 1,2,4-triazoles in regard to other small-ring azoles has been reviewed recently by Schofield, Grimmett, and Keene.[3]

Bladin reported the preparation of derivatives of *s*-triazole (**1**) in 1885,[4] and soon thereafter Pellizzari obtained the parent ring system from the reaction of formylhydrazine with formamide.[5] This and related reactions, which gave low and variable yields of *s*-triazole (**1**), have been reviewed.[1] Later the condensation of hydrazine sulfate with formamide was reported to give a 53% yield of *s*-triazole (**1**).[6] Ainsworth and Jones observed that a large quantity of ammonia was evolved in the reaction of hydrazine with formamide, and to prevent the loss of ammonia, the intermediate N,N'-di-formylhydrazine was reacted with excess ammonia in a pressure vessel to give a 70 to 80% yield of s-triazole (**1**).[7] A further improvement in the yield of s-triazole (**1**) resulted from the work of Grundmann and Rätz, who obtained a 95% yield of s-triazole (**1**) from the interaction of s-triazine (**4**) with hydrazine hydrochloride.[8] Apparently, the intermediate amidrazone (**5**) was initially formed, which was postulated to react with another molecule of s-triazine (**4**) to give s-triazole (**1**). However, the acid-catalyzed self-condensation of amidrazones is well documented,[9-11] and s-triazole (**1**)

might be formed via intermediate **6**. With hydrazine rather than its hydrochloride, s-triazine (**4**) reacted to give 1,2-diformylhydrazine dihydrazone.[12]

References

1. K. T. Potts, *Chem. Rev.*, **61, 87** (1961).
2. J. H. Boyer, "Heterocyclic Compounds," R. C. Elderfield, Ed., John Wiley, New York, 1961, p. 425.
3. K. Schofield, M. R. Grimmett, and B. R. T. Keene, "The Azoles," Cambridge U. P., Cambridge, 1976.
4. J. A. Bladin, *Berichte*, **18**, 1544 (1885).
5. G. Pellizzari, *Gazz. Chim. Ital.*, **24**, 222 (1894).
6. M. Sekiya and S. Ishikawa, *Yakugaku Zasshi*, **78**, 549 (1958); *Chem. Abstr.*, **52**, 17244f.
7. C. Ainsworth and R. G. Jones, *J. Amer. Chem. Soc.* **77**, 621 (1955).
8. C. Grundmann and R. Rätz, *J. Org. Chem.*, **21**, 1037 (1956).
9. D. D. Libman and R. Slack, *J. Chem. Soc.*, 2253 (1956).
10. H. Behringer and H. J. Fischer, *Berichte*, **95**, 2546 (1962).
11. A. Spassov, E. Golovinsky and G. Demirov, *Chem. Ber.*, **99**, 3734 (1966).
12. C. Grundmann and A. Kreutzberger, *J. Amer. Chem. Soc.*, **79**, 2839 (1952).

CHAPTER 1

Alkyl- or Aryl-Monosubstituted 1,2,4-Triazoles

1.1. 1-Alkyl- or Aryl-Substituted 1H-1,2,4-Triazoles

The reaction of either 1-substituted hydrazines or the corresponding formyl derivatives with formamide (Pellizzari reaction) at high temperatures (>200°) produced both 1-alkyl- and 1-aryl-1,2,4-triazoles (**1.1-1**, **-2**). The products are generally isolated in low yields because the separation of by-products is often difficult.[*,1] Both 1-methyl- and 1-phenyltriazole are formed in high yields (>80%) via an amidrazone intermediate (**1.1-3**) in the acid-catalyzed condensation of the appropriate hydrazine with *s*-triazine in refluxing ethanol.[2] An amidrazone intermediate was also involved in the reaction of the imino ether (**1.1-4**) with methyl hydrazine in ether to give a

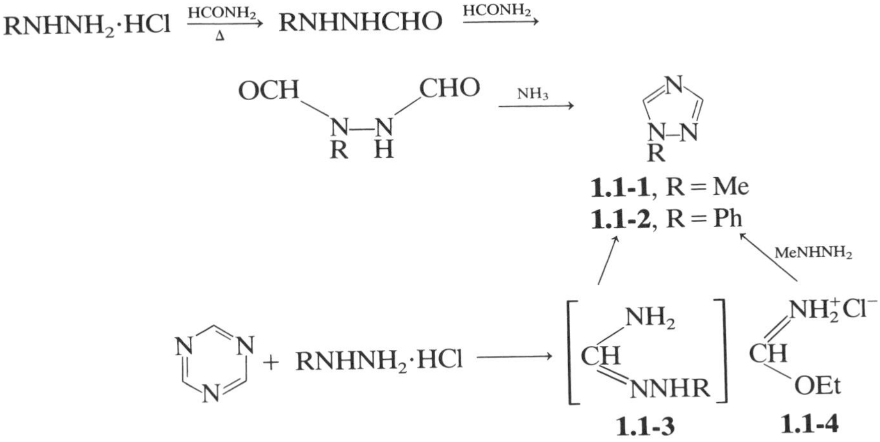

RNHNH2·HCl
HCONH2
Δ
RNHNHCHO
HCONH2
OCH
CHO
N—N
R H
NH3
N
N–N
R
1.1-1, R = Me
1.1-2, R = Ph
MeNHNH2
N
N
N
+ RNHNH2·HCl
NH2
CH
NNHR
NH2+Cl−
CH
OEt
1.1-3
1.1-4

mixture of a dihydrotetrazine (38%) and 1-methyl-1,2,4-triazole (**1.1-1**) (27%).[3] In a related reaction, treatment of 2,4,6-trichloro-*s*-triazine (**1.1-5**)

* G. Pellizzari and A. Soldi, *Gazz. Chim. Ital.*, **35,** 373 (1905).

with dimethylformamide gave the reactive [(dimethylaminomethylene)-amino]dimethylammonium intermediate (**1.1-6**), which was condensed with substituted hydrazines to give good yields of triazoles (e.g., **1.1-2**, **1.1-7**, **1.1-8**).[4,5]

1.1-2, R = Ph (77%)
1.1-7, R = HOCH$_2$CH$_2$ (62%)
1.1-8, R = Et$_2$NCH$_2$CH$_2$ (50%)

Another important method, the alkylation of the anion of *s*-triazole, can be used with a variety of both simple and complex alkyl halides to give 1-alkyl-1,2,4-triazoles (e.g., **1.1-1**, **1.1-9**, **1.1-10**).[6,7] In addition, the silver salt

1.1-1, R = Me (78%)
1.1-9, R = PhtNCH$_2$ (41%)
1.1-10, R = EtO$_2$CCH$_2$ (81%)

of *s*-triazole was alkylated with butyl iodide in refluxing benzene to give 1-butyl-1,2,4-triazole,[8] and the condensation of the trimethylsilyl derivative of *s*-triazole with 2,3,5-tri-*O*-benzoyl-D-ribofuranosyl bromide in acetonitrile gave the *O*-benzoyl derivative of 1-(β-D-ribofuranosyl)-1,2,4-triazole (**1.1-12**).[9] The latter was also prepared by the reductive deamination of 3-amino-1-(β-D-ribofuranosyl)-1,2,4-triazole (**1.1-11**) with nitrous acid in the

1.1-11   **1.1-12**

presence of hypophosphorous acid. A new method for the alkylation of s-triazole used a trialkyl phosphate as the alkylating reagent.[10,11] For example, a neat mixture of s-triazole and trimethyl phosphate was heated at 85° to give a 54% yield of 1-methyl-1,2,4-triazole (**1.1-1**). Also, 1-vinyl-1,2,4-triazole has been reported to result from the vinylation of s-triazole with acetylene at elevated temperature and pressure.[12]

In these reactions, substitution occurred mainly at N-1 to give 1-alkyl triazoles because of the greater nucleophilicity of the N-N linkage in comparison with that of N-4.[13,14] Procedures have been developed to completely eliminate the formation of 4-alkyl-1,2,4-triazoles. Benzylation of 4-amino-1,2,4-triazole gave 4-amino-1-benzyl-1,2,4-triazolium bromide (**1.1-13**), which was deaminated with nitric acid to give 1-benzyl-1,2,4-triazole (**1.1-14**).[15,16] Also, the imine (**1.1-15**) was cleaved either with zinc dust in acetic acid or with phosphorus oxychloride in dioxane to give **1.1-14**.[17]

| **1.1-13** | **1.1-14** | **1.1-15** |

The alkylation of s-triazole with activated aryl halides has also been accomplished. Treatment of s-triazole with 1-chloro-2-nitrobenzene in pyridine containing potassium carbonate and copper (II) oxide (Ullmann reaction) gave a 68% yield of 1-(2-nitrophenyl)-1,2,4-triazole (**1.1-16**).[18] Low yields, however, were obtained with a number of other aryl halides. In the alkylation of s-triazole with 2-chloropyridine in an autoclave at 190°, a 61% yield of 1-(2-pyridinyl)-1,2,4-triazole (**1.1-17**) was obtained.[19] In contrast, the alkylation of s-triazole with 1-fluoro-4-nitrobenzene at 180° gave a mixture of the 1- and 4-(4-nitrophenyl)-1,2,4-triazoles.[20] Another method for the preparation of 1-aryl-1,2,4-triazoles (e.g., **1.1-2**) is illustrated by the treatment of the triazolinone (**1.1-18**) with phosphorus (V) sulfide to remove

| **1.1-16** | **1.1-17** | **1.1-18** |

the oxo function.* Apparently, this reaction is limited to N-aryltriazoles and N-unsubstituted triazoles since the removal of both the oxo and 1-methyl groups in a 1-methyltriazolinone has been reported.†

---

* O. Widmann, *Berichte*, **26**, 2612 (1893).
† G. Young and W. H. Oates, *J. Chem. Soc.*, **79**, 659 (1901).

References

1. **57:** 3982a	2. **52:** 364c	3. **78:** 4224j	4. **55:** 14478a
5. **78:** 72019w	6. **49:** 1710i	7. **50:** 1785c	8. **68:** 29645d
9. **73:** 66847v	10. **80:** 47869g	11. **84:** 121737p	12. **83:** P114414g
13. **49:** 11630g	14. **73:** 45425r	15. **76:** 14436c	16. **83:** 164183b
17. **78:** 43384f	18. **72:** 55340c	19. **77:** 152071b	20. **68:** 29646e

1.2. 3-Alkyl- or Aryl-Substituted s-Triazoles

The formation of 3-[(phthalimido)methyl]-s-triazole (**1.2-2**) and related compounds was accomplished by the condensation of the amidrazone **1.2-1** with ethyl orthoformate.[1-3] A higher yield of this triazole was obtained when the *N*-1 formyl derivative (**1.2-3**) of the amidrazone was heated 5 to 10°

$$\text{PhtNCH}_2\text{C}\begin{array}{c}\diagup\text{NH}\\\diagdown\text{NHNH}_2\end{array} \xrightarrow[50\%]{(\text{EtO})_3\text{CH}} \text{PhtNCH}_2\text{ (triazole)}$$

1.2-1   **1.2-2**

80% ↘ Δ

$$\text{PhtNCH}_2\text{C}\begin{array}{c}\diagup\text{NH}\\\diagdown\text{NHNHCHO}\end{array}$$

1.2-3

above the melting point in the absence of a solvent. This method has also been used for the preparation of s-triazoles containing perhalo alkyl[4] and silyl alkyl[5] side chains. In contrast, 1-benzimidoyl-2-formylhydrazine (**1.2-4**)

$$\text{PhC}\begin{array}{c}\diagup\text{NH}\\\diagdown\text{NHNHCHO}\end{array} \xrightarrow[18\%]{\Delta} \text{Ph (triazole)}$$

1.2-4   **1.2-5**

was converted to 3-phenyl-s-triazole (**1.2-5**) in poor yield[6] by heating above its melting point. In addition to these methods, amidrazones have been converted with both formic acid[7] and s-triazine[8] to 3-substituted s-triazoles.

A number of 3-alkyl- and 3-aryl-s-triazoles have been obtained from triazoles by rearrangement, by removal of a functional group from the 5-position, and by modification of a functional group at the 5-position. Rearrangement of the amidine (**1.2-6**) with acetic anhydride at room temperature gave a 65% yield of the 1-formyl derivative (**1.2-10**), which was probably formed via **1.2-7** and **1.2-8**.[9,10] Pyrolysis of **1.2-10** gave both 3-phenyl-s-triazole (**1.2-5**) and **1.2-9** in high yields.

In the removal of a thione group from an *s*-triazole, oxidation of the triazoline-5-thione **1.2-11** with nitric acid gave 3-methyl-*s*-triazole (**1.2-14**) in good yield.[11] Also, this type of reaction was successful in the preparation of triazoles containing 3-alkyl groups substituted with phthalimido and (ethoxycarbonyl)amino moieties.[12,13] Triazoline-5-thiones substituted with 3-aryl[14] and 3-heteroaryl[15] moieties are also desulfured in good yields by this method, as illustrated by the conversion of **1.2-12** to **1.2-15**. In addition, treatment of 3-aryltriazoline-5-thiones (e.g., **1.2-13**) with Raney nickel in refluxing alcohol readily gave 3-aryl-*s*-triazoles (e.g., **1.2-5**) in good yields.[16] In the preparation of **1.2-5**, a slightly lower yield was obtained by

1.2-11, R = Me   **1.2-14**, R = Me
1.2-12, R = 2-MeOC$_6$H$_4$ **1.2-15**, R = 2-MeOC$_6$H$_4$
1.2-13, R = Ph   **1.2-5**, R = Ph

the use of hydrogen peroxide rather than Raney nickel.[16,17] Also, **1.2-5** has been prepared by heating 3-phenyltriazolin-5-one with P$_4$S$_{10}$.*

The amino group in 3-alkyl-5-amino-*s*-triazoles (e.g., **1.2-16**) can be diazotized followed by reductive removal of the diazo group with hypophosphorus acid to give 3-alkyl-*s*-triazoles (e.g., **1.2-17**).[18,19] Diazotiza-

1.2-16   **1.2-17**

*G. Young, *J. Chem. Soc.*, **87**, 625 (1905).

tion must be carried out in the absence of hydrochloric acid to avoid preferential displacement of the diazo group by chloride instead of hydrogen.[19] The Gomberg-Bachmann arylation reaction in the triazole series was demonstrated by the conversion of the hydroxide of 3-diazotriazole in benzene to 3-phenyl-s-triazole (**1.2-5**).[20] In the diazotization of 4-amino-3-(2-aminophenyl)triazole (**1.2-18**), the 4-amino group was removed, and the 2-amino group of the phenyl moiety was transformed to an azido group to give **1.2-19**.[21] Apparently, the initially formed diazophenyl group underwent

1.2-18   **1.2-19**

coupling with the 4-amino group, followed by ring opening of the resulting cyclic structure to give the observed product.

In the modification of functional groups on the triazole ring, 3-methyl-s-triazole (**1.2-12**) was shown to condense with 5-nitrofurfural to give a 28%

1.2-12   **1.2-20**

yield of **1.2-20**.[22] Further, the methyl ester (**1.2-21**) was shown to react with phenyl lithium to give a 62% yield of **1.2-22**.[23]

1.2-21   **1.2-22**

A number of other ring systems have been transformed to 3-substituted s-triazoles. Treatment of 4,6-dihydrazinopyrimidine (**1.2-23**) with hydrazine resulted in the formation of 3-methyl-s-triazole (**1.2-12**).[24,25] This ring contraction reaction is believed to involve C-2, N-3, C-4, and C-5 of **1.2-23**. Ring contraction was also observed in the reaction of the 4-aminopyridazine (**1.2-24**) with potassium amide in liquid ammonia to give 3-(2-cyanomethyl)-s-triazole (**1.2-27**).[26] An unusual rearrangement was observed in derivatives of 1-aminoadenine as illustrated by the conversion of **1.2-25** with base to give **1.2-26**.[27,28] This reaction involved hydrolytic

1.2-23   **1.2-12**   **1.2-24**

1.2-25   **1.2-26**   **1.2-27**

cleavage of the 1,2-bond of **1.2-25**, migration of a formyl group, and dehydrative ring closure. The hydrazinolysis of several bicyclic triazole systems has been investigated.[29] For example, treatment of the condensed triazole (**1.2-28**) with hydrazine cleaved the six-membered ring at the N—N linkage, followed by cyclization of the resulting intermediate to give **1.2-29** (58%).[30] Similarly, cleavage of the N—N bond of the six-membered ring of **1.2-30** with alkoxide gave 3-(cyanovinyl)-s-triazole (**1.2-33**) in 80% yield.[31] The tricyclic condensed triazole (**1.2-31**) also underwent N—N bond cleavage in the presence of base to give 3-(2-cyanophenyl)-s-triazole (**1.2-32**) in 40% yield.[32]

1.2-28   **1.2-29**   **1.2-30**

1.2-31   **1.2-32**   **1.2-33**

In the treatment of the tricyclic condensed triazole (**1.2-34**) with base, the pyrimidine ring was opened to give **1.2-35**.[33] Similarly, treatment of **1.2-36**

1.2-34   **1.2-35**

with hydrochloric acid opened the pyrimidine ring and resulted in the formation of **1.2-37** in good yield.[34] On oxidation of **1.2-39** with potassium

1.2-36   **1.2-37**

permanganate, 3-(2-carboxyphenyl)-s-triazole (**1.2-38**) was formed.[35,36] Derivatives of this tricyclic ring system were also cleaved with base as demonstrated by the conversion of **1.2-40** to **1.2-41** in 56% yield.[37]

1.2-38   **1.2-39**, R = H   **1.2-41**
              1.2-40, R = NH₂

References

1. **61:** 655h	2. **69:** 18568a	3. **75:** 20356x	4. **81:** 49647c
5. **80:** 96082d	6. **54:** 9898c	7. **83:** 22239a	8. **79:** 18620t
9. **71:** 38867s	10. **71:** 91386k	11. **50:** 1784b	12. **49:** 13978h
13. **48:** 12739c	14. **79:** 115500m	15. **70:** 3966a	16. **44:** 2514i
17. **49:** 10937f	18. **48:** 1343d	19. **43:** 7018d	20. **78:** 72019w
21. **77:** 88421m	22. **80:** 10273g	23. **80:** 133389g	24. **68:** 68940r
25. **73:** 87861k	26. **83:** 9953j	27. **81:** 169748h	28. **83:** 131874z
29. **85:** 46578c	30. **82:** 31304z	31. **78:** 136179h	32. **72:** 12671u
33. **73:** 120582v	34. **63:** 18084h	35. **76:** 72456v	36. **76:** 85193g
37. **76:** 72459y			

1.3. 4-Alkyl- or Aryl-Substituted 4H-1,2,4-Triazoles

The interaction of hydrazine, N,N-dimethylformamide, and thionyl chloride gave N,N-dimethylformamide azine hydrochloride (**1.3-1**), which underwent transamination reactions with both aliphatic and aromatic amines to give 4-alkyl- and 4-aryl-4H-1,2,4-triazoles (e.g., **1.3-2** and **1.3-3**),[1,2]

The thermally induced condensation of N,N'-diformylhydrazine (**1.3-4**) with aryl and heteroaryl primary amines is a method that has been used extensively for the preparation of 4-substituted triazoles. This procedure was successful for aniline and its derivatives,[3,4] and amines derived from pyridine,[5] quinoline,[5] and pyrazole,[6] but was unsuccessful with benzylamine.[3] In a related reaction, a good yield was claimed for the preparation of **1.3-2** by the interaction of N-formylhydrazine (**1.3-5**) with butyl amine in the presence of ethyl orthoformate.[7] Apparently, amidrazone intermediates are involved in these reactions as well as the reaction of diphenylformamidine with **1.3-5** at 150° to give **1.3-3**, which was probably formed via **1.3-6**.[8] In the condensation of diethyl (2-pyridylamino)methyl-

enemalonate (**1.3-7**) with hydrazine, the initially formed amidrazone (**1.3-9**) was transformed via **1.3-10** and **1.3-12** to 4-(2-pyridyl)-4H-1,2,4-triazole (**1.3-11**).[9] This reaction was limited to those (aminomethylene)malonates in which the N had at least some double bond character. In the condensation of **1.3-8** with hydrazine, aniline was eliminated in the initial reaction, and the major product was a pyrazole.

Although the alkylation of s-triazole under basic conditions gave mixtures of 1- and 4-alkyltriazoles in which the 1-alkyltriazole was the major product, a procedure has been developed to accomplish exclusive alkylation at N-4. Treatment of 1-acetyl-1H-1,2,4-triazole (**1.3-13**) with trimethyloxonium fluoroborate in nitromethane gave **1.3-14**, which was not isolated but hydrolyzed with methanol to give 4-methyl-4H-1,2,4-triazole (**1.3-15**) in 88% yield.[10] Arylation of s-triazole has also been accomplished, but the product was a mixture of 1- and 4-aryltriazoles.[11,12]

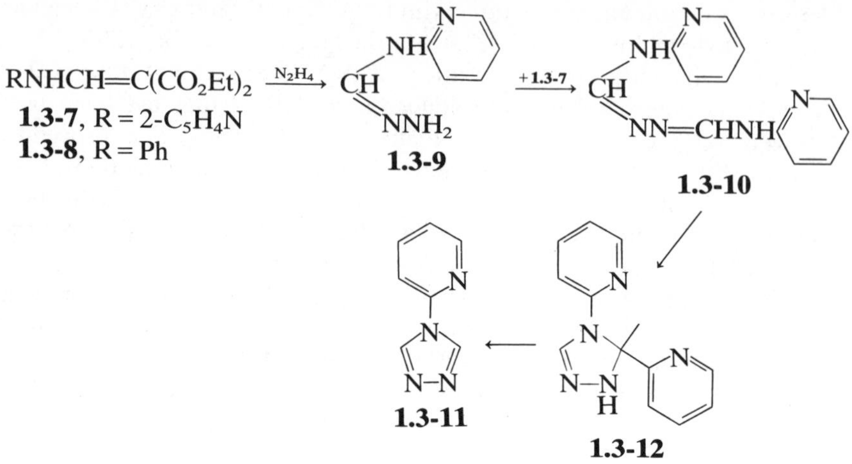

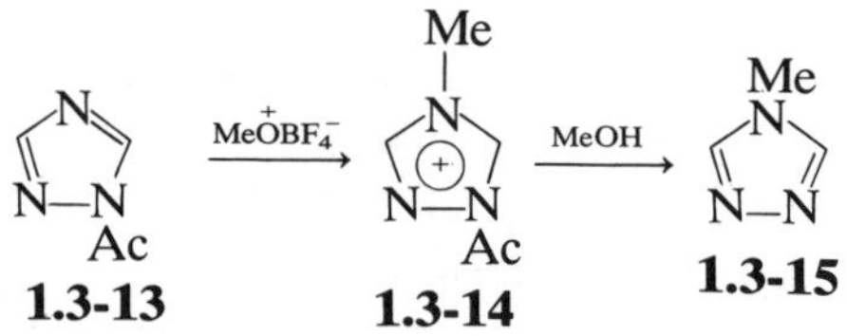

RNHCH=C(CO$_2$Et)$_2$ $\xrightarrow{\text{N}_2\text{H}_4}$ [structure 1.3-9] $\xrightarrow{+1.3\text{-}7}$ [structure 1.3-10]

1.3-7, R = 2-C$_5$H$_4$N
1.3-8, R = Ph

1.3-9

1.3-10

[structures 1.3-11, 1.3-12]

1.3-11

1.3-12

[structures 1.3-13, 1.3-14, 1.3-15]

$\xrightarrow{\text{Me\overset{+}{O}BF}_4^-}$ $\xrightarrow{\text{MeOH}}$

1.3-13   **1.3-14**   **1.3-15**

A convenient route for the preparation of 4-substituted 4H-1,2,4-triazoles involves the oxidative removal of the thioxo group of Δ^2-triazoline-5-thiones, as illustrated for the conversion of **1.3-16** to **1.3-3**[13] with hydrogen peroxide in acetic acid and of **1.3-17** to **1.3-15** with Raney nickel.[14]

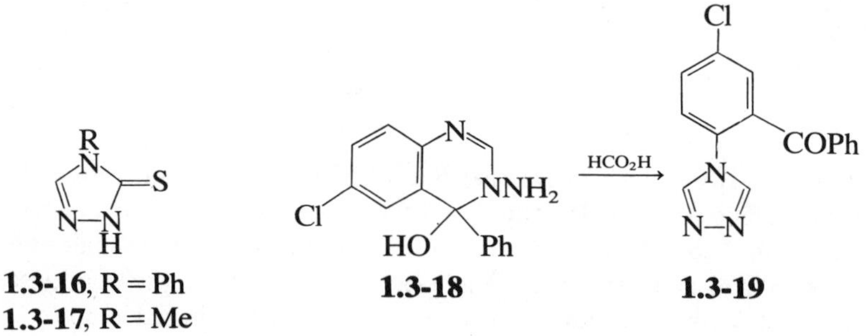

[structures 1.3-16/1.3-17, 1.3-18, 1.3-19]

$\xrightarrow{\text{HCO}_2\text{H}}$

1.3-16, R = Ph   **1.3-18**   **1.3-19**
1.3-17, R = Me

Interestingly, treatment of the quinazoline derivative **1.3-18** with formic acid gave **1.3-19**.[15,16]

References

1. **67:** 90738g	2. **82:** 156003g	3. **57:** 3982a	4. **78:** 72019w
5. **48:** 12092c	6. **54:** 2332i	7. **76:** 113224p	8. **50:** 966i
9. **73:** 66512g	10. **73:** 45425r	11. **68:** 29646e	12. **72:** 55340c
13. **53:** 21904h	14. **68:** 29645d	15. **74:** 125579e	16. **83:** 193188n

TABLE 1. ALKYL- OR ARYL-MONOSUBSTITUTED 1,2,4-TRIAZOLES

Compound	Reference

1.1. 1-Alkyl- or Aryl-substituted 1H-1,2,4-triazoles

Compound	Reference
Parent (1H-1,2,4-Triazole)	**49:** 12451e, **50:** 1785c, **51:** 14751h, **51:** 18009a, **52:** 364c, **55:** 4337i, **75:** 35894v
Parent, 3,5-d_2	**83:** 113229p
Parent, 1,3,5-d_3	**79:** 11610f
Parent, 1-[(4-chlorophenyl)diphenylstannyl]-	**77:** P101899v, **79:** P78965u
Parent, 1-(difluorophosphinylidene)-	**82:** 140033f
Parent, 1-(tributylgermyl)-	**64:** 17632a
Parent, 1-(tributylstannyl)-	**60:** 6860e, **77:** P101899v, **78:** P159868w, **79:** P78965u, **81:** P91733c, **84:** P100878v
Parent, 1-(tricyclohexylstannyl)-	**78:** P159868w, **79:** P78965u, **81:** P91733c, **84:** P100878v
Parent, 1-(tricyclopentylstannyl)-	**81:** P91733c, **84:** P100878v
Parent, 1-(trimethylsilyl)-	**55:** 5484d, **69:** 43249e
Parent, 1-trimethylstannyl-	**60:** 6860g
Parent, 1-(triphenylgermyl)-	**84:** 135765h
Parent, 1-(triphenylstannyl)-	**77:** P101899v, **79:** 78965u
Parent, 1-[tris(tert-butyl)stannyl]-	**81:** P91733c
Parent, 1-[tris(4-chlorophenyl)stannyl]-	**79:** P78965u
Parent, 1-[tris(1,1-dimethylethyl)stannyl]-	**84:** P100878v
Parent, 1-[tris(isopropyl)stannyl]-	**81:** P91733c
Parent, 1-[tris(1-methylethyl)stannyl]-	**84:** P100878v
-1-acetamide	**50:** 1785f
-1-acetamide, N-[5-[(4-[2,4-bis(1,1-dimethylpropyl)-phenoxy]-1-oxobutyl)amino]-2-chlorophenyl]-α-(2,2-dimethyl-1-oxopropyl)-	**85:** P102316e
-1-acetamide, N-[2-chloro-5-([(3-pentadecylphenoxy)-acetyl]amino)phenyl]-α-(2,2-dimethyl-1-oxopropyl)-	**85:** P102316e
-1-acetamide, α-(2,2-dimethyl-1-oxopropyl)-N-[2-(hexa-decyloxy)-5-[(methylamino)-sulfonyl]phenyl]-	**85:** P151754e
1-(4'-acetamido-2,2'-disulfostilben-4-yl)-	**74:** P43528y
1-(7-acetamido-2-oxo-2H-1-benzopyran-3-yl)-	**82:** P18628p
-1-acetic acid	**50:** 1785e
-1-acetic acid, α,α-diphenyl-, 1-methyl-2-(4-morpholinyl)ethyl ester	**76:** P59636j
-1-acetic acid, ethyl ester	**83:** 97150y
-1-acetic acid, hydrazide	**83:** 97150y
-1-acetic acid, [(5-nitro-2-furanyl)methylene]-hydrazide	**83:** 97150y
1-(2-acetylphenyl)-	**69:** 59162g, **72:** 55340c
1-(3-acetylphenyl)-	**72:** 55340c
1-(4-acetylphenyl)-	**72:** 55340c
1-(4-acetylphenyl)-, semicarbazone	**72:** 55340c
1-(3-amino-4-chlorophenacyl)-	**78:** P148956x
1-[2-(4-amino-3-chlorophenoxy)ethyl]-	**78:** P148956x

13

TABLE 1 (*Continued*)

Compound	Reference

1.1. 1-Alkyl- or Aryl-substituted 1H-1,2,4-triazoles (*Continued*)

Compound	Reference
1-(2-[(4-amino-3-chlorophenyl)sulfonyl]ethyl)-	**78:** P148956x
1-(2-[(4-amino-2,5-dichlorophenylsulfonyl)amino]ethyl)	**78:** P148956x
1-(2-aminoethyl)	**50:** 1785c, **80:** 65b
1-[2-([(6-amino-2-naphthalenyl)sulfonyl]amino)ethyl]-	**78:** P148956x
1-(7-amino-2-oxo-2H-1-benzopyran-3-yl)-	**72:** P22599r, **73:** P100073b
1-(2-aminophenyl)-	**78:** 72019w
1-(*p*-aminophenyl)-	**57:** 3982a, **81:** 62940n
1-(3-aminopropyl)-	**50:** 1785i
1-[4'-[(4-anilino-6-morpholino-*s*-triazin-2-yl)amino]- 2,2'-disulfostilben-4-yl]-	**74:** P43528y
1-[4'-(*p*-anisamido)-2,2'-disulfostilben-4-yl]-	**74:** P43528y
1-(1,3,2-benzodioxaphosphol-2-yl)-	**85:** 21797k
1-[4-[2-(2-benzofuranyl)ethenyl]phenyl]-	**84:** P91665u
1-[2-[2-benzoyl-2-(4-chlorophenoxy)]ethyl]⁻	**83:** P43336v
1-(4-benzoyl-1,3-dimethyl-1H-pyrazol-5-yl)-	**82:** P140121h
1-[2-(2-benzoyl-2-phenoxy)ethyl]-	**83:** P43336v
1-benzyl	**76:** 14436c
1-benzyl-	**50:** 1785b, **64:** 19596g, **78:** 43384f
1-[1,1'-biphenyl]-2-yl-	**78:** 72019w
1-[1,1'-biphenyl]-4-yl-	**78:** 72019w
1-[(1,1'-biphenyl)-4-yl(2-chlorophenyl)methyl]-	**85:** P160095t
1-[4-(2-[1,1'-biphenyl]-4-ylethenyl)phenyl]-	**84:** P91665u
1-[α-4-biphenylyl-α-(1-methylimidazol-2-yl)benzyl]-	**75:** P76798k
1-[(1,1'-biphenyl)-4-yl(1-methyl-1H-imidazol- 5-yl)phenylmethyl]-	**84:** P169551a
1-[1-([1,1'-biphenyl]-2-yloxy)-3,3-dimethyl-]- oxobutyl]-	**80:** P146169k
1-[1-([1,1'-biphenyl]-4-yloxy)-3,3-dimethyl-2- oxobutyl]-	**79:** P105257y, **80:** P146169k, **83:** P179070m, **83:** P189336s, **83:** P206289y
1-([1,1'-biphenyl]-4-ylphenylmethyl)-	**85:** P160095t
1-[1-[(3-bromo]1,1'-biphenyl]-4-yl)oxy]-3,3- dimethyl-2-oxobut-1-yl]-	**83:** P179070m
1-[1-(4-bromo-2-chlorophenoxy)-3,3-dimethyl-2- oxobut-1-yl]-	**83:** P179070m
1-[4-(bromomethyl)phenyl]-	**84:** 91665u
1-([2-(3-bromo-4-methylphenyl)-1,3-dioxolan- 2-yl]methyl)-	**85:** P94368f
1-([2-(4-bromo-2-methylphenyl)-1,3-dioxolan- 2-yl]methyl)-	**85:** P94368f
1-[1-(4-bromophenoxy)-3,3-dimethyl-2-oxobutyl]-	**80:** P146169k, **83:** P179070m, **83:** P206289y, **83:** P189336s
1-[2-(4-bromophenoxy)-4,4-dimethyl-3-oxo-1-pentyl]-	**82:** P170966e, **83:** P43336v, **83:** P43337w
1-([2-(2-bromophenyl)-1,3-dioxolan-2-yl]methyl)-	**85:** P94368f
1-([2-(3-bromophenyl)-1,3-dioxolan-2-yl]methyl)-	**85:** P94368f

TABLE 1 (*Continued*)

Compound	Reference

1.1. 1-Alkyl- or Aryl-substituted 1H-1,2,4-triazoles (*Continued*)

Compound	Reference
1-([2-(4-bromophenyl)-1,3-dioxolan-2-yl]methyl)-	**85:** P94368f
1-[9-(butoxycarbonyl)-9-fluorenyl]-	**75:** P110320k
1-[3-(p-butoxyphenyl)coumarin-7-yl]-	**71:** P14180m
1-butyl-	**57:** 3982a, **64:** 19596g, **78:** 43384f, **80:** 47869g
1-sec-butyl-	**57:** 3982a
1-tert-butyl-	**68:** 29645d
1-([4-butyl-2-(2,4-dichlorophenyl)-1,3-dioxolan-2-yl]methyl)-	**85:** P94368f
1-[p-tert-butyl-α-(1-methylimidazol-2-yl)-α-phenylbenzyl]-	**75:** P76798k
1-(8-caffeinyl)-	**60:** 13236g
1-(4′-carbamoyl-2-sulfostilben-4-yl)-	**74:** P43528y
1-(4-carbamoylphenyl)-	**72:** 55340c
1-[p-(carboxyamino)phenyl]-, diester with 2,2′-(p-phenylenedioxy)diethanol	**72:** P100715s
1-[p-(carboxyamino)phenyl]-, phenyl ester	**72:** P100715s
1-[2-(4-carboxymethoxy- -2,3-dichlorobenzoyl)butyl]-	**75:** P49083s
1-[1-(4-carboxyphenoxy)-3,3-dimethyl-2-oxobut-1-yl]-	**83:** P189336s
1-[1-(4-carboxyphenoxy)-2-hydroxy-3,3-dimethyl-1-butyl]-	**83:** P189336s
1-(p-carboxyphenyl)methyl-	**24:** 368[5]
1-(2-[2-(4-chlorobenzoyl)-2-(2,4-dichlorophenoxy)]ethyl)-	**83:** P43336v
1-(2-[2-(4-chlorobenzoyl)-2-(4-chlorophenoxy)]ethyl)-	**83:** P43336v
1-(2-chlorobenzyl)-	**76:** 14436c
1-(4-chlorobenzyl)-	**57:** 3982a, **76:** 14436c
1-[1-[(3-chloro[1,1′-biphenyl]-4-yl)oxy]-3,3-dimethyl-2-oxobut-1-yl]-	**83:** P179070m
1-[4-chloro-α-(2,4-dichlorophenoxy)phenacyl]-	**79:** P105257y, **80:** P146169k
1-[1-(4-chloro-3,5-dimethylphenoxy)-3,3-dimethyl-2-oxobutyl]-	**80:** P146169k
1-(m-chloro-α,α-diphenylbenzyl)-	**74:** P100062t
1-(o-chloro-α,α-diphenylbenzyl)-	**74:** 100062t, **74:** 112048f, **74:** P125697s, **74:** P125698t, **76:** P59632e
1-(p-chloro-α,α-diphenylbenzyl)-	**74:** P100062t
1-(2-chloroethyl)-	**62:** 13270g, **83:** P164183b
1-[2-chloro-9-(methoxycarbonyl)-9-fluorenyl]-	**75:** P76791c, **75:** P110320k
1-([2-(4-chloro-2-methoxyphenyl)-1,3-dioxolan-2-yl]methyl)-	**85:** P94368f
1-([2-(2-chloro-4-methoxyphenyl)-1,3-dioxolan-2-yl]methyl)-	**85:** P94368f
1-[p-chloro-α-(1-methylimidazol-2-yl)-α-phenylbenzyl]-	**75:** P76798k
1-[p-chloro-α-(5-methyl-3-isoxazolyl)-α-phenylbenzyl]-	**75:** P76798k, **76:** P59632e
1-[1-(4-chloro-2-methylphenoxy)-3,3-dimethyl-2-oxobutyl]-	**79:** P105257y, **80:** P146169k

TABLE 1 (Continued)

Compound	Reference

1.1. 1-Alkyl- or Aryl-substituted 1H-1,2,4-triazoles (Continued)

Compound	Reference
1-[1-(4-chloro-3-methylphenoxy)-3,3,-dimethyl-2-oxobutyl]-	80: P146169k
1-[1-(5-chloro-2-methylphenoxy)-3,3-dimethyl-2-oxobutyl]-	80: P146169k, 83: P206289y
1-[2-(4-chloro-2-methylphenoxy)-4,4-dimethyl-3-oxo-1-pentyl]-	82: P170966e, 83: P43336v, 83: P43337w, 83: P43338x, 83: P58837h
1-[[2-(4-chloro-2-methylphenyl)-1,3-dioxolan-2-yl]methyl]-	85: P94368f
1-[7-(4-chloro-3-methylpyrazol-1-yl)-2-oxo-2H-1-benzopyran-3-yl]-	74: P127556u
1-[1-(3-chloro-4-nitrophenoxy)-3,3-dimethyl-2-oxobut-1-yl]-	83: P179070m
1-(4-chloro-2-nitrophenyl)-	72: P79056r
1-(2-[([(2-chloro-4-nitrophenyl)azo]phenyl)ethyl amino]ethyl)-	82: P126600g
1-[1-(2-chlorophenoxy)-3,3-dimethyl-2-oxobutyl]-	79: P105257y, 83: P189336s, 83: P206289y
1-[1-(3-chlorophenoxy)-3,3-dimethyl-2-oxo-1-butyl]-	82: P156325p, 83: P179070m
1-[1-(4-chlorophenoxy)-3,3-dimethyl-2-oxobutyl]-	79: P105257y, 80: P146169k, 83: P179070m, 83: P189336s, 83: P206289y
1-[1-(4-chlorophenoxy)-3,3-dimethyl-2-oxobutyl]-, oxime	80: P146169k, 83: P189336s
1-[2-(2-chlorophenoxy)-4,4-dimethyl-3-oxo-1-pentyl]-	82: P170966e, 83: P43336v, 83: P43337w
1-[2-(3-chlorophenoxy)-4,4-dimethyl-3-oxo-1-pentyl]-	82: P170966e, 83: P43336v
1-[2-(4-chlorophenoxy)-4,4-dimethyl-3-oxo-1-pentyl]-	82: P170966e, 83: P43336v, 83: P43337w
1-[1-(4-chlorophenoxy)-3,3-dimethyl-1-phenyl-2-oxobutyl]-	79: P105257y, 80: P146169k
1-[1-(4-chlorophenoxy)-2-oxopropyl]-	80: P146169k, 82: P156325p
1-[α-(4-chlorophenoxy)phenacyl]-	82: P156325p
1-(p-chlorophenyl)-	57: 3982a
1-[5-(4-chlorophenyl)-10,11-dihydro-5H-dibenzo[a,d]-cyclohepten-5-yl]-	85: P58083a
1-([2-(2-chlorophenyl)-1,3-dioxolan-2-yl]methyl)-	85: P94368f
1-([2-(3-chlorophenyl)-1,3-dioxolan-2-yl]methyl)-	85: P94368f
1-([2-(4-chlorophenyl)-1,3-dioxolan-2-yl]methyl)-	85: P94368f
1-[1-(2-chlorophenyl)-2-nitroethyl]-	81: P91539u
1-[1-(4-chlorophenyl)-2-nitroethyl]-	81: P91539u
1-(2-[[p-[3-(p-chlorophenyl)-2-pyrazolin-1-yl]phenyl)-sulfonyl]ethyl)-	72: P91491m
1-([2-(5-chloro-2-thienyl)-1,3-dioxolan-2-yl]methyl)-	85: P94368f
1-(o-cyano-α,α-diphenylbenzyl)-	74: P100062t
1-(m-cyano-α,α-diphenylbenzyl)-	74: P100062t
1-(p-cyano-α,α-diphenylbenzyl)-	74: P100062t

TABLE 1 (*Continued*)

Compound	Reference
1.1. 1-Alkyl- or Aryl-substituted 1H-1,2,4-triazoles (*Continued*)	

Compound	Reference
1-(7-cyano-2-oxo-2H-1-benzopyran-3yl)-	**73:** P57178a
1-(*o*-cyanophenyl)-	**72:** 55340c
1-(*m*-cyanophenyl)-	**72:** 55340c
1-(*p*-cyanophenyl)	**72:** 55340c
1-[(cyclohexylcarbonyl)(2,4-dichlorophenoxy)methyl]-	**79:** P105257y
1-[1-(2-cyclohexylphenoxy)-3,3-dimethyl- 2-oxobut-1-yl]-	**83:** P179070m
1-[1-(4-cyclohexylphenoxy)-3,3-dimethyl- 2-oxobut-1-yl]-	**83:** P179070m
1-[(cyclopentyl)(2,4-dichlorophenoxy)methyl]-	**79:** P105257y
1-[[2-(2,4-dibromophenyl)-1,3-dioxolan-2-yl]methyl]-	**85:** P94368f
1-(1-[(3,5-dichloro[1,1'-biphenyl]-4-yl)oxy]-3,3- dimethyl-2-oxobut-1-yl)-	**83:** P179070m
1-(2,4-dichloro-α,α-diphenylbenzyl)-	**74:** P100062t
1-[1-(2,4-dichlorophenoxy)-3,3-dimethyl-2-oxobutyl]-	**79:** P105257y, **80:** P146169k, **83:** P179070m, **83:** P189336s, **83:** P206289y
1-[1-(2,5-dichlorophenoxy)-3,3-dimethyl-2-oxobutyl]-	**79:** P105257y, **80:** P146169k
1-[1-(2,6-dichlorophenoxy)-3,3-dimethyl-2-oxobutyl]-	**79:** P105257y, **80:** P146169k, **83:** P206289y
1-[1-(2,4-dichlorophenoxy)-3,3-dimethyl-2- oxobut-1-yl]-	**83:** P189336s
1-[1-(3,4-dichlorophenoxy)-3,3-dimethyl-2- oxobut-1-yl]-	**83:** P179070m
1-[2-(2,4-dichlorophenoxy)-4,4-dimethyl-3- oxo-1-pentyl]-	**82:** P170966e, **83:** P43336v, **83:** P43337w
1-[2-(2,5-dichlorophenoxy)-4,4-dimethyl-3- oxo-1-pentyl]-	**82:** P170966e, **83:** P43336v, **83:** P43337w
1-[1-(2,4-dichlorophenoxy)-3,3-dimethyl-2- oxo-1-phenylbutyl]-	**80:** P146169k, **83:** P189336s
1-[1-(2,4-dichlorophenoxy)-3-methyl-2-oxobutyl]-	**79:** P105257y
1-[1-(2,4-dichlorophenoxy)-2-oxopropyl]-	**80:** P146169k, **83:** P189336s
1-[α-(2,6-dichlorophenoxy)phenacyl]-	**79:** P105257y, **80:** P146169k
1-([2-(2,3-dichlorophenyl)-1,3-dioxolan-2-yl]methyl)-	**85:** P94368f
1-([2-(2,4-dichlorophenyl)-1,3-dioxolan-2-yl]methyl)-	**85:** P94368f
1-([2-(2,5-dichlorophenyl)-1,3-dioxolan-2-yl]methyl)-	**85:** P94368f
1-([2-(3,4-dichlorophenyl)-1,3-dioxolan-2-yl]methyl)-	**85:** P94368f
1-([2-(3,5-dichlorophenyl)-1,3-dioxolan-2-yl]methyl)-	**85:** P94368f
1-([2-(2,4-dichlorophenyl)-4-ethyl-1,3-dioxolan-2-yl]- methyl)-	**85:** P94368f
1-([2-(2,4-dichlorophenyl)-4-heptyl-1,3-dioxolan- 2-yl]methyl)-	**85:** P94368f
1-([2-(2,4-dichlorophenyl)-4-hexyl-1,3-dioxolan- 2-yl]methyl)-	**85:** P94368f
1-([2-(2,4-dichlorophenyl)-4-methyl-1,3-dioxolan- 2-yl]methyl)-	**85:** P94368f
1-[1-(2,6-dichlorophenyl)-2-nitroethyl]-	**81:** P91539u

TABLE 1 (*Continued*)

Compound	Reference

1.1. 1-Alkyl- or Aryl-substituted 1H-1,2,4-triazoles (*Continued*)

Compound	Reference
1-([2-(2,4-dichlorophenyl)-4-octyl-1,3-dioxolan- 2-yl]methyl)-	**85:** P94368f
1-([2-(2,4-dichlorophenyl)-4-pentyl-1,3-dioxolan- 2-yl]methyl)-	**85:** P94368f
1-([2-(2,4-dichlorophenyl)-4-propyl-1,3-dioxolan- 2-yl]methyl)-	**85:** P94368f
1-[2-(diethylamino)ethyl]-	**55:** 14478d, **80:** 65b
1-(10,11-dihydro-5H-dibenzo[*a,d*]cyclohepten-5-yl)-	**85:** P58084b
1-(2,3-dihydro-3,3-dimethyl-2,5-dihydroxynaphtho- [1,2-b]furan-4-yl)-	**78:** 84321b
1-(5,6-dihydro-4H-1,3-thiazin-2-yl)-	**80:** P27263b
1-(4,5-dihydro-2-thiazolyl)-	**80:** P27263b
1-[2,2-dihydroxy-3,3-dimethyl-1-(pentachloro- phenoxy)butyl]-	**80:** P146169k
1-(1,4-dihydroxynaphthalen-2-yl)-	**78:** 84321b
1-[(2-(dimethylamino)ethyl]-	**80:** 65b
1-(1-[4-(1,1-dimethylethyl)phenoxyl]-3,3-dimethyl- 2-oxobutyl)-	**79:** P105257y, **80:** P146169k, **83:** P189336s
1-[4[2-(3,3-dimethyl-3H-2-indoyl)vinylamino]phenyl]-, 1-methyl-3H-indolium perchlorate	**73:** 26583z
1-(4-[2-(3,3-dimethyl-3H-2-indoyl)vinyl]phenyl)-, 1-methyl-3H-indolium iodide	**73:** 26583z
1-[3,3-dimethyl-1-[4-(1-methylethyl)phenoxy]- 2-oxobutyl]-	**79:** P105257y
1-[3,3-dimethyl-1-(2-methyl-5-nitrophenoxy)- 2-oxobut-1-yl]-	**83:** P179070m
1-[3,3-dimethyl-1-(4-methylphenoxy)-2-oxobutyl]-	**79:** P105257y, **83:** P206289y
1-[4,4-dimethyl-2-(4-methylphenoxy)-3-oxo-1-pentyl]-	**82:** P170966e, **83:** P43336v, **83:** P43337w
1-[3,3-dimethyl-1-(4-nitrophenoxy)-2-oxobut-1-yl]-	**83:** P189336s
1-[3,3-dimethyl-1-(4-nitrophenoxy)-2-oxobutyl]-	**79:** P105257y, **80:** P146169k, **83:** P179070m, **83:** P206289y
1-[3,3-dimethyl-1-(4-nitrophenoxy)-2-oxobutyl]-, oxime	**80:** P146169k, **83:** P189336s
1-[4,4-dimethyl-2-(4-nitrophenoxy)-3-oxo-1-pentyl]-	**82:** P170966e, **83:** P43336v, **83:** P43337w
1-[3,3-dimethyl-2-oxo-1-phenoxybutyl]-	**79:** P105257y, **80:** P146169k, **83:** P206289y
1-(4,4-dimethyl-3-oxo-2-phenoxy-1-pentyl)-	**82:** P170966e, **83:** P43336v
1-[3,3-dimethyl-2-oxo-1-(4-phenoxyphenoxy)-but-1-yl]-	**83:** P179070m
1-[3,3-dimethyl-2-oxo-1-phenoxy-1-phenyl-butyl]-	**79:** P105257y
1-[3,3-dimethyl-2-oxo-1-(2,4,6-trichlorophenoxy)- but-1-yl]-	**83:** P179070m
1-[3,3-dimethyl-2-oxo-1-(2,4,5-trichlorophenoxy- 1-butyl]-	**83:** P189336s
1-[3,3-dimethyl-2-oxo-1-(2,4,6-trichlorophenoxy)- 1-butyl]-	**83:** P189336s
1-[1-(2,4-dimethylphenoxy)-3,3-dimethyl- 2-oxobut-1-yl]-	**83:** P179070m

18

TABLE 1 (*Continued*)

Compound	Reference

1.1. 1-Alkyl- or Aryl-substituted 1H-1,2,4-triazoles (*Continued*)

1-[1-(2,3-dimethylphenoxy)-3,3-dimethyl-2-oxobutyl]-	**79:** P105257y, **80:** P146169k, **83:** P206289y
1-[1-(3,4-dimethylphenoxy)-3,3-dimethyl-2-oxobutyl]-	**79:** P105257y, **80:** P146169k, **83:** P206289y
1-[2-(2,3-dimethylphenoxy)-4,4-dimethyl-3-oxo-1-pentyl]-	**82:** P170966e, **83:** P43336v, **83:** P43337w
1-[2-(2,4-dimethylphenoxy)-4,4-dimethyl-3-oxo-1-pentyl]-	**82:** P170966e, **83:** P43336v, **83:** P43337w
1-[2-(3,4-dimethylphenoxy)-4,4-dimethyl-3-oxo-1-pentyl]-	**82:** P170966e, **83:** P43336v, **83:** P43337w
1-[2,2-dimethyl-3-oxo-4-[4-(trifluoromethyl)-phenoxy]butyl]-	**79:** P105257y
1-[3,3-dimethyl-2-oxo-1-[3-(trifluoromethyl)-phenoxy]-butyl]-	**80:** P146169k, **83:** P189336s
1-[7-(3,5-dimethylpyrazol-1-yl)-2-oxo-2H-1-benzopyran-3-yl]-	**72:** P22599r
1-(2,4-dinitrophenyl)-	**65:** 18448g, **68:** 29646e
1-(1,3,2-dioxaphospholan-2-yl)-	**85:** 21797k
1-(1,4-dioxonaphthalen-2-yl)-	**78:** 84321b
1-[diphenyl[4-(phenylthio)phenyl]methyl]-	**84:** P169551a
1-(diphenylphosphino)-	**85:** 78185p
1-[α,α-diphenyl-m-(trifluoromethyl)benzyl]-	**74:** P100062t
1-dodecyl-	**35:** 3598[8]
-1-ethanol	**50:** 1785g, **55:** 14478d, **85:** 178509r
-1-ethanol, 4-aminobenzoate ester	**78:** P148956x
-1-ethanol, α-benzyl-α-(tert-butyl)-β-(4-chlorophenoxy)-	**82:** P72997w, **82:** P156325p
-1-ethanol, β-([1,1'-biphenyl]-4-yloxy)-α-(1,1-dimethylethyl)-	**84:** P85617p
-1-ethanol, α,β-bis(4-chlorophenyl)-	**84:** P150636k
-1-ethanol, β-(4-bromophenoxy)-α-(tert-butyl)-	**82:** P72997w, **82:** P156325p, **83:** P189336s
-1-ethanol, β-(4-bromophenoxy)-α-(1,1-dimethylethyl)-	**84:** P85617p
-1-ethanol, α-(tert-butyl)-β-([1,1'-biphenyl]-4-yloxy)-	**82:** P72997w, **82:** P156325p
-1-ethanol, α-(tert-butyl)-β-[4-(tert-butyl)phenoxy]-	**82:** P72997w, **82:** P156325p, **83:** P189336s
-1-ethanol, α-tert-butyl-β-(4-chloro-2-methylphenoxy)-	**82:** P72997w, **82:** P156325p
-1-ethanol, α-(tert-butyl)-β-(2-chlorophenoxy)-	**82:** P72997w, **82:** P156325p
-1-ethanol, α-(tert-butyl)-β-(3-chlorophenoxy)-	**82:** P72997w, **82:** P156325p
-1-ethanol, α-(tert-butyl)-β-(4-chlorophenoxy)-	**82:** P72997w, **82:** P156325p, **83:** P189336s
-1-ethanol, α-(tert-butyl)-β-(4-chlorophenoxy)-α-methyl-	**82:** P72997w, **82:** P156325p
-1-ethanol, α-tert-butyl-β-(4-cyclohexylphenoxy)-	**83:** P189336s
-1-ethanol, α-(tert-butyl)-β-(2,4-dichlorophenoxy)-	**82:** P72997w, **82:** P156325p
-1-ethanol, α-(tert-butyl)-β-(2,4-dichlorophenoxy)-α-methyl-	**82:** P72997w, **82:** P156325p
-1-ethanol, α-(tert-butyl)-β-(3,4-dimethylphenoxy)-	**82:** P72997w, **82:** P156325p

TABLE 1 (Continued)

Compound	Reference

1.1. 1-Alkyl- or Aryl-substituted 1H-1,2,4-triazoles (Continued)

Compound	Reference
-1-ethanol, α-(tert-butyl)-β-(4-fluorophenoxy)-	**82:** P72997w, **82:** P156325p, **83:** P189336s
-1-ethanol, α-(tert-butyl)-β-(4-methylphenoxy)-	**82:** P72997w, **82:** P156325p
-1-ethanol, α-(tert-butyl)-β-(4-nitrophenoxy)-	**82:** P72997w, **82:** P156325p
-1-ethanol, α-(tert-butyl)-β-phenoxy-	**82:** P72997w, **82:** P156325p
-1-ethanol, α-tert-butyl-β-phenoxy-	**83:** P189336s
-1-ethanol, α-(tert-butyl)-β-(2,4,5-trichlorophenoxy)-	**82:** P72997w, **82:** P156325p
-1-ethanol, β-(4-chlorophenoxy)-α-(1,1-dimethylethyl)-	**84:** P85617p
-1-ethanol, β-(4-chlorophenoxy)-α-methyl-	**82:** P72997w, **82:** P156325p
-1-ethanol, β-(4-chlorophenoxy)-α-methyl-α-phenyl-	**84:** P169551a
-1-ethanol, β-(4-chlorophenoxy)-α-phenyl-	**82:** P72997w, **82:** P156325p, **83:** P189336s, **84:** P85617p
-1-ethanol, α-(4-chlorophenyl)-β-methyl-	**84:** P150636k
-1-ethanol, α-(4-chlorophenyl)-β-phenyl-	**84:** P150636k
-1-ethanol, β-(4-chlorophenyl)-α-phenyl-	**84:** P150636k
-1-ethanol, β-[(3,5-dichloro[1,1'-biphenyl]-4-yl)oxy]-α-(1,1-dimethylethyl)-	**84:** P169551a
-1-ethanol, β-(2,4-dichlorophenoxy)-α-(1,1-dimethylethyl)-	**82:** P72997w, **82:** P156325p
-1-ethanol, β-(2,4-dichlorophenoxy)-α-(1,1-dimethylethyl)-α-methyl-	**84:** P169551a
-1-ethanol, α-(2,4-dichlorophenyl)-	**84:** P150636k
-1-ethanol, α-(1,1-dimethylethyl)-β-[4-(1,1-dimethylethyl)phenoxy]-	**84:** P169551a
-1-ethanol, α-(1,1-dimethylethyl)-β-(3,4-dimethylphenoxy)-	**84:** P169551a
-1-ethanol, α-(1,1-dimethylethyl)-β-(4-fluorophenoxy)-	**84:** P85617p, **84:** P169551a
-1-ethanol, α-(1,1-dimethylethyl)-β-(4-nitrophenyl)-	**84:** P85617p
1-(2-ethoxy-1,4-dioxonaphthalen-3-yl)-	**78:** 84321b
1-[9-(ethoxycarbonyl)-9-fluorenyl]-	**75:** P76791c, **75:** P110320k
1-[3-(p-ethoxyphenyl)coumarin-7-yl]-	**71:** P14180m
1-ethyl-	**68:** 29645d, **80:** 47869g, **84:** 121737p
1-[7-[(1-ethylacetonylidene)hydrazino]-2-oxo-2H-1-benzopyran-3-yl]-, 7-oxime	**73:** P100073b, **74:** P14197j, **82:** P172621n
1-[2-[ethyl(3-methylphenyl)amino]ethyl]-	**82:** P126600g
1-[2-[ethyl(phenyl)amino]ethyl]-	**82:** P126600g
1-[[2-(9H-fluoren-2-yl)-1,3-dioxolan-2-yl]methyl]-	**85:** P94368f
1-(p-fluoro-α,α-diphenylbenzyl)-	**74:** P100062t
1-[p-fluoro-α-(3-methyl-5-isoxazolyl)-α-phenylbenzyl]-	**75:** P76798k, **76:** P59632e
1-[p-fluoro-α-(5-methyl-3-isoxazolyl)-α-phenylbenzyl]-	**75:** P76798k
1-[1-(3-fluorophenoxy)-3,3-dimethyl-2-oxo-1-butyl]-	**82:** P156325p
1-[1-(4-fluorophenoxy)-3,3-dimethyl-2-oxobutyl]-	**80:** P146169k, **83:** P179070m, **83:** P206289y
1-[2-(4-fluorophenoxy)-4,4-dimethyl-3-oxo-1-pentyl]-	**82:** P170966e, **83:** P43336v, **83:** P43337w
1-[[2-(4-fluorophenyl)-1,3-dioxolan-2-yl]methyl]-	**85:** P94368f

TABLE 1 (*Continued*)

Compound	Reference

1.1. 1-Alkyl- or Aryl-substituted 1H̲-1,2,4-triazoles (*Continued*)

Compound	Reference
1-(4-formylphenyl)-	**73:** 26583z
1-(7-hydrazino-2-oxo-2H̲-1-benzopyran-3-yl)-	**72:** P22599r
3-(3-hydroxybenzo[*b*]thiophen-2-yl)-	**81:** P49560u
1-(hydroxyethyl)-	**82:** P170962a
1-[4-[(2-hydroxy-5-methylphenyl)azo]phenyl]-	**79:** 54855g
1-[(2-hydroxy-2-phenyl)ethyl]-	**69:** P77265x
1-[[2-(4-iodophenyl)-1,3-dioxolan-2-yl]methyl]-	**85:** P94368f
1-isobutyl-	**57:** 3982a
1-(*p*-isocyanatophenyl)-	**72:** P100715s
1-(2-isopropoxy-1,4-dioxonaphthalen-3-yl)-	**81:** P8428p
1-[9-(isopropoxycarbonyl)-9-fluorenyl]-	**75:** P76791c, **75:** P110320k
1-isopropyl-	**57:** 3982a
-1-methanol, α-(trichloromethyl)-	**83:** P193337k
-1-methanol, α-(trichloromethyl)-, butylcarbamate ester	**83:** P193337k
-1-methanol, α-(trichloromethyl)-, tert-butylcarbamate ester	**83:** P193337k
-1-methanol, α-(trichloromethyl)-, (2-chloroethyl)carbamate ester	**83:** P193337k
-1-methanol, α-(trichloromethyl)-, cyclohexylcarbamate ester	**83:** P193337k
-1-methanol, α-(trichloromethyl)-, dimethylcarbamate ester	**83:** P193337k
-1-methanol, α-(trichloromethyl)-, ethylcarbamate ester	**83:** P193337k
-1-methanol, α-(trichloromethyl)-, isopropylcarbamate ester	**83:** P193337k
-1-methanol, α-(trichloromethyl)-, methoxymethylcarbamate ester	**83:** P193337k
-1-methanol, α-(trichloromethyl)-, (2-methoxymethyl)carbamate ester	**83:** P193337k
-1-methanol, α-(trichloromethyl)-, methylcarbamate ester	**83:** P193337k
-1-methanol, α-(trichloromethyl)-, (2-propenyl)carbamate ester	**83:** P193337k
-1-methanol, α-(trichloromethyl)-, propylcarbamate ester	**83:** P193337k
1-(2-methoxy-1,4-dioxonaphthalen-3-yl)-	**78:** 84321b
1-[9-(methoxycarbonyl)-9-fluorenyl]-	**75:** P76791c, **75:** P110320k
1-[(4-methoxycarbonyl)phenoxy]-3,3-dimethyl-2-oxobut-1-yl	**83:** P206289y
1-[1-(2-methoxyphenoxy)-3,3-dimethyl-2-oxobut-1-yl]-	**83:** P179070m
1-[3-(*p*-methoxyphenyl)coumarin-7-yl]-	**71:** P14180m
1-[[2-(2-methoxyphenyl)-1,3-dioxolan-2-yl]methyl]-	**85:** P94368f
1-[[2-(3-methoxyphenyl)-1,3-dioxolan-2-yl]methyl]-	**85:** P94368f
1-[[2-(4-methoxyphenyl)-1,3-dioxolan-2-yl]methyl]-	**85:** P94368f
1-methyl-	**49:** 1711d, **52:** 364e,

TABLE 1   (*Continued*)

Compound	Reference

1.1. 1-Alkyl- or Aryl-substituted 1H-1,2,4-triazoles   (*Continued*)

1-methyl- (*continued*)	**64:** 19596g, **68:** 29645d, **73:** 45425r, **78:** 4224j, **80:** 47869g, **83:** 27047w
1-[(3'-methyl[1,1'-biphenyl]-4-yl)benzyl]-	**85:** P160095t
1-[(3-methyl-5-isoxazolyl)diphenylmethyl]-	**75:** P76798k
1-[(3-methyl-5-isoxazolyl)[4-(methylthio)-phenyl]benzyl]-	**75:** P76798k, **76:** P59632e
1-[α-(5-methyl-3-isoxazolyl)-*p*-(methylthio)-α-phenylbenzyl]-	**75:** P76798k, **76:** P59632e
1-[*m*-methyl-α-(1-methylimidazol-2-yl)-α-phenylbenzyl]-	**75:** P76798k
1-[[2-(2-methylphenyl)-1,3-dioxolan-2-yl]methyl]-	**85:** 94368f
1-[[2-(3-methylphenyl)-1,3-dioxolan-2-yl]methyl]-	**85:** P94368f
1-[[2-(4-methylphenyl)-1,3-dioxolan-2-yl]methyl]-	**85:** P94368f
1-[1-(4-methylphenyl)-2-nitroethyl]-	**81:** P91539u
1-[7-(3-methylpyrazol-1-yl)-2-oxo-2H-1-benzopyran-3-yl]-	**83:** 105984r
1-[7-(5-methylpyrazol-1-yl)-2-oxo-2H-1-benzopyran-3-yl]-	**72:** P22599r
1-[2-(morpholine)ethyl]-	**80:** 65b
1-[[2-(2-naphthalenyl)-1,3-dioxolan-2-yl]methyl]-	**85:** P94368f
1-[4-[2-(1-naphthalenyl)ethenyl]phenyl]-	**84:** P91665u
1-[4-[2-(2-naphthalenyl)ethenyl]phenyl]-	**84:** P91665u
1-[4'-(2H-naphtho[1,2-*d*]triazol-2-yl)-2,2'-disulfostilben-4-yl]-	**74:** P43528y
1-[7-(2H-naphtho[1,2-*d*]triazol-2-yl)-2-oxo-2H-1-benzopyran-3-yl)-	**74:** P65595v
1-[3-[(1,4-naphthoquinon-2-yl)amino]propyl]-	**71:** 2032b
1-(2-naphthylmethyl)-	**57:** 3982a
1-(2-nitrophenyl)-	**72:** 55340c, **72:** P79056r
1-(3-nitrophenyl)-	**72:** 55340c
1-(4-nitrophenyl)-	**57:** 3982a, **65:** 705a, **68:** 29646e, **72:** 55340c, **72:** P79056r, **75:** 35894v, **78:** 72019w, **82:** 139570x
1-[[2-(3-nitrophenyl)-1,3-dioxolan-2-yl]methyl]-	**85:** P94368f
1-[[2-(4-nitrophenyl)-1,3-dioxolan-2-yl]methyl]-	**85:** P94368f
1-(2-nitro-1-phenylethyl)-	**81:** P91539u
1-(3-nitro-2-pyridinyl)-	**72:** P79056r
1-[2-nitro-1-[2-trifluoromethyl)phenyl]ethyl]-	**81:** P91539u
1-[2-oxo-7-(5-phenylpyrazol-1-yl)-2H-1-benzopyran-3-yl]-	**72:** P22599r
1-[α-(phenacyl)benzyl]-	**50:** 3418a
1-phenethyl-	**57:** 3982a
1-[α-(phenoxy)phenacyl]-	**79:** P105257y, **80:** P146169k,
1-phenyl-	**57:** 3982a, **62:** 6476c, **83:** P206289y, **84:** 90078t

TABLE 1 (*Continued*)

Compound	Reference

1.1. 1-Alkyl- or Aryl-substituted 1H-1,2,4-triazoles (*Continued*)

1-[4-[(5-phenyl-2H-benzotriazol-2-yl)vinylene]phenyl]- **74:** P100066x, **75:** P110997m
1-(3-phenylcoumarin-7-yl)- **71:** P14180m
1-[(2-phenyl-1,3-dioxolan-2-yl)methyl]- **85:** P94368f
1-[7-(3-phenylpyrazol-1-yl)coumarin-3-yl]- **72:** P22599r
1-[2-(phthalimido)ethyl]- **50:** 1785h
1-(phthalimido)methyl- **50:** 1785h
1-[3-(phthalimido)propyl]- **50:** 1785i
1-picryl- **68:** 105111u
1-[2-(piperidino)ethyl]- **80:** 65b
-1-propanol, α-tert-butyl-β-(4-chloro-2-methylphenoxy)- **83:** P43338x, **83:** P58837h
-1-propanol, α-tert-butyl-β-(2-chlorophenoxy)- **83:** P43338x, **83:** P58837h
-1-propanol, α-tert-butyl-β-(3-chlorophenoxy)- **83:** P43338x, **83:** P58837h
-1-propanol, α-tert-butyl-β-(4-chlorophenoxy)- **83:** P43338x, **83:** P58837h
-1-propanol, α-tert-butyl-β-(4-chlorophenoxy)-α-methyl- **83:** P43338x, **83:** P58837h
-1-propanol, α-tert-butyl-β-(2,4-dichlorophenoxy)- **83:** P43338x, **83:** P58837h
-1-propanol, α-tert-butyl-β-(2,5-dichlorophenoxy)- **83:** P43338x, **83:** P58837h
-1-propanol, α-tert-butyl-β-(2,3-dimethylphenoxy)- **83:** P43338x, **83:** P58837h
-1-propanol, α-tert-butyl-β-(4-fluorophenoxy)- **83:** P43338x, **83:** P58837h
-1-propanol, α-tert-butyl-β-phenoxy- **83:** P43338x, **83:** P58837h
-1-propanol, β-(2-chlorophenoxy)-α- **84:** P169551a
  (1,1-dimethylethyl)-
-1-propanol, β-(4-chlorophenoxy)-α- **84:** P169551a
  (1,1-dimethylethyl)-α-methyl-
-1-propanol, α-(1,1-dimethylethyl)-β- **84:** P169551a
  (2,3-dimethylphenoxy)-
-1-propanol, α-(1,1-dimethylethyl)-β- **84:** P169551a
  (4-fluorophenoxy)-
-1-propionamide, N-(3β-hydroxyandrostan-6β-yl)-, **71:** 113155p
  acetate ester
-1-propionic acid **50:** 3418a
1-propionic acid, ethyl ester **71:** 113155p
1-(2-propoxy-1,4-dioxonaphthalen-3-yl)- **78:** 84321b
1-[3-(p-propoxyphenyl)coumarin-7-yl]- **71:** P14180m
1-propyl- **57:** 3982a
1-[7-(pyrazol-1-yl)-2-oxo-2H-1-benzopyran-3-yl]- **72:** P22599r
1-[pyridin-2,6-diyl]bis- **77:** 152071b
1-(2-pyridinyl)- **72:** 55340c, **77:** 152071b
1-(3-pyridinyl)- **72:** 55340c
1-(4-pyridinyl)- **72:** 55340c
1-[2-(1-pyrrolidinyl)ethyl]- **80:** 65b
1-β-D-ribofuranosyl- **73:** 66847v
1-β-D-ribofuranosyl-, 2′, 3′, 5′-tribenzoate **73:** 66847v
1-[4′-(6-sulfo-2H-naphtho[1,2-d]triazol-2-yl)-2,2′- **74:** P43528y
  disulfostilben-4-yl]-
1-[4′-(7-sulfo-2H-naphtho[1,2-d]triazol-2-yl)-2,2′- **74:** P43528y
  disulfostilben-4-yl]-
1,1′-[terephthaloylbis(imino-p-phenylene)]bis- **72:** P100715s

23

TABLE 1   (*Continued*)

Compound	Reference

1.1. 1-Alkyl- or Aryl-substituted 1$\underline{H}$-1,2,4-triazoles   (*Continued*)

Compound	Reference
1-(2-thiazolyl)-	**46:** 5581g
1-[[2-(2-thienyl)-1,3-dioxolan-2-yl]methyl]-	**85:** P94368f
1-*p*-tolyl-	**62:** 6476c
1-[3-(*p*-tolyl)coumarin-7-yl]-	**71:** P14180m
1-(2,3,5-tri-*O*-benzoyl-β-D-ribofuranosyl)-	**85:** 193010q
1-[1-(2,4,5-trichlorophenyl)-3,3-dimethyl-2-oxo-1-butyl]-	**82:** P156325p
1-[[2-(2,3,4-trichlorophenyl)-1,3-dioxolan-2-yl]methyl]-	**85:** P94368f
1-[[2-(2,4,5-trichlorophenyl)-1,3-dioxolan-2-yl]methyl]-	**85:** P94368f
1-(triphenylmethyl)-	**74:** 100062t, **74:** 112048f, **74:** 125697s, **74:** P125698t, **76:** P59632e
1-[tris(*p*-chlorophenyl)methyl]-	**74:** P100062t
1-trityl-	**57:** 807f, **74:** P100062t
1-vinyl-	**59:** 8880c, **63:** 7119b, **83:** P114414g, **83:** P114416j

1.2. 3-Alkyl- or Aryl-substituted *s*-triazoles

Compound	Reference
-3-acetamide	**49:** 13979d
-3-acetamide, *N,N*-dimethyl-	**49:** 13979d
-3-acetamide, *N*-methyl-	**49:** 13979d
3-[2-(acetamido)ethyl]-	**48:** 12739c
-3-acetic acid, ethyl ester	**49:** 13979c, **71:** 38867s
-3-acetonitrile	**50:** 16789e, **73:** 87861k, **83:** 9953j
-3-alanine	**50:** 1784b
3-[2-[(amidino)thio]ethyl]-	**75:** P20399p, **79:** P137155e
3-(4-amino-1-benzylimidazol-5-yl)-	**63:** 18084h
3-(4-aminobutyl)-	**76:** P72514n, **85:** P104199t
3-(2-amino-5-chlorophenyl)-	**82:** P57724m, **82:** P171103h
3-(2-amino-6-chlorophenyl)-	**77:** 88421m
3-(1-aminoethyl)-	**49:** 13980b
3-(2-aminoethyl)-	**48:** 12739c, **78:** P136290n
3-[[(2-aminoethyl)thio]methyl]-	**77:** P164704y
3-(4-aminoimidazol-5-yl)-	**81:** 169748h
3-[5(or 4)-aminoimidazol-4(or 5)-yl]-	**63:** 18084h
3-(aminomethyl)-	**49:** 13980b, **61:** 656d, **85:** 5552w, **85:** 46578c
3-(2-amino-1-methylethyl)-	**49:** 13980b
3-(5-amino-1-methyl-1$\underline{H}$-imidazole-4-yl)-	**81:** 169748h, **83:** 131874z
3-(*m*-aminophenyl)-	**65:** 705a,
3-(*o*-aminophenyl)-	**65:** 705a, **73:** 120582v, **77:** 88421m
3-(*p*-aminophenyl)-	**64:** 5072h, **65:** 705a

TABLE 1 (*Continued*)

Compound	Reference

1.2. 3-Alkyl- or Aryl-substituted *s*-triazoles (*Continued*)

Compound	Reference
3-(2-aminopropyl)-	**49:** 13980b
3-(3-aminopropyl)-	**49:** 13980b, **78:** P136290n
3-(2-amino-3-pyridinyl)-	**81:** 91461n
3-(3-amino-2-pyridinyl)-	**81:** 91461n
3-(5-amino-1-β-D-ribofuranosyl-1$\underline{H}$-imidazole-4-yl)-	**81:** 169748h, **83:** 131874z
3-(4-amino-7-β-D-ribofuranosyl-7$\underline{H}$-pyrrolo[2,3-$\underline{1}$]-pyrimidin-5-yl)	**83:** 22239a
3-(2-azidophenyl)-	**77:** 88421m
3-[2-(benzamido)ethyl]-	**48:** 12740e
3-[(benzamido)methyl]-	**71:** 38867s, **71:** P124441e
3-[[4-(2-benzoyl-4-chlorophenyl)-5-[(phenylmethoxy)-methyl]-1,3(2$\underline{H}$)-dioxo-1$\underline{H}$-isoindol-2-yl]methyl]-	**83:** P79295v
3-benzyl-	**69:** 18568a, **71:** 38867s, **71:** P124441e, **54:** 9898f
3-[2-(benzylamino)ethyl]-	**48:** 12739c
3-benzyl-4-benzyloxy-	**77:** 139906h, **80:** 47060m
3-benzyl-4-hydroxy-	**74:** 111967t, **74:** 141651c
3-benzyl-, 4-oxide	**74:** 111182h
3-(*p*-bromophenyl)-4-hydroxy-	**74:** 141651c
3-(2-butoxyphenyl)-	**79:** 115500m
3-butyl-	**69:** 18568a
3-[2-(carboxymethyl)phenyl]-	**76:** 72459y
3-(2-carboxyphenyl)-	**76:** 72456v, **76:** 85193g
3-(chlorodiphenylmethyl)-	**80:** 133389g
3-(2-chloroethyl)-	**49:** 13979f
3-(chloromethyl)-	**50:** 1784g, **57:** 5903i, **83:** 9953j
3-(2-chlorophenyl)-	**82:** 170243s
3-(3-chlorophenyl)-	**81:** P13524w
3-(*p*-chlorophenyl)-	**44:** 2515e, **71:** 38867s, **71:** P124441e
3-(2-cyanoethenyl)-	**78:** 136179h
3-(*o*-cyanophenyl)-	**72:** 12671u
3-(2,2-dichloro-1,1-difluoroethyl)-	**81:** 49647c
3-(3,4-dichlorophenyl)-	**81:** P13524w
3-[2-(diethylamino)ethyl]-	**49:** 13980a
3-(difluoronitromethyl)-	**59:** 3927e
3-(4,5-dihydro-1$\underline{H}$-pyrazol-3-yl)-	**78:** 29701a
3-[2-(dimethylamino)ethyl]-	**49:** 13979f, **75:** 96997c
3-[(dimethylamino)methyl]-	**82:** P140121h
3-(2,2-dimethylpropyl)-	**78:** 77189m
3-[4-(1,3(2$\underline{H}$)-dioxo-1$\underline{H}$-isoindol-2-yl)butyl]-	**76:** P72514n
3-(2-ethoxyethyl)-	**49:** 13979h, **79:** P137155e
3-(ethoxymethyl)-	**50:** 1785a
3-(2-ethoxyphenyl)-	**79:** 115500m
3-ethyl-	**43:** 7019c, **68:** 68940r, **69:** 18568a, **71:** 38867s

TABLE 1 (*Continued*)

Compound	Reference

1.2. 3-Alkyl- or Aryl-substituted *s*-triazoles (*Continued*)

Compound	Reference
3-[2-(ethylamino)ethyl]-	**49:** 13980a
3-(2-furyl)-	**52:** 16341c, **71:** P124441e, **73:** 87861k
3-(2-guanidinoethyl)-	**75:** P118317k, **71:** 38867s, **78:** P136290n
3-(3-guanidinopropyl)-	**75:** P118317k, **78:** P136290n
3-heptadecyl-	**55:** P26807a
4-hydroxy-3-(*p*-methoxyphenyl)-	**74:** 141651c
4-hydroxy-3-phenyl-	**74:** 111967t, **74:** 141651c
3-isopropyl-	**69:** 18568a, **71:** 38867s, **71:** P124441e
3-[2-(isopropylamino)ethyl]-	**48:** 12739c
-3-methanol	**50:** 1784b, **57:** 5903i
-3-methanol, α,α-diphenyl-	**80:** 133389g
-3-methanol, α-phenyl-	**68:** 105162m
3-(methoxymethyl)-	**50:** 1785a, **71:** 38867s, **71:** P124441e
3-(2-methoxyphenyl)-	**79:** 115500m
3-(*p*-methoxyphenyl)-	**44:** 2515e
4-methoxy-3-phenyl-	**77:** 139906h, **80:** 47060m
4-(methoxy-d_3)-3-phenyl-	**80:** 47060m
3-methyl-	**48:** 1344a, **50:** 1784d, **65:** 8898b **68:** 29645d, **68:** 68940r, **69:** 18568a **70:** 77876t, **71:** 38867s, **73:** 87861k, **79:** 18620t **80:** 96082d, **83:** 114298x, **85:** 46578c
3-[2-(methylamino)ethyl]-	**49:** 13979f, **75:** 96997c
3-[2-(3-methylbutoxy)phenyl]-	**79:** 115500m
3-[2-(1-methylethoxy)phenyl]-	**79:** 115500m
3-[2-(2-methylpropoxy)phenyl]-	**79:** 115500m
3-(2-methylpropyl)-	**78:** 77189m
3-[4-[(*N*-methylthiocarbamoyl)amino]butyl]-	**76:** P72514n
3-[[2-[(*N*-methylthiocarbamoyl)amino]ethylthio]methyl]-	**77:** P164704y
3-(2-naphthyl)-	**71:** 38867s, **71:** P124441e
3-[2-(5-nitro-2-furanyl)ethenyl]-	**80:** 10273g
3-(5-nitro-2-furyl)-	**64:** 19596c, **70:** P87037j, **75:** 20356x
3-(*m*-nitrophenyl)-	**65:** 705a
3-(*o*-nitrophenyl)-	**65:** 705a, **78:** 72019w
3-(*p*-nitrophenyl)-	**65:** 705a, **70:** 3966a, **78:** 72019w
-3-nonanoic acid	**63:** 1887g
3-pentyl-	**57:** 7253i
3-[2-(pentyloxy)phenyl]-	**79:** 115500m
3-phenoxymethyl-	**50:** 1784i

TABLE 1   (*Continued*)

Compound	Reference

1.2. 3-Alkyl- or Aryl-substituted *s*-triazoles   (*Continued*)

Compound	Reference
3-phenyl-	**22:** 4123[7], **48:** 4525b, **65:** 8898b, **71:** 38867s, **71:** 91386k, **71:** P124441e, **74:** 3559c, **74:** P100066x, **78:** 72019w, **83:** 114298x **84:** 135602c
3-phenyl-, 4-oxide	**74:** 111182h, **77:** 139906h, **80:** 47060m
3-phenyl-1-(tributylstannyl)-	**78:** 29956n
3-[1-(phthalimido)ethyl]-	**49:** 13979a
3-[2-(phthalimido)ethyl]-	**48:** 12740a
3-[(phthalimido)methyl]-	**49:** 13979a, **61:** 656b
3-[2-(phthalimido)propyl]-	**49:** 13979b
3-[3-(phthalimido)propyl]-	**49:** 13979b
-3-propanoic acid	**69:** 16431b
-3-propanoic acid, α-amino-	**74:** 39709r
-3-propionic acid, α-amino-	**66:** 73426r
-3-propionic acid, α-carboxy-α-formamido-, diethyl ester	**50:** P12116c
3-(2-propoxyphenyl)-	**79:** 115500m
3-propyl-	**69:** 18568a
3-(1H̲-pyrazol-3-yl)-	**82:** 31304z
3-(3-pyridyl)-	**70:** 3966a, **71:** 38867s, **71:** P124441e
3-(4-pyridyl)-	**49:** 10938a, **62:** 11804g, **70:** 3966a, **71:** 38867s, **71:** P124441e
3-*m*-totyl-	**71:** P124441e
3-*p*-tolyl-	**71:** 38867s
3-(trifluoromethyl)-	**85:** 138543r
3-[2-(trimethylsilyl)ethyl]-	**80:** 96082d
3-[2-(ureido)ethyl]-	**48:** 12740d

1.3. 4-Alkyl- or Aryl-substituted 4H̲-1,2,4-triazoles

Compound	Reference
-4-acetic acid, α-[(3-carboxy-2-hydroxy-1-naphthyl)-methylene]-, δ-lactone, ethyl ester	**59:** P15420h
-4-acetic acid, α-[(2-hydroxy-1-naphthyl)methylene]-, δ-lactone	**59:** P15420h
4-(*p*-aminophenyl)-	**65:** 705a
4-(*o*-aminophenyl)-	**78:** 72019w
4-(4-aminophenyl)-	**81:** 62940n
4-(1-anilino-2,2,2-trichloro-1-ethyl)-	**79:** 5280g
4-(anilinomethyl)-	**79:** 5280g
4-(2-benzothiazolyl)-	**76:** P113224p
4-(2-benzoyl-4-chlorophenyl)-	**74:** 125579e, **79:** 92176u, **83:** 193188n

27

TABLE 1 (*Continued*)

Compound	Reference

1.3. 4-Alkyl- or Aryl-substituted 4H̲-1,2,4-triazoles (*Continued*)

Compound	Reference
4-benzyl-	**57:** 3982a, **67:** 90738g, **72:** P100713q, **76:** P113224p, **82:** P150485u
4-[1,1'-biphenyl]-2-yl-	**78:** 72019w
4-[1,1'-biphenyl]-3-yl-	**78:** 72019w
4-[1,1'-biphenyl]-4-yl-	**78:** 72019w
4-(p-bromophenyl)-	**57:** 3982a
4-butyl-	**67:** 90738g, **72:** P100713q, **76:** P113224p, **82:** P150485u
4-(sec-butyl)-	**76:** P113224p
4-(tert-butyl)-	**76:** P113224p
4-(3-carboxy-4-hydroxyphenyl)-	**81:** P63478e
4-(4-chloro-2-benzothiazolyl)-	**76:** P113224p
4-[1-(4-chlorobenzamido)-2-oxo-2-(2-thienyl)ethyl]-	**77:** 101459b
4-(p-chlorobenzyl)-	**72:** P100713q, **76:** P113224p
4-[(4'-chloro[1,1'-biphenyl]-4-yl)phenylmethyl]-	**85:** P160095t
4-[2-(4-chlorophenoxy)ethyl]-	**76:** P113224p
4-(m-chlorophenyl)-	**57:** 3982c
4-(o-chlorophenyl)-	**57:** 3982a
4-(p-chlorophenyl)-	**57:** 3982a
4-[2-[(4-chlorophenyl)thio]ethyl]-	**76:** P113224p
4-cyclopropyl-	**76:** P113224p
4-decyl-	**72:** P100713q, **76:** P113224p, **82:** P150485u
4-[2-(2,4-dichlorophenoxy)ethyl]-	**76:** P113224p
4-(2,4-dichlorophenyl)-	**57:** 3982a, **72:** P100713q
4-(3,4-dichlorophenyl)-	**72:** P100713q, **76:** P113224p
4-[(2,4-dichlorophenyl)methyl]-	**72:** P100713q
4-[(3,4-dichlorophenyl)methyl]-	**72:** P100713q, **76:** P113224p
4-[2-[(3,4-dichlorophenyl)thio]ethyl]-	**76:** P113224p
[(2-diethylamino)ethyl]-	**76:** P113224p
4-(2,3-dihydro-3,3-dimethyl-2,5-dihydroxynaphtho-[1,2-b]furan-4-yl)-	**78:** 84321b
4-(1,1-dimethyl-2-propynyl)-	**76:** P113224p
4-dodecyl-	**72:** P100713q, **82:** P150485u
4-(2-ethoxyethyl)-	**72:** P100713q, **76:** P113224p, **82:** P150485u
4-(p-ethoxyphenyl)-	**50:** 967b
4-(3-ethoxypropyl)-	**72:** P100713q, **76:** P113224p, **82:** P150485u
4-ethyl-	**79:** 78696g
4-(2-ethylhexyl)-	**72:** P100713q, **76:** P113224p, **82:** P150485u
4-hexyl-	**72:** P100713q, **76:** P113224p, **82:** P150485u
4-[4-[(2-hydroxy-5-methylphenyl)azo]phenyl]-	**79:** 54855g
4-[4-[(4-hydroxyphenyl)azo]phenyl]-	**79:** 54855g

TABLE 1 (*Continued*)

Compound	Reference

1.3. 4-Alkyl- or Aryl-substituted 4H-1,2,4-triazoles (*Continued*)

Compound	Reference
4-isobutyl-	**72:** P100713q, **82:** P150485u
4-isopropyl-	**73:** 45425r
4-(*m*-methoxyphenyl)-	**57:** 3982c
4-(*o*-methoxyphenyl)-	**57:** 3982a
4-(*p*-methoxyphenyl)-	**57:** 3982a
4-[(4-methoxyphenyl)methyl]-	**76:** P113224p
4-[2-[(4-methoxyphenyl)thio]ethyl]-	**76:** P113224p
4-(3-methoxypropyl)-	**72:** P100713q, **76:** P113224p, **82:** P150485u
4-methyl-	**65:** 12205a, **68:** 29645d, **73:** 45425r
4-(3-methylbutyl)-	**72:** P100713q, **76:** P113224p, **82:** P150485u
4-[3-(1-methylethoxy)propyl]-	**72:** P100713q, **76:** P113224p, **82:** P150485u
4-(3-methyl-1-phenylpyrazol-5-yl)-	**54:** 2333e
4-(2-nitrophenyl)-	**72:** 55340c
4-(3-nitrophenyl)-	**76:** P113224p
4-(4-nitrophenyl)-	**68:** 29646e, **57:** 3982a
4-octyl-	**72:** P100713q, **76:** P113224p, **82:** P150485u
4-pentyl-	**76:** P113224p
4-(2-phenoxyethyl)-	**76:** P113224p
4-phenyl-	**23:** 836[3], **53:** 21904i, **72:** P100713q
4-(2-phenylethyl)-	**76:** P113224p
4-[2-(phenylimino-3-thiazolidinyl]-	**74:** 87901u
4-[5(or 3)-phenylpyrazol-3(or 5)-yl]-	**54:** 2333e
-4-propanoic acid, β-methyl-, ethyl ester	**76:** P113224p
4-propanol	**72:** P100713q, **76:** P113224p
4-propyl-	**72:** P100713q, **76:** 113224p, **82:** P150485u
4-(2-propynyl)-	**76:** P113224p
4-pyrazinyl-	**73:** 66512g
4-(2-pyridyl)-	**72:** 55340c, **48:** 12092c
4-(3-pyridyl)-	**72:** 55340c, **48:** 12092c
4-(4-pyridyl)-	**72:** 55340c, **76:** P113224P
4-(2-pyrimidinyl)-	**76:** P113224p
4-(2-thiazolyl)-	**76:** P113224p
4-*m*-tolyl-	**57:** 3982c
4-*o*-tolyl-	**57:** 3982a
4-*p*-tolyl-	**57:** 3982a
4-(2,4,6-trimethylphenyl)-	**82:** 156003g

CHAPTER 2

Alkyl- or Aryl-Disubstituted 1,2,4-Triazoles

2.1. 1,3-Alkyl- or Aryl-Disubstituted 1H-1,2,4-Triazoles

The cyclization of amidrazone hydrochlorides with formic acid provided an unambiguous method for the synthesis of 1,3-disubstituted triazoles. For example, both the 1-methyl- (**2.1-1**) and 1-phenyl- (**2.1-2**) amidrazones are

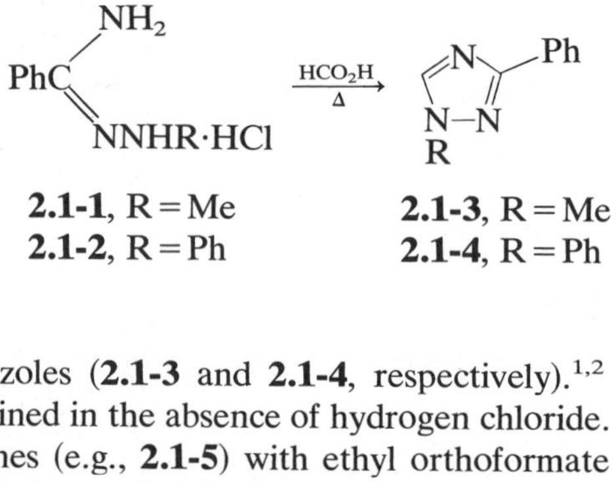

2.1-1, R = Me   **2.1-3**, R = Me
2.1-2, R = Ph   **2.1-4**, R = Ph

converted to triazoles (**2.1-3** and **2.1-4**, respectively).[1,2] Lower yields of products are obtained in the absence of hydrogen chloride. Also, the cyclization of amidrazones (e.g., **2.1-5**) with ethyl orthoformate to give 1-aryl-3-[(benzyloxycarbonyl)amino]alkyl triazoles (e.g., **2.1-6**) has been reported.[3–6]

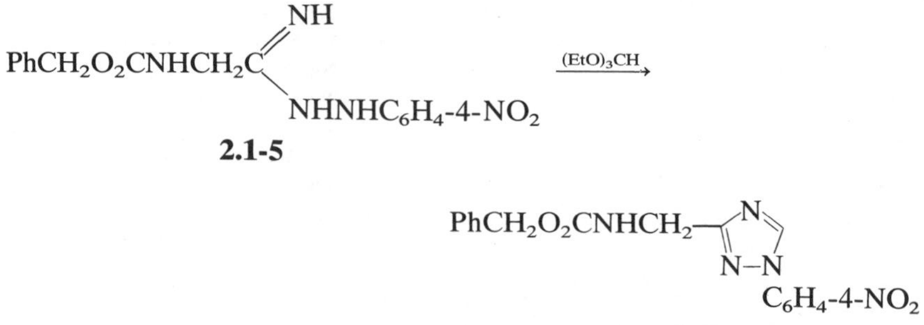

2.1-5

2.1-6

In more recent work, good yields of triazoles were obtained by the cyclization of amidrazones with s-triazine.[7]

Although the tritylation of 3-methyl-s-triazole appeared to give a good yield of 3-methyl-1-trityltriazole, it has been demonstrated that the alkylation of the sodium salt of **2.1-7** gave a mixture of the 1,3- (**2.1-8**) and 1,5- (**2.1-9**) isomers in the ratio of 1:2.[8] Alkylation of 3-phenyl-s-triazole

$$PhtNCH_2CH_2\text{-triazole(H)} \xrightarrow{MeI} PhtNCH_2CH_2\text{-triazole(Me)} + PhtNCH_2CH_2\text{-triazole(Me)}$$

2.1-7   **2.1-8**   **2.1-9**

(**2.1-10**) with methyl iodide, however, gave a mixture of 1,3- (**2.1-3**) and 1,5- (**2.1-11**) isomers in the ratio of 2:1 (route a).[2] In contrast, treatment of

$$Ph\text{-triazole(H)} \;\; \underset{\underset{b}{CH_2N_2}}{\overset{\overset{a}{MeI}}{\longrightarrow}} \;\; Ph\text{-triazole(Me)} + Ph\text{-triazole(Me)}$$

2.1-10   **2.1-3**   **2.1-11**

(a) 56%   (a) 28%
(b) 34%   (b) 56%

2.1-10 with diazomethane gave a mixture containing mainly the 1,5-isomer (**2.1-11**) (route b). The formation of a carbon-carbon bond was reported to occur in the reaction of 3-iodo-1-methyltriazole (**2.1-12**) with 2-methyl-3-butyn-2-ol in the presence of copper (I) iodide to give **2.1-13**, which was converted with base to 3-ethynyl-1-methyltriazole (**2.1-16**).[9]

$$\text{triazole(Me)-I} + HC{\equiv}CC(Me)_2OH \xrightarrow{CuI} \text{triazole(Me)-}C{\equiv}CC(Me)_2OH$$

2.1-12   **2.1-13**

$\Big| OH^-$

$$Me\text{-triazole(Me)-}NH_2 \xrightarrow[EtOH]{HNO_2} Me\text{-triazole(Me)} \qquad \text{triazole(Me)-}C{\equiv}CH$$

2.1-14   **2.1-15**   **2.1-16**

The removal of a functional group from a triazole system is a common method for the preparation of 1,3-disubstituted triazoles. Alcohol reduction of the diazotriazole resulting from diazotization of 5-amino-1,3-dimethyltriazole (**2.1-14**) gave a 36% yield of 1,3-dimethyltriazole (**2.1-15**).[10] In addition, the removal of the thioxo group from triazolines-5-thiones is often used to prepare 1,3-disubstituted triazoles. For example, treatment of either **2.1-17** with nitric acid[8,11] or **2.1-19** with Raney nickel[12,13] gave the desulfurated triazoles, **2.1-18** and **2.1-3**, respectively. The oxidative removal of the thione group has also been effected with hydrogen peroxide.[14]

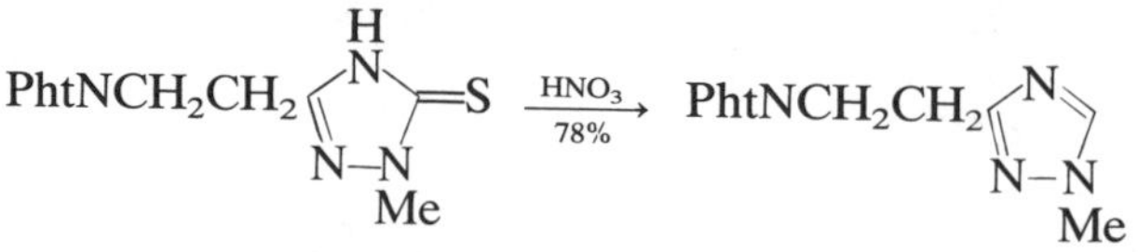

2.1-17   **2.1-18**

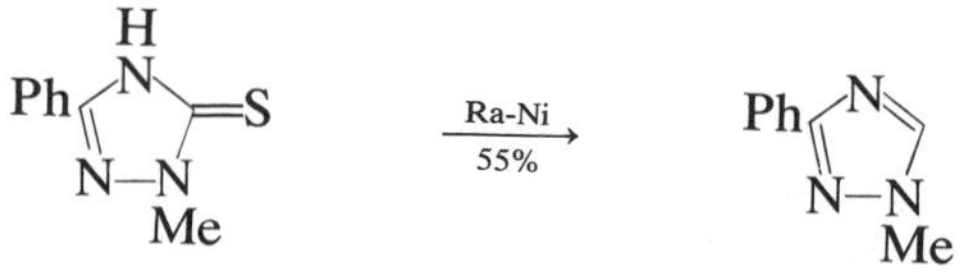

2.1-19   **2.1-3**

An interesting intramolecular rearrangement was observed in N-(isoxazolyl)formamidines (e.g., **2.1-20**), which on heating underwent rupture of the N—O bond of the isoxazole ring followed by N—N bond formation between the resulting electron-deficient nitrogen and an amidine nitrogen to give 1-aryl-3-acetonyltriazoles (e.g., **2.1-21**).[15] In the hydrazinolysis of 4-

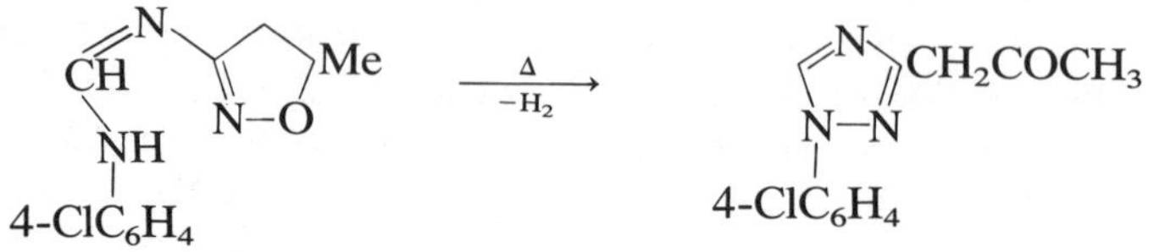

2.1-20   **2.1-21**

chloroquinazoline (**2.1-22**) with phenyl hydrazine, the intermediate 4-hydrazinoquinazoline (**2.1-23**) underwent ring opening at the 2-position of the pyrimidine ring, followed by recyclization of the resulting intermediate to give a 25% yield of **2.1-24**.[16]

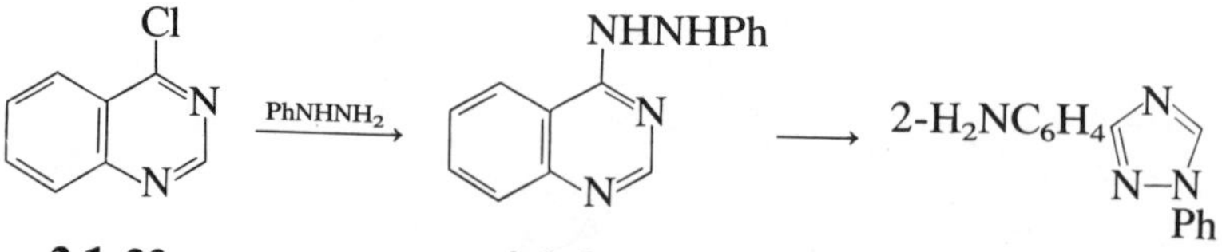

2.1-22   **2.1-23**   **2.1-24**

References

1. **49:** 3950c   2. **49:** 11630g   3. **64:** 19458a   4. **68:** 105162m
5. **75:** 118255p   6. **84:** 90081p   7. **79:** 18620t   8. **50:** 1785c
9. **83:** 28165v   10. **81:** 25613n   11. **68:** 29645d   12. **66:** 37843r
13. **77:** 164587n   14. **84:** 59312r   15. **60:** 13238g   16. **77:** 88421m

2.2. 1,5-Alkyl- or Aryl-Disubstituted 1H-1,2,4-Triazoles

The condensation of *N*-formylbenzamide (**2.2-1**) with 1-substituted hydrazines [Einhorn Brunner reaction], usually under acidic conditions, provided a direct route for the synthesis of 1-alkyl- (e.g., **2.2-3**)[1] and 1-aryl- (e.g., **2.2-4**)[2-4] 5-phenyltriazoles in good yields. The orientation of the substitutents in these products indicated the initial formation of a 1-substituted amidrazone intermediate (**2.2-2**).

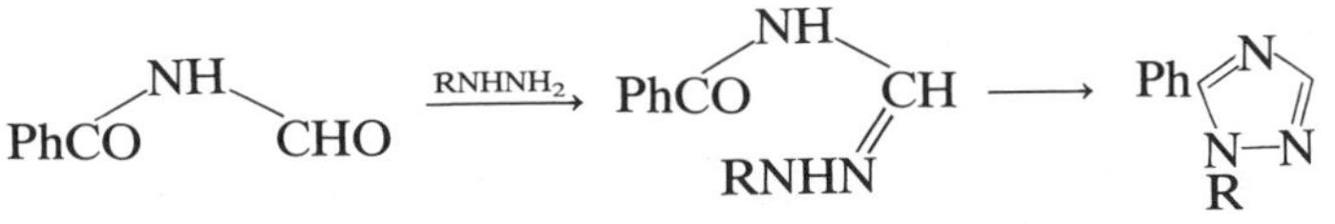

2.2-1	**2.2-2**	**2.2-3**, R = Me
		2.2-4, R = Ph

A variety of convenient methods have been developed for the preparation of 1,5-disubstituted Δ^4-triazolin-3-ones (see Section 13.2), which are potential intermediates for the synthesis of 1,5-disubstituted triazoles. For example, removal of the oxo function of the triazolin-3-one (**2.2-5**) with P_4S_{10}

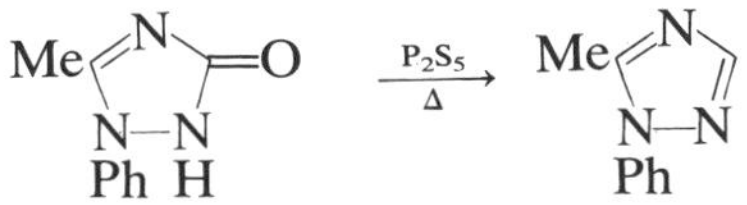

2.2-5   **2.2-6**

gave **2.2-6** (see Section 1.1 for limitation).[5] In addition, other preformed triazoles have been used as intermediates. Decarboxylation of the triazole-3-carboxylic acid (**2.2-7**) gave **2.2-4**,[6] and the ring opening of the tricyclic condensed triazole (**2.2-8**) gave **2.2-9**.[7] Lithiation of 1-(2-pyridyl)triazole

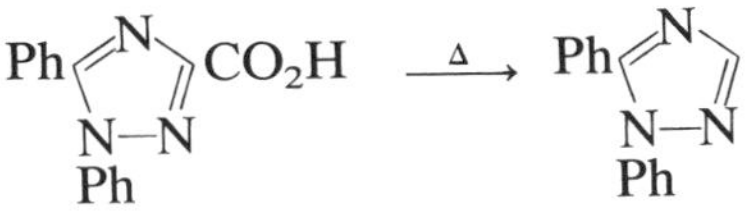

2.2-7   **2.2-4**

2.2-8   **2.2-9**

(**2.2-10**) with butyl lithium gave **2.2-11**, which underwent oxidative coupling in the presence of copper (II) chloride to give the 3,3'-bis(triazole) (**2.2-12**) (see Section 21.1).[8]

2.2-10   **2.2-11**   **2.2-12**

The alkylation of 3-substituted *s*-triazoles to give mixtures of 1,3- and 1,5-disubstituted triazoles has been discussed in Section 2.1. In addition, 2-methyl-3-butyn-2-ol was alkylated with 5-iodo-1-methyltriazole (**2.2-13**) in

2.2-13   **2.2-14**   **2.2-15**

the presence of copper (I) chloride to give **2.2-14**, which was converted with base to 5-ethynyl-1-methyltriazole (**2.2-15**).[9]

References

1. **49:** 11630g 2. **46:** 11147a 3. **47:** 4335a 4. **69:** 35596a
5. **57:** 3982a   6. **51:** 12079a 7. **67:** 43155e 8. **77:** 152071b
9. **83:** 28165v

2.3. 3,4-Alkyl- or Aryl-Disubstituted 4H-1,2,4-Triazoles

Several methods have been developed for the preparation of 3,4-disubstituted triazoles by intermolecular cyclization of two reactants. The condensation of diarylamidrazones (e.g., **2.3-3** and **2.3-4**) with purified ethyl formate, either at reflux or by heating in a sealed tube, provided a route to 3,4-diaryltriazoles (e.g., **2.3-1** and **2.3-2**).[1] Although the products are obtained in good yields, the formation of two types of by-products was observed. When the ethyl formate was contaminated with formic acid, self-condensation of **2.3-3** to give **2.3-5** resulted in the formation of 3,4,5-triphenyltriazole (**2.3-8**). In the preparation of **2.3-2**, self-condensation of **2.3-4** with the elimination of aniline gave **2.3-6**, which resulted in the formation of the tetrazine (**2.3-7**).

2.3-1, R = Ph   **2.3-3**, R = Ph   **2.3-5**
2.3-2, R = 2-C₅H₄N   **2.3-4**, R = 2-C₅H₄N

2.3-6   **2.3-7**   **2.3-8**

Treatment of the imino ether **2.3-9** with methyl- and benzylamines produced the 3,4-dialkyltriazoles (**2.3-12** and **2.3-13**), presumably formed via the 1-formylamidrazones (**2.3-10** and **2.3-11**, respectively).[2] These products were intermediates for the preparation of the corresponding 3-(aminomethyl)triazoles.

2.3-9

2.3-10, R = Me
2.3-11, R = PhCH₂

2.3-12, R = Me
2.3-13, R = PhCH₂

In a related reaction in which the 3-aryl group is derived from a hydrazide, treatment of the imino chloride (**2.3-14**) with **2.3-15** in the presence of triethylamine gave **2.3-17**.[3] In another procedure the intermediate corresponding to **2.3-16** was isolated. Reaction of the amidine (**2.3-18**) with

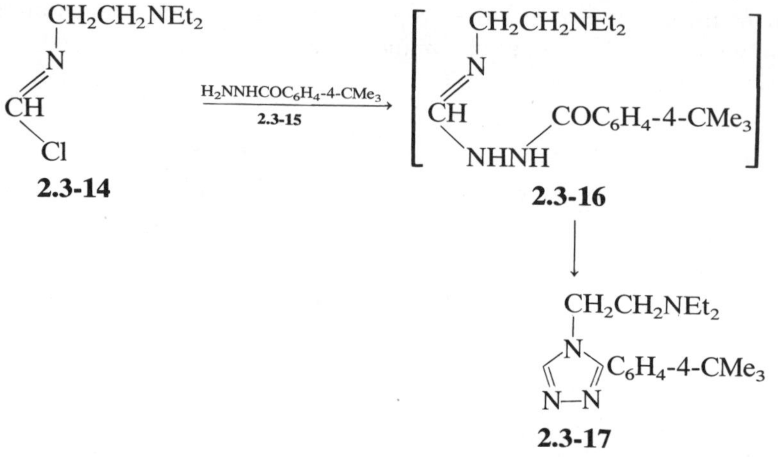
CH2CH2NEt2
N
CH
Cl
2.3-14
H2NNHCOC6H4-4-CMe3
2.3-15
CH2CH2NEt2
N
CH
COC6H4-4-CMe3
NHNH
2.3-16
CH2CH2NEt2
N
C6H4-4-CMe3
N—N
2.3-17

benzylhydrazide at 160° for 5 minutes gave **2.3-19**, which was heated at 250° for a short period of time to give **2.3-1**.[4] Similarly, treatment of **2.3-18** with an alkyl hydrazide gave **2.3-20**, which was converted to the 3-(cyanomethyl)triazole **2.3-21**. This type of acylamidrazone has also been cyclized in refluxing diglyme containing pyridine.[5,6] In recent work the

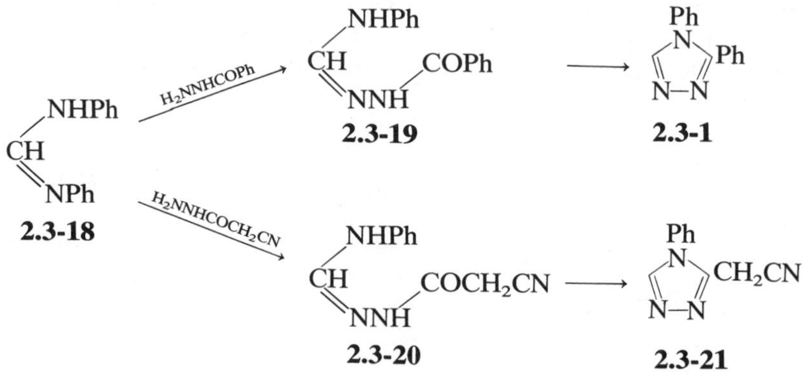
NHPh
CH
NPh
2.3-18
H2NNHCOPh
H2NNHCOCH2CN
NHPh
CH
COPh
NNH
2.3-19
Ph
N
Ph
N—N
2.3-1
NHPh
CH
COCH2CN
NNH
2.3-20
Ph
N
CH2CN
N—N
2.3-21

acylamidrazones have been generated by the successive treatment of an arylamine with triethylorthoformate and a hydrazide.[7]

The benzooxazin-4-ones (e.g., **2.3-22**) condensed with hydrazides to give quinazolin-4-ones (e.g., **2.3-23**), which undergo ring opening under basic conditions to give acylamidrazone intermediates followed by ring closure to give 3,4-disubstituted triazoles (e.g., **2.3-24**).[8] In related work, the benzodiazepines fused to an *s*-triazolo ring were found to be potent central

2.3-22   **2.3-23**   **2.3-24**

nervous system agents. These benzodiazepines are prepared from 4-(2-benzoylphenyl)triazole intermediates,[9] and a number of reports have appeared on the preparation of the latter. Although several variations in the procedure are possible,[10,11] the best method appears to involve the reaction of a 2-(acylamino)benzophenone (e.g., **2.3-25**) with hydrazine to give a dihydroquinazoline (e.g., **2.3-26**), followed by treatment of the latter with formic acid to effect formylation, ring opening, and ring closure to give a 4-(2-benzoylphenyl)triazole (e.g., **2.3-27**).[12] The type of compound has also been prepared by periodate oxidation of s-triazolo[4,3-a]quinolines[13-15] and

2.3-25   **2.3-26**   **2.3-27**

by chromic acid oxidation of the benzyl group of 4-(2-benzylphenyl)-triazoles.[7]

The preparation of 3,4-disubstituted triazoles from preformed triazoles has been reported and includes the saponification and decarboxylation of

2.3-28   **2.3-1**

2.3-28 to give **2.3-1**[16,17] and the oxidative decarboxylation of a 3-(cyanomethyl) group of a triazole with potassium permanganate.[8] The desulfurization of triazoline-5-thiones to give 3,4-disubstituted triazoles has been effected with both nitric acid[18,19] and hydrogen peroxide.[20] In the desulfurization of **2.3-29** with Raney nickel, both the sulfur and chlorine groups were simultaneously removed to give **2.3-30**.[21]

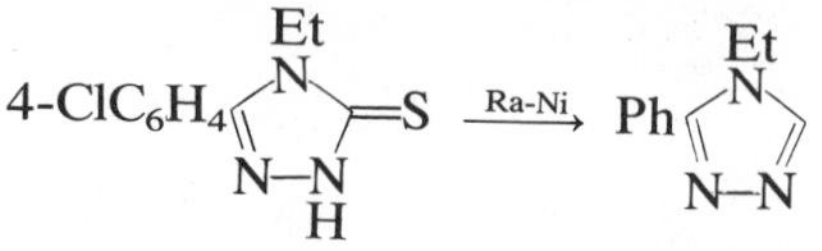

$$4\text{-ClC}_6\text{H}_4 \overset{\text{Et}}{\underset{\text{H}}{\overset{N}{\underset{N-N}{}}}}{=}S \xrightarrow{\text{Ra-Ni}} Ph \overset{\text{Et}}{\underset{N-N}{\overset{N}{}}}$$

2.3-29   **2.3-30**

References

1. **66:** 46412b
5. **71:** 70607t
9. **74:** 125579e
13. **77:** 126707u
17. **73:** 66517n
21. **74:** 111968u

2. **62:** 1647a
6. **78:** P136354m
10. **77:** 88557k
14. **77:** 126708v
18. **68:** 29645d

3. **62:** 9124b
7. **85:** 56539e
11. **79:** 92176u
15. **80:** 83083e
19. **81:** 49624t

4. **54:** 2332i
8. **72:** 43550h
12. **83:** 193188n
16. **54:** 21063g
20. **77:** 34439e

2.4. 3,5-Alkyl- or Aryl-Disubstituted s-Triazoles

The Pellizzari reaction, the thermal condensation of an acid hydrazide (e.g., **2.4-1**) with an amide (e.g., **2.4-2**), was one of the first methods used for the synthesis of 3,5-disubstituted-s-triazoles (e.g., **2.4-4**).* The low yields of triazoles usually obtained were attributed to transamination reactions between the acid hydrazide and the amide, which often lead to mixtures of triazoles.[1] The reaction of aryl thioamides (e.g., **2.4-3**) with aryl hydrazides (e.g., **2.4-1**) occurred at lower temperatures and gave higher yields of triazoles.[2–4]

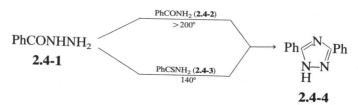

In another method developed early in work in the preparation of triazoles, 3,5-dimethyl-s-triazole (**2.4-6**) was obtained in low yield by the reaction of the s-diacylhydrazine (**2.4-5**) with ammonia in the presence of a dehydrating agent.† The use of excesss ammonium chloride and a higher

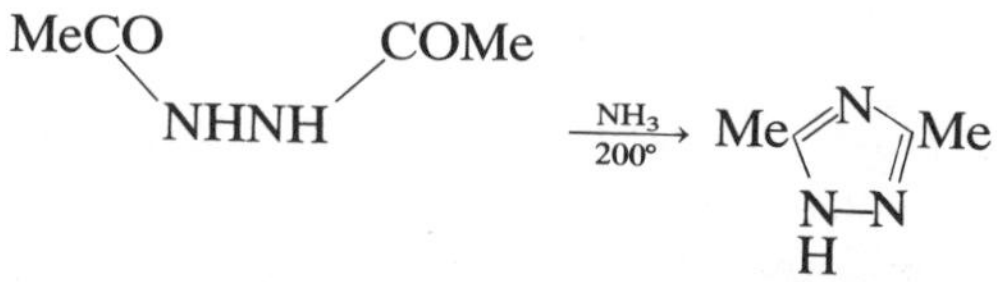

$$\underset{\textbf{2.4-5}}{\overset{\text{MeCO}\quad\quad\text{COMe}}{\underset{\text{NHNH}}{}}} \xrightarrow[200°]{\text{NH}_3} \underset{\textbf{2.4-6}}{Me\overset{N}{\underset{N-N}{\overset{}{}}}Me}$$

* G. Pellizzari, *Gazz. Chim. Ital.*, **41,** 20 (1911).
† R. Stolle, *Berichte*, **32,** 797 (1899).

reaction temperature resulted in higher yields with the s-diacylhydrazine (**2.4-8**) to give **2.4-7**, but not with the related s-diacylhydrazine (**2.4-9**) to give **2.4-10**.[5] Another reaction of this type involved the condensation of

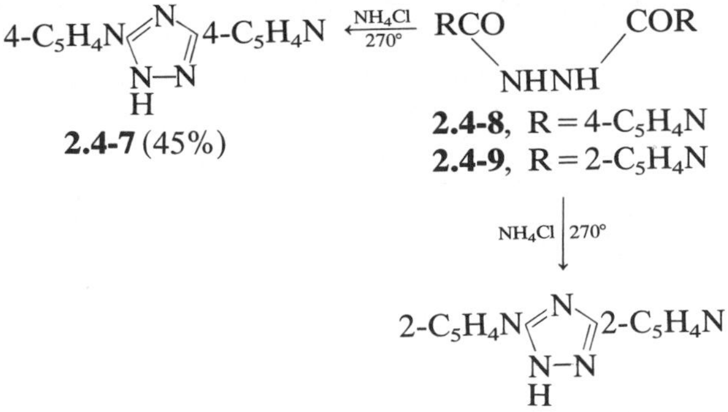

2.4-7 (45%)

2.4-8, R = 4-C$_5$H$_4$N
2.4-9, R = 2-C$_5$H$_4$N

2.4-10 (5%)

semicarbazide with the diacylamine (**2.4-11**) to give the triazole (**2.4-12**).[6] Monothio analogs of diacylamines also react readily with hydrazine to give good yields of 3,5-disubstituted triazoles.[7]

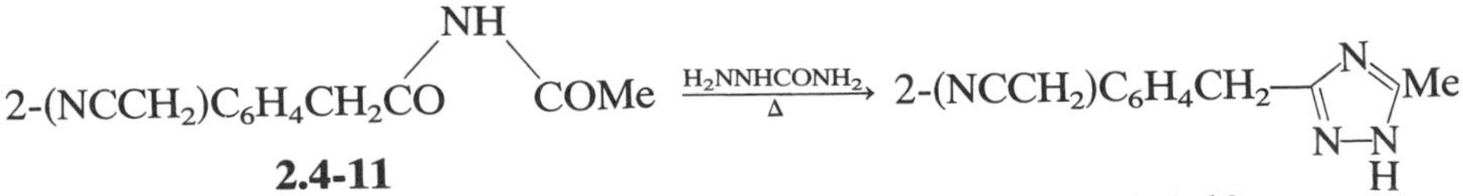

2.4-11

2.4-12

Considerable improvement in these types of condensations resulted from the discovery that equimolar amounts of an aryl nitrile (e.g., benzonitrile) and the aryl sulfonate salt of an aryl hydrazide (e.g., **2.4-13**) undergo fusion at 200° to give good yields of 3,5-diaryl-s-triazoles (e.g., **2.4-15**).[8,9] Apparently, acylamidrazones (e.g., **2.4-14**), which are known to dehydrate readily

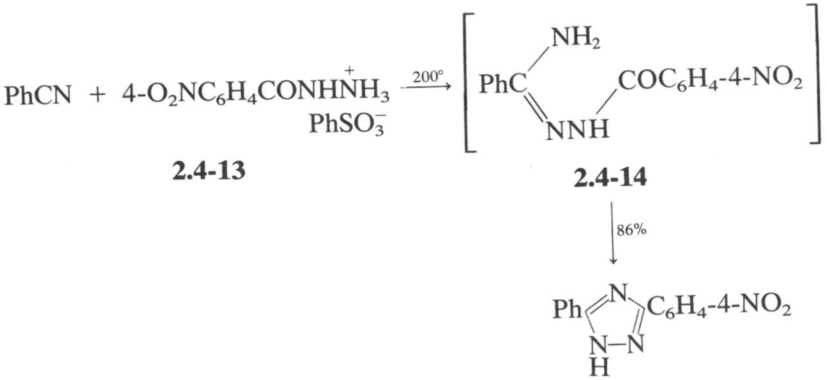

2.4-13

2.4-14

2.4-15

to give triazoles (discussed later in this section), are intermediates. Although the fusion of alkyl nitriles with the benzenesulfonate salts of aryl hydrazides to give 3-alkyl-5-aryl-s-triazoles was unsuccessful, these s-triazoles (e.g., **2.4-18**) were obtained by the fusion of aryl nitriles (e.g., **2.4-16**) with alkyl hydrazides (e.g., **2.4-17**), either as the benzenesulfonate salt or in the presence of an equimolar amount of benzene sulfonic acid.[1] The lower yields of 3-alkyl-5-aryl-s-triazoles result from the simultaneous formation of 3,5-diaryl-s-triazoles (e.g., **2.4-19**) and the aryl amide (e.g., **2.4-20**) corresponding to the nitrile. The by-products were shown to be formed by the

$$2\text{-MeC}_6\text{H}_4\text{CN} \xrightarrow[\text{PhSO}_3\text{H}]{\text{H}_2\text{NNHCOMe (2.4-17)}} 2\text{-MeC}_6\text{H}_4 \overset{\displaystyle \begin{array}{c} N \\ N-N \\ H \end{array}}{\diagdown} \text{Me} +$$

2.4-16   **2.4-18** (50%)

$$2\text{-MeC}_6\text{H}_4 \overset{\displaystyle \begin{array}{c} N \\ N-N \\ H \end{array}}{\diagdown} \text{C}_6\text{H}_4\text{-2-Me} + 2\text{-MeC}_6\text{H}_4\text{CONH}_2$$

2.4-19   **2.4-20**

hydration of the nitrile during the reaction followed by a transamination reaction between the resulting aryl amide (e.g., **2.4-20**) and the alkyl hydrazide (e.g., **2.4-17**) to give an aryl hydrazide (e.g., **2.4-21**). The

$$2\text{-MeC}_6\text{H}_4\text{CONH}_2 + \text{H}_2\text{NNHCOMe} \longrightarrow 2\text{-MeC}_6\text{H}_4\text{CONHNH}_2 + \text{H}_2\text{NCOMe}$$

2.4-20   **2.4-17**   **2.4-21**

interaction of the latter with either the aryl nitrile or amide provided the 3,5-diaryl-s-triazole by-product. This method has been described as unsatisfactory for the preparation of 3,5-dialkyl-s-triazoles from aliphatic nitriles and alkyl hydrazides.[1]

Weidinger modified the Potts procedure by using the free acid hydrazide rather than its benzenesulfonic acid salt.[10] In this method the acid hydrazide was reacted with an excess of nitrile in a sealed vessel at 200 to 230° to give good yields of 3,5-diaryl- and 3-alkyl-5-aryl-s-triazoles. Under these conditions alkyl nitriles and hydrazides undergo condensation to give low yields of 3,5-dialkyl-s-triazoles. For example, acethydrazide (**2.4-17**) and excess isobutyl nitrile (**2.4-22**) at 200° for 1 hr gave a 17% yield of **2.4-23**. In

$$\text{Me}_2\text{CHCN} + \text{H}_2\text{NNHCOMe} \longrightarrow \text{Me}_2\text{CH} \overset{\displaystyle \begin{array}{c} N \\ N-N \\ H \end{array}}{\diagdown} \text{Me}$$

2.4-22   **2.4-17**   **2.4-23**

contrast, the reaction of the complex formed from zinc chloride and hydrazine with propionitrile at 140° was reported in a patent to give an 86% yield of 3,5-diethyl-1,2,4-triazole.[11]

The direct cyclization of amidrazones to 3,5-disubstituted s-triazoles can be effected with anhydrides and ortho esters but not satisfactorily with aliphatic carboxylic acids. For example, treatment of **2.4-24** with refluxing acetic anhydride gave a 72% yield of an N-acetylated triazole, which was readily hydrolyzed in hot water to give **2.4-25**.[12,13] In contrast, treatment of

2.4-24   **2.2-25**

2.4-24 with acetic acid in the presence of several drops of nitric acid gave a mixture of **2.4-25** and the corresponding 1,3,4-oxadiazole.[12] The ortho ester method was used for the synthesis of the 3-[(benzyloxycarbonyl)amino]-methyl-s-triazole (**2.4-27**) from **2.4-26**.[14] The cyclization of **2.4-26** with ethyl orthobenzoate gave the corresponding 3-phenyl derivative (**2.4-29**), which has also been prepared by oxidative ring closure of the benzylidene

2.4-26   **2.4-27**

2.4-28   **2.4-29**

derivative (**2.4-28**) with palladium in refluxing mesitylene.[15,16] Presumably, the type of reaction involves the initial formation of a triazoline intermediate.[17] Another type of methylene amidrazone (**2.4-31**) was generated by the interaction of 1-(α-chlorobenzylidene)-2-(ethoxycarbonyl)hydrazine (**2.4-30**) with benzamidine.[18] The *in situ* ring closure of **2.4-31** provided a high yield of 3,5-diphenyl-s-triazole (**2.4-4**). In addition, the 1,2-bis(imidoyl)hydrazine (**2.4-32**) was cyclized in refluxing formic acid to give a 91% yield of **2.4-33**.[19]

The self-condensation of amidrazones has been observed and can lead to both triazoles and tetrazines. (Perfluoroacetimidoyl)hydrazine (**2.4-34**) underwent slow conversion to 1,2-bis(perfluoroacetimidoyl)hydrazine (**2.4-35**) at room temperature. The structure of **2.4-35** was confirmed by conversion of the hydrochloride to the triazole (**2.4-36**) in a sealed tube at 140°.[20] In

another example, the hydrochloride of the amidrazone (**2.4-37**) was heated in a sealed tube at 130° to give a mixture of **2.4-9** and **2.4-40**, presumably formed via **2.4-38** and **2.4-39**, respectively.[2] Intermediates like **2.4-39**[21] and **2.4-40** (see the following), however, have been converted to s-triazoles. In the transformation of **2.4-41** in refluxing acetic acid to give a high yield (90%) of **2.4-15**,[22] the tetrahydrotetrazine **2.4-42** might be formed initially. Rearrangement of **2.4-42** gave the 4-aminotriazole (**2.4-43**), which was followed by elimination of 4-nitroaniline to give the 3,5-disubstituted triazole (**2.4-15**).

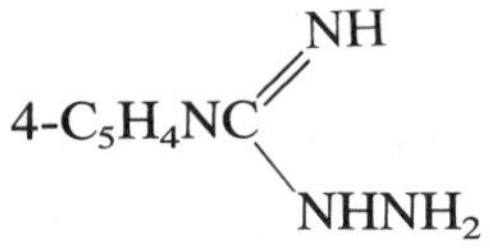

2.4-37

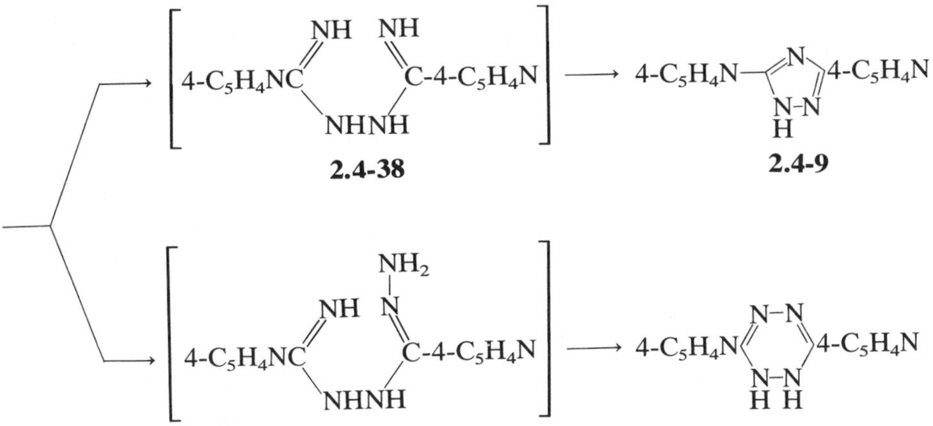

2.4-38 → **2.4-9**

2.4-39 → **2.4-40**

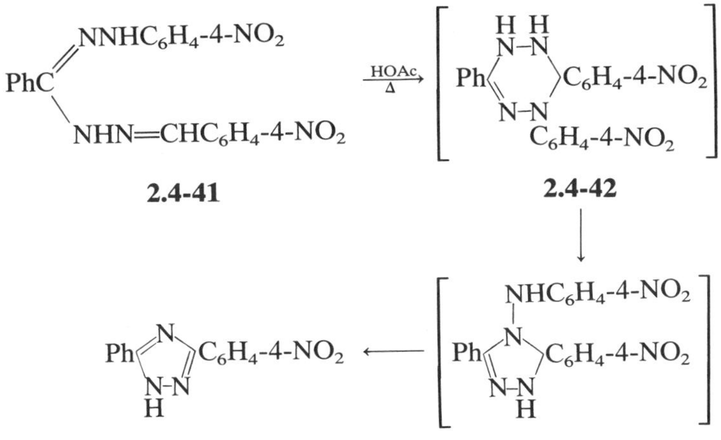

2.4-41 $\xrightarrow[\Delta]{\text{HOAc}}$ **2.4-42**

2.4-43

2.4-15 ←

An excellent method for the preparation of 3,5-diaryl-s-triazoles in good yields (74 to 94%) was developed by Postovskii and Vereshchagina.[23] This method is illustrated by the reaction of the imino ether (**2.4-44**) with acethydrazide (**2.4-17**) in refluxing ethanol to give the acylamidrazone (**2.4-45**), which after isolation, was heated 5 to 10° above the melting point to give 3,5-dimethyl-s-triazole (**2.4-6**). In addition to 3,5-dialkyl-s-

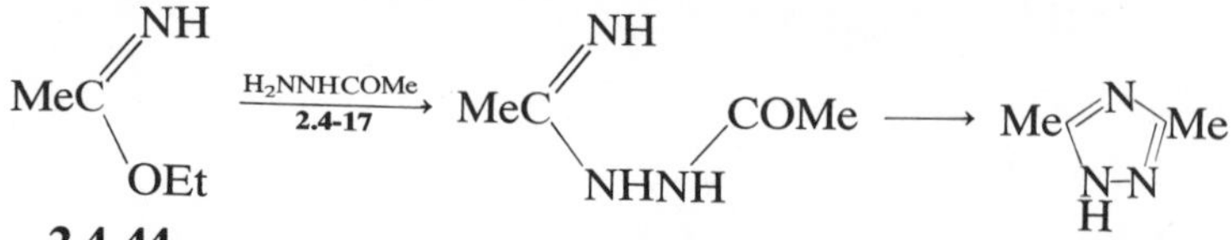

$$\text{MeC}\underset{\text{OEt}}{\overset{\text{NH}}{\|}} \quad \xrightarrow[\textbf{2.4-17}]{\text{H}_2\text{NNHCOMe}} \quad \text{MeC}\underset{\text{NHNH}}{\overset{\text{NH}}{\|}}\text{COMe} \quad \longrightarrow \quad \text{Me}\overset{\text{N}}{\underset{\text{N-N}}{\diagdown}}\text{Me}$$

2.4-44　　　　　　　　**2.4-45**　　　　　　　　**2.4-6**

triazoles,[24] s-triazoles containing either long-chain alkyl groups[25] or substituted alkyl groups[26] (e.g., **2.4-46** to **2.4-48**) are readily prepared. Also, the

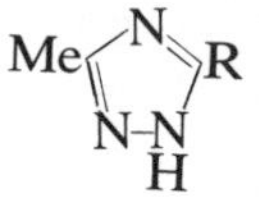

2.4-46, R = CH₂OMe

2.4-46, $R = CH_2OMe$
2.4-47, $R = CH_2CN$
2.4-48, $R = CH(OMe)_2$

thermal dehydration of acylamidrazones either as a melt or in a high-boiling solvent is the method of choice for the synthesis of 3,5-disubstituted s-triazoles containing alkyl-aryl,[27,28] alkyl-heteroaryl,[29] aryl-aryl,[30–32] aryl-heteroaryl,[33–35] and heteroaryl-heteroaryl[36] (e.g., **2.4-9**, **2.4-10**, **2.4-49**) substituents. Bistriazoles connected by an aliphatic or aryl bridge are also prepared by this method (see Section 21.1).[24]

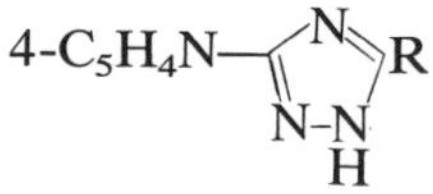

2.4-9, $R = 4\text{-}C_5H_4N$
2.4-10, $R = 2\text{-}C_5H_4N$
2.4-49, $R = 3\text{-}C_5H_4N$

　　In solution, the cyclization of acylamidrazones (e.g., **2.4-51**) was found to give exclusively s-triazoles (e.g., **2.4-50**) under basic conditions, whereas ammonia was eliminated with formation of oxadiazoles (e.g., **2.4-52**) under acidic conditions.[37,38]. An exception to oxadiazole formation was the dehydration of **2.4-53** in polyphosphoric acid to give **2.4-54**.[39]

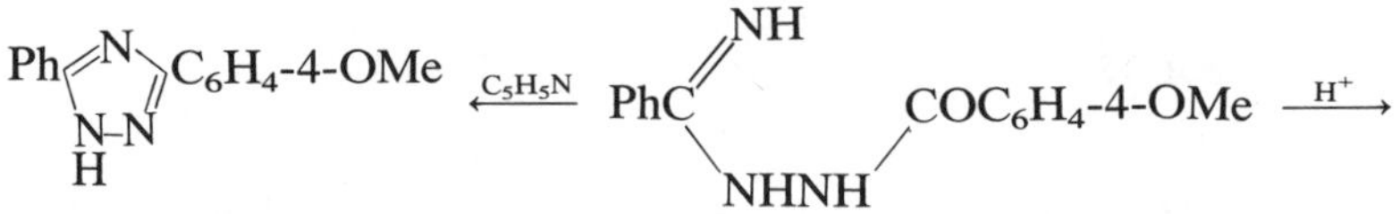

2.4-50　　　　　　　　**2.4-51**

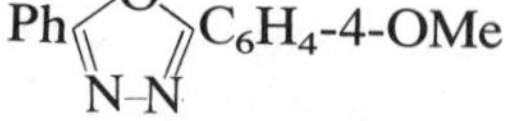

2.4-52

2.4-53   **2.4-54**

The ring carbons of 2,5-bis(perfluoroalkyl)oxadiazoles (e.g., **2.4-55**) are readily attacked by ammonia resulting in opening of the ring to give 1-acylamidrazones (e.g., **2.4-56**), which are dehydrated in the presence of phosphorus pentaoxide at 60 to 70° to give 3,5-bis(perfluoroalkyl)-s-triazoles (e.g., **2.4-57**).[40] Thermal cyclization of this type of 1-acylamidrazones has been reported to occur at 172 to 197°.[41]

2.4-55   **2.4-56**   **2.4-57** (73%)

In contrast to the 1-acylamidrazones, the 3-acylamidrazones (e.g., **2.4-58**) are cyclized in hydrochloric acid to give 3,5-diaryl- and 3-alkyl-5-aryl-s-triazoles (e.g., **2.4-59**).[42] The formation of 3,5-diphenyl-s-triazole (**2.4-4**)

2.4-58   **2.4-59**

from **2.4-60** and hydrazine apparently involved the displacement of the cyano group to give the 3-acylamidrazone (**2.4-61**).[43] Presumably, 3-

2.4-60   **2.4-61**   **2.4-4**

acylamidrazones (**2.4-64**) are also intermediates in the conversion of the 1,3-benzoxazines (**2.4-62** and **2.4-63**) with hydrazine to the 3,5-diaryl-s-triazoles (**2.4-66**).[44] Furthermore, the hydrazinolysis of 2-oxazolin-4-ones (e.g., **2.4-65**) proceeded via 3-acylamidrazones (e.g., **2.4-67**) to give good yields of s-triazoles (e.g., **2.4-68**).[45]

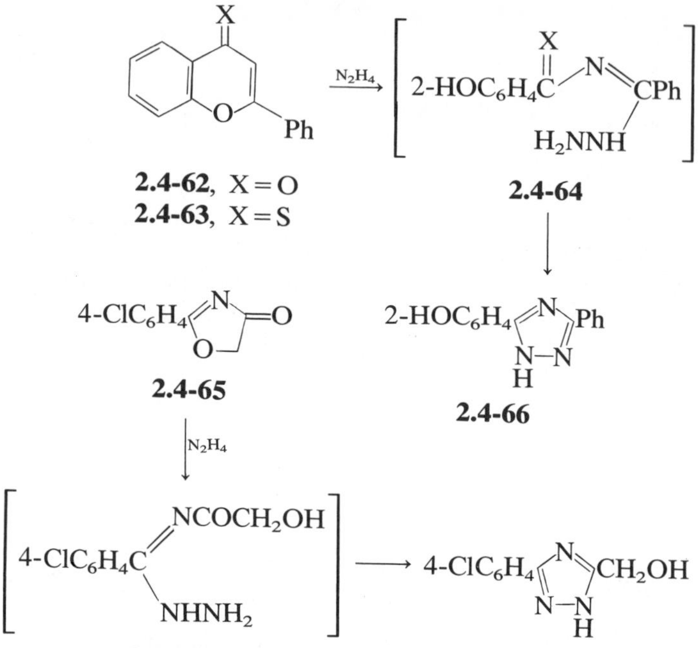

2.4-62, X = O
2.4-63, X = S

2.4-64

2.4-65

2.4-66

2.4-67

2.4-68

The 1-acylamidrazone (**2.4-69**) underwent transamination with hydroxylamine to give **2.4-70**, which was dehydrated in base to give the 4-hydroxytriazole (**2.4-71**).[46] Spectral data indicated that this type of compound existed mainly as the tautomeric N-oxide (e.g., **2.4-72**).[47]

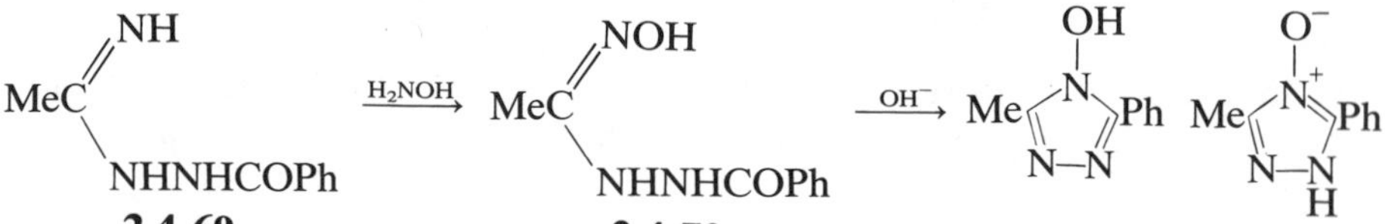

2.4-69   **2.4-70**   **2.4-71**   **2.4-72**

Other methods of theoretical interest for the preparation of s-triazoles include the Chichibabin reaction of aryl azines (e.g., **2.4-73**)[48] to give 3,5-diaryltriazoles (e.g., **2.4-74**), and the oxidative dimerization of the sodium derivatives of benzamidines (**2.4-75**) with cupric chloride to give 3,5-diphenyltriazole (**2.4-4**).[49] In addition, aryl azines (e.g., **2.4-73**) are converted to triazoles (e.g., **2.4-74**) with potassium *tert*-butoxide in refluxing toluene.[50] This transformation was believed to involve cleavage of the azine to a benzonitrile, which underwent addition to another azine molecule to give the triazole product.

TABLE 2 (*Continued*)

Compound	Reference

2.1. 1,3-Alkyl- or Aryl-Disubstituted 1H-1,2,4-Triazoles (*Continued*)

Compound	Reference
1-(1,2-diphenyletheny)-3-phenyl-	**82:** 156237m
3-[ethoxy(2-methoxyphenyl)methyl]-1-phenyl-	**84:** 90081p
3-(ethoxyphenylmethyl)-1-phenyl-	**84:** 90081p
3-ethynyl-1-methyl-	**83:** 28165v
3-(2-hydroxy-2-methyl-3-butyn-4-yl)-1-methyl-	**83:** 28165v
-3-methanol, α-(2-chlorophenyl)-1-phenyl-	**84:** 90081p
-3-methanol, α,1-diphenyl-	**68:** 105162m, **84:** 90081p
-3-methanol, 1-phenyl-	**63:** 7001c
3-[methoxy(2-methoxyphenyl)methyl]-1-phenyl-	**84:** 90081p
3-(methoxyphenylmethyl)-1-phenyl-	**84:** 90081p
1-[3-(4-methoxyphenyl)-2(2H)-oxo-1-benzopyran-7-yl]-3-methyl-	**83:** 105984r
3-methyl-1-(4-nitrophenyl)-	**68:** 29646e
1-methyl-3-phenyl-	**49:** 11630i, **66:** 37843r, **77:** 164587n, **79:** 18620t, **81:** 37521u
3-methyl-1-phenyl-	**57:** 3982a, **68:** 29646e, **79:** 92113w
1-methyl-3-phenyl-, 4-oxide	**77:** 139906h
3-methyl-1-trityl-	**57:** 807f
1-(*p*-nitrophenyl)-3-phenyl-	**61:** 5626e
3-(*p*-nitrophenyl)-1-phenyl-	**61:** 5626d
3-phenyl-1-(phenylmethyl)-	**84:** 59312r
1-phenyl-3-(trifluoromethyl)-	**85:** 138543r

2.2. 1,5-Alkyl- or Aryl-Disubstituted 1H-1,2,4-Triazoles

Compound	Reference
5-(2-aminoethyl)-1-methyl-	**50:** 1786a
5-(5-aminoimidazol-4-yl)-1-benzyl-	**67:** 43155e
1-benzyl-3-phenyl-, 4-oxide	**77:** 139906h
5-difluoronitromethyl-1 (or 4)-methyl-	**59:** 3927e
5-(2,6-difluorophenyl)-1-phenyl-	**69:** 35596a
1,5-dimethyl-	**49:** 15916b, **68:** 29645d
1-(2,4-dinitrophenyl)-5-methyl-	**68:** 29646e
1,5-diphenyl-	**46:** 11147b, **47:** 4335c, **51:** 12079c, **82:** 156003g
1-ethyl-5-methyl-	**85:** 46578c
1-[7-(4-ethyl-5-methyl-2H-1,2,3-triazol-2-yl)-2-oxo-2H-1-benzopyran-3-yl]-5-ethyl-	**73:** P100073b
5-ethynyl-1-methyl-	**83:** 28165v
5-(2-hydroxy-2-methyl-3-butyn-4-yl)-1-methyl-	**83:** 28165v
-5-methanol, α,α-diphenyl-1-(2-pyridinyl)-	**77:** 152071b
1-[3-(*p*-methoxyphenyl)coumarin-7-yl]-5-phenyl-	**71:** P14180m
5-methyl-1-(*p*-nitrophenyl)-	**68:** 29646e
1-methyl-1-phenyl-	**49:** 11631a, **74:** 3559c
5-methyl-1-phenyl-	**57:** 3982a, **83:** 114298x
5-phenyl-1-[3-(*p*-tolyl)coumarin-7-yl]-	**71:** P14180m
1-(2-pyridinyl)-5-[1-(2-pyridinyl)-1,2,4-triazol-5-yl]-	**77:** 152071b

TABLE 2   (*Continued*)

Compound	Reference
2.3. 3,4-Alkyl- or Aryl-Disubstituted 4H-1,2,4-Triazoles	
-3-acetic acid, 4-(2-pyridinyl)-	**83:** 9958q
-3-acetonitrile, α-[(p-anilinophenyl)imino]-4-phenyl-	**63:** P15026b
-3-acetonitrile, α-[[p-(dimethylamino)phenyl]imino]- 4-phenyl-	**63:** P15026b
-3-acetonitrile, 4-(p-ethoxyphenyl)-	**50:** 967a
-3-acetonitrile, 4-phenyl-	**50:** 967a, **54:** 2333i
4-[2-(acetyl(benzyl)amino]-4-chlorophenyl]- 3-methyl-	**80:** P59944r
4-[2-(α-aminobenzyl)-4-chlorophenyl]- 3-(hydroxymethyl)-	**83:** 193188n
4-[2-(α-aminobenzyl)-4-chlorophenyl]-3-methyl-	**80:** P59944r, **83:** 193188n
4-(3-amino-4-bromo-2-naphthyl)-3-methyl-	**66:** 55438s
4-(2-amino-3,5-dibromophenyl)-3-methyl-	**66:** 55438s
3-(2-aminoethyl)-4-methyl-	**50:** 1786b, **85:** P21373u
3-(2-aminoethyl)-4-phenyl-	**50:** 1786c
3-[[(2-aminoethyl)thio]methyl]-4-methyl-	**77:** P164704y
3-(aminomethyl)-4-benzyl-	**62:** 1647f
3-(aminomethyl)-4-[3-(2-chlorobenzoyl)- 5-ethyl-2-thienyl]-	**80:** P133443v
3-(aminomethyl)-4-methyl-	**62:** 1647f
4-(3-amino-2-naphthyl)-3-methyl-	**66:** 55438s
4-(o-aminophenyl)-3-methyl-	**66:** 10878f, **66:** 55438s
3-(o-aminophenyl)-4-phenyl-	**59:** 2804c
3-[(benzamido)methyl]-4-(4-chlorophenyl)- ·	**81:** 49624t
3-[(benzamido)methyl]-4-phenyl	**81:** 49624t
4-(2-benzoyl-4-chlorophenyl)-3-[[(benzyloxy- carbonyl)amino]methyl]-	**83:** 193188n
4-(2-benzoyl-4-chlorophenyl)-3-(chloromethyl)-	**77:** P88557k, **79:** 92176u, **82:** P31330e, **83:** P43339y, **83:** P58895a
4-(2-benzoyl-4-chlorophenyl)-3-ethyl-	**82:** P156401k
4-(2-benzoyl-4-chlorophenyl)-3-[(formyloxy)methyl]-	**83:** 193188n
4-(2-benzoyl-4-chlorophenyl)-3-(hydroxymethyl)-	**81:** P152287j, **83:** 193188n
4-(2-benzoyl-4-chlorophenyl)-3-(hydroxymethyl)-, oxime	**83:** 193188n
4-(2-benzoyl-4-chlorophenyl)-3-methyl-	**77:** P126707u, **77:** P126708v, **78:** P136354m, **78:** P159694m, **80:** P83083e, **82:** P156401k
4-(2-benzoyl-4-chlorophenyl)-3-methyl-, oxime	**80:** P59944r, **83:** 193188n
4-(2-benzoyl-4-chlorophenyl)-3-[(phthalimino)methyl]-	**81:** P152287j
4-(2-benzoyl-4-chlorophenyl)-3-(4-pyridinyl)-	**78:** P136354m
4-(2-benzoyl-4-nitrophenyl)-3-(chloromethyl)-	**77:** P88557k, **83:** 43339y
4-(2-benzoylphenyl)-3-(chloromethyl)-	**77:** P88557k, **82:** P31330e, **83:** P43339y, **83:** P58895a
4-[2-benzoyl-4-(trifluoromethyl)phenyl]-3-(chloromethyl)-	**82:** P31330e, **83:** P43339y, **83:** P58895a
4-[2-benzyl-5-chloro-phenyl]-3-methyl-	**78:** P136354m
4-[2-benzyl-5-chloro-phenyl]-3-(4-pyridinyl)-	**78:** P136354m
3-benzyl-4-methyl-	**73:** 66517n

TABLE 2 (*Continued*)

Compound	Reference
2.3. 3,4-Alkyl- or Aryl-Disubstituted 4H-1,2,4-Triazoles (*Continued*)	
4-[4-bromo-2-(2-pyridinylcarbonyl)phenyl]-3-methyl-	**83:** 193188n
3-(*p̄*-tert-butylphenyl)-4-[2-(diethylamino)ethyl]-	**62:** P568b, **62:** 11829h
4-(*o*-carboxyphenyl)-3-methyl-	**72:** 43550h
4-[3-(2-chlorobenzoyl)-5-ethyl-2-thienyl]- 3-(chloromethyl)-	**80:** P133443v, **81:** P152238v
4-[3-(2-chlorobenzoyl)-5-ethyl-2-thienyl]- 3-[(dimethylamino)methyl]-	**80:** P133443v
4-[2-(4-chlorobenzyl)phenyl]-3-methyl	**82:** P156401k
4-[4-chloro-2-benzylphenyl]-3-methyl-	**78:** P136302t
4-[4-chloro-2-benzylphenyl]-3-(4-pyridinyl)-	**78:** P136302t

4-[4-chloro-2-(2-chlorobenzoyl)phenyl]-
  3-(chloromethyl)-
	77: P88557k, **83:** P43339y,
	  83: P58895a,

4-[4-chloro-2-(2-chlorobenzoyl)phenyl]-3-methyl-	**81:** P152289m, **82:** P156401k
4-[4-chloro-2-[(2-chlorophenyl)methyl]phenyl] 3-methyl-	**82:** P156401k, **85:** 56539e

4-[4-chloro-2-(2-fluorobenzoyl)phenyl]-
  3-(chloromethyl)-
	77: P88557k, **83:** P43339y,
	  83: P58895a

4-[4-chloro-2-(fluorobenzoyl)phenyl]-3-methyl	**83:** 193188n
4-[4-chloro-2-(α-hydroxybenzyl)phenyl]-3-methyl-	**81:** P49689t
4-[5-chloro-2-(4-methoxybenzoyl)phenyl]- 3-(chloromethyl)-	**82:** P31330e
4-[4-chloro-2-(4-methoxyphenyl)phenyl]-	**77:** P88557k, **83:** P43339y, **83:** P58895a
3-(*p*-chlorophenyl)-4-ethyl-	**74:** 111968u
4-[4-chloro-2-(phenylmethyl)phenyl]-3-ethyl-	**82:** P156401k
4-[4-chloro-2-(phenylmethyl)phenyl]-3-methyl-	**85:** 56539e
4-(4-chlorophenyl)-3-(trifluoromethyl)-	**85:** 138543r
4-cyclohexyl-3-ethyl-	**26:** P258[4], **26:** P3516[6]
3-(3,5-di-tert-butyl-2-hydroxyphenyl)-4-phenyl-	**71:** P70607t
3-(2,2-dichloro-1,1-difluoroethyl)-4- [tris(4-chlorophenyl)methyl]-	**81:** 49647c
3-(3,4-dihydro-2,4-dimethyl-5(2H)-oxo-3-thioxo- 1,2,4-triazin-6-yl)-4-methyl-	**83:** 114345k
3,4-dimethyl-	**68:** 29645d, **81:** 37521u
3-[*p*-(dimethylamino)phenyl]-4-phenyl-	**66:** 46412b
4-[2-(1,1-dimethylethy)phenyl]-3-(trifluoromethyl)-	**85:** 138543r

3,4-diphenyl-	**44:** 2516g, **54:** 2333h,
	  54: 21064g, **57:** 12473c,
	  66: 46412b, **71:** 81268j,
	  77: P164704y, **85:** P21373u,
	  85: 33015s

3-(ethoxymethyl)-4-methyl-	**85:** 138543r
4-(4-ethoxyphenyl)-3-(trifluoromethyl)-	**85:** 46578c
4-ethyl-3-methyl-	**74:** 111968u
4-ethyl-3-phenyl-	**74:** 111968u
4-ethyl-3-(*p*-sulfamoylphenyl)-	**71:** P70607t
3-(2-hydroxy-3-methylphenyl)-4-phenyl-	**71:** P70607t
3-(*o*-hydroxyphenyl)-4-phenyl-	**85:** P33015s
-3-methanol, 4-methyl-	

TABLE 2 (*Continued*

Compound	Reference

2.3. 3,4-Alkyl- or Aryl-Disubstituted 4H̲-1,2,4-Triazoles (*Continued*)

Compound	Reference
4-(2-methoxyphenyl)-3-(trifluoromethyl)-	**85:** 138543r
3-methyl-4-(3-methylcyclohexyl)-	**26:** P3516[6]
3-methyl-4-(*p*-nitrophenyl)-	**68:** 29646e
3-[2-(2-methyl-10H̲-phenothiazin-10-yl)ethyl]-4-phenyl-	**77:** 34439e
3-methyl-4-phenyl-	**49:** 15916b, **66:** 55438s
4-methyl-3-phenyl-	**44:** 2516i, **49:** 15916b
4-(2-methylphenyl)-3-(trifluoromethyl)-	**85:** 138543r
3-methyl-4-(2-pyridinyl)-	**83:** 9958q
3-methyl-4-[2-(2-pyridinylcarbonyl)phenyl]-	**83:** 193188n
3-methyl-4-*o*-tolyl-	**72:** 43550h
3-(*p*-nitrophenyl)-4-phenyl-	**66:** 46412b
4-(*p*-nitrophenyl)-3-phenyl-	**66:** 46412b
4-phenyl-3-(2-pyridyl)-	**66:** 46412b
4-phenyl-3-(4-pyridyl)-	**67:** 37867x
3-phenyl-4-*p*-tolyl-	**66:** 46412b
4-phenyl-3-(trifluoromethyl)-	**85:** 138543r
3-(4-pyridyl)-4-*m*-tolyl-	**67:** 37867x
3-(4-pyridyl)-4-*o*-tolyl-	**67:** 37867x
3-(4-pyridyl)-4-*p*-tolyl-	**67:** 37867x

2.4. 3,5-Alkyl- or Aryl-Disubstituted *s*-Triazoles

Compound	Reference
-3-acetonitrile, 5-benzyl-	**58:** 2448a
-3-acetonitrile, 5-methyl-	**58:** 2448a
-3-acetonitrile, 5-(5-nitro-2-furyl)-	**64:** 19596c
-3-acetonitrile, 5-phenyl-	**58:** 2448a
3-(2-amino-5-chlorophenyl)-5-methyl-	**82:** P57724m, **82:** P171103h
3-[4-[(2-amino-3,6-disulfo-2-naphthalenyl)azo]phenyl]-5-[4-[(1-amino-4-sulfo-2-naphthalenyl)azo]phenyl]-	**83:** P116930w
3-[4-[(1-amino-8-hydroxy-3,6-disulfonaphthalen-2-yl)azo]phenyl]-5-[4-[(2,4-diaminophenyl)azo]phenyl]-	**83:** P195215t
3-[4-[(1-amino-8-hydroxy-7-[(4-nitrophenyl)azo]-3,6-disulfonaphthalen-2-yl)azo]phenyl]-5-[4-[[2,4-diamino-3-[(4-nitrophenyl)azo]-phenyl]azo]phenyl]-	**83:** P195215t
3-[4-[4-amino-5-hydroxy-6-(phenylazo)-2,7-disulfo-3-naphthalenylazo]phenyl]-5-[4-[(2,4-diaminophenyl)azo]phenyl]-	**81:** P122761s **83:** P81197h
3-[3-[(1-amino-8-hydroxy-7-(phenylazo)-3,6-disulfonaphthalen-2-yl)azo]phenyl]-5-[3-[(4-hydroxyphenyl)azo]phenyl]-	**83:** P81197h
5-[4-[(7-amino-1-hydroxy-3-sulfo-2-naphthalenyl)azo]phenyl]-3-[4-[(3-carboxy-4-hydroxyphenyl)azo]phenyl]-	**83:** P116930w
3-(aminomethyl)-5-benzyl-4-hydroxy-	**74:** 141651c
3-(aminomethyl)-4-hydroxy-5-phenyl-	**74:** 141651c
3-(aminomethyl)-5-phenyl-	**61:** 656d
3-[m-(1-aminophenoxy)phenyl]-5-(m-aminophenyl)	**72:** 68047n, **78:** P30495t
3-[4-(4-aminophenoxy)phenyl]-5-(3-aminophenyl)-	**78:** P30495t
3-(3-aminophenyl)-5-(4-aminophenyl)-	**83:** P97302z

TABLE 2   (*Continued*)

Compound	Reference

2.4. 3,5-Alkyl- or Aryl-Disubstituted *s*-Triazoles   (*Continued*)

Compound	Reference
3-(3-aminophenyl)-5-[3-[(4-aminophenyl)sulfonyl]phenyl]-	**72:** 68047n, **78:** P30495t
3-(*o*-aminophenyl)-5-phenyl-	**73:** 120582v
3-(2-aminophenyl)-5-(2-pyridyl)-	**76:** 14998n, **76:** 99614g
3-(4-aminophenyl)-5-(4-pyridyl)-	**68:** 105162m
3-(2-amino-3-pyridinyl)-5-methyl-	**81:** 91461n
3-(3-amino-2-pyridinyl)-5-methyl-	**81:** 91461n
3-[(1-azabicyclo[2,2,2]octan-3-idene)methylene]-5-(4-pyridinyl)-	**79:** 66307t
3-[(benzamido)methyl]-5-benzyl-4-hydroxy-	**74:** 141651c
3-[(benzamido)methyl]-4-hydroxy-5-phenyl-	**74:** 141651c
3-[(benzamido)methyl]-5-(4-pyridyl)-	**70:** 114407p
3-(2-benzamidophenyl)-5-phenyl-	**80:** 15236g
3-[2-(benzamido)phenyl]-5-(2-pyridyl)-	**76:** 14998n, **76:** 99614g
3-(1H-benzimidazol-2-yl)-5-methyl-	**79:** 92107x
3-(2-benzoxazolyl)-5-(*p*-methoxyphenyl)-	**75:** 118255p, **77:** P164709d
5-benzyl-3-(7-diethylamino-2-imino-2H-1-benzopyran-3-yl)-	**83:** P207574z
3-benzyl-5-(dimethoxymethyl)-	**58:** 2447h
3-benzyl-5-ethyl-	**55:** 14441a
3-benzyl-4-hydroxy-5-methyl-	**74:** 111967t, **74:** 141651c
3-benzyl-5-(methoxymethyl)-	**58:** 2448a
3-benzyl-5-methyl-	**54:** 9898e, **67:** 11793v
3-benzyl-5-methyl-, 4-oxide	**74:** 111182h
3-benzyl-5-(*p*-nitrophenyl)-	**49:** 13227c
3-benzyl-5-phenyl-	**54:** 9898f, **62:** 6392c
3-benzyl-5-phenyl-1-(trimethylsilyl)-	**68:** 69071v
3-benzyl-5-(3-pyridyl)-	**68:** 105162m
3-benzyl-5-*p*-tolyl-	**63:** 11541b
3,3'-[([1,1'-biphenyl]-4,4'-dicarboxamido)bis-2,1-phenylene]bis[5-phenyl-	**80:** 15236g, **80:** 59912d
3,5-bis(4-amidinophenyl)-	**82:** 156162h
3,5-bis(3-amino-4-chlorophenyl)-	**83:** P97302z
3,5-bis[[(2-aminoethyl)thio]methyl]-	**77:** P164704y
3,5-bis[4-[(4-amino-5-hydroxy-1,3-disulfonaphthalen-6-yl)azo]phenyl]-	**83:** P116930w
3,5-bis(4-amino-2-methylphenyl)-	**83:** P97302z
3,5-bis(3-aminophenyl)-	**82:** P59896t, **83:** P81197h
3,5-bis(*p*-aminophenyl)-	**56:** P11111i, **83:** P179068s
3,5-bis(2-benzoxazolyl)-	**77:** P164709d
3,5-bis[4-(2-[1,1'-biphenyl]-4-ylethenyl)phenyl]-	**68:** P21961h, **80:** P146980t, **80:** P146981u
3,5-bis(4-bromophenyl)-	**82:** 156162h
3,5-bis[2-(butylthio)ethyl]-	**80:** P62968g
3,5-bis(*p*-carboxyphenyl)-	**53:** P1748g
3,5-bis[4-[[1-(4-chloro-2,5-dimethoxyanilino)-1,3-dioxo-2-butyl]azo]phenyl]-	**82:** P59896t
3,5-bis[3[[3-(4-chloro-2,5-dimethoxyphenylcarbamoyl)-2-hydroxy-1-naphthalenyl]azo]phenyl]-	**82:** P59896t

TABLE 2 (*Continued*)

Compound	Reference
2.4. 3,5-Alkyl- or Aryl-Disubstituted *s*-Triazoles (*Continued*)	
3,5-bis(chloromethyl)-	**70:** 115076s
3,5-bis(4-chloro-3-nitrophenyl)-	**83:** P97302z, **84:** P150633g
3,5-bis(4-chlorophenyl)-	**83:** P97302z, **84:** P150633g
3,5-bis(4-cyanophenyl)-	**82:** 156162h
3,5-bis(2,2-dichloro-1,1-difluoroethyl)-	**81:** 49647c
3,5-bis[4-[[4-[(2,5-dichlorophenyl)azo]-3-hydroxynaphthan-2-ylcarbonyl]amino]phenyl]-	**74:** P77385q, **76:** P60936p
3,5-bis[2-(diethylamino)ethyl]-	**80:** P62968g
3,5-bis[(diethylamino)methyl]-	**74:** P141811e
3,5-bis[*p*-(diethylamino)phenyl]-	**64:** PC16886c, **74:** P149220g, **79:** P47838a, **81:** P8424j
3,5-bis[2,6-dimethyl-4-pyridinyl)-	**77:** P5487p
3,5-bis[*p*-(dipropylamino)phenyl]-	**61:** P2637g, **64:** PC16886c
3,5-bis[2-(dodecylthio)ethyl]-	**81:** P105517z, **82:** P170962a
3,5-bis[4-[(3-ethoxycarbonyl-5-oxo-1-phenyl-4-1<u>H</u>-pyrazolyl)azo]phenyl]-	**82:** P59896t
3,5-bis[3-[[3-(2-ethoxyphenylcarbamoyl)-2-hydroxy-1-naphthalenyl]azo]phenyl]-	**82:** P59896t
3,5-bis(heptafluoropropyl)-	**57:** 12471a, **67:** 21871h
3,5-bis[2-(2(hydroxyethylthio)ethyl]-	**80:** P62968g
3,5-bis(*p*-methoxyphenyl)-	**62:** 14661g, **84:** 90123d
3-[bis(4-methoxyphenyl)methylene]-5-phenyl-	**84:** 4866u
3,5-bis[4-[[1-(2-methylanilino)-1,3-dioxo-2-butyl]azo]phenyl]-	**82:** P59896t
3,5-bis(3-methyl-4-nitrophenyl)-	**83:** P97302z, **84:** P150633g
3,5-bis(3-methylphenyl)-	**83:** P97302z, **84:** P150633g
3,5-bis(4-methylphenyl)-	**84:** 105503d
3,5-bis(2-methyl-4-pyridinyl)-	**77:** P5487p
3,5-bis[[2-[(*N*-methylthiocarbamoyl)amino]ethylthio]methyl]-	**77:** P164704y
3,5-bis(*p*-nitrophenyl)-	**49:** 13227c, **80:** P62968g, **85:** P123928w
3,5-bis[2-(octylthio)ethyl]-	**80:** P62968g
3,5-bis(pentafluoroethyl)-	**57:** 12471a, **64:** 12662e
3,5-bis(pentafluorophenyl)-	**68:** P39628q
3,5-bis(2-phenoxyethyl)-	**80:** P62968g
3,5-bis(2-pyridinyl)-	**74:** 13103p
3,5-bis(3-pyridinyl)-	**68:** 114391n, **77:** P5487p, **83:** 188199n
3,5-bis[4-pyridinyl]-, pyridine *N*-oxides	**77:** P5487p
3,5-bis(4-pyridyl)-	**68:** 114391n, **83:** 188199n
3,5-bis(4-pyrimidinyl)-	**77:** P5480f
3,5-bis(trifluoromethyl)-	**57:** 12471a, **64:** 2081c
3,5-bis[3-trifluoromethyl)phenyl]-	**84:** 105503d
3-(2-bromoacetamido-5-chlorophenyl)-5-methyl-	**82:** P171103h
3-[2-(bromomethyl)phenyl]-5-(3-chlorophenyl)-	**83:** P206286v
3-[2-(bromomethyl)phenyl]-5-phenyl-	**83:** P206286v
3-(4-bromophenyl)-5-phenyl-	**83:** 58722s

TABLE 2   (*Continued*)

Compound	Reference

2.4. 3,5-Alkyl- or Aryl-Disubstituted *s*-Triazoles   (*Continued*)

Compound	Reference
3-butyl-5-isopropyl-	**56:** 10132d, **59:** 6387a
3-(2-carboxybenzyl)-5-methyl-	**81:** 105408q, **81:** 37521u
3-(2-carboxybenzyl)-5-phenyl-	**81:** 37521u
3-(2-carboxybenzyl)-5-(2-pyridinyl)-	**83:** 43245q
3-(*o*-carboxyphenyl)-5-methyl-	**76:** 85193g
3-(*o*-carboxyphenyl)-5-(trifluoromethyl)-	**74:** 53671d
4-(4-chlorobenzyloxy)-3-methyl-5-phenyl-	**77:** 139906h
3-(4-chlorobenzyl)-5-(4-pyridyl)-	**68:** 105162m
3-[4-[[2-chloro-5-[[[2-chloro-5-(trifluoromethyl)-phenyl]amino]carbonyl]phenyl]azo]-3-hydroxy-naphthan-2-ylcarbonyl]amino]phenyl]-5-phenyl-	**74:** P77385q, **76:** P60936p
3-(chlorodiphenylmethyl)-5-phenyl-	**80:** 133389g, **84:** 4866u
5-[5-chloro-2-(hydroxymethyl)phenyl]-3-(4-nitrophenyl)-	**77:** 61916q
3-(chloromethyl)-5-phenyl-	**73:** 56038t
3-(*p*-chlorophenyl)-4-hydroxy-5-methyl-	**74:** 141651c
3-(4-chlorophenyl)-5-[(2-hydroxymethyl)phenyl]-	**78:** P159619r
3-(*p*-chlorophenyl)-5-methyl-, 4-oxide	**74:** 111182h
3-(3-chlorophenyl)-5-(1-methyl-4-pyridinio)-, iodide	**77:** P5480f
3-(4-chlorophenyl)-5-(1-methyl-4-pyridinio)-, iodide	**77:** P5480f
3-(3-chlorophenyl)-5-(4-pyridinyl)-	**77:** P5480f, **83:** 188199n
3-(4-chlorophenyl)-5-(4-pyridinyl)-	**77:** P5480f, **83:** 188199n
3-(4-chlorophenyl)-5-(trifluoromethyl)-	**85:** 138543r
3-(4-chloro-3-sulfamoylphenyl)-5-(4-pyridinyl)-	**77:** P5480f
3-(2-cyanoethenyl)-5-methyl-	**78:** 136179h
3-(2-cyanoethenyl)-5-phenyl-	**78:** 136179h
3-(2-cyanoethenyl)-5-(trifluoromethyl)-	**78:** 136179h
3-(α-cyano-2-methylbenzyl)-5-methyl-	**69:** 43772v
3-[2-(cyanomethyl)benzyl]-5-methyl-	**69:** 43772v
3-(cyanomethyl)-5-(2-pyridinyl)-	**84:** 59317w
3-(cyanomethyl)-5-(3-pyridinyl)-	**84:** 59317w
3-(cyanomethyl)-5-(4-pyridinyl)-	**84:** 59317w
3-(*o*-cyanophenyl)-5-methyl-	**72:** 12671u
3-cyclohexyl-5-phenyl-	**56:** 10132d, **59:** 6387a
3,5-dibenzyl-	**54:** 9898f, **58:** 2449a, **85:** 123928w
3-[4-[[4-[(2,5-dichlorophenyl)azo]-3-hydroxy-2-naphthalenylcarbonyl]amino]phenyl]-5-phenyl-	**76:** P60936p
3-(3,4-dichlorophenyl)-5-[(2-hydroxymethyl)phenyl]-	**78:** P135625p
3-(3,4-dichlorphenyl)-5-(4-pyridinyl)-	**77:** P5480f, **83:** 188199n
3-(3,5-dichlorphenyl)-5-(4-pyridinyl)-	**77:** P5480f, **83:** 188199n
3-(3,4-dichlorophenyl)-5-(trifluoromethyl)-	**85:** 138543r
3-(4,5-dichloro-3-sulfamoylphenyl)-5-(4-pyridinyl)-	**77:** P5480f
3,5-diethyl-	**21:** 3200[9], **21:** 3201[1], **64:** 4919f
3-(7-diethylamino-2-imino-2$\underline{H}$-1-benzopyran-3-yl)-5-methyl-	**83:** P207574z
3-(7-diethylamino-2-imino-2$\underline{H}$-1-benzopyran-3-yl)-5-phenyl-	**83:** P207574z

TABLE 2 (*Continued*)

Compound	Reference

2.4. 3,5-Alkyl- or Aryl-Disubstituted s-Triazoles (*Continued*)

3,5-di-2-furanyl-	**64:** 5072h, **85:** 177373e
3,5-di-3-furanyl-	**85:** 177373e
3,5-diheptadecyl-	**85:** P123928w
3,5-diheptyl-	**83:** P114414g
3,5-diisopropyl-	**21:** 3201[1]
-3,5-dimethanol	**70:** 115076s, **78:** P136305w,
	85: 21373u
-3,5-dimethanol, α,α'-dimethyl-	**70:** 115076s
-3,5-dimethanol, α,α'-dimethyl-α,α'-	**78:** 84378a
bis(4-methylphenyl)-	
-3,5-dimethanol, dinitrate (ester)	**70:** 115076s
3-(dimethoxymethyl)-5-methyl-	**58:** 2447h
3-(dimethoxymethyl)-5-(p-nitrophenyl)-	**58:** 2447h
3-(dimethoxymethyl)-5-phenyl-	**58:** 2447h
3-(dimethoxymethyl)-5-p-tolyl-	**58:** 2447h
3-(3,5-dimethoxyphenyl)-5-(4-pyridinyl)-	**77:** P5480f, **83:** 188199n
3,5-dimethyl-	**49:** 1710i, **64:** 4919f,
	74: P141811e, **77:** 5410h,
	78: P136305w, **80:** 96082d
	85: P178509r
3-[α-(dimethylamino)diphenylmethyl]-5-phenyl-	**84:** 4866u
3-(2,6-dimethyl-4-pyridinyl)-5-(4-pyridinyl)-	**77:** P5487p
3,5-dimethyl-1-(trimethylsilyl)-	**80:** 96082d
3,5-di-2-naphthalenyl-	**78:** P136305w
3,5-di-m-tolyl-	**55:** 14441a
3,5-di-o-tolyl-	**59:** 11494a
3,5-di-p-tolyl-	**31:** 3054[1], **57:** 5909f,
	62: 14661f
3,5-diphenyl-	**49:** 1710i, **49:** 11631c,
	49: 13227b, **51:** 6607g,
	56: 8708i, **57:** 16599h,
	58: 2449a, **58:** 5661h,
	59: 6387b, **59:** 11494a,
	62: 16232h, **63:** 11541b,
	67: 73570d, **69:** 96601v,
	72: 100864q, **74:** 125574z,
	77: 34426y, **78:** 72019w
	78: P136305w, **81:** 129811s,
	83: 58722s, **83:** P179068s,
	84: 135602c
3-(diphenylmethyl)-5-methyl-	**28:** 5458[5]
3-(diphenylmethyl)-5-phenyl-	**24:** 4032[7], **84:** 59312r
5-(diphenylmethyl)-3-(2-pyridinyl)-	**83:** 43245q
3,5-diphenyl-1-(trimethylsilyl)-	**68:** 69071v
3,5-dipropyl-	**68:** P2902y
3,5-di-2-pyridyl-	**62:** 9122a
3,5-di-4-pyridyl-	**52:** 6345g, **52:** 20151h
3,5-diselenophene-2-yl-	**80:** 82840n, **85:** 177373e

TABLE 2   (*Continued*)

Compound	Reference

2.4. 3,5-Alkyl- or Aryl-Disubstituted s-Triazoles   (*Continued*)

Compound	Reference
3,5-diselenophene-3-yl-	**80:** 82840n, **85:** 177373e
3,5-di-2-thienyl-	**80:** 82840n, **85:** 177373e
3,5-di-3-thienyl-	**80:** 82840n, **85:** 177373e
3,5-diundecyl-	**78:** P136305w, **83:** P179068s, **85:** P123928w
3-ethyl-5-(5-nitro-2-furyl)-	**64:** 19596c, **70:** P87037j
3-ethyl-5-phenyl-	**55:** 14441a, **63:** 11541b, **78:** 84327h
3-ethyl-5-o-tolyl-	**55:** 14441a
3-(2-furanyl)-5-(2-propenyl)-	**79:** 87327q
3-(2-furanyl)-5-(4-pyridinyl)-	**77:** P5480f, **83:** 188199n
3-(2-furanyl)-5-selenophene-2-yl-	**85:** 177373e
3-(2-furanyl)-5-(2-thienyl)-	**80:** 82840n, **85:** 177373e
3-(2-furanyl)-5-(3-thienyl)-	**80:** 82840n, **85:** 177373e
3-heptadecyl-5-methyl-	**62:** 6392e
3-(heptafluoropropyl)-5-phenyl-	**60:** P5513a, **64:** P646b
5-(3-hydroxybenzo[b]thiophen-2-yl)-3-methyl-	**81:** P49560u
5-(3-hydroxybenzo[b]thiophen-2-yl)-3-(4-pyridinyl)-	**81:** P49560u
4-hydroxy-3,5-dimethyl-	**76:** 71575w
4-hydroxy-3-(p-methoxyphenyl)-5-methyl-	**74:** 111967t, **74:** 141651c
4-hydroxy-3-methyl-5-phenyl-	**74:** 111967t, **74:** 141651c
3-[(o-hydroxymethyl)phenyl]-5-methyl-	**77:** 61916q
3-[(o-hydroxymethyl)phenyl]-5-(4-nitrophenyl)-	**77:** 61916q
3-[(o-hydroxymethyl)phenyl]-5-phenyl-	**77:** 61916q
3-(4-hydroxyphenyl)-5-methyl-	**80:** 96082d
3-(o-hydroxyphenyl)-5-phenyl-	**51:** 17924b
3-(2-hydroxyphenyl)-5-(2-pyridyl)-	**68:** 105162m
3-(2-hydroxyphenyl)-5-(3-pyridyl)-	**70:** 114407p
3-(2-hydroxyphenyl)-5-(4-pyridyl)-	**68:** 105162m
3-(3-hydroxyphenyl)-5-(4-pyridyl)-	**68:** 105162m
3-(4-hydroxyphenyl)-5-(2-pyridyl)-	**68:** 105162m
3-[4-[[3-hydroxy-4-[(2,4,5-trichlorophenyl)azo]-naphthan-2-ylcarbonyl]amino]phenyl]-5-p-tolyl-	**74:** P77385q, **76:** P60936p
3-isobutyl-5-isopropyl-	**55:** P13450a
3-isopropyl-5-methyl-	**56:** 10132d
3-isopropyl-5-(5-nitro-2-furyl)-	**64:** 19596c
3-(2-mercaptophenyl)-5-methyl-	**69:** 52074q
-3-methanol, α,α-bis[4-(dimethylamino)phenyl]-5-phenyl-	**84:** 59312r
-3-methanol, α,α-bis(4-methoxyphenyl)-5-phenyl-	**84:** 59312r
-3-methanol, α,5-diphenyl-	**73:** 56038t
-3-methanol, 5-(o-ethoxyphenyl)-	**73:** 56038t
-3-methanol, 5-[2-(5-nitro-2-furyl)vinyl]-	**63:** P4306f
-3-methanol, α-(nitromethyl)-5-phenyl-	**79:** 66307t
-3-methanol, 5-(p-nitrophenyl)-	**74:** 125572x
-3-methanol, 5-phenyl-	**73:** 56038t, **74:** 141651c
-3-methanol, 5-(4-pyridinyl)-, N-oxide	**79:** 66307t
-3-methanol, 5-(3-pyridyl)-	**75:** 140792h
-3-methanol, 5-p-tolyl-	**74:** 125572x

59

TABLE 2 (*Continued*)

Compound	Reference
2.4. 3,5-Alkyl- or Aryl-Disubstituted *s*-Triazoles (*Continued*)	
-3-methanol, α,α,5-triphenyl-	**24:** 4032[7], **80:** 133389g, **84:** 59312r
3-(methoxydiphenylmethyl)-5-phenyl-	**80:** 133389g, **84:** 4866u
3-(methoxymethyl)-5-methyl-	**58:** 2448a
3-(methoxymethyl)-5-phenyl-	**58:** 2448a, **62:** 16232h
3-(4-methoxyphenyl)-5-methyl-	**74:** 111967t
3-(4-methoxyphenyl)-5-(4-pyridinyl)-	**77:** P5480f
5-methyl-3-(1-methyl-1H-benzimidazo)-2-yl)-	**79:** 92107x
3-methyl-5-(1-methyl-5-nitroimidazol-2-yl)-	**73:** P25480b, **85:** P5644c
3-methyl-5-[4-(1-methyl)pyridinio]-, chloride	**71:** 91398r
3-methyl-5-(2-methylphenyl)-	**77:** 61916q
3-methyl-5-(5-nitro-2-furyl)-	**64:** 19596c, **70:** P87037j, **76:** 126882b
3-methyl-5-nonyl-	**62:** 6392d
3-methyl-5-phenyl-	**23:** 836[3], **49:** 11631c, **55:** 14441a, **56:** 10132d, **59:** 6387a, **63:** 11541b, **71:** P124441e, **84:** 135602c
3-methyl-5-phenyl-, 4-oxide	**74:** 111182h
3-(2-methylphenyl)-5-phenyl-	**85:** P155072d
3-(4-methylphenyl)-5-(41pyridinyl)-	**77:** P5487p
3-methyl-5-propyl-	**80:** 96082d
5-methyl-3-(2-pyridinyl)-	**72:** 43572s
3-(2-methyl-4-pyridinyl)-5-(4-pyridinyl)-	**77:** P5487p
3-methyl-5-*o*-tolyl-	**55:** 14441a
3-methyl-5-*p*-tolyl-	**63:** 11541b
3-methyl-5-*p*-tolyl-, 4-oxide	**74:** 111182h
3-methyl-5-[2-(trimethylsilyl)ethyl]-	**80:** 96082d
3-methyl-5-undecyl-	**62:** 6392d
3-(2-naphthalenyl)-5-(4-pyridinyl)-	**77:** P5480f
3-(1-naphthyl)-5-(*p*-nitrophenyl)-	**49:** 13227c
3-(1-naphthyl)-5-phenyl-	**49:** 13227c
3-(5-nitro-2-furyl)-5-propyl-	**64:** 19596c
3-[*m*-(*p*-nitrophenoxy)phenyl]-5-(*m*-nitrophenyl)-	**72:** 68047n
3-[4-(4-nitrophenoxy)phenyl]-5-(3-nitrophenyl)-	**78:** P30495t
3-(3-nitrophenyl)-5-(4-nitrophenyl)-	**83:** P97302z, **84:** P150633g
3-(*p*-nitrophenyl)-5-phenyl-	**49:** 13227c, **59:** 6387b, **56:** 10132e, **62:** 16232h, **83:** 58722s
3-(*p*-nitrophenyl)-5-(4-pyridyl)-	**68:** 105162m
5-(2-nitrophenyl)-3-(2-pyridyl)-	**68:** 105162m, **76:** 14998n, **76:** 99614g
3-(*p*-nitrophenyl)-5-*p*-tolyl-	**49:** 13227c
3-[(1(2H)-oxo-3,4-dihydronaphthalen-2-idene)-methylene]-5-phenyl-	**79:** 66307t
3-[(1(2H)-oxo-3,4-dihydronaphthalen-2-idene)-methylene]-5-(4-pyridinyl)-,	**79:** 66307t
3-[(1(2H)-oxo-3,4-dihydronaphthalen-2-idene)-methylene]-5-(4-pyridinyl)-, *N*-oxide	**79:** 66307t

TABLE 2   (*Continued*)

Compound	Reference
2.4. 3,5-Alkyl- or Aryl-Disubstituted s-Triazoles (*Continued*)	
3-(2-oxo-3-phenyl-2H̲-1-benzopyran-7-yl)- 5-(2-pyridinyl)-	**74:** P65592s
3-(2-oxo-3-*p*-tolyl-2H̲-1-benzopyran-7-yl)-5-phenyl-	**74:** P65592s
5-phenyl-3-(4-phenyl-2-pyridinyl)-	**74:** 13103p
3-phenyl-5-propyl-	**55:** 14441a
5-phenyl-3-(2-pyridinyl)-	**72:** 43572s
3-phenyl-5-(4-pyridinyl)-, *N*-oxide	**79:** 66307t
3-phenyl-5-(4-pyridyl)-	**59:** 6387b, **83:** 188199n
3-phenyl-5-*m*-tolyl-	**56:** 10132e, **59:** 6387b
3-phenyl-5-*o*-tolyl-	**49:** 13227c
3-phenyl-5-*p*-tolyl-	**49:** 13227c, **57:** 5909e, **58:** 2449a, **63:** 11541b
3-phenyl-5-(trifluoromethyl)-	**85:** 138543r
3-propyl-5-*o*-tolyl-	**55:** 14441a
3-pyrazinyl)-5-(2-pyridinyl)-	**77:** P5480f
3-pyrazinyl-5-(4-pyridinyl)-	**77:** P5480f, **83:** 188199n
3-(4-pyridazinyl)-5-(4-pyridinyl)-	**77:** P5480f, **83:** 188199n
3,3'-[(2,6-pyridinedicarboxamido)bis-2,1- phenylene]bis[5-phenyl-	**80:** 15236g, **80:** 59912d
3-(2-pyridinyl)-5-(3-pyridinyl)-	**77:** P5487p
3-(2-pyridinyl)-5-(4-pyridinyl)-	**77:** P5487p, **83:** 188199n
3-(4-pyridinyl)-5-[(2-pyridinyl)methyl]-	**77:** P5480f
3-(4-pyridinyl)-5-[(3-pyridinyl)methyl]-	**77:** P5480f
3-(4-pyridinyl)-5-(2-pyrimidinyl)-	**77:** P5480f, **83:** 188199n
3-(4-pyridinyl)-5-(4-pyrimidinyl)-	**77:** P5480f, **83:** 188199n
3-(4-pyridinyl)-5-(6-quinolinyl)-	**77:** P5480f, **83:** 188199n
3-(4-pyridinyl)-5-(4-sulfamoyl)-	**77:** P5480f, **83:** 188199n
3-(4-pyridinyl)-5-(2-thienyl)-	**77:** P5480f, **83:** 188199n
3-(3-pyridyl)-5-(4-pyridyl)-	**68:** 114391n
3-(selenophene-2-yl)-5-(selenophene-3-yl)-	**85:** 177373e
3-selenophene-2-yl-5-(2-thienyl)-	**80:** 82840n, **85:** 177373e
3-selenophene-3-yl-5-(2-thienyl)-	**80:** 82840n, **85:** 177373e
3-(2-thienyl)-5-(3-thienyl)-	**80:** 82840n, **85:** 177373e

CHAPTER 3

Alkyl- or Aryl-Trisubstituted 1,2,4-Triazoles

3.1. 1,3,5-Alkyl- or Aryl-Trisubstituted 1H-1,2,4-Triazoles

The condensation of diacylamines with hydrazine or its monoalkyl or aryl derivatives to give triazoles has been named the Einhorn–Brunner reaction.[*,1,2] This reaction is catalyzed by weak acids (acetic acid, acidic buffers, pyridine hydrochloride), but is inhibited by strong acids, which increases the yields of by-products.[3] Satisfactory yields of triazoles are obtained at moderate temperatures when the acyl moieties of the diacylamine are derived from aliphatic or aryl acids or from a combination of both.[3–7] The 3-substituent in triazoles formed from unsymmetrical diacylamine is derived from the acyl moiety corresponding to the stronger acid, indicating that 3-acylamidrazones are intermediates (e.g., **3.1-2**).[3] For example, the condensation of the hydrochloride of phenylhydrazine (**3.1-1**) and *N*-acetylbenzamide under a variety of conditions gave **3.1-3** and none of **3.1-5**.[3] However, the latter is preferentially formed by the reaction of **3.1-1** with *N*-acetylthiobenzamide, in which the thio group is initially displaced to give **3.1-4**.[8] In a related

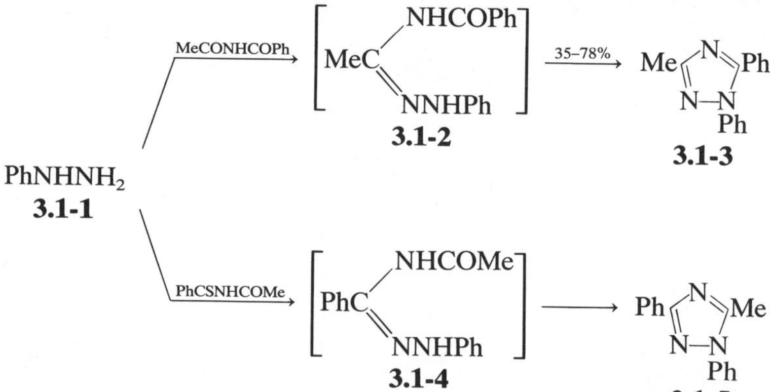

MeCONHCOPh
NHCOPh
MeC
NNHPh
35–78%
Me
N
Ph
N—N
Ph
PhNHNH$_2$
PhCSNHCOMe
NHCOMe
PhC
NNHPh
Ph
N
Me
N—N
Ph

3.1-1 **3.1-2** **3.1-3** **3.1-4** **3.1-5**

* A. Einhorn, E. Bischkopff, C. Ladisch, T. Mauermayer, G. Schupps, and B. Szelinski, *Justus Liebigs Ann. Chem.*, **343,** 207 (1905).

reaction the alkylation of phenylhydrazine with the polychloroazaalkene **3.1-6** afforded an excellent yield of the 3,5-bis(trichloromethyl)triazole (**3.1-7**) (92%).[9] The polychloroazaalkene (**3.1-8**) has also been used to prepare triazoles.[10]

Reaction of the imino ether (**3.1-9**) with methylhydrazine provided the 1-methylamidrazone, 1-benzimidozyl-2-methylhydrazine (**3.1-10**), rather than the 2-methylamidrazone (**3.1-12**). Cyclization of **3.1-10** with acetic anhydride gave the 1,5-dimethyltriazole (**3.1-11**), not identical with the 1,3-dimethyltriazole (**3.1-13**) obtained by the reaction of *N*-acetylbenzamide with methylhydrazine (Einhorn–Brunner reaction).[5] Amidrazones containing a phenyl substituent in place of the methyl of **3.1-10** have also been cyclized with acetic anhydride to give triazoles.[11,12]

In the preparation of the 3,5-di(4-pyridyl)triazole (**3.1-16**) (37%), the 1-acylamidrazone (**3.1-15**) was generated *in situ* by the condensation of the imino ether (**3.1-14**) with isonicotinoylhydrazine.[13]

Compounds (e.g., **3.1-18**) similar to the 3-acylamidrazones, proposed as intermediates for the Einhorn–Brunner reaction, have been isolated by treatment of the acyl derivatives of imino ethers (**3.1-17**) with *N*-methyl-*N*-phenylhydrazine.[14] Although these products (**3.1-18**) cannot undergo cyclization to triazoles, the reaction of **3.1-17** with phenylhydrazine occurred instantaneously and exothermally to give a series of 5-alkyl- and 5-aryl-1,3-diphenyl-1H-1,2,4-triazoles (**3.1-20** to **3.1-22**). The results indicated that the 3-acylamidrazones intermediates (**3.1-19**) underwent dehydration

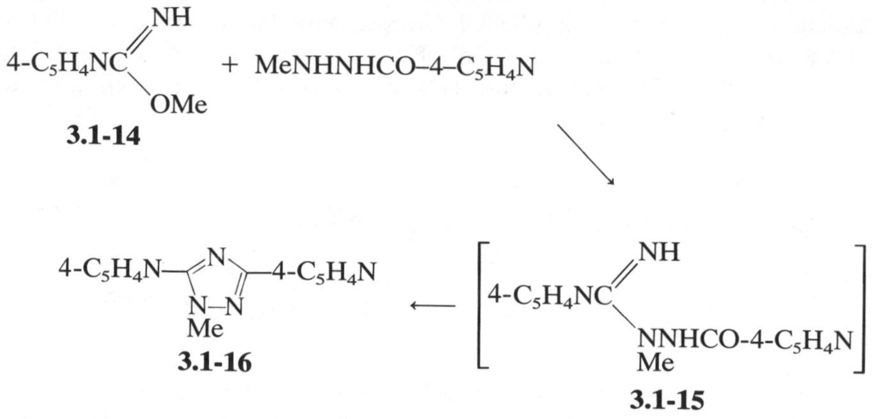

$$4\text{-}C_5H_4NC\underset{\text{OMe}}{\overset{\text{NH}}{<}} \quad + \quad MeNHNHCO\text{-}4\text{-}C_5H_4N$$

3.1-14

$$4\text{-}C_5H_4N\underset{\underset{\text{Me}}{N-N}}{\overset{N}{\diagdown}}4\text{-}C_5H_4N \quad \longleftarrow \quad \left[4\text{-}C_5H_4NC\underset{\underset{\text{Me}}{NNHCO\text{-}4\text{-}C_5H_4N}}{\overset{NH}{<}} \right]$$

3.1-16   **3.1-15**

readily.[15–18] Many of these triazoles (**3.1-20–21**) were also prepared by the condensation of the amidrazone (**3.1-23**) with aldehydes to give the triazolines (**3.1-24**), followed by oxidation of the latter with potassium

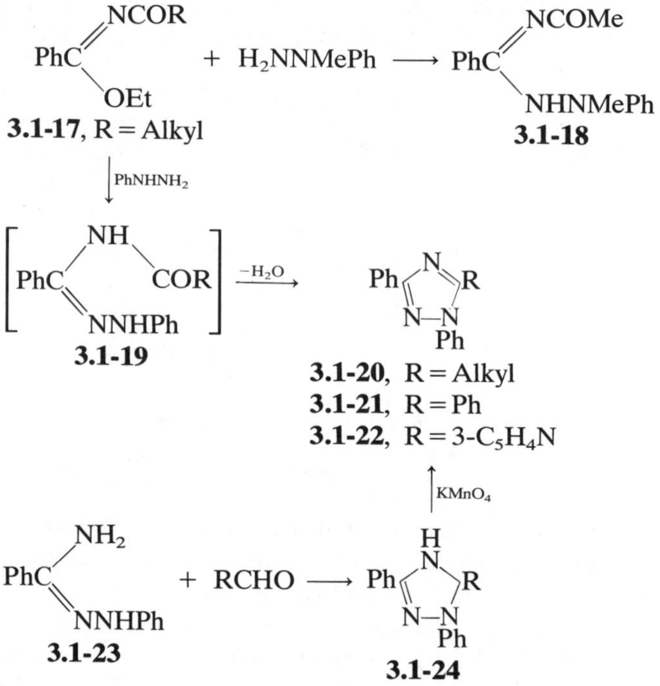

$$PhC\underset{\text{OEt}}{\overset{NCOR}{<}} \quad + \quad H_2NNMePh \quad \longrightarrow \quad PhC\underset{\text{NHNMePh}}{\overset{NCOMe}{<}}$$

3.1-17, R = Alkyl   **3.1-18**

$$\downarrow PhNHNH_2$$

$$\left[PhC\underset{\text{NNHPh}}{\overset{NH}{<}}COR \right] \overset{-H_2O}{\longrightarrow} \quad Ph\underset{\underset{\text{Ph}}{N-N}}{\overset{N}{\diagup}}R$$

3.1-19

3.1-20, R = Alkyl
3.1-21, R = Ph
3.1-22, R = 3-C$_5$H$_4$N

$$\uparrow KMnO_4$$

$$PhC\underset{\text{NNHPh}}{\overset{NH_2}{<}} \quad + \quad RCHO \quad \longrightarrow \quad Ph\underset{\underset{\text{Ph}}{N-N}}{\overset{\overset{H}{N}}{\diagup}}R$$

3.1-23   **3.1-24**

permanganate in acetic acid.[19] In some reactions of this type, the inter-mediate triazoline is oxidized spontaneously to the triazole.[20] In addition, amidrazones might be intermediates in the conversion of a number of ring systems to triazoles. Treatment of 4-arylazo-2-phenyloxazolin-5-ones (e.g., **3.1-25**) with excess phenylmagnesium bromide gave good yields of the triazolemethanols (**3.1-27**), presumably formed via the 3-acylamidrazone intermediates (**3.1-26**).[21]

In the reaction of the 2-oxazolin-4-one **3.1-28** with phenylhydrazine, a quantitative yield of 5-(hydroxymethyl)triazole (**3.1-30**) was formed, presumably via **3.1-29**.[22] Apparently, the 3-acylamidrazone (**3.1-32**) was an intermediate in the reaction of benzoyl derivative of 3-amino-1,5-diphenylpyrazole (**3.1-31**) with phenylhydrazine to give **3.1-33**.[23]

Imino derivatives of the 3-acylamidrazones (**3.1-35**) are possible intermediates in the reaction of arylhydrazines with 4,6-diaryl-1,2,3,5-oxathiadiazine 2,2-dioxides (e.g., **3.1-34**) under mild conditions to give good yields of 1,3,5-triaryltriazoles (e.g., **3.1-36**).[24]

Huisgen has developed a method for the preparation of 1,3,5-triaryl-, 1-alkyl-3,5-diaryl-, and 5-alkyl- and 5-substituted alkyl-1,3-diaryl-1H-1,2,4-triazoles in good yields from nitrilimine intermediates (e.g., **3.1-38**).[25–27] The latter are generated by the thermal decomposition of 2,5-disubstituted

3.1-34 → H$_2$NNHC$_6$H$_4$-4-NO$_2$, 80° → **3.1-35** → −SO$_3$, −NH$_3$ → **3.1-36** (91%)

tetrazoles and undergo addition reactions with a variety of compounds containing multiple bonds. For example, heating a mixture of 2,5-diphenyltetrazole (**3.1-37**) and 4-chlorobenzonitrile at 160° gave **3.1-39**.[26] Similarly, a mixture of **3.1-37** and phenoxyacetonitrile gave an 80% yield of the 5-phenoxymethyltriazole (**3.1-40**). In addition, nitrilimine intermediates are generated by the thermolysis of the 1,2,4-oxadiazolin-5-one (**3.1-41**)[28]

3.1-37 $\xrightarrow{160°}_{-N_2}$ [PhC≡N⁺N⁻Ph] **3.1-38**

3.1-39 (52%)　　　　**3.1-40**

and the 1,3,4-oxadiazoline (**3.1-42**).[29] The formation of nitrilimine intermediates at lower temperatures was realized by the dehydrochlorination of

3.1-41　　　　**3.1-42**

α-chlorobenzylidenephenylhydrazones either with triethylamine[26,27] or with a Lewis acid such as aluminum chloride.[30] Treatment of **3.1-43** with triethylamine in benzene at 80° in the presence of the aldoxime (**3.1-44**) gave **3.1-45** formed via the triazoline (**3.1-46**).[27] The 1,3-addition of each azomethine linkage of **3.1-47** to **3.1-38** gave the triphenyltriazole (**3.1-21**) and the corresponding dihydrotriazole (**3.1-49**), which results from cleavage of the N—N bond of the intermediate (**3.1-48**).[27]

The 1,3,5-trisubstituted triazoles can be prepared from several other ring systems.[31,32] For example, a high yield of **3.1-51** was obtained by heating the 1H-2,3-benzoxazine **3.1-50** in aqueous acid.[33]

As might be expected from the relative nucleophilicities of hydrazine and ammonia, the alkylation of 3,5-disubstituted-s-triazoles in which the 3,5-substituents are identical gave mainly 1-alkyl- rather than 4-alkyltriazoles. This was demonstrated by methylation of 3,5-dimethyl-s-triazole (**3.1-52**) with either diazomethane in methanol or methyl iodide in methanolic sodium methoxide to give the 1-methyltriazole (**3.1-53**) (81%) and none of the isomeric 4-methyl compound (**3.1-55**).[7,34,35] However, ethylation of **3.1-52** with ethyl iodide in sodium ethoxide gave **3.1-54** (58%) and a small

3.1-52 → **3.1-53**, R = Me   **3.1-55**, R = Me
              3.1-54, R = Et   **3.1-56**, R = Et

amount of **3.1-56** (1%). In the alkylation of 3-methyl-5-phenyl-s-triazoles (**3.1-57**)[34] and other unsymmetrical 3,5-disubstituted s-triazole,[36,37] substitution occurred only in the N—N linkage, but the structure of the product depended upon the nature of the alkylation reagent. Methylation of **3.1-57** with methyl iodide in the presence of sodium methoxide gave only **3.1-11** (90%), whereas treatment of **3.1-57** with diazomethane gave a mixture of

3.1-57   **3.1-11**   **3.1-13**

3.1-11 (10%) and **3.1-13** (37%). These results are attributed to the lower electron density in the nitrogen adjacent to the phenyl group in the reaction of diazomethane with the neutral molecule of **3.1-57** and to the greater electron density on the nitrogen adjacent to the methyl group in the reaction of methyl iodide with the anion of **3.1-57**. In addition to the reagents described previously, 3,5-disubstituted triazoles have been alkylated with epoxides,[38,39] ethylene carbonate,[40] and acetylene.[41] The 3,5-diethynyltriazole (**3.1-60**) was reported to result from the interaction of the 3,5-diiodotriazole (**3.1-58**) with 2-hydroxy-2-methyl-3-butyne to give

3.1-58   **3.1-59**   **3.1-60**

3.1-61   **3.1-62**

3.1-59 followed by base cleavage of the side chains of **3.1-59**.[42] In addition, the lithio derivative of **3.1-61** condensed with benzophenone to give **3.1-62**.[43]

References

1. **9:** 210	2. **9:** 3058	3. **47:** 4335a	4. **11:** 1412
5. **49:** 11630g	6. **49:** 15915i	7. **76:** 85759w	8. **54:** 15371d
9. **66:** 10907m	10. **85:** 21300t	11. **79:** 66285j	12. **79:** 146129b
13. **83:** 188199n	14. **61:** 13232d	15. **69:** 96601v	16. **70:** 28872e
17. **74:** 125574z	18. **76:** 34177x	19. **61:** 7007a	20. **84:** 90081p
21. **55:** 24729e	22. **73:** 56038t	23. **59:** 7530e	24. **59:** 11493d
25. **54:** 515a	26. **57:** 11198d	27. **62:** 13147e	28. **69:** 36039h
29. **81:** 104454q	30. **80:** 82818m	31. **81:** 37521u	32. **82:** 72909u
33. **77:** 61916q	34. **49:** 1710i	35. **68:** 29645d	36. **64:** 19596c
37. **77:** 139906h	38. **69:** 77265x	39. **82:** 170962a	40. **81:** 105517z
41. **83:** 114414g	42. **83:** 28165v	43. **84:** 59312r	

3.2. 3,4,5-Alkyl- or Aryl-Trisubstituted 4H-1,2,4-Triazoles

The condensation of diaroylhydrazines and amines in the presence of a dehydration agent such as phosphorus pentoxide or anhydrous zinc chloride was one of the first methods used to prepare 3,4,5-trisubstituted triazoles. Under these conditions, however, this method has little preparative value.[1,2] In an improved procedure developed by Klingsberg[3] for the preparation of

$$PhCO \quad COPh \xrightarrow[\text{3.2-2}]{4\text{-}ClC_6H_4N=PNHC_6H_4\text{-}4\text{-}Cl}$$

$$\underset{\text{NHNH}}{\diagdown \diagup}$$

3.2-1

$$Ph \underset{N-N}{\overset{4\text{-}ClC_6H_4}{\overset{|}{\underset{N}{\diagup}}}} Ph$$

3.2-3

triaryltriazoles, arylamines are treated with phosphorus trichloride to give N,N'-diarylphosphenimidous amides (**3.2-2**), which condensed readily with diaroylhydrazines (**3.2-1**) in refluxing 2-dichlorobenzene to give high yields of triazoles (**3.2-3**). Apparently this reaction is limited to the preparation of triaryltriazoles since no alkyl triazoles were obtained when either the dihydrazide or the phosphorus adduct contained an alkyl group. Higher yields of triazoles were also obtained by the chlorodehydroxylation of a diaroylhydrazine with phosphorus pentachloride to give a dichloroaldazine followed by a reaction of the latter with an amine.*† For example, the dichloro-

---

* R. Stollé and K. Thomä, *J. Prakt. Chem.*, **73**, 288 (1906).
† R. Stollé and A. Bamback, *J. Prakt. Chem.*, **74**, 13 (1906).

$$2\text{-}(C_{14}H_7O_2)C \overset{Cl\ Cl}{\underset{N\ N}{\diagdown}} CPh \xrightarrow{PhNH_2} 2\text{-}(C_{14}H_7O_2) \overset{Ph}{\underset{N-N}{\diagdown}} Ph$$

3.2-4   **3.2-5**

aldazine (**3.2-4**), prepared from 1-(2-anthraquinonylcarbonyl)-2-benzoylhydrazine, was reacted with aniline at 175° to give **3.2-5**.[1] In addition to the 3,4,5-triaryltriazoles,[4-8] this method has been used to prepare a series of 4-alkyl-3,5-diaryltriazoles.[9-11]

Although a variety of methods have been investigated for the preparation of 3,4,5-trisubstituted-1,2,4-4$\underline{H}$-triazole, most of these procedures involve some type of amidrazone intermediate. The condensation of the amidrazone **3.2-6** with 1-diethylamino-1-propyne and ethoxyacetylene, respectively, gave the 3-ethyl-[12] and 3-methyltriazoles[13] (**3.2-7** and **3.2-8**). The aroylation

$$PhC \overset{NHPh}{\underset{NNH_2}{\diagdown}} \quad \overset{Et_2NC\equiv CMe}{\underset{EtOC\equiv CH}{\diagup}} \quad Ph \overset{Ph}{\underset{N-N}{\diagdown}} R$$

3.2-6   **3.2-7,** R = Et
                                             3.2-8, R = Me

of the amidrazone (**3.2-9**) with 4-nitrobenzoyl chloride gave the 1-aroylamidrazone (**3.2-10**), which underwent intramolecular cyclization to

$$2\text{-}C_5H_4NC \overset{NHPh}{\underset{NNH_2}{\diagdown}} \longrightarrow 2\text{-}C_5H_4NC \overset{NHPh}{\underset{NNHCOC_6H_4\text{-}4\text{-}NO_2}{\diagdown}}$$

3.2-9   **3.2-10**

$$2\text{-}C_5H_4N \overset{Ph}{\underset{N-N}{\diagdown}} C_6H_4\text{-}4\text{-}NO_2$$

3.2-11

give **3.2-11**.[14,15] Usually the 1-acylamidrazones are dehydrated by heat alone, but an alkaline medium can also be used satisfactorily.[16] In the conversion of **3.2-12** to **3.2-13**, heat and the presence of phosphorus pentaoxide was required for dehydration.[17] The most common method for the *in situ* formation of 1-acylamidrazones involves the interaction of a hydrazide with an alkyl-[18] or arylimidoyl chloride.[19] For example, the

$$\text{C}_3\text{F}_7\text{C}\overset{\text{NHMe}}{\underset{\text{NNHCOPh}}{\Big\langle}} \quad\xrightarrow[\Delta]{\text{P}_2\text{O}_5}\quad \text{C}_3\text{F}_7\text{—}\underset{\text{N—N}}{\overset{\text{Me}}{N}}\text{—Ph}$$

3.2-12   **3.2-13**

amidrazones (**3.6-16** and **3.6-18**, respectively), are apparently formed in the reaction of either *N*-phenylbenzimidoyl chloride (**3.2-14**) with 2-anthra-quinonecarboxylic acid hydrazide (**3.2-15**) or *N*-phenyl-2-anthraquinone-carboximidoyl chloride (**3.2-17**) with benzhydrazide to give the same

$$\text{PhC}\overset{\text{NPh}}{\underset{\text{Cl}}{\Big\langle}} \quad\xrightarrow[\text{3.2-15}]{\text{H}_2\text{NNHCO-2-(C}_{14}\text{H}_7\text{O}_2)}\quad \left[\text{PhC}\overset{\text{NPh}}{\underset{\text{NHNHCO-2-(C}_{14}\text{H}_7\text{O}_2)}{\Big\langle}}\right]$$

3.2-14   **3.2-16**

$$2\text{-(C}_{14}\text{H}_7\text{O}_2)\text{—}\underset{\text{N—N}}{\overset{\text{Ph}}{N}}\text{—Ph}$$

3.2-5

$$2\text{-(C}_{14}\text{H}_7\text{O}_2)\text{C}\overset{\text{NPh}}{\underset{\text{Cl}}{\Big\langle}} \quad\xrightarrow{\text{H}_2\text{NNHCOPh}}\quad \left[2\text{-(C}_{14}\text{H}_7\text{O}_2)\text{C}\overset{\text{NPh}}{\underset{\text{NHNHCOPh}}{\Big\langle}}\right]$$

3.2-17   **3.2-18**

triazole (**3.2-5**).[1] Similar procedures were used in the reaction of *N*-alkyl-benzimidoyl chlorides with both alkyl- and arylhydrazides to give 3,4-dialkyl-5-aryl- and 4-alkyl-3,5-diaryl-1,2,4-4$\underline{H}$-triazoles.[9,10,20]

A unique method for the preparation of 1-aroylamidrazones for conversion to 3,4-dialkyl-5-aryltriazole was developed by Westermann. A trans-amination reaction between benzhydrazide and the hydrochloride of the imino ether (**3.2-19**) in ethanol provided the *N*-substituted imino ether (**3.2-20**), which on treatment with methylamine gave the triazole (**3.2-21**) formed via the amidrazone intermediate (**3.2-22**).[21]

$$\text{PhCH}_2\text{O}_2\text{CNHCH}_2\text{C}\overset{\text{OEt}}{\underset{\text{NH}_2^+\text{Cl}^-}{\Big\langle}} \quad\xrightarrow{\text{H}_2\text{NNHCOPh}}\quad \text{PhCH}_2\text{O}_2\text{CNHCH}_2\text{C}\overset{\text{OEt}}{\underset{\text{NNHCOPh}}{\Big\langle}}$$

3.2-19   **3.2-20**

$$\Big\downarrow \text{MeNH}_2$$

$$\text{PhCH}_2\text{O}_2\text{CNHCH}_2\text{—}\underset{\text{N—N}}{\overset{\text{Me}}{N}}\text{—Ph} \quad\xleftarrow{79\%}\quad \left[\text{PhCH}_2\text{O}_2\text{CNHCH}_2\text{C}\overset{\text{NHMe}}{\underset{\text{NNHCOPh}}{\Big\langle}}\right]$$

3.2-21   **3.2-22**

Reaction of 2,5-diphenyl-1,3,4-oxadiazole (**3.2-23**) with equimolar amounts of aryl amines at elevated temperatures (270 to 310°) gave triaryltriazoles (**3.2-25**), apparently formed via 1-aroylamidrazone intermediates (**3.2-24**).[22,23] Good yields were obtained for a series of substituted anilines,

with the exception of 2-anisidine, which gave a 29% yield of the triazole. Similarly, 2,5-dimethyl-1,3,4-oxadazole was the starting material for the preparation of 3,4,5-trimethyl-1,2,4-4$\underline{H}$-triazole and some 3,5-dimethyl-4-aryltriazoles in moderate yields.[24–26]

In contrast to ammonia, which reacted in a 1:1 ratio with 2,5-bis(perfluoromethyl)-1,3,4-oxadiazole (**3.2-26**) (see Section 2.4) to give 1-aroylamidrazones, treatment of **3.2-26** with methylamine at room temperature gave 1,2-bis(N-methylperfluoroacetylimidoyl)hydrazine (**3.2-27**).[27] Ring closure of the latter was effected with phosphorus pentaoxide at 85° to give 3,5-bis(perfluoromethyl)-4-methyl-1,2,4-4$\underline{H}$-triazole (**3.2-28**). Good yields

of the perfluoroethyl and perfluoropropyl triazoles were obtained by the same procedure. Intermediates (**3.2-32**) of the same type as **3.2-27** were indicated by the obtainment of 3,4,5-triaryl (e.g., **3.2-31**) and 4-alkyl-3,5-diaryltriazoles either from amidrazones (e.g., **3.2-6**) in the presence of formic acid[28] or from amidrazones generated via **3.2-30** by the thermal decomposition of 5-phenyltetrazole (**3.2-29**) in the presence of aryl amines.[29]

Many of the triaryltriazoles (e.g., **3.2-11**) described above have also been prepared in good yields by oxidation of the benzylidene derivatives of amidrazones (**3.2-33**) with mercuric oxide or silver oxide in refluxing xylene.[14] In an unusual reaction the hydrazones derived from amidrazones and a β-keto acid ester or β-diketone were converted to 3-alkyl-4,5-diaryltriazoles. For example, when **3.2-34** was heated in the melted state or

in xylene, cyclization occurred with cleavage of the α,β-C—C bond of the carbonyl moiety to give the triazole (**3.2-8**) in good yield.[13,30] In a related reaction the methyleneamidrazone (**3.2-35**) was converted to 3,4,5-triphenyl-1,2,4-4H-triazole (**3.2-32**) in refluxing dimethylformamide.[31,32]

In his work with tetrazoles, Huisgen discovered an elegant method for the preparation of a series of 3,4,5-triaryl- and several 3-alkyl-4,5-diaryltriazoles. This method is illustrated by the interaction of 5-phenyltetrazole (**3.2-29**) with the imino chloride (**3.2-36**) in refluxing pyridine to give the intermediate (**3.2-37**), which underwent intramolecular cyclization with loss of nitrogen to give **3.2-3** (77%).[33,34] In related reactions both the oxazolium and thiazolium compounds (**3.2-38** and **3.2-39**) are converted with diethyl azodicarboxylate to **3.2-40** in good yields.[35,36]

3.2-29 + **3.2-36** $\xrightarrow{\underset{\Delta}{C_5H_5N}}$ **3.2-37**

$\xrightarrow{-N_2}$

3.2-3 ⟵ **3.2-37 intermediate**

3.2-38, X = O
3.2-39, X = S $\xrightarrow{EtO_2CN=NCO_2Et}$ **3.2-40**

In the conversion of other ring systems to triazoles, the quinazolin-4-one (**3.2-41**) was rearranged in base to give **3.2-42**.[18] Similarly, the dihydro-quinazoline (**3.2-43**) was rearranged in formic acid to give **3.2-44**.[37] The latter is an important intermediate in the synthesis of benzodiazepines (central nervous system agents) and has also been prepared by a route involving the electrophilic addition of formaldehyde to a triazole (e.g., **3.2-45**) to give a hydroxymethyl derivative (e.g., **3.2-46**).[38,39] In addition, the triazole (**3.2-48**) has been shown to undergo the Mannich reaction with pyrrolidine to give **3.2-47**.[40]

3.2-41 $\xrightarrow{OH^-}$ **3.2-42**

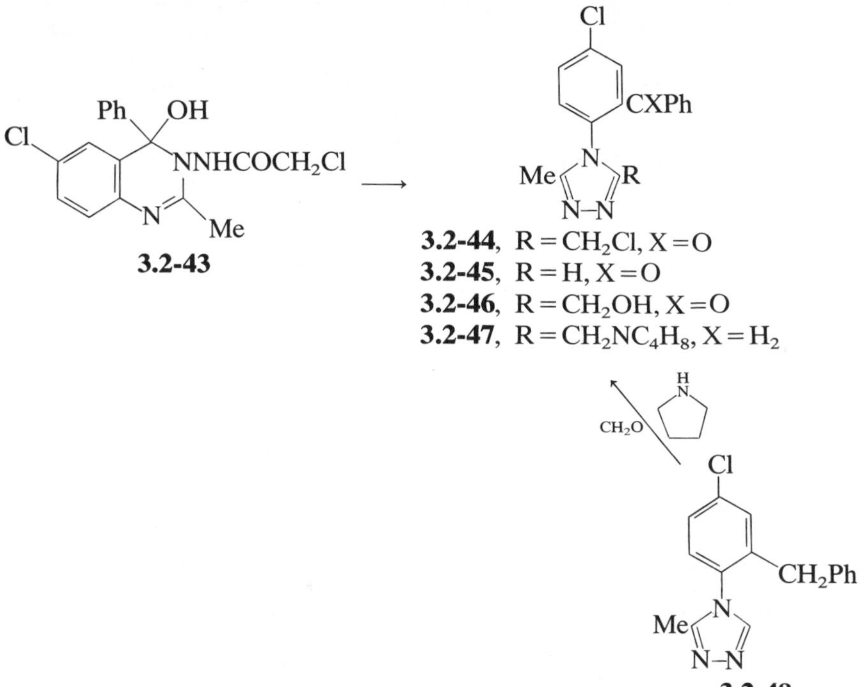

3.2-43

3.2-44, R = CH$_2$Cl, X = O
3.2-45, R = H, X = O
3.2-46, R = CH$_2$OH, X = O
3.2-47, R = CH$_2$NC$_4$H$_8$, X = H$_2$

3.2-48

References

1. **53:** 9200b	2. **68:** 39628q	3. **53:** 16121a	4. **73:** 130941a
5. **75:** 118255p	6. **77:** 101478g	7. **83:** 206172e	8. **84:** 17236p
9. **62:** 567d	10. **62:** 11829c	11. **77:** 34431w	12. **73:** 109744d
13. **74:** 87898y	14. **60:** 4131c	15. **69:** 19082z	16. **71:** 81268j
17. **70:** 77879w	18. **72:** 43550h	19. **72:** 68047n	20. **76:** 85759w
21. **62:** 1647a	22. **67:** 100076f	23. **72:** 44284m	24. **51:** 16162i
25. **54:** 7695e	26. **69:** 106627z	27. **57:** 12470i	28. **66:** 46412b
29. **57:** 5909c	30. **66:** 46376t	31. **54:** 21063g	32. **69:** 51807a
33. **53:** 5253e	34. **55:** 6473g	35. **75:** 20254n	36. **75:** 35845e
37. **79:** 92176u	38. **77:** 126708v	39. **81:** 152289m	40. **85:** 56539e

TABLE 3.   ALKYL- OR ARYL-TRISUBSTITUTED 1,2,4-TRIAZOLES

Compound	Reference

3.1. 1,3,5-Alkyl- or Aryl-Trisubstituted 1H-1,2,4-triazoles

Compound	Reference
1-acetamide, N-[5-[[4-[2,4-bis(1,1-dimethylpropyl)-phenoxy]-1-oxobutyl]amino]-2-chlorophenyl]-α-(2,2-dimethyl-1-oxopropyl)-5-methyl-3-phenyl-	**85:** P102316e
-1-acetamide, N,N-dimethyl-3,5-dipyrazinyl-	**77:** P5480f
1-(2-acetamidophenyl)-5-methyl-3-phenyl-	**79:** 146129b
5-acetic acid, 1,3-diphenyl-	**85:** P94369g
5-acetic acid, 1,3-diphenyl-, ethyl ester	**85:** 94369g
-5-acetonitrile, 3-(p-formylphenyl)-1-phenyl-, chlorophenylhydrazone	**74:** P118381f, **75:** P93013t
1-[2'-(acetylamino)-1,1'-biphenyl-2-yl]-5-methyl-3-phenyl	**79:** 66285j
1-[2'-[acetyl(methyl)amino]-1,1'-biphenyl-2-yl]-5-methyl-3-phenyl-	**79:** 66285j
-3-acrylic acid, 1,5-diphenyl-	**55:** 2625d, **57:** 807e
-3-acrylic acid, 1-(p-methoxyphenyl)-5-phenyl-	**55:** 2625d, **57:** 807e
-3-acrylic acid, 5-phenyl-1-p-tolyl-	**55:** 2625d, **57:** 807e
1-(4-amino-2-chlorophenyl)-3,5-dimethyl-	**80:** P49275j
1-(5-amino-2-chlorophenyl)-3,5-dimethyl-	**80:** P49275j
1-(m-aminophenyl)-3,5-dimethyl-	**24:** 2130[5]
1-(o-aminophenyl)-3,5-dimethyl-	**24:** 2130[5]
1-(p-aminophenyl)-3,5-dimethyl-	**24:** 2130[5]
3-(3-aminophenyl-1-methyl-5-(2-methylphenyl)-	**78:** P159619r
1-[2-(2-aminophenyl)phenyl]-5-methyl-3-phenyl-	**82:** 72909u
3-(anilinomethyl)-1,5-diphenyl-	**57:** 806f
1-(benzamidophenyl)-3,5-diethyl-	**21:** 3200[9]
1-(benzamidophenyl)-3,5-dimethyl-	**21:** 3200[8]
3-(1,3-benzodioxol-5-yl)-5-[(2-hydroxymethyl)phenyl]-1-methyl-	**78:** P135625p
3-(1,3-benzodioxol-5-yl)-1-methyl-5-(2-methylphenyl)-	**78:** P159619r
1-benzyl-3,5-dimethyl-	**64:** 19596g
5-benzyl-1,3-diphenyl-	**80:** 82818m
1-benzyl-3,5-di-4-pyridinyl-	**83:** P164184c
3-(5-benzylidine-4-oxoimidazolidin-2-yl)-1-(2-methylphenyl)-5-phenyl-	**83:** 10007k
3-(4-benzylidene-5(4H)-oxo-2-oxazolyl)-1,5-diphenyl-	**83:** 10007k
3-(4-benzylidene-5(4H)-oxo-2-oxazolyl)-1-(2-methylphenyl)-5-phenyl-	**83:** 10007k
3-(4-benzylidene-5(4H)-oxo-2-oxazolyl)-1-(3-methylphenyl)-5-phenyl-	**83:** 10007k
3-(4-benzylidene-5(4H)-oxo-2-oxazolyl)-1-(4-methylphenyl)-5-phenyl-	**83:** 10007k
3-benzyl-1-phenyl-5-propenyl-	**58:** 2447h
3,5-bis[4-(2-[1,1'-biphenyl]-4-ylethenyl)phenyl]-1-phenyl-	**68:** P21961h, **80:** P146981u
3,5-bis(3-chlorophenyl)-1-methyl-	**78:** P159619r
3,5-bis(p-chlorophenyl)-1-methyl-	**57:** 11199i
3,5-bis[2-(dodecylthio)ethyl]-1-(2-hydroxyethyl)-	**81:** P105517z, **82:** 170962a

TABLE 3 (*Continued*)

Compound	Reference

3.1. 1,3,5-Alkyl- or Aryl-Trisubstituted 1H̲-1,2,4-triazoles (*Continued*)

Compound	Reference
3,5-bis(2-hydroxy-2-methyl-3-butyn-4-yl)-1-methyl-	**83:** 28165v
3,5-bis(3-methoxyphenyl)-1-[(3-methoxyphenyl)methyl]-	**84:** 90123d
3,5-bis(4-methoxyphenyl)-1-[(4-methoxyphenyl)methyl]-	**84:** 90123d
3,5-bis(4-methylphenyl)-1-phenyl-	**80:** P146981u
3,5-bis(4-pyridinyl)-1-methyl-	**77:** P5487p, **83:** 188199n
3,5-bis(4-pyridinyl)-1-methyl-, pyridine *N*-oxides	**77:** P5487p
5-[2-(bromomethyl)phenyl]-3-(3-chlorophenyl)-1-methyl-	**78:** P159619r
5-[2-(bromomethyl)phenyl]-1-methyl-3-(3-methylphenyl)-	**78:** P159619r
5-[2-(bromomethyl)phenyl]-1-methyl-3-(4-nitrophenyl)-	**77:** 61916q, **78:** P159619r
5-[2-(bromomethyl)phenyl]-1-methyl-3-phenyl-	**78:** P159619r
1-(*p*-bromophenyl)-3,5-dimethyl-	**21:** 3200^8, **21:** 3620^4
1-butyl-3,5-dicyclohexyl-	**85:** P178509r
5-butyl-1,3-diphenyl-	**61:** 13232e
5-[2-[[(carbamoyl)oxy]methyl]phenyl]-3-(2-chlorophenyl)-1-methyl-	**78:** P159619r
5-[2-[[(carbamoyl)oxy]methyl]phenyl]-3-(3-chlorophenyl)-1-methyl	**78:** P159619r
5-[2-[[(carbamoyl)oxy]methyl]phenyl]-3-(4-chlorophenyl)-1-methyl-	**78:** P159619r
5-[2-[[(carbamoyl)oxy]methyl]phenyl]-1-methyl-3-(2-methylphenyl)-	**78:** P159619r
5-[2-[[(carabamoyl)oxy]methyl]phenyl]-1-methyl-3-(3-methylphenyl)-	**78:** P159619r
5-[2-[[(carbamoyl)oxy]methyl]phenyl]-1-methyl-3-(4-methylphenyl)-	**78:** P159619r
5-(2-carboxybenzyl)-1,3-dimethyl-	**81:** 3752u
3-(2-carboxybenzyl)-1,5-dimethyl-	**81:** 37521u
5-(2-carboxybenzyl)-1-methyl-3-phenyl-	**81:** 37521u
5-(2-carboxybenzyl)-1-methyl-3-phenyl-, methyl ester	**81:** 37521u
3-(2-carboxybenzyl)-1-methyl-5-phenyl-	**81:** 37521u
5-(carboxymethyl)-1,3-diphenyl-, ethyl ester	**80:** 82818m
1-(*m*-carboxyphenyl)-3,5-dimethyl-	**24:** 368^{3,4,5}
1-(*o*-carboxyphenyl)-3,5-dimethyl-	**24:** 368^{3,45}
1-(*p*-carboxyphenyl)-3,5-dimethyl-	**24:** 368^{3,45}
1-(4-carboxyphenyl)-5-(hydroxymethyl)-3-phenyl-	**73:** 56038t
1-(4-chlorobenzyl)-5-methyl-3-phenyl-	**77:** 139906h
1-(3-chlorobenzyl)-3,5-di-4-pyridinyl-	**83:** P164184c
1-(4-chlorobenzyl)-3,5-di-4-pyridinyl-	**83:** P164184c
1-(2-chloroethyl)-3,5-diethyl-	**81:** P120637g
1-(2-chloroethyl)-3,5-diheptyl-	**81:** P120637g
1-(2-chloroethyl)-3,5-dimethyl-	**81:** P120637g
1-(2-chloroethyl)-3,5-diphenyl-	**81:** P120637g, **83:** P164183b
5-(2-chloroethyl)-1,3-diphenyl-	**80:** 82818m
1-(2-chloroethyl)-3,5-diundecyl-	**81:** P120637g
5-[5-chloro-2-(hydroxymethyl)phenyl]-1-methyl-3-(4-nitrophenyl)-	**77:** 61916q, **78:** P135625p
5-[5-chloro-2-(hydroxymethyl)phenyl]-1-methyl-3-phenyl-	**78:** P135625p

77

TABLE 3 (*Continued*)

Compound	Reference

3.1. 1,3,5-Alkyl- or Aryl-Trisubstituted 1H-1,2,4-triazoles (*Continued*)

Compound	Reference
1-[2-chloro-4-(7-methoxy-2-oxo-2H-1-benzopyran-3-yl)phenyl]-3,5-dimethyl-	**80:** P49275j
1-[2-chloro-5-(7-methoxy-2-oxo-2H-1-benzopyran-3-yl)phenyl]-3,5-dimethyl-	**80:** P49275j
5-(chloromethyl)-1,3-diphenyl-	**73:** 56038t
3-(chloromethyl)-1,5-diphenyl-	**73:** 56038t
5-(chloromethyl)-1-methyl-3-phenyl-	**73:** 56038t
3-(4-chlorophenyl)-5-(dichloromethyl)-1-phenyl-	**85:** 21300t
3-(3-chlorophenyl)-5-(2,6-dimethoxyphenyl)-1-methyl-	**78:** P159619r
1-(p-chlorophenyl)-3,5-dimethyl-	**51:** P899e
3-(3-chlorophenyl)-5-[(2-dimethylaminomethyl)phenyl]-1-methyl-	**78:** P159619r
3-(3-chlorophenyl)-5-(2,4-dimethylphenyl)-1-methyl-	**78:** P159619r
1-(p-chlorophenyl)-3,5-diphenyl-	**67:** 116333y
3-(p-chlorophenyl)-1,5-diphenyl-	**46:** 8634e
5-(p-chlorophenyl)-1,3-diphenyl-	**57:** 11198g. **62:** 13147h, **80:** 82818m
3-(2-chlorophenyl)-5-[2-hydroxymethyl)phenyl]-1-methyl-	**78:** P159619r
3-(3-chlorophenyl)-5-[(2-hydroxymethyl)phenyl]-1-methyl-	**78:** P159619r
3-(p-chlorophenyl)-5-[3-(p-methoxyphenyl)-2-oxo-2H-1-benzopyran-7-yl]-1-methyl-	**74:** P65592s
5-(2-chlorophenyl)-1-methyl-3-(2-methylphenyl)-	**78:** P159519r
5-(3-chlorophenyl)-1-methyl-3-(2-methylphenyl)-	**78:** P159619r
5-(4-chlorophenyl)-1-methyl-3-(2-methylphenyl)-	**78:** P159619r
3-(3-chlorophenyl)-1-methyl-5-(2-methylphenyl)-	**78:** P159619r
3-(p-chlorophenyl)-1-methyl-5-(2-oxo-3-phenyl-2H-1-benzopyran-7-yl)-	**74:** P65592s
5-(p-chlorophenyl)-1-methyl-3-phenyl-	**54:** 515c
3-(p-chlorophenyl)-5-[2-oxo-3-(pyrazol-1-yl)-2H-1-benzopyran-7-yl]-1-methyl-	**74:** P65592s
1-(2-cyanoethyl)-3,5-di(3-pyridinyl)-	**84:** P74275z
1-(2-cyanoethyl)-3,5-di(4-pyridinyl)-	**84:** P74275z
1-(p-cyanophenyl)-3,5-dimethyl-	**21:** 3620[5]
1-cyclohexyl-3,5-diphenyl-	**72:** 90379u
3,5-dibenzyl-1-(2,4-dinitrophenyl)-	**61:** 4515f
3,5-dibenzyl-1-(p-nitrophenyl)-	**61:** 4515f
1-(2,4-dibromophenyl)-3,5-diphenyl-	**32:** 7456[9]
5-(dichloromethyl)-1,3-diphenyl-	**85:** 21300t
5-(dichloromethyl)-3-(4-methoxyphenyl)-1-phenyl-	**85:** 21300t
5-(dichloromethyl)-1-(4-methylphenyl)-3-phenyl-	**85:** 21300t
5-(dichloromethyl)-3-(4-methylphenyl)-1-phenyl-	**85:** 21300t
5-(dichloromethyl)-3-(4-nitrophenyl)-1-phenyl-	**85:** 21300t
5-(dichloromethyl)-1-phenyl-3-(trichloromethyl)-	**85:** 21300t
1[2-(diethylamino)ethyl]-3,5-di-4-pyridinyl-	**83:** P164184c, **84:** P177435b
3,5-diethyl-1-(2-hydroxyethyl)-	**81:** P120637g
3,5-diethyl-1-(p-nitrophenyl)-	**61:** 4515f

78

TABLE 3 (Continued)

Compound	Reference

3.1. 1,3,5-Alkyl- or Aryl-Trisubstituted 1H-1,2,4-triazoles (Continued)

3,5-diethyl-1-phenyl-	**21:** 3201[1]
3,5-diethyl-1-propyl-	**85:** 46523f
3,5-diethynyl-1-methyl-	**83:** 28165v
3,5-diheptyl-1-(2-hydroxyethyl)-	**81:** P120637g
1-(2,3-dihydroxy-1-propyl)-3,5-di-4-pyridinyl-	**83:** P164184c
3-(3,4-dimethoxyphenyl)-5-[(2-hydroxymethyl)phenyl]-1-methyl	**78:** P135625p
3-(3,5-dimethoxyphenyl)-1-methyl-5-(2-methylphenyl)-	**78:** P159619r
5-[p-(dimethylamino)phenyl]-1,3-diphenyl-	**62:** 13147h
5-[(2-dimethylaminomethyl)phenyl]-1-methyl-3-(3-methylphenyl)-	**78:** P159619r
5-[(2-dimethylaminomethyl)phenyl]-1-methyl-3-phenyl-	**78:** P159619r
3,5-dimethyl-1-(1-naphthyl)-	**21:** 3201[1]
3,5-dimethyl-1-(2-naphthyl)-	**21:** 3201[1]
1,5-dimethyl-3-(5-nitro-2-furyl)-	**64:** 19596c
3,5-dimethyl-1-(nitro-1-naphthyl)-	**21:** 3200[9]
3,5-dimethyl-1-(nitro-2-naphthyl)-	**21:** 3200[9]
3,5-dimethyl-1-(m-nitrophenyl)-	**24:** 2130[5]
3,5-dimethyl-1-(o-nitrophenyl)-	**24:** 2130[5]
3,5-dimethyl-1-(p-nitrophenyl)-	**24:** 2130[5], **68:** 29646e
3,5-dimethyl-1-(1-oxo-9-octadecenyl)-	**85:** P178509r
1,3-dimethyl-5-phenyl-	**49:** 11631b, **55:** 8394a
1,5-dimethyl-3-phenyl-	**49:** 11630i, **55:** 8394a
3,5-dimethyl-1-phenyl-	**47:** 4335e, **49:** 6928c
3,5-dimethyl-1-(phenylmethyl)-	**85:** P178509r
5-(2,4-dimethylphenyl)-1-methyl-3-(3-methylphenyl)-	**78:** P159619r
3,5-dimethyl-1-(2-propenyl)-	**85:** P178509r
3,5-dimethyl-1-m-tolyl-	**24:** 368[3,4,5], **23:** 3226[6,7]
3,5-dimethyl-1-o-tolyl-	**24:** 368[3,4,5], **23:** 3226[6,7]
3,5-dimethyl-1-p-tolyl-	**24:** 368[3,4,5], **23:** 3226[6,7]
3,5-dimethyl-1-[3-(p-tolyl)coumarin-7-yl]-	**71:** P14180m
1-(2,4-dinitrophenyl)-3,5-diethyl-	**61:** 4515f
1-(2,4-dinitrophenyl)-3,5-dimethyl-	**68:** 29646e
1-(2,4-dinitrophenyl)-3,5-diphenyl-	**59:** 11494c
5-(diphenylmethyl)-1,3-diphenyl-	**80:** 82818m, **80:** 133389g
5-(diphenylmethyl)-3-phenyl-1-(phenylmethyl)-	**84:** 59312r
1,5-diphenyl-3-phenacyl-,	**59:** 7530e
1,5-diphenyl-3-phenacyl-, phenylhydrazone	**59:** 7530e
1,5-diphenyl-3-phenacyl-, oxime	**59:** 7530e
1,3-diphenyl-5-propenyl-	**58:** 2447h
1,3-diphenyl-5-propyl-	**61:** 13232e, **80:** 82818m
1,3-diphenyl-5-o-tolyl-	**57:** 11199a
1,3-diphenyl-5-p-tolyl-	**57:** 11198i
1,5-diphenyl-3-p-tolyl-	**69:** 36039h
3,5-diphenyl-1-(2,4,6-trichlorophenyl)-	**67:** 116333y
1-(4-dipropylsulfamoyl)phenyl)-3, 5-di-4-pyridinyl	**83:** P164184c
3,5-di-4-pyridinyl-1-(3-pyridinylmethyl)-	**83:** P164184c
3,5-di-4-pyridinyl-1-(4-pyridinylmethyl)-	**83:** P164184c

TABLE 3 (*Continued*)

Compound	Reference

3.1. 1,3,5-Alkyl- or Aryl-Trisubstituted 1H-1,2,4-triazoles (*Continued*)

Compound	Reference
-3,3′-(dithiodi-*o*-phenylene)bis[5-methyl-1-phenyl-	**69:** 52047q
1-dodecyl-3,5-dimethyl-	**85:** P178509r
-1-ethanol, 3,5-diphenyl-	**85:** P178509r
-1-ethanol, 3,5-di-4-pyridinyl-	**83:** P164184c, **84:** P74275z, **85:** P177435b
1-ethenyl-3,5-diethyl-	**81:** P120637g, **83:** P114414g
1-ethenyl-3,5-diheptyl-	**81:** P120637g, **83:** P114414g
1-ethenyl-3,5-dimethyl-	**81:** P106281e, **83:** P114414g
1-ethenyl-3,5-diphenyl-	**81:** P120637g, **83:** P114414g
1-ethenyl-3,5-diundecyl-	**81:** P120637g, **83:** P114414g
1-ethyl-3,5-dimethyl-	**49:** 1711c, **76:** 85759w, **80:** 107696k
1-ethyl-3,5-diphenyl-	**76:** 85759w
5-ethyl-1,3-diphenyl-	**61:** 13232e, **62:** 13147f, **80:** 82818m
3-ethyl-5-methyl-1-phenyl-	**47:** 4335b
5-ethyl-3-methyl-1-phenyl-	**47:** 4335b
3-(3-fluorophenyl)-5-[(2-hydroxymethyl)phenyl]-1-methyl-	**78:** P159619r
3-(2-furanyl)-5-[(2-hydroxymethyl)phenyl]-1-methyl-	**81:** P105518a
3-(2-furanyl)-1-methyl-5-(2-methylphenyl)-	**81:** P105518a
1-β-D-glucopyranosyl-5-(4-pyridinyl)-3-(4-pyrimidinyl)-	**82:** 25627r
3-heptadecyl-5-methyl-1-phenyl-	**43:** P7730c
1-[[2-hydroxy-2-(*p*-chlorophenoxy)methyl]ethyl]-3,5-dimethyl-	**69:** P77265x
1-(2-hydroxyethyl)-3,5-dimethyl-	**81:** P105517z, **82:** P170962a
1-(2-hydroxyethyl)-3,5-diphenyl-	**81:** P105517z, **82:** P170962a
1-(2-hydroxyethyl)-3,5-diundecyl-	**81:** P105517z, **82:** P170962a
5-[1-(hydroxymethyl)-3-hydroxybutyl]-1,3-diphenyl-	**76:** 34177x
5-[1-(hydroxymethyl)-3-hydroxypropyl]-1,3-diphenyl-	**76:** 34177x
5-[(*o*-hydroxymethyl)phenyl]-1,3-dimethyl-	**77:** 61916q
3-[(*o*-hydroxymethyl)phenyl]-1,5-dimethyl-	**77:** 61916q
5-[2-(hydroxymethyl)phenyl]-3-(1H-indol-3-yl)-1-methyl-	**81:** P105518a
5-[(2-hydroxymethyl)phenyl]-3-(3-methoxyphenyl)-1-methyl-	**78:** P159619r
5-[2-(hydroxymethyl)phenyl]-1-methyl-3-(5-methyl-2-furanyl)-	**81:** P105518a
5-[2-(hydroxymethyl)phenyl]-1-methyl-3-(6-methyl-2-pyridinyl)	**81:** P105518a
5-[2-(hydroxymethyl)phenyl]-1-methyl-3-(1-methyl-1H-pyrrol-2-yl)-	**81:** P105518a
5-[2-(hydroxymethyl)phenyl]-1-methyl-3-(5-methyl-2-thienyl)-	**81:** P105518a
5-[(2-hydroxymethyl)phenyl]-1-methyl-3-(2-methylphenyl)-	**78:** P159619r
5-[(2-hydroxymethyl)phenyl]-1-methyl-3-(3-methylphenyl)-	**78:** P159619r
5-[(2-hydroxymethyl)phenyl]-1-methyl-3-(4-methylphenyl)-	**78:** P159619r
5-[(2-hydroxymethyl)phenyl]-1-methyl-3-(2-nitrophenyl)-	**78:** P159619r
5-[(*o*-hydroxymethyl)phenyl]-1-methyl-3-(4-nitrophenyl)-	**77:** 61916q, **78:** P135625p

TABLE 3 (*Continued*)

Compound	Reference

3.1. 1,3,5-Alkyl- or Aryl-Trisubstituted 1H̲-1,2,4-triazoles (*Continued*)

Compound	Reference
5-[(*o*-hydroxymethyl)phenyl]-1-methyl-3-(4-nitrophenyl)-acetate (ester)	**77:** 61916q, **78:** P159619r
5-[(*o*-hydroxymethyl)phenyl]-1-methyl-3-phenyl-	**77:** 61916q, **78:** P159619r
3-[(*o*-hydroxymethyl)phenyl]-1-methyl-5-phenyl-	**77:** 61916q, **78:** P159619r
5-[2-hydroxymethyl)phenyl]-1-methyl-3-(2-pyridinyl)-	**81:** P105518a
5-[2-hydroxymethyl)phenyl]-1-methyl-3-(3-pyridinyl)-	**81:** P105518a
5-[2-hydroxymethyl)phenyl]-1-methyl-3-(4-pyridinyl)-	**81:** P105518a
5-[2-hydroxymethyl)phenyl]-1-methyl-3-(1H̲-pyrrol-2-yl)	**81:** P105518a
5-[2-hydroxymethyl)phenyl]-1-methyl-3-(2-thienyl)-	**81:** P105518a
5-[(2-hydroxymethyl)phenyl]-1-methyl-3-(3,4,5-trimethoxy-phenyl)-	**78:** P135625p
1-[[2-hydroxy-2-(phenoxymethyl)]ethyl]-3,5-dimethyl-	**69:** P77265x
1-[(2-hydroxy-2-phenyl)ethyl]-3,5-dimethyl-	**69:** P77265x
1-[[2-hydroxy-2-(*o*-tolyloxy)methyl]ethyl]-3,5-dimethyl-	**69:** P77265x
3-(1H̲-indol-3-yl)-1-methyl-5-(2-methylphenyl)-	**81:** P105518a
5-isopropyl-1-[(*O,O*-dimethylphosphorodithio)methyl]-3-methyl	**79:** P74954s
5-isopropyl-1,3-diphenyl-	**62:** 13147g
1-mesityl-3,5-diphenyl-	**69:** 36039h
1-mesityl-5-(*p*-nitrophenyl)-3-phenyl-	**69:** 36039h
-5-methanol, 1-benxyl-α,α-bis[4-(dimethylamino)phenyl]-3-phenyl-	**84:** 59312r
-5-methanol, 1-benzyl-α,α-bis(4-methoxyphenyl)-3-phenyl-	**84:** 59312r
-5-methanol, 1-benzyl-3-(*p*-chlorophenyl)-	**73:** 56038t
-5-methanol, 1-benzyl-3-phenyl-	**73:** 56038t, **84:** 59312r
-5-methanol, 1-benzyl α,α,3-triphenyl-	**80:** 133389g, **84:** 59312r
-3-methanol, 1-(*p*-bromophenyl)-5-phenyl-	**55:** 2625h, **57:** 806i
-3-methanol, 5-(*p*-bromophenyl)-1-phenyl-	**57:** 806i
-5-methanol, 1-(*p*-bromophenyl)-3-phenyl-	**73:** 56038t
-3-methanol, 1-[4-chloro-2-(hydroxyphenylmethyl)phenyl]-5-[(dimethylamino)methyl]-	**84:** P121844w
-3-methanol, 1-[4-chloro-2-(hydroxyphenylmethyl)phenyl]-5-[(dimethylamino)methyl]-, 3-acetate	**84:** P121844w
-3-methanol, 1-(*p*-chlorophenyl)-5-methyl-	**58:** 2447h
-3-methanol, α-(2-chlorophenyl)-5-methyl-1-phenyl-	**84:** 90081p
-5-methanol, 3-(*p*-chlorophenyl)-1-phenyl-	**73:** 56038t
-3-methanol, α-(2-chlorophenyl)-1-phenyl-5-propyl-	**84:** 90081p
-3-methanol, 5-(2,6-dichlorophenyl)-α,1-diphenyl-	**84:** 90081p
-3-methanol, 1,5-diphenyl-	**55:** 2625i, **57:** 806i, **57:** 11191f
-5-methanol, 1,3-diphenyl-	**73:** 56038t
-3-methanol, 1,5-diphenyl-α,α-di-*p*-tolyl-	**55:** 24729g
-3-methanol, α,1-diphenyl-5-propyl-	**84:** 90081p
-3-methanol, α,1-diphenyl-5-(2-pyridinyl)-	**84:** 90081p
-3-methanol, 5-ethyl-α,1-diphenyl-	**84:** 90081p
-3-methanol, 5-ethyl-α-(2-methoxyphenyl)-1-phenyl-	**84:** 90081p
-3-methanol, 5-(2-furanyl)-α,1-diphenyl-	**84:** 90081p
-3-methanol, 5-(4-hydroxy-3-methoxyphenyl)-α,1-diphenyl-	**84:** 90081p

TABLE 3 (*Continued*)

Compound	Reference

3.1. 1,3,5-Alkyl- or Aryl-Trisubstituted 1H̱-1,2,4-triazoles (*Continued*)

Compound	Reference
-3-methanol, 5-(2-hydroxyphenyl)-α,1-diphenyl-	**84:** 90081p
-3-methanol, 5-(4-hydroxyphenyl)-α,1-diphenyl-	**84:** 90081p
-3-methanol, 5-(4-methoxyphenyl)-α,1-diphenyl-	**84:** 90081p
-3-methanol, 1-(*p*-methoxyphenyl)-5-phenyl-	**55:** 2625h, **57:** 806i
-3-methanol, α-(2-methoxyphenyl)-1-phenyl-5-propyl-	**84:** 90081p
-3-methanol, 5-methyl-α,1-diphenyl-	**84:** 90081p
-5-methanol, 1-methyl-3-phenyl-	**73:** 56038t
-3-methanol, 1-(4-methylphenyl)-α,5-diphenyl-	**84:** 90081p
-3-methanol, 5-(4-methylphenyl)-α,1-diphenyl-	**84:** 90081p
-5-methanol, 1-(*p*-nitrophenyl)-3-phenyl-	**73:** 56038t
-3-methanol, 5-phenyl-1-*m*-tolyl-	**55:** 2625h, **57:** 806i
-3-methanol, 5-phenyl-1-*o*-tolyl-	**55:** 2625h, **57:** 806i
-3-methanol, 5-phenyl-1-*p*-tolyl-	**55:** 2625h, **57:** 806h
-3-methanol, α,α,1,5-tetraphenyl-	**55:** 24729f
-3-methanol, α,1,5-triphenyl-	**84:** 90081p
-3-methanol, α,α,5-triphenyl-1-*o*-tolyl-	**55:** 24729g
-3-methanol, α,α,5-triphenyl-1-*p*-tolyl-	**55:** 24729g
3-[5-(2-methoxybenzylidene)-4-oxoimidazolinin-2-yl]-1,5-diphenyl-	**83:** 10007k
5-(2-methoxyethyl)-1,3-diphenyl-	**80:** 82818m
5-(methoxymethyl)-1,3-diphenyl-	**57:** 11199d, **80:** 82818m
1[3-(*p*-methoxyphenyl)coumarin-7-yl]-3,5-dimethyl-	**71:** P14180m
5-(*p*-methoxyphenyl)-1,3-diphenyl-	**62:** 13147h
3-(3-methoxyphenyl)-1-methyl-5-(2-methylphenyl)-	**78:** P159619r
3-(*p*-methoxyphenyl)-1-(*p*-nitrophenyl)-5-phenyl-	**67:** 116333y
5-[3-(*p*-methoxyphenyl)-1-oxo-2H̱-1-benzopyran-7-yl]-1-methyl-3-phenyl-	**74:** P65592s
3-[4-(2-methylbenzylidene)-5-(4H̱-oxo-2-oxazolyl]-1,5-diphenyl-	**83:** 10007k
3-[4-(4-methylbenzylidene)-5(4H̱)-oxo-2-oxazolyl]-1,5-diphenyl-	**83:** 10007k
1-methyl-3,5-bis(2-methylphenyl)-	**78:** P159619r
1-methyl-3,5-bis(3-methylphenyl)-	**78:** P159619r
1-methyl-3,5-bis(trichloromethyl)-	**66:** 10907m
1-methyl-3,5-diphenyl-	**49:** 1711d, **49:** 11630i, **54:** 515c, **57:** 11199g
3-methyl-1,5-diphenyl-	**47:** 4335c, **49:** 15916b, **57:** 11199g, **79:** 31989q
5-methyl-1,3-diphenyl-	**44:** 8904b, **47:** 1625f, **47:** 4335e, **54:** 15371i, **57:** 11199b, **61:** 13232e, **62:** 13148c, **70:** 28872e, **80:** 82818m
1-methyl-3-(5-methyl-2-furanyl)-5-(2-methylphenyl)-	**81:** P105518a
1-methyl-3-(2-methylphenyl)-5-(3-methylphenyl)-	**78:** P159619r
1-methyl-3-(2-methylphenyl)-5-(4-methylphenyl)-	**78:** P159619r
1-methyl-5-(2-methylphenyl)-3-(4-methylphenyl)-	**78:** P159619r
1-methyl-5-(2-methylphenyl)-3-(6-methyl-2-pyridinyl)-	**81:** P105518a

TABLE 3   (*Continued*)

Compound	Reference

3.1. 1,3,5-Alkyl- or Aryl-Trisubstituted 1H-1,2,4-triazoles   (*Continued*)

Compound	Reference
1-methyl-5-(2-methylphenyl)-3-(1-methyl-1H-pyrrol-2-yl)-	**81:** P105518a
1-methyl-5-(2-methylphenyl)-3-(5-methyl-2-thienyl)-	**81:** P105518a
1-methyl-3-(2-methylphenyl)-5-phenyl-	**78:** P159619r
1-methyl-5-(2-methylphenyl)-3-phenyl-	**77:** 619916q, **78:** P159619r
1-methyl-5-(2-methylphenyl)-3-(3-pyridinyl)-	**81:** P105518a
1-methyl-5-(2-methylphenyl)-3-(4-pyridinyl)-	**81:** P105518a
1-methyl-5-(2-methylphenyl)-3-(2-thienyl)-	**81:** P105518a
1-methyl-5-(2-methylphenyl)-3-[3-trifluoromethyl)- phenyl]-	**78:** P159619r
1-methyl-5-(2-methylphenyl)-3-(3,4,5-trimethoxy- phenyl)-	**78:** P159619r
5-methyl-1-(*p*-nitrophenyl)-3-phenyl-	**61:** 13232f
1-methyl-5-(2-oxo-3-(pyrazol-1-yl)-2H-1-benzopyran- 7-yl]-3-phenyl-	**74:** 65592s
5-(5-methyl-2-oxotetrahydrofuryl)-1,3-diphenyl-	**76:** 34177x
1-methyl-5-(2-oxo-3-*p*-tolyl-2H-1-benzopyran-7-yl)-3- phenyl-	**74:** P65592s
5-(3-methylphenyl)-1,3-diphenyl-	**80:** 82818m
3-(4-methylphenyl)-1,5-diphenyl-, 2-oxide	**81:** 169484u
1-(3-methylphenyl)-3-(4-oxo-4H-3,1-benzoxazin-2-yl)-5- phenyl	**83:** 10007k
1-(4-methylphenyl)-3-(4-oxo-4H-3,1-benzoxazin-2-yl)-5- phenyl-	**83:** 10007k
3-methyl-1-phenyl-5-styryl-	**49:** 15916b
1-methyl-3-phenyl-5-(2,4,6-trimethoxyphenyl)-	**80:** 82270b
1-[2-(4-morpholino)ethyl]-3,5-di-4-pyridinyl-	**83:** P164184c
1-(2-naphthyl)-α,α,5-triphenyl-	**55:** 24729g
1-(4-nitrobenzyl)-3,5-di-4-pyridinyl-	**83:** P164184c
1-(*p*-nitrophenyl)-3,5-diphenyl-	**59:** 11494d, **67:** 116333y
3-(4-nitrophenyl)-1,5-diphenyl-	**81:** 104454q
5-(*p*-nitrophenyl)-1,3-diphenyl-	**57:** 11198i, **62:** 13148a
3-(4-oxo-4H-3,1-benzoxazin-2-yl)-1,5-diphenyl-	**83:** 10007k
5-(2-oxotetrahydrofuryl)-1,3-diphenyl-	**76:** 34177x
5-pentadecyl-1,3-diphenyl-	**61:** 13232e
5-(phenoxymethyl)-1,3-diphenyl-	**57:** 11199e
1-phenyl-3,5-bis[4-(2-phenylethenyl)phenyl]-	**68:** P21961h, **80:** P146981u
1-phenyl-3,5-bis[*p*-(*p*-phenylstyryl)phenyl]-	**67:** 21858j
1-phenyl-3,5-bis(*p*-styrylphenyl)-	**67:** 21858j
1-phenyl-3,5-bis(trichloromethyl)-	**66:** 10907m
1-phenyl-3,5-di-*p*-tolyl-	**59:** 11494b, **67:** 21858j
1-(2-phenylethyl)-3,5-di-4-pyridinyl-	**83:** P164184c
1-phenyl-5-propenyl-3-*p*-tolyl-	**58:** 2447h
1-[2-(1-piperidinyl)ethyl-3,5-di-4-pyridinyl-	**83:** P164184c
-1-propanamide, 3,5-diethyl-	**85:** P178509r
-1-propanenitrile, 3,5-dimethyl-	**85:** P178509r
-1-propanenitrile, 3,5-di-3-pyridinyl-	**83:** P164184c, **85:** P177435b
-1-propanenitrile, 3,5-di-4-pyridinyl-	**83:** P164184c, **85:** P177435b
-1-propanoic acid, 3,5-bis(benzyl)-	**85:** P178509r

TABLE 3 (*Continued*)

Compound	Reference

3.1. 1,3,5-Alkyl- or Aryl-Trisubstituted 1H-1,2,4-triazoles (*Continued*)

Compound	Reference
-1-propanoic acid, 3,5-bis(isopropyl)-	**85:** P178509r
-1-propanoic acid, 3,5-dicyclohexyl-, methyl ester	**85:** P178509r
-1-propanoic acid, 3,5-diethyl-	**85:** P178509r
-1-propanoic acid, 3,5-di-4-pyridinyl-	**83:** P164184c, **84:** P74275z, **85:** P177435b
-1-propanol, 3,5-di-4-pyridinyl-	**83:** P164184c, **84:** P74275z, **85:** P177435b
-5-propanol, γ,γ-dinitro-3-(4-nitrophenyl)-1-phenyl), nitrate ester	**76:** 34177x
1-(2-propenyl)-3,5-diundecyl-	**85:** P178509r
1,3,5-triethyl-	**76:** 85759w
1,3,5-trimethyl-	**44:** 8904c, **49:** 1711b, **68:** 29645d
1,3,5-triphenyl-	**32:** 7456[8], **54:** 515c, **57:** 11198f, **59:** 11494c, **61:** 13232e, **62:** 13147g, **67:** 43097n, **69:** 96601v, **70:** 28872e, **74:** 125574z, **79:** 31989q, **80:** 82818m
1,3,5-triphenyl-, 2-oxide	**81:** 169484u

3.2. 3,4,5-Alkyl- or Aryl-Trisubstituted 4H-1,2,4-triazoles

Compound	Reference
3-(*o*-acetamidophenyl)-4-(*o*-carboxyphenyl)-5-methyl-	**72:** 43550h
4-[(4-acetamido)phenyl]-3,5-diphenyl	**83:** 206172e
-3-acetic acid, 4,5-diphenyl-	**71:** 81268j
-3-acetonitrile, 5-methyl-4-*o*-tolyl-	**72:** 43550h
-3-acrylic acid, 4,5-diphenyl-	**71:** 81268j
-3-acrylic acid, β-methyl-4,5-diphenyl-	**71:** 81268j
3-[[(aminocarbonyl)oxy]methyl]-5-(4-chlorophenyl)-4-phenyl-	**83:** P179072p
3-[[(aminocarbonyl)oxy]methyl]-5-(3,4-dichlorophenyl)-4-phenyl-	**83:** P179072p
3-[[(aminocarbonyl)oxy]methyl]-5-(3,4-dimethoxyphenyl)-4-phenyl-	**83:** P179072p
3-[[(aminocarbonyl)oxy]methyl]-4,5-diphenyl-	**83:** P179072p
3-[[(aminocarbonyl)oxy]methyl-5-(3-fluorophenyl)-4-phenyl-	**83:** P179072p
3-[[(aminocarbonyl)oxy]methyl-5-(4-fluorophenyl)-4-phenyl-	**83:** P179072p
3-[[(aminocarbonyl)oxy]methyl-5-(2-methoxyphenyl)-4-phenyl-	**83:** P179072p
3-[[(aminocarbonyl)oxy]methyl-5-(4-methoxyphenyl)-4-phenyl-	**83:** P179072p
3-[[(aminocarbonyl)oxy]methyl-5-(3-methylphenyl)-4-phenyl-	**83:** P179072p
3-[[(aminocarbonyl)oxy]methyl-5-(4-methylphenyl)-4-phenyl-	**83:** P179072p

TABLE 3   (*Continued*)

Compound	Reference

3.2. 3,4,5-Alkyl- or Aryl-Trisubstituted 4H-1,2,4-triazoles   (*Continued*)

Compound	Reference
3-[[(aminocarbonyl)oxy]methyl-4-phenyl-5-[3-(tri-fluoromethyl)phenyl]-	**83:** P179072p
3-[[(aminocarbonyl)oxy]methyl-4-phenyl-5-(3,4,5-trimethoxyphenyl)-	**83:** P179072p
4-[2-[amino(2-chlorophenyl)methyl]-4-chlorophenyl]-3-[(dimethylamino)methyl]-5-methyl-	**84:** 121845x
4-(2-amino-3,5-dibromophenyl)-3,5-dimethyl-	**66:** 55438s
4-(2-aminoethyl)-3-(*p-tert*-butylphenyl)-5-phenyl-	**62:** 568e
4-(2-aminoethyl)-3,5-diphenyl-	**62:** P568d, **62:** P11830b
3-[4-[(1-amino-8-hydroxy-3,6-disulfonaphthalen-2-yl)-azo]phenyl]-5-[4-[(2,4-diamino-5-nitrophenyl)azo]phenyl]-4-methyl-	**83:** P195215t
3-[4[[1-amino-8-hydroxy-7-(phenylazo)-3,6-disulfonaph-thalen-2-yl]azo]phenyl]-5-4-[(2,4-diaminophenyl)azo]-phenyl]-4-methyl-	**82:** P172596h, **83:** P195211p
3-[3-[(1-amino-8-hydroxy-7-(phenylazo)-3,6-disulfonaph-thalen-2-yl)azo]phenyl]-5-[3-[(4-hydroxyphenyl)azo]-phenyl]-4-methyl-	**82:** P172596h
3-(aminomethyl)-4-benzyl-5-phenyl-	**62:** 1647f
3-(aminomethyl)-4-[3-(2-chlorobenzoyl)-4,5-dimethyl-2-thienyl]-5-methyl-	**80:** P133443v
3-(aminomethyl)-4-[3-(2-chlorobenzoyl)-5-ethyl-2-thienyl]-5-methyl-	**80:** P133443v
3-(aminomethyl)-4-[3-(2-chlorobenzoyl)-5-methyl-2-thienyl]-5-methyl-	**80:** P133443v
3-(aminomethyl)-4-[4-chloro-2-(α-hydroxybenzyl)phenyl]-5-methyl-	**81:** P49689t
3-(aminomethyl)-4-methyl-5-phenyl-	**62:** 1647f
3-[*m*-(*p*-aminophenoxy)phenyl]-5-(*m*-aminophenyl)-4-methyl-	**72:** 68047n
3-[*m*-(1-aminophenoxy)phenyl]-5-(*m*-aminophenyl)-4-phenyl-	**72:** 68047n
3-[*p*-(*p*-aminophenoxy)phenyl]-5-(*m*-aminophenyl)-4-phenyl-	**72:** 68047n
3-[*p*-(*p*-aminophenoxy)phenyl]-5-(*p*-aminophenyl)-4-phenyl-	**72:** 68047n
3-(*o*-aminophenyl)-4-benzyl-5-(phenylethnyl)-	**77:** 152103p
3-(*o*-aminophenyl)-5-[2-(4-chlorophenyl)ethenyl]-4-methyl-	**77:** 152103p
4-(*o*-aminophenyl)-3,5-dimethyl-	**66:** 55438s
3-(*o*-aminophenyl)-4,5-diphenyl-	**54:** P20217g
4-(2-aminophenyl)-3,5-diphenyl-	**83:** 206172e
4-(3-aminophenyl)-3,5-diphenyl-	**83:** 206172e
4-(4-aminophenyl)-3,5-diphenyl-	**83:** 206172e
3-(*o*-aminophenyl)-4-methyl-5-(2-phenylethenyl)-	**77:** 152103p
3-(2-amino-α,α,α-trifluoro-*p*-tolyl)-4-5-diphenyl-	**54:** P20217h
4-*p*-anisyl-3-benzyl-5-phenyl-	**25:** 295^2
4-*p*-anisyl-3,5-diphenyl-	**25:** 295^3

TABLE 3   (*Continued*)

Compound	Reference
3.2. 3,4,5-Alkyl- or Aryl-Trisubstituted 4H-1,2,4-triazoles (*Continued*)	
5-(2-benzoyl-4-chlorophenyl)-3,5-bis(hydroxymethyl)-	**81:** P152287j
4-(2-benzoyl-4-chlorophenyl)-3,5-bis(phthalimidomethyl)-	**81:** P152287j
4-(2-benzoyl-4-chlorophenyl)-3-bromo-5-[(dimethyl-amino)methyl]-	**83:** P79302v
4-(2-benzoyl-4-chlorophenyl)-5-(bromomethyl)-3-[(phthalimino)methyl]-	**81:** P152287j
4-(2-benzoyl-4-chlorophenyl)-3-(chloromethyl)-5-(2-ethoxyethyl)-	**83:** P79295v
4-(2-benzoyl-4-chlorophenyl)-3-(chloromethyl)-5-(ethoxymethyl)-	**83:** P79295v
4-(2-benzoyl-4-chlorophenyl)-3-(chloromethyl)-5-(methoxymethyl)-	**83:** P79295v
4-(2-benzoyl-4-chlorophenyl)-3-(chloromethyl)-5-methyl-	**77:** P88557k, **77:** P126708v **78:** P159694m, **79:** 92176u, **80:** P14934w, **82:** P31330e, **83:** P43339y, **83:** P58895a
4-(2-benzoyl-4-chlorophenyl)-3-(chloromethyl)-5-[(phenylmethoxy)methyl]-	**83:** P79295v
4-(2-benzoyl-4-chlorophenyl)-3-(chloromethyl)-3-[(phthalimino)methyl]-	**81:** P152287j
4-(2-benzoyl-4-chlorophenyl)-3,5-dimethyl-	**79:** 92176u
4-(2-benzoyl-4-chlorophenyl)-3-[(dimethylamino)methyl]-5-methyl-	**77:** P126708v, **81:** P49689t, **81:** P152289m
4-(2-benzoyl-4-chlorophenyl)-5-[(dimethylamino)methyl]-3-[(phthalimino)methyl]-	**81:** P152287j
4-(2-benzoyl-4-chlorophenyl)-5-(2-furanyl)-3-[(dimethylamino)methyl]-	**80:** P37116v
4-(2-benzoyl-4-chlorophenyl)-3-(hydroxymethyl)-5-methyl-	**77:** P126708v, **78:** P136354m, **78:** P159694m, **80:** P14934w, **80:** P83083e, **81:** P49689t, **81:** P152289m
4-(2-benzoyl-4-chlorophenyl)-5-(hydroxymethyl)-3-[phthalimino)methyl]-	**81:** P152287j
4-(2-benzoyl-4-chlorophenyl)-3-(iodomethyl)-5-methyl-	**77:** P126708v, **78:** P159694m, **80:** 14934w, **82:** P31330e
4-(2-benzoyl-4-chlorophenyl)-5-(iodomethyl)-3-[(phthalimino)methyl]-	**81:** P152287j
4-(2-benzoyl-4-chlorophenyl)-3-methyl-5-[(methylamino)-methyl]-	**77:** P126708v, **80:** P83083e
4-(2-benzoyl-4-chlorophenyl)-3-methyl-5-[(4-methyl-1-piperazinyl)methyl]-	**81:** P152289m
4-(2-benzoyl-4-chlorophenyl)-3-methyl-5-[[(methyl-sulfonyl)oxy]methyl]-	**81:** P152289m
4-(2-benzoyl)-4-chlorophenyl)-3-methyl-5-(4-morpholinylmethyl)-	**78:** P136302t, **80:** P14934w, **81:** P49689t, **81:** P152289m
4-(2-benzoyl-4-chlorophenyl)-5-methyl-3-[(phthalimino)-methyl]-	**81:** P152289m
4-(2-benzoyl-4-chlorophenyl)-3-methyl-5-(1-piperdinyl-methyl)-	**78:** P136302t, **80:** P14934w

TABLE 3 (*Continued*)

Compound	Reference

3.2. 3,4,5-Alkyl- or Aryl-Trisubstituted 4H-1,2,4-triazoles (*Continued*)

Compound	Reference
4-(2-benzoyl-4-chlorophenyl)-3-methyl-5-(1-pyrrolidinyl-methyl)-	**78:** P136302t, **80:** P14934w, **81:** P49689t, **81:** P152289m
4-(3-benzoyl-5-ethyl-2-thienyl)-3-[dimethylamino)methyl]-5-(hydroxymethyl)-	**80:** P133443v
4-(2-benzoyl-4-nitrophenyl)-3-(chloromethyl)-5-methyl-	**77:** P88557k, **82:** P31330e, **83:** P43339y, **83:** P58895a
3-(2-benzoxazolyl)-5-(4-methoxyphenyl)-4-phenyl-	**75:** 118255p, **77:** P164709d
3-[[(benzoxycarbonyl)amino]methyl]-4-(2-benzoyl-4-chlorophenyl)-5-methyl-	**82:** P31364u
5-benzyl-4-(3-carboxy-4-methylthiophen-2-yl)-3-methyl-	**83:** 178980w
5-benzyl-4-(3-carboxy-4,5,6,7-tetrahydrobenzo[b]-thiophen-2-yl)-3-methyl-	**83:** 178980w
3-benzyl-4-[2-diethylamino)ethyl]-5-phenyl-	**62:** 568b
3-benzyl-4,5-diphenyl-	**25:** 295[2]
4-benzyl-3,5-diphenyl-	**57:** 5910g, **73:** 130941a
4-(2-benzyl-4-methylphenyl-3-(chloromethyl)-5-methyl-	**77:** P88557k, **82:** P31330e, **83:** P58895a
3-benzyl-5-phenyl-4-*m*-tolyl-	**25:** 295[2]
4-benzyl-5-(4-pyridyl)-3-[(*p*-toluenesulfonamido)methyl]-	**69:** 19082z
4-(4-biphenylyl)-3,5-diphenyl-	**73:** 66511f
3,5-bis(3-amino-4-anilinophenyl)-4-phenyl-	**71:** P62234d, **72:** 44284m
3,5-bis[3-amino-4-(*p*-phenoxyanilino)phenyl]-4-(*p*-phenoxyphenyl)-	**72:** 44284m
3,5-bis[*m*-(*p*-aminophenoxy)phenyl]-4-phenyl-	**72:** 68047n
3,5-bis[*p*-(*p*-aminophenoxy)phenyl]-4-phenyl-	**72:** 68047n
3,5-bis(3-aminophenyl)-4-methyl-	**82:** P157819q, **83:** P195211p
3,5-bis(4-aminophenyl)-4-methyl-	**82:** P172595h, **83:** P195211p, **83:** P195215t
3,5-bis(*m*-aminophenyl)-4-phenyl-	**64:** 8175c, **66:** 46705z
3,5-bis(4-aminophenyl)-4-phenyl-	**82:** P157819q
3,5-bis(4-anilino-3-nitrophenyl)-4-phenyl-	**72:** 44284m
3,5-bis(*o*-chlorophenyl)-4-[2-(diethylamino)ethyl]-	**62:** P567d, **62:** P11829e
3,4-bis(4-chlorophenyl)-5-(trifluoromethyl)-	**85:** 138543r
3,5-bis[*p*-diethylamino)phenyl]-4-(2-hydroxyethyl)-	**71:** P35023c
3,5-bis[*p*-(dimethylamino)phenyl]-4-phenyl-	**60:** 4131f
3,5-bis[4-[(3-ethoxycarbonyl-5-oxo-1-phenyl-1H-pyrazol-4-yl)azo]phenyl]-4-phenyl-	**82:** P157819q
3,5-bis(heptafluoropropyl)-4-methyl-	**57:** 12471a
3,5-bis(2-hydroxyphenyl)-4-(4-methylphenyl)-	**77:** 76506e
3,4-bis(4-methylphenyl)-5-phenyl-	**80:** P146980t, **80:** P146981u, **83:** P29887u
3,5-bis(4-methylphenyl-4-phenyl-	**80:** P146980t, **80:** P146981u, **83.** P29887u
3,5-bis[3-nitro-4-(*p*-phenoxyanilino)phenyl]-4-(*p*-phenoxyphenyl)-	**72:** 44284m
3,5-bis[*m*-(*p*-nitrophenoxy)phenyl]-4-phenyl-	**72:** 68047n
3,5-bis(*m*-nitrophenyl)-4-phenyl-	**64:** 8175c, **66:** 46705z
3,5-bis(pentafluorophenyl)-4-phenyl-	**68:** P39628q
4-[3-(2-bromobenzoyl)-5-ethyl-2-thienyl]-3-	**80:** P133443v

87

TABLE 3 (*Continued*)

Compound	Reference

3.2. 3,4,5-Alkyl- or Aryl-Trisubstituted 4H̲-1,2,4-triazoles (*Continued*)

Compound	Reference
[(dimethylamino)methyl]-5-methyl-3-(bromomethyl)-4-[2-(2-chlorobenzoyl)phenyl]-5-methyl-	**80:** P14934w
3-(bromomethyl)-4-[4-chloro-2-(2-chlorobenzoyl)phenyl]-5-methyl-	**80:** P14934w
4-(*p*-bromophenyl)-3,5-diethyl-	**69:** 106627z
3-(*p*-bromophenyl)-4-[2-(diethylamino)ethyl]-5-phenyl-	**62:** P568b, **62:** P11829h
4-(*o*-bromophenyl)-3,5-diphenyl-	**73:** 130941a
4-(*m*-bromophenyl)-3,5-diphenyl-	**73:** 130941a
4-(*p*-bromophenyl)-3,5-diphenyl-	**73:** 130941a
4-(butoxypyridyl)-3,5-dimethyl-	**27:** P4541[3]
4-[2-(butylamino)ethyl]-3,5-diphenyl-	**62:** P568d, **62:** P11830a
4-sec-butyl-3,5-bis(*p*-nitrophenyl)-	**69:** 51807a
3-butyl-4-[2-(diethylamino)ethyl]-5-phenyl-	**62:** P568a, **62:** P11829g
3-(*p*-tert-butylphenyl)-4-[2-(diethylamino)ethyl]-5-methyl-	**62:** P568b, **62:** P11829h
3-(*p*-tert-butylphenyl)-4-[2-diethylamino)ethyl]-5-phenyl-	**62:** P568b, **62:** P11829h
3-(*p*-tert-butylphenyl)-4-[3-(diethylamino)propyl]-5-phenyl-	**62:** P567d, **62:** P11829f
3-(*p*-tert-butylphenyl)-4-[2-(dimethylamino)ethyl]-5-phenyl-	**62:** P567d, **62:** P11829f
3-(*p*-tert-butylphenyl)-4-[3-(dimethylamino)propyl]-5-phenyl-	**62:** P567d, **62:** P11829f
-3-butyric acid, 4,5-diphenyl-	**71:** 81268j
3-(carbamoylmethyl)-5-(2-furyl)-4-methyl-	**74:** P100068z
3-(2-carboxybenzyl)-4,5-dimethyl-	**81:** 37521u
4-(3-carboxy-4-methylthiophen-2-yl)-3-(cyanomethyl)-5- methyl-	**83:** 178980w
4-(3-carboxy-4-methylthiophen-2-yl)-3,5-dimethyl-	**83:** 178980w
4-(3-carboxy-4-methylthiophen-2-yl)-3-methyl-5-phenyl-	**83:** 178980w
4-(3-carboxy-4-methylthiophen-2-yl)-3-methyl-5-(4-pyridinyl)-	**83:** 178980w
4-(*o*-carboxyphenyl)-3,5-bis(*p*-chlorophenyl)-	**23:** 836[2]
4-(*o*-carboxyphenyl)-3-(*p*-chlorophenyl)-5-phenyl-	**23:** 836[1,2], **24:** 2748[1]
4-(*o*-carboxyphenyl)-3-(cyanomethyl)-5-methyl-	**72:** 43550h
4-(*o*-carboxyphenyl)-3-(*α*-cyano-*p*-nitrostyryl)-5-methyl-	**72:** 43550h
4-(*o*-carboxyphenyl)-3-[(diethylamino)methyl]-5-methyl	**72:** 43550h
4-(*o*-carboxyphenyl)-3,5-dimethyl-	**72:** 43550h
3-(*o*-carboxyphenyl)-4,5-diphenyl-	**71:** 81268j
4-(*o*-carboxyphenyl)-3,5-diphenyl-	**23:** 835[9], **83:** 206172e
4-(3-carboxyphenyl)-3,5-diphenyl-	**83:** 206172e
4-(4-carboxyphenyl)-3,5-diphenyl-	**83:** 206172e
4-(*o*-carboxyphenyl)-3-methyl-5-(nitromethyl)-	**72:** 43550h
4-(*o*-carboxyphenyl)-3-methyl-5-(phenoxymethyl)-	**72:** 43550h
4-(*o*-carboxyphenyl)-3-methyl-5-phenyl-	**23:** 836[1], **72:** 43550h
4-(4-carboxy-1-phenyl-1H̲-pyrazol-5-yl)-3-(cyanomethyl)-5-methyl-	**80:** P70813t
4-(4-carboxy-1-phenyl-1H̲-pyrazol-5-yl)-3-(2-furanyl)-5-methyl-	**80:** P70813t

TABLE 3 (Continued)

Compound	Reference
3.2. 3,4,5-Alkyl- or Aryl-Trisubstituted 4H-1,2,4-triazoles (Continued)	
4-(4-carboxy-1-phenyl-1H-pyrazol-5-yl)-3-(1H-indol-2-yl)-5-methyl-	**80:** P70813t
4-(4-carboxy-1-phenyl-1H-pyrazol-5-yl)-3-(1H-indol-3-yl)-5-methyl-	**80:** P70813t
4-(4-carboxy-1-phenyl-1H-pyrazol-5-yl)-5-isopropyl-3-methyl	**80:** P70813t
4-(4-carboxy-1-phenyl-1H-pyrazol-5-yl)-3-methyl-5-(1-naphthalenyl)-	**80:** P70813t
4-(4-carboxy-1-phenyl-1H-pyrazol-5-yl)-3-methyl-5-(1-naphthalenylmethyl)-	**80:** P70813t
4-(4-carboxy-1-phenyl-1H-pyrazol-5-yl)-3-methyl-5-(4-pyridinyl)-	**80:** P70813t
4-(3-carboxy-4,5,6,7-tetrahydrobenzo[b]thiophen-2-yl)-3-(cyanomethyl)-5-methyl-	**83:** 178980w
4-(3-carboxy-4,5,6,7-tetrahydrobenzo[b]thiophen-2-yl)-3,5-dimethyl-	**83:** 178980w
4-(3-carboxy-4,5,6,7-tetrahydrobenzo[b]thiophen-2-yl)-3-methyl-5-phenyl-	**83:** 178980w
4-(3-carboxy-4,5,6,7-tetrahydrobenzo[b]thiophen-2-yl)-3-methyl-5-(4-pyridinyl)-	**83:** 178980w
4-[3-(2-chlorobenzoyl)-4,5-dimethyl-2-thienyl]-3-(chloromethyl)-5-methyl-	**80:** P133443v
4-[3-(2-chlorobenzoyl)-4,5-dimethyl-2-thienyl]-3-[(dimethylamino)methyl]-5-methyl-	**80:** P133443v
4-[3-(2-chlorobenzoyl)-5-ethyl-2-thienyl]-3-(chloromethyl)-5-methyl-	**80:** P133443v, **81:** P152238u
4-[3-(2-chlorobenzoyl)-5-ethyl-2-thienyl]-3-[(dimethylamino)methyl]-5-(hydroxymethyl)-	**80:** P133443v
4-[3-(2-chlorobenzoyl)-5-ethyl-2-thienyl]-3-[(dimethylamino)methyl]-5-methyl-	**80:** P133443v
4-[3-(2-chlorobenzoyl)-5-ethyl-2-thienyl]-3-(hydroxymethyl)-5-methyl-	**81:** P152238u
4-[5-chloro-2-(α-hydroxybenzyl)phenyl]-3-methyl-5-(1-pyrrolidinylmethyl)-	**81:** P49689t
3-(chloromethyl)-4-[5-ethyl-3-(2-fluorobenzoyl)-2-thienyl]-5-methyl-	**80:** P133443v
3-(o-chlorophenyl)-4-[2-(diethylamino)ethyl]-5-phenyl-	**62:** P567d, **62:** P11829e
3-(m-chlorophenyl)-4-[2-(diethylamino)ethyl]-5-phenyl-	**62:** P568b, **62:** P11829h
4-(p-chlorophenyl)-3,5-dimethyl-	**54:** 7696c
3-(p-chlorophenyl)-4,5-diphenyl-	**55:** 6474a
4-(o-chlorophenyl)-3,5-diphenyl-	**73:** 130941a
4-(m-chlorophenyl)-3,5-diphenyl-	**73:** 130941a
4-(p-chlorophenyl)-3,5-diphenyl-	**53:** 16121b, **55:** 6474a, **56:** 1388g, **57:** 5910f, **67:** 100076f, **73:** 130941a
3-(p-chlorophenyl)-5-isopropyl-4-phenyl-	**55:** 6474b
3-(p-chlorophenyl)-5-(p-methoxyphenyl)-4-phenyl-	**55:** 6474b
4-(p-chlorophenyl)-3-(p-methoxyphenyl)-5-phenyl-	**55:** 6474b
5-(2-chlorophenyl)-3-[[(methylaminocarbonyl)oxy]-	**83:** P179072p

TABLE 3 (*Continued*)

Compound	Reference

3.2. 3,4,5-Alkyl- or Aryl-Trisubstituted 4H-1,2,4-triazoles (*Continued*)

Compound	Reference
methyl]-4-phenyl-	
5-(4-chlorophenyl)-3-[[(methylaminocarbonyl)oxy]-methyl]-4-phenyl-	**83:** P179072p
4-[4-chloro-2-(phenylmethyl)phenyl]-3-methyl-5-(1-pyrrolidinylmethyl)-	**85:** 56539e
3-(*p*-chlorophenyl)-4-phenyl-5-*p*-tolyl-	**55:** 6474b
4-(*p*-chlorophenyl)-3-phenyl-5-*p*-tolyl-	**55:** 6474b
3-(cyanomethyl)-4-(*o*-carboxyphenyl)-5-phenyl-	**72:** 43550h
3-(cyanomethyl)-4-(*o*-methoxycarbonylphenyl)-5-methyl-	**72:** 43550h
4-(3-cyanophenyl)-3,5-diphenyl-	**83:** 206172e
3-(cyclohexylamino)-4,5-diphenyl-	**74:** 87898y
4-[3-(2-chlorobenzoyl)-5-ethyl-2-thienyl]-3-methyl-5-[(methylamino)methyl]-	**80:** P133443v
4-[3-(2-chlorobenzoyl)-5-ethyl-2-thienyl]-3-methyl-5-[[[(4-methylphenyl)sulfonyl]-oxy]methyl]-	**80:** P133443v
4-[3-(2-chlorobenzoyl)-5-ethyl-2-thienyl]-3-methyl-5-[(4-methyl-1-piperazinyl)methyl]-	**80:** P133443v
4-[3-(2-chlorobenzoyl)-5-ethyl-2-thienyl]-3-methyl-5-(1-pyrrolidinylmethyl)-	**80:** P133443v
4-[3-(2-chlorobenzoyl)-5-methyl-2-thienyl]-3-(chloromethyl)-5-methyl-	**80:** P133443v
4-[3-(2-chlorobenzoyl)-5-methyl-2-thienyl]-3-(hydroxymethyl)-5-methyl-	**80:** P133443v, **81:** P152238u
4-[3-(2-chlorobenzoyl)-5-methyl-2-thienyl]-3-methyl-5-(4-morpholinylmethyl)-	**80:** P133443v
4-[2-(2-chlorobenzoyl)phenyl]-3-[(dimethylamino)-methyl]-5-methyl-	**80:** P14934w
4-[3-(2-chlorobenzoyl)-4,5,6,7-tetrahydrobenzo[*b*]thien-2-yl]-3-(chloromethyl)-5-methyl-	**80:** P133443v
4-[3-(2-chlorobenzoyl)-4,5,6,7-tetrahydrobenzo[*b*]thien-2-yl]-3-[(dimethylamino)methyl]-5-methyl-	**80:** P133443v
4-[4-chloro-2-benzylphenyl]-3-methyl-5-(1-pyrrolidinylmethyl)-	**78:** P136302t
4-[4-chloro-2-(2-chlorobenzoyl)]-3-[(dimethylamino)-methyl]-5-(4-pyridinyl)-	**80:** P37116v
4-[4-chloro-2-(2-chlorobenzoyl)phenyl]-3-(chloromethyl)-5-cyclopropyl	**80:** P14934w
4-[4-chloro-2-(2-chlorobenzoyl)phenyl]-3-cyclopropyl-5-[(dimethylamino)methyl]-	**80:** P37116v
4-[4-chloro-2-(2-chlorobenzoyl)phenyl]-3-[(dimethylamino)methyl]-5-methyl-	**80:** P14934w
4-[4-chloro-2-(2-chlorobenzoyl)phenyl]-3-(hydroxymethyl)-5-methyl-	**81:** P152289m
4-[4-chloro-2-(2-chlorobenzoyl)phenyl]-5-methyl-3-[(phthalimino)methyl]-	**81:** P152289m
4-[4-chloro-2-(2-chlorobenzoyl)phenyl]-3-methyl-5-(1-pyrrolidinylmethyl)-	**80:** P14934w
3-cyclohexyl-4-[2-(diethylamino)ethyl]-5-phenyl-	**62:** P568b, **62:** P11829h
4-cyclohexyl-3,5-dimethyl-	**26:** P3516[6]

TABLE 3 (*Continued*)

Compound	Reference

3.2. 3,4,5-Alkyl- or Aryl-Trisubstituted 4H-1,2,4-triazoles (*Continued*)

Compound	Reference
4-cyclohexyl-3,5-bis(*p*-nitrophenyl)-	**69:** 51807a
5-(3,4-dichlorophenyl)-3-[[(methylaminocarbonyl)oxy]-methyl]-4-phenyl-	**83:** P179072p
4-[2-(diethylamino)ethyl]-3,5-diphenyl-	**62:** P567d, **62:** 11829e, **77:** 34431w
4-[2-(diethylamino)ethyl]-3-(*p*-methoxyphenyl)-5-phenyl-	**62:** P568b, **62:** P11829h
4-[2-(diethylamino)ethyl]-3-methyl-5-phenyl-	**62:** P568d, **62:** P11830a
4-[2-(diethylamino)ethyl]-3-phenyl-5-*p*-tolyl-	**62:** P568b, **62:** 11829h
4-[3-(diethylamino)propyl]-3,5-diphenyl-	**62:** P567d, **62:** P11829e
3,5-diethyl-4-(*o*-methoxyphenyl)-	**69:** 106627z
3,5-diethyl-4-(*p*-methoxyphenyl)-	**69:** 106627z
3,5-diethyl-4-phenyl-	**69:** 106627z
3,5-diethyl-4-*o*-tolyl-	**69:** 106627z
3,5-diethyl-4-*m*-tolyl-	**69:** 106627z
3,5-diethyl-4-*p*-tolyl-	**69:** 106627z
-3,5-dimethanol, α,α′,4-triphenyl-	**66:** 94962c
4-(3,4-dimethoxyphenyl)-3-[[(methylaminocarbonyl)oxy]-methyl]-4-phenyl-	**83:** P179072p
4-[2-(dimethylamino)ethyl]-3,5-diphenyl-	**62:** P567d, **62:** P11829e
3-[(dimethylamino)methyl]-4-[5-ethyl-3-(4-fluorobenzoyl)-2-thienyl]-5-methyl-	**80:** P133443v
3-[(dimethylamino)methyl]-4-[3-(2-fluorobenzoyl)-5-ethyl-2-thienyl]-5-methyl-	**80:** P133443v
3-[*p*-(dimethylamino)phenyl]-5-methyl-4-phenyl-	**66:** 46376t
3-[*p*-(dimethylamino)phenyl]-4-(*p*-nitrophenyl)-5-phenyl-	**53:** 5253f, **55:** 6474b
3-[*p*-(dimethylamino)phenyl]-5-(*p*-nitrophenyl)-4-phenyl-	**55:** 6474b, **60:** 4131f
4-[3-(dimethylamino)propyl]-3,5-diphenyl-	**62:** P567d, **62:** 11829e
3,4-dimethyl-5-phenyl-	**25:** P4557[2], **26:** P2469[5], **55:** 8394a
3,5-dimethyl-4-phenyl-	**23:** 836[3], **49:** 15915i, **66:** 55438s, **78:** 58327p
3,5-dimethyl-4-*o*-tolyl-	**51:** P16163b
3,5-dimethyl-4-*m*-tolyl-	**51:** P16163b
3,5-dipentyl-4-phenyl-	**69:** 106627z
4,5-diphenyl-3-(diphenylmethyl)-	**24:** 4032[4], **28:** 5457[9]
3-(diphenylmethyl)-4,5-diphenyl-	**74:** 87898y
5-(diphenylmethyl)-4-(4-methylphenyl)-3-(2-pyridinyl)-	**83:** 43245q
3-(diphenylmethyl)-5-phenyl-4-benzyl-	**84:** 59312r
3,4-diphenyl-5-(*p*-3-phenyl-1,3,4-oxadiazol-2-ylphenyl)-	**55:** 6474c
3,5-diphenyl-4-propyl-	**77:** 34431w
3,5-diphenyl-4-[2-(1-pyrrolidinyl)ethyl]-	**62:** P568d
3,5-diphenyl-4-(2-pyrrol-1-ylethyl)-	**62:** P11830a
3,4-diphenyl-5-*p*-tolyl	**55:** 6474b
3,5-diphenyl-4-*o*-tolyl-	**53:** 16121b, **67:** 100076f, **73:** 130941a
3,5-diphenyl-4-*m*-tolyl-	**25:** 295[2], **53:** 16121b, **67:** 100076f **73:** 130941a
3,5-diphenyl-4-*p*-tolyl-	**25:** 295[2], **53:** 1612b, **66:** 46412b, **67:** 100076f, **73:** 130941a

TABLE 3 (*Continued*)

Compound	Reference

3.2. 3,4,5-Alkyl- or Aryl-Trisubstituted 4H-1,2,4-triazoles (*Continued*)

Compound	Reference
3,5-diphenyl-4-(α,α,α-trifluoro-*m*-tolyl)-	**73:** 130941a
3,5-dipropyl-4-*p*-tolyl-	**69:** 106627z
-4-ethanol, 3,5-bis[*p*-(diethylamino)phenyl]-	**61:** P2637g, **70:** P92277s
-4-ethanol, 3,5-diphenyl-	**62:** P568c, **62:** P11829h
4-[2-(ethoxycarbonyl)phenyl]-3,5-diphenyl-	**83:** 206172e
4[3-(ethoxycarbonyl)phenyl]-3,5-diphenyl-	**83:** 206172e
4-[4-(ethoxycarbonyl)phenyl]-3,5-diphenyl-	**83:** 206172e
4-(*p*-ethoxyphenyl)-3,5-diethyl-	**69:** 106627z
4-(*o*-ethoxyphenyl)-3,5-diphenyl-	**73:** 130941a
4-(*m*-ethoxyphenyl)-3,5-diphenyl-	**73:** 130941a
4-(*p*-ethoxyphenyl)-3,5-diphenyl-	**25:** 295[3], **67:** 100076f
4-[2-(ethylamino)ethyl]-3,5-diphenyl-	**62:** P568d, **62:** P11830a
4-ethyl-3,5-bis(4-diethylaminophenyl)-	**83:** P35680n
4-ethyl-3,5-dimethyl-	**49:** 1711c
3-ethyl-4,5-diphenyl-	**73:** 109744d, **74:** 87898y
4-ethyl-3,5-diphenyl-	**76:** 85759w, **77:** 34431w
3-ethyl-5-methyl-4-(4-methylcyclohexyl)-	**26:** P3516[6]
3-ethyl-5-methyl-4-phenyl-	**26:** P2469[5], **26:** P3263[9], **78:** 58327p
4-ethyl-5-(4-pyridyl)-3-[(*p*-toluenesulfonamido)methyl]-	**69:** 19082z
4-(*p*-fluorophenyl)-3,5-diphenyl-	**73:** 130941a
5-(3-fluorophenyl)-3-[[(methylaminocarbonyl)oxy]-methyl]-4-phenyl-	**83:** P179072p
5-(4-fluorophenyl)-3-[[(methylaminocarbonyl)oxy]-methyl]-4-phenyl-	**83:** P179072p
3-(heptafluoropropyl)-4-methyl-5-phenyl-	**70:** 77879w, **77:** 126515e
-3-hexanoic acid, 4,5-diphenyl-	**71:** 81268j
4-hexyl-3,5-diphenyl-	**77:** 34431w
4-[4-(hydrazinocarbonyl)phenyl]-3,5-diphenyl	**83:** 206172e
4-[2-(α-hydroxybenzyl)phenyl]-3-methyl-5-(1-pyrrolidinylmethyl)-	**81:** P49689t
4-isopropyl-3,5-bis(*p*-nitrophenyl)-	**69:** 51807a
3-isopropyl-5-methyl-4-xylyl-	**25:** P2442[9]
4-(1-naphthyl)-3,5-diphenyl-	**67:** 100076f
4-(2-naphthyl)-3,5-diphenyl-	**53:** 16121b
4-(1-naphthyl)-3,5-dipropyl-	**69:** 106627z
3-[*m*-(*p*-nitrophenoxy)phenyl]-5-(*m*-nitrophenyl-4-phenyl-	**72:** 68047n
3-(*p*-nitrophenyl)-4,5-diphenyl-	**55:** 6474a
4-(2-nitrophenyl)-3,5-diphenyl-	**73:** 130941a, **83:** 206172e
4-(*m*-nitrophenyl)-3,5-diphenyl-	**73:** 130941a, **83:** 206172e
4-(*p*-nitrophenyl)-3,5-diphenyl-	**55:** 6474a, **66:** 46412b
3-(*p*-nitrophenyl)-4-phenyl-5-*p*-tolyl-	**55:** 6474b
-3-nonanoic acid, 4,5-diphenyl-	**71:** 81268j
4-mesityl-3,5-diphenyl-	**73:** 66511f
-3-methanol, 4-[2-[amino(2-chlorophenyl)methyl]-4-chlorophenyl]-5-methyl-	**84:** P121845x
-3-methanol, 4-(3-bromophenyl)-5-phenyl-	**84:** P164789z

92

TABLE 3 (*Continued*)

Compound	Reference

3.2. 3,4,5-Alkyl- or Aryl-Trisubstituted 4H̱-1,2,4-triazoles (*Continued*)

Compound	Reference
-3-methanol, 4-[4-chloro-2-(hydroxyphenylmethyl)-phenyl]-5-[(dimethylamino)methyl]-	**85:** P46692k
-3-methanol, 4-(4-chlorophenyl)-5-phenyl-	**84:** P164789z
-3-methanol, 5-(2-chlorophenyl)-4-phenyl-	**83:** P179072p
-3-methanol, 5-(4-chlorophenyl)-4-phenyl-	**83:** P179072p
-3-methanol, 5-(3,4-dichlorophenyl)-4-phenyl-	**83:** P179072p
-3-methanol, 4-(3,5-dimethoxyphenyl)-5-phenyl-	**84:** P164789z
-3-methanol, 5-(3,4-dimethoxyphenyl)-4-phenyl-	**83:** P179072p
-3-methanol, 4-(3,5-dimethylphenyl)-5-phenyl-	**84:** P164789z
-3-methanol, 4,5-diphenyl-	**83:** P179072p
-3-methanol, 4-(4-fluorophenyl)-5-phenyl-	**84:** P164789z
-3-methanol, 5-(4-fluorophenyl)-4-phenyl-	**83:** P179072p
-3-methanol, 4-(2-hydroxyphenyl)-5-phenyl-	**84:** P164789z
-3-methanol, 4-(4-hydroxyphenyl)-5-phenyl-	**84:** P164789z
-3-methanol, 4-(3-methoxyphenyl)-5-phenyl-	**84:** P164789z
-3-methanol, 4-(4-methoxyphenyl)-5-phenyl-	**84:** P164789z
-3-methanol, 5-(2-methoxyphenyl)-4-phenyl-	**83:** P179072p
-3-methanol, 5-(4-methoxyphenyl)-4-phenyl-	**83:** P179072p
-3-methanol, 4-(3-methylphenyl)-5-phenyl-	**84:** P164789z
-3-methanol, 4-(4-methylphenyl)-5-phenyl-	**84:** P164789z
-3-methanol, 5-(4-methylphenyl)-4-phenyl-	**83:** P179072p
-3-methanol, 5-(3-methylphenyl)-4-phenyl-	**83:** P179072p
-3-methanol, 5-phenyl-4-[2-(phenylmethoxy)phenyl]-	**84:** P164789z
-3-methanol, 5-phenyl-4-[4-(phenylmethoxy)phenyl]-	**84:** P164789z
-3-methanol, 4-phenyl-5-[3-trifluoromethyl)phenyl]-	**83:** 179072p
-3-methanol, 5-phenyl-4-[3-(trifluoromethyl)phenyl]-	**84:** P164789z
-3-methanol, 5-phenyl-4-[4-(trifluoromethyl)phenyl]-	**84:** P164789z
-3-methanol, 4-phenyl-5-(3,4,5-trimethoxyphenyl)-	**83:** P179072p
-3-methanol, 5-phenyl-4-(3,4,5-trimethoxyphenyl)-	**84:** P164789z
-3-methanol, α,α,4,5-tetraphenyl-	**24:** 4032^5
-3-methanol, α,4,5-triphenyl-	**66:** 94962c
3-[[(methoxycarbamoyl)oxy]methyl]-5-(4-methoxy-phenyl)-4-phenyl-	**83:** P179072p
4-(*o*-methoxycarbonylphenyl)-3,5-dimethyl-	**72:** 43550h
4-(*p*-methoxyphenyl)-3,5-dimethyl-	**54:** 7696c, **67:** 100076f
4-(*p*-methoxyphenyl)-5-(*p*-nitrophenyl)-4-phenyl-	**55:** 6474b
4-(*p*-methoxyphenyl)-4,5-diphenyl-	**55:** 6474b
4-(*o*-methoxyphenyl)-3,5-diphenyl-	**67:** 100076f, **73:** 130941a
4-(*m*-methoxyphenyl)-3,5-diphenyl-	**73:** 130941a
4-(*p*-methoxyphenyl)-3,5-diphenyl-	**73:** 130941a
4-(*p*-methoxyphenyl)-3,5-dipropyl-	**69:** 106627z
5-(2-methoxyphenyl)-3-[[methylaminocarbonyl)oxy]-methyl]-4-phenyl-	**83:** P179072p
5-(4-methoxyphenyl)-3-[[(phenoxycarbonyl)oxy]methyl]-4-phenyl-	**83:** P179072p
3-[[(methylaminocarbonyl)oxy]methyl]-4,5-diphenyl-	**83:** P179072p
3-[[(methylaminocarbonyl)oxy]methyl]-5-(3-methyl-phenyl)-4-phenyl-	**83:** P179072p

93

TABLE 3 (*Continued*)

Compound	Reference

3.2. 3,4,5-Alkyl- or Aryl-Trisubstituted 4H-1,2,4-triazoles (*Continued*)

3-[[(methylaminocarbonyl)oxy]methyl]-5-(4-methyl-phenyl)-4-phenyl-	**83:** P179072p
3-[[(methylaminocarbonyl)oxy]methyl]-4-phenyl-5-[3-(trifluoromethyl)phenyl]-	**83:** P179072p
3-[[(methylaminocarbonyl)oxy]methyl]-4-phenyl-5-(3,4,5-trimethoxyphenyl)-	**83:** P179072p
4-methyl-3,5-bis[3-[[1-(2-methylphenylcarbamoyl)-aceton-1-yl]azo]phenyl]-	**82:** P157819q
4-methyl-3,5-bis(pentafluoroethyl)-	**57:** 12471a
4-methyl-3,5-bis(trifluoromethyl)-	**57:** 12471a
4-(3-methylbutyl)-3,5-diphenyl-	**77:** 34431w
3-methyl-4,5-diphenyl-	**66:** 46376t **74:** 87898y
4-methyl-3,5-diphenyl-	**49:** 1711d, **55:** 6474a, **75:** 20254n, **75:** 35845e, **77:** 34431w
4-methyl-3-[o-methyldithio)phenyl]-5-phenyl-	**75:** 76665q
4-(4-methylpentyl)-3,5-diphenyl-	**77:** 34431w
4-(2-methylphenyl)3,5-bis(4-methylphenyl)-	**84:** 17236p
4-(3-methylphenyl)-3,5-bis(4-methylphenyl)-	**84:** 17236p
3-(2-methylphenyl)4,5-diphenyl-	**84:** 17236p
3-(4-methylphenyl)-4,5-diphenyl-	**84:** 17236p
5-methyl-4-phenyl-3-(diphenylmethyl)-	**28:** 5458²
3-(2-methylphenyl)-5-(4-methylphenyl)-4-phenyl-	**84:** 17236p
1-methyl-3-phenyl-5-(2-oxo-3-phenyl-2H-1-benzopyran-7-yl)-	**74:** P65592s
3-methyl-5-phenyl-4-[o-(phenylcarbamyl)phenyl]-	**23:** 836²
4-methyl-3-phenyl-5-(2-thienyl)-	**75:** 35845e
4-methyl-5-(4-pyridyl)-3-[(p-toluenesulfonamido)methyl]-	**69:** 19082z
3-methyl-5-phenyl-4-o-tolyl-	**23:** 836³
3-methyl-5-phenyl-4-p-tolyl-	**23:** 836³
3-methyl-4-phenyl-5-(trifluoromethyl)-	**85:** 138543r
4-(p-nitrophenyl)-3,5-diphenyl-	**73:** 130941a, **83:** 206172e
3-phenyl-4,5-bis[4-(2-phenylethenyl)phenyl]-	**68:** P21961h, **80:** P146981u
4-phenyl-3,5-bis[4-(2-phenylethenyl)phenyl]-	**68:** P21961h, **80:** P146981u
3-phenyl-4,5-bis[p-(p-phenylstyryl)phenyl]-	**68:** P21961h, **80:** P146980t, **80:** P146981u
4-phenyl-3,5-bis[p-(p-phenylstyryl)phenyl]-	**67:** 21858j
3-phenyl-4,5-bis(p-styrylphenyl)-	**67:** 21858j
4-phenyl-3,5-bis(p-styrylphenyl)-	**67:** 21858j
4-phenyl-3,5-dipropyl-	**69:** 106627z
3-phenyl-4,5-di-p-tolyl-	**67:** 21858j
4-phenyl-3,5-di-p-tolyl-	**67:** 21858j
3-phenyl-4-(2-pyridinyl)-5-(tribromomethyl)-	**83:** 9958q
4-phenyl-5-(4-pyridyl)-3-[(p-toluenesulfonamido)methyl]-	**69:** 19082z
-3-propionic acid, 4,5-diphenyl-	**71:** 81268j
-3-propionic acid, β-methylene-4,5-diphenyl-	**71:** 81268j
4-(2-pyridinyl)-3,5-diphenyl-	**83:** 206172e
4-(3-pyridinyl-3,5-diphenyl-	**83:** 206172e

TABLE 3 (*Continued*)

Compound	Reference

3.2. 3,4,5-Alkyl- or Aryl-Trisubstituted 4H̲-1,2,4-triazoles (*Continued*)

Compound	Reference
4-(4-pyridinyl)-3,5-diphenyl-	**83:** 206172e
3,4,5-trimethyl-	**27:** P4541[3], **54:** 7696c
3,4,5-triphenyl-	**53:** 5253g, **53:** 16121b,
	54: 21064a, **55:** 6473i,
	57: 5910d, **66:** 46412b,
	67: 100076f, **77:** 101478g
3,4,5-tris[p-(p-chlorostyryl)phenyl]-	**68:** P21961h, **80:** P146981u
3,4,5-tris[4-[2-(4-methoxyphenyl)ethenyl]phenyl]-	**68:** P21961h, **80:** P146981u
3,4,5-tris[p-(p-methoxystyryl)phenyl]-	**67:** 21858j
3,4,5-tris(2-methylphenyl)-	**84:** 17236p
3,4,5-tris(4-methylphenyl)-	**80:** P146980t, **80:** P146981u,
	83: P29887u, **84:** 17236p
3,4,5-tris[4-[2-(1-naphthalenyl)ethenyl]phenyl]-	**68:** P21961h, **80:** P146980t,
	80: 146981u
3,4,5-tris[4-[2-(2-naphthalenyl)ethenyl]phenyl]-	**80:** P146980t, **80:** P146981u
3,4,5-tris[p-[2-(1-naphthyl)vinyl]phenyl]-	**67:** 21858j
3,4,5-tris[p-[2-(2-naphthyl)vinyl]phenyl]-	**67:** 21858j
3,4,5-tris(pentafluorophenyl)-	**68:** P39628q
3,4,5-tris[4-(2-phenylethenyl)phenyl]-	**68:** P21961h, **80:** P146981u
3,4,5-tris[p-(p-phenylstyryl)phenyl]-	**67:** 21858j, **68:** P21961h,
	80: P146980t, **80:** P146981u
3,4,5-tris(p-styrylphenyl)-	**67:** 21858j
3,4,5-tri-p-tolyl-	**67:** 21858j
-3-valeric acid, 4,5-diphenyl-	**71:** 81268j

CHAPTER 4

1,2,4-Triazolecarboxylic Acids and Their Functional Derivatives

Amidrazones and their derivatives are key intermediates for the synthesis of triazolecarboxylic acids. The condensation of the amidrazone (**4.1-1**) with ethyl orthoformate gave a good yield of the triazolecarboxylate (**4.1-2**).[1] The

introduction of the carboxylate group via the condensing reagent was demonstrated by the acylation of the amidrazone (**4.1-3**) with the acid chloride (**4.1-4**) to give the 1,3-diphenyltriazole-5-carboxylate (**4.1-5**).[2] Also this compound was prepared in higher yield by treatment of ethyl cyano-formate with the nitrilimine (**4.1-7**), which was generated *in situ* from **4.1-6**.[2] The isomeric ethyl 3,4-diphenyltriazole-5-carboxylate (**4.1-9**) was prepared in low yield (19%) by the condensation of the amidrazone (**4.1-8**) with diethyl oxalate.[3,4] In this reaction at 150°, a 34% yield of 3,4,5-tri-phenyltriazole (**4.1-10**) resulted from self-condensation of the amidrazone.[4]

96

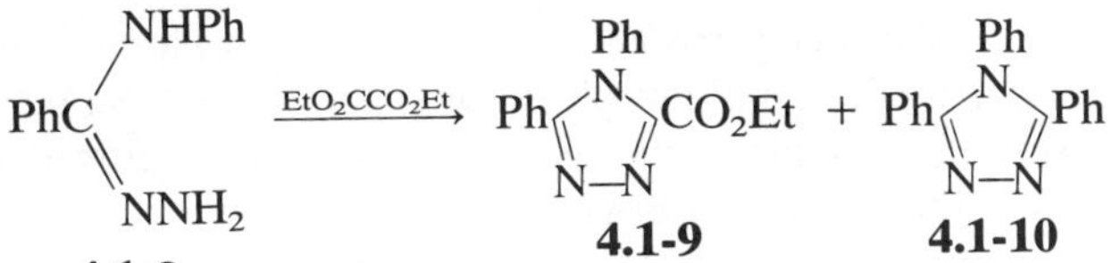

4.1-8   **4.1-9**   **4.1-10**

The 1-acetylamidrazone (**4.1-11**) has been dehydrated thermally[5] and in refluxing acetic anhydride[6] to give ethyl 3-methyltriazole-5-carboxylate (**4.1-12**). Also, the 1-acylamidrazone (**4.1-16**), generated either from the *S*-

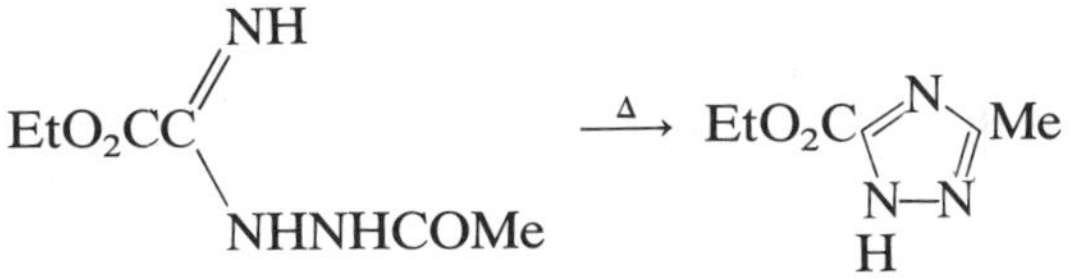

4.1-11   **4.1-12**

methylthioamide (**4.1-13**) and **4.1-14** or from the iminoether (**4.1-15**) and methylamine, was converted in good yield to the triazole-5-carboxamide (**4.1-17**).[7]

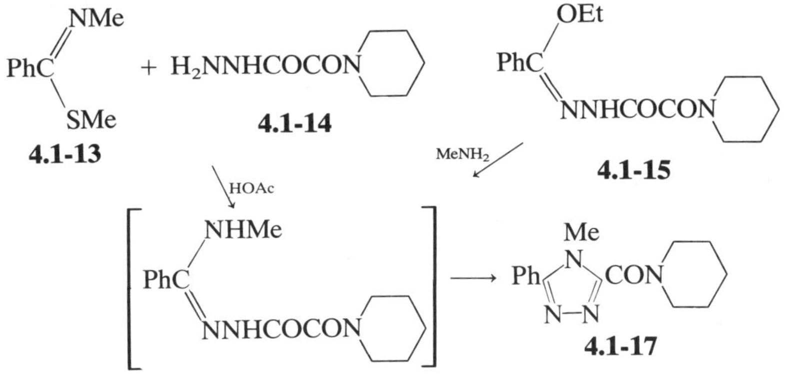

4.1-13   **4.1-14**   **4.1-15**

4.1-16   **4.1-17**

The conversion of 4-phenylazo-2-phenyloxazolin-5-one (**4.1-18**) with methanolic potassium hydroxide to 1,5-diphenyl-1H-1,2,4-triazole-3-carboxylic acid (**4.1-23**) was discovered by Sawdey, who postulated that the rearrangement proceeded via the acyclic (**4.1-19**) and triazole (**4.1-24**) esters, followed by saponification of the latter.[8] In the reaction of **4.1-18** with methanolic ammonium hydroxide, Sawdey suggested that the amide (**4.1-25**) also was formed from the ester (**4.1-24**). However, the rearrangement of **4.1-18** to **4.1-23**, presumably via the acyclic acid intermediate (**4.1-20**), was successful in a mixture of HCl in acetic acid.[9] Rearrangements of this type have also been carried out in acetic acid alone[10] and in acetic acid containing a catalytic amount of sodium acetate.[11] These results were in contrast to reports that no rearrangement of **4.1-18** occurred in aqueous acid.[12,13] In addition, Browne and Polya found that treatment of **4.1-18** with aniline to give the anilide (**4.1-26**) was preceded by the formation of the acyclic amide (**4.1-21**).[12,13] In the reaction of azooxazolin-5-ones with aryl amides, N-aroyltriazole-3-carboxamides were isolated.[14] Furthermore, reaction of a series of 4-arylazo-2-phenyloxazolin-5-ones with hydrazine gave acyclic hydrazides (e.g., **4.1-22**), which were cyclized with aqueous base to give the corresponding triazolecarboxylic acid hydrazides (e.g., **4.1-27**) except for those hydrazides containing bulky groups (e.g., 2-MeC$_6$H$_4$, 2-C$_{10}$H$_7$).[13] Treatment of **4.1-18** and other 4-(arylazo)oxazolin-5-ones with aryl thiols gave the triazole thiol esters (e.g., **4.1-28**).[15] In addition, the 5-

4.1-18

$\longrightarrow$

4.1-19, R = MeO
4.1-20, R = HO
4.1-21, R = PhNH
4.1-22, R = H$_2$NNH

$\longrightarrow$

4.1-23, R = HO
4.1-24, R = MeO
4.1-25, R = NH$_2$
4.1-26, R = PhNH
4.1-27, R = H$_2$NNH
4.1-28, R = PhS

amino-2-phenyloxazole (**4.1-29**) was treated with p-tolyl diazonium tetrafluoroborate to give the salt, **4.1-30**, which was rearranged in refluxing N,N-dimethylformamide (DMF) to give a 90% yield of **4.1-31**.[16] The

4.1-29   **4.1-30**   **4.1-31**

thiazolin-5-one analogs of **4.1-18** also undergo rearrangement with the elimination of the ring sulfur.[17] A similar rearrangement was observed for 2-methyl-4-(phenylazo)imidazol-5-(1H)-one (**4.1-32**), which was converted to the carboxylic acid (**4.1-33**) in 2% NaOH and to a mixture of **4.1-33** and the corresponding amide (**4.1-34**) in 0.12 N NaOH.[18,19]

Acylamidrazones similar to those postulated as intermediates in the preceding reactions have been prepared by other methods. Reaction of α-(acetylamino)malonic acid monoethyl ester (**4.1-35**) with aryl diazonium salts resulted in C—C bond cleavage to give 3-acyl-1-arylamidrazones (e.g.,

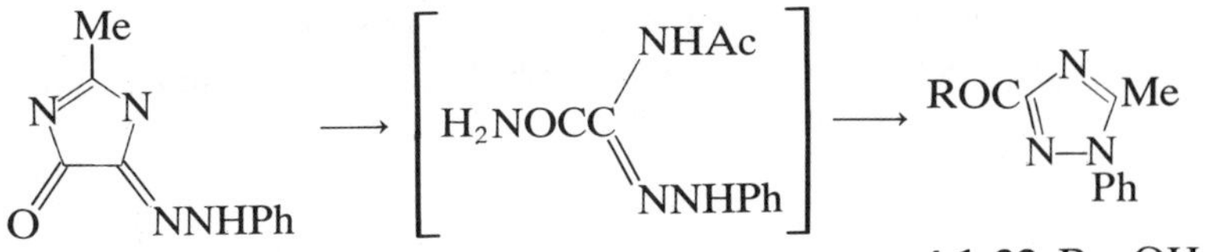

4.1-32

4.1-33, R = OH
4.1-34, R = NH$_2$

4.1-36), which were cyclized in refluxing acetic anhydride to give excellent yields of 1-aryl-5-methyltriazole-3-carboxylic acid esters (e.g., **4.1-37**).[20–23]

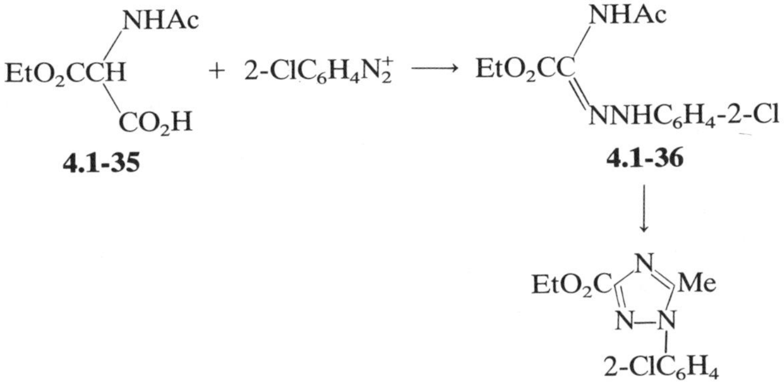

4.1-35   + 2-ClC$_6$H$_4$N$_2^+$ ⟶ **4.1-36**

4.1-37

Also, this method was successful with both α-(formylamino)- and α-(benzoylamino)malonic acid monoethyl esters to give 1-aryl- and 1,5-diaryl-triazole-3-carboxylic acid esters, respectively.[20,24] Other amidrazone intermediates were formed by loss of a *C*-acetyl moiety when ethyl α-(acetylamino)acetoacetate (**4.1-38**) was treated with aryl diazonium salts.[25] This route is illustrated by reaction of the ester (**4.1-38**) with diazotized 4-methoxyaniline to give the 3-acylamidrazone (**4.1-39**), which was treated with aqueous alkali to give **4.1-40** and with concentrated ammonium hydroxide to give **4.1-41**. Carbon–carbon bond cleavage was also observed in

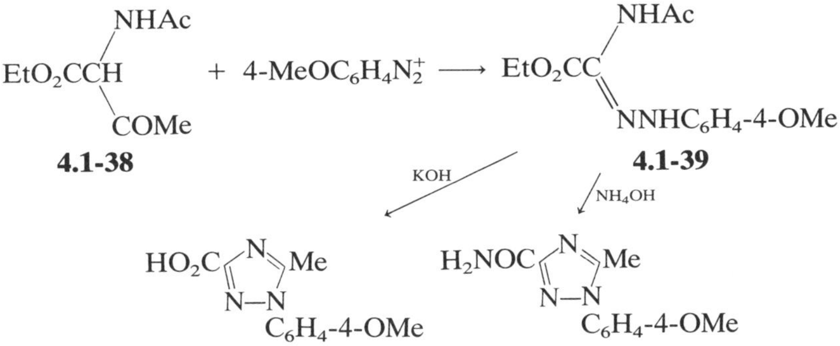

4.1-38   + 4-MeOC$_6$H$_4$N$_2^+$ ⟶ **4.1-39**

KOH   NH$_4$OH

4.1-40   **4.1-41**

the cyclization step when either **4.1-42** or **4.1-44** was converted to the 4,5-diphenyltriazole-3-carboxylic acid ester (**4.1-43**).[26] Another unusual method

involved the intramolecular cycloaddition of the isocyanoacetate (**4.1-45**) to give a 46% yield of **4.1-46**.[27]

In related reactions, triazole-3-carbonitriles were prepared from α-(acetylamino)cyanoacetic acid and aryl diazonium salts. For example, treatment of **4.1-47** with diazotized 4-nitroaniline gave the amidrazone (**4.1-48**), which was cyclized with hot acetic anhydride to give **4.1-49**.[24] An inter-

mediate similar to **4.1-48** resulted from the displacement of the diethylamino group of **4.1-50** with phenylhydrazine, followed by cyclization to give **4.1-51** (70%).[28] In contrast, the cyclization of 1-acylamidrazones unsubstituted on the hydrazino moiety with an aryl group gave lower yields of

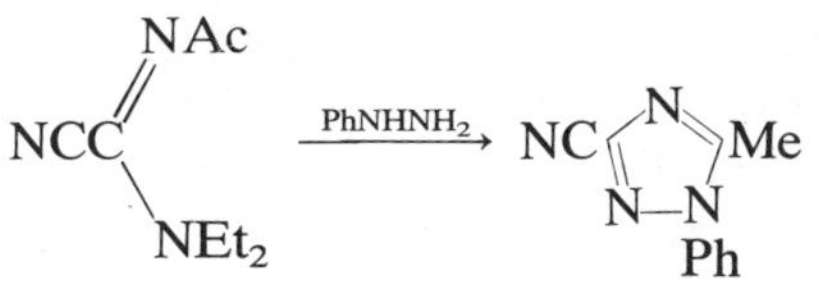

NAc
‖
NCC
  |
  NEt$_2$

$\xrightarrow{\text{PhNHNH}_2}$

NC — triazole ring with Me, N—N, Ph

4.1-50   **4.1-51**

triazole-3-carbonitriles, as demonstrated by the conversion of **4.1-52** to **4.1-54**.[29] With the 1-benzoyl derivative (**4.1-53**) one experiment gave a low yield of the triazole (**4.1-55**), and another gave the oxadiazole (**4.1-56**).[29] In

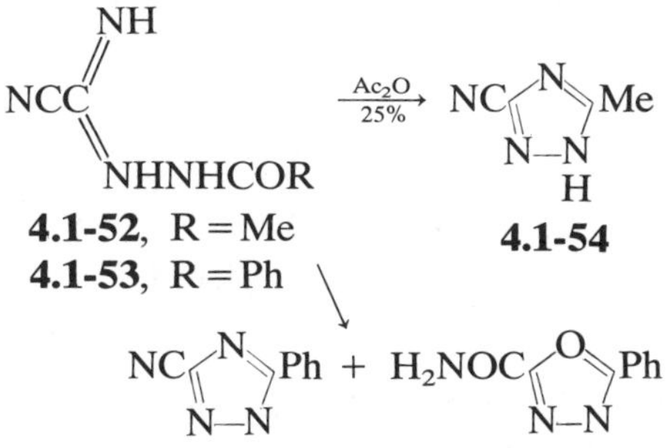

NH
‖
NCC
  ‖
  NHNHCOR

$\xrightarrow[25\%]{\text{Ac}_2\text{O}}$

NC — triazole ring with Me, N—N, H

4.1-52, R = Me
4.1-53, R = Ph   **4.1-54**

NC — triazole ring with Ph, N—N   +   H$_2$NOC — oxadiazole ring with Ph, N—N

4.1-55   **4.1-56**

addition, triazole-3-carbonitrile (**4.1-58**) has been reported to result from dehydration of triazole-5-carboxamide with phosphorus oxychloride,[30] from nucleophilic displacement of the chloro group of 3-chlorotriazole with cyanide in N,N-dimethylformamide,[31] and from the acid-catalyzed cyclization of cyanoformimidic acid hydrazide (**4.1-57**) with triethyl orthoformate.[32]

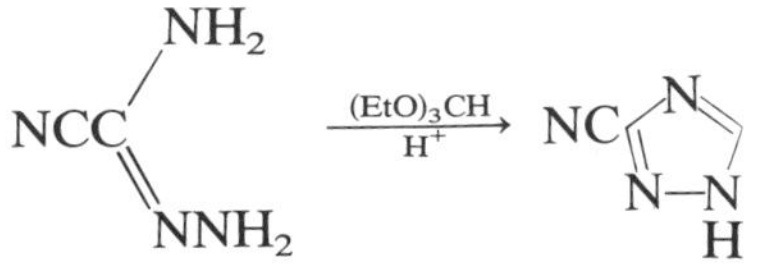

NH$_2$
  |
NCC
  ‖
  NNH$_2$

$\xrightarrow[\text{H}^+]{\text{(EtO)}_3\text{CH}}$

NC — triazole ring with N, N—N, H

4.1-57   **4.1-58** (57%)

A number of triazolecarboxylic acids and their derivatives have been prepared by N-alkylation and N-acylation of preformed triazoles. The acid-catalyzed fusion of methyl triazole-3-carboxylate (**4.1-59**) with 1,2,3,5-tetra-O-acetyl-β-D-ribofuranose gave a 78% yield of the ribo derivative substituted at N-1 (**4.1-60**) and a 7% yield of the ribo derivative substituted at N-2 (**4.1-62**).[33] In contrast to these yields of isomers, treatment of the trimethylsilyl derivative of **4.1-59** with 2,3,5-tri-O-acetyl-D-ribofuranosyl bromide in acetonitrile at room temperature gave a 51% yield of **4.1-60** and a 46% yield of **4.1-62**. In the ribosylation of triazole-3-carbonitrile (**4.1-58**) both procedures gave an 80% yield of **4.1-61** and only a 6% yield of the isomeric ribo derivative (**4.1-63**).[32] These procedures have also been used for the preparation of nucleosides (e.g., ribo- and arabinofuranose) derived from the methyl esters of 3-amino-, 3-chloro-, 3-methyl-, and 3-nitrotriazole-5-carboxylic acids.[34,35] Of interest 1-β-D-ribofuranosyl-1,2,4-triazole-3-carboxamide (virazole) showed broad spectrum antiviral activity against both DNA and RNA viruses.[33] Also, the 4-thio-β-D-ribofuranosyl analog of virazole has been prepared.[36]

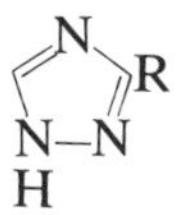

4.1-59, R = CO₂Me
4.1-58, R = CN

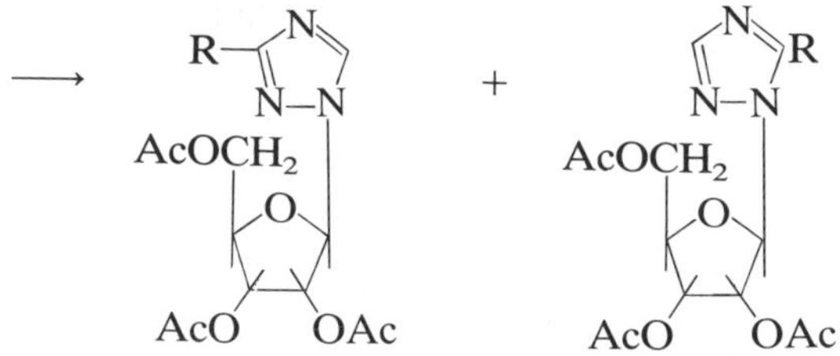

4.1-60, R = MeO₂C   **4.1-62**, R = CO₂Me
4.1-61, R = NC   **4.1-63**, R = CN

Derivatives of triazole-1-carboxylic acid are readily formed by the interaction of s-triazole with ethyl chloroformate, methyl isocyanate, dimethylcarbamoyl chloride, cyanamide, 2-methoxyphenylcyanate, and thiobenzoyl chloride to give **4.1-64**,[37,38] **4.1-65**,[39] **4.1-66**,[40] **4.1-67**,[41] **4.1-68**,[42] and **4.1-69**,[43] respectively. The latter has been reported to be a powerful thioacylating reagent. Similarly, the tert-butyl ester (**4.1-70**) has been used to block the amino group of amino acids.[44] In the reaction of s-triazole with cyanogen bromide, the reaction conditions can be modified to give either triazole-1-carboxamide (**4.1-71**) or the bitriazole (**4.1-73**) (see Section 21.1).[45] The peroxycarbamic acid (**4.1-72**), generated *in situ* from 1,1'-carbonylbis(1,2,4-triazole) (**4.1-74**) and hydrogen peroxide, epoxidizes alkenes at rates some 200 times faster than peroxybenzoic acid.[46]

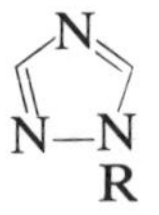

4.1-64, R = EtO₂C
4.1-65, R = MeNHCO
4.1-66, R = Me₂NCO
4.1-67, R = H₂N(HN═)C
4.1-68, R = 2-MeOC₆H₄(HN═)C
4.1-69, R = PhCS
4.1-70, R = Me₃COCO
4.1-71, R = H₂NCO
4.1-72, R = CO₃H

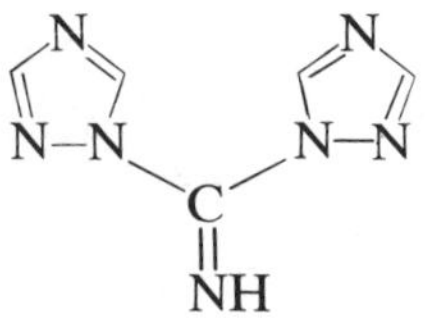

4.1-73

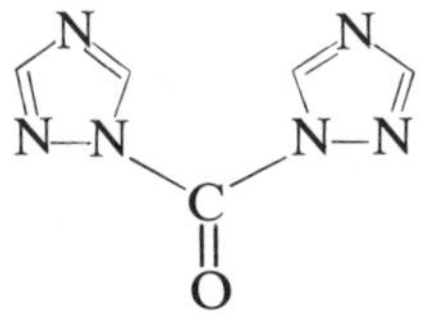

4.1-74

In other interconversions of triazoles, the low melting form of the oxime (4.1-75) was rearranged in polyphosphoric acid to give the amide (4.1-76) in 70% yield.[47] Diazotization of 3-aminotriazole-5-carboxylic acid (4.1-77)

4.1-75   **4.1-76**

produced the 3-diazotriazole intermediate (4.1-78), which was reduced with methanol to give triazole-3-carboxylic acid (4.1-79)[48] and reacted with copper (I) chloride and potassium cyanide to give 3-cyanotriazole-5-carboxylic acid (4.1-80).[49] The latter was decarboxylated with heat to give triazole-3-carbonitrile (4.1-58),[49] which can be converted to the thioamide (4.1-81) with hydrogen sulfide in ethanol.[30]

4.1-77   **4.1-78**   **4.1-79**

4.1-80   **4.1-58**   **4.1-81**

The triazole ring is stable to oxidizing conditions and many types of C̲-carbon derivatives can be converted to triazolecarboxylic acids. For example, the 3-acyl group of the 5-methyltriazole (4.1-82) was preferentially oxidized with potassium permanganate to give 5-methyl-1-phenyl-triazole-3-carboxylic acid (4.1-33).[50] Also, 3-hydroxymethyl),[51] 3-acetyl-,[52] 3-acetonyl,[53] and 3-phenacyltriazoles[54] as well as 3-methyl-s-triazole[55–57] are easily converted to triazole-3-carboxylic acids with potassium permanganate in neutral or alkaline medium. Under acidic conditions, it has been observed that N-aryl substituents are removed.* The carbonation of s-triazole in the presence of CdF$_2$ in an autoclave has been reported to give either 4.1-79 or 4.1-83 depending upon the reaction temperature.[58]

4.1-82   **4.1-33**

---

* A. Andreocci, *Gazz. Chim. Ital.*, **29**, 1 (1899).

4.1-79 (43.7%)   **4.1-83** (60%)

The condensed ring systems, s-triazolo[4,3-a]pyridine (**4.1-84**) and the isomeric s-triazolo[1,5-a]pyridine (**4.1-86**), are readily oxidized with aqueous potassium permanganate at room temperatures to give s-triazole-3-carboxylic acid (**4.1-79**).[59,60] The pyridine ring rather than the methyl group of **4.1-87** was oxidized to give 3-methyltriazole-5-carboxylic acid (**4.1-85**).[61]

4.1-84   **4.1-79**, R = H   **4.1-86**, R = H
                  4.1-85, R = Me   **4.1-87**, R = Me

References

1. **62:** 13147e	2. **57:** 11198d	3. **71:** 81268j	4. **74:** 87898y
5. **78:** 136177f	6. **83:** 178256q	7. **73:** 66517n	8. **51:** 12079a
9. **59:** 2815h	10. **83:** 10007k	11. **82:** 16754c	12. **55:** 2625d
13. **57:** 805h	14. **85:** 177322n	15. **55:** 24729e	16. **81:** 105370w
17. **85:** 177325r	18. **67:** 100068e	19. **71:** 70546x	20. **55:** 22300h
21. **58:** 2447h	22. **84:** P31151p	23. **85:** 5701u	24. **57:** 15096c
25. **60:** 4133a	26. **54:** 21063g	27. **83:** 193193k	28. **70:** 28872e
29. **56:** 7307a	30. **63:** 13243f	31. **71:** 81375s	32. **80:** 10254b
33. **78:** 12030h	34. **80:** 121258a	35. **83:** 114784c	36. **85:** 191t
37. **55:** 5484b	38. **71:** 38867s	39. **76:** 85792b	40. **67:** 54140x
41. **69:** 59251b	42. **62:** 5222f	43. **79:** 78696g	44. **79:** 5560y
45. **79:** 105150h	46. **82:** 111099b	47. **68:** 105162m	48. **64:** 5074f
49. **72:** 111375g	50. **53:** 17020e	51. **75:** 140792h	52. **54:** 510e
53. **60:** 13238g	54. **59:** 7530e	55. **57:** 5903d	56. **70:** 77876t
57. **74:** 99951y	58. **54:** 1552e	59. **64:** 5072e	60. **64:** 5072d
61. **65:** 8898a			

TABLE 4.   1,2,4-TRIAZOLECARBOXYLIC ACIDS AND THEIR FUNCTIONAL DE-
RIVATIVES

Compound	Reference
3-[[[7-[(α-aminophenylacetyl)amino]-2-carboxy-8-oxo- 5-thia-1-azabicyclo[4·2·0]oct-2-en-3-yl]methylthio] carbonyl]-	**82:** 11679x
1-(2-benzoyl-4-chlorophenyl)-3-cyano-5-[(dimethylamino)- methyl]-	**84:** P121844w
1-(3-benzoyl-4,5-dimethyl-2-thienyl)-5-(chloromethyl)- 3-[(1-piperidinyl)carbonyl]-	**83:** P206340h
1-(2-benzoylphenyl)-5-(chloromethyl)-4-methyl-3- (piperazinocarbonyl)-	**81:** P169567y
5-benzyl-4-butyl-3-[(1-piperidino)carbonyl]-	**70:** P87820j, **73:** 66517n
5-benzyl-4-cyclohexyl-3-[(1-piperidino)carbonyl]-	**70:** P87820j, **73:** 66517n
5-benzyl-4-ethyl-3-[(1-piperidino)carbonyl]-	**70:** P87820j, **73:** 66517n
5-benzyl-4-methyl-3-[(1-piperidino)carbonyl]-	**70:** P87820j, **73:** 66517n
5-benzyl-3-[(1-piperidino)carbonyl]-4-propyl-	**70:** P87820j, **73:** 66517n
4-butyl-5-phenyl-3-[(1-piperidino)carbonyl]-	**70:** P87820j, **73:** 66517n
-3-carbohydrazonamide, 1-β-D-ribofuranosyl-	**80:** 10254b
-3-carbonitrile	**57:** 15096c, **63:** 13243g, **72:** 111375g, **80:** 10254b
-3-carbonitrile, 1-(p-chlorophenyl)-5-methyl-	**57:** 15096g
-3-carbonitrile, 1-(2-deoxy-β-D-erythro-pentofuranosyl)-, di(4-methyl benzoate ester)	**83:** 114784c
-3-carbonitrile, 1,5-diphenyl-	**70:** 28872e
-3-carbonitrile, 1-(2-methoxy-4-nitrophenyl)-5-methyl-	**57:** 15096g
-3-carbonitrile, 1-(p-methoxyphenyl)-5-methyl-	**57:** 15096g
-3-carbonitrile, 5-methyl-	**56:** 7307g
-3-carbonitrile, 5-methyl-1-(p-nitrophenyl)-	**57:** 15096g
-3-carbonitrile, 5-methyl-1-(4-nitro-o-tolyl)-	**57:** 15096g
-3-carbonitrile, 5-methyl-1-phenyl-	**57:** 15096g **70:** 28872e
-3-carbonitrile, 5-phenyl-	**56:** 7307g
-3-carbonitrile, 1-(2,3,4-tri-O-acetyl-α-L- lyxopyranosyl)-	**83:** 114784c
-3-carbonitrile, 1-(2,3,5-tri-O-acetyl-β-D- ribofuranosyl)-	**78:** P84766a, **80:** 10254b
-5-carbonitrile, 1-(2,3,5-tri-O-acetyl-β-D- ribofuranosyl)-	**80:** 10254b
-3-carbonyl azide	**85:** P123930r
-3-carbonyl azide, 1,5-diphenyl-	**55:** 2625g, **57:** 806f
-3-carbonyl azide, 1-(p-methoxyphenyl)-5-phenyl-	**55:** 2625g, **57:** 806g
-3-carbonyl chloride	**85:** P123930r
-3-carbonyl chloride, 1-(2-benzoyl-4-chlorophenyl)- 5-(chloromethyl)-	**84:** P44206k
-3-carbonyl chloride, 1-[4-chloro-2-(2-chlorobenzoyl)- phenyl]-5-(chloromethyl)-	**84:** P105667k
-3-carbonyl chloride, 1-[4-chloro-2-(2-chlorobenzoyl)- phenyl]-5-(4-morpholinylmethyl)-	**84:** P121844w
-3-carbonyl chloride, 1,5-diphenyl-	**83:** 10007k
-1-carboperoxoic acid	**82:** 111099b

TABLE 4   (*Continued*)

Compound	Reference
-3-carbothioamide, 1-(2-deoxy-α-D-erythro-pentofuranosyl)-	**83:** 114784c
-3-carbothioamide, 1-(2-deoxy-β-D-erythro-pentofuranosyl)-	**83:** 114784c
-3-carbothioamide, 1-(2-deoxy-α-D-erythro-pentofuranosyl)-, di(4-methyl benzoate ester)	**83:** 114784c
-3-carbothioamide, 1-(2-deoxy-β-D-erythro-pentofuranosyl)-, di(4-methyl benzoate ester)	**83:** 114784c
-3-carbothioamide, 1-β-D-ribofuranosyl-	**78:** P84766a, **80:** 10254b
-5-carbothioamide, 1-β-D-ribofuranosyl-	**80:** 10254b
-3-carbothioamide, 1-(tetrahydro-2H-pyran-2-yl)-	**84:** P4966b
-5-carbothioamide, 1-(2,3,5-tri-O-acetyl-β-D-ribofuranosyl)-	**80:** 10254b
-1-carbothioic acid, 2,2-diphenylhydrazide	**67:** 99764v
-5-carboxamide, 1-α-L-arabinopyranosyl-	**83:** 114784c
-3-carboxamide, 1-β-D-arabinofuranosyl-	**78:** P84766a, **83:** 114784c
-3-carboxamide, 5-(azidomethyl)-1-(2-benzoyl-4-chlorophenyl)-N,N-diethyl-	**84:** P105667k
-3-carboxamide, 5-(azidomethyl)-1-[4-chloro-2-(2-chlorobenzoyl)phenyl]-	**84:** P105667k
-3-carboxamide, 1-(2-benzoyl-4-chlorophenyl)-5-(chloromethyl)-	**81:** P169567y. **84:** P121844w
-3-carboxamide, 1-(2-benzoyl-4-chlorophenyl)-5-(chloromethyl)-N,N-diethyl-	**79:** P126535c, **84:** P44206k
-3-carboxamide, 1-(2-benzoyl-4-chlorophenyl)-5-(chloromethyl)-N,N-dimethyl-	**81:** P169567y, **84:** P44206k
-3-carboxamide, 1-(2-benzoyl-4-chlorophenyl)-5-(chloromethyl)-N-(2-hydroxyethyl)-	**81:** P169567y
-3-carboxamide, 1-(2-benzoyl-4-chlorophenyl)-5-[(1,3-dihydro-1,3-dioxo-2H-isoindol-2-yl)methyl]-N,N-diethyl-	**79:** P126535c, **84:** P44206k
-3-carboxamide, 1-(2-benzoyl-4-chlorophenyl)-5-[(1,3-dihydro:1,3-dioxo-2H-isoindol-2-yl)methyl]-N,N-dimethyl-	**84:** P44206k
-3-carboxamide, 1-(2-benzoyl-4-chlorophenyl)-5-[(dimethylamino)methyl]-	**84:** P121844w
-3-carboxamide, 4-(2-benzoyl-4-chlorophenyl)-5-[(dimethylamino)methyl]-	**85:** P46692k
-3-carboxamide, 4-(2-benzoyl-4-chlorophenyl)-5-[(dimethylamino)methyl]-N-(hydroxymethyl)-	**85:** P46692k
-3-carboxamide, 1-(2-benzoyl-4-chlorophenyl)-5-(4-morpholinylmethyl)-	**84:** P121844w
-3-carboxamide, 1-(2-benzoyl-4-chlorophenyl)-5-(1-piperidinylmethyl)-	**84:** P121844w
-3-carboxamide, 1-(2-benzoyl-4-chlorophenyl)-5-(1-pyrrolidinylmethyl)-	**84:** P121844w
-1-carboxamide	**76:** 85792b, **79:** 105150h
-3-carboxamide	**63:** 13243g, **82:** P4534t
-3-carboxamide, 1-(5-O-acetyl-β-D-ribofuranosyl)	**84:** P44603n
-3-carboxamide, N-[2-(acetylamino)phenyl]-5-(2-chlorophenyl)-1-(4-methylphenyl)-	**85:** 177322n

TABLE 4   (*Continued*)

Compound	Reference
-3-carboxamide, *N*-[2-(acetylamino)phenyl]-5-(2-chlorophenyl)-1-phenyl-	**85:** 177322n
-3-carboxamide, *N*-[2-(acetylamino)phenyl]-5-(4-methoxyphenyl)-1-(4-methylphenyl)-	**85:** 177322n
-3-carboxamide, *N*-[2-(acetylamino)phenyl]-5-(4-methoxyphenyl)-1-phenyl-	**85:** 177322n
-3-carboxamide, 1-[6-aminocarbonyl-5,6-dideoxy-β-D-ribo-hexfuranosyl]-	**81:** 114436z
-3-carboxamide, *N*-[1-(aminocarbonyl)-2-phenylethenyl]-1,5-diphenyl-	**83:** 10007k
-3-carboxamide, *N*-(2-aminophenyl)-5-(2-chlorophenyl)-1-(4-methylphenyl)-	**85:** 177322n
-3-carboxamide, *N*-(2-aminophenyl)-5-(2-chlorophenyl)-1-(4-methylphenyl)-	**85:** 177322n
-3-carboxamide, *N*-(2-aminophenyl)-5-(2-chlorophenyl)-1-phenyl-	**85:** 177322n
-3-carboxamide, *N*-(2-aminophenyl)-5-(4-methoxyphenyl)-1-(4-methylphenyl)-	**85:** 177322n
-3-carboxamide, *N*-(2-aminophenyl)-5-(4-methoxyphenyl)-1-phenyl-	**85:** 177322n
-3-carboxamide, *N*-(4-aminophenyl)-5-(4-methoxyphenyl)-1-phenyl-	**85:** 177322n
-3-carboxamide, 1-α-D-arabinofuranosyl-	**83:** 114784c
-5-carboxamide, 1-β-D-arabinofuranosyl-	**83:** 114784c
-3-carboxamide, 1-α-L-arabinopyranosyl-	**83:** 114784c
-3-carboxamide, 1-(2-benzoyl-4-methylphenyl)-5-(chloromethyl)-	**81:** P169567y
-3-carboxamide, 1-(2-benzoyl-4-nitrophenyl)-5-[(dimethylamino)methyl]-	**84:** P121844w
-3-carboxamide, 1-(5-*O*-benzoyl-β-D-ribofuranosyl)-	**84:** P44603n
-3-carboxamide, 5-benzyl-*N*,*N*-diethyl-4-methyl-	**73:** 66517n
-5-carboxamide, 1-benzyl-*N*-methyl-	**77:** 139909m
-1-carboxamide, 3-benzyl-*N*-phenyl-, 4-oxide	**77:** 139906h
-3-carboxamide, *N*,5-bis(4-chlorophenyl)-1-(4-methylphenyl)-	**82:** 16754c, **85:** 177325r
-3-carboxamide, 1,5-bis(4-chlorophenyl)-*N*-4-methylphenyl)-	**82:** 16754c
-3-carboxamide, *N*,5-bis(4-chlorophenyl)-1-(4-nitrophenyl)-	**82:** 16754c
-3-carboxamide, 1,5-bis(4-chlorophenyl)-*N*-phenyl)-	**82:** 16754c
-3-carboxamide, *N*,5-bis(4-chlorophenyl)-1-phenyl-	**82:** 16754c, **85:** 177325r
-3-carboxamide, 1-[6-[bis(phenylmethoxy)phosphinyl]-5,6-dideoxy-2,3-*O*-(isopropylidene)-β-D-ribo-hexofuranosyl]-	**81:** 114436z
-3-carboxamide, 5-(4-bromophenyl)-*N*,1-bis(4-methylphenyl)-	**85:** 177325r
-3-carboxamide, 5-(4-bromophenyl)-*N*-(4-chlorophenyl)-1-(4-methylphenyl)-	**85:** 177325r
-3-carboxamide, 5-(4-bromophenyl)-*N*-(4-chlorophenyl)-1-phenyl-	**85:** 177325r
-3-carboxamide, 5-(4-bromophenyl)-*N*,1-diphenyl-	**85:** 177325r

TABLE 4 (*Continued*)

Compound	Reference
-3-carboxamide, 1-(*p*-bromophenyl)-5-phenyl-	**57:** 806g
-3-carboxamide, 1-[4-bromo-2-(2-pyridinylcarbonyl)-phenyl]-5-(chloromethyl)-	**84:** P31151p, **84:** P150671t
-3-carboxamide, 1-[4-bromo-2-(2-pyridinylcarbonyl)-phenyl]-5-(chloromethyl)-*N*-methyl-	**84:** P31151p, **84:** P150671t
-4-carboxamide, *N*-butyl-3-(2,2-dichloro-1,1-difluoroethyl)-	**81:** 49647c
-3-carboxamide, 1-[6-carboxy-5,6-dideoxy-β-D-ribo-hexfuranosyl]-	**81:** 114436z
-3-carboxamide, 1-[3-(2-chlorobenzoyl)-4-ethyl-2-thienyl]-5-(chloromethyl)-	**84:** P44197h
-3-carboxamide, 1-[3-(2-chlorobenzoyl)-4-ethyl-2-thienyl]-5-(chloromethyl)-*N*-methyl-	**84:** P44197h
-3-carboxamide, 1-[3-(2-chlorobenzoyl)-5-ethyl-2-thienyl]-5-(chloromethyl)-*N*-methyl-	**83:** P206340h
-3-carboxamide, 1-[3-(2-chlorobenzoyl)-5-ethyl-2-thienyl]-5-[(1,3-dihydro-1,3-dioxo-2H̲-isoindol-2-yl)-methyl]-*N*-methyl-	**83:** P206340h
-3-carboxamide, 5-(*p*-chlorobenzyl)-*N*,*N*-diethyl-4-methyl-	**70:** P87820j, **73:** 66517n
-3-carboxamide, 1-[4-chloro-2-(2-chlorobenzoyl)phenyl]-5-(chloromethyl)-	**81:** P169567y
-3-carboxamide, 1-[4-chloro-2-(2-chlorobenzoyl)phenyl]-5-(chloromethyl)-	**84:** P105667k
-3-carboxamide, 1-[4-chloro-2-(2-chlorobenzoyl)phenyl]-5-(chloromethyl)-*N*,*N*-diethyl-	**84:** P44206k
-3-carboxamide, 1-[4-chloro-2-(2-chlorobenzoyl)phenyl]-5-(chloromethyl)-*N*,*N*-dimethyl-	**84:** P44206k
-3-carboxamide, 1-[4-chloro-2-(2-chlorobenzoyl)phenyl]-5-[(1,3-dihydro-1,3-dioxo-2H̲-isoindol-2-yl)methyl]-*N*,*N*-diethyl-	**84:** P44206k
-3-carboxamide, 1-[4-chloro-2-(2-chlorobenzoyl)phenyl]-5-[(1,3-dihydro-1,3-dioxo-2H̲-isoindol-2-yl)methyl]-*N*,*N*-dimethyl-	**84:** P44206k
-3-carboxamide, 1-[4-chloro-2-(2-chlorobenzoyl)phenyl]-5-[(dimethylamino)methyl]-	**84:** P121844w
-3-carboxamide, 1-[4-chloro-2-(2-chlorobenzoyl)phenyl]-5-[(4-methyl-1-piperazinyl)methyl]-	**84:** P121844w
-3-carboxamide, 1-[4-chloro-2-(2-chlorobenzoyl)phenyl]-5-(4-morpholinylmethyl)-	**84:** P121844w
-1-carboxamide, *N*-(2-chloroethyl)-	**76:** 85792b
-3-carboxamide, 1-[4-chloro-2-(2-fluorobenzoyl)-phenyl]-5-(chloromethyl)-*N*,*N*-dimethyl-	**84:** P44206k
-3-carboxamide, 1-[4-chloro-2-(2-fluorobenzoyl)-phenyl]-5-[(1,3-dihydro-1,3-dioxo-2H-isoindol-2-yl)-methyl]-*N*,*N*-dimethyl-	**84:** P44206k
-3-carboxamide, 1-[4-chloro-2-(2-fluorobenzoyl)-phenyl]-5-[(dimethylamino)methyl]-*N*,*N*-dimethyl-	**84:** P121844w
-3-carboxamide, 1-[4-chloro-2-(hydroxyphenylmethyl)-phenyl]-5-(chloromethyl)-	**84:** P121844w
-3-carboxamide, 1-[4-chloro-2-(hydroxyphenylmethyl)-phenyl]-5-[(dimethylamino)methyl]-	**84:** P121844w

TABLE 4 (*Continued*)

Compound	Reference
-3-carboxamide, 5-(chloromethyl)-1-[2-(2-pyridinyl-carbonyl)phenyl]-	**84:** P31151p, **84:** P15067t
-3-carboxamide, 5-[(*p*-chlorophenoxy)methyl]-*N*,*N*-diethyl-	**70:** P87820j, **73:** 66517n
-3-carboxamide, 5-(4-chlorophenyl)-*N*,1-bis(4-methylphenyl)-	**82:** 16754c, **85:** 177325r
-3-carboxamide, *N*-(2-chlorophenyl)-5-(4-chlorophenyl)-1(4-methylphenyl)-	**82:** 16754c
-3-carboxamide, *N*-(3-chlorophenyl)-5-(4-chlorophenyl)-1-(4-methylphenyl)-	**82:** 16754c
-3-carboxamide, *N*-(2-chlorophenyl)-5-(4-chlorophenyl)-1-phenyl-	**82:** 16754c
-3-carboxamide, *N*-(3-chlorophenyl)-5-(4-chlorophenyl)-1-phenyl-	**82:** 16754c
-3-carboxamide, 5-(*p*-chlorophenyl)-*N*,*N*-diethyl-	**70:** P87820j, **73:** 66517n
-3-carboxamide, 5-(4-chlorophenyl)-*N*,1-diphenyl-	**82:** 16754c, **85:** 177325r
-3-carboxamide, 5-(2-chlorophenyl)-*N*-formyl-1-(4-methylphenyl)-	**85:** 177322n
-3-carboxamide, 5-(2-chlorophenyl)-*N*-formyl-1-phenyl-	**85:** 177322n
-3-carboxamide, 5-(2-chlorophenyl)-*N*-(2-hydroxy-benzoyl)-1-(4-methylphenyl)-	**85:** 177322n
-3-carboxamide, 5-(2-chlorophenyl)-*N*-(2-hydroxy-benzoyl)-1-phenyl-	**85:** 177322n
-3-carboxamide, 5-(4-chlorophenyl)-*N*-(3-methylphenyl)-1-(4-methylphenyl)-	**82:** 16754c
-3-carboxamide, 5-(2-chlorophenyl)-1-(4-methylphenyl)-*N*-[(4-methylphenyl)sulfonyl]-	**85:** 177322n
-3-carboxamide, 5-(4-chlorophenyl)-*N*-(4-methylphenyl)-1-(4-nitrophenyl)-	**82:** 16754c
-3-carboxamide, 5-(2-chlorophenyl)-1-(4-methylphenyl)-*N*-[4-(4-oxo-2-phenyl-3-thiazolidinyl)phenyl]-	**85:** 177322n
-3-carboxamide, 5-(4-chlorophenyl)-*N*-(2-methylphenyl)-1-phenyl-	**82:** 16754c
-3-carboxamide, 5-(4-chlorophenyl)-*N*-(4-methylphenyl)-1-phenyl-	**82:** 16754c, **85:** 177325r
-3-carboxamide, 5-(4-chlorophenyl)-*N*-(3-methylphenyl)-1-phenyl-	**82:** 16754c
-3-carboxamide, 5-(4-chlorophenyl)-1-(4-methylphenyl)-*N*-phenyl-	**82:** 16754c, **85:** 177325r
-3-carboxamide, 5-(2-chlorophenyl)-1-(4-methylphenyl)-*N*-[2-[(phenylmethylene)amino]phenyl]-	**85:** 177322n
-3-carboxamide, 5-(2-chlorophenyl)-1-(4-methylphenyl)-*N*-[4-[(phenylmethylene)amino]phenyl]-	**85:** 177322n
-3-carboxamide, 5-(2-chlorophenyl)-1-(4-methylphenyl)-*N*-2-pyridinyl-	**85:** 177322n
-3-carboxamide, 5-(2-chlorophenyl)-*N*-[(4-methylphenyl)-sulfonyl]-1-phenyl-	**85:** 177322n
-3-carboxamide, 5-(4-chlorophenyl)-1-(4-nitrophenyl)-*N*-phenyl-	**82:** 16754c

TABLE 4   (*Continued*)

Compound	Reference
-3-carboxamide, 5-(2-chlorophenyl)-1-phenyl-*N*-[2-[(phenylmethylene)amino]-phenyl]-	
-3-carboxamide, 5-(2-chlorophenyl)-1-phenyl-*N*-2-pyridinyl-	**85:** 177322n
-4-carboxamide, 3-cyano-	**31:** 7849[4]
-3-carboxamide, 1-(6-deoxy-β-D-erythro-hex-5-en-2-ulofuranosyl]-	**84:** 180518r
-3-carboxamide, 1-(2-deoxy-α-D-erythro-pentofuranosyl)-	**78:** P84766a, **83:** 114874c
-3-carboxamide, 1-(2-deoxy-β-D-erythro-pentofuranosyl)-	**78:** P84766a, **83:** 114874c
-3-carboxamide, 1-(3-deoxy-β-D-erythro-pentofuranosyl)-	**83:** 114784c
-3-carboxamide, 1-[5,6-dideoxy-6-(diphenoxyphosphinyl)-2,3-*O*-(isopropylidene)-β-D-ribo-hex-5-enofuranoxyl]-	**81:** 114436z
-3-carboxamide, 1-[5,6-dideoxy-6-(diphenoxyphosphinyl)-2,3,-*O*-(isopropylidene)-β-D-ribo-hexofuranoxyl]-	**81:** 114436z
-3-carboxamide, 1-[5,6-dideoxy-6-ethoxycarbonyl-2,3-*O*-(isopropylidene)-β-D-ribo-hex-5-enofuranosyl]-	**81:** 114436z
-3-carboxamide, 1-[5,6-dideoxy-6-ethoxycarbonyl-2,3-*O*-(isopropylidene)-β-D-ribo-hexfuranosyl]-	**81:** 114436z
-3-carboxamide, 1-[5,6-dideoxy-2,3-*O*-(isopropylidene)-6-(methoxycarbonyl)-β-D-ribo-hexfuranosyl]-	**81:** 114436z
-3-carboxamide, 1-[5,6-dideoxy-2,3-*O*-(1-methylethylidene)-6-phosphono-β-D-ribo-hexofuranosyl]-	**81:** 114436z
-3-carboxamide, 1-(5,6-dideoxy-6-phosphono-β-D-ribo-hexofuranosyl)-	**81:** 114436z
-3-carboxamide, *N,N*-diethyl-5-methyl-	**70:** P87820j, **73:** 66517n
-3-carboxamide, *N,N*-diethyl-4-methyl-5-phenyl-	**70:** P87820j
-3-carboxamide, *N,N*-diethyl-5-(phenoxymethyl)-	**70:** P87820j, **73:** 66517n
-3-carboxamide, *N,N*-diethyl-5-(3-pyridyl)-	**70:** P87820j, **73:** 66517n
-3-carboxamide, *N,N*-diethyl-5-(4-pyridyl)-	**70:** P87820j
-3-carboxamide, 5-[(1,3-dihydro-1,3-dioxo-2H-isoindol-2-yl)methyl]-1-[2-(2-pyridinylcarbonyl)phenyl]-	**84:** P31151p
-1-carboxamide, *N,N*-dimethyl-	**67:** PC54140x, **84:** 174980u
-1-carboxamide, 3,5-dimethyl-	**21:** 3200[5]
-3-carboxamide, *N*,5-dimethyl-	**83:** 178256q
-1-carboxamide, *N,N*-dimethyl-3,5-di-4-pyridinyl-	**77:** P5487p, **84:** P184927w, **85:** P177430w
-1-carboxamide, *N,N*-dimethyl-3-(4-pyridyl)-	**67:** PC54140x
-3-carboxamide, 1,5-diphenyl-	**51:** 12079c, **57:** 806g
-3-carboxamide, 1-[6-(1,3-diphenyl-2-imidazolidinyl)-tetrahydro-2,2-dimethylfuro[3,4-*d*]-1,3-dioxol-4-yl]-	**81:** 114436z
-3-carboxamide, 1,5-diphenyl-*N*-[2-phenyl-1-[(phenylamino)carbonyl]ethenyl]-	**83:** 10007k
-3-carboxamide, 1-(1-ethoxyethyl)-	**84:** P4966b
-3-carboxamide, *N*-ethyl-1,5-diphenyl-	**57:** 806g
-3-carboxamide, *N*-formyl-5-(4-methoxyphenyl)-1-(4-methylphenyl)-	**85:** 177322n
-3-carboxamide, *N*-formyl-5-(4-methoxyphenyl)-1-phenyl-	**85:** 177322n
-4-carboxamide, *N*-heptadecyl-	**65:** P2275a
-3-carboxamide, *N*-(2-hydroxybenzoyl)-5-(4-methoxyphenyl)-1-(4-methylphenyl)-	**85:** 177322n

TABLE 4 (*Continued*)

Compound	Reference
-3-carboxamide, *N*-(2-hydroxybenzoyl)-5-(4-methoxyphenyl)-1-phenyl-	**85:** 177322n
-3-carboxamide, *N*-hydroxy-1-β-D-ribofuranosyl-	**78:** 12030h
-3-carboxamide, 1-[2,3-*O*-(isopropylidene)-β-D-ribofuranosyl]-	**81:** 114436z
-3-carboxamide, 1-[2,3-*O*-(isopropylidene)-β-D-ribo-pentodialdo-1,4-furanosyl]-	**81:** 114436z
-3-carboxamide, 1-α-L-lyxopyranosyl-	**83:** 114784c
-3-carboxamide, 1-(*p*-methoxyphenyl)-5-methyl-	**60:** 4133g
-3-carboxamide, *N*-[2-(2-methoxyphenyl)-1-[[2-methylphenyl)amino]carbonyl]ethenyl-1,5-diphenyl-	**83:** 10007k
-3-carboxamide, *N*-[2-(2-methoxyphenyl)-1-[[(3-methylphenyl)amino]carbonyl]ethenyl]-1,5-diphenyl-	**83:** 10007k
-3-carboxamide, *N*-[2-(2-methoxyphenyl)-1-[[4-methylphenyl)amino]carbonyl]-ethenyl]-1,5-diphenyl-	**83:** 10007k
-3-carboxamide, 5-(4-methoxyphenyl)-1-(4-methylphenyl)-*N*-[(methylphenyl)sulfonyl]-	**85:** 177322n
-3-carboxamide, 5-(4-methoxyphenyl)-1-(4-methylphenyl)-*N*-[2-[(phenylmethylene)amino]phenyl]-	**85:** 177322n
-3-carboxamide, 5-(4-methoxyphenyl)-1-(4-methylphenyl)-*N*-2-pyridinyl-	**85:** 177322n
-3-carboxamide, 5-(4-methoxyphenyl)-*N*-[(4-methylphenyl)sulfonyl]-1-phenyl-	**85:** 177322n
-3-carboxamide, 5-(4-methoxyphenyl)-*N*-(2-nitrophenyl)-1-phenyl-	**85:** 177322n
-3-carboxamide, 5-(4-methoxyphenyl)-*N*-[4-(4-oxo-2-phenyl-3-thiazolidinyl)phenyl]-1-phenyl-	**85:** 177322n
-3-carboxamide, 1-(*p*-methoxyphenyl)-5-phenyl-	**57:** 806g
-3-carboxamide, *N*-[2-(2-methoxyphenyl)-1-[(phenylamino)carbonyl]ethenyl]-1,5-diphenyl-	**83:** 10007k
-3-carboxamide, 5-(4-methoxyphenyl)-1-phenyl-*N*-[2-[(phenylmethylene)amino]-phenyl]-	**85:** 177322n
-3-carboxamide, 5-(4-methoxyphenyl)-1-phenyl-*N*-2-pyridinyl-	**85:** 177322n
-1-carboxamide, *N*-methyl	**76:** 85792b
-3-carboxamide, 5-methyl-	**74:** 99951y, **83:** 178256q
-3-carboxamide, 1-[2,3-*O*-(1-methylethylidene)-β-D-ribofuranosyl]-	**84:** P44603n
-3-carboxamide, 1-[2,3-*O*-(1-methylethylidene)-4-thio-β-D-ribofuranosyl]-	**85:** 191t
-1-carboxamide, *N*-methyl-3-(5-nitro-2-furyl)-	**64:** 19596c
-3-carboxamide, 5-methyl-1-(*p*-nitrophenyl)-	**60:** 4133h
-3-carboxamide, 5-methyl-1-phenyl-	**60:** 4133g, **67:** 100068e
-3-carboxamide, *N*-[1-[[(4-methylphenyl)amino]carbonyl]-2-phenylethenyl]-1,5-diphenyl-	**83:** 10007k
-3-carboxamide, 5-methyl-1-β-D-ribofuranosyl-	**80:** 121258a
-5-carboxamide, 3-methyl-1-β-D-ribofuranosyl-	**80:** 121258a

TABLE 4   (*Continued*)

Compound	Reference
-4-carboxamide, *N*-octadecyl-	**65:** PC2275a
-3-carboxamide, *N*-[4-[(phenylmethylene)amino]-phenyl]-5-(4-methoxyphenyl)-1-phenyl-	**85:** 177322n
-3-carboxamide, 5-phenyl-1-*p*-tolyl-	**57:** 806d
-3-carboxamide, 1-[2,3,*O*-phosphinico-β-D-ribofuranosyl)-	**84:** P44603n
-3-carboxamide, 1-(3,5-*O*-phosphinico-β-D-ribofuranosyl)-	**78:** P84756a
-3-carboxamide, 1-(2-*O*-phosphono-β-D-ribofuranosyl)-	**84:** P44603n
-3-carboxamide, 1-(3-*O*-phosphono-β-D-ribofuranosyl)-	**84:** P44603n
-3-carboxamide, 1-(5-*O*-phosphono-β-D-ribofuranosyl)-	**78:** P84766a
-3-carboxamide, 1-β-D-psicofuranosyl-	**84:** 180518r
-3-carboxamide, 1-α-D-ribofuranosyl-	**83:** 114784c
-3-carboxamide, 1-β-D-ribofuranosyl-	**78:** P84766a, **80:** 10254b, **82:** P4534t, **82:** P73409t, **82:** P73413q, **84:** P44603n
-5-carboxamide, 1-β-D-ribofuranosyl-	**78:** 12030h, **80:** 121258a
-3-carboxamide, 1-β-D-ribofuranosyl-, 5'-(dihydrogen phosphate)	**79:** 38876j,
-3-carboxamide, 1-(tetrahydro-2-furanyl)-	**84:** P4966b
-3-carboxamide, 1-(tetrahydro-2H-pyran-2-yl)-	**84:** P4966b
-1-carboxamide, *N*,*N*,3,5-tetramethyl-	**67:** P54140x
-3-carboxamide, 1-(4-thio-β-D-ribofuranosyl)-	**85:** 191t
-3-carboxamide, 1-(2,3,5-tri-*O*-acetyl-β-D-ribofuranosyl)-	**78:** P84766a
-1-carboxamide, *N*,*N*,3-trimethyl-	**67:** PC54140x
-3-carboxamide, *N*,1,5-triphenyl-	**82:** 16754c
-5-carboxamide, 1-[2,3,5-tris-*O*-benzyl-β-D-arabinofuranosyl]-	**83:** 114784c
-3-carboxamide, 1-[2,3,5-tris-*O*-benzyl-α-D-ribofuranosyl]-	**83:** 114784c
-3-carboxamide, *N*,1,5-tris(4-chlorophenyl)-	**82:** 16754c
-3-carboxamide, 1-[2,3,5-tris-*O*-(phenylmethyl)-β-D-arabinofuranosyl)]-	**78:** P84766a, **83:** 114784c
-3-carboxamide, 1-β-D-xylofuranosyl-	**78:** P84766a, **83:** 114784c
-1-carboxamidine	**68:** 105162m
-1-carboxanilide	**43:** 8365g, **52:** 16341f
-3-carboxanilide, 1,5-diphenyl-	**57:** 806e
-3-carboxanilide, 1-phenyl-	**79:** 105150h
-1-carboximidamide, *N*-butyl-	**79:** 105150h
-1-carboximidamide, *N*'-(3-chloro-4-methoxyphenyl)-*N*,*N*-dimethyl-	**82:** P72999y
-1-carboximidamide, *N*'-(3-chloro-4-methylphenyl)-*N*,*N*-dimethyl-	**82:** P72999y
-1-carboximidamide, *N*'-(4-chloro-2-methylphenyl)-*N*,*N*-dimethyl-	**82:** P72999y
-1-carboximidamide, *N*-(4-chlorophenyl)-	**79:** 105150h
-1-carboximidamide, *N*'-(4-chlorophenyl)-*N*,*N*-dimethyl-	**82:** P72999y
-1-carboximidamide, *N*'-[4-chloro-2-(trifluoromethyl)phenyl]-*N*,*N*-dimethyl-	**82:** P72999y

TABLE 4   (*Continued*)

Compound	Reference
-1-carboximidamide, *N'*-(2,4-dichlorophenyl)-*N*,*N*-dimethyl-	**82:** P72999y
-1-carboximidamide, *N'*-(2,5-dichlorophenyl)-*N*,*N*-dimethyl-	**82:** P72999y
-1-carboximidamide, *N'*-(2,6-dichlorophenyl)-*N*,*N*-dimethyl-	**82:** P72999y
-1-carboximidamide, *N'*-(3,4-dichlorophenyl)-*N*,*N*-dimethyl-	**82:** P72999y
-1-carboximidamide, *N'*-(3,5-dichlorophenyl)-*N*,*N*-dimethyl-	**82:** P72999y
-4-carboximidamide, *N'*-(2,5-dichlorophenyl)-*N*,*N*-dimethyl-	**82:** P72999y
-1-carboximidamide, *N*,*N*-diethyl-*N'*-phenyl-	**82:** P72999y
-1-carboximidamide, *N*,*N*-dimethyl-*N'*-phenyl-	**82:** P72999y
-1-carboximidamide, *N*,*N*-dimethyl-*N'*-(2,4,6-trichlorophenyl)-	**82:** P72999y
-1-carboximidamide, *N*,*N*-dimethyl-*N'*-[3-(trifluoromethyl)phenyl]-	**82:** P72999y
	82: P72999y
-1-carboximidamide, *N*-ethyl-	**79:** 105150h
-3-carboximidamide, *N*-hydroxy-1-β-D-ribofuranosyl-	**78:** P84766a, **80:** 10254b
-1-carboximidamide, *N*-(2-methoxyphenyl)-	**79:** 105150h
-1-carboximidamide, *N*-(4-methoxyphenyl)-	**79:** 105150h
-1-carboximidamide, *N*-(4-methylphenyl)-	**79:** 105150h
-1-carboximidamide, *N*-phenyl-	**79:** 105150h
-1-carboximidamide, *N*-2-propenyl-	**79:** 105150h
-3-carboximidamide, 1-β-D-ribofuranosyl-	**78:** P84766a, **80:** 10254b
-1-carboximidic acid, ethyl ester	**79:** 105150h
-1-carboximidic acid, isopropyl ester	**79:** 105150h
-1-carboximidic acid, *o*-methoxyphenyl ester	**62:** 5222f
-1-carboximidic acid, 2-phenylhydrazide	**79:** 105150h
3-[(2-carboxyanilino)carbonyl]-1,5-diphenyl-	**83:** 10007k
3-[(2-carboxyanilino)carbonyl]-1-(2-methylphenyl)-5-phenyl-	**83:** 10007k
3-[(2-carboxyanilino)carbonyl]-1-(3-methylphenyl)-5-phenyl-	**83:** 10007k
3-[(2-carboxyanilino)carbonyl]-1-(4-methylphenyl)-5-phenyl-	**83:** 10007k
-3-carboxylic acid	**50:** 1784d, **64:** 5072d, **64:** 5074f, **69:** 27360d, **70:** 77876t, **85:** 5552w
-3-carboxylic acid, 1-(6-*O*-acetyl-1,3,4-tri-*O*-benzoyl-β-D-psicofuranosyl)-, methyl ester	**84:** 180518r
-3-carboxylic acid, 1-(*p*-acetylphenyl)-, methyl ester	**57:** 15096e
-3-carboxylic acid, 5-(azidomethyl)-1-(2-benzoyl-4-chlorophenyl)-	**78:** P4289j, **85:** P5701u
-3-carboxylic acid, 5-(azidomethyl)-1-[4-chloro-2-(2-fluorobenzoyl)phenyl]-, ethyl ester	**78:** P4289j, **85:** P21507r
-3-carboxylic acid, 1-(2-benzoyl-4-chlorophenyl)-5-chloromethyl)-	**77:** P101690v, **78:** P124642r, **78:** P124646v, **79:** P126535c, **81:** P169567y, **82:** P43479y **84:** P44206k, **85:** P5701u

TABLE 4   (*Continued*)

Compound	Reference
-3-carboxylic acid, 1-(2-benzoyl-4-chlorophenyl)-5-(chloromethyl)-, ethyl ester	**81:** P169567y
-3-carboxylic acid, 1-(2-benzoyl-4-chlorophenyl)-5-(chloromethyl)-, methyl ester	**84:** P121844w
-3-carboxylic acid, 1-(2-benzoyl-4-chlorophenyl)-5-[(dimethylamino)methyl]-, methyl ester	**84:** P121844w
-3-carboxylic acid, 1-(2-benzoyl-4-chlorophenyl)-5-(iodomethyl)-, methyl ester	**84:** P121844w
-3-carboxylic acid, 1-(2-benzoyl-4-chlorophenyl)-5-methyl-, ethyl ester	**73:** 77155u
-3-carboxylic acid, 1-(2-benzoyl-4-chlorophenyl)-5-(4-morpholinylmethyl)-, methyl ester	**84:** P121844w
-3-carboxylic acid, 1-(2-benzoyl-4-chlorophenyl)-5-(1-piperidinylmethyl)-, methyl ester	**84:** P121844w
-3-carboxylic acid, 1-(2-benzoyl-4-chlorophenyl)-5-(1-pyrrolidinylmethyl)-, methyl ester	**84:** P121844w
-3-carboxylic acid, 1-(3-benzoyl-4,5-dimethyl-2-thienyl)-5-(chloromethyl)-	**83:** P206340h, **84:** P44197h
-3-carboxylic acid, 1-(3-benzoyl-4,5-dimethyl-2-thienyl)-5-(chloromethyl)-, ethyl ester	**83:** P206340h, **84:** P135730t
-3-carboxylic acid, 1-(*p*-bromophenyl)-5-phenyl-, methyl ester	**55:** 2625h, **57:** 806g
-3-carboxylic acid, 5-(*p*-bromophenyl)-1-phenyl-, methyl ester	**57:** 806h
-3-carboxylic acid, 1-[4-bromo-2-(2-pyridinylcarbonyl)-phenyl]-5-(chloromethyl)-	**84:** P31151p, **84:** P150671t
-3-carboxylic acid, 1-[4-bromo-2-(2-pyridinylcarbonyl)-phenyl]-5-(chloromethyl)-, ethyl ester	**84:** P31151p, **84:** P150671t
-1-carboxylic acid, *tert*-butyl ester	**79:** 5560y
-3-carboxylic acid, 1(or 2)-(*o*-carboxyphenyl)-	**24:** 368[4,5,6]
-3-carboxylic acid, 1(or 2)-(*m*-carboxyphenyl)-	**24:** 368[4,5,6]
-3-carboxylic acid, 1(or 2)-(*p*-carboxyphenyl)-	**24:** 368[4,5,6]
-3-carboxylic acid, 1-(*p*-carboxyphenyl)-5-methyl-, diethyl ester	**55:** 22301b
-3-carboxylic acid, 1-[2-(2-chlorobenzoyl)-4-chloro-phenyl]-5-(chloromethyl)-	**77:** P101690v, **78:** 4289j, **78:** P124642r, **78:** P124646v, **79:** P126535c, **81:** P169567y, **84:** P44206k, **85:** P5701u
-3-carboxylic acid, 1-[3-(2-chlorobenzoyl)-4-ethyl-2-thienyl]-5-(chloromethyl)-, anhydride with ethyl hydrogen carbonate	**84:** P44197h
-3-carboxylic acid, 1-[3-(2-chlorobenzoyl)-5-ethyl-2-thienyl]-5-(chloromethyl)	**83:** P206340h, **84:** P44197h, **84:** P135730t
-3-carboxylic acid, 1-[3-(2-chlorobenzoyl)-5-ethyl-2-thienyl]-5-(chloromethyl)-, ethyl ester	**83:** P206340h, **84:** P44197h, **84:** P135730t
-3-carboxylic acid, 1-[4-chloro-2-(2-chlorobenzoyl)-phenyl]-5-(chloromethyl)-, ethyl ester	**78:** P124646v, **81:** P169567y
-3-carboxylic acid, 1-[4-chloro-2-(2-chlorobenzoyl)-phenyl]-5-[(dimethylamino)methyl]-	**84:** P121844w
-3-carboxylic acid, 1-[4-chloro-2-(2-chlorobenzoyl)-phenyl]-5-[(dimethylamino)methyl]-, methyl ester	**84:** P121844w

TABLE 4 (*Continued*)

Compound	Reference
-3-carboxylic acid, 1-(2-benzoyl-4-methylphenyl)-5-(chloromethyl)-	**81:** P169567y
-3-carboxylic acid, 1-(2-benzoyl-4-methylphenyl)-5-(chloromethyl)-, ethyl ester	**81:** P169567y
-3-carboxylic acid, 1-(2-benzoyl-4-nitrophenyl)-5-(chloromethyl)-, ethyl ester	**79:** P126535c
-3-carboxylic acid, 1-(2-benzoyl-4-nitrophenyl)-5-[(dimethylamino)methyl]-, ethyl ester	**84:** P121844w
-3-carboxylic acid, 1-(2-benzoylphenyl)-5-(chloromethyl)-	**81:** P169567y
-3-carboxylic acid, 1-(2-benzoylphenyl)-5-(chloromethyl)-, ethyl ester	**81:** P169567y
-3-carboxylic acid, 1-[2-benzoyl-4-(trifluoromethyl)-phenyl]-5-(chloromethyl)-, ethyl ester	**79:** P126535c
-5-carboxylic acid, 1-benzyl-	**77:** 139909m
-3-carboxylic acid, 5-benzyl-1-(*p*-carboxyphenyl)-, ethyl 3-methyl ester	**55:** 22301b, **57:** P8585f
-3-carboxylic acid, benzylidenehydrazide	**63:** 13243h
-3-carboxylic acid, 5-benzyl-1-phenyl-	**62:** 13148h, **67:** 100068e
-3-carboxylic acid, 5-benzyl-1-phenyl-, hydrazide	**67:** 100068e
-3-carboxylic acid, 1-[1-[[[5-[[4-[2,4-bis(1,1-dimethylpropyl)phenoxy]-1-oxobutyl]amino]-2-chloro-phenyl]amino]carbonyl]-3,3-dimethyl-2-oxobutyl]-, 2-ethylbutyl ester	**85:** P102316e
-3-carboxylic acid, 1-[1-[[[5-[[4-[2,4-bis(1,1-dimethylpropyl)phenoxy]-1-oxobutyl]amino]-2-chloro-phenyl]amino]carbonyl]-3,3-dimethyl-2-oxobutyl]-, ethyl ester	**85:** P102316e
-3-carboxylic acid, 1-[1-[[[5-[[4-[2,4-bis(1,1-dimethylpropyl)phenoxy]-1-oxobutyl]amino]-2-chloro-phenyl]amino]carbonyl]-3,3-dimethyl-2-oxobutyl]-, 2-methylpropyl ester	**85:** P102316e
-3-carboxylic acid, 1-(*p*-bromophenyl)-5-phenyl-	**55:** 2625h, **57:** 806g
-3-carboxylic acid, 1(*p*-bromophenyl)-5-phenyl-, hydrazide	**56:** 9421c
-3-carboxylic acid, 5-(*p*-bromophenyl)-1-phenyl-, hydrazide	**57:** 806h
-3-carboxylic acid, 1-[4-chloro-2-(2-chlorobenzoyl)-phenyl]-5-[(4-methyl-1-piperazinyl)methyl]-, methyl ester	**84:** P121844w
-3-carboxylic acid, 1-[4-chloro-2-(2-chlorobenzoyl)-phenyl]-5-(4-morpholinylmethyl)-	**84:** P121844w
-3-carboxylic acid, 1-[4-chloro-2-(2-chlorobenzoyl)-phenyl]-5-(4-morpholinylmethyl)-, methyl ester	**84:** P121844w
-3-carboxylic acid, 1-[4-chloro-2-(2-fluorobenzoyl)-phenyl]-5-(chloromethyl)-	**84:** P44206k, **85:** P5701u
-3-carboxylic acid, 1-[4-chloro-2-(2-fluorobenzoyl)-phenyl]-5-(chloromethyl)-, ethyl ester	**84:** P164874y, **85:** P5701u, **85:** P21507r
-3-carboxylic acid, 1-[4-chloro-2-(2-fluorobenzoyl)-phenyl]-5-[(1,3-dihydro-1,3-dioxo-2H-isoindol-2-yl)-methyl]-, ethyl ester	**78:** P4289j, **85:** P21507r

TABLE 4   *(Continued)*

Compound	Reference
-3-carboxylic acid, 1-[4-chloro-2-(2-fluorobenzoyl)-phenyl]-5-(iodomethyl)-, ethyl ester	**78:** P4289j, **85:** P21507r
-3-carboxylic acid, 1-[4-chloro-2-(4-methoxybenzoyl)-phenyl]-5-(chloromethyl)-	**81:** P169567y
-3-carboxylic acid, 1-[4-chloro-2-(4-methoxybenzoyl)-phenyl]-5-(chloromethyl)-, ethyl ester	**81:** P169567y
-3-carboxylic acid, 5-(chloromethyl)-1-[4-chloro-2-(2-fluorobenzoyl)phenyl]-	**78:** P124642r, **78:** P124646v, **78:** P4289j, **82:** P43479y
-3-carboxylic acid, 5-(chloromethyl)-1-[4-chloro-2-(2-fluorobenzoyl)phenyl]-, ethyl ester	**78:** P124646v, **78:** P4289j, **79:** P126535c, **81:** P43479y
-3-carboxylic acid, 5-(chloromethyl)-1-[2-(2-fluoro-benzoyl)-4-nitrophenyl]-, ethyl ester	**79:** P126535c
-3-carboxylic acid, 5-(chloromethyl)-1-[2-(2-pyridinyl-carbonyl)phenyl]-	**84:** P31151p, **84:** P150671t
-3-carboxylic acid, 5-(chloromethyl)-1-[2-(2-pyridinyl-carbonyl)phenyl]-, ethyl ester	**84:** P31151p, **84:** P150671t
-3-carboxylic acid, 1-(*m*-chlorophenyl)-	**60:** 13239c
-3-carboxylic acid, 1-(*p*-chlorophenyl)-	**54:** 510i
-3-carboxylic acid, 1-(*p*-chlorophenyl)-, methyl ester	**57:** 15096d, **83:** 193193k
-3-carboxylic acid, 1-(*o*-chlorophenyl)-5-methyl-, ethyl ester	**55:** 22301b
-3-carboxylic acid, 1-(*p*-chlorophenyl)-5-methyl-, ethyl ester	**58:** 2447h
-3-carboxylic acid, 1-[2-deoxy-3,5-bis-*O*-(4-methyl-benzoyl)-α-D-erythro-pentofuranosyl]-, methyl ester	**78:** P84766a, **83:** 114784c
-3-carboxylic acid, 1-(2,5-di-*O*-benzoyl-3-deoxy-β-D-erythro-pentofuranosyl)-, methyl ester	**83:** 114784c
-4-carboxylic acid, 3-(2,2-dichloro-1,1-difluoroethyl)-, phenyl ester	**81:** 49647c
-3-carboxylic acid, 1-[1-[[(2,4-dichlorophenyl)amino]-carbonyl]-3,3-dimethyl-2-oxobutyl]-, 2-ethylbutyl ester	**85:** P102316e
-3-carboxylic acid, 1-[2-dioxy-3,5-bis-*O*-(4-methyl-benzoyl)-β-D-erythro-pentofuranosyl]-, methyl ester	**78:** P84766a, **83:** 114784c
-3-carboxylic acid, 1,5-diphenyl-	**51:** 12079b, **55:** 2625h, **57:** 806f, **59:** 7530h, **71:** 70546x
-3-carboxylic acid, 4,5-diphenyl-	**71:** 81268j
-3-carboxylic acid, 1,3-diphenyl-	**76:** 34177x
-3-carboxylic acid, 1,5-diphenyl-, ethyl ester	**81:** 105370w
-3-carboxylic acid, 4,5-diphenyl-, ethyl ester	**71:** 81268j, **74:** 87898y
-5-carboxylic acid, 1,3-diphenyl-, ethyl ester	**57:** 11199e, **62:** 13147g
-3-carboxylic acid, 1,5-diphenyl-, hydrazide	**57:** 806c, **59:** 2816a
-3-carboxylic acid, 1,5-diphenyl-, methyl ester	**51:** 12079b, **55:** 2625h, **57:** 806f
-3-carboxylic acid, 4,5-diphenyl-, methyl ester	**54:** 21064f
-1-carboxylic acid, 3,5-di-4-pyridyl-, benzylidene-hydrazide	**64:** 687f, **68:** 114391n
-1-carboxylic acid, 3,5-di-4-pyridyl-, cinnamylidene-hydrazide	**64:** 687f, **68:** 114391n
-1-carboxylic acid, 3,5-di-4-pyridyl-, hydrazide	**64:** 687f, **68:** 114391n

TABLE 4 (*Continued*)

Compound	Reference
-3-carboxylic acid, 1-(*p*-ethoxyphenyl)-	**54:** 510i
-1-carboxylic acid, ethyl ester	**52:** 16341f, **55:** 5484h, **78:** P28517h
-3-carboxylic acid, ethyl ester	**50:** 1784d, **57:** 5903i
-3-carboxylic acid, hydrazide	**63:** 13243g, **70:** 77876t
-3-carboxylic acid, (*p*-methoxybenzylidene)hydrazide	**63:** 13243h
-3-carboxylic acid, 1-(2-methoxy-4-nitrophenyl)-5-methyl-, ethyl ester	**55:** 22301b
-3-carboxylic acid, 1-(*p*-methoxyphenyl)-	**60:** 13239c
-3-carboxylic acid, 1-(*p*-methoxyphenyl)-5-methyl-	**60:** 4133f
-3-carboxylic acid, 1-(2-methoxyphenyl)-, methyl ester	**83:** 193193k
-3-carboxylic acid, 1-(4-methoxyphenyl)-, methyl ester	**83:** 193193k
-3-carboxylic acid, 1-(*p*-methoxyphenyl)-5-phenyl-	**55:** 2625h, **57:** 806g
-3-carboxylic acid, 1-(*p*-methoxyphenyl)-5-phenyl-, hydrazide	**57:** 806g
-3-carboxylic acid, 1-(*p*-methoxyphenyl)-5-phenyl-, methyl ester	**55:** 2625h
-3-carboxylic acid, 5-(*o*-methoxystyryl)-1-phenyl-	**71:** 70546x
-3-carboxylic acid, 5-(*o*-methoxystyryl)-1-phenyl-, methyl ester	**71:** 70546x
-3-carboxylic acid, 5-methyl-	**65:** 8898a, **74:** 99951y
-1-carboxylic acid, 5-methyl ester	**78:** P28517h
-3-carboxylic acid, methyl ester	**63:** 13243g, **64:** 5074f **70:** 77876t, **82:** P4534t
-3-carboxylic acid, 5-methyl-, ethyl ester	**78:** 136177f, **83:** 178256q
-3-carboxylic acid, 5-methyl-, hydrazide	**74:** 99951y
-3-carboxylic acid, 1-(5-methyl-3-isoxazolyl)-	**60:** 13239c
-3-carboxylic acid, 5-methyl-, methyl ester	**74:** 99951y, **78:** 136177f
-3-carboxylic acid, 5-methyl-1-(*p*-nitrophenyl)-	**62:** 13148h, **67:** 100068e
-3-carboxylic acid, 5-methyl-1-(*p*-nitrophenyl)-, ethyl ester	**55:** 22301b, **57:** 15096h, **60:** 4133c, **67:** 100068e
-3-carboxylic acid, 5-methyl-1-(4-nitro-*o*-tolyl)-, ethyl ester	**55:** 22301b,
-3-carboxylic acid, 1-methyl-5-phenyl-	**74:** 3559c
-3-carboxylic acid, 5-methyl-1-phenyl-	**53:** 17020e, **60:** 4133e, **62:** 13148h, **67:** 100068e
-3-carboxylic acid, 1-methyl-5-phenyl-, ethyl ester	**74:** 3559c
-3-carboxylic acid, 1-(4-methylphenyl)-, methyl ester	**83:** 193193k
-3-carboxylic acid, 5-methyl-1-phenyl-, methyl ester	**53:** 17020e
-3-carboxylic acid, 1-(2-naphthyl)-5-phenyl-	**55:** 2625h, **57:** 806h
-3-carboxylic acid, 1-(2-naphthyl)-5-phenyl-, methyl ester	**55:** 2625h
-3-carboxylic acid, 1-(*p*-nitrophenyl)-5-phenyl-, hydrazide	**56:** 9421c, **57:** 806h
-3-carboxylic acid, 1-(*p*-nitrophenyl)-5-phenyl-, methyl ester	**55:** 22301b
-3-carboxylic acid, 1-phenyl-	**54:** 510i, **60:** 13239c **71:** 70546x
-3-carboxylic acid, 5-phenyl-	**65:** 8898a, **74:** 3559c, **75:** 140792h

117

TABLE 4   (*Continued*)

Compound	Reference
-3-carboxylic acid, 1-phenyl-, ethyl ester	**62:** 13148c
-1-carboxylic acid, 3-phenyl-, ethyl ester	**71:** 38867s
-3-carboxylic acid, 5-phenyl-, ethyl ester	**74:** 3559c
-3-carboxylic acid, 1-phenyl ester, 4-oxide	**62:** 13148c
-3-carboxylic acid, 1-phenyl-, methyl ester	**60:** 13239c, **83:** 193193k
-3-carboxylic acid, 1-phenyl-5-(4-phenyl-1,3-butadienyl)-	**71:** 70546x
-3-carboxylic acid, 1-phenyl-5-(4-phenyl-1,3-butadienyl)-, methyl ester	**71:** 70546x
-3-carboxylic acid, 1-phenyl-5-styryl-	**71:** 70546x
-3-carboxylic acid, 1-phenyl-5-styryl-, methyl ester	**71:** 70546x
-3-carboxylic acid, 5-phenyl-1-*o*-tolyl-	**55:** 2625H, **57:** 806g
-3-carboxylic acid, 5-phenyl-1-*m*-tolyl-	**55:** 2625h, **57:** 806g
-3-carboxylic acid, 5-phenyl-1-*p*-tolyl-	**55:** 2625h, **57:** 806c
-3-carboxylic acid, 5-phenyl-1-*o*-tolyl-, hydrazide	**57:** 806g
-3-carboxylic acid, 5-phenyl-1-*p*-tolyl-, hydrazide	**57:** 806c
-3-carboxylic acid, 5-phenyl-1-*o*-tolyl-, methyl ester	**55:** 2625h, **57:** 806g
-3-carboxylic acid, 5-phenyl-1-*m*-tolyl-, methyl ester	**55:** 2625h, **57:** 806g
-3-carboxylic acid, 5-phenyl-1-*p*-tolyl-, methyl ester	**57:** 806b
-3-carboxylic acid, 5-(2-pyridinyl)-	**84:** 59317w
-3-carboxylic acid, 5-(3-pyridinyl)-	**84:** 59317w
-3-carboxylic acid, 5-(4-pyridinyl)-	**84:** 59317w
-3-carboxylic acid, 1-(2-pyridyl)-	**60:** 13239c
-1-carboxylic acid, 3-(3-pyridyl)-5-(4-pyridyl)-, hydrazide	**68:** 114391n
-3-carboxylic acid, 1-β-D-ribofuranosyl-	**78:** 12030h
-3-carboxylic acid, 1-β-D-ribofuranosyl-, hydrazide	**78:** 12030h
-5-carboxylic acid, 1-β-D-ribofuranosyl-, hydrazide	**78:** 12030h
-3-carboxylic acid, 1-β-D-ribofuranosyl-, methyl ester	**78:** 12030h, **78:** P84766a
-5-carboxylic acid, 1-β-D-ribofuranosyl-, methyl ester	**78:** 12030h
-3-carboxylic acid, 1-(tetrahydro-2H-pyran-2-yl)-, methyl ester	**84:** P4966b
-3-carboxylic acid, 1-(2,3,4-tri-*O*-acetyl-α-L-arabinopyranosyl)-, methyl ester	**83:** 114784c
-5-carboxylic acid, 1-(2,3,4-tri-*O*-acetyl-α-L-arabinopyranosyl)-, methyl ester	**83:** 114784c
-3-carboxylic acid, 1-(2,3,5-tri-*O*-acetyl-β-D-ribofuranosyl)-	**82:** P73413q
-3-carboxylic acid, 1-(2,3,5-tri-*O*-acetyl-β-D-ribofuranosyl)-, methyl ester	**78:** 12030h, **78:** P84766a, **80:** 121258a, **82:** P58077q
-5-carboxylic acid, 1-(2,3,5-tri-*O*-acetyl-β-D-ribofuranoxyl)-, methyl ester	**78:** 12030h
-3-carboxylic acid, 1-(2,3,5-tri-*O*-acetyl-β-D-xylofuranosyl)-, methyl ester	**83:** 114784c
-3-carboxylic acid, 1-(1,3,4-tri-*O*-benzoyl-6-chloro-6-deoxy-β-D-psicofuranosyl)-, methyl ester	**84:** 180518r
-3-carboxylic acid, 1-(1,3,4-tri-*O*-benzoyl-6-deoxy-β-D-*erythro*-hex-5-en-2-ulofuranosyl)-	**84:** 180518r
-3-carboxylic acid, 1-(1,3,4-tri-*O*-benzoyl-6-deoxy-6-iodo-β-D-psicofuranosyl)-, methyl ester	**84:** 180518r
-5-carboxylic acid, 1-(1,3,4-tri-*O*-benzoyl-6-deoxy-6-iodo-β-D-psicofuranosyl)-, methyl ester	**84:** 180518r

TABLE 4 (*Continued*)

Compound	Reference
-3-carboxylic acid, 1-(2,3,5-tri-*O*-benzoyl-β-D-ribofuranosyl)-	**82:** P73413q
-3-carboxylic acid, 1-(2,3,5-tri-*O*-benzoyl-β-D-ribofuranosyl)-, methyl ester	**78:** 12030h, **78:** P84766a, **82:** P73409t
-5-carboxylic acid, 1-(2,3,5-tri-*O*-benzoyl-β-D-ribofuranosyl)-, methyl ester	**78:** 12030h, **78:** P84766a
-3-carboxylic acid, 1-(trimethylsilyl)-, methyl ester	**78:** P84766a, **82:** P73409t, **83:** 114784c, **84:** 180518r
-3-carboxylic acid, 1-[2,3,5-tris-*O*-benzyl-β-D-arabinofuranosyl]-, methyl ester	**83:** 114784c
-5-carboxylic acid, 1-[2,3,5-tris-*O*-benzyl-β-D-arabinofuranosyl]-, methyl ester	**83:** 114784c
-3-carboxylic acid, veratrylidenehydrazide	**63:** 13243h
3-[[(carboxymethyl)amino]carbonyl]-1,5-diphenyl-	**83:** 10007k
3-[[(carboxymethyl)amino]carbonyl]-1-(2-methylphenyl-5-phenyl-	**83:** 10007k
3-[[(carboxymethyl)amino]carbonyl]-1-(3-methylphenyl)-5-phenyl-	**83:** 10007k
3-[[(carboxymethyl)amino]carbonyl]-1-(4-methylphenyl)-5-phenyl-	**83:** 10007k
5-(*p*-chlorobenzyl)-4-methyl-3-[(4-morpholino)carbonyl]-	**70:** P87820j, **73:** 66517n
5-(*p*-chlorobenzyl)-4-methyl-3-[(1-piperidino)carbonyl]-	**70:** P87820j, **73:** 66517n
1-[4-chloro-2-(2-chlorobenzoyl)phenyl]-3-cyano-5-[(dimethylamino)methyl]-	**84:** 121844w
1-[4-chloro-2-(4-methoxybenzoyl)phenyl]-5-(chloro-methyl)-3-(morpholin-4-ylcarbonyl)-	**81:** P169567y
3-[(*p*-chlorophenoxy)methyl]-5-[(4-morpholino)carbonyl]-	**70:** P87820j, **73:** 66517n
3-[(*p*-chlorophenoxy)methyl]-5-[(1-piperidinyl)carbonyl]-	**70:** P87820j, **73:** 66517n
3-[(*p*-chlorophenoxy)methyl]-5-[(1-pyrrolidino)carbonyl]-	**70:** P87820j, **73:** 66517n
3-(*p*-chlorophenyl)-5-[(1-piperidinyl)carbonyl]-	**70:** P87820j, **73:** 66517n
3-(*p*-chlorophenyl)-5-[(1-pyrrolidino)carbonyl]-	**70:** P87820j, **73:** 66517n
3-cyano-	**85:** 191t
3-cyano-1-(tetrahydro-2H-pyran-2-yl	**84:** P4966b
3-cyano-1-(2,3,5-tri-*O*-acetyl-4-thio-β-D-ribofuranosyl)-	**85:** 191t
4,5-dibenzyl-3-[(1-piperidino)carbonyl]-	**70:** P87820j, **73:** 66517n
-3,5-dicarboxylic acid	**54:** P1552g
-3,5-dicarboxylic acid, 1-(*p*-nitrophenyl)-, diethyl ester	**67:** 116333y
3-(2-furyl)-5-[(1-piperidinyl)carbonyl]-	**70:** P87820j, **73:** 66517n
3-(2-furyl)-5-[(1-pyrrolidino)carbonyl]-	**70:** P87820j, **73:** 66517n
3-methyl-5-[(4-morpholino)carbonyl]-	**70:** P87820j, **73:** 66517n
4-(methyl-3-[(4-morpholinyl)carbonyl]-5-phenyl-	**73:** 66517n
1-(4-methylphenyl)-3-(4-morpholinylcarbonyl)-5-phenyl-	**81:** 105370w
1-(4-methylphenyl)-5-(4-nitrophenyl)-3-(1-piperidinylcarbonyl)-	**81:** 105370w
4-methyl-5-phenyl-3-[(1-piperidino)carbonyl]-	**70:** P87820j, **73:** 66517n
4-methyl-5-phenyl-3-[(1-pyrrolidino)carbonyl]-	**70:** P87820j, **73:** 66517n
3-methyl-5-[(1-piperidinyl)carbonyl]-	**70:** P87820j, **73:** 66517n
3-methyl-5-[(1-pyrrolidino)carbonyl]-	**70:** P87820j, **73:** 66517n
3-[(4-morpholino)carbonyl]-1,5-diphenyl-	**81:** 105370w
3-[(4-morpholino)carbonyl]-5-(phenoxymethyl)-	**70:** P87820j, **73:** 66517n

119

TABLE 4 (*Continued*)

Compound	Reference
3-[(4-morpholino)carbonyl]-5-phenyl-	**70:** P87820j, **73:** 66517n
3-[(4-morpholino)carbonyl]-5-(3-pyridyl)-	**70:** P87820j, **73:** 66517n
3-[(4-morpholino)carbonyl]-5-(4-pyridyl)-	**70:** P87820j
5-(4-nitrophenyl)-3-(1-piperidinylcarbonyl-1-phenyl-	**81:** 105370w
3-(phenoxymethyl)-5-[(1-piperidinyl)carbonyl]-	**70:** P87820j, **73:** 66517n
3-(phenoxymethyl)-5-[(1-pyrrolidino)carbonyl]-	**70:** P87820j, **73:** 66517n
3-phenyl-5-[(1-pyrrolidino)carbonyl]-	**70:** P87820j, **73:** 66517n
1-(phenylthioxomethyl)-	**79:** 78696g
3-[(1-piperidino)carbonyl]-5-phenyl-	**70:** P87820j, **73:** 66517n
3-[(1-piperidino)carbonyl]-5-(3-pyridyl)-	**70:** P87820j, **73:** 66517n
3-[(1-piperidino)carbonyl]-5-(4-pyridyl)-	**70:** P87820j, **73:** 66517n
3-(3-pyridyl)-5-[(1-pyrrolidino)carbonyl]-	**70:** P87820j, **73:** 66517n
3-(4-pyridyl)-5-[(1-pyrrolidino)carbonyl]-	**70:** P87820j
-3-thiocarboxamide	**63:** 13243g, **82:** P4534t
-1-thiocarboxamide, *N*-benzyl-	**76:** 85792b
-1-thiocarboxamide, *N,N*-dimethyl-	**67:** P54140x
-1-thiocarboxamide, *N*-methyl-	**76:** 85792b
-3-thiocarboxylic acid, 1,5-diphenyl-	**55:** 24729h
-3-thiocarboxylic acid, 1-(2-naphthyl)-5-phenyl-, S-*p*-tolyl ester	**55:** 24729i
-3-thiocarboxylic acid, 5-phenyl-1-*o*-tolyl-, S-*p*-tolyl ester	**55:** 24729i
-3-thiocarboxylic acid, 5-phenyl-1-*p*-tolyl-, S-*p*-tolyl ester	**55:** 24729i

CHAPTER 5

Acyl-1,2,4-Triazoles

A study of the nuclear magnetic resonance spectra of some model triazoles indicated that N-acetyltriazole is best represented as 1-acetyl-1H-1,2,4-triazole (**5.1-1**).[1] Triazoles substituted at the 4-position have a plane of symmetry, and the 4-substituent influence the chemical shifts of the 3- and 5-CH to the same degree. In the unsymmetrical 1-acetyltriazole (**5.1-1**), the carbonyl function has a greater deshielding effect on the adjacent 5-CH proton than on the 3-CH proton.

Treatment of s-triazole with acetyl chloride in dry benzene provided the low-melting 1-acetyl derivative (**5.1-1**), which was purified by sublination at 25°.[2,3] The 1-trimethylsilyl derivative (**5.1-2**) of s-triazole was reported to acylate at a rapid rate, and reaction of **5.1-2** with acetyl chloride in benzene gave a good yield of **5.1-1**.[4]

The rate of hydrolysis of 1-acetyltriazole was compared to that of other N-acetyl five-membered ring heterocyclics and was found to increase as the number of N atoms of the ring increased: 1(or 2)-acetyltetrazole > 1-acetyl-1H-1,2,4-triazole > 1-acetylimidazole ≫ 1-acetylpyrole.[2,5,6]

A number of N-benzoyl derivatives of s-triazole (e.g., **5.1-4**) were readily obtained by the reaction of the appropriate benzoyl chloride with 1,1-carbonylbis(1H-1,2,4-triazole) (**5.1-3**).[7] The 1-benzoyltriazole (**5.1-6**) was reported to be formed directly in a cyclization reaction when the N-acetylimino ether (**5.1-5**) was treated with benzhydrazide.[8] Similarly, the condensation of either **5.1-7** or **5.1-8** with refluxing acetic anhydride gave the 1-acetyltriazole (**5.1-9**) in good yield.[9] Of interest, the conversion of **5.1-10** to **5.1-13** with acetic anhydride was reported to involve N-acylation of the ring to give **5.1-11**, formation of the bicyclic system, **5.1-12**, by intramolecular cyclization, and opening of the original triazole ring.[10]

5.1-3   **5.1-4**

5.1-5   **5.1-6**

5.1-7, R = H   **5.1-9**
5.1-8, R = Ac

5.1-10   **5.1-11**   **5.1-12**   **5.1-13**

A variety of reagents have been investigated for the conversion of preformed triazole derivatives to triazole-3-carboxaldehydes.[11] One of the most successful routes involved the oxidation of N-substituted triazole-3-methanols.[11-14] For example, treatment of 1,5-diphenyl-1H-1,2,4-triazole-3-methanol (**5.1-14**) with lead tetraacetate in refluxing anhydrous benzene gave a 50% yield of the 1,5-diphenyltriazole-3-carboxaldehyde (**5.1-15**), which was purified via Girard's reagent (H$_2$NNHCOCH$_2$N$^+$Me$_3$Cl$^-$).[11] Related N-substituted triazole-3-carboxaldehydes were obtained by the same procedure in 10 to 70% yields. In other methods investigated for the preparation of this type of aldehyde, the oxime, semicarbazone, and 2,4-dinitrophenylhydrazone derivatives of the triazole-3-carboxaldehyde were isolated, but not the free aldehyde.[11] The aldehyde derivatives were prepared from the reaction mixtures obtained in the oxidation of **5.1-14** with *N*-bromosuccinimide, the reaction of the triazole-3-carboxylic acid hydrazide (**5.1-16**) with potassium ferricyanide and ammonia, and the decomposition of the benzenesulfonyl derivative (**5.1-17**) with sodium carbonate in hot

Ph⟨N–N⟩CH₂OH —Pb(OAc)₄→ Ph⟨N–N⟩CHO   Ph⟨N–N⟩CONHNHR
   |Ph |Ph |Ph

5.1-14   **5.1-15**   **5.1-16**, R = H
                                          5.1-17, R = O₂SPh

ethylene glycol in the presence of powdered glass (McFadyen–Stevens reaction). Furthermore, the conversion of the carbonyl derivatives to the free aldehydes was unsuccessful. Formation of the aldehyde was not observed when the 5-propenyltriazole (**5.1-18**) was treated with either preoxyformic acid or selenium dioxide.[13] Hydrolysis of the acetals of *N*-unsubstituted triazoles (e.g., **5.1-19** and **5.1-20**) produced syrups from which derivatives of the aldehydes can be prepared.[13] The absence of a free carbonyl group in 5-aryltriazole-3-carboxaldehydes (e.g., **5.1-22**) was attributed to hemiaminal (e.g., **5.1-21**) formation,[15] which was later confirmed by mass spectrometry.[16] The preparation of **5.1-21** also resulted from oxidation of **5.1-23** with manganese dioxide and of **5.1-24** with sodium metaperiodate.[15] It is of interest that the hemiaminals with nonaromatic

Ph⟨N–N⟩CH=CHMe   R⟨N–N⟩CH(OMe)₂
   |Ph |H

5.1-18   **5.1-19**, R = Me
                        5.1-20, R = Ph

5.1-21   ←   [Ph⟨N–N⟩CHO] **5.1-22**
                        |H

5.1-23 Ph⟨N–N⟩CH₂OH   **5.1-24** Ph⟨N–N⟩CH(OH)–CH(OH)⟨N–N⟩Ph
               |H

5-substituents are less stable and that 1,2,4-triazole-3-carboxaldehyde itself existed predominantly in the carbonyl form.[17] In related work, phenyl triazol-3-yl ketones (e.g., **5.1-26**) were prepared by neutral permanganate oxidation of the corresponding 3-benzyltriazoles (e.g., **5.1-25**).[18]

5.1-25   **5.1-26**

High yields of 3-acetyl-5-methyl-1-phenyl-1$\underline{H}$-1,2,4-triazole (**5.1-28**) and the corresponding 1-(4-methoxyphenyl)- and 1-(4-nitrophenyl) derivatives were obtained by cyclization of the appropriate acylamidrazones (e.g., **5.1-27**) in warm concentrated ammonium hydroxide.[19]

5.1-27   **5.1-28**

Similarly, the amidrazone (**5.1-29**) was converted to the keto acid (**5.1-30**). In addition, this type of compound has been prepared by thermal dehydration of 3-acylamidrazones.[20]

5.1-29   **5.1-30**

According to Fusco, reaction of 4-phenyl-4$\underline{H}$-1,2,4-triazole (**5.1-31**) with benzoyl chloride at elevated temperatures gave either the 3-benzoyl or the 3,5-dibenzoyltriazoles (**5.1-34** and **5.1-36**) depending upon the reaction temperatures.[21] In addition, reaction of 3,4-diphenyltriazole (**5.1-32**) with benzoyl chloride gave 3-benzoyl-4,5-diphenyl-4$\underline{H}$-1,2,4-triazole (**5.1-35**). These benzoyltriazoles are postulated to result from the rearrangement of the initially formed quaternarized N-benzoyltriazole (e.g., **5.1-33** and **5.1-37**).

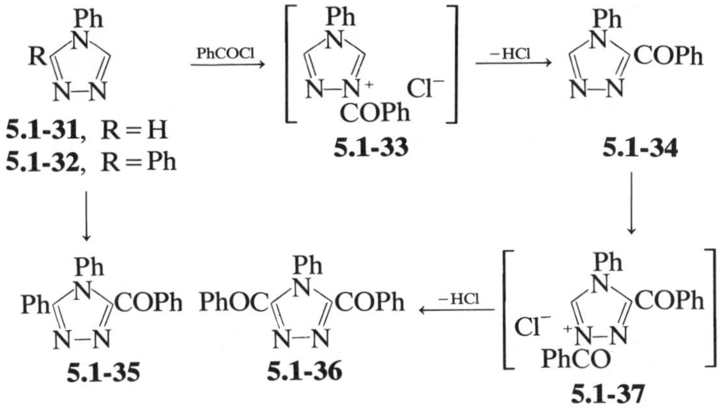

5.1-31, R = H
5.1-32, R = Ph

5.1-33

5.1-34

5.1-35   **5.1-36**   **5.1-37**

References

1. **57:** 12472f   2. **51:** 8732g   3. **73:** 45425r   4. **55:** 5484b
5. **51:** 10504i   6. **52:** 4297g   7. **57:** 5907i   8. **61:** 13232d
9. **76:** 126882b   10. **71:** 91386k   11. **57:** 805h   12. **55:** 2625b
13. **58:** 2447h   14. **83:** 193188n   15. **72:** 132636h   16. **79:** 66307t
17. **74:** 125572x   18. **68:** 105162m   19. **60:** 4133a   20. **83:** 193182f
21. **66:** 94962c

TABLE 5.   ACYL-1,2,4-TRIAZOLES

Compound	Reference
-1-acetamide, α-(2,2-dimethyl-1-oxopropyl)-N-[2-(hexa-decyloxy)-5-[(methylamino)sulfonyl]phenyl]-5-[3-(trifluoromethyl)benzoyl]-	**85:** P151754e
1-acetyl-	**51:** 8732h, **55:** 5484g, **57:** 12472f, **73:** 45425r
4-acetyl-	**51:** 8732h
1-acetyl-3-[2-[(acetyloxy)methyl]phenyl]-5-methyl-	**77:** 61916q
3-acetyl-1-(2,4-dichlorophenyl)-5-methyl-	**31:** 688[3]
1-acetyl-3,5-dimethyl-	**49:** 15916a
1-acetyl-3,5-diphenyl-	**49:** 13227b
1-acetyl-3,5-dipyrazinyl-	**77:** P5480f
1-acetyl-3,5-di-4-pyridinyl-	**77:** P5487p, **85:** P177430w
1-acetyl-5-ethyl-3-(5-nitro-2-furanyl)-	**76:** 126882b
1-[(acetylhydrazino)methylene]-3-phenyl	**71:** 91386k
1-acetyl-3-[2-(hydroxymethyl)phenyl]-5-methyl-	**77:** 61916q
4-acetyl-3-methyl-5-(2-naphthyl)-	**29:** 1817[7]
1-acetyl-5-methyl-3-(5-nitro-2-furanyl)-	**76:** 126882b
3-acetyl-5-methyl-1-(p-nitrophenyl)-	**60:** 4134a
3-acetyl-5-methyl-1-phenyl-	**60:** 4134a
1-acetyl-3-(5-nitro-2-furyl)-	**64:** 19596c
1-acetyl-3-phenyl-	**66:** 104601a, **71:** 38867s
1-acetyl-3-phenyl-5-p-tolyl-	**49:** 13227c

TABLE 5   (*Continued*)

Compound	Reference
3-aminobenzoyl-5-benzyl	**59:** 11494f
4-(2-amino-3-phenylpropionyl)-	**82:** 43697t
4-(2-aminopropionyl)-	**82:** 43697t
1-*p*-anisoyl-	**57:** 5908f
4-*m*-anisoyl-	**52:** 4297h
4-*p*-anisoyl-	**52:** 4297h
1-benzoyl-	**57:** 5908f
3-benzoyl-	**68:** 105162m
4-benzoyl-	**52:** 4297h
3-benzoyl-5-(*o*-carboxyphenyl)-	**76:** 85193g
3-benzoyl-, (2,4-dinitrophenyl)hydrazone	**68:** 105162m
1-benzoyl-5-methyl-3-phenyl-	**61:** 13232f
3-benzoyl-5-(*m*-nitrophenyl)-	**68:** 105162m
3-benzoyl-5-(*p*-nitrophenyl)-	**68:** 105162m
3-benzoyl-1-phenyl-	**68:** 105162m
3-benzoyl-5-phenyl-	**68:** 105162m
3-benzoyl-1-phenyl-, (2,4-dinitrophenyl)hydrazone	**68:** 105162m
3-benzoyl-5-phenyl-, (2,4-dinitrophenyl)hydrazone	**68:** 105162m
3-benzoyl-1-phenyl-, oxime	**68:** 105162m
3-benzoyl-1-phenyl-, thiosemicarbazone	**68:** 105162m
3-benzoyl-5-phenyl-, thiosemicarbazone	**68:** 105162m
3-benzoyl-5-(3-pyridyl)-	**68:** 105162m
3-benzoyl-5-(4-pyridyl)-	**68:** 105162m
3-benzoyl-5-(4-pyridyl)-, (2,4-dinitrophenyl)hydrazone	**68:** 105162m
3,5-bis(*p*-chlorobenzoyl)-4-phenyl-	**66:** 94962c
3,5-bis(*p*-nitrobenzoyl)-4-phenyl-	**66:** 94962c
4-[2-[(tert-butoxycarbonyl)amino]acetyl]-	**82:** 43697t
4-[4-[(tert-butoxycarbonyl)amino]-3-methylbutyryl]-	**82:** 43697t
4-[5-[(tert-butoxycarbonyl)amino]-3-methylvaleryl]-	**82:** 43697t
4-[2-[(tert-butoxycarbonyl)amino]-3-phenylpropionyl]-	**82:** 43697t
4-[2-[(tert-butoxycarbonyl)amino]propionyl]-	**82:** 43697t
1-butyryl-3,5-di-4-pyridinyl-	**77:** P5487p, **85:** P177430w
1-[(butyrylhydrazino)methylene]-3-phenyl-	**71:** 91386k
1-butyryl-3-phenyl-	**71:** 38867s
1,1'-carbonylbis-	**79:** 5560y
-1-carbothioamide, *N,N*-dimethyl-	**78:** P28517h
-3-carboxaldehyde	**57:** 805h, **74:** 125572x
-3-carboxaldehyde, 4-[2-[amino(2-chlorophenyl)methyl]-4-chlorophenyl]-5-methyl-, dimethylhydrazone	**84:** 121845x
-3-carboxaldehyde, 4-[2-[amino(2-chlorophenyl)methyl]-4-chlorophenyl]-5-methyl-, methylhydrazone	**84:** 121845x
-3-carboxaldehyde, 4-(2-benzoyl-4-chlorophenyl)-	**83:** 193188n
-3-carboxaldehyde, 1-(2-benzoyl-4-chlorophenyl)-5-[(dimethylamino)methyl]-	**84:** P121844w
-3-carboxaldehyde, 4-(2-benzoyl-4-chlorophenyl)-5-[(dimethylamino)methyl]-	**85:** P46692k
-3-carboxaldehyde, 4-(2-benzoyl-4-chlorophenyl)-5-methyl-	**77:** P126707u, **80:** P59944r, **81:** P49689t, **84:** P31146r
-3-carboxaldehyde, 4-(2-benzoyl-4-chlorophenyl)-5-methyl-, *O*-acetyloxime	**84:** P31084u

126

TABLE 5   (*Continued*)

Compound	Reference
-3-carboxaldehyde, 4-(2-benzoyl-4-chlorophenyl)-5-methyl-, 3-(*O*-acetyloxime)	**81:** P49689t, **85:** 56539e
-3-carboxaldehyde, 4-(2-benzoyl-4-chlorophenyl)-5-methyl-, oxime	**84:** P31084u, **85:** 56539e
-3-carboxaldehyde, 4-(2-benzoyl-4-chlorophenyl)-5-methyl-, 3-oxime	**77:** P126708v, **78:** P159694m, **81:** P49689t
-3-carboxaldehyde, 5-benzyl-, dimethyl acetal	**58:** 2447h
-3-carboxaldehyde, 5-benzyl-, (2,4-dinitrophenyl)-hydrazone	**58** 2447h
-3-carboxaldehyde, 5-benzyl-, thiosemicarbazone	**74:** 125572x
-3-carboxaldehyde, 5-(*p*-bromophenyl)-1-phenyl-	**57:** 807d
-3-carboxaldehyde, 1-(*p*-bromophenyl)-5-phenyl-	**55:** 2625d, **57:** 807d
-3-carboxaldehyde, 1-(*p*-bromophenyl)-5-phenyl-, (2,4-dinitrophenyl)hydrazone	**57:** 807d
-3-carboxaldehyde, 1-(*p*-bromophenyl)-5-phenyl-, oxime	**57:** 807d
-3-carboxaldehyde, 1-(*p*-bromophenyl)-5-phenyl-, semicarbazone	**57:** 807d
-3-carboxaldehyde, 5-(*p*-chlorophenyl)-	**72:** 132636h
-3-carboxaldehyde, 5-(*p*-chlorophenyl)-, dimethyl acetal	**74:** 125572x
-3-carboxaldehyde, 5-(*p*-chlorophenyl)-, (2,4-dinitrophenyl)hydrazone	**74:** 125572x
-3-carboxaldehyde, 1-(*p*-chlorophenyl)-5-methyl-	**58:** 2447h
-3-carboxaldehyde, 1-(*p*-chlorophenyl)-5-methyl-, (2,4-dinitrophenyl)hydrazone	**58:** 2447h
-3-carboxaldehyde, diethyl acetal	**74:** 125572x, **75:** 140792h
-3-carboxaldehyde, dimethyl acetal	**74:** 125572x
-3-carboxaldehyde, (2,4-dinitrophenyl)hydrazone	**74:** 125572x
-3-carboxaldehyde, 1,5-diphenyl-	**55:** 2625c, **57:** 807a
-3-carboxaldehyde, 1,5-diphenyl-, (2,4-dinitrophenyl)-hydrazone	**57:** 807a
-3-carboxaldehyde, 1,5-diphenyl-, oxime	**57:** 807c
-3-carboxaldehyde, 1,5-diphenyl-, semicarbazone	**57:** 807c
-3-carboxaldehyde, 5-ethyl-, azine	**74:** 125572x
-3-carboxaldehyde, 5-ethyl-, (2,4-dinitrophenyl)-hydrazone	**74:** 125572x
-3-carboxaldehyde, 1-(*p*-methoxyphenyl)-5-phenyl-	**55:** 2625d, **57:** 807c
-3-carboxaldehyde, 1-(*p*-methoxyphenyl)-5-phenyl-, (2,4-dinitrophenyl)hydrazone	**57:** 807c
-3-carboxaldehyde, 5-methyl-, diethyl acetal	**74:** 125572x
-3-carboxaldehyde, 5-methyl-, dimethyl acetal	**58:** 2447h
-3-carboxaldehyde, 5-methyl-, (2,4-dinitrophenyl)-hydrazone	**58:** 2447h
-3-carboxaldehyde, 5-methyl-, thiosemicarbazone	**74:** 125572x
-3-carboxaldehyde, 5-(*p*-nitrophenyl)-	**72:** 132636h
-3-carboxaldehyde, 5-(*p*-nitrophenyl)-, dimethyl acetal	**58:** 2447h
-3-carboxaldehyde, 5-(*p*-nitrophenyl)-, (2,4-dinitrophenyl)hydrazone	**58:** 2447h, **74:** 125572x
-3-carboxaldehyde, 5-phenyl-	**72:** 132636h
-3-carboxaldehyde, 5-phenyl-, diethyl acetal	**74:** 125572x
-3-carboxaldehyde, 5-phenyl-, dimethyl acetal	**58:** 2447h

127

TABLE 5 (*Continued*)

Compound	Reference
-3-carboxaldehyde, 5-phenyl-, (2,4-dinitrophenyl)-hydrazone	**58:** 2447h
-3-carboxaldehyde, 5-phenyl-, oxime	**74:** 125572x
-3-carboxaldehyde, 5-phenyl-, semicarbazone	**74:** 125572x
-3-carboxaldehyde, 5-phenyl-, thiosemicarbazone	**74:** 125572x
-3-carboxaldehyde, 5-phenyl-1-*o*-tolyl-	**55:** 2625d, **57:** 807c
-3-carboxaldehyde, 5-phenyl-1-*m*-tolyl-	**55:** 2625d, **57:** 807c
-3-carboxaldehyde, 5-phenyl-1-*o*-tolyl-, (2,4-dinitrophenyl)hydrazone	**57:** 807c
-3-carboxaldehyde, 5-phenyl-1-*p*-tolyl-	**55:** 2625d, **57:** 806i
-3-carboxaldehyde, 5-phenyl-1-*p*-tolyl-, (2,4-dinitrophenyl)hydrazone	**57:** 807a
-3-carboxaldehyde, 5-phenyl-1-*p*-tolyl-, semicarbazone	**57:** 806i
-3-carboxaldehyde, 5-(4-pyridinyl)-	**79:** 66307t
-3-carboxaldehyde, 5-(4-pyridinyl)-, *N*-oxide	**79:** 66307t
-3-carboxaldehyde, 5-(4-pyridinyl)-, thiosemicarbazone, *N*-oxide	**79:** 66307t
-3-carboxaldehyde, 5-[3(or 4)-pyridyl]-	**72:** 132636h
-3-carboxaldehyde, 5-(3-pyridyl)-, dimethyl acetal	**74:** 125572x
-3-carboxaldehyde, 5-(4-pyridyl)-, dimethyl acetal	**74:** 125572x
-3-carboxaldehyde, 5-(2-pyridyl)-, (2,4-dinitrophenyl)-hydrazone	**74:** 125572x
-3-carboxaldehyde, 5-(3-pyridyl)-, (2,4-dinitrophenyl)-hydrazone	**74:** 125572x
-3-carboxaldehyde, 5-(4-pyridyl)-, (2,4-dinitrophenyl)-hydrazone	**74:** 125572x
-3-carboxaldehyde, 5-(4-pyridyl)-, thiosemicarbazone	**75:** 140792h
-3-carboxaldehyde, 5-*p*-tolyl-	**72:** 132636h
-3-carboxaldehyde, 5-*p*-tolyl-, dimethyl acetal	**58:** 2447h
-3-carboxaldehyde, 5-*p*-tolyl-, (2,4-dinitrophenyl)hydrazone	**58:** 2447h
-1-carboxamide, *N,N*-dimethyl-	**78:** P28517h
5-(2-carboxybenzoyl)-1,3-dimethyl-	**81:** 37521u
3-(2-carboxybenzoyl)-1,5-dimethyl-	**81:** 37521u
3-(2-carboxybenzoyl)-4,5-dimethyl-	**81:** 37521u
3-(2-carboxybenzoyl)-1-methyl-5-phenyl-	**81:** 37521u
5-(2-carboxybenzoyl)-1-methyl-5-phenyl-	**81:** 37521u
3-(2-carboxybenzoyl)-5-phenyl-	**81:** 37521u
4-*m*-chlorobenzoyl-	**52:** 4297h
4-*p*-chlorobenzoyl-	**52:** 4297h
3-(4-chlorobenzoyl)-5-(4-pyridyl)-	**68:** 105162m
1-cinnamoyl-	**57:** 5908e
1-crotonoyl-	**57:** 5908f
3,5-dibenzoyl-4-phenyl-	**66:** 94962c
3,5-dibenzoyl-4-phenyl-, dioxime	**66:** 94962c
4-(*p*-dimethylaminobenzoyl)-	**52:** 4297h
3,3′-(dithiodi-*o*-phenylene)bis[1-acetyl-5-methyl-	**69:** 52047q
5-ethyl-3-(5-nitro-2-furanyl)-1-propionyl-	**76:** 126882b
1(or 4)-isobutyryl-	**51:** 10505b
-5-methanol, 1-acetyl-3-(*p*-chlorophenyl)-, acetate ester	**74:** 141651c

TABLE 5 (*Continued*)

Compound	Reference
-5-methanol, 1-acetyl-3-phenyl-, acetate ester	**74:** 141651c
5-methyl-1-(3-methylphenyl)-3-(4-nitrobenzoyl)-	**83:** 193182f
5-methyl-3-(4-nitrobenzoyl)-1-(4-nitrophenyl)-	**83:** 193182f
5-methyl-3-(4-nitrobenzoyl)-1-phenyl-	**83:** 193182f
5-methyl-3-(5-nitro-2-furanyl)-1-propionyl-	**76:** 126882b
5-methyl-3-(1-oxo-4-carboxybutyl)-1-phenyl-	**53:** 17021a, **60:** 4134a
1-*p*-nitrobenzoyl-	**57:** 5908f
4-*m*-nitrobenzoyl-	**52:** 4297h
4-*p*-nitrobenzoyl-	**52:** 4297h
3-(5-nitro-2-furanyl)-1-propionyl-	**77:** 67602f
1-[(3-nitrophenyl)acetyl]-	**84:** P121506n
3-phenyl-1-propionyl-	**71:** 38867s
1(or 4)-pivaloyl-	**51:** 10505b
1(or 4)-propionyl-	**51:** 10505b
1-[(propionylhydrazino)methylene]-3-phenyl	**71:** 91386k
1-*p*-toluoyl-	**57:** 5908f
4-*p*-toluoyl-	**52:** 4297h

CHAPTER 6

Amino-1,2,4-Triazoles

6.1a. 3(or 5)-Amino(unsubstituted)-s-Triazoles

The preparation of 3-aminotriazole (**6.1a-3**) via *N*-(formylamino)-guanidine (**6.1a-2**) was readily accomplished by heating a mixture of an aminoguanidine (**6.1a-1**) salt (H$_2$CO$_3$, HCl, HNO$_3$, H$_2$SO$_4$) and formic acid in toluene and separating the water formed in the reaction.[1,2] Pure 3-aminotriazole (**6.1a-3**) has also been claimed to result from the interaction of an equimolar mixture of cyanamide and hydrazine with formic acid,[3] and from the reaction of free *N*-aminoguanidine with *s*-triazine at 210°.[4]

In addition to formic acid, extensive work has been reported on the condensation of aliphatic carboxylic acids with *N*-aminoguanidine (**6.1a-1**) to give 3-amino-5-alkyl-[1,5] and 5-substituted alkyltriazoles.[6,7] The latter included the preparation of a series of 3-amino-5-aryloxymethyltriazoles[7] from aryloxyacetic acid and of 3-amino-5-(hydroxymethyl)triazoles (**6.1a-4**) from 70% glycolic acid.[6] Also, treatment of *N*-aminoguanidine (**6.1a-1**) with γ-butyrolactone in refluxing pyridine gave a 70% yield of 3-amino-5-(3-hydroxypropyl)triazole (**6.1a-5**).[8] A similar procedure was used for the

6.1a-6   **6.1a-7**

conversion of **6.1a-6** to the block C-nucleoside (**6.1a-7**).[9,10] The homo C-nucleoside (**6.1a-9**) was prepared by treatment of the ester (**6.1a-8**) with N-aminoguanidine (**6.1a-1**).[11] In addition, treatment of the 1,3-oxazolidinone

6.1a-8   **6.1a-9**

(**6.1a-10**) with excess hydrazine hydrate gave a fair yield of **6.1a-12**, presumably formed via the N-acyl-N′-aminoguanidine (**6.1a-11**).[12]

6.1a-10   **6.1a-11**

6.1a-12

The conversion of N-(acylamino)guanidines to 3-amino-5-alkyltriazoles has been effected by azeotropic removal of water in hot toluene[13] and by treatment with dilute base either at room temperature[14,15] or at reflux.[16,17] Excellent yields of perfluoroalkyl 3-aminotriazoles (**6.1a-14**) were obtained by the thermal dehydration of the hydrazides of perfluoro carboxylic acids (**6.1a-13**).[18] In contrast, the cyclization of this type of hydrazide in a basic medium was unsuccessful.

$$\underset{\textbf{6.1a-13}}{\overset{\displaystyle NH}{\underset{NHNHCOCF_3}{\parallel}}{H_2NC}} \longrightarrow \underset{\textbf{6.1a-14}}{H_2N\text{—triazole—}CF_3}$$

In the reaction of hydrazides with 2-methyl-2-thiopseudourea (**6.1a-15**) under mild conditions, the intermediate N-(acylamino)guanidines (**6.1a-16**) are usually not isolated but undergo cyclization to triazoles as demonstrated by the reaction of (benzamido)acethydrazide with **6.1a-15** to give 3-amino-5-(benzamidomethyl)triazole (**6.1a-17**).[19] An intermediate similar to

$$\underset{\textbf{6.1a-15}}{\overset{NH}{\underset{SMe}{\parallel}}{H_2NC}} + \underset{PhCONH}{H_2NNHCOCH_2} \overset{\Delta}{\longrightarrow} \left[\underset{\textbf{6.1a-16}}{\overset{NH}{\underset{NHNH}{\parallel}}{H_2NC}}\ COCH_2NHCOPh \right]$$

$$\downarrow 58\%$$

$$\underset{\textbf{6.1a-17}}{H_2N\text{—triazole—}CH_2NHCOPh}$$

6.1a-16 in which the side-chain amino group contained a phthaloyl blocking group was also cyclized with aqueous base to give 3-amino-5-[(phthalimido)methyl]triazole.[17] Good yields of 3-amino-5-alkyltriazoles resulted from the condensation of imino ethers with N-aminoguanidine in refluxing pyridine or tripropylamine. For example, equimolar amounts of ethyl propionimidate hydrochloride (**6.1a-18**) and N-aminoguanidine (**6.1a-1**) nitrate were refluxed for 2 hr in pyridine to give 3-amino-5-ethyltriazole (**6.1a-19**).[20,21] Other methods include the preparation of 3-amino-5-methyl-

$$\underset{\textbf{6.1a-1}}{\overset{NH}{\underset{NHNH_2}{\parallel}}{H_2NC}} + \underset{\textbf{6.1a-18}}{\overset{NH}{\underset{}{\parallel}}{EtOCEt}} \overset{86\%}{\longrightarrow} \underset{\textbf{6.1a-19}}{H_2N\text{—triazole—}Et}$$

triazole (**6.1a-21**) by C—C bond cleavage of **6.1a-20** in hot water containing sodium carbonate[22] and by addition of hydrazine with cooling to ethyl N-cyanoacetimidate (**6.1a-22**).[23,24] In the latter type of reaction with alkyl hydrazines, mixtures of 3-aminotriazoles were obtained.[24]

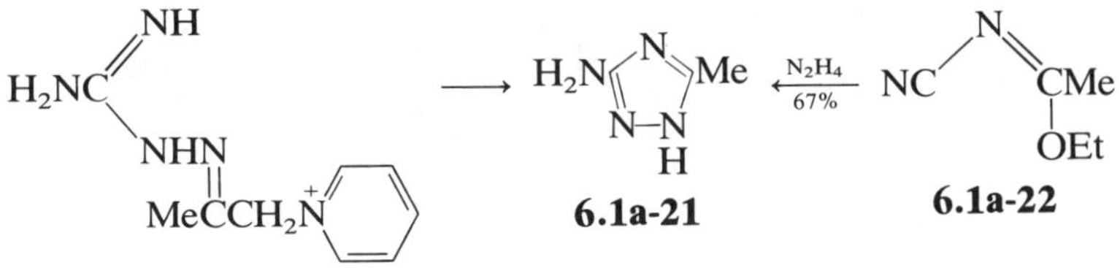

6.1a-20 **6.1a-21** **6.1a-22**

The 3-amino-5-aryltriazoles are prepared via similar types of intermediates as described in the preceding, including the cyclization of N-(aroylamino)guanidines either in hot base or fusion of the solid at temperatures greater than 100°.[16,25-27] Treatment of aroylhydrazines (**6.1a-23**) with cyanamide in refluxing acetic acid gave 3-amino-5-aryltriazoles (**6.1a-24**)

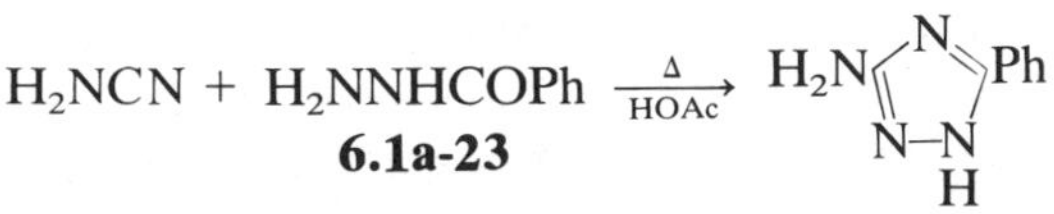

$$H_2NCN + H_2NNHCOPh$$

6.1a-23

6.1a-24

directly.[16] The compounds (**6.1a-27**) prepared by this method were also obtained via N-aroyl-N'-aminoguanidines (**6.1a-26**) by reduction of N-aroyl-N'-nitroguanidines (**6.1a-25**) with zinc dust in acetic acid.[16] The N-aroyl-N'-aminoguanidine (**6.1a-29**) is also an intermediate in the reaction of

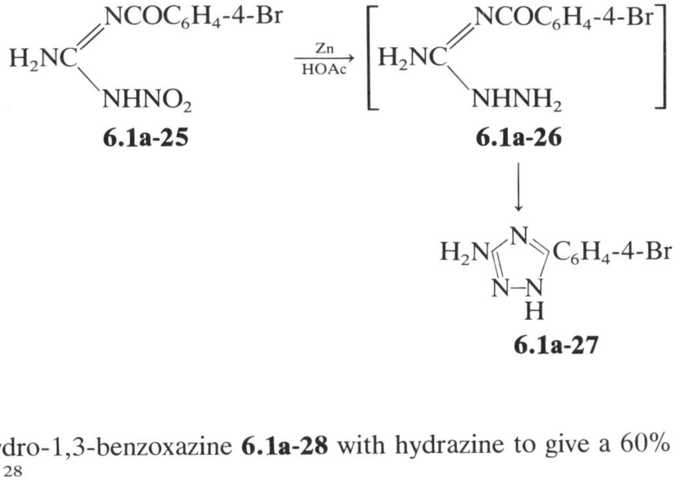

6.1a-25 **6.1a-26**

6.1a-27

the dihydro-1,3-benzoxazine **6.1a-28** with hydrazine to give a 60% yield of **6.1a-30**.[28]

6.1a-28 → **6.1a-29** → **6.1a-30**

The 5-(2-furyl)triazole (**6.1a-34**) was prepared by the condensation of the imidate (**6.1a-31**) with *N*-aminoguanidine (**6.1a-1**) to give **6.1a-32**, followed by cyclization of the latter either in hot nitrobenzene[29] or in hot propylene glycol.[30] The same product (**6.1a-34**) was also prepared by treatment of the amidrazone (**6.1a-33**) with cyanogen bromide in refluxing methanol.[31]

6.1a-1 + **6.1a-31** → **6.1a-32**

6.1a-33 $\xrightarrow{\text{BrCN}}$ **6.1a-34**

The hydrazinolysis of the isomeric triazolopyrimidines (**6.1a-35** and **6.1a-36**) gave a mixture of 3,5-dimethylpyrazole and 3-amino-1,2,4-triazole (**6.1a-3**).[32]

6.1a-35   **6.1a-36**

References

1. **51:** 13934g   2. **53:** 17151e   3. **76:** 153796h   4. **51:** 14751b
5. **60:** 2951h   6. **53:** 2264c   7. **58:** 13965f   8. **69:** 27342z
9. **79:** 66255z   10. **83:** 43657a   11. **82:** 98308y   12. **81:** 136067g
13. **66:** 46427k   14. **64:** 12685d   15. **67:** 32651n   16. **55:** 23507a
17. **70:** 77877u   18. **72:** 78954v   19. **53:** 11354d   20. **55:** 25991f
21. **59:** 6386f   22. **51:** 8085e   23. **59:** 3927g   24. **81:** 25613n
25. **54:** 9937a   26. **70:** 96755u   27. **77:** 126518h   28. **83:** 97141w
29. **63:** 611e   30. **65:** 18595e   31. **75:** 20356x   32. **85:** 94307k

6.1b. 3(or 5)-Amino(substituted)-s-Triazoles

The simple 3-substituted amino-s-triazoles (e.g., **6.1b-1** and -**3**) are easily prepared in high yields by amination of 3-bromo- or 3-chloro-s-triazole (**6.1b-2**) with alkyl- and arylamines at temperatures greater than 130°.[1-3]

6.1b-1   **6.1b-2**, X = Br, Cl   **6.1b-3**

The cyclization of *N*-alkyl-*N'*-aminoguanidines (e.g., **6.1b-4**) with hot formic acid gave 4-substituted-3-aminotriazoles (e.g., **6.1b-5**) containing a minor amount of the corresponding isomeric 3-substituted aminotriazoles (e.g., **6.1b-6**).[4-6] Similarly, reaction of benzoyl chloride with *N*-amino-*N'*-

6.1b-4   **6.1b-5** (44%)   **6.1b-6** (5%)

ethylguanidine (**6.1b-7**) in aqueous sodium hydroxide gave a poor yield of a mixture of triazoles (**6.1b-10** and **6.1b-11**).[7] Presumably **6.1b-10** was formed from **6.1b-8**, and **6.1b-11** from **6.1b-9**. Excellent yields of 3-arylamino-5-alkyltriazoles were produced by the cyclization of preformed *N*-aryl-*N'*-(acylamino)guanidines with aqueous sodium hydroxide.[8,9] This method is illustrated by the conversion of **6.1b-12** to **6.1b-13**, in which cyclization occurred between the carbonyl function and the unsubstituted

6.1b-7 → (PhCOCl / NaOH) → **6.1b-8** + PhCONHC(NHEt)=NNHCOPh **6.1b-9**

6.1b-10　　　　**6.1b-11**

amino group. A similar procedure was used in the treatment of N-(acetyl-amino)-N',N'-dimethylguanidine with base to give 3-(dimethylamino)-5-methyltriazole.[10] In addition, the cyclization of a series of N-(furoylamino)-N'-alkylguanidines (e.g., **6.1b-14**) in base gave 3-alkylamino-5-furyltriazoles (e.g., **6.1b-15**).[11] In contrast, dehydration of **6.1b-14** in refluxing N,N-di-methylformamide gave a mixture of **6.1b-15** and **6.1b-16**. However, several

6.1b-12 → (NaOH, Δ) → **6.1b-13**

3-substituted amino-5-aryltriazoles have been prepared by heating the corresponding N-aryl-N'-(aroylamino)guanidines above their melting point.[10]

6.1b-14 → (NaOH, Δ) → **6.1b-15**　　**6.1b-16**

The reaction of aliphatic 2-cyanohydrazides[12] or the isomeric 5-substituted 2-amino-1,3,4-oxadiazoles[13,14] with primary aliphatic amines lead to the *in situ* formation of N-(acylamino)guanidines, which were cyclized to give 3-alkylamino-5-alkyltriazoles in 30–70% yield.[12] For exam-ple, 2-cyanobutyrohydrazide (**6.1b-17**) and isopropylamine gave a 53% yield of **6.1b-19** formed via **6.1b-18**. In the reaction of 2-cyanoacethydra-zide (**6.1b-20**) with aniline, a mixture of a triazole (**6.1b-23**) and a triazol-inone (**6.1b-24**) were obtained. Apparently the triazolinone (**6.1b-24**) re-

6.1b-17 → **6.1b-18** → **6.1b-19**

sulted from the cyclization of **6.1b-22**, which was formed by the hydrolysis of **6.1b-21**. Also the triazolinone was the major product when the reaction was carried out in the presence of excess 20% hydrochloric acid. In the presence of a small amount of hydrochloric acid, however, a 40% yield of the triazole (**6.1b-23**) was obtained. Under these conditions the preparation of other 3-anilinotriazoles from 2-cyanohydrazides and aniline was also successful.

6.1b-20, **6.1b-21**, **6.1b-22**, **6.1b-23**, **6.1b-24**

The condensation of 5-alkyl- and 5-aryl-2-amino-1,3,4-oxadiazoles (e.g., **6.1b-25**) with primary alkyl and aryl amines in the presence of 20% hydrochloric acid provided variable yields (5 to 52%) of 3-substituted amino-5-alkyl- or aryltriazoles (e.g., **6.1b-26**).[10,15] The *in situ* formation of N-(acylamino)guanidines was also involved in a number of other preparations of 3-substituted amino-s-triazoles. Treatment of the hydrazide

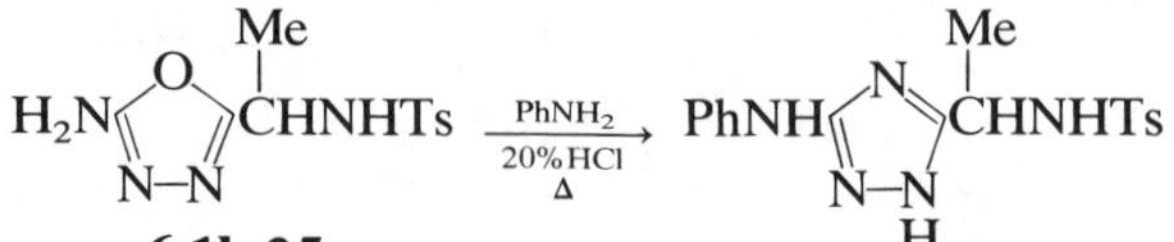

$$H_2N\overset{O}{\underset{N-N}{\diamond}}\overset{Me}{\underset{}{CHNHTs}} \xrightarrow[\Delta]{\underset{20\%\,HCl}{PhNH_2}} PhNH\overset{Me}{\underset{\underset{H}{N-N}}{\diamond}}CHNHTs$$

6.1b-25   **6.1b-26**

(**6.1b-27**) with diallyl cyanamide in the presence of pyridine hydrochloride at 175° gave a 23% yield of **6.1b-28**).[16,17] In contrast, replacement of the

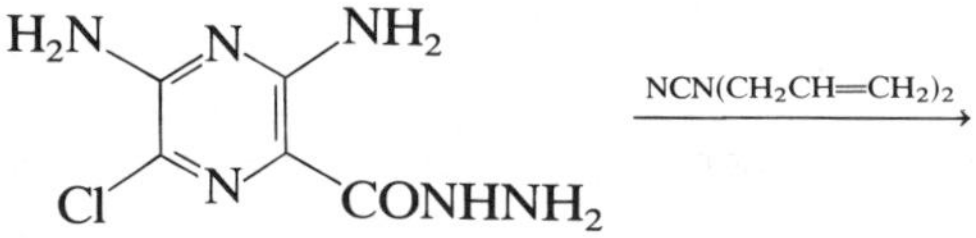

$$\underset{\underset{\textbf{6.1b-27}}{Cl\diagdown}}{H_2N}\diagdown\diagup NH_2 \quad\quad \xrightarrow{NCN(CH_2CH=CH_2)_2}$$

(with CONHNH₂ substituent)

$$H_2N\diagdown\diagup NH_2$$
$$Cl \quad N(CH_2CH=CH_2)_2$$

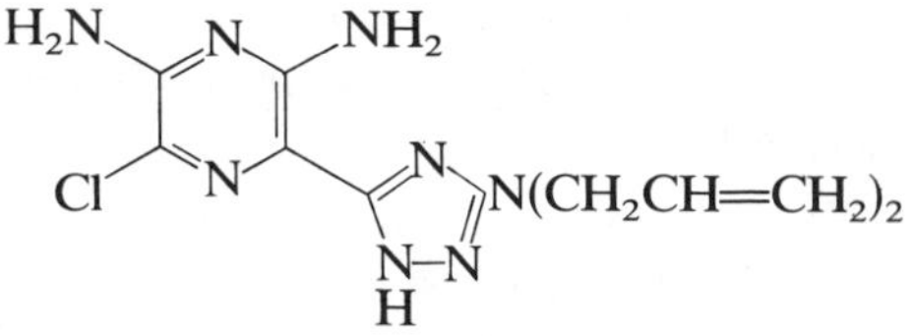

6.1b-28

cyano group of C-cyanoamidines (e.g., **6.1b-29**) with hydrazine resulted in the formation of good yields of 3-substituted amino-s-triazoles [e.g., **6.1b-30** (65%)].[18] In the cyclization of N-(aroylamino)biguanides (e.g.,

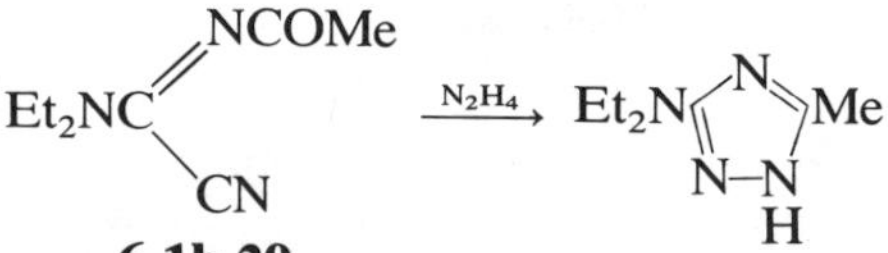

$$Et_2NC\overset{NCOMe}{\underset{CN}{\diagup}} \xrightarrow{N_2H_4} Et_2N\overset{N}{\underset{\underset{H}{N-N}}{\diamond}}Me$$

6.1b-29   **6.1b-30**

6.1b-31) with base at room temperature, good yields of 5-aryl-3-guanidi-notriazoles [e.g., **6.1b-33** (60%)] were obtained.[19,20] Another mode of cyclization was observed in the reaction of the tautomer (**6.1b-32**) of **6.1b-31** with formic acid, which gave the s-triazine (**6.1b-34**) (65%).

Presumably, the related N-acyl-N'-aminoguanidines are intermediates in the reaction of 3-aryl-1-benzoylthioureas (e.g., **6.1b-35**) with hydrazine in hot toluene to give 3-arylamino-5-phenyltriazoles (e.g., **6.1b-36**).[21] Under similar conditions the conversion of 3-alkyl-1-benzoylthioureas to triazoles was unsuccessful. In contrast, the 1,3-dibenzoyl-S-methylisothiuronium

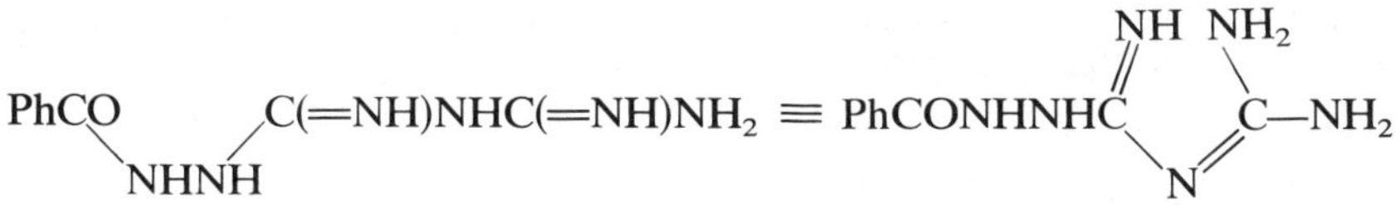

6.1b-31 ≡ **6.1b-32**

6.1b-33 **6.1b-34**

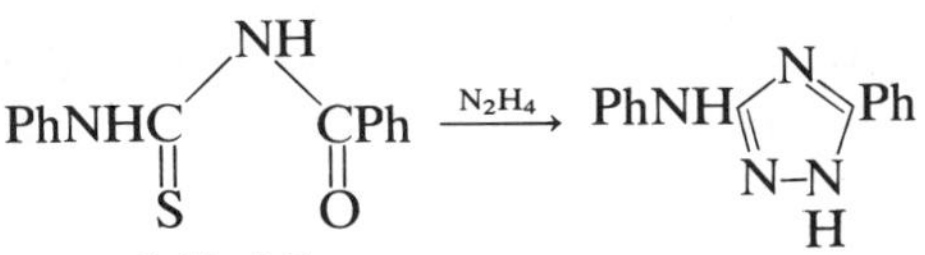

6.1b-35 **6.1b-36**

compound (**6.1b-37**) and hydrazine hydrate in refluxing ethanol provided an 85% yield of the 3-benzamido-s-triazole (**6.1b-38**).[22] Similarly, treatment of

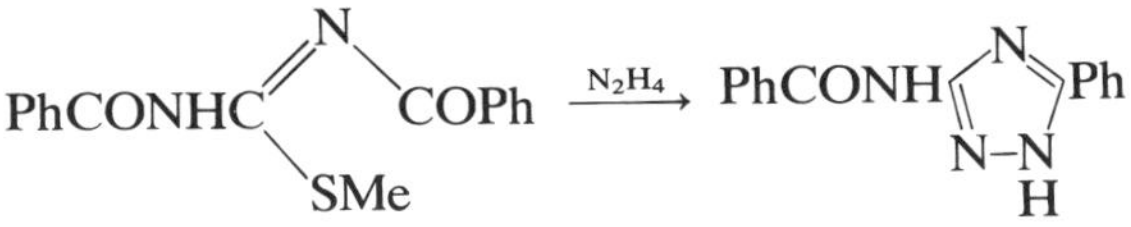

6.1b-37 **6.1b-38**

the 1,3-oxazolin-4-one (**6.1b-39**) with hydrazine gave a good yield of **6.1b-41**, presumably formed via the intermediate, N-acyl-N'-aminoguanidine (**6.1b-40**).[23,24] In an unusual method, the 3-anilino-s-triazole (**6.1b-23**)

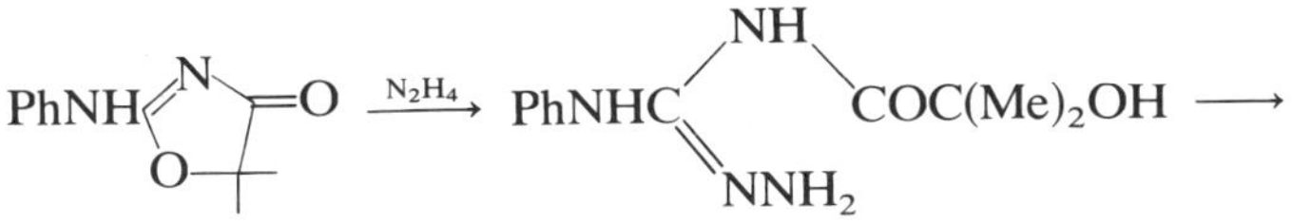

6.1b-39 **6.1b-40**

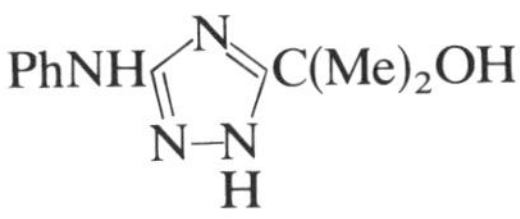

6.1b-41

was prepared in 35% yield by the thermal dehydration of the *N*-carbamoyl-amidrazone (**6.1b-42**) at 250°.[25]

6.1b-42   **6.1b-23**

The cyclization of aliphatic and aromatic *N*-(acylamino)-*N*′-(arylamino)-guanidines in aqueous potassium hydroxide containing alcohol usually gave good yields (30 to 95%) of 3-(2-arylhydrazino)triazoles, although air oxidation of the hydrazino group to give an azo group has been observed.[26,27] In addition, the 4-aminotriazole resulting from ring closure via the hydrazino moieties was a by-product in some instances. For example, the cyclization of **6.1b-43** in base gave a mixture of **6.1b-44–46** in the ratio 2.2 : 1 : 1.[26] Furthermore, 4-aminotriazoles (e.g., **6.1b-46**) are the major product in the

6.1b-43   **6.1b-44**

6.1b-45   **6.1b-46**

dehydration of this type of guanidine derivative (e.g., **6.1b-43**) in refluxing butanol.[26,28] Similarly, *N*-(benzamido)-*N*′-(benzylamino)guanidine (**6.1b-47**) was converted with base to give a 69% yield of **6.1b-48** and a small amount of the 4-aminotriazole (**6.1b-49**).[29] As described in the preceding for the cyclization of **6.1b-43**, dehydration of **6.1b-47** in refluxing butanol gave as the major product a 26% yield of **6.1b-49**.[29]

6.1b-47   **6.1b-48**   **6.1b-49**

The air oxidation of the hydrazino group mentioned above was unobserved in the cyclization of N,N'-di(benzoylamino)guanidine (**6.1b-50**) to give the 3-(2-benzoylhydrazino)triazole (**6.1b-51**).[28] Hydrolysis of the ben-

zoyl group to give the corresponding 3-hydrazinotriazole was effected with refluxing 50% hydrochloric acid. Similarly, the N-benzamido-N-methylguanidine (**6.1b-52**) was dehydrated either in the presence of base or in refluxing butanol to give good yields of the 3-(2-benzoyl-1-methylhydrazino)-s-triazole (**6.1b-53**).[30] When one of the acyl functions was an ethoxycarbonyl group as in **6.1b-54**, cyclization of the latter in boiling water gave a

mixture containing **6.1b-55** (40%) and the 4-amino-s-triazole **6.1b-56** (25%).[31-33] The ethoxycarbonyl group was involved in the cyclization of **6.1b-54** in the presence of base to give a mixture of two triazolin-5-ones.[31]

An attractive route to 3-hydrazinotriazoles involved with cyclization of N-(acylamino)-N'-ureidoguanidines (e.g., **6.1b-57**) with dilute base. This route provided good yields (59 to 85%) of both 5-alkyl- and aryl-3-(semicarbazido)triazoles (e.g., **6.1b-60**), which were hydrolyzed to the 3-hydrazinotriazoles (e.g., **6.1b-58**) in 41 to 92% yields with aqueous sulfuric acid.[34,35] The 4-ureidotriazoles (e.g., **6.1b-59**) were also isolated in low yields from some of the reaction mixtures. The cyclization of **6.1b-61** and similar guanidine derivatives in strong base gave the 3-hydrazinotriazole (**6.1b-62**) directly; however, the triazolin-5-ones (**6.1b-63** and **6.1b-64**) were produced

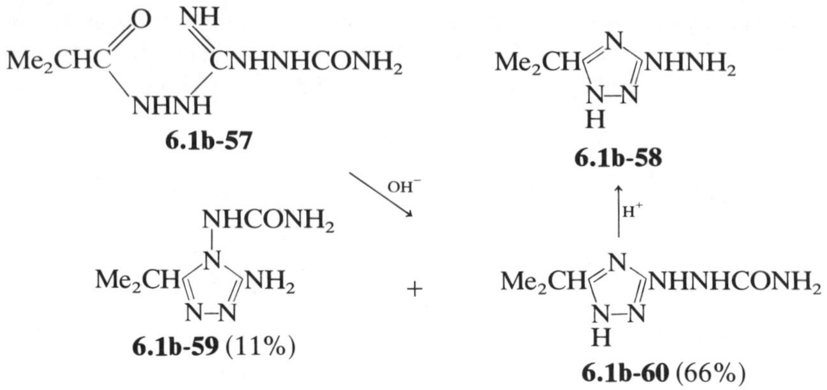

$$\underset{\textbf{6.1b-57}}{\text{Me}_2\text{CHC}\overset{\text{O}}{\underset{\text{NHNH}}{\diagdown}}\overset{\text{NH}}{\underset{}{\text{CNHNHCONH}_2}}}$$

$$\underset{\textbf{6.1b-58}}{\text{Me}_2\text{CH}\overset{\text{N}}{\underset{\text{N–N}}{\diagup}}\text{NHNH}_2}$$

OH⁻

H⁺

$$\underset{\textbf{6.1b-59}\,(11\%)}{\text{Me}_2\text{CH}\overset{\overset{\displaystyle\text{NHCONH}_2}{|}}{\underset{\text{N–N}}{\diagup}}\text{NH}_2}$$

+

$$\underset{\textbf{6.1b-60}\,(66\%)}{\text{Me}_2\text{CH}\overset{\text{N}}{\underset{\text{N–N}}{\diagup}}\text{NHNHCONH}_2}$$

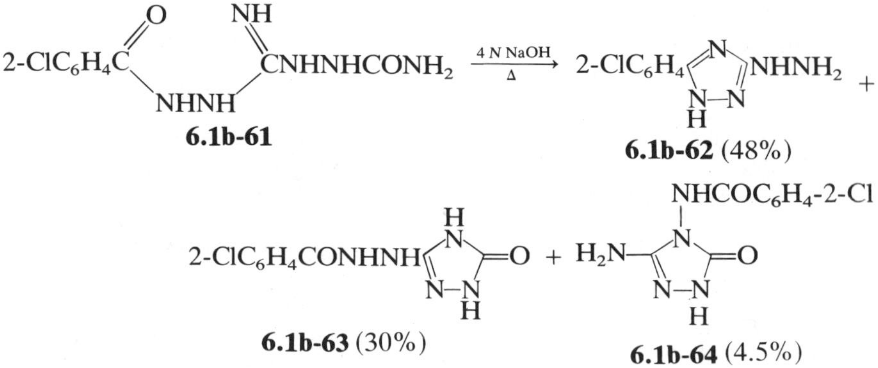

$$\underset{\textbf{6.1b-61}}{\text{2-ClC}_6\text{H}_4\text{C}\overset{\text{O}}{\underset{\text{NHNH}}{\diagdown}}\overset{\text{NH}}{\text{CNHNHCONH}_2}}$$ $\xrightarrow[\Delta]{4\,N\,\text{NaOH}}$ $\underset{\textbf{6.1b-62}\,(48\%)}{\text{2-ClC}_6\text{H}_4\overset{\text{N}}{\underset{\text{N–N}}{\diagup}}\text{NHNH}_2}$ +

$$\underset{\textbf{6.1b-63}\,(30\%)}{\text{2-ClC}_6\text{H}_4\text{CONHNH}\overset{\text{H}}{\underset{\text{N–N}}{\diagup}}\text{=O}}$$ + $\underset{\textbf{6.1b-64}\,(4.5\%)}{\text{H}_2\text{N}\overset{\overset{\displaystyle\text{NHCOC}_6\text{H}_4\text{-2-Cl}}{|}}{\underset{\text{N–N}}{\diagup}}\text{=O}}$

as by-products. In the reaction of *N*-(benzoylamino)-*N'*-(thioureido)guanidine (**6.1b-65**) with base, a mixture of **6.1b-66** and the triazoline-5-thione (**6.1b-67**) was obtained, with **6.1b-66** being the major product.[35] In contrast, cyclization of **6.1b-65** in boiling water gave the thione (**6.1b-67**) as the major product.[36]

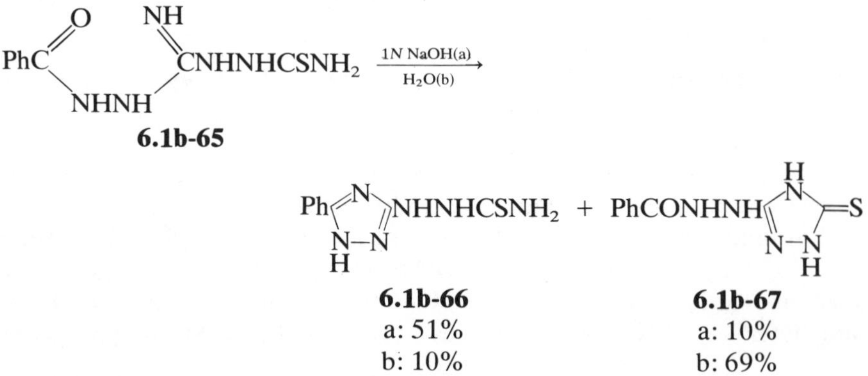

$$\underset{\textbf{6.1b-65}}{\text{PhC}\overset{\text{O}}{\underset{\text{NHNH}}{\diagdown}}\overset{\text{NH}}{\text{CNHNHCSNH}_2}}$$ $\xrightarrow[\text{H}_2\text{O(b)}]{1\,N\,\text{NaOH(a)}}$

$$\underset{\substack{\textbf{6.1b-66}\\ \text{a: 51\%}\\ \text{b: 10\%}}}{\text{Ph}\overset{\text{N}}{\underset{\text{N–N}}{\diagup}}\text{NHNHCSNH}_2}$$ + $\underset{\substack{\textbf{6.1b-67}\\ \text{a: 10\%}\\ \text{b: 69\%}}}{\text{PhCONHNH}\overset{\text{H}}{\underset{\text{N–N}}{\diagup}}\text{=S}}$

Many of the 3-substituted amino-s-triazoles are prepared by the acylation, alkylation, and condensation of preformed 3-amino- and 3-hydrazino-s-triazoles with an appropriate reagent. Although a number of reports have claimed products resulting from the acylation of the amino group of 3-amino-s-triazole (**6.1b-69**), it has been demonstrated that **6.1b-69** acetylates preferentially to give 1-acetyl-5-aminotriazole (**6.1b-68**), whereas 3-acetamido-s-triazole (**6.1b-70**) acetylates preferentially to give 1-acetyl-3-acetamido-s-triazole (**6.1b-71**).[37] On hydrolysis of **6.1b-71**, the ring acetyl group is preferentially removed.

In the methylation of **6.1b-69** and related 3-(substituted amino)triazoles, substitution occurred only at the ring nitrogens;[38,39] however, alkylation of the amino group has been reported to occur with activated aryl halides.[40–42]

The products resulting from the addition of isocyanates and isothiocyanates to **6.1b-69** depend upon the reaction conditions and the nature of the reactants. For example, treatment of **6.1b-69** with methyl isocyanate at low temperature gave **6.1b-72**, which rearranged near room temperature to give **6.1b-73**.[43] The latter was also formed directly by treatment of **6.1b-69** with methyl isocyanate in refluxing pyridine.[44] In contrast, under similar conditions **6.1b-69** and methyl isothiocyanate gave only the thioureido derivative (**6.1b-74**).[44]

Substituted aminotriazoles were also derived from 3-diazo-s-triazole (**6.1b-76**), which coupled with nitromethane to give **6.1b-75**,[45] and with the hydrazone **6.1b-77** to give the formazanyl derivative **6.1b-78**.[46]

References

1. **68:** 95761q	2. **71:** 81375s	3. **84:** 17241m	4. **59:** 3918f
5. **60:** 5484g	6. **66:** 85757n	7. **55:** 8393d	8. **55:** 24728c
9. **77:** 126518h	10. **59:** 12789f	11. **63:** 18071a	12. **53:** 9197b
13. **55:** 8387d	14. **71:** 101781b	15. **57:** 3423i	16. **70:** 96755u
17. **72:** 90516m	18. **70:** 28872e	19. **60:** 1752h	20. **60:** 1753f
21. **64:** 9727d	22. **77:** 48344g	23. **81:** 136067g	24. **83:** 97141w
25. **81:** 37516w	26. **59:** 12788b	27. **70:** 115073p	28. **63:** 13275d
29. **70:** 115080p	30. **76:** 14437d	31. **72:** 78953u	32. **73:** 87864p
33. **76:** 99568v	34. **64:** 3557c	35. **67:** 43756b	36. **66:** 104959y
37. **74:** 22770f	38. **79:** 105154n	39. **84:** 17241m	40. **68:** 105111u
41. **69:** 27451j	42. **72:** 55458x	43. **79:** 42421z	44. **76:** 85792b
45. **77:** 126523f	46. **77:** 152042t		

6.2. 3(or 5)-Amino-1H-1,2,4-Triazoles

The cyclization of N-aminoguanidine derivatives to give 1-substituted aminotriazoles has been directed toward the preparation of both 1-alkyl- and aryl-3-amino- or 5-amino-1,2,4-1H-triazoles. Treatment of N-anilino-N'-phenylguanidine (**6.2-1**) with formic acid gave 3-anilino-1-phenyl-triazole (**6.2-2**) in good yield.[1] The preparation of **6.2-2** by another route has been claimed,[2] but the product of this reaction was later identified as the iminotriazoline **6.2-3**.[1] The preparation of a 1-alkyl-3-aminotriazole is il-

lustrated by the cyclization of the aminoguanidine (**6.2-4**) with formic acid to give the 1-piperonyl derivative (**6.2-5**) in 47% yield.[3] Similarly, the

condensation of N-(phenethylamino)guanidine (**6.2-6**) with the methyl ester (**6.2-7**) in refluxing isopropanol gave a 60% yield of (**6.2-8**).[4,5]

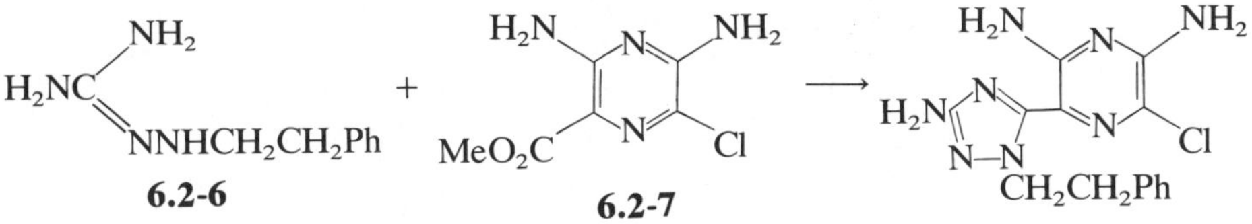

6.2-6 + **6.2-7** → **6.2-8**

In the cyclization of *N*-amino-*N*-methylguanidine (**6.2-9**) with formic acid, a good yield of 5-amino-1-methyltriazole (**6.2-11**) was obtained.[6-8] Treatment of **6.2-9** with acetic and propionic acids also gave high yields of the 3-methyl and 3-ethyl derivatives of **6.2-11**.[7] In addition, the formyl-

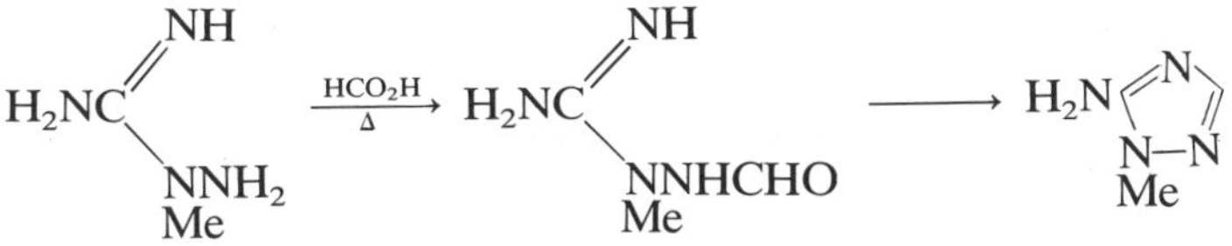

6.2-9   **6.2-10**   **6.2-11**

amino intermediate (**6.2-10**) has been isolated and heated above its melting point to give **6.2-11**.[9] Thermal dehydration of *N*-(aroylamino)-*N*-methyl-guanidines (e.g., **6.2-12**) by the melt procedure also gave good yields of 5-aminotriazoles (e.g., **6.2-13**),[9] however, the cyclization of the amidinoguani-

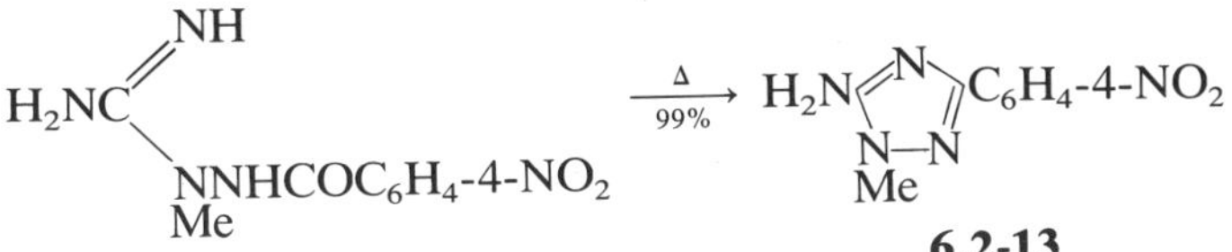

6.2-12   **6.2-13**

dine derivative (**6.2-14**) in refluxing nitrobenzene gave a poor yield of **6.2-15**.[10]

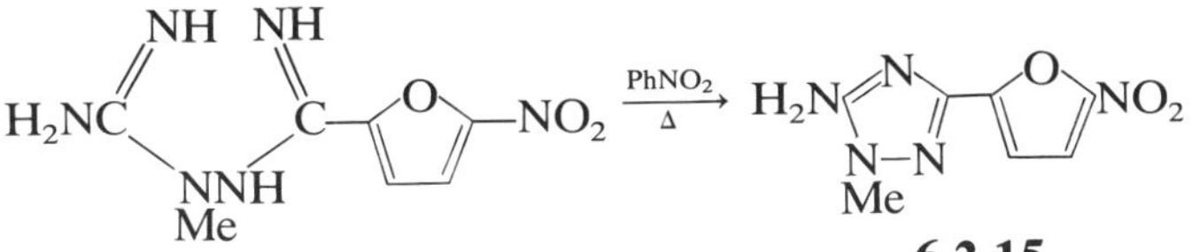

6.2-14   **6.2-15**

Another method for the preparation of 1-(substituted)-5-aminotriazoles involves the cyclization of *N*-(acylamino)guanidines in a basic medium. For example, **6.2-10** was converted to **6.2-11**[9] and **6.2-16** to **6.2-17**.[11] The *N*-(acylamino)guanidines (e.g., **6.2-19**) can also be generated *in situ* and

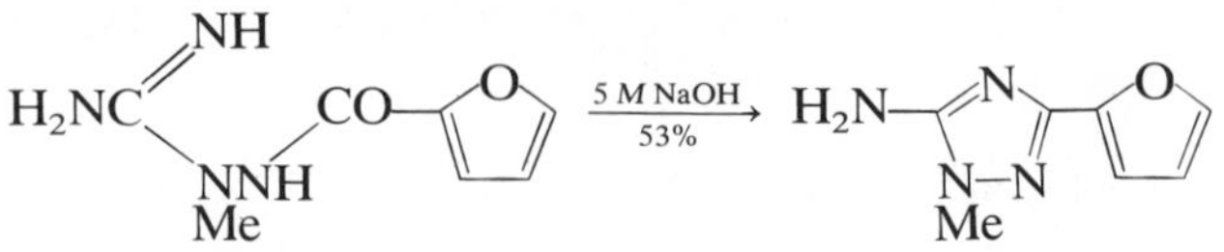

6.2-16   **6.2-17**

converted directly to triazoles, as demonstrated by the condensation of the hydrazide (**6.2-18**) with cyanamide to give **6.2-20**.[12] In addition, *N*-(2-

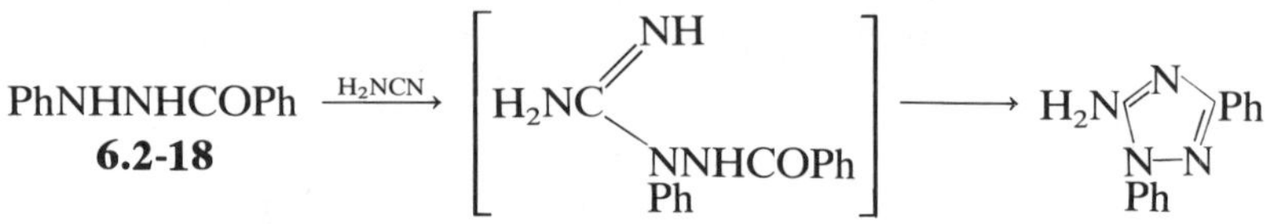

PhNHNHCOPh   →H₂NCN→
6.2-18

6.2-19   **6.2-20**

hydroxybenzoyl)-*N′*-anilinoguanidine (**6.2-21**) was cyclized with aqueous base to give a 68% yield of the 3-amino-1-phenyltriazole (**6.2-22**).[13] Also,

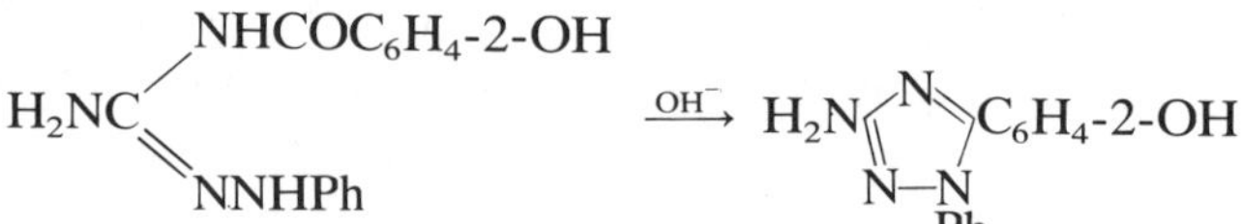

6.2-21

6.2-22

the *N*-formylguanidine (**6.2-24**) was probably an intermediate in the catalytic hydrogenation of the 3-(phenylazo)-1,2,4-oxadiazole (**6.2-23**) to give 3-amino-1-(4-chlorophenyl)triazole (**6.2-25**) in 68% yield.[14]

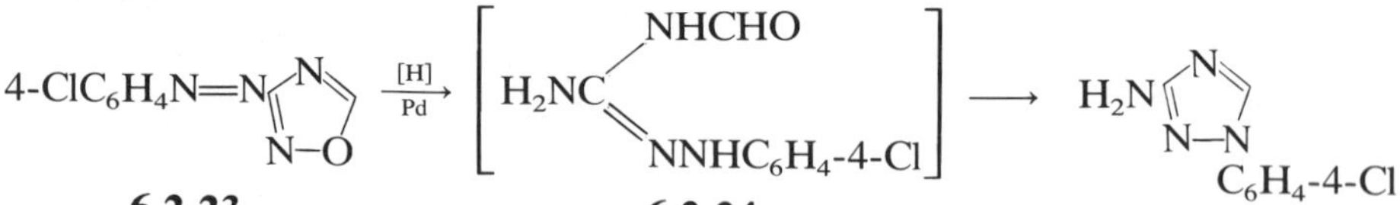

6.2-23   **6.2-24**

6.2-25

   The more complex *N,N′*-di(carbonylamino)guanidine derivatives can undergo different modes of cyclization to triazoles, the direction of cyclization depending upon the reaction conditions and the nature of the carbonyl substituents. For example, the diaminoguanidine (**6.2-26**) is thermally dehydrated in water or butanol to give a 76% yield of the hydrazinotriazole (**6.2-27**).[15] In contrast, treatment of **6.2-26** with base gave the triazolin-5-one (**6.2-28**) (69%). Contrary to these results, treatment of **6.2-29** with base

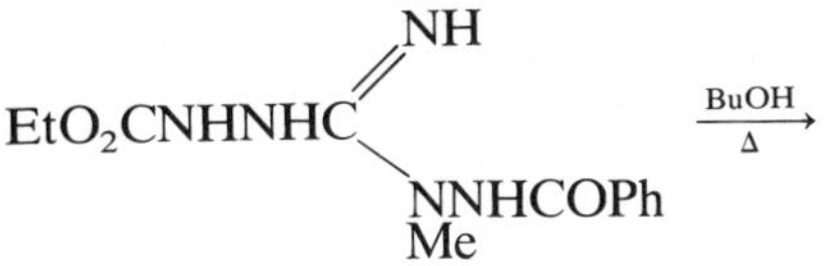

EtO$_2$CNHNHC$\overset{\text{NH}}{\underset{\text{NNHCOPh}}{\parallel}}$
Me

6.2-26

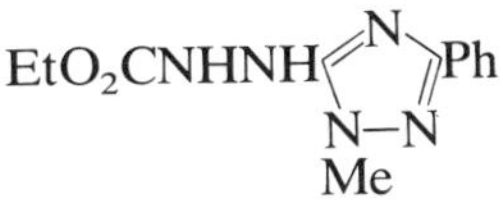

$\xrightarrow[\Delta]{\text{BuOH}}$

EtO$_2$CNHNH$\overset{\text{N}}{\underset{\text{N}-\text{N}}{}}$Ph
Me

6.2-27

PhCONH(Me)N$\overset{\text{H}}{\underset{\text{N}-\text{N}}{}}$=O
H

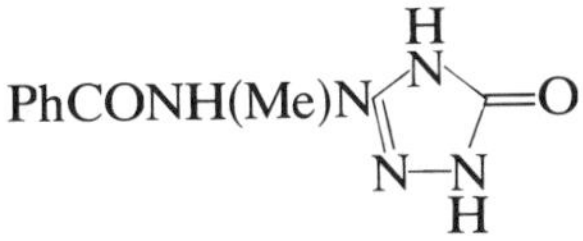

6.2-28

provided a 50% yield of the 5-(1-semicarbazido)triazole (**6.2-30**), which was cleaved by acid to give the 5-hydrazinotriazole (**6.2-31**).[16]

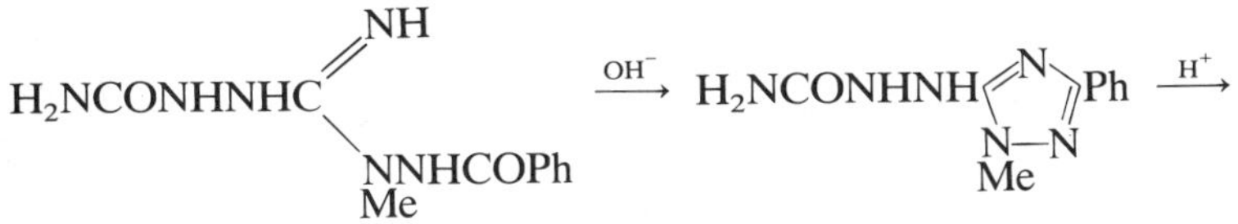

H$_2$NCONHNHC$\overset{\text{NH}}{\underset{\text{NNHCOPh}}{\parallel}}$
Me

6.2-29

$\xrightarrow{\text{OH}^-}$ H$_2$NCONHNH$\overset{\text{N}}{\underset{\text{N}-\text{N}}{}}$Ph $\xrightarrow{\text{H}^+}$
Me

6.2-30

H$_2$NNH$\overset{\text{N}}{\underset{\text{N}-\text{N}}{}}$Ph
Me

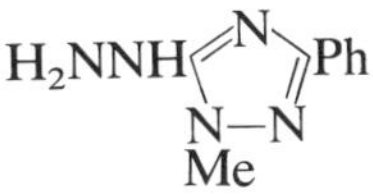

6.2-31

In the thermal cyclization of the amidrazone derivative (**6.2-32**), it was found that water rather than aniline was eliminated to give a good yield of **6.2-33**.[17] The N-cyano amidrazones are implicated in several conversions to

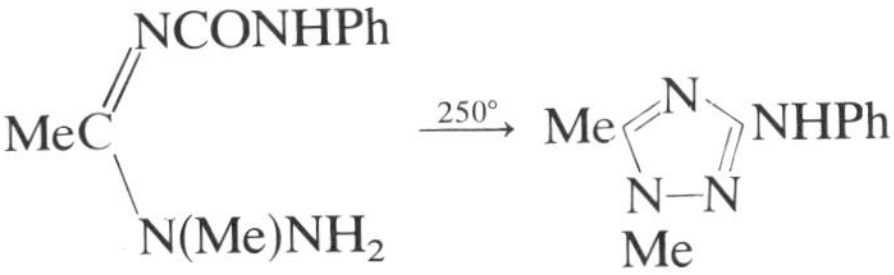

MeC$\overset{\text{NCONHPh}}{\underset{\text{N(Me)NH}_2}{\parallel}}$

6.2-32

$\xrightarrow{250°}$ Me$\overset{\text{N}}{\underset{\text{N}-\text{N}}{}}$NHPh
Me

6.2-33

give aminotriazoles in good yields under mild conditions. For example, the reaction of phenylhydrazine with ethoxymethylene cyanamide (**6.2-34**) gave **6.2-36** presumably formed via **6.2-35**.[18] However, treatment of **6.2-34** with methyl hydrazine produced a mixture of the 1-methyl-5-amino- and 1-methyl-3-aminotriazoles (**6.2-11** and **6.2-37**), respectively, in which **6.2-11** was the major product.[19] Also it was suggested that **6.2-39** was an intermediate in the treatment of the 1,3,4-oxadiazolium salt (**6.2-38**) with cyanamide in the presence of triethylamine to give **6.2-40**.[20] In this reaction the benzoyl group of the intermediate was recovered as ethyl benzoate.

Nitrilimines, generated *in situ* from benzhydrazonic acid chlorides, are known to undergo cycloaddition reactions with aryl nitriles (see Section 3.1). In addition, treatment of the nitrilimine **6.2-41** with cyanamide gave the

aminotriazole (**6.2-42**).[21] Also cyanohydrazine was implicated as an intermediate in the reaction of the nitrilimine (**6.2-43**) with thiosemicarbazide to give a 36% yield of **6.2-44**.[22]

Several amino-1,2,4-1H-triazoles have been prepared by the thermal rearrangement of other ring systems. The free radical, 1,3,5-triphenylverdazyl (**6.2-45**), was thermally disproportionated at 200° to give a mixture of **6.2-46**, **6.2-47**, and aniline.[23] Disproportionation of **6.2-45** at 80° produced only **6.2-46**, indicating that **6.2-47** and aniline are formed from **6.2-46** at

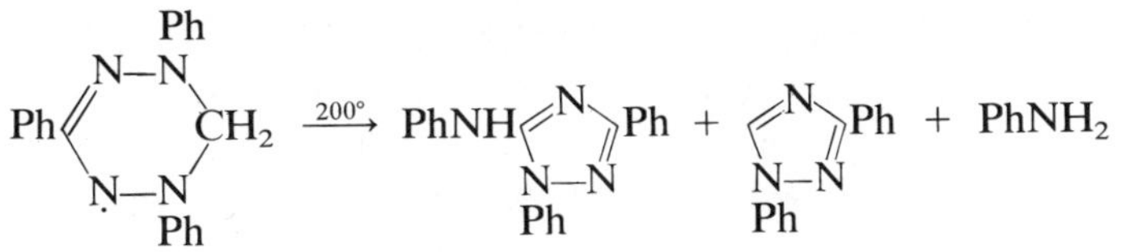

6.2-45   **6.2-46**   **6.2-47**

200°. Another rearrangement reaction was encountered in the treatment of the 1,2,4-oxadiazole (**6.2-48**) either with base or heat to give the 3-aceta-midotriazole (**6.2-49**).[24]

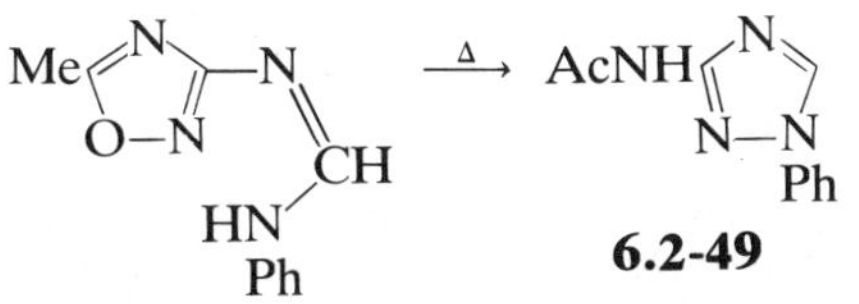

6.2-48   **6.2-49**

The conversion of preformed triazoles to 3-amino-1,2,4-1H-triazoles is illustrated by the displacement of the nitro group of **6.2-50** with methyl-amine at 75° to give **6.2-51**.[8] The same product was prepared from **6.2-52**

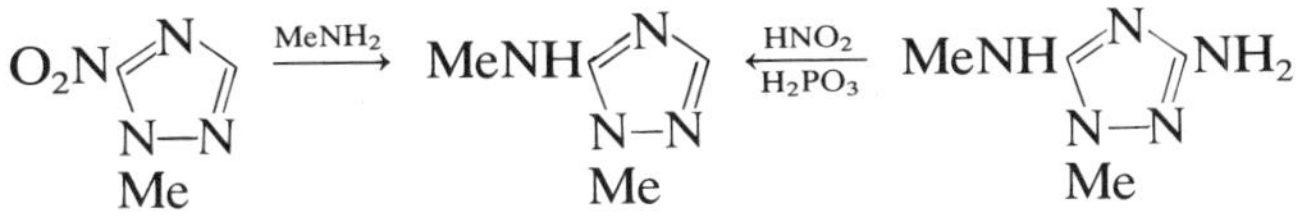

6.2-50   **6.2-51**   **6.2-52**

via reduction of the corresponding 3-diazotriazole with hypophosphorous acid. In addition, the nitro group of **6.2-53** was reduced with 85% hydrazine to give an 88% yield of 3-amino-1-β-D-ribofuranosyl-1,2,4-triazole

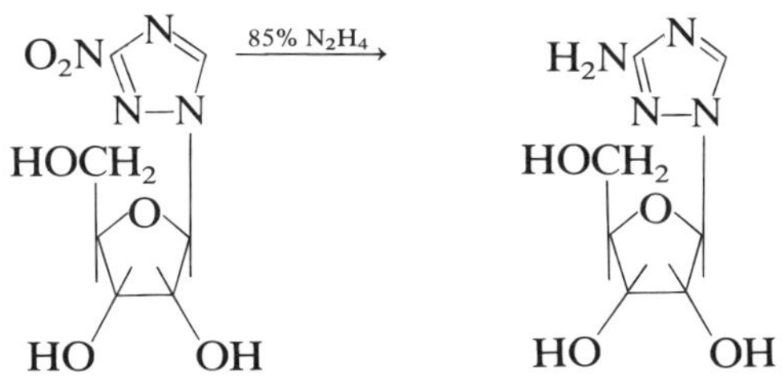

6.2-53   **6.2-54**

(**6.2-54**).[25] Also, the 5-hydrazinotriazole (**6.2-56**) was reported to result from the displacement of the bromo group of **6.2-55** with hydrazine.[26]

$$Br\overset{N}{\underset{\underset{Ph}{N-N}}{\diagup}} \xrightarrow{N_2H_4} H_2NNH\overset{N}{\underset{\underset{Ph}{N-N}}{\diagup}}$$

6.2-55   **6.2-56** (94%)

In the alkylation of 3-aminotriazoles, substitution can occur at any of the ring nitrogens, and mixtures of aminotriazoles are usually obtained. The alkylation of 3-aminotriazole (**6.2-57**) in the absence of base was reported to give a mixture of 3-amino-1-methyl-, 5-amino-1-methyl-, and 3-amino-4-methyltriazoles (**6.2-37**, **6.2-11**, and **6.2-58**) in yields of 25, 30, and 45%, respectively.[27,28] In the presence of base the yields were 40, 55, and 5%, respectively. A mixture of three products was also observed to result from the ribosylation of **6.2-57**.[29] The addition of acetylene to **6.2-57** in the presence of base at 180° was reported to give mainly 3-amino-1-vinyl-1,2,4-1H-triazole (**6.2-59**)[30,31] and a small amount of the corresponding 4-vinyl isomer. Only the 1-methyl-5-aminotriazole (**6.2-61**) was isolated in the methylation of **6.2-60** in aqueous alcoholic potassium hydroxide.[9,10] How-

$$H_2N\overset{N}{\underset{\underset{H}{N-N}}{\diagup}} \xrightarrow{MeI} H_2N\overset{N}{\underset{\underset{Me}{N-N}}{\diagup}} + H_2N\overset{N}{\underset{\underset{Me}{N-N}}{\diagup}} + H_2N\overset{\overset{Me}{N}}{\underset{N-N}{\diagup}}$$

6.2-57 $\searrow^{HC\equiv CH}$ **6.2-37**   **6.2-11**   **6.2-58**

$$H_2N\overset{N}{\underset{\underset{CH=CH_2}{N-N}}{\diagup}} \qquad H_2N\overset{N}{\underset{\underset{H}{N-N}}{\diagup}}Ph \xrightarrow[34\%]{Me_2SO_4} H_2N\overset{N}{\underset{\underset{Me}{N-N}}{\diagup}}Ph$$

6.2-59   **6.2-60**   **6.2-61**

ever, treatment of **6.2-62** with methyl iodide under basic conditions gave a mixture containing about equal amount of **6.2-63** and **6.2-17** and trace amounts of di- and trimethylated products.[32,33] In the ethylation of **6.2-64,** the resulting mixture of isomers was separated to give a 33% yield of the 1-ethyl-5-amino derivative (**6.2-65**) and a 13% yield of the 1-ethyl-3-amino derivative (**6.2-66**), indicating that alkylation on the nitrogen adjacent to the amino group was favored.[34] A metabolite of **6.2-57**, the 1-substituted 3-aminotriazole (**6.2-67**) is formed in plants[35,36] and *Escherichia coli*,[37] but not in *Saccharomyces cerevisiae* (yeast).[38]

$$H_2N\overset{N}{\underset{\underset{H}{N-N}}{\diagup}}\overset{O}{\diagdown} \longrightarrow H_2N\overset{N}{\underset{\underset{Me}{N-N}}{\diagup}}\overset{O}{\diagdown} + H_2N\overset{N}{\underset{\underset{Me}{N-N}}{\diagup}}\overset{O}{\diagdown}$$

6.2-62   **6.2-63**   **6.2-17**

6.2-64   **6.2-65**   **6.2-66**   **6.2-67**

References

1. **63:** 2975c	2. **55:** 1645g	3. **74:** 99946a	4. **70:** 96755u
5. **72:** P90516m	6. **59:** 12790f	7. **60:** 9264f	8. **72:** 111380e
9. **58:** 10202c	10. **63:** 611e	11. **63:** 18071a	12. **79:** 18648h
13. **83:** 97141w	14. **80:** 133347s	15. **73:** 14772x	16. **76:** 14437d
17. **81:** 37516w	18. **60:** 520h	19. **81:** 25613n	20. **76:** 14444d
21. **80:** 59602c	22. **84:** 59334z	23. **76:** 113185b	24. **74:** 125578d
25. **73:** 66847v	26. **70:** 37729r	27. **79:** 105154n	28. **84:** 17241m
29. **85:** 193010q	30. **79:** 66253x	31. **80:** 82823j	32. **63:** 18071g
33. **79:** 87327q	34. **55:** 8393d	35. **59:** 6729f	36. **80:** 67324d
37. **62:** 10857b	38. **64:** 4190a		

6.3. 3-Amino-4$\underline{H}$-1,2,4-Triazoles

The amidrazone derivatives similar to **6.3-1** are obtained readily, and the
S-methylation of **6.3-1** under basic conditions gave **6.3-2**, which was heated
in base to form the 3-amino-4$\underline{H}$-triazole (**6.3-3**) in good yield.[1] A series of
4-(substituted pyridinyl)-3-amino triazoles was prepared by the same
method.[2] In contrast, the cyclization of **6.3-2** under acidic conditions resulted in the elimination of aniline to give 3-(methylthio)triazole (**6.3-4**).[3] In

a one-step reaction the amidrazone (**6.3-5**) was condensed with N-(dichloro-
methylene)benzamide to give **6.3-6**.[3]

$$PhC\underset{NNH_2}{\overset{NHPh}{|}} + Cl_2C{=}NCOPh \longrightarrow Ph\underset{N-N}{\overset{N-Ph}{\diagup}}NHCOPh$$

6.3-5   **6.3-6**

The cyclization of *N*-amino-*N'*-alkyl- or arylguanidines with carboxylic acids are well-documented reactions[4–7] and are illustrated by the conversion of **6.3-7** to **6.3-8** with acetic acid[8] and of **6.3-9** to **6.3-3** with formic acid.[9] In

$$H_2NC\underset{NNH_3^+}{\overset{NHMe}{|}} \xrightarrow[\Delta]{MeCO_2H} H_2N\underset{N-N}{\overset{Me}{\diagup}}Me$$

6.3-7   **6.3-8** (80%)

some studies the isolation of a minor amount of the isomeric 3-substituted amino-*s*-triazole was reported.[5,6]

$$H_2NC\underset{NNH_3^+}{\overset{NHPh}{|}} \xrightarrow{HCO_2H} H_2N\underset{N-N}{\overset{Ph}{\diagup}}$$

6.3-9   **6.3-3** (75%)

The cyclization of *N*-(acylamino)-*N'*-alkyl- or arylguanidines in base usually provided 3-substituted aminotriazoles as the major product; however, as previously described, base treatment of the product obtained from the reaction of **6.3-10** with benzoyl chloride gave a mixture of **6.3-11** and **6.3-12**, with the latter being the major product (see Section 6.1b).[10,11] In

$$H_2NC\underset{NNH_2}{\overset{NHEt}{|}} \xrightarrow[OH]{PhCOCl} PhCONH\underset{N-N}{\overset{Et}{\diagup}}Ph + EtNH\underset{\underset{H}{N-N}}{\overset{N}{\diagup}}Ph$$

6.3-10   **6.3-11**   **6.3-12**

contrast, *N*-substituted *N'*-(acylamino)guanidines are converted on heating either in a neutral solvent or in an aliphatic carboxylic acid to give mainly 3-amino-4*H*-triazoles.[12,13] For example, the 4-(4-chlorophenyl)triazole (**6.3-14**) was obtained in good yield by heating **6.3-13** in water containing ethanol. The isomeric 3-(4-chloroanilino)triazole (**6.3-15**) was also formed, but was readily separated from the major product (**6.3-14**).[12] In addition, both dimethyl sulfoxide and *N,N*-dimethylformamide have been used as solvents in this type of reaction to give 3-amino-4*H*-triazoles.[14,15]

$$\underset{\textbf{6.3-13}}{\overset{\displaystyle \text{NHC}_6\text{H}_4\text{-4-Cl}}{\underset{\displaystyle \text{NNHCOMe}}{\text{H}_2\text{NC}}}} \xrightarrow[\Delta]{\text{H}_2\text{O}} \underset{\textbf{6.3-14}}{\text{H}_2\text{N}\!\!\overset{\text{C}_6\text{H}_4\text{-4-Cl}}{\underset{\text{N}-\text{N}}{\diagdown}}\!\!\text{Me}} \; + \; \underset{\textbf{6.3-15}}{4\text{-ClC}_6\text{H}_4\text{NH}\!\!\overset{\text{N}}{\underset{\text{N}-\text{N}}{\diagdown}}\!\!\text{Me}}$$

In related work acetylation of **6.3-16** was postulated to give **6.3-17**, which in base formed the 4H-triazole (**6.3-18**).[16] The cyclization of N-(acetyl-

$$\underset{\textbf{6.3-16 (X = O, NH)}}{\overset{\displaystyle \text{NHPh}}{\underset{\displaystyle \text{NNHC(X)NH}_2}{\text{PhNHC}}}} \xrightarrow{\text{Ac}_2\text{O}} \left[\underset{\textbf{6.3-17 (X = O, NH)}}{\overset{\displaystyle \text{NHPh}}{\underset{\overset{\displaystyle \text{NNCOMe}}{\displaystyle \text{C(X)NH}_2}}{\text{PhNHC}}}} \right] \xrightarrow{\text{OH}^-} \underset{\textbf{6.3-18}}{\text{PhNH}\!\!\overset{\text{Ph}}{\underset{\text{N}-\text{N}}{\diagdown}}\!\!\text{Me}}$$

amino)-N'-amino-N''-methylguanidine (**6.3-19**) with base gave a 3,4-diamino-4H-triazole (see Section 7.2), but treatment of the benzylidene derivative (**6.3-20**) with base gave the 3-hydrazino-4H-triazole (**6.3-21**).[17]

$$\underset{\textbf{6.3-19}}{\overset{\displaystyle \text{NHMe}}{\underset{\displaystyle \text{NNHCOMe}}{\text{H}_2\text{NNHC}}}} \xrightarrow{\text{PhCHO}} \underset{\textbf{6.3-20}}{\overset{\displaystyle \text{NHMe}}{\underset{\displaystyle \text{NNHCOMe}}{\text{PhCH}=\text{NNHC}}}} \xrightarrow{\text{OH}^-}$$

$$\underset{\textbf{6.3-21}}{\text{PhCH}=\text{NNH}\!\!\overset{\text{Me}}{\underset{\text{N}-\text{N}}{\diagdown}}\!\!\text{Me}}$$

The chloride of an N-acylaminoguanidine (e.g., **6.3-23**) was shown to be an intermediate in the reaction of 1,1,4-trichlorodiazabutadienes (e.g., **6.3-22**) with excess aniline under mild conditions to give 3-anilino-4H-triazoles (e.g., **6.3-24**) in high yields.[18]

$$\underset{\textbf{6.3-22}}{\overset{\displaystyle \text{Cl}}{\underset{\displaystyle \text{NN}=\text{CCl}_2}{4\text{-ClC}_6\text{H}_4\text{C}}}} \xrightarrow{\text{PhNH}_2} \left[\underset{\textbf{6.3-23}}{\overset{\displaystyle \text{Cl NHPh}}{\underset{\displaystyle \text{NNH}}{4\text{-ClC}_6\text{H}_4\text{C} \qquad \text{C}=\text{NPh}}}} \right] \longrightarrow$$

$$\underset{\textbf{6.3-24}}{4\text{-ClC}_6\text{H}_4\!\!\overset{\text{Ph}}{\underset{\text{N}-\text{N}}{\diagdown}}\!\!\text{NHPh}}$$

The 1-benzoylsemicarbazides (**6.3-25** and **6.3-26**) are converted with aniline in the presence of anhydrous zinc chloride at 190° to the 3-anilino-4-phenyl-4H-triazole (**6.3-27**).[19] The same product (**6.3-27**) and a triazoline-

6.3-25, R = H
6.3-26, R = Ph

6.3-27

5-thione are obtained as a mixture in the reaction of aniline with 1-benzoyl-4-phenylthiosemicarbazide (**6.3-28**).[20,21] Presumably, the N-(benzamido)-guanidine (**6.3-29**) is an intermediate in the formation of **6.3-27** from **6.3-25**, **6.3-26**, and **6.3-28**.

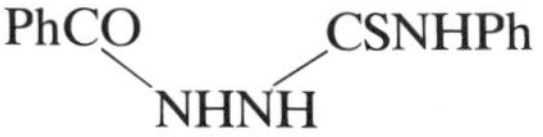

6.3-28

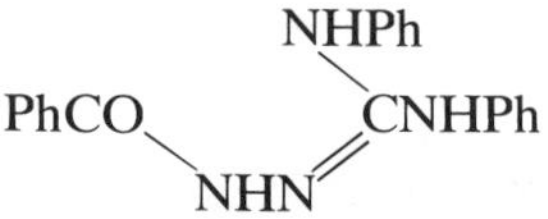

6.3-29

In addition to the alkylation of 3-aminotriazoles to give mixtures containing 4-substituted 4H-triazoles as described in Section 6.2, the results of the reaction of 3-amino-1,2,4-triazole (**6.3-30**) with the dimethyl acetal of bromoacetaldehyde are of interest.[22] Under some conditions all of the possible ring N-alkylated products were observed; but in the presence of equivalent amounts of triazole, alkylating reagent, and sodium ethoxide in refluxing ethanol, a 3:1 mixture of **6.3-31** and **6.3-32** was isolated in 50% yield. In contrast, in the absence of base only **6.3-33** was isolated in 15%

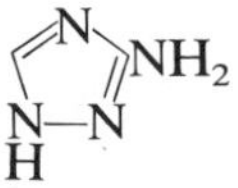

6.3-30 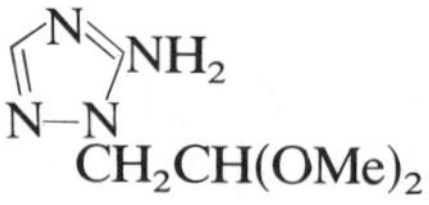   **6.3-31**

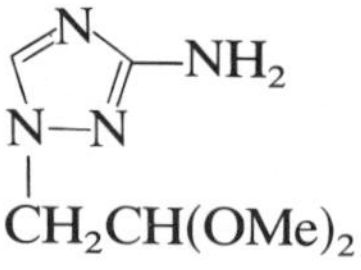

6.3-32

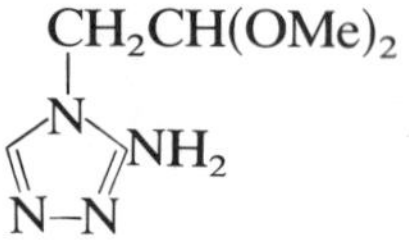

6.3-33

yield. Other reactions of preformed systems to give 4H-triazoles include the hydrazinolysis of triazolo[1,5-a]pyrimidines (e.g., **6.3-34**) to give ring

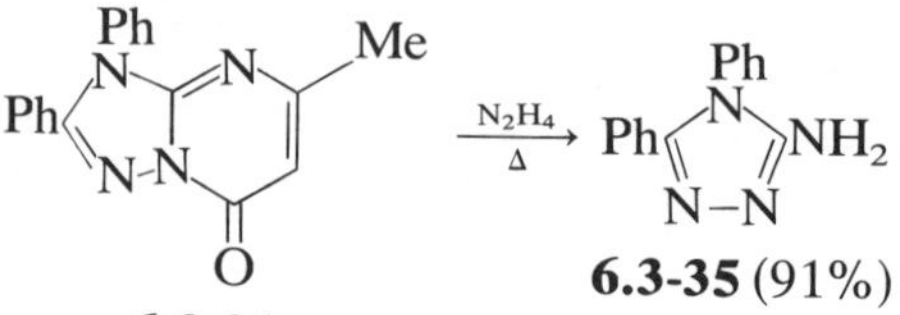

6.3-34

6.3-35 (91%)

opened products (e.g., **6.3-35**),[23,24] and the desulfuration of **6.3-36** with bromine to give **6.3-37**.[25]

6.3-36   **6.2-37**

References

1. **51:** 17894e	2. **58:** 5703f	3. **73:** 98876v	4. **55:** 10452h
5. **59:** 3918f	6. **66:** 85757n	7. **67:** 100157h	8. **60:** 9264f
9. **63:** 7012g	10. **55:** 8393d	11. **77:** 126518h	12. **55:** 23504e
13. **63:** 18071a	14. **70:** 96755u	15. **79:** 87327q	16. **57:** 12472h
17. **59:** 10050a	18. **77:** 101478g	19. **68:** 59500v	20. **51:** 5095b
21. **57:** 8561c	22. **85:** 45813p	23. **73:** 56050r	24. **81:** 4188n
25. **80:** 14879g			

6.4. 4-Amino-4H-1,2,4-Triazoles

The reaction of ethyl formate with hydrazine gave formhydrazide, which was heated at 150 to 200° to effect self-condensation to give 4-aminotriazole (**6.4-1**).[1,2] Similarly, the interaction of formic acid and hydrazine gave a good yield of **6.4-1**.[3–5] In addition, 4-aminotriazole was prepared by the reaction of hydrazine with s-triazine (**6.4-2**),[6] thionooxamic acid (**6.4-3**),[7]

OCHNHNH$_2$ $\xrightarrow{\Delta}$

6.4-1 (80%)   **6.4-2**   H$_2$NCSCO$_2$H
                                            6.4-3

and carbon monoxide;[8] numerous other methods have been described.[1] Other carboxylic acids and their functional derivatives have also been used to prepare alkyl and substituted alkyl 4-aminotriazoles. For example, acetic acid and hydrazine hydrate at a high temperature gave 3,5-dimethyl-4-aminotriazole (**6.4-4**).[3,4,9] A similar route was used to prepare 3,5-diethyl- and 3,5-dipropyl-4-aminotriazoles, but this method was unsuccessful for the preparation of 3,5-di(isopropyl)-4-aminotriazole from isobutyric acid.[3] Similar reactions have been reported for the preparation of 4-amino-3,5-bis-(aminoalkyl)triazoles,[10–12] which is illustrated by the reaction of glycine with hydrazine to give **6.1-5**.[10,13]

$$ \text{MeCO}_2\text{H} \xrightarrow[\Delta]{\text{N}_2\text{H}_4} \underset{\substack{\text{N}-\text{N} \\ \textbf{6.4-4} \ (75\%)}}{\text{Me} \overset{\overset{\displaystyle \text{NH}_2}{|}}{\diagdown}\text{Me}} \qquad \text{H}_2\text{NCH}_2\text{CO}_2\text{H} \xrightarrow{\text{N}_2\text{H}_4} \underset{\substack{\text{N}-\text{N} \\ \textbf{6.4-5}}}{\text{H}_2\text{NCH}_2 \overset{\overset{\displaystyle \text{NH}_2}{|}}{\diagdown}\text{CH}_2\text{NH}_2} $$

Fusion of glycolic acid hydrazide by gradually raising the temperature to 175° gave a good yield of 4-amino-3,5-bis(hydroxymethyl)triazole (**6.4-6**),[14] a compound that was earlier identified incorrectly by Curtius and Schwan (1895) as the eight-membered ring heterocyclic (**6.4-7**). A violent decomposition was reported to occur when this fusion reaction was attempted at

$$ \text{HOCH}_2\text{CONHNH}_2 \longrightarrow \underset{\substack{\text{N}-\text{N} \\ \textbf{6.4-6}}}{\text{HOCH}_2 \overset{\overset{\displaystyle \text{NH}_2}{|}}{\diagdown}\text{CH}_2\text{OH}} $$

6.4-7

higher temperatures.[14] In the condensation of 4-hydroxybutyric acid hydrazide with hydrazine hydrate at 180°, the major product was **6.4-8**, but intramolecular cyclization of the hydrazide also gave the *N*-amino-2-pyrrolidone (**6.4-9**) as a by-product.[15] A variety of 4-amino-3,5-bis(hydroxyalkyl)-triazoles have been prepared by treatment of hydrazine with esters,[16,17]

$$ \text{HO(CH}_2)_3\text{CONHNH}_2 \xrightarrow{\text{N}_2\text{H}_4} \underset{\substack{\text{N}-\text{N} \\ \textbf{6.4-8}}}{\text{HO(CH}_2)_3 \overset{\overset{\displaystyle \text{NH}_2}{|}}{\diagdown}\text{(CH}_2)_3\text{OH}} + \underset{\textbf{6.4-9}}{\overset{\displaystyle \text{NH}_2}{\diagdown}\text{O}} $$

lactones,[15] and nitriles.[14,16] In the condensation of aliphatic,[18–20] aryl,[21] and heteroaryl[22,23] nitriles with hydrazine, higher yields of 3,5-disubstituted 4-aminotriazoles were obtained when the reaction mixture contained sulfur or a sulfur-containing compound. For example, the 4-aminotriazole (**6.4-10**), substituted at the 3,5-positions with different substituents, was prepared by the reaction of equimolar amounts of acetonitrile and propionitrile with a mixture of hydrazine and sulfur.[18] This type of reaction apparently involves the initial formation of a dihydrotetrazine, which undergoes rearrangement to a 4-aminotriazole (see the following).

$$ \text{MeCN} + \text{EtCN} + \text{H}_2\text{NNH}_2\cdot\text{H}_2\text{O} \xrightarrow[\Delta]{\text{S}} \underset{\substack{\text{N}-\text{N} \\ \textbf{6.4-10}}}{\text{Me} \overset{\overset{\displaystyle \text{NH}_2}{|}}{\diagdown}\text{Et}} $$

The preparation of 4-amino-3,5-diaryltriazoles (e.g., **6.4-12**) are usually effected by the acidic rearrangement of 1,2-dihydrotetrazines (e.g., **6.4-11**), which are obtained by the reaction of equimolar amounts of aryl nitriles and hydrazine in refluxing benzene or toluene.[24] In one experiment prolonged

$$RC_6H_4CN + N_2H_4 \longrightarrow RC_6H_4 \underset{\substack{N-N \\ H\ H}}{\overset{N-N}{\diagdown\diagup}} C_6H_4R \xrightarrow{HCl} RC_6H_4 \underset{N-N}{\overset{\overset{\displaystyle NH_2}{\mid}}{N}} C_6H_4R$$

6.4-11   **6.4-12**

heating of benzonitrile and hydrazine without solvent gave a mixture of the 1,2-dihydrotetrazine (**6.4-13**) and the 4-aminotriazole (**6.4-14**).[25] Also the

$$PhCN + 93\%\ N_2H_4 \xrightarrow[64\ hrs]{100°} Ph \underset{\substack{N-N \\ H\ H}}{\overset{N-N}{\diagdown\diagup}} Ph\ +\ Ph \underset{N-N}{\overset{\overset{\displaystyle NH_2}{\mid}}{N}} Ph$$

6.4-13 (22%)   **6.4-14** (36%)

reaction of aryl nitriles with excess hydrazine gave good yields of 4-amino-triazoles directly; however, in some of these reactions hydrazine reacted with the substituents on the aryl ring causing the reduction of nitro groups and the demethylation of methoxy groups.[24] Dihydrotetrazines substituted with heteroaromatic groups are readily rearranged to give 4-aminotriazoles substituted with 2-pyridinyl,[26,27] 3-pyridinyl,[26] 4-pyridinyl,[26,28] 2-thiazolyl,[29] and pyranyl[30] groups.

In the rearrangement of **6.4-15** to **6.4-16**, a considerable amount of the diaroylhydrazine (**6.4-17**) was obtained as a by-product.[28] A lower yield of **6.4-16** (30%) was obtained by the reaction of the diaroylhydrazine (**6.4-17**)

$$4\text{-}C_5H_4N \underset{\substack{N-N \\ H\ H}}{\overset{N-N}{\diagdown\diagup}} 4\text{-}C_5H_4N \xrightarrow[\Delta]{2\ N\ HCl} 4\text{-}C_5H_4N \underset{N-N}{\overset{\overset{\displaystyle NH_2}{\mid}}{N}} 4\text{-}C_5H_4N$$

6.4-15   **6.4-16** (59%)

$$4\text{-}C_5H_4NC \overset{\displaystyle O}{\diagup} \quad \overset{\displaystyle O}{\diagdown} C\text{-}4\text{-}C_5H_4N$$
$$\underset{NHNH}{}$$

6.4-17 (40%)

with hydrazine at high temperatures.[31] Similarly the 3,5-di(2-furyl) compound (**6.4-18**) was obtained in low yield (18%) from the corresponding diacylhydrazine;[31] however, the 3,5-diphenyl compound (**6.4-14**) was prepared in good yield (58%) from 1,2-dibenzoylhydrazine.[3] In addition, **6.4-14** was obtained by heating the benzenesulfonate salt of benzhydrazide.[32]

6.4-18

The monoaryl 4-(benzamido)triazole (**6.4-19**) was obtained as its benzenesulfonate salt by refluxing benzhydrazide benzenesulfonate in N,N-dimethylformamide.[33,34] The reaction mixture also contained the benzhydrazidine (**6.4-20**) and 2,5-diphenyl-1,3,4-oxadiazole. Treatment of **6.4-19** with dilute sodium hydroxide opened the triazole ring to give **6.4-20**, which was reconverted to **6.4-19** on reaction with benzenesulfonic acid. In contrast, treatment of **6.4-19** with strong sodium hydroxide gave the parent 4-benzamido base (**6.4-22**). Hydrolysis of the benzamido group of **6.4-19** to give the 4-aminotriazole (**6.4-21**) was effected with hydrochloric acid.

The imino ether (**6.4-23**) reacted with acethydrazide in hot ethanol to give the 4-acetamidotriazole (**6.4-26**), presumably formed via the intermediate (**6.4-24**).[35] Under acidic conditions **6.4-25** was converted directly to the 4-aminotriazole (**6.4-27**) in 56% yield.[36] Similarly, the monoacyl derivative (**6.4-28**) was cyclized in acetic acid to give **6.4-29**.[37,38] In addition, oxidative

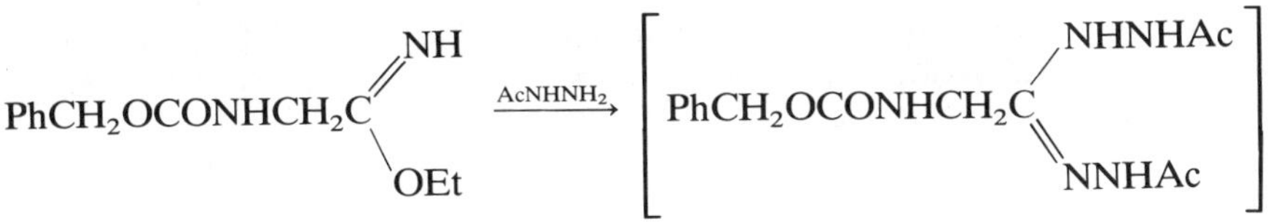

PhCH$_2$OCONHCH$_2$C(=NH)(OEt)

6.4-23

$\xrightarrow{\text{AcNHNH}_2}$

$\left[\text{PhCH}_2\text{OCONHCH}_2\text{C}(\text{NHNHAc})(=\text{NNHAc}) \right]$

6.4-24

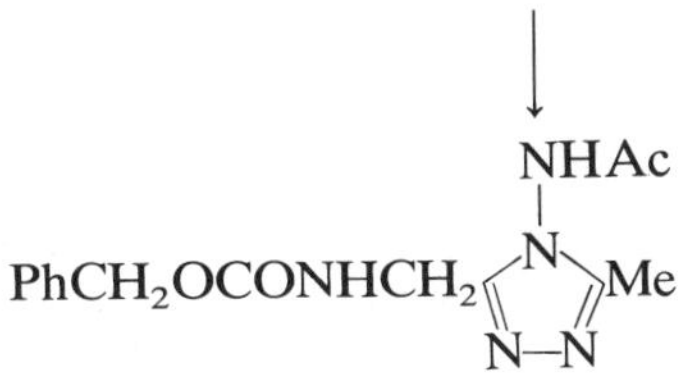

CH(NHNHCOCH$_2$Ph)(=NNHCOCH$_2$Ph)

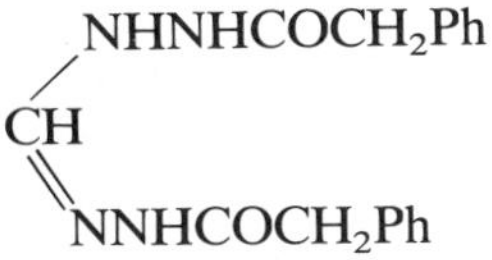

6.4-25

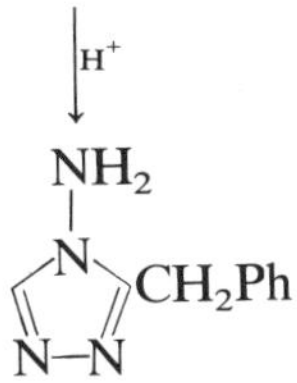

PhCH$_2$OCONHCH$_2$—triazole(NHAc)(Me)

6.4-26

triazole NH$_2$, CH$_2$Ph

6.4-27

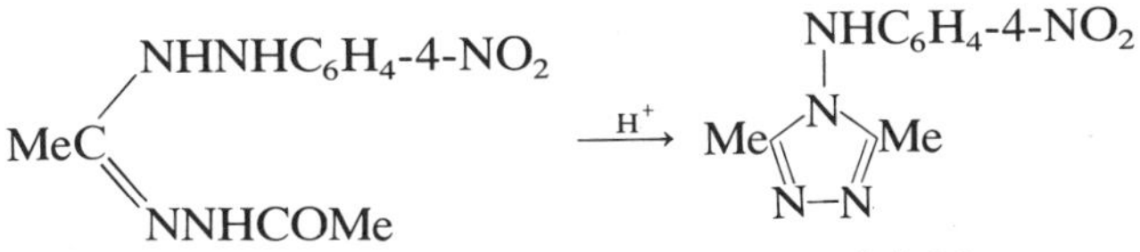

MeC(NHNHC$_6$H$_4$-4-NO$_2$)(=NNHCOMe)

6.4-28

$\xrightarrow{\text{H}^+}$

Me—triazole(NHC$_6$H$_4$-4-NO$_2$)—Me

6.4-29

cyclization of **6.4-30** was effected with mercuric oxide to give **6.4-31**.[39] The preparation of N-unsubstituted hydrazidines has been described, and the cyclization of the hydrochlorides of these hydrazidines with triethyl ortho-formate provided high yields of 3-substituted 4-aminotriazoles.[40]

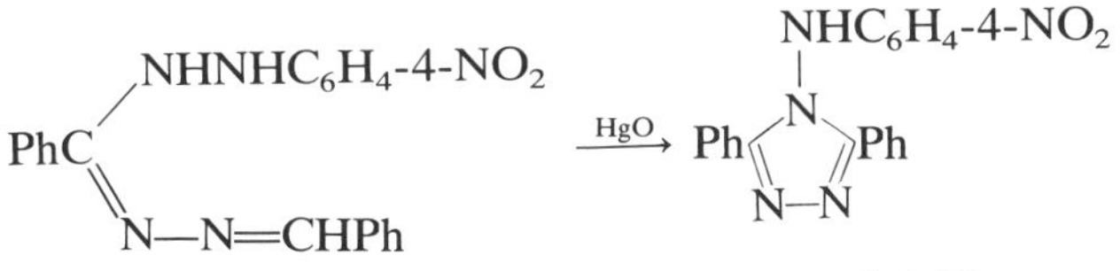

PhC(NHNHC$_6$H$_4$-4-NO$_2$)(N—N=CHPh)

6.4-30

$\xrightarrow{\text{HgO}}$

Ph—triazole(NHC$_6$H$_4$-4-NO$_2$)—Ph

6.4-31

Perfluoroalkyl triazoles were prepared by a number of routes as demonstrated by the condensation of perfluoroacetonitrile with hydrazine in a sealed tube at room temperature to give **6.4-32**,[41] by the rearrangement of

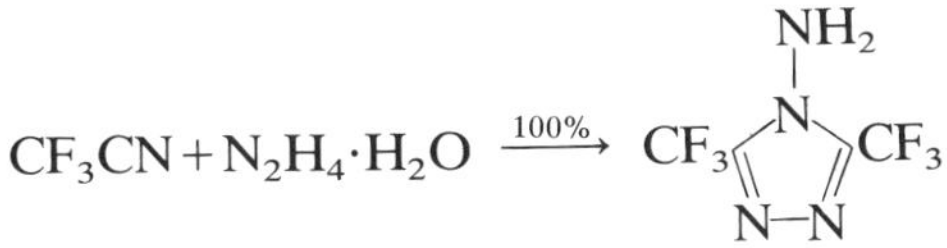

CF$_3$CN + N$_2$H$_4$·H$_2$O $\xrightarrow{100\%}$ CF$_3$—triazole(NH$_2$)—CF$_3$

6.4-32

the dihydrotetrazine (**6.4-33**) with acid to give **6.4-34**, and by reaction of the oxadiazole (**6.4-36**) with hydrazine to give **6.4-35** formed via the intermediate (**6.4-37**).[42]

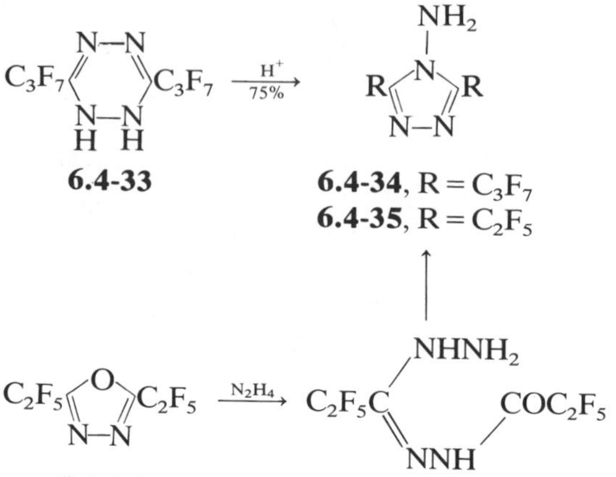

6.4-33

6.4-34, R = C_3F_7
6.4-35, R = C_2F_5

6.4-36

6.4-37

The conversion of bicyclic ring systems containing a condensed triazole ring to 4-aminotriazoles has been observed in the reaction of **6.4-38** with hydrazine to give **6.4-39**[43] and of **6.4-40** with aqueous base to give **6.4-41**.[44]

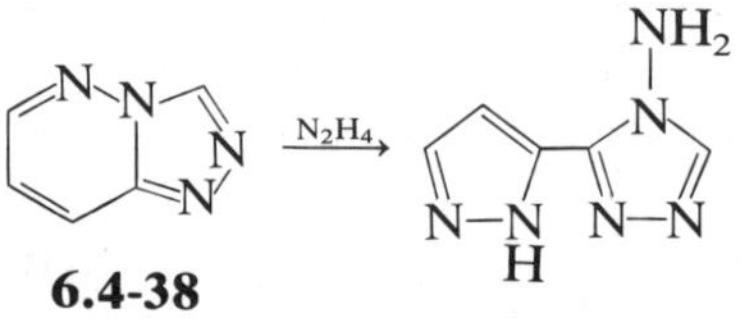

6.4-38

6.4-39

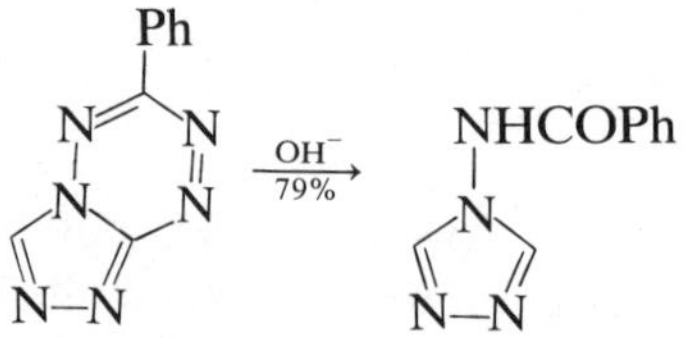

6.4-40

6.4-41

The preparation of 4-aminotriazoles from 1,3,4-oxa- and thiadiazoles and hydrazine has also been reported.[45–47] In the reaction of 4-chloroquinazoline (**6.4-42**) with hydrazine to give a mixture of **6.4-43** and **6.4-44**, the 2-carbon of the quinazoline becomes the 5-carbon of the triazole ring.[48] Also, the formation of 4-amino-3,5-diphenyltriazole (**6.4-14**) by the decomposition of

6.4-42 $\xrightarrow[150°]{N_2H_4}$ 2-H$_2$NC$_6$H$_4$ **6.4-43** (66%) + 2-H$_2$NC$_6$H$_4$ **6.4-44** (22%)

other heteroaromatic ring systems is illustrated by the reaction of the oxo-pyrimidine **6.4-45** with hydrazine[49] and by the thermolysis of 5-phenyltetra-zole (**6.4-46**) in a variety of solvents.[25]

6.4-45 $\xrightarrow[75\%]{N_2H_4}$ Ph **6.4-14** Ph $\xleftarrow{\Delta}$ Ph **6.4-46**

Triazoles substituted at the 1-position with an amino group have been reported to result from the thermolysis of 2-(trimethylsilyl)-5-phenyltetra-zole (**6.4-47**).[50] When **6.4-47** was heated at 190 to 230°, nitrogen was evolved to give the nitrilimine (**6.4-48**), two molecules of which were combined to give the 4-aminotriazole (**6.4-49**). The latter was postulated to rearrange under the conditions of the reaction to give the 1-aminotriazole (**6.4-50**). In later work, Ettenhuber and Ruehlmann claimed that the pro-duct of this reaction was the 4-aminotriazole (**6.4-49**).[51]

6.4-47 $\xrightarrow{-N_2}$ [PhC≡N⁺N⁻SiMe$_3$] **6.4-48** $\longrightarrow$ Ph **6.4-49** Ph $\longrightarrow$ Ph **6.4-50** Ph

The amino group of 4-aminotriazoles undergoes most of the reactions expected of an unsymmetrical disubstituted hydrazine including alkylation,[52] acylation,[53-55] Schiff base formation,[56-58] and oxidative removal of the amino group.[9] Also, 4-benzylideneaminotriazoles (e.g., **6.4-51**) eliminated benzonitrile in the presence of sodium glycolate.[59] Other reactions of 4-amino-1,2,4-triazole (**6.4-1**) include condensation with 2,5-diethoxytetra-hydrofuran to give **6.4-52**,[60] the addition of styrene in the presence of lead tetraacetate to give the aziridinyltriazole (**6.4-53**),[61] and the base-catalyzed addition of alkyl- and arylnitriles to give amidinotriazoles (e.g., **6.4-55**).[62]

The 4-amidinotriazoles can also be prepared from **6.4-1** and imino ethers[62] or by the successive treatment of **6.4-1** with triethyl orthoformate and a primary amine.[63] Becker and coworkers have shown that treatment of 4-amidinotriazoles (e.g., **6.4-55**) with acylation reagents resulted in the formation of 3-substituted triazoles (e.g., **6.4-54**).[62,64]

References

1. **38:** 5828[3]	2. **61:** 14662g	3. **48:** 7006c	4. **55:** 27282f
5. **76:** P3867t	6. **51:** 14751b	7. **53:** 11398g	8. **44:** 2941i
9. **77:** 5410h	10. **57:** 5911h	11. **58:** 8071d	12. **59:** 10065h
13. **84:** 90156s	14. **54:** 18484g	15. **59:** 1624b	16. **55:** 18778e
17. **59:** 15278h	18. **77:** 152130v	19. **78:** 29776d	20. **85:** P192740r
21. **81:** 120640c	22. **80:** 82840n	23. **85:** 177373e	24. **60:** 14508e
25. **57:** 5909c	26. **54:** 24790h	27. **62:** 9122a	28. **51:** 1957e
29. **55:** 6495d	30. **77:** 34469q	31. **64:** 5072h	32. **55:** 14440f
33. **57:** 3436i	34. **58:** 7924d	35. **62:** 1647a	36. **77:** 139906h
37. **82:** 170800w	38. **85:** 159991a	39. **83:** 58722s	40. **85:** 77541q
41. **55:** 6474c	42. **64:** 12662e	43. **78:** 29701a	44. **76:** 126949d
45. **73:** 66511f	46. **76:** 59542a	47. **85:** 123823h	48. **77:** 88421m
49. **60:** 12006c	50. **60:** 5536c	51. **68:** 69071v	52. **68:** 105111u
53. **72:** 21667z	54. **73:** 130946f	55. **76:** 140662m	56. **74:** 13074e
57. **76:** 14220c	58. **77:** 114143u	59. **68:** 58906v	60. **83:** 131395u
61. **77:** 101467c	62. **71:** 38867s	63. **76:** 113138p	64. **71:** 91386k

TABLE 6. AMINO-1,2,4-TRIAZOLES

Compound	Reference

6.1. 3(or 5)-Amino-s-triazoles

6.1a. 3(*or* 5)-*Amino*(*unsubstituted*)-*s-triazoles*

3-acetonyl-5-amino- — **60:** 4141f

3-(2-acetylamino-3-pyridinyl)-5-amino- — **69:** 27374m

3-amino- — **41:** 755g, **51:** 14752c,
53: 17151f, **71:** P124443g,
76: P153796h, **85:** 94307k

3-amino-5-(*p*-aminobenzyl)- — **58:** P13965g

3-amino-5-(2-amino-5-chloro-3-pyrazinyl)- — **70:** 96755u, **72:** P90516m

3-amino-5-(2-aminoethyl)- — **53:** 11354g, **83:** 126268a

3-amino-5-(aminomethyl)- — **53:** 11354f, **64:** P12685e,
70: 77877u

3-amino-5-(α-aminophenethyl)- — **53:** 11354h

3-amino-5-[3-amino-5-(trifluoromethyl)-2-pyrazinyl]- — **70:** 96755u, **72:** P90516m

3-amino-5-amyl- — **40:** P369^2

5-amino-, 3-(2,5-anhydro-1-deoxy-D-allitol derivative) — **82:** 98308y

5-amino-, 3-(2,5-anhydro-1-deoxy-3,4-O-(isopropylidene)-D-allitol derivative) — **82:** 98308y

5-amino-, 3-(1,4-anhydro-2,3-O-isopropylidene-D-
ribitol derivative) — **83:** 43657a

5-amino-, 3-(1,4-anhydro-D-ribitol derivative) — **83:** 43657a

3-amino-5-(2-benzothiazolyl)- — **77:** 126518h

3-amino-5-benzyl- — **55:** 16523a, **59:** 11494f

3-amino-5-[2-(benzylmethylamino)ethyl]- — **66:** P46427k

3-amino-5-[(benzylmethylamino)methyl]-, — **66:** P46427k

3-amino-5-[*p*-(benzyloxy)phenyl]- — **54:** 9937i

3-amino-5-[([1,1'-biphenyl]-4-yloxy)methyl]- — **81:** P65189k

5-amino-1-[bis(dimethylamino)phosphinyl]-3-phenyl- — **68:** P21922w

5-amino-3-[2-[3,5-bis(tert-butyl)-4-hydroxyphenyl]ethyl]- — **79:** P19662v

3-amino-5-[2-bromo-5-(*N*-methylsulfamoyl)phenyl]- — **76:** P99678f, **77:** P164710x

3-amino-5-[4-bromo-3-(*N*-methylsulfamoyl)phenyl]- — **76:** P99678f, **77:** P164710x

3-amino-5-(*p*-bromophenyl)- — **55:** 23507c, **81:** P63636e

3-amino-5-(4-bromo-3-sulfamoylphenyl- — **76:** P99678f, **77:** P164710x

3-amino-5-butyl- — **55:** 16523a

3-amino-5-(*p-tert*-butylphenyl)- — **69:** P11964t

3-amino-5-(*p*-chlorobenzyl)- — **58:** P13965g

3-amino-5-(2-chlorophenyl)- — **81:** P63636e

3-amino-5-(*p*-chlorophenyl)- — **44:** 6855i

3-amino-5-(4-chloro-3-sulfamoylphenyl)- — **76:** P99678f, **77:** P164710x

3-amino-5-[[(4-chloro-*o*-tolyl)oxy]methyl]- — **58:** P13965g

3-amino-5-cyclohexyl- — **55:** 25991h, **56:** 10132e,
59: 6386h

3-amino-5-[(cyclohexylmethylamino)methyl]- — **66:** P46427k

3-amino-5-[3-(cyclohexylmethylamino)propyl]- — **66:** P46427k

3-amino-5-[2-(diallylamino)ethyl]- — **66:** P46427k

3-amino-5-[(diallylamino)methyl]- — **66:** P46427k

3-amino-5-(2,6-diamino-5-chloro-3-pyrazinyl)- — **70:** 96755u, **72:** P90516m

TABLE 6 (*Continued*)

Compound	Reference

6.1. 3(or 5)-Amino-*s*-triazoles (*Continued*)

6.1a. 3(*or 5*)-*Amino(unsubstituted)-s-triazoles* (*Continued*)

Compound	Reference
3-amino-5-[2-(dibutylamino)ethyl]-	**66:** P46427k
3-amino-5-[(dibutylamino)methyl]-	**66:** P46427k
3-amino-5-[(2,4-dichlorophenoxy)methyl]-	**58:** P13965g
3-amino-5-[2-(diethylamino)ethyl]-	**66:** P46427k
3-amino-5-[(diethylamino)methyl]-	**66:** P46427k
3-amino-5-[3-(diethylamino)propyl]-	**66:** P46427k
3-amino-5-(4,5-dihydro-1H̲-pyrazol-3-yl)-	**78:** 29701a
3-amino-5-[1,2-dihydroxy-5-(hydroxymethyl)-3-cyclopentyl]-	**79:** 66255z
3-amino-5-[2-(dimethylamino)ethyl]-	**66:** P46427k
3-amino-5-[(dimethylamino)methyl]-	**66:** P46427k
3-amino-5-[3-(dimethylamino)propyl]-	**66:** P46427k
3-amino-5-[(dipentylamino)methyl]-	**66:** P46427k
3-amino-5-[2-(dipropylamino)ethyl]-	**66:** P46427k
3-amino-5-[(dipropylamino)methyl]-	**66:** P46427k
3-amino-5-(2-ethoxyethyl)-	**43:** 7019c
3-amino-5-ethyl-	**23:** 3470[9], **56:** 10132e, **59:** 6386h
3-amino-5-(2-furyl)-	**63:** 18071g
3-amino-5-heptadecyl-	**40:** P369[2]
3-amino-5-(heptafluoropropyl)-	**72:** 78954v
3-amino-5-heptyl-	**55:** 16523a, **60:** P2952c, **75:** P98576p
3-amino-5-hexyl-	**49:** 15916g
5-amino-3-[(hydroxyacetoxy)methyl]-	**82:** 57642h
5-amino-3-(1-hydroxy-1-methylethyl)-	**81:** 136067g
5-amino-3-(1-hydroxy-2-methylpropyl)-	**81:** 136067g
3-amino-4-(*o*-hydroxyphenyl)-	**49:** 14744h, **83:** 97141w
5-amino-3-(3-hydroxypropyl)-	**69:** 27342z, **79:** P20314q
5-amino-3-(3-hydroxypropyl)-, hydrogen sulfate (ester)	**69:** 27342z
5-amino-3-(3-hydroxypropyl)-, sulfate (2:1) (ester)	**69:** 27342z
3-amino-5-(*p*-isopropylphenyl)-	**80:** 107585y
3-amino-5-isobutyl-	**23:** 3471[1]
3-amino-5-[4-methoxy-3-(*N*-methylsulfamoyl)phenyl]-	**76:** P99678f, **77:** P164710x
3-amino-5-(*p*-methoxyphenyl)-	**44:** 6855i, **81:** P63636e
3-amino-5-methyl-	**48:** 1276c, **49:** 6240d, **56:** 10132e, **59:** 3928a, **64:** 5072h, **65:** 12205a, **81:** 25613n
3-amino-5-[(N-methylanilino)methyl]-	**66:** P46427k
3-amino-5-[4-methyl-3-(*N*-methylsulfamoyl)phenyl]-	**76:** P99678f, **77:** P164710x
3-amino-5-(1-methyl-5-nitroimidazol-2-yl)-	**73:** P25480b
3-amino-5-(1-methyl-5-nitro-1H̲-imidazol-2-yl)-	**85:** P5644c
3-amino-5-(4-methylphenyl)-	**80:** 107585y, **81:** P63636e
3-amino-5-[3-(*N*-methylsulfamoyl)phenyl]-	**76:** P99678f, **77:** P164710x

TABLE 6 (*Continued*)

Compound	Reference

6.1. 3(or 5)-Amino-*s*-triazoles (*Continued*)

6.1a. 3(*or* 5)-*Amino(unsubstituted)-s-triazoles* (*Continued*)

Compound	Reference
3-amino-5-(1-naphthyl)-	**61:** 7403h
3-amino-5-[(2-naphthyloxy)methyl]-	**58:** P13965g
3-amino-5-(*p*-nitrobenzyl)-	**58:** P13965g
3-amino-5-(5-nitro-2-furyl)-	**63:** P611g, **65:** PC20139a, **65:** P18596a, **75:** 20356x, **79:** 87327q
3-amino-5-(*m*-nitrophenyl)-	**55:** 23507c
3-amino-5-(*p*-nitrophenyl)-	**55:** 23507c, **56:** 10132e, **59:** 6386h
3-amino-5-nonyl-	**71:** 124342y
3-amino-5-octyl-	**54:** 21498h
5-amino-2-[(2(3H)-oxo-5-phenyl-3-furanidene-α-phenacylmethyl]-	**69:** 27342z
3-amino-5-pentadecyl-	**54:** 21498h
3-amino-5-(pentafluoroethyl)-	**72:** 78954v
3-amino-5-pentyl-	**48:** 1276c, **48:** 13687d, **55:** 23507c, **70:** 11640e
3-amino-5-(phenoxymethyl)-	**58:** P13965g, **81:** P65180a
3-amino-5-phenyl-	**44:** 6855e, **48:** 1276d, **49:** 15916g, **55:** 23507c, **56:** 10132e, **64:** 5072h, **70:** 115073p
5-amino-3-(2-phenylphenacyl)-	**69:** 27342z
3-amino-5-[(phthalimido)methyl]-	**70:** 77877u
3-amino-5-piperonyl-	**67:** 32651n
3-amino-5-(3-propionylphenyl)-	**69:** 27342z
3-amino-5-(3-propionylphenyl)-, (2,4-dinitrophenyl)-hydrazone	**69:** 27342z
3-amino-5-propyl-	**51:** P13934h
3-amino-5-(3-pyridyl)-	**72:** P43682c
3-amino-5-(4-pyridyl)-	**53:** 19169d, **77:** 70181y
3-amino-5-[2-(1-pyrrolidinyl)ethyl]-	**66:** P46427k
3-amino-5-(1-pyrrolidinylmethyl)-	**66:** P46427k
3-amino-5-[3-(1-pyrrolidinyl)propyl]-	**66:** P46427k
3-amino-5-styryl-	**55:** 16523a, **60:** P2952c
5-amino-3-[tetrahydro-4-(hydroxymethyl)-2,2-dimethyl-4H-cyclopenta-1,3-dioxolyl]	**79:** 66255z
3-amino-5-[tetrafluoro-2-(trifluoromethoxy)ethyl]-	**72:** 78954v
3-amino-5-(2-thienyl)-	**56:** 10132e, **59:** 6386h, **64:** 5072h
3-amino-5-*m*-tolyl-	**56:** 10132e, **59:** 6386h
3-amino-5-(trifluoromethyl)-	**72:** 78954v
3-amino-5-undecyl-	**55:** 16523a, **60:** P2952c
-3-methanol, 5-amino-	**73:** P25482d
-3-methanol, 5-amino-, glycolate	**53:** P2264c

TABLE 6 (*Continued*)

Compound	Reference

6.1. 3(or 5)-Amino-*s*-triazoles   (*Continued*)

6.1b. 3(*or* 5)-*Amino*(*substituted*)-*s*-*Triazoles*

3-acetamido-   **38:** 970[4], **53:** 21901i,
                                                     54: 1535b, **55:** 3565i,
                                                     60: 13243a

3-(*p*-acetamidobenzenesulfonamido)-   **70:** 47368m
3-(*p*-acetamidobenzenesulfonamido)-5-(*p*-chlorophenyl)-   **70:** 47368m
3-(*p*-acetamidobenzenesulfonamido)-5-phenyl-   **70:** 47368m
3-acetamido-5-(*p*-bromophenyl)-   **59:** 12790f
3-acetamido-5-methyl-   **49:** 15917a
3-(acetamido)methyl-5-anilino   **57:** 3423i
3-acetamido-5-(5-nitro-2-furyl)-   **65:** PC20139a, **79:** 87327q
3-acetamido-5-phenyl-   **55:** 3566a, **58:** 519d,
                                                     59: 12790d

3-[*p*-(acetamido)phenyl]-5-ureido-   **64:** 5072h
3-(2-acetoacetamido)-   **72:** P67723t
3-[(1-acetyl-2-benzoyl)hydrazino]-5-phenyl-   **73:** 35293g, **77:** 87362f
3-[[1-acetyl-2-(4-bromobenzoyl)]hydrazino]-5-phenyl-   **73:** 35293g, **77:** 87362f
3-[[1-acetyl-2-(4-chlorobenzoyl)]hydrazino]-5-phenyl-   **73:** 35293g, **77:** 87362f
3-[[1-acetyl-2-(*p*-isopropylbenzoyl)]hydrazino]-5-phenyl-   **73:** 35293g, **77:** 87362f
3-[[-1-acetyl-2-(4-methylbenzoyl)]hydrazino]-5-phenyl   **73:** 35293g, **77:** 87362f
3-[[1-acetyl-]-(4-nitrobenzoyl)]hydrazino]-5-phenyl-   **73:** 35293g, **77:** 87362f
3-(N^4-acetylsulfanilamido)-   **36:** 6511[5], **38:** 730[2], **38:** 2938[4]
3-(allylamino)-   **59:** 10048d, **68:** 95761q,
                                                     70: P28922w, **71:** P81375s,
                                                     71: P124443g

3-(allylamino)-5-(2-furyl)-   **63:** 18071c
3,3'-(*N,N'*-amidino)bis[triazole]   **71:** P90233w
3-[2-(*N*-aminoamidino)hydrazino]-5-phenyl-   **68:** 78203s
3-(*o*-aminobenzamido)-   **45:** 629h
3-(4-aminobenzenesulfonamido)-   **81:** P27329y, **83:** 173146u
3-(5-amino-3-phenyl-1H̲-pyrazol-1-yl)   **83:** 10006j
3-anilino-   **62:** 11803c, **63:** 7012h,
                                                     70: P28922w, **71:** P81375s,
                                                     71: P124443g

3-anilino-5-[1-(benzamido)ethyl]-   **57:** 3423i
3-anilino-5-[2-(benzamido)ethyl]-   **57:** 3423i
3-anilino-5-benzyl-   **53:** 9197e, **55:** 24728d
3-anilino-1-[bis(dimethylamino)phosphinylidyne]-   **77:** P126641t
3-(α-anilino-4-bromobenzylidenehydrazino)-5-phenyl- **82:** 57055u
3-anilino-5-ethyl-   **55:** 24728d, **53:** 9197d
5-anilino-3-(1-hydroxy-1-methylethyl)-   **81:** 136067g
5-anilino-3-(2-hydroxyphenyl)-   **83:** 97141w
3-anilino-5-isobutyl-   **55:** 24728d
3-anilino-5-isopropyl-   **71:** 91382f
5-anilino-3-isopropyl-1-[bis(dimethylamino)phosphinyl]-   **76:** 42580p
3-anilino-5-methyl-   **53:** 9197c, **71:** 91382f,
                                                     81: 37516w

TABLE 6 (*Continued*)

Compound	Reference

6.1. 3(or 5)-Amino-*s*-triazoles   (*Continued*)

6.1b. 3(*or* 5)-*Amino*(*substituted*)-*s*-*Triazoles*   (*Continued*)

Compound	Reference
3-anilino-5-pentyl-	**55:** 24728d
3-anilino-5-phenyl-	**55:** 24728d, **59:** 12790b, **64:** 9727e
3-anilino-5-propyl-	**55:** 24728d
3-anilino-5-[1-(4-toluenesulfonamido)ethyl]-	**57:** 3423i
3-anilino-5-[2-(4-toluenesulfonamido)ethyl]-	**57:** 3423i
3-anilino-5-(4-toluenesulfonamido)methyl-	**57:** 3423i
3-*o*-anisamido-	**55:** P2990e
3-(*p*-anisamido)-	**70:** 11640e
3-*p*-anisidino-	**68:** 95761q
3-*p*-anisidino-5-methyl-	**71:** 91382f
3-[2-[*N*-(2-anisoylamino)amidino]hydrazino]-5-phenyl-	**68:** 78203s
3-benzamido-	**53:** 21901i, **64:** P8197a
3-[(benzamido)acetamido]-	**70:** 11640e
3-[(benzamido)acetamido]-5-(ethoxymethyl)-	**70:** 11640e
3-[2-[*N*-(benzamido)amidino]hydrazino]-	**68:** 78203s
3-[2-[*N*-(benzamido)amidino]hydrazino]-5-phenyl-	**68:** 78203s
3-benzamido-5-benzyl-	**59:** 11494f
3-benzamido-5-(*p*-bromophenyl)-	**59:** ,12790f
3-benzamido-5-ethyl-	**59:** 6227e, **59:** 12791e
3-[(6-benzamidohexanoyl)amino]-	**70:** 11640e
3-[(α-benzamido)hydrocinnamamido]-	**70:** 11640e
3-benzamido-5-methyl-	**59:** 12790f
3-benzamido-5-(5-nitro-2-furanyl)-	**79:** 87327q
3-benzamido-5-(*p*-nitrophenyl)-	**59:** 12790f
3-benzamido-5-pentyl-	**59:** 6227e, **59:** 12791e
3-benzamido-5-phenyl-	**58:** 519c
3-benzamido-5-phenyl-	**77:** 48344g
3-(benzenesulfonamido)-	**70:** 47368m
3-(benzenesulfonamido)-5-(*p*-chlorophenyl)-	**70:** 47368m
3-(benzenesulfonamido)-5-methyl	**70:** 47368m
3-(benzenesulfonamido)-5-phenyl-	**70:** 47368m
5-(1,3-benzodioxol-5-yl)-3-(2-benzoyl-1-methylhydrazino)-	**76:** 14437d
3-[2-(1,3-benzodioxole-5-carbonyl)-1-methylhydrazino]-5-(3,4-dimethoxyphenyl)-	**76:** 14437d
3-[2-(1,3-benzodioxole-5-carbonyl)-1-methylhydrazino]-5-phenyl-	**76:** 14437d
5-(1,3-benzodioxol-5-yl)-3-[(4-nitrobenzylidene)-1-methylhydrazino]-	**76:** 14437d
3-(2-benzothiazolyl)-5-methylamino-	**77:** 126518h
3-(2-benzothiazolyl)-5-(2-propenylamino)-	**77:** 126518h
3-(2-benzoyl-1-methylhydrazino)-5-(3,4-dimethoxyphenyl)-	**76:** 14437d
3-(2-benzoyl-1-methylhydrazino)-5-(3,5-dimethoxyphenyl)-	**76:** 14437d
3-(2-benzoyl-1-methylhydrazino)-5-phenyl-	**76:** 14437d
5-(2-benzoyl-1-phenylhydrazino)-3-phenyl-	**82:** 125326k

TABLE 6 (*Continued*)

Compound	Reference

6.1. 3(or 5)-Amino-*s*-triazoles (*Continued*)

6.1b. 3(*or* 5)-*Amino*(*substituted*)-*s*-*Triazoles* (*Continued*)

Compound	Reference
3-(benzylamino)-	**55:** 10452h, **64:** 4935f, **71:** P81375s, **71:** P124443g
5-benzylamino-3-(2-hydroxyphenyl)-	**83:** 97141w
3-(benzylamino)-5-phenyl-	**59:** 12790a
3-benzyl-5-[2-(*p*-chlorophenyl)hydrazino]-	**59:** 12789d
3-benzyl-5-[2-(2,4-dichlorophenyl)hydrazino]-	**59:** 12789d
3-(4-benzyl-1,2-dihydro-5-hydroxy-3-oxo-2-phenyl-3H-pyrazol-1-yl)-5-phenyl-	**83:** 146905j
3-[(α-benzyl-1,3-dioxo-2-isoindolinyl)acetamido]-	**70:** 11640e
3-(4-benzyl-3,5-dioxo-1-phenyl-2-pyrazolidinyl)-5-phenyl-	**82:** 4183c
3-(4-benzyl-3,5-dioxo-1-pyrazolidinyl)-5-phenyl-	**82:** 4183c
3-(2-benzylhydrazino)-5-butyl-	**70:** 115080p
3-(2-benzylhydrazino)-5-ethyl-	**70:** 115080p
3-(2-benzylhydrazino)-5-isobutyl-	**70:** 115080p
3-(2-benzylhydrazino)-5-isopropyl-	**70:** 115080p
3-(2-benzylhydrazino)-5-phenyl-	**70:** 115080p
3-(2-benzylhydrazino)-5-propyl-	**70:** 115080p
3-(benzylideneamino)-5-phenyl-	**58:** 519d
3-[2-(benzylidenecarbazoyl)hydrazino]-5-(2-chlorophenyl)-	**76:** 99568v
3-[2-(benzylidenecarbazoyl)hydrazino]-5-phenyl-	**76:** 99568v
3-[(benzylidene)hydrazino]-5-butyl-	**70:** 115080p
3-[(benzylidene)hydrazino]-5-ethyl-	**70:** 115080p
3-(benzylidenehydrazino)-5-phenyl-	**77:** 87362f, **82:** 4183c
3-(4-benzyl-4-methyl-3,5-dioxo-1-pyrazolidinyl)-5-phenyl-	**82:** 4183c
3-[[(benzyloxycarbonyl)amino]acetamido]-	**70:** 11640e
3-[[2-(benzyloxycarbonylamino)-3-phenylpropionyl]-amino]-	**68:** 78612z
3-benzyl-5-(2-phenylhydrazino)-	**59:** 12789d
3-benzyl-5-[2-(3-phenyl-*s*-triazol-5-yl)hydrazino]-	**70:** 115073p
5-benzyl-3-[(4-pyridinecarbonyl)amino]-	**81:** P63636e
3-[(*N*-benzylthiocarbamoyl)amino]-	**76:** 85792b
3-[[(benzylthio)carbonimidoyl]amino]-5-methyl-	**78:** 147921v
3-[[(benzylthio)carbonimidoyl]amino]-5-phenyl-	**78:** 147921v
3-[[(benzylthio)carbonothioyl]amino]-	**78:** P148008q, **82:** P43429g
5-[(benzylthio)methyl]-3-(2-hydroxybenzamido)-	**77:** P115460g
3-[[[(benzylthio)(methylthio)]methylene]amino]-	**78:** P148008q
3-biguanido-	**69:** P52142s
3-[2-[*N*-[(2-biphenylcarbonyl)amino]amidino]-hydrazino]-	**68:** 78203s
3-[[[bis(benzylthio)]methylene]amino]	**82:** P43429g
3-[[[bis(benzylthio)]methylene]amino]-	**78:** P148008q
3-[[[bis(dodecylthio)]methylene]amino]-	**78:** P148008q
3-[[bis[2-(ethylthio)ethylthio]methylene]amino]-	**78:** P148008q, **82:** P43429g

TABLE 6 (*Continued*)

Compound	Reference

6.1. 3(or 5)-Amino-*s*-triazoles (*Continued*)

6.1b. 3(*or* 5)-*Amino*(*substituted*)-*s*-*Triazoles* (*Continued*)

Compound	Reference
3-[[[bis(ethylthio)]methylene]amino]-	**78:** P148008q, **82:** P43429g
3-[[[3,4-bis(4-methoxyphenyl)-2-furanyl]methylene]-amino]-	**83:** 43117z
3-[[[bis(methylthio)]methylene]amino]-	**82:** P43429g
3-[[[bis(2-propenylthio)]methylene]amino]-	**78:** P148008q, **82:** P43429g
3-(*p*-bromobenzamido)-	**70:** 11640e
3-(*p*-bromobenzamido)-5-(ethoxymethyl)-	**70:** 11640e
3-(*p*-bromobenzamido)-5-isopropyl-	**70:** 11640e
3-[2-(4-bromobenzoyl)hydrazino]-5-phenyl-	**82:** 57055u
3-(α-bromobenzylidenehydrazino)-5-phenyl-	**82:** 57055u
3-[(3-bromobenzylidene)hydrazino]-5-phenyl-	**82:** 57055u
3-[(4-bromobenzylidene)hydrazino]-5-phenyl-	**77:** 87362f, **82:** 57055u
3-[(2-bromobutyryl)amino]-	**81:** 147547q
3-[(3-bromobutyryl)amino]-	**81:** 147547q
3-(α-bromo-4-chlorobenzylidenehydrazino)-5-phenyl-	**82:** 57055u
3-(4-bromo-α-chlorobenzylidenehydrazino)-5-phenyl-	**82:** 57055u
3-[[(3-bromo-4-ethoxyphenyl)thiocarbamoyl]amino]-	**74:** 141216q
3-(2-bromo-5-fluorobenzenesulfonamido)-	**75:** 118258s
3-(2-bromo-5-fluorobenzenesulfonamido)-5-ethyl-	**75:** 118258s
3-(2-bromo-5-fluorobenzenesulfonamido)-5-methyl-	**75:** 118258s
3-(bromo-5-fluorobenzenesulfonamido)-5-propyl-	**75:** 118258s
3-(α-bromo-4-isopropylbenzylidenehydrazino)-5-phenyl-	**82:** 57055u
3-(α-bromo-4-methylbenzylidenehydrazino)-5-phenyl-	**82:** 57055u
3-(4-bromonaphthalimino)-	**73:** P67700d, **83:** P116968q
3-(α-bromo-3-nitrobenzylidenehydrazino)-5-phenyl-	**82:** 57055u
3-(α-bromo-4-nitrobenzylidenehydrazino)-5-phenyl-	**82:** 57055u
5-(4-bromophenyl)-3-[(4-pyridinecarbonyl)amino]-	**81:** P63636e
3-[(2-bromopropionyl)amino]-	**81:** 25611k
3-[[3-[(2-butoxycarbonylethyl)thio]propionyl]amino]-	**72:** P67723t
3-(5-butoxy-2-hydroxybenzamido)-	**77:** P115460g
3-[1-(*p*-butoxyphenyl)thioureido]-5-methyl-	**70:** 77881r
3-butylamino-5-methyl-	**53:** 9197g
3-butylamino-5-propyl-	**53:** 9197g
3-[6-[[3-(butylamino)propyl]amino]-1,3(2H)-dioxo-1H-benz[*de*]isoquinolin-2(3H)-yl]-	**83:** P61703s
3-(4-butyl-3,5-dioxo-1-phenyl-2-pyrazolidinyl)-5-phenyl-	**82:** 4183c
3-[(α-butyl-1,3-dioxo-2-isoindolinyl)acetamido]-	**70:** 11640e
3-(4-butyl-3,5-dioxo-1-pyrazolidinyl)-5-phenyl-	**82:** 4183c
3-butyl-5-hydrazino-	**64:** P3557g
3-(4-butyl-4-hydroxy-3,5-dioxopyrazolidin-1-yl)-5-phenyl-	**82:** 111992n
3-(4-butyl-4-methyl-3,5-dioxo-1-pyrazolidinyl)-5-phenyl-	**82:** 4183c
3-[(butylthiocarbamoyl)amino]-	**83:** 53854b
butyrylamino-	**38:** 970[4]

TABLE 6 (*Continued*)

Compound	Reference

6.1. 3(or 5)-Amino-*s*-triazoles (*Continued*)

6.1b. 3(*or* 5)-*Amino*(*substituted*)-*s*-*Triazoles* (*Continued*)

Compound	Reference
5-butyrylamino-3-methyl-	**38:** 970[5]
-3-carbamic acid, 4-chloro-2-butynyl ester	**60:** P6852d, **61:** P4364f
-3-carbamic acid, 5-phenyl-	**48:** 1276a
-3-carbamimidic acid, 3-chlorophenyl ester	**79:** P78436x
3-[2-(carbazoyl)hydrazino]-5-(2-hydroxyphenyl)-	**76:** 99568v
3-[2-(carbazoyl)hydrazino]-5-phenyl-	**76:** 99568v
3-[2-(2-carboxybutyryl)-2-phenylhydrazino]-5-phenyl-	**82:** 111992n
3-[2-(carboxy)(cyclohexyl)acetyl]hydrazino]-5-phenyl-	**82:** 4183c
3-[2-(2-carboxyheptanoyl)-2-phenylhydrazino]-5-phenyl-	**82:** 111992n
3-[2-(2-carboxyhexanoyl)-2-phenylhydrazino]-5-phenyl-	**82:** 111992n
3-[2-(2-carboxy-2-hydroxybutyryl)hydrazino]-5-phenyl-	**82:** 111992n
3-[2-(α-carboxy-α-hydroxycyclohexylacetyl)hydrazino]-5-phenyl-	**82:** 111992n
3-[2-(α-carboxy-α-hydroxycyclohexylacetyl)-2-phenylhydrazino]-5-phenyl-	**82:** 111992n
3-[2-(2-carboxy-2-hydroxyheptanoyl)hydrazino]-5-phenyl-	**82:** 111992n
3-[2-(2-carboxy-2-hydroxyheptanoyl)-2-phenylhydrazino]-5-phenyl-	**82:** 111992n
3-[2-(2-carboxy-2-hydroxyhexanoyl)hydrazino]-5-phenyl-	**82:** 111992n
3-[2-(2-carboxy-2-hydroxyhexanoyl)-2-phenylhydrazino]-5-phenyl-	**82:** 111992n
3-[2-(2-carboxy-2-hydroxy-3-phenylpropionyl)hydrazino]-5-phenyl-	**82:** 111992n
3-[2-(2-carboxy-2-hydroxy-3-phenylpropionyl)-2-phenylhydrazino]-5-phenyl-	**82:** 111992n
3-[(2-carboxy-2-hydroxyvaleryl)hydrazino]-5-phenyl-	**82:** 111992n
3-[2-(2-carboxy-2-hydroxyvaleryl)-2-phenylhydrazino]-5-phenyl-	**82:** 111992n
3-[(carboxymethyl)amino]-, hydrazide	**72:** P67723t
3-[2-(2-carboxy-3-phenylpropionyl)-2-phenylhydrazino]-5-phenyl-	**82:** 111992n
3-[2-(2-carboxyvaleryl)-2-phenylhydrazino]-5-phenyl-	**82:** 111992n
3-(2-chloroacetamido)-	**55:** 27048e, **81:** 25611k
3-(*p*-chloroanilino)-	**68:** 95761q
3-*p*-chloroanilino-5-ethyl-	**55:** 24728d
3-*p*-chloroanilino-5-isobutyl-	**55:** 24728e
3-(*o*-chloroanilino)-5-methyl-	**71:** 91382f
3-*m*-chloroanilino-5-methyl-	**55:** 24728d
3-*p*-chloroanilino-5-methyl-	**55:** 24728d
3-(*p*-chloroanilino)-5-phenyl-	**55:** 24728e, **64:** 9727e
3-*m*-chloroanilino-5-propyl-	**55:** 24728e
3-*p*-chloroanilino-5-propyl-	**55:** 24728e
3-(2-chlorobenzamido)-	**81:** 147547q
3-*p*-chlorobenzamido-	**55:** P2990e
3-[2-[*N*-(2-chlorobenzamido)amidino]hydrazino]-5-phenyl]	**68:** 78203s

TABLE 6   (Continued)

Compound	Reference

6.1. 3(or 5)-Amino-s-triazoles   (Continued)

6.1b. 3(or 5)-Amino(substituted)-s-Triazoles   (Continued)

Compound	Reference
3-[2-[N-(3-chlorobenzamido)amidino]hydrazino]-5-phenyl-	**68:** 78203s
3-[2-[N-chlorobenzamido)amidino]hydrazino]-5-phenyl-	**68:** 78203s
3-(4-chlorobenzamido)-5-(3-pyridinyl)-	**81:** P63636e
3-(4-chlorobenzamido)-5-(4-pyridinyl)-	**81:** P63636e
3-[2-(4-chlorobenzoyl)hydrazino]-5-phenyl-	**77:** 87362f, **82:** 57055u
3-[(o-chlorobenzyl)amino]-	**66:** 85757n
3-[(p-chlorobenzyl)amino]-	**66:** 85757n
3-[(p-chlorobenzylidene)amino]-	**64:** P8196d
3-[(3-chlorobenzylidene)hydrazino]-5-phenyl-	**82:** 57055u
3-[(4-chlorobenzylidene)hydrazino]-5-phenyl-	**77:** 87362f, **82:** 57055u
3-(6-chloro-1,3(2H)-dioxo-1H-benz[de]-isoquinolin-2(3H)-yl)-	**83:** P61703s
3-[[4-chloro-6-(ethylamino)-1,3,5-triazin-2-yl]amino]-	**83:** P179133j
3-(4-chloro-2-fluorobenzenesulfonamido)-	**75:** 118258s
3-(4-chloro-3-fluorobenzenesulfonamido)-	**75:** 118258s
3-(5-chloro-2-fluorobenzenesulfonamido)-	**75:** 118258s
3-(4-chloro-2-fluorobenzenesulfonamido)-5-ethyl-	**75:** 118258s
3-(4-chloro-3-fluorobenzenesulfonamido)-5-ethyl-	**75:** 118258s
3-(5-chloro-2-fluorobenzenesulfonamido)-5-ethyl-	**75:** 118258s
3-(4-chloro-2-fluorobenzenesulfonamido)-5-methyl-	**75:** 118258s
3-(4-chloro-3-fluorobenzenesulfonamido)-5-methyl-	**75:** 118258s
3-(5-chloro-2-fluorobenzenesulfonamido)-5-methyl-	**75:** 118258s
3-(4-chloro-2-fluorobenzenesulfonamido)-5-propyl-	**75:** 118258s
3-(4-chloro-3-fluorobenzenesulfonamido)-5-propyl-	**75:** 118258s
3-(5-chloro-2-fluorobenzenesulfonamido)-5-propyl-	**75:** 118258s
3-(5-chloro-2-hydroxybenzamido)-	**77:** P115460g, **78:** P147974q
3-[[(4-chlorophenyl)carbamoyl]amino]-	**82:** 156189x
5-(2-chlorophenyl)-3-[2-(ethoxycarbonyl)hydrazino]-	**72:** 78953u
3-(o-chlorophenyl)-5-hydrazino-	**67:** 43756b, **77:** 164610q
3-(m-chlorophenyl)-5-hydrazino-	**67:** 43756b
3-(p-chlorophenyl)-5-hydrazino-	**67:** 43756b
3-[2-(p-chlorophenyl)hydrazino]-5-isobutyl-	**59:** 12789d
3-[2-(p-chlorophenyl)hydrazino]-5-phenyl-	**59:** 12789d
3-(p-chlorophenyl)-5-(p-methoxybenzenesulfonamido)-	**70:** 47368m
3-(p-chlorophenyl)-5-(2-naphthylamino)-	**64:** 9727e
3-(o-chlorophenyl)-5-[2-(3-phenyl-s-triazol-5-yl)-hydrazino]-	**70:** 115073p
3-(p-chlorophenyl)-5-[2-(3-phenyl-s-triazol-5-yl)-hydrazino]-	**70:** 115073p
5-(2-chlorophenyl)-3-[4-pyridinecarbonyl)amino]-	**81:** P63636e
5-(4-chlorophenyl)-3-[(4-pyridinecarbonyl)amino]-	**81:** P63636e
3-(2-chlorophenyl)-5-semicarbazido	**66:** 104959y
3-(p-chlorophenyl)-5-sulfanilamido-	**70:** 47368m
3-(p-chlorophenyl)-5-(p-toluenesulfonamido)-	**70:** 47368m
3-[2-(4-chloro-o-tolyl)oxy]acetamido]-	**56:** P12907h

TABLE 6 (*Continued*)

Compound	Reference

6.1. 3(or 5)-Amino-*s*-triazoles (*Continued*)

6.1b. 3(*or* 5)-*Amino*(*substituted*)-*s*-*Triazoles* (*Continued*)

Compound	Reference
3-(cinnamamido)-	**70:** 11640e
3-(cinnamylideneamino)-	**71:** P38973y
3-(2-cyanoacetamido)-	**61:** 3105f
3-(2-cyanoacetamido)-5-ethyl-	**61:** 3105f
3-(2-cyanoacetamido)-5-hexyl-	**61:** 3105f
3-(2-cyanoacetamido)-5-isobutyl-	**61:** 3105f
3-(2-cyanoacetamido)-5-methyl-	**61:** 3105f
3-(2-cyanoacetamido)-5-propyl-	**61:** 3105f
3-[[(cyclohexanecarbonyl)amino]acetamido]-	**70:** 11640e
3-(cyclohexylamino)-	**53:** 6216a, **68:** 95761q
3-[(cyclohexylcarbonyl)amino]-	**70:** 11640e
3-(4-cyclohexyl-3,5-dioxo-1-phenyl-2-pyrazolidinyl)-5-phenyl-	**82:** 4183c
3-(4-cyclohexyl-3,5-dioxo-1-pyraziolidinyl)-5-phenyl-	**82:** 4183c
3-(4-cyclohexyl-4-hydroxy-3,5-dioxo-1-phenyl-pyrazolidin-2-yl)-5-phenyl-	**82:** 111992n
3-(4-cyclohexyl-4-hydroxy-3,5-dioxopyrazolidin-1-yl)-5-phenyl-	**82:** 111992n
3-[(2-cyclohexyloxy)acetamido]-	**70:** 11640e
3-(3,5-diamino-6-chloropyrazinecarbonyl)amino]	**75:** P20438a
3-(2,6-diamino-5-chloro-3-pyrazinyl)-5-anilino-	**70:** 96755u, **72:** P90516m
3-(2,6-diamino-5-chloro-3-pyrazinyl)-5-(tert-butylamino)-	**72:** P90516m
3-(2,6-diamino-3-chloro-5-pyrazinyl)-5-(diallylamino)-	**70:** 96755u, **72:** P90516m
3-(2,6-diamino-3-chloro-5-pyrazinyl)-5-(dibutylamino)-	**72:** P90516m
3-(2,6-diamino-3-chloro-5-pyrazinyl)-5-(methylamino)-	**70:** 96755u, **72:** P90516m
5-(1,2-dibenzoylhydrazino)-3-phenyl-	**82:** 125326k
3-(α,3-dibromobenzylidenehydrazino)-5-phenyl-	**82:** 57055u
3-(α,4-dibromobenzylidenehydrazino)-5-phenyl-	**82:** 57055u
3-(4,4-dibutyl-3,5-dioxo-1-pyrazolidinyl)-5-phenyl-	**82:** 4183c, **83:** 146905j
3-(2,6-dicarboxy-1,4-dihydro-4-oxo-1-pyridinyl)-	**83:** 193226y
3-(2,2-dichloroacetamido)-	**57:** 16607c
3-(2,2-dichloroacetamido)-5-(5-nitro-2-furyl)-	**65:** PC20139a, **79:** 87327q
3-(2,4-dichlorobenzylideneamino)-	**55:** P5855h
3-(2,4-dichloro-6-methylbenzamido)-	**81:** P73392m
3-[2-(2,4-dichlorophenoxy)acetamido]-	**56:** P12907h
3-[2-(2,4-dichlorophenyl)hydrazino]-5-isobutyl-	**59:** 12789c
3-[(1-(3,4-dichlorophenyl)-3-methyl-2,4-dioxoimidazolidin-5-yl]amino-	**77:** P152186t
3-[(2,2-dichloropropionyl)amino]-	**77:** 30239s
3-(diethylamino)-5-(2-furyl)-	**63:** 18071f
3-(diethylamino)-5-methyl-	**70:** 28872e
3-(diethylamino)-5-phenyl-	**70:** 28872e
3-(4,4-diethyl-3,5-dioxo-1-phenylpyrazolidin-2-yl-5-phenyl-	**82:** 125326k

TABLE 6 (*Continued*)

Compound	Reference

6.1. 3(or 5)-Amino-*s*-triazoles (*Continued*)

6.1b. 3(*or* 5)-*Amino*(*substituted*)-*s*-*Triazoles* (*Continued*)

Compound	Reference
3-(4,4-diethyl-3,5-dioxo-1-pyrazolidinyl)-5-phenyl-	**82:** 4183c
3-[(*N,N*-diethylthiocarbamoyl)amino]-	**76:** 101078f
3-[[*O,O*-(2,2-diethyltrimethylenedioxo)phosphinyl]amino]-	**73:** P109799a
3-(8,13-dihydro-8,13-dioxodinaphtho[2,1-*b*: 2′,3′-*d*]-furan-6-carboxamido)-	**72:** 3301w
3-(9,10-dihydro-9,10-ethanoanthracene-11,12-dicarboximido)-	**69:** P67151t
3-[(3,4-dihydro-2-methyl-4-oxo-2H-1,2-benzothiazin-3-ylcarbonyl)amino]-5-phenyl-, *S,S*-dioxide	**73:** P120647v
3-(2,4-dihydroxybenzamido)-	**77:** P115460g
3-(2,5-dihydroxybenzamido)-	**77:** P115460g
3-(2,5-dihydroxybenzamido)-5-ethyl	**78:** P147974q
3-(dimethoxyamino)-	**68:** 95761q
3-[2-(3,4-dimethoxybenzoyl)-1-methylhydrazino]-5-phenyl-	**76:** 14437d
3-[(2,4-dimethoxybenzylidene)amino]-	**64:** P8196g
5-(2,4-dimethoxyphenyl)-3-[(4-nitrobenzylidene)-1-methylhydrazino]-	**76:** 14437d
5-(3,4-dimethoxyphenyl)-3-[(4-nitrobenzylidene)-1-methylhydrazino]-	**76:** 14437d
5-(3,5-dimethoxyphenyl)-3-[(4-nitrobenzylidene)-1-methylhydrazino]-	**76:** 14437d
3-(dimethylamino)-	**64:** P8197c, **70:** P28922w, **71:** P81375s, **71:** P124443g
3-[[*p*-(dimethylamino)benzylidene]amino]-	**64:** P8196f
3-[[*p*-(dimethylamino)benzylidene]amino]-5-phenyl-	**58:** 519e
3-[2-[[(4-dimethylamino)benzylidene]carbazoyl]-hydrazino]-5-phenyl-	**76:** 99568v
3-dimethylamino-5-methyl-	**59:** 12787h, **84:** 17241m
3-(dimethylamino)-5-(*m*-nitrophenyl)-	**59:** 12790d
3-dimethylamino-5-phenyl-	**45:** 1122f, **59:** 12790d
3-[[[4-(dimethylamino)phenyl]methylene]amino]-	**77:** P146201c
3-[(dimethylcarbamoyl)amino]-	**83:** P164195g
3-(4,4-dimethyl-3,5-dioxo-1-pyrazolidinyl)-5-phenyl-	**82:** 4183c
3-(3,5-dimethylpyrazol-1-yl)-	**59:** 10049d, **63:** 11545c
3-(3,5-dimethylpyrazol-1-yl)-5-methyl-	**62:** 11805b
3-(2,5-dimethylpyrrol-1-yl)-	**54:** 19641i, **73:** 87717t
3-[[*O,O*-(2,2-dimethyltrimethylenedioxo)phos-phinothioyl]amino]-	**73:** P109799a
3-[[*O,O*-(2,2-dimethyltrimethylenedioxo)phosphinyl]-amino]-	**73:** P109799a
3-(3,5-dinitrobenzamido)-	**61:** P14690h
3-(3,4-dinitro-2-toluamido)-	**69:** P27451j
3-(3,5-dioxo-4,4-dipentyl-1-pyrazolidinyl)-5-phenyl-	**82:** 4183c
3-[(1,3-dioxo-2-isoindolinyl)acetamido]-	**70:** 11640e

TABLE 6   (Continued)

Compound	Reference

6.1. 3(or 5)-Amino-s-triazoles   (Continued)

6.1b. 3(or 5)-Amino(substituted)-s-Triazoles   (Continued)

3-[(1,3-dioxo-2-isoindolinyl)acetamido]- 5-(ethoxymethyl)-	**70:** 11640e
3-[(1,3-dioxo-2-isoindolinyl)acetamido]-5-isopropyl-	**70:** 11640e
3-[(1,3-dioxo-2-isoindolinyl)acetamido]-5-pentyl-	**70:** 11640e
3-[3-(1,3-dioxo-2-isoindolinyl)propionamido]-	**70:** 11640e
3-[3-(1,3-dioxo-2-isoindolinyl)propionamido]- 5-(ethoxymethyl)-	**70:** 11640e
3-[3-(1,3-dioxo-2-isoindolinyl)propionamido]-5-pentyl-	**70:** 11640e
3-(3,5-dioxo-4-pentyl-1-phenyl-2-pyrazolidinyl-5-phenyl-	**82;** 4183c
3-(3,5-dioxo-4-pentyl-1-pyrazolidinyl)-5-phenyl-	**82:** 4183c
3-(3,5-dioxo-1-phenyl-4-propyl-2-pyrazolidinyl)- 5-phenyl-	**82:** 4183c
3-[(2,6-dioxo-4-piperidineacetyl)amino]-	**68:** P68975f
3-(3,5-dioxo-4-propyl-1-pyrazolidinyl)-5-phenyl-	**82:** 4183c
3-(2,5-dioxo-1H-pyrrol-1-yl)-	**77:** 28922r
3-(4,4-diphenyl-3,5-dioxo-1-pyrazolidinyl)-5-phenyl-	**83:** 146905j
3-(1,3-diphenyl-5-formazanyl)-	**77:** 152042t
3-[[(3,4-diphenyl-2-furanyl)methylene]amino]-	**83:** 43117z
3,3'-[3,3'-dithiobis(6-hydroxybenzamido)]bis[5-methyl-	**77:** P115460g
-3-dithiocarbamic acid, 2-mercaptoethanesulfonic acid ester	**64:** P14100f
3-[3-[[2-(dodecoxycarbonyl)ethyl]thio]propionylamino]-	**81:** P14321q
3-(p-ethoxybenzylamino)-	**53:** 6216a
3-(ethoxycarbonylamino)-	**79:** 42421z
3-[[(ethoxycarbonylamino)thioacetyl]amino]-5-methyl-	**78:** 136237a
3-[[(ethoxycarbonylamino)thioacetyl]amino]-5-phenyl	**78:** 136237a
3-[N-[(ethoxycarbonyl)carbamoyl]amino]-	**76:** 14500u, **79:** 42421z
3-[[(1-ethoxycarbonyl)ethylidene]hydrazino]-	**76:** 113183z
3-[[(1-ethoxycarbonyl)ethylidene]hydrazino]- 5-(mercaptomethyl)-	**76:** 113183z
3-[[(1-ethoxycarbonyl)ethylidene]hydrazino]-5-methyl-	**76:** 113183z
3-[2-(ethoxycarbonyl)hydrazino]-5-(2-hydroxyphenyl)-	**72:** 78953u
3-[2-(ethoxycarbonyl)hydrazino]-5-phenyl-	**70:** P68377d, **72:** 78953u
3-[N-[(ethoxycarbonyl)thiocarbamoyl]amino]-	**76:** 14500u, **76:** 85792b, **78:** 136237a
5-(2-ethoxyethyl)-3-(2-hydroxybenzamido)-	**77:** P115460g
3-(ethoxymethyl)-5-ureido-	**70:** 11640e
3-[(ethoxyoxalyl)amino]-	**78:** 72011n
3-[[(p-ethoxyphenyl)thiocarbamoyl]amino]-	**74:** 141216q
3-[1-(p-ethoxyphenyl)thioureido]-	**70:** 77881r
3-[1-(p-ethoxyphenyl)thioureido]-5-methyl-	**70:** 77881r
3-[1-(p-ethoxyphenyl)thioureido]-5-pentyl-	**70:** 77881r
3-[1-(p-ethoxyphenyl)thioureido]-5-propyl-	**70:** 77881r
3-ethylamino-	**54:** 9937f, **59:** 3918h
3-(ethylamino)-5-(2-furyl)-	**63:** 18071c

TABLE 6 (*Continued*)

Compound	Reference

6.1. 3(or 5)-Amino-*s*-triazoles (*Continued*)

6.1b. 3(*or* 5)-*Amino*(*substituted*)-*s*-*Triazoles* (*Continued*)

Compound	Reference
3-(ethylamino)-5-(5-nitro-2-furyl)-	**65:** PC20138h, **79:** 87327q
3-ethylamino-5-phenyl-	**55:** 8393f
3-[(ethylcarbamoyl)amino]-	**83:** P164195g
3-[(1-ethyl-1,4-dihydro-7-methyl-4-oxo-1,8-naphthyridine-3-carbonyl)amino]-	**74:** 125500x
3-(4-ethyl-3,5-dioxo-1-phenyl-2-pyrazolidinyl)-5-phenyl-	**82:** 4183c
3-(4-ethyl-3,5-dioxo-1-pyrazolidinyl)-5-phenyl-	**82:** 4183c
5-ethyl-3-(4-fluorobenzenesulfonamido)-	**75:** 118258s
3-ethyl-5-(4-fluoro-3-methylbenzenesulfonamido)-	**75:** 118258s
5-ethyl-3-(4-fluoro-2-methylbenzenesulfonamido)-	**75:** 118258s
5-ethyl-3-(5-fluoro-2-methylbenzenesulfonamido)-	**75:** 118258s
3-ethyl-5-(4-fluoro-1-naphthalenesulfonamido)-	**75:** 118258s
3-ethyl-5-(2-hydroxybenzamido)-	**77:** P115460g
3-ethyl-5-(2-mercaptobenzamido)-	**77:** P115460g
3-ethyl-5-(*N*-phenylbenzamido)-	**62:** 11803g
3-(4-ethylpyrazol-1-yl)-	**72:** P3491h
3-[(ethylthiocarbamoyl)amino]-	**83:** 53854b
3-[[(ethylthio)carbonothioyl]amino]-	**78:** P148008q, **82:** P43429g
3-[[(ethylthio)carbonyl]amino]-	**76:** 81641d
3-[[(ethylthio)carbonyl]amino]-5-phenyl-	**76:** 81641d
3-[[[2-(ethylthio)ethylthio]carbonothioyl]amino]-	**78:** P148008q, **82:** P43429g
3-ethyl-5-*p*-toluidino-	**55:** 24728d
5-ethyl-3-[3-(trifluoromethyl)benzenesulfonamido]-	**75:** 118258s
3-(*p*-fluorobenzenesulfonamido)-	**75:** 118258s
3-(4-fluorobenzenesulfonamido)-5-methyl-	**75:** 118258s
3-(4-fluoro-benzenesulfonamido)-5-propyl-	**75:** 118258s
3-(4-fluoro-2-methylbenzenesulfonamido)-	**75:** 118258s
3-(4-fluoro-3-methylbenzenesulfonamido)-	**75:** 118258s
3-(5-fluoro-2-methylbenzenesulfonamido)-	**75:** 118258s
3-(4-fluoro-2-methylbenzenesulfonamido)-5-methyl-	**75:** 118258s
3-(4-fluoro-3-methylbenzenesulfonamido)-5-methyl-	**75:** 118258s
3-(5-fluoro-2-methylbenzenesulfonamido)-5-methyl-	**75:** 118258s
3-(4-fluoro-2-methylbenzenesulfonamido)-5-propyl-	**75:** 118258s
3-(4-fluoro-3-methylbenzenesulfonamido)-5-propyl-	**75:** 118258s
3-(5-fluoro-2-methylbenzenesulfonamido)-5-propyl-	**75:** 118258s
3-(4-fluoro-1-naphthalenesulfonamido)-	**75:** 118258s
3-(4-fluoro-1-naphthalenesulfonamido)-5-methyl-	**75:** 118258s
3-(4-fluoro-1-naphthalenesulfonamido)-5-propyl-	**75:** 118258s
3-formamido-	**79:** 105154n
3-(formamido)-5-(5-nitro-2-furanyl)-	**79:** 87327q
3-(5-formanido-3-phenyl-1H-pyrazol-1-yl)-	**83:** 10006j
3-[[*N*-formyl(methylthio)carbonimidoyl]amino]-5-methyl-	**78:** 147921v
3-[2-[(2-furanylmethylene)carbazoyl]hydrazino]-5-phenyl-	**76:** 99568v
5-(2-furanyl)-3-[*N*-(2-propenyl)acetamido]-	**79:** 87327q
3-furfurylamino-	**53:** 6216a

TABLE 6 (*Continued*)

Compound	Reference

6.1. 3(or 5)-Amino-s-triazoles (*Continued*)

6.1b. 3(*or 5*)-*Amino*(*substituted*)-*s-Triazoles* (*Continued*)

Compound	Reference
3-furfurylideneamino-	**53:** P14125d, **59:** 11392b
3-(2-furoylamino)-	**70:** 11640e
3-(2-furyl)-5-(4,5-dihydro-3-phenyl-1,2,4-oxaeiazol-4-yl)-	**77:** 152067e
3-(2-furyl)-5-(methylamino)-	**63:** 18071c
3-(2-furyl)-5-(propylamino)-	**63:** 18071c
5-glycolamido-	**54:** 1536d
3-guanidino-	**69:** P52142s
3-guanidino-5-(4-nitrophenyl)	**60:** 1752h
3-guanidino-5-phenyl	**60:** 1752h
3-guanidino-5-(3-pyridinyl)	**60:** 1752h
3-guanidino-5-(4-pyridinyl)	**60:** 1752h
3-(heptylideneamino)-	**71:** P38973y
3-hydrazino-	**48:** 5182b, **70:** P28922w, **71:** P124443g, **71:** P81375s, **77:** 164610q
5-hydrazino-3-(4,5-dihydro-1H̱-pyrazol-3-yl)-	**78:** 29701a
3-hydrazino-5-(2-hydroxyphenyl)-	**70:** P68377d
3-hydrazino-5-isobutyl-	**67:** 43756b
3-hydrazino-5-isopentyl-	**67:** 43756b
3-hydrazino-5-isopropyl-	**67:** 43756b
3-hydrazino-5-(*m*-methoxyphenyl)-	**67:** 43756b, **77:** 164610q
3-hydrazino-5-(*p*-methoxyphenyl)-	**67:** 43756b
3-hydrazino-5-methyl-	**48:** 1344b
3-hydrazino-5-(2-methylphenyl)-	**77:** 164610q
3-hydrazino-5-(3-methylphenyl)-	**77:** 164610q
3-hydrazino-5-(4-methylphenyl)-	**77:** 164610q
3-hydrazino-5-pentyl-	**67:** 43756b
3-hydrazino-5-phenyl-	**63:** P13275f, **67:** 43756b **70:** P68377d
3-hydrazino-5-propyl-	**67:** 43756b
3-hydrazino-5-*m*-tolyl-	**67:** 43756b
3-hydrazino-5-*o*-tolyl-	**67:** 43756b
3-hydrazino-5-*p*-tolyl-	**67:** 43756b
3-(2-hydroxybenzamido)-	**78:** P147974q
3-(2-hydroxybenzamido)-5-[2-[3,5-bis(tert-butyl)-4-hydroxyphenyl]ethyl]	**81:** 121645v
3-(2-hydroxybenzamido)-5-methyl-	**77:** P115460g, **78:** P147974q
3-(2-hydroxybenzamido)-5-octyl-	**77:** P115460g
3-(2-hydroxybenzamido)-5-(phenoxymethyl)-	**77:** P115460g, **78:** P147974q
3-(2-hydroxybenzamido)-5-(2-pyridinyl)-	**77:** P115460g
3-(2-hydroxybenzamido)-5-(1,1,3,3-tetramethylbutyl)-	**81:** 121645v
3-[(*o*-hydroxybenzylidene)amino]-	**72:** P67723t
3-(4-hydroxy-3,5-dioxo-4-propylpyrazolidin-1-yl)-5-phenyl-	**82:** 111992n
3-[2-(2-hydroxyethoxycarbonyl)hydrazino]-5-phenyl-	**72:** 78953u

TABLE 6 (*Continued*)

Compound	Reference

6.1. 3(or 5)-Amino-s-triazoles (*Continued*)

6.1b. 3(*or* 5)-*Amino(substituted)-s-Triazoles* (*Continued*)

Compound	Reference
3-[6-[(2-hydroxyethyl)amino]-1,3(2H)-dioxo-1H- benz[de]isoquinolin-2(3H)-yl]-	**83:** P61703s
4-hydroxy-3-(hydroxyamino)-	**82:** 132249r
3-[hydroxymethyl)amino]-	**72:** P67723t
3-(2-hydroxy-5-methylbenzamido)-	**77:** P115460g
3-[(4-hydroxy-2-methyl-2H-1,2-benzothiazine-3- carbonyl)amino]-S,S-dioxide	**79:** 87336s
3-[(4-hydroxy-2-methyl-2H-1,2-benzothiazine-3- carbonyl)amino]-5-phenyl-, S,S-dioxide	**79:** 87336s
3-[(3-hydroxy-2-naphthalenecarbonyl)amino]-	**72:** 3301w, **78:** P147974q
3-(1-hydroxy-2-naphthalenecarboxamido)-5-methyl-	**77:** P115460g
3-[(hydroxyoxalyl)amino]-	**78:** 72011n
3-(4-hydroxy-4-pentyl-3,5-dioxopyrazolidin-1-yl)- 5-phenyl-	**82:** 111992n
3-(o-hydroxyphenyl)-5-(3,5-dimethylpyrazol-1-yl)-	**71:** 91375f
3-(o-hydroxyphenyl)-5-(3,5-diphenyl-2-pyrazolin-1-yl)-	**71:** 91375f
3-(o-hydroxyphenyl)-5-(3,5-diphenylpyrazol-1-yl)-	**71:** 91375f
3-(o-hydroxyphenyl)-5-[3-ethoxycarbonyl)-5-oxo- 2-pyrazolin-1-yl]-	**71:** 91375f
5-(o-hydroxyphenyl)-3-[2-(methoxycarbonyl)hydrazino]-	**73:** 87864p
3-(2-hydroxyphenyl)-5-(N-methylanilino)-	**83:** 97141w
3-(o-hydroxyphenyl)-5-[3-(p-nitrophenyl)-5-oxo- 2-pyrazolin-1-yl]-	**71:** 91375f
3-(o-hydroxyphenyl)-5-[3-(o-octadecylphenyl)-5-oxo- 2-pyrazolin-1-yl]-	**71:** 91375f
3-(o-hydroxyphenyl)-5-(5-oxo-3-undecyl-2-pyrazolin-1-yl)-	**71:** 91375f
3-(o-hydroxyphenyl)-5-(3-phenyl-5-oxo-2-pyrazolin-1-yl)-	**71:** 91375f
3-[(2-hydroxypropionyl)amino]-	**81:** 25611k
3-[(2-hydroxy)thiobenzamido]-5-methyl-	**77:** P115460g
3-[(2-hydroxy)thiobenzamido]-5-phenyl-	**77:** P115460g
3-(3-hydroxy-2,4,6-trinitroanilino)-	**82:** 142281j
3-(4-imidazolecarboxamido)-	**72:** P67724u
3-o-iodobenzamido-	**45:** 629h
3-(2-isobutylhydrazino)-5-isopropyl-	**63:** P13275f
3-isobutyl-5-(2-phenylhydrazino)-	**59:** 12789c
3-isobutyl-5-m-toluidino-	**55:** 24728d
3-isobutyl-5-p-toluidino-	**55:** 24728d
(1,3-isoindoline)bis[3-imino-	**69:** P60044b
3-[(isonicotinoyl)amino]-	**70:** 11640e
3-isopropylamino-5-propyl-	**53:** 9197f
3-[(4-isopropylbenzylidene)hydrazino]-5-phenyl-	**77:** 87362f, **82:** 57055u
3-[(isopropylidene)hydrazino]-	**76:** 113183z
3-[[(isopropylthio)carbonyl]amino]-5-phenyl-	**76:** 81641d
3-isopropyl-5-ureido-	**70:** 11640e

TABLE 6   (*Continued*)

Compound	Reference

6.1. 3(or 5)-Amino-*s*-triazoles   (*Continued*)

6.1b. 3(*or* 5)-*Amino*(*substituted*)-*s*-*Triazoles*   (*Continued*)

Compound	Reference
3-[2-[*N*-(isovalerylamino)amindino]hydrazino]-5-phenyl-	**68:** 78203s
3-(2-mercaptoacetamido)-	**72:** 67723t, **72:** P79854z, **77:** P6570x
3-(2-mercaptoacetamido)-, S-ester with *O*,*O*-di-Me phosphorodithioate	**55:** 27048h
3-[(mercaptocarbonothioyl)amino]-	**78:** P148008q, **82:** P43429g
3-[(3-mercaptopropionyl)amino]-	**72:** P67723t
3-[(2-mercapto)thioacetamido]-	**72:** P67723t
3-(2-methoxyacetamido)-	**54:** P4623b
3-(2-methoxybenzamido)-	**81:** 121645v
3-[(*o*-methoxybenzyl)amino]-	**53:** 6216a
3-[(*p*-methoxybenzyl)amino]-,	**53:** 6216a
3-[(*p*-methoxybenzylidene)amino]-	**64:** P8196f
3-[(4-methoxybenzylidene)hydrazino]-5-phenyl-	**77:** 87362f, **82:** 57055u
3-(*p*-methoxybenzenesulfonamido)-	**70:** 47368m
3-(*p*-methoxybenzenesulfonamido)-5-methyl-	**70:** 47368m
3-(*p*-methoxybenzenesulfonamido)-5-phenyl-	**70:** 47368m
3-(4-methoxybenzylideneamino)-	**78:** 72011n
3-[2-[(4-methoxybenzylidene)carbazoyl]hydrazino]-5-phenyl-	**76:** 99568v
3-[2-[[2-(methoxycarbonyl)ethyl]thio]benzamido]-	**77:** P115460g
3-(4-methoxynaphthalimino)-	**73:** P67700d
3-[(2-methoxyphenyl)amino]-1-[bis(dimethylamino)-phosphinylidyne]-	**77:** P126641t
5-[(2-methoxyphenyl)amino]-1-[bis(dimethylamino)-phosphinylidyne]-	**77:** P126641t
3-(*o*-methoxyphenyl)-5-[2-(3-phenyl-*s*-triazol-5-yl)-hydrazino]-	**70:** 115073p
3-(*p*-methoxyphenyl)-5-piperidino-	**44:** 6856a
5-(4-methoxyphenyl)-3-[(4-pyridinecarbonyl)amino]-	**81:** P63636e
3-methylamino-	**53:** 6216a, **59:** 3918h, **59:** 10051a, **70:** P28922w, **71:** P81375s, **71:** P124443g, **79:** 105154n, **85:** 5552w
3-(methylamino)-1-[bis(dimethylamino)phosphinylidyne]-	**77:** P126641t
5-(methylamino)-1-[bis(dimethylamino)phosphinylidyne]-	**77:** P126641t
3-methylamino-5-(5-nitro-2-furanyl)-	**79:** 87327q
3-methylamino-5-phenyl-	**81:** P63636e
3-(2-methylbenzamido)-	**76:** P24217w
3-(2-methylbenzamido)-5-(4-pyridinyl)-	**81:** P63636e
3-[3-[2-[(4-methylbenzoyl)oxy]ethylthio]propinylamino]-	**81:** P14321q
3-[(4-methylbenzylidene)hydrazino]-5-phenyl-	**77:** 87362f, **82:** 57055u
3-[(α-methyl-1,3-dioxo-2-isoindolinyl)acetamido]-	**70:** 11640e
3-(4-methyl-3,5-dioxo-4-pentyl-1-pyrazolidinyl)-5-phenyl-	**82:** 4183c
3-(1-methyl-2,4-dioxo-3-phenylimidazolidin-5-yl)amino-	**77:** P152186t
3-(3-methyl-2,4-dioxo-1-phenylimidazolidin-5-yl)amino-	**77:** P152186t

TABLE 6   (*Continued*)

Compound	Reference

6.1. 3(or 5)-Amino-*s*-triazoles   (*Continued*)

6.1b. 3(*or* 5)-*Amino*(*substituted*)-*s*-*Triazoles*   (*Continued*)

3-(1-methylhydrazino)-5-phenyl-	**45:** 1122f
3-(1-methylhydrazino)-5-phenyl-, anisaldehyde hydrazone	**45:** 1122f
3-(1-methylhydrazino)-5-phenyl-, benzaldehyde hydrazone	**45:** 1122f
3-methyl-5-methylamino-	**84:** 17241m
3-[1methyl-3-(3-methylphenyl)-2,4-dioxoimidazolidin-5-yl]amino-	**77:** P152186t
5-methyl-3-[[(methylthio)carbonimidoyl]amino]-	**78:** 147921v
3-[1-methyl-(4-nitrobenzylidene)hydrazino]-5-phenyl-	**76:** 14437d
5-methyl-3-[(1-nitroethylidene)hydrazino]-	**77:** 126523f
3-methyl-5-[(5-nitrofurfurylidene)amino]-	**58:** 9071g
3-methyl-5-(*p*-nitrophenylsulfonamido)-	**38:** 2938[3]
5-methyl-3-[(1-nitropropylidene)hydrazino]-	**77:** 126523f
3-(5-methyl-5-octadecyl-1,3,2-dioxaphosphorinan-2-yl)-amino-, *P*-oxide	**77:** P166993r
3-methyl-5-[1-[*p*-(pentyloxy)phenyl]thioureido]-	**70:** 103962q
3-[(α-methylphenethyl)amino]-	**53:** 6216a
3-methyl-5-(*N*-phenylbenzamido)-	**62:** 11803g
3-(3-methyl-1-phenyl-5-formazanyl)-	**77:** 152042t
5-(4-methylphenyl)-3-[(3-pyridinecarbonyl)amino]-	**81:** P63636e
5-(4-methylphenyl)-3-[(4-pyridinecarbonyl)amino]-	**81:** P63636e
3-(2-methyl-5-phenylpyrrol-1-yl)-	**54:** 19641i
5-methyl-3-[(4-phenyl-2-thiazolyl)amino]-	**78:** 147921v
3-[[(4-methylphenyl)thiocarbamoyl]amino]-	**82:** 156189x
3-methyl-5-[1-(phenyl)thioureido]-	**70:** 77881r
3-methyl-5-[1-(*p*-propoxyphenyl)thioureido]-	**70:** 77881r
3-methyl-5-propionylamino-	**38:** 970[5]
3-methyl-5-propylamino-	**53:** 9197f
3-[methyl(4-pyridinecarbonyl)amino]-5-phenyl-	**81:** P63636e
3-methyl-5-sulfanilamido-	**36:** 428[1], **38:** 2938[3], **38:** 6322[3]
3-[[(methylthio)carbonimidoyl]amino]-5-phenyl-	**78:** 147921v
3-[[(methylthio)carbonothioyl]amino]-	**78:** P148008q, **82:** P43429g
3-[[[(methylthio)(2-propenylthio)]methylene]amino]-	**78:** P148008q, **82:** P43429g
3-[[[(methylthio)(2-propynylthio)]methylene]amino]-	**78:** P148008q, **82:** P43429g
3-[1-(methyl)thioureido]-	**70:** 77881r, **76:** 85792b, **82:** 156189x
3-[1-(methyl)thioureido]-5-methyl-	**70:** 77881r
3-methyl-5-*p*-toluidino-	**55:** 24728d
3-(methylureylene)-	**81:** 105467h
3-(4-morpholino)-	**68:** 95761q, **70:** P28922w, **71:** P81375s **71:** P124443g
3-(4-*m*-morpholinecarboxamidino)-5-phenyl	**60:** 1753f
3-(4-*m*-morpholinecarboxamidino)-5-(4-pyridinyl)	**60:** 1753f
3-[α-(4-morpholinyl)-4-nitrobenzylidenehydrazino]-5-phenyl	**82:** 57055u
5-(4-morpholinyl)-1-(*p*-nitrophenyl)-	**40:** 3444[3]

179

TABLE 6 (*Continued*)

Compound	Reference

6.1. 3(or 5)-Amino-*s*-triazoles (*Continued*)

6.1b. 3(*or* 5)-*Amino(substituted)-s-Triazoles* (*Continued*)

Compound	Reference
3-[[(1-naphthabenyl)carbamoyl]amino]-	**82:** 156189x
3-(naphthalimido)-	**69:** 2896v
3-(1-naphthylamino)-5-phenyl-	**64:** 9727e
3-(1-naphthyl)-5-(2-naphthylamino)-	**64:** 9727e
3-[(nicotinoyl)amino]-	**70:** 11640e
3-*o*-nitrobenzamido-	**45:** 629h
3-(*p*-nitrobenzamido)-	**59:** 12790f **77:** 341b
5-(*p*-nitrobenzamido)-3-phenyl-	**58:** 10202f
3-[(*p*-nitrobenzylidene)amino]-	**64:** P8196e
3-[(*p*-nitrobenzylidene)amino]-5-phenyl-	**70:** 115073p
3-[2-[(2-nitrobenzylidene)carbazoyl]hydrazino]-5-phenyl-	**76:** 99568v
3-[2-[(3-nitrobenzylidene)carbazoyl]hydrazino]-5-phenyl-	**76:** 99568v
3-[2-[(4-nitrobenzylidene)carbazoyl]hydrazino]-5-phenyl-	**76:** 99568v
3-[(3-nitrobenzylidene)hydrazino]-5-phenyl-	**77:** 87362f, **82:** 57055u
3-[(4-nitrobenzylidene)hydrazino]-5-phenyl-	**68:** 78203s, **77:** 87362f, **82:** 57055u
3-[[(4-nitrobenzylthio)carbonimidoyl]amino-5-phenyl-	**78:** 147921v
3-[(5-nitro-3,4-diphenyl-2-furanyl)methylene]amino-	**84:** 17043y
3-[(1-nitroethylidene)hydrazino]-	**77:** 126523f
5-(5-nitro-2-furanyl)-3-(2-propenyl)amino-	**79:** 87327q
5-(5-nitro-2-furanyl)-3-propylamino-	**79:** 87327q
3-[(5-nitrofurfurylidene)amino]-	**58:** 9071g, **77:** 146070j
3-(5-nitro-2-furyl)-5-(propylamino)-	**65:** PC20138h
3-[(nitromethylene)hydrazino]-	**77:** 126523f
3-[(1-nitro-2-oximidoethylidene)hydrazino]-	**71:** 112866j
5-(*p*-nitrophenylsulfonamido)-	**38:** 2938^5
3-[(1-nitropropylidene)hydrazino]-	**77:** 126523f
3-(4-nitrosalicylamido)-	**47:** 3820a
3-(5-octadecyl-1,3,2-dioxaphosphorinan-2-yl)amino-, P-sulfide	**77:** P166993r
3-(7-oxabicyclo[2.2.1]heptane-2,3-dicarboximideo)-	**69:** 2896v
3-[(5-oxo-2-pyrrolidinecarbonyl)amino]-	**72:** P67724u
3-[(2,2,3,3,3-pentachloropropionyl)amino]-	**72:** P43185t, **73:** P14233x
3-(2,2,3,4,4-pentamethyl-1-phosphetanyl)amino-, P-oxide	**80:** 144999g
5-pentyl-3-[1-(5-pentyl-*s*-triazol-3-yl)thioureido]-	**70:** 103962q
3-pentyl-5-ureido-	**70:** 11640e
3-phenethylamino-	**53:** 6216a
3-[2-(2-phenylacetamido)acetamido]-	**70:** 11640e
3-(*N*-phenylbenzamido)-5-propyl-	**62:** 11803g
3-phenyl-5-(*N*-phenylbenzamido)-	**59:** 12791f
3-[(phenylcarbamoyl)amino]-	**82:** 156189x
5-phenyl-3-[2-[*N*-(phenylacetylamino)amidino]hydrazino]-	**68:** 78203s
5-phenyl-3-[2-(phenylacetyl)hydrazino]-	**82:** 4183c
3-phenyl-5-(2-phenylhydrazino)-	**59:** 12789d, **82:** 4183c
3-phenyl-5-[(4-phenyl-2-thiazolyl)amino]-	**64:** 9727e

TABLE 6   (Continued)

. Compound	Reference

6.1. 3(or 5)-Amino-s-triazoles   (Continued)

6.1b. 3(or 5)-Amino(substituted)-s-Triazoles   (Continued)

3-phenyl-5-[1-(phenyl)thioureido]-	**70:** 103962q
3-phenyl-5-[2-(3-phenyl-s-triazol-5-yl)hydrazino]-	**70:** 115073p
3-phenyl-5-piperidino-	**44:** 6855f
3-(3-phenylpyrazol-1-yl)-	**72:** P3491h
3-(4-phenylpyrazol-1-yl)-	**72:** P3491h
5-phenyl-3-[(2-pyridinecarbonyl)amino]-	**81:** P63636e
5-phenyl-3-[(3-pyridinecarbonyl)amino]-	**81:** P63636e
5-phenyl-3-[(4-pyridinecarbonyl)amino]-	**81:** P63636e
5-phenyl-3-[2-[N-(2-salicyloylamino)amidino]- hydrazino]-	**68:** 78203s
3-phenyl-5-semicarbazido	**66:** 104959y
3-phenyl-5-sulfanilamido-	**70:** 47368m
3-[2-(3-phenyl-1,2,4-thiadiazol-5-yl)hydrazino]- 5-phenyl-	**70:** 115073p
3-[(phenylthiocarbamoyl)amino]-	**74:** 141216q
3-[[(phenylthio)carbonyl]amino]-	**76:** 81641d
3-[1-(phenyl)thioureido]-	**70:** 77881r, **82:** 156189x
3-phenyl-5-(p-toluenesulfonamido)-	**70:** 47368m
3-phenyl-5-m-toluidino-	**55:** 24728d
3-phenyl-5-p-toluidino-	**55:** 24728d
3-phenyl-5-ureido-	**64:** 5437d
5-phenyl-3-ureido-, O-methyl-	**73:** 77206m
5-phenyl-3-[2-[N-(valerylamino)amidino]hydrazino]-	**68:** 78203s
3-phenyl-5-(veratrylideneamino)-	**58:** 519d
3-[3-(phthalidyl)amino]-	**75:** 76355p
3-(picolinamido-	**70:** 11640e
3-(1-piperidino)-	**68:** 95761q, **70:** P28922w, **71:** P81375s **71:** P124443g
3-(piperonylamido)-	**70:** 11640e
3-(piperonylideneamino)-	**64:** P8196h
-3-propanol, 5-[(2-hydroxyethyl)amino]-	**71:** 101781b
3-[[(2-propenyl)thiocarbamoyl]amino]-	**82:** 15689x
3-[[(2-propenylthio)carbonothioyl]amino]-	**78:** P148008q, **82:** P43429g
3-propionylamino-	**38:** 970⁴
3-(propylamino)-	**60:** 5484g
3-propyl-5-propylamino-	**53:** 9197f, **70:** P28922w, **71:** P81375s, **71:** P124443g,
3-[[(propylthio)carbonyl]amino]-5-phenyl-	**76:** 81641d
3-propyl-5-m-toluidino-	**55:** 24728d
3-propyl-5-p-toluidino-	**55:** 24728d
5-propyl-3-[3-(trifluoromethyl)benzenesulfonamido]- 5-propyl-	**75:** 118258s
3-pyrazol-1-yl-	**72:** P3491h
3-[(3-pyridinecarbonyl)amino]-5-(3-pyridinyl)-	**81:** P63636e
3-[(3-pyridinecarbonyl)amino]-5-(4-pyridinyl)-	**81:** P63636e

TABLE 6 (Continued)

Compound	Reference

6.1. 3(or 5)-Amino-s-triazoles   (Continued)

6.1b. 3(or 5)-Amino(substituted)-s-Triazoles   (Continued)

Compound	Reference
3-[(4-pyridinecarbonyl)amino]-5-(4-pyridinyl)-	**81:** P63636e
5-(2-pyridinylamino)-3-(2-hydroxyphenyl)-	**83:** 97141w
3-[[[5-(3-pyridinyl)-2-furanyl]methylene]amino]-	**78:** P4136g, **83:** 206077c
	85c 46328w
3-[(2-pyridinylthiocarbonyl)amino]-	**83:** 147414k
3-[[(2-pyridyl)methylene]amino]-	**71:** P38973y
3-(2-pyrolidinecarboxamido)-	**72:** P67724u
3-[(1H-pyrrole-3-carbonyl)amino]-5-methyl-	**77:** 88381y
3-[(rifamycin-3-imino)methyl]-	**77:** 147920y
3-p-sulfamoylanilino-	**41:** P2607a
3-sulfanilamido-	**37:** 1402[4]
3-(2-thiazolecarboxamido)	**72:** P67724u
3-[4-thiocarbamoyl)anilino]-	**81:** P27330s
3-(p-toluenesulfonamido)-	**70:** 47368m
3-(p-toluenesulfonamido)-5-methyl-	**70:** 47368m
3-p-toluidino-	**68:** 95761q, **70:** P28922w,
	71: P81375s,**71:** P124443g
3-(1,2,4-triazol-4-yl)-	**60:** 15861f
3-[3-(s-triazol-3-ylamino)but-2-enoyl]amino]-	**71:** P66072x, **75:** P157035m
3-(2,2,2-trichloroacetamido)-	**54:** P4623b
3-[(2,2,2-trichloro-1-hydroxy)ethylamino]-	**68:** 104067x
3-[[(2,2,2-trichloro-1-hydroxy)ethyl]amino]-, 3-[14]C	**74:** 142323j
3-[(2,2,2-trichloro-1-hydroxy)ethylamino]-5-methyl	**68:** 104067x
3-[2-(2,4,5-trichlorophenoxy)acetamido]-	**56:** P12907h
3-[(2,2,3-trichloropropionyl)amino]-	**77:** 30239s
3-(2,2,2-trifluoroacetamido)-	**74:** 22770f
3-[3-(trifluoromethyl)benzenesulfonamido]-	**75:** 118258s
3-[3-(trifluoromethyl)benzenesulfonamido]-5-methyl-	**75:** 118258s
3-[(p-trifluoromethylphenyl)ureylene]-	**71:** P91056c
3-(trifluoromethylsulfonamido)-	**71:** P49571s, **72:** P121183g
3-(2,4,6-trinitroanilino)-	**68:** 105111u, **72:** P55458x,
	84: 92352b
3-(2,4,6-triphenyl-1-pyridinio)-, perchlorate	**82:** 97303e
3-ureido-	**40:** P4227[8]
3-veratrylamino-	**53:** 6216a

6.2. 3(or 5)-Amino-1H-1,2,4-Triazoles

Compound	Reference
-1-acetamide, 5-(acetylamino)-α-(2,2-dimethyl-1-oxopropyl)-N-[2-(hexadecyloxy)phenyl]-	**85:** P151754e
-1-acetamide, 5-(acetylamino)-α-(2,2-dimethyl-1-oxopropyl)-N-[2-(hexadecyloxy)-5-[(methylamino)-sulfonyl]phenyl]-	**85:** P151754e
-1-acetamide, 3-(acetylamino)-N-(2-methoxyphenyl)-α-[2-(tetradecyloxy)benzoyl]-	**85:** P151754e

TABLE 6 (*Continued*)

Compound	Reference

6.2. 3(or 5)-Amino-1H̲-1,2,4-Triazoles (*Continued*)

-1-acetamide, 3-amino-N-[5-[[4-[2,4-bis(1,1-dimethyl-propyl)phenoxy]-1-oxobutyl]amino]-2-chlorophenyl]-α-(2,2-dimethyl-1-oxopropyl)-　　**85:** P102316e

-1-acetamide, 5-(benzoylamino)-α-(2,2-dimethyl-1-oxopropyl)-N-[2-(hexadecyloxy)-5-[(methylamino)-sulfonyl]phenyl]-　　**85:** P151754e

-1-acetamide, N-[5-[[4-[2,4-bis(1,1-dimethylpropyl)-phenoxy]-1-oxobutyl]amino]-2-chlorophenyl]-α-(2,2-dimethyl-1-oxopropyl)-3-(4-morpholinyl)-　　**85:** P102316e

-1-acetamide, N-[5-[[4-[2,4-bis(1,1-dimethylpropyl)-phenoxy]-1-oxobutyl]amino]-2-chlorophenyl]-α-(2,2-dimethyl-1-oxopropyl)-3-(phenylamino)-　　**85** P102316e

-1-acetamide, N-[2-chloro-5-[[(3-pentadecylphenoxy)-acetyl]amino]phenyl]-α-(2,2-dimethyl-1-oxopropyl)-3-(phenylamino)-　　**85:** P102316e

-1-acetamide, α-(2,2-dimethyl-1-oxopropyl)-N-[2-(hexadecyloxy)-5-[(methylamino)sulfonyl]phenyl]-5-[[[3-(trifluoromethyl)phenyl]sulfonyl]amino]-　　**85:** *P*151754e

5-acetamido-1,3-dimethyl-　　**60:** 9246g

5-acetamido-1-methyl-　　**60:** 9264g

5-acetamido-1-methyl-3-(5-nitro-2-furyl)-　　**65:** P20138h, **79:** 87327q

3-acetamido-1-methyl-5-(5-nitro-2-furyl)-　　**79:** 87327q

3-acetamido-1-(*p*-nitrophenyl)-　　**74:** 125578d

3-acetamido-1-phenyl-　　**74:** 125578d

3-acetamido-1-*p*-tolyl-　　**74:** 125578d

3-acetonyl-5-amino-1-methyl-　　**60:** 4141f

5-(N-acetylanilino)-1,3-diphenyl-　　**76:** 113185b

-1-acrylic acid, 3-amino-　　**73:** 130967p

-1-acrylic acid, 5-amino-, ethyl ester　　**66:** 85757n

-1-acrylic acid, 3-amino-, methyl ester　　**73:** 130967p

3-amino-1-(2-amino-2-carboxyethyl)-　　**59:** 6729f, **62:** 10857b, **64:** 4190a, **80:** 67324d

5-amino-1-(2-aminophenyl)-　　**81:** 25598m, **83:** 178946q

3-amino-1-benzyl-　　**55:** 10453d, **73:** P36546d, **81:** 49629y

5-amino-1-benzyl-　　**55:** 10453a

3-amino-1-benzyl-5-methyl-　　**73:** P36546d, **73:** P46632t, **76:** P25296h

3-amino-1-benzyl-5-phenyl-　　**55:** 8393g, **73:** P36546d, **73:** P46632t, **76:** P25296h

5-amino-1,3-bis(4-chlorophenyl)-　　**80:** 59602c

5-amino-1-[bis(dimethylamino)phosphinyl]-　　**73:** P120632m

5-amino-1-[bis(dimethylamino)phosphinyl]-3-phenyl-　　**61:** 16717c

5-amino-3-(*p*-bromophenyl)-1-methyl-　　**58:** 10202e

3-amino-5-[4-chloro-3-(N-methylsulfamoyl)phenyl]-1-methyl-　　**76:** P99678f, **77:** P164710x

3-amino-1-(4-chlorophenyl)-　　**80:** 133347s

TABLE 6   (*Continued*)

Compound	Reference

6.2. 3(or 5)-Amino-1H-1,2,4-Triazoles   (*Continued*)

5-amino-1-(4-chlorophenyl)-3-phenyl-	**79:** 18648h
5-amino-1-(2,4-dibromophenyl)-3-phenyl-	**84:** 59334z
5-amino-1-(2,2-dicarboxyvinyl)-	**43:** P52i, **44:** P66b
5-amino-1-(2,2-dicarboxyvinyl)-3-methyl-	**43:** P53a, **44:** P66b
5-amino-1-(2,2-dicarboxyvinyl)-3-propyl-	**44:** P66b
3-amino-5-(2,6-diamino-5-chloro-3-pyrazinyl)-1-methyl-	**72:** P90516m
3-amino-5-(2,6-diamino-5-chloro-3-pyrazinyl)-1-phenethyl-	**70:** 96755u, **72:** P90516m
3-amino-1,5-dibenzyl-	**76:** P25296h
3-amino-1-(1,2-dibromoethyl)-	**79:** 66253x, **80:** 82823j
3-amino-1-(1,2-dicarboxyvinylene)-	**75:** 140782e
3-amino-1-(2,2-dimethoxyethyl)-	**85:** 45813p
5-amino-1-(2,2-dimethoxyethyl)-	**85:** 45813p
3-amino-1,5-dimethyl-	**76:** 14444d, **81:** 25613n, **84:** 17241m
5-amino-1,3-dimethyl-	**59:** 10052a, **60:** 9264g, **84:** 17241m
5-amino-1-(4,6-dimethyl-2-pyrimidinyl)-3-methyl-	**78:** 147898t
3-amino-1-(2,4-dinitrophenyl)-	**83:** 178946q
5-amino-1-(2,4-dinitrophenyl)-	**83:** 178946q
5-amino-1,3-diphenyl-	**79:** 18648h
3-amino-1-ethenyl-	**79:** 66253x, **80:** 82823j
3-amino-1-ethyl-	**79:** 66253x, **80:** 82823j, **85:** 186070f
3(or 5)-amino-1-ethyl-5(or 3)-phenyl-	**55:** 8393g
5-amino-3-ethyl-1-methyl-	**59:** 10052a, **60:** 9264g
3-amino-5-(2-furyl)-1-methyl-	**63:** 18071g
3-amino-1-glucosyl-	**56:** 15949e, **57:** 3703d
3-amino-1-methyl-	**79:** 105154n
5-amino-1-methyl-	**58:** 10202e, **59:** 10052a, **59:** 12790f, **65:** 12205b, **60:** 9264g, **79:** 105154n
5-amino-3-methyl-1-[2,3-*O*-(1-methylethylidene)-β-D-ribofuranosyl]-	**82:** P171367x
3-amino-1-methyl-5-(5-nitro-2-furyl)-	**79:** 87327q, **81:** 163330u
5-amino-1-methyl-3-(5-nitro-2-furyl)-	**63:** P612c, **65:** P20139a, **79:** 87327q
3-amino-5-methyl-1-(4-nitrophenyl)-	**76:** 14444d
5-amino-1-methyl-3-(*p*-nitrophenyl)-	**58:** 10202e
5-amino-1-methyl-3-phenyl-	**58:** 10102e, **59:** 12791c, **66:** 104601a
3-amino-1-(2-nitrophenyl)-	**83:** 178946q
3-amino-1-(4-nitrophenyl)-	**80:** 133347s
5-amino-1-(2-nitrophenyl)-	**81:** 25598m, **83:** 178946q
5-amino-1-phenyl-	**60:** 520h
3-amino-1-phenyl-5-(2-hydroxyphenyl)-	**83:** 97141w
3-amino-1-piperonyl-	**74:** 99946a

184

TABLE 6 (*Continued*)

Compound	Reference

6.2. 3(or 5)-Amino-1H̲-1,2,4-Triazoles (*Continued*)

Compound	Reference
3-amino-1-β-D-ribofuranosyl-	**73:** 66847v, **85:** 193010q
5-amino-1-β-D-ribofuranosyl-	**85:** 193010q
3-amino-1-β-D-ribofuranosyl-, 2′,3′,5′-tribenzoate	**73:** 66847v
3-amino-1-(4,4,4-trichloro-3-hydroxy-1-oxobutyl)-	**76:** 140733k
3-amino-1-triethylstannyl-	**60:** 6860f
3-amino-1-(trimethylsilyl)-	**85:** 193010q
3-anilino-1-[bis(dimethylamido)phosphoryl]-	**65:** 7027b
3-anilino-1-[bis(dimethylamido)phosphoryl]-5-ethyl-	**65:** 7027b
3-anilino-1-[bis(dimethylamido)phosphoryl]-5-isopropyl-	**65:** 7027b
3-anilino-1-[bis(dimethylamido)phosphoryl]-5-methyl-	**65:** 7027b
3-anilino-1-[bis(dimethylamido)phosphoryl]-5-pentyl-	**65:** 7027b
3-anilino-1-[bis(dimethylamido)phosphoryl]-5-propyl-	**65:** 7027b
5-anilino-3-butyl-1-[bis(dimethylamido)phosphoryl]-	**65:** 7027b
5-anilino-1,3-diphenyl-	**76:** 113185b
3-anilino-1-phenyl-	**45:** 6211i, **55:** 1646a, **63:** 2975e
5-benzamido-1-(4-chlorophenyl)-3-phenyl-	**79:** 18648h
5-benzamido-1,3-diphenyl-	**79:** 18648h
5-benzamido-1-methyl-	**59:** 12790g
5-benzamido-1-methyl-3-phenyl-	**49:** 15917a, **58:** 10202f, **59:** 12791h
3-benzamido-1-phenyl-	**74:** 125578d
3-benzamido-1-*p*-tolyl-	**74:** 125578d
3-(2-benzoylhydrazino)-1-methyl-5-phenyl	**82:** 125326k
5-(2-benzoylhydrazino)-1-methyl-3-phenyl-	**82:** 125326k
5-benzylamino-1-[bis(dimethylamino)phosphinyl]-	**73:** P120632m, **77:** P126641t
5,5′-[2,2′-(benzylmethylmalonyl)bishydrazino]bis-[1-methyl-3-phenyl-	**82:** 125326k
5-[2-(benzyloxycarbonylacetyl)hydrazino]-	**82:** 125326k
5-[(*N*-benzylthiocarbamoyl)amino]-1-(diethoxymethyl)-	**76:** 85792b
5-[2-(butoxycarbonylacetyl)hydrazino]-1-methyl-3-phenyl-	**82:** 125326k
5,5′-[2,2′-(butylmethylmalonyl)bishydrazino]bis-[1-methyl-3-phenyl-	**82:** 125326k
5-[2-(2-carboxy-2-butylhexanoyl)hydrazino]-1-methyl-3-phenyl-	**82:** 125326k
5-[2-(2-carboxy-2-methylhexanoyl)hydrazino]-1-methyl-3-phenyl-	**82:** 125326k
5-[(2-carboxy-2-methyl-3-phenylpropionyl)hydrazino]-1-methyl-3-phenyl-	**82:** 125326k
5-[2-(2-carboxy-2-methylpropiohyl)hydrazino]-1-methyl-3-phenyl-	**82:** 125326k
5-[2-(2-carboxy-3-phenylpropionyl)hydrazino]-1-methyl-3-phenyl-	**82:** 125326k
3-*p*-chloroanilino-1-(*p*-chlorophenyl)-	**45:** 6211i
3-(*p*-chlorophenyl)-5-[2-(ethoxycarbonyl)hydrazino]-1-methyl-	**73:** 14772x
1-(4-chlorophenyl)-5-(*N*-ethyl-*N*¹-guanidino)-3-phenyl-	**79:** 18648h

TABLE 6 (*Continued*)

Compound	Reference

6.2. 3(or 5)-Amino-1H-1,2,4-Triazoles (*Continued*)

Compound	Reference
1-(4-chlorophenyl)-5-guanidino-3-phenyl-	**79:** 18648h
3-(4-chlorophenyl)-5-hydrazino-1-methyl-	**76:** 14437d
1-(4-chlorophenyl)-5-(N-methyl-N^1-guanidino)-3-phenyl-	**79:** 18648h
3-(4-chlorophenyl)-1-methyl-5-(1-semicarbazido)-	**76:** 14437d
5-(3-chloropropionamido)-1-methyl-3-(5-nitro-2-furyl)-	**70:** P96798k
5-[2-(cyclohexoxycarbonylacetyl)hydrazino]-1-methyl- 3-phenyl-	**82:** 125326k
3-(diacetylamino)-1-methyl-5-(5-nitro-2-furanyl)-	**79:** 87327q
5-(diacetylamino)-1-methyl-3-(5-nitro-2-furanyl)-	**79:** 87327q
5-(1,2-dibenzoylhydrazino)-1-methyl-3-phenyl-	**82:** 125326k
1-(2,4-dibromophenyl)-5-hydrazino-3-phenyl-	**84:** 59334z
5-(4,4-diethyl-3,5-dioxopyrazolidin-1-yl)-1-methyl- 3-phenyl-	**82:** 125326k
5,5'-[2.2'-(diethylmalonyl)bishydrazino]bis[1-methyl- 3-phenyl-	**82:** 125326k
1,3-dimethyl-5-amino	**80:** 70754z, **81:** 25613n, **81:** 49629y
3-(dimethylamino)-1-methyl-	**78:** 4224j
5-(dimethylamino)-1-methyl-	**72:** 111380e
1,5-dimethyl-3-dimethylamino-	**84:** 17241m
1,3-dimethyl-5-dimethylamino-	**84:** 17241m
1,3-dimethyl-(5-methylamino)-	**81:** 25613n
1,5-dimethyl-3-methylamino-	**84:** 17241m
1,3-diphenyl-5-hydrazino-	**84:** 59334z
-1-ethanol, 5-amino-3-(5-nitro-2-furyl)-	**63:** P612d
-1-ethanol, 5-amino-α-(p-nitrophenyl)-	**66:** 75958q
1-ethenyl-3-dimethylamino-	**79:** 66253x
5-[2-(ethoxycarbonylacetyl)hydrazino]-1-methyl- 3-phenyl-	**82:** 125326k
5-[2-(ethoxycarbonyl)hydrazino]-	**73:** 14772x
5-[2-(ethoxycarbonyl)hydrazino]-1-ethyl-3-phenyl	**73:** 14772x
5-[2-(ethoxycarbonyl)hydrazino]-1-methyl-3- (*m*-nitrophenyl)-	**73:** 14772x
5-[3-(ethoxycarbonyl)thioureido]-1-methyl-	**78:** 136237a
5-[3-ethoxycarbonyl)thioureido]-1-methyl-3-phenyl-	**78:** 136237a
5-(N-ethyl-N^1-guanidino)-1,3-diphenyl-	**79:** 18648h
5-(N-ethyl-N^1-guanidino)-1-(4-nitrophenyl)-3-phenyl-	**79:** 18648h
3-[ethyl(methyl)amino]-1-methyl-	**78:** 4224j
1-ethyl-3-phenyl-5-(1-semicarbazido)-	**76:** 14437d
3-[ethyl(methyl)amino]-1-methyl-	**78:** 4224j
3-(2-furyl)-1-methyl-5-(methylamino)-	**63:** 18071e
5-guanidino-1,3-diphenyl-	**79:** 18648h
5-guanidino-1-(4-nitrophenyl)-3-phenyl-	**79:** 18648h
5-hydrazino-1-methyl-3-phenyl-	**76:** 14437d
5-hydrazino-1-phenyl-	**70:** 37729r
5-[(hydroxymethyl)amino]-1-methyl-3-(5-nitro-2-furyl)	**71:** P12444h, **79:** 87327q
1-(hydroxymethyl)-3-[di(hydroxymethyl)amino]-	**72:** P67723t

186

TABLE 6 (*Continued*)

Compound	Reference

6.2. 3(or 5)-Amino-1H-1,2,4-Triazoles   (*Continued*)

Compound	Reference
1-(hydroxymethyl)-3-[1-(hydroxymethyl)imidazole- 4-carboxamido]-	**72:** P67724u
5-(methylamino)-1-methyl-3-(5-nitro-2-furanyl)-	**79:** 87327q
3-(methylamino)-1-methyl-5-(5-nitro-2-furanyl)-	**79:** 87327q
5-methylamino-1-(2-methylpropyl)-	**82:** 43271z
5-(N-methylanilino)-1,3-diphenyl-	**76:** 113185b
1-methyl-3-dimethylamino-	**78:** 4224j
1-methyl-5-(3,5-dioxo-4-phenylpyrazolidin-1-yl)- 3-phenyl-	**82:** 125326k
5-(N-methyl-N^1-guanidino)-1,3-diphenyl-	**79:** 18648h
5-(N-methyl-N^2-guanidino)-1-(4-nitrophenyl)-3-phenyl-	**79:** 18648h
1-methyl-5-(methylamino)-	**72:** 111380e
1-methyl-5-[2-(α-methylphenylacetyl)hydrazino]- 3-phenyl-	**82:** 125326k
3-methyl-5-(2-naphthalenylamino)-1-phenyl-	**81:** 37516w
1-methyl-5-(p-nitrobenzamido)-	**59:** 12790h
1-methyl-5-[2-(pentoxycarbonylacetyl)hydrazino]- 3-phenyl-	**82:** 125326k
1-methyl-3-phenyl-5-[2-(phenylacetyl)hydrazino]-	**82:** 125326k
1-methyl-3-phenyl-5-[2-(propoxycarbonylacetyl)- hydrazino]-	**82:** 125326k
1-methyl-3-phenyl-5-(1-semicarbazido)-	**76:** 14437d
1-methyl-3-phenyl-5-thioureido-	**78:** 136237a
1-methyl-5-thioureido-	**78:** 136237a
5-[(2-methylpropyl)amino]-1-methyl-	**82:** 43271z
5-[(4-nitrobenzylidene)hydrazino]-	**76:** 14437d
1-phenyl-3-(N-phenylacetamido)-	**55:** 1646a, **63:** 2976a
1-picryl-5-(2,4,6-trinitroanilino)-	**68:** 10511u
-1-propanoic acid, 3-(2-hydroxybenzamido)-, methyl ester	**81:** 121645v
3-p-toluidino-1-p-tolyl-	**45:** 6211i

6.3. 3(or 5)-Amino-4H-1,2,4-Triazoles

Compound	Reference
3-[p-(acetamido)benzenesulfonamido]-4-(2-pyridyl)-	**58:** P5703h
3-[p-(acetamido)benzenesulfonamido]-4-(3-pyridyl)-	**58:** P5704a
3-[p-(acetamido)benzenesulfonamido]-4-(4-pyridyl)-	**58:** P5704a
3-[p-(acetamido)benzenesulfonamido]-4-(2-pyrimidinyl)-	**58:** P5704b
3-acetamido-4,5-dimethyl-	**60:** 9264h
3-acetamido-4-ethyl-5-(5-nitro-2-furanyl)-	**79:** 87327q
3-acetamido-4-methyl-5-(5-nitro-2-furyl)-	**65:** P20138h, **74:** P100068z, **79:** 87327q
3-acetonitrile, 5-hydrazino-4-methyl-	**85:** 21243b
4-allyl-3-amino-	**59:** 10048d
3-amino-5-(2-benzothiazolyl)-4-ethyl-	**77:** 126518h
3-amino-5-(2-benzothiazolyl)-4-(2-propenyl)-	**77:** 126518h
3-amino-4-benzyl-	**55:** 10453b, **64:** 4935f
3-amino-4-benzyl-5-phenyl-	**48:** 1276b

TABLE 6   (*Continued*)

Compound	Reference

6.3. 3(or 5)-Amino-4<u>H</u>-1,2,4-Triazoles   (*Continued*)

Compound	Reference
3-amino-4-(*o*-chlorobenzyl)-	**66:** 85757n
3-amino-4-(*p*-chlorobenzyl)-	**66:** 85757n
3-amino-4-(*p*-chlorophenyl)-5-ethyl-	**55:** 23504h
3-amino-4-(*p*-chlorophenyl)-5-isobutyl-	**55:** 23504h
3-amino-4-(*p*-chlorophenyl)-5-methyl-	**55:** 23504h
3-amino-4-(*p*-chlorophenyl)-5-propyl-	**55:** 23504h, **59:** 11473h
3-(2-amino-5-chloro-3-pyrazinyl)-5-(dimethylamino)-4-methyl-	**70:** 96755u, **72:** P90516m
3-(2-amino-5-chloro-3-pyrazinyl)-4-methyl-5-(methylamino)-	**70:** 96755u, **72:** P90516m
3-amino-4-(1,2-dibromoethyl)-	**79:** 66253x
3-amino-4-(2,2-diethoxyethyl)-	**85:** 45813p
3-amino-4-(2,2-dimethoxyethyl)-	**85:** 45813p
3-amino-4-(2,5-dimethoxyphenyl)-	**58:** P5703g
3-amino-4,5-dimethyl-	**60:** 9264g, **81:** 49629y, **84:** 17241m
3-amino-4,5-diphenyl-	**57:** 12473c, **73:** 56050r
3-amino-4-ethenyl-	**79:** 66253x
3-amino-4-ethyl-	**59:** 3918h, **79:** 66253x
3-amino-4-ethyl-5-(5-nitro-2-furanyl)-	**79:** 87327q
3-amino-5-ethyl-4-methyl-	**60:** 9264h
3-amino-4-ethyl-5-phenyl-	**55:** 8393f
3-amino-5-ethyl-4-*p*-tolyl-	**55:** 23504h
3-amino-5-(2-furyl)-4-methyl-	**63:** 18071dh, **74:** P100068z
3-amino-5-isobutyl-4-*p*-tolyl-	**55:** 23504h
3-amino-4-methyl-	**59:** 3918h, **60:** 9264g, **79:** 105154n
3-amino-4-methyl-5-(5-nitro-2-furyl)-	**65:** PC20139a, **74:** P100068z, **79:** 87327q
3-amino-4-methyl-5-phenyl-	**81:** P63636e
3-amino-5-methyl-4-*p*-tolyl-	**55:** 23504g
3-amino-4-phenethyl-	**66:** 85757n, **67:** P100157h
3-amino-4-phenyl-	**63:** 7012h
3-amino-4-phenyl-5-propyl-	**55:** 23504g
3-amino-4-propyl-	**60:** 5484g
3-amino-5-propyl-4-*p*-tolyl-	**55:** 23504h
3-amino-4-β-D-ribofuranosyl-	**85:** 193010q
3-amino-4-(1,2,5-tri-*O*-benzoyl-β-D-ribofuranosyl)-	**85:** 193010q
3-amino-4-(4,4,4-trichloro-3-hydroxy-1-oxobutyl)-	**76:** 140733k
3-amino-4-[2-(2,6-xylyloxy)ethyl]-	**64:** 4935f
3-anilino-4-(4-bromophenyl)-	**80:** 14879g
3-anilino-5-(4-bromophenyl)-4-phenyl-	**75:** 35895w, **77:** 101478g
3-anilino-5-(3-chlorophenyl)-4-phenyl-	**77:** 101475d
3-anilino-5-(*p*-chlorophenyl)-4-phenyl-	**75:** 35895w, **77:** 101478g
3-anilino-5-*p*-cumenyl-4-phenyl-	**75:** 35895w, **77:** 101478g
3-anilino-4,5-diphenyl-	**51:** 5095b, **53:** 21326d, **68:** 59500v

TABLE 6 (*Continued*)

Compound	Reference

6.3. 3(or 5)-Amino-4H̲-1,2,4-Triazoles (*Continued*)

Compound	Reference
3-anilino-5-methyl-4-phenyl-	**57:** 12472i
3-anilino-5-(4-nitrophenyl)-4-phenyl-	**75:** 35895w, **77:** 101478g
3-anilino-4-phenyl	**80:** 14879g
3-anilino-4-phenyl-5-(4-pyridyl)-	**68:** 59500v
3-anilino-4-phenyl-5-*p*-tolyl-	**51:** 5095c
3-(1-anthraquinonylamino)-4-methyl-	**69:** P37104n
3-benzamido-4,5-diphenyl-	**73:** 98876v
3-benzamido-4-ethyl-5-phenyl-	**55:** 8393f
3-(benzylideneamino)-5-(2-furyl)-4-methyl-	**63:** 18071d
3-(benzylidenehydrazino)-4-(*p*-bromophenyl)-5-propyl-	**59:** 11474b
3-(benzylidenehydrazino)-4-(*p*-chlorophenyl)-5-propyl-	**59:** 11473h
3-(benzylidenehydrazino)-4-phenyl-5-propyl-	**59:** 11474b
3-(benzylidenehydrazino)-5-propyl-4-*p*-tolyl-	**59:** 11474b
3-(4-bromoanilino)-4-(4-bromophenyl)-	**80:** 14879g
3-(4-bromoanilino)-4-phenyl	**80:** 14879g
5-(4-bromophenyl)-3-(4-morpholinyl)-4-phenyl-	**75:** 35895w, **77:** 101478g
3-(chloroacetamido)-4-methyl-5-(5-nitro-2-furanyl)-	**79:** 87327q
3-(*m*-chloroanilino)-4-(*m*-chlorophenyl)-5-phenyl-	**68:** 59500v
5-(*p*-chlorophenyl)-3-(4-morpholinyl)-4-phenyl-	**75:** 35895w, **77:** 101478g
5-*p*-cumenyl-3-(4-morpholinyl)-4-phenyl-	**75:** 35895w, **77:** 101478g
4-cyclohexyl-3-(cyclohexylamino)-5-phenyl-	**68:** 59500v
3-(diacetylamino)-4-methyl-5-(5-nitro-2-furanyl)-	**79:** 87327q
3-[(1,2-dihydro-3-oxo-3H̲-pyrazo)-5-yl)amino]- 4-β-D-ribofuranosyl-	**81:** 4188n
3-(dimethylamino)-4,5-dimethyl-	**60:** 9265c
3-(dimethylamino)-5-ethyl-4-methyl-	**60:** 9265c
3-(dimethylamino)-5-(2-furyl)-4-methyl-	**63:** 18071d
3-(dimethylamino)-4-methyl-	**60:** 9265c
3,4-dimethyl-5-(methylamino)-	**60:** 9265a
3-(3,5-dimethylpyrazol-1-yl)-4-phenyl-	**64:** 8174a
3-(1,3-diphenyl-5-formazano)-4-(2-pyridinyl)-	**72:** 131663c
3,5-di-*p*-toluidino-4-*p*-tolyl-	**69:** 10398z
3,4-diphenyl-5-*N*-phenylacetamido-	**51:** 5095b
4,5-di-*p*-tolyl-3-*N*-*p*-tolylacetamido-	**51:** 5095c
3-[(3-ethoxycarbonyl)thioureido]-4-methyl-	**78:** 136237a
4-ethenyl-3-dimethylamino-	**79:** 66253x, **74:** P100068z
4-ethyl-3-(ethylamino)-5-(2-furyl)-	**63:** 18071d
4-ethyl-3-(ethylamino)-5-(5-nitro-2-furyl)-	**74:** P100068z
3-ethyl-5-hydrazino-4-methyl-	**59:** 10051b
3-ethyl-4-methyl-5-(methylamino)-	**60:** 9265a
3-(2-furyl)-4-methyl-5-(methylamino)-	**63:** 18071d
3-hydrazino-4,5-dimethyl-	**59:** 10051b
3-[(*o*-hydroxybenzylidene)amino]-4,5-diphenyl-	**73:** 56050r
3-[(hydroxymethyl)amino]-4-methyl-5- (5-nitro-2-furyl)-	**71:** P124444h, **79:** 87327q
4-isopropyl-3-isopropylamino-	**81:** 91439m
4-methyl-3-(methylamino)-	**60:** 9264h

TABLE 6 (*Continued*)

Compound	Reference

6.3. 3(or 5)-Amino-4$\underline{H}$-1,2,4-Triazoles (*Continued*)

Compound	Reference
4-methyl-5-phenyl-3-[(4-pyridinecarbonyl)amino]-	**81:** P63636e
3-(4-morpholinyl)-5-(4-nitrophenyl)-4-phenyl-	**75:** 35895w, **77:** 101478g
4-phenyl-3-(*N*-phenylacetamido)-5-(4-pyridyl)-	**57:** 8561d
4-(2-pyridyl)-3-sulfanilamido-S^{35}-	**58:** P5703h
4-(3-pyridyl)-3-sulfanilamido-S^{35}-	**58:** P5704a
4-(4-pyridyl)-3-sulfanilamido-S^{35}-	**58:** P5704a
3-(4-pyridyl)-5-*p*-toluidino-4-*m*-tolyl-	**57:** 8561d
3-(4-pyridyl)-5-*p*-toluidino-4-*p*-tolyl-	**57:** 8561d
3-*p*-toluidino-4,5-di-*p*-tolyl-	**51:** 5095c

6.4. 4-Amino-4$\underline{H}$-1,2,4-Triazoles

Compound	Reference
4-(acetamidino)-	**71:** 38867s
4-acetamido-3,5-bis(heptafluoropropyl)-	**64:** 12662f
4-acetamido-3,5-bis(λ-hydroxyheptadecyl)-, diacetate	**28:** 4418^9
4-acetamido-3,5-bis(pentafluoroethyl)-	**64:** 12662f
4-acetamido-3-(carbobenzyloxyamino)methyl-5-methyl	**62:** 1647a
4-acetamido-3,5-diethyl-,	**65:** 12205ab
4-acetamido-3,5-diheptadecyl-	**28:** 4419^1
4-acetamido-3,5-dimethyl-	**56:** 472e
4-acetamido-3,5-diphenyl-	**50:** 6189c
4-[[*p*-(acetamido)phenyl]amino]-	**76:** 99572s
4-[(acetimidoyl)amino]-3,5-dimethyl-	**76:** 113138p
4-(acetoacetamido)-3,5-dimethyl-	**73:** 130946f
3-(1-acetyl-4,5-dihydro-1$\underline{H}$-pyrazol-3-yl)-4- (diacetylamino)-	**78:** 29701a
3-(1-acetyl-4,5-dihydro-1$\underline{H}$-pyrazol-3-yl)-4- (diacetylamino)-5-phenyl-	**78:** 29701a
4-(2-acetyl-3,3-dimethyl-1-aziridinyl)-3,5-diphenyl-	**77:** 101467c
4-[N^4-acetylsulfanilamido]-	**37:** 1402^4

4-amino-	**38:** 5828^3, **45:** P8561a,
	   53: 11399a, **71:** 38872q,
	   76: P3867t

4-amino-3-(2-amino-4-chlorophenyl)-	**77:** 88421m
4-amino-3-(2-amino-5-chlorophenyl)-	**77:** 88421m
4-amino-3-(2-amino-4-chlorophenyl)-5-methyl-	**77:** 88421m
4-amino-3-(2-amino-3,5-dichlorophenyl)-	**77:** 88421m
4-amino-3-(2-amino-5-methylphenyl)-	**77:** 88421m
4-amino-3-(2-aminophenyl)-	**77:** 88421m
4-amino-3-(2-aminophenyl)-5-benzyl-	**77:** 88421m
4-amino-3-(2-aminophenyl)-5-ethyl-	**77:** 88421m
4-amino-3-(2-aminophenyl)-5-isopropyl-	**77:** 88421m
4-amino-3-(2-aminophenyl)-5-methyl-	**77:** 88421m
4-amino-3-(2-aminophenyl)-5-phenyl-	**77:** 88421m
4-amino-3-(2-aminophenyl)-5-propyl-	**77:** 88421m
4-(4-aminobenzenesulfonamido)-	**81:** P27329y
4-amino-3-benzyl-	**77:** 139906h

TABLE 6   (Continued)

Compound	Reference

6.4. 4-Amino-4H-1,2,4-Triazoles   (Continued)

Compound	Reference
4-[(m-aminobenzylidene)amino]-	**76:** 99572s
4-amino-5-benzyl-3-phenyl-	**77:** 152130v
4-amino-3,5-bis(α-aminobenzyl)-	**84:** P90156s
4-amino-3,5-bis[4-([1-(2-aminocarbonyl-5-methylanilino)- 1,3-dioxo-2-butyl]azo)phenyl]-	**82:** P59896t
4-amino-3,5-bis(4-aminocyclohexyl)-	**58:** P8071e, **67:** P65353v
4-amino-3,5-bis(2-aminoethyl)-	**81:** P25678n, **85:** P192740r
4-amino-3,5-bis(aminomethyl)-	**57:** 5911i, **84:** P90156s
4-amino-3,5-bis(5-aminopentyl)-	**58:** P8071e, **72:** P112771v
4-amino-3,5-bis(o-aminophenyl)-	**60:** 14508g
4-amino-3,5-bis(m-aminophenyl)-	**60:** 14508g, **81:** P120640c, **81:** P25678n,
4-amino-3,5-bis(p-aminophenyl)-	**60:** 14508g, **75:** P93013t, **82:** P59896t
4-amino-3,5-bis(3-aminopropyl)-	**57:** 5911i, **58:** P8071d, **59:** P10065h, **69:** P11393f, **72:** P112771v
4-amino-3,5-bis(11-aminoundecyl)-	**69:** P11393f, **72:** P112771v
4-amino-3,5-bis[2-(benzyloxy)ethyl]-	**70:** 57739x
4-amino-3,5-bis[3-([1-(4-chloro-2,5-dimethoxyanilino)- 1,3-dioxo-2-butyl]azo)phenyl]-	**82:** P59896t
4-amino-3,5-bis[4-([(4-chloro-2,4-dimethoxyphenyl- carbamoyl)-2-hydroxy-1-naphthalenyl]azo)phenyl]-	**82:** P59896t
4-amino-3,5-bis(1-chloroethyl)-	**70:** 115076s
4-amino-3,5-bis(chloromethyl)-	**70:** 115076s
4-amino-3,5-bis(o-chlorophenyl)-	**60:** 14508g
4-amino-3,5-bis(p-chlorophenyl)-	**60:** 14508g, **81:** 120640c, **81:** P25678n
4-amino-3,5-bis(μ-Δ²-cyclopentenyldodecyl)-	**23:** 114[8]
4-amino-3,5-bis[2-(diethylmethylsilyl)ethyl]-	**84:** 121930w
4-amino-3,5-bis(3,4-dihydro-2H-pyran-6-yl)-	**77:** 34469q
4-amino-3,5-bis(diphenylmethyl)-	**24:** 4031[4]
4-amino-3,5-bis[4-([3-(2-ethoxyphenylcarbamoyl)- 2-hydroxy-1-naphthalenyl]azo)phenyl]-	**82:** P59896t
4-amino-3,5-bis(2-formylthiophene-4-yl)-, dihydrazone	**80:** 82840n
4-amino-3,5-bis(heptafluoropropyl)-	**64:** 12662f
4-amino-3,5-bis(λ-hydroxyheptadecyl)-	**28:** 4418[8]
4-amino-3,5-bis(2-hydroxyphenyl)-	**81:** P120640c
4-amino-3,5-bis(p-methoxybenzyl)-	**55:** 6495c
4-amino-3,5-bis(m-methoxyphenyl)-	**60:** 14508g
4-amino-3,5-bis[3-[[1-(2-methylanilino)-1,3-dioxo- 2-butyl]azo]phenyl]-	**82:** P59896t
4-amino-3,5-bis(4-methylphenyl)-	**78:** P29776d, **81:** P25678n
4-amino-3,5-bis(pentafluoroethyl)-	**55:** 6474f, **64:** 12662f
4-amino-3,5-bis(2-thiazolyl)-	**55:** 6495d
4-amino-3,5-bis(trifluoromethyl)-	**55:** 6474f

191

TABLE 6 (Continued)

Compound	Reference

6.4. 4-Amino-4H-1,2,4-Triazoles (Continued)

4-amino-3,5-bis[2-(trimethylsilyl)ethyl]-	**84:** 121930w
4-amino-3,5-di(2-aminoethyl)-	**79:** P32461y
4-amino-3,5-dibenzyl-	**68:** 69071v, **79:** 115418r,
	81: P25678n, **81:** P120540c
4-amino-3,5-diethyl-	**48:** 7006d, **55:** 27282f,
	61: 4515f, **71:** 38872q,
	78: P29776d, **81:** P25678n,
	85: 21220s
4-amino-3,5-di-2-furyl-	**64:** 5072h, **85:** 177373e
4-amino-3,5-di-3-furyl-	**85:** 177373e
4-amino-3,5-diheptadecyl-	**28:** 4419[1], **81:** P25678n,
	85: 192740r
4-amino-3,5-diheptyl-	**85:** P147920p
4-amino-3-(4,5-dihydro-1H-pyrazol-3-yl)-	**78:** 29701a
4-amino-3-(4,5-dihydro-1H-pyrazol-3-yl)-5-phenyl-	**78:** 29701a
4-amino-3,5-di-(1-hydroxyethyl)-	**70:** 57739x
4-amino-3,5-diisoamyl-	**24:** 3216[2]
4-amino-3,5-diisobutyl-	**24:** 3216[2]
4-amino-3,5-diisohexyl-	**24:** 3216[6]
4-amino-3,5-diisopropyl-	**48:** 7006e, **69:** 83184m,
	71: 56483t
4-amino-3,5-dimethyl-	**28:** 2679[7], **29:** 5088[7],
	48: 7006d, **55:** 27282f,
	61: 4515f, **61:** 14662h,
	69: 86504b, **70:** 11687a,
	71: 38872q **77:** 5410h,
	77: 75194q, **78:** P29776d
	81: P25678n, **85:** 21220s,
	85: 77541q, **85:** 123823h
4-amino-3,5-di-2-naphthalenyl-	**76:** 59542a, **85:** 123878e
4-amino-3,5-dinonyl-	**49:** 2427a
4-amino-3,5-diphenyl-	**48:** 7006d, **55:** 14440g,
	57: 5909c, **60:** 12006d,
	60: 14508e, **66:** 85767r,
	78: P29776d, **81:** P25678n,
	81: P120640c
4-amino-3-diphenylmethyl-5-methyl-	**28:** 5458[2]
4-amino-3-diphenylmethyl-5-phenyl-	**24:** 4032[6]
4-amino-3,5-dipropyl-	**48:** 7006d, **65:** 12205b
4-amino-3,5-di-2-pyridyl-	**54:** 24791c, **62:** 9122a
4-amino-3,5-di-3-pyridyl-	**54:** 24791c, **79:** 115418r
4-amino-3,5-di-4-pyridyl-	**51:** 1957e, **54:** 24791c,
	77: 152130v
4-amino-3,5-diselenophene-2-yl-	**80:** 82840n, **85:** 177373e
4-amino-3,5-diselenophene-3-yl-	**80:** 82840n, **85:** 177373e
4-amino-3,5-di-2-thienyl-	**80:** 82840n, **85:** 177373e
4-amino-3,5-di-3-thienyl-	**80:** 82840n, **85:** 177373e

TABLE 6 (*Continued*)

Compound	Reference

6.4. 4-Amino-4H-1,2,4-Triazoles (*Continued*)

4-amino-3,5-di-*o*-tolyl-	**60:** 14508f
4-amino-3,5-di-*m*-tolyl-	**50:** 6189d
4-amino-3,5-di-*p*-tolyl-	**50:** 6189d
4-amino-3,5-diundecyl-	**67:** P44809h, **78:** P29776d, **81:** P25678n, **85:** P63073b
4-amino-3-ethyl-	**85:** 77541q
4-amino-3-ethyl-5-methyl-	**77:** 152130v
4-amino-3-(heptafluoropropyl)-5-trifluoromethyl-	**64:** 12662f
4-amino-3-(α-hydroxy-α-methyl-4-methylbenzyl)-5-methyl-	**77:** 152130v
4-amino-3-(α-hydroxy-α-methyl-4-methylbenzyl)-5-phenyl-	**77:** 152130v
4-amino-3-isopropyl-	**85:** 77541q
4-amino-3-methyl-	**72:** 66873e, **85:** 77541q
4-amino-3-[(4-methylbenzenesulfonamido)methyl]-5-(4-pyridinyl)-	**77:** 114314a
4-amino-3-methyl-5-phenyl-	**77:** 139906h
4-amino-3-phenyl-	**48:** 4525a, **57:** 3437a, **77:** 139906h
4-amino-3-propyl-	**85:** 77541q
4-anilino-3,5-dimethyl-	**85:** 21220s
4-anilino-3,5-diphenyl-	**82:** 43275d
4-anilino-3-[(4-methylbenzenesulfonamido)methyl]-5-(4-pyridinyl)-	**77:** 114314a
4-[(9(10H)-anthracenylidene)amino]-	**76:** 99572s
4-[(9-anthracenylmethylene)amino]-	**61:** 13232g, **76:** 99572s
4-(7-azabicyclo[4.1.0]heptan-7-yl)-3,5-diphenyl-	**77:** 101467c
4-(6-azabicyclo[3.1.0]hexan-6-yl)-3,5-diphenyl-	**77** x 101467c
4-benzamidino	**71:** 38867s, **76:** 113138p
4-benzamido-	**54:** 7959f, **75:** P129817z, **76:** 126949d
4-benzamido-3,5-diphenyl-	**55:** 14440g, **73:** 66511f
4-[[(benzamido)methyl]amidino]-	**71:** 38867s
4-benzamido-3-phenyl-	**58:** 7924e, **57:** 3436i
4-(benzenecarboximidamido)-3,5-diethyl-	**76:** 113138p
4-(benzenecarboximidamido)-3,5-dimethyl-	**76:** 113138p
4-(benzenecarboximidamido)-3,5-diphenyl-	**76:** 113138p
4-(benzenecarboximidamido)-3-methyl-	**76:** 113138p
4-(benzenecarboximidamido)-3-phenyl-	**76:** 113138p
4-(benzenesulfonamido)-	**75:** 35893u, **80:** 81776j
4-[(1,3-benzodioxol-5-ylmethylene)amino]-	**76:** 99572s, **85:** 42100t
4-(*N*-benzylacetamido)-	**70:** 87681q
4-(*N*-benzylacetamido)-3,5-dimethyl-	**70:** 87681q
4-(benzylamino)-	**64:** 19596h, **70:** 87681q, **75:** 110242m
4-benzylamino-3,5-bis[(4-methylbenzenesulfonamido)methyl]-	**77:** 114314a

TABLE 6 (*Continued*)

Compound	Reference

6.4. 4-Amino-4H̲-1,2,4-Triazoles (*Continued*)

Compound	Reference
4-(benzylamino)-3,5-dimethyl-	**70:** 87681q
4-(benzylamino)-3-methyl-	**72:** 66873e
4-[[(*N*-benzyl)benzeneacetimidoyl]amino]-	**76:** 113138p
4-[(*N*-benzyl)benzenecarboximidamido]-	**76:** 113138p
4-[(*N*-benzylformimidoyl)amino]-	**76:** 113138p
4-(benzylideneamino)-	**44:** 2942a, **61:** 13232g, **76:** 99572s, **85:** 138489c
4-(benzylideneamino)-3,5-bis(4-methoxyphenyl)-	**85:** 123878c
4-(benzylideneamino)-3,5-dimethyl-	**61:** 13232g
4-(benzylideneamino)-3,5-diphenyl-	**47:** 8035h, **74:** 13074e, **85:** 123878e
4-(benzylideneamino)-3-methyl-	**72:** 66873e
4-[2-(benzyloxy)-*N*,2-dimethylpropionamido]-	**75:** P129817z
4-(2-(benzyloxy)-2-ethylbutyramido)-	**75:** P129817z
4-[2-(benzyloxy)-2-ethyl-*N*-methylbutyramido]-	**75:** P129817z
4-[2-(benzyloxy)-2-methylpropionamido]-	**75:** P129817z
4-[2-(benzylthio)-2-ethylbutyramido]-	**75:** P129817z
4-[2-(benzylthio)-2-ethyl-*N*-methylbutyramido]-	**75:** P129817z
4-[*N*-[(benzylthio)iminocarbonyl]anilino]-	**76:** 85755s
4-[([1,1'-biphenyl]-4-ylmethylene)amino]-	**76:** 99572s
3,5-bis(acetonyl)-4-amino-	**75:** 129725t
3,5-bis(acetonyl)-4-(benzylideneamino)-	**75:** 129725t
3,5-bis(acetonyl)-4-(2-butenylideneamino)-	**75:** 129725t
3,5-bis(acetonyl)-4-(butylideneamino)-	**75:** 129725t
3,5-bis(acetonyl)-4-(cinnamylideneamino)-	**75:** 129725t
3,5-bis(acetonyl)-4-[([(ethoxycarbonylmethyl)thio]-thioxomethyl)amino]-	**81:** 3897f
3,5-bis(acetonyl)-4-(ethylideneamino)-	**75:** 129725t
3,5-bis(acetonyl)-4-([2-hydroxyethylthio)thioxomethyl]-amino)-	**81:** 3897f
3,5-bis(acetonyl)-4-[(3-methoxysalicylidene)amino]-	**75:** 129725t
3,5-bis(acetonyl)-4-[(phenylthiocarbamoyl)amino]-	**75:** 129725t
3,5-bis(acetonyl)-4-(salicylideneamino)-	**75:** 129725t
3,5-bis(acetonyl)-4-[[(3-sulfopropylthio)thioxomethyl]-amino]	**80:** 36682q
3,5-bis(tert-butyl)-4-[(4-chlorophenyl)ureylene]-	**82:** P12285c
4-[[4-[bis(2-chloroethyl)amino]-2-methylbenzylidene]-amino]-	**73:** 35288j
4-(3,4-bis(ethoxycarbonyl)-2,5-dimethyl-1H̲-pyrrol-1-yl)-	**83:** 131395u
4-[2-[bis(isopropyl)amino]acetamido]-3,5-dimethyl-	**76:** 140662m
3,5-bis(isopropyl)-4-[(4-isopropylphenyl)ureylene]-	**82:** P12285c
3,5-bis(isopropyl-4-([3-(trifluoromethyl)phenyl]-ureylene)	**82:** P12285c
4-[bis(*p*-methoxyphenylsulfonyl)amino]-	**75:** 35893u
4-[bis(phenylsulfonyl)amino]-	**75:** 35893u, **80:** 81776j
4-[bis(*p*-toluenesulfonyl)amino]-	**75:** 35893u, **80:** 81776j
4-[bis(trimethylsilyl)amino]-3,5-diphenyl-	**68:** 69071v

194

TABLE 6 *(Continued)*

Compound	Reference

6.4. 4-Amino-4H-1,2,4-Triazoles *(Continued)*

Compound	Reference
4-[([2,2'-bithiophen]-5-ylmethylene)amino]-	**78:** 147712c
4-(4-bromoanilino)-3-[(4-methylbenzenesulfonamido)- methyl]-5-(4-pyridinyl)-	**77:** 114314a
4-[(*o*-bromobenzylidene)amino]-	**76:** 99572s
4-[(*m*-bromobenzylidene)amino]-	**76:** 99572s
4-[(*p*-bromobenzylidene)amino]-	**61:** 13232g, **75:** 14233p, **75:** 29276d, **85:** 42100t
4-[([4-(2-bromoethoxy)phenyl]methylene)amino]-	**76:** 99572s
4-[(5-bromofurfurylidene)amino]-	**56:** 2440f, **58:** 9071e
4-([3-(5-bromo-2-furyl)allylidene]amino)-	**58:** 9071h
4-([(5-bromo-1H-indol-3-yl)methylene]amino)-	**76:** 99572s
4-(2-bromo-4-nitroanilino)-3,5-diphenyl-	**83:** 58722s
4-(2-bromo-4-nitroanilino)-3-(3-nitrophenyl)-5-phenyl-	**83:** 58722s
4-([1-(3-bromophenyl)ethylidene]amino)-	**76:** 99572s
4-[[3-(4-bromophenyl)-2-propenylidene]amino]-	**76:** 99572s
4-[(4-bromophenyl)ureylene]-	**82:** P12285c
4-(2-butenylideneamino)-3,5-diphenyl-	**74:** 13074e
4-(*p*-butoxybenzamido)-	**61:** 8298e
4-(4-butoxybenzeneethanimidamido)-	**81:** 105402h
4-(4-tert-butoxybenzeneethanimidamido)-	**81:** 105402h
4-(*p*-butoxybenzenesulfonamido)-	**72:** 21667z
4-[α-(butoxycarbonylamino)benzylideneamino]-	**74:** 125651x
4-[α-(butoxycarbonylamino)-4-chlorobenzylideneamino]-	**74:** 125651x
4-[α-(butoxycarbonylamino)-2-chlorophene- thylideneamino]-	**74:** 125651x
4-[α-(butoxycarbonylamino)-4-chlorophene- thylideneamino]-	**74:** 125651x
4-[1-(butoxycarbonylamino)ethylideneamino]-	**74:** 125651x
4-[α-(butoxycarbonylamino)-4-methoxybenzyl- ideneamino]-	**74:** 125651x
4-[α-(butoxycarbonylamino)-4-methylbenzyl- ideneamino]-	**74:** 125651x
4-[(butoxycarbonylamino)methyleneamino]-	**74:** 125651x
4-[[α-butoxycarbonylamino)naphthalen-1- ylmethylene]amino]-	**74:** 125651x
4-[α-(butoxycarbonylamino)phenethylideneamino]-	**74:** 125651x
4-[1-(butoxycarbonylamino)propylideneamino]-	**74:** 125651x
4-[α-(butoxycarbonylamino)-2,4,6-trimethylphene- thylideneamino]-	**74:** 125651x
4-[(2-butoxyphenyl)ureylene]-	**82:** P12285c
4-(butylamino)-	**70:** 87681q
4-(butylamino)-3-methyl-	**72:** 66873e
4-[(*p*-tert-butylbenzylidene)amino]-	**68:** 58906v
3-butyl-4-[[2-fluorophenyl)thiocarbamoyl]amino]-	**82:** P12285c
4-([(butyl)(2-fluorophenyl)thiocarbamoyl]amino)-	**82:** P12285c
4-([N-butyl-N-(2-fluorophenylcarbamoyl)]amino)-	**82:** P12285c
4-[(butylidene)amino]-	**76:** 99572s

195

TABLE 6   (Continued)

Compound	Reference

6.4. 4-Amino-4H-1,2,4-Triazoles   (Continued)

Compound	Reference
4-(N-butyl-2-phenylacetamido)-	**75:** P129817z
4-[[butyl(phenyl)carbamoyl]amino]-	**82:** P12285c
4-[N-(butylthiocarbamoyl)amino]-	**74:** 87901u
4-[[2-(butylthio)phenyl]ureylene]-	**82:** P12285c
4-[1-(3-butyl-2-thio)ureido]-	**64:** 6644b
-4-carbamic acid, ethyl ester	**47:** 497a
-4-carbamodithioic acid, 3,5-bis(acetonyl)-	**80:** 36682q, **81:** 3897f
4-[(carbamoyl)amino]-	**76:** 85755s
4-(o-carboxybenzamido)-3,5-dimethyl-	**51:** P4440f
4-[(o-carboxybenzyl)amino]-	**76:** 99572s
4-[(p-carboxybenzyl)amino]-	**76:** 99572s
4-[[(2-carboxy-5-furyl)methyl]amino]-, methyl ester	**77:** 114143u
4-[(3-carboxy-4-hydroxybenzyl)amino]-	**76:** 99572s
4-[(4-carboxymethoxybenzyl)amino]-	**76:** 99572s
4-[5-(carboxymethyl)-3-(p-chlorophenyl)-4-oxothiazolidin-2-ylimino]-	**74:** 87901u
4-[5-(carboxymethyl)-3-(p-ethoxyphenyl)-4-oxothiazolidin-2-ylimino]-	**74:** 87901u
4-[5-(carboxymethyl)-3-(o-methoxyphenyl)-4-oxothiazolidin-2-ylimino]-	**74:** 87901u
4-[(5-carboxymethyl-4-oxo-3-phenylthiazolidinyl-2-idene)amino]-	**74:** 87901u
4-[5-(carboxymethyl)-4-oxo-3-(2,4-xylyl)thiazolidin-2-ylimino]-	**74:** 87901u
4-(2-chloroacetamido)-3,5-dimethyl-	**76:** 140662m
4-(4-chloroanilino)-3-[(4-methylbenzenesulfonamido)methyl]-5-(4-pyridinyl)-	**77:** 114314a
4-(p-chlorobenzamidino)-	**71:** 38867s
4-(4-chlorobenzamido)-	**76:** 126949d
4-(p-chlorobenzemesulfonamido)-	**75:** 35893u
4-[(p-chlorobenzyl)amino]-	**70:** 87681q
4-[(p-chlorobenzyl)amino]-3-methyl-	**72:** 66873e
4-[(o-chlorobenzylidene)amino]-	**75:** 14233p, **76:** 99572s
4-[(m-chlorobenzylidene)amino]-	**75:** 14233p, **76:** 99572s
4-[(p-chlorobenzylidene)amino]-	**61:** 13232g, **76:** 99572s
4-[(o-chlorobenzylidene)amino]-3,5-diphenyl-	**74:** 13074e
4-[(m-chlorobenzylidene)amino]-3,5-diphenyl-	**74:** 13074e
4-[(p-chlorobenzylidene)amino]-3,5-diphenyl-	**74:** 13074e
4-[([2-chloro-4-(dimethylamino)phenyl]methylene)amino]-	**76:** 99572s, **85:** 42100t
4-(4-chloro-α-[(2-ethoxyethoxy)carbonylamino]benzylideneamino)-	**74:** 125651x
4-[(4-chloro-3-fluorophenyl)ureylene]-	**82:** P12285c
4-[(5-chlorofurfurylidene)amino]-	**58:** 9071f
4-([3-(5-chloro-2-furyl)allylidene]amino]-	**58:** 9071h
4-(3-chloro-4-methoxybenzeneethanimidamido)-	**81:** 105402h
4-([[(2-chloro-5-nitrophenyl)methylene]amino)-	**85:** 42100t

TABLE 6 (Continued)

Compound	Reference

6.4. 4-Amino-4H-1,2,4-Triazoles (Continued)

Compound	Reference
4-([(2-chloro-6-nitrophenyl)methylene]amino)-	**76:** 99572s.
4-([(4-chloro-3-nitrophenyl)methylene]amino)-	**76:** 99572s
4-([(5-chloro-2-nitrophenyl)methylene]amino)-	**76:** 99572s
4-([N-chlorophenyl)carbamoyl]amino)-	**76:** 85755s
4-([1-(4-chlorophenyl)ethylidene]amino)-	**76:** 99572s
4-(2-[(p-chlorophenyl)imino]-3-thiazolidinyl)-	**74:** 87901u
4-([(4-chlorophenyl)methylene]amino)-	**85:** 42100t
1-([2-(4-chlorophenyl)-3-(5-nitro-2-furanyl)-2-propenylidene]amino)-	**81:** P68558r
4-([N-(4-chlorophenyl)thiocarbamoyl]amino)-	**74:** 87901u
4-(4-chloro-3-sulfamoylbenzamido)-	**58:** P1470b, **61:** 1786a, **69:** P10473v
4-(cinnamylideneamino)-3,5-diphenyl-	**74:** 13074e
4-(2-cyanoacetamido)-	**74:** 141669q
4-[(p-cyanobenzyl)amino]-	**76:** 99572s
4-[(4-cyanophenyl)ureylene]-	**82:** P12285c
4-[(3-cyclohexen-1-ylmethylene)amino]-	**76:** 99572s
4-[(cyclohexylmethylene)amino]-	**76:** 99572s
4-[(cyclohexylmethylene)amino]-3,5-diphenyl-	**74:** 13074e
4-[1-(3-cyclohexyl)-2-thioureido]-	**64:** 6644b
3,5-diacetic acid, 4-amino-, dihydrazide	**85:** 21243b
4-(diacetylamino)-	**51:** 14751h
4-(diacetylamino)-3,5-diphenyl-	**50:** 6189c
4-(diallylamino)-	**64:** 19596g
4-(dibenzylamino)-	**64:** 19596g
4-(dibenzylamino)-3,5-dimethyl-	**64:** 19596g
3,5-dibenzyl-4-[bis(trimethylsilyl)amino]-	**68:** 69071v
4-(2,4-dibromoanilino)-3,5-diphenyl-	**83:** 58722s
4-[(3,5-dibromo-2-hydroxybenzyl)amino]-	**76:** 99572s
4-[[1-(2,4-dibromophenyl)ethylidene]amino]-	**76:** 99572s
-3,5-dibutanol, 4-amino-	**59:** 1624b, **69:** 83184m, **71:** 56483t
3,5-dibutyl-4-[(2-nitrophenyl)ureylene]-	**82:** P12285c
4-[3-(1,2-dichloro-1-carboxy-1-propenyl)amino]-	**76:** 99572s
4([(2,4-dichlorophenyl)methylene]amino)-	**76:** 99572s, **85:** 42100t
4-([(2,6-dichlorophenyl)methylene]amino)-	**64:** PC11132g, **67:** PC11327w, **74:** 13074e, **85:** 42100t
4-([(3,4-dichlorophenyl)methylene]amino)-	**76:** 99572s, **85:** 42100t
-3,5-diethanol, 4-amino-	**55:** P18778f, **59:** 15279a, **71:** 38872q, **78:** P29776d, **81:** P25678n
-3,5-diethanol, 4-[(2,2-dinitropropyl)amino]-	**59:** 15279b
4-(3,4-diethoxycarbonyl-2,5-dimethyl-1H-pyrrol-1-yl)-	**83:** 131395u
4-[2-(diethylamino)acetamido]-3,5-dimethyl-	**76:** 140662m
4-[([4-[2-(diethylamino)ethoxy]phenyl]methylene)amino]-	**76:** 99572s
4-[([4-(diethylamino)phenyl]methylene)amino]-	**76:** 99572s
3,5-diethyl-4-[(2-fluorophenyl)ureylene]-	**82:** P12285c

TABLE 6 *(Continued)*

Compound	Reference

6.4. 4-Amino-4H̲-1,2,4-Triazoles *(Continued)*

Compound	Reference
4-([(3,4-dihydro-2,5-dimethyl-2H̲-pyran-2-yl)methylene]amino)-	**76:** 99572s
3-(4,5-dihydro-1H̲-pyrazol-3-yl)-4-(isopropylideneamino)-	**78:** 29701a
4-[(2,5-dihydroxybenzyl)amino]-	**76:** 99572s
4-[(3,4-dihydroxybenzyl)amino]-	**76:** 99572s
4-[[3,4-(dihydroxy)benzylidene]amino)-	**48:** P10468a
4-[(2,5-dihydroxyphenethyl)amino]-	**76:** 99572s
-3,5-dimethanol, 4-amino-	**54:** 18484g, **55:** P18778f, **59:** 15279a, **71:** 38872q
-3,5-dimethanol, 4-amino-α,α′-diethyl-α,α′-diphenyl-	**78:** 84378a
-3,5-dimethanol, 4-amino-α,α′-dimethyl-	**59:** 15279a, **71:** 38872q
-3,5-dimethanol, 4-amino-α,α′-dimethyl-α,α′-bis(4-methylphenyl)-	**78:** 84378a
-3,5-dimethanol, 4-amino-α,α′-dimethyl-α-α′-bis(4-nitrophenyl)-	**78:** 84378a
-3,5-dimethanol, 4-amino-α,α′-dimethyl-α,α′-diphenyl-	**78:** 84378a
-3,5-dimethanol, 4-amino-α,α,α′,α′-tetramethyl)	**81:** P25678n, **85:** P192740r
-3,5-dimethanol, 4-(benzylideneamino)-	**61:** 13232g
-3,5-dimethanol, 4-[(2,2-dinitropropyl)amino]-	**59:** 15279b
-3,5-dimethanol, 4-[(2,2-dinitropropyl)amino]-α,α′-dimethyl-	**59:** 15279b
4-([(2,3-dimethoxyphenyl)methylene]amino)-	**76:** 99572s
4-([(2,4-dimethoxyphenyl)methylene]amino)-	**76:** 99572s, **85:** 42100t
4-([(2,5-dimethoxyphenyl)methylene]amino)-	**76:** 99572s
4-([(3,4-dimethoxyphenyl)methylene]amino)-	**85:** 42100t
4-([(3,5-dimethoxyphenyl)methylene]amino)-	**76:** 99572s
4-([(4,5-dimethoxy-2-nitrophenyl)methylene]amino)-	**76:** 99572s
4-([p-(dimethylamino)benzyl]amino)-	**64:** 19596h
4-([(p-dimethylamino)benzylidene]amino)-	**76:** 99572s
4-([p-(dimethylamino)benzylidene]amino)-3-phenyl-	**48:** 4525a
4-(dimethylamino)-3-[(4-methylbenzenesulfonamido)methyl]-5-(4-pyridinyl)-	**77:** 114314a
4-[([4-(dimethylamino)phenyl]methylene)amino]-	**85:** 42100t
4-[(3-[4-(dimethylamino)phenyl]-2-propenylidene)-amino]-	**76:** 99572s
4-[(3,5-dimethylbenzylidene)amino]-	**68:** 58906v
4-(N,3-dimethylbutyramido)-	**75:** P129817z
4-(N,N′-dimethylformamidino)-	**71:** 91386k
4-(N,α-dimethylhydratropamido)-	**75:** P129817z
3,5-dimethyl-4-methylamino-	**85:** 123823h
3,5-dimethyl-4-([(1-methyl-5-nitro-1H̲-imidazol-2-yl)-methylene]amino)-	**82:** P4260a
3,5-dimethyl-4-([(N-methyl-N-(phenylthiocarbamoyl)]-amino)-	**82:** P12285c
3,5-dimethyl-4-(4-nitroanilino)-	**82:** 170800w
3,5-dimethyl-4-[(5-nitrofurfurylidene)amino]-	**56:** 14194a, **58:** 9071g, **61:** 14663a

TABLE 6   (*Continued*)

Compound	Reference

6.4. 4-Amino-4H-1,2,4-Triazoles   (*Continued*)

Compound	Reference
4-(2,6-dimethyl-4-(1H)-oxo-1-pyridinyl)-	**83:** 131395u
4-(N,3-dimethyl-3-phenylbutyramido)-	**75:** P129817z
3,5-dimethyl-4-[(phenylthiocarbamoyl)amino]-	**75:** 129725t
3,5-dimethyl-4-(phenylureylene)-	**82:** P12285c
4-(2,5-dimethyl-1H-pyrrol-1-yl)-	**83:** 131395u
3,5-dimethyl-4-(xanthen-9-ylamino)-	**73:** P120633n
4-[[(2,4-dinitrophenyl)methylene]amino]-	**76:** 99572s
4-[[(2,6-dinitrophenyl)methylene]amino]-	**76:** 99572s
4-[(2,2-dinitropropyl)amino]-	**55:** 27282g, **65:** 12205b
4-[(2,2-dinitropropyl)amino]-3,5-diethyl-	**55:** 27282g
4-[(2,2-dinitropropyl)amino]-3,5-dimethyl-	**55:** 27282g, **65:** 12205b
4-[(2,2-dinitropropyl)amino]-3,5-dipropyl-	**65:** 12205b
4-(2,5-dioxo-1-pyrrolidinyl)-	**81:** 49309u, **82:** P124906n
3,5-diphenyl-4-(S,S-dimethylsulfonimidoyl)-	**77:** 101467c
3,5-diphenyl-4-(2-phenyl-1-aziridinyl)-	**77:** 101467c
3,5-diphenyl-4-(salicylideneamino)-	**74:** 13074e
3,5-diphenyl-4-(2-thenylideneamino)-	**74:** 13074e
3,5-diphenyl-4-[(2,4,6-trimethoxybenzylidene)amino]-	**74:** 13074e
3,5-diphenyl-4-[(2,4,6-trimethylbenzylidene)amino]-	**74:** 13074e
-3,5-dipropanol, 4-amino-	**55:** P18778f, **59:** 1624b, **69:** 83184m, **71:** 56483t
-3,5-dipropanol, 4-amino-, dibenzoate	**59:** 1624c
-3,5-dipropanol, 4-benzamido-, dibenzoate	**59:** 1624c
4-(dipropylamino)-	**64:** 19596g
4-[2-(dipropylamino)acetamido]-3,5-dimethyl-	**76:** 140662m
4-(p-ethoxybenzamido)-	**61:** 8298d
4-(4-ethoxybenzeneethanimidamido)-	**81:** 105402h
4-(p-ethoxybenzenesulfonamido)-	**72:** 21667z
4-[(ethoxycarbonyl)amino]-	**82:** P12285c
4-[[(ethoxycarbonyl)methyl]amidino]-	**71:** 38867s, **74:** 141669q
4-[α-[(2-ethoxyethoxy)carbonylamino]benzylideneamino]-	**74:** 125651x
4-(α-[(2-ethoxyethoxy)carbonylamino]-4-methoxy-benzylideneamino)-	**74:** 125651x
4-(α-[(2-ethoxyethoxy)carbonylamino]-4-methyl-benzylideneamino)-	**74:** 125651x
4-[(α-[(2-ethoxyethoxy)carbonylamino]naphthalen-1-ylmethylene)amino]-	**74:** 125651x
4-[(3-ethoxy-4-hydroxybenzyl)amino]-	**76:** 99572s
4-[(ethoxymalonyl)amino]-	**74:** 141669q
4-[(ethoxyoxalyl)amino]-	**83:** 172416v
4-([(2-ethoxyphenyl)methylene]amino)-	**76:** 99572s
4-([(4-ethoxyphenyl)methylene]amino)-	**76:** 99572s
4-([(p-ethoxyphenyl)thiocarbamoyl]amino)-	**74:** 87901u
4-(ethylamino)-	**70:** 87681q
4-(ethylamino)-3,5-dimethyl-	**70:** 87681q, **85:** 123823h
3-ethyl-5-hexyl-4-(phenylrueylene)-	**82:** P12285c
4-(ethylideneamino)-	**75:** 129725t
3-ethyl-5-methyl-4-(4-nitroanilino)-	**82:** 170800w

TABLE 6 (*Continued*)

Compound	Reference

6.4. 4-Amino-4H-1,2,4-Triazoles (*Continued*)

Compound	Reference
3-ethyl-4-(4-nitroanilino)-5-(4-nitrophenyl)-	**82:** 170800w
3-ethyl-4-(phenylureylene)-	**82:** P12285c
4-[(2-fluoro-4-methylphenyl)ureylene]-	**82:** P12285c
4-([[(4-fluoro-2-nitrophenyl)methylene]amino)-	**76:** 99572s
4-[(4-fluoro-3-nitrophenyl)ureylene]-	**82:** P12285c
4-([(*p*-fluorophenyl)benzylidene]amino)-	**76:** 99572s
4-([2-fluorophenyl)methyl)carbamoyl]amino)-	**82:** P12285c
4-([(2-fluorophenyl)methylene]amino)-	**76:** 99572s
4-([(3-fluorophenyl)methylene]amino)-	**76:** 99572s
4-[(2-fluorophenyl)ureylene]-	**82:** P12285c
4-[(4-fluorophenyl)ureylene]-	**82:** P12285c
4-[(2-fluorophenyl)ureylene]-3,5-dimethyl-	**82:** P12285c
4-[(2-fluorophenyl)ureylene]-3-(2-methylpropyl)-	**82:** P12285c
4-(2-furamidino)-	**71:** 38867s
4-[(2-furanylmethylene)amino]	**76:** 99572s, **85:** 42100t
4-[[3-(2-furanyl)-2-propenylidene]amino]-	**76:** 99572s
4-(furfurylideneamino)-	**58:** 9071f
4-(furfurylideneamino)-3,5-diphenyl-	**74:** 13074e
4-([3-(2-furyl)allylidene]amino)-	**58:** 9071g
4-[(heptylidene)amino]-	**76:** 99572s
4-[1-(3-hexyl-2-thio)ureido]-	**64:** 6644b
4-[(2-hydroxybenzyl)amino]-3,5-diheptyl-	**83:** P80456e
4-[(*o*-hydroxybenzylidene)amino]-	**75:** 14233p, **75:** 29276d, **76:** 99572s
4-[(*m*-hydroxybenzylidene)amino]-	**75:** 14233p, **75:** 29276d, **76:** 99572s
4-[(4-hydroxybenzylidene)amino]-	**75:** 14233p, **75:** 29276d, **76:** 99572s
4-[(2-hydroxy-5-chlorobenzyl)amino]-	**76:** 99572s
4-[([1-(2-hydroxyethyl)-5-nitro-1H-imidazol-2-yl]-methyl)amino]-	**82:** P4260a
4-([5-hydroxy-3-(hydroxymethyl)-5-methyl-4-pyridinyl-methyl]amino)-, 5-dihydrogen phosphate	**72:** 74965b
4-[(4-hydroxy-3-methoxy-5-nitrobenzyl)amino]-	**76:** 99572s
4-[(2-hydroxy-5-methylbenzyl)amino]-	**76:** 99572s
4-([(5-hydroxymethyl-2-furyl)methylene]amino)-	**76:** 99572s
4-([(2-hydroxy-1-naphthalenyl)methyl]amino)-3,5-diphenyl-	**83:** P80456e
4-([(1-hydroxy-2-naphthalenyl)methylene]amino)-	**76:** 99572s
4-[(2-hydroxy-3-nitrobenzyl)amino]-	**76:** 99572s
4-[(2-hydroxy-5-nitrobenzyl)amino]-	**76:** 99572s
4-[(5-hydroxy-2-nitrobenzyl)amino]-	**76:** 99572s
4-[(2-hydroxyphenethyl)amino]-	**76:** 99572s
4-[(1H-indol-3-ylmethylene)amino]-	**76:** 99572s
4-[(5-iodofurfurylidene)amino]-	**58:** 9071f
4-([3-(5-iodo-2-furyl)allylidene]amino)-	**58:** 9071h
4-([(2-iodophenyl)methylene]amino)-	**76:** 99572s

TABLE 6   (Continued)

Compound	Reference

6.4.   4-Amino-4H-1,2,4-Triazoles   (Continued)

Compound	Reference
4-([(3-iodophenyl)methylene]amino)-	**76:** 99572s
4-([(4-iodophenyl)methylene]amino)-	**76:** 99572s
4-[(4-iodophenyl)ureylene]-	**82:** P12285c
4-(p-isobutoxybenzamido)-	**61:** 8298e
4-(isonicotinamidino)-	**71:** 38867s
4-[p-(isopentyloxy)benzamido]-	**61:** 8298e
4-[p-(isopropoxy)benzamido]-	**61:** 8298e
4-(4-isopropoxybenzeneethanimidamido)-	**81:** 105402h
4-[(4-isopropylbenzylidene)amino]-	**75:** 14233p, **75:** 29276d
4-[(N-isopropylformimidoyl)amino]-	**76:** 113138p
4-[[(isopropyl)(2-methoxyphenyl)carbamoyl]amino]-	**82:** P12285c
4-[[N-isopropyl-N-(phenylcarbamoyl)]amino]-	**82:** P12285c
4-[(4-isopropylphenyl)ureylene]-	**82:** P12285c
-3-methanol, 4-amino-5-ethyl-α-methyl-	**71:** 56483t
-3-methanol, 4-amino-5-methyl-	**71:** 38872q
4-(2-methoxyacetamidino)-	**71:** 38867s
4-(4-methoxybenzeneethanimidamido)-	**81:** 105402h
4-(p-methoxybenzenesulfonamido)-	**72:** 21667z, **75:** 35893u
4-[(p-methoxybenzyl)amino]-	**64:** 19596h
4-[(p-methoxybenzylidene)amino]-	**61:** 13232g
4-[(o-methoxybenzylidene)amino]-3,5-diphenyl-	**74:** 13074e
4-[(m-methoxybenzylidene)amino]-3,5-diphenyl-	**74:** 13074e
4-[(p-methoxybenzylidene)amino]-3,5-diphenyl-	**74:** 13074e
4-[[(p-methoxyphenyl)benzylidene]amino]-	**76:** 99572s
4-[2-(4-methoxyphenyl)-2,2-dimethylacetyl]amino-	**76:** P34258z
4-[2-[(p-methoxyphenyl)imino]-3-thiazolidinyl]-	**74:** 87901u
4-[[(2-methoxyphenyl)methylene]amino]-	**76:** 99572s **85:** 42100t
4-[[(3-methoxyphenyl)methylene]amino]-	**85:** 42100t
4-[[(4-methoxyphenyl)methylene]amino]-	**85:** 42100t
4-[[(o-methoxyphenyl)thiocarbamoyl]amino]-	**74:** 87901u
4-[(4-methoxyphenyl)thiocarbamoyl]amino]-	**74:** 87901u
4-[(2-methoxyphenyl)ureylene]-	**82:** P12285c
4-[(4-methoxyphenyl)ureylene]-	**82:** P12285c
4-[(2-methoxyphenyl)ureylene]-3-methyl-	**82:** P12285c
4-[(2-methoxyphenyl)ureylene]-3-octyl-	**82:** P12285c
4-(methylamino)-	**70:** 87681q
4-(methylamino)-3-[(methylbenzenesulfonamido)- methyl]-5-(4-pyridinyl)-	**77:** 114314a
4-(4-methylamino)-3-[(methylbenzenesulfonamido)- methyl]-5-(4-pyridinyl)-	**77:** 114314a
4-(N-methylbenzamido)-	**75:** P129817z
4-[(o-methylbenzylidene)amino]-3,5-diphenyl-	**74:** 13074e
4-[(m-methylbenzylidene)amino]-3,5-diphenyl-	**74:** 13074e
4-[(p-methylbenzylidene)amino]-3,5-diphenyl-	**74:** 13074e
4-(3-methylbutyramido)-	**75:** P129817z
4-[(N-methylcarbamoyl)amino]-	**76:** 85755s
4-(2-methyl-4,6-diphenyl-1-pyridinio-, perchlorate	**83:** 131395u

TABLE 6 (*Continued*)

Compound	Reference

6.4. 4-Amino-4H-1,2,4-Triazoles (*Continued*)

4-[2-methyl-4,6-di-*p*-tolyl-1-pyridinio]-, perchlorate	**75:** P13530w
4-(methyleneamino)-	**59:** 15276h
4-[([4-(1-methylethyl)phenyl]methylene)amino]-	**85:** 42100t
4-(*N*-methyl-*N*-[(2-fluorophenyl)carbamoyl]amino)	**82:** P12285c
4-[(*N*-methylforminidoyl)amino]-	**76:** 113138p
4-[(5-methylfurfurylidene)amino]-	**58:** 9071f, **59:** 15276h
4-([3-(5-methyl-2-furyl)allylidene]amino)-	**58:** 9071g
4-(α-methylhydratropamido)-	**75:** P129817z
4-(β-methylhydrocinnamamido)-	**75:** P129817z
4-[*N*-methyl-*N*-[(methyl)(phenyl)carbamoyl]amino]-	**82:** P12285c
4-(*N*-methyl-1-naphthaleneacetamido)-	**75:** P129817z
3-methyl-4-(2-nitroanilino)-	**85:** 159991a
3-methyl-4-(2-nitroanilino)-5-(2-nitrophenyl)-	**82:** 170800w
3-methyl-4-(2-nitroanilino)-5-(3-nitrophenyl)-	**82:** 170800w
3-methyl-4-(2-nitroanilino)-5-(4-nitrophenyl)-	**82:** 170800w
3-methyl-4-(2-nitroanilino)-5-phenyl-	**82:** 170800w
4-[(α-methyl-*p*-nitrobenzylidene)amino]-	**61:** 8784a
4-([(1-methyl-5-nitro-1H-imidazol-2-yl)methylene]- amino)-	**82:** P4260a
4-(*N*-methyl-2-phenylacetamido)-	**75:** P129817z
4-(2-methyl-3-phenyl-1-aziridinyl)-3,5-diphenyl-, cis-	**77:** 101467c
4-(2-methyl-3-phenyl-1-aziridinyl)-3,5-diphenyl-, trans-	**77:** 101467c
4-(*N*-methyl-2-phenylbutyramido)-	**75:** P129817z
4-(*N*-methyl-4-phenylbutyramido)-	**75:** P129817z
4-([(methyl)(phenyl)carbamoyl]amino)-	**82:** P12285c
4-(*N*-methyl-2-phenylhexanamido)-	**75:** P129817z
4-([(2-methylphenyl)methylene]amino)-	**76:** 99572s
4-([(3-methylphenyl)methylene]amino)-	**76:** 99572s
4-([(4-methylphenyl)methylene]amino)-	**61:** 13232g, **76:** 99572s, **85:** 42100t
4-[(2-methyl-3-phenyl-2-propenylidene)amino]-	**76:** 99572s
4-(2-methylpropionamidino)-	**71:** 38867s
4-([(6-methyl-2-pyridinyl)methylene]amino)-	**76:** 99572s
4-([(1-methyl-1H-pyrrol-2-yl)methylene]amino)-	**76:** 99572s
4-(2-[3-(3-methyl-4-thiazolin-2-ylidene)propenyl)- 4,6-di-*p*-tolyl-1-pyridinio]-, perchlorate	**75:** P13530w
4-([(5-methyl-2-thienyl)methylene]amino)-	**76:** 99572s
4-[[2-(methylthio)phenyl)]ureylene]-	**82:** P12285c
4-[1-(3-methyl-2-thio)ureido]-	**64:** 6644b
4-(methylxanthen-9-ylamino)-	**73:** P120633n
4-(1-naphthaleneacetamido)-	**75:** P129817z
4-[(1-naphthalenylmethylene)amino]-	**85:** 42100t
4-[(2-naphthalenylmethylene)amino]-	**76:** 99572s, **85:** 42100t
4-(2-naphthamidino)-	**71:** 38867s
4-[(1-naphthylmethyl)amino]-	**70:** 87681q

TABLE 6 (*Continued*)

Compound	Reference

6.4. 4-Amino-4H-1,2,4-Triazoles (*Continued*)

Compound	Reference
4-[(1-naphthylmethylene)amino]-3,5-diphenyl-	**74:** 13074e
4-[(2-naphthylmethylene)amino]-3,5-diphenyl-	**74:** 13074e
4-(nicotinamidino)-	**71:** 38867s
4-(4-nitroanilino)-3,5-diphenyl-	**82:** 43275d, **83:** 58722s
4-(4-nitroanilino)-3-4(-nitrophenyl)-5-phenyl-	**82:** 43275d
4-[(o-nitrobenzylidene)amino]-3,5-diphenyl-	**74:** 13074e
4-[(m--nitrobenzylidene)amino]-3,5-diphenyl-	**74:** 13074e
4-[(p-nitrobenzylidene)amino]-3,5-diphenyl-	**74:** 13074e
4-[([5′-nitro(2,2′-bithiophen)-5-yl]methylene)amino]-	**78:** 147712c
4-[3-(5-nitro-2-furanyl)-2-propenylidene]amino	**69:** P77107x
4-(5-nitrofurfurylideneamino)-	**50:** 972e, **61:** 14663a, **76:** 99572s
4-[3-(5-nitro-2-furyl)allylideneamino]-	**50:** 972f, **67:** 97910r
4-([1-(3-nitrophenyl)ethylidene]amino)-	**76:** 99572s
4-([(2-nitrophenyl)methylene]amino)-	**76:** 99572s, **85:** 42100t
4-([(2-nitrophenyl)methylene]amino)-	**61:** 8784a, **75:** 14233p, **75:** 29276d, **85:** 42100t
4-[(4-nitrophenyl)methylene]amino-	**61:** 8784b, **76:** 99572s, **77:** 159501a, **85:** 42100t
4-[(2-nitrophenyl)-3-phenylamino]-	**85:** 159991a
4-([3-(2-nitrophenyl)-2-propenylidene]amino)-	**76:**·99572s
4-(p-nitrophenylsulfonamido)-	**39:** P2382[8]
4-[(2-nitrophenyl)ureylene]-	**82:** P12285c
4-[(3-nitrophenyl)ureylene]-	**82:** P12285c
4-[(5-nitro-2-thenylidene)amino]-	**61:** 8784a, **76:** 14220c, **76:** 99572s
4-([(4-nitro-2-thienyl)methylene]amino)-	**76:** 99572s
4-[p-(pentyloxy)benzamido]-	**61:** 8298e
4-(2-phenylacetamidino)-	**71:** 38867s
4-(phenylacetamido)-	**75:** P129817z
4-(2-phenylbutyramido)-	**75:** P129817z
4-(4-phenylbutyramido)-	**75:** P129817z
4-[(N-phenylcarbamoyl)amino]-	**76:** 85755s, **82:** P12285c
4-[(1-phenylethylidene)amino]-	**76:** 25193x
4-[(N-phenylforminidoyl)amino]-	**76:** 113138p
4-(2-phenylhexanamido)-	**75:** P129817z
4-([2-(phenylmethylene)heptiylidene]amino)-	**76:** 99572s
4-[([1-(phenylmethyl)-1H-indol-3-yl]methylene)amino]-	**76:** 99572s
4-([N-(5-phenyl-1,3,4-oxadiazol-2-yl)carbamoyl]amino)-	**72:** 43579z
4-[(3-phenyl-2-propenylidene)amino]-	**76:** 99572s, **85:** 42100t
4-(2-phenyl-N-propylacetamido)-	**75:** P129817z
4-[(3-phenyl-2-propynylidene)amino]-	**76:** 99572s
4-[(α-phenyl)tetrahydrofuran-2-ylbutyramido]-	**75:** P129817z
4-[N-[2-phenyl-4-(tetrahydrofuran-2-yl)butyryl]-N-methylamino]-	**75:** P129817z
4-(2-phenyl-N-(tetrahydrofurfuryl)acetamido)-	**75:** P129817z
4-[(N-phenylthiocarbamoyl)amino]-	**75:** 129725t, **76:** 85755s

TABLE 6 (*Continued*)

Compound	Reference

6.4. 4-Amino-4H-1,2,4-Triazoles (*Continued*)

Compound	Reference
5-[1-(3-phenyl-2-thio)ureido]-	**64:** 6644a
4-[(phenyl)ureylene]-3-hexyl-	**82:** P12285c
4-[[p-(phthalimido)phenyl]sulfonamido]-	**39:** 2741[9]
4-(picolinamidino)-	**71:** 38867s
4-(propionamidino)-	**71:** 38867s
4-(p-propoxybenzamido)-	**61:** 8298d
4-(4-propoxybenzeneethanimidamido)-	**81:** 105402h
4-(p-propoxybenzenesulfonamido)-	**72:** 21667z
4-[(propylidene)amino]-	**76:** 99572s
4-[(1-pyrenylmethylene)amino]-	**61:** 13232g
4-[(2-pyridinylmethylene)amino]-	**76:** 99572s
4-[(3-pyridinylmethylene)amino]-	**76:** 99572s
4-[(4-pyridinylmethylene)amino]-	**76:** 99572s
4-(1H-pyrrol-1-yl)-	**83:** 131395u
4-[(1H-pyrrol-2-ylmethylene)amino]-	**76:** 99572s
4-[(rifamycin)-3-methyleneamino]-	**77:** 135025x
4-sulfanilamido-	**37:** 1402[2], **39:** 2742[1]
4-[(o-sulfobenzyl)amino]-	**76:** 99572s
4-[(tetradecylidene)amino]-	**76:** 99572s
4-[N-(tetrahydrofurfuryl)-1-naphthaleneacetamido]-	**75:** P129817z
4-(2-thenylideneamino)-	**61:** 13232g, **76:** 14220c, **76:** 99572s
4-[1-(2-thio-3-o-tolyl)ureido]-	**64:** 6644b
4-[1-(2-thio-3-m-tolyl)ureido]-	**64:** 6644b
4-[1-(2-thio-3-p-tolyl)ureido]-	**64:** 6644b
4-(p-toluamidino)-	**71:** 38867s
4-(p-toluenesulfonamido)-	**75:** 35893u, **80:** 81776j
4-(2,2,2-trichloroacetamido)-	**81:** P135150s
4-[(2,2,2-trichloro-1-hydroxy)ethylamino]-	**68:** 104067x, **77:** P5482h
4-[(2,2,2-trichloroethylidene)amino]-	**76:** 99572s, **81:** P135150s
4-([3-(trifluoromethyl)phenyl]ureylene)-	**82:** P12285c
4-[(2,4,6-trihydroxybenzyl)amino]-	**76:** 99572s
4-([(p-4,4-trimethylamino)benzyl]amino)-, iodide	**76:** 25193x
4-(2,4,6-trimethyl-1-pyridinio)-, perchlorate	**82:** 140030c, **83:** 131395u
4-(2,4,6-trinitroanilino)-	**68:** 105111u
4-(2,4,6-triphenyl-1-pyridinio)-, perchlorate	**82:** 140030c, **83:** 131395u
4-[(N-xanthen-9-yl)acetamido]-	**73:** P120633n
4-(xanthen-9-ylamino)-	**73:** P120633n
4-([(2,4-xylyl)thiocarbamoyl]amino)-	**74:** 87901u
4-([(2,6-xylyl)thiocarbamoyl]amino)-	**74:** 87901u

CHAPTER 7

Diamino- and Triamino-1,2,4-Triazoles

7.1. 3,5-Diamino-1,2,4-Triazoles

The preparation of a variety of 1-aryl-3,5-diaminotriazoles (1-arylguan-azoles) was accomplished in good yields by the interaction of the hydrochlorides of aryl hydrazines with dicyandiamide (**7.1-1**) in refluxing water, as shown for the preparation of the 1-phenyltriazole (**7.1-2**).[1-4] In the condensation of **7.1-1** with hydrazine dihydrochloride, a 2:1 molar ratio of the reactants gave the 4,4'-bistriazole (**7.1-3**).[5] Variations of this method include the interaction of **7.1-4** with hydrazine to give **7.1-5** (78%),[6] and the

interaction of the alkaline salts of dicyanamide (**7.1-6**) with hydrazine and its alkyl and aryl derivatives to give 3,5-diamino-s-triazole (guanazole) (**7.1-7**) and its 1-substituted derivatives (e.g., **7.1-8**).[7,8] Similarly, **7.1-2**,

7.1-7, and **7.1-8** were obtained by the reaction of *N*-cyanoamidino azide (**7.1-9**) with the appropriate hydrazine.[8]

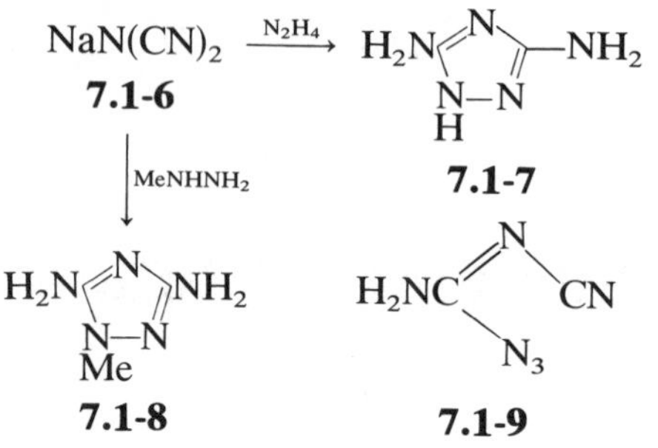

$$NaN(CN)_2 \xrightarrow{N_2H_4}$$

7.1-6

| MeNHNH$_2$

7.1-7

7.1-8

7.1-9

The cyclization of the *N*-(thiocarbamoyl)-*N'*-aminoguanidine derivatives (e.g., **7.1-10** and **7.1-11**) with the elimination of hydrogen sulfide was effected in a basic medium to give fair yields of 3,5-diaminotriazoles (e.g., **7.1-14** and **7.1-15**).[9] Several related intermediates such as **7.1-12** were also cyclized in 50% acetic acid; however, this method was unsuccessful for the cyclization of those compounds containing *N*-alkyl groups (e.g., **7.1-10**). In addition, Kurzer has determined the structures of the products resulting from the cyclization of the *N*-(phenylthiocarbamoyl) derivative (**7.1-13**) under both acidic and basic conditions. In hot aqueous sodium hydroxide, hydrogen sulfide was eliminated to give **7.1-16**, and in hot hydrochloric acid, aniline was eliminated to give the 3-aminotriazoline-5-thione (**7.1-17**).[10] In contrast to the results described previously, both **7.1-16** and **7.1-17** were obtained when **7.1-13** was treated with aqueous acetic acid. Also treatment of the thiadiazole (**7.1-18**) with boiling ethanolic hydrochloric acid gave sulfur (85%) and the triazole (**7.1-16**) (80%).[11]

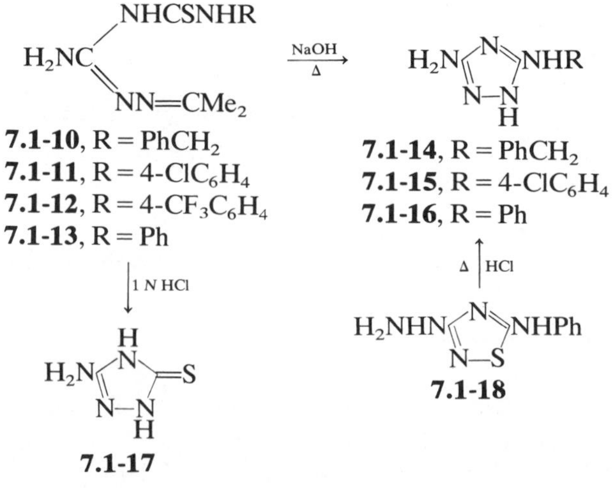

7.1-10, R = PhCH$_2$
7.1-11, R = 4-ClC$_6$H$_4$
7.1-12, R = 4-CF$_3$C$_6$H$_4$
7.1-13, R = Ph

7.1-14, R = PhCH$_2$
7.1-15, R = 4-ClC$_6$H$_4$
7.1-16, R = Ph

7.1-18

7.1-17

Different modes of cyclization were also observed with the N-(thiocarbamoylamino)guanidine (**7.1-20**).[12] In contrast to **7.1-13**, treatment of **7.1-20** with base resulted in the elimination of ammonia to give the thione (**7.1-22**). Under acidic conditions **7.1-20** was converted to the thiadiazole (**7.1-23**).[12] However, treatment of the S-benzyl derivative (**7.1-21**) with base

gave a good yield of the triazole (**7.1-16**). In a related reaction treatment of the S-benzyl compound (**7.1-24**) with hot ethanolic hydrazine give 3,5-bis-(anilino)-1,2,4-triazole (**7.1-25**).[13]

A number of 3,5-diaminotriazoles have been prepared from carbodiimides. Treatment of N,N'-dicyclohexylcarbodiimide (**7.1-26**) with hydrazine in hot N,N-dimethylformamide gave a 75% yield of **7.1-27**.[14] The related triazole (**7.1-28**) resulted from the interaction of N,N-diphenylcarbodiimide with the S-methyl derivative (**7.1-29**) of 4-phenylthiosemicarbazide.[15]

The condensation of N-aminoguanidine and N,N'-diphenylcarbodiimide gave two products, the monoadduct (**7.1-30**) and the triazoline (**7.1-33**), the latter resulting from the diadduct (**7.1-31**) by the elimination of aniline.[16] When **7.1-30** was pyrolyzed both **7.1-32** (48%) and **7.1-16** (32%) were formed, **7.1-32** by the elimination of ammonia and **7.1-16** by the elimination of aniline. Treatment of **7.1-33** with base gave the 4H̲-1.2.4-triazole (**7.1-28**), which could also be obtained directly from N-aminoguanidine and N,N'-diphenylcarbodiimide without the isolation of **7.1-33** in 60% yield. In addition, the reaction of **7.1-34** with N,N'-diphenylcarbodiimide gave **7.1-35**, which was hydrolyzed and cyclized in acid to give a 90% yield of **7.1-16**.[17]

In the condensation of N,N'-diaminoguanidine with excess N,N'-diphenylcarbodiimide in hot dimethylformamide, the major product was the 4H̲-1,2,4-triazole (**7.1-28**) (72%), probably formed via the diadduct (**7.1-36**)[18] and the triazoline (**7.1-38**). The formation of a small amount of **7.1-32** (22%) possibly occurred by the double cyclization and cleavage of the triadduct (**7.1-37**) to give both **7.1-32** and **7.1-28**. A good yield of **7.1-28** (85%) resulted from the reaction of N,N'-diamino-N''-phenylguanidine with N,N'-diphenylcarbodiimide to give **7.1-39** followed by cyclization.[19]

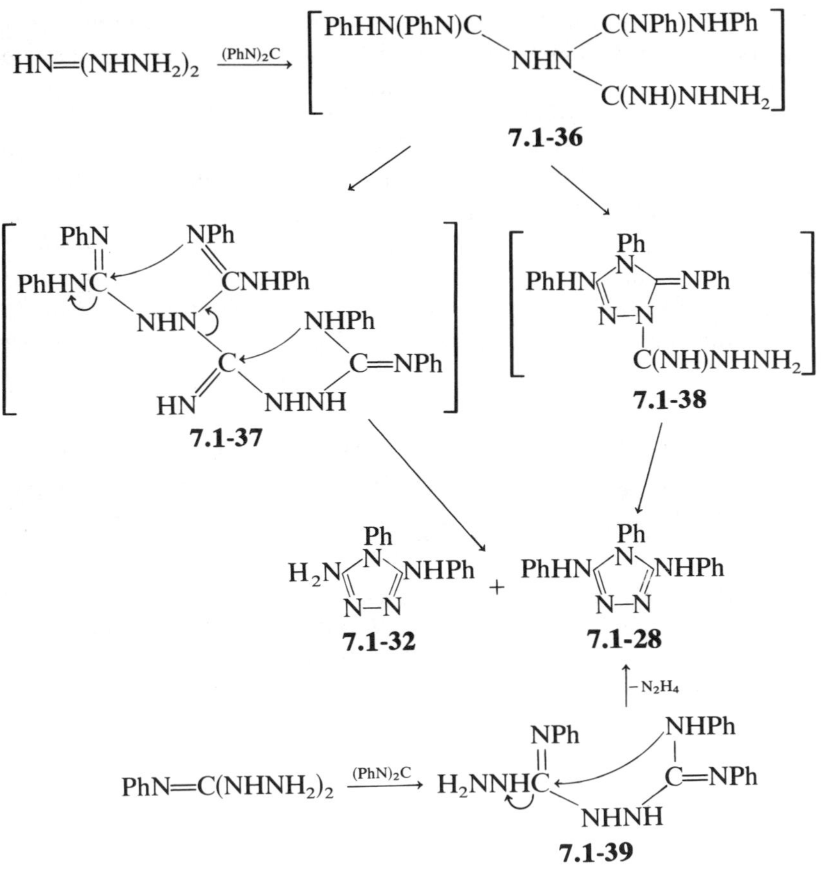

$$HN{=}(NHNH_2)_2 \xrightarrow{(PhN)_2C} \left[\begin{array}{c} PhHN(PhN)C{\diagdown}C(NPh)NHPh \\ NHN \\ C(NH)NHNH_2 \end{array}\right]$$

7.1-36

$$\left[\begin{array}{c} PhN NPh \\ \| \\ PhHNC CNHPh \\ NHN NHPh \\ C C{=}NPh \\ HN NHNH \end{array}\right]$$

7.1-37

$$\left[\begin{array}{c} Ph \\ N \\ PhHN{-}\langle \rangle{=}NPh \\ N{-}N \\ C(NH)NHNH_2 \end{array}\right]$$

7.1-38

$$\underset{\textbf{7.1-32}}{H_2N{-}\!\!\begin{array}{c} Ph \\ N \\ \langle \rangle \\ N{-}N \end{array}\!\!{-}NHPh} \quad + \quad \underset{\textbf{7.1-28}}{PhHN{-}\!\!\begin{array}{c} Ph \\ N \\ \langle \rangle \\ N{-}N \end{array}\!\!{-}NHPh}$$

$$\uparrow {-}N_2H_4$$

$$PhN{=}C(NHNH_2)_2 \xrightarrow{(PhN)_2C} \begin{array}{c} NPh NHPh \\ \| \\ H_2NNHC C{=}NPh \\ NHNH \end{array}$$

7.1-39

In contrast to these results, the blocked N,N'-diaminoguanidine (**7.1-40**) reacted with N,N'-diphenylcarbodiimide to give **7.1-41**, which was cyclized in hot hydrochloric acid to give the 5-hydrazino-s-triazole (**7.1-43**).[18] Also, **7.1-43** was formed by the cyclization of **7.1-42** in either aqueous base or dimethylformamide.[20] In a acidic medium, **7.1-42** was converted to a triazoline-5-thione.

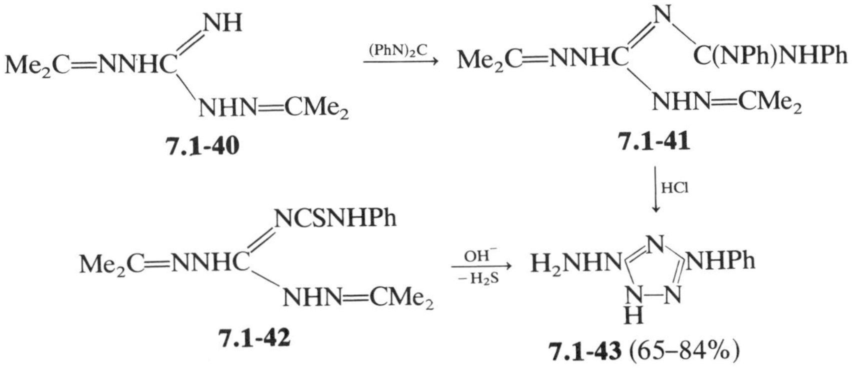

$$\underset{\textbf{7.1-40}}{Me_2C{=}NNHC\!\!\begin{array}{c} NH \\ \diagup \\ \diagdown \\ NHN{=}CMe_2 \end{array}} \xrightarrow{(PhN)_2C} \underset{\textbf{7.1-41}}{Me_2C{=}NNHC\!\!\begin{array}{c} N \\ \diagup \diagdown C(NPh)NHPh \\ \diagdown \\ NHN{=}CMe_2 \end{array}}$$

$$\Big\downarrow \text{HCl}$$

$$\underset{\textbf{7.1-42}}{Me_2C{=}NNHC\!\!\begin{array}{c} NCSNHPh \\ \diagup \\ \diagdown \\ NHN{=}CMe_2 \end{array}} \xrightarrow[-H_2S]{OH^-} \underset{\textbf{7.1-43 }(65–84\%)}{H_2NHN{-}\!\!\begin{array}{c} N \\ \langle \rangle \\ N{-}N \\ H \end{array}\!\!{-}NHPh}$$

Recently, the use of *S*-alkylated derivatives of dithiobiurets for the preparation of 3,5-diaminotriazoles has been investigated. Preliminary work showed that treatment of the 4-*S*-methyl derivative (**7.1-44**) of 1-phenyl-dithiobiuret with hydrazine gave 3-amino-5-anilinotriazole (**7.1-16**) with the elimination of hydrogen sulfide,[21,22] whereas treatment of the 2-ethyl derivative (**7.1-46**) with hydrazine gave the triazoline-5-thione (**7.1-47**) with the elimination of ammonia.[21] Further work on 1,5-*N*,*N*-disubstituted 2-methylthio derivatives of dithiobiuret indicated that *N*-5 alkyl groups favored amine elimination to give triazoline-5-thiones, while *N*-5 aryl groups favored hydrogen sulfide elimination to give 3,5-diaminotriazoles.[23] In addition, *S*,*S*-dimethyl derivatives (e.g., **7.1-45**) of dithiobiuret and hydrazine

gave good yields of 3,5-diaminotriazoles (e.g., **7.1-16**).[24] Similarly, treatment of the 1,2,4-dithiazolium bromide **7.1-48** with hydrazine gave 3,5-bis-(dimethylamino)-1,2,4-triazole (**7.1-50**).[25] In reactions related to those described previously, the chloro groups of **7.1-49** were displaced with phenyl hydrazine to give **7.1-51** (93%).[26,27] Also, **7.1-51** was prepared by amination of **7.1-52** with ammonia.[28]

Several 3,5-diaminotriazoles have been prepared from preformed triazoles. Both 3,5-dichloro- and dibromo-1,2,4-triazoles are aminated with ammonia, aliphatic, and aryl amines at 180°, which is illustrated by the conversion of **7.1-53** to 3,5-diamino-1,2,4-triazole (**7.1-7**).[29,30] In addition,

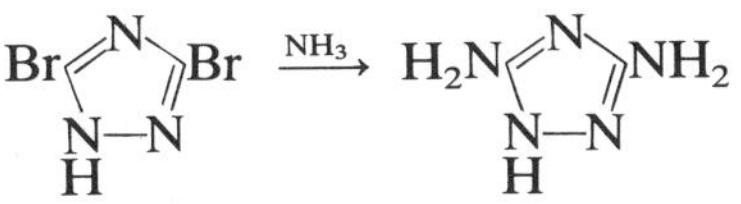

7.1-53   **7.1-7**

catalytic hydrogenation of the nitrotriazole (**7.1-54**) gave **7.1-55**,[31] and the reductive alkylation of **7.1-16** with benzaldehyde gave **7.1-56**.[9]

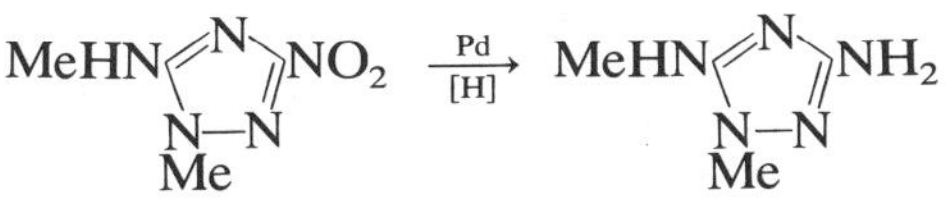

7.1-54   **7.1-55**

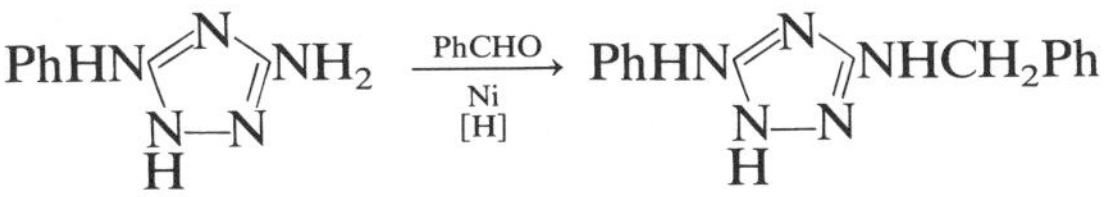

7.1-16   **7.1-56**

References

1. **53:** 3198b	2. **58:** 5568c	3. **74:** 99946a	4. **77:** 164610q
5. **78:** 29676w	6. **82:** 89356m	7. **59:** 8589e	8. **60:** 10674d
9. **77:** 56420w	10. **55:** 2623g	11. **59:** 10052c	12. **56:** 12877h
13. **59:** 2822a	14. **72:** 55345h	15. **81:** 91439m	16. **57:** 12472d
17. **62:** 9124b	18. **63:** 7012g	19. **66:** 115650s	20. **63:** 11545b
21. **68:** 12914t	22. **79:** 126446z	23. **78:** 111227j	24. **70:** 68272r
25. **76:** 149945p	26. **78:** 97563g	27. **80:** 3472t	28. **84:** 30981x
29. **68:** 95761q	30. **70:** 28922w	31. **72:** 111380e	

7.2. Other Diamino- and Triamino-1,2,4-Thiazoles

The interaction of a variety of 5-alkyl-2-amino-1,3,4-oxadiazoles (e.g., **7.2-1**) with refluxing hydrazine hydrate provided a simple method for the preparation of 5-alkyl-3,4-diaminotriazoles (e.g., **7.2-2**).[1-4] The yields are usually poor (<50%) because of side reactions to give semicarbazide, carbohydrazide,[2] and the triazolin-5-one isomeric with the 1,3,4-oxadiazole.[4] In addition, the preparation of 5-aryl-3,4-diaminotriazoles by this

$$7.2\text{-}1 \xrightarrow{N_2H_4} 7.2\text{-}2 \,(28\%)$$

method was unsuccessful.[2] The latter were obtained, however, in four steps from 2-hydrazino-1,3,4-oxadiazoles by a pathway that included the cyclization of **7.2-3** to the bicyclic system (**7.2-4**) followed by treatment of **7.2-4** with aniline to give **7.2-5**.[5]

The conversion of *N*-(acylamino)-*N'*-(arylamino)guanidines to 5-substituted-3-(2-arylhydrazino)-1,2,4-thiazoles in a basic medium was described in Section 6.1b. In a neutral solvent the same *N*-(arylamino)guanidines are dehydrated to give good yields of 3,4-diaminotriazoles and only a minor amount of the isomeric 3-hydrazinotriazoles.[6,7] For example, a refluxing solution of **7.2-6** in butanol gave 3-amino-4-anilino-5-phenyl-4H-1,2,4-triazole (**7.2-7**). This method can also be used to prepare the corresponding

5-alkyltriazoles. In a similar reaction the *N,N'*-(diacylamino)guanidine (**7.2-8**) was cyclized in hot butanol to give a 56% yield of 4-[(2-chlorobenzoyl)-amino]triazole (**7.2-9**) and a 23% yield of 3-[2-(2-chlorobenzoyl)hydrazino]triazole (**7.2-10**).[8] When one of the acyl groups of **7.2-8** is replaced by an carboethoxy group, cyclization might give either a mixture of triazolin-5-ones (basic conditions) or a mixture of 3-hydrazino- and 3,4-diaminotriazoles (neutral conditions).[9] The *in situ* preparation of the acyl intermediate

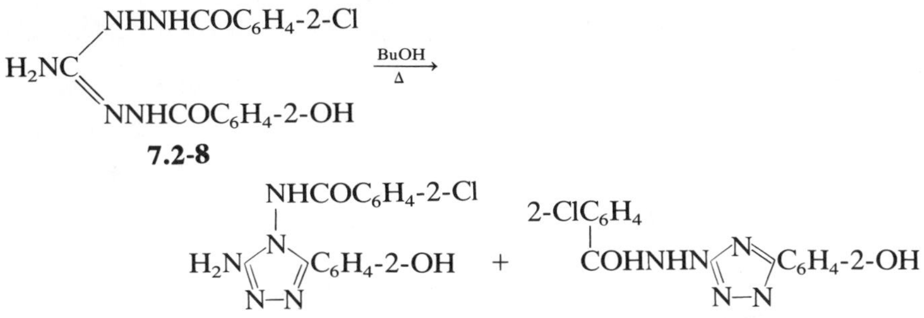

NHNHCOC$_6$H$_4$-2-Cl

H$_2$NC

NNHCOC$_6$H$_4$-2-OH

$\xrightarrow[\Delta]{\text{BuOH}}$

7.2-8

NHCOC$_6$H$_4$-2-Cl

H$_2$N⟨N⟩C$_6$H$_4$-2-OH
      N—N

7.2-9

+

2-ClC$_6$H$_4$

COHNHN⟨N⟩C$_6$H$_4$-2-OH
       N—N

7.2-10

was demonstrated by the condensation of N,N'-diamino-N''-methylguan-idine (**7.2-12**) with hot 85% formic acid to give 4-amino-3-(methylamino)-4$\underline{H}$-1,2,4-triazole (**7.2-11**)[10] and by the cyclization of N,N'-diaminoguan-idine (**7.2-13**) hydroiodide with anhydrous formic acid to give **7.2-14**.[11]

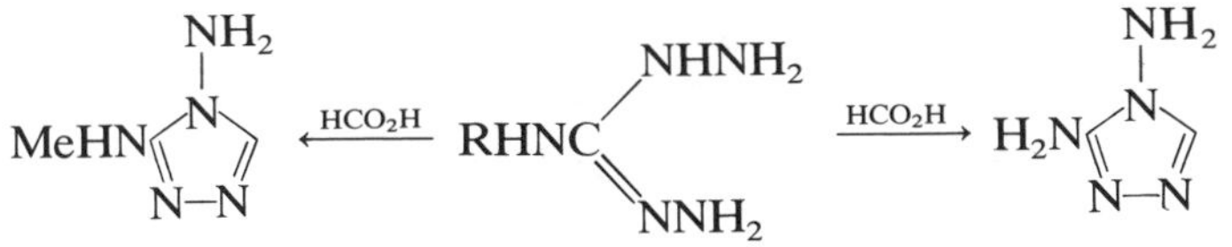

NH$_2$

MeHN⟨N⟩
     N—N

7.2-11

$\xleftarrow{\text{HCO}_2\text{H}}$

NHNH$_2$

RHNC

NNH$_2$

7.2-12, R = Me
7.2-13, R = H

$\xrightarrow{\text{HCO}_2\text{H}}$

NH$_2$

H$_2$N⟨N⟩
     N—N

7.2-14

Another reaction leading to 3,4-diaminotriazoles involved the condensation of glycolic acid cyanohydrazide (**7.2-15**) with hydrazine to give **7.2-16**.[12]

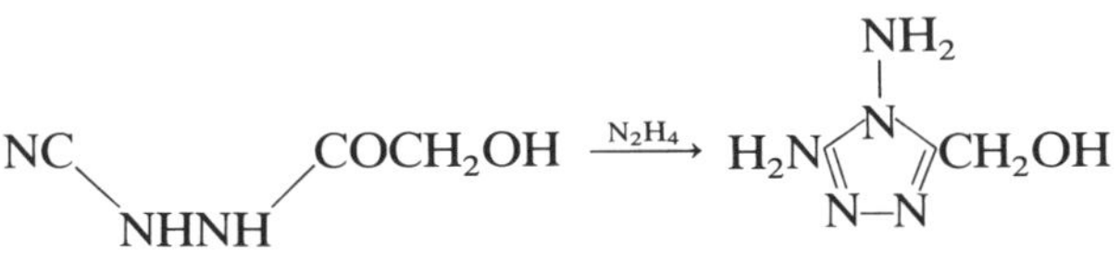

NC

NHNH

COCH$_2$OH

$\xrightarrow{\text{N}_2\text{H}_4}$

NH$_2$

H$_2$N⟨N⟩CH$_2$OH
     N—N

7.2-15   **7.2-16**

The preparation of 4-amino-3-hydrazino-1,2,4-triazole (**7.2-17**) and its 5-alkyl derivatives was accomplished in good yields by the reaction of N,N',N''-triaminoguanidine (**7.2-18**) with aliphatic carboxylic acids.[13–15] Al-though the interaction of **7.2-18** and benzoic acid to give the 5-phenyltria-zole (**7.2-19**) was unsuccessful,[14] treatment of **7.2-18** with benzoyl chloride in refluxing pyridine gave a tribenzoyl triazole, which was hydrolyzed with hydrochloric acid to give **7.2-19**.[16]

7.2-17   **7.2-18**   **7.2-19**

The reaction of hydrazine either with cyanogen bromide or dimethyl-cyanamide readily provided 3,4,5-triamino-1,2,4-triazole (guanazine **7.2-22**).[17,18] In the former reaction, the intermediate cyanohydrazide (**7.2-20**) is dimerized to give **7.2-22**. Apparently, **7.2-20** is also involved in the treatment of thiosemicarbazide with lead (II) oxide to give **7.2-22** in 74% yield.[19] A tautomer of **7.2-20**, the aminocarbodiimide (**7.2-21**), has been shown to be an intermediate in the conversion of **7.2-23** with base to **7.2-22**.[20]

The formation of 3,4-diaminotriazoles from bicyclic systems is illustrated by the hydrazinolysis of **7.2-24** to give **7.2-25** in 90% yield.[21,22] In condensation reactions of 4-amino-3-hydrazinotriazoles with aldehydes, reaction occurs initially at the 3-hydrazino group.[23-25] In contrast, activated esters have been reported to preferentially acylate the 4-amino group of 3,4,5-tri-amino-1,2,4-triazole.[26]

7.2-24

References

1. **55:** 22335b	2. **59:** 2820h	3. **64:** 4935f	4. **76:** 126953a
5. **76:** 85758y	6. **59:** 12788b	7. **70:** 115080p	8. **63:** 13275d
9. **72:** 78953u	10. **59:** 10050a	11. **77:** 19615r	12. **70:** 115076s
13. **59:** 10048f	14. **62:** 11805b	15. **66:** 94965f	16. **65:** 704d
17. **60:** 2951c	18. **63:** 1783b	19. **83:** 164137q	20. **72:** 66185g
21. **76:** 140711b	22. **78:** 29701a	23. **73:** 35288j	24. **78:** 92411w
25. **85:** 21230v	26. **77:** 34570r		

TABLE 7.   DIAMINO- AND TRIAMINO-1,2,4-TRIAZOLES

Compound	Reference

7.1.   3,5-Diamino-1,2,4-triazoles

3-acetamido-5-amino-	**55:** 3565i
3-acetamido-5-anilino-	**77:** 56420u
3-(o-acetamido)anilino-5-amino-	**40:** P1285[8]
3-(m-acetamido)anilino-5-amino-	**40:** P1285[8]
3-(p-acetamido)anilino-5-amino-	**40:** P1285[8]
3(or 5)-(p-acetamidoanilino)-5(or 3)-amino-1-methyl	**40:** P1285[8]
3-acetamido-5-(N-phenylacetamido)-	**56:** 12880a
-1-acrylic acid, 3,5-diamino-α-cyano-	**52:** 4607c
5-allylamino-3-amino-1-phenyl-	**23:** 1640[1]
3-(or 5)-(β-allylthiocarbamido)-5(or 3)-amino-1-phenyl-	**23:** 2178[1]
3-(β-allylthiocarbamido)-5-anilino-	**23:** 2177[9]
3-(β-allylthiocarbamido)-5-anilino-1-phenyl-	**23:** 2177[9]
3-amino-5-anilino-	**56:** 12879f, **59:** 10052d, **62:** 9124c
3-amino-5-anilino-1-methyl-	**77:** 56420u
3-amino-5-anilino-4-phenyl-	**57:** 12472i, **63:** 7013a
3-amino-5-(benzylamino)-	**73:** P25482d, **77:** 56420u
3-amino-5-(benzylidene)amino-	**28:** 2714[7], **47:** 12371h
3-amino-5-(benzylidene)hydrazino-	**48:** 2053d
5-amino-3-[2,2-bis(2-chloroethyl)hydrazino]-1-(2-chloroethyl)-	**64:** 12534b
5-amino-1-bis(dimethylamino)phosphinyl-3-(4-morpholinyl)-	**80:** P82992p
5-amino-1-bis(dimethylamino)phosphinyl-3-(1-piperidinyl)-	**80:** P82992p
3-amino-5-(carboxymethylamino)-, hydrazide	**76:** 46484h, **82:** 98356m
3-amino-5-(p-chloroanilino)-	**68:** 12914t, **77:** 56420u, **85:** 46603g
3-amino-5-(3-chloroanilino)-	**77:** 56420u
3-amino-5-(4-cyanoanilino)-	**68:** 12914t
3-amino-5-(2,4-dibromoanilino)-	**68:** 12914t
3-amino-5-(2,4-dichloroanilino)-	**68:** 12914t
3-amino-5-(2,5-dichloroanilino)-	**68:** 12914t
3-amino-5-(3,4-dichloroanilino)-	**79:** 126446z
3-amino-5-(dimethylamino)-	**60:** 10674d
5-amino-3-(dimethylamino)-1-bis(dimethylamino)-phosphinyl-	**80:** P82992p
3-amino-5-(2,5-dimethylpyrrol-1-yl)-	**52:** 4607a
5-amino-3-[(ethoxyoxalyl)amino]-	**78:** 72011n
3-amino-5-(4-fluoroanilino)-	**77:** 56420u
5-amino-3-(guanidino)-	**76:** P3866s
3-amino-5-hydrazino-	**28:** 2714[3], **85:** 192674x
3-amino-5-(o-hydroxybenzylidene)amino-	**47:** 12371h
5-amino-3-[(hydroxyoxalyl)amino]-	**78:** 72011n
5-amino-3-[(2-hydroxynaphthylidene)amino]-1-phenyl-	**47:** 12371i
5-amino-3-[(2-hydroxypropionyl)amino]-	**81:** 25611k
3-amino-5-(p-methoxyanilino)-	**68:** 12914t, **77:** 56420u
3-amino-5-(3-methylanilino)-	**77:** 56420u

TABLE 7 (Continued)

Compound	Reference

7.1. 3,5-Diamino-1,2,4-triazoles (Continued)

Compound	Reference
3-amino-5-(4-methylanilino)-	**85:** 46603g
3-amino-1-methyl-5-dimethylamino	**72:** 111380e
3-amino-1-methyl-5-methylamino-	**72:** 111380e
3-amino-5-(4-morpholinyl)-	**80:** P82992p, **80:** P95962d
3-amino-5-[(1-naphthalenyl)amino]-	**77:** 56420u
3-amino-5-(2-naphthalenylamino)-	**77:** 56420u
5-amino-3-[(o-nitrobenzylidene)amino]-1-phenyl-	**47:** 12371i
5-amino-3-(4-nitro-2H-1,2,3-triazol-2-yl)-1-phenyl-	**61:** P1873h
3-amino-5-(2-phenethyl)amino-	**77:** 56420u
3(or 5)-amino-1-phenyl-5(or 3)-(β-phenylthiocarbamido)-	**23:** 2177⁹
3-amino-5-(1-piperidinyl)-	**80:** P82992p
3-amino-5-[p-(propionylamino)anilino]-	**40:** 1285⁸
3-amino-5-[p-(sulfamoyl)anilino]-	**39:** P3536⁵, **39:** P5553²
3-amino-5-(o-toluidino)-	**68:** 12914t, **77:** 56420u
3-amino-5-p-toluidino-	**55:** 2624e, **56:** 12879h, **62:** 9124c, **68:** 12914t
3-amino-5-(p-toluidino)-4-p-tolyl-	**63:** 7013a
3-amino-5-(4H-1,2,4-triazol-4-yl)-	**60:** 15861e, **72:** 121448x
5-amino-3-[(2,2,2-trichloro-1-hydroxy)ethylamino]-	**68:** 104067x
3-amino-5-[(tricyclo[3.3.1.1³,⁷]dec-1-yl)amino]-	**77:** 56420u
3-amino-5-(m-trifluoromethylanilino)-	**68:** 12914t
3-amino-5-[p-(trifluoromethyl)anilino]-	**77:** 56420u
3-amino-5-(2,4,6-trinitroanilino)-	**68:** 105111u
5-anilino-3-benzamido-1-phenyl-	**23:** 2177⁹
3-anilino-5-(benzylamino)-	**77:** 56420u
3-(N-anilino-N′,N′-dimethylguanidino)-5-(dimethylamino)-1-phenyl-	**80:** 3472t
3-anilino-5-(3,5-dimethylpyrazol-1-yl)-	**63:** 7012h, **63:** 11545c
3-anilino-5-(diphenylamino)-	**78:** 111227j
3-anilino-5-(N-ethylanilino)-	**78:** 111227j
3-anilino-5-hydrazino-	**63:** 7012h, **63:** 11545c
3-anilino-5-(N-methylanilino)-	**78:** 111227j
5-anilino-1-phenyl-3-(β-phenylthiocarbamido)-	**23:** 2177⁹
3-anilino-5-(β-phenylthiocarbamido)-	**23:** 2177⁹
3-anilino-5-p-toluidino-	**62:** 9124c
3-(4-anisidino)-5-(diphenylamino)-	**78:** 111227j
3-benzamido-5-anilino-	**77:** 56420u
3-(benzylamino)-5-(propylamino)-	**73:** P25482d
3-(benzylidene)amino-5-(benzylidene)hydrazino	**28:** 2714²
3-(benzylidene)amino-5-hydrazino-	**48:** 2053c
3-(benzylidene)hydrazino-5-(α-hydroxybenzylamino)-	**28:** 2714³
4,4′-bi[3,5-diamino-	**78:** 29676w
3,5-bis[(acetyl)acetamido]-1-phenyl-	**41:** P42i
3,5-[bis(3-amino-1H-isoindol-1-ylidene)amino]-	**79:** 93420f
3,5-bis(anilino)-	**70:** P28922w, **71:** P81375s, **71:** P124443g
3,5-bis(anilino)-4-phenyl-	**81:** 91439m
3,5-bis(benzamido)-	**28:** 2714⁷, **81:** 77848n

TABLE 7 (*Continued*)

Compound	Reference

7.1. 3,5-Diamino-1,2,4-triazoles (*Continued*)

Compound	Reference
3,5-bis[(benzoyl)acetamido]-1-phenyl-	**41:** P43a
3,5-bis[(benzylidene)amino]-	**47:** 12371h, **60:** 10674d
3,5-bis[(benzylidene)hydrazino]-	**48:** 2053d
3,5-bis(*p*-bromoanilino)-4-(*p*-bromophenyl)-	**69:** 10398z
3,5-bis[(2-carboxy-1-methylvinyl)amino]-1-phenyl-	**46:** 4534d
3,5-bis(5-carboxyphthalimido)-	**74:** P141817m, **77:** P115048d
3,5-bis(2-chloroacetamido)-	**81:** 25611k
3,5-bis[(1,3-dicarboxybenzoyl)amino]-	**74:** P88677n
3,5-bis(dimethylamino)-	**68:** 95761q, **70:** P28922w, **71:** P124443g
3,5-bis(dimethylamino)-1-methyl-	**76:** 149945p, **84:** 30981x
3,5-bis(dimethylamino)-1-phenyl-	**78:** 97563g, **80:** P146175j, **84:** 30981x
3,5-bis-[4-(dimethylamino)phenyl]-4-methyl-	**81:** P44127y
3,5-bis(ethylamino)-	**73:** P25482d
3,5-bis[(2-formylhydrazono)methylamino]-	**48:** 12092e
3,5-bis[(*o*-hydroxybenzylidene)amino]-	**47:** 12371i
3,5-bis[(*o*-hydroxybenzylidene)amino]-1-phenyl-	**47:** 12371i
3,5-bis(methylamino)-	**81:** 25613n
3,5-bis(3,4-methylenedioxybenzylideneamino)-1-phenyl-	**47:** 12371i
3,5-bis(4-morpholino)-	**68:** 95761q, **70:** P28922w, **71:** P81375s, **71:** P124443g
3,5-bis(phthalimido)-	**77:** P115048d
3,5-bis(1-piperidino)-	**68:** 95761q, **70:** P28922w, **71:** P81375s, **71:** P124443g
3,5-bis(propylamino)-	**73:** P25482d
3,5-bis(2,4,6-trinitroanilino)-	**68:** 105111u
3-*p*-bromoanilino-5-isopropylamino-	**44:** 7850b
3-butylamino-5-*p*-chloroanilino-	**44:** 7850b
3-chloroamino-5-chloroimino-	**28:** 2714^8
3-*p*-chloroanilino-5-ethylamino-	**44:** 7850b
3-(3-chloroanilino)-5-(*N*-ethylanilino)-	**78:** 111227j
3-*p*-chloroanilino-5-isopropylamino-	**44:** 7850a
3-*p*-chloroanilino-5-methylamino-	**44:** 7850a
3-(3-chloroanilino)-5-(*N*-methylanilino)-	**78:** 111227j
3-(4-chloroanilino)-5-(*N*-methylanilino)-	**78:** 111227j
3-(4-cyanoanilino)-5-(diphenylamino)-	**78:** 111227j
3-(2-cyanoanilino)-5-(*N*-methylanilino)-	**78:** 111227j
3-(4-cyanoanilino)-5-(*N*-methylanilino)-	**78:** 111227j
4-cyclohexyl-3,5-bis[(cyclohexyl)amino]-	**72:** 55345h, **81:** 91439m
3,5-diacetamido-	**28:** 2714^7, **55:** 3565h
3,5-di(allylamino)-	**68:** 95761q, **70:** P28922w, **71:** P81375s, **71:** P124443g
3,5-diamino-	**48:** 2050b, **48:** P8268i, **59:** 8589e, **68:** 95761g, **70:** P28922w, **71:** P81375s, **71:** 124443g, **72:** P79052m
3,5-diamino-1-(4-biphenylyl)-	**51:** 17453c

TABLE 7   (*Continued*)

Compound	Reference

7.1. 3,5-Diamino-1,2,4-triazoles   (*Continued*)

Compound	Reference
3,5-diamino-1-(*p*-bromophenyl)-	**48:** 2719i, **53:** 3198c
3,5-diamino-1-(*p*-carboxyphenyl)-	**58:** 5568c
3,5-diamino-1-(*o*-chlorophenyl)-	**55:** P15943i
3,5-diamino-1-(*m*-chlorophenyl)-	**53:** 3198b
3,5-diamino-1-(*p*-chlorophenyl)-	**48:** 675d, **53:** 3198b
3,5-diamino-1-[*p*-(*p*-chlorophenylthio)phenyl]-	**53:** 3198d
3,5-diamino-1-(3-chloro-*p*-tolyl)-	**53:** 3198c, **51:** 17453c
3,5-diamino-1-[5-(2-chlorophenyl)-*s*-triazol-3-yl]-	**77:** 164610q
3,5-diamino-1-(2,6-diamino-4-pyrimidyl)-	**37:** P1131[1]
3,5-diamino-1-(4,6-diamino-2-pyrimidyl)-	**37:** P1131[1]
3,5-diamino-1-(2,4-dichlorophenyl)-	**48:** 675d
3,5-diamino-1-(3,4-dichlorophenyl)-	**48:** 675d
3,5-diamino-1-(3,4-diiodophenyl)-	**58:** 5568c
3,5-diamino-1-(*o*-ethoxyphenyl)-	**73:** P77251x
3,5-diamino-1-(*p*-fluorophenyl)-	**53:** 3198b
3,5-diamino-1-(*m*-iodophenyl)-	**58:** 5568c
3,5-diamino-1-(*p*-iodophenyl)-	**53:** 3198c, **58:** 5568c
3,5-diamino-1-isonicotinoyl-	**53:** 19169d
3,5-diamino-1-(*p*-methoxyphenyl)-	**73:** P77251x
3,5-diamino-1-[5-(2-methoxyphenyl)-*s*-triazol-3-yl]-	**77:** 164610q
3,5-diamino-1-methyl-	**60:** 10674d, **78:** 120168t
3,5-diamino-1-[5-(2-methylphenyl)-*s*-triazol-3-yl]-	**77:** 164610q
3,5-diamino-1-[5-(3-methylphenyl)-*s*-triazol-3-yl]-	**77:** 164610q
3,5-diamino-1-[5-(4-methylphenyl)-*s*-triazol-3-yl]-	**77:** 164610q
3,5-diamino-1-(2-naphthyl)-	**53:** 3198d,
3,5-diamino-1-(*o*-nitrophenyl)-	**55:** P15943i
3,5-diamino-1-phenyl-	**47:** 12371a, **53:** 3198b, **60:** 10674d
3,5-diamino-4-phenyl-	**68:** 12914t, **70:** 68272r
3,5-diamino-1-piperonyl-	**74:** 99946a
3,5-diamino-1-*p*-tolyl-	**47:** 12371g, **53:** 3198c
3,5-diamino-1-[*p*-(*p*-tolyloxy)phenyl]-	**53:** 3198c
3,5-diamino-1-[*p*-(*p*-tolylthio)phenyl]-	**53:** 3198d
3,5-diamino-1-(*s*-triazol-3-yl)-	**77:** 164610q
3,5-diamino-1-(5-*s*-triazol-3-yl)-	**77:** 164610q
3,5-dianilino-	**32:** 3399[5], **59:** 2822a, **62:** 9124c
3,5-dianilino-1-(*p*-bromophenyl)-	**32:** 3399[5]
3,5-dianilino-1-phenyl-	**32:** 3399[5]
3,5-dianilino-4-phenyl-	**57:** 12472i, **63:** 7013a, **66:** 115650s
3,5-dianilino-1-*o*-tolyl-	**32:** 3399[5]
3,5-dianilino-1-*p*-tolyl-	**32:** 3399[5]
3,5-dibenzylamino-	**70:** P28922w, **71:** P81375s, **71:** P124443g
3,5-di(cyclohexylamino)-	**68:** 95761q
3,5-di(methylamino)-	**68:** 95761q, **70:** P28922w, **71:** P81375s

TABLE 7 (*Continued*)

Compound	Reference

7.1. 3,5-Diamino-1,2,4-triazoles (*Continued*)

Compound	Reference
3-(3,5-dimethylpyrazol-1-yl)-5-*p*-toluidino-	**63:** 7012h
3,5-disulfonamido-4-phenyl-	**45:** 4671d
3,5-di-*p*-toluidino-	**70:** P28922w, **71:** P81375s, **71:** P124443g
3,5-di-*p*-toluidino-4-(*p*-tolyl)-	**63:** 7013a
3,5-diureido-	**54:** P6376e
3-ethylamino-5-(diphenylamino)-	**78:** 111227j
3-(*N*-ethylanilino)-5-[2-(trifluoromethyl)anilino]-	**78:** 111227j
3-(*N*-ethylanilino)-5-[3-(trifluoromethyl)anilino]-	**78:** 111227j
3-hydrazino-5-*p*-toluidino-	**63:** 7012h
-4-methanol, 3,5-diamino-α-methyl-	**81:** 3897f
-4-methanol, 3,5-diamino-α-phenyl-	**81:** 3897f
3-(*N*-methylanilino)-5-(2-methylanilino)-	**78:** 111227j
3-(*N*-methylanilino)-5-[2-(trifluoromethyl)anilino]-	**78:** 111227j
3-(*N*-methylanilino)-5-[3-(trifluoromethyl)anilino]-	**78:** 111227j
4-(4-methylphenyl)-3,5-bis(*p*-toluidino)-	**69:** 10398z
3-(4-nitroanilino)-5-(diphenylamino)-	**78:** 111227j
4-phenyl-3,5-bis(*N*-phenylacetamido)-	**57:** 12472i

7.2. Other Diamino- and Triamino-1,2,4-triazoles (*Continued*)

Compound	Reference
4-acetamido-3-amino-	**58:** 9051g
4-acetamido-3-amino-5-benzyl-	**58:** 9051g
4-acetamido-3-amino-5-methyl-	**58:** 9051g
4-acetamido-3-amino-5-propyl-	**58:** 9051g
5-(1-acetyl-4,5-dihydro-1H-pyrazol-3-yl)-4-(diacetyl-amino)-3-(1,2,2-triacetylhydrazino)-	**78:** 29701a
4-amino-3-(15-acetamido-13-isopropylpodocarpa-8,11,13-trien-7-idenehydrazino)-	**73:** 75210j
4-amino-3-[(aminocarbonylmethyl)amino]-5-benzyl-	**73:** 87862m
4-amino-3-(15-amino-13-isopropylpodocarpa-8,11,13-trien-7-idenehydrazino)-	**73:** 75210j
4-amino-3-anilino-	**66:** 115650s
3-amino-4-anilino-5-benzyl-	**59:** 12789b
4-amino-3-anilino-5-benzyl-	**71:** 91382f
4-amino-3-anilino-5-butyl-	**71:** 91382f
4-amino-3-anilino-5-ethyl-	**71:** 91382f
4-amino-5-anilino-3-(*o*-hydroxyphenyl)-	**71:** 91382f
3-amino-4-anilino-5-isobutyl-	**59:** 12788h
4-amino-3-anilino-5-isobutyl-	**71:** 91382f
4-amino-3-anilino-5-isopropyl-	**71:** 91382f
4-amino-3-anilino-5-methyl-	**71:** 91382f
3-amino-4-anilino-5-phenyl-	**59:** 12789b
4-amino-3-anilino-5-phenyl-	**45:** 1122b
4-amino-5-anilino-3-(4-pyridyl)-	**71:** 91382f
4-amino-3-*p*-anisidino-5-methyl-	**71:** 91382f
4-amino-3-benzamido-	**77:** 341b
3-amino-4-benzamido-5-benzyl-	**58:** 9051h

TABLE 7 (*Continued*)

Compound	Reference

7.2. Other Diamino- and Triamino-1,2,4-triazoles (*Continued*)

Compound	Reference
3(or 4)-amino-4(or 3)-benzamido-5-phenyl-	**44:** 6856e
3-amino-4-(benzylamino)-5-butyl-	**70:** 115080p
3-amino-4-(benzylamino)-5-isobutyl-	**70:** 115080p
3-amino-4-(benzylamino)-5-isopropyl-	**70:** 115080p
3-amino-4-(benzylamino)-5-phenyl-	**70:** 115080p
3-amino-4-(benzylamino)-5-propyl-	**70:** 115080p
4-amino-5-benzyl-3-[(carboxymethyl)amino]-	**73:** 87862m
3-amino-5-benzyl-4-(*p*-chloroanilino)-	**59:** 12789b
3(or 4)-amino-4(or 3)-(benzylidene)amino-	**48:** 5182b, **59:** 2821a
4-amino-5-benzyl-3-[(ethoxycarbonylmethyl)amino]-	**73:** 87862m
4-amino-5-benzyl-3-[(ethoxyoxalyl)amino)-	**73:** 87862m
3-amino-5-benzyl-4-[(3-oxobutyryl)amino]-	**76:** 126953a
3-amino-4,5-bis(benzylideneamino)-	**63:** 1783c
4-amino-3-[[4-[bis(2-chloroethyl)amino]benzylidene]- hydrazino]-	**73:** 35288j
4-amino-3-[[4-[bis(2-chloroethyl)amino]benzylidene]- hydrazino]-5-ethyl-	**73:** 35288j
4-amino-3-[[*p*-[bis(2-chloroethyl)amino]benzylidene]- hydrazino]-5-methyl-	**73:** 35288j
4-amino-3-[[4-[bis(2-chloroethyl)amino]-2-chloro- benzylidene]hydrazino]-	**73:** 35288j
4-amino-3-[[4-(bis(2-chloroethyl)amino]-2,5-dimethy- oxybenzylidene]hydrazino]-	**73:** 35288j
4-amino-3-[[4-[bis(2-chloroethyl)amino]-3-ethoxy- benzylidene]hydrazino]-	**73:** 35288j
4-amino-3-[[4-[bis(2-chloroethyl)amino]-5-ethoxy-2- methylbenzylidene]hydrazino]-	**73:** 35288j
4-amino-3-[[4-[bis(2-chloroethyl)amino]-2-methoxy- benzylidene]hydrazino]-	**73:** 35288j
4-amino-3-[[4-[bis(2-chloroethyl)amino]-3-methoxy- benzylidene]hydrazino]	**73:** 35288j
4-amino-3-[[4-[bis(2-chloroethyl)amino]-2-methyl- benzylidene]hydrazino]-5-methyl-	**73:** 35288j
4-amino-3-[(carboxymethyl)amino]-5-ethyl-	**73:** 87862m
4-amino-3-[*N*-(carboxymethyl)anilino]-5-methyl-	**73:** 87862m
3-amino-4-(*p*-chloroanilino)-5-isobutyl-	**59:** 12789b
4-amino-3-(*o*-chloroanilino)-5-methyl-	**71:** 91382f
3-amino-4-(*p*-chloroanilino)-5-phenyl-	**59:** 12789b
3-amino-4-(*o*-chlorobenzamido)-5-(*o*-hydroxyphenyl)-	**63:** P13275g
3(or 4)-amino-4(or 3)-(*p*-chlorobenzylideneamino)-	**48:** 5182b
3-amino-4-[(4-chlorobenzylidene)amino]-5- (4-chlorophenyl)-	**83:** 126029y
4-amino-3,5-dianilino-	**49:** 2446h
4-amino-3-(2,6-dichloroanilino)-	**85:** 21230v
4-amino-3-(2,4-dichlorobenzylidenehydrazino)-	**82:** P170960y
4-amino-3-[1-[(2,6-dichlorophenyl)methyl]hydrazino]-	**85:** 21230v
4-amino-3,5-dihydrazino-	**47:** 9863a, **63:** P4305g, **68:** P106539b

TABLE 7 (*Continued*)

Compound	Reference

7.2. Other Diamino- and Triamino-1,2,4-triazoles (*Continued*)

Compound	Reference
4-amino-5-(4,5-dihydro-1H-pyrazol-3-yl)-3-(isopropylidenehydrazino)-	**78:** 29701a
4-amino-3-(3,5-dimethylpyrazol-1-yl)-	**59:** 10049d
4-amino-3-(3,5-dimethylpyrazol-1-yl)-5-methyl-	**62:** 11805b
4-amino-3-[(2,5-dimethylpyrrol-1-yl)amino]-	**59:** 10050a
4-amino-3-[(ethoxyoxalyl)amino]-5-ethyl-	**73:** 87862m
4-amino-3-[(ethoxyoxalyl)amino]-5-methyl-	**73:** 87862m
4-amino-3-[(ethoxyoxalyl)amino]-5-phenyl-	**73:** 87862m
3-amino-4-(ethoxycarbonyl)amino)-5-(*o*-hydroxyphenyl)-	**72:** 78953u
3-amino-4-[(ethoxycarbonyl)amino]-5-isobutyl-	**72:** 78953u
3-amino-4-[(ethoxycarbonyl)amino]-5-isopentyl-	**72:** 78953u
3-amino-4-[(ethoxycarbonyl)amino]-5-phenyl-	**72:** 78953u
4-amino-3-[(ethoxycarbonylmethyl)amino]-	**73:** 87862m
4-amino-3-[(ethoxycarbonylmethyl)amino]-5-methyl-	**73:** 87862m
4-amino-3-[*N*-(ethoxycarbonylmethyl)anilino]-5-benzyl-	**73:** 87862m
4-amino-3-[*N*-(ethoxycarbonylmethyl)anilino]-5-ethyl-	**73:** 87862m
4-amino-3-[*N*-(ethoxycarbonylmethyl)anilino]-5-methyl-	**73:** 87862m
4-amino-3-ethyl-5-hydrazino-	**62:** 11805b
4-amino-5-ethyl-3-hydrazino-, 2,6-dichlorobenzaldehyde hydrazone	**82:** P170960y
4-amino-3-ethyl-5-(methylamino)-	**59:** 10050h
4-amino-3-hydrazino-	**59:** 10048f, **64:** 12534b, **66:** 94965f, **82:** P170960y
4-amino-3-hydrazino-, [2-(1,3-benzodioxol-5-yl)-1-methylethylidene) hydrazone	**85:** 21230v
4-amino-3-hydrazino-, 2-chlorobenzaldehyde hydrazone	**82:** P170960y
4-amino-3-hydrazino-, [1-(4-chlorophenyl)ethylidene] hydrazone	**85:** 21230v
4-amino-3-hydrazino-, 2,6-dichlorobenzaldehyde hydrazone	**82:** P170960y
4-amino-3-hydrazino-, 2,6-dimethylbenzaldehyde hydrazone	**82:** P170960y
4-amino-3-hydrazino-5-methyl-	**62:** 11805b, **80:** P59693h, **82:** P155791a, **82:** P170960y
4-amino-3-hydrazino-, 2-methylbenzaldehyde hydrazone	**82:** P170960y
4-amino-3-hydrazino-5-methyl-, 2,6-dichlorobenzaldehyde hydrazone	**82:** P170960y
4-amino-3-hydrazino-5-phenyl-	**65:** 704d
4-amino-3-hydrazino-, (phenyl-2-pyridinylmethylene) hydrazone	**85:** 21230v
4-amino-3-hydrazino-, 2,4,6-trimethylbenzaldehyde hydrazone	**82:** P170960y
3-amino-5-isobutyl-4-ureido-	**67:** 43756b
3-amino-5-isopentyl-4-ureido-	**67:** 43756b
3-amino-5-isopropyl-4-ureido-	**67:** 43756b
3(or 4)-amino-4(or 3)-(*p*-methoxybenzylideneamino)-	**48:** 5182b
4-amino-3-(methylamino)-	**59:** 10050f
4-amino-3-methylamino-5-phenyl-	**45:** 1122e

TABLE 7 (*Continued*)

Compound	Reference

7.2. Other Diamino- and Triamino-1,2,4-triazoles (*Continued*)

Compound	Reference
4-amino-3-methyl-5-(methylamino)-	**59:** 10050g
3-amino-5-methyl-4-[(5-nitrofurfurylidene)amino]-	**56:** 14194a
4-amino-3-methyl-5-*p*-toluidino-	**71:** 91382f
5-amino-1-methyl-3-(4<u>H</u>-1,2,4-triazol-4-yl)-	**72:** 121448x
4-amino-3-[[3-(5-nitro-2-furanyl)-2-propenylidene]-hydrazino]-	**78:** 92411w
3-amino-4-[(5-nitrofurfurylidene)amino]-	**56:** 14194a, **61:** 14663a
3-amino-5-pentyl-4-ureido-	**67:** 43756b
5-amino-1-phenyl-3-(4<u>H</u>-1,2,4-triazol-4-yl)-	**72:** 121448x
3-amino-5-propyl-4-ureido-	**67:** 43756b
3-anilino-4-benzamido-5-(4-nitrophenyl)-	**76:** 85758v
4-anilino-3-benzyl-5-(benzylideneamino)-	**59:** 12789b
4-anilino-5-benzyl-3-[(carboxymethyl)amino]-	**73:** 87862m
4-anilino-5-benzyl-3-[(ethoxycarbonylmethyl)amino]-	**73:** 87862m
3-anilino-5-benzyl-4-[2-[1-(ethoxycarbonyl)-1-propenyl]-amino-	**71:** 91382f
3-anilino-4-(benzylideneamino)-5-ethyl-	**71:** 91382f
4-anilino-3-(benzylideneamino)-5-isobutyl-	**59:** 12789a
4-anilino-3-(benzylideneamino)-5-phenyl-	**59:** 12789b
3-anilino-5-benzyl-4-[(*o*-nitrobenzylidene)amino]-	**71:** 91382f
3-anilino-5-butyl-4-[(*o*-hydroxylbenzylidene)amino]-	**71:** 91382f
4-anilino-3-[(carboxymethyl)amino]-5-isobutyl-	**73:** 87862m
4-anilino-3-[(ethoxycarbonylmethyl)amino]-5-isobutyl-	**73:** 87862m
3-anilino-4-[2-[1-(ethoxycarbonyl)-1-propenyl]amino]-5-ethyl-	**71:** 91382f
3-anilino-4-[2-[1-(ethoxycarbonyl)-1-propenyl]amino]-5-isobutyl-	**71:** 91382f
3-anilino-4-[2-[1-(ethoxycarbonyl)-1-propenyl]amino]-5-methyl-	**71:** 91382f
3-anilino-5-ethyl-4-(formamido)-	**71:** 91382f
3-anilino-4-(formamido)-5-isobutyl-	**71:** 91382f
3-anilino-4-formamido-5-methyl-	**71:** 91382f
3-anilino-4-[(*o*-hydroxybenzylidene)amino]-5-methyl-	**71:** 91382f
3-anilino-5-isobutyl-4-[(*o*-nitrobenzylidene)amino]-	**71:** 91382f
3-*p*-anisidino-4-(benzylideneamino)-5-methyl-	**71:** 91382f
3-*p*-anisidino-4-[2-[1-(ethoxycarbonyl)-1-propenyl]-amino]-5-methyl-	**71:** 91382f
4-(benzylamino)-3-butyl-5-[(*p*-nitrobenzylidene)amino]-	**70:** 115080p
4-(benzylamino)-3-isobutyl-5-[(*p*-nitrobenzylidene)amino]-	**70:** 115080p
4-(benzylamino)-3-isopropyl-5-[(*p*-nitrobenzylidene)amino]-	**70:** 115080p
4-(benzylamino)-3-[(*p*-nitrobenzylidene)amino]-5-phenyl-	**70:** 115080p
4-(benzylamino)-3-[(*p*-nitrobenzylidene)amino]-5-propyl-	**70:** 115080p
5-benzyl-3-[(ethoxyoxalyl)amino]-4-(salicylideneamino)-	**73:** 87862m
4-(benzylideneamino)-3-(*o*-chloroanilino)-5-methyl-	**71:** 91382f
4-(benzylideamino)-3-(3,5-dimethylpyrazol-1-yl)-	**59:** 10049d
4-(benzylideamino)-3-(3,5-dimethylpyrazol-1-yl)-5-methyl-	**62:** 11805b
4-(benzylideneamino)-3-[(2,5-dimethylpyrrol-1-yl)amino]-	**59:** 10050a

222

TABLE 7 (*Continued*)

Compound	Reference

7.2. Other Diamino- and Triamino-1,2,4-triazoles (*Continued*)

Compound	Reference
4-(benzylideneamino)-3-ethyl-5-(methylamino)-	**59:** 10050h
4-(benzylideneamino)-3-(*o*-hydroxyanilino)-5-isopropyl-	**71:** 91382f
4-(benzylideneamino)-3-(methylamino)-	**59:** 10050f
4-(benzylideneamino)-3-methyl-5-(methylamino)-	**59:** 10050g
4-(benzylideneamino)-3-methyl-5-*p*-toluidino-	**71:** 91382f
3,4-bis(benzylideneamino)-5-isobutyl-	**67:** 43756b
3,4-bis(benzylideneamino)-5-isopentyl-	**67:** 43756b
3,4-bis(benzylideneamino)-5-isopropyl-	**59:** 2821b, **67:** 43756b
3,4-bis(benzylideneamino)-5-methyl-	**48:** 1344a, **59:** 2821b
3,4-bis(benzylideneamino)-5-pentyl-	**59:** 2821b, **67:** 43756b
3,4-bis(benzylideneamino)-5-phenyl-	**44:** 6856c
3,4-bis(benzylideneamino)-5-propyl-	**59:** 2821b, **67:** 43756b
4-[[[4-[bis(2-chloroethyl)amino]benzylidene]amino]-3-[[(5-nitro-2-furanyl)methylene]hydrazino]-	**78:** 92412x
4-[[[4-[bis(2-chloroethyl)amino]benzylidene]amino]-3-[[3-(5-nitro-2-furanyl)-2-propenylidene]hydrazino]-	**78:** 92412x
3-[[4-[bis(2-chloroethyl)amino]benzylidene]hydrazino]-4-[[2-(5-nitro-2-furanyl)-2-propenylidene]amino]-	**78:** 92412x
3,4-bis(*p*-methoxybenzylideneamino)-5-phenyl-	**44:** 6856c
-4-carbamic acid, 3-amino-5-(*o*-hydroxyphenyl)-, ethyl ester	**70:** P68377d, **72:** 78953u
-4-carbamic acid, 3-amino-5-(*o*-hydroxyphenyl)-, methyl ester	**73:** 87864p
-4-carbamic acid, 3-amino-5-phenyl-, ethyl ester	**70:** P68377d, **72:** 78953u
3,4-diacetamido-	**58:** 9052a
3,4-diacetamido-5-benzyl-	**58:** 9052d
3,4-diacetamido-5-methyl-	**58:** 9052b
3,4-diacetamido-5-propyl-	**58:** 9052d
3,4-diamino-	**48:** 5182b, **59:** 2821a, **61:** 14662h, **77:** 19615r, **83:** 193255g
3,4-diamino-5-benzyl-	**50:** 9396g, **59:** 2821b, **76:** 126953a
3,5-diamino-4-benzylideneamino-	**63:** 1783c
3,4-diamino-5-butyl-	**55:** 22335c
3,4-diamino-5-(3-chlorophenyl)-	**76:** 140711b
3,4-diamino-5-(*p*-chlorophenyl)-	**45:** 1122e
3,5-diamino-4-[(3,5-diamino-6-chloropyrazine-carbonyl)amino]-	**77:** P34570r
4-[(2,6-dichlorobenzylidene)amino]-3,5-bis[2,6-dichlorobenzylidene)hydrazino]-	**80:** P59693h, **82:** P155791a
3,4-diamino-5-(4,5-dihydro-1H-pyrazol-3-yl)-	**78:** 29701a
5-ethyl-3-[(ethoxyoxalyl)amino]-4-(salicylideneamino)-	**73:** 87862m
3,4-diamino-5-ethyl-	**50:** 9396f, **59:** 2821b
3,4-diamino-5-heptyl-	**55:** 22335c
4,5-diamino-3-(*o*-hydroxyphenyl)-	**70:** P68377d
3,4-diamino-5-isobutyl-	**50:** 9396f, **59:** 2821b
3,4-diamino-5-isopentyl-	**67:** 43756b

223

TABLE 7 (*Continued*)

Compound	Reference
7.2. Other Diamino- and Triamino-1,2,4-triazoles (*Continued*)	

Compound	Reference
3,4-diamino-5-isopropyl-	**59:** 2821b
3,4-diamino-5-(*p*-methoxyphenyl)-	**45:** 1122b
3,4-diamino-5-methyl-	**48:** 598g, **59:** 2821b,
	82: 57659u
3,4-diamino-5-nonyl-	**55:** 22335c
3,4-diamino-5-pentyl-	**55:** 22335c, **59:** 2821b,
	67: 43756b
3,4-diamino-5-phenethyl-	**50:** 9396g, **59:** 2821b
3,4-diamino-5-phenyl-	**44:** 6856a, **50:** 9396g
	70: P68377d, **76:** 140711b
3,4-diamino-5-piperonyl-	**67:** 32651n
3,4-diamino-5-propyl-	**59:** 2821b
3,5-dianilino-,	**59:** 2822a
3,4-dibenzamido-5-methyl-	**58:** 9052c
4-[(2,6-dichlorobenzylidene)amino]-3-[(2,6-dichloro-benzylidene)hydrazino]-5-methyl-	**80:** P59693h, **82:** P155791a
5-(4,5-dihydro-1H̲-pyrazol-3-yl)-4-(isopropylideneamino)-3-(isopropylidenehydrazino)-	**78:** 29701a
-3-methanol, 4,5-diamino-	**70:** 115076s
5-methyl-3-[(ethoxyoxalyl)amino]-4-(salicylideneamino)-	**73:** 87862m
4-[[3-(5-nitro-2-furanyl)-2-propenylidene]amino]-3-[[3-(5-nitro-2-furanyl)-2-propenylidene]hydrazino]-	**78:** 92411w
3,4,5-triamino-	**60:** 2951c, **63:** 1783b,
	64: 4935f, **83:** 164137q
3,4,5-tri(benzylideneamino)-	**72:** 66185g

Azido-, Azo-, Diazo-, Diazoamino (Triazeno-)-, Nitramino-, Nitrosamino-, and Nitro-1,2,4-Triazoles Containing One or More of These Functional Groups

A large number of the azotriazoles were prepared by the diazotization and coupling of 3-amino-1,2,4-triazole with electron-rich aromatic systems. For example, the interaction of **8.1-2** with N,N-diethylaniline occurred readily to give a 36% yield of 3-[[4-(diethylamino)phenyl]azo]-1,2,4-triazole (**8.1-1**).[1,2] In general, low yields of azotriazoles were obtained in the coupling of **8.1-2** with either phenol ethers or aromatic hydrocarbons.[3] The azotriazoles prepared by this route are usually converted to cationic dyes, which result from quaternarization of the triazole ring with alkylating reagents [see Section 22.3]. In addition to these couplers, the anions of nitroalkanes react with **8.1-2** to give azotriazoles (e.g., **8.1-3**).[4]

8.1-1   **8.1-2**   **8.1-3**

Azotriazoles have also been prepared from 2-amino-1,3,4-oxadiazoles by a three-step procedure.[5-7] This method is illustrated by the reaction of phenylhydrazine hydrochloride with 2-amino-5-isobutyl-1,3,4-oxadiazole (**8.1-4**) to give the N,N'-diaminoguanidine (**8.1-5**), which is oxidized by oxygen in strong base to give the formazan (**8.1-7**), followed by cyclization of the latter to give **8.1-6**. In the absence of alkali, 3-(2-arylhydrazino)trio-

iso-Bu $\overset{O}{\diagup}$ NH$_2$ $\xrightarrow{\text{PhNHNH}_2 \cdot \text{HCl}}$ iso-BuCO $\diagdown$ C(NH)NHNHPh

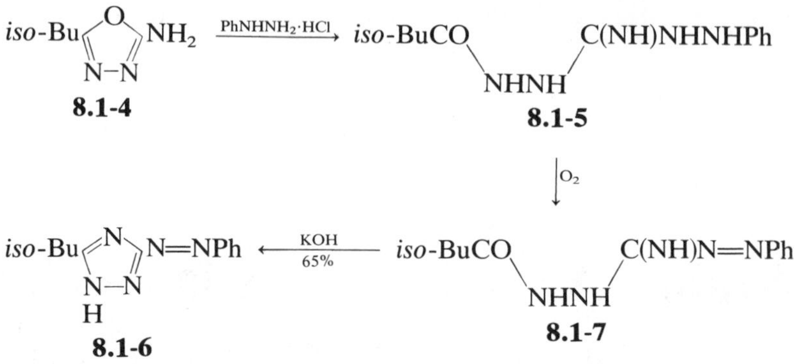

N–N   NHNH

8.1-4   **8.1-5**

$\Big\downarrow$ O$_2$

iso-Bu $\overset{N}{\diagup}$ N=NPh $\xleftarrow[\text{65\%}]{\text{KOH}}$ iso-BuCO $\diagdown$ C(NH)N=NPh

N–N   NHNH

H

8.1-6   **8.1-7**

zoles (e.g., **8.1-8**) can be prepared, which are oxidized to azotriazoles as demonstrated by the treatment of **8.1-8** with ethanolic iodine to give **8.1-9**. The latter was quantitatively reduced to **8.1-8** with ethanolic ammonium sulfide.

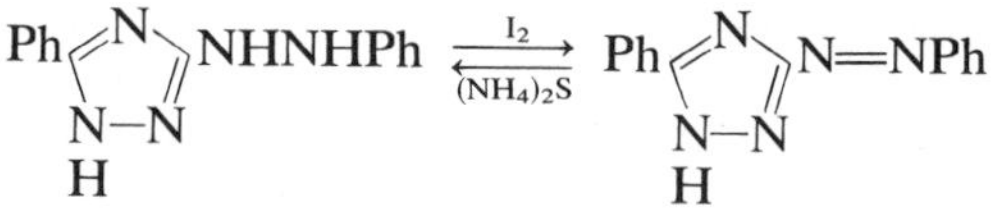

Ph $\overset{N}{\diagup}$ NHNHPh $\underset{(NH_4)_2S}{\overset{I_2}{\rightleftarrows}}$ Ph $\overset{N}{\diagup}$ N=NPh

N–N   N–N

H   H

8.1-8   **8.1-9**

The oxidative coupling of 5-amino-1-methyl-1,2,4-triazole (**8.1-10**) to give 5,5′-azobis(1-methyl-1,2,4-triazole) (**8.1-11**) was effected either with potassium permanganate in base[8,9] or with sodium nitrite in 60% hydrobromic acid.[10] The dissimilarity in melting points of **8.1-11** obtained by these

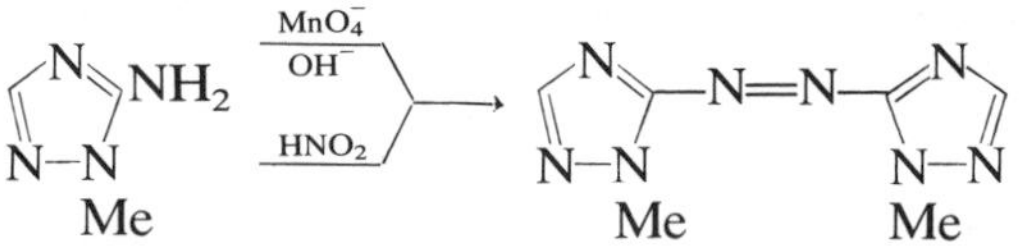

$\overset{N}{\diagup}$ NH$_2$ $\overset{\overset{MnO_4^-}{OH^-}}{\underset{HNO_2}{\diagup}}$ $\overset{N}{\diagup}$ –N=N– $\overset{N}{\diagdown}$

N–N   N–N   N–N

Me   Me   Me

8.1-10   **8.1-11**

two methods suggested that one of the samples was contaminated with an impurity. The 4,4′-azobis(3,5-dimethyl-1,2,4-triazole) (**8.1-13**) was reported to result from oxidation of **8.1-12** with acidified potassium bromate.[11]

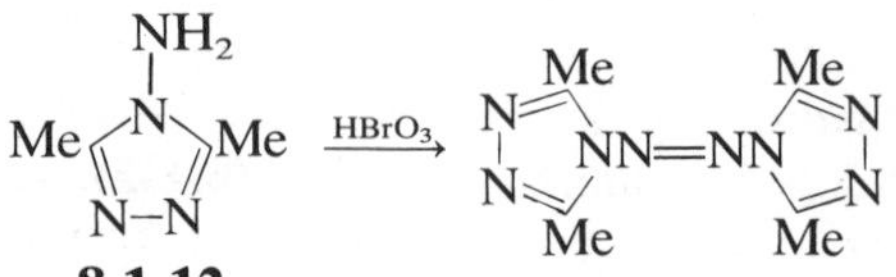

NH$_2$

Me $\overset{N}{\diagup}$ Me $\xrightarrow{\text{HBrO}_3}$ $\overset{Me}{\underset{N}{\diagup}}$ N–N=N–N $\overset{Me}{\diagdown}$

N–N   Me   Me

8.1-12   **8.1-13**

An investigation of the nitrosation of 4,5-disubstituted 3-aminotriazoles showed that either triazenes or nitrosamines could be isolated depending on the concentration of hydrochloric acid.[12] For example, nitrosation of the 3-aminotriazole (**8.1-15**) in 10% hydrochloric acid at 20° gave the triazene (**8.1-14**) (35%) and in 18% hydrochloric acid at 0° the nitrosamine (**8.1-17**) (65%). Reduction of the triazene (**8.1-14**) with zinc dust in aqueous acetic acid gave the expected mixture of the corresponding 3-amino- and 3-hydrazino-triazoles (**8.1-15** and **8.1-16**, respectively). Also, reduction of **8.1-17** with zinc dust and acetic acid gave **8.1-16**, and the sodium salt of (**8.1-17**) was shown to couple with N,N-dimethylaniline in the presence of sulfuric acid to give the azotriazole (**8.1-20**) (66%). In the nitrosation of **8.1-18**, the 4-N-nitroso derivative (**8.1-19**) was obtained in 60% yield.[13] Cleavage of the N—N bond at the 4-position of **8.1-19** was effected with sodium hydroxide to give 1,2,4-triazole.[14] Both deamination and nitrosation occured in the conversion of

8.1-14   **8.1-15**, R = H   **8.1-17**
                    8.1-16, R = NH$_2$

8.1-18   **8.1-19**   **8.1-20**

8.1-21 to **8.1-22** with nitrous acid.[15] In addition, treatment of 3-amino-5-nitrosamino triazole (**8.1-23**) with concentrated hydrochloric acid gave a

8.1-21   **8.1-22**

solution of the diazonium compound (**8.1-24**), which was coupled with 3,5-diaminotriazole to give the triazene (**8.1-25**) and was neutralized with base to give poly(3-diazoamino-1,2,4-triazole) (**8.1-26**).[16] Further reaction of the dimer (**8.1-25**) with **8.1-24** was shown to give the trimer (**8.1-27**). The sodium salt of 4-nitroaminotriazole (**8.1-29**) was prepared in 90% yield by nitration of 4-aminotriazole (**8.1-28**) with ethyl nitrate in ethanolic sodium ethoxide.[17] Also, 3,5-dimethyl-4-nitroaminotriazole was prepared from the

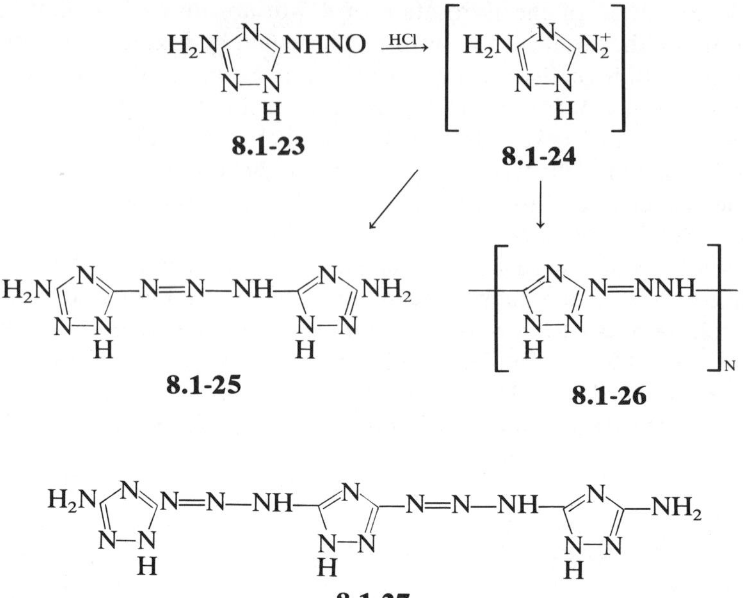

8.1-23   **8.1-24**

8.1-25   **8.1-26**

8.1-27

4-aminotriazole and nitryl tetrafluoroborate.[18] In addition, a series of 3-nitroamino-5-alkyl and aryl-1,2,4-triazoles (e.g., **8.1-31**) were prepared in 44 to 99% yields by dehydration of N-(acylamino)-N'-nitroguanidines (e.g., **8.1-30**) with aqueous sodium hydroxide.[19,20]

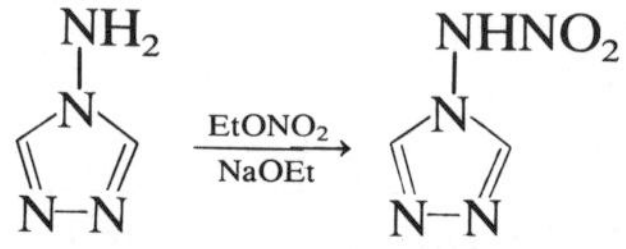

8.1-28   **8.1-29**

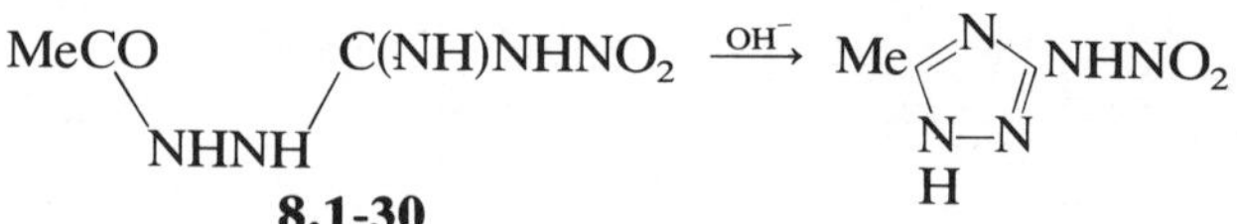

8.1-30   **8.1-31**

The addition of a nitric acid solution of 3-diazo-1,2,4-triazole (**8.1-2**) to an aqueous solution of sodium nitrite gave, after the addition of acetic acid and warming to 50°, a 58% yield of 3-nitro-1,2,4-triazole (**8.1-33**).[21] This nitro compound has also been obtained by the addition of a solution of 3-amino-1,2,4-triazole (**8.1-32**) in 10% sulfuric acid to a 10% aqueous solution of sodium nitrile at 45°.[22] Alkyl-substituted 3-aminotriazoles were also converted to the corresponding 3-nitrotriazoles by this procedure.[22,23]

8.1-32 → [8.1-2] → 8.1-33

The direct nitration of **8.1-34** gave the *N*-nitro compound (**8.1-35**), which rearranged to **8.1-36** on heating at 120°.[24] Oxidation of an amino group to a nitro group is also possible, as demonstrated by treatment of **8.1-32** with perfluoroacetic acid to give a 45% yield of **8.1-33**.[25] The methylation of **8.1-33** with dimethylsulfate under alkaline conditions gave a 66% yield of **8.1-37**, whereas treatment of **8.1-33** with diazomethane in a mixture of dioxane and ether gave a 76% yield of **8.1-37** and a 24% yield of **8.1-38**.[23]

$4\text{-}O_2NC_6H_4$ — N — (ring) **8.1-34** ⟶ $4\text{-}O_2NC_6H_4$ — N — (ring, NO₂) **8.1-35** ⟶ $4\text{-}O_2NC_6H_4$ — N — NO₂ **8.1-36**

8.1-37   **8.1-38**

In the diazotization of 3,4,5-triamino-1,2,4-triazole (**8.1-39**) in the presence of copper (II) nitrate, the 4-amino group was lost, and both the 3- and 5-amino groups were replaced with nitro groups to give 3,5-dinitro-1,2,4-triazole (**8.1-40**).[26] Also, **8.1-40** was obtained from 3,5-diaminotriazole (**8.1-41**).[22,27] The addition of nitroalkenes and epoxides to **8.1-40** and the alkylation of the silver or sodium salt of **8.1-40** has been reported to

H_2N — N(NH₂) — NH₂ **8.1-39** $\xrightarrow{\text{NaNO}_2}_{\text{Cu(NO}_3)_2}$ O_2N — N — NO₂ **8.1-40** $\xleftarrow{\text{NaNO}_2}_{\text{H}_2\text{SO}_4}$ H_2N — N — NH₂ **8.1-41**

give 1-alkyl-3,5-dinitro-1,2,4-triazoles.[27–31] The 1-methyl derivative (**8.1-43**) contains an activated 5-nitro group, as shown by displacement of this group with ammonium hydroxide at 80° to give **8.1-42**.[32] The same compound was obtained in 86% yield by reduction of **8.1-43** with 25%

hydrazine hydrate.[33] In contrast, treatment of **8.1-43** with 95% hydrazine hydrate displaced the 5-nitro group to give a 70% yield of the corresponding 5-hydrazinotriazole **8.1-44**.

Azidotriazoles such as the 5-phenyl derivative **8.1-46** are prepared in high yields by the nitrosation of the corresponding 3-hydrazinotriazoles (e.g., **8.1-45**).[34] The preparation of 3-azidotriazoles (e.g., **8.1-46**) substituted with various groups has also been accomplished by diazotization of 3-aminotriazoles (e.g., **8.1-47**) followed by treatment of the resulting 3-diazotriazole with sodium azide.[35,36] In reactions directed toward the preparation of the triazolotriazole (**8.1-49**) by heating **8.1-48** in 95% ethanol, the product was identified as the tautomeric azido isomer (**8.1-46**), indicating that the azido compound was thermodynamically more stable than the bicyclic compound.[35,36] Similar intermediates (**8.1-51** and **8.1-49**) were involved in the diazotization of **8.1-50** to give a 84% yield of **8.1-46**.[37]

References

1. **76:** 85792b
2. **79:** 54855g
3. **54:** 12118h
4. **77:** 126523f
5. **59:** 12788b
6. **68:** 78203s
7. **70:** 115073p
8. **59:** 12790g
9. **79:** 18648h
10. **67:** 72992f
11. **82:** 4178e
12. **59:** 11473d
13. **70:** 87681q
14. **73:** 87860j
15. **71:** 91382f
16. **62:** 4024d
17. **80:** 47913s
18. **85:** 123823h
19. **58:** 518g
20. **79:** 78692c
21. **71:** 112866j
22. **72:** 111383h
23. **72:** 111384j
24. **76:** 99569w
25. **73:** 66847v
26. **60:** 2951c
27. **58:** 10220g
28. **56:** 4776e
29. **81:** 77930h
30. **82:** 43269e
31. **84:** 30967x
32. **72:** 111380e
33. **74:** 76376a
34. **67:** 43756b
35. **75:** 97926r
36. **81:** 25613n
37. **72:** 3440r

TABLE 8. AZIDO-, AZO-, NITRAMINO-, NITROSAMINO-, DIAZOAMINO-TRIAZENO), AND NITRO-1,2,4-TRIAZOLES CONTAINING ONE OR MORE OF THESE FUNCTIONAL GROUPS

Compound	Reference
1(or 4)-allyl-3,5-dinitro	**58:** P10221a
4-amino-3-azido-5-methyl-	**69:** 110617p
3-[[4-amino-3-[2-(4-methylmorpholinium-4-yl)ethoxy]phenyl]azo]-	**71:** P82637r
3-(aminonaphthylazo)-5-ethyl-	**23:** 3470[9]
3-amino-5-nitrosoamino-	**28:** 2714[9]
3-[3-(5-amino-s-triazol-3-yl)-1-triazeno]-5-[3-(5-amino-s-triazol-3-yl)-2-triazeno]-	**62:** 4024d
3-azido-	**69:** 110617p
3-azido-5-(p-bromophenyl)-	**75:** 97926r
3-azido-5-(p-chlorophenyl)-	**75:** 97926r
3-azido-5-p-cumenyl-	**75:** 97926r
3-azido-5-(4,5-dihydro-1H-pyrazol-3-yl)-	**78:** 29701a
5-azido-1,3-dimethyl-	**81:** 25613n
3-azido-5-[p-(dimethylamino)phenylazo]-	**28:** 2714[5]
3-azido-5-ethyl-	**69:** 110617p
3-azido-5-(2-hydroxy-1-naphthylazo)-	**28:** 2714[5]
3-azido-1-methyl-	**82:** 30784u
3-azido-4-methyl-	**82:** 30784u
3-azido-5-methyl-	**69:** 110617p, **81:** 25613n, **82:** 30522g
5-azido-1-methyl-	**82:** 30784u
5-azido-1-methyl-3-nitro-	**74:** 76376a
3-azido-5-nitro-	**82:** 30784u
3-azido-5-(3-nitrophenyl)-	**82:** 30784u
3-azido-5-(p-nitrophenyl)-	**75:** 97926r
3-azido-5-nitrosoamino-	**28:** 2714[4]
3-azido-5-phenyl-	**67:** 43756b, **72:** 3440r **75:** 97926r
3-azido-5-p-tolyl-	**75:** 97926r
3,3'-azobis-	**59:** 6227c, **78:** P130578r

TABLE 8 (Continued)

Compound	Reference
5,5'-azobis[1-(4-chlorophenyl)-3-phenyl-	**79:** 18648h
4,4'-azobis[3,5-dimethyl-	**82:** 4178e
4,4'-azobis[3,5-diphenyl-	**82:** 4178e
5,5'-azobis[1,3-diphenyl-	**79:** 18648h
3,3'-azobis[5-methyl-	**78:** P130578r
5,5'-azobis[1-methyl-	**59:** 12790g, **67:** 72992f
5,5'-azobis[1-(4-nitrophenyl)-3-phenyl-	**79:** 18648h
3,3'-azobis[5-phenyl-	**59:** 12790g, **70:** 115073p
5,5'-azoxybis[1-methyl-3-nitro-	**74:** 76376a
3-[[N-(benzamido)amidino]azo]-5-phenyl	**68:** 78203s
3-[[4-[2-(benzoyloxy)ethyl(ethyl)amino]phenyl]azo]-	**83:** P165797s
3-benzyl-5-[(p-chlorophenyl)azo]-	**59:** 12789e
3-benzyl-5-[(2,4-dichlorophenyl)azo]-	**59:** 12789e
1-benzyl-3-[[p-(diethylamino)phenyl]azo]-	**73:** P36546d
3-[[4-[benzyl(ethyl)amino]phenyl]azo]-	**80:** P61056w, **83:** P81203g
5-[[4-(N-benzyl-N-methylamino)phenyl]azo]-1-methyl-	**81:** P37390a
1-benzyl-3-[(1-methyl-2-phenyl-1H-indol-3-yl)azo]-	**73:** P46632t
1-benzyl-5-[(1-methyl-2-phenyl-1H-indol-3-yl)azol]-	**80:** P28461b
4-(benzylnitrosamino)-	**70:** 87681q
4-(benzylnitrosamino)-3,5-dimethyl-	**70:** 87681q
3-benzyl-5-(N-nitrosoanilino)-	**71:** 91382f
3-benzyl-5-(phenylazo)-	**59:** 12789e
5-benzyl-5'-phenyl-3,3'-azobis-	**70:** 115073p
5-[([1,1'-biphenyl]-4-yloxy)methyl]-3-[[4-(diethyl-amino)phenyl]azo]-	**81:** P65189k
5-[[4-[N-[2-([1,1'-biphenyl]-4-yloxypropyl]methyl-amino]phenyl]azo]-1-methyl-	**81:** P65189k
3-[[2,6-bis(butylamino)-3-cyano-4-methyl-5-pyridinyl]-azo]-	**83:** P99180g
3,5-bis[5-chloro-1,2-dimethyl-1H-indol-3-yl)azo]-	**83:** P181082k
3,5-bis[5-chloro-1-ethyl-2-phenyl-1H-indol-3-yl)azo]-	**83:** P181082k
3,5-bis-[[4-(diethylamino)phenyl]azo]-	**81:** 137569j, **81:** P171335q
3,5-bis[(1,2-dimethyl-1H-indol-3-yl)azo]-	**83:** P181082k
3,5-bis[(2-ethyl-1H-indol-3-yl)azo]-	**83:** P181082k
3,5-bis[(1-ethyl-5-methoxy-2-phenyl-1H-indol-3-yl)azo]-	**83:** P181082k
3,5-bis[[2-hydroxy-4-(diethylamino)phenyl]azo]-	**81:** P65182c
3,5-bis[(p-hydroxyphenyl)azo]-	**28:** 2714[6]
3,5-bis[(2-methyl-1H-indol-3-yl)azo]-	**83:** P181082k
3,5-bis[[1-methyl-2-(4-methylphenyl)-1H-indol-3-yl]-azo]-	**83:** P181082k
3,5-bis-[(1-methyl-2-phenyl-1H-indol-3-yl)azo]-	**81:** P79380c, **81:** P93041z, **83:** P181082k
3,5-bis(nitrosoamino)-	**28:** 2714[2]
3,5-bis[(2-phenyl-1H-indol-3-yl)azo]-	**83:** P181082k
3-[[4-[N-[2([1,1'-bisphenyl]-4-yloxy)propyl]methyl-amino]phenyl]azo]-	**81:** P65189k
3,5-bis[(1,2,5-trimethyl-1H-indol-3-yl)azo]-	**83:** P181082k
4-(p-bromophenyl)-3-[[p-(dimethylamino)phenyl]azo]-5-propyl-	**59:** 11474c
4-(p-bromophenyl)-3-nitrosamino-5-propyl-	**59:** 11473e

TABLE 8 (*Continued*)

Compound	Reference
5-butyl-5'-phenyl-3,3'-azobis-	**70:** 115073p
3-[4-[(5-carboxy-1,3-dihydro-2H-isoindol-2-yl)- phenyl]azo]-	**78:** P73641e
3-[[4-[(2-carboxyethyl)ethylamino]phenyl]azo]-	**80:** P84685q
3-[(3-carboxy-4-hydroxyphenyl)azo]-	**81:** 25609r
3-[[N-(2-chlorobenzamido)amidino]azo]-5-phenyl	**68:** 78203s
3-[[N-(4-chlorobenzamido)amidino]azo]-5-phenyl-	**68:** 78203s
4-[(p-chlorobenzyl)nitrosamino]-	**73:** 87860j
3-[[2-chloro-4-(dimethylamino)phenyl]azo]-5-methyl-	**72:** P121541x
3-[[4-[(2-chloroethyl)methylamino]phenyl]azo]-	**83:** P81203g
3-(o-chloro-N-nitrosoanilino)-5-methyl-	**71:** 91382f
3-[(p-chlorophenyl)azo]-5-isobutyl-	**59:** 12789e
3-[(p-chlorophenyl)azo]-5-phenyl-	**59:** 12789b
4-(p-chlorophenyl)-3-[(p-(dimethylamino)phenyl]azo]- 5-ethyl-	**59:** 11474c
4-(p-chlorophenyl)-3-[[p-(dimethylamino)phenyl]azo]- 5-propyl-	**59:** 11474c
4-(p-chlorophenyl)-3-ethyl-5-nitrosamino-	**59:** 11473f
5-(3-chlorophenyl)-3-nitramino-	**79:** 78692c
4-(p-chlorophenyl)-3-nitrosamino-5-propyl-	**59:** 11473f
5-(m-chlorophenyl)-5'-phenyl-3,3'-azobis-	**70:** 115073p
5-(o-chlorophenyl)-5'-phenyl-3,3'-azobis-	**70:** 115073p
5-(p-chlorophenyl)-5'-phenyl-3,3'-azobis-	**70:** 115073p
3-[(3-cyano-1,2-dihydro-6-hydroxy-1,4-dimethyl-2-oxo- 5-pyridinyl)azo]-	**73:** P16264g
3-[(3-cyano-2,6-dihydroxy-4-methyl-5-pyridinyl)azo]-	**73:** P100007h
3-[[4-[(2-cyanoethyl)amino]-3-(phenylethyl)phenyl]azo]-	**78:** P148964y
3-[(3-cyano-1-ethyl-1,2-dihydro-6-hydroxy-4-methyl- 2-oxo-5-pyridinyl)azo]-	**81:** P79384g
3-[4-[N-(cyanoethyl)-N-ethylanilino]azo]-	**72:** P121541x
5-[[p-[(2-cyanoethyl)methylamino]phenyl]azo]-3- (1-methylpyridinium-3-yl)-	**72:** P45020j
3-[[2-(cyanoethyl)-2-methyl-1H-indol-3-yl]azo]-	**81:** P51113u
3-[[4-[(2-cyanoethyl)3,3,4,4,4-pentafluorobutyl)amino]- phenyl]azo]-	**81:** P93049h
3-[[1-(2-cyanoethyl)-2-phenyl-1H-indol-3-yl]azo]	**81:** P154525j
3-[[1-(2-cyanoethyl)-2-phenyl-1H-indol-3-yl]azo]-4- 2-hydroxypropyl)-	**81:** P154525j
3-[[3-cyano-4-methyl-2,6-bis(methylamino)-5- pyridinyl]azo]-	**80:** P84684p
3-[(2,6-diamino-3-cyano-4-methyl-5-pyridinyl)azo]-	**83:** P99180g
3-[(1,3-diamino-4-isoquinolinyl)azo]-	**83:** P81207m
3,5-diazido-	**82:** 30784u
3-diazo-	**83:** 9910t, **83:** 178931f, **84:** 135600a
3,3'-(diazoamino)bis[5-amino-	**62:** 4024c, **70:** P96802g
3,3'-(diazoamino)bis[4-(p-bromophenyl)-5-propyl-	**59:** 11473g
3,3'-(diazoamino)bis[4-(p-chlorophenyl)-5-ethyl-	**59:** 11473g
3,3'-(diazoamino)bis[4-(p-chlorophenyl)-5-propyl-	**59:** 11473g
3,3'-(diazoamino)bis[1,5-dimethyl-	**81:** 25613n

TABLE 8 (*Continued*)

Compound	Reference
3,3'-(diazoamino)bis[4-phenyl-5-propyl-	**59:** 11473g
3,3'-(diazoamino)bis[5-propyl-4-*p*-tolyl-	**59:** 11473g
3-diazo-5-methyl-	**73:** 45420k
3-diazo-5-(*m*-nitrophenyl)-	**73:** 45420k
3-diazo-5-(*p*-nitrophenyl)-	**73:** 45420k
3-diazo-5-(*p*-nitrophenyl)-, hydroxide, inner salt	**73:** 45420k
3-diazo-5-phenyl-, chloride	**73:** 14771w, **79:** 78698j
3-diazo-5-phenyl-, tetrafluoroborate(1-)	**73:** 45420k
1-(2,3-dibromopropyl)-3,5-dinitro-	**58:** P10221a
3-[[4-(dibutylamino)phenyl]azo]-	**79:** P93432m
3-[(2,4-dichlorophenyl)azo]-5-isobutyl-	**59:** 12789d
3-[[*p*-(diethylamino)phenyl]azo]-	**53:** P1734d, **54:** P3970a, **68:** P115683w
3-[[4-(diethylamino)phenyl]azo]-4-methyl-	**79:** 67690z
3-[[4-(diethylamino)phenyl]azo]-5-(phenoxymethyl)-	**81:** P65180a
3-[(2,4-dihydroxyphenyl)azo]-	**81:** 25609r
1-(2,3-dihydroxypropyl)-3,5-dinitro-	**81:** P77930h
3-[[4-(dimethylamino)phenyl]azo]-	**81:** 25609r
3-[[(4-dimethylamino)phenyl]azo]-1-methyl-	**76:** P60914e
3-[[(*p*-dimethylamino)phenyl]azo]-4-methyl-	**77:** 90026e
3-[[*p*-(dimethylamino)phenyl]azo]-4-phenyl-5-propyl-	**59:** 11474c
3-[[*p*-(dimethylamino)phenyl]azo]-5-propyl-4-*p*-tolyl-	**59:** 11474c
3,5-dimethyl-4-nitramino-	**85:** 123823h
1,3-dimethyl-5-nitro-	**72:** 111384j
1,5-dimethyl-3-nitro-	**72:** 111384j
3,4-dimethyl-5-nitro-	**72:** 111384j
3,5-dinitramino-	**48:** 2050d
3,5-dinitro-	**58:** P10220h, **60:** 2951c, **72:** 111383h, **80:** 26672x
5,5'-dinitro-3,3'-bis[	**72:** 111383h
3-[(1,1-dinitroethyl)azo]-	**77:** 126523f
3-[(1,1-dinitroethyl)azo]-5-methyl-	**77:** 126523f
3,5-dinitro-4(or 1)-(2-nitroethyl)-	**56:** P4776f
3,5-dinitro-4(or 1)-(β-nitrophenethyl)-	**56:** P4776f
3,5-dinitro-4(or 1)-(2-nitropropyl)-	**56:** P4776f
1-(2,4-dinitrophenyl)-3-nitro-	**74:** 53664d
3,5-dinitro-1-(2-propenyl)-	**82:** 43269e, **82:** P43428f, **85:** 78052t
3-[(1,1-dinitropropyl)azo]-	**77:** 126523f
3-[(1,1-dinitropropyl)azo]-5-methyl-	**77:** 126523f
-1-ethanol, 3(or 5)-[[4-[bis(phenylmethyl)amino]-phenyl]azo]-	**85:** P79672g
-1-ethanol, 3(or 5)-[[4-[bis(phenylmethyl)amino]-phenyl]azo]-α-ethyl-	**85:** P22741z
-1-ethanol, α-(butoxymethyl)-3(or 5)-[(5-chloro-1-methyl-2-phenyl-1H-indol-3-yl)azo]-	**85:** P22740y
-1-ethanol, α-(butoxymethyl)-3(or 5)-[[4-(diethylamino)-phenyl]azo]-	**85:** P22741z
-1-ethanol, 3(or 5)-[[4-[(2-chloroethyl)ethylamino]-phenyl]azo]-α-methyl-	**85:** P22741z

TABLE 8 (*Continued*)

Compound	Reference
-1-ethanol, α-(chloromethyl)-3(or 5)-[[4-[ethyl(phenyl-methyl)amino]phenyl]azo]-	**85:** P22741z
-1-ethanol, 3(or 5)-[[4-(dibutylamino)phenyl]azo]-α-(phenoxymethyl)-	**85:** P22741z
-1-ethanol, 3(or 5)-[[4-(diethylamino)-2-methyl-phenyl]azo]-α-methyl-	**85:** P22741z
-1-ethanol, 3(or 5)-[[4-(diethylamino)-2-methylphenyl]-azo]-α-(phenoxymethyl)-	**85:** P22741z
-1-ethanol, 3(or 5)-[[4-(diethylamino)phenyl]azo]-α-ethyl-	**85:** P22741z
-1-ethanol, 3(or 5)-[[4-(diethylamino)phenyl]azo]-α-methyl-	**85:** P22741z
-1-ethanol, 3(or 5)-[[4-(diethylamino)phenyl]azo]-α-phenyl-	**85:** P22741z
-1-ethanol, 3(or 5)-[(1,2-dimethyl-1H-indol-3-yl)azo]-α-(phenoxymethyl)-	**85:** P22740y
-1-ethanol, 3,5-dinitro	**84:** 30967x
-1-ethanol, 3,5-dinitro-α-(nitromethyl)-	**85:** 78052t
-1-ethanol, 3,5-dinitro-α-(nitromethyl)-, acetate (ester)	**85:** 78052t
-1-ethanol, 3,5-dinitro-α-(nitromethyl)-, nitrate (ester)	**85:** 78052t
-1-ethanol, α-ethyl-3(or 5)-[(1-ethyl-2-phenyl-1H-indol-3-yl)azo]-	**85:** P22740y
-1-ethanol, α-ethyl-3(or 5)-[[ethyl(phenylmethyl)-amino]phenyl]azo]-	**85:** P22741z
-1-ethanol, 3(or 5)-[(1-ethyl-2-phenyl-1H-indol-3-yl)azo]-	**85:** P144687g
-1-ethanol, 3(or 5)-[(1-ethyl-2-phenyl-1H-indol-3-yl)azo]-α-(methoxymethyl)-	**85:** P22740y
-1-ethanol, 3(or 5)-[(1-ethyl-2-phenyl-1H-indol-3-yl)azo]-α-(phenoxymethyl)-	**85:** P22740y
-1-ethanol, 3(or 5)-[[4-[ethyl(phenylmethyl)amino]-2-methylphenyl]azo]-α-methyl-	**85:** P79672g
-1-ethanol, 3(or 5)-[[4-[ethyl(phenylmethyl)amino]-phenyl]azo]-, acetate (ester)	**85:** P22745d
-1-ethanol, 3(or 5)-[[4-[ethyl(phenylmethyl)amino]-phenyl]azo]-α-methyl-	**85:** P22741z
-1-ethanol, 3(or 5)-[[4-[ethyl(phenylmethyl)amino]-phenyl]azo]-α-(phenoxymethyl)-	**85:** P22741z
-1-ethanol, α-(methoxymethyl)-3,5-dinitro-	**84:** 30967x
-1-ethanol, α-(methoxymethyl)-3(or 5)-[[4-(methylphenylamino)phenyl]azo]-	**85:** P22741z
-1-ethanol, α-methyl-3,5-dinitro-	**84:** 30967x, **85:** 159994d
-1-ethanol, α-methyl-3,5-dinitro-α-(1-nitroethyl)-	**84:** 105501b
-1-ethanol, α-methyl-3(or 5)-[(1-methyl-2-phenyl-1H-indol-3-yl)azo]-	**85:** P22740y
-1-ethanol, α-methyl-3(or 5)-[[4-(4-morpholinyl)-phenyl]azo]-	**85:** P22741z
-1-ethanol, α-methyl-3-nitro	**85:** 159994d
-1-ethanol, 3(or 5)-[(1-methyl-2-phenyl-1H-indol-3-yl)azo]-α-(phenoxymethyl)-	**85:** P22740y

TABLE 8   (*Continued*)

Compound	Reference
3-(*p*-ethoxyphenylazo)-5-methyl-	**54:** 12119i
3-[[4-[ethyl[1,2-bis(methoxycarbonyl)ethyl]amino]- 2-methylphenyl-azo]-	**83:** P181099w
1(or 4)-ethyl-3,5-dinitro-	**58:** P10221a
3-ethyl-5-nitramino-	**58:** 519a, **72:** P43692f
5-ethyl-3-nitramino-	**79:** 78692c
3-ethyl-5-nitro-	**72:** 111383h, **72:** 12648s
3-[ethyl(nitro)amino]-5-(5-nitro-2-furanyl)-	**79:** 87327q
3-[ethyl(nitroso)amino]-5-(5-nitro-2-furanyl)-	**79:** 87327q
3-ethyl-5-(*N*-nitrosoanilino)-	**71:** 91382f
3-[4-[ethyl(2-phenoxysulfonylethyl)amino]phenylazo]-	**83:** P81198j
3-[(1-ethyl-2-phenyl-1H-indol-3-yl)azo]-	**80:** P28454b
3-[[4-[ethyl(phenylmethyl)amino]phenyl]azo]-4- (2-hydroxypropyl)-	**80:** P61056w
3-[[*p*-[ethyl[2-*N*-propylacetamido)ethyl]amino]- phenyl]azo]-4-methyl-	**70:** P97943r
3-[(10-hydroxyanthracene-9-yl)azo]-	**81:** 25609r
3-[(2-hydroxy-3,5-dimethylphenyl)azo]-	**79:** 54855g
3-[(4-hydroxy-2,5-dimethylphenyl)azo]-	**79:** 54855g
3-[(4-hydroxy-3,5-dimethylphenyl)azo]-	**79:** 54855g
1-(2-hydroxyethyl)-3,5-dinitro-	**81:** P77930h
1-(1-hydroxy-3-methoxy-2-propyl)-3,5-dinitro-	**81:** P77930h
3-[(2-hydroxy-5-methylphenyl)azo]-	**79:** 54855g, **81:** 25609r
3-[(4-hydroxy-3-methylphenyl)azo]-	**79:** 54855g
3-[(2-hydroxy-5-methylphenyl)azo]-5-methyl-	**82:** 67667j
3-[(2-hydroxynaphthalen-1-yl)azo]-	**81:** 25609r, **83:** 9910t
3-[(4-hydroxynaphthalen-1-yl)azo]-	**81:** 25609r
3-(2-hydroxy-1-naphthylazo)-5-methyl-	**48:** 1344a, **50:** 9396e
3-[(4-hydroxyphenyl)azo]-	**79:** 54855g, **81:** 25609r
3-isobutyl-5-(*N*-nitrosoanilino)-	**71:** 91382f
3-isobutyl-5-(phenylazo)-	**59:** 12789c
5-isopropyl-3-nitramino-	**79:** 78692c
3-isopropyl-5-(*N*-nitrosoanilino)-	**71:** 91382f
3-[[*N*-(isovalerylamino)amidino]azo]-5-phenyl-	**68:** 78203s
3-mesitylazo-	**54:** 12119i
3-mesitylazo-5-methyl-	**54:** 12120a
4-[(*p*-methoxybenzyl)nitrosamino]-	**73:** 87860j
5-(*o*-methoxyphenyl)-5'-phenyl-3,3'-azobis-	**70:** 115073p
1-(2-methyl-2-butenyl)-3,5-dinitro-	**84:** 105501b
1-(3-methyl-2-butenyl)-3,5-dinitro-	**82:** 43269e
1-methyl-3,5-dinitro-	**72:** 111384j
1(or 4)-methyl-3,5-dinitro-	**62:** P7770e
3-methyl-5-[(1-methyl-1-nitroethyl)azo]-	**77:** 126523f
3-methyl-5-(methylnitrosamino)-	**59:** 10051a, **60:** 9264h
1-methyl-5-(methylnitrosamino)-3-nitro-	**74:** 76376a
3-[[4-[(methyl)[2-(*N*-methyl-*N*-phenylsulfamoyl)ethyl]- amino]phenyl]azo]-	**82:** P100064k
1-methyl-5-nitramino-	**59:** 12791c
3-methyl-5-nitramino-	**46:** 6088g, **58:** 519e, **72:** P43692f
5-methyl-3-nitramino-	**79:** 78692c

TABLE 8 (*Continued*)

Compound	Reference
1-methyl-3-nitramino-5-(5-nitro-2-furanyl)-	**81:** 163330u
1-methyl-3-nitro-	**72:** 111380e, **72:** 111384j
1-methyl-5-nitro-	**72:** 111383h
3-methyl-5-nitro-	**71:** 112866j, **72:** 12648s,
	72: 111383h, **80:** 26672x
4-methyl-3-nitro-	**72:** 111383h
4-methyl-3-(nitroamino)-5-(5-nitro-2-furanyl)-	**79:** 87327q, **81:** 163330u
1-(2-methyl-3-nitro-2-butenyl)-3,5-dinitro-	**84:** 105501b
3-[(1-methyl-1-nitroethyl)azo]-	**77:** 126523f
1-methyl-3-nitro-5-nitrosamino-	**72:** 111380e
1-methyl-5-nitro-3-phenyl-	**77:** 74519n
3-(methylnitrosamino)-	**59:** 10051a
3-[methyl(nitroso)amino]-5-(5-nitro-2-furanyl)-	**79:** 87327q
3-methyl-5-(*N*-nitrosoanilino)-	**71:** 91382f
3-methyl-5-(*N*-nitroso-*p*-anisidino)-	**71:** 91382f
3-methyl-5-(*N*-nitroso-*p*-toluidino)-	**71:** 91382f
3-[(1-methyl-2-phenyl-1 H̲-indol-3-yl)azo]-	**81:** P65203k, **83:** P81203g
5-methyl-1-phenyl-3-phenylazo-	**42:** 1232h
1-(2-methyl-2-propenyl)-3,5-dinitro-	**82:** 43269e
3-methyl-5-[(trinitromethyl)azo]-	**77:** 126523f
3-[[4-(4-morpholinyl)phenyl]azo]-	**72:** P121541x
3-nitramino-	**46:** 6088e, **58:** 519a,
	72: P43692f
5-(nitramino)-1-methyl-3-(5-nitro-2-furanyl)-	**81:** 163330u
3-(nitramino)-5-(5-nitro-2-furanyl)-	**81:** 163330u
3-nitramino-5-(2-nitrophenyl)-	**79:** 78692c
3-nitramino-5-(3-nitrophenyl)-	**79:** 78692c
3-nitramino-5-(4-nitrophenyl)-	**79:** 78692c
3-nitramino-5-phenyl-	**58:** 519a, **79:** 78692c
3-nitro-	**70:** 77876t, **71:** 112866j,
	72: 111383h, **73:** 66847v,
	80: 26672x, **84:** 90089x
4-nitroamino-	**80:** 47913s
1-nitro-3-(4-nitrophenyl)-	**76:** 99569w
3-nitro-5-(3-nitrophenyl)-	**72:** 111383h
3-nitro-5-(4-nitrophenyl)-	**72:** 111383h, **76:** 99569w
1-nitro-3-phenyl-	**76:** 99569w
3-nitro-5-phenyl-	**72:** 111383h, **76:** 99569w
3-nitro-5-propyl-	**72:** 111383h
3-nitro-1-β-D-ribofuranosyl-	**73:** 66847v, **80:** 121258a
3-nitro-1-β-D-ribofuranosyl-, 2′,3′,5′-tribenzoate	**73:** 66847v
3-nitrosamino-4-phenyl-5-propyl-	**59:** 11473e
3-nitrosamino-5-propyl-4-*p*-tolyl-	**59:** 11473e
3-[[*N*-(phenylacetylamino)amidino]azo]-5-phenyl-	**68:** 78203s
3-[(2-phenyl-1H̲-indol-3-yl)azo]-	**83:** P29853e
3-phenyl-5-(phenylazo)-	**59:** 12789c
5-phenyl-3-[(3-phenyl-1,2,4-thiadiazol-5-yl)azo]-	**70:** 115073p
5-phenyl-3-[(4-pyridyl)azo]-	**70:** 115073p
5-phenyl-3-[[*N*-(salicyloylamino)amidino]azo]-	**68:** 78203s
poly(3-diazoamino)-	**62:** 4024d

TABLE 8 (*Continued*)

Compound	Reference
-1-propanamide, *N*-butyl-3(or 5)-[[4-benzyl(ethyl)amino]-phenyl]azo]-5(or 3)-phenyl-	**85:** P22745d
-1-propanamide, *N*-tert-butyl-3(or 5)-[(1-methyl-2-phenyl-1H-indol-3-yl)azo]-	**85:** P7255a
-1-propanamide, 3(or 5)-[(1-ethyl-2-phenyl-1H-indol-3-yl)azo]-	**85:** P7255a
-1-propanamide, 3(or 5)-[[4-[ethyl(phenylmethyl)amino]-phenyl]azo]-	**85:** P22745d
-1-propanenitrile, 3(or 5)-[[4[bis(phenylmethyl)amino]-	**85:** P22745d
-1-propanenitrile, 3(or 5)-[[4-[(2-chloroethyl)-ethylamino]phenyl]azo]-	**85:** P22745d
-1-propanenitrile, 3(or 5)-[(5-chloro-1-methyl-2-phenyl-1H-indol-3-yl)azo]-	**85:** P7255a
-1-propanenitrile, 3(or 5)-[[4-[(2-cyanoethyl)methyl-amino]phenyl]azo]-	**85:** P22745d
-1-propanenitrile, 3(or 5)-[[4-[(2-cyanoethyl)(phenyl-methyl)amino]phenyl]azo]-	**85:** P22745d
-1-propanenitrile, 3(or 5)-[[4-(dibutylamino)phenyl]azo]-	**85:** P22745d
-1-propanenitrile, 3(or 5)-[[4-(diethylamino)-2-ethoxyphenyl]azo]-5(or 3)-phenyl-	**85:** P22745d
-1-propanenitrile, 3(or 5)-[[4-(diethylamino)-2-methylphenyl]azo]-	**85:** P22745d
-1-propanenitrile, 3(or 5)-[[4-(dimethylamino)-1-naphthalenyl]azo]-	**85:** P22745d
-1-propanenitrile, 3(or 5)-[[4-(diethylamino)-phenyl]azo]-	**85:** P22745d
-1-propanenitrile, 3(or 5)-[(1,2-dimethyl-1H-indol-3-yl)azo]-	**85:** P7255a
-1-propanenitrile, 3(or 5)-[[4-[ethyl(2-methoxyethyl)-amino]phenyl]azo]-	**85:** P22745d
-1-propanenitrile, 3(or 5)-[[4-[ethyl(2-phenoxyethyl)-amino]phenyl]azo]-	**85:** P22745d
-1-propanenitrile, 3(or 5)-[[4-(ethylphenylamino)-phenyl]azo]-	**85:** P22745d
-1-propanenitrile, 3(or 5)-[(1-ethyl-2-phenyl-1H-indol-3-yl)azo]-	**85:** P7255a
-1-propanenitrile, 3(or 5)-[[4-[ethyl(phenylmethyl)-amino]phenyl]azo]-	**85:** P22745d
-1-propanenitrile, 3-methyl-5-[(2-phenyl-H-indol-3-yl)azo]-	**85:** P7255a
-1-propanenitrile, 3(or 5)-[(1-methyl-2-phenyl-1H-indol-3-yl)azo]-	**85:** P7255a
-1-propanenitrile, 3(or 5)-[[4-(4-morpholinyl)-phenyl]azo]-	**85:** P22745d
-1-propanoic acid, 3(or 5)-[[4-(diethylamino)phenyl]-azo]-, methyl ester	**85:** P22745d
3-[(trinitromethyl)azo]-	**77:** 126523f

CHAPTER 9

O-Substituted Oxy-1,2,4-Triazoles

9.1. Alkyl-, Aryl-, and Acyl- Mono- and Di-*O*-Substituted 1,2,4-Triazoles

Alcoholysis of a variety of 5-alkyl- and aryl-2-amino-1,3,4-oxadiazoles (e.g., **9.1-1** to **9.1-3**) in the presence of potassium hydroxide gave good yields of alkoxytriazoles (e.g., **9.1-7** to **9.1-9**), presumably formed via the corresponding imino ether intermediates (e.g., **9.1-4** to **9.1-6**).[1-6] Although isopropyl alcohol gave a lower yield than propyl alcohol (27 vs. 39%), Gehlen has reported that this rearrangement was successful with both straight- and branched-chain C-1 to C-5 alcohols[7] including glycol.[8]

9.1-1, R = Me
9.1-2, R = PhCONHCH$_2$
9.1-3, R = Ph

9.1-4, R = Me
9.1-5, R = PhCONHCH$_2$
9.1-6, R = Ph

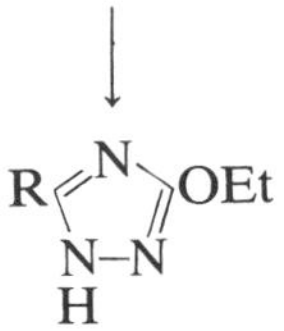

9.1-7, R = Me (41%)
9.1-8, R = PhCONHCH$_2$ (84%)
9.1-9, R = Ph (81%)

Aryl ethers of 1,2,4-triazoles have been prepared by the 1,3-dipolar addition of nitrilimines to aryl cyanates.[9] For example, refluxing a benzene solution of **9.1-10** in the presence of triethylamine generated the nitrilimine **9.1-11**, which reacted with phenyl cyanate to give **9.1-12** in 48% yield. In a related reaction treatment of **9.1-13** with hydrazine gave a 94% yield of 3-phenoxy-5-phenyl-1,2,4-triazole.[10] The 3-phenoxy-1,2,4-triazole (**9.1-15**)

PhCCl $\xrightarrow{\text{Et}_3\text{N}}$ [PhĊ] $\xrightarrow{\text{PhOCN}}$
‖   ‖
NNHPh   NN̄Ph

9.1-10   **9.1-11**

Ph⟨triazole⟩OPh   PhCO⟨⟩COPh

9.1-12   **9.1-13**

resulted from *N—N* bond formation when **9.1-14** was heated in acetic acid.[11]

4-O$_2$NC$_6$H$_4$C⟨⟩COPh $\xrightarrow{\text{HOAc}}$ 4-O$_2$NC$_6$H$_4$⟨triazole⟩OPh

9.1-14   **9.1-15**

Nucleophilic displacement of the halogen of halotriazoles occurred readily, as demonstrated by treatment of **9.1-16** with methoxide and phenoxide to give **9.1-17** and **9.1-19**, respectively.[12–15] Although the disilver salt of **9.1-20** was alkylated with ethyl iodide to give **9.1-21**,[16] the sodium salts of 3-methyl-, 1,3-dimethyl-, 3,4-dimethyl-, 1-methyl-3-phenyl-, 2-methyl-3-phenyl-, and 4-methyl-3-phenyl-1,2,4-triazolin-5-ones were alkylated with

⟨triazole⟩Cl $\xrightarrow[59\%]{\text{MeO}^-}$ ⟨triazole⟩OMe $\xleftarrow{\text{OH}^-}$ ⟨triazolium⟩OMe   I$^-$

9.1-16   **9.1-17**   **9.1-18**

⟨triazole⟩OPh   O=⟨triazolinone⟩=O $\xrightarrow{\text{EtI}}$ EtO⟨triazole⟩OEt

9.1-19   **9.1-20**   **9.1-21**

methyl iodide to give *N*-alkylated products.[15–17] In contrast, 2-methyl-3-phenyl-1,2,4-triazolin-5-one (**9.1-22**) was *O*-methylated with diazomethane in dimethyl sulfoxide to give **9.1-23**,[17] and 2,3-dimethyl-1,2,4-triazolin-5-one (**9.1-24**) was *O*-methylated with both diazomethane and methyl iodide-base to give mainly **9.1-25**.[15] In other methods, the 1-methyl-1,2,4-triazinium iodide (**9.1-18**) underwent ring contraction in base to give **9.1-17**.[18] The triazole ethers are hydrolyzed with strong hydrochloric acid to the corresponding triazolin-5-ones (see Section 13).[14,19] The potassium salts of the latter (e.g., **9.1-26**) have been reported to *O*-acylate with dimethylcarbamyl chloride in 2-butanone to give triazole esters (e.g., **9.1-27**).[20–22]

References

1. **57:** 3425c	2. **69:** 67293r	3. **71:** 91400k	4. **73:** 14770v
5. **74:** 53661a	6. **81:** 13442t	7. **71:** 101781b	8. **68:** 78201q
9. **64:** 8171e	10. **74:** 100065w	11. **81:** 136062b	12. **57:** 12473b
13. **71:** 81375s	14. **70:** 28922w	15. **84:** 17237q	16. **66:** 28303n
17. **85:** 159988e	18. **76:** 59576q	19. **70:** 68268u	20. **78:** 4261u
21. **83:** 114420f	22. **82:** 156321j		

9.2. *O*-(1,2,4-Triazole) Derivatives of Phosphorus Acids

The diverse biological activities of organophosphorus compounds, which act by the irreversibly phosphorylation of cholinesterase in which the organo moiety is the leaving group, prompted the synthesis of triazoles substituted with a variety of derivatives of phosphorus acids. These compounds, useful as insecticides, acaricides, fungicides, and nematocides, are prepared by *O*-acylation of 1,2,4-triazolin-5-ones with phosphorus chlorides in acetone or 2-butanone in the presence of an acid acceptor. Typical examples are the interaction of the triazolin-5-ones (**9.2-1**, **9.2-4**, and **9.2-7**) with the chlorides (**9.2-2**, **9.2-5**, and **9.2-8**) to give the corresponding phosphono-thioate (**9.2-3**),[1] phosphorothioate (**9.2-6**),[2–4] and phosphorodithioate (**9.2-9**).[5] Related types of compounds are described in Section 12.6.

$$\text{9.2-1} + \text{EtP(S)(OEt)Cl} \quad \text{9.2-2} \longrightarrow \text{9.2-3}$$

$$\text{9.2-4} + \text{(EtO)}_2\text{P(S)Cl} \quad \text{9.2-5} \longrightarrow \text{9.2-6}$$

$$\text{9.2-7} + \text{MeSP(S)OEt)Cl} \quad \text{9.2-8} \longrightarrow \text{9.2-9}$$

References

1. **83:** 206281q 2. **77:** P152192s 3. **79:** 28414t 4. **79:** 66363h
5. **81:** 13521t

TABLE 9. *O*-SUBSTITUTED OXY-1,2,4-TRIAZOLES

Compound	Reference
9.1. Alkyl, Aryl, and Acyl Mono- and Di-*O*-substituted 1,2,4-Triazoles	
-1-acetamide, *N*-[5-[[4-[2,4-bis(1,1-dimethylpropyl)- phenoxy]-1-oxobutyl]amino]-2-chlorophenyl]-α- (2,2-dimethyl-1-oxopropyl)-5-ethoxy-3-phenyl-	**85:** P102316e
-1-acetamide, α-(2,2-dimethyl-1-oxopropyl)-*N*-[2- (hexadecyloxy)-5-[(methylamino)sulfonyl]phenyl]-5- (2-hydroxyethoxy)-3-phenyl-	**85:** P151754e
3-(2-acetamidoethyl)-5-(4-hydroxybutoxy)-	**68:** 78201q
3-(2-acetamidoethyl)-5-(2-hydroxyethoxy)-	**68:** 78201q
5-(4-aminophenyl)-3-(2-hydroxyethoxy)-	**68:** 78201q
3-(1H̲-benzotriazol-1-ylmethyl)-5-ethoxy-	**73:** 14770v
3-(1H̲-benzotriazol-1-ylmethyl)-5-methoxy-	**73:** 14770v
3-(1H̲-benzotriazol-1-ylmethyl)-5-propoxy-	**73:** 14770v
3-benzyl-5-(2-bromoethoxy)-	**68:** 78201q
3-benzyl-5-butoxy-	**57:** 3425e, **71:** 91400k
3-benzyl-5-(2-chloroethoxy)-	**68:** 78201q
3-benzyl-5-ethoxy-	**57:** 3425e, **63:** P13275d, **71:** 91400k

TABLE 9 (*Continued*)

Compound	Reference

9.1. Alkyl, Aryl, and Acyl Mono- and Di-*O*-substituted 1,2,4-Triazoles (*Continued*)

Compound	Reference
5-benzyl-3-(4-hydroxybutoxy)-	**68:** 78201q
5-benzyl-3-(2-hydroxyethoxy)-	**68:** 78201q
3-benzyl-5-(2-iodoethoxy)-	**68:** 78201q
3-benzyl-5-isobutoxy-	**57:** 3425e, **71:** 91400k
3-benzyl-5-isopropoxy-	**57:** 3425e, **71:** 91400k
3-benzyl-5-methoxy-	**57:** 3425e, **71:** 91400k
5-benzyl-3-(2-methoxyethoxy)-	**81:** 13442t
3-(benzyloxy)-	**70:** P28922w, **71:** P81375s, **71:** P124443g
5-benzyloxy-3-benzyl	**81:** 13442t
3-benzyloxy-1-isopropyl-	**79:** P66364j
5-benzyloxy-3-(4-methoxyphenyl)-	**81:** 13442t
5-benzyloxy-3-methyl-	**81:** 13442t
5-benzyloxy-3-phenyl-	**81:** 13442t
3-benzyl-5-propoxy-	**57:** 3425e, **71:** 91400k
1,5-bis(isopropyl)-3-[(*N*,*N*-dimethylcarbamoyl)oxy]-	**78:** P4261u
3,5-bis[(trimethylsilyl)oxy]-	**83:** 97822u
3-(2-bromoethoxy)-5-(*p*-chlorophenyl)-	**68:** 78201q
3-(2-bromoethoxy)-5-(*p*-methoxyphenyl)-	**68:** 78201q
3-(2-bromoethoxy)-5-methyl-	**68:** 78201q
3-(2-bromoethoxy)-5-phenethyl-	**68:** 78201q
3-(2-bromoethoxy)-5-phenyl-	**68:** 78201q
3-butoxy-5-(3,4-dimethoxyphenyl)-	**69:** 96593u
3-butoxy-5-phenyl-	**71:** 91400k
5-butyl-3-(2-hydroxyethoxy)-	**68:** 78201q
3-butyl-5-(2-methoxyethoxy)-	**81:** 13442t
1-butyl-3-[(methylsulfonyl)oxy]-	**82:** P156321
3-(2-chloroethoxy)-5-(*p*-chlorophenyl)-	**68:** 78201q
3-(2-chloroethoxy)-5-phenyl-	**68:** 78201q
3-(*p*-chlorophenoxy)-5-[[(4-chloro-*o*-tolyl)oxy]methyl]-	**74:** P100065w
3-[(*p*-chlorophenoxy)methyl]-5-ethoxy-	**70:** 68268u
3-[(*o*-chlorophenoxy)methyl]-5-methoxy-	**70:** 68268u
3-[(*p*-chlorophenoxy)methyl]-5-methoxy-	**70:** 68268u
3-(p-chlorophenyl)-5-ethoxy-	**44:** 2518b
5-(4-chlorophenyl)-3-(4-hydroxybutoxy)-	**68:** 78201q
5-(2-chlorophenyl)-3-(2-hydroxyethoxy)-	**68:** 78201q
5-(4-chlorophenyl)-3-(2-hydroxyethoxy)-	**68:** 78201q
3-(*p*-chlorophenyl)-5-(2-iodoethoxy)-	**68:** 78201q
1-(3-chlorophenyl)-5-methyl-3-[(methylsulfonyl)oxy]-	**82:** P156321j
1-(4-chlorophenyl)-5-methyl-3-[(methylsulfonyl)oxy]-	**82:** P156321j
1-(2-chlorophenyl)-3-[(methylsulfonyl)oxy]-	**82:** P156321j
1-(2-cyanoethyl)-3-[(dimethylcarbamoyl)oxy]-	**83:** P114420f
1-(2-cyanoethyl)-5-methyl-3-[(dimethylcarbamoyl)oxy]-	**83:** P114420f
1-cyclopentyl-3-[(*N*,*N*-dimethylcarbamoyl)oxy]-	**78:** P4261u
3-[(2,4-dichlorophenoxy)methyl]-5-methoxy-	**70:** 68268u
3-(2,5-dichlorophenyl)-5-(*p*-tolyloxy)-	**74:** P100065w
3,5-diethoxy-1-phenyl-	**66:** 28303n

TABLE 9   (*Continued*)

Compound	Reference

9.1. Alkyl, Aryl, and Acyl Mono- and Di-*O*-substituted 1,2,4-Triazoles   (*Continued*)

Compound	Reference
3-[(*N*,*N*-diethylcarbamoyl)oxy]-1-isopropyl-	**78:** P4261u
3-[(diisobutylamino)methyl]-5-ethoxy-	**69:** 67293r
3-(3,4-dimethoxyphenyl)-5-ethoxy-	**69:** 96593u
3-(3,4-dimethoxyphenyl)-5-methoxy-	**69:** 96593u
3-(3,4-dimethoxyphenyl)-5-propoxy-	**69:** 96593u
3,5-dimethoxy-1-β-D-ribofuranosyl-, 2′,3′,5′-tribenzoate	**73:** 66847v
3-[(*N*,*N*-dimethylcarbamoyl)oxy]-5-ethyl-1-phenyl-	**48:** P187h
5-[(*N*,*N*-dimethylcarbamoyl)oxy]-3-ethyl-1-phenyl-	**48:** P187h
3-[(*N*,*N*-dimethylcarbamoyl)oxy]-1-isopropyl-	**78:** P4261u
5-(*N*,*N*-dimethylcarbamoyl)oxy-1-isopropyl-3-methyl-	**49:** P5532h
5-[(*N*,*N*-dimethylcarbamoyl)oxy]-3-methyl-1-phenyl-	**49:** P5532g
3-[(*N*,*N*-dimethylcarbamoyl)oxy]-1-(1-methylpropyl)-	**78:** P4261u
3-[(*N*,*N*-dimethylcarbamoyl)oxy]-1-(2-methylpropyl)-	**78:** P4261u
3-[(dimethylcarbamoyl)oxy]-1-phenyl-	**48:** P187h, **49:** 15916b
5-[(dimethylcarbamoyl)oxy]-1-phenyl-	**48:** P187h
3-[(dimethylcarbamoyl)oxy]-1-phenyl-5-propyl-	**48:** P187h
5-[(dimethylcarbamoyl]-1-phenyl-3-propyl-	**48:** P187h
3-[(*N*,*N*-dimethylcarbamoyl)oxy]-1-propyl-	**78:** P4261u
3-[(*N*,*N*-dimethylcarbamoyl)oxy]-1-(1,2,2-trimethylpropyl)-	**78:** P4261u
3,5-diphenoxy-	**73:** 56073a
1-(diphenylmethyl)-3-[(methylsulfonyl)oxy]-	**82:** P156321j
1,3-diphenyl-5-(*p*-tolyloxy)-	**64:** 8171e
3-ethoxy-5-ethyl-	**63:** P13275d, **71:** 91400k
3-ethoxy-5-(2-furyl)-	**73:** 14770v
5-ethoxy-3-(9-hydroxy-9-fluorenyl)-	**74:** 53661a
5-ethoxy-1-isopropyl-3-[(*N*,*N*-dimethylcarbamoyl)oxo]-	**78:** P4261u
3-ethoxy-5-(*p*-methoxyphenyl)-	**44:** 2518b
3-ethoxy-5-methyl-	**57:** 3425g
3-ethoxy-5-phenyl-	**44:** 2518b, **57:** 3425f, **71:** 91400k
4-ethoxy-3-phenyl-	**80:** 47060m
3-ethoxy-5-propyl-	**71:** 91400k
3-ethoxy-5-(2-thienyl)-	**73:** 14770v
1-ethyl-3-[(*N*,*N*-dimethylcarbamoyl)oxy]-	**78:** P4261u
5-ethyl-3-(2-hydroxyethoxy)-	**68:** 78201q
5-ethyl-1-isopropyl-3-[(*N*,*N*-dimethylcarbamoyl)oxy]-	**78:** P4261u
1-ethyl-5-methyl-3-[(*N*,*N*-dimethylcarbamoyl)oxy]-	**78:** P4261u
1-ethyl-5-methyl-3-[(methylsulfonyl)oxy]-	**82:** P156321j
1-ethyl-3-[(methylsulfonyl)oxy]-	**82:** P156321j
3-ethyl-5-propoxy-	**71:** 91400k
1-(1-ethylpropyl)-3-[(*N*,*N*-dimethylcarbamoyl)oxy]-	**78:** P4261u
3-(2-furanyl)-5-methoxy-4-methyl-	**79:** 87327q
3-(2-furyl)-5-propoxy-	**73:** 14770v
1-hexayl-3-[(*N*,*N*-dimethylcarbamoyl)oxy]-	**78:** P4261u
3-(4-hydroxybutoxy)-5-(4-methoxyphenyl)-	**68:** 78201q

TABLE 9 (*Continued*)

Compound	Reference

9.1. Alkyl, Aryl, and Acyl Mono- and Di-O-substituted 1,2,4-Triazoles (*Continued*)

Compound	Reference
3-(4-hydroxybutoxy)-5-methyl-	**68:** 78201q
3-(4-hydroxybutoxy)-5-phenethyl-	**68:** 78201q
3-(4-hydroxybutoxy)-5-phenyl-	**68:** 78201q
3-(2-hydroxyethoxy)-5-(4-hydroxyphenyl)-	**68:** 78201q
3-(2-hydroxyethoxy)-5-isobutyl-	**68:** 78201q
3-(2-hydroxyethoxy)-5-(4-methoxyphenyl)-	**68:** 78201q
3-(2-hydroxyethoxy-5-methyl-	**68:** 78201q
3-(2-hydroxyethoxy)-5-phenethyl-	**68:** 78201q
3-(2-hydroxyethoxy)-5-phenyl-	**68:** 78201q
3-(2-hydroxyethoxy)-5-propyl-	**74:** 53661a
3-(9-hydroxy-9-fluorenyl)-5-propoxy-	**68:** 78201q
3-(2-iodoethoxy)-5-(*p*-methoxyphenyl)-	**68:** 78201q
3-(2-iodoethoxy)-5-methyl-	**68:** 78201q
3-(2-iodoethoxy)-5-phenethyl-	**68:** 78201q
3-(2-iodoethoxy)-5-phenyl-	**71:** 91400k
3-isobutoxy-5-phenyl-	**57:** 3425f, **71:** 91400k
3-isopropoxy-5-phenyl-	**82:** P156321j
1-isopropyl-3-[(butylsulfonyl)oxy]-	**82:** P156321j
1-isopropyl-3-[(isopropylsulfonyl)oxy]-	**78:** P4261u
1-isopropyl-3-[(*N*-methoxymethylcarbamoyl)oxy]-	**78:** P4261u
1-isopropyl-5-methyl-3-[(*N,N*-dimethylcarbamoyl)oxy]-	**82:** P156321j
1-isopropyl-5-methyl-3-[(methylsulfonyl)oxy]-	**82:** P156321j
1-isopropyl-3-[(methylsulfonyl)oxy]-	**74:** 53661a
-3-methanol, 5-(benzyloxy)-α,α-diphenyl-	**71:** 101781b
-3-methanol, 5-butoxy-α,α-dimethyl-	**74:** 53661a
-3-methanol, 5-butoxy-α,α-diphenyl-	**71:** 101781b
-3-methanol, 5-butoxy-α-phenyl-	**71:** 101781b
-3-methanol, α,α-dimethyl-5-(pentyloxy)-	**71:** 101781b
-3-methanol, α,α-dimethyl-5-propoxy-	**74:** 53661a
-3-methanol, α,α-diphenyl-5-propoxy-	**71:** 101781b
-3-methanol, 5-ethoxy-α,α-dimethyl-	**74:** 53661a
-3-methanol, 5-ethoxy-α,α-diphenyl-	**71:** 101781b
-3-methanol, 5-ethoxy-α-phenyl-	**74:** 53661a
-3-methanol, 5-(hexyloxy)-α,α-diphenyl-	**71:** 101781b
-3-methanol, 5-isobutoxy-α,α-dimethyl-	**74:** 53661a
-3-methanol, 5-isobutoxy-α,α-diphenyl-	**71:** 101781b
-3-methanol, 5-isobutoxy-α-phenyl-	**71:** 101781b
-3-methanol, 5-(isopentyloxy)-α,α-dimethyl-	**74:** 53661a
-3-methanol, 5-(isopentyloxy)-α,α-diphenyl-	**71:** 101781b
-3-methanol, 5-isopropoxy-α,α-dimethyl-	**74:** 53661a
-3-methanol, 5-isopropoxy-α,α-diphenyl-	**74:** 53661a
-3-methanol, 5-methoxy-α,α-diphenyl-	**74:** 53661a
-3-methanol, 5-(2-methoxyethoxy)-α,α-diphenyl-	**71:** 101781b
-3-methanol, 5-methoxy-α-phenyl-	**71:** 101781b
-3-methanol, α-phenyl-5-propoxy-	**74:** 53661a
-3-methanol, 5-(pentyloxy)-α,α-diphenyl-	

TABLE 9   (*Continued*)

Compound	Reference

9.1. Alkyl, Aryl, and Acyl Mono- and Di-*O*-substituted 1,2,4-Triazoles   (*Continued*)

Compound	Reference
3-methoxy-	**64:** 4919f, **70:** P28922w, **70:** 87692u, **71:** P81375s, **71:** P124443g
3-methoxy-1,5-dimethyl-	**76:** 59576q, **84:** 17237q
3-methoxy-4,5-dimethyl-	**84:** 17237q
5-methoxy-1,3-dimethyl-	**84:** 17237q
3-methoxy-4,5-diphenyl-	**58:** 4569h
3-(2-methoxyethoxy)-5-phenyl-	**81:** 13442t
3-methoxy-5-[(*p*-methoxyphenoxy)methyl]-	**70:** 68268u
3-methoxy-5-(*p*-methoxyphenyl)-	**44:** 2518b
3-methoxy-1-methyl-	**76:** 59576q
3-methoxy-5-methyl-	**84:** 17237q
3-methoxy-4-methyl-5-(5-nitro-2-furanyl)-	**79:** 87327q
3-methoxy-4-methyl-5-phenyl-	**79:** 78749b, **85:** 159988e
3-methoxy-1-methyl-5-phenyl-	**79:** 78749b, **85:** 159988e
5-methoxy-1-methyl-3-phenyl-	**79:** 78749b, **85:** 159988e
3-methoxy-5-[(1-naphthyloxy)methyl]-	**70:** 68268u
3-methoxy-5-(phenoxymethyl)-	**70:** 68268u
3-methoxy-5-phenyl-	**57:** 3425f, **71:** 91400k, **83:** 163373b
3-methoxy-5-[(phenylthio)methyl]-	**70:** 68268u
3-methoxy-5-[(*m*-tolyloxy)methyl]-	**70:** 68268u
3-methoxy-5-[(*o*-tolyloxy)methyl]-	**70:** 68268u
3-methoxy-5-(3,4,5-trimethoxyphenyl)-	**69:** 96593u
3-methoxy-5-[(3,4-xylyloxy)methyl]-	**70:** 68268u
1-(1-methylbutyl)-3-[(*N*,*N*-dimethylcarbamoyl)oxy]-	**78:** P4261u
3-methyl-5-propoxy-	**71:** 91400k
3-[(methylsulfonyl)oxy]-1-phenyl-	**82:** P156321j
3-[(methylsulfonyl)oxy]-1-(1-phenylethyl)-	**82:** P156321j
3-[(methylsulfonyl)oxy]-1-propyl-,	**82:** P156321j
3-[(methylsulfonyl)oxy]-1-[3-(trifluoromethyl)phenyl]-,	**82:** P156321j
5-(*p*-nitrophenoxy)-1,3-diphenyl-	**64:** 8171e
3-(4-nitrophenyl)-5-phenoxy-	**81:** 136062b
3-phenoxy-	**70:** P28922w, **71:** P81375s, **71:** P124443g
3-phenoxy-4,5-diphenyl-	**57:** 12473c
5-phenoxy-1,3-diphenyl-	**64:** 8171e
3-phenoxy-5-phenyl-	**74:** P100065w
5-phenyl-3-[[(5-phenyl-1,3,4-oxadiazol-2-yl)-carbamoyl]oxy]-	**80:** 108456u
3-phenyl-5-propoxy-	**71:** 91400k
-1-propanenitrile, 5-methyl-3-[(methylsulfonyl)oxy]-	**82:** P156321j
-1-propanenitrile, 3-[(methylsulfonyl)oxy]-	**82:** P156321j
3-propoxy-5-propyl-	**71:** 91400k
3-propoxy-5-(2-thienyl)-	**73:** 14770v
3-[[(3,3,3-trichloro-1-propen-1-yl)sulfonyl]oxy]-1-isopropyl-	**82:** P156321j

TABLE 9 (*Continued*)

Compound	Reference

9.2. O-(1,2,4-Triazole) Derivatives of Phosphorus Acids

9.2a. *P-Substituted-O-ethyl-O-(substituted 1,2,4-Triazol-3-yl)phosphonothioates*[a]

Compound	Reference
—, 1-[4-chloro-3-(trifluoromethyl)phenyl]-5-propyl-, O-methyl-	**79:** P42515h
ethyl-, 1-benzyl-	**79:** P42515h
ethyl-, 1,5-bis(isopropyl)-	**77:** P48632z
ethyl, 1-(4-bromophenyl)-5-methyl-	**77:** P48632z, **79:** P42515h
ethyl-, 1-butyl-	**77:** P48632z
ethyl-, 1-(4-chloro-2-methylphenyl)-	**79:** P42515h
ethyl-, 1-(2-chlorophenyl)-	**77:** P48632z
ethyl-, 1-(3-chlorophenyl)-	**77:** P48632z
ethyl-, 1-(4-chlorophenyl)-	**77:** P48632z
ethyl-, 1-(3-chlorophenyl)-5-isopropyl-	**77:** P48632z
ethyl-, 1-(2-chlorophenyl)-5-methyl-	**77:** P48632z
ethyl-, 1-(3-chlorophenyl)-5-methyl-	**77:** P48632z
ethyl-, 1-(4-chlorophenyl)-5-methyl-	**77:** P48632z
ethyl-, 1-(2-cyanoethyl)-5-methyl-	**83:** P206281q
ethyl-, 1-cyclopentyl-	**77:** P48632z
ethyl-, 1-(2,5-dichlorophenyl)-	**77:** P48632z
ethyl-, 1-(2,5-dichlorophenyl)-5-methyl-	**77:** P48632z
ethyl-, 1-(diphenylmethyl)-	**79:** P42515h
ethyl-, 1-ethyl-	**77:** P48632z
ethyl-, 5-ethyl-1-isopropyl-	**77:** P48632z
ethyl-, 1-ethyl-5-methyl-	**77:** P48632z
ethyl-, 1-ethyl-5-phenyl-	**77:** P48632z
ethyl-, 5-ethyl-1-phenyl-	**77:** P48632z
ethyl-, 1-(1-ethylpropyl)-	**79:** P42515h
ethyl-, 1-(4-fluorophenyl)-	**79:** P42515h
ethyl-, 1-(4-fluorophenyl)-5-methyl-	**77:** P48632z
ethyl-, 1-hexyl-	**77:** P48632z
ethyl-, 1-*sec*-hexyl-	**77:** P48632z, **79:** P28415u
ethyl-, 1-isopropyl-	**77:** P48632z
ethyl-, 1-isopropyl-5-methyl-	**77:** P48632z
ethyl-, 1-(2-methylbutyl)-	**79:** P42515h
ethyl-, 5-methyl-1-(4-methylphenyl)-	**77:** P48632z
ethyl-, 1-methyl-3-phenyl-	**77:** P48632z
ethyl-, 5-methyl-1-phenyl-	**77:** P48632z
ethyl-, 1-(1-methylpropyl)-	**77:** P48632z
ethyl-, 1-(2-methylpropyl)-	**79:** P42515h
ethyl-, 1-sec-neohexyl-	**79:** P42515h
ethyl-, 1-(2-nitrophenyl)-	**79:** P42515h
ethyl-, 1-(3-nitrophenyl)-	**77:** P48632z
ethyl-, 1-(4-nitrophenyl)-	**77:** P48632z
ethyl-, 1-sec-pentyl-	**79:** P42515h
ethyl-, 1-phenyl-	**77:** P48632z
ethyl-, 1-(1-phenylethyl)-	
ethyl-, 1-phenyl-, O-methyl-	

TABLE 9 (*Continued*)

Compound	Reference

9.2. *O*-(1,2,4-Triazole) Derivatives of Phosphorus Acids (*Continued*)

9.2a. *P-Substituted-O-ethyl-O-(substituted* 1,2,4-Triazol-3-yl)phosphonothioates[a] (*Continued*)

ethyl-, 1-propyl-	**77:** P48632z
ethyl-, 4-[3-(trifluoromethyl)phenyl]-	**79:** P42515h
methyl-, 1-[4-chloro-3-(trifluoromethyl)phenyl]-,	**79:** P42515h
O-propyl-	
methyl-, 1-isopropyl-, O-propyl	**77:** P48632z
methyl-, 5-methyl-1-isopropyl-, O-propyl	**77:** P48632z
methyl-, 5-methyl-1-phenyl-, O-propyl	**77:** P48632z
methyl-, 1-[3-(trifluoromethyl)phenyl]-	**79:** P42515h
methyl-, 1-[3-(trifluoromethyl)phenyl]-, O-propyl-	**79:** P42515h
phenyl-, 1-ethyl-	**77:** P48632z
phenyl-, 1-(4-methylphenyl)-	**79:** P42515h

9.2b. *O,O-Diethyl-O-(substituted* 1,2,4-Triazol-3-yl)phosphorates[b]

1-(2-cyanoethyl)-	**83:** P206281q
1-isopropyl-, O,O-dimethyl-	**77:** P34528h
1-phenyl-	**71:** P101861c

9.2c. *O,O-Diethyl-O-(substituted* 1,2,4-Triazol-3-yl)phosphorothioates[b]

1-benzyl-	**79:** P18723d
5-benzyl-2-methyl-	**75:** P88615m, **77:** P152192s
5-benzyl-2-phenyl-	**75:** P88615m, **77:** P152192s
1,5-bis(isopropyl)-	**77:** P34528h
1-(4-bromophenyl)-5-methyl-	**77:** P126637w
1-butyl-	**77:** P34528h
1-butyl-, O,O-dimethyl	**77:** P34528h
1-(4-chloro-2-methylphenyl)-	**79:** P18723d
1-(m-chlorophenyl)-	**71:** P101861c
1-(p-chlorophenyl)-	**71:** P101861c
1-(3-chlorophenyl)-5-isopropyl-	**77:** P126637w
1-(3-chlorophenyl)-5-methyl-	**77:** P126637w
1-(4-chlorophenyl)-5-methyl-	**77:** P126637w
1-[4-chloro-3-(trifluoromethyl)phenyl]-	**79:** P42513f
1-(2-cyanoethyl)-	**83:** P206281q
1-(2-cyanoethyl)-5-methyl-	**83:** P206281q
1-cyclohexyl-	**77:** P34528h, **79:** P28414t
1-cyclohexyl-, O,O-dimethyl	**77:** P34528h
1-cyclopentyl-	**77:** P34528h, **79:** P28414t
1-cyclopentyl-, O,O-dimethyl	**77:** P34528h
1-(2,5-dichlorophenyl)-5-methyl-	**77:** P126637w
1-(3,4-dichlorophenyl)-5-methyl-	**77:** P126637w
2,5-dimethyl-	**75:** P88615m, **77:** P152192s, **82:** 107475d

TABLE 9 (*Continued*)

Compound	Reference

9.2. *O*-(1,2,4-Triazole) Derivatives of Phosphorus Acids (*Continued*)

9.2c. *O,O-Diethyl-O-(substituted* 1,2,4-*Triazol*-3-*yl)phosphorothioates*[b] (*Continued*)

1-(diphenylmethyl)-	**79:** P18723d
1-ethyl-	**77:** P34528h, **79:** P28414t
1-ethyl-, *O,O*-dimethyl	**77:** P34528h, **79:** P28414t
5-ethyl-1-isopropyl-	**77:** P34528h
1-ethyl-5-methyl-	**77:** P34528h
3-ethyl-1-methyl-	**79:** P66363h
1-ethyl-5-phenyl-	**77:** P34528h
3-ethyl-1-phenyl-	**79:** P66363h
5-ethyl-1-phenyl-	**77:** P126637w
1-ethyl-5-phenyl-, *O,O*-dimethyl	**77:** P34528h
5-ethyl-1-phenyl-, *O,O*-dimethyl	**77:** P126637w
1-(4-fluorophenyl)-	**79:** P42513f
1-hexyl-	**77:** P34528h
1-sec-hexyl-	**77:** P34528h
1-hexyl-, *O,O*-dimethyl	**77:** P34528h
1-isopropyl-	**77:** P34528h, **79:** P28414t, **82:** 107475d
1-isopropyl-, *O,O*-bis(isopropyl)	**77:** P34528h
1-isopropyl-, *O,O*-dibutyl	**77:** P34528h
1-isopropyl-, *O,O*-dimethyl	**77:** P34528h, **79:** P28414t
1-isopropyl-5-methyl-	**77:** P34528h, **79:** P28414t, **82:** 107475d
5-isopropyl-1-methyl-	**77:** P34528h
1-methyl-	**77:** P34528h
2-methyl-	**77:** P152192s
1-(2-methylbutyl)-	**77:** P34528h
1-(2-methylbutyl)-, *O,O*-dimethyl	**77:** P34528h
2-methyl-5-isopropyl-	**75:** P88615m, **77:** P152192s, **82:** 107475d
2-methyl-4-(4-methylphenyl)-	**75:** P88615m, **77:** P152192s
2-methyl-5-phenyl-	**75:** P88615m, **77:** P152192s
5-methyl-1-phenyl-	**77:** P126637w
5-methyl-2-phenyl-	**75:** P88615m, **77:** P152192s
1-(2-methylpropyl)-	**77:** P34528h
1-(2-methylpropyl)-, *O,O*-dimethyl	**77:** P34528h
5-methyl-1-[3-(trifluoromethyl)phenyl]-	**77:** P126637w
1-neohexyl-	**77:** P34528h
1-(2-nitrophenyl)-	**79:** P18723d
1-(3-nitrophenyl)-	**79:** P18723d
1-(4-nitrophenyl)-	**79:** P18723d
1-sec-pentyl-	**77:** P34528h
1-(pentafluorophenyl)-	**79:** P42513f
1-phenyl-	**71:** P101861c
1-phenyl, *O,O*-dimethyl	**71:** P101861c
1-(1-phenylethyl)-	**79:** P18723d

TABLE 9 (Continued)

Compound	Reference

9.2. O-(1,2,4-Triazole) Derivatives of Phosphorus Acids (Continued)

9.2c. O,O-Diethyl-O-(substituted 1,2,4-Triazol-3-yl)phosphorothioates[b] (Continued)

Compound	Reference
1-propyl-	**77:** P34528h, **79:** P28414t
1-propyl-, O,O-dimethyl	**77:** P34528h
1-[2-(trifluoromethyl)phenyl]-	**79:** P42513f
1-[3-(trifluoromethyl)phenyl]-	**79:** P42513f

9.2d. S-Alkyl-O-ethyl-O-(substituted 1,2,4-Triazol-3-yl)phosphorodithionates[c]

Compound	Reference
S-methyl-, 1-isopropyl-	**81:** P13520s
S-methyl-, 1-phenyl-	**81:** P13521t
S-propyl-, 1-butyl-	**81:** P13520s
S-propyl-, 1-(4-chlorophenyl)-	**81:** P13521t
S-propyl-, 1-(2-cyanoethyl)-	**83:** P206281q
S-propyl-, 1-(2-cyanoethyl)-5-methyl-	**83:** P206281q
S-propyl-, 1-(3,5-dichlorophenyl)-	**81:** P13521t
S-propyl-, 1-ethyl-5-methyl-	**81:** P13520s
S-propyl-, 1-isopropyl-	**81:** P13520s
S-propyl-, 1-(methylpropyl)-	**81:** P13520s
S-propyl-, 1-(2-methylpropyl)-	**81:** P13520s
S-propyl-, 1-phenyl-	**81:** P13521t
S-propyl-, 1-[3-(trifluoromethyl)phenyl]-	**79:** P42513f

[a] Listed as P-substituent, triazole substituent(s), O-alkyl if not ethyl.
[b] Listed as triazole substituent(s), O,O-dialkyl if not ethyl.
[c] Listed as S-substituent, triazole substituent.

CHAPTER 10

S-Substituted Thio-, Sulfinyl-, or Sulfonyl-1,2,4-Triazoles, 3,3′-Dithobis[1,2,4-Triazoles], and 1,2,4-Triazolesulfonic Acids and Their Functional Derivatives

10.1. Mono- and Di-S-Substituted Thio-1,2,4-Triazoles

The alkyl-, aryl-, and acylthio-1,2,4-triazoles are composed of a large number of compounds in which the S-substituent can be derived from a variety of both simple and complex functions. Several of the S-substituted thiotriazoles have shown biological activity: **10.1-1** against *Mycobacterium tuberculosis*,[1,2] **10.1-2** as an anticoccidial agent in chickens,[3] and the

10.1-1   **10.1-2**

cephalosporin (5-thia-1-azabicyclo[4.2.0]oct-2-ene ring system) derivative (**10.1-3**) as an antibacterial agent.[4] A 3-(2,4-dichlorophenoxymethyl)-5-(ethylthio)-1,2,4-triazole was also reported to have good fungicidal activity against *Aspergillus niger*.[5] In addition, the alkylthio-1,2,4-thiazoles have been used to stabilize photographic emulsions[6,7] (e.g., **10.1-4**) and plastics (e.g., **10.1-5**).[8]

The alkylation of preformed alkyl- and aryl-1,2,4-triazoline-5-thiones in a basic medium has been effected with a variety of reagents: PhCH₂Cl with **10.1-6**,[5,9–11] PhOCH₂Cl with **10.1-7**,[12] EtO₂CCH₂Cl with **10.1-7**,[13]

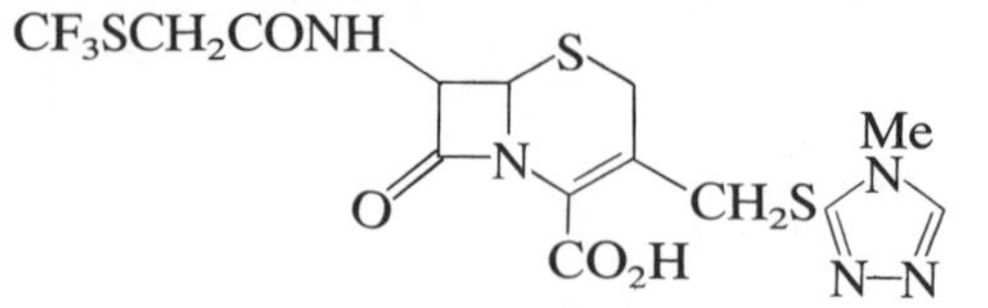

10.1-3

10.1-4

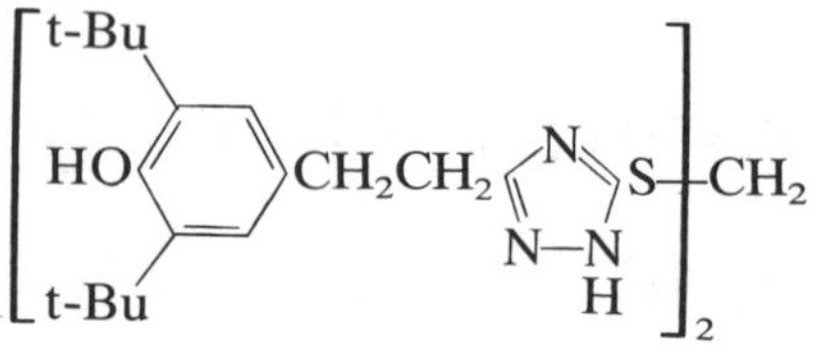

10.1-5

H_2NCOCH_2Cl with **10.1-8**,[14] $Me_2NCH_2CH_2Cl$ with **10.1-9**,[15] $PhCOCH_2Br$ with **10.1-10**,[15,16] and the cyclic sulfonyl $O(CH_2)_3SO_2$ with **10.1-6**[17] to give good yields of S-substituted thio-1,2,4-triazoles (**10.1-12** to **10.1-18**). This type of reaction was also accomplished with several heterocyclic systems substituted with chloromethyl groups.[18,19] In one example, 5-(hydroxymethyl)uracil was reacted with **10.1-11** in 1 N HCl to give a 59% yield of **10.1-19**, presumably formed via 5-(chloromethyl)uracil.[20] In addition, heterocyclics containing activated halogen groups such as 2-bromo-5-nitro-1,3,4-thiadiazole alkylate triazoline-5-thiones to give the corresponding S-substituted products.[21-23] Compound **10.1-17** underwent intramolecular cyclization to the N-1 of the triazole ring when heated in polyphosphoric acid to give a condensed bicyclic thiazolotriazole, and related dihydrothiazolotriazoles were obtained by treatment of **10.1-10** with 1,2-dibromoethane.[16] In contrast, **10.1-6** was reported to react with 1,2-dibromoethane (and other dihaloalkanes) to give a 66% yield of 3,3'-(1,2-ethanedithio)bis(triazole)[24] (see Section 21.1). A reaction that involved hydrolysis of the thione group

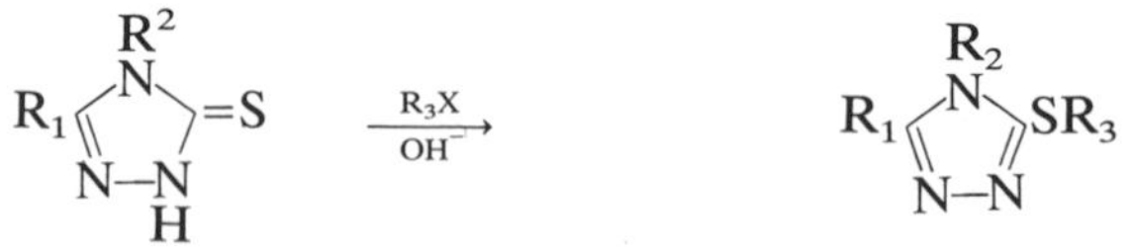

10.1-6, $R_1 = R_2 = H$
10.1-7, $R_1 = Me, R_2 = Ph$
10.1-8, $R_1 = H, R_2 = Ph$
10.1-9, $R_1 = H, R_2 = 4\text{-}C_5H_4N$
10.1-10, $R_1 = Ph, R_2 = H$
10.1-11, $R_1 = Me, R_2 = H$

10.1-12, $R_1 = R_2 = H, R_3 = CH_2Ph$
10.1-13, $R_1 = Me, R_2 = Ph, R_3 = CH_2OPh$
10.1-14, $R_1 = Me, R_2 = Ph, R_3 = CH_2CO_2Et$
10.1-15, $R_1 = H, R_2 = Ph, R_3 = CH_2CONH_2$
10.1-16, $R_1 = H, R_2 = 4\text{-}C_5H_4N, R_3 = CH_2CH_2NM$
10.1-17, $R_1 = Ph, R_2 = H, R_3 = CH_2COPh$
10.1-18, $R_1 = R_2 = H, R_3 = (CH_2)_3SO_3H$
10.1-19, $R_1 = Me, R_2 = H, R_3 = $ (uracil-5-yl)methyl

resulted from treatment of **10.1-20** with ethylene oxide in acetic acid to give the 1-(2-hydroxyethyl)-1,2,4-triazolin-5-one (**10.1-23**).[25] The intermediate (**10.1-21**) was independently prepared from **10.1-20** and chloroethanol in ethanolic sodium hydroxide and converted to **10.1-23** with ethylene oxide in acetic acid, indicating that **10.1-22** was also an intermediate. In the alkylation of **10.1-7** with methyl p-toluenesulfonate, the product was incorrectly

described as a 1,3,4-thiadiazole rather than the p-toluenesulfonate salt of **10.1-24**.[26] Although these reactions demonstrate the greater nucleophilicity of sulfur over that of a ring nitrogen, a second alkylation of **10.1-25** with methyl iodide to give a mixture of the corresponding 1-, 2-, and 4-methyl derivatives[27] suggested that minor amounts of N-alkylated products are obtained in most alkylations of triazoline-5-thiones.[9,28] Also, alkylation with diazomethane gave predominantly (methylthio)triazoles.[29]

Several examples of reactions that give N rather than S substitution have been reported. In the addition of formaldehyde to the thione (**10.1-8**), the product was assigned to structure **10.1-26** rather than **10.1-27**.[30] Another

exception to S-substitution occurred in the reaction of **10.1-8** with 3-chloropropionitrile under basic conditions to give the N-1 substituted triazoline-5-thione (**10.1-28**).[31] The same product resulted from the cyanoethylation of **10.1-8** with acrylonitrile in the presence of triethylamine.[32,33] Similarly, 1-phenyl-1,2,4-triazoline-5-thione (**10.1-29**) gave only the corresponding N-4 cyanoethyl derivative, whereas 2-phenyl-1,2,4-triazoline-5-thione (**10.1-30**) gave only the corresponding S-substituted cyanoethyl derivative.[32]

10.1-28   **10.1-29**   **10.1-30**

The preparation of simple alkyl and aryl thio-1,2,4-triazoles (**10.1-32** to **10.1-34**) has also been effected by displacement of the chloro group of 3-chloro-1,2,4-triazole (**10.1-31**) at high temperatures with methyl mercaptan, benzyl mercaptan, and thiophenol.[34] In addition, the oxidation of the triazolinethione (**10.1-11**) with the iodine-potassium iodide reagent apparently

10.1-31   **10.1-32**, R = Me (70%)
10.1-33, R = CH$_2$Ph (90%)
10.1-34, R = Ph (82%)

generated the iodide intermediate (**10.1-35**), which reacted *in situ* with indole to give the electrophilic substitution product, **10.1-36**.[35] The usefulness of this approach is limited by the simultaneous formation of the

10.1-11   **10.1-35**   **10.1-36** (52%)

disulfide of the triazole as a by-product. No difficulties were encountered, however, in the removal of the 4-amino group of 4-amino-1,2,4-triazoles with nitrous acid to give alkylthio-1,2,4-triazoles, as illustrated by the conversion of **10.1-37** to **10.1-38**.[36,37]

10.1-37   **10.1-38**

Several methods have been described for the direct synthesis of *S*-substituted thiotriazoles by cyclization reactions. Treatment of the *S*-methyl derivative of 3-thiosemicarbazide (**10.1-40**) and the related compounds, **10.1-41** and **10.1-42**, with refluxing formic acid gave formyl intermediates that were converted *in situ* to give good yields of the methylthio-1,2,4-triazoles (**10.1-32**, **10.1-39**, and **10.1-43**, respectively).[38] The isolated 1-acyl or

10.1-32, R$_1$ = H   **10.1-40**, R$_1$ = R$_2$ = H   **10.1-43**
10.1-39, R$_1$ = Me   **10.1-41**, R$_1$ = Me, R$_2$ = H
                              10.1-42, R$_1$ = H, R$_2$ = Me

1-aroyl S-methyl-3-thiosemicarbazides are readily dehydrated in the presence of either an acid or base catalyst[38,39] or in the absence of a catalyst,[40] as demonstrated by the conversion of **10.1-44** to **10.1-45** on recrystallization from water.[41] Compound **10.1-45** was the precursor of **10.1-46**, formed from **10.1-45** and hydrobromic acid at room temperature. In the cyclization

10.1-45, R$_1$ = PhCH$_2$O$_2$C (77%)
10.1-46, R$_1$ = H

of **10.1-47** in refluxing alcohol to give **10.1-24**, a minor amount of the oxadiazole (**10.1-48**) was also formed.[42] In addition, oxadiazole formation

10.1-47   **10.1-24**   **10.1-48**

was observed when a 1-aroyl-S-methyl-3-thiosemicarbazide was refluxed in ethanol.[43] Although treatment of **10.1-50** with a refluxing mixture of methyl iodide and sodium hydroxide gave the 3-(methylthio)-1,2,4-triazole (**10.1-49**), treatment of **10.1-50** with dimethyl sulfate in the presence of excess 10% sodium hydroxide gave the oxadiazole **10.1-51**.[44] Ring closure

10.1-49   **10.1-50**   **10.1-51**

of S-methylated 1,4-diacyl-3-thiosemicarbazides resulted in the removal of the 1-acyl function[45] as shown by the alkylation of 4-benzoyl-1-ethoxycarbonyl-3-thiosemicarbazide (**10.1-52**) with methyl iodide to give **10.1-53**, which on treatment with hot base gave **10.1-54**.[46]

10.1-52   **10.1-53**   **10.1-54**

In a variation of the Einhorm–Brunner reaction, condensation of phenyl hydrazine with methyl N-acetyldithiocarbamate (**10.1-55**) gave the 3-(methylthio)-1,2,4-triazole (**10.1-57**), formed via intermediate **10.1-56**.[47]

10.1-55   **10.1-56**   **10.1-57**

When the substituted amidine hydroiodide (**10.1-58**) was refluxed in water, ammonia was liberated to give a 15% yield of **10.1-60**.[48] In the corresponding furan derivative (**10.1-59**), cyclization produced a 53% yield of **10.1-61**.

10.1-58, X = S
10.1-59, X = O

10.1-60, X = S
10.1-61, X = O

In addition, 3-(methylthio)-1,2,4-triazoles have been formed by ring contraction in base of triazine derivatives: **10.1-62** to give **10.1-63**[49] and **10.1-64** to give **10.1-65**.[50]

10.1-62   **10.1-63**

10.1-64   **10.1-65**

The *S*-acylation of 1,2,4-triazolinethiones can be effected under mild conditions. For example, the 5-thiocyanato-1,2,4-triazole (**10.1-67**) was prepared from the thione (**10.1-66**) and cyanogen bromide under basic conditions.[51–53] Similarly, both an isocyanate and an isothiocyanate reacted

10.1-66   **10.1-67**

with **10.1-69** to give the *S*-substituted products, **10.1-68** and **10.1-70**, respectively.[54] In the reaction of the sodium salt (**10.1-72**) with benzoyl

10.1-68   **10.1-70**

chloride, the *S*-benzoyl derivative (**10.1-71**) was obtained at room temperature and the *N*-benzoyl derivative (**10.1-73**) at reflux in benzene.[55] Reactions of this type were also observed when **10.1-74** was treated with the tetraacetate of α-D-glucopyranoxyl bromide. In the absence of a metal ion

10.1-71   **10.1-72**   **10.1-73**

the *S*-substituted product (**10.1-75**) was obtained. In contrast the mercury (II) salt (**10.1-76**) reacted with the bromide to give the *N*-substituted product (**10.1-77**), which was also obtained by heating **10.1-75** in the presence of mercury (II).[56]

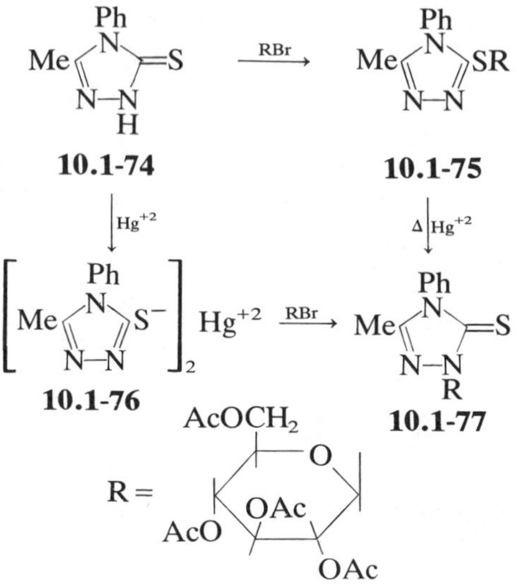

10.1-74   **10.1-75**

10.1-76   **10.1-77**

AcOCH₂

R =

References

1. **74:** 53660z	2. **79:** 105155p	3. **83:** 126969e	4. **83:** 109119k
5. **82:** 125327m	6. **81:** 129840a	7. **81:** 144229b	8. **79:** 19662v
9. **57:** 804i	10. **82:** 72886j	11. **84:** 30975y	12. **82:** 140028h
13. **61:** 4337h	14. **63:** 13242f	15. **63:** 10497b	16. **81:** 136112t
17. **80:** 47911q	18. **78:** 159355b	19. **83:** 114412e	20. **70:** 68301z
21. **74:** 14185d	22. **83:** 164088z	23. **84:** 164664e	24. **71:** 124344a
25. **76:** 126876c	26. **54:** 5630e	27. **76:** 34180t	28. **78:** 124513z
29. **85:** 159988e	30. **75:** 35891s	31. **63:** 13242b	32. **74:** 141644c
33. **82:** 43276e	34. **71:** 81375s	35. **75:** 19520q	36. **55:** 23507d
37. **57:** 12471f	38. **55:** 23508e	39. **61:** 9490f	40. **66:** 115714r
41. **66:** 2536b	42. **53:** 11355h	43. **71:** 49853k	44. **75:** 76692w
45. **53:** 10033i	46. **66:** 115651t	47. **54:** 7695e	48. **69:** 77177v
49. **76:** 59576q	50. **73:** 66517n	51. **61:** 4338e	52. **68:** 59501w
53. **68:** 21884k	54. **81:** 49647c	55. **81:** 91429h	56. **80:** 146458d

10.2. 3,3′-Dithiobis[1,2,4-triazoles], Sulfinyl- or Sulfonyl-1,2,4-triazoles, and 1,2,4-Triazolesulfonic Acids and Their Functional Derivatives

10.2a. 3,3′-Dithiobis[1,2,4-triazoles]

Oxidation of triazoline-5-thione (**10.2a-1**) with hydrogen peroxide[1] and a series of 3,4-diayltriazoline-5-thiones (e.g., **10.2a-2**) with iodine[2] gave the corresponding disulfides (**10.2a-4** and **10.2a-5**). In the oxidation of **10.2a-7**

with bromine, the cinnamoyl group was removed to give a 85% yield of
10.2a-5.[3] The electrolytic oxidation of triazoline-5-thiones to disulfides in
5% hydrochloric acid in the presence of a platinum-iridium electrode at a
potential of 0.7 to 1.0 volts has been demonstrated.[4] In addition, treatment
of triazoline-5-thiones with ozone gave the corresponding disulfides.[4] Also,
a mixture of sodium nitrite and sodium hydrogen sulfate was used as the
oxidant in the conversion of **10.2a-3** to **10.2a-6**.[5] The resulting disulfides are
readily reduced chemically (e.g., Sn or Zn) in an acidic medium to give the
parent triazoline-5-thione.[2,6]

10.2a-1, $R_1 = R_2 = H$ **10.2a-4**, $R_1 = R_2 = H$ **10.2a-7**
10.2a-2, $R_1 = R_2 = Ph$ **10.2a-5**, $R_1 = R_2 = Ph$
10.2a-3, $R_1 = 4\text{-}C_5H_4N$, **10.2a-6**, $R_1 = 4\text{-}C_5H_4N$,
    $R_2 = 4\text{-}MeC_6H_4$ $R_2 = 4\text{-}MeC_6H_4$

References

1. **51:** 12079f   2. **54:** 8794f   3. **81:** 91429h   4. **74:** 3557a
5. **63:** 2967c   6. **64:** 8174a

10.2b. *S*-Substituted Sulfinyl- or Sulfonyl-1,2,4-Triazoles

Oxidation of several 4-aryl-3-(alkylthio)-1,2,4-triazoles (e.g., **10.2b-1**)
with potassium permanganate gave sulfonyl-1,2,4-triazoles (sulfones) (e.g.,
10.2b-2) in yields ranging from 15 to 75%.[1,2] Under the same conditions,
oxidation of **10.2b-3** resulted in decarboxylation to give an 82% yield of

10.2b-1   **10.2b-2**   **10.2b-3**

10.2b-2. Similarly, 1-aryl-3-(alkylthio)- and 1-aryl-, 3-aryl-, and 3,4-diaryl-
5-(alkylthio)-1,2,4-triazoles are oxidized either with potassium permanganate or hydrogen peroxide in acetic acid to give good yields of alkylsulfonyl-
1,2,4-triazoles.[3-6] The conversion of **10.2b-4** either to **10.2b-5** (74%) or
10.2b-6 (68%) with peroxyphthalic acid in ether demonstrated the readiness

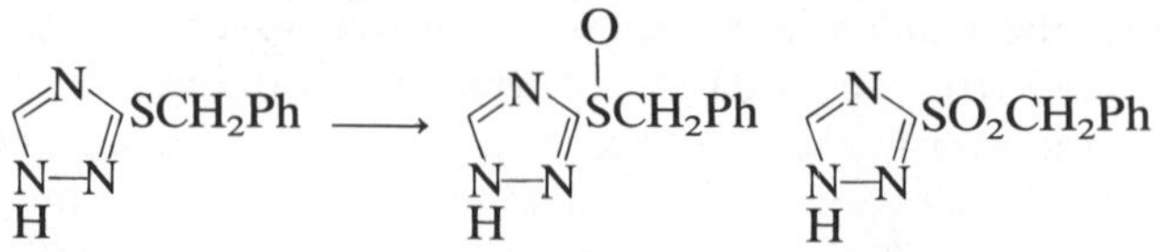

10.2b-4   **10.2b-5**   **10.2b-6**

in which the (alkylthio)triazoles are oxidized.[7] In addition, several alkylsulfonyl- and arylsulfonyl-1,2,4-triazoles (e.g., **10.2b-8**) were prepared by the cyclization of amidrazones (e.g., **10.2b-7**) with orthoesters.[8] Also, the base-

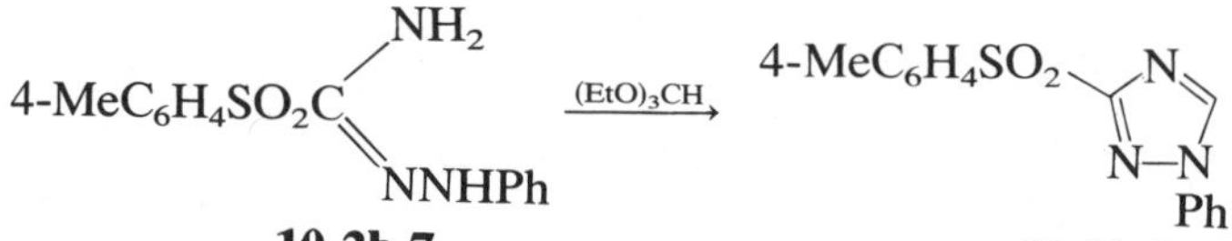

10.2b-7   **10.2b-8**

promoted cycloaddition of *p*-tosylmethyl isocyanide (**10.2b-9**) to phenyl diazonium tetrafluoroborate gave a mixture of **10.2b-10** and **10.2b-11**, in which **10.2b-10** was the major product.[9]

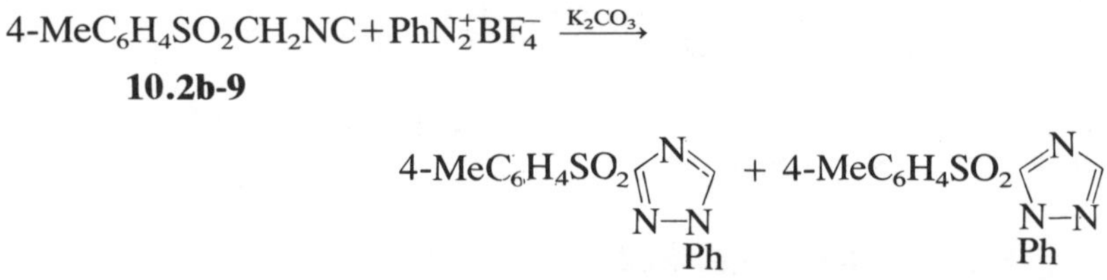

$4\text{-MeC}_6\text{H}_4\text{SO}_2\text{CH}_2\text{NC} + \text{PhN}_2^+\text{BF}_4^- \xrightarrow{\text{K}_2\text{CO}_3}$

10.2b-9

10.2b-10   **10.2b-11**

Hydrolysis of the alkylsulfonyl group occurred readily under basic conditions.[2–4] In one investigation it was shown that displacement of the sulfonyl group of **10.2b-2** in aqueous sodium hydroxide to give the corresponding triazolin-5-one was 90% complete in 3 hr at 80°, whereas the methylthio group of **10.2b-1** was stable in refluxing base.[10]

The condensation of *p*-toluenesulfonhydrazide (**10.2b-12**) with benzyl cyanide was reported to give the 1-(*p*-toluenesulfonyl)-1,2,4-triazole (**10.2b-13**), a reaction that is similar to the condensation of an acid hydrazide with a nitrile (Section 2.4).[11] Also, a series of 1-(arylsulfonyl)-1,2,4-triazoles (e.g., **10.2b-15**) were prepared by reaction of 1,2,4-triazoles (e.g.,

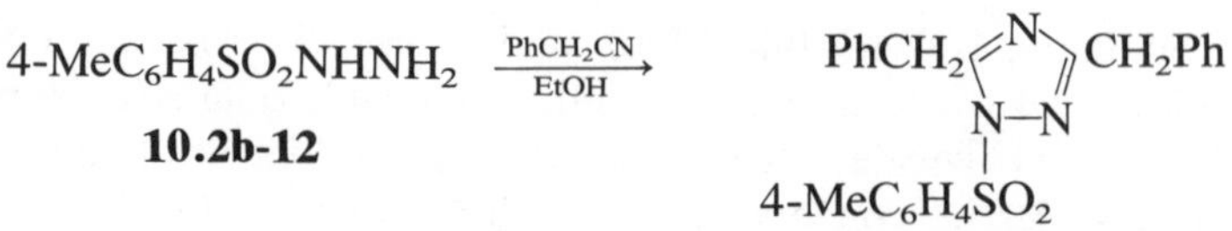

10.2b-12

10.2b-13 (63%)

10.2b-14) with benzenesulfonyl chlorides.[12] The 1-(mesitylsulfonyl)triazole (**10.2b-16**) has been used as a condensation reagent for the conversion of nucleotides (e.g., trinucleotides) to polynucleotides (e.g., hexanucleotides).[13–16]

Me $\overset{N}{\underset{N-N}{\|}}$ Me $\xrightarrow[\text{Et}_3\text{N}]{\text{PhSO}_2\text{Cl}}$ Me $\overset{N}{\underset{N-N}{\|}}$ Me $\overset{N}{\underset{N-N}{\|}}$

H   PhSO$_2$ 2,4,6-(Me)$_3$C$_6$H$_2$SO$_2$

10.2b-14   **10.2b-15**   **10.2b-16**

References

1. **63**: 13242f	2. **76**: 34182v	3. **71**: 49853k	4. **74**: 111968u
5. **74**: 141644c	6. **82**: 125327n	7. **74**: 3557a	8. **83**: 27827a
9. **84**: 90078t	10. **63**: 13243b	11. **61**: 4515f	12. **66**: 55489j
13. **81**: 91849v	14. **84**: 17652w	15. **84**: 105990k	16. **85**: 94637t

10.2c. 1,2,4-Triazolesulfonic Acids and Their Functional Derivatives

Oxidation of triazoline-5-thiones (e.g., **10.2c-1**) by heating with aqueous potassium permanganate gave the corresponding triazole-5-sulfonic acids (e.g., **10.2c-4**).[1] With this oxidant and also alkaline hydrogen peroxide, difficulties were encountered in removing residual matter from the more soluble sulfonic acids. For these compounds the triazoline-5-thione (e.g., **10.2c-2**) was oxidized with aqueous chlorine to give the corresponding sulfonyl chloride (e.g., **10.2c-8**), which was hydrolyzed with ethanol to give the sulfonic acid (e.g., **10.2c-5**).[1–3] Also, the conversion of triazole disulfides to triazolesulfonic acids with ozone has been demonstrated.[2] The chlorine oxidation of triazoline-5-thiones has several limitations including partial dethiation to give the parent triazole[2] and also, in some instances, the corresponding chlorotriazole.[4] With some 1- and 4-substituted triazoline-5-thiones, the formation of the parent triazole was the major product of the reaction.[2] In contrast, the 4-substituted triazoline-5-thiones (**10.2c-13** and **10.2c-14**) were converted in high yields to the sulfonic acids (**10.2c-6** and **10.2c-7**) with alkaline hydrogen peroxide.[5]

The oxidation of triazoline-5-thiones to sulfonic acids with aqueous bromine was unsatisfactory, many of these reactions giving a high yield of the parent triazole.[1,2] In addition, the isolation of sulfonic acids from triazoline-5-thiones and hydrogen peroxide under acidic conditions was unsuccessful. In contrast, both 3-(4-nitrophenyl)triazoline-5-thione (**10.2c-3**) and the 5-(benzylthio)triazole (**10.2c-12**) were oxidized under acidic conditions to give the corresponding sulfonyl chlorides (**10.2c-9** and **10.2c-10**).[6,7]

Although the benzenesulfonyl chlorides are more stable than the triazolesulfonyl chlorides, treatment of the latter with aqueous ammonium hydroxide (e.g., **10.2c-9**) and liquid ammonia (e.g., **10.2c-10**) readily gave the sulfonamide derivatives (e.g., **10.2c-15** and **10.2c-16**).[2,8] Reaction of sulfonyl chlorides with hydrazine, however, resulted in reductive cleavage of the sulfonyl function to give the parent triazole.[2] In other work, hydrazinolysis of **10.2c-11** and related compounds reduced the sulfonyl moiety to give triazoline-5-thiones.[9] Of interest, reaction of the sulfonamide (**10.2c-17**) with a mixture of nitric and sulfuric acids gave the *N*-nitro sulfonamide (**10.2c-18**).[10]

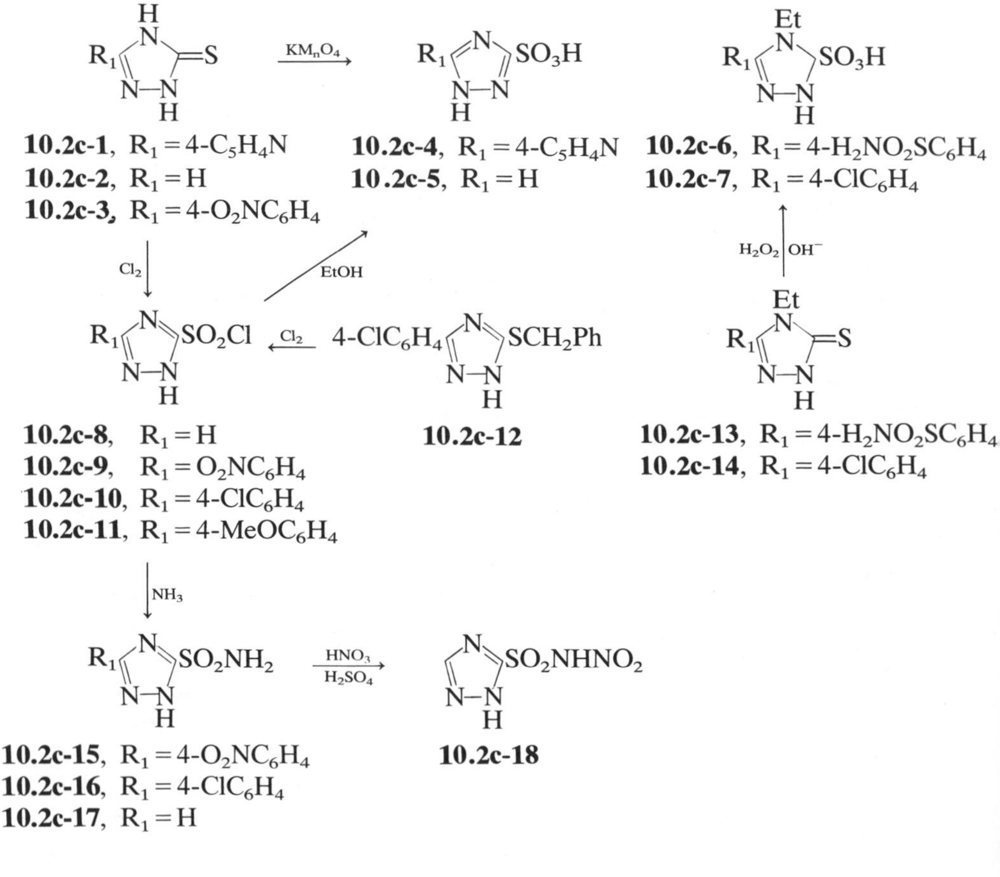
KM$_n$O$_4$
Cl$_2$
EtOH
Cl$_2$
H$_2$O$_2$ OH$^-$
NH$_3$
HNO$_3$
H$_2$SO$_4$
4-ClC$_6$H$_4$
SCH$_2$Ph
SO$_2$Cl
SO$_3$H
SO$_2$NH$_2$
SO$_2$NHNO$_2$
Et
10.2c-1, R$_1$ = 4-C$_5$H$_4$N
10.2c-2, R$_1$ = H
10.2c-3, R$_1$ = 4-O$_2$NC$_6$H$_4$
10.2c-4, R$_1$ = 4-C$_5$H$_4$N
10.2c-5, R$_1$ = H
10.2c-6, R$_1$ = 4-H$_2$NO$_2$SC$_6$H$_4$
10.2c-7, R$_1$ = 4-ClC$_6$H$_4$
10.2c-8, R$_1$ = H
10.2c-9, R$_1$ = O$_2$NC$_6$H$_4$
10.2c-10, R$_1$ = 4-ClC$_6$H$_4$
10.2c-11, R$_1$ = 4-MeOC$_6$H$_4$
10.2c-12
10.2c-13, R$_1$ = 4-H$_2$NO$_2$SC$_6$H$_4$
10.2c-14, R$_1$ = 4-ClC$_6$H$_4$
10.2c-15, R$_1$ = 4-O$_2$NC$_6$H$_4$
10.2c-16, R$_1$ = 4-ClC$_6$H$_4$
10.2c-17, R$_1$ = H
10.2c-18

References

1. **66:** 104948u
2. **74:** 3557a
3. **74:** 99946a
4. **84:** 17238r
5. **74:** 111968u
6. **64:** 5072h
7. **51:** 2874h
8. **85:** P33028y
9. **78:** 43372a
10. **59:** 12826f

TABLE 10. S-SUBSTITUTED THIO, SULFINYL, OR SULFONYL-1,2,4-TRIAZOLES, 3,3′-DITHIOBIS[1,2,4-TRIAZOLES], AND 1,2,4-TRIAZOLESULFONIC ACIDS AND THEIR FUNCTIONAL DERIVATIVES

Compound	Reference

10.1. Mono- and Di-S-substituted Thio-1,2,4-triazoles

Compound	Reference
1-acetamide, N-[5-[[4-[2,4-bis(1,1-dimethylpropyl)-phenoxy]-1-oxobutyl]amino]-2-chlorophenyl]-α-(2,2-dimethyl-1-oxopropyl)-3-(methylthio)-	**85:** P102316e
4-acetamide, α-(2,2-dimethyl-1-oxopropyl)-N-[2-(hexadecyloxy)-5-[(methylamino)sulfonyl]phenyl]-3-methyl-5-(methylthio)-	**85:** P151754e
3-(p-acetamidophenyl)-5-(benzylthio)-	**47:** 131f
3-(p-acetamidophenyl)-5-(methylthio)-	**47:** 131i
-1-acetic acid, 3-(2-methyl-4-quinolinyl)-5-(methylthio)-, methyl ester	**84:** 17287f
-1-acetic acid, 5-(2-methyl-4-quinolinyl)-3-(methylthio)-, methyl ester	**84:** 17287f
3-[(acetonyl)thio]-	**80:** P37118x
3-[(acetonyl)thio]-5-methyl-	**74:** 42320n, **80:** P37118x
3-[(acetonyl)thio]-5-methyl-4-phenyl-	**82:** 140028h
3-[[[3-(acetoxymethyl)-2-carboxy-8-oxo-5-thia-1-azabicyclo[4.2.0]oct-2-en-7-yl]carbamoyl]-methylthio]-	**71:** P91505y
3-[2-[[2-acetylamino-4-[[2-(acetylamino)ethyl]-ethylamino]phenyl]azo]phenylthio]-	**79:** P43665u
3-[2-[3-acetylamino-4-[(2-cyano-4,6-dinitrophenyl)-azo]-N-ethylanilino]ethylthio]-	**79:** P43665u
3,3′-(2-[(2-acetylamino-4-cyclohexylamino-5-methyl-phenyl)azo]-4-cyanophenylene-1,3-dithio)bis-	**79:** P43665u
3-(2-[(2-acetylamino-4-cyclohexylamino-5-methyl-phenyl)azo]-4-cyanophenylthio)-	**79:** P43665u
3,3′-[2-([2-acetylamino-4-(diethylamino)phenyl]azo)-4-bromophenylene-1,3-dithio]bis	**79:** P43665u
3-[2-([2-acetylamino-5-(diethylamino)phenyl]azo)-4-cyanophenylthio]-	**79:** P43665u
3-([4-([2-acetylamino)-4-(diethylamino)phenyl]azo)-5-ethoxycarbonylphenyl]thio)-	**79:** P43665u
3-(2-[(2-acetylamino-4-(diethylamino)phenyl)-azo]-5-methylsulfonylphenylthio)-	**79:** P43665u
3-[2-(3-acetylamino-N-ethylanilino)ethylthio]-	**79:** P43665u
3-[2[3-(acetylamino)-N-ethyl-4-([4-(phenylazo)phenyl]-azo)anilino]ethylthio]-	**79:** P43665u
3-[2-(7-acetylamino-1-ethyl-1,2,3,4-tetrahydro-2,2,4-trimethyl-6-quinolinyl)-4-(methylsulfonyl)phenylthio]-	**79:** P43665u
5-[([3(acetyloxy)methyl]-2-carboxy-8-oxo-5-thia-1-azabicyclo[4.2.0]oct-2-en-7-yl)carbamoylmethylthio]-1-methyl-	**79:** 61983a
4-allyl-3-(carboxymethylthio)-5-(2-chlorophenyl)-	**68:** 105106w
4-allyl-3-(carboxymethylthio)-5-(o-hydroxyphenyl)-	**68:** 105106w
4-allyl-3-(carboxymethylthio)-5-(m-chlorophenyl)-	**68:** 105106w
4-allyl-3-(carboxymethylthio)-5-(p-chlorophenyl)-	**68:** 105106w
4-allyl-3-(carboxymethylthio)-5-(p-ethoxyphenyl)-	**68:** 105106w

TABLE 10 (*Continued*)

Compound	Reference
10.1. Mono- and Di-S-substituted Thio-1,2,4-triazoles (*Continued*)	
4-allyl-3-(carboxymethylthio)-5-(*p*-methoxyphenyl)-	**68:** 105106w
4-allyl-3-(carboxymethylthio)-5-(*m*-nitrophenyl)-	**68:** 105106w
4-allyl-3-(carboxymethylthio)-5-(4-pyridyl)-	**68:** 105106w
4-allyl-3-(carboxymethylthio)-5-(3,4,5-tri- methoxyphenyl)-	**68:** 105106w
3-[(1-amino-4-anthraquinonyl)thio]-	**74:** P14185d
3-[(1-amino-4-anthraquinonyl)thio]-1-(3-chloropropyl)-	**74:** P14185d
3-[(1-amino-4-anthraquinonyl)thio]-5-phenyl-	**74:** P14185d
3-[(1-amino-2-bromo-4-anthraquinonyl)thio]-	**74:** P14185d
3-[(1-amino-2-bromo-4-anthraquinonyl)thio]-1-ethyl-	**74:** P14185d
3-[[7-[(aminocarbonyl)amino]-2-carboxy-8-oxo-5-thia- 1-azabicyclo[4.2.0]oct-2-en-3-yl]methylthio]-4-methyl-	**79:** P53343b
3-[(7-amino-2-carboxy-8-oxo-5-thia-1-azabicyclo- [4.2.0]oct-2-en-3-yl)methylthio]-	**82:** P57711e
3-[(7-amino-2-carboxy-8-oxo-5-thia-1-azabicyclo- [4.2.0]oct-2-en-3-yl)methylthio]-4,5-dimethyl-	**77:** P140109g
3-[(7-amino-2-carboxy-8-oxo-5-thia-1-azabicyclo- [4.2.0]oct-2-en-3-yl)methylthio]-4-4-methyl-	**79:** P53343b, **82:** P57711e
3-[(7-amino-2-carboxy-8-oxo-5-thia-1-azabicyclo- [4.2.0]oct-2-en-3-yl)methylthio]-5-methyl-	**80:** 83012f, **82:** P57711e
3-[(7-amino-2-carboxy-8-oxo-5-thia-1-azabicyclo- [4.2.0]oct-2-en-3-yl)methylthio]-4-methyl-5- (trifluoromethyl)-	**81:** 68556p, **83:** P209412u
3-(2-aminoethyl)-5-(methylthio)-	**48:** 12739g
3-[4-(1-amino-4-hydroxyanthraquinon-2-yl)oxy]- benzylthio]-	**74:** P14185d
3-[2-[2-(1-amino-4-hydroxyanthraquinon-2-yloxy)- ethoxy]ethylthio]-	**74:** P14185d
3-[(1-amino-4-hydroxy-2-anthraquinonyl)thio]-	**74:** P14185d
3-[(1-amino-2-methoxy-4-anthraquinonyl)thio]-	**74:** P14185d
3-(aminomethyl)-5-(methylthio)-	**66:** 2536b
3-[(7-[([2-(aminomethyl)phenyl]acetyl)amino]- 2-carboxy-8-oxo-5-thia-1-azabicyclo[4.2.0]- oct-2-en-3-yl)methylthio]-	**83:** P28252w
3-[3-amino-4-(nitrophenyl)thio]-	**82:** P31324f
3-[(7-[(α-amino-4-oxazolylactyl)amino]-2-carboxy-8- oxo-5-thia-1-azabicyclo[4.2.0]oct-2-ene-3-yl)- methylthio]-5-methyl-	**78:** P97680t
3-[(7-[(α-aminophenylacetyl)amino]-2-carboxy-8-oxo- 5-thia-1-azabicyclo[4.2.0]oct-2-en-3-yl)methylthio]-	**77:** P140109g, **83:** P209412u
3-[(7-[(α-aminophenylacetyl)amino]-2-carboxy-8-oxo- 5-thia-1-azabicyclo[4.2.0]oct-2-en-3-yl)methylthio]-	**82:** P170991j
3-[(7-[(α-aminophenylacetyl)amino]-2-carboxy-8-oxo- 5-thia-1-azabicyclo[4.2.0]oct-2-en-3-yl)methylthio]- 4,5-dimethyl-	**77:** P140109g, **81:** P68556p, **83:** P209412u
3-[(7-[(α-aminophenylacetyl)amino]-2-carboxy-8-oxo- 5-thia-1-azabicyclo[4.2.0]oct-2-en-3-yl)methylthio]- 4-ethyl-	**77:** P140109g

TABLE 10 (*Continued*)

Compound	Reference
10.1. Mono- and Di-S-substituted Thio-1,2,4-triazoles (*Continued*)	
3-[(7-[(α-aminophenylacetyl)amino]-2-carboxy-8-oxo-5-thia-1-azabicyclo[4.2.0]oct-2-en-3-yl)methylthio]-1-methyl-	**77:** P140109g
3-[(7-[(α-aminophenylacetyl)amino]-2-carboxy-8-oxo-5-thia-1-azabicyclo[4.2.0]oct-2-en-3-yl)methylthio]-4-methyl-	**77:** P140109g, **83:** P209412u
3-[(7-[(α-aminophenylacetyl)amino]-2-carboxy-8-oxo-5-thia-1-azabicyclo[4.2.0]oct-2-en-3-yl)methylthio]-5-methyl-	**77:** P140109g, **83:** P209412u
3-[(7-[(α-aminophenylacetyl)amino]-2-carboxy-8-oxo-5-thia-1-azabicyclo[4.2.0]oct-2-en-3-yl)methylthio]-5-methyl-5-(trifluoromethyl)-	**77:** P140109g, **83:** P209412u
3-(4-aminophenyl)-1-methyl-3-(methylthio)-	**83:** 163373b
3-(4-aminophenyl)-1-methyl-5-(methylthio)-	**83:** 163373b
3-(4-aminophenyl)-4-methyl-5-(methylthio)-	**83:** 163373b
3-(4-aminophenyl)-5-(methylthio)-	**83:** 163373b
3-(o-aminophenyl)-5-(methylthio)-4-phenyl-	**63:** 13256h
3-(2-aminophenylthio)-	**79:** P43665u
3-([1-amino-2-(1-piperidinylsulfonyl)-4-anthraquinonyl]thio)-	**74:** P14185d
3-[(7-[(4-aminopyridinoacetyl)amino]-2-carboxy-8-oxo-5-thia-1-azabicyclo-[4.2.0]oct-2-en-3-yl)-methylthio]-5-methyl-, hydroxide, inner salt	**82:** P31344n
1-(2-[2-([4-amino-1-(p-toluenesulfonamido)-3-anthraquinonyl]oxo)ethoxy]ethyl)-3-(ethylthio)-	**74:** P14185d
3-[(anilinocarbonyl)thio]-5-(2,2-dichloro-1,1-difluoroethyl)-4-phenyl-	**81:** 49647c
3-(4-anisoylthio)-4-butyl-	**72:** P100713q
3-[(1-anthraquinonyl)thio]-	**74:** P14185d
3-[2-[3-benzamido-N-(2-cyanoethyl)anilino]ethylthio]-	**79:** P43665u
5-[2-(1H-benzimidazol-2-yl)ethyl]-3-[(carboxymethyl)-thio]-4-cyclohexyl-, hydrazide	**77:** 147663s
5-[2-(1H-benzimidazol-2-yl)ethyl]-3-[(carboxymethyl)-thio]-4-(2-methoxyphenyl)-, hydrazide	**77:** 147663s
5-[2-(1H-benzimidazol-2-yl)ethyl]-3-[(carboxymethyl)-thio]-4-(4-methoxyphenyl)-, hydrazide	**77:** 147663s
5-[2-(1H-benzimidazol-2-yl)ethyl]-3-[(carboxymethyl)-thio]-4-(3-methylphenyl)-, hydrazide	**77:** 147663s
5-[2-(1H-benzimidazol-2-yl)ethyl]-3-[(carboxymethyl)-thio]-4-phenyl-, hydrazide	**77:** 147663s
5-[2-(1H-benzimidazol-2-yl)ethyl]-3-[(carboxymethyl)-thio]-4-(2-propenyl)-, hydrazide	**77:** 147663s
5-[2-(1H-benzimidazol-2-yl)ethyl]-4-cyclohexyl-3-[(ethoxycarbonylmethyl)thio]-	**77:** 147663s
5-[2-(1H-benzimidazol-2-yl]-3-[(ethoxycarbonyl-methyl)thio]-4-(2-methoxyphenyl)-	**77:** 147663s
5-[2-(1H-benzimidazol-2-yl]-3-[(ethoxycarbonyl-methyl)thio]-4-(4-methoxyphenyl)-	**77:** 147663s

TABLE 10 (*Continued*)

Compound	Reference
10.1. Mono- and Di-S-substituted Thio-1,2,4-triazoles (*Continued*)	
5-[2-(1H̲-benzimidazol-2-yl]-3-[(ethoxycarbonyl- methyl)thio]-4-(3-methylphenyl)-	**77:** 147663s
5-[2-(1H̲-benzimidazol-2-yl]-3-[(ethoxycarbonyl- methyl)thio]-4-phenyl-	**77:** 147663s
5-[2-(1H̲-benzimidazol-2-yl)ethyl]-3-[(ethoxycarbonyl- methyl)thio]-4-(2-propenyl)-	**77:** 147663s
3-(2-benzothiazolyl)-1-benzyl-5-(methylthio)-	**77:** 126518h
3-(2-benzothiazolyl)-4-benzyl-5-(methylthio)-	**77:** 126518h
3-(2-benzothiazolyl)-4-ethyl-5-(methylthio)-	**77:** 126518h
5-(2-benzothiazolyl)-4-ethyl-3-[(phenacyl)thio]-	**82:** 139995q
3-(2-benzothiazolyl)-4-methyl-5-(methylthio)-	**77:** 126518h
3-(2-benzothiazolyl)-5-(methylthio)-	**77:** 126518h
3-(2-benzothiazolyl)-5-(methylthio)-4-phenyl-	**77:** 126518h
3-(2-benzothiazolyl)-5-(methylthio)-4-(2-propenyl)-	**77:** 126518h
3-[benzoylthio]-4,5-diphenyl-	**81:** 91429h
1-benzyl-3-(benzylthio)-	**57:** 805e
1-benzyl-5-(benzylthio)-	**74:** 13070a
3-[[(N-benzylcarbamoyl)methyl]thio]-	**72:** 31712z
3-[[(N-benzylcarbamoyl)methyl]thio]-5-(o-ethoxyphenyl)-	**71:** 49853k
5-benzyl-3-[(2-carboxyethyl)thio]-4-phenyl-	**83:** 164091v
5-benzyl-3-[(2-carboxyethyl)thio]-4-(2-propenyl)-	**83:** 164091v
5-benzyl-3-[(carboxymethyl)thio]-4-phenyl-	**83:** 164091v
5-benzyl-3-[(carboxymethyl)thio]-4-phenyl-, 2-(chloroacetyl)hydrazide	**79:** 126404j
5-benzyl-3-[(carboxymethyl)thio]-4-phenyl-, 2-(cyanoacetyl)hydrazide	**79:** 126404j
5-benzyl-3-[(carboxymethyl)thio]-4-phenyl-, hydrazide	**79:** 126404j
5-benzyl-3-[(carboxymethyl)thio]-4-(2-propenyl)-	**83:** 164091v
3-benzyl-5-[([(3,5-dimethyl-1H̲-pyrazol-1-yl)carbonyl]- methylene)thio]-4-phenyl-	**76:** 153661k
5-benzyl-3-[(methoxycarbonylmethyl)thio]-4-phenyl-	**79:** 126404j
3-benzyl-5-(methylthio)-	**53:** 10034d
3-(benzylthio)-	**57:** 805d, **70:** P28922w, **71:** P81375s, **71:** P124344a, **71:** P124443g
5-(benzylthio)-4-(4-chlorophenyl)-3-(3-fluorophenyl)-	**82:** 72886j
5-(benzylthio)-3-(4-chlorophenyl)-4-(4-iodophenyl)-	**78:** 159528k
5-(benzylthio)-3-(4-chlorophenyl)-4-(4-methylphenyl)-	**78:** 159528k
5-(benzylthio)-3-[(2,4-dichlorophenoxy)methyl]- 4-(4-iodophenyl)-	**82:** 125327m
5-(benzylthio)-3-[(2,4-dichlorophenoxy)methyl]- 4-(2-methylphenyl)-5-	**82:** 125327m
5-(benzylthio)-3-[(2,4-dichlorophenoxy)methyl]- 4-(3-methylphenyl)-	**82:** 125327m
5-(benzylthio)-3-[(2,4-dichlorophenoxy)methyl]- 4-(4-methylphenyl)-	**82:** 125327m
5-(benzylthio)-1,3-dimethyl-	**55:** 23507h
3-(benzylthio)-5-ethyl-	**57:** 805e

TABLE 10 (*Continued*)

Compound	Reference
10.1. Mono- and Di-S-substituted Thio-1,2,4-triazoles (*Continued*)	
5-(benzylthio)-3-(3-fluorophenyl)-4-(4-fluorophenyl)-	**82:** 72886j
3-(benzylthio)-1(?)-(hydroxymethyl)-	**76:** 30510c
3-(benzylthio)-5-methyl-	**57:** 805e
5-(benzylthio)-1-methyl-	**55:** 23507g
3-(benzylthio)-5-methyl-1-phenyl-	**51:** P899e
3-(benzylthio)-1(?)-(4-morpholinomethyl)-	**76:** 30510c
3-(benzylthio)-1-phenyl-	**74:** 141644c
3-(benzylthio)-4-phenyl-	**63:** 13242g
3-(benzylthio)-5-phenyl-	**65:** 13691g, **66:** 115651t
5-(benzylthio)-1-phenyl-	**74:** 141644c
3-(benzylthio)-1(?)-(1-piperidinomethyl)-	**76:** 30510c
3,5-bis(benzylthio)-	**57:** 12472b
3,5-bis(methylthio)-	**44:** 6416h
3,5-bis(methylthio)-1-phenyl-	**84:** 135551k
3,5-bis(methylthio)-4-phenyl-	**44:** 6416i
1-bis(dimethylamino)phosphinyl-3-(isopropylthio)-	**80:** P82992p
1-bis(dimethylamino)phosphinyl-3-(methylthio)-	**80:** P82992p
3-([bis(dimethylamino)phosphinyl]thio)-	**60:** P14514b
3-([bis(dimethylamino)phosphinyl]thio)-5-pentyl-	**60:** P14514b
3,5-bis(methylthio)-4-phenyl-	**79:** 42423b
3-[(2-bromobenzoyl)thio]-4-butyl-	**72:** P100713q
3-[(4-bromobenzoyl)thio]-4-butyl-	**72:** P100713q
3-(4-bromobenzylthio)-1(?)-(hydroxymethyl)-	**76:** 30510c
5-(5-bromo-2-hydroxyphenyl)-4-ethyl-3-(thiocyanato)-	**68:** 59501w
5-(5-bromo-2-hydroxyphenyl)-4-hexyl-3-(thiocyanato)-	**68:** 59501w
3-[(5-bromo-2-hydroxy)phenyl]-5-methylthio-	**66:** P115714r
5-(5-bromo-2-hydroxyphenyl)-4-pentyl-3-(thiocyanato)-	**68:** 59501w
5-(5-bromo-2-hydroxyphenyl)-4-propyl-3-(thiocyanato)-	**68:** 59501w
3-[(4-bromophenacyl)thio]-	**77:** 126492v
3-[(4-bromophenacyl)thio]-5-(4-methoxyphenyl)-	**81:** 136112t
3-[(4-bromophenacyl)thio]-5-methyl-	**77:** 126492v
3-[(4-bromophenacyl)thio]-5-(2-methylphenyl)-	**81:** 136112t
3-[(4-bromophenacyl)thio]-5-phenyl-	**81:** 136112t
3-([(N-p-bromophenylcarbamoyl)methyl]thio)-	**72:** 31712z
3-([(4-bromophenyl)methyl]thio)-	**76:** 30510c
4-(p-bromophenyl)-3-(methylthio)-5-phenyl-	**75:** 76692w
3-[(2-bromo-2-propenyl)thio]-5-(1,1-dimethylethyl)-	**84:** P135677f
5-(5-bromosalicyl)-3-(carboxymethylthio)-4-cyclohexyl-	**68:** 105106w
3-(2-butenylthio)-5-(1,1-dimethylethyl)-	**84:** P135677f
3-((7-[(α-[[2-(tert-butoxycarbonyl)amino]-4-(methylthio)butryl]oxy)phenylacetylamino)-2-carboxy-8-oxo-5-thia-1-azabicyclo[4.2.0]oct-2-en-3-yl]methylthio)-5-methyl-	**77:** P140109g
3-(p-butoxyphenyl)-5-(butylthio)-	**52:** 12852b
3-(p-butoxyphenyl)-5-(ethylthio)-	**52:** 12852b
3-(p-butoxyphenyl)-5-(isobutylthio)-	**52:** 12852b
3-(p-butoxyphenyl)-5-(isopentylthio)-	**52:** 12852b
3-(p-butoxyphenyl)-5-(isopropylthio)-	**52:** 12852b

TABLE 10   (*Continued*)

Compound	Reference

10.1. Mono- and Di-S-substituted Thio-1,2,4-triazoles   (*Continued*)

Compound	Reference
3-([[(4-butoxyphenyl)methyl]thio)-	**84:** 30975y
3-(*p*-butoxyphenyl)-5-(methylthio)-	**52:** 12852b
3-([[(4-butoxyphenyl)methyl]thio)-5-ethyl-	**84:** 30975y
3-([[(4-butoxyphenyl)methyl]thio)-5-(4-methoxyphenyl)-	**84:** 30975y
3-([[(4-butoxyphenyl)methyl]thio)-5-methyl-	**84:** 30975y
3-(*p*-butoxyphenyl)-5-(penthylthio)-	**52:** 12852b
3-(*p*-butoxyphenyl)-5-(propylthio)-	**52:** 12852b

3-([7-([[(butylamino)carbonyl]amino)-2-carboxy-8-oxo-
   5-thia-1-azabicyclo[4.2.0]oct-2-en-3-yl]methylthio)-
   5-methyl-
	80: P83012f

3-([[(*N*-butylcarbamoyl)methyl]thio)-	**72:** 31712z
4-butyl-3-[(carbamoylmethyl)thio]-	**72:** P100713q
3-([[(*N*-butylcarbamoyl)methyl]thio)-5-(*o*-ethoxyphenyl)-	**71:** 49853k
4-butyl-3-(carboxymethylthio)-5-(*m*-chlorophenyl)-	**68:** 105106w
4-butyl-3-(carboxymethylthio)-5-(*p*-chlorophenyl)-	**68:** 105106w
4-butyl-3-(carboxymethylthio)-5-(*p*-methoxyphenyl)-	**68:** 105106w
4-butyl-3-(carboxymethylthio)-5-(*m*-nitrophenyl)-	**68:** 105106w
4-butyl-3-(carboxymethylthio)-5-phenyl-	**68:** 105106w
4-butyl-3-(carboxymethylthio)-5-(3,4,5-trimethoxyphenyl)-	**68:** 105106w
4-butyl-3-[(5-chloro-2-hydroxy)phenyl]-5-methylthio-	**66:** P115714r
4-butyl-3-[(3,5-dichloro-2-hydroxy)phenyl]-5-methylthio-	**66:** P115714r
4-butyl-3-[(2,4-dinitrophenyl)thio]-	**72:** P100713q
4-butyl-3-[(2-ethoxyethyl)thio]-	**72:** P100713q
3-butyl-5-([[(4-ethoxyphenyl)methyl]thio)-	**84:** 30975y
4-butyl-3-(2-furoylthio)-	**72:** P100713q
4-butyl-3-[(*p*-methoxybenzyl)thio]-	**72:** P100713q
4-butyl-3-[(4-methoxyphenacyl)thio]-	**72:** P100713q
4-butyl-3-(methylthio)-	**72:** P100713q
4-butyl-3-(methylthio)-5-[2-(5-nitro-2-furyl)vinyl]-	**61:** 9490h
4-butyl-3-[(3-nitrobenzoyl)thio]-	**72:** P100713q
4-butyl-3-[(2-phenoxyethyl)thio]-	**72:** P100713q
3-(butylthio)-	**71:** 124344a, **76:** 30510c
3-(butylthio)-5-(*p*-chlorophenyl)-4-ethyl-	**74:** 111968u
3-(butylthio)-5-(*p*-ethoxyphenyl)-	**52:** 12852a
3-(butylthio)-5-(2-furyl)-	**52:** 16341e
3-(butylthio)-5-(*p*-isobutoxyphenyl)-	**52:** 12852c
3-(butylthio)-5-[*p*-isopentyloxyphenyl]-	**52:** 12852c
3-(butylthio)-5-(*p*-isopropoxyphenyl)-	**52:** 12852b
3-(butylthio)-5-(*p*-methoxyphenyl)-	**52:** 12852a
3-(butylthio)-5-(*p*-pentyloxyphenyl)-	**52:** 12852c
3-(butylthio)-5-(*p*-propoxyphenyl)-	**52:** 12852b
5-(butylthio)-3-(2-pyridinyl)-	**74:** P87996d
4-butyl-3-[(2,2,2-trichloro-1-hydroxyethyl)thio]-	**72:** P100713q
3-([[(4-carbamoylanilino)carbonylmethyl]thio)-	**72:** 31711y
3-[(carbamoylmethyl)thio]-	**71:** 124346c
3-[(carbamoylmethyl)thio]-5-(2-furyl)-	**74:** 53660z
3,3'-[carbonylbis(thio)]bis[1-phenyl-	**74:** P81774h, **79:** P59979k
3-[(carbonylmethyl)thio]-5-(2-furyl)-, hydrazide	**74:** 53660z

TABLE 10 (*Continued*)

Compound	Reference

10.1. Mono- and Di-S-substituted Thio-1,2,4-triazoles (*Continued*)

Compound	Reference
3-[[[(4-carboxyanilino)carbonyl]methyl]thio]-	**72:** 31711y
3-[[(4-carboxyanilino)carbonylmethyl]thio]-, hydrazide	**72:** 31711y
3-(2-carboxybenzyl)-5-(methylthio)-	**77:** 164601n
3-(2-carboxybenzyl)-5-(methylthio)-, 2-phenylhydrazide	**77:** 164601n
3-([2-carboxy-7-([5-carboxy-5-([(2-chloroethoxy)-carbonyl]amino)-1-oxopentyl]amino)-8-oxo-5-thia-1-azabicyclo[4.2.0]oct-2-en-3-yl]methylthio-	**82:** P57711e
3-([2-carboxy-7-([5-carboxy-5-([(2-chloroethoxy)-carbonyl]amino)-1-oxopentyl]amino)-8-oxo-5-thia-1-azabicyclo[4.2.0]oct-2-en-3-yl]methylthio)-4-methyl-	**82:** P57711e
3-([2-carboxy-7-([5-carboxy-5-([(2-chloroethoxy)-carbonyl]amino)-1-oxopentyl]amino)-8-oxo-5-thia-1-azabicyclo[4.2.0]oct-2-en-3-yl]methylthio)-5-methyl-	**82:** P57711e
3-[([2-carboxy-7-([[2-chloroethyl)amino]carbonyl)-amino]-8-oxo-5-thia-1-azabicyclo[4.2.0]oct-2-en-3-yl)methylthio]-	**80:** P83012f
3-[(2-carboxy-7-[([2-chloroethyl)amino]carbonyl)-amino]-8-oxo-5-thia-1-azabicyclo[4.2.0]oct-2-en-3-yl)methylthio]-5-methyl-	**80:** P83012f
3-[(2-carboxy-7-[(cyanoacetyl)amino]-8-oxo-5-thia-1-azabicyclo[4.2.0]oct-2-en-3-yl)-methylthio]-5-methyl-	**79:** P78824x
3-([(2-carboxy-3,2-dimethyl-7-oxo-4-thia-1-azabicyclo-[3.3.0]hept-6-yl)carbamoyl(phenyl)methyl]-carbamoylmethylthio)-	**80:** P82954c
3-[(*N*-[7-(2-carboxy-4,4-dimethyl-8-oxo-5-thia-1-azabicyclo[4.2.0]oct-2-enyl)carbamoyl]methylene)thio]-	**76:** P99686g, **77:** P114397e, **81:** P37556j, **83:** P114441p
3-[(2-carboxy-7-[(1,3-dioxybutyl)amino]-8-oxo-5-thia-1-azabicyclo[4.2.0]oct-2-en-3-yl)methylthio]-	**83:** P79262g
3-([[(1-carboxyethyl)carbamoyl]methylthio)-	**72:** 31711y
3-([2-carboxy-7-([(ethylsulfonyl)acetyl]amino)-8-oxo-5-thia-1-azabicyclo[4.2.0]oct-2-en-3-yl]methylthio)-5-methyl-	**80:** P48019s
3-[(2-carboxyethyl)thio]-4,5-diphenyl-	**82:** 43276e
3-[(2-carboxyethyl)thio]-1-phenyl-	**74:** 141644c
5-[(2-carboxyethyl)thio]-1-phenyl	**74:** 141644c
3-[(2-carboxyethyl)thio]-5-phenyl-4-(2-propenyl)-	**82:** 43276e
3-[(2-carboxy-7-[(α-hydroxyphenylacetyl)amino]-8-oxo-5-thia-1-azabicyclo[4.2.0]oct-2-en-3-yl)methylthio]-1-methyl-	**77:** P140109g
3-[(2-carboxy-7-[(α-hydroxyphenylacetyl)amino]-8-oxo-5-thia-1-azabicyclo[4.2.0]oct-2-en-3-yl)methylthio]-2-methyl-	**83:** P209412u
3-[(2-carboxy-7-[(α-hydroxyphenylacetyl)amino]-8-oxo-5-thia-1-azabicyclo[4.2.0]oct-2-en-3-yl)methylthio]-4-methyl-	**77:** P140109g, **83:** P209412u
3-[(2-carboxy-7-[(α-hydroxyphenylacetyl)amino]-8-oxo-5-thia-1-azabicyclo[4.2.0]oct-2-en-3-yl)methylthio]-5-methyl-	**77:** P140109g, **83:** P209412u

TABLE 10 (*Continued*)

Compound	Reference

10.1. Mono- and Di-S-substituted Thio-1,2,4-triazoles (*Continued*)

3-[(2-carboxy-7-[(α-hydroxyphenylacetyl)amino]-8-oxo-
5-thia-1-azabicyclo[4.2.0]oct-2-en-3-yl)methylthio]-
4-methyl-5-(trifluoromethyl)-
77: P140109g, **81:** P68556p,
83: P209412u

3-[(2-carboxy-7-[(α-hydroxy-5-isothiazolyacetyl)amino]-
8-oxo-5-thia-1-azabicyclo[4.2.0]oct-2-ene-3-yl)-
methylthio]-
78: P97680t

5-[(2-carboxy-7-[(5-hydroxy-3-oxadiazolioacetyl)amino]-
8-oxo-5-thia-1-azabicyclo[4.2.0]oct-2-en-3-yl)methyl-
thio]-1,3-dimethyl-, hydroxide, inner salt
71: P124458r

5-[(2-carboxy-7-[(5-hydroxy-3-oxadiazolioacetyl)amino]-
8-oxo-5-thia-1-azabicyclo[4.2.0]oct-2-en-3-yl)methyl-
thio]-4,5-dimethyl-, hydroxide, inner salt
71: P124458r

5-[(2-carboxy-7-[(5-hydroxy-3-oxadiazolioacetyl)amino]-
8-oxo-5-thia-1-azabicyclo[4.2.0]oct-2-en-3-yl)methyl-
thio]-1-methyl-, hydroxide, inner salt
71: P124458r

3-[(2-carboxy-7-[(5-hydroxy-3-oxadiazolioacetyl)amino]-
8-oxo-5-thia-1-azabicyclo[4.2.0]oct-2-en-3-yl)methyl-
thio]-4-methyl-, hydroxide, inner salt
71: P124458r

3-[(2-carboxy-7-[(5-hydroxy-3-oxadiazolioacetyl)amino]-
8-oxo-5-thia-1-azabicyclo[4.2.0]oct-2-en-3-yl)methyl-
thio]-4-methyl-5-phenyl-, hydroxide, inner salt
71: P124458r

3-[(2-carboxy-7-[(5-hydroxy-3-oxadiazolioacetyl)amino]-
8-oxo-5-thia-1-azabicyclo[4.2.0]oct-2-en-3-yl)methyl-
thio]-5-methyl-4-phenyl-, hydroxide, inner salt
71: P124458r

5-[(2-carboxy-7-[(5-hydroxy-3-oxadiazolioacetyl)amino]-
8-oxo-5-thia-1-azabicyclo[4.2.0]oct-2-en-3-yl)methyl-
thio]-1-phenyl-, hydroxide, inner salt
71: P124458r

3-[(2-carboxy-7-[[(5-hydroxy-3-oxadiazolio)methyl]-
carbamoylamino]-8-oxo-5-thia-1-azabicyclo-
[4.2.0]oct-2-ene-3-yl)methylthio]-1-methyl-,
hydroxide,
71: P124458r, **73:** P3923w

3-[(2-carboxy-7-[(α-hydroxyphenylacetyl)amino]-8-
oxo-5-thia-1-azabicyclo[4.2.0]oct-2-en-3-yl)-
methylthio]-
77: P140109g, **83:** P209412u

3-[(2-carboxy-7-[(α-hydroxyphenylacetyl)amino]-8-
oxo-5-thia-1-azabicyclo[4.2.0]oct-2-en-3-yl)-
methylthio]-4,5-dimethyl-
77: P140109g, **83:** P209412u

3-[(2-carboxy-7-[(α-hydroxyphenylacetyl)amino]-8-
oxo-5-thia-1-azabicyclo[4.2.0]oct-2-en-3-yl)-
methylthio]-4-ethyl-
77: P140109g

3-[([(carboxymethyl)amino]carbonylmethyl)thio]-
72: 31711y

3-[(2-carboxy-3-methoxy-8-oxo-5-thia-1-azabicyclo-
[4.2.0]oct-2-en-7-yl)carbamoylmethylthio]-
80: P83019p

3-([(1-carboxy-3-methylbutyl)carbamoyl]methylthio)-
72: 31711y

3-[(2-carboxy-7-[[([methyl(1-oxo-3-phenyl-2-
propenyl)amino]carbonyl)amino]phenylacetyl)amino]-
8-oxo-5-thia-1-azabicyclo[4.2.0]oct-2-en-3-yl)-
methylthio]-
82: P170991j

TABLE 10 (*Continued*)

Compound	Reference

10.1. Mono- and Di-S-substituted Thio-1,2,4-triazoles (*Continued*)

Compound	Reference
3-([(1-carboxy-2-methylpropyl)carbamoyl]methylthio)-	**72:** 31711y
3-([2-carboxy-7-([(methylsulfonyl)acetyl]amino)-8-oxo-5-thia-1-azabicyclo[4.2.0]oct-2-en-3-yl]methylthio)-4-methyl-	**80:** P48019s
3-([2-carboxy-7-([(methylsulfonyl)acetyl]amino)-8-oxo-5-thia-1-azabicyclo[4.2.0]oct-2-en-3-yl]-methylthio)-4-methyl-	**83:** P10106s
3-([2-carboxy-7-([(methylsulfonyl)acetyl]amino)-8-oxo-5-thia-1-azabicyclo[4.2.0]oct-2-en-3-yl]methylthio)-5-methyl-	**80:** P48019s
3-[(carboxymethyl)thio]-	**71:** 124346c, **76:** 30510c, **78:** P130586s
3-[(carboxymethyl)thio]-, benzylidenehydrazide	**71:** 124346c
3-[(carboxymethyl)thio]-, hydrazide	**71:** 124346c, **76:** 30510c
3-[(carboxymethyl)thio]-, (*p*-nitrobenzylidene)hydrazide	**71:** 124346c
3-([2-carboxy-7-([(methylthio)acetyl]amino)-8-oxo-5-thia-1-azabicyclo[4.2.0]oct-2-en-3-yl]methylthio)-5-methyl-	**83:** P164206m
3-[(carboxymethyl)thio]-5-(2-butoxyphenyl)-	**79:** 115500m
3-[(carboxymethyl)thio]-5-(4-chlorophenyl)-4-(4-iodophenyl)-	**78:** 159528k
3-(carboxymethylthio)-5-(*m*-chlorophenyl)-4-isopropyl-	**68:** 105106w
3-(carboxymethylthio)-5-(*p*-chlorophenyl)-4-isopropyl-	**68:** 105106w
3-[(carboxymethyl)thio]-5-(4-chlorophenyl)-4-(4-methylphenyl)-	**78:** 159528k
3-(carboxymethylthio)-5-(*p*-chlorophenyl)-4-phenyl-	**68:** 105106w
3-(carboxymethylthio)-4-cyclohexyl-5-(*p*-methoxyphenyl)-	**68:** 105106w
3-(carboxymethylthio)-4-cyclohexyl-5-phenyl-	**68:** 105106w
3-(carboxymethylthio)-5-(2,4-dichlorophenyl)-	**68:** 105106w
3-[carboxymethyl)thio]-4,5-diphenyl-	**82:** 43276e
3-[(carboxymethyl)thio]-4,5-diphenyl-, 2-(chloroacetyl)hydrazide	**79:** 126404j
3-[(carboxymethyl)thio]-4,5-diphenyl-, 2-(cyanoacetyl)hydrazide	**79:** 126404j
3-[(carboxymethyl)thio]-4,5-diphenyl-, hydrazide	**79:** 126404j
3-[(carboxymethyl)thio]-4,5-diphenyl-, 2-(phenoxyacetyl)hydrazide	**79:** 126404j
3-[(carboxymethyl)thio]-5-(*o*-ethoxyphenyl)-	**71:** 49853k
3-[(carboxymethyl)thio]-5-(*o*-ethoxyphenyl)-, hydrazide	**71:** 49853k
3-[(carboxymethyl)thio]-5-(*o*-ethoxyphenyl)-4-ethyl-	**71:** 49853k
3-[(carboxymethyl)thio]-5-(*o*-ethoxyphenyl)-4-phenyl-	**71:** 49853k
5-[(carboxymethyl)thio]-1-ethyl-3-phenyl-	**73:** 45427t
3-[(carboxymethyl)thio]-5-(2-furyl)-	**74:** 53660z
3-[(carboxymethyl)thio-5-(2-furyl)-, (*N*-allylthiocarbamoyl)hydrazide	**74:** 53660z
3-[(carboxymethyl)thio-5-(2-furyl)-, benzylidenehydrazide	**74:** 53660z
3-[(carboxymethyl)thio-5-(2-furyl)-, cinnamylidenehydrazide	**74:** 53660z

TABLE 10 (*Continued*)

Compound	Reference

10.1. Mono- and Di-S-substituted Thio-1,2,4-triazoles (*Continued*)

Compound	Reference
3-[(carboxymethyl)thio-5-(2-furyl)-, ethylidenehydrazide	**74:** 53660z
3-[(carboxymethyl)thio-5-(2-furyl)-, furfurylidenehydrazide	**74:** 53660z
3-[(carboxymethyl)thio-5-(2-furyl)-, hydrazide, hydrazone with L-arabinose	**74:** 53660z
3-[(carboxymethyl)thio-5-(2-furyl)-, hydrazide, hydrazone with D-mannose	**74:** 53660z
3-[(carboxymethyl)thio-5-(2-furyl)-, isopropylidenehydrazide	**74:** 53660z
3-[(carboxymethyl)thio-5-(2-furyl)-, (α-methylbenzylidene)hydrazide	**74:** 53660z
3-[(carboxymethyl)thio-5-(2-furyl)-, methylenehydrazide	**74:** 53660z
3-[(carboxymethyl)thio-5-(2-furyl)-, piperonylidenehydrazide	**74:** 53660z
3-[(carboxymethyl)thio-5-(2-furyl)-, propylidenehydrazide	**74:** 53660z
3-[(carboxymethyl)thio-5-(2-furyl)-, (3-pyridylmethylene)hydrazide	**74:** 53660z
3-[(carboxymethyl)thio-5-(2-furyl)-, (4-pyridylmethylene)hydrazide	**74:** 53660z
3-[(carboxymethyl)thio-5-(2-furyl)-, (2-pyrrolylmethylene)hydrazide	**74:** 53660z
3-[(carboxymethyl)thio-5-(2-furyl)-, salicylidenehydrazide	**74:** 53660z
3-[(carboxymethyl)thio-5-(2-furyl)-, 2-thienylidenehydrazide	**74:** 53660z
3-[(carboxymethyl)thio-5-(2-furyl)-, vanillylidenehydrazide	**74:** 53660z
3-[(carboxymethyl)thio-5-(2-furyl)-, veratrylidenehydrazide	**74:** 53660z
3-(carboxymethylthio)-5-(*o*-hydroxyphenyl)-4-phenyl-	**68:** 105106w
3-(carboxymethylthio)-4-isopentyl-5-(*m*-nitrophenyl)-	**68:** 105106w
3-[(carboxymethyl)thio]-, isopropylidenehydrazide	**71:** 12346c
3-(carboxymethylthio)-4-isopropyl-5-(*p*-methoxyphenyl)-	**68:** 105106w
3-(carboxymethylthio)-4-isopropyl-5-(*m*-nitrophenyl)-	**68:** 105106w
3-(carboxymethylthio)-4-isopropyl-5-phenyl-	**68:** 105106w
3-[(carboxymethyl)thio]-5-(2-methoxyphenyl)-	**79:** 115500m
3-(carboxymethylthio)-4-(*p*-methoxyphenyl)-5-phenyl-	**68:** 105106w
3-(carboxymethylthio)-4-(*p*-methoxyphenyl)-4-phenyl-	**68:** 105106w
3-[(carboxymethyl)thio]-5-[2-(1-methylethoxy)phenyl]-	**79:** 115500m
3-[(carboxymethyl)thio]-5-(4-methyl-3-furazanyl)-,	**79:** 105155p
3-[(carboxymethyl)thio]-5-(4-methyl-3-furazanyl)-, hydrazide	**79:** 105155p
3-[(carboxymethyl)thio]-5-(4-methyl-3-furazanyl)-, hydrazide, N-oxide	**79:** 105155p
5-[(carboxymethyl)thio]-1-methyl-3-phenyl-	**73:** 45427t

TABLE 10 (*Continued*)

Compound	Reference
10.1. Mono- and Di-S-substituted Thio-1,2,4-triazoles (*Continued*)	
3-[(carboxymethyl)thio-5-[2-(2-methylpropoxy)phenyl]-	**79:** 115500m
3-(carboxymethylthio)-5-methyl-	**44:** P9287d
3-[(carboxymethyl)thio]-5-(phenoxymethyl)-4-phenyl-, 2-(chloroacetyl)hydrazide	**79:** 126404j
3-[(carboxymethyl)thio]-5-(phenoxymethyl)-4-phenyl-, 2-(cyanoacetyl)hydrazide	**79:** 126404j
3-[(carboxymethyl)thio]-5-(phenoxymethyl)-4-phenyl-, hydrazide	**79:** 126404j
3-(carboxymethylthio)-4-phenyl-, benzylidenehydrazide	**68:** 95763s
3-[(carboxymethyl)thio]-4-phenyl-, (*o*-chlorobenzylidene)hydrazide	**71:** 124310m
3-(carboxymethylthio)-4-phenyl-, cinnamylidenehydrazide	**68:** 95763s
3-(carboxymethylthio)-4-phenyl-, (*p*-methoxybenzylidene)hydrazide	**68:** 95763s
3-[(carboxymethyl)thio-4-phenyl-, (*o*-nitrobenzylidene)hydrazide	**71:** 124310m
3-[(carboxymethyl)thio]-4-phenyl-, (*m*-nitrobenzylidene)hydrazide	**71:** 124310m
3-(carboxymethylthio)-4-phenyl-, salicylidenehydrazide	**68:** 95763s
3-[(carboxymethyl)thio]-4-phenyl-, vanillylidenehydrazide	**71:** 124310m
3-(carboxymethylthio)-5-phenyl-4-(2-propenyl)-	**68:** 105106w, **82:** 43276e
3-(carboxymethylthio)-4-phenyl-5-(4-pyridyl)-	**68:** 105106w
3-(carboxymethylthio)-4-phenyl-5-(3,4,5-trimethoxyphenyl)-	**68:** 105106w
3-[(carboxymethyl)thio]-5-(2-propoxyphenyl)-	**79:** 115500m
3-(carboxymethylthio)-4-*o*-tolyl-, benzylidene-hydrazide	**68:** 95763s
3-[(carboxymethyl)thio]-4-*o*-tolyl-, (*o*-chlorobenzylidene)hydrazide	**71:** 124310m
3-(carboxymethylthio)-4-*o*-tolyl-, cinnamylidene-hydrazide	**68:** 95763s
3-(carboxymethylthio)-4-*o*-tolyl-, (*p*-methoxybenzyl-idene)hydrazide	**68:** 95763s
3-[(carboxymethyl)thio]-4-*o*-tolyl-, (*o*-nitrobenzyl-idene)hydrazide	**71:** 124310m
3-[(carboxymethyl)thio]-4-*o*-tolyl-, (*m*-nitrobenzyl-idene)hydrazide	**71:** 124310m
3-(carboxymethylthio)-4-*o*-tolyl-, salicylidenehydrazide	**68:** 95763s
3-[(carboxymethyl)thio]-4-*o*-tolyl-, vanillylidene hydrazide	**71:** 124310m
3-([2-carboxy-7-([α-([4-(methylthio)-2-[(trifluoro-acetyl)amino]butyryl]oxy)phenylacetyl]amino)-8-oxo-5-thia-1-azabicyclo[4.2.0]oct-2-en-3-yl]-methylthio)-5-methyl-	**77:** P140109g
3-[(2-carboxy-8-oxo-7-[((1-phenylethyl)amino]-carbonyl)amino]-5-thia-1-azabicyclo[4.2.0]oct-2-en-3-yl)methylthio]-	**80:** P83012f

TABLE 10 (*Continued*)

Compound	Reference
10.1. Mono- and Di-S-substituted Thio-1,2,4-triazoles (*Continued*)	
3-[(2-carboxy-8-oxo-7-[([(2-phenylethyl)amino]-carbonyl)amino]-5-thia-1-azabicyclo[4.2.0]oct-2-en-3-yl)methylthio]-5-methyl-	**80:** P83012f
3-([2-carboxy-8-oxo-7-([(2-propenylamino)carbonyl]-amino)-5-thia-1-azabicyclo[4.2.0]oct-2-en-3-yl]-methylio)-	**80:** P83012f
3-([2-carboxy-8-oxo-7-([(2-propenylamino)carbonyl]-amino)-5-thia-1-azabicyclo[4.2.0]oct-2-en-3-yl]-methylthio)-5-methyl-	**80:** P83012f
3-([2-carboxy-8-oxo-7-([(propylamino)carbonyl]amino)-5-thia-1-azabicyclo[4.2.0]oct-2-en-3-yl]methylthio)-4-methyl-	**80:** P83012f
3-([2-carboxy-8-oxo-7-([(propylsulfonyl)acetyl]amino)-5-thia-1-azabicyclo[4.2.0]oct-2-en-3-yl]methylthio)-5-methyl-	**80:** P48019s
3-[(2-carboxy-8-oxo-7-[(2-thienylacetyl)amino]-5-thia-1-azabicyclo[4.2.0]oct-2-ene-3-yl)methylthio]-4,5-dimethyl-	**79:** 61983a
3-[(2-carboxy-8-oxo-7-[([(trifluoromethyl)thio]acetyl)-amino]-5-thia-1-azabicyclo[4.2.0]oct-2-en-3-yl)-methylthio]-4-ethyl-	**83:** 109119k
3-[[2-carboxy-8-oxo-7-[([(trifluoromethyl)thio]acetyl)-amino]-5-thia-1-azabicyclo[4.2.0]oct-2-en-3-yl]-methylthio]-4-methyl-	**83:** 109119k
3-[(2-carboxy-8-oxo-7-[([(trifluoromethyl)thio]acetyl)-amino]-5-thia-1-azabicyclo[4.2.0]oct-2-en-3-yl)-methylthio]-5-methyl-	**80:** P120970q, **83:** 109119k
3-[([(2-carboxyphenyl)amino]carbonylmethyl)thio]-	**72:** 31711y
3-([(1-carboxy-2-phenylethyl)carbamoyl]methylthio)-	**72:** 31711y
3-[2-(3-chloro-*N*-ethylanilino)ethylthio]-	**79:** P43665u
3-[(2-chloroethyl)thio]-5-(4-chlorophenyl)-4-(4-methylphenyl)-	**78:** 159528k
3-[(2-chloroethyl)thio]-5-(1,1-dimethylethyl)-	**84:** P135677f
3-[(5-chloro-2-hydroxy)phenyl]-4-ethyl-5-methylthio-	**66:** P115714r
3-[(4-chlorophenacyl)thio]-	**77:** 126492v
3-[(4-chlorophenacyl)thio]-5-methyl-	**77:** 126492v
4-(4-chlorophenyl)-3,5-bis(methylthio)-	**79:** 42423b
3-([2-(4-[(4-chlorophenyl)azo]-*N*-ethyl-*m*-toluidino)ethyl]thio)-	**79:** P43665u
3-([(*N*-*o*-chlorophenylcarbamoyl)methyl]thio)-	**72:** 31712z
3-([(*N*-*m*-chlorophenylcarbamoyl)methyl]thio)-	**72:** 31712z
3-([(*N*-*p*-chlorophenylcarbamoyl)methyl]thio)-	**72:** 31712z
3-(*p*-chlorophenyl)-4-ethyl-5-(ethylthio)-	**74:** 111968u
3-(*p*-chlorophenyl)-4-ethyl-5-(methylthio)-	**74:** 111968u
3-(*p*-chlorophenyl)-4-ethyl-5-(propylthio)-	**74:** 111968u
3-(4-chlorophenyl)-5-(ethylthio)-	**83:** P126969e
4-(4-chlorophenyl)-3-(ethylthio)-5-(3-fluorophenyl)-	**82:** 72886j

TABLE 10 (*Continued*)

Compound	Reference
10.1. Mono- and Di-S-substituted Thio-1,2,4-triazoles (*Continued*)	
3-(4-chlorophenyl)-5-(ethylthio)-4-(4-iodophenyl)-	**78:** 159528k
3-(4-chlorophenyl)-5-(ethylthio)-4-(4-methylphenyl)-	**78:** 159528k
3-(4-chlorophenyl)-4-(4-iodophenyl)-5-(propylthio)-	**78:** 159528k
3-(4-chlorophenyl)-1-methyl-5-(methylthio)-	**83:** 163373b
3-(4-chlorophenyl)-4-methyl-5-(methylthio)-	**83:** 163373b
5-(4-chlorophenyl)-1-methyl-3-(methylthio)-	**83:** 163373b
3-(4-chlorophenyl)-4-(4-methylphenyl)-5-(propylthio)-	**78:** 159528k
3-(4-chlorophenyl)-5-(methylthio)-	**44:** 6855g, **83:** 163373b
4-(*p*-chlorophenyl)-3-(methylthio)-	**63:** 13242g
3-(*p*-chlorophenyl)-5-(methylthio)-4-*p*-tolyl	**75:** 76692w
3-[(2-chloro-2-propenyl)thio]-5-(1,1-dimethylethyl)-	**84:** P135677f
3-[(2-chloro-2-propenyl)thio]-5-(1-methylethyl)-	**84:** P135677f
3-[2-[*N*-cyanoethyl)anilino]ethyl]amino]-	**79:** P43665u
1-(2-cyanoethyl)-3-(2-[3-benzamido-*N*-(2-cyanotheyl)- 4([2,4-bis(methylsulfonyl)phenyl)azo)anilino]ethylthio)-	**79:** P43665u
3-[(2-cyanoethyl)thio]-1-phenyl-	**74:** 141644c
3-([2-cyano-3-nitro-5-(trifluoromethyl)phenyl]thio)-	**83:** P178845f
4-cyclohexyl-3-[(5-nitro-2-thiazolyl)thio]-5-phenyl-	**78:** P4286f
3-([(2,2-diacetylhydrazino)carbonyl]methylthio)- 5-(2-furyl)-	**74:** 53660z
3-[(4,5-diamino-1,8-dihydroxy-2-anthraquinonyl)thio]-	**74:** P14185d
3-[(3,4-diaminophenyl)thio]-	**82:** P33124f
3-[(3,5-dibromo-2-hydroxy)phenyl]-5-methylthio-	**66:** P115714r
3-([(dibutylamino)thiocarbonyl]thio)-	**69:** P11257q
5-(2,2-dichloro-1,1-difluoroethyl)-3- ([(3,4-dichloroanilino)carbonyl]thio)-4-phenyl-	**81:** 49647c
3-(2,2-dichloro-1,1-difluoroethyl)-4- [(trichloromethyl)thio]-	**81:** 49647c
3-[(2,4-dichlorophenoxy)methyl]-4-(4-ethoxyphenyl)- 5-(ethylthio)-	**82:** 125327m
3-[(2,4-dichlorophenoxy)methyl]-5-(ethylthio)-4- (2-methylphenyl)-	**82:** 125327m
3-[(2,4-dichlorophenoxy)methyl]-5-(ethylthio)-4- (3-methylphenyl)-	**82:** 125327m
3-[(2,4-dichlorophenoxy)methyl]-5-(ethylthio)-4- (4-methylphenyl)-	**82:** 125327m
3-[(2,4-dichlorophenoxy)methyl]-4-(2-methylphenyl)- 5-(propylthio)-	**82:** 125327m
3-[(2,4-dichlorophenoxy)methyl]-4-(3-methylphenyl)- 5-(propylthio)-	**82:** 125327m
3-[(2,2-diethoxyethyl)thio]-5-methyl-	**82:** P171068a
3-[(diethoxyphosphinothioyl)thio]-4-methyl-	**72:** P132735q
3-[(diethoxyphosphinothioyl)thio]-4-phenyl-	**72:** P132735q
3-[(diethoxyphosphinyl)thio]-4-methyl-	**72:** P132735q
3-[(diethoxyphosphinyl)thio]-4-phenyl-	**72:** P132735q
3-([(diethylamino)thiocarbonyl]thio)-	**69:** P11257q, **71:** P82422s
3-([(*N*,*N*-diethylcarbamoyl)methyl]thio)-	**72:** 31712z
3-(3,4-dihydro-2,4-dimethyl-5(2<u>H</u>)-oxo-3-thioxo-1,2,4- triazin-6-yl)-4-methyl-5-(methylthio)	**83:** 114345k

TABLE 10 (*Continued*)

Compound	Reference
10.1. Mono- and Di-S-substituted Thio-1,2,4-triazoles (*Continued*)	
3-(3,4-dihydro-5(2H̲)-oxo-3-thioxo-1,2,4-triazin-6-yl)-4-methyl-5-(methylthio)-	**83:** 114345k
3-([(6,7-dihydroxy-1,2,3,4-tetrahydroisoquinolin-1-yl)-methyl]thio)-5-methyl-	**82:** P171068a
3-([(2,5-dihydroxy-4-(1,1,3,3-tetramethylbutyl)-phenyl]thio)-5-pentyl-4-phenyl-	**82:** P9971t, **83:** 50701p
3-([(N,N-diisobutylcarbamoyl)methyl]thio)-	**72:** 31712z
3-[(dimethoxyphosphinyl)thio]-4-phenyl-	**72:** P132735q
3-(1,1-dimethylethyl)-5-(ethenylthio)-	**84:** P135677f
3-(1,1-dimethylethyl)-5-(ethylthio)-	**84:** P135677f
3-(1,1-dimethylethyl)-5-(methylthio)-	**84:** P135677f
3-(1,1-dimethylethyl)-5-(2-propynylthio)-	**84:** P135677f
1,3-dimethyl-5-(methylthio)-	**78:** 124513z
1,5-dimethyl-3-(methylthio)-	**76:** 59576q, **78:** 124513z
3,4-dimethyl-5-(methylthio)-	**54:** 7696b, **55:** 23509e, **78:** 124513z
3-[([(3,5-dimethyl-1H̲-pyrazol-2-yl)carbonyl]methyl)-thio]-4,5-diphenyl-	**76:** 153661k, **79:** 126404j
3-[(2,4-dinitrophenyl)thio]-4,5-diphenyl-	**81:** P97737d
5-[(2,4-dinitrophenyl)thio]-1,3-diphenyl-	**77:** 164587n
3-[(2,4-dinitrophenyl)thio]-4-ethyl-5-pentyl-	**81:** P97737d
3-[(2,4-dinitrophenyl)thio]-5-ethyl-4-phenyl-	**81:** P97737d
5-[(2,4-dinitrophenyl)thio]-1-methyl-3-phenyl-	**52:** 18380a, **77:** 164587n
5-(2,4-dinitrophenylthio)-3-methyl-1-phenyl-	**52:** 18380a
3-[(2,4-dinitrophenyl)thio]-5-nonyl-4-phenyl-	**81:** P97737d
3-[(2,4-dinitrophenyl)thio]-5-pentyl-4-phenyl-	**81:** P97737d
3-[(2,4-dinitrophenyl)thio]-4-phenyl-	**72:** P100713q
3-[(2,4-dinitrophenyl)thio]-4-phenyl-5-tridecyl-	**81:** P97737d
3-[(2,5-dinitrophenyl)thio]-4-phenyl-5-undecyl-	**81:** P97737d
4,5-diphenyl-3-(ethylthio)-	**25:** 2119[9]
4,5-diphenyl-3-(methylthio)-	**25:** 2119[8]
4,5-diphenyl-3-[(1-oxo-3-phenyl-2-propenyl)thio]	**81:** 91429h
3,4-diphenyl-5-(2-propenylthio)-	**85:** 159997g
1-(4,5-diphenyl-2-thiazolyl)-5-methyl-3-(methylthio)-	**46:** 5581g
3-([(dipropylamino)thiocarbonyl]thio)-	**69:** P11257q, **71:** P82422s
3-([(N,N-dipropylcarbamoyl)methyl]thio)-	**72:** 31712z
3-[([3-(N-dodecylsulfamoyl)phenyl]-carbonylmethyl)thio]-	**72:** P56720P
3-(dodecylthio)-5-methyl-	**67:** P54140x
3-([(4-ethoxycarbonylanilino)carbonylmethyl]thio)-	**72:** 31711y
3-[(3-ethoxycarbonyl-2,4-dimethyl-5-pyrrolyl)thio]-	**73:** 14605v
3-[(3-ethoxycarbonyl-5-methoxybenzofuranyl-2-methyl)thio]-	**78:** 159355b
3-[(ethoxycarbonylmethyl)thio]-5-(o-ethoxyphenyl)-	**71:** 49853k
3-[(ethoxycarbonylmethyl)thio]-5-ethyl-4-(2-mercapto-4,4,6-trimethyl-1(4H̲)-pyrimidinyl)-	**74:** 141645d
5-[(ethoxycarbonylmethyl)thio]-1-ethyl-3-phenyl-	**73:** 45427t

TABLE 10   (*Continued*)

Compound	Reference
10.1. Mono- and Di-S-substituted Thio-1,2,4-triazoles (*Continued*)	
3-[(ethoxycarbonylmethyl)thio]-5-(2-furyl)-	**74:** 53660z
3-[(ethoxycarbonylmethyl)thio]-5-(4-methyl- 3-furazanyl)-	**79:** 105155p
3-[(ethoxycarbonylmethyl)thio]-5-(4-methyl- 3-furazanyl)-, N-oxide	**79:** 105155p
3-[(ethoxycarbonylmethyl)thio]-5-methyl-4-phenyl-	**61:** 4337h
3-[(ethoxycarbonylmethyl)thio]-4-phenyl-	**61:** 4337h
3-[(ethoxycarbonylmethyl)thio]-4-*o*-tolyl]-	**61:** 4337h
3-[(ethoxycarbonyl)thio]-4,5-dimethyl-	**62:** P16259e, **77:** P171214s
3-[(ethoxycarbonyl)thio]-4-methyl-	**77:** P171214s
3-[(ethoxycarbonyl)thio]-1-phenyl	**62:** P16259e, **69:** P112206c
3-[(2-ethoxyethyl)thio]-	**79:** P92235n, **84:** P175174w
3-(*o*-ethoxyphenyl)-4-ethyl-5-(methylthio)-	**71:** 49853k
3-(*o*-ethoxyphenyl)-5-(ethylthio)-	**71:** 49853k
3-(*p*-ethoxyphenyl)-5-(ethylthio)-	**52:** 12852a
3-(*p*-ethoxyphenyl)-5-(isobutylthio)-	**52:** 12852a
3-(*p*-ethoxyphenyl)-5-(isopentylthio)-	**52:** 12852b
3-(*p*-ethoxyphenyl)-5-(isopropylthio)-	**52:** 12852a
3-([(4-ethoxyphenyl)methyl]thio)-	**84:** 30975y
3-(*o*-ethoxyphenyl)-5-(methylthio)-	**71:** 49853k
3-(*p*-ethoxyphenyl)-3-(methylthio)-	**63:** 13242f
3-(*p*-ethoxyphenyl)-5-(methylthio)-	**52:** 12852a
3-([(4-ethoxyphenyl)methyl]thio)-5-ethyl-	**84:** 30975y
3-([(4-ethoxyphenyl)methyl]thio)-5-methyl-	**84:** 30975y
3-(*o*-ethoxyphenyl)-5-(methylthio)-4-phenyl-	**71:** 49853k
3-(*p*-ethoxyphenyl)-5-(pentylthio)-	**52:** 12852b
3-(*p*-ethoxyphenyl)-5-(propylthio)-	**52:** 12852a
3-(*o*-ethoxyphenyl)-5-([(5-thioxo-Δ^2-1,3,4-oxadiazolin- 2-yl)methyl]thio)-	**71:** 49853k
3-[2-(*N*-ethylanilino)ethylthio]-	**79:** P43665u
3-([(*N*-ethylanilino)thiocarbonyl]thio)-	**69:** P11257q, **71:** P8422s
3-([(*N*-ethylcarbamoyl)methyl]thio)-	**72:** 31712z
3,3'-[*N*,*N*'-ethylenebis(carbamoylmethylthio)]bis]-	**72:** 31712z
3-ethyl-5-([(4-methoxyphenyl)methyl]thio)-	**84:** 30975y
3-[2-(*N*-ethyl-3-methylanilino)ethylthio]-	**79:** P43665u
3-ethyl-5-[([4-(1-methylethoxy)phenyl]methyl)thio]-	**84:** 30975y
3-ethyl-4-methyl-5-(methylthio)-	**55:** 23509e
3-ethyl-3-methyl-5-(methylthio)-	**54:** 7696b
4-ethyl-3-(methylthio)-5-[2-(5-nitro-2-furyl)vinyl]-	**61:** 9490h
3-(2-[*N*-ethyl-3-methyl-4-([4-(phenylazo)phenyl]azo)- anilino]ethylthio)-	**79:** P43665u
3-ethyl-5-[([4-(2-methylpropoxy)phenyl]methyl)thio]-	**84:** 30975y
3-ethyl-5-(methylthio)-	**55:** 23508d
4-ethyl-5-(methylthio)-3-(*p*-sulfamoylphenyl)-	**74:** 111968u
3-ethyl-5-[(5-nitro-2-thiazolyl)thio]-	**78:** P4286f
4-ethyl-3-[(5-nitro-2-thiazolyl)thio]-5-phenyl-	**78:** P4286f
3-[(2-[ethyl(phenyl)amino]ethyl)thio]-	**79:** P43665u

TABLE 10 (*Continued*)

Compound	Reference
10.1. Mono- and Di-S-substituted Thio-1,2,4-triazoles (*Continued*)	
4-ethyl-5-(propylthio)-3-(*p*-sulfamoylphenyl)-	**74:** 111968u
1-ethyl-3-[2-[1,2,3,4-tetrahydro-2,2,4,7-tetramethyl-6-[[2,4-bis(methylsulfonyl)phenyl]azo]-1-quinolinyl-ethylthio]-	**79:** P43665u
3-(ethylthio)-	**67:** 64315x, **71:** 124344a
3-(ethylthio)-4,5-dimethyl-	**54:** 7696c
5-(ethylthio)-1,3-dimethyl-	**52:** 18380a
3-(ethylthio)-4,5-diphenyl-	**82:** 43276e
3-(ethylthio)-5-(2-furyl)-	**52:** 16341d
3-(ethylthio)-5-(*p*-isobutoxyphenyl)-	**52:** 12852b
3-(ethylthio)-5-[*p*-(isopentyloxy)phenyl]-	**52:** 12852c
3-(ethylthio)-5-(*p*-isopropoxyphenyl)-	**52:** 12852b
3-(ethylthio)-5-(*p*-methoxyphenyl)-	**52:** 12852a
3-(ethylthio)-5-(1-methylethyl)-	**84:** P135677f
3-(ethylthio)-5-methyl-1-phenyl-	**51:** P899e, **54:** 7695h
3-(ethylthio)-5-methyl-4-phenyl-	**51:** P16163a, **54:** 7696b
5-(ethylthio)-3-methyl-1-phenyl-	**52:** 18380a
3-(ethylthio)-5-(1-methylpropyl)-	**84:** P135677f
3-(ethylthio)-5-(*p*-pentyloxyphenyl)-	**52:** 12852c
3-(ethylthio)-5-(phenoxymethyl)-4-phenyl-	**70:** 47369n
3-(ethylthio)-5-phenyl-	**25:** 2119[8]
3-(ethylthio)-5-phenyl-1-*o*-tolyl-	**25:** 2119[9]
3-(ethylthio)-5-*p*-propoxyphenyl-	**52:** 12852b
3-(ethylthio)-5-phenyl-1-*p*-tolyl-	**25:** 2119[9]
3-[3-(2-formylhydrazino)-5(2H̲)-oxo-1,2,4-triazin-6-yl]-4-methyl-5-(methylthio)-	**83:** 114345k
3-(2-furyl)-5-[(2-furylmethyl)thio]-	**84:** 4868w
5-(2-furyl)-3-([(hydroxamino)carbonyl]methylthio)-	**74:** 53660z
3-(2-furyl)-5-(hexylthio)-	**52:** 16341e
3-(2-furyl)-5-(isobutylthio)-	**52:** 16341e
3-(2-furyl)-5-(isopentylthio)-	**52:** 16341e
3-(2-furyl)-5-(isopropylthio)-	**52:** 16341e
3-(2-furyl)-4-methyl-5-(methylthio)-	**79:** 87327q
3-(2-furyl)-5-(methylthio)-	**52:** 16341d
3-[(2-furylmethyl)thio]-5-(5-nitro-2-furyl)-	**84:** 4868w
3-(2-furyl)-5-([(4-nitro-2-furyl)methyl]thio)-	**84:** 4868w
3-(2-furyl)-5-(pentylthio)-	**52:** 16341e
3-(2-furyl)-5-(propylthio)-	**52:** 16341e
3-(3-hydrazino-5(4H̲)-oxo-1,2,4-triazin-6-yl)-4-methyl-5-methylthio)-	**83:** 114345k
3-[(2-hydroxyethyl)thio]-1-phenyl-	**76:** 126876c
3-[(2-hydroxyethyl)thio]-4-phenyl-	**76:** 126876c
1(?)-(hydroxymethyl)-3-[(methoxycarbonylmethyl)thio]-	**76:** 30510c
1(?)-(hydroxymethyl)-3-(methylthio)-	**76:** 30510c
1(?)-(hydroxymethyl)-3-(4-nitrobenzylthio)-	**76:** 30510c
1(?)-(hydroxymethyl)-3-[(4-nitrophenyl)thio]-	**76:** 30510c
3-[4-hydroxy-3-([2-(tetradecyloxy)anilino]carbonyl)-1-naphthalenylthio]-	**79:** P137937m

TABLE 10 (*Continued*)

Compound	Reference

10.1. Mono- and Di-S-substituted Thio-1,2,4-triazoles (*Continued*)

Compound	Reference
3-(3-indolylthio)-5-methyl-	**75:** 19520q
3-(*p*-isobutoxyphenyl)-5-(isobutylthio)-	**52:** 12852c
3-(*p*-isobutoxyphenyl)-5-(isopentylthio)-	**52:** 12852c
3-(*p*-isobutoxyphenyl)-5-(isopropylthio)-	**52:** 12852c
3-(*p*-isobutoxyphenyl)-5-(methylthio)-	**52:** 12852b
3-(*p*-isobutoxyphenyl)-5-(pentylthio)-	**52:** 12852c
3-(*p*-isobutoxyphenyl)-5-(propylthio)-	**52:** 12852c
3-[[(*N*-isobutylcarbamoyl)methyl]thio]-	**72:** 31712z
3-(isobutylthio)-5-[*p*-(isopentyloxy)phenyl]-	**52:** 12852c
3-(isobutylthio)-5-(*p*-isopropoxyphenyl)-	**52:** 12852b
3-(isobutylthio)-5-(*p*-methoxyphenyl)-	**52:** 12852a
3-(isobutylthio)-5-(*p*-pentyloxyphenyl)-	**52:** 12852c
3-(isobutylthio)-5-(*p*-propoxyphenyl)-	**52:** 12852b
3-[*p*-(isopentyloxy)phenyl]-5-(isopentylthio)-	**52:** 12852c
3-[*p*-(isopentyloxy)phenyl]-5-(isopropylthio)-	**52:** 12852c
3-[*p*-(isopentyloxy)phenyl]-5-(methylthio)-	**52:** 12852c
3-[*p*-(isopentyloxy)phenyl]-5-(pentylthio)-	**52:** 12852c
3-[*p*-(isopentyloxy)phenyl]-5-(propylthio)-	**52:** 12852c
3-(isopentylthio)-5-(*p*-isopropoxyphenyl)-	**52:** 12852b
3-(isopentylthio)-5-(*p*-methoxyphenyl)-	**52:** 12852a
3-(isopentylthio)-5-(*p*-pentyloxyphenyl)-	**52:** 12852c
3-(isopentylthio)-5-(*p*-propoxyphenyl)-	**52:** 12852b
3-(*p*-isopropoxyphenyl)-5-(isopropylthio)-	**52:** 12852b
3-(*p*-isopropoxyphenyl)-5-(methylthio)-	**52:** 12852b
3-(*p*-isopropoxyphenyl)-5-(pentylthio)-	**52:** 12852b
3-(*p*-isopropoxyphenyl)-5-(propylthio)-	**52:** 12852b
3-(isopropylthio)-5-(*p*-methoxyphenyl)-	**52:** 12852a
3-(isopropylthio)-5-(*p*-pentyloxyphenyl)-	**52:** 12852c
3-(isopropylthio)-5-(*p*-propoxyphenyl)-	**52:** 12852b
-1-methanol, 5-(benzylthio)-	**75:** 35891s
3-[(*p*-methoxybenzyl)thio]-	**68:** 49563d
3-[(*p*-methoxybenzoyl)thio]-	**72:** P100713q
3-[(2-[(methoxycarbonyl)amino]-1H- benzimidazol-5-yl)thio]-	**82:** P31324f
3-([(methoxycarbonyl)methyl]thio)-	**71:** 124346c
3-[(methoxycarbonylmethyl)thio]-	**76:** 30510c
3-[(methoxycarbonylmethyl)thio]-4,5-diphenyl-	**79:** 126404j
3-[(methoxycarbonylmethyl)thio]-5-(phenoxymethyl)- 4-phenyl-	**79:** 126404j
3-(4-methoxyphenyl)-5-([(4-methoxyphenyl)methyl]thio)-	**84:** 30975y
3-(4-methoxyphenyl)-1-methyl-5-(methylthio)-	**83:** 163373b
3-(4-methoxyphenyl)-4-methyl-5-(methylthio)-	**83:** 163373b
5-(4-methoxyphenyl)-1-methyl-3-(methylthio)-	**83:** 163373b
5-(4-methoxyphenyl)-3-[(4-methylphenacyl)thio]-	**81:** 136112t
3-([(4-methoxyphenyl)methyl]thio)-	**84:** 30975y
3-(4-methoxyphenyl)-5-(methylthio)-	**44:** 2515h **44:** 6855g, **52:** 12852a, **83:** 163373b
3-([(4-methoxyphenyl)methyl]thio)-5-methyl-	**84:** 30975y

TABLE 10 (*Continued*)

Compound	Reference

10.1. Mono- and Di-S-substituted Thio-1,2,4-triazoles (*Continued*)

Compound	Reference
5-(4-methoxyphenyl)-3-[(4-nitrophenacyl)thio]-	**81:** 136112t
3-(*p*-methoxyphenyl)-5-(pentylthio)-	**52:** 12852a
5-(4-methoxyphenyl)-3-[(phenacyl)thio]-	**81:** 136112t
5-(4-methoxyphenyl)-3-[(4-phenylphenacyl)thio]-	**81:** 136112t
3-(*p*-methoxyphenyl)-5-(propylthio)-	**52:** 12852a
3,3'-[3-methyl-*N*,*N*-anilinobis(2-ethylthio)]bis-	**79:** P43665u
3-([(*N*-methylcarbamoyl)methyl]thio)-	**72:** 31712z
3-methyl-5-[(2,4-dioxo-3-pentyl)thio]-	**80:** P37118x
3-methyl-5-([[(2(1H),4(3H)-dioxopyrimidin-5-yl)methyl]-thio)-	**70:** 68301z
3,3'-(methylenedithio)bis[5-[2-[3,5-bis(*tert*-butyl)-4-hydroxyphenyl]ethyl]-	**79:** P19662v
3-[[[4-(1-methylethoxy)phenyl]methyl]thio]-	**84:** 30975y
3-(1-methylethyl)-5-(methylthio)-	**84:** P135677f
3-(1-methylethyl)-5-(propylthio)-	**84:** P135677f
3-(1-methylethyl)-5-(2-propynylthio)-	**84:** P135677f
3-methyl-5-[([4-(3-methylbutoxy)phenyl]methyl)thio]-	**84:** 30975y
3-methyl-5-[([4-(1-methylethoxy)phenyl]methyl)thio]-	**84:** 30975y
5-methyl-3-[2-(3-methyl-*N*-ethylanilino)ethylthio]-	**79:** P43665u
3-methyl-5-([(1-methyl-5-nitro-1H-imidazol-2-yl)-methyl]thio)-	**83:** P114412e
3-methyl-5-[([4-(2-methylpropoxy)phenyl]methyl)thio]-	**84:** 30975y
5-methyl-1-(4-methyl-2-thiazolyl)-3-(methylthio)-	**46:** 5582a
1-methyl-3-(methylthio)-	**76:** 59576q
1-methyl-5-(methylthio)-	**55:** 23509a, **83:** 163373b
3-methyl-5-(methylthio)-	**55:** 23508dg, **78:** 124513z
4-methyl-3-(methylthio)-	**55:** 23509d, **70:** 11686z, **73:** 66517n
4-methyl-3-(methylthio)-5-(5-nitro-2-furanyl)-	**79:** 87327q
4-methyl-3-(methylthio)-5-[2-nitro-2-furyl)vinyl]-	**61:** 9490h
1-methyl-3-(methylthio)-5-(4-nitrophenyl)-	**83:** 163373b
1-methyl-5-(methylthio)-3-(4-nitrophenyl)-	**83:** 163373b
4-methyl-3-(methylthio)-5-(4-nitrophenyl)-	**83:** 163373b
4-methyl-5-(methylthio)-3-(4(1H)-oxo-[1,2,4]triazola-[5,1-c][1,2,4]triazin-3-yl)-	**83:** 114345k
4-methyl-5-(methylthio)-3-(5(1H)-oxo-[1,2,4]triazola-[3,4-c][1,2,4]triazin-6-yl)-	**83:** 114345k
1-methyl-3-(methylthio)-5-phenyl-	**76:** 59576q, **79:** 78749b, **85:** 159988e
1-methyl-5-(methylthio)-3-phenyl-	**79:** 787449b, **85:** 159988e
3-methyl-5-(methylthio)-1-phenyl-	**54:** 7695g, **79:** 92113w
3-methyl-5-(methylthio)-4-phenyl-	**53:** 11356h, **54:** 5630e, **54:** 7696b, **78:** P36272q
4-methyl-3-(methylthio)-5-phenyl-	**44:** 2517g, **45:** 1122e, **55:** 17009i, **85:** 159988e
5-methyl-3-(methylthio)-1-phenyl-	**54:** 7695g
5-methyl-3-(methylthio)-1-(4-phenyl-2-thiazolyl)-	**46:** 5582a
1-methyl-3-(methylthio)-5-(4-pyridinyl)-	**83:** 163373b

280

TABLE 10 (*Continued*)

Compound	Reference
10.1. Mono- and Di-S-sub ;tituted Thio-1,2,4-triazoles (*Continued*)	
1-methyl-5-(methylthio)-3-(4-pyridinyl)-	**83:** 163373b
4-methyl-5-(methylthio)-3-(4-pyridinyl)-	**83:** 163373b
1-methyl-3-(methylthio)-5-(2-pyridyl)-	**76:** 34180t
1-methyl-5-(methylthio)-3-(2-pyridyl)-	**76:** 34180t
4-methyl-5-(methylthio)-3-(2-pyridyl)-	**76:** 34180t
5-methyl-3-(methylthio)-1-(2-thiazolyl)-	**46:** 5582a
1-methyl-5-(methylthio)-3-(trifluoromethyl)-	**80:** P82990m
4-methyl-3-(methylthio)-5-(trifluoromethyl)-	**82:** P72995u
3-[[(1-methyl-5-nitro-1H-imidazol-2-yl)methyl]thio]-	**83:** P114412e
3-[(3-methyl-4-nitro-5-isothiazolyl)thio]-	**84:** 164664e
3-methyl-5-[(5-nitro-2-thiazolyl)thio]-	**78:** P4286f
4-methyl-3-[(5-nitro-2-thiazolyl)thio]-	**78:** P4286f
3-methyl-5-[(2-oxo-3-butyl)thio]-	**80:** P37118x
3-methyl-5-[([(4-(pentyloxy)phenyl]methyl)thio]-	**84:** 30975y
3-[(4-methylphenacyl)thio]-5-(2-methylphenyl)-	**81:** 136112t
5-methyl-3-[(phenacyl)thio]-4-phenyl-	**82:** 140028h
3-methyl-5-[(phenoxymethyl)thio]-4-phenyl-	**72:** 78951s
4-(4-methylphenyl)-3,5-bis(methylthio)-	**79:** 42423b
5-(2-methylphenyl)-3-[(4-nitrophenacyl)thio]-	**81:** 136112t
5-(2-methylphenyl)-3-[(phenacyl)thio]-	**81:** 136112t
5-(2-methylphenyl)-3-[(4-phenylphenacyl)thio]-	**81:** 136112t
3-methyl-5-([(4-propoxyphenyl)methyl]thio)-	**84:** 30975y
3-(1-methylpropyl)-5-(methylthio)-	**84:** P135677f
3-(1-methylpropyl)-5-(propylthio)-	**84:** P135677f
3-(1-methylpropyl)-5-(2-propynylthio)-	**84:** P135677f
3-[(2-methylpropyl)thio]-	**78:** P57044g
3-(methylthio)-	**51:** 17895a, **70:** P28922w, **71:** P81375s, **71:** P124344a, **71:** P124443g
3-(methylthio)-4,5-diphenyl-	**75:** 76692w
5-(methylthio)-1,3-diphenyl-	**80:** 82818m
3-[(methylthio)methyl]-5-[(3-sulfopropyl)thio]-	**83:** 186250y
3-(methylthio)-5-(5-nitro-2-furyl)-	**69:** 77177v
3-(methylthio)-5-[2-(5-nitro-2-furyl)vinyl]-	**61:** 9490h
3-(methylthio)-5-[2-(5-nitro-2-furyl)vinyl]-4-propyl-	**61:** 9490h
3-(methylthio)-5-(4-nitrophenyl)-	**83:** 163373b
3-(methylthio)-5-(*p*-nitrophenyl)-4-*p*-tolyl-	**75:** 76692w
3-(methylthio)-5-(5-nitro-2-thienyl)-	**69:** 77177v
3-[3-(methylthio)-5(2H)-oxo-1,2,4-triazin-6-yl]- 4-methyl-5-(methylthio)-	**83:** 114345k
3-(methylthio)-5-(*p*-pentyloxyphenyl)-	**52:** 12852c
3-(methylthio)-1-phenyl-	**45:** 6212a, **54:** 7695h
3-(methylthio)-4-phenyl-	**53:** 21905a
3-(methylthio)-5-phenyl-	**25:** 2119[8], **44:** 2515g, **45:** 1122f, **48:** 4525a, **53:** 10034c, **65:** 13691g, **83:** 163373b
3-(methylthio)-5-phenyl-1-*p*-tolyl-	**25:** 2119[9], **25:** 2120[1]

281

TABLE 10 (*Continued*)

Compound	Reference
10.1. Mono- and Di-S-substituted Thio-1,2,4-triazoles (*Continued*)	
3-(methylthio)-5-phenyl-1-4-*o*-tolyl-	**25:** 2119[9], **25:** 2120[1]
3-(methylthio)-5-phenyl-4-*p*-tolyl-	**75:** 76692w
3-(methylthio)-5-*p*-propoxyphenyl-	**52:** 12852b
5-(methylthio)-3-(4-pyridyl)-	**83:** 163373b
3-(methylthio)-4-(3-pyridyl)-	**76:** 34182v
3-(methylthio)-5-(2-pyridyl)-	**76:** 34180t
5-methyl-3-[(thiocyanato)methylthio]-	**74:** 125575a
3-(methylthio)-5-(3,4,5-trimethoxyphenyl)-	**67:** P54140x
3-(methylthio)-5-undecyl-	**67:** P54140x
3-(1-naphthylmethylthio)-5-phenyl-	**51:** 15699i
3-[(1-nitro-4-anthraquinonyl)thio]-	**74:** P14185d
3-[(*m*-nitrobenzoyl)thio]-	**72:** P100713q
3-[(*p*-nitrobenzyl)thio]-4-phenyl-	**63:** 13242g
3-[(4-nitrophenacyl)thio]-5-phenyl-	**81:** 136112t
3-[[(N-*p*-nitrophenylcarbamoyl)methyl]thio]-	**72:** 31712z
3-[[(N-*o*-nitrophenylcarbamoyl)methyl]thio]-	**72:** 31712z
3-[[(4-nitrophenyl)methyl]thio]-	**76:** 30510c
5-(4-nitrophenyl)-3-[(4-nitrophenacyl)thio]-	**81:** 136112t
3-[(2-nitrophenyl)thio]-	**79:** P43665u
3-[(4-nitrophenyl)thio]-	**76:** 30510c
3-[(2-nitro-1,3,4-thiodiazol-5-yl)thio]-	**83:** 164088z
3-[(2-oxo-3-butyl)thio]-	**80:** P37118x
3-(*p*-pentyloxyphenyl)-5-(pentylthio)-	**52:** 12852c
3-(*p*-pentyloxyphenyl)-5-(propylthio)-	**52:** 12852c
3-(pentylthio)-5-*p*-propoxyphenyl-	**52:** 12852b
3-[(phenacyl)thio]-	**75:** 5441c, **77:** 126492v
3-(phenacylthio)-5-methyl-	**74:** 42320n, **77:** 126492v
3-[(phenacyl)thio]-5-methyl-	**77:** 126492v
3-[(phenoxycarbonyl)thio]-1-phenyl-	**77:** P171214s
3-[(phenoxymethyl)-5-[([(3,5-dimethyl-1H-pyrazol-1-yl)-carbonyl]methylene)thio]-4-phenyl-	**76:** 153661k
3-([(N-phenylcarbamoyl)methyl]thio)-	**72:** 31712z
4-phenyl-3-(phenylmethyl)-5-(2-propenylthio)-	**85:** 159997g
5-phenyl-3-[(phenylphenacyl)thio]-	**81:** 136112t
1-phenyl-3-(2-propynylthio)-	**81:** P129840a, **81:** P144229b
3-(phenylthio)-	**70:** P28922w, **71:** P81375s, **71:** P124443g
1-phenyl-3-thiocyanato-	**61:** 4338g
3-phenyl-5-thiocyanato-	**61:** 4388g
5-phenyl-3-[(thiocyanato)methylthio]-	**74:** 125575a
1-phenyl-3-[(trimethylammonio)methylthio]-, chloride	**72:** P56720p
3-(2-propenylthio)-	**78:** P57044g
-1-propionitrile, 3-[(4-amino-3-methoxy-1-anthraquinonyl)thio]-	**74:** P14185d
3-([(4-propoxyphenyl)methyl]thio)-	**84:** 30975y
3-([(4-propoxyphenyl)methyl]thio)-5-propyl-	**84:** 30975y
3-*p*-propoxyphenyl-5-(propylthio)-	**52:** 12852b

TABLE 10 (*Continued*)

Compound	Reference
10.1. Mono- and Di-S-substituted Thio-1,2,4-triazoles (*Continued*)	
3-(propylthio)-	**71:** 124344a, **76:** P72528v, **79:** P92235n, **84:** P175174w
3-(2-propynylthio)-	**81:** P129840a, **81:** P144229b
3-(4-pyridyl)-5-(2-pyridylthio)-	**63:** 10497b
3-[(3-sulfopropyl)thio]-	**80:** 47911q
3-[(2,3,4,6-tetra-*O*-acetyl-β-D-glucopyrano-1-yl)thio]-5-methyl-4-phenyl-	**80:** 146458d
3-[(2-[1,2,3,4-tetrahydro-6-([2,4-bis(methylsulfonyl)-phenyl]azo)-2,2,4,7-tetramethyl-1-quinolinyl]ethyl)thio]-	**79:** P43665u
3-[2-(1,2,3,4-tetrahydro-2,2,4,7-tetramethyl-1-quinolinyl)ethylthio]-	**79:** P43665u
3-(2-thenylthio)-	**60:** 9263c
3-[(4-thiazolylmethyl)thio]-	**60:** 9263c
3-[(thioacetyl)thio]-	**78:** P57044g
3-[(thiobenzoyl)thio]-	**78:** P57044g
3-(thiocyanato)-	**68:** 21884k, **75:** 35361u
3-([(*N*-*p*-tolylcarbamoyl)methyl]thio)-	**72:** 31712z
3-([(*N*-*o*-tolylcarbamoyl)methyl]thio)-	**72:** 31712z
3-[(2,3,4-tri-*O*-acetyl-β-D-xylopyranos-1-yl)thio]-5-methyl-4-phenyl-	**80:** 146458d
1-[(trichloromethyl)thio]-	**79:** 78696g
4-[(trichloromethyl)thio]-	**79:** 78696g

10.2. 3,3'-Dithiobis[1,2,4-triazoles], Sulfinyl- or Sulfonyl-1,2,4-triazoles, and 1,2,4-Triazolesulfonic Acids and Their Functional Derivatives

10.2a. 3,3'-Dithiobis[1,2,4-triazoles]	
3,3'-dithiobis-	**51:** P3669c, **51:** 12079f, **67:** P54140x
3,3'-dithiobis[5-(*p*-acetamidophenyl)-	**47:** 131g
3,3'-dithiobis[4-allyl-5-(*m*-nitrophenyl)-	**22:** 4123[9]
3,3'-dithiobis[4-allyl-5-phenyl-	**22:** 4123[8]
3,3'-dithiobis[5-(2-chloro-4-nitrophenyl)-4-phenyl-	**54:** 8795b
3,3'-dithiobis[5-(*p*-chlorophenyl)-4-phenyl-	**54:** 8795b
3,3'-dithiobis[5-(3,5-dimethylpyrazol-1-yl)-4-phenyl-	**64:** 8174a
3,3'-dithiobis[4,5-diphenyl-	**22:** 4123[8], **54:** 8795b, **81:** 91429h
3,3'-dithiobis[4-ethyl-5-phenyl-	**22:** 4123[8]
3,3'-dithiobis[4-methyl-	**82:** 43271z
3,3'-dithiobis[5-(*p*-nitrophenyl)-4-phenyl-	**54:** 8795b
3,3'-dithiobis[5-(*o*-nitrophenyl)-4-(2,4-xylyl)-	**22:** 4123[7]
3,3'-dithiobis[4-phenyl-5-benzyl-	**83:** 164091v
3,3'-dithiobis[4-phenyl-5-(4-pyridyl)-	**63:** 2967c
3,3'-dithiobis[5-phenyl-4-*p*-tolyl-	**22:** 4123[8]
3,3'-dithiobis[5-(4-pyridyl)-4-*m*-tolyl-	**63:** 2967d
3,3'-dithiobis[5-(4-pyridyl)-4-*o*-tolyl-	**63:** 2967d
3,3'-dithiobis[5-(4-pyridyl)-4-*p*-tolyl-	**63:** 2967d

TABLE 10 *(Continued)*

Compound	Reference
10.2. 3,3′-Dithiobis[1,2,4-triazoles], Sulfinyl- or Sulfonyl-1,2,4-triazoles, and 1,2,4-Triazolesulfonic Acids and Their Functional Derivatives *(Continued)*	

10.2b. *S*-Substituted Sulfinyl- or Sulfonyl-1,2,4-triazoles

Compound	Reference
-1-acetamide, *N*-[5-[[4-[2,4-bis(1,1-dimethylpropyl)-phenoxy]-1-oxobutyl]amino]-2-chlorophenyl]-α-(2,2-dimethyl-1-oxopropyl)-3-(methylsulfonyl)-	**85:** P102316e
-1-acetamide, *N*-[2-chloro-5-[[(3-pentadecylphenoxy)-acetyl]amino]phenyl]-α-(2,2-dimethyl-1-oxopropyl)-3-(methylsulfonyl)-	**85:** P102316e
3-(2-benzothiazolyl)-1-benzyl-5-(methylsulfinyl)-	**77:** 126518h
5-(2-benzothiazolyl)-1-benzyl-3-(methylsulfinyl)-	**77:** 126518h
3-(2-benzothiazolyl)-4-benzyl-5-(methylsulfonyl)-	**77:** 126518h
3-(2-benzothiazolyl)-4-ethyl-5-(methylsulfonyl)-	**77:** 126518h
3-(2-benzothiazolyl)-4-methyl-5-(methylsulfinyl)-	**77:** 126518h
3-(2-benzothiazolyl)-5-(methylsulfonyl)-	**77:** 126518h
3-(2-benzothiazolyl)-5-(methylsulfonyl)-4-phenyl-	**77:** 126518h
3-(2-benzothiazolyl)-5-(methylsulfonyl)-4-(2-propenyl)-	**77:** 126518h
3-(benzylsulfonyl)-	**76:** 30510c
5-(benzylsulfonyl)-3-(4-chlorophenyl)-4-(4-methylphenyl)-	**78:** 159528k
5-(benzylsulfonyl)-3-[(2,4-dichlorophenoxy)methyl]-4-(4-iodophenyl)-	**78:** 159528k
5-(benzylsulfonyl)-3-[(2,4-dichlorophenoxy)methyl]-4-(2-methylphenyl)-	**78:** 159528k
5-(benzylsulfonyl)-3-[(2,4-dichlorophenoxy)methyl]-4-(3-methylphenyl)-	**78:** 159528k
3-benzylsulfonyl-1-(?)-(hydroxymethyl)-	**76:** 30510c
3-(benzylsulfonyl)-5-methyl-1-phenyl-	**83:** 27827a
3-(benzylsulfonyl)-1-phenyl-	**83:** 27827a
3-(benzylsulfonyl)-4-phenyl-	**63:** 13242h
3-[(4-bromobenzyl)sulfonyl]-	**76:** 30510c
3-(butylsulfonyl)-5-(*p*-chlorophenyl)-4-ethyl-	**74:** 111968u
3-[(2-carboxyethyl)sulfonyl]-1-phenyl-	**74:** 141644c
5-[(2-carboxyethyl)sulfonyl]-1-phenyl-	**74:** 141644c
3-[(2-chloroethyl)sulfonyl]-	**65:** PC5404g
3-(*p*-chlorophenyl)-4-ethyl-5-(ethylsulfonyl)-	**74:** 111968u
3-(*p*-chlorophenyl)-4-ethyl-5-(methylsulfonyl)-	**74:** 111968u
3-(*p*-chlorophenyl)-4-ethyl-5-(propylsulfonyl)-	**74:** 111968u
3-(4-chlorophenyl)-5-(ethylsulfonyl)-4-(4-iodophenyl)-	**78:** 159528k
4-(*p*-chlorophenyl)-3-(methylsulfonyl)-	**63:** 13242h, **63:** 13243b
1-[(*p*-chlorophenyl)sulfonyl]-	**66:** PC55489j
3-[(2-cyanoethyl)sulfonyl]-1-phenyl-	**74:** 141644c
3,5-dibenzyl-1-(*p*-tolylsulfonyl)-	**61:** 4515f
3-[(2,4-dichlorophenoxy)methyl]-5-(ethylsulfonyl)-4-(2-methylphenyl)-	**82:** 125327m
1-[1-(dimethylamino)-5-naphthalenylsulfonyl]-	**79:** 91273m
1-[(2,3-dimethylphenyl)sulfonyl]-	**84:** 17652w
3,5-dimethyl-1-(phenylsulfonyl)-	**66:** PC55489j
3,5-dimethyl-1-(2,4-xylylsulfonyl)-	**66:** PC55489j
1-[(2,4-dinitrophenyl)sulfonyl]-	**84:** 17652w

TABLE 10 (Continued)

Compound	Reference

10.2b. S-Substituted Sulfinyl- or Sulfonyl-1,2,4-triazoles (Continued)

Compound	Reference
3-[[(ethoxycarbonyl)methyl]-5-(o-ethoxyphenyl)-	**71:** 49853k
3-[(2-ethoxyethyl)sulfonyl]-	**79:** P92235n, **84:** P175174w
3-(o-ethoxyphenyl)-4-ethyl-5-(methylsulfonyl)-	**71:** 49853k
3-(o-ethoxyphenyl)-5-(ethylsulfonyl)-	**71:** 49853k
3-(o-ethoxyphenyl)-5-(methylsulfonyl)-	**71:** 49853k
4-(p-ethoxyphenyl)-3-(methylsulfonyl)-	**63:** 13242h, **63:** 13243b
3-(o-ethoxyphenyl)-5-(methylsulfonyl)-4-phenyl-	**71:** 49853k
4-ethyl-5-(ethylsulfonyl)-3-(p-sulfamoylphenyl)-	**74:** 111968u
4-ethyl-5-(methylsulfonyl)-3-(p-sulfamoylphenyl)-	**74:** 111968u
4-ethyl-5-(propylsulfonyl)-3-(p-sulfamoylphenyl)-	**74:** 111968u
1-[(p-fluorophenyl)sulfonyl]-	**66:** PC55489j
1-[(p-fluorophenyl)sulfonyl]-3,5-dimethyl-	**66:** PC55489j
3-(2-furanyl)-4-methyl-5-(methylsulfinyl)-	**79:** 87327q
3-(2-furanyl)-4-methyl-5-(methylsulfonyl)-	**79:** 87327q
1-(1H-imidazol-4-ylsulfonyl)-	**84:** 136515p
-1-methanol, 5-(bensulfonyl)-	**75:** 35891s
3-[[2-[(methoxycarbonyl)amino]-1H-benzimidazol-5-yl]sulfinyl]-	**82:** P31324f
3-[(methoxycarbonylmethyl)sulfonyl]-	**76:** 30510c
3-[(2-methoxyethyl)sulfonyl]-	**79:** P92235n, **84:** P175174w
1-(4-methoxyphenyl)-3-[(4-methylphenyl)sulfonyl]-	**84:** 90078t
1-(4-methoxyphenyl)-5-[(4-methylphenyl)sulfonyl]-	**84:** 90078t
1-[(p-methoxyphenyl)sulfonyl]-	**66:** PC55489j
4-[(p-methoxyphenyl)sulfonyl]-	**66:** PC55489j
1-[(p-methoxyphenyl)sulfonyl]-3,5-dimethyl-	**66:** PC55489j
5-methyl-3-[(4-methylphenyl)sulfonyl]-1-phenyl-	**83:** 27827a
4-methyl-3-(methylsulfinyl)-5-(5-nitro-2-furanyl)-	**79:** 87327q
3-methyl-5-(methylsulfonyl)-	**67:** P54140x
4-methyl-3-(methylsulfonyl)-5-(5-nitro-2-furanyl)-	**79:** 87327q
1-[(4-methylphenyl)sulfonyl]-	**84:** 17652w
3-[(4-methylphenyl)sulfonyl]-1-(1-naphthalenyl)-	**84:** 90078t
3-[(4-methylphenyl)sulfonyl]-1-phenyl-	**84:** 90078t
5-[(4-methylphenyl)sulfonyl]-1-phenyl-	**84:** 90078t
5-methyl-1-phenyl-3-(tricyclo[3.3.1.1^{3,7}]dec-1-ylsulfonyl)-	**83:** 27827a
3-[(2-methylpropyl)sulfonyl]-	**79:** P92235n
3-(methylsulfonyl)-	**63:** 13243c
3-(methylsulfonyl)-5-[2-(5-nitro-2-furyl)vinyl]-	**61:** 9490h
3-(methylsulfonyl)-4-phenyl-	**63:** 13243g, **63:** 13243b
3-(methylsulfonyl)-5-phenyl-	**44:** 2515h
3-(methylsulfonyl)-4-(3-pyridyl)-	**76:** 34182v
5-methyl-3-[(thiocyanato)methylsulfonyl]-	**74:** 125575a
3-[(4-nitrobenzyl)sulfonyl]-	**76:** 30510c
3-[(p-nitrobenzyl)sulfonyl]-4-phenyl-	**63:** 13242h
1-[(2-nitrophenyl)sulfonyl]-	**84:** 17652w
1-[(4-nitrophenyl)sulfonyl]-	**84:** 17652w
3-[(4-nitrophenyl)sulfonyl]-	**76:** 30510c
3-(p-nitrophenylsulfonyl)-5-phenyl-	**51:** 15699i
3-phenyl-5-p-sulfanilyl-	**51:** 15699i

285

TABLE 10 (*Continued*)

Compound	Reference

10.2b. *S*-Substituted Sulfinyl- or Sulfonyl-1,2,4-triazoles (*Continued*)

Compound	Reference
1-(phenylsulfonyl)-	**66:** PC55489j, **84:** 17652w
1-(phenylsulfonyl)-3.5-di-4-pyridinyl-	**77:** P5487p, **85:** P177430w
5-phenyl-3-[(thiocyanato)methylsulfonyl]-	**74:** 125575a
1-phenyl-3-(tricyclo[3.3.1.1^{3,7}]dec-1-ylsulfonyl)-	**83:** 27827a
3-(1-piperidinylsulfonyl)-	**79:** P92235n
3-(propylsulfonyl)-	**76:** P72528v, **84:** P175174w
1-(*p*-tolylsulfonyl)-	**66:** PC55489j
1-[(2,4,6-trimethylphenyl)sulfonyl]-	**81:** 91849v, **84:** 105990k, **85:** 94637t
1-[[2,4,6-tris(1-methylethyl)phenyl]sulfonyl]-	**81:** 91849v, **84:** 17652w
1-(2,3-xylylsulfonyl)-	**66:** PC55489j

10.2c. 1,2,4-Triazolesulfonic Acids and Their Functional Derivatives

Compound	Reference
-3,5-disulfonyl chloride, 4-phenyl-	**45:** 4671b
-3-sulfonamide	**45:** 4671d, **53:** 5268c
-3-sulfonamide, 5-(2-aminoethyl)-	**48:** 12739i
-3-sulfonamide, 5-(*p*-aminophenyl)-	**64:** 5072h
-3-sulfonamide, 5-(2-bromophenyl)-	**78:** 43372a
-3-sulfonamide, 5-(4-bromophenyl)-	**78:** 43372a
-3-sulfonamide, 5-tert-butyl	**85:** P33028y
-3-sulfonamide, 5-tert-butyl-*N*,*N*-diethyl-	**85:** P33028y
-3-sulfonamide, 5-tert-butyl-*N*,*N*-dimethyl-	**85:** P33028y
-3-sulfonamide, 5-tert-butyl-*N*-ethyl-	**85:** P33028y
-3-sulfonamide, 5-tert-butyl-*N*-methyl-	**85:** P33028y
-3-sulfonamide, 5-(*p*-chlorophenyl)-	**51:** P2875a
-3-sulfonamide, 5-(*p*-chlorophenyl)-4-isopropyl-	**72:** 30148h
-3-sulfonamide, 5-(*o*-chlorophenyl)-4-phenyl-	**72:** 30148h
-3-sulfonamide, 1-(5-chloropiperonyl)-	**74:** 99946a
-3-sulfonamide, *N*,*N*-dimethyl-5-isopropyl-	**85:** P33028y
-3-sulfonamide, *N*,*N*-di-2-propenyl-	**79:** P92235n
-3-sulfonamide, 5-(2-methoxyphenyl)-	**78:** 43372a
-3-sulfonamide, N-nitro-	**59:** P12827b
-3-sulfonamide, 5-(*p*-nitrophenyl)-	**64:** 5072h
-3-sulfonamide, 5-(2-phthalimidoethyl)-	**48:** 12739h
-3-sulfonamide, 1-piperonyl-	**74:** 99946a
-3-sulfonamide, 5-(4-pyridyl)-	**64:** 5072h
-3-sulfonic acid	**66:** 104948u, **74:** 3557a
-3-sulfonic acid, 5-(5-bromo-2-methoxyphenyl)-	**85:** 159998h
-3-sulfonic acid, 5-(*p*-chlorophenyl)-4-ethyl-	**74:** 111968u
-3-sulfonic acid, 4-ethyl-5-(*p*-sulfamoylphenyl)-	**74:** 111968u
-3-sulfonic acid, 5-methyl-	**66:** 104948u
-3-sulfonic acid, 1-phenyl-	**66:** 104948u
-3-sulfonic acid, 5-phenyl-	**66:** 104948u

TABLE 10 (*Continued*)

Compound	Reference
10.2c. 1,2,4-Triazolesulfonic Acids and Their Functional Derivatives (*Continued*)	
-3-sulfonic acid, 5-(3-pyridyl)-	**66:** 104948u
-3-sulfonic acid, 5-(4-pyridyl)-	**66:** 104948u
-3-sulfonyl azide, 3-cyclopropyl-1-[(dimethylamino)- carbonyl]-	**85:** P33028y
-3-sulfonyl azide, 1-[(dimethylamino)carbonyl]- 3-tert-butyl-	**85:** P33028y
-3-sulfonyl chloride	**45:** 4671b
-3-sulfonyl chloride, 5-(2-bromophenyl)-	**78:** 43372a
-3-sulfonyl chloride, 5-(4-bromophenyl)-	**78:** 43372a
-3-sulfonyl chloride, 5-tert-butyl-	**85:** P33028y
-3-sulfonyl chloride, 5-(*p*-chlorophenyl)-	**51:** P2874i
-3-sulfonyl chloride, 5-isopropyl-	**85:** P33028y
-3-sulfonyl chloride, 5-(2-methoxyphenyl)-	**78:** 43372a

CHAPTER 11

Mono-, Di-, and Trihalo-1,2,4-Triazoles

The synthesis of halotriazoles from noncyclic reactants has been limited to one method. The interaction of monosubstituted alkyl- and arylhydrazines (e.g., **11.1-1** and **11.1-2**) with 1,1,1,3,3-pentachloro-2-azapropene (**11.1-3**) in a two-phase system of aqueous sodium hydroxide and methylene chloride gave the 1-substituted 3,5-dichloro-1,2,4-triazoles (**11.1-4** and **11.1-5**).[1]

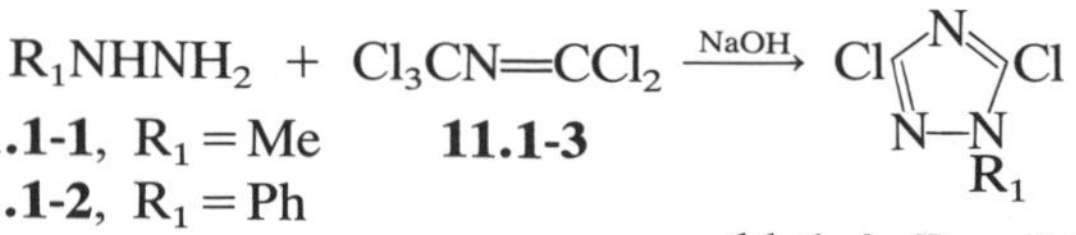

$$R_1NHNH_2 \ + \ Cl_3CN{=}CCl_2 \ \xrightarrow{\text{NaOH}} \ \text{Cl} \underset{\underset{R_1}{N-N}}{\overset{N}{\diagup\!\!\diagdown}} \text{Cl}$$

11.1-1, $R_1 =$ Me   **11.1-3**
11.1-2, $R_1 =$ Ph

11.1-4, $R_1 =$ Me (26%)
11.1-5, $R_1 =$ Ph (60%)

A common method for the preparation of halotriazoles including N-alkyl and N-aryl derivatives involves the diazotization of 3-amino-1,2,4-triazoles and treatment of the resulting 3-diazotriazole with a halogen acid.[2-5] For example, addition of hydrobromic acid to a mixture of 3-amino-1,2,4-triazole (**11.1-9**) and sodium nitrite gave 3-bromo-1,2,4-triazole (**11.1-6**) in 85% yield.[6] Similarly, 3-amino-5-chloro- and 3,5-diamino-1,2,4-triazoles (**11.1-10** and **11.1-11**) gave the bromotriazoles **11.1-12** (60%) and **11.1-13** (89%), respectively. The addition of potassium iodide to an acidified solution of 3-diazotriazole gave a 62% yield of 3-iodo-1,2,4-triazole (**11.1-7**),[7] and ultraviolet irradiation of 3-diazotriazole in fluoboric acid gave 3-fluoro-1,2,4-triazole (**11.1-8**).[8]

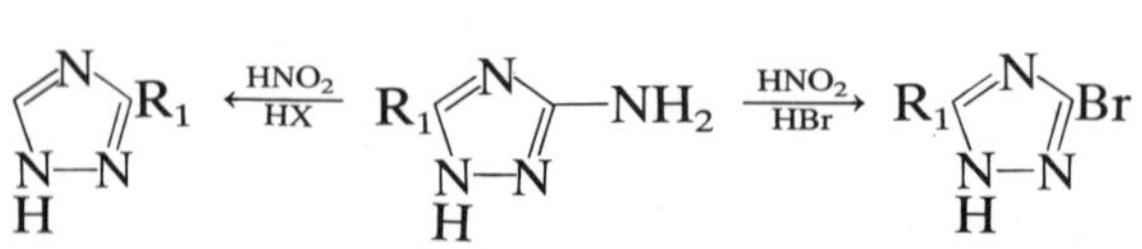

$$\underset{\underset{H}{N-N}}{\overset{N}{\diagup\!\!\diagdown}} R_1 \xleftarrow[\text{HX}]{\text{HNO}_2} \ R_1 \underset{\underset{H}{N-N}}{\overset{N}{\diagup\!\!\diagdown}} {-}NH_2 \xrightarrow[\text{HBr}]{\text{HNO}_2} \ R_1 \underset{\underset{H}{N-N}}{\overset{N}{\diagup\!\!\diagdown}} Br$$

11.1-6, $R_1 =$ Br   **11.1-9**, $R_1 =$ H   **11.1-12**, $R_1 =$ Cl
11.1-7, $R_1 =$ I   **11.1-10**, $R_1 =$ Cl   **11.1-13**, $R_1 =$ Br
11.1-8, $R_1 =$ F   **11.1-11**, $R_1 =$ NH$_2$

288

Electrophilic substitution of 1,2,4-triazole and its 3-alkyl and 3-aryl derivatives with either chlorine or bromine gave good yields of halotriazoles under mild conditions. N-Chlorination of 11.1-14 in an aqueous solution containing an equilmolar amount of potassium hydrogen carbonate gave a N-chlorotriazole, presumably 11.1-15, which was thermally rearranged to 11.1-17 in dichloroethane.[9] In both water and *tert*-butylalcohol, this rearrangement has been described as a second order reaction.[10] The chlorination of 11.1-14 in the presence of two molar equivalents of aqueous sodium hydroxide was reported to give 1,3-dichloro-1,2,4-triazole (11.1-18), which on treatment with excess sodium hydrogen carbonate also gave 3-chlorotriazole (11.1-17).[11] Further reaction of 11.1-18 with chlorine replaced the remaining hydrogen to give 1,3,5-trichloro-1,2,4-triazole (11.1-22).

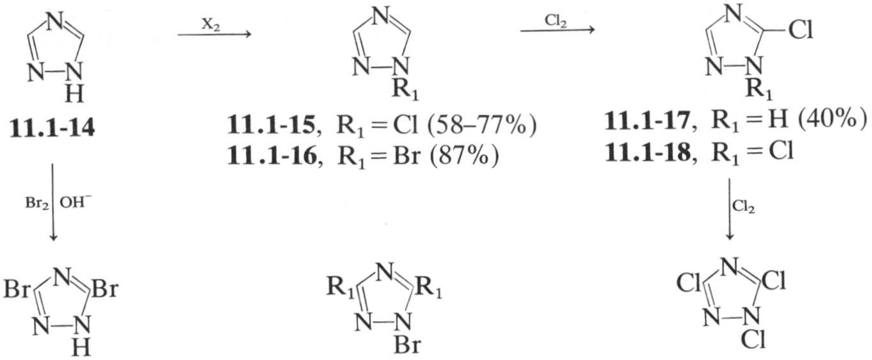

11.1-14

11.1-15, R_1 = Cl (58–77%)
11.1-16, R_1 = Br (87%)

11.1-17, R_1 = H (40%)
11.1-18, R_1 = Cl

11.1-19

11.1-20, R_1 = Me
11.1-21, R_1 = Ph

11.1-22

The N-bromotriazole (11.1-16) was prepared from an aqueous solution of the potassium salt of 11.1-14 and bromine at 0°. In addition, 1-halo derivatives (e.g., 11.1-20 and 11.1-21) were obtained on treatment of the corresponding 3,5-disubstituted-1,2,4-triazoles with bromine.[6] In the bromination of 11.1-14 with two molar equivalents of bromine and three molar equivalents of aqueous sodium hydroxide, 3,5-dibromo-1,2,4-triazole (11.1-19) was obtained directly in high yield.[6,9] Similarly, 3-ethyl-1,2,4-triazole (11.1-23) was converted to 5-ethyl-3-bromotriazole (11.1-25) in 73% yield. In contrast to the bromination of 11.1-24 to give 11.1-26 (99%),[6] the chlorination of 11.1-24 gave only a N-chloro derivative.[11] Both 4-substituted and 3,4-disubstituted-1,2,4-triazoles brominate on carbon in water via

11.1-23, R_1 = Et
11.1-24, R_1 = Ph

11.1-25, R_1 = Et
11.1-26, R_1 = Ph

N-bromotriazolium intermediates. For example, treatment of 4-phenyltria-zole (**11.1-27**) with bromine gave **11.1-28**, which rearranged at 80° to give **11.1-29**.[12] Under similar conditions some 1-alkyl- and aryltriazoles gave only the corresponding hydrobromides.

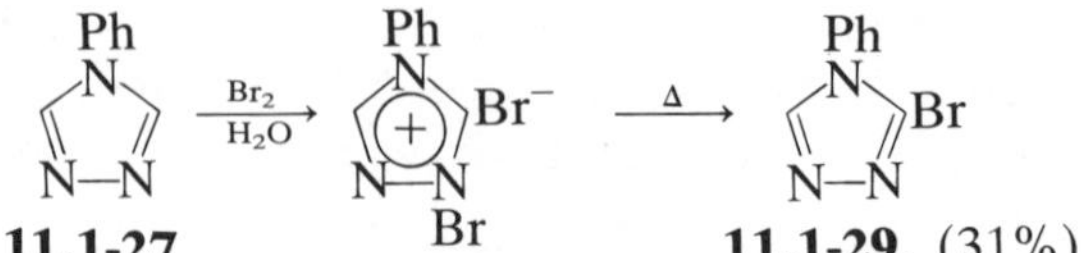

Ph
N
‖ ⟩   Br₂/H₂O →
N—N
11.1-27

Ph
N
(+) ⟩ Br⁻   Δ →
N—N
Br
11.1-28, (60%)

Ph
N
‖ ⟩ Br
N—N
11.1-29, (31%)

A high yield of the *N*-iodotriazole (**11.1-30**) resulted from treatment of 1,2,4-triazole with iodine chloride in either ethanol[13] or an alkaline medium.[9,13] Although thermal rearrangement of **11.1-30** to 3-iodo-triazole (**11.1-31**) was unsuccessful,[9] heating a mixture of **11.1-30** with 1,2,4-triazole gave **11.1-31**.[13]

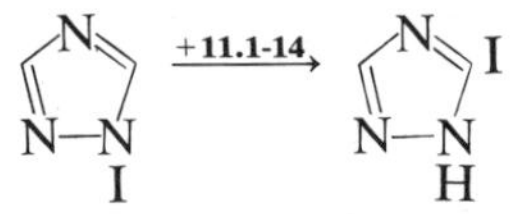

N
‖ ⟩   +11.1-14 →
N—N
I
11.1-30

N
‖ ⟩ I
N—N
H
11.1-31

Although the yields are usually low, *N*-alkyl- and aryl-1,2,4-triazoles (e.g., **11.1-32**) are converted to C-bromo derivatives (e.g., **11.1-33**) with *N*-bromosuccinimide in refluxing chloroform.[12,14] Similarly, 1,3,5-tribromotria-

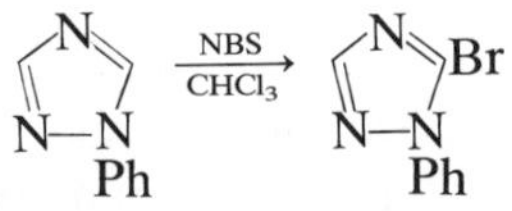

N
‖ ⟩   NBS/CHCl₃ →
N—N
Ph
11.1-32

N
‖ ⟩ Br
N—N
Ph
11.1-33

zole (**11.1-35**) has been used as a halogenating agent, as demonstrated by the conversion of a melt of **11.1-35** and 1-methyl-1,2,4-triazole (**11.1-34**) to **11.1-36**.[13,15]

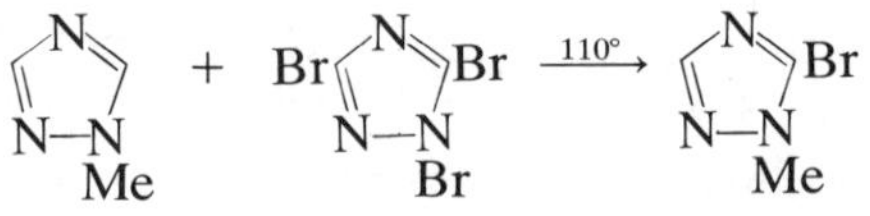

N
‖ ⟩   +
N—N
Me
11.1-34

N
Br‖ ⟩ Br   110° →
N—N
Br
11.1-35

N
‖ ⟩ Br
N—N
Me
11.1-36

Chlorodehydroxylation of triazoline-5-ones (e.g., **11.1-37**) to give chloro-triazoles (e.g., **11.1-38**) has been accomplished with phosphorus oxychloride at reflux,[12,16] and a mixture of this reagent either with *N*,*N*-dimethylform-amide[17] or phosphorus pentachloride[18–20] in a sealed tube at 150 to 160°.

Also, **11.1-38** has been prepared by the oxidative chlorination of the tria-
zoline-5-thione (**11.1-39**) in chloroform.[21] Hydrolysis of the chloro group of

11.1-38 to give **11.1-37** occurred readily in mineral acids but not in aqueous
sodium hydroxide.[22] In contrast, 5-bromo-1-phenyl-1,2,4-triazole (**11.1-33**)
was converted to **11.1-40** with refluxing base and to **11.1-41** with hydrazine.[14]

Kinetic studies showed that 5-bromo-1-methyl-1,2,4-triazole was more
reactive toward nucleophilic substitution with ethanolic piperidine than 3-
bromo-4-methyl-1,2,4-triazole and that both of these compounds were
appreciably less reactive that 5-bromo-1-methyltetrazole but more reactive
than either 2- or 5-bromo-1-methylimidazole.[23]

The nitrotriazoles possess an activated nitro group, which is readily
displaced by halogen acids to give the corresponding halotriazoles. The type
of reaction is illustrated by the conversion of **11.1-42** to **11.1-17** in refluxing
concentrated hydrochloric acid.[24] In the reaction of 1-methyl-3,5-dinitro-

1,2,4-triazole (**11.1-43**) with hydrochloric and hydrobromic acids, respectively, the nitro group adjacent to the *N*-methyl substitutent was displaced
to give good yields of **11.1-44** and **11.1-45**. Fluorotriazoles are also prepared
by this method, as shown by the conversion of **11.1-42** to **11.1-8** at 150° in
liquid hydrofluoric acid.[25]

Three isomers are possible in the alkylation of 3-halotriazoles, and the formation of each, regardless of the nature of the alkylating agent, has been observed.[12,26] The yield, however, of each isomer is dependent upon the nature of the alkylating agent. For example, treatment of **11.1-6** with methyliodide in methanolic sodium methoxide gave a 60% yield of product, which was composed of **11.1-46** (56%), **11.1-36** (34%), and **11.1-47** (10%). In the reaction of **11.1-6** with diazomethane in methanol, a 70% yield of product consisting of **11.1-46** (50%), **11.1-36** (45%), and **11.1-47** (5%) was obtained. In contrast, alkylation of **11.1-6** with methyl sulfate in methanol in the absence of base gave a 70% yield of product, which was composed of **11.1-46** (5%), **11.1-36** (10%), and **11.1-47** (85%).

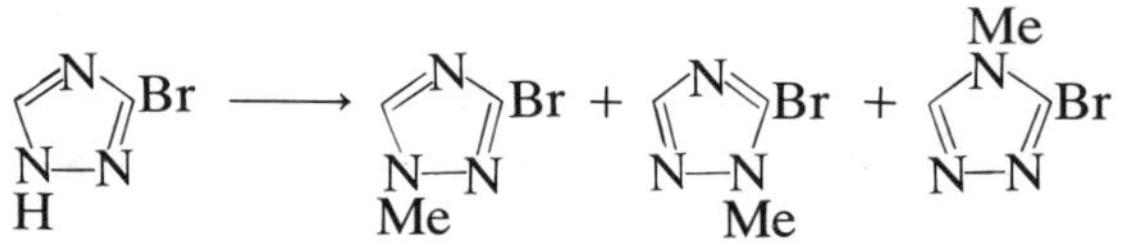

11.1-6   **11.1-46**   **11.1-36**   **11.1-47**

References

1. **66:** 10907m
2. **55:** 8393d
3. **57:** 7253h
4. **59:** 11474a
5. **84:** 180140m
6. **67:** 82167e
7. **78:** 16143v
8. **83:** 178919h
9. **72:** 121456y
10. **78:** 84327h
11. **67:** 64314w
12. **83:** 114298x
13. **71:** 91399s
14. **70:** 37729r
15. **72:** 3437v
16. **57:** 808f
17. **56:** 10133e
18. **57:** 12473b
19. **59:** 2803g
20. **81:** 49647c
21. **84:** 17238r
22. **57:** 12473b
23. **67:** 72992f
24. **74:** 99948c
25. **80:** 14898n
26. **84:** 31078v

TABLE 11   MONO-, DI-, AND TRIHALO-1,2,4-TRIAZOLES

Compound	Reference
-1-acetamide, N-[5-[[4-[2,4-bis(1,1-dimethylpropyl) phenoxy]-1-oxobutyl]amino]-2-chlorophenyl]-3-bromo-α-(2,2-dimethyl-1-oxopropyl)-	**85:** P102316e
-1-acetamide, N-[5-[[4-[2,4-bis(1,1-dimethylpropyl)- phenoxy]-1-oxobutyl]amino]-2-chlorophenyl]-3-chloro-α-(2,2-dimethyl-1-oxopropyl)-	**85:** P102316e
-1-acetamide, N-[5-[[4-[2,4-bis(1,1-dimethylpropyl)- phenoxy]-1-oxobutyl]amino]-2-chlorophenyl]-3-chloro-α-(2,2-dimethyl-1-oxopropyl)-5-methyl-	**85:** P102316e
-1-acetamide, N-[5-[[4-[2,4-bis(1,1-dimethylpropyl)- phenoxy]-1-oxobutyl]amino]-2-chlorophenyl]-3,5-dichloro-α-(2,2-dimethyl-1-oxopropyl)-	**85:** P102316e
-1-acetamide, 3-chloro-α-(2,2-dimethyl-1-oxopropyl)-N-(5-[([2-(hexadecycloxy)phenyl]amino)-sulfonyl]-2-methoxyphenyl)-	**85:** P102316e
3-(o-aminophenyl)-5-chloro-4-phenyl-	**59:** 2804c
4-(2-benzoyl-4-chlorophenyl)-3-bromo-5-(bromomethyl)-	**83:** 193188n
4-(2-benzoyl-4-chlorophenyl)-3-bromo-5-(chloromethyl)-	**83:** P79302v
4-(2-benzoyl-4-chlorophenyl)-3-bromo-5-[(formyloxy)methyl]-	**83:** 193188n
4-(2-benzoyl-4-chlorophenyl)-3-bromo-5-(hydroxymethyl)-	**83:** 193188n
4-(2-benzoyl-4-chlorophenyl)-3-bromo-	**82:** P156401k, **83:** 193188n
4-(2-benzoyl-4-methylphenyl)-3-chloro-5-methyl-	**83:** P43339y
3-benzyl-5-bromo-4-hydroxy-	**74:** 141651c
3-benzyl-5-bromo-, 4-oxide	**74:** 111182h
1-(1-[((1,1'-biphenyl)-4-yloxy]-3,3-dimethyl-2-oxo-1-butyl)-3-chloro-	**82:** P170967f
1-(1-[((1,1'-biphenyl)-4-yloxy]-3,3-dimethyl-2-oxo-1-butyl)-3,5-dibromo-	**82:** P170967f
1(or 4)-bromo	**70:** P11703c, **72:** 121456y
3-bromo-	**51:** 16159i, **67:** 82167e, **74:** 99948c
5-bromo-1-[1-(4-bromophenoxy)-3,3-dimethyl-2-oxo-1-butyl]-	**82:** P170967f
3-bromo-5-(p-bromophenyl)-	**67:** 82167e
3-bromo-1(or 4)-chloro-	**70:** P11703c, **71:** 91399s, **72:** 121456y
3-bromo-5-chloro-	**67:** 82167e, **72:** 121456y
3-bromo-1-[1-(4-chlorophenoxy)-3,3-dimethyl-2-oxo-1-butyl]-	**82:** P170967f
5-bromo-1-[1-(4-chlorophenoxy)-3,3-dimethyl-2-oxo-1-butyl]-	**82:** P170967f
3-bromo-1(or 4), 5-dichloro-	**70:** P11703, **72:** 1211456y
5-bromo-1-[1-(2,4-dichlorophenoxy)-3,3-dimethyl-2-oxo-1-butyl]-	**82:** P170967f
1-bromo-3,5-diethyl-	**67:** 82167e
1-bromo-3,5-dimethyl-	**67:** 82167e

TABLE 11 (*Continued*)

Compound	Reference
3-bromo-1,5-dimethyl-	**83:** 114298x
3-bromo-4,5-dimethyl-	**83:** 114298x
5-bromo-1,3-dimethyl-	**83:** 114298x, **84:** 17237q
5-bromo-1-[2,2-dimethyl-1-(4-nitrophenoxy)-2-oxo- 1-butyl]-	**82:** P170967f
5-bromo-1-[3,3-dimethyl-1-(2,4,5-trichlorophenoxy)- 2-oxo-1-butyl]-	**82:** P170967f
1-bromo-3,5-diphenyl-	**67:** 82167e
3-bromo-5-ethyl-	**67:** 82167e
3-bromo-1-ethyl-5-phenyl-	**55:** 8393i
3-bromo-4-ethyl-5-phenyl-	**55:** 8393i
5-bromo-1-ethyl-3-phenyl-	**55:** 8393i
1-bromo-3-fluoro-	**70:** P11703c
3-bromo-5-fluoro-	**80:** 14898n
5-bromo-1-[1-(4-fluorophenoxy)-3,3-dimethyl-2- oxo-1-butyl]-	**82:** P170967f
5-bromo-3-(hydroxymethyl)-	**70:** 114407p
3-bromo-4-hydroxy-5-phenyl-	**74:** 111967t, **74:** 141651c
3-bromo-5-iodo-	**67:** 82167e
3-bromo-5-methoxy-	**67:** 82167e
3-bromo-1-methyl-	**83:** 114298x
3-bromo-4-methyl-	**67:** 72992f, **83:** 114298x
3-bromo-5-methyl-	**67:** 82167e, **72:** 121456y
5-bromo-1-methyl-	**67:** 16500g, **67:** 72992f **83:** 146700p
3-bromo-5-methyl-1-phenyl-	**83:** 114298x
3-bromo-5-methyl-4-phenyl-	**83:** 114298x
5-bromo-3-methyl-1-phenyl-	**83:** 114298x
3-bromo-5-methyl-4-[2-(2-pyridinylcarbonyl)phenyl]-	**83:** 193188n
3-bromo-1-phenyl-	**83:** 114298x
3-bromo-5-phenyl-	**67:** 82167e, **83:** 114298x
5-bromo-1-phenyl-	**70:** 37729r, **83:** 114298x
3-(2-bromophenyl)-5-chloro-	**84:** 180140m
4-(*p*-bromophenyl)-3-chloro-5-propyl-	**59:** 11474a
3-bromo-5-phenyl-, 4-oxide	**74:** 111182h, **77:** 139906h
1-butyl-3-chloro-	**84:** P31078v
1-butyl-5-chloro-	**84:** P31078v
3-butyl-5-chloro-	**82:** 97518d
4-butyl-3-chloro-	**84:** P31078v
1(or 4)-chloro-	**70:** P11703c, **72:** 121456y
3-chloro-	**51:** 16159i, **67:** 64314w, **69:** P96734r, **70:** 77876t, **71:** P81374r, **72:** 121456y, **74:** 99948c, **78:** 84327h
5-chloro-1-[bis(dimethylamino)phosphinyl]-	**73:** P120632m
3-chloro-4-[3-(2-chlorobenzoyl)-5-methyl-2-thienyl]- 5-methyl-	**81:** P152238u
3-chloro-1-[1-(4-chlorophenoxy(-3,3-dimethyl-2-oxo- 1-butyl]-	**82:** P170967f
3-chloro-4-(*p*-chlorophenyl)-5-ethyl-	**59:** 11474a
3-chloro-4-(*p*-chlorophenyl)-5-propyl-	**59:** 11474a

TABLE 11 (*Continued*)

Compound	Reference
3-chloro-5-(-chloro-2-thienyl)-4-phenyl-	**83:** 163180m, **84:** 17238r
3-chloro-5-(2,2-dichloro-1,1-difluoroethyl)-	**81:** 49647c
3-chloro-5-(2,2-dichloro-1,1-difluoroethyl)-4-(triphenylmethyl)-	**81:** 49647c
3-chloro-1-[1-(2,4-dichlorophenoxy)-3,3-dimethyl-2-oxo-1-butyl]-	**82:** P170967f
3-chloro-1-[1-(2,6-dichlorophenyl)-2-nitroethyl]-	**81:** P91539u
1-chloro-3,5-diethyl-	**78:** 84327h
1-chloro-3,5-diheptadecyl-	**70:** P11703c
1-chloro-3,5-dimethyl-	**70:** P11703c, **78:** 84327h
3-chloro-1,5-dimethyl-	**83:** 114298x
3-chloro-4,5-dimethyl-	**83:** 114298x
5-chloro-1,3-dimethyl-	**83:** 114298x
3-chloro-1-[3,3-dimethyl-1-(4-nitrophenoxy)-2-oxo-1-butyl]-	**82:** P170967f
1-chloro-3,5-diphenyl-	**67:** 64314w, **78:** 84327h
3-chloro-4,5-diphenyl-	**57:** 12473b, **83:** 163180m, **84:** 17238r
5-chloro-1,3-diphenyl-	**66:** 10907m
1-chloro-3-ethyl-	**70:** P11703c
1-chloro-5-ethyl-	**78:** 84327h
3-chloro-5-ethyl-	**43:** 7019c, **78:** 84327h
1-chloro-5-ethyl-3-phenyl-	**78:** 84327h
3-chloro-5-isopropyl-	**82:** 97518d
1-chloro-3-methyl-	**70:** P11703c
1-chloro-5-methyl-	**78:** 84327h
3-chloro-1-methyl-	**83:** 114298x
3-chloro-4-methyl-	**83:** 114298x
3-chloro-5-methyl-	**67:** 64314w, **78:** 84327h
5-chloro-1-methyl-	**83:** 114298x
1-chloro-5-methyl-3-phenyl-	**78:** 84327h
3-chloro-4-methyl-5-phenyl-	**79:** 78749b
3-chloro-5-(2-methylphenyl)-	**84:** 180140m
3-chloro-5-(3-methylphenyl)-	**84:** 180140m
3-chloro-5-(4-methylphenyl)-	**84:** 180140m
5-chloro-3-methyl-1-phenyl-	**56:** 10134b
5-chloro-1-methyl-3-(4H-1,2,4-triazol-4-yl)-	**72:** 121448x
3-chloro-5-(5-nitro-2-furanyl)-	**79:** 87327q
3-chloro-5-(2-nitrophenyl)-	**84:** 180140m
3-chloro-5-(3-nitrophenyl)-	**84:** 180140m
3-chloro-5-(4-nitrophenyl)-	**84:** 180140m
3-chloro-4-phenyl-	**56:** 2440c, **57:** 809a
3-chloro-5-phenyl-	**57:** 7254a
1-chloro-3-phenyl-	**67:** 64314w, **70:** P11703c
3-chloro-1-phenyl-	**45:** 6212a, **75:** 110244p
3-chloro-5-phenyl-	**57:** 7254a, **74:** 141651c, **84:** 180140m
3-chloro-4-phenyl-5-propyl-	**59:** 11474a
3-chloro-4-phenyl-5-*p*-tolyl-	**51:** 5095c
5-chloro-1-phenyl-3-(4H-1,2,4-triazol-4-yl)-	**72:** 121448x
3-chloro-5-(4H-1,2,4-triazol-4-yl)-	**72:** 121448x

TABLE 11 (*Continued*)

Compound	Reference
3-chloro-1-trityl-	**74:** P100062t
3,5-dibenzyl-1-bromo-	**70:** P11703c
3,5-dibenzyl-1-chloro-	**70:** P11703c
3,5-dibenzyl-4-chloro-	**72:** 121456y
1,3-dibromo-	**63:** 4281g
3,5-dibromo-	**67:** 82167e, **69:** P96734r, **70:** P47462n, **71:** P81374r **72:** 121456y
3,5-dibromo-1-(or 4)-chloro-	**70:** P11703c, **71:** 91399s, **72:** 121456y
3,5-dibromo-1-[1-(2-,4-dichlorophenoxy)-3,3-dimethyl-2-oxo-1-butyl]-	**82:** P170967f
3,5-dibromo-1-[3,3-dimethyl-1-(4-nitrophenoxy)-2-oxo-1-butyl]-	**82:** P170967f
3,5-dibromo-1-fluoro-	**70:** P11703c
3,5-dibromo-4-iodo-	**72:** 121456y
3,5-dibromo-1-methyl-	**71:** 91399s, **83:** 114298x
3,5-dibromo-4-methyl-	**83:** 114298x
3,5-dibromo-4-phenyl-	**83:** 114298x
1(or 4),3-dichloro-	**67:** 64314w, **69:** P96734r, **70:** P47462n, **71:** P81374r, **72:** 3437v
3,4-dichloro-	**69:** P96734r, **70:** P47462n, **71:** P81374r
3,5-dichloro-	**28:** 2714[5], **69:** P96734r, **70:** P47462n, **71:** P81374r
3,5-dichloro-1-(*o*-chlorophenyl)-	**65:** P8926d
3,5-dichloro-1-(*p*-chlorophenyl)-	**66:** 10907m
1,3-dichloro-5-ethyl	**78:** 84327h
3,5-dichloro-1-methyl-	**66:** 10907m
1,3-dichloro-5-methyl-	**67:** 64314w, **78:** 84327h
3,5-dichloro-1-(*m*-nitrophenyl)-	**65:** P8926d
3,5-dichloro-1-phenyl-	**66:** 10907m
3,5-dichloro-1-(tetrahydro-3-theinyl)-, *S*,*S*-dioxide	**66:** 10907m
3,5-diiodo-	**73:** 66513h
3,5-diiodo-1-methyl-	**83:** 28165v
3-fluoro-	**80:** 14898n, **83:** 178919h
1(or 4)-iodo-	**70:** P11703c, **71:** 91399s, **78:** 16143v
3-iodo-	**64:** 4919f, **78:** 16143v
4-iodo-	**72:** 121456y
1-iodo-5-methyl-	**71:** 91399s
3-iodo-1-methyl-	**83:** 28165v
3-iodo-5-methyl-	**71:** 91399s, **83:** 114298x
5-iodo-1-methyl-	**83:** 28165v
1,3,5-tribromo-	**67:** 82167e, **70:** P11703c
3,4,5-tribromo-	**69:** P96734r, **70:** P47462n, **71:** P81374r **72:** 121456y
1,3,5-trichloro-	**67:** 64314w

CHAPTER 12

1,2,4-Triazoles Containing More Than One Representative Function

12.1. Mono- and Diamino- 1,2,4-Triazolecarboxylic Acids and Their Functional Derivatives

A high yield (80%) of 3-amino-1,2,4-triazole-5-carboxylic acid (**12.1-7**) was obtained by slowly adding **12.1-1** to a refluxing aqueous solution of oxalic acid to give **12.1-3**, which was cyclized in an alkaline medium.[1,2] Similarly, the reaction of *N*-amino-*N*-methylguanidine (**12.1-2**) with oxalic acid gave **12.1-4**, which was converted to 1-methyl-5-amino-1,2,4-triazole-3-carboxylic acid (**12.1-8**) in 57% yield.[1] Esterification of **12.1-7** with methanolic hydrogen chloride gave the methyl ester (**12.1-5**), and hydrazinolysis of the latter gave the acid hydrazide (**12.1-6**),[1] Glycosylation of the trimethylsilyl derivative of (**12.1-5**) with 2,3,5-tri-*O*-benzoyl-D-ribofuranosyl bromide followed by deblocking with ethanolic ammonia gave a good yield of the 1-substituted 3-aminotriazole (**12.1-9**).[3] The isomeric 5-aminotriazole (**12.1-10**) was prepared in excellent yield by amination of the 5-chlorotriazole (**12.1-11**) with liquid ammonia in a pressure vessel at 100°.

The 5-hydrazino-1,2,4-triazole-3-carboxylic acid derivative (**12.1-15**) was formed from phenylcyanamide (**12.1-12**) and **12.1-13** in refluxing benzene containing triethylamine.[4] The initial product (**12.1-14**) reacted further with **12.1-13** to give **12.1-17**, which underwent rearrangement to **12.1-15** via the spiran (**12.1-16**).

The reaction of 3-amino-1,2,4-triazole (**12.1-22**) with potassium cyanate in dilute hydrochloric acid and with phenyl isothiocyanate in 95% ethanol has been reported to give the ring-substituted derivatives **12.1-20** and **12.1-21**, respectively, and minor amounts of the ureas and thioureas resulting from substitution on the amino group.[5] However, in the reaction of 3-aminotriazole with potassium cyanate, the structure of the product was shown to be **12.1-18**, based on chemical and spectroscopic evidence.[6] In addition, Driscoll used proton magnetic resonance spectra to assign the structure of products from this type of reaction and found that the site of substitution was dependent upon both the nature of the reactants and the

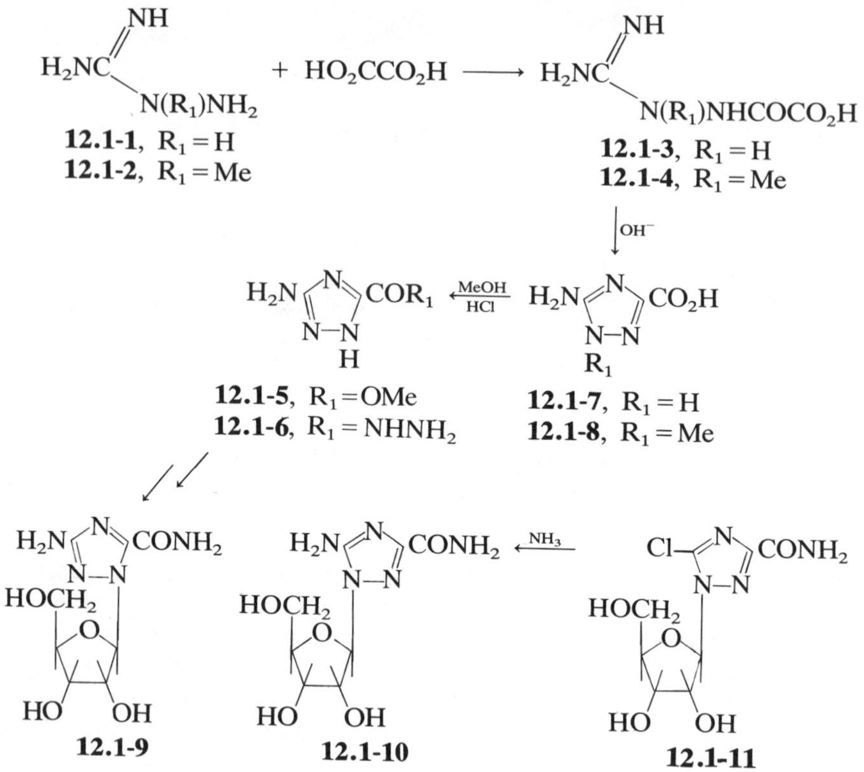

12.1-1, $R_1 = H$
12.1-2, $R_1 = Me$

12.1-3, $R_1 = H$
12.1-4, $R_1 = Me$

12.1-5, $R_1 = OMe$
12.1-6, $R_1 = NHNH_2$

12.1-7, $R_1 = H$
12.1-8, $R_1 = Me$

12.1-9

12.1-10

12.1-11

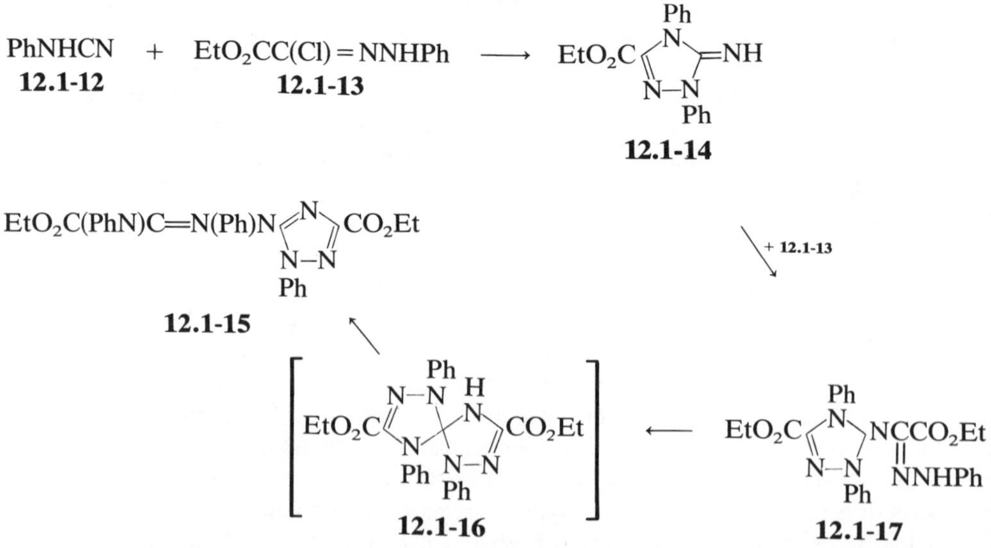

PhNHCN + $EtO_2CC(Cl)=NNHPh$ $\longrightarrow$ EtO_2C
12.1-12   **12.1-13**

12.1-14

$EtO_2C(PhN)C=N(Ph)N$ triazole CO_2Et

12.1-15

+ **12.1-13**

12.1-16

EtO_2C triazole $NCCO_2Et$

12.1-17

conditions.[7] In contrast to a patent claim,[5] reaction of **12.1-22** with methyl isothiocyanate in dimethylformamide gave a 64% yield of a 2:1 mixture of **12.1-23** and **12.1-24**. In refluxing pyridine these reactants gave the 3-(thioureido)triazole (**12.1-25**), whereas under the same conditions **12.1-22** and methyl isocyanate gave 1-(methylcarbamoyl)-5-aminotriazole (**12.1-26**). Treatment of **12.1-22** with ethyl isothiocyanatoformate in a mixture of dimethylformamide and acetone gave **12.1-27** in 27% yield at room temperature and the 5-(thioureido)triazole (**12.1-28**) in 72% yield at reflux. In earlier work the structure of compound **12.1-28** was misassigned,[8] no doubt because the rearrangement of **12.1-27** to **12.1-28** in dimethylformamide was rapid at room temperature.[9] Both the 3-methyl and 3-phenyl derivatives of **12.1-27** have been prepared.[10] In addition, the product from **12.1-22** and phenylisothiocyanate in ethanol at room temperature was assigned the structure of **12.1-19** in later work.[11]

The reaction of **12.1-22** with ethyl chloroformate in pyridine at room temperature gave the N-1-substituted 5-aminotriazole **12.1-29**, whereas at $-8°$ in dimethylformamide, the product was a mixture of **12.1-29** and **12.1-30**.[9] Both **12.1-29** and **12.1-30** were rearranged in refluxing pyridine to give the 5-ureidotriazole (**12.1-31**) with **12.1-29** being an intermediate in the conversion of **12.1-30**. In the reaction of **12.1-22** and its 3-alkyl and 3-aryl derivatives with 1-guanyl-1,2,3-benzotriazole hydrochloride in hot ethanol, the 1-guanyl-5-aminotriazoles (e.g., **12.1-32**) were obtained in 36 to 62% yield.[12] The product from the interaction of phenyl isothiocyanate with 3,5-diamino-1,2,4-triazole (**12.1-33**) was assigned to the structure resulting from addition to the 4-position to give **12.1-34**.[13] In contrast, acylation of 5-amino-3-(dimethylamino)-1,2,4-triazole with a carbamoyl chloride was reported to give the corresponding 1-substituted triazole.[14]

12.1-22   12.1-29   12.1-30   12.1-31

12.1-32   12.1-33   12.1-34

Acylation of 3-aminotriazole (**12.1-22**) with dimethylcarbamoyl chloride in pyridine gave a product that has been assigned both to structure **12.1-35**[5] and **12.1-36**.[15] However, acylation of the sodium salt of **12.1-22** with the carbamoyl chloride in tetrahydrofuran was shown to give a mixture of **12.1-35** and **12.1-36**, from which **12.1-36** was obtained pure by column chromatography.[16] A pure sample of **12.1-35** was obtained by alkylation of the 4-methoxybenzylidene derivative of **12.1-22** followed by removal of the amino-blocking group with 2,4-dinitrophenylhydrazine. Also, alkylation of the 3-(acylamino) derivatives of **12.1-22** with dimethylcarbamoyl chloride gave 1-substituted 3-(acylamino)-1,2,4-triazoles.[16–18]

12.1-22   12.1-35   12.1-36

References

1. **63:** 13243f   2. **67:** 64306v   3. **80:** 121258a   4. **63:** 7001c
5. **59:** 8759b   6. **63:** 1786f   7. **76:** 85792b   8. **69:** 67388a
9. **79:** 42421z   10. **78:** 136237a   11. **70:** 77881r   12. **73:** 87897b
13. **81:** 3897f   14. **80:** 95962d   15. **82:** 156320h   16. **64:** 8196c
17. **65:** 725f   18. **74:** 3636a

12.2. S-Substituted Thio, Sulfinyl-, Sulfonyl-, and Acylthio-1,2,4-Triazolecarboxylic Acids and Their Functional Derivatives

The direct synthesis of ethyl 5-(benzylthio)-1-phenyl-1,2,4-triazole-3-carboxylate (**12.2-2**) was accomplished by the rearrangement of 4-(phenylazo)-2-benzylthio-2-thiazolin-5-one (**12.2-1**) in refluxing ethanolic sulfuric acid.[1] Similarly, treatment of **12.2-1** with aryl amines (and phenyl-hydrazine) such as aniline gave 5-(benzylthio)-1,2,4-triazole-3-carboxamide derivatives (e.g., **12.2-3**). In contrast, a 1,2,4-triazole-5-thione was obtained when **12.2-1** was treated with ethanolic potassium hydroxide. Oxidation of **12.2-3** with hydrogen peroxide in acetic acid occurred readily at room temperature to give the 5-(benzylsulfonyl)triazole (**12.2-4**). The 4-substituted

3-(methylthio)-1,2,4-triazole-5-carboxamide (**12.2-6**) was obtained in 52% yield by the cyclization of **12.2-5** in refluxing sodium ethoxide in ethanol.[2]

The acylation of 3-(methylthio)-1,2,4-triazole (**12.2-8**) with *N,N*-dimethylcarbamoyl chloride in tetrahydrofuran in the presence of triethylamine was reported to give *N,N*-dimethyl-3-(methylthio)-1,2,4-triazole-1-carboxamide (**12.2-7**).[3] In the acylation of **12.2-9** and related compounds, the position of the entering *N,N*-dimethylcarbamoyl moiety was uncertain. In contrast, acylation of 3-(methylthio)-1,2,4-triazoles containing bulky 5-substituents (e.g., **12.2-10**) was claimed to give 5-(methylthio)-1,2,4-triazole-1-carboxamides (e.g., **12.2-11**).

12.2-7	**12.2-8**, $R_1 = H$	**12.2-11**
	12.2-9, $R_1 = Et$	
	12.2-10, $R_1 = 3,4,5\text{-}(MeO)_3C_6H_2$	

The diacylation of triazoline-5-thione (**12.2-12**) with dimethylcarbamoyl chloride in pyridine gave the triazole (**12.2-13**).[4,5] A similar substitution pattern was also observed in the reaction of **12.2-14** with dipropylcarbamoyl chloride to give the 3-(propylsulfonyl)triazole-1-carboxamide (**12.2-15**).[6,7]

| **12.2-12** | **12.2-13** | **12.2-14** | **12.2-15** |

References

1. **75**: 98505q   2. **73**: 66517n   3. **67**: 54140x   4. **77**: 34529j
5. **81**: 164763z   6. **76**: 72528v   7. **79**: 92235n

12.3. Miscellaneous 1,2,4-Triazolecarboxylic Acids and Their Functional Derivatives

The rearrangement of 4-(arylazo)-2-alkoxy-2-thiazolin-5-ones in the presence of amines can provide one of three types of 1,2,4-triazoles, depending upon the nature of both the amine and the 2-alkoxy group of the thiazoline. Treatment of the 2-ethoxythiazoline (**12.3-1**) with an aliphatic amine (e.g., $PhCH_2NH_2$) gave the triazolin-5-one (**12.3-9**), whereas the 2-benzyloxythiazoline (**12.3-2**) with the same amine gave the triazoline-5-thione (**12.3-8**)[1,2] (see Sections 15.4 and 16.4). To account for the formation of

12.3-9 and an alkyl mercaptan, intermediate **12.3-5** was proposed to result from **12.3-4** by an O to S migration of the alkyl group. In contrast, the reaction of 2-alkoxythiazolines (e.g., **12.3-1** and **12.3-3**) with aryl amines (e.g., aniline) gave 5-alkoxy-1,2,4-triazole-3-carboxamide derivatives (e.g., **12.3-6** and **12.3-7**).[2,3] This reaction is similar to the transformation of the corresponding 2-(alkylthio)thiazolines with aryl amines to 5-(alkylthio)-triazoles (see Section 12.2). Both phenyl hydrazine and glycine were also found to convert 2-alkoxythiazolines to 5-alkoxytriazoles.[1,2] Other workers found that O-acyl (e.g., **12.3-10**) rather than N-acyl triazole derivatives were prepared by acylation of 2-alkyl- and aryl-1,2,4-triazolin-5-ones with methanesulfonyl chloride.[4]

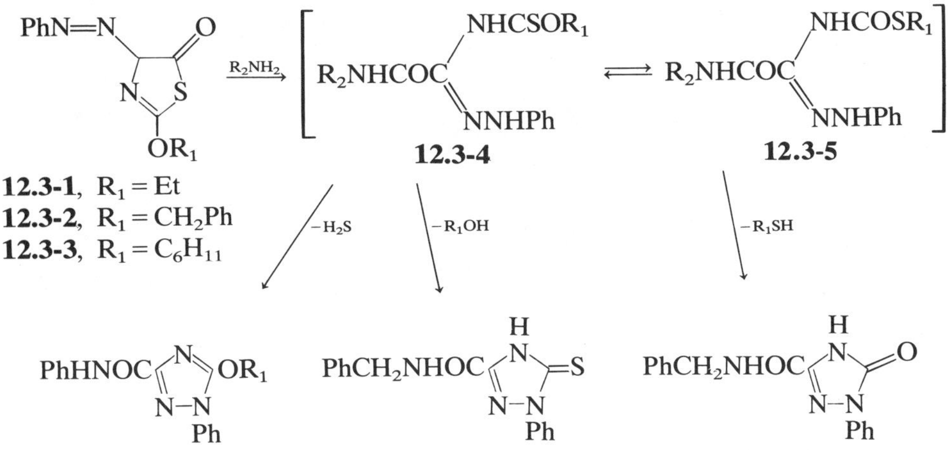

12.3-1, R_1 = Et
12.3-2, R_1 = CH_2Ph
12.3-3, R_1 = C_6H_{11}

12.3-4

12.3-5

12.3-6, R_1 = Et
12.3-7, R_1 = C_6H_{11}

12.3-8

12.3-9

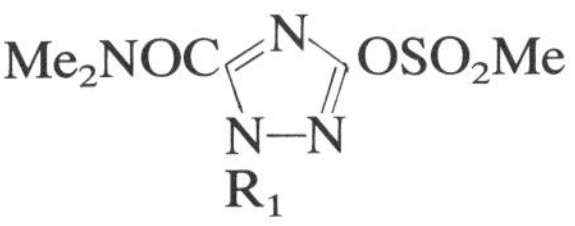

12.3-10 (R_1 = alkyl or aryl)

Diazotization of 3-amino-1,2,4-triazole-5-carboxylic acid (**12.3-11**) gave the diazotriazole (**12.3-12**) which underwent both addition and displacement reactions. The coupling of **12.3-12** with 3-diethylaminoacrolein was accompanied by hydrolysis of the diethylamine group to give the azomalonaldehyde derivative (**12.3-13**).[5] Also, **12.3-12** coupled with N,N-dimethylaniline to give the azotriazole (**12.3-14**),[6-8] and condensed with di- and trinitroalkanes to give **12.3-15**.[9] Catalytic hydrogenation of **12.3-16** with Raney nickel in N,N-dimethylformamide resulted in the reduction of the nitro group but not the azo function.[10]

The intermediate 3-diazotriazole (**12.3-12**) was treated with a mixture of aqueous copper (II) nitrate and sodium nitrate to give 3-nitro-1,2,4-triazole-5-carboxylic acid (**12.3-17**).[11] This compound was also prepared directly from **12.3-11** by nitrosation of the latter in a mixture of acetic and sulfuric acids.[12] Alkylation of the methyl ester of **12.3-17** with dimethyl sulfate gave

a 51% yield of the 1-methyl-3-nitrotriazole (**12.3-18**).[13] Similarly, the fusion of the methyl ester of **12.3-17** with 1,2,3,5-tetra-O-acetyl-β-D-ribofuranose gave a 77% yield of a nucleoside, which was identified as the 3-nitro-1-ribofuranosyl-1,2,4-triazole (**12.3-19**).[14]

The versatile diazotriazole intermediate (**12.3-12**) was also reacted with concentrated hydrochloric acid to give a mixture of the 3-chlorotriazoles (**12.3-20** and **12.3-21**), from which **12.3-20** was separated via its potassium salt.[11] The same products were obtained in the reaction of the 3-nitro-triazole (**12.3-17**) with a mixture of phosphorus oxychloride and phosphorus

pentachloride. In a similar reaction, the methyl ester of **12.3-17** was treated with liquid hydrofluoric acid at 100° to give a 70% yield of the 3-fluoro-triazole (**12.3-22**).[15] Fusion of the methyl ester of the 3-chlorotriazole (**12.3-20**) with 1,2,3,5-tetra-*O*-acetyl-β-D-ribofuranose in the presence of an acid catalyst gave a mixture of the 1- and 2-ribofuranosyl-1,2,4-triazole-5-carboxylic acid derivatives (**12.3-23** and **12.3-24**).[14] The mixture was

separated by silica gel chromatography to give a 50% yield of **12.3-23** and a 36% yield of **12.3-24**. Other halotriazolecarboxylic acid were prepared from preformed triazoles by *N*-acylation or *N*-halogenation. Treatment of the 3-chlorotriazole (**12.3-25**) with ethyl chloroformate was claimed to give the triazole-4-carboxate (**12.3-26**),[16] whereas the 3-chlorotriazoles (**12.3-21**

and **12.3-27**) were reported to react with dimethylcarbamoyl chloride to give **12.3-28** and **12.3-29**.[17] In contrast, acylation of the 5-amino-3-(methyl-thio)- and 3-(methylsulfonyl)-1,2,4-triazoles (**12.3-30** and **12.3-31**) with dimethylcarbamoyl chloride gave the 5-amino-1,2,4-triazole-1-carbox-amides (**12.3-32** and **12.3-33**).[18] The *N*-bromotriazolecarboxylic acid

$$R_1 \underset{\underset{H}{N-N}}{\overset{N}{\diagdown}} Cl \quad \xrightarrow[\text{Et}_3\text{N}]{\text{Me}_2\text{NCOCl}} \quad R_1 \underset{\underset{\text{Me}_2\text{NCO}}{N-N}}{\overset{N}{\diagdown}} Cl$$

12.3-21, $R_1 = H$ **12.3-28**, $R_1 = H$
12.3-27, $R_1 = Cl$ **12.3-29**, $R_1 = Cl$

(**12.3-34**) was assigned the structure of the product resulting from the bromination of 1,2,4-triazole-5-carboxylic acid.[19]

$$R_1 \underset{\underset{H}{N-N}}{\overset{N}{\diagdown}} NH_2 \quad \xrightarrow{\text{Me}_2\text{NCOCl}} \quad R_1 \underset{\underset{\text{CONMe}_2}{N-N}}{\overset{N}{\diagdown}} NH_2 \qquad \underset{\underset{Br}{N-N}}{\overset{N}{\diagdown}} CO_2H$$

12.3-30, R = MeS   **12.3-32**, $R_1 = MeS$ **12.3-34**
12.3-31, R = MeSO$_2$ **12.3-33**, $R_1 = MeSO_2$

Oxidation of 3,5-dimethyl-1,2,4-triazole (**12.3-35**) with potassium permanganate in 1% potassium hydroxide at 70° gave an 85% yield of triazole-3,5-dicarboxylic acid (**12.3-36**), isolated as the mono-potassium salt.[20] This dicarboxylic acid was also prepared by oxidation of the corresponding 4-aminotriazole (**12.3-37**).[21] Similarly, 1,3,5- and 3,4,5-trialkyl-1,2,4-triazoles, respectively, were converted to 1- and 4-alkyl derivatives of **12.3-36**.[22] The reaction of the diazotriazole (**12.3-12**) with copper (I)

$$Me \underset{\underset{H}{N-N}}{\overset{N}{\diagdown}} Me \xrightarrow{\text{KMnO}_4} HO_2C \underset{\underset{H}{N-N}}{\overset{N}{\diagdown}} CO_2H \xleftarrow{\text{KMnO}_4} Me \underset{\underset{N-N}{}}{\overset{N}{\diagdown}} Me$$

                                                NH$_2$

12.3-35   **12.3-36**   **12.3-37**

cyanide gave the 3-cyanotriazole (**12.3-38**), which was readily decarboxylated when heated above its melting point to give 3-cyano-1,2,4-triazole.[20] In addition, **12.3-38** was converted to the monocarboxamide (**12.3-39**) in hydrochloric acid and to the dimethyl ester (**12.3-40**) in methanolic hydrogen chloride.

$$HO_2C \underset{\underset{H}{N-N}}{\overset{N}{\diagdown}} N_2^+ \xrightarrow{\text{CuCN}} HO_2C \underset{\underset{H}{N-N}}{\overset{N}{\diagdown}} CN \xrightarrow{\text{HCl}} R_1O_2C \underset{\underset{H}{N-N}}{\overset{N}{\diagdown}} COR_2$$

12.3-12   **12.3-38**   **12.3-39**, $R_1 = H$, $R_2 = NH_2$
                                                        12.3-40, $R_1 = Me$, $R_2 = OMe$

The complex 1,2,4-triazole-1-carboxylate (**12.3-45**) was prepared via intermediates **12.3-43** and **12.3-46** by the addition of two equivalents of the aryl cyanate (**12.3-42**) to the quasi 1,3-dipole (**12.3-41**).[23] Acid hydrolysis of **12.3-45** gave the aminotriazole (**12.3-44**), which was identified by chemical synthesis.

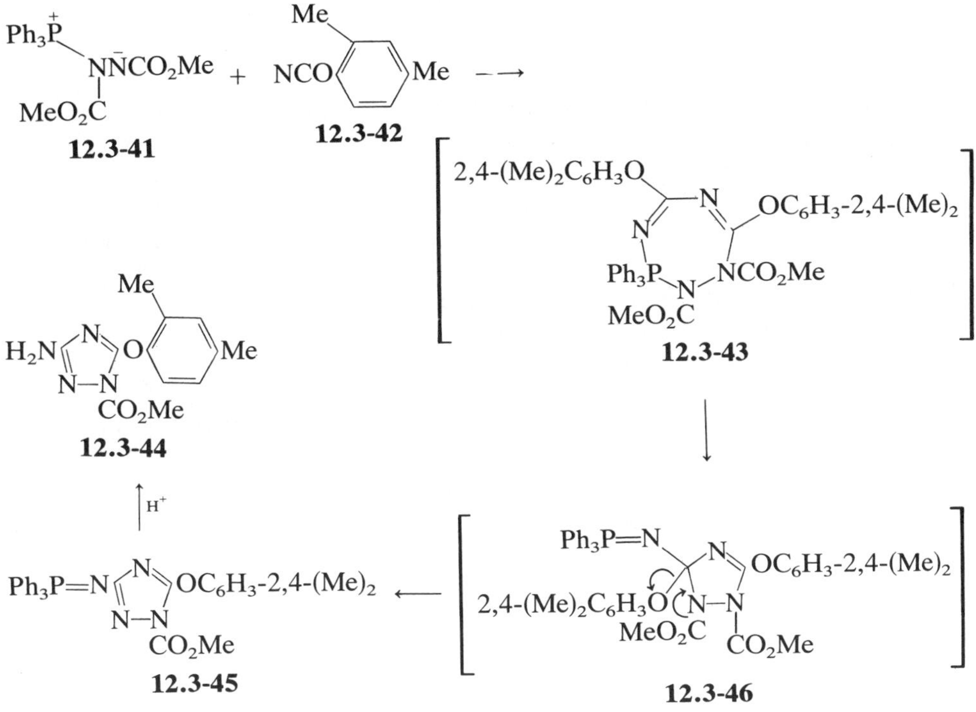

12.3-41

12.3-42

12.3-43

12.3-44

12.3-45

12.3-46

References

1. **76:** 59533y
2. **78:** 136152u
3. **75:** 98505q
4. **82:** 156321j
5. **60:** 9383e
6. **55:** 11862h
7. **57:** 6072h
8. **72:** 56717t
9. **77:** 126523f
10. **61:** 14829a
11. **70:** 77876t
12. **72:** 111383h
13. **72:** 111384j
14. **80:** 121258a
15. **80:** 14898n
16. **81:** 49647c
17. **67:** 54140x
18. **80:** 95962d
19. **71:** 91399s
20. **72:** 111375g
21. **76:** 3766j
22. **73:** 120638t
23. **61:** 61474v

12.4. Miscellaneous Acyl-1,2,4-Triazoles

Acylation of 1,2,4-triazoles containing a 3-amino function occurred readily. For example, acetylation of 3-amino-1,2,4-triazole (**12.4-1**) with acetyl chloride either in tetrahydrofuran at room temperature or in acetonitrile in the presence of 2,4,6-trimethylpyridine at 70° gave a product in which the

$$R_1 \overbrace{\underset{N-N}{\underset{H}{\parallel}}}^{N} NH_2$$

12.4-1, $R_1 = H$
12.4-2, $R_1 = Me$
12.4-3, $R_1 = Ph$

$$\overbrace{\underset{N-N}{\underset{R_1CO}{\parallel}}}^{N} NH_2$$

12.4-4, $R_1 = Me$
12.4-5, $R_1 = Ph$

acetyl group was substituted on a ring nitrogen rather than on the amino group.[1,2] The structure of the product has been assigned both to 1-acetyl-3-amino-1,2,4-triazole (**12.4-4**)[1] and to 4-acetyl-3-amino-1,2,4-triazole (**12.4-6**).[3] Another isomer, 1-acetyl-5-amino-1,2,4-triazole (**12.4-8**), is also possible, and this compound was identified as the correct structure of the product by ^{1}H-NMR (nuclear magnetic resonance) spectroscopy.[4,5] Acetylation at this position was attributed to the higher election density at the N of the N—N linkage adjacent to the amino group. Similarly, in the benzoylation of **12.4-1** with benzoyl chloride in tetrahydrofuran, the product was considered to be 3-amino-1-benzoyl- (**12.4-5**), 3-amino-4-benzoyl- (**12.4-7**),[3,6] or 5-amino-1-benzoyl-1,2,4-triazole (**12.4-9**).[7] The determination of the dipole moments of a series of triazoles indicated that the structure of this benzoylated product was **12.4-9**.[7] These results also suggested that benzoylation of **12.4-2** and **12.4-3** gave **12.4-10** and **12.4-11**, respectively.

$$R_1CO \overbrace{\underset{N-N}{\parallel}}^{N} NH_2$$

12.4-6, $R_1 = Me$
12.4-7, $R_1 = Ph$

$$R_1 \overbrace{\underset{N-N}{\underset{COR_2}{\parallel}}}^{N} NH_2$$

12.4-8, $R_1 = H$, $R_2 = Me$
12.4-9, $R_1 = H$, $R_2 = Ph$
12.4-10, $R_1 = Me$, $R_2 = Ph$
12.4-11, $R_1 = R_2 = Ph$

Acetylation of 3,5-diamino-1,2,4-triazole (**12.4-12**) with acetyl chloride under mild conditions also gave a ring-acetylated product, most likely **12.4-13**.[2] Migration of the acyl function from a ring nitrogen to an amino group is a characteristic reaction of ring-acetylated aminotriazoles, as shown by melting **12.4-13** and **12.4-8** to give **12.4-14**[1,2] and **12.4-15**,[2] respectively. In contrast to the site of substitution in **12.4-8**, further acetylation of either **12.4-8** or **12.4-15** gave the 1-acetyl-3-acetamidotriazole (**12.4-16**).[4] In this conversion the acetyl group of **12.4-8** migrated to the adjacent ring N to form the thermodynamically more stable product, **12.4-16**.

$$H_2N\text{-triazole-}NH_2 \xrightarrow{AcCl} H_2N\text{-triazole-}NH_2 \quad \text{triazole-}NH_2$$

12.4-12 (N–N, H) **12.4-13** (N–N, Ac) **12.4-8** (N–N, Ac)

$$R_1\text{-triazole-}NHAc \xrightarrow{AcCl} \text{triazole-}NHAc$$

12.4-14, $R_1 = NH_2$ **12.4-16**
12.4-15, $R_1 = H$

The simultaneous introduction of more than one acyl group can be accomplished readily as demonstrated by heating a mixture of **12.4-18** and benzoyl anhydride to give a product assigned structure **12.4-17**,[8] and refluxing a solution of **12.4-12** and acetic anhydride to give **12.4-19**.[2] Of

$$Ph\text{-triazole-}NHCOPh \xleftarrow{(PhCO)_2O} R_1\text{-triazole-}NH_2 \xrightarrow{Ac_2O} AcHN\text{-triazole-}NHAc$$

(COPh)   (H)   (Ac)
12.4-17   **12.4-18**, $R_1 = Ph$ **12.4-19**
       12.4-12, $R_1 = NH_2$

interest, **12.4-8**, **12.4-15**, and **12.4-16** reacted with trifluoroacetic anhydride to give **12.4-20**, which was identified by hydrolysis of the acetyl group to give **12.4-21**.[4] The latter was acetylated with acetic anhydride to give **12.4-20**, but was recovered unchanged after treatment with trifluoroacetic anhydride.

$$\text{triazole-}NHCOCF_3 \underset{Ac_2O}{\overset{H_2O}{\rightleftharpoons}} \text{triazole-}NHCOCF_3$$

(Ac)   (H)
12.4-20   **12.4-21**

Benzoylation of the 3-hydrazinotriazoles (**12.4-22** and **12.4-23**) with benzoyl chloride in pyridine gave the 1-benzoyl-5-hydrazinotriazoles (**12.4-24** and **12.4-25**, respectively).[9] The structural assignments were based on the ^{1}H-NMR spectra of the products. The ring benzoyl group of these compounds migrated to the hydrazino group in refluxing pyridine containing a catalytic amount of silver nitrate.[10]

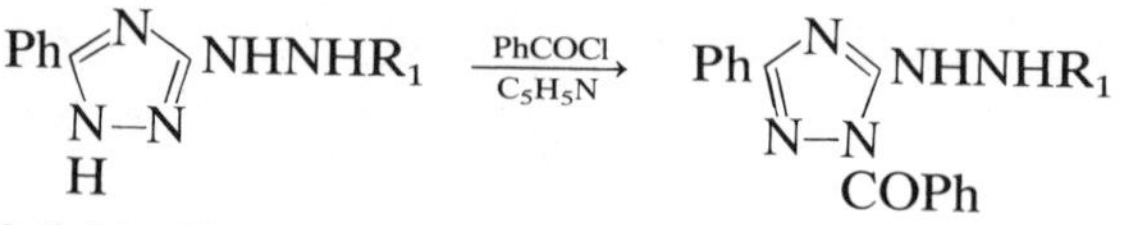

$$\text{Ph}\underset{\underset{\text{H}}{N-N}}{\overset{N}{\diagdown}}\text{NHNHR}_1 \xrightarrow[\text{C}_5\text{H}_5\text{N}]{\text{PhCOCl}} \text{Ph}\underset{\underset{\text{COPh}}{N-N}}{\overset{N}{\diagdown}}\text{NHNHR}_1$$

12.4-22, R$_1$ = Ph   **12.4-24**, R$_1$ = Ph
12.4-23, R$_1$ = PhCO   **12.4-25**, R$_1$ = PhCO

Refluxing the triazolin-5-one (**12.4-26**) in acetic anhydride was reported to result in both *N*- and *O*-acetylation to give 1-acetyl-3-acetoxy-1,2,4-triazole-5-carboxylic acid (**12.4-27**).[11] In other work a good yield of the 5-

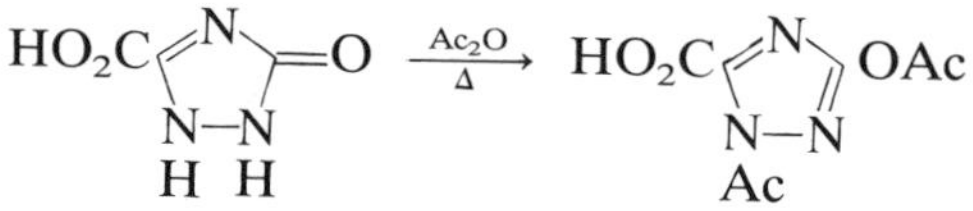

$$\text{HO}_2\text{C}\underset{\underset{\text{H H}}{N-N}}{\overset{N}{\diagdown}}=\text{O} \xrightarrow[\Delta]{\text{Ac}_2\text{O}} \text{HO}_2\text{C}\underset{\underset{\text{Ac}}{N-N}}{\overset{N}{\diagdown}}\text{OAc}$$

12.4-26   **12.4-27**

butyryltriazole (**12.4-29**) was obtained by cleavage of the pyrazine ring of the bicyclic triazole (**12.4-28**) under acidic conditions.[12] The keto triazole (**12.4-31**) was prepared by dehydration of **12.4-30** in refluxing acetic acid.[13]

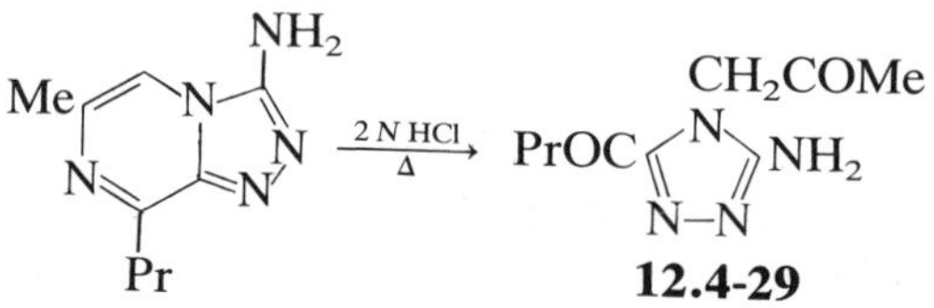

$$\xrightarrow[\Delta]{\text{2 N HCl}} \text{PrOC}\underset{\underset{}{N-N}}{\overset{N}{\diagdown}}\text{NH}_2$$

CH$_2$COMe

12.4-28   **12.4-29**

$$\text{MeCOC}\begin{matrix}\text{NHNHC}_6\text{H}_4\text{-4-NO}_2\\ \| \\ \text{NNHCOMe}\end{matrix} \xrightarrow[\Delta]{\text{HOAc}} \text{MeOC}\underset{N-N}{\overset{N}{\diagdown}}\text{Me}$$

NHC$_6$H$_4$-4-NO$_2$

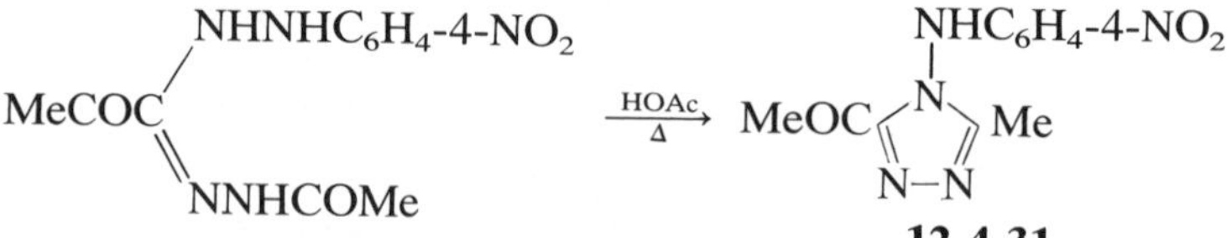

12.4-30   **12.4-31**

References

1. **53:** 21901h   2. **55:** 3565g   3. **59:** 6227e   4. **74:** 22770f
5. **77:** 101474c   6. **59:** 12791e   7. **66:** 104601a   8. **58:** 519b
9. **82:** 4183c   10. **82:** 125326k   11. **65:** 7169g   12. **76:** 58494z
13. **83:** 9917a

12.5. Amino-*S*-Substituted Thio-, Sulfinyl-, or Sulfonyl-1,2,4-Triazoles

Cyclization of the *S*-benzyl derivative of 2,5-dithiobiurea (**12.5-1**) with aqueous sodium hydroxide resulted in the elimination of one of the benzyl-thio group to give 3-amino-5-(benzylthio)-1,2,4-triazole (**12.5-2**).[1] In a

related conversion, the 2,5-dithiobiurea (**12.5-3**) was treated with methyl iodide in 10% aqueous sodium hydroxide to give **12.5-4** directly.[2] The

formation of the *S,S*-dimethyl derivative of **12.5-3** before ring closure was indicated by the recovery of **12.5-3** from aqueous base. Similarly, methyla-tion of the thiocarbohydrazide derivative (**12.5-5**) in the presence of base at room temperature gave a good yield of **12.5-7** formed via the *S,S',S"*-trimethyl compound (**12.5-6**).[3] The structure of **12.5-7** was confirmed by its synthesis from the 3-hydrazinotriazoline-5-thione (**12.5-8**).[4]

Diphenylcarbodiimide was found to add to the 4-position of the S-benzyl derivative of isopropylidene-3-thiosemicarbazone (**12.5-9**) to give **12.5-10**, which was cyclized in mineral acids by hydrolysis of the isopropylidene group and the elimination of aniline to give **12.5-11**.[5] Different modes of

12.5-9   **12.5-10**

12.5-11

cyclization were observed for the 1-amidino S-benzyl compound (**12.5-14**). In the reaction of **12.5-14** with refluxing aqueous sodium hydroxide, benzyl mercaptan was eliminated to give **12.5-12** (78%), whereas in refluxing aniline both **12.5-12** (35%) and 3-amino-5-(benzylthio)-4-phenyl-1,2,4-triazole (**12.5-13**), (32%) were obtained.[6] When the p-toluenesulfonate salt of **12.5-14** was refluxed in aniline, S-debenzylation occurred to give the triazoline-5-thione (**12.5-15**). The condensation of the S-methyl derivative

12.5-12   **12.5-13**

12.5-14   **12.5-15**

of thiocarbohydrazide (**12.5-16**) with refluxing formic acid gave directly a good yield of 4-amino-5-(methylthio)-1,2,4-triazole (**12.5-17**).[7] An elegant

12.5-16   **12.5-17**

and practical method for the synthesis of 5-amino-3-(methylthio)triazoles (**12.5-19** and **12.5-20**) resulted from the addition of the appropriate hydrazine to a cold solution of dimethyl cyanodithioimidocarbonate (**12.5-18**) in acetonitrile.[8]

$$\underset{\textbf{12.5-18}}{\overset{\displaystyle NCN}{\underset{\displaystyle SMe}{MeSC}}} \quad + \quad R_1NHNH_2 \quad \xrightarrow{\text{MeCN}} \quad \underset{\underset{R_1}{N-N}}{MeS\overset{N}{\diagdown}NH_2}$$

12.5-19, $R_1 = H$
12.5-20, $R_1 = Me$

Alkylation of preformed 3-amino-1,2,4-triazoline-5-thiones is an important and the most common method used for the preparation of 5-alkylthio-3-amino-1,2,4-triazoles.[9-16] For example, treatment of **12.5-21** and **12.5-22** with benzyl chloride in a refluxing mixture of alcohol and aqueous sodium hydroxide gave **12.5-23** (82%) and **12.5-24** (74%), respectively.[17] Although

$$\underset{\underset{H}{N-N}}{\overset{R_1}{H_2N\overset{N}{\diagdown}}}=S \quad \xrightarrow[\text{NaOH}]{\text{PhCH}_2\text{Cl}} \quad \underset{N-N}{\overset{R_1}{H_2N\overset{N}{\diagdown}SCH_2Ph}}$$

12.5-21, $R_1 = C_6H_{11}$ **12.5-23**, $R_1 = C_6H_{11}$
12.5-22, $R_1 = 3\text{-}CF_3C_6H_4$ **12.5-24**, $R_1 = 3\text{-}CF_3C_6H_4$

the methylation of the 3-hydrazinotriazole (**12.5-25**) gave only an intractable gum,[4] methylation of the isopropylidene derivative (**12.5-26**) proceeded without difficulty to give the *S*-methylthio derivative (**12.5-27**).[18] Similarly,

$$\underset{\underset{H}{N-N}}{\overset{Ph}{H_2NHN\overset{N}{\diagdown}}}=S \longrightarrow \underset{\underset{H}{N-N}}{\overset{Ph}{Me_2C=NNH-\overset{N}{\diagdown}}}=S \longrightarrow \underset{N-N}{\overset{Ph}{Me_2C=NHN\overset{N}{\diagdown}SMe}}$$

12.5-25   **12.5-26**   **12.5-27**

4-amino- (e.g., **12.5-28**)[7,19] and 3,4-diamino-1,2,4-triazoline-5-thiones (e.g., **12.4-29**)[20,21] are readily converted to the corresponding alkylthio derivatives (e.g., **12.5-30** and **12.5-31**). Of interest, the 4-formamidotriazole (**12.4-32**) was prepared in 38% yield from a mixture of the thione, formalin, and piperidine in refluxing ethanol.[22] Also methylation of 4-amino-1,2,4-triazolidine-3,5-dithione (**12.5-33**) gave a 71% yield of the 3,5-bis(methylthio)triazole (**12.5-34**).[23,24]

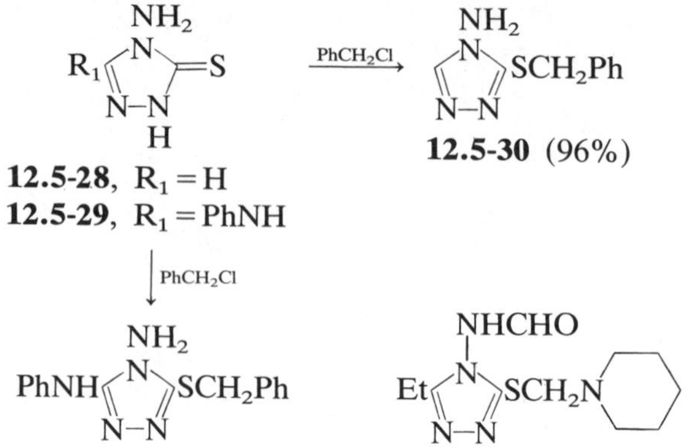

NH₂ structure with R₁, S, PhCH₂Cl → 12.5-30 (96%)

12.5-28, R₁ = H
12.5-29, R₁ = PhNH

| PhCH₂Cl

12.5-31 (70%)

12.5-32

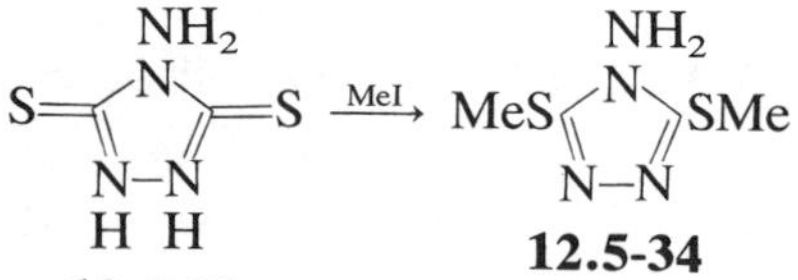

12.5-33 MeI → **12.5-34**

The acid-catalyzed condensation of the 4-aminotriazole (**12.5-35**) with 5-nitrofurfural gave (**12.5-36**),[25] and the deamination of **12.5-37** with nitrous acid gave **12.5-38**.[20] In addition, amination of chelidonic acid with

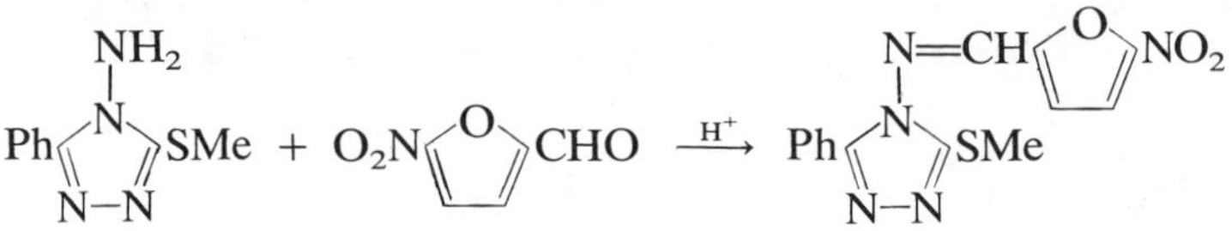

12.5-35 + O₂N—furan—CHO →ᴴ⁺ **12.5-36**

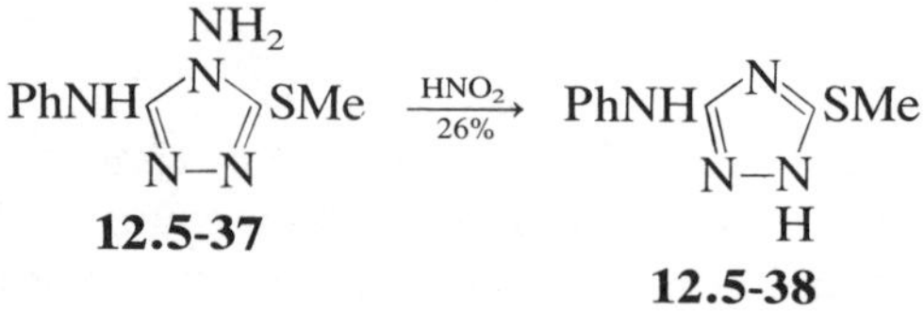

12.5-37 →(HNO₂, 26%) **12.5-38**

12.5-39 gave **12.5-40**,[26] and acylation of **12.5-41** was reported to give 1-substituted 5-aminotriazoles derivatives (e.g., **12.5-42**).[27,28]

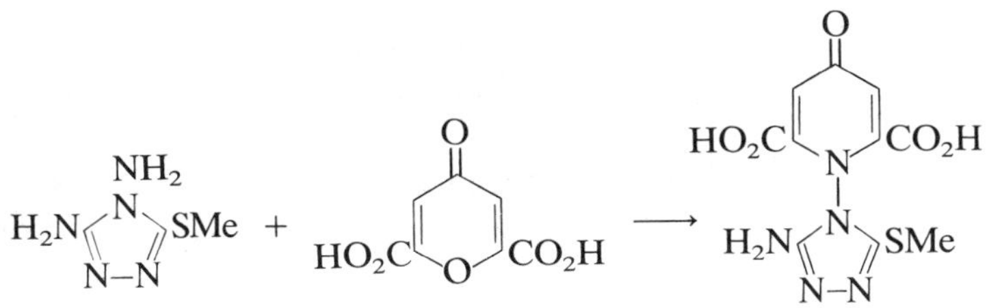

12.5-39

12.5-40

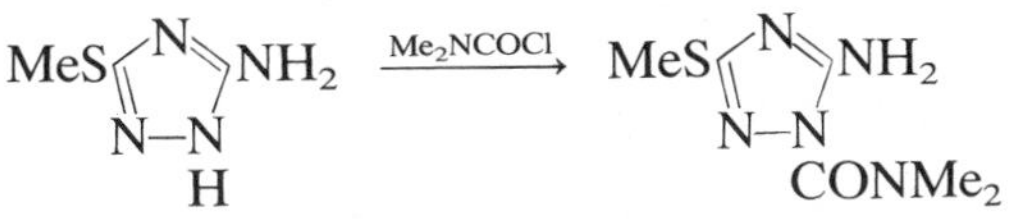

12.5-41 **12.5-42**

Oxidation of the 5-(methylthio)-1,2,4-triazole (**12.5-44**) with 30% hydrogen peroxide in warm acetic acid gave a 78% yield of the 5-(methylsulfinyl)-1,2,4-triazole (**12.5-43**).[10] With a large excess of hydrogen peroxide, **12.5-44** was converted to the 5-(methylsulfonyl)-1,2,4-triazole (**12.5-45**) (75%). The stability of the methylsulfonyl group of **12.5-45** was indicated by the unsuccessful displacement of this group with ethanolic hydrazine at 120°.

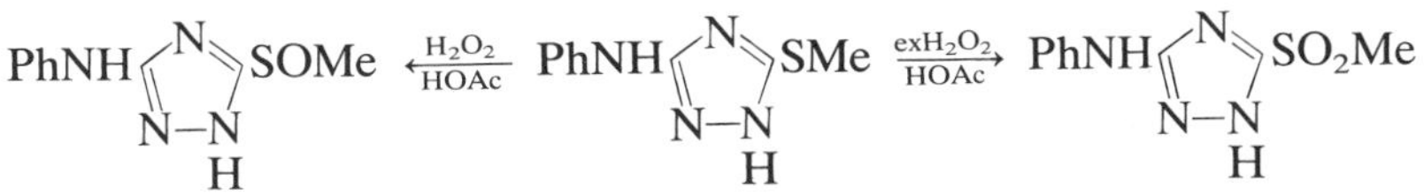

12.5-43 **12.5-44** **12.5-45**

In the oxidation of the triazolopyrimidine (**12.5-46**) with methanolic chlorine, the methylthio group was converted to a methylsulfonyl group and the condensed pyrimidine ring was cleaved to give 3-amino-5-(methylsulfonyl)-1,2,4-triazole (**12.5-47**).[29]

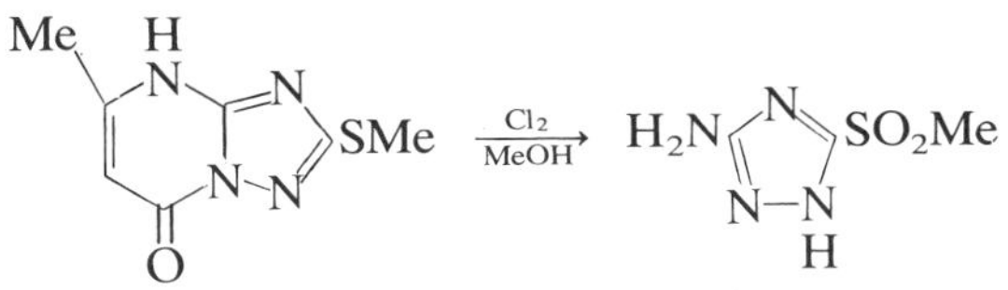

12.5-46 **12.5-47**

Reaction of 3-amino-5-phenyl-1,2,4-triazole (**12.5-48**) with *p*-toluenesulfonyl chloride in aqueous base at 60° was reported to give a 44% yield of the sulfonamide (**12.5-49**) and only a 12% yield of 5-amino-3-phenyl-1-(*p*-toluenesulfonyl)-1,2,4-triazole (**12.5-50**).[30]

$$\text{Ph} \underset{\underset{H}{N-N}}{\overset{N}{\diagup}} \text{NH}_2 \xrightarrow[\text{OH}^-]{4\text{-MeC}_6\text{H}_4\text{SO}_2\text{Cl}} \text{Ph} \underset{\underset{H}{N-N}}{\overset{N}{\diagup}} \text{NHSO}_2\text{C}_6\text{H}_4\text{-4-Me} + \text{Ph} \underset{\underset{SO_2C_6H_4-4Me}{N-N}}{\overset{N}{\diagup}} \text{NH}_2$$

12.5-48   **12.5-49**   **12.5-50**

References

1. **54:** 19695f	2. **71:** 91395n	3. **64:** 8173f	4. **69:** 77151g
5. **62:** 9124b	6. **56:** 12877h	7. **55:** 23507d	8. **81:** 25613n
9. **55:** 2624d	10. **63:** 7012g	11. **63:** 11545b	12. **66:** 2536b
13. **66:** 115650s	14. **80:** 14879g	15. **84:** 4865t	16. **84:** 135551k
17. **60:** 1754b	18. **64:** 8174a	19. **57:** 804i	20. **69:** 77170n
21. **72:** 55347k	22. **74:** 141645d	23. **57:** 12471f	24. **61:** 10676a
25. **61:** 14662g	26. **83:** 193226y	27. **80:** 82992p	28. **80:** 95962d
29. **57:** 16613g	30. **58:** 519c		

12.6. Miscellaneous O-(1,2,4-Triazole) Derivatives of Phosphorus Acids

In addition to O-triazole derivatives of phosphorus acids in which the triazole moiety was substituted with alkyl or aryl groups (see Section 9.2), a large number of compounds have been prepared in which the triazole moiety was substituted with carboxy,[1] amino,[2] alkoxy,[3] alkyl and arylthio,[4,5] and halo[6] groups. In general, triazolin-5-ones was O-alkylated with a phosphorus acid chloride in 2-butanone containing potassium carbonate. An example is the synthesis of the phosphonothioate (**12.6-3**) by reaction of **12.6-1** with the thiophosphonate (**12.6-2**).[7]

$$\text{NCCH}_2\text{S} \underset{\underset{Ph\ H}{N-N}}{\overset{N}{\diagup}} =O \xrightarrow[\textbf{12.6-2}]{\text{ClPS(Me)(OMe)}} \text{NCCH}_2\text{S} \underset{\underset{Ph}{N-N}}{\overset{N}{\diagup}} \text{OPS(Me)(OMe)}$$

12.6-1   **12.6-3**

References

1. **77:** 140089a	2. **80:** 82651b	3. **79:** 66363h	4. **79:** 56362g
5. **81:** 91540n	6. **79:** 66364j	7. **83:** 206285u	

12.7. Miscellaneous 1,2,4-Triazoles

The synthesis of triazafulvenes (e.g., **12.7-2**) was accomplished by treatment of a 5-(chlorodiphenylmethyl)triazoles (e.g., **12.7-1**) with triethylamine in tetrahydrofuran at $-78°$.[1] These reactive 5-(diphenylmethylene)triazoles were converted to dimers on warming to $30°$ and to 5-(methoxydiphenyl-methyl)triazoles (e.g., **12.7-3**) in the presence of methanol.

Aryl cyanates (e.g., **12.7-5**) were added to the terminal nitrogen of benzenesulfonyl hydrazide (**12.7-4**) to give (aryloxy)iminomethyl derivatives (e.g., **12.7-6**).[2] Further reaction of **12.7-6** with 4-chlorophenyl cyanate (**12.7-7**) gave the 5-amino-1,2,4-triazole (**12.7-8**) (41%), formed via the elimination of 4-chlorophenol from the intermediate **12.7-9**. The direction

$$\text{PhSO}_2\text{NHNH}_2 \; + \; 1\text{-C}_{10}\text{H}_7\text{OCN} \longrightarrow 1\text{-C}_{10}\text{H}_7\text{OC(NH)NHNHSO}_2\text{Ph}$$

12.7-4   **12.7-5**   **12.7-6**

of cyclization in this adduct was dependent upon the nature of the leaving group as demonstrated by the interaction of **12.7-10** and 1-naphthyl cyanate (**12.7-5**) to give **12.7-11**, from which 4-chlorophenol was again eliminated to give the isomeric 3-amino-1,2,4-triazole (**12.7-12**) (33%). A simple method

for the preparation of 3-amino-5-methoxy-1,2,4-triazole (**12.7-14**) in 79% yield involved the condensation of dimethyl *N*-cyanoimidocarbonate

(**12.7-13**) with hydrazine.[3] Treatment of **12.7-14** in aqueous sulfuric acid successively with sodium nitrite and sodium azide gave the 3-azidotriazole (**12.7-15**) (80%). Similar reactions were used in the conversion of the 5-

12.7-13   **12.7-14**   **12.7-15**

(methylthio)triazole (**12.7-16**) to **12.7-17** and **12.7-18**.[3,4] In the patent literature a large number of 3-substituted 5-O-acyltriazoles (e.g., **12.7-20**) have been prepared from triazolin-5-ones (e.g., **12.7-19**).[5-7]

12.7-16, $R_1 = NH_2$ **12.7-19**   **12.7-20**
12.7-17, $R_1 = N_3$
12.7-18, $R_1 = Cl$

Diazotization of the 3-aminotriazole (**12.7-21**) in the presence of sodium azide gave the corresponding 3-azidotriazole (**12.7-22**) (72%).[3] Also,

12.7-21   **12.7-22**

diazotriazole intermediates were involved in the coupling of 3-amino-5-(nitrosoamino)-1,2,4-triazole (**12.7-23**) with N-benzyl-N-ethylaniline to give the azotriazole (**12.7-24**).[8] In the nitrosation of **12.7-25** it was possible

12.7-23

to convert the hydrazino group to an azido group without cleavage of the 4-amino group to give **12.7-26**.[9,10] The amino group of **12.7-26** condensed normally with benzaldehyde to give **12.7-27**.

12.7-25   **12.7-26**   **12.7-27**

The bromination of 3-nitrotriazole (**12.7-28**) was slow (~20 hr), but the reaction gave a high yield of 3-bromo-5-nitrotriazole (**12.7-29**).[11] Both the fusion of **12.7-29** with 1,2,3,5-tetra-O-acetyl-β-D-ribofuranose[11] and the alkylation of **12.7-29** with epoxides[12] has been shown to give 1-substituted 5-bromo-3-nitrotriazoles (e.g., **12.7-30**). The bromo group of these products was readily displaced by alkoxide at room temperature to give 5-alkoxytriazoles (e.g., **12.7-31**).

12.7-28   **12.7-29**

12.7-30   **12.7-31**

The discovery that 3-amino- and 3,5-diamino-1,2,4-triazoles are converted with a mixture of sodium nitrite and sulfuric acid to the corresponding 3-nitro- and 3,5-dinitro-1,2,4-triazoles provided a number of valuable intermediates (see Section 8). The major product from treatment of **12.7-32** in dioxane with hydrazine was dependent upon the concentration of hydrazine.[13] With 25% hydrazine hydrate, reduction of one of the nitro groups to give an 86% yield of the 5-amino-3-nitrotriazole (**12.7-33**) was observed. In contrast, with 95 to 99% hydrazine hydrate a mixture of **12.7-33** and predominantly **12.7-34** (70%) was obtained. Also, amination of **12.7-32** with ammonium hydroxide provided **12.7-33**,[14] although the hydrazine method was reported to be more convenient. The nitro group of **12.7-33** was displaced in concentrated hydrochloric acid in the presence of urea, which consumed the generated nitrous acid, to give the 3-chlorotriazole (**12.7-37**).[15] When this reaction was performed in the absence of urea, the liberated nitrous acid diazotized the amino group of **12.7-33**, and the resulting diazo intermediate reacted further with chloride to give the by-product, **12.7-38**, in higher yields. Similarly, the 3,5-dinitrotriazole (**12.7-32**) reacted readily with either hydrochloric or hydrobromic acids to give the corresponding 5-halo-3-nitrotriazoles (**12.7-35** and **12.7-36**). The activated halogen of **12.7-35** and **12.7-36** was displaced by aliphatic amines to give 5-amino-3-nitrotriazoles (e.g., **12.7-39**),[15] which were also prepared directly from **12.7-32**.[14] In addition, the 5-nitro groups of **12.7-32** and the halo group of either **12.7-35** or **12.7-36** were displaced with alkoxides to give 5-alkoxy- and 5-aryloxytriazoles (e.g., **12.7-40**) (51%).[15,16] When **12.7-32** was treated with aqueous triethylamine or **12.7-35** with 10% aqueous sodium hydroxide, the product was the triazolin-5-one (**12.7-41**).[14,15]

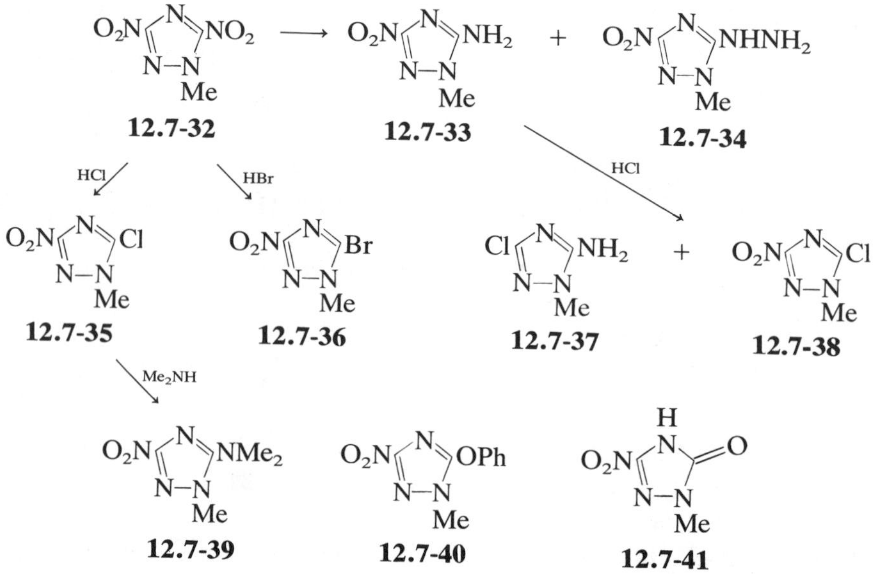

$$O_2N \underset{\underset{Me}{N-N}}{\overset{N}{\diagup\diagdown}} NO_2 \longrightarrow O_2N \underset{\underset{Me}{N-N}}{\overset{N}{\diagup\diagdown}} NH_2 \; + \; O_2N \underset{\underset{Me}{N-N}}{\overset{N}{\diagup\diagdown}} NHNH_2$$

12.7-32   **12.7-33**   **12.7-34**

HCl   HBr   HCl

$$O_2N \underset{\underset{Me}{N-N}}{\overset{N}{\diagup\diagdown}} Cl \qquad O_2N \underset{\underset{Me}{N-N}}{\overset{N}{\diagup\diagdown}} Br \qquad Cl \underset{\underset{Me}{N-N}}{\overset{N}{\diagup\diagdown}} NH_2 \; + \; O_2N \underset{\underset{Me}{N-N}}{\overset{N}{\diagup\diagdown}} Cl$$

12.7-35   **12.7-36**   **12.7-37**   **12.7-38**

Me₂NH

$$O_2N \underset{\underset{Me}{N-N}}{\overset{N}{\diagup\diagdown}} NMe_2 \qquad O_2N \underset{\underset{Me}{N-N}}{\overset{N}{\diagup\diagdown}} OPh \qquad O_2N \underset{\underset{Me}{N-N}}{\overset{\overset{H}{N}}{\diagup\diagdown}} O$$

12.7-39   **12.7-40**   **12.7-41**

References

1. **80:** 133389g
2. **64:** 15886g
3. **81:** 25613n
4. **67:** 82167e
5. **81:** 13523v
6. **82:** 16845h
7. **82:** 156321j
8. **75:** 130778u
9. **65:** 704d
10. **66:** 37902j
11. **73:** 66847v
12. **81:** 120541w
13. **74:** 76376a
14. **72:** 111380e
15. **74:** 99948c
16 **73:** 45419s

TABLE 12. 1,2,4-TRIAZOLES CONTAINING MORE THAN ONE REPRESENTATIVE
FUNCTION

Compound	Reference

12.1. Mono- and Diamino-1,2,4-Triazolecarboxylic Acids and Their Functional Derivatives

Compound	Reference
3-acetamido-1-(1-pyrrolidinylcarbonyl)-	**64:** P8197b
5-amino-1-[(ethoxycarbonyl)carbonyl]-	**79:** 42421z
5-amino-1-[*N*-(ethoxycarbonyl)thiocarbamoyl]-	**69:** P67388a, **76:** 85792b, **78:** 136237a
5-amino-1-[*N*-(ethoxycarbonyl)thiocarbamoyl]-3-(4-nitrophenyl)-	**69:** P67388a
5-amino-1-[*N*-(ethoxycarbonyl)thiocarbamoyl]-3-phenyl-	**69:** P67388a, **78:** 136237a
3-amino-1-(1-pyrrolidinylcarbonyl)-	**64:** P8197b
-1-carbamic acid, 3,5-diamino-	**39:** P1255[1]
-1-carbothioamide, 5-amino-*N*-methyl-	**70:** 77881r, **76:** 85792b
-4-carbothioamide, 3,5-diamino-*N*-phenyl-	**81:** 3897f
-1-carbothioic acid, 5-amino-, *O*-pentyl ester	**70:** 47370f
-1-carboxamide, 3-acetamido-*N*-butyl-*N*-methyl-	**64:** P8197c
-1-carboxamide, 3-acetamido-*N,N*-dimethyl-	**64:** P8196h
-1-carboxamide, 5-(acetylamino)-*N,N*-diethyl-	**82:** P156320h
-1-carboxamide, 5-(acetylamino)-*N,N*-dimethyl-	**82:** P156320h
-1-carboxamide, 3-amino-	**59:** P8759c, **59:** P8759b
-1-carboxamide, 5-amino-	**63:** 1786f
-1-carboxamide, 5-amino-*N*-(2-chloroethyl)-	**76:** 85792b
-1-carboxamide, 5-amino-*N,N*-diethyl-	**82:** P156320h
-1-carboxamide, 3-amino-*N,N*-dimethyl-	**59:** P8759c, **64:** P8196d
-1-carboxamide, 5-amino-*N,N*-dimethyl-	**64:** P8196d, **82:** P156320h
-1-carboxamide, 5-amino-*N*,3-dimethyl-	**70:** 77881r
-1-carboxamide, 5-amino-3-(dimethylamino)-*N,N*-dimethyl-	**80:** P95962d
-1-carboxamide, 5-amino-3-(dimethylamino)-*N*-methoxy-*N*-methyl-	**80:** P95962d
-1-carboxamide, 5-amino-3-(dimethylamino)-*N*-methyl-*N*-2-propenyl-	**80:** P95962d
-1-carboxamide, 5-amino-3-(dimethylamino)-*N*-methyl-*N*-propyl-	**80:** P95962d
-1-carboxamide, 5-amino-*N,N*-dimethyl-3-(4-morpholinyl)-	**80:** P95962d
-1-carboxamide, 5-amino-*N*-methoxy-*N*-methyl-3-(4-morpholinyl)-	**80:** P95962d
-1-carboxamide, 3-amino-*N*-methyl-	**76:** 85792b, **79:** 42421z
-1-carboxamide, 3-amino-5-methyl-	**59:** P8759c
-1-carboxamide, 5-amino-*N*-methyl-	**63:** 1786f
-1-carboxamide, 5-amino-*N*-methyl-3-(4-morpholinyl)-*N*-2-propenyl-	**80:** P95962d
-1-carboxamide, 5-amino-*N*-methyl-3-(5-nitro-2-furyl)-	**65:** P18595e, **79:** 87327q
-3-carboxamide, 5-amino-1-β-D-ribofuranosyl-	**80:** 121258a
-5-carboxamide, 3-amino-1-β-D-ribofuranosyl-	**80:** 121258a
-1-carboxamide, 3-benzamido-*N,N*-dimethyl-	**64:** P8197a
-1-carboxamide, *N*-sec-butyl-3-formamido-*N*-methyl-	**65:** PC725g, **74:** P3636a
-1-carboxamide, 3-butyramido-*N,N*-dimethyl-	**64:** P8197a
-1-carboxamide, 3-(2-chloroacetamido)-*N,N*-dimethyl-	**64:** P8197d

TABLE 12 (*Continued*)

Compound	Reference
12.1. Mono- and Diamino-1,2,4-Triazolecarboxylic Acids and Their Functional Derivatives (*Continued*)	
-1-carboxamide, 3-[(*p*-chlorobenzylidene)amino]-*N*,*N*-dimethyl-	**64:** P8196d
-1-carboxamide, 5-[(*p*-chlorobenzylidene)amino]-*N*,*N*-dimethyl-	**64:** P8196d
-1-carboxamide, 3-cyclopropanecarboxamido-*N*,*N*-dimethyl-	**64:** P8197e
-1-carboxamide, 3,5-diamino-	**39:** P1255[4]
-1-carboxamide, 3-(2,2-dichloroacetamido)-*N*,*N*-dimethyl-	**64:** P8197e
-1-carboxamide, 3-[(2,4-dimethoxybenzylidene)amino]-*N*,*N*-dimethyl-	**64:** P8196h
-1-carboxamide, 3-([*p*-(dimethylamino)benzylidene]-amino)-*N*,*N*-dimethyl-	**64:** P8196e
-1-carboxamide, 5-([*p*-(dimethylamino)benzylidene]-amino)-*N*,*N*-dimethyl-	**64:** P8196f
-1-carboxamide, 3-(dimethylamino)-*N*,*N*-dimethyl-	**64:** P8197c
-1-carboxamide, *N*,*N*-dimethyl-3-(*N*-methylacetamido)-	**64:** P8197e
-1-carboxamide, *N*,*N*-dimethyl-3-[(*p*-nitrobenzylidene)-amino]-	**64:** P8196e
-1-carboxamide, *N*,*N*-dimethyl-3-(piperonylideneamino)-	**64:** P8196h
-1-carboxamide, *N*,*N*-dimethyl-3-propionamido-	**64:** P8197d
-1-carboxamide, *N*,*N*-dimethyl-5-[(propionyl)amino]-	**82:** P156320h
-1-carboxamide, *N*,*N*-dimethyl-3-succinimido-	**64:** P8197c
-1-carboxamide, *N*,*N*-dimethyl-3-(2,2,2-trichloroacet-amido)-	**64:** P8197e
-1-carboxamide, *N*,*N*-dimethyl-3-(2,2,2-trifluoroacet-amido)-	**64:** P8197d
-1-carboxamide, *N*-ethyl-3-formamido-*N*-methyl-	**65:** P725g, **74:** P3636a
-1-carboxamide, 3-formamido-*N*,*N*-dimethyl-	**65:** P725f, **74:** P3636a
-1-carboxamide, 3-formamido-*N*-isopropyl-*N*-methyl-	**65:** P725g, **74:** P3636a
-1-carboxamide, 5-formamido-*N*-methyl-	**63:** 1786f
-1-carboxamide, 3-[(*p*-methoxybenzylidene)amino]-*N*,*N*-dimethyl-	**64:** P8196g
-1-carboxamide, 5-amino-	**73:** 87897b
-1-carboxamidine, 5-amino-3-ethyl-	**73:** 87897b
-1-carboxamidine, 5-amino-3-(*p*-methoxyphenyl)-	**73:** 87897b
-1-carboxamidine, 5-amino-3-methyl-	**73:** 87897b
-1-carboxamidine, 5-amino-3-phenyl-	**73:** 87897b
-1-carboxamidine, 5-amino-3-propyl-	**73:** 87897b
-1-carboxamidine, 3,5-diamino-	**39:** P1255[1], **43:** P3843d
-1-carboximidamide, 5-amino-3-methyl-	**73:** 87897b
-3-carboxylic acid, 5-(acetylamino)-1-(1-[2-(hexa-decyloxy)benzoyl]-2-[(2-methoxyphenyl)amino]-2-oxoethyl)-, methyl ester	**85:** P151754e
-3-carboxylic acid, 5-amino-	**63:** 13244a, **64:** 5437d, **67:** 64306v, **78:** 16143v, **83:** 193189p
-3-carboxylic acid, 5-amino-, benzylidenehydrazide	**63:** 13244a
-3-carboxylic acid, 5-amino-1-[bis(dimethylamino)-phosphinyl]-	**55:** 16523c

TABLE 12   (*Continued*)

Compound	Reference

12.1. Mono- and Diamino-1,2,4-Triazolecarboxylic Acids and Their Functional Derivatives (*Continued*)

Compound	Reference
-1-carboxylic acid, 3-amino-, ethyl ester	**79:** 42421z
-1-carboxylic acid, 5-amino-, ethyl ester	**79:** 42421z
-3-carboxylic acid, 5-amino-, hydrazide	**63:** 13244a
-3-carboxylic acid, 5-amino-, isopropyl ester	**60:** P2952c
-3-carboxylic acid, 5-amino-1-methyl-	**63:** 13244b
-3-carboxylic acid, 5-amino-, methyl ester	**63:** 13244a, **84:** 180140m, **85:** 77387u
-3-carboxylic acid, 5-amino-, pentyl ester	**60:** P2952c
-5-carboxylic acid, 3-amino-1-β-D-ribofuranosyl-, methyl ester	**80:** 121258a
-5-carboxylic acid, 3-amino-1-(2,3,5-tri-O-benzoyl-β-D-ribofuranosyl)-, methyl ester	**80:** 121258a
-3-carboxylic acid, 5-(2-[(1-carboxy-N-phenyl-formimidoyl)-1-phenyl]hydrazino)-1-phenyl-, diethyl ester	**63:** 7001c
-3-carboxylic acid, 5,5'-(vinylenebis[(3-sulfo-p-phenylene)amino(6-[bis(2-hydroxyethyl)amino]-s-triazine-4,2-diyl)amino])di-	**75:** P65304v
-3-carboxylic acid, 5,5'-(vinylenebis[(3-sulfo-p-phenylene)amino(6-morpholino-s-triazine-4,2-diyl)-amino])di-	**75:** P65304v
-1-thiocarbamic acid, 3,5-diamino-	**39:** 1255[1]
-1-thiocarboxamide, N-allyl-3-amino-	**59:** P8759d
-1-thiocarboxamide, 5-amino-N-benzyl-	**76:** 85792b
-1-thiocarboxamide, 3-amino-N-methyl-	**59:** P8759d
-1-thiocarboxamide, 5-amino-3-methyl-	**70:** 47370f
-1-thiocarboxamide, 5-amino-3-phenyl-	**70:** 47370f
-1-thiocarboxanilide, 3-amino-	**59:** P8759d
-1-thiocarboxanilide, 5-amino-	**70:** 77881r
-1-thiocarboxanilide, 3-amino-3'-chloro-	**59:** P8759d
-1-thiocarboxanilide, 3-amino-4'-chloro-	**59:** P8759d

12.2 S-Substituted Thio-, Sulfinyl-, Sulfonyl-, and Acylthio-1,2,4-Triazolecarboxylic Acids and Their Functional Derivatives

Compound	Reference
-1-carboxamide, 5-(aminosulfonyl)-3-tert-butyl-N,N-dimethyl-	**85:** P33028y
-1-carboxamide, 3-(benzylthio)-N,N,5-trimethyl-	**67:** P54140x
-1-carboxamide, 3-([bis(2-ethoxyethyl)amino]sulfonyl)-N,N-diethyl-	**79:** P92235n
-1-carboxamide, 3-([bis(2-ethoxyethyl)amino]sulfonyl)-N,N-di-2-propenyl-	**79:** P92235n
-1-carboxamide, 3-([bis(2-methyl-2-propenyl)amino]-sulfonyl)-N,N-di-2-propenyl-	**79:** P92235n
-1-carboxamide, 5-[(2-bromo-2-propenyl)thio]-3-(1,1-dimethylethyl)-N,N-dimethyl-	**84:** P135677f
-1-carboxamide, 5-[(2-bromo-2-propenyl)thio]-N,N-dimethyl-3-(1-methylethyl)-	**84:** P135677f

TABLE 12   (*Continued*)

Compound	Reference
12.2 *S*-Substituted Thio-, Sulfinyl-, Sulfonyl-, and Acylthio-1,2,4-Triazole-carboxylic Acids and Their Functional Derivatives (*Continued*)	
-1-carboxamide, 5-(2-butenylthio)-3-(1,1-dimethylethyl)-*N*,*N*-dimethyl-	**84:** P135677f
-1-carboxamide, 3-(2-butenylthio)-*N*,*N*-di-2-propenyl-	**76:** P72528v
-1-carboxamide, 3-(2-butenylthio)-*N*,*N*-dipropyl-	**76:** P72528v
-1-carboxamide, *N*-(2-butoxyethyl)-3-[(2-ethoxyethyl)-sulfonyl]-*N*-2-propenyl-	**79:** P92235n, **84:** P175174w
-1-carboxamide, 3-([(2-butoxyethyl)-2-propenylamino]-sulfonyl)-*N*,*N*-di-2-propenyl-	**79:** P92235n
-1-carboxamide, 3-[(2-butoxyethyl)sulfonyl]-*N*,*N*-di-2-propenyl-	**79:** P92235n, **84:** P175174w
-1-carboxamide, 3-tert-butyl-5-[(butylmethylamino)-sulfonyl]-*N*,*N*-dimethyl-	**85:** P33028y
-1-carboxamide, *N*-butyl-3-[(butylmethylamino)-sulfonyl]-*N*-ethyl-	**79:** P92235n
-1-carboxamide, *N*-butyl-3-[(butylmethylamino)-sulfonyl]-*N*-methyl-	**79:** P92235n
-1-carboxamide, *N*-butyl-3-(butylsulfonyl)-*N*-(2-ethoxyethyl)-	**79:** P92235n, **84:** P175174w
-1-carboxamide, *N*-butyl-3-(butylsulfonyl)-*N*-(2-methoxyethyl)-	**79:** P92235n, **84:** P175174w
-1-carboxamide, 3-tert-butyl-5-[(diethylamino)sulfonyl]-*N*,*N*-dimethyl-	**85:** P33028y
-1-carboxamide, *N*-butyl-3-[(diethylamino)sulfonyl]-*N*-ethyl-	**79:** P92235n
-1-carboxamide, *N*-butyl-3-[(diethylamino)sulfonyl]-*N*-methyl-	**79:** P92235n
-1-carboxamide, 3-tert-butyl-5-[(diethylamino)sulfonyl]-*N*-methyl-	**85:** P33028y
-1-carboxamide, 3-tert-butyl-5-[(dimethylamino)sulfonyl]-*N*,*N*-dimethyl-	**85:** P33028y
-1-carboxamide, 3-tert-butyl-5-[(dimethylamino)sulfonyl]-*N*-methyl-	**85:** P33028y
-1-carboxamide, *N*-butyl-3-[(dimethylamino)sulfonyl]-*N*-methyl-	**79:** P92235n
-1-carboxamide, 3-tert-butyl-5-[(3,3-dimethyl-2-isoxazolidinyl)sulfonyl]-*N*,*N*-dimethyl-	**85:** P33028y
-1-carboxamide, 3-tert-butyl-*N*,*N*-dimethyl-5-[(methylamino)sulfonyl]-	**85:** P33028y
-1-carboxamide, 3-tert-butyl-*N*,*N*-dimethyl-5-([[(1-methylethyl)(phenylmethyl)amino]sulfonyl)-	**85:** P33028y
-1-carboxamide, *N*-butyl-*N*,3(or 5)-dimethyl-5(or 3)-(methylthio)-	**67:** P54140x
-1-carboxamide, 3-tert-butyl-*N*,*N*-dimethyl-5-(4-morpholinylsulfonyl)-	**85:** P33028y
-1-carboxamide, 3-tert-butyl-*N*,*N*-dimethyl-5-[(phenylamino)sulfonyl]-	**85:** P33028y
-1-carboxamide, 3-tert-butyl-*N*,*N*-dimethyl-5-(1-piperidinylsulfonyl)-	**85:** P33028y

TABLE 12 (*Continued*)

Compound	Reference

12.2. *S*-Substituted Thio-, Sulfinyl-, Sulfonyl-, and Acylthio-1,2,4-Triazole-carboxylic Acids and Their Functional Derivatives (*Continued*)

Compound	Reference
-1-carboxamide, 3-tert-butyl-*N*,*N*-dimethyl-5-[(propylamino)sulfonyl]-	**85:** P33028y
-1-carboxamide, 3-tert-butyl-5-[(2,5-dimethyl-1-pyrrolidinyl)sulfonyl]-*N*,*N*-dimethyl-	**85:** P33028y
-1-carboxamide, *N*-butyl-3-[(di-2-propenylamino)-sulfonyl]-*N*-ethyl-	**79:** P92235n
-1-carboxamide, *N*-butyl-3-[(di-2-propenylamino)-sulfonyl]-*N*-methyl-	**79:** P92235n
-1-carboxamide, *N*-butyl-3-[(di-2-propenylamino)-sulfonyl]-*N*-2-propenyl-	**79:** P92235n
-1-carboxamide, *N*-butyl-*N*-(2-ethoxyethyl)-3-(ethylsulfonyl)-	**79:** P92235n, **84:** P175174w
-1-carboxamide, *N*-butyl-*N*-(2-ethoxyethyl)-3-(propylsulfonyl)-	**79:** P92235n, **84:** P175174w
-1-carboxamide, *N*-butyl-3-[(2-ethoxyethyl)sulfonyl]-*N*-ethyl-	**79:** P92235n, **84:** P175174w
-1-carboxamide, *N*-butyl-3-[(2-ethoxyethyl)sulfonyl]-*N*-(2-methoxyethyl)-	**79:** P92235n, **84:** P175174w
-1-carboxamide, *N*-butyl-3-[(2-ethoxyethyl)sulfonyl]-*N*-2-propenyl-	**79:** P92235n, **84:** P175174w
-1-carboxamide, 3-[(butylethylamino)sulfonyl]-*N*,*N*-diethyl-	**79:** P92235n
-1-carboxamide, 3-[(butylethylamino)sulfonyl]-*N*,*N*-di-2-propenyl-	**79:** P92235n
-1-carboxamide, 3-[(butylethylamino)sulfonyl]-*N*-ethyl-*N*-(1-methylethyl)-	**79:** P92235n
-1-carboxamide, 3-[(butylethylamino)sulfonyl]-*N*-2-propenyl-*N*-propyl-	**79:** P92235n
-1-carboxamide, *N*-butyl-*N*-ethyl-3-[(ethylmethylamino)-sulfonyl]-	**79:** P92235n
-1-carboxamide, *N*-butyl-*N*-ethyl-3-[(ethylpropylamino)-sulfonyl]-	**79:** P92235n
-1-carboxamide, *N*-butyl-*N*-ethyl-3-(ethylsulfinyl)-	**76:** P72528v
-1-carboxamide, *N*-butyl-*N*-ethyl-3-(ethylsulfonyl)-	**76:** P72528v
-1-carboxamide, *N*-butyl-*N*-ethyl-3-(ethylthio)-	**76:** P72528v
-1-carboxamide, *N*-butyl-*N*-ethyl-3-([(4-fluorophenyl)-methylamino]sulfonyl)-	**79:** P92235n
-1-carboxamide, *N*-butyl-*N*-ethyl-3-[(2-methoxyethyl)-sulfonyl]-	**79:** P92235n, **84:** P175174w
-1-carboxamide, *N*-butyl-*N*-ethyl-3-[(3-methoxypropyl)-sulfonyl]-	**79:** P92235n, **84:** P175174w
-1-carboxamide, *N*-butyl-*N*-ethyl-3-([methyl(1-methyl-propyl)amino]sulfonyl)-	**79:** P92235n
-1-carboxamide, *N*-butyl-*N*-ethyl-3[(methyl-2-propenyl-amino)sulfonyl]-	**79:** P92235n
-1-carboxamide, *N*-butyl-*N*-ethyl-3-(1-piperidinyl-sulfonyl)-	**79:** P92235n

TABLE 12 (*Continued*)

Compound	Reference
12.2. *S*-Substituted Thio-, Sulfinyl-, Sulfonyl-, and Acylthio-1,2,4-Triazole-carboxylic Acids and Their Functional Derivatives (*Continued*)	
-1-carboxamide, *N*-butyl-*N*-ethyl-3-[(2-propenylpropyl-amino)sulfonyl]-	**79:** P92235n
-1-carboxamide, *N*-butyl-*N*-ethyl-3-[(2-propoxyethyl)-sulfonyl]-	**79:** P92235n, **84:** P175174w
-1-carboxamide, *N*-butyl-3-(ethylsulfinyl)-*N*-2-propenyl-	**76:** P72528v
-1-carboxamide, *N*-butyl-3-(ethylsulfonyl)-*N*-(2-methoxyethyl)-	**79:** P92235n, **84:** P175174w
-1-carboxamide, *N*-butyl-3-(ethylsulfonyl)-*N*-2-propenyl-	**76:** P72528v
-1-carboxamide, *N*-butyl-3-(ethylthio)-*N*-2-propenyl-	**76:** P72528v
-1-carboxamide, *N*-butyl-3-(ethylthio)-*N*-propyl-	**76:** P72528v
-1-carboxamide, *N*-butyl-3-[[(4-fluorophenyl)methyl-amino]sulfonyl]-*N*-methyl-	**79:** P92235n
-1-carboxamide, *N*-butyl-*N*-(2-methoxyethyl)-3-(propylsulfonyl)-	**79:** P92235n, **84:** P175174w
-1-carboxamide, *N*-butyl-3-[(2-methoxyethyl)sulfonyl]-*N*-2-propenyl-	**79:** P92235n, **84:** P175174w
-1-carboxamide, 3-[(butylmethylamino)sulfonyl]-*N*-(2-chloro-2-propenyl)-*N*-propyl-	**79:** P92235n
-1-carboxamide, 3-[(butylmethylamino)sulfonyl]-*N*-cyclohexyl-*N*-methyl-	**79:** P92235n
-1-carboxamide, 3-[(butylmethylamino)sulfonyl]-*N*,*N*-diethyl-	**79:** P92235n
-1-carboxamide, 3-[(butylmethylamino)sulfonyl]-*N*,*N*-di-2-propenyl-	**79:** P92235n
-1-carboxamide, 3-[(butylmethylamino)sulfonyl]-*N*-(2-ethoxyethyl)-*N*-ethyl-	**79:** P92235n
-1-carboxamide, 3-[(butylmethylamino)sulfonyl]-*N*-(2-ethoxyethyl)-*N*-propyl-	**79:** P92235n
-1-carboxamide, 3-[(butylmethylamino)sulfonyl]-*N*-ethyl-*N*-(3-methylbutyl)-	**79:** P92235n
-1-carboxamide, 3-[(butylmethylamino)sulfonyl]-*N*-ethyl-*N*-(1-methylpropyl)-	**79:** P92235n
-1-carboxamide, 3-[(butylmethylamino)sulfonyl]-*N*-ethyl-*N*-pentyl-	**79:** P92235n
-1-carboxamide, 3-[(butylmethylamino)sulfonyl]-*N*-ethyl-*N*-2-propenyl-	**79:** P92235n
-1-carboxamide, 3-[(butylmethylamino)sulfonyl]-*N*-ethyl-*N*-2-propynyl-	**79:** P92235n
-1-carboxamide, 3-[(butylmethylamino)sulfonyl]-*N*-hexyl-*N*-methyl-	**79:** P92235n
-1-carboxamide, 3-[(butylmethylamino)sulfonyl]-*N*-hexyl-*N*-2-propenyl-	**79:** P92235n
-1-carboxamide, 3-[(butylmethylamino)sulfonyl]-*N*-(2-methoxyethyl)-*N*-propyl-	**79:** P92235n
-1-carboxamide, 3-[(butylmethylamino)sulfonyl]-*N*-(1-methylethyl)-*N*-2-propenyl-	**79:** P92235n
-1-carboxamide, 3-[(butylmethylamino)sulfonyl]-*N*-methyl-*N*-2-propenyl-	**79:** P92235n

TABLE 12 (*Continued*)

Compound	Reference
12.2. *S*-Substituted Thio-, Sulfinyl-, Sulfonyl-, and Acylthio-1,2,4-Triazole-carboxylic Acids and Their Functional Derivatives (*Continued*)	
-1-carboxamide, 3-[(butylmethylamino)sulfonyl]-N-methyl-N-propyl-	**79:** P92235n
-1-carboxamide, 3-[(butylmethylamino)sulfonyl]-N-2-propenyl-N-propyl-	**79:** P92235n
-1-carboxamide, 3-[(butylmethylamino)sulfonyl]-N-propyl-N-2-propynyl-	**79:** P92235n
-1-carboxamide, N-butyl-N-methyl-3-(1-piperidinylsulfonyl)-	**79:** P92235n
-1-carboxamide, N-butyl-N-methyl-3-(1-pyrrolidinylsulfonyl)-	**79:** P92235n
-1-carboxamide, 3-[(butylpropylamino)sulfonyl]-N,N-di-2-propenyl-	**79:** P92235n
-1-carboxamide, 3-(butylsulfinyl)-N,N-diethyl-	**76:** P72528v
-1-carboxamide, 3-(butylsulfinyl)-N,N-di-2-propenyl-	**76:** P72528v
-1-carboxamide, 3-(butylsulfinyl)-N,N-dipropyl-	**76:** P72528v, **84:** P175174w
-1-carboxamide, 3-(butylsulfonyl)-N-(2-chloro-2-propenyl)-N-ethyl-	**79:** P92235n, **84:** P175174w
-1-carboxamide, 3-(butylsulfonyl)-N-(2-chloro-2-propenyl)-N-2-propenyl-	**79:** P92235n, **84:** P175174w
-1-carboxamide, 3-(butylsulfonyl)-N-(2-chloro-2-propenyl-N-propyl-	**79:** P92235n, **84:** P175174w
-1-carboxamide, 3-(butylsulfonyl)-N-cyclopropyl-N-ethyl-	**79:** P92235n, **84:** P175174w
-1-carboxamide, 3-(butylsulfonyl)-N-cyclopropyl-N-propyl-	**79:** P92235n, **84:** P175174w
-1-carboxamide, 3-(butylsulfonyl)-N,N-diethyl-	**76:** P72528v
-1-carboxamide, 3-(butylsulfonyl)-N,N-di-2-propenyl-	**76:** P72528v
-1-carboxamide, 3-(butylsulfonyl)-N,N-dipropyl-	**76:** P72528v
-1-carboxamide, 3-(butylsulfonyl)-N-(2-ethoxyethyl)-N-ethyl-	**79:** P92235n, **84:** P175174w
-1-carboxamide, 3-(butylsulfonyl)-N-(2-ethoxyethyl)-N-2-propenyl-	**79:** P92235n, **84:** P175174w
-1-carboxamide, 3-(butylsulfonyl)-N-(2-ethoxyethyl)-N-propyl-	**79:** P92235n, **84:** P175174w
-1-carboxamide, 3-(butylsulfonyl)-N-(3-ethoxypropyl)-N-ethyl-	**79:** P92235n, **84:** P175174w
-1-carboxamide, 3-(butylsulfonyl)-N-(3-ethoxypropyl)-N-2-propenyl-	**79:** P92235n, **84:** P175174w
-1-carboxamide, 3-(butylsulfonyl)-N-ethyl-N-(2-methoxyethyl)-	**79:** P92235n, **84:** P175174w
-1-carboxamide, 3-(butylsulfonyl)-N-ethyl-N-(3-methoxypropyl)-	**84:** P175174w
-1-carboxamide, 3-(butylsulfonyl)-N-ethyl-N-[2-(1-methylethoxy)ethyl]-	**79:** P92235n, **84:** P175174w
-1-carboxamide, 3-(butylsulfonyl)-N-ethyl-N-(1-methyl-2-propoxyethyl)-	**79:** P92235n, **84:** P175174w
-1-carboxamide, 3-(butylsulfonyl)-N-ethyl-N-(2-propoxyethyl)-	**79:** P92235n, **84:** P175174w

TABLE 12 (*Continued*)

Compound	Reference
12.2. *S*-Substituted Thio-, Sulfinyl-, Sulfonyl-, and Acylthio-1,2,4-Triazolecarboxylic Acids and Their Functional Derivatives (*Continued*)	
-1-carboxamide, 3-(butylsulfonyl)-*N*-(2-methoxyethyl)-*N*-2-propenyl-	**79:** P92235n, **84:** P175174w
-1-carboxamide, 3-(butylsulfonyl)-*N*-(2-methoxyethyl)-*N*-propyl-	**79:** P92235n, **84:** P175174w
-1-carboxamide, 3-(butylsulfonyl)-*N*-ethyl-*N*-(3-methoxypropyl)-	**79:** P92235n
-1-carboxamide, 3-(butylthio)-*N*,*N*-diethyl-	**76:** P72528v
-1-carboxamide, 3-(butylthio)-*N*,*N*-di-2-propenyl-	**76:** P72528v
-1-carboxamide, 3-(butylthio)-*N*,*N*-dipropyl-	**76:** P72528v
-1-carboxamide, 3-[(1-chlorobutyl)sulfinyl]-*N*,*N*-dipropyl-	**79:** P92235n, **84:** P175174w
-1-carboxamide, *N*-(2-chloroethyl)-3-[(diethylamino)sulfonyl]-*N*-propyl-	**79:** P92235n
-1-carboxamide, *N*-(2-chloroethyl)-*N*-ethyl-3-(ethylsulfonyl)-	**79:** P92235n, **84:** P175174w
-1-carboxamide, *N*-(2-chloroethyl)-3-(ethylsulfonyl)-*N*-propyl-	**79:** P92235n, **84:** P175174w
-1-carboxamide, *N*-(2-chloroethyl)-3-(methylsulfonyl)-*N*-propyl-	**79:** P92235n, **84:** P175174w
-1-carboxamide, *N*-(chloromethyl)-3-[(2-ethoxyethyl)sulfonyl]-*N*-ethyl-	**79:** P92235n, **84:** P175174w
-1-carboxamide, *N*-(5-chloropentyl)-3-[(diethylamino)sulfonyl]-*N*-ethyl-	**79:** P92235n
-1-carboxamide, 3-([(4-chlorophenyl)ethylamino]sulfonyl)-*N*,*N*-diethyl-	**79:** P92235n
-1-carboxamide, 3-([(4-chlorophenyl)ethylamino]sulfonyl)-*N*,*N*-di-2-propenyl-	**79:** P92235n
-1-carboxamide, 3-([(4-chlorophenyl)methylamino]sulfonyl)-*N*,*N*-diethyl-	**79:** P92235n
-1-carboxamide, 3-([(4-chlorophenyl)methylamino]sulfonyl)-*N*,*N*-di-2-propenyl-	**79:** P92235n
-1-carboxamide, *N*-(2-chloro-2-propenyl)-3-[(diethylamino)sulfonyl]-*N*-ethyl-	**79:** P92235n
-1-carboxamide, *N*-(2-chloro-2-propenyl)-3-[(diethylamino)sulfonyl]-*N*-2-propenyl-	**79:** P92235n
-1-carboxamide, *N*-(2-chloro-2-propenyl)-3-[(diethylamino)sulfonyl]-*N*-propyl-	**79:** P92235n
-1-carboxamide, *N*-(2-chloro-2-propenyl)-3-[(dimethylamino)sulfonyl]-*N*-2-propenyl-	**79:** P92235n
-1-carboxamide, *N*-(2-chloro-2-propenyl)-3-[(di-2-propenylamino)sulfonyl]-*N*-ethyl-	**79:** P92235n
-1-carboxamide, *N*-(2-chloro-2-propenyl)-3-[(di-2-propenylamino)sulfonyl]-*N*-propyl-	**79:** P92235n
-1-carboxamide, *N*-(2-chloro-2-propenyl)-3-[(2-ethoxyethyl)sulfonyl]-*N*-ethyl-	**79:** P92235n, **84:** P175174w
-1-carboxamide, *N*-(2-chloro-2-propenyl)-3-[(2-ethoxyethyl)sulfonyl]-*N*-2-propenyl-	**79:** P92235n, **84:** P175174w
-1-carboxamide, *N*-(2-chloro-2-propenyl)-3-[(2-ethoxyethyl)sulfonyl]-*N*-propyl-	**79:** P92235n, **84:** P175174w

TABLE 12 (*Continued*)

Compound	Reference

12.2. *S*-Substituted Thio-, Sulfinyl-, Sulfonyl-, and Acylthio-1,2,4-Triazole-
carboxylic Acids and Their Functional Derivatives (*Continued*)

Compound	Reference
-1-carboxamide, *N*-(2-chloro-2-propenyl)-*N*-ethyl-3-(ethylsulfonyl)-	**79:** P92235n, **84:** P175174w
-1-carboxamide, *N*-(2-chloro-2-propenyl)-*N*-ethyl-3-([[(4-fluorophenyl)methylamino]sulfonyl)-	**79:** P92235n
-1-carboxamide, *N*-(2-chloro-2-propenyl)-*N*-ethyl-3-(isopropylsulfonyl)-	**79:** P92235n
-1-carboxamide, *N*-(2-chloro-2-propenyl)-*N*-ethyl-3-[(2-methoxyethyl)sulfonyl]-	**79:** P92235n, **84:** P175174w
-1-carboxamide, *N*-(2-chloro-2-propenyl)-*N*-ethyl-3-[(1-methylethyl)sulfonyl]-	**84:** P175174w
-1-carboxamide, *N*-(2-chloro-2-propenyl)-*N*-ethyl-3-[(1-methylpropyl)sulfonyl]-	**79:** P92235n, **84:** P175174w
-1-carboxamide, *N*-(2-chloro-2-propenyl)-*N*-ethyl-3-[(2-methylpropyl)sulfonyl]-	**79:** P92235n, **84:** P175174w
-1-carboxamide, *N*-(2-chloro-2-propenyl)-*N*-ethyl-3-(propylsulfonyl)-	**79:** P92235n, **84:** P175174w
-1-carboxamide, *N*-(2-chloro-2-propenyl)-3-(ethylsulfonyl)-*N*-2-propenyl-	**79:** P92235n, **84:** P175174w
-1-carboxamide, *N*-(2-chloro-2-propenyl)-3-(ethylsulfonyl)-*N*-propyl-	**79:** P92235n, **84:** P175174w
-1-carboxamide, *N*-(2-chloro-2-propenyl)-3-(isopropylsulfonyl)-*N*-2-propenyl-	**79:** P92235n
-1-carboxamide, *N*-(2-chloro-2-propenyl)-3-[(2-methoxyethyl)sulfonyl]-*N*-2-propenyl-	**79:** P92235n, **84:** P175174w
-1-carboxamide, *N*-(2-chloro-2-propenyl)-3-[(2-methoxyethyl)sulfonyl]-*N*-propyl-	**79:** P92235n, **84:** P175174w
-1-carboxamide, *N*-(2-chloro-2-propenyl)-3-[(1-methylethyl)sulfonyl]-*N*-2-propenyl-	**84:** P175174w
-1-carboxamide, *N*-(2-chloro-2-propenyl)-3-[(1-methylethyl)sulfonyl]-*N*-propyl-	**79:** P92235n, **84:** P175174w
-1-carboxamide, *N*-(2-chloro-2-propenyl)-3-[(methyl-2-propenylamino)sulfonyl]-*N*-propyl-	**79:** P92235n
-1-carboxamide, *N*-(2-chloro-2-propenyl)-3-[(1-methylpropyl)sulfonyl]-*N*-2-propenyl-	**79:** P92235n, **84:** P175174w
-1-carboxamide, *N*-(2-chloro-2-propenyl)-3-[(2-methylpropyl)sulfonyl]-*N*-2-propenyl-	**79:** P92235n, **84:** P175174w
-1-carboxamide, *N*-(2-chloro-2-propenyl)-3-[(1-methylpropyl)sulfonyl]-*N*-propyl-	**79:** P92235n, **84:** P175174w
-1-carboxamide, *N*-(2-chloro-2-propenyl)-3-[(2-methylpropyl)sulfonyl]-*N*-propyl-	**79:** P92235n, **84:** P175174w
-1-carboxamide, *N*-(2-chloro-2-propenyl)-*N*-2-propenyl-3-(propylsulfonyl)-	**79:** P92235n, **84:** P175174w
-1-carboxamide, 3-[[(2-chloro-2-propenyl)propylamino[sulfonyl]-*N,N*-di-2-propenyl-	**79:** P92235n
-1-carboxamide, *N*-(2-chloro-2-propenyl)-*N*-propyl-3-(propylsulfonyl)-	**79:** P92235n, **84:** P175174w
-1-carboxamide, 5-[(2-chloro-2-propenyl)thio]-3-(1,1-dimethylethyl)-*N,N*-dimethyl-	**84:** P135677f

TABLE 12 (*Continued*)

Compound	Reference

12.2. *S*-Substituted Thio-, Sulfinyl-, Sulfonyl-, and Acylthio-1,2,4-Triazole-
carboxylic Acids and Their Functional Derivatives (*Continued*)

Compound	Reference
-1-carboxamide, 5-[(2-chloro-2-propenyl)thio]-*N*,*N*-dimethyl-3-(1-methylethyl)-	**84:** P135677f
-1-carboxamide, 5-[(2-chloro-2-propenyl)thio]-*N*,*N*-dimethyl-3-(1-methylpropyl)-	**84:** P135677f
-1-carboxamide, *N*-cyclohexyl-3-[(diethylamino)-sulfonyl]-*N*-methyl-	**79:** P92235n
-1-carboxamide, *N*-cyclohexyl-3-[(diethylamino)-sulfonyl]-*N*-2-propenyl-	**79:** P92235n
-1-carboxamide, *N*-cyclohexyl-3-[(di-2-propenylamino)-sulfonyl]-*N*-methyl-	**79:** P92235n
-1-carboxamide, *N*-cyclohexyl-*N*-ethyl-3-(propylsulfonyl)-	**79:** P92235n, **84:** P175174w
-1-carboxamide, *N*-cyclohexyl-3-([(4-fluorophenyl)-methylamino]sulfonyl)-*N*-methyl-	**79:** P92235n
-1-carboxamide, 3-[(cyclohexylmethylamino)sulfonyl]-*N*,*N*-diethyl-	**79:** P92235n
-1-carboxamide, 3-[(cyclohexylmethylamino)sulfonyl]-*N*,*N*-di-2-propenyl-	**79:** P92235n
-1-carboxamide, *N*-cyclohexyl-*N*-methyl-3-[(methyl-2-propenylamino)sulfonyl]-	**79:** P92235n
-1-carboxamide, *N*-cyclohexyl-*N*-methyl-3-(1-piperidinyl-sulfonyl)-	**79:** P92235n
-1-carboxamide, *N*-cyclohexyl-*N*-methyl-3-(1-pyrrolidinyl-sulfonyl)-	**79:** P92235n
-1-carboxamide, *N*-cyclopropyl-3-[(diethylamino)-sulfonyl]-*N*-ethyl-	**79:** P92235n
-1-carboxamide, 3-cyclopropyl-5-[(dimethylamino)-sulfonyl]-*N*,*N*-dimethyl-	**85:** P33028y
-1-carboxamide, *N*-cyclopropyl-3-[(2-ethoxyethyl)-sulfonyl]-*N*-ethyl-	**79:** P92235n, **84:** P175174w
-1-carboxamide, 3-cyclopropyl-5-[(ethylamino)sulfonyl]-*N*,*N*-dimethyl-	**85:** P33028y
-1-carboxamide, *N*-cyclopropyl-*N*-ethyl-3-(ethylsulfonyl)-	**79:** P92235n, **84:** P175174w
-1-carboxamide, *N*-cyclopropyl-*N*-ethyl-3-(isopropylsulfonyl)-	**79:** P92235n
-1-carboxamide, *N*-cyclopropyl-*N*-ethyl-3-[(1-methylethyl)sulfonyl]-	**84:** P175174w
-1-carboxamide, *N*-cyclopropyl-3-(ethylsulfonyl)-*N*-propyl-	**79:** P92235n
-1-carboxamide, *N*-cyclopropyl-3-(isopropylsulfonyl)-*N*-propyl-	**79:** P92235n
-1-carboxamide, *N*-cyclopropyl-3-[(1-methylethyl)-sulfonyl]-*N*-propyl-	**84:** P175174w
-1-carboxamide, *N*-cyclopropyl-*N*-propyl-3-(propylsulfonyl)-	**79:** P92235n, **84:** P175174w
-1-carboxamide, 3-[(dibutylamino)sulfonyl]-*N*,*N*-diethyl-	**79:** P92235n
-1-carboxamide, 3-[(dibutylamino)sulfonyl]-*N*,*N*-di-2-propenyl-	**79:** P92235n

TABLE 12 (*Continued*)

Compound	Reference
12.2. *S*-Substituted Thio-, Sulfinyl-, Sulfonyl-, and Acylthio-1,2,4-Triazole- carboxylic Acids and Their Functional Derivatives (*Continued*)	
-1-carboxamide, *N*-(2,3-dichloro-2-propenyl)- 3-[(diethylamino)sulfonyl]-*N*-ethyl-	**79:** P92235n
-1-carboxamide, *N*-(2,3-dichloro-2-propenyl)- 3-[(2-ethoxyethyl)sulfonyl]-*N*-ethyl-	**79:** P92235n, **84:** P175174w
-1-carboxamide, 3-[[(2,3-dichloro-2-propenyl)- ethylamino]-*N,N*-di-2-propenyl-	**79:** P92235n
-1-carboxamide, 3-[[2-(diethylamino)ethyl]thio]-*N,N,* 5-trimethyl-	**67:** P54140x
-1-carboxamide, 3-[(diethylamino)sulfonyl]-*N,N*- bis(2-methyl-2-propenyl)-	**79:** P92235n
-1-carboxamide, 3-[(diethylamino)sulfonyl]- *N,N*-diethyl-	**79:** P92235n
-1-carboxamide, 5-[(diethylamino)sulfonyl]-*N,N*- dimethyl-3-(1-methylethyl)-	**85:** P33028y
-1-carboxamide, 3-[(diethylamino)sulfonyl]-*N,N*- di-2-propenyl-	**79:** P92235n
-1-carboxamide, 3-[(diethylamino)sulfonyl]-*N*- (4-ethoxybutyl)-*N*-ethyl-	**84:** 4425z
-1-carboxamide, 3-[(diethylamino)sulfonyl]-*N*- (2-ethoxyethyl)-*N*-ethyl-	**79:** P92235n
-1-carboxamide, 3-[(diethylamino)sulfonyl]-*N*- (2-ethoxyethyl)-*N*-propyl-	**79:** P92235n
-1-carboxamide, 3-[(diethylamino)sulfonyl]-*N*- (5-ethoxypentyl)-*N*-ethyl-	**84:** 4425z
-1-carboxamide, 3-[(diethylamino)sulfonyl]-*N*-ethyl-*N*- hexyl-	**79:** P92235n
-1-carboxamide, 3-[(diethylamino)sulfonyl]-*N*-ethyl-*N*- (2-methoxyethyl)-	**79:** P92235n
-1-carboxamide, 3-[(diethylamino)sulfonyl]-*N*-ethyl-*N*- (3-methoxypropyl)-	**79:** P92235n
-1-carboxamide, 3-[(diethylamino)sulfonyl]-*N*-ethyl-*N*- methyl-	**79:** P92235n
-1-carboxamide, 3-[(diethylamino)sulfonyl]-*N*-ethyl-*N*- (3-methylbutyl)-	**79:** P92235n
-1-carboxamide, 3-[(diethylamino)sulfonyl]-*N*-ethyl-*N*- (1-methylethyl)-	**79:** P92235n
-1-carboxamide, 3-[(diethylamino)sulfonyl]-*N*-ethyl-*N*- (1-methylpropyl)-	**79:** P92235n
-1-carboxamide, 3-[(diethylamino)sulfonyl]-*N*-ethyl-*N*- (2-methylpropyl)-	**79:** P92235n
-1-carboxamide, 3-[(diethylamino)sulfonyl]-*N*-ethyl-*N*- pentyl-	**79:** P92235n
-1-carboxamide, 3-[(diethylamino)sulfonyl]-*N*-ethyl-*N*- 2-propenyl-	**79:** P92235n
-1-carboxamide, 3-[(diethylamino)sulfonyl]-*N*-ethyl-*N*- [2-(2-propenyloxy)ethyl]-	**79:** P92235n
-1-carboxamide, 3-[(diethylamino)sulfonyl]-*N*-ethyl-*N*- 2-propynyl-	**79:** P92235n

TABLE 12   (*Continued*)

Compound	Reference
12.2. *S*-Substituted Thio-, Sulfinyl-, Sulfonyl-, and Acylthio-1,2,4-Triazole-carboxylic Acids and Their Functional Derivatives (*Continued*)	
-1-carboxamide, 3-[(diethylamino)sulfonyl]-*N*-hexyl-*N*-2-propenyl-	**79:** P92235n
-1-carboxamide, 3-[(diethylamino)sulfonyl]-*N*-(2-methoxyethyl)-*N*-propyl-	**79:** P92235n
-1-carboxamide, 3-[(diethylamino)sulfonyl]-*N*-(methoxymethyl)-*N*-propyl-	**79:** P92235n
-1-carboxamide, 3-[(diethylamino)sulfonyl]-*N*-(1-methylethyl)-*N*-2-propenyl-	**79:** P92235n
-1-carboxamide, 3-[(diethylamino)sulfonyl]-*N*-methyl-*N*-pentyl-	**79:** P92235n
-1-carboxamide, 3-[(diethylamino)sulfonyl]-*N*-methyl-*N*-2-propenyl-	**79:** P92235n
-1-carboxamide, 3-[(diethylamino)sulfonyl]-*N*-methyl-*N*-propyl-	**79:** P92235n
-1-carboxamide, 3-[(diethylamino)sulfonyl]-*N*-(2-methylpropyl)-*N*-2-propenyl-	**79:** P92235n
-1-carboxamide, 3-[(diethylamino)sulfonyl]-*N*-pentyl-*N*-2-propenyl-	**79:** P92235n
-1-carboxamide, 3-[(diethylamino)sulfonyl]-*N*-2-propenyl-*N*-propyl-	**79:** P92235n
-1-carboxamide, 3-[(diethylamino)sulfonyl]-*N*-2-propynyl-*N*-propyl-	**79:** P92235n
-1-carboxamide, *N*,*N*-diethyl-3-[[ethyl(4-fluorophenyl)-amino]sulfonyl]-	**79:** P92235n
-1-carboxamide, *N*,*N*-diethyl-3-[[ethyl(2-methoxyethyl)-amino]sulfonyl]-	**79:** P92235n
-1-carboxamide, *N*,*N*-diethyl-3-[(ethylmethylamino)-sulfonyl]-	**79:** P92235n
-1-carboxamide, *N*,*N*-diethyl-3-[(ethylpentylamino)-sulfonyl]-	**79:** P92235n
-1-carboxamide, *N*,*N*-diethyl-3-[(ethyl-2-propenyl-amino)sulfonyl]-	**79:** P92235n
-1-carboxamide, *N*,*N*-diethyl-3-[(ethylpropylamino)-sulfonyl]-	**79:** P92235n
-1-carboxamide, *N*,*N*-diethyl-3-(ethylsulfonyl)-	**76:** P72528v
-1-carboxamide, *N*,*N*-diethyl-3-(ethylthio)-	**76:** P72528v
-1-carboxamide, *N*,*N*-diethyl-3-([(4-fluorophenyl)-methylamino]sulfonyl)-	**79:** P92235n
-1-carboxamide, *N*,*N*-diethyl-3-[(hexylmethylamino)-sulfonyl]-	**79:** P92235n
-1-carboxamide, *N*,*N*-diethyl-3-[(hexylpropylamino)-sulfonyl]-	**79:** P92235n
-1-carboxamide, *N*,*N*-diethyl-3-(isopropylthio)-	**76:** P72528v
-1-carboxamide, *N*,*N*-diethyl-3-[(2-methoxyethyl)-sulfonyl]-	**79:** P92235n, **84:** P175174w
-1-carboxamide, *N*,*N*-diethyl-3-[[(1-methylethyl)-propylamino]sulfonyl]-	**79:** P92235n
-1-carboxamide, *N*,*N*-diethyl-3-[(1-methylethyl)-sulfinyl]-	**76:** P72528v

TABLE 12   (*Continued*)

Compound	Reference
12.2. *S*-Substituted Thio-, Sulfinyl-, Sulfonyl-, and Acylthio-1,2,4-Triazole-carboxylic Acids and Their Functional Derivatives (*Continued*)	
-1-carboxamide, *N,N*-diethyl-3-[(1-methylethyl)-sulfonyl]-	**76:** P72528v
-1-carboxamide, *N,N*-diethyl-3-([methyl(1-methyl-propyl)amino]sulfonyl)-	**79:** P92235n
-1-carboxamide, *N,N*-diethyl-3-([methyl(2-methyl-propyl)amino]sulfonyl)-	**79:** P92235n
-1-carboxamide, *N,N*-diethyl-3(or 5)-methyl-5(or 3)-methylthio)-	**67:** P54140x
-1-carboxamide, *N,N*-diethyl-3-[(methylpentylamino)-sulfonyl]-	**79:** P92235n
-1-carboxamide, *N,N*-diethyl-3-[(methyl-2-propenyl-amino)sulfonyl]-	**79:** P92235n
-1-carboxamide, *N,N*-diethyl-3-[(2-methylpropyl)-sulfonyl]-	**76:** P72528v
-1-carboxamide, *N,N*-diethyl-3-[(1-methylpropyl)thio]-	**76:** P72528v
-1-carboxamide, *N,N*-diethyl-3-[(2-methylpropyl)thio]-	**76:** P72528v
-1-carboxamide, *N,N*-diethyl-3-(methylsulfonyl)-	**76:** P72528v
-1-carboxamide, *N,N*-diethyl-3-(4-morpholinylsulfonyl)-	**79:** P92235n
-1-carboxamide, *N,N*-diethyl-3-[(pentylpropylamino)-sulfonyl]-	**79:** P92235n
-1-carboxamide, *N,N*-diethyl-3-(pentylsulfonyl)-	**76:** P72528v
-1-carboxamide, *N,N*-diethyl-3-(pentylthio)-	**76:** P72528v
-1-carboxamide, *N,N*-diethyl-3-(1-peperidinylsulfonyl)-	**79:** P92235n
-1-carboxamide, *N,N*-diethyl-3-[[2-(2-propenyloxy)-ethyl]thio]-	**79:** P92235n, **84:** P175174w
-1-carboxamide, *N,N*-diethyl-3-[(2-propenylpropyl-amino)sulfonyl]-	**79:** P92235n
-1-carboxamide, *N,N*-diethyl-3-[(2-propoxyethyl)-sulfonyl]-	**79:** P92235n, **84:** P175174w
-1-carboxamide, *N,N*-diethyl-3-(propylsulfinyl)-	**76:** P72528v
-1-carboxamide, *N,N*-diethyl-3-(propylsulfonyl)-	**76:** P72528v
-1-carboxamide, *N,N*-diethyl-3-(propylthio)-	**76:** P72528v
-1-carboxamide, *N,N*-diethyl-3-(1-pyrrolidinylsulfonyl)-	**79:** P92235n
-1-carboxamide, 3-[(difluoromethyl)thio]-*N,N*-dimethyl-	**67:** P54140x
-1-carboxamide, 5-[(dimethylamino)sulfonyl]-*N,N*-dimethyl-3-propyl-	**85:** P33028y
-1-carboxamide, 3-[(dimethylamino)sulfonyl]-*N,N*-di-2-propenyl-	**79:** P92235n
-1-carboxamide, 3-[(dimethylamino)sulfonyl]-*N*-ethyl-*N*-methyl-	**79:** P92235n
-1-carboxamide, 3-[(dimethylamino)sulfonyl]-*N*-ethyl-*N*-(1-methylethyl)-	**79:** P92235n
-1-carboxamide, 3-[(dimethylamino)sulfonyl]-*N*-ethyl-*N*-2-propenyl-	**79:** P92235n
-1-carboxamide, 3-[(dimethylamino)sulfonyl]-*N*-(2-methylpropyl)-*N*-2-propenyl-	**79:** P92235n
-1-carboxamide, 5-(1,1-dimethylethyl)-*N,N*-dimethyl-3-(methylthio)-	**84:** P135677f

TABLE 12 (*Continued*)

Compound	Reference
12.2. *S*-Substituted Thio-, Sulfinyl-, Sulfonyl-, and Acylthio-1,2,4-Triazole-carboxylic Acids and Their Functional Derivatives (*Continued*)	
-1-carboxamide, 3-(1,1-dimethylethyl)-*N*,*N*-dimethyl-5-(2-propynylthio)-	**84:** P135677f
-1-carboxamide, 3-(1,1-dimethylethyl)-5-(ethenylthio)-*N*,*N*-dimethyl-	**84:** P135677f
-1-carboxamide, 3-(1,1-dimethylethyl)-*N*-ethyl-*N*-methyl-5-(2-propynylthio)-	**84:** P135677f
-1-carboxamide, 3-(1,1-dimethylethyl)-5-(ethylthio)-*N*,*N*-dimethyl-	**84:** P135677f
-1-carboxamide, *N*,*N*-dimethyl-3-(1-methylethyl)-5-(methylthio)-	**84:** P135677f
-1-carboxamide, *N*,*N*-dimethyl-3-(1-methylethyl)-5-(propylthio)-	**84:** P135677f
-1-carboxamide, *N*,*N*-dimethyl-3-(1-methylethyl)-5-(2-propynylthio)-	**84:** P135677f
-1-carboxamide, *N*,*N*-dimethyl-3-(1-methylpropyl)-5-(methylthio)-	**84:** P135677f
-1-carboxamide, *N*,*N*-dimethyl-3-(1-methylpropyl)-5-(propylthio)-	**84:** P135677f
-1-carboxamide, *N*,*N*-dimethyl-3-(1-methylpropyl)-5-(2-propynylthio)-	**84:** P135677f
-1-carboxamide, *N*,*N*-dimethyl-3-(methylthio)-	**67:** P54140x
-1-carboxamide, *N*,*N*-dimethyl-5-(methylthio)-3-phenyl-	**67:** P54140x
-1-carboxamide, *N*,*N*-dimethyl-5-(methylthio)-3-(3,4,5-trimethoxy-phenyl)-	**67:** P54140x
-1-carboxamide, *N*,*N*-dimethyl-5-(methylthio)-3-undecyl-	**67:** P54140x
-1-carboxamide, 3-[(2,6-dimethyl-4-morpholinyl)-sulfonyl]-*N*,*N*-diethyl-	**79:** P92235n
-1-carboxamide, 3-[(2,6-dimethyl-4-morpholinyl)-sulfonyl]-*N*,*N*-di-2-propenyl-	**79:** P92235n
-1-carboxamide, 3-[(di-2-propenylamino)sulfonyl]-*N*,*N*-bis(2-methyl-2-propenyl)-	**79:** P92235n
-1-carboxamide, 3-[(di-2-propenylamino)sulfonyl]-*N*,*N*-diethyl-	**79:** P92235n
-1-carboxamide, 3-[(di-2-propenylamino)sulfonyl]-*N*,*N*-di-2-propenyl-	**79:** P92235n
-1-carboxamide, 3-[(di-2-propenylamino)sulfonyl]-*N*-(2-ethoxyethyl)-*N*-ethyl-	**79:** P92235n
-1-carboxamide, 3-[(di-2-propenylamino)sulfonyl]-*N*-ethyl-*N*-(2-methoxyethyl)-	**79:** P92235n
-1-carboxamide, 3-[(di-2-propenylamino)sulfonyl]-*N*-ethyl-*N*-(3-methoxypropyl)-	**79:** P92235n
-1-carboxamide, 3-[(di-2-propenylamino)sulfonyl]-*N*-ethyl-*N*-methyl-	**79:** P92235n
-1-carboxamide, 3-[(di-2-propenylamino)sulfonyl]-*N*-ethyl-*N*-(3-methylbutyl)-	**79:** P92235n
-1-carboxamide, 3-[(di-2-propenylamino)sulfonyl]-*N*-ethyl-*N*-(1-methylethyl)-	**79:** P92235n

TABLE 12 (*Continued*)

Compound	Reference

12.2. *S*-Substituted Thio-, Sulfinyl-, Sulfonyl-, and Acylthio-1,2,4-Triazole-carboxylic Acids and Their Functional Derivatives (*Continued*)

Compound	Reference
-1-carboxamide, 3-[(di-2-propenylamino)sulfonyl]-N-ethyl-N-(1-methylpropyl)-	**79:** P92235n
-1-carboxamide, 3-[(di-2-propenylamino)sulfonyl]-N-ethyl-N-(2-methylpropyl)-	**79:** P92235n
-1-carboxamide, 3-[(di-2-propenylamino)sulfonyl]-N-ethyl-N-pentyl-	**79:** P92235n
-1-carboxamide, 3-[(di-2-propenylamino)sulfonyl]-N-ethyl-N-2-propenyl-	**79:** P92235n
-1-carboxamide, 3-[(di-2-propenylamino)sulfonyl]-N-ethyl-N-propyl-	**79:** P92235n
-1-carboxamide, 3-[(di-2-propenylamino)sulfonyl]-N-ethyl-N-2-propynyl-	**79:** P92235n
-1-carboxamide, 3-[(di-2-propenylamino)sulfonyl]-N-hexyl-N-2-propenyl-	**79:** P92235n
-1-carboxamide, 3-[(di-2-propenylamino)sulfonyl]-N-(2-methoxyethyl)-N-propyl-	**79:** P92235n
-1-carboxamide, 3-[(di-2-propenylamino)sulfonyl]-N-(1-methylethyl)-N-2-propenyl-	**79:** P92235n
-1-carboxamide, 3-[(di-2-propenylamino)sulfonyl]-N-methyl-N-pentyl-	**79:** P92235n
-1-carboxamide, 3-[(di-2-propenylamino)sulfonyl]-N-methyl-N-2-propenyl-	**79:** P92235n
-1-carboxamide, 3-[(di-2-propenylamino)sulfonyl]-N-methyl-N-propyl-	**79:** P92235n
-1-carboxamide, 3-[(di-2-propenylamino)sulfonyl]-N-(2-methylpropyl)-N-2-propenyl-	**79:** P92235n
-1-carboxamide, 3-[(di-2-propenylamino)sulfonyl]-N-propyl-N-2-propynyl-	**79:** P92235n
-1-carboxamide, 3-[(di-2-propenylamino)sulfonyl]-N-pentyl-N-2-propenyl-	**79:** P92235n
-1-carboxamide, N,N-di-2-propenyl-3-([2-(2-propenyloxy)ethyl]thio)-	**79:** P92235n, **84:** P175174w
-1-carboxamide, N,N-di-2-propenyl-3-[(2-propenyl-propylamino)sulfonyl]-	**79:** P92235n
-1-carboxamide, N,N-di-2-propenyl-3-(2-propenylthio)-	**76:** P72528v
-1-carboxamide, N,N-di-2-propenyl-3-[(2-propoxy-ethyl)sulfonyl]-	**79:** P92235n, **84:** P175174w
-1-carboxamide, N,N-di-2-propenyl-3-[(propyl-2-propynylamino)sulfonyl]-	**79:** P92235n
-1-carboxamide, N,N-di-2-propenyl-3-(propylsulfinyl)-	**76:** P72528v
-1-carboxamide, N,N-di-2-propenyl-3-(propylsulfonyl)-	**76:** P72528v
-1-carboxamide, N,N-di-2-propenyl-3-(propylthio)-	**76:** P72528v
-1-carboxamide, N,N-di-2-propenyl-3-(1-pyrrolidinyl-sulfonyl)-	**79:** P92235n
-1-carboxamide, 3-[(dipropylamino)sulfonyl]-N,N-diethyl-	**79:** P92235n
-1-carboxamide, 3-[(dipropylamino)sulfonyl]-N,N-di-2-propenyl-	**79:** P92235n

TABLE 12 (*Continued*)

Compound	Reference

-1-carboxamide, *N,N*-dipropyl-3-(propylsulfinyl)-	**76:** P72528v
-1-carboxamide, *N,N*-dipropyl-3-(propylsulfonyl)-	**76:** P72528v
-1-carboxamide, *N,N*-dipropyl-3-(propylthio)-	**76:** P72528v
-1-carboxamide, 3,3'-(dithiodi-*o*-phenylene)bis-(5-methylthio-	**69:** 52047q
-1-carboxamide, 3-(dodecylthio)-*N,N*,5-trimethyl-	**67:** P54140x
-1-carboxamide, 5-(ethenylthio)-*N,N*-dimethyl-3-(1-methylethyl)-	**84:** P135677f
-1-carboxamide, 5-(ethenylthio)-*N,N*-dimethyl-3-(1-methylpropyl)-	**84:** P135677f
-1-carboxamide, *N*-(4-ethoxybutyl)-*N*-ethyl-3-(propylsulfonyl)-	**84:** 4425z
-1-carboxamide, *N*-(2-ethoxyethyl)-3-[(2-ethoxyethyl)-sulfonyl]-*N*-propyl-	**79:** P92235n, **84:** P175174w
-1-carboxamide, *N*-(2-ethoxyethyl)-*N*-ethyl-3-(ethylsulfonyl)-	**79:** P92235n, **84:** P175174w
-1-carboxamide, *N*-(2-ethoxyethyl)-*N*-ethyl-3-(isopropylsulfonyl-	**79:** P92235n
-1-carboxamide, *N*-(2-ethoxyethyl)-*N*-ethyl-3-[(2-methoxyethyl)sulfonyl]-	**79:** P92235n, **84:** P175174w
-1-carboxamide, *N*-(2-ethoxyethyl)-*N*-ethyl-3-[(1-methylethyl)sulfonyl]-	**84:** P175174w
-1-carboxamide, *N*-(2-ethoxyethyl)-*N*-ethyl-3-[(2-methylpropyl)sulfonyl]-	**79:** P92235n, **84:** P175174w
-1-carboxamide, *N*-(2-ethoxyethyl)-*N*-ethyl-3-(pentylsulfonyl)-	**79:** P92235n, **84:** P175174w
-1-carboxamide, *N*-(2-ethoxyethyl)-*N*-ethyl-3-(1-piperidinylsulfonyl)-	**79:** P92235n
-1-carboxamide, *N*-(2-ethoxyethyl)-*N*-ethyl-3-(propylsulfonyl)-	**79:** P92235n, **84:** P175174w
-1-carboxamide, *N*-(2-ethoxyethyl)-3-(ethylsulfonyl)-*N*-2-propenyl-	**79:** P92235n, **84:** P175174w
-1-carboxamide, *N*-(2-ethoxyethyl)-3-(ethylsulfonyl)-*N*-propyl-	**79:** P92235n, **84:** P175174w
-1-carboxamide, *N*-(2-ethoxyethyl)-3-(isopropyl-sulfonyl)-*N*-propyl-	**79:** P92235n
-1-carboxamide, *N*-(2-ethoxyethyl)-3-[(2-methoxyethyl)-sulfonyl-*N*-propyl-	**79:** P92235n, **84:** P175174w
-1-carboxamide, *N*-(2-ethoxyethyl)-3-[(1-methylethyl)-sulfonyl]-*N*-propyl-	**84:** P175174w
-1-carboxamide, *N*-(2-ethoxyethyl)-3-[(2-methylpropyl)-sulfonyl]-*N*-propyl-	**79:** P92235n, **84:** P175174w
-1-carboxamide, *N*-(2-ethoxyethyl)-3-(pentylsulfonyl)-*N*-propyl-	**79:** P92235n, **84:** P175174w
-1-carboxamide, *N*-(2-ethoxyethyl)-*N*-2-propenyl-3-(propylsulfonyl)-	**79:** P92235n, **84:** P175175w
-1-carboxamide, *N*-(2-ethoxyethyl)-*N*-propyl-3-(propylsulfonyl)-	**79:** P92235n, **84:** P175174w

TABLE 12 (*Continued*)

Compound	Reference
12.2. *S*-Substituted Thio-, Sulfinyl-, Sulfonyl-, and Acylthio-1,2,4-Triazole-carboxylic Acids and Their Functional Derivatives (*Continued*)	
-1-carboxamide, 3-[(2-ethoxyethyl)sulfonyl]-N,N-bis(2-methyl-2-propenyl)-	**79:** P92235n, **84:** P175174w
-1-carboxamide, 3-[(2-ethoxyethyl)sulfonyl]-N,N-di-2-propenyl-	**79:** P92235n, **84:** P175174w
-1-carboxamide, 3-[(2-ethoxyethyl)sulfonyl]-N,N-dipropyl-	**79:** P92235n, **84:** P175174w
-1-carboxamide, 3-[(2-ethoxyethyl)sulfonyl]-N-ethyl-N-hexyl-	**79:** P92235n, **84:** P175174w
-1-carboxamide, 3-[(2-ethoxyethyl)sulfonyl]-N-ethyl-N-isopropyl-	**79:** P92235n
-1-carboxamide, 3-[(2-ethoxyethyl)sulfonyl]-N-ethyl-N-(2-methoxyethyl)-	**79:** P92235n, **84:** P175174w
-1-carboxamide, 3-[(2-ethoxyethyl)sulfonyl]-N-ethyl-N-(3-methylbutyl)-	**79:** P92235n, **84:** P175174w
-1-carboxamide, 3-[(2-ethoxyethyl)sulfonyl]-N-ethyl-N-[2-(1-methylethoxy)ethyl]-	**79:** P92235n, **84:** P175174w
-1-carboxamide, 3-[(2-ethoxyethyl)sulfonyl]-N-ethyl-N-(1-methylethyl)-	**84:** P175174w
-1-carboxamide, 3-[(2-ethoxyethyl)sulfonyl]-N-ethyl-N-(2-methylpropyl)-	**79:** P92235n, **84:** P175174w
-1-carboxamide, 3-[(2-ethoxyethyl)sulfonyl]-N-ethyl-N-pentyl-	**79:** P92235n, **84:** P175174w
-1-carboxamide, 3-[(2-ethoxyethyl)sulfonyl]-N-ethyl-N-(2-propoxyethyl)-	**79:** P92235n, **84:** P175174w
-1-carboxamide, 3-[(2-ethoxyethyl)sulfonyl]-N-ethyl-N-2-propynyl-	**79:** P92235n, **84:** P175174w
-1-carboxamide, 3-[(2-ethoxyethyl)sulfonyl]-N-hexyl-N-2-propenyl-	**79:** P92235n, **84:** P175174w
-1-carboxamide, 3-[(2-ethoxyethyl)sulfonyl]-N-hexyl-N-propyl-	**79:** P92235n, **84:** P175174w
-1-carboxamide, 3-[(2-ethoxyethyl)sulfonyl]-N-isopropyl-N-2-propenyl-	**79:** P92235n
-1-carboxamide, 3-[(2-ethoxyethyl)sulfonyl]-N-isopropyl-N-propyl-	**79:** P92235n
-1-carboxamide, 3-[(2-ethoxyethyl)sulfonyl]-N-(2-methoxyethyl)-N-2-propenyl-	**79:** P92235n, **84:** P175174w
-1-carboxamide, 3-[(2-ethoxyethyl)sulfonyl]-N-(2-methoxyethyl)-N-propyl-	**79:** P92235n, **84:** P175174w
-1-carboxamide, 3-[(2-ethoxyethyl)sulfonyl]-N-(1-methylethyl)-N-2-propenyl-	**84:** P175174w
-1-carboxamide, 3-[(2-ethoxyethyl)sulfonyl]-N-(1-methylethyl)-N-propyl-	**84:** P175174w
-1-carboxamide, 3-[(2-ethoxyethyl)sulfonyl]-N-(2-methylpropyl)-N-2-propenyl)	**79:** P92235n, **84:** P175174w
-1-carboxamide, 3-[(2-ethoxyethyl)sulfonyl]-N-(1-methylpropyl)-N-propyl-	**79:** P92235n, **84:** P175174w
-1-carboxamide, 3-[(2-ethoxyethyl)sulfonyl]-N-pentyl-N-2-propenyl-	**79:** P92235n, **84:** P175174w

TABLE 12 (*Continued*)

Compound	Reference
12.2. *S*-Substituted Thio-, Sulfinyl-, Sulfonyl-, and Acylthio-1,2,4-Triazole-carboxylic Acids and Their Functional Derivatives (*Continued*)	
-1-carboxamide, 3-[(2-ethoxyethyl)sulfonyl]-N-2-propenyl-N-propyl-	**79:** P92235n, **84:** P175174w
-1-carboxamide, 3-[(2-ethoxyethyl)sulfonyl]-N-propyl-N-2-propenyl-	**79:** P92235n
-1-carboxamide, 3-[(2-ethoxyethyl)sulfonyl]-N-propyl-N-2-propynyl-	**84:** P175174w
-1-carboxamide, N-(5-ethoxypentyl)-N-ethyl-3-(propylsulfonyl)-	**84:** 4425z
-1-carboxamide, N-(3-ethoxypropyl)-N-ethyl-3-(ethylsulfonyl)-	**79:** P92235n, **84:** P175174w
-1-carboxamide, N-(3-ethoxypropyl)-N-ethyl-3-[(1-methylethyl)sulfonyl]-	**79:** P92235n, **84:** P175174w
-1-carboxamide, N-(3-ethoxypropyl)-N-ethyl-3-[(2-methylpropyl)sulfonyl]-	**79:** P92235n, **84:** P175174w
-1-carboxamide, N-(3-ethoxypropyl)-N-ethyl-3-(pentylsulfonyl)-	**79:** P92235n, **84:** P175174w
-1-carboxamide, N-(3-ethoxypropyl)-N-ethyl-3-(propylsulfonyl)-	**79:** P92235n, **84:** P175174w
-1-carboxamide, N-(3-ethoxypropyl)-3-(ethylsulfonyl)-N-2-propenyl-	**79:** P92235n, **84:** P175174w
-1-carboxamide, N-(3-ethoxypropyl)-N-2-propenyl-3-(propylsulfonyl)-	**79:** P92235n, **84:** P175174w
-1-carboxamide, 3-[(3-ethoxypropyl)sulfonyl]-N,N-diethyl-	**79:** P92235n, **84:** P175174w
-1-carboxamide, 3-[(3-ethoxypropyl)sulfonyl]-N,N-di-2-propenyl-	**79:** P92235n, **84:** P175174w
-1-carboxamide, 3(or 5)-ethyl-N,N-dimethyl-5(or 3)-(methylthio)-	**67:** P54140x
-1-carboxamide, N-ethyl-3-(ethylsulfinyl)-N-hexyl-	**76:** P72528v
-1-carboxamide, N-ethyl-3-(ethylsulfinyl)-N-(3-methylbutyl)-	**76:** P72528v
-1-carboxamide, N-ethyl-3-(ethylsulfinyl)-N-(2-methylpropyl)-	**76:** P72528v
-1-carboxamide, N-ethyl-3-(ethylsulfinyl)-N-2-propenyl-	**76:** P72528v
-1-carboxamide, N-ethyl-3-(ethylsulfonyl)-N-hexyl-	**76:** P72528v
-1-carboxamide, N-ethyl-3-(ethylsulfonyl)-N-(2-methoxyethyl)-	**79:** P92235n, **84:** P175174w
-1-carboxamide, N-ethyl-3-(ethylsulfonyl)-N-(3-methoxypropyl)-	**79:** P92235n, **84:** P175174w
-1-carboxamide, N-ethyl-3-(ethylsulfonyl)-N-(3-methylbutyl)-	**76:** P72528v
-1-carboxamide, N-ethyl-3-(ethylsulfonyl)-N-[2-(1-methylethoxy)ethyl]-	**79:** P92235n, **84:** P175174w
-1-carboxamide, N-ethyl-3-(ethylsulfonyl)-N-(2-methylpropyl)-	**76:** P72528v
-1-carboxamide, N-ethyl-3-(ethylsulfonyl)-N-2-propenyl-	**76:** P72528v

TABLE 12 (*Continued*)

Compound	Reference
12.2. *S*-Substituted Thio-, Sulfinyl-, Sulfonyl-, and Acylthio-1,2,4-Triazole-carboxylic Acids and Their Functional Derivatives (*Continued*)	
-1-carboxamide, *N*-ethyl-3-(ethylsulfonyl)-*N*-(2-propoxyethyl)-	**79:** P92235n, **84:** P175174w
-1-carboxamide, *N*-ethyl-3-(ethylsulfonyl)-*N*-propyl-	**76:** P72528v
-1-carboxamide, *N*-ethyl-3-(ethylthio)-*N*-hexyl-	**76:** P72528v
-1-carboxamide, *N*-ethyl-3-(ethylthio)-*N*-(3-methylbutyl)-	**76:** P72528v
-1-carboxamide, *N*-ethyl-3-(ethylthio)-*N*-(2-methylpropyl)-	**76:** P72528v
-1-carboxamide, *N*-ethyl-3-(ethylthio)-*N*-pentyl-	**76:** P72528v
-1-carboxamide, *N*-ethyl-3-(ethylthio)-*N*-2-propenyl-	**76:** P72528v
-1-carboxamide, *N*-ethyl-3-(ethylthio)-*N*-propyl-	**76:** P72528v
-1-carboxamide, *N*-ethyl-3-([(4-fluorophenyl)methylamino]sulfonyl)-*N*-(1-methylpropyl)-	**79:** P92235n
-1-carboxamide, *N*-ethyl-3-([(4-fluorophenyl)methylamino]sulfonyl)-*N*-pentyl-	**79:** P92235n
-1-carboxamide, *N*-ethyl-3-([(4-fluorophenyl)methylamino]sulfonyl)-*N*-2-propenyl-	**79:** P92235n
-1-carboxamide, *N*-ethyl-3-([(4-fluorophenyl)methylamino]sulfonyl)-*N*-propyl-	**79:** P92235n
-1-carboxamide, 3-[(ethylhexylamino)sulfonyl]-*N*,*N*-di-2-propenyl-	**79:** P92235n
-1-carboxamide, *N*-ethyl-*N*-hexyl-3-[(2-methoxyethyl)-sulfonyl]-	**79:** P92235n, **84:** P175174w
-1-carboxamide, *N*-ethyl-3-[(hexyl-2-propenylamino)-sulfonyl]-*N*-propyl-	**79:** P92235n
-1-carboxamide, 3-[[ethyl(2-methoxyethyl)amino]-sulfonyl]-*N*,*N*-di-2-propenyl-	**79:** P92235n
-1-carboxamide, *N*-ethyl-*N*-(2-methoxyethyl)-3-[(2-methoxyethyl)sulfonyl]-	**79:** P92235n, **84:** P175174w
-1-carboxamide, *N*-ethyl-*N*-(2-methoxyethyl)-3-[(1-methylethyl)sulfonyl]-	**79:** P92235n, **84:** P175174w
-1-carboxamide, *N*-ethyl-*N*-(2-methoxyethyl)-3-[(2-methylpropyl)sulfonyl]-	**79:** P92235n, **84:** P175174w
-1-carboxamide, *N*-ethyl-3-([(2-methoxyethyl)propylamino]sulfonyl)-*N*-2-propenyl-	**79:** P92235n
-1-carboxamide, *N*-ethyl-*N*-(2-methoxyethyl)-3-(propylsulfonyl)-	**79:** P92235n, **84:** P175174w
-1-carboxamide, *N*-ethyl-3-[(2-methoxyethyl)sulfonyl]-*N*-(3-methoxypropyl)-	**79:** P92235n, **84:** P175174w
-1-carboxamide, *N*-ethyl-3-[(2-methoxyethyl)sulfonyl]-*N*-(1-methylethyl)-	**79:** P92235n, **84:** P175174w
-1-carboxamide, *N*-ethyl-3-[(2-methoxyethyl)sulfonyl]-*N*-(2-methylpropyl)-	**79:** P92235n, **84:** P175174w
-1-carboxamide, *N*-ethyl-3-[(2-methoxyethyl)sulfonyl]-*N*-pentyl-	**79:** P92235n, **84:** P175174w
-1-carboxamide, *N*-ethyl-3-[(2-methoxyethyl)sulfonyl]-*N*-2-propenyl-	**79:** P92235n, **84:** P175174w
-1-carboxamide, *N*-ethyl-3-[(2-methoxyethyl)sulfonyl]-*N*-2-propynyl-	**79:** P92235n, **84:** P175174w

TABLE 12 *(Continued)*

Compound	Reference
12.2. *S*-Substituted Thio-, Sulfinyl-, Sulfonyl-, and Acylthio-1,2,4-Triazole-carboxylic Acids and Their Functional Derivatives *(Continued)*	
-1-carboxamide, *N*-ethyl-*N*-(3-methoxypropyl)-3-(propylsulfonyl)-	**79:** P92235n, **84:** P175174w
-1-carboxamide, 3-[(ethylmethylamino)sulfonyl]-*N,N*-di-2-propenyl-	**79:** P92235n
-1-carboxamide, 3-[(ethylmethylamino)sulfonyl]-*N*-2-propenyl-*N*-propyl-	**79:** P92235n
-1-carboxamide, *N*-ethyl-*N*-(3-methylbutyl)-3-(1-piperidinylsulfonyl)-	**79:** P92235n
-1-carboxamide, *N*-ethyl-*N*-[2-(1-methylethoxy)ethyl]-3-(propylsulfonyl)-	**79:** P92235n, **84:** P175174w
-1-carboxamide, *N*-ethyl-*N*-(1-methylpropyl)-3-(1-piperidinylsulfonyl)-	**79:** P92235n
-1-carboxamide, *N*-ethyl-3-[(methyl-2-propenylamino)sulfonyl]-*N*-2-propenyl-	**79:** P92235n
-1-carboxamide, *N*-ethyl-*N*-(1-methyl-2-propoxyethyl)-3-(propylsulfonyl)-	**79:** P92235n, **84:** P175174w
-1-carboxamide, *N*-ethyl-*N*-methyl-3-(1-pyrrolidinyl-sulfonyl)-	**79:** P92235n
-1-carboxamide, *N*-ethyl-*N*-pentyl-3-(1-piperidinyl-sulfonyl)-	**79:** P92235n
-1-carboxamide, 3-[(ethylpentylamino)sulfonyl]-*N,N*-di-2-propenyl-	**79:** P92235n
-1-carboxamide, *N*-ethyl-3-(1-piperidinylsulfonyl)-*N*-2-propenyl-	**79:** P92235n
-1-carboxamide, *N*-ethyl-3-(1-piperidinylsulfonyl)-*N*-2-propynyl-	**79:** P92235n
-1-carboxamide, 3-[(ethyl-2-propenylamino)sulfonyl]-*N,N*-di-2-propenyl-	**79:** P92235n
-1-carboxamide, 3-[(ethyl[2-(2-propenyloxy)ethyl]-amino)sulfonyl]-*N,N*-di-2-propenyl-	**79:** P92235n
-1-carboxamide, *N*-ethyl-3-[(2-propenylpropylamino)-sulfonyl-*N*-2-propenyl-	**79:** P92235n
-1-carboxamide, *N*-ethyl-*N*-2-propenyl-3-(1-pyrrolidinylsulfonyl)-	**79:** P92235n
-1-carboxamide, *N*-ethyl-*N*-(2-propoxyethyl)-3-(propylsulfonyl)-	**79:** P92235n, **84:** P175174w
-1-carboxamide, 3-[(ethylpropylamino)sulfonyl]-*N,N*-di-2-propenyl-	**79:** P92235n
-1-carboxamide, *N*-ethyl-*N*-propyl-3-(propylsulfonyl)-	**84:** 174980u, **85:** 29406c
-4-carboxamide, *N*-ethyl-*N*-propyl-3-(propylsulfonyl)-	**84:** 174980u
-1-carboxamide, *N*-ethyl-*N*-propyl-5-(propylsulfonyl)-	**84:** 174980u
-1-carboxamide, 3-(ethylsulfinyl)-*N,N*-di-2-propenyl-	**76:** P72528v
-1-carboxamide, 3-(ethylsulfinyl)-*N*-hexyl-*N*-2-propenyl-	**76:** P72528v
-1-carboxamide, 3-(ethylsulfinyl)-*N*-(2-methylpropyl)-*N*-2-propenyl-	**76:** P72528v
-1-carboxamide, 3-(ethylsulfinyl)-*N*-pentyl-*N*-2-propenyl-	**76:** P72528v

TABLE 12   (*Continued*)

Compound	Reference
12.2. *S*-Substituted Thio-, Sulfinyl-, Sulfonyl-, and Acylthio-1,2,4-Triazolecarboxylic Acids and Their Functional Derivatives (*Continued*)	
-1-carboxamide, 3-(ethylsulfinyl)-*N*-2-propenyl-*N*-propyl-	**76:** P72528v
-1-carboxamide, 3-(ethylsulfonyl)-*N*,*N*-bis(2-methyl-2-propenyl)-	**76:** P72528v
-1-carboxamide, 3-(ethylsulfonyl)-*N*,*N*-di-2-propenyl-	**76:** P72528v
-1-carboxamide, 3-(ethylsulfonyl)-*N*,*N*-dipropyl-	**76:** P72528v
-1-carboxamide, 3-(ethylsulfonyl)-*N*-hexyl-*N*-2-propenyl-	**76:** P72528v
-1-carboxamide, 3-(ethylsulfonyl)-*N*-hexyl-*N*-propyl-	**76:** P72528v
-1-carboxamide, 3-(ethylsulfonyl)-*N*-isopropyl-*N*-propyl-	**76:** P72528v
-1-carboxamide, 3-(ethylsulfonyl)-*N*-(2-methoxyethyl)-*N*-2-propenyl-	**79:** P92235n, **84:** P175174w
-1-carboxamide, 3-(ethylsulfonyl)-*N*-(2-methoxyethyl)-*N*-propyl-	**79:** P92235n, **84:** P175174w
-1-carboxamide, 3-(ethylsulfonyl)-*N*-(1-methylpropyl)-*N*-propyl-	**76:** P72528v
-1-carboxamide, 3-(ethylsulfonyl)-*N*-(2-methylpropyl)-*N*-propyl-	**76:** P72528v
-1-carboxamide, 3-(ethylsulfonyl)-*N*-pentyl-*N*-2-propenyl-	**76:** P72528v
-1-carboxamide, 3-(ethylsulfonyl)-*N*-pentyl-*N*-propyl-	**76:** P72528v
-1-carboxamide, 3-(ethylsulfonyl)-*N*-2-propenyl-*N*-propyl-	**76:** P72528v
-1-carboxamide, 3-(ethylsulfonyl)-*N*-propyl-*N*-2-propynyl-	**76:** P72528v
-1-carboxamide, 3-(ethylthio)-*N*,*N*-bis(2-methyl-2-propenyl)-	**76:** P72528v
-1-carboxamide, 5-(ethylthio)-*N*,*N*-dimethyl-3-(1-methylethyl)-	**84:** P135677f
-1-carboxamide, 5-(ethylthio)-*N*,*N*-dimethyl-3-(1-methylpropyl)-	**84:** P135677f
-1-carboxamide, 3-(ethylthio)-*N*,*N*-di-2-propenyl-	**76:** P72528v
-1-carboxamide, 3-(ethylthio)-*N*,*N*-dipropyl-	**76:** P72528v
-1-carboxamide, 3-(ethylthio)-*N*-hexyl-*N*-2-propenyl-	**76:** P72528v
-1-carboxamide, 3-(ethylthio)-*N*-(2-methylpropyl)-*N*-2-propenyl-	**76:** P72528v
-1-carboxamide, 3-(ethylthio)-*N*-(2-methylpropyl)-*N*-propyl-	**76:** P72528v
-1-carboxamide, 3-(ethylthio)-*N*-pentyl-*N*-propyl-	**76:** P72528v
-1-carboxamide, 3-(ethylthio)-*N*-pentyl-*N*-2-propenyl-	**76:** P72528v
-1-carboxamide, 3-(ethylthio)-*N*-2-propenyl-*N*-propyl-	**76:** P72528v
-1-carboxamide, 3-(ethylthio)-*N*-propyl-*N*-2-propynyl-	**76:** P72528v
-1-carboxamide, 3(or 5)-(ethylthio)-*N*,*N*,5(or 3)-trimethyl-	**67:** P54140x
-1-carboxamide, 3-([(4-fluorophenyl)methylamino]-sulfonyl)-*N*,*N*-di-2-propenyl-	**79:** P92235n

TABLE 12 (*Continued*)

Compound	Reference

12.2. *S*-Substituted Thio-, Sulfinyl-, Sulfonyl-, and Acylthio-1,2,4-Triazole-
carboxylic Acids and Their Functional Derivatives (*Continued*)

Compound	Reference
-1-carboxamide, 3-([(4-fluorophenyl)methylamino]-sulfonyl)-*N*-hexyl-*N*-methyl-	**79:** P92235n
-1-carboxamide, 3-([(4-fluorophenyl)methylamino]-sulfonyl)-*N*-(2-methoxyethyl)-*N*-propyl-	**79:** P92235n
-1-carboxamide, 3-([(4-fluorophenyl)methylamino]-sulfonyl)-*N*-methyl-*N*-2-propenyl-	**79:** P92235n
-1-carboxamide, 3-([(4-fluorophenyl)methylamino]-sulfonyl)-*N*-methyl-*N*-propyl-	**79:** P92235n
-1-carboxamide, 3-([(4-fluorophenyl)methylamino]-sulfonyl)-*N*-(2-methylpropyl)-*N*-2-propenyl-	**79:** P92235n
-1-carboxamide, 3-([(4-fluorophenyl)methylamino]-sulfonyl)-*N*-2-propenyl-*N*-propyl-	**79:** P92235n
-1-carboxamide, 3-[(hexahydro-1H̲-azepin-1-yl)-sulfonyl-*N*,*N*-di-2-propenyl-	**79:** P92235n
-1-carboxamide, 3-[(hexahydro-1(2H̲)-azocinyl)-sulfonyl]-*N*,*N*-di-2-propenyl-	**79:** P92235n
-1-carboxamide, *N*-hexyl-3-[(2-methoxyethyl)-sulfonyl]-*N*-2-propenyl-	**79:** P92235n, **84:** P175174w
-1-carboxamide, 3-[(hexylmethylamino)sulfonyl]-*N*,*N*-di-2-propenyl-	**79:** P92235n
-1-carboxamide, *N*-hexyl-*N*-methyl-3-(1-piperidinyl-sulfonyl)-	**79:** P92235n
-1-carboxamide, 3-[(hexyl-2-propenylamino)sulfonyl]-*N*,*N*-di-2-propenyl-	**79:** P92235n
-1-carboxamide, 3-[(hexylpropylamino)sulfonyl]-*N*,*N*-di-2-propenyl-	**79:** P92235n
-1-carboxamide, *N*-isopropyl-3-[(2-methoxyethyl)-sulfonyl]-*N*-2-propenyl-	**79:** P92235n
-1-carboxamide, *N*-isopropyl-3-[(2-methoxyethyl)-sulfonyl]-*N*-propyl-	**79:** P92235n
-1-carboxamide, *N*-isopropyl-3-[(methyl-2-propenyl-amino)sulfonyl]-*N*-2-propenyl-	**79:** P92235n
-1-carboxamide, 3-(isopropylsulfonyl)-*N*,*N*-di-2-propenyl-	**76:** P72528v
-1-carboxamide, *N*-(2-methoxyethyl)-3[(2-methoxyethyl)-sulfonyl]-*N*-propyl-	**79:** P92235n, **84:** P175174w
-1-carboxamide, *N*-(2-methoxyethyl)-3-[(1-methylethyl)-sulfonyl]-*N*-propyl-	**79:** P92235n, **84:** P175174w
-1-carboxamide, *N*-(2-methoxyethyl)-3-[(1-methyl-propyl)sulfonyl]-*N*-propyl-	**79:** P92235n, **84:** P175174w
-1-carboxamide, *N*-(2-methoxyethyl)-3-[(2-methyl-propyl)sulfonyl]-*N*-propyl-	**79:** P92235n, **84:** P175174w
-1-carboxamide, *N*-(2-methoxyethyl)-3-(pentylsulfonyl)-*N*-propyl-	**79:** P92235n, **84:** P175174w
-1-carboxamide, *N*-(2-methoxyethyl)-3-(1-piperidinyl-sulfonyl)-*N*-propyl-	**79:** P92235n
-1-carboxamide, *N*-(2-methoxyethyl)-*N*-2-propenyl-3-(propylsulfonyl)-	**79:** P92235n, **84:** P175174w

TABLE 12 *(Continued)*

Compound	Reference
12.2. *S*-Substituted Thio-, Sulfinyl-, Sulfonyl-, and Acylthio-1,2,4-Triazole-carboxylic Acids and Their Functional Derivatives *(Continued)*	
-1-carboxamide, 3-([(2-methoxyethyl)propylamino]-sulfonyl)-*N,N*-di-2-propenyl-	**79:** P92235n
-1-carboxamide, *N*-(2-methoxyethyl)-*N*-propyl-3-(propylsulfonyl)-	**79:** P92235n, **84:** P175174w
-1-carboxamide, 3-[(2-methoxyethyl)sulfinyl]-*N,N*-di-2-propenyl-	**79:** P92235n, **84:** P175174w
-1-carboxamide, 3-[(2-methoxyethyl)sulfonyl]-*N,N*-bis(2-methyl-2-propenyl)-	**79:** P92235n, **84:** P175174w
-1-carboxamide, 3-[(2-methoxyethyl)sulfonyl]-*N,N*-di-2-propenyl-	**79:** P92235n, **84:** P175174w
-1-carboxamide, 3-[(2-methoxyethyl)sulfonyl]-*N,N*-dipropyl-	**79:** P92235n, **84:** P175174w
-1-carboxamide, 3-[(2-methoxyethyl)sulfonyl]-*N*-(1-methylethyl)-*N*-2-propenyl-	**84:** P175174w
-1-carboxamide, 3-[(2-methoxyethyl)sulfonyl]-*N*-(1-methylethyl)-*N*-propyl-	**84:** P175174w
-1-carboxamide, 3-[(2-methoxyethyl)sulfonyl]-*N*-(2-methyl propyl)-*N*-2-propenyl-	**79:** P92235n, **84:** P175174w
-1-carboxamide, 3-[(2-methoxyethyl)sulfonyl]-*N*-(1-methylpropyl)-*N*-propyl-	**79:** P92235n, **84:** P175174w
-1-carboxamide, 3-[(2-methoxyethyl)sulfonyl]-*N*-pentyl-*N*-2-propenyl-	**79:** P92235n, **84:** P175174w
-1-carboxamide, 3-[(2-methoxyethyl)sulfonyl]-*N*-2-propenyl-*N*-propyl-	**79:** P92235n, **84:** P175174w
-1-carboxamide, 3-[(2-methoxyethyl)sulfonyl]-*N*-propyl-*N*-2-propynyl-	**79:** P92235n, **84:** P175174w
-1-carboxamide, *N*-(methoxymethyl)-3-[(1-methylpropyl)sulfonyl]-*N*-propyl-	**79:** P92235n, **84:** P175174w
-1-carboxamide, *N*-(methoxymethyl)-*N*-propyl-3-(propylsulfonyl)-	**79:** P92235n, **84:** P175174w
-1-carboxamide, 3-[(3-methoxypropyl)sulfonyl]-*N*-2-propenyl-*N*-propyl-	**79:** P92235n, **84:** P175174w
-1-carboxamide, 3-([2-(1-methylethoxy)ethyl]-sulfonyl)-*N*-2-propenyl-*N*-propyl-	**79:** P92235n, **84:** P175174w
-1-carboxamide, *N*-(1-methylethyl)-3-[(methyl-2-propenylamino)sulfonyl]-*N*-2-propenyl-	**79:** P92235n
-1-carboxamide, 3-([(1-methylethyl)propylamino]-sulfonyl)-*N,N*-di-2-propenyl-	**79:** P92235n
-1-carboxamide, 3-[(1-methylethyl)sulfinyl]-*N,N*-di-2-propenyl-	**76:** P72528v
-1-carboxamide, 3-[(1-methylethyl)sulfinyl]-*N,N*-dipropyl-	**76:** P72528v
-1-carboxamide, 3-[(1-methylethyl)sulfonyl]-*N,N*-dipropyl-	**76:** P72528v
-1-carboxamide, 3-[(1-methylethyl)thio]-*N,N*-di-2-propenyl-	**76:** P71528v
-1-carboxamide, 3-([methyl(1-methylpropyl)amino]-sulfonyl)-*N,N*-di-2-propenyl-	**79:** P92235n

TABLE 12 (*Continued*)

Compound	Reference
12.2. *S*-Substituted Thio-, Sulfinyl-, Sulfonyl-, and Acylthio-1,2,4-Triazole-carboxylic Acids and Their Functional Derivatives (*Continued*)	
-1-carboxamide, 3-([methyl(2-methylpropyl)amino]-sulfonyl)-*N*,*N*-di-2-propenyl-	**79:** P92235n
-1-carboxamide, 3-([methyl(1-methylpropyl)amino]-sulfonyl)-*N*-2-propenyl-*N*-propyl-	**79:** P92235n
-1-carboxamide, 3-[(methylpentylamino)sulfonyl]-*N*,*N*-di-2-propenyl-	**79:** P92235n
-1-carboxamide, 3-[(4-methyl-1-piperidinyl)sulfonyl]-*N*,*N*-di-2-propenyl-	**79:** P92235n
-1-carboxamide, *N*-methyl-3-(1-piperidinylsulfonyl)-*N*-2-propenyl-	**79:** P92235n
-1-carboxamide, *N*-methyl-3-(1-piperidinylsulfonyl)-*N*-propyl-	**79:** P92235n
-1-carboxamide, 3-[(methyl-2-propenylamino)sulfonyl]-*N*,*N*-di-2-propenyl-	**79:** P92235n
-1-carboxamide, 3-[(methyl-2-propenylamino)sulfonyl]-*N*-2-propenyl-*N*-propyl-	**79:** P92235n
-1-carboxamide, 3-[(methyl-2-propenylamino)sulfonyl]-*N*-propyl-*N*-2-propynyl-	**79:** P92235n
-1-carboxamide, 3-[(1-methylpropyl)sulfinyl]-*N*,*N*-di-2-propenyl-	**76:** P72528v
-1-carboxamide, 3-[(2-methylpropyl)sulfinyl]-*N*,*N*-di-2-propenyl-	**76:** P72528v
-1-carboxamide, 3-[(1-methylpropyl)sulfinyl]-*N*,*N*-dipropyl-	**76:** P72528v
-1-carboxamide, 3-[(2-methylpropyl)sulfinyl]-*N*,*N*-dipropyl-	**76:** P72528v
-1-carboxamide, 3-[(1-methylpropyl)sulfonyl]-*N*,*N*-di-2-propenyl-	**76:** P72528v
-1-carboxamide, 3-[(2-methylpropyl)sulfonyl]-*N*,*N*-di-2-propenyl-	**76:** P72528v
-1-carboxamide, 3-[(1-methylpropyl)sulfonyl]-*N*,*N*-dipropyl-	**76:** P72528v
-1-carboxamide, 3-[(2-methylpropyl)sulfonyl]-*N*,*N*-dipropyl-	**76:** P72528v
-1-carboxamide, 3-[(1-methylpropyl)thio]-*N*,*N*-di-2-propenyl-	**76:** P72528v
-1-carboxamide, 3-[(2-methylpropyl)thio]-*N*,*N*-di-2-propenyl-	**76:** P72528v
-1-carboxamide, 3-[(1-methylpropyl)thio]-*N*,*N*-dipropyl-	**76:** P72528v
-1-carboxamide, 3-[(2-methylpropyl)thio]-*N*,*N*-dipropyl-	**76:** P72528v
-1-carboxamide, 3-(methylsulfonyl)-*N*,*N*-di-2-propenyl-	**76:** P72528v
-1-carboxamide, 3-(methylsulfonyl)-*N*,*N*-dipropyl-	**76:** P72528v
-1-carboxamide, 3-[(pentylpropylamino)sulfonyl]-*N*,*N*-di-2-propenyl-	**79:** P92235n
-1-carboxamide, 3-pentylsulfinyl)-*N*,*N*-di-2-propenyl-	**76:** P72528v

TABLE 12   (*Continued*)

Compound	Reference
12.2. *S*-Substituted Thio-, Sulfinyl-, Sulfonyl-, and Acylthio-1,2,4-Triazole-carboxylic Acids and Their Functional Derivatives (*Continued*)	
-1-carboxamide, 3-(pentylsulfinyl)-*N,N*-dipropyl-	**76**: P72528v
-1-carboxamide, 3-(pentylsulfonyl)-*N,N*-di-2-propenyl-	**76**: P72528v
-1-carboxamide, 3-(pentylsulfonyl)-*N,N*-dipropyl-	**76**: P72528v
-1-carboxamide, 3-(pentylthio)-*N,N*-di-2-propenyl-	**76**: P72528v
-1-carboxamide, 3-(pentylthio)-*N,N*-dipropyl-	**76**: P72528v
-1-carboxamide, 3-(1-piperidinylsulfonyl)-*N,N*-di-2-propenyl-	**79**: P92235n
-1-carboxamide, 3-(1-piperidinylsulfonyl)-*N*-2-propenyl-*N*-propyl-	**79**: P92235n
-1-carboxamide, 3-(1-piperidinylsulfonyl)-*N*-propyl-*N*-2-propynyl-	**79**: P92235n
-1-carboxamide, *N*-2-propenyl-*N*-propyl-3-(1-pyrrolidinylsulfonyl)-	**79**: P92235n
-1-carboxamide, 3-(2-propenylthio)-*N,N*-dipropyl-	**76**: P72528v
-1-carboxamide, 3-(2-propenylthio)-*N*-propyl-*N*-2-propynyl-	**76**: P72528v
-1-carboxamide, *N,N,*3(or 5)-trimethyl-5(or 3)-(methylsulfonyl)-	**67**: P54140x
-1-carboxamide, *N,N,*3(or 5)-trimethyl-5(or 3)-(methylthio)-	**67**: P54140x
-1-carboxamide, *N,N,*5-trimethyl-3-[(*p*-nitrophenyl)thio]-	**67**: P54140x
-3-carboxanilide, 5-(benzylsulfonyl)-1-phenyl	**75**: 98505q
-3-carboxanilide, 5-(benzylsulfonyl)-1-*o*-tolyl-	**75**: 98505q
-3-carboxanilide, 5-(benzylsulfonyl)-1-*p*-tolyl-	**75**: 98505q
-3-carboxanilide, 5-(benzylthio)-1-phenyl-	**75**: 98505q
-3-carboxanilide, 5-(benzylthio)-1-*o*-tolyl-	**75**: 98505q
-3-carboxanilide, 5-(benzylthio)-1-*p*-tolyl-	**75**: 98505q
-3-carboxylic acid, 5-(benzylthio)-1-phenyl-	**75**: 98505q
-3-carboxylic acid, 5-(benzylthio)-1-phenyl-, ethyl ester	**75**: 98505q
-3-carboxylic acid, 5-(benzylthio)-1-phenyl-, 2-phenylhydrazide	**75**: 98505q
-3-carboxylic acid, 5-(benzylthio)-1-*o*-tolyl-	**75**: 98505q
-3-carboxylic acid, 5-(benzylthio)-1-*p*-tolyl-	**75**: 98505q
-3-carboxylic acid, 5-(benzylthio)-1-*o*-tolyl-, ethyl ester	**75**: 98505q
-3-carboxylic acid, 5-(benzylthio)-1-*o*-tolyl-, 2-phenylhydrazide	**75**: 98505q
-3-carboxylic acid, 5-(benzylthio)-1-*p*-tolyl-, 2-phenylhydrazide	**75**: 98505q
-3-carboxylic acid, 5-benzyloxy-1-phenyl-, 2-phenylhydrazide	**76**: 59533y
-4-carboxylic acid, 3-[(ethoxycarbonyl)thio]-, ethyl ester	**77**: P171214s
-5-carboxylic acid, 1-methyl-3-nitro-, methyl ester	**72**: 111384j
-1-carboxylic acid, 3-(methylthio)-5-phenyl-, ethyl ester	**66**: 115651t
-3-carboxy-*o*-toluidide, 5-(benzylthio)-1-*o*-tolyl-	**75**: 98505q
1-(*N,N*-dimethylcarbamoyl)-3-[(*N,N*-dimethylcarbamoyl)thio]-	**77**: P34529j, **81**: P164763z

TABLE 12 (*Continued*)

Compound	Reference

TABLE 12 (*Continued*)

Compound	Reference

12.3. Miscellaneous 1,2,4-Triazolecarboxylic Acids and Their Functional Derivatives
(*Continued*)

Compound	Reference
-5-carboxamide, 3-[(diethoxyphosphinothioyl)oxy]-1-phenyl-	**77:** P140089a
-5-carboxamide, 3-[(dimethoxyphosphinothioyl)oxy]-1-phenyl-	**77:** P140089a
-5-carboxamide, *N*,*N*-dimethyl-3-[[ethoxy(propylthio)phosphinothioyl]oxy]-1-phenyl-,	**81:** P63640b
-5-carboxamide, *N*,*N*-dimethyl-3-[(methylsulfonyl)oxy]-1-phenyl-	**82:** P156321j
-3-carboxamide, *N*, 1-diphenyl-5-(2-phenylethoxy)-	**78:** 136152u
-5-carboxamide, 3-([ethoxy(ethyl)phosphinothioyl]oxy)-*N*,*N*-dimethyl-1-phenyl-	**77:** P140089a
-5-carboxamide, 3-([ethoxy(ethyl)phosphinothioyl]oxy)-*N*-methyl-1-phenyl-	**77:** P140089a
-5-carboxamide, 3-([ethoxy(methylthio)phosphinothioyl]oxy)-1-phenyl-,	**81:** P63640b
-1-carboxamide, 3-(ethylsulfonyl)-*N*-(2-methylpropyl)-*N*-2-propenyl-	**76:** P72528v
-5-carboxamide, 3-nitro-1-β-D-ribofuranosyl-	**80:** 121258a
-3-carboxamide, 5-(pentyloxy)-*N*, 1-diphenyl-	**78:** 136152u
-1-carboxanilide, 3-amino-5-(methylthio)-	**23:** 1639[9]
-1-carboxanilide, 5-amino-3-methylthio-	**70:** 77881r
-3-carboxanilide, 5-ethoxy-1-phenyl-	**75:** 98505q
-3-carboxanilide, 5-ethoxy-1-*o*-tolyl-	**75:** 98505q
-3-carboxanilide, 5-ethoxy-1-*p*-tolyl-	**75:** 98505q
-3-carboxanilide, 5-isopropoxy-1-*o*-tolyl-	**75:** 98505q
-3-carboxanilide, 5-isopropoxy-1-*p*-tolyl-	**75:** 98505q
-3-carboxanilide, 5-isopropoxy-1-phenyl-	**79:** 152590m
5-carboxy-3-diazonio-	**56:** 7307g
-3-carboxylic acid, 1-acetyl-5-methyl-	**67:** P44841n
-3-carboxylic acid, 5-[(2-amino-6-[(4-[(6-anilino-1-hydroxy-3-sulfo-2-naphthyl)azo]-3,3′-dimethoxy-4-biphenylyl)azo]-5-hydroxy-7-sulfo-1-naphthyl)azo]-	
-3-carboxylic acid, 5-([2-amino-6-([4-(*p*-[(3-carboxy-4-hydroxyphenyl)azo]benzamido)-2-methoxyphenyl]azo)-5-hydroxy-7-sulfo-1-naphthyl]azo)-	**60:** P6963c
-3-carboxylic acid, 5-([2-amino-6-([4-(*p*-[(3-carboxy-4-hydroxyphenyl)azo]benzamido)-6-methoxy-*m*-tolyl]-azo)-5-hydroxy-7-sulfo-1-naphthyl]azo)-	**60:** P6963c
-3-carboxylic acid, 5-(2-amino-6-[4′-(3-carboxy-4-hydroxyphenylazo)-3,3′-dimethoxy-4-biphenylylazo]-5-hydroxy-7-sulfo-1-naphthylazo)-	**52:** P6803i
-3-carboxylic acid, 5-[(8-amino-1-hydroxy-3,6-disulfo-2-naphthalenyl)azo]-	**76:** 80560q, **77:** 109080s, **78:** 131702p
-3-carboxylic acid, 5-[(1-[(*p*-aminophenyl)carbamoyl]-acetonyl)azo]-	**61:** P14829b
-3-carboxylic acid, 5-[(4-amino-2,5-xyly)azo]-	**57:** P6073a
-1-carboxylic acid, 3-amino-5-(2,4-xylyloxy)-, methyl ester	**71:** 61474v
-3-carboxylic acid, 5-azido-	**82:** 30784u

347

TABLE 12 (*Continued*)

Compound	Reference

12.3. Miscellaneous 1,2,4-Triazolecarboxylic Acids and Their Functional Derivatives (*Continued*)

Compound	Reference
-3-carboxylic acid, 5-azido-, ethyl ester	**82:** 30784u
-5-carboxylic acid, 3-azido-1-methyl-	**82:** 30784u
-5-carboxylic acid, 3-azido-1-methyl-, methyl ester	**82:** 30784u
-3-carboxylic acid, 5-azido-1-methyl-, methyl ester	**82:** 30784u
-3-carboxylic acid, 5-azido-, propyl ester	**82:** 30784u
-3-carboxylic acid, 5-benzyloxy-1-*o*-tolyl-, 2-phenylhydrazide	**76:** 59533y
-3-carboxylic acid, 5-benzyloxy-1-*p*-tolyl-, 2-phenylhydrazide	**76:** 59533y
-3-carboxylic acid, 1-bromo-, methyl ester	**71:** 91399s
-3-carboxylic acid, 5-carbamoyl-	**72:** 111375g
-3-carboxylic acid, 5-carbamoyl-, methyl ester	**72:** 111375g
-3-carboxylic acid, 5-[(3-carboxy-4-hydroxyphenyl)azo]-	**72:** P56717t
-3-carboxylic acid, 5-[(3-carboxy-4-hydroxyphenyl)azo]-, 5-methyl ester	**72:** P56717t
-3-carboxylic acid, 5-[(3-carboxy-4-methoxyphenyl)azo]-	**72:** P56717t
-3-carboxylic acid, 5-[(3-carboxy-4-methoxyphenyl)azo]-, dimethyl ester	**72:** P56717t
-3-carboxylic acid, 5-chloro-	**70:** 77876t
-4-carboxylic acid, 3-chloro-5-(2,2-dichloro-1,1-difluoroethyl)-, ethyl ester	**81:** 49647c
-3-carboxylic acid, 5-chloro-, methyl ester	**70:** 77876t, **84:** 180140m
-3-carboxylic acid, 5-chloro-1-(*p*-nitrophenyl)-, ethyl ester	**40:** 3444[2]
-5-carboxylic acid, 1-(3-chlorophenyl)-3-[(diethoxyphosphinothioyl)oxy]-, ethyl ester	**77:** P140089a
-5-carboxylic acid, 3-chloro-1-(2,3,5-tri-*O*-acetyl-β-D-ribofuranosyl)-, methyl ester	**80:** 121258a
-3-carboxylic acid, 5-chloro-1-(2,3,5-tri-*O*-acetyl-β-D-ribofuranosyl)-, methyl ester	**80:** 121258a
-3-carboxylic acid, 5-cyano-	**72:** 111375g
-3-carboxylic acid, 5-[[cyano(1,3,3-trimethyl-2-indolinylidene)methyl]azo]-	**61:** P750f
-3-carboxylic acid, 5-(cyclohexyloxy)-1-phenyl-, 2-phenylhydrazide	**78:** 136152u
-3-carboxylic acid, 3-diazo-, hydroxide	**42:** 6336a
-5-carboxylic acid, 3-diazo-, methyl ester	**73:** 45420k
-5-carboxylic acid, 3-[(diethoxyphosphinyl)oxy]-1-phenyl-, ethyl ester	**77:** P140089a
-5-carboxylic acid, 3-[diethoxyphosphinothioyl)oxy]-1-isopropyl-, ethyl ester	**77:** P140089a
-5-carboxylic acid, 3-[diethoxyphosphinothioyl)oxy]-1-phenyl-, ethyl ester	**77:** P140089a
-5-carboxylic acid, 3-[diethoxyphosphinothioyl)oxy]-1-phenyl-, methyl ester	**77:** P140089a
-3-carboxylic acid, 5-[[2-(diethylamino)-6-hydroxyphenyl]azo]-	**81:** P72295v
-3-carboxylic acid, 5-[(diformylmethyl)azo]-	**60:** 9383h

TABLE 12   (*Continued*)

Compound	Reference

12.3. Miscellaneous 1,2,4-Triazolecarboxylic Acids and Their Functional Derivatives (*Continued*)

Compound	Reference
-3-carboxylic acid, 5-[(1,8-dihydroxy-3,6-disulfo-2-naphthalenyl)azo]-	**76:** 67703u, **83:** 52878a
-5-carboxylic acid, 3-[(dimethoxyphosphinyl)oxy]-1-phenyl-, ethyl ester	**77:** P140089a
-5-carboxylic acid, 3-[(dimethoxyphosphinothioyl)oxy]-1-phenyl-, ethyl ester	**77:** P140089a
-3-carboxylic acid, 5-([*p*-(dimethylamino)phenyl]azo)-	**55:** P11863a, **57:** P6072i
-3-carboxylic acid, 5-[(1,1-dinitroethyl)azo]-	**77:** 126523f
-3-carboxylic acid, 5-[(1,1-dinitroethyl)azo]-, methyl ester	**77:** 126523f
-5-carboxylic acid, 3-[(ethoxyethylphosphinothioyl)oxy]-1-isopropyl-, ethyl ester	**77:** P140089a
-5-carboxylic acid, 3-[(ethoxyethylphosphinothioyl)oxy]-1-phenyl-, ethyl ester	**77:** P140089a
-5-carboxylic acid, 3-([ethoxy(methylthio)-phosphinothioyl]oxy)-1-phenyl-, ethyl ester	**81:** P63640b
-3-carboxylic acid, 5-ethoxy-1-phenyl-, 2-phenylhydrazide	**76:** 59533y
-5-carboxylic acid, 3-([ethoxy(propylthio)-phosphinothioyl]oxy)-1-phenyl-, ethyl ester	**81:** P63640b
-3-carboxylic acid, 5-ethoxy-1-*p*-tolyl-, 2-phenylhydrazide	**76:** 59533y
-3-carboxylic acid, 5-fluoro-, methyl ester	**80:** 14898n
-3-carboxylic acid, 5-[1-hydroxy-(4,8-disulfo-2-naphthyl)azo]-	**60:** P14668e, **60:** P14667b, **61:** P1982c
-3-carboxylic acid, 5-([2-hydroxy-3-([(2-methoxyphenyl)-amino]carbonyl)-1-naphthalenyl]azo)-	**76:** 80560q
-3-carboxylic acid, 5-([2-hydroxy-3-([(4-methoxyphenyl)-amino]carbonyl)-1-naphthalenyl]azo)-	**76:** 80560q, **79:** 152590m
-3-carboxylic acid, 1-methyl-5-nitro-	**72:** 111383h, **77:** 74519n
-3-carboxylic acid, 1-methyl-5-nitro-, methyl ester	**72:** 111384j
-3-carboxylic acid, 5-([3-methyl-5-oxo-1-(4-sulfo-*o*-tolyl)-2-pyrazolin-4-yl]azo)-	**61:** P1982c
-5-carboxylic acid, 3-[(methylpropoxyphosphinothioyl)-oxy]-1-phenyl-, ethyl ester	**77:** P140089a
-3-carboxylic acid, 5-nitro-	**70:** 77876t, **72:** 111383h
-3-carboxylic acid, 5-nitro-, hydrazide	**70:** 77876t
-3-carboxylic acid, 5-nitro-, methyl ester	**70:** 77876t, **72:** 111383h
-5-carboxylic acid, 3-nitro-1-(2,3,5-tri-*O*-acetyl-β-D-ribofuranosyl)-, methyl ester	**80:** 121258a
-3-carboxylic acid, 5-(pentyloxy)-1-phenyl-, 2-phenylhydrazide	**78:** 136152u
-3-carboxylic acid, 1-phenyl-5-(2-phenylethoxy)-, 2-phenylhydrazide	**78:** 136152u
-3-carboxylic acid, 5-[(trinitromethyl)azo]-	**77:** 126523f
-3-carboxylic acid, 5-[(trinitromethyl)azo]-, methyl ester	**77:** 126523f
-1-carboxylic acid, 3-[(triphenylphosphoranylidene)-amino]-5-(2,4-xylyloxy)-, methyl ester	**71:** 61474v

TABLE 12 (*Continued*)

Compound	Reference

12.3. Miscellaneous 1,2,4-Triazolecarboxylic Acids and Their Functional Derivatives (*Continued*)

Compound	Reference
3-[*N*-(carboxymethyl)carbamoyl]-5-ethoxy-1-(2-methylphenyl)-	**76:** 59533y
3-[*N*-(carboxymethyl)carbamoyl]-5-ethoxy-1-(4-methylphenyl)-	**76:** 59533y
3-[*N*-(carboxymethyl)carbamoyl]-5-ethoxy-1-phenyl-	**76:** 59533y
3-[*N*-(carboxymethyl)carbamoyl]-1-(4-methylphenyl)-5-(phenylmethoxy)-	**76:** 59533y
3-[*N*-(carboxymethyl)carbamoyl]-1-phenyl-5-phenylmethoxy)-	**76:** 59533y
-3-carboxy-*o*-toluidide, 5-ethoxy-1-phenyl-	**75:** 98505q
-3-carboxy-*o*-toluidide, 5-ethoxy-1-*p*-tolyl	**75:** 98505q
-3-carboxy-*o*-toluidide, 5-isopropoxy-1-phenyl-	**75:** 98505q
-3-carboxy-*o*-toluidide, 5-isopropoxy-1-*p*-tolyl-	**75:** 98505q
5-chloro-3-cyano-	**84:** 180140m
5-cyano-1-isopropyl-3-[(*N,N*-dimethylcarbamoyl)oxo]-	**78:** P4261u
-3-diazonium, 5-carboxy-, hydroxide, inner salt	**75:** 129025q
-3,5-dicarboxylic acid	**72:** 111375g, **76:**3766j
-4,5-dicarboxylic acid	**80:** 14221m
-3,5-dicarboxylic acid, dimethyl ester	**72:** 111375g
-3,5-dicarboxylic acid, 1-methyl-, dimethyl ester	**76:** 3766j
-1-thiocarboxamide, 5-amino-*N*-benzoyl-	**70:** 47370f
-1-thiocarboxamide, 5-amino-*N*-benzoyl-3-methyl-	**70:** 47370f
-1-thiocarboxamide, 5-amino-*N*-benzoyl-3-phenyl-	**70:** 47370f

12.4. Miscellaneous Acyl-1,2,4-triazoles

Compound	Reference
3-acetamido-4-(acetonyl)-5-butyryl-	**76:** 58494z
3-acetamido-1(or 2)-acetyl-	**38:** 970[4], **55:** 3565i
3-acetamido-1-acetyl-5-amino-	**55:** 3565h
5-acetamido-1(or 4)-acetyl-3-methyl-	**38:** 970[4]
5-acetamido-1(or 2 or 4)-acetyl-3-phenyl-	**55:** 3566a, **58:** 519d, **59:** 12791e
5-acetamido-1-(1-piperidinylcarbonyl)-	**82:** P156320h
4-(acetonyl)-3-amino-5-butyryl-	**76:** 58494z
4-(acetonyl)-3-(diacetylamino)-5-butyryl-	**76:** 58494z
3-acetoxy-1-acetyl-	**65:** 7169h
1(or 4)-acetyl-3-amino-	**53:** 21901i, **55:** 3565i, **59:** 6227e
1-acetyl-5-amino-	**77:** 101474c
1-acetyl-5-amino-3-[14]C-,	**77:** 101474c
4-acetyl-3-amino-5-anilino-	**62:** 9124c
1-acetyl-3-[(4-amino-1-anthraquinonyl)thio]-	**74:** P14185d
4-acetyl-3-amino-5-(*p*-bromophenyl)-	**59:** 12790e
4-acetyl-3-amino-5-(*p*-methoxyphenyl)-	**59:** 6227e
1-acetyl-5-amino-3-(5-nitro-2-furyl)-	**75:** 20356x
1(or 2 or 4)-acetyl-5-amino-3-phenyl-	**55:** 3566b, **58:** 519d, **59:** 6227e, **59:** 12791e, **63:** 6817f

TABLE 12 (*Continued*)

Compound	Reference

12.4. Miscellaneous Acyl-1,2,4-triazoles (*Continued*)

Compound	Reference
1-acetyl-3-anilino-5-ethyl-	**62:** 11803h
1-acetyl-3-anilino-5-methyl-	**62:** 11803h
4-acetyl-3-(benzylthio)-5-phenyl-	**66:** 115651t
4-acetyl-3,5-bis(*p*-hydroxyphenylazo)-, diacetate	**28:** 2714[6]
4-acetyl-3-(*p*-bromophenyl)-5-(dimethylamino)-	**59:** 6227e
1-acetyl-3,5-diamino-	**55:** 3565i
1-acetyl-3-[2,3-dihydro-1H-cyclopenta[*b*]quinolin-9-yl)-5-(methylthio)-	**84:** 17287f
4-acetyl-3-(dimethylamino)-5-phenyl-	**59:** 6227e
4-acetyl-3-(3,5-dimethylpyrazol-1-yl)-	**59:** 10049e
1-acetyl-3-(2,3-dimethyl-4-quinolinyl)-5-(methylthio)-	**84:** 17287f
1-acetyl-3-[2-ethyl-4-quinolinyl]-5-(methylthio)-	**84:** 17287f
3-acetyl-5-methyl-4-(4-nitroanilino)-	**83:** 9917a
1-acetyl-5-(2-methyl-3-phenyl-4-quinolinyl)-3-(methylthio)-	**84:** 17287f
1-acetyl-3-(2-methyl-4-quinolinyl)-5-(methylthio)-	**84:** 17287f
1-acetyl-3-(methylthio)-5-(1,2,3,4-tetrahydro-9-acridinyl)-	**84:** 17287f
3-acetyl-4-(4-nitroanilino-5-(4-nitrophenyl)-	**83:** 9917a
1-acetyl-5-(2-oxo-1,2-diphenylethyl)amino-	**84:** 59321t
1-acetyl-3-(2,2,2-trifluoroacetamido)	**74:** 22770f
1-acetyl-3-(2,4,6-trinitroanilino)-	**68:** 105111u
4-acetyl-3-(2,4,6)-trinitroanilino)-	**68:** 10511u
5-amino-1(or 2 to 4)-benzoyl-	**53:** 21901i, **59:** 6227e, **59:** 12791e, **66:** 104601a
3-amino-4-benzoyl-5-(*p*-bromophenyl)-	**59:** 12790f
3-amino-4-benzoyl-5-ethyl-	**59:** 6227e, **59:** 12791e
5-amino-1(or 4)-benzoyl-3-methyl-	**59:** 12790e, **63:** 6817f, **66:** 104601a
5-amino-1-benzoyl-3-(5-nitro-2-furanyl)-	**79** 87327q
3-amino-4-benzoyl-5-(*p*-nitrophenyl)-	**59:** 12790f
3-amino-4-benzoyl-5-pentyl-	**59:** 6227e
5-amino-1(or 4)-benzoyl-3-phenyl-	**58:** 519b, **59:** 6227e, **59:** 12791e, **63:** 6817f, **66:** 104601a
5-amino-1-(4-chlorobenzoyl)-3-(3-pyridinyl)-	**81:** P63636e
5-amino-1-(4-chlorobenzoyl)-3-(4-pyridinyl)-	**81:** P63636e
5-amino-1-decanoyl-	**77:** 101474c
5-amino-1-dodecanoyl-	**77:** 101474c
5-amino-1-[(ethoxycarbonylamino)thioacetyl]-3-methyl-	**78:** 136237a
5-amino-1-heptanoyl-	**77:** 101474c
5-amino-1-(2-methylbenzoyl)-3-(4-pyridinyl)-	**81:** P63636e
3-amino-4-(*p*-nitrobenzoyl)-5-phenyl-	**58:** 10202f
5-amino-1-nonanoyl-	**77:** 101474c
3-anilino-1-benzoyl-5-ethyl-	**62:** 11803e
3-anilino-1-benzoyl-5-methyl-	**62:** 11803e
3-anilino-1-benzoyl-3-phenyl-	**59:** 12790b

TABLE 12 (*Continued*)

Compound	Reference

12.4. Miscellaneous Acyl-1,2,4-triazoles (*Continued*)

3-anilino-1-benzoyl-5-propyl-	**62:** 11803e
3-benzamido-5-benzoyl)	**57:** 5897d
5-benzamido-1(or 4)-benzoyl-3-phenyl-	**58:** 519d, **59:** 12791e
1-benzoyl-5-[2-(benzoyl)hydrazino]-3-phenyl-	**82:** 4183c
1(or 4)-benzoyl-5-(benzylamino)-3-phenyl-	**59:** 6227e, **59:** 12790a
1(or 4)-benzoyl-5-(dimethylamino)-3-phenyl-	**59:** 6227e, **59:** 12790d, **59:** 12791e
1-benzoyl-5-ethyl-3-(*N*-phenylbenzamido)-	**62:** 11803g
1-benzoyl-3-(*N*-phenylbenzamido)-5-propyl-	**62:** 11803h
1-benzoyl-3-phenyl-5-[2-(phenyl)hydrazino]-	**82:** 4183c
3,5-bis(acetamido)-1-acetyl-	**55:** 3565h
3,5-bis(anilino)-1-benzoyl-	**32:** 3399[6]
5-butyramido-1(or 2)-butyryl-	**38:** 970[4]
5-butyramido-1(or 4) -butyryl-3-methyl-	**38:** 970[5]
3,5-diamino-1-(*p*-nitrobenzoyl)-	**60:** 1753e
5-(dimethylamino))-3-methyl-1-(*p*-nitrobenzoyl)-	**59:** 12789h
1-(2-ethylbutyryl)-3-phenyl-5-[2-(phenylacetyl)hydrazino]-	**82:** 4183c
3-methyl-5-propionamido-1(or 4)1propionyl-	**38:** 970[5]
5-propionamido-1(or 2)-propionyl-	**38:** 970[4]

12.5. Amino-*S*-Substituted Thio-, Sulfinyl- or Sulfonyl-1,2,4-triazoles

3-acetamido-5-(benzylthio)-	**77:** 56420u
1-(*N*-acetylsulfanilyl)-5-amino-3-methyl-	**36:** 5793[3]
3-(β-allylthiòcarbamido)-5-(benzylthio]-	**23:** 2178[2]
3-(β-allylthiocarbamido)-5-(benzylthio)-1-phenyl-	**23:** 2178[2]
5-(β-allylthiocarbamido)-3-(benzylthio)-1-phenyl-	**23:** 2178[2]
3-(β-allylthiocarbamido)-5-methylthio)-1-phenyl-	**23:** 2178[2]
4-amino-3-anilino-5-(benzylthio)-	**69:** 77170n
4-amino-3-anilino-5-(methylthio)-	**64:** 8173f, **69:** 77170n
4-amino-3-(benzoylthio)-5-ethyl-	**74:** 141645d
3-amino-5-(benzylthio)-	**54:** 19695f, **55:** 2624d, **62:** 9124d, **79:** 5307w
4-amino-3-(benzylthio)-	**57:** 805b
5-amino-3-(benzylthio)-1-bis(dimethylamino)phosphinyl-	**80:** P82992p
3-amino-5-(benzylthio)-4-(*p*-bromophenyl)-	**60:** 1755a
3-amino-5-(benzylthio)-4-(*p*-chlorophenyl)-	**60:** 1755a
3-amino-5-(benzylthio)-4-cyclohexyl-	**60:** 1755a
4-amino-3-(benzylthio)-5-ethyl-	**57:** 805b, **74:** 141645d
3-amino-5-(benzylthio)-4-isobutyl-	**60:** 1755a
3-amino-5-(benzylthio)-4-(*p*-methoxyphenyl)-	**60:** 1755a
4-amino-3-(benzylthio)-5-methyl-	**57:** 805b
3-amino-5-(benzylthio)-4-phenyl-	**56:** 12879g
4-amino-3-(benzylthio)-5-phenyl-	**60:** 526c
3-amino-5-(benzylthio)-4-(α,α,α-trifluoro-*m*-tolyl)-	**60:** 1755a
4-amino-3,5-bis(benzylthio)-	**56:** 14293i, **56:** 15513b, **57:** 12472a, **69:** 83184m, **71:** 56483t
4-amino-3,5-bis(butylthio)-	**80:** 82830j

TABLE 12 (*Continued*)

Compound	Reference

12.5.   Amino-*S*-Substituted Thio-, Sulfinyl- or Sulfonyl-1,2,4-triazoles   (*Continued*)

Compound	Reference
5-amino-1-bis(dimethylamino)phosphinyl- 3-(methylsulfonyl)-	**80:** P82992p
5-amino-1-bis(dimethylamino)phosphinyl-3-(propylthio)-	**80:** P82992p
4-amino-3,5-bis(ethylthio)-	**80:** 82830j
4-amino-3,5-bis(methylthio)	**56:** 15513b, **57:** 12472a, **61:** 10676a, **71:** 21509d, **79:** 31993m
4-amino-3,5-bis(propylthio)-	**80:** 82830j
4-amino-3-([(4-bromophenyl)methyl]thio)-5-methyl-	**84:** 17271w
4-amino-3-([(4-butoxyphenyl)methyl]thio)-	**84:** 30975y
4-amino-3-([(4-butoxyphenyl)methyl]thio)-5-butyl-	**84:** 30975y
4-amino-3-([(4-butoxyphenyl)methyl]thio)-5-ethyl-	**84:** 30975y
4-amino-3-([(4-butoxyphenyl)methyl]thio)- 5-(4-methoxyphenyl)-	**84:** 30975y
4-amino-3-([(4-butoxyphenyl)methyl]thio)-5-methyl-	**84:** 30975y
4-amino-3-([(4-butoxyphenyl)methyl]thio)-5-methyl- *N*-(phenylmethylene)-	**84:** 30975y
4-amino-3-butyl-5-([(4-ethoxyphenyl)methyl]thio)-	**84:** 30975y
4-amino-3-butyl-5-([(4-methoxyphenyl)methyl]thio)-	**84:** 30975y
4-amino-3-(carboxymethylthio)-5-phenyl-	**48:** 4525c
3-amino-5-[(*p*-chlorobenzyl)thio]-	**55:** 2624d
3-amino-5-[(*p*-chlorobenzyl)thio]-4-methyl-	**56:** 12879a
4-amino-5-(4-chlorophenyl)-3-([α-(ethoxycarbonyl)- benzyl]thio)-	**78:** 72075m
4-amino-5-(4-chlorophenyl)-3-([α-(ethoxycarbonyl)- α-methylbenzyl]thio)-	**78:** 72075m
4-amino-5-(2-chlorophenyl)-3-[(ethoxycarbonyl- methyl)thio]-	**78:** 72075m
4-amino-5-(4-chlorophenyl)-3-[(ethoxycarbonyl- methyl)thio]-	**78:** 72075m
3-amino-5-([(4-chlorophenyl)methyl]thio)-	**79:** 5307w, **80:** P83040p
3-amino-4-(2,6-dicarboxy-1,4-dihydro-4-oxo- 1-pyridinyl)-5-(methylthio)-	**83:** 193226y
4-amino-3,5-di(carboxymethylthio)-	**69:** 83184m, **71:** 56483t
4-amino-3-([(2,6-dichlorophenyl)methyl]thio)-	**85:** 21230v
4-amino-3-([2-(diethylamino)ethyl]thio)-5-ethyl-	**74:** 141645d
4-amino-3-([α-(ethoxycarbonyl)benzyl]thio)- 5-(3-methylphenyl)-	**78:** 72075m
4-amino-3-([α-(ethoxycarbonyl)benzyl]thio)- 5-(4-methylphenyl)-	**78:** 72075m
4-amino-3-([α-(ethoxycarbonyl)benzyl]thio)-5-phenyl-	**78:** 72075m
4-amino-3-([α-(ethoxycarbonyl)-α-methylbenzyl]thio)- 5-(4-methylphenyl)-	**78:** 72075m
4-amino-3-([α-(ethoxycarbonyl)-α-methylbenzyl]thio)- 5-phenyl-	**78:** 72075m
4-amino-3-[(ethoxycarbonylmethyl)thio]- 5-(3-methylphenyl)-	**78:** 72075m
4-amino-3-[(ethoxycarbonylmethyl)thio]- 5-(4-methylphenyl)-	**78:** 72075m

TABLE 12   (*Continued*)

Compound	Reference
12.5. Amino-S-Substituted Thio-, Sulfinyl- or Sulfonyl-1,2,4-triazoles (*Continued*)	
4-amino-3-[(ethoxycarbonylmethyl)thio]-5-(4-pyridinyl)-	**78:** 72075m
4-amino-3-([(4-ethoxyphenyl)methyl]thio)-	**84:** 30975y
4-amino-3-([(4-ethoxyphenyl)methyl]thio)-5-ethyl-	**84:** 30975y
4-amino-3-([(4-ethoxyphenyl)methyl]thio)-5-(4-methoxy-phenyl)-	**84:** 30975y
4-amino-3-([(4-ethoxyphenyl)methyl]thio)-5-methyl-	**84:** 30975y
4-amino-3-([(4-ethoxyphenyl)methyl]thio)-5-propyl-	**84:** 30975y
4-amino-3-ethyl-5-([(4-methoxyphenyl)methyl]thio)-	**84:** 30975y
4-amino-3-ethyl-5-[([4-(1-methylethoxy)phenyl]-methyl)thio]-	**84:** 30975y
4-amino-3-ethyl-5-[([4-(2-methylpropoxy)phenyl]-methyl)thio]-	**84:** 30975y
4-amino-3-ethyl-5-(methylthio)-	**55:** 23508d
4-amino-5-ethyl-3-([2-(4-morpholinyl)ethyl]thio)-	**74:** 141645d
4-amino-5-ethyl-3-(phenacylthio)-	**72:** 3470a
4-amino-3-ethyl-5-([(4-propoxyphenyl)methyl]thio)-	**84:** 30975y
4-amino-3-ethyl-5-(2-propynylthio)-	**74:** 141645d
3-amino-5-(ethylthio)-	**80:** P82992p
4-amino-3-(ethylthio)-5-(methylthio)-	**80:** 82830j
3-amino-5-(isopropylthio)-	**80:** P82992p
4-amino-3-(4-methoxyphenyl)-5-([(4-methoxyphenyl)-methyl]thio)-	**84:** 30975y
4-amino-3-([(4-methoxyphenyl)methyl]thio)-	**84:** 30975y
4-amino-5-(p-methoxyphenyl)-3-(methylthio)-	**48:** 4525a
4-amino-3-([(4-methoxyphenyl)methyl]thio)-5-methyl-	**84:** 30975y
4-amino-3-[([4-(3-methylbutoxy)phenyl]methyl)thio]-	**84:** 30975y
4-amino-3-[([4-(3-methylethoxy)phenyl]methyl)thio]-	**84:** 30975y
4-amino-3-(3-methyl-5-isoxazolyl)-5-(methylthio)-	**78:** 43419w
4-amino-3-(5-methyl-3-isoxazolyl)-5-(methylthio)-	**78:** 43419w
4-amino-3-methyl-5-[([4-(2-methylpropoxy)phenyl]-methyl)thio]-	**84:** 30975y
4-amino-3-methyl-5-[([4-(1-methylethoxy)phenyl]-methyl)thio]-	**84:** 30975y
4-amino-3-methyl-5-[([4-(2-methylpropoxy)phenyl]-methyl)thio]-	**84:** 30975y
3-amino-1-methyl-3-(methylthio)-	**81:** 25613n
4-amino-3-methyl-5-(methylthio)-	**48:** 4525a, **55:** 23508b
3-amino-5-([(1-methyl-5-nitro-1H-imidazol-2-yl)-methyl]thio)-	**83:** P114412e
3-amino-1-methyl-5-[(phenylmethyl)thio]-	**77:** 56420u
4-amino-3-[([4-(2-methylpropoxy)phenyl]methyl)thio]-	**84:** 30975y
4-amino-5-methyl-3-([(4-propoxyphenyl)methyl]thio)-	**84:** 30975y
4-amino-3-(5-methyl-1H-pyrazol-3-yl)-5-(methylthio)-	**78:** 136235y
3-amino-5-(methylsulfonyl)-	**57:** 16613g
3-amino-5-(methylsulfonyl)-1-phenyl-	**23:** 2178³
4-amino-3-(methylsulfonyl)-5-phenyl-	**48:** 4524i
3-amino-5-(methylthio)-	**59:** 1623b, **81:** 25613n
4-amino-3-(methylthio)-	**55:** 23507e

TABLE 12 (*Continued*)

Compound	Reference

12.5. Amino-*S*-Substituted Thio-, Sulfinyl- or Sulfonyl-1,2,4-triazoles (*Continued*)

Compound	Reference
3-amino-5-(methylthio)-1-phenyl-	**84:** 135551k
4-amino-3-(methylthio)-5-phenyl-	**48:** 4524i, **61:** 14662h,
	78: 43419w
5-amino-3-(methylthio)-1-phenyl-	**23:** 2178[2]
4-amino-3-(methylthio)-5-(5-phenyl-3-isoxazolyl)-	**78:** 43419w
4-amino-5-(methylthio)-3-(3-pyridazinyl)-	**68:** 105174s
4-amino-3-(methylthio)-5-*p*-toluidino-	**69:** 77170n
3-amino-5-[[(4-nitrophenyl)methyl]thio]-	**79:** 5307w
3-amino-5-[(5-nitro-2-thiazolyl)thio]-4-phenyl-	**78:** P4286f
4-amino-3-[([4-(pentyloxy)phenyl]methyl)thio]-	**84:** 30975y
3-amino-5-[(phenylmethyl)sulfonyl]-	**84:** 130811k
3-amino-5-phenyl-4-(*p*-tolylsulfonyl)-	**59:** 6227e
5-amino-3-phenyl-1-(*p*-tolylsulfonyl)-	**58:** 519c
4-amino-3-([(4-propoxyphenyl)methyl]thio)-	**84:** 30975y
4-amino-3-([(4-propoxyphenyl)methyl]thio)-5-propyl-	**84:** 30975y
3-amino-5-(propylthio)-	**80:** P95962d
3-amino-5-(2-propynylthio)-	**81:** P129840a, **81:** P144229b
4-amino-3-([2-(2,2,2-trichloroethoxycarbonyl)-	**82:** P43443g
3-methyl-8-oxo-7-[(phenylacetyl)amino]-5-thia-	
1-azabicyclo[4·2·0]oct-2-en-4yl]thio)-5-phenyl-,	
5-*S*-oxide	
4-anilino-3-benzylsulfonyl-5-phenyl-	**84:** 4865t
3-anilino-5-(benzylthio)-	**62:** 9124c, **79:** 42423b
3-anilino-5-(benzylthio)-4-(2,3-diphenylguanidino)-	**69:** 77170n
3-anilino-5-(benzylthio)-4-phenyl-	**21:** 2900[5], **70:** 3963x
4-anilino-3-(benzylthio)-5-phenyl-	**84:** 4865t
3-anilino-4-(4-bromophenyl)-5-(methylthio)-	**80:** 14879g
3-anilino-5-[(ethoxycarbonyl)thio]-4-phenyl-	**77:** P171214s
3-anilino-5-(ethylthio)-4-phenyl-	**49:** 2447a
4-anilino-3-(ethylthio)-5-phenyl-	**84:** 4865t
4-anilino-3-(ethylthio)-5-*p*-toluidino-	**72:** 55347k
3-anilino-5-(methylsulfinyl)-	**63:** 7012h
3-anilino-5-(methylsulfonyl)-	**63:** 7012h
3-anilino-5-(methylsulfonyl)-phenyl-	**80:** 14879g
4-anilino-3-(methylsulfonyl)-5-phenyl-	**84:** 4865t
3-anilino-5-(methylthio)-4-(2,3-diphenylguanidino)-	**69:** 77170n
3-anilino-5-(methylthio)-4-phenyl-	**21:** 2900[5], **49:** 2446h,
	66: 115650s, **71:** 91395n
3-*p*-anisidino-5-(benzylthio)-4-(*p*-methoxyphenyl)-,	**70:** 3963x
4-benzamido-3-(methylthio)-5-phenyl-	**48:** 4524i
5-benzamido-3-(methylthio)-1-phenyl-	**23:** 2178[2]
3-(benzoylthio)-5-ethyl-4-(isopropylideneamino)-	**74:** 141645d
4-(benzylideneamino)-3-(benzylthio)-	**74:** 13070a
4-(benzylideneamino)-3-(benzylthio)-5-methyl-	**57:** 805b
4-(benzylideneamino)-3-(benzylthio)-5-phenyl-	**60:** 526c
4-(benzylideneamino)-3,5-bis(benzylthio)-	**57:** 12472b
4-(benzylideneamino)-3,5-bis(methylthio)-	**56:** 15513b
4-(benzylideneamino)-3-([(4-butoxyphenyl)methyl]thio)-	**84:** 30975y

TABLE 12 (Continued)

Compound	Reference
12.5. Amino-S-Substituted Thio-, Sulfinyl- or Sulfonyl-1,2,4-triazoles (Continued)	
4-(benzylideneamino)-3-[(4-methoxybenzyl)thio]-	**84:** 30975y
4-(benzylideneamino)-3-[(4-methoxybenzyl)thio]-5-methyl-	**84:** 30975y
4-(benzylideneamino)-3-methyl-5-[4-(2-methylpropoxy)-benzylthio]-	**84:** 30975y
4-(benzylideneamino)-3-(methylthio)-	**55:** 23508a
3-(benzylthio)-5-(p-chloroanilino)-4-(p-chlorophenyl)-	**70:** 3963x
3-(benzylthio)-5-dibenzoylamino-	**23:** 2178[2]
3-(benzylthio)-5-(3,5-dimethylpyrazol-1-yl)-4-phenyl-	**64:** 8174a
3-(benzylthio)-5-(methylamino)-4-phenyl-,	**70:** 3963x
3-(benzylthio)-4-methyl-5-(methylamino)-,	**70:** 3963x
3-(benzylthio)-1-phenyl-5-(β-phenylthiocarbamido)-	**23:** 2178[1,2]
5-(benzylthio)-1-phenyl-3-(β-phenylthiocarbamido)-	**23:** 2178[1,2]
3-(benzylthio)-5-(β-phenylthiocarbamido)-	**23:** 2178[2]
3-(benzylthio)-5-p-toluidino-	**62:** 9124c
3-(benzylthio)-5-p-toluidino-4-(2,3-di-p-tolylguanidino)-	**69:** 77170n
3-(benzylthio)-5-o-toluidino-4-o-tolyl-,	**70:** 3963x
3-(benzylthio)-5-p-toluidino-4-p-tolyl-,	**70:** 3963x, **79:** 42423b
3,4-bis(benzylideneamino)-5-(methylthio)-	**48:** 5182a
3,4-bis(p-chlorobenzylideneamino)-5-(methylthio)-	**48:** 5182b
3,4-bis(p-methoxybenzylideneamino)-5-(methylthio)-	**48:** 5182a
3-(p-bromoanilino)-4-(p-bromophenyl)-5-(methylthio)-	**71:** 91395n, **80:** 14879g
3-(4-bromoanilino)-5-(methylsulfonyl)-4-phenyl-	**80:** 14879g
3-(4-bromoanilino)-5-(methylthio)-4-phenyl-	**80:** 14879g
3-(butylthio)-4-(1,4-dihydro-2-mercapto-4,4,6-trimethyl-1-pyrimidinyl)-	**74:** 141645d
3,4-diamino-5-(methylthio)-	**48:** 5182a, **82:** 125368a, **83:** 193226y
3,5-diamino-1-(phenylsulfonyl)-	**55:** P15944a
3,4-dianilino-5-(benzylthio)-	**72:** 55347k
3,4-dianilino-5-(methylthio)-	**72:** 55347k
4-(p-dimethylaminobenzylideneamino)-3-(methylthio)-5-phenyl-	**48:** 4524i
3-(3,5-dimethylpyrazol-1-yl)-5-(methylthio)-	**63:** 11545c
3,3'-dithiobis [5-amino-	**55:** 2624e
3,3'-dithiobis[5-anilino-	**64:** 8173e
3-ethyl-4-formamido-5-[(piperidinomethyl)thio]-	**74:** 141645d
3-hydrazino-5-(methylthio)-	**48:** 5182b
3-(methylamino)-5-(methylthio)-	**66:** 2536b
4-methyl-3-(1-methyl-2-propenyl)amino]-5-(methylthio)-	**76:** 10155n, **77:** 109606t
3-(methylthio)-4-[(5-nitrofurfurylidene)amino]-5-phenyl-	**61:** 14663a
5-(methylthio)-1-phenyl-3-(β-phenylthiocarbamido)-S-methyl derivative	**23:** 2178[2]
5-(methylthio)-4-phenyl-3-(1-thiosemicarbazido)-, S-methyl derivative	**69:** 77151g
5-(methylthio)-1-phenyl-3-tolylsulfonamido-	**23:** 2178[3]
3-(methylthio)-5-p-toluidino-4-p-tolyl-	**71:** 91395n
3-(methylthio)-5-p-toluidino-4-(2,3-di-p-tolylguanidino)-	**69:** 77170n
3-methylthio-5-p-toluidine-4-p-tolyl-	**21:** 2900[6]

TABLE 12 (*Continued*)

Compound	Reference

12.6. Miscellaneous *O*-(1,2,4-triazole) Derivatives of Phosphorous Acids

12.6a. *Miscellaneous P-Substituted-O-Ethyl-O-(Substituted 1,2,4-Triazol-3-yl)phosphono-thioates*[a]

Compound	Reference
ethyl-, 1-butyl-5-chloro-	**79:** P66364j
ethyl-, 5-chloro-1-ethyl-	**79:** P66364j
ethyl-, 5-chloro-1-isopropyl-	**79:** P66364j
ethyl-, 5-chloro-1-(1-methylpropyl)-	**79:** P66364j
ethyl-, 5-chloro-1-phenyl-	**79:** 66364j
ethyl-, 5-[(cyanomethyl)thio]-1-phenyl-	**83:** P206285u
ethyl-, 5-(dimethylamino)-1-methyl	**80:** P82651b
ethyl-, 1-ethyl-5-(methylthio)-	**81:** P120638h
ethyl-, 1-isopropyl-5-(isopropylthio)-	**79:** P66362g
ethyl-, 1-isopropyl-5-methoxy-	**79:** P66363h
ethyl-, 1-isopropyl-5-(methylamino)-	**80:** P82651b
ethyl-, 1-isopropyl-5-(methylthio)-	**79:** P66362g
ethyl-, 5-(isopropylthio)-1-methyl-	**79:** P66362g
ethyl-, 1-methyl-5-(methylthio)-	**79:** P66362g
ethyl-, 1-methyl-5-(propylthio)-	**79:** P66362g
ethyl-, 5-(methylthio)-1-phenyl-	**79:** P66362g
ethyl-, 5-(methylthio)-1-(1-phenylethyl)-	**79:** P66362g
methyl-, 5-bromo-1-isopropyl-, *O*-propyl-	**79:** P66364j
methyl-, 5-chloro-1-isopropyl-	**79:** P66364j
methyl-, 5-chloro-1-(2-methylpropyl)-	**79:** P66364j
phenyl-, 5-chloro-1-isopropyl-	**79:** P66364j
phenyl-, 1-isopropyl-5-(methylamino)-	**80:** P82651b
phenyl-, 1-isopropyl-5-(methylthio)-	**81:** P120638h
phenyl-, 5-methoxy-1-methyl-	**81:** P120638h

12.6b. *Miscellaneous N-Substituted-O-Ethyl-O-(Substituted-1,2,4-Triazol-3-yl)phosphoro-amidothioates*[b]

Compound	Reference
-,5-chloro-1-isopropyl-	**79:** P66364j
dimethyl-, 5-chloro-1-isopropyl-	**79:** P66364j
isopropyl-, 5-chloro-1-isopropyl-	**79:** P66364j
isopropyl-, 5-chloro-1-phenyl-	**79:** P66364j
isopropyl-, 5-(dimethylamino)-1-methyl-	**80:** P82651b
isopropyl-, 1-isopropyl-5-(methylthio)-	**81:** P120638h
methyl-, 1-benzyl-5-chloro-	**79:** P66364j
methyl-, 5-chloro-1-isopropyl-	**79:** P66364j
methyl-, 5-chloro-1-phenyl-	**79:** P66364j
methyl-, 5-chloro-1-(1-phenylethyl)-	**79:** P66364j
methyl-, 5-chloro-1-(1,2,2-trimethylpropyl)-	**79:** P66364j
methyl-, 5-(dimethylamino)-1-isopropyl-	**80:** P82651b
methyl-, 1-methyl-5-[(2-methyl-2-propenyl)thio]-	**81:** P91536r

12.6c. *Miscellaneous Diethyl (Substituted 1,2,4-Triazol-3-yl)phosphorates*[c]

Compound	Reference
5-chloro-1-isopropyl-	**79:** P66364j
5-chloro-1-isopropyl-, *O,O*-dimethyl-	**79:** P66364j
5-(dimethylamino)-1-methyl-	**80:** P82651b
1-methyl-5-(isopropylthio)-	**79:** P66362g

TABLE 12 (*Continued*)

Compound	Reference

12.6. Miscellaneous *O*-(1,2,4-triazole) Derivatives of Phosphorous Acids (*Continued*)

12.6d. Miscellaneous O,O-Diethyl-O-(Substituted 1,2,4-Triazol-3-yl)phosphorothionates[c]

Compound	Reference
5-amino-1-isopropyl-	**80:** P82651b
5-[benzyl(methyl)amino]-1-methyl-	**80:** P82651b
1-(benzylthio)-5-(methylthio)-	**79:** P66362g
5-bromo-1-isopropyl-	**79:** P66364j
5-bromo-1-isopropyl-, *O,O*-dimethyl-	**79:** P66364j
5-([(4-bromophenyl)methyl]thio)-1-methyl-	**81:** P91540n
5-(2-butenylthio)-1-methyl-	**81:** P91536r
1-butyl-5-chloro-	**79:** P66364j
5-(butylthio)-1-phenyl-	**79:** P66362g
5-chloro-1-benzyl-	**79:** P66364j
5-chloro-1-benzyl-, *O,O*-dimethyl-	**79:** P66364j
5-[(4-chloro-2-buteneyl)thio]-1-methyl-	**81:** P91536r
5-chloro-1-cyclopentyl-	**79:** P66364j
5-chloro-1-ethyl-	**79:** P66364j
5-chloro-1-ethyl-, *O,O*-dimethyl	**79:** P66364j
5-chloro-1-isopropyl-	**79:** P66364j, **82:** 107475d
5-chloro-1-isopropyl-, *O,O*-dimethyl-	**79:** P66364j
5-chloro-1-isopropyl-, *O*-ethyl-*O*-methyl-	**79:** P66364j
5-chloro-1-isopropyl-, *O*-ethyl-*S*-propyl-	**79:** P66364j
5-chloro-1-(1-methylpropyl)-	**79:** P66364j
5-chloro-1-(2-methylpropyl)-	**79:** P66364j
5-chloro-1-(1-methylpropyl)-, *O,O*-dimethyl	**79:** P66364j
5-chloro-1-(2-methylpropyl)-, *O,O*-dimethyl	**79:** P66364j
5-chloro-1-phenyl-	**79:** P66364j
5-chloro-1-phenyl-, *O,O*-dimethyl	**79:** P66364j
5-chloro-1-(1-phenylethyl)-	**79:** P66364j
5-chloro-1-(1-phenylethyl)-, *O,O*-dimethyl-	**79:** P66364j
5-([(4-chlorophenyl)methyl]thio)-1-cyclohexyl-	**81:** P91540n
5-([(4-chlorophenyl)methyl]thio)-1-ethyl-	**81:** P91540n
5-([(4-chlorophenyl)methyl]thio)-1-isopropyl-	**81:** P91540n
5-([(4-chlorophenyl)methyl]thio)-1-methyl-	**81:** P91540n
5-chloro-1-(1,2,2-trimethylpropyl)-	**79:** P66364j
5-chloro-1-(1,2,2-trimethylpropyl)-, *O,O*-dimethyl-	**79:** P66364j
1-(2-cyanoethyl)-5-(ethylthio)-	**83:** P206281q
1-(2-cyanoethyl)-5-(methylthio)-	**83:** P206281q
5-[(2-cyanoethyl)thio]-1-phenyl-	**83:** P206285u
5-cyano-1-isopropyl-	**82:** 107475d
5-[(cyanomethyl)thio]-1-ethyl-	**81:** P120638h
5-[(cyanomethyl)thio]-1-phenyl-	**83:** P206285u
5-[(cyanomethyl)thio]-1-phenyl-, *O,O*-dimethyl-	**83:** P206285u
1-cyclobutyl-5-[(2,3-dichloro-2-propenyl)thio]-	**81:** P91536r
1-cyclopentyl-5-(dimethylamino)-	**80:** P82651b
5-([(2,4-dichlorophenyl)methyl]thio)-1-methyl-	**81:** P91540n
5-([(2,6-dichlorophenyl)methyl]thio)-1-methyl-	**81:** P91540n
5-([(3,4-dichlorophenyl)methyl]thio)-1-methyl-	**81:** P91540n
5-[(3,3-dichloro-2-propenyl)thio]-1-ethyl-	**81:** P91536r
5-(dimethylamino)-1-ethyl-	**80:** P82651b
5-(dimethylamino)-1-isopropyl-	**80:** P82651b

TABLE 12 (*Continued*)

Compound	Reference

12.6. Miscellaneous *O*-(1,2,4-triazole) Derivatives of Phosphorous Acids (*Continued*)

12.6*d*. *Miscellaneous O,O-Diethyl-O-(Substituted 1,2,4-Triazol-3-yl)phosphorothionates[c]* (*Continued*)

Compound	Reference
5-(dimethylamino)-1-isopropyl-, *O,O*-dimethyl-	**80:** P82651b
5-(dimethylamino)-1-methyl-	**80:** P82651b
5-(dimethylamino)-1-methyl-, *O,O*-dimethyl	**80:** P82651b
5-(dimethylamino)-1-methyl-, *O*-ethyl-*S*-propyl-	**80:** P82651b
5-[(diphenylmethyl)thio]-1-methyl-	**81:** P91540n
5-([(3-fluorophenyl)methyl]thio)-1-methyl-	**81:** P91540n
5-([(4-fluorophenyl)methyl]thio)-1-methyl-	**81:** P91540n
5-ethoxy-1-isopropyl-	**82:** 107475d
5-ethoxy-1-methyl-	**81:** P120638h
1-ethyl-5-(ethylthio)-	**79:** P66362g
1-ethyl-5-(methylamino)-	**80:** P82651b
1-ethyl-5-(methylthio)-	**79:** P66362g, **81:** P120638h, **82:** 107475d
1-ethyl-5-(methylthio)-, *O,O*-dimethyl-	**81:** P120638h
1-ethyl-5-(2-propenylthio)-	**81:** P120638h
5-(ethylthio)-1-isopropyl-	**79:** P66362g, **82:** 107475d
5-(ethylthio)-1-phenyl-	**79:** P66362g
1-isopropyl-5-(di-2-propenylamino)-	**80:** P82651b
1-isopropyl-5-(isopropylthio)-	**79:** P66362g
1-isopropyl-5-(isopropylthio)-, *O,O*-dimethyl-	**79:** P66362g
1-isopropyl-5-(methylamino)-	**80:** P82651b
1-isopropyl-5-(methylamino)-, *O,O*-dimethyl-	**80:** P82651b
1-isopropyl-5-(methylthio)-	**79:** P66362g, **81:** P120638h, **82:** 107475d
1-isopropyl-5-(methylthio)-, *O*-ethyl-*O*-propyl-	**81:** P120638h
1-isopropyl-5-(methylthio)-, *O*-methyl-*O*-propyl-	**81:** P120638h
1-isopropyl-5-(4-morpholinyl)-	**80:** P82651b
1-isopropyl-5-(phenylthio)-	**79:** P66362g
1-isopropyl-5-(2-propenylthio)-	**81:** P91536r
1-isopropyl-5-(1-pyrroldinyl)-	**80:** P82651b
5-(isopropylthio)-	**79:** P66362g
5-(isopropylthio)-1-phenyl-	**79:** P66362g
5-methoxy-1-isopropyl-	**79:** P66363h
5-methoxy-1-methyl-	**81:** P120638h
5-[(methylamino)carbonyl]-1-phenyl-	**77:** P140089a
1-methyl-5-(benzylthio)-	**79:** P66362g
1-methyl-5-(isopropylthio)-	**79:** P66362g
1-methyl-5-(methylphenylamino)-	**80:** P82651b
1-methyl-5-([(2-methylphenyl)methyl]thio)-	**81:** P91540n
1-methyl-5-([(3-methylphenyl)methyl]thio)-	**81:** P91540n
1-methyl-5-[(2-methylpropyl)thio]-	**79:** P66362g
1-methyl-5-(methylthio)-	**79:** 66362g, **81:** P120638h, **82:** 107475d
1-methyl-5-(methylthio)-, *O,O*-dimethyl-	**81:** P120638h
1-methyl-5-([(2-nitrophenyl)methyl]thio)-	**81:** P91540n
1-methyl-5-([(4-nitrophenyl)methyl]thio)-	**81:** P91540n

TABLE 12   (*Continued*)

Compound	Reference

12.6. Miscellaneous *O*-(1,2,4-triazole) Derivatives of Phosphorous Acids   (*Continued*)

12.6d. Miscellaneous *O,O*-Diethyl-*O*-(Substituted 1,2,4-Triazol-3-yl)phosphorothionates[c]
    (*Continued*)

Compound	Reference
1-methyl-3-phenoxy-	**79:** P66363h
1-methyl-5-[(1-phenylethyl)thio]-	**81:** P91540n
1-methyl-5-(2-propenylthio)-	**81:** P91535q, **81:** P91536r
1-methylpropyl)-5-(methylthio)-	**82:** 107475d
5-(methylsulfinyl)-1-phenyl-	**79:** P66362g
5-(methylsulfonyl)-1-phenyl-	**79:** P66362g
3-(methylthio)-1-phenyl-	**79:** 53334z
5-(methylthio)-1-phenyl-	**79:** P66362g
5-(methylthio)-1-(1-phenylethyl)-	**79:** P66362g, **82:** 107475d
5-(methylthio)-1-propyl-	**79:** P66362g, **82:** 107475d

12.6e. *Miscellaneous S-Alkyl-O-Ethyl-O-(Substituted 1,2,4-Triazol-3-yl)phosphoro-
    dithionates*[d]

Compound	Reference
S-propyl-, 5-chloro-1-benzyl-	**79:** P66364j
S-propyl-, 5-chloro-1-isopropyl-	**79:** P66364j
S-propyl-, 5-chloro-1-(1-phenylethyl)-	**79:** P66364j
S-propyl-, 1-(3-chlorophenyl)-5-(methylthio)-	**79:** P66362g
S-propyl-, 5-([(2-chlorophenyl)methyl]thio)-1-methyl-	**81:** P91540n
S-propyl-, 5-chloro-1-(1,2,2-trimethylpropyl)-	**79:** P66364j
S-propyl-, 5-(dimethylamino)-1-isopropyl-	**80:** P82651b
S-propyl-, 5-(dimethylamino)-1-methyl-	**80:** P82651b
S-propyl-, 1-ethyl-5-(ethylthio)-	**79:** P66362g
S-propyl-, 1-ethyl-5-(methylthio)-	**79:** P66362g
S-propyl-, 1-isopropyl-5-methoxy-	**79:** P66363h
S-propyl, 1-isopropyl-5-(methylamino)-	**80:** P82651b
S-propyl-, 1-isopropyl-5-(methylthio)-	**79:** P66362g
S-propyl, 5-(isopropylthio)-1-phenyl-	**79:** P66362g
S-propyl-, 1-methyl-5-[(2-methylpropyl)thio]-	**79:** P66362g
S-propyl-, 1-methyl-5-(methylthio)-	**79:** P66362g
S-propyl-, 5-(methylthio)-1-phenyl-	**79:** P66362g
S-propyl-, 5-(methylthio)-1-(1-phenylethyl)-	**79:** P66362g
S-propyl-, 5-(methylthio)-1-propyl-	**79:** P66362g

[a] Listed as *P*-substituent, triazole substituent, *O*-alkyl if not ethyl.
[b] Listed as *N*-substituent, triazole substituent.
[c] Listed as triazole substituent, *O,O*-dialkyl if not ethyl.
[d] Listed as *S*-substituent, triazole substituent, *O*-alkyl if not ethyl.

12.7. Miscellaneous 1,2,4-triazoles

Compound	Reference
5-(acetamido)-1-methyl-3-nitro-	**72:** 111380e
3-acetamido-5-nitro-	**59:** 13277e
1-acetonyl-5-bromo-3-nitro-	**81:** 120541w
4-acetyl-3,5-dibromo-	**72:** 121456y
4-amino-3-azido-	**66:** 37902j, **69:** 110617p
4-amino-3-azido-5-(4,5-dihydro-1H-pyrazol-3-yl)-	**78:** 29701a
4-amino-3-azido-5-ethyl-	**69:** 110617p

TABLE 12 (*Continued*)

Compound	Reference

12.7. Miscellaneous 1,2,4-triazoles (*Continued*)

Compound	Reference
4-amino-3-azido-5-methyl-	**66:** 37902j
4-amino-3-azido-5-phenyl-	**69:** 110617p
5-amino-3-([p-(benzylethylamino)phenyl]azo)-	**75:** P130778u
4-amino-3,5-bis(benzyloxy)-	**71:** 113323s
5-amino-3-chloro-1-methyl-	**74:** 99948c
5-amino-3-(p-chlorophenoxy)-1-(phenylsulfonyl)-	**67:** P90816f
5-amino-3-([p-(dimethylamino)phenyl]azo)-	**75:** P130778u
3-amino-5-(p-hydroxyphenylazo)-	**28:** 2714[6]
3-amino-5-methoxy-	**81:** 25613n
5-amino-3-[(4-[(2-methoxycarbonylethyl)ethylamino]-phenyl)azo]-1-phenyl-	**80:** P84685q
5-amino-1-methyl-3-nitro-	**69:** P59245m, **72:** 111380e
3-amino-5-[(1-methyl-2-phenyl-1H̲-indol-3-yl)azo]-	**79:** P147415d
3-amino-5-(1-naphthyloxy)-1-(phenylsulfonyl)-	**64:** 15887h, **68:** P114607n
5-amino-3-(1-naphthyloxy)-1-(phenylsulfonyl)-	**64:** 15887h
3-amino-5-nitro-	**84:** 145644g
3-amino-5-nitrosamino-	**70:** P96802g
3-amino-5-phenoxy-1-(phenylsulfonyl)-	**64:** 15888a, **68:** P114607n
5-amino-3-phenoxy-1-(phenylsulfonyl)-	**64:** 15888a
3-amino-1-(phenylsulfonyl)-5-(p-tolyloxy)-	**64:** 15888a, **68:** P114607n
5-amino-1-(phenylsulfonyl)-3-(p-tolyloxy)-	**64:** 15888a
3-amino-1-(phenylsulfonyl)-5-(2,4-xylyloxy)-	**64:** 15888a, **68:** P114607n
5-amino-1-(phenylsulfonyl)-3-(2,4-xylyloxy)-	**64:** 15887h
3-amino-5-(2,4-xylyloxy)-	**71:** 61474v
3-azido-4-[(o-azidobenzylidene)amino]-5-methyl-	**65:** 704d
3-azido-4-(benzylideneamino)-	**65:** 704d
3-azido-4-(benzylideneamino)-5-ethyl-	**65:** 704d
3-azido-4-(benzylideneamino)-5-methyl-	**62:** 11805b, **65:** 704d
3-azido-4-(benzylideneamino)-5-phenyl-,	**65:** 704d
3-azido-5-bromo-	**82:** 30784u
3-azido-4-[(p-chlorobenzylidene)amino]-	**65:** 704d
3-azido-4-[(p-chlorobenzylidene)amino]-5-methyl-	**65:** 704d
3-azido-5-ethyl-4-[(p-methoxybenzylidene)amino]-	**65:** 704d
3-azido-5-ethyl-4-[(p-methylbenzylidene)amino]-	**65:** 704d
3-azido-5-ethyl-4-[(p-nitrobenzylidene)amino]-	**65:** 704d
3-azido-5-methoxy-	**81:** 25613n
3-azido-4-[(p-methoxybenzylidene)amino]-	**65:** 704d
3-azido-4-[(p-methoxybenzylidene)amino]-5-methyl-	**65:** 704d
3-azido-4-[(p-methylbenzylidene)amino]-	**65:** 704d
3-azido-5-methyl-4-[(p-methylbenzylidene)amino]-	**65:** 704d
5-azido-1-methyl-3-(methylthio)-	**81:** 25613n
3-azido-5-methyl-4-[(o-nitrobenzylidene)amino]-	**65:** 704d
3-azido-5-methyl-4-[(p-nitrobenzylidene)amino]-	**65:** 704d
3-azido-5-(methylthio)-	**81:** 25613n
3-azido-4-[(p-nitrobenzylidene)amino]-	**65:** 704d
3-azido-4-[(p-nitrobenzylidene)amino]-5-phenyl-	**65:** 704d
5-(1-aziridinyl)-1-methyl-3-nitro-	**72:** 111380e
4-benzamido-3-chloro-5-phenyl-	**81:** 3549u
5-(benzylidenehydrazino)-1-methyl-3-nitro-	**74:** 76376a

TABLE 12 (*Continued*)

Compound	Reference
12.7. Miscellaneous 1,2,4-triazoles (*Continued*)	
5-(benzyloxy)-3-nitro-1-β-D-ribofuranosyl-	**73:** 66847v
3-(*p*-bromoanilino)-4-(*p*-bromophenyl)-5-ethoxy-	**66:** 115651t
5-(4-bromobenzylthio)-1-methyl-3-[(methylsulfonyl)oxy]-	**82:** P156321j
5-bromo-1-[2-(chloromethyl)-2-hydroxyethyl]-3-nitro	**81:** 120541w
5-bromo-1-(2,3-dihydroxypropyl)-3-nitro-	**81:** 120541w
5-bromo-1-(2-hydroxyethyl)-3-nitro-	**81:** 120541w
5-bromo-1-[2-hydroxy-2-(methoxymethyl)ethyl]-3-nitro-	**81:** 120541w
5-bromo-1-(2-hydroxy-2-methylethyl)-3-nitro-	**81:** 120541w
5-bromo-1-methyl-3-nitro-	**74:** 99948c
3-bromo-5-(methylthio)-	**67:** 82167e
1-bromo-3-nitro-	**71:** 91399s
3-bromo-5-nitro-	**73:** 66847v
5-bromo-3-nitro-1-β-D-ribofuranosyl-, 2′,3′,5′-triacetate	**73:** 66847v
5-(2-butenylthio)-1-methyl-3-[(methylsulfonyl)oxy]-	**82:** P156321j
1-butyl-3-[(methylsulfonyl)oxy]-5-(methylthio)-	**82:** P156321j
5-[(4-chloro-2-butenyl)thio]-1-methyl-3-[(methylsulfonyl)oxy]-	**82:** P156321j
3-chloro-5-(2,2-dichloro-1,1-difluoroethyl)-3-[(trichloromethyl)thio]-	**81:** 49647c
5-[(2-chloroethyl)amino]-1-methyl-3-nitro-	**72:** 111380e
5-chloro-1-ethyl-3-[(methylsulfonyl)oxy]-	**82:** P156321j
5-chloro-1-isopropyl-3-[(methylsulfonyl)oxy]-	**82:** P156321j
3-chloro-1-methyl-5-nitro-	**77:** 74519n
5-chloro-1-methyl-3-nitro-	**74:** 99948c
5-chloro-1-(1-methylpropyl)-3-[(methylsulfonyl)oxy]-	**82:** P156321j
3-[(chloromethylsulfonyl)oxy]-1-methyl-5-(2-propenylthio)-	**82:** P156321j
5-chloro-3-[(methylsulfonyl)oxy]-1-neohexyl-	**82:** P156321j
5-chloro-3-[(methylsulfonyl)oxy]-1-(1-phenylethyl)-	**82:** P156321j
5-chloro-3-[(methylsulfonyl)oxy]-1-propyl-,	**82:** P156321j
3-chloro-5-(methylthio)-	**81:** 25613n
5-chloro-3-nitro-1-(2-nitroethyl)-	**74:** 99948c
5-chloro-3-([(3,3,3-trichloro-1-propen-1-yl)sulfonyl]oxy)-1-(1-methylpropyl)-	**82:** P156321j
5-(cyanomethylthio)-1-methyl-3-[(methylsulfonyl)oxy]-	**82:** P156321j
5-(2,4-dichlorobenzylthio)-1-methyl-3-[(methylsulfonyl)oxy]-	**82:** P156321j
5-(3,4-dichlorobenzylthio)-1-methyl-3-[(methylsulfonyl)oxy]-	**82:** P156321j
5-(diethylamino)-1-methyl-3-nitro-	**72:** 111380e
5-(dimethylamino)-3-azido-	**81:** 25613n
5-(dimethylamino)-1-isopropyl-3-[(methylsulfonyl)oxy]-	**82:** P156321j
5-(dimethylamino)-1-methyl-3-nitro-	**72:** 111380e
3-[(dimethylcarbamoyl)oxy]-1-ethyl-5-(methylthio)-	**81:** P13523v
3-[(dimethylcarbamoyl)oxy]-1-isopropyl-5-(methylthio)-	**81:** P13523v
3-[(dimethylcarbamoyl)oxy]-4-isopropyl-5-(methylthio)-	**81:** P13523v
3-[(dimethylcarbamoyl)oxy]-5-(isopropylthio)-4-methyl-	**81:** P13523v
3-[(dimethylcarbamoyl)oxy]-4-methyl-5-(ethylthio)-	**81:** P13523v

TABLE 12 (*Continued*)

Compound	Reference
12.7. Miscellaneous 1,2,4-triazoles (*Continued*)	
3-[(dimethylcarbamoyl)oxy]-1-methyl-5-(methylthio)-	**81:** P13523v, **82:** P16845h
3-[(dimethylcarbamoyl)oxy]-4-methyl-5-(methylthio)-	**81:** P13523v
5-[(dimethylcarbamoyl)oxy]-1-methyl-3-(methylthio)-	**81:** P13523v
3-(diphenylmethylene)-5-phenyl-	**78:** P71045q, **80:** 133389g, **84:** 4866u
3,3′-diselenobis-	**77:** P146189e, **80:** P89539e
5,5′-diselenobis[1,3-dimethyl-	**77:** P146189e
3,3′-diselenobis[5-methyl-	**77:** P48469b, **80:** P89539e
3,3′-diselenobis[5-phenyl-	**77:** P146189e
5-(dodecylthio)-1-methyl-3-[(methylsulfonyl)oxy]-	**82:** P156321j
-1-ethanol, 5-bromo-α-methyl-3-nitro-	**85:** 159994d
-1-ethanol, 5-chloro-3-nitro-α-(nitromethyl)-	**85:** 78052t
5-ethoxy-1-methyl-3-nitro-	**70:** P96800e, **73:** 45419s
3,3′-(ethylenedithio)bis[5-(3-sulfopropylthio)-	**75:** P103658y, **76:** P66326m, **83:** 186250y
1-ethyl-5-(ethylthio)-3-[(methylsulfonyl)oxyl]-	**82:** P156321j
5-ethyl-1-(methylsulfonyl)-3-[(methylsulfonyl)oxy]-	**82:** P156321j
1-ethyl-3-[(methylsulfonyl)oxy]-5-(methylthio)-,	**82:** P156321j
5-[[2-(ethylthio)ethyl]thio]-3-[(3-sulfopropyl)thio]-	**73:** 115028q
5-(ethylthio)-1-isopropyl-3-[(methylsulfonyl)oxy]-	**82:** P156321j
5-(ethylthio)-3-[(methylsulfonyl)oxy]-1-phenyl-	**82:** P156321j
5-(3-fluorobenzylthio)-1-methyl-3-[(methylsulfonyl)oxy]-	**82:** P156321j
3-(2-furancarboxamido)-5-nitro-	**80:** 10273g
5-(hexylthio)-1-methyl-3-[(methylsulfonyl)oxy]-	**82:** P156321j
5-hydrazino-1-methyl-3-nitro-	**74:** 76376a
5-(hydroxyamino)-1-methyl-3-nitro-	**74:** 76376a
3-isocyanato-	**85:** P123930r
4-(isopropylidenamino)-3-(isopropylidenehydrazino)- 5-(3-methyl-2-[(isopropylidenehydrazino)methyl]-1- oxo-2-butenyl)-	**78:** 29701a
1-isopropyl-3-[(methylsulfonyl)oxy]- 5-[(methylsulfonyl)thio]-	**82:** P156321j
5-(isopropylthio)-1-methyl-3-[(methylsulfonyl)oxy]-	**82:** P156321j
5-methoxy-1-methyl-3-nitro-	**70:** P96800e, **73:** 45419s
5-methoxy-3-nitro-1-β-D-ribofuranosyl-	**73:** 66847v
1-methyl-3-[(dimethylcarbamoyl)oxy]-5-(isopropylthio)-	**82:** P16845h
1-methyl-3-[(dimethylcarbamoyl)oxy]-5-(pentylthio)-	**82:** P16845h
1-methyl-3-[(dimethylcarbamoyl)oxy]-5-(2-propenylthio)-	**82:** P16845h
1-methyl-3-[(dimethylcarbamoyl)oxy]-5-(propylthio)-	**82:** P16845h
1-methyl-5-(methylamino)-3-nitro-	**72:** 111380e
1-methyl-5-(3-methylbenzylthio)-3-[(methylsulfonyl)oxy]-	**82:** P156321j
1-methyl-5-(1-methylhydrazino)-3-nitro-	**74:** 76376a
1-methyl-5-(1-methylhydrazino)-3-nitro-, benzaldehyde hydrazone	**74:** 76376a
1-methyl-5-[(2-methylphenyl)thio]-3- [(methylsulfonyl)oxy]-	**82:** P156321j
1-methyl-5-[(2-methyl-2-propenyl)thio]- 3-[(methylsulfonyl)oxy]-	**82:** P156321j

TABLE 12 (*Continued*)

Compound	Reference

12.7. Miscellaneous 1,2,4-triazoles (*Continued*)

Compound	Reference
1-methyl-5-[(2-methylpropyl)thio]-3-[(methylsulfonyl)oxy]-	**82:** P156321j
1-methyl-3-[(methylsulfonyl)oxy]-5-(methylthio)-	**82:** P156321j
1-methyl-3-[(methylsulfonyl)oxy]-5-([2-(methylthio)ethyl]thio)-	**82:** P156321j
1-methyl-3-[(methylsulfonyl)oxy]-5-(3-nitrobenzylthio)-	**82:** P156321j
1-methyl-3-[(methylsulfonyl)oxy]-5-(pentylthio)-	**82:** P156321j
1-methyl-3-[(methylsulfonyl)oxy]-5-[(1-phenylethyl)thio]-,	**82:** P156321j
1-methyl-3-[(methylsulfonyl)oxy]-5-(2-propenylthio)-	**82:** P156321j
1-methyl-3-[(methylsulfonyl)oxy]-5-(propylthio)-	**82:** P156321j
1-methyl-3-nitro-5-(*m*-nitrophenoxy)-	**73:** 45419s
1-methyl-3-nitro-5-(*p*-nitrophenoxy)-	**70:** P96800e, **73:** 45419s
1-methyl-3-nitro-5-phenoxy-	**70:** P96800e, **73:** 45419s
1-methyl-3-nitro-5-propoxy-	**73:** 45419s
1-methyl-3-nitro-5-(vinylamino)-	**72:** 111380e
3-[(methylsulfonyl)oxy]-5-(methylthio)-1-phenyl-	**82:** P156321j
3-[(methylsulfonyl)oxy]-5-(methylthio)-1-propyl-	**82:** P156321j
-3-selenenic acid	**77:** P48469b
-3-selenenic acid, 5-methyl-	**77:** P48469b
5,5'-(1-triazene-1,3-diyl)bis[1-methyl-3-(methylthio)-	**81:** 25613n

CHAPTER 13

Alkyl or Aryl-1,2,4-Triazolin-5-ones

13.1. Alkyl- or Aryl-Monsubstituted Δ^2(or Δ^3)-1,2,4-Triazolin-5-ones

s-Triazolin-3-one. Although the reaction of thiosemicarbazide and ethyl orthoformate lead to the formation of 2-amino-1,2,4-thiadiazole,[1] the condensation of semicarbazide hydrochloride with this reagent at reflux gave triazolin-5-one (**13.1-1**) instead of 2-amino-1,3,4-oxadiazole.[2] In addition reaction of ethyl ethoxymethylenecarbamate with hydrazine afforded a 60% yield of **13.1-1**.[3] Also, a low yield of **13.1-1** was obtained by treatment of semicarbazide with *s*-triazine.[4]

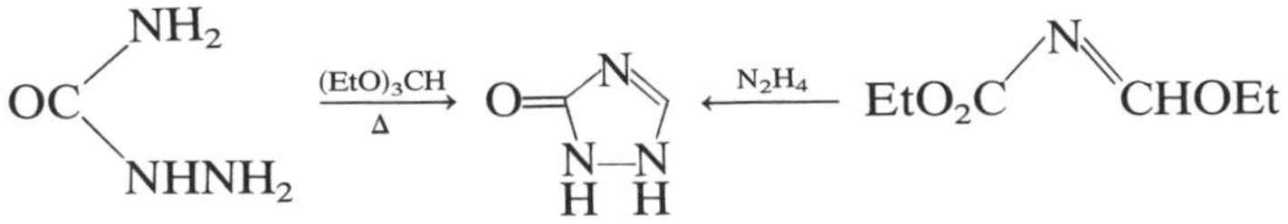

13.1-1

The interaction of 1-alkyl semicarbazide hydrochlorides with refluxing ethyl orthoformate provided good yields of 2-alkyltriazolin-5-ones, as illustrated for the preparation of the 2-methyl derivative **13.1-2**.[5,6] The 1-

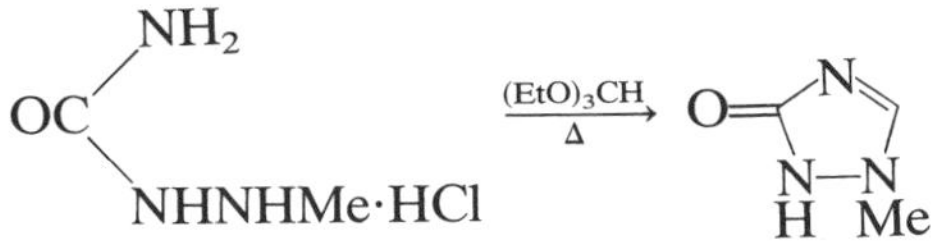

13.1-2 (81%)

substituted triazolin-5-ones (e.g., **13.1-3**) are readily obtained by dehydration of 1-formyl-2-arylsemicarbazides with dilute, hot base.[7,8] Similarly, 1-acetyl semicarbazide was converted to 3-methyltriazolin-5-one in good yield.[9,10] In addition, the 4-alkyl- and 4-arylsemicarbazides are readily converted to triazolin-5-ones. When 4-methylsemicarbazide (**13.1-4**) hydrochloride was heated with ethyl orthoformate, the 4-methyltriazolinone

365

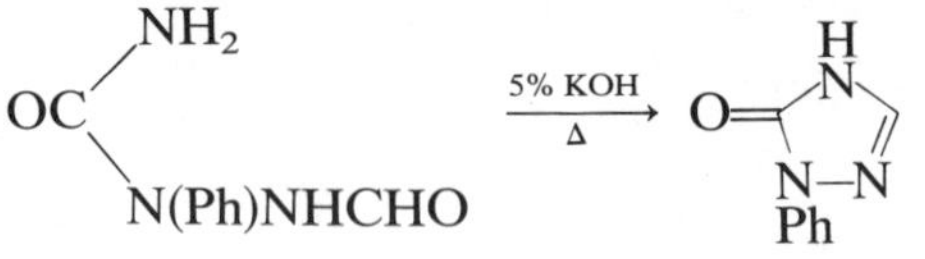

13.1-3 (60%)

(**13.1-6**) was obtained in 76% yield.[5] Similarly, the 4-phenyl derivative (**13.1-7**) was prepared in good yield from the condensation of 4-phenylsemicarbazide (**13.1-5**) with ethyl formate in the presence of base.[11,12]

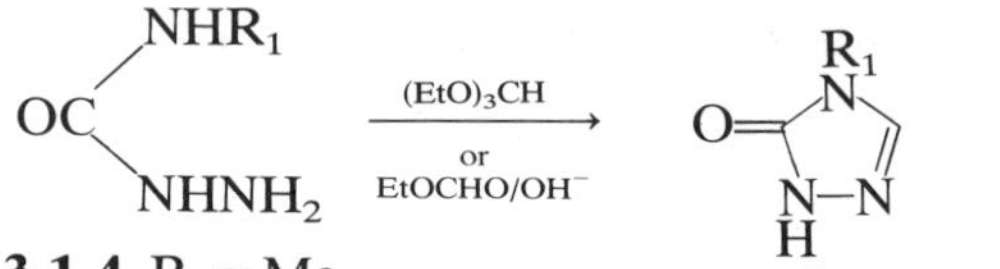

13.1-4, R$_1$ = Me
13.1-5, R$_1$ = Ph

13.1-6, R$_1$ = Me
13.1-7, R$_1$ = Ph

In the conversion of 1,4-dibenzoylsemicarbazide (**13.1-8**) with base to give 3-phenyltriazolin-5-one (**13.1-11**), ^{14}C labeling indicated that cyclization was preceded by hydrolysis of the 4- benzoyl group to give intermediate **13.1-10**.[13] Also, the transformation of **13.1-9** to **13.1-11** (93%) with base presumably occurred via **13.1-10**;[14] however, the reaction of **13.1-12** with hydrazine hydrate to give **13.1-11** (57%) probably occurred via intermediate **13.1-13**.[14,15]

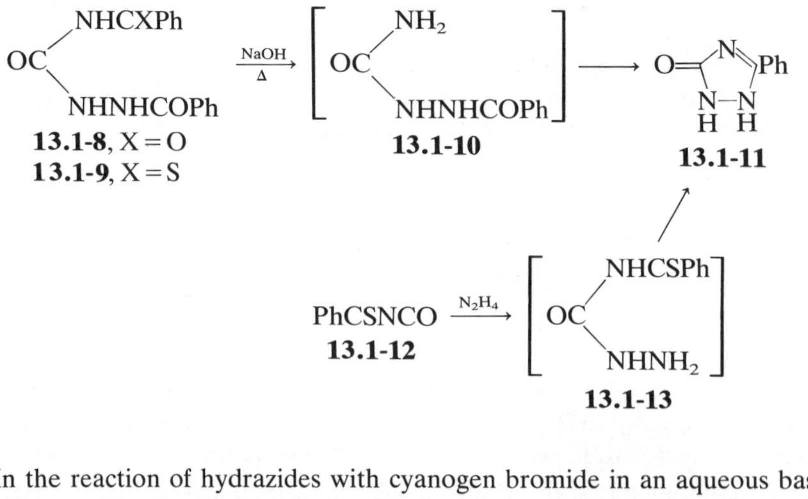

13.1-8, X = O
13.1-9, X = S

13.1-10

13.1-11

PhCSNCO
13.1-12

13.1-13

In the reaction of hydrazides with cyanogen bromide in an aqueous basic medium to give 3-substituted triazolin-5-ones (e.g., **13.1-16**), the latter are formed via 2-amino-1,3,4-oxadiazole (e.g., **13.1-14**) and 1-acylsemicarbazide

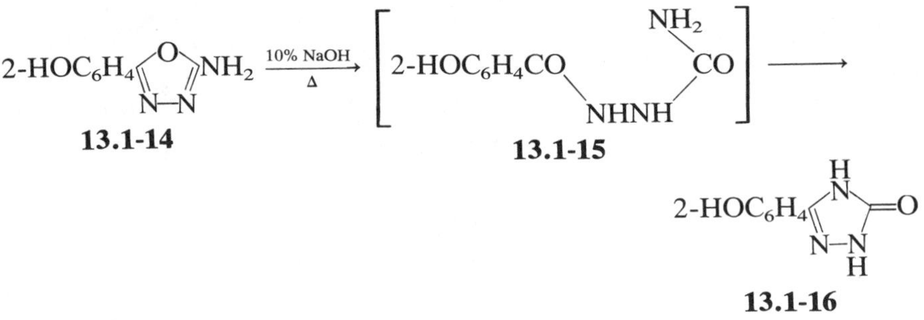

13.1-14   **13.1-15**

13.1-16

(e.g., **13.1-15**) intermediates.[16–20] In contrast, treatment of the 1,3,4-oxadiazole (**13.1-17**) with ethanolic potassium hydroxide gave the 3-ethoxytriazole (**13.1-19**), probably formed via the O-alkyl-1-acylsemicarbazide (**13.1-18**).[21,22] Hydrolysis of **13.1-19** to the triazolin-5-one (**13.1-20**) was effected with 20% hydrochloric acid.

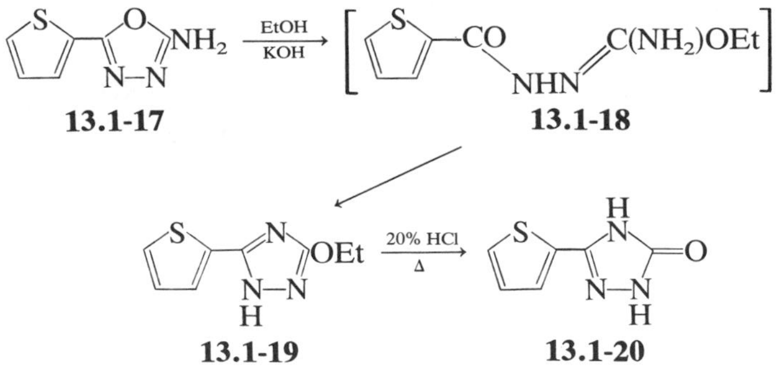

13.1-17   **13.1-18**

13.1-19   **13.1-20**

The synthesis of 3-substituted triazolin-5-ones (e.g., **13.1-11**) was accomplished by the cyclization of amidrazone intermediates (**13.1-22** to **13.1-24**), which were prepared by the condensation of thiobenzamide (**13.1-21**) with ethyl carbazate[23,24] and by the reaction of ethyl imidates (e.g., **13.1-25**) with ethyl carbazate,[25–28] N,N'-dicarboxyhydrazine,[29] and semicarbazide.[30–32] For example, the condensation of **13.1-29** and semicarbazide in acetic acid gave a 27% yield of **13.1-31**,[33] whereas the cyclization of the isolated intermediate (**13.1-30**) in nitrobenzene gave a 78% yield of **13.1-31**.[34] Amidrazone intermediates are also involved in the treatment of the N-cyanoamidine (**13.1-26**) with hydrazine in aqueous methanol, but this reaction provided a poor yield of **13.1-11**.[35] In contrast, reaction of N-(ethoxycarbonyl)amidines (e.g., **13.1-27**) with hydrazine readily gave 3-alkyl- and 3-aryltriazolinones (e.g., **13.1-11**).[36] Also, triazolin-5-ones were prepared by the reaction of **13.1-28** with hydrazine.[37] Furthermore, several 4-alkyl- and 4-aryltriazolin-5-ones (e.g., **13.1-7**) have been obtained by the cyclization of N-(carboethoxy) amidrazones (e.g., **13.1-32**) with dilute base.[24,25]

PhCSNH$_2$ $\longrightarrow$ $\left[\text{PhC} \begin{array}{c} \text{NH}_2 \\ \diagdown \\ \text{NNHCOR}_1 \end{array} \right]$ $\longleftarrow$ PhC$\begin{array}{c} \text{NH} \\ \diagup \\ \text{OEt} \end{array}$

13.1-21

13.1-22, R$_1$ = OEt
13.1-23, R$_1$ = OH
13.1-24, R$_1$ = NH$_2$

13.1-25

PhC$\begin{array}{c} \text{NHR}_1 \\ \diagdown \\ \text{NH} \end{array}$ $\xrightarrow{\text{N}_2\text{H}_4}$ Ph$\begin{array}{c} \text{N} \\ \diagup\quad\diagdown \\ \text{N—N} \\ \text{H H} \end{array}$O   ArC$\begin{array}{c} \text{NHCO}_2\text{Et} \\ \diagup \\ \text{S} \end{array}$

13.1-26, R$_1$ = CN
13.1-27, R$_1$ = CO$_2$Et

13.1-11

13.1-28

O$_2$N$\begin{array}{c}\text{O}\end{array}$—C$\begin{array}{c} \text{NH} \\ \diagup \\ \text{OEt} \end{array}$ $\xrightarrow{\text{H}_2\text{NNHCONH}_2}$ O$_2$N$\begin{array}{c}\text{O}\end{array}$—C$\begin{array}{c} \text{NH NH}_2 \\ \diagup\qquad | \\ \text{NHNH}\quad\text{CO} \end{array}$ $\longrightarrow$

13.1-29   **13.1-30**

O$_2$N$\begin{array}{c}\text{O}\end{array}$$\begin{array}{c} \text{N} \\ \diagup\quad\diagdown \\ \text{N—N} \\ \text{H H} \end{array}$O

13.1-31

CH$\begin{array}{c} \text{NHPh} \\ \diagup \\ \text{NNHCO}_2\text{Et} \end{array}$ $\xrightarrow[\Delta]{\text{OH}^-}$ $\begin{array}{c} \text{Ph} \\ \text{N} \\ \diagup\quad\diagdown \\ \text{N—N} \\ \text{H} \end{array}$O

13.1-32   **13.1-7**

The reaction of thiazolidin-2-ones with hydrazine and its derivatives have been extensively investigated and shown to give open-ringed products, either 4-acylsemicarbazide or amidrazones, which are further transformed to 1,2,4-triazolin-5-ones.[38,39] For example, the reaction of **13.1-33** with hydrazine gave **13.1-35**, which underwent ring opening at the CO—S linkage and reclosure on N to give **13.1-36**.[38,40] Similarly, treatment of

X$\begin{array}{c} \text{H} \\ \text{N} \\ | \quad\diagdown \\ \quad\quad \\ \text{S} \end{array}$O $\xrightarrow{\text{N}_2\text{H}_4}$ H$_2$NHN$\begin{array}{c} \text{N} \\ \diagup\quad\diagdown \\ \quad\quad \\ \text{S} \end{array}$O $\longrightarrow$ O$\begin{array}{c} \text{N} \\ \diagdown\quad \\ \diagup\quad\diagdown \\ \text{N—N} \\ \text{H H} \end{array}CH_2$SH

13.1-33, X = S   **13.1-35**   **13.1-36**
13.1-34, X = NH

13.1-34 with hydrazine gave **13.1-36**.[39] This procedure is also applicable for the preparation of triazolin-5-ones from thiazolidin-2-ones containing a variety of hydrocarbon substituents.[41–43]. In addition, the selenazolidin-2-one (**13.1-37**) was converted to **13.1-38** with hydrazine hydrate.[44] A differ-

ent pathway was encountered in the reaction of thiazolidine-2,4-dione (**13.1-39**) with hydrazine, which gave the 4-acylsemicarbazide (**13.1-40**).

The latter was dehydrated in hydrochloric acid to give **13.1-36**.[45] Presumably, the 1-acylsemicarbazide (**13.1-42**) was an intermediate in the rearrangement of **13.1-41** with base to give **13.1-43**.[46]

Several elegant but undeveloped methods for the synthesis of 1,2,4-triazolin-5-ones have been described. These included the condensation of the trichloromethyl group of **13.1-44** with ethanolic semicarbazide to give **13.1-45**,[47] the nitrosation of **13.1-46** to give **13.1-50** formed via **13.1-47**

and **13.1-48**,[48,49] and the interaction of the carbamoyl chloride (**13.1-49**) with hydrazine to give **13.1-11**.[50] The readily available benzylidenesemicarbazones (e.g., **13.1-51**) are converted with bromine in anhydrous acetic acid to give good yields of 1,2,4-triazolin-5-ones (e.g., **13.1-53**).[51,52] In the presence of water or sodium acetate, however, the intermediate brominated product (**13.1-52**) was converted to a 2-amino-1,3,4-oxadiazole. The latter was postulated to result from the conversion of **13.1-52** to a nitrilimine intermediate, whereas the triazolin-5-one was formed from the ionization of **13.1-52** to give a carbonium ion intermediate.

In addition, triazolin-5-ones have been prepared from preformed triazoles by the displacement with hydroxide of bromo[53] and methylthio[54] groups. Although nitration of 1- and 4-substituted triazolin-5-ones gave the corresponding 3-nitro derivatives, nitration of 2-substituted triazolin-5-ones was unsuccessful.[55] Also, the 3-aryl s-triazolin-5-ones are oxidized with lead tetraacetate to give the corresponding s-triazol-5-ones, which form Diels-Alder adducts in the presence of 1,3-dienes.[56]

References

1. **50:** 13886b	2. **54:** 22602f	3. **60:** 520h	4. **51:** 14751b
5. **63:** 16340e	6. **67:** 32651n	7. **58:** 6827g	8. **61:** 10673g
9. **63:** 16339d	10. **84:** 17237q	11. **56:** 2440b	12. **57:** 808f
13. **52:** 16363g	14. **64:** 15870g	15. **69:** 52075x	16. **54:** 24680e
17. **55:** 23509g	18. **63:** 13275d	19. **68:** 105101r	20. **69:** 2912x
21. **70:** 68268u	22. **73:** 14770v	23. **56:** 4748g	24. **57:** 16598e
25. **57:** 2229d	26. **61:** 14675c	27. **81:** 49647c	28. **84:** 150568q
29. **64:** 11119b	30. **59:** 6386f	31. **63:** 611e	32. **83:** 43245q
33. **69:** 77177v	34. **72:** 100607h	35. **59:** 7531a	36. **63:** 11541a
37. **85:** 143036g	38. **71:** 70540r	39. **75:** 35882q	40. **76:** 3765h
41. **76:** 34220f	42. **80:** 3422b	43. **81:** 136067g	44. **79:** 115502p
45. **74:** 87899z	46. **59:** 3930f	47. **66:** 46372p	48. **71:** 21770g
49. **71:** 21830b	50. **75:** 118283w	51. **75:** 76693x	52. **77:** 87502b
53. **70:** 37729r	54. **76:** 34182v	55. **70:** 87692u	56. **74:** 87894u

13.2. Alkyl- or Aryl-Disubstituted Δ^2(or Δ^3)-1,2,4-Triazolin-5-ones

Many types of disubstituted triazolin-5-ones have been prepared from N-substituted semicarbazides, semicarbazones, and their acyl derivatives. The cyclization of both 1,2-dimethylsemicarbazide and its hydrochloride to give 1,2-dimethyltriazolin-5-one (**13.2-1**) was effected with refluxing ethyl orthoformate, but the reaction was completed in a shorter period of time with the hydrochloride.[1] Similarly, the acid-catalyzed condensation of 2,4-dimethylsemicarbazide with refluxing ethyl orthoformate gave a 84% yield of 1,4-dimethytriazolin-5-one (**13.2-2**).

The preparation of 3-substituted 1-phenyltriazolin-5-ones from 1-acyl-2-phenylsemicarbazides provided only low yields ($<36\%$) of products.[2,3] For example, treatment of both **13.2-3** and **13.2-4** with dilute potassium hydroxide at reflux gave a 36% yield of **13.2-5** and a 6% yield of **13.2-6**, respectively. In a related cyclization, the imino ether (**13.2-7**) was treated with refluxing phosphorus oxychloride to give **13.2-8**.[4,5] In addition, the benzylidene semicarbazone (**13.2-9**) was readily cyclized with bromine in acetic acid to give a 78% yield of the 1,3-disubstituted triazolin-5-one (**13.2-10**).[6]

$$OC\overset{NH_2}{\underset{N(Ph)NCOR_1}{}} \xrightarrow[\Delta]{5\% KOH} O{=}\overset{N}{\underset{\underset{Ph}{N-N}}{}}R_1$$

13.2-3, $R_1 = Me$ **13.2-5**, $R_1 = Me$
13.2-4, $R_1 = Ph$ **13.2-6**, $R_1 = Ph$

$$O_2N\text{-furan-}CH{=}CHC\overset{OEt}{\underset{NN(Me)CONH_2}{}} \xrightarrow[\Delta]{POCl_3} O_2N\text{-furan-}CH{=}CH\text{-}\overset{N}{\underset{\underset{H\ Me}{N-N}}{}}{=}O$$

13.2-7   **13.2-8**

$$OC\overset{NH_2}{\underset{N(Me)N=CHC_6H_4\text{-}4\text{-}Cl}{}} \xrightarrow[HOAc]{Br_2} O{=}\overset{N}{\underset{\underset{Me\ H}{N-N}}{}}C_6H_4\text{-}4\text{-}Cl$$

13.2-9   **13.2-10**

Although a low yield of 1,4-diphenyltriazolin-5-one (**13.2-12**) was obtained by dehydration of **13.2-11** either thermally or in refluxing formic acid,[7] treatment of **13.2-11** with dilute base gave a good yield of **13.2-12**.[2] The cyclization of the semicarbazone (**13.2-13**) in refluxing formic acid gave **13.2-14**, which was also obtained by ring contraction of 5-bromo-1,3-dibenzyl-6-azauracil (**13.2-15**) in the presence of sodium ethoxide.[8]

$$OC\overset{NHPh}{\underset{N(Ph)NHCHO}{}} \xrightarrow{5\% KOH} O{=}\overset{\overset{Ph}{N}}{\underset{\underset{Ph}{N-N}}{}}$$

13.2-11   **13.2-12**

$$OC\overset{NHCH_2Ph}{\underset{N(CH_2Ph)N=CHPh}{}} \xrightarrow{HCO_2H} O{=}\overset{\overset{CH_2Ph}{N}}{\underset{\underset{PhCH_2}{N-N}}{}} \xleftarrow{EtO^-} PhCH_2N\text{-azauracil-}Br$$

13.2-13   **13.2-14**   **13.2-15**

The elimination of water from 4-acyl-1-phenyl semicarbazides (e.g., **13.2-16**) to produce 3-alkyl- and aryl-2-phenyltriazolin-5-ones (e.g., **13.2-17**) has been effected both with dilute base[9,10] and polyphosphoric

13.2-16   **13.2-17**

acid.[10] Also, the 2-(4-nitrophenyl) derivative of **13.2-17** was prepared by the polyphosphoric acid method, which, however, was unsuccessful for the preparation of the 2-(2,4-dinitrophenyl) compound. The synthesis of bis(triazolin-5-ones) (e.g., **13.2-19**) was effected by either alkaline or acidic dehydration of the appropriate bis(semicarbazide) (e.g., **13.2-18**) (see Section 21.2).[11] As described in the following, the 2,3-disubstituted triazolin-5-ones are obtained as by-products in the preparation of 1,3-disubstituted triazolinones from the reaction of phenylhydrazine with imines and with thiobenzoyl isocyanate.[12,13]

13.2-18   **13.2-19**

The thermal dehydration of 4-aryl-1-benzoylsemicarbazides (e.g., **13.2-20**) in the presence of anhydrous zinc chloride has been reported to give 3,4-diaryltriazolin-5-ones (e.g., **13.2-21**).[14] Lower yields (15 to 30%) of

13.2-20   **13.2-21**   **13.2-22**

this type compound were obtained in the intramolecular cyclization of 4-aryl-1-benzoylsemicarbazides (e.g., **13.2-20**) with dilute base.[2,3] In contrast, good yields have been obtained in the conversion of 1-acyl-4-aryl semicarbazides with base to triazolin-5-ones, as illustrated by the preparation of **13.2-24** from **13.2-23**.[2,3,15,16] Similarly, the conversion of **13.2-25** to **13.2-26** was reported,[17,18] and **13.2-27** was cyclized to give a 53% yield of **13.2-28** with base.[19]

13.2-23   **13.2-24**   **13.2-25**   **13.2-26**

$$\underset{\textbf{13.2-27}}{OC\overset{NHC_6H_3\text{-}3,4\text{-}Cl_2}{\underset{NHNHCOCF_2CHCl_2}{\Big\langle}}} \quad \xrightarrow{OH^-} \quad \underset{\textbf{13.2-28}}{O=\underset{\underset{H}{N-N}}{\overset{\overset{C_6H_3\text{-}3,4\text{-}Cl_2}{N}}{\Big\langle}}CF_2CHCl_2}$$

The oxidation of 4-aryl thiosemicarbazones gave the expected 1,3,4-thiadiazolines and also low yields of triazolin-5-ones.[20–22] This observation led to the investigation of the oxidation of 4-aryl semicarbazones, which produced high yields of triazolin-5-ones. For example, the semicarbazone (**13.2-22**) was treated with potassium hexacyanoferrate (III) in a basic medium to give **13.2-21** in 96% yield. Although 4-arylsemicarbazones derived from aliphatic, aryl, and heteroaryl aldehydes were smoothly oxidized to triazolin-5-ones, this procedure was unsuccessful for the cyclization of those semicarbazones either unsubstituted at the 4-position or containing a 4-alkyl group, except for 4-benzylsemicarbazones.[23]

Semicarbazide derivatives are often formed *in situ* and converted without isolation to disubstituted triazolin-5-ones. A high yield of 1,3-diphenyl-1,2,4-triazolin-5-one (**13.2-6**) and a small amount of the isomeric 2,3-diphenyl-1,2,4-triazolin-5-one (**13.2-17**) were obtained in the reaction of thiobenzoylisocyanate (**13.2-29**) with phenylhydrazine.[12,24] Presumably the minor product (**13.2-17**) is formed from the expected intermediate **13.2-30**, whereas the major product (**13.2-6**) is generated from **13.2-31** formed by displacement of the thioxo group of **13.2-30** with phenyl hydrazine.[25] The

$$PhCSNCO + H_2NNHPh \longrightarrow O=\underset{\underset{H\;\;Ph}{N-N}}{\overset{N}{\Big\langle}}Ph \;+\; O=\underset{\underset{Ph\;\;H}{N-N}}{\overset{N}{\Big\langle}}Ph$$

13.2-29   **13.2-17**   **13.2-6**

$$\left[OC\overset{NHCSPh}{\underset{NHNHPh}{\Big\langle}}\right] \xrightarrow{PhNHNH_2} \left[\underset{PhHNNH}{\overset{N}{PhNHNHCO\;\;\;CPh}}\right]$$

$$\textbf{13.2-30} \qquad\qquad\qquad\qquad\qquad \textbf{13.2-31}$$

(–H₂S) applied to the left path.

interaction of **13.2-29** with the benzylidene hydrazone (**13.2-32**) to give **13.2-33** followed by acid hydrolysis of the latter also gave **13.2-6**.[26,27] In contrast, the acetonylidene hydrazone (**13.2-34**) reacted with **13.2-29** to give a 19% yield of **13.2-36**, the isomerization product of the initially formed acylated hydrazone **13.2-35**, and a 60% yield of **13.2-37**, the (4+2) cycloadduct between **13.2-29** and **13.2-34**. On treatment with acid **13.2-37** was converted via **13.2-30** in 69% yield to **13.2-17**. In addition, benzoyl

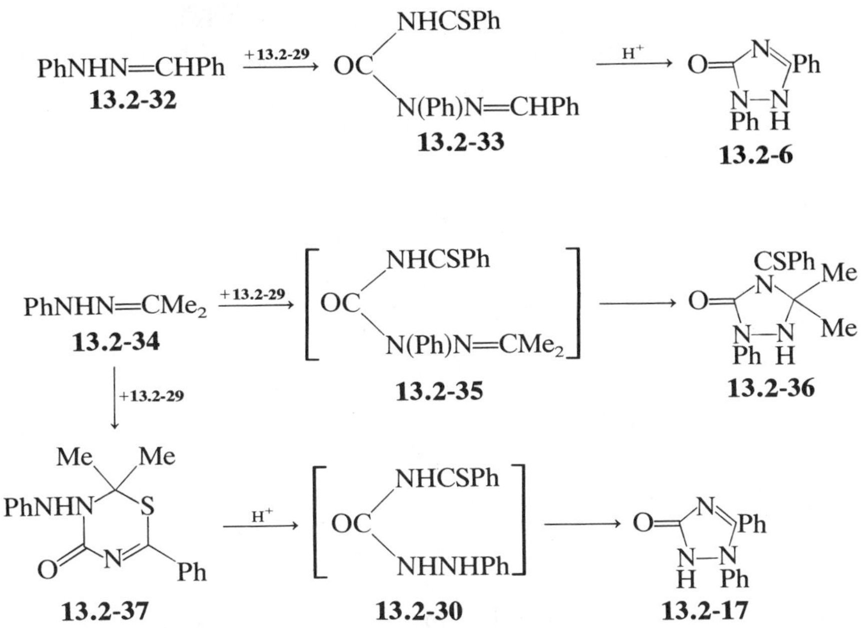

PhNHN=CHPh $\xrightarrow{+13.2-29}$ 13.2-33 $\xrightarrow{H^+}$ 13.2-6
13.2-32

PhNHN=CMe$_2$ $\xrightarrow{+13.2-29}$ [13.2-35] $\longrightarrow$ 13.2-36
13.2-34

$\downarrow$ +13.2-29

13.2-37 $\xrightarrow{H^+}$ [13.2-30] $\longrightarrow$ 13.2-17

isocyanate (**13.2-38**) reacted with **13.2-32** to give **13.2-39** and with **13.2-34** to give via **13.2-40** the triazolidin-5-one (**13.2-41**). However, both **13.2-39** and **13.2-41** were converted with acid to **13.2-6**. Although the reaction of both **13.2-29** and **13.2-38** with benzylhydrazine in benzene appeared to proceed via **13.2-43** to give good yields of the 1-benzyltriazolin-5-one (**13.2-44**), the condensation of **13.2-38** with 4-chlorophenylhydrazine gave the 4-acylsemicarbazide **13.2-42**, which was dehydrated in 15% hydrochloric acid to give the 2-(4-chlorophenyl)triazolin-5-one (**13.2-45**).[24,28] In contrast, **13.2-29** and 4-chlorophenylhydrazine gave a 3:1 mixture of **13.2-46** and **13.2-45**.

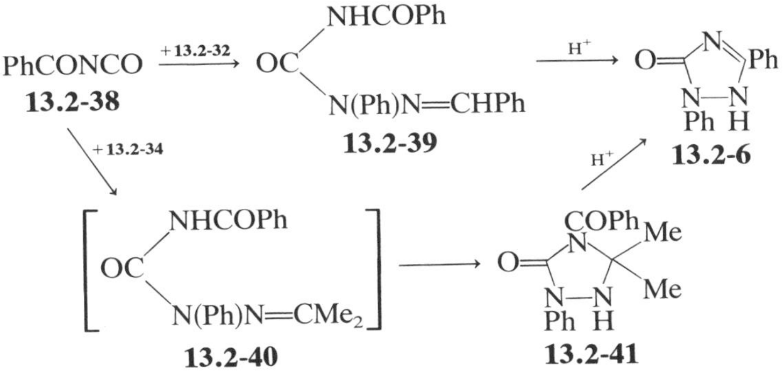

PhCONCO $\xrightarrow{+13.2-32}$ 13.2-39 $\xrightarrow{H^+}$ 13.2-6
13.2-38

$\searrow$ +13.2-34

[13.2-40] $\longrightarrow$ 13.2-41 $\xrightarrow{H^+}$ 13.2-6

PhCONCO   PhCSNCO
13.2-38   **13.2-29**

4-ClC$_6$H$_4$NHNH$_2$ /   \ PhCH$_2$NHNH$_2$ / PhCH$_2$NHNH$_2$

NHCOPh   NHCXPh
OC   [OC   ]   →   O=⟨triazole⟩Ph
    NHNHC$_6$H$_4$-4-Cl   N(CH$_2$Ph)NH$_2$ PhCH$_2$

13.2-42   **13.2-43**   **13.2-44**

15% HCl

O=⟨triazole⟩Ph   O=⟨triazole⟩Ph
  H   C$_6$H$_4$-4-Cl   N—N
13.2-45   4-ClC$_6$H$_4$
                                  13.2-46

Another common method for the synthesis of disubstituted 1,2,4-triazolin-5-ones begins with an amidrazone. The thermal intramolecular cyclization of the hydrochloride of the 1-acylamidrazone (**13.2-47**) gave a 73% yield of the 1-methyltriazolinone (**13.2-48**).[29,30] Although the dehydra-

NH$_2$
O$_2$N⟨furan⟩—C $\xrightarrow[\Delta]{PhNO_2}$ O$_2$N⟨furan⟩—⟨triazolinone⟩=O
              NN(Me)CONH$_2$ N—N
                                                                   H   Me
13.2-47   **13.2-48**

tion of the 4-acylamidrazone (**13.2-49**) gave a 70% yield of **13.2-5**, thermal cyclization of those amidrazones unsubstituted on the hydrazino moiety gave mixtures of 3-aminotriazoles and triazolin-5-ones.[31] Also, 3,4-disubstituted

NHCONHPh
MeC $\xrightarrow{200-250°}$ Me⟨triazolinone⟩=O
     NNHPh   N—N
13.2-49   H   Ph
                                                13.2-5

triazolin-5-ones have been prepared directly from amidrazones and isocyanates at high temperatures.[32] The acylamidrazones like **13.2-50** and **13.2-53** are readily converted either by hot, dilute base[33–37] or by heating above their melting points[33,38,39] to 3-alkyl-4-aryl- and 3,4-diaryltriazoles-5-ones such as **13.2-52** and **13.2-51**. Similar arylamidrazone intermediates are

$$\text{13.2-50} \xrightarrow[\Delta]{10\% \text{ NaOH}} \text{13.2-51, } R_1 = 2\text{-}C_5H_4N$$

13.2-51, $R_1 = 2$-C$_5$H$_4$N
13.2-52, $R_1 = $ Et

15.2-53

generated in the reaction of imino ethers (e.g., **13.2-54**) with excess aliphatic amines at 150° to give 3,4-dialkyltriazolinones (e.g., **13.2-55**).[40,41] In addi-

$$\text{13.2-54} \xrightarrow[\Delta]{BuNH_2} \text{13.2-55 (62\%)}$$

tion, the amidrazone (**13.2-56**) was postulated as the intermediate in the fusion of 1-acetyl-2-phenylhydrazine with urea at high temperatures to give

$$\text{PhNHNHCOMe} \xrightarrow{H_2NCONH_2} [\text{13.2-56}] \longrightarrow \text{13.2-5}$$

a good yield of the 3-methyltriazolin-5-one (**13.2-5**).[42,43] Some 1,3-diaryltriazolin-5-ones (e.g., **13.2-59**) were prepared via amidrazones (e.g., **13.2-58**) by the condensation of phenyl hydrazine with acylimino ethers (e.g., **13.2-57**) under mild conditions.[44] The substitution pattern in the

$$\text{13.2-57} \xrightarrow{PhNHNH_2} [\text{13.2-58}] \longrightarrow \text{13.2-59}$$

products also indicated that amidrazone intermediates were formed in the following type of reaction. The condensation of phenylhydrazine with imines (e.g., **13.2-60**) in an inert solvent containing a dehydration agent (P_2O_5) gave good yields of 3-alkyl-1-phenyltriazolin-5-ones (e.g., **13.2-63**) as the major product apparently formed via **13.2-61**.[45] In the example shown, a low yield of the isomeric 2,3-distributed triazolin-5-one (**13.2-64**) was isolated, which probably resulted from the 4-acyl-1-phenylsemicarbazide (**13.2-62**).[13,46] Also, similar reactions have been carried out thermally in the

13.2-63 (70%)   **13.2-64**

absence of solvents.[47,48] Only mild conditions are required for the cycloaddition of thioamides and hydrazines, as illustrated by the condensation of **13.2-65** with methylhydrazine in ethanol to give **13.2-66**.[49]

Some unique methods for the preparation of triazolin-5-ones have been reported. The addition of the imino sulfurane ylide (**13.2-68**) to **13.2-67** gave a 55% yield of **13.2-70**, presumably formed via **13.2-69** and **13.2-71**.[50]

The hydrazone (**13.2-73**) was reported to result from the addition of phenylhydrazine to the acetylene (**13.2-72**). Isomerization of **13.2-73** occurred in chloroform to give **13.2-74** (71%) under mild conditions.[51] In

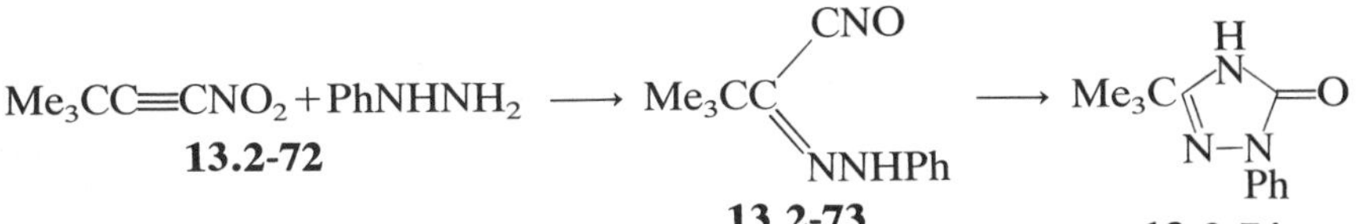

$$Me_3CC\equiv CNO_2 + PhNHNH_2 \longrightarrow$$

13.2-72

Me₃CC with CNO above and NNHPh below

13.2-73

$\longrightarrow$ Me₃C ring structure with N, H, N—N, Ph, =O

13.2-74

addition, the nitrosation of **13.2-75** gave the azide (**13.2-76**), which readily lost nitrogen and underwent rearrangement to form **13.2-70**.[52,53]

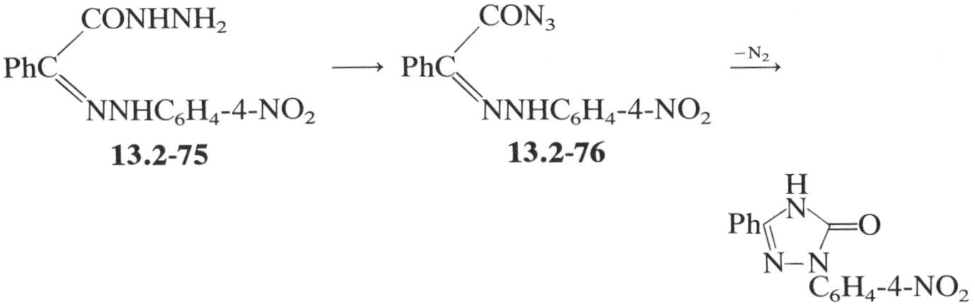

PhC with CONHNH₂ above and NNHC₆H₄-4-NO₂ below

13.2-75

$\longrightarrow$ PhC with CON₃ above and NNHC₆H₄-4-NO₂ below

13.2-76

$\xrightarrow{-N_2}$

Ph ring structure with N, H, N—N, C₆H₄-4-NO₂, =O

13.2-70

Several methods have been investigated for the preparation of disubstituted triazolin-5-ones from other ring systems. These include the reaction of the thiazolidones (**13.2-77**) with methyl hydrazine to give good yields of **13.2-78**,[54] the rearrangement of mesoionic **13.2-79** in base to give

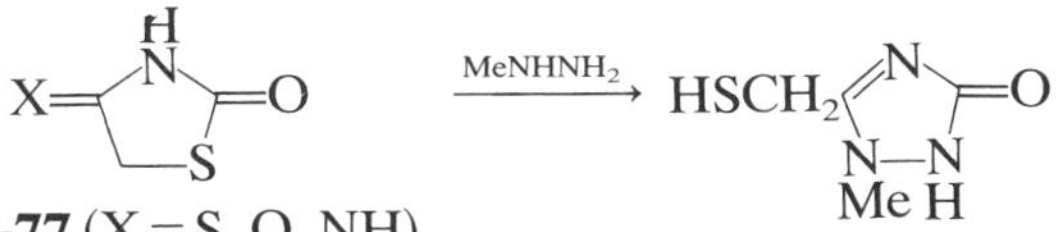

X= ring structure with N, H, =O, S

13.2-77 (X = S, O, NH)

$\xrightarrow{MeNHNH_2}$

HSCH₂ ring structure with N, N—N, Me H, =O

13.2-78

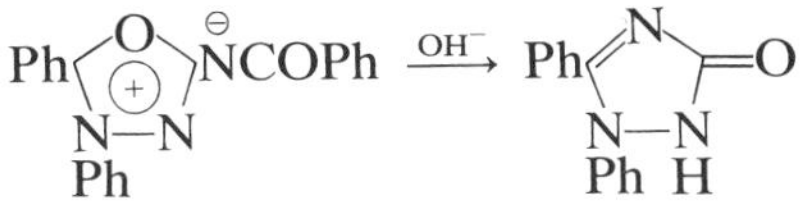

Ph ring structure with O, ⊖, NCOPh, N—N, Ph, ⊕

13.2-79

$\xrightarrow{OH^-}$

Ph ring structure with N, N—N, Ph H, =O

13.2-17

13.2-17,[55] and the condensation of **13.2-80** with phenylhydrazine to give **13.2-81**.[56,57] The latter method has also been used in the preparation of 3,4-disubstituted triazolin-5-ones.[58] The formation of triazolin-5-ones from preformed triazoles includes the rearrangement of the ether (**13.2-82**) with heat to give **13.2-83**[59] and the displacement of the methylsulfinyl or sulfonyl

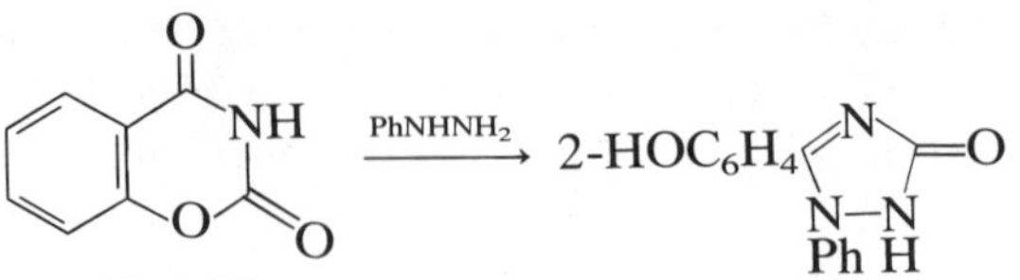

13.2-80   **13.2-81**

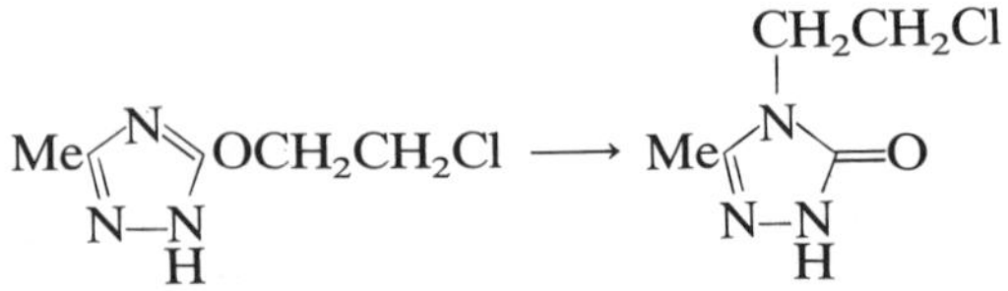

13.2-82   **13.2-83**

group of **13.2-84** with base to give **13.2-85**.[60–62] In addition, the ring nitrogen adjacent to the oxo function in triazolin-5-ones is favored toward

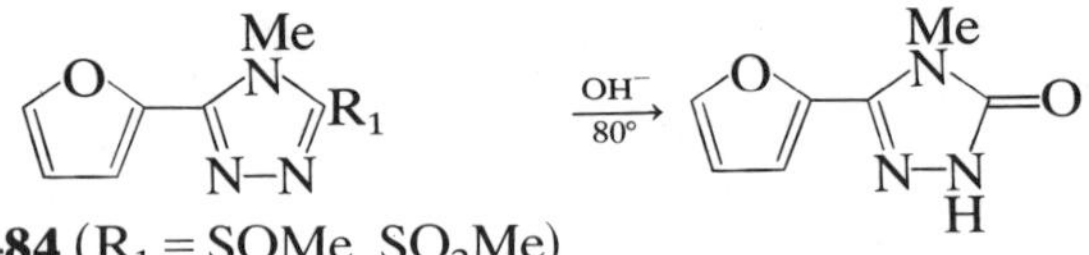

13.2-84 (R$_1$ = SOMe, SO$_2$Me)

13.2-85

substitution in alkylation reactions.[63] For example, treatment of **13.2-86** with acrylonitrile in the presence of sodium methoxide gave **13.2-87**,[64] and reaction of **13.2-86** with ethylene oxide in acetic acid gave **13.2-88**.[65] The latter was also formed by reaction of the triazolin-5-thione (**13.2-89**) with ethylene oxide in acetic acid. Of interest, 1,4-dimethyl-1,2,4-triazolin-5-one

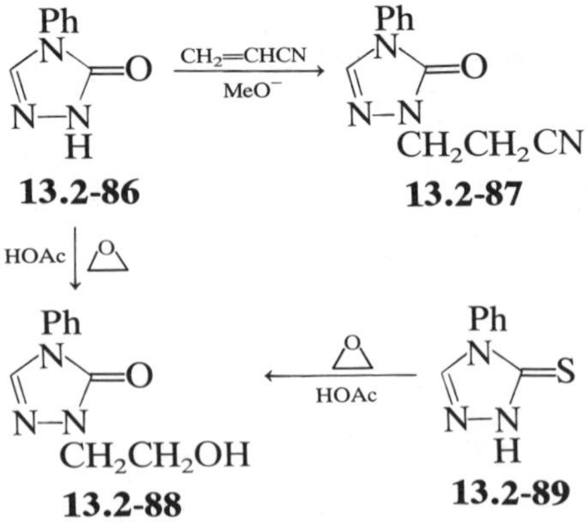

13.2-86   **13.2-87**

13.2-88   **13.2-89**

(**13.2-2**), but not the isomeric 1,2-dimethyl-1,2,4-triazolin-5-one (**13.2-1**), underwent nitration to give a 65% yield of the corresponding 3-nitrotriazolin-5-one (**13.2-90**).[66] In addition, 3-aryltriazolin-5-ones, but not the corresponding 3-alkyl compounds, were oxidized with lead tetraacetate to give triazol-5-ones, which reacted with 1,3-dienes to form Diels–Alder adducts.[67]

$$\text{13.2-2} \xrightarrow{\text{HNO}_3} \text{13.2-90}$$

13.2-2   **13.2-90**

References

1. **63:** 16340e	2. **58:** 6827g	3. **61:** 10673g	4. **65:** P7188c
5. **68:** 39627p	6. **77:** 87502b	7. **63:** 2975c	8. **76:** 4159a
9. **51:** 10506b	10. **52:** 11863e	11. **69:** 52073v	12. **64:** 15870g
13. **65:** 10579g	14. **71:** 49864q	15. **56:** 2440b	16. **57:** 808h
17. **80:** 82989t	18. **82:** 72995u	19. **81:** 49647c	20. **54:** 7695d
21. **57:** 16600d	22. **58:** 4569g	23. **75:** 88541j	24. **69:** 52075x
25. **79:** 91736q	26. **82:** 42707j	27. **85:** 46601e	28. **79:** 78693d
29. **63:** 611e	30. **72:** 100607h	31. **81:** 37516w	32. **69:** 67340d
33. **55:** 10420b	34. **57:** P2229d	35. **57:** 16598e	36. **62:** 14661b
37. **70:** 28924y	38. **56:** 4748g	39. **56:** 8708e	40. **55:** 27284a
41. **84:** 150568q	42. **73:** 77210h	43. **73:** 77253z	44. **63:** 11541a
45. **57:** 4651d	46. **67:** 90815e	47. **61:** 5640a	48. **63:** 2965h
49. **85:** 143036g	50. **81:** 63263f	51. **79:** 66260x	52. **71:** 21770g
53. **71:** 21830b	54. **76:** 3765h	55. **80:** 27175z	56. **77:** 126508e
57. **83:** 97141w	58. **84:** 105498f	59. **68:** 78201q	60. **71:** 49853k
61. **74:** 111968u	62. **79:** 87327q	63. **85:** 159988e	64. **75:** 20303c
65. **76:** 126876c	66. **70:** 87692u	67. **74:** 87894u	

13.3. Alkyl- or Aryl-Trisubstituted Δ^2(or Δ^1, Δ^3)-1,2,4-Triazolin-5-ones

An excellent method for the synthesis of 1,2,3-triaryltriazolin-5-ones resulted from the reaction of 1,2-diphenylhydrazines with benzoyl isocyanate to give 4-benzoylsemicarbazides, followed by cyclization of the latter in hydrochloric acid.[1] This method is illustrated by the combination of **13.3-1** with **13.3-2** to give **13.3-4** and with **13.3-3** to give **13.3-5**, acylation occurring exclusively on the most basic nitrogen of the hydrazine. Treatment of **13.3-4** with hydrochloric acid gave **13.3-7** in 84% yield, and **13.3-5** gave **13.3-6** in 74% yield. Similarly, 1,2-diethylhydrazine was converted in low yield to the corresponding 1,2-diethyltriazolin-5-one.[2] In the acylation of 1,2-diphenylhydrazine (**13.3-9**) with thiobenzoyl isocyanate (**13.3-8**) in a refluxing mixture of methylcyclohexane and tetrahydrofuran, 1,2,3-triphenyltriazolin-5-one (**13.3-11**) was formed by the elimination of hydrogen sulfide from the intermediate semicarbazide **13.3-10**.[1,3] In the reaction of **13.3-8** with 1,2-diphenylhydrazines containing dissimilar phenyl

PhCONCO + 4-R₁C₆H₄NHNHPh ⟶

$$PhCONCO + 4\text{-}R_1C_6H_4NHNHPh \longrightarrow$$

13.3-1   **13.3-2**, $R_1 = Cl$
        13.3-3, $R_1 = MeO$

substituents, acylation occurred perferentially but not exclusively on the most basic nitrogen of the hydrazine. For example, the combination of **13.3-8** and **13.3-2** in xylene at 95° gave a 52% yield of **13.3-7** and a 23% yield of **13.3-15**, formed from the intermediate semicarbazides (**13.3-13** and **13.3-14**, respectively).[1] Apparently, the rate of cyclization of **13.3-14** to **13.3-15** is faster than the reversible conversion of **13.3-14** to the thermodynamic more stable isomer (**13.3-13**). In contrast, reaction of **13.3-8** with **13.3-3** proceeded via **13.3-12** to give only **13.3-6**.

The cyclization of the 1-valeryl semicarbazide (13.3-17) to give 13.3-16 was unsuccessful presumably because the carbonyl moiety cannot undergo enolization.[4] This explanation is supported by the conversion of the 1-valerylsemicarbazide (13.3-18) with dilute base to the trisubstituted triazolinone (13.3-19).[4,5] Treatment of 1,2-dimethyl semicarbazide

(13.3-20) with acetic anhydride, however, gave the triazolin-5-one (13.3-21).[6]

The reaction of the amidrazone (13.3-22) with phenyl isocyanate in hot acetonitrile gave directly a 92% yield of 1,3,4-triphenyltriazolinone (13.3-23).[7,8] The same product was formed in good yield via the nitrilimine

(13.3-25) in the reaction of 2,5-diphenyltetrazole (13.3-24) with phenylisocyanate in the presence of acidic aluminum oxide.[9] The addition of the oxygen rather than the nitrogen of the isocyanate to the electrophilic carbon of 13.3-25 resulted in the obtainment of the oxadiazoline (13.3-26) as a by-product. Good yields of 4-aryl-1,2-dimethyltriazolin-5-ones (e.g.,

13.3-28) have been reported to result from the interaction of N-ethoxycarbonylthioamides (e.g., **13.3-27**) with 1,2-dimethylhydrazine.[10]

The formation of trisubstituted triazolin-5-ones by rearrangement reactions has been observed in several instances. The reaction of the carbohydroxamoyl chloride (**13.3-29**) with dicyclohexylcarbodiimide to give **13.3-32** in 40% yield was postulated to occur via the rearrangement of the 1,2,4-oxadiazoline (**13.3-30**).[11] A reaction sequence involving the migration of two functional groups was discovered in the oxidation of benzophenone 2-phenylsemicarbazone (**13.3-31**) with chromyl acetate to give the azo compound (**13.3-33**), which underwent ring closure and migration of a phenyl group to give a 90% yield of **13.3-11**.[12] Azo compounds like **13.3-33** are also generated by the thermolysis of the carbonyl azide derivative of hydrazones such as **13.3-34**.[1]

The formation of trisubstituted triazolin-5-ones from other ring systems includes the reaction of the benzoxazine (**13.3-35**) with 1,2-dimethylhydrazine to give **13.3-36**,[13,14] the reaction of the diaziridinone

13.3-35   **13.3-36**

(**13.3-37**) with benzonitrile to give a mixture containing **13.3-38** and its dealkylated products (**13.3-39** and **13.3-40**),[15] and the alkylation of **13.3-41** followed by ring opening to give **13.3-42**.[16]

13.3-37

13.3-38, $R_1 = R_2 = Me_3C$
13.3-39, $R_1 = H$, $R_2 = Me_3C$
13.3-40, $R_1 = Me_3C$, $R_2 = H$

13.3-41   **13.3-42**

The conversion of preformed triazole systems to trisubstituted triazolin-5-ones includes the alkylation of 1,3-disubstituted triazolin-5-ones at N-4[17] and 3,4-disubstituted triazolin-5-ones at N-1.[18-20] In addition, the oxidation of the triazolin-5-thione (**13.3-43**) with mercuric acetate gave **13.3-44**.[21,22] A similar compound was also prepared by oxidation of a semicarbazone of acetone with lead tetraacetate.[23]

13.3-43   **13.3-44**

References

1. **79:** 91736q	2. **81:** 169457n	3. **64:** 15870g	4. **65:** 10579g
5. **61:** 10673g	6. **84:** 17237q	7. **63:** 7000e	8. **69:** 67340d
9. **61:** 654h	10. **85:** 143036g	11. **70:** 106436q	12. **71:** 38870n
13. **77:** 126508e	14. **83:** 97141w	15. **81:** 13434s	16. **79:** 78811r
17. **85:** 159988e	18. **75:** 20303c	19. **81:** 13553e	20. **81:** 163330u
21. **72:** 55339j	22. **72:** 78951s	23. **70:** 106024w	

TABLE 13.   ALKYL- OR ARYL-1,2,4-TRIAZOLIN-5-ONES

Compound	Reference
13.1 Alkyl- or Aryl-Monosubstituted Δ^2(or Δ^3)-1,2,4-Triazolin-5-ones	
Parent	**54:** 22602h, **60:** 521a, **62:** 14437d, **64:** 8174c, **65:** 7169h, **70:** P28922w, **71:** P81375s, **71:** P124443g
-1-acetamide, α-(2,2-dimethyl-1-oxopropyl)-N-[2-(hexadecyloxy)-5-[(methylamino)sulfonyl]phenyl]-	**85:** P151754e
3-(5-amino-1-benzylimidazol-4-yl)-	**67:** 43155e
3-(1-aminoethyl)-	**57:** 3426a
3-(2-aminoethyl)-	**57:** 3426a
3-(aminomethyl)-	**57:** 3426a
3-(D-arabino-1,2,3,4-tetrahydroxybutyl)-	**68:** 105101r
3-(2-benzimidazolyl)-	**66:** 46372p
3-(1H-benzotriazol-1-ylmethyl)-	**73:** 14770v
4-(2-benzoyl-4-chlorophenyl)-	**82:** 25638v
3-benzyl	**37:** 2738[1], **37:** 3092[3], **44:** 6855c, **61:** 14675h, **62:** 14437d, **63:** 11541d **74:** 111967t
3-(1-benzyl-4-[(2-hydroxy-1-naphthyl)azo]-imidazol-4-yl)-	**67:** 43155e
2-benzyl-	**79:** P18723d, **79:** P66364j
2-(4-bromophenyl)-	**79:** P18723d
3-(3-bromophenyl)-	**84:** 4861p
3-(p-bromophenyl)-	**75:** 76693x, **77:** 87502b
3-butyl-	**58:** 6827h, **62:** 14437d, **84:** 150568q
3-tert-butyl-	**62:** 14437d, **84:** 150568q
3-(4-tert-butylphenyl)-	**85:** 143036g
2-(p-carboxyphenyl)-	**50:** 8651a
3-(p-chlorobenzyl)-	**63:** 11541d
4-[4-chloro-2-(2-hydroxybenzoyl)phenyl]-	**82:** 25638v
4-[4-chloro-2-(4-hydroxybenzoyl)phenyl]-	**82:** 25638v
2-(4-chloro-2-methylphenyl)-	**79:** P18723d, **79:** P42515h
3-[(o-chlorophenoxy)methyl]-	**70:** 68268u
3-[(p-chlorophenoxy)methyl]-	**70:** 68268u
2-(2-chlorophenyl)-	**77:** P48632z
2-(p-chlorophenyl)-	**71:** 101441r

TABLE 13 (*Continued*)

Compound	Reference

13.1. Alkyl- or Aryl-Monosubstituted Δ^2(or Δ^3)-1,2,4-Triazolin-5-ones (*Continued*)

3-(*o*-chlorophenyl)-	**47:** 3929a
3-(*p*-chlorophenyl)-	**44:** 2518c, **75:** 76693x,
	85: 143036g
4-(*p*-chlorophenyl)-	**56:** 2440c, **57:** 809b,
	57: 16598i
2-[4-chloro-3-(trifluoromethyl)phenyl]-	**79:** P42513f
2-cyclohexyl-	**77:** P34528h
4-cyclohexyl-	**74:** 3557a
2-cyclopentyl-	**77:** P34528h
3-[(dibutylamino)methyl]-	**69:** 67293r
3-(2,2-dichloro-1,1-difluoroethyl)-	**81:** 49647c
3-[(2,4-dichlorophenoxy)methyl]-	**70:** 68268u
2-(2,5-dichlorophenyl)-	**77:** P48632z
2-(3,5-dichlorophenyl)-	**81:** P13521t
3-[(diethylamino)methyl]-	**69:** 67293r
3-[1-(diisobutylamino)ethyl]-	**69:** 67293r
3-[(diisobutylamino)methyl]-	**69:** 67293r
3-(3,5-dimethoxyphenyl)-	**69:** 96593u
3-(3,4-dimethyl-4-pyrazolyl)-	**32:** 9088²
1-(2,4-dinitrophenyl)-	**71:** 21770g
2-(diphenylmethyl)-	**79:** P18723d, **82:** P156321j
3-[(dipropylamino)methyl]-	**69:** 67293r
3-(*o*-ethoxyphenyl)-	**69:** 2912x
3-(4-ethoxyphenyl)-	**85:** 143036g
4-(*p*-ethoxyphenyl)-	**56:** 2440c, **57:** 809b,
	57: 16599a
1-ethyl-	**63:** 16340g
2-ethyl-	**63:** 16340f, **79:** P66364j,
	82: P156321j
3-ethyl-	**62:** 14437d, **84:** 150568q,
	85: 143036g
3-(α-ethylbenzyl)-	**57:** P2229g
4-(α-ethylbenzyl)-	**57:** P2229h
3-(4-ethylphenyl)-	**85:** 143036g
2-(4-fluorophenyl)-	**79:** P42513f
3-(2-furyl)-	**73:** 14770v
3-(D-galacto-1,2,3,5-pentahydroxypentyl)-	**68:** 105101r
3-(D-gluco-1,2,3,4,5-pentahydroxypentyl)-	**68:** 105101r
3-(D-glycero-D-gluco-1,2,3,4,5,6-hexahydroxyhexyl)-	**68:** 105101r
3-heptyl-	**62:** 14437d
2-hexyl-	**77:** P34528h
2-sec-hexyl-	**77:** P34528h
3-hexyl-	**48:** 5184g, **62:** 14437d
3-(hydroxymethyl)-	**71:** 70540r, **74:** 87899z,
	76: 34220f
3-(1-hydroxy-1-methylethyl)-	**81:** 136067g
3-(*o*-hydroxyphenyl)-	**62:** 14437d, **69:** 2912x,
	83: 97141w

387

TABLE 13 (*Continued*)

Compound	Reference

13.1. Alkyl- or Aryl-Monosubstituted Δ^2(or Δ^3)-1,2,4-Triazolin-5-ones (*Continued*)

3-(*m*-hydroxyphenyl)-	**62:** 14437d
3-(*p*-hydroxyphenyl)-	**62:** 14437d
3-(1H̲-indol-3-yl)-	**85:** 143036g
3-isobutyl-	**62:** 14437d
3-isopentyl-	**62:** 14437d
2-(isopropyl)-	**77:** P34528h, **22:** P48632z
3-isopropyl-	**44:** 6855c, **62:** 14437d
3-(4-isopropylphenyl)-	**85:** 143036g
3-(D-lyxo-1,2,3,4-tetrahydroxybutyl)-	**68:** 105101r
3-(mercaptomethyl)-	**71:** 70540r, **72:** P12733r, **74:** 87899z, **75:** 35882q, **76:** 3765h, **76:** 34220f
3-(1-mercaptopropyl)-	**76:** 34220f
3-(*p*-methoxybenzyl)-	**37:** 2738³, **37:** 3092³
3-[*p*-methoxyphenoxy)methyl]-	**70:** 68268u
3-(*o*-methoxyphenyl)-	**69:** 2912x
3-(*p*-methoxyphenyl)-	**44:** 2518c **75:** 76693x, **77:** 87502b **85:** 143036g
1-methyl-	**63:** 16340ag, **83:** 28165v
2-methyl-	**83:** 28165v
3-methyl-	**50:** 6437i, **58:** 6827h, **61:** 14675g, **62:** 14437d, **63:** 16339g, **81:** 37516w **84:** 17237q, **84:** 150568q
4-methyl-	**63:** 16340g
2-(2-methylbutyl)-	**77:** P34528h
3-(3-methylbutyl)-	**84:** 150568q
3-[α-(2-methylhydrazino)-*p*-tolyl]-	**62:** P14688h
3-(1-methyl-5-nitroimidazol-2-yl)-	**73:** P25480b, **85:** P5644c
3-(4-methyl-3-nitrophenyl)-	**84:** 4861p
3-(4-methylphenyl)-	**84:** 4861p, **85:** 143036g
2-(1-methylpropyl)-	**77:** P34528h
2-(2-methylpropyl)-	**77:** P34528h
3-(2-naphthalenyl)-	**84:** 4861p
2-(2-naphthyl)-	**63:** 5633a
3-[(1-naphthyloxy)methyl]-	**70:** 68268u
2-(*p*-nitrobenzyl)-	**70:** 87692u
3-(5-nitro-2-furyl)-	**63:** P611f, **72:** 100607h
1-(*p*-nitrophenyl)-	**71:** 21770g
2-(2-nitrophenyl)-	**79:** P18723d
2-(3-nitrophenyl)-	**79:** P18723d
2-(4-nitrophenyl)-	**70:** 87692u
3-(*o*-nitrophenyl)-	**71:** 21830b
3-(3-nitrophenyl)-	**84:** 4861p
3-(4-nitrophenyl)-	**70:** 87692u, **84:** 4861p
3-(5-nitro-2-thienyl)-	**69:** 77177v
2-(pentafluorophenyl)-	**79:** P42513f
2-sec-pentyl-	**77:** P34528h
3-pentyl-	**62:** 14437d

TABLE 13 (*Continued*)

Compound	Reference
13.1. Alkyl- or Aryl-Monosubstituted Δ²(or Δ³)-1,2,4-Triazolin-5-ones *(Continued)*	

Compound	Reference
3-phenethyl-	**37:** 27838[3], **37:** 3092[3], **44:** 6855c, **62:** 14437d
3-(phenoxymethyl)-	**70:** 68268u
1-phenyl-	**53:** 8053a, **60:** 521a, **61:** 10674f, **70:** 37729r, **71:** 21770g, **73:** 77210h, **84:** 90078t
2-phenyl-	**50:** 8651a, **63:** 16336a **71:** 101441r, **84:** 90078t
3-phenyl-	**37:** 2738[2], **44:** 2518b, **47:** 8752d, **50:** 6437i, **54:** 24680i, **55:** 23509h, **56:** 4748h, **57:** 16599g, **59:** 3930h, **59:** 6386g, **59:** 7531h, **61:** 14675g, **62:** 14437d, **63:** 11541d, **64:** 11119c, **64:** 15871g, **68:** 39599f, **69:** 52075x, **75:** 118283w, **77:** 139992h, **75:** 76693x, **77:** 152070a, **79:** 78749b, **84:** 4861p, **84:** 105494b
3-phenyl-3-[14]C	**52:** 16364a
4-phenyl-	**56:** 2440b, **57:** 808h, **57:** 16598f, **61:** 10673h, **62:** 14437d, **76:** 85755s
2-(1-phenylethyl)-	**79:** P18723d, **79:** P66364j, **82:** P156321j
3-[1-(1-phenyliminoethyl)acetonyl]-	**32:** 9088[3]
3-[(phenylthio)methyl]-	**70:** 68268u
3-(piperidinomethyl)-	**56:** 4748h, **57:** 16599h
2-piperonyl-	**67:** 32651n
3-piperonyl-	**67:** 32651n
-2-propanenitrile	**82:** P156321j, **83:** P114420f, **83:** P206281q
2-propyl-	**77:** P34528h
3-propyl-	**58:** 6827h, **62:** 14437d, **84:** 150568q
3-(2-pyridinyl)-	**83:** 43245q
3-(4-pyridyl)-	**49:** 10937i **62:** 14437d
4-(2-pyridyl)-	**65:** 16961b
4-(3-pyridyl)-	**76:** 34182v
3-(1H-pyrrol-2-yl)-	**85:** 143036g
3-(selenylmethyl)-	**79:** 115502p
3-(2-thienyl)-	**73:** 14770v, **85:** 143036g
2-*p*-tolyl-	**71:** 101441r, **79:** P18723d
3-*p*-tolyl-	**63:** 11541d, **75:** 76693x, **77:** 87502b
4-*p*-tolyl-	**61:** 10674a

TABLE 13 (*Continued*)

Compound	Reference

13.1. Alkyl- or Aryl-Monosubstituted Δ^2(or Δ^3)-1,2,4-Triazolin-5-ones (*Continued*)

3-[(*o*-tolyloxy)methyl]-	**70:** 68268u
3-[(*m*-tolyloxy)methyl]-	**70:** 68268u
2-[2-(trifluoromethyl)phenyl]-	**79:** P42513f
2-[3-(trifluoromethyl)phenyl]-	**79:** P42513f
2-(1,2,2-trimethylpropyl)-	**79:** P66364j
3-[(3,4-xylyloxy)methyl]-	**70:** 68268u

13.2. Alkyl- or Aryl-Disubstituted Δ^2(or Δ^3)-1,2,4-Triazolin-5-ones

-1-acetamide, *N*-[2-chloro-5-[(4-[2,4-bis(1,1-dimethylpropyl)phenoxy]-1-oxobutyl)amino]phenyl]-α-(2,2-dimethyl-1-oxopropyl)-3-phenyl-	**83:** P155710u
4-acetamide, α-(2,2-dimethyl-1-oxopropyl)-*N*-[2-(hexadecycloxy)-5-[(methylamino)sulfonyl]phenyl]-	**85:** P151754e
-3-acetic acid, 1-phenyl	**30:** 4860[1]
3-acetonyl-	**30:** 4860[3]
3-(1-acetylacetonyl)-	**30:** 4860[3]
3-(1-acetylacetonyl)-1-phenyl-	**32:** 9087[9]
3-(1-allyl-3-butenyl)-1-phenyl-	**61:** 5640b
3-(2-benzothiazolyl)-4-ethyl-	**77:** 126518h
3-(2-benzothiazolyl)-4-methyl-	**77:** 126518h
3-(2-benzothiazolyl)-4-phenyl-	**77:** 126518h
4-(2-benzoyl-4-chlorophenyl)-3-[(methylamino)methyl]-	**79:** P78808v, **79:** P78811r
4-(2-benzoyl-4,6-dibromophenyl)-3-[1-(methylamino)-propyl]-	**79:** P78811r
3-benzyl-4-(2-bromoethyl)-	**68:** 78201q
4-benzyl-3-(*o*-carboxyphenyl)-	**75:** 88541j
3-benzyl-4-(2-chloroethyl)-	**68:** 78201q
3-benzyl-4-(2-diethylaminoethyl)-	**55:** 27284c, **57:** 16599d
3-benzyl-4-(2-diethylaminopropyl)-	**55:** 27284c
3-benzyl-4-[3-(diethylamino)propyl-	**57:** 16599d
3-benzyl-4-[3-(dimethylamino)propyl-	**57:** 16599d
3-benzyl-4-ethyl-	**61:** 10674b
3-benzyl-4-(2-hydroxyethyl)-	**68:** 78201q
3-benzyl-4-(2-hydroxyethyl)-, acetate ester	**68:** 78201q
3-benzyl-4-(2-iodoethyl)-	**68:** 78201q
4-benzyl-3-(*p*-methoxyphenyl)-	**75:** 88541j
4-benzyl-3-methyl-	**55:** 27284c, **57:** 16599c
4-benzyl-3-(*m*-nitrophenyl)-	**75:** 88541j
4-benzyl-3-(*p*-nitrophenyl)-	**75:** 88541j
1-benzyl-3-phenyl	**69:** 52075x, **77:** 139909m
3-benzyl-1-phenyl-	**63:** 11541d, **69:** 52073v
3-benzyl-2-phenyl-	**69:** 52073v
3-benzyl-4-phenyl-	**55:** 27284c, **57:** 16599d, **61:** 10674b
4-benzyl-3-phenyl-	**75:** 75910d, **75:** 88541j
3,4-bis(*o*-chlorophenyl)-	**75:** 88541j
3,4-bis(*p*-chlorophenyl)-	**58:** 4570a

TABLE 13 (*Continued*)

Compound	Reference

13.2. Alkyl- or Aryl-Disubstituted Δ^2(or Δ^3)-1,2,4-Triazolin-5-ones (*Continued*)

Compound	Reference
3,4-bis(*p*-methoxyphenyl)-	**75:** 88541j
4-(2-bromoethyl)-3-(*p*-chlorophenyl)-	**68:** 78201q
4-(2-bromoethyl)-3-(*p*-methoxyphenyl)-	**68:** 78201q
4-(2-bromoethyl)-3-methyl-	**68:** 78201q
3-(1-bromoethyl)-2-phenyl-	**79:** 78693d
4-(2-bromoethyl)-3-phenyl-	**68:** 78201q
3-(bromoethyl)-1-phenyl-	**57:** 4651f
4-(*p*-bromophenyl)-3-(*o*-chlorophenyl)-	**75:** 88541j
4-(*p*-bromophenyl)-3-(*p*-chlorophenyl)-	**75:** 88541j
4-(*p*-bromophenyl)-3-(2,4-dichlorophenyl)-	**75:** 88541j
4-(*p*-bromophenyl)-3-(*p*-methoxyphenyl)-	**75:** 88541j
1-(*p*-bromophenyl)-3-methyl-	**75:** 110244p
2-(4-bromophenyl)-3-methyl-	**77:** P48632z, **77:** P126637w
3-(4-bromophenyl)-1-methyl-	**77:** 87502b
4-(*p*-bromophenyl)-3-(*m*-nitrophenyl)-	**75:** 88541j
4-(*p*-bromophenyl)-3-(*p*-nitrophenyl)-	**75:** 88541j
3-(*p*-bromophenyl)-1-phenyl-	**21:** 743^3
4-(*p*-bromophenyl)-3-phenyl-	**75:** 88541j
3-(*p*-bromophenyl)-4-*p*-tolyl-	**58:** 4570a, **59:** 10053a
4-(*p*-bromophenyl)-3-*p*-tolyl-	**75:** 88541j
3,3'-(1,4-butanediyl)bis[1-phenyl-	**81:** 37516w
2-butyl-	**77:** P34528h
3-tert-butyl-4-cyclohexyl-	**57:** P2229h
1-tert-butyl-3-ethenyl-	**84:** P59486a
1-tert-butyl-3-ethyl-	**84:** P59486a
4-butyl-3-methyl-	**55:** 27284c, **57:** 16599c
3-tert-butyl-1-(4-methylphenyl)-	**79:** 66260x
3-tert-butyl-1-(4-nitrophenyl)-	**79:** 66260x
1-tert-butyl-3-phenyl	**81:** 13434s, **84:** P59486a
3-butyl-1-phenyl-	**65:** 10579g, **73:** 77210h
3-tert-butyl-1-phenyl	**73:** 77210h, **79:** 66260x
3-butyl-2-phenyl-	**65:** 10579h
3-butyl-4-phenyl-	**56:** 4748i, **57:** 16599i, **58:** 6827h, **61:** 10674a
3-(4-tert-butylphenyl)-1-methyl-	**85:** 143036g
3-(4-tert-butylphenyl)-1-phenyl-	**85:** 143036g
3-(*o*-carboxyphenyl)-4-(*p*-chlorophenyl)-	**75:** 88541j
3-(*o*-carboxyphenyl)-4-(*p*-methoxyphenyl)-	**75:** 88541j
3-(*o*-carboxyphenyl)-4-phenyl-	**75:** 88541j
4-[4-chloro-2-(2-chlorobenzoyl)phenyl]-3-[(ethylmethylamino)methyl]-	**79:** P78811r
4-[4-chloro-2-(2-chlorobenzoyl)phenyl]-3-[(methylamino)methyl]-	**79:** P78811r
4-[2-(2-chlorobenzoyl)-4-nitrophenyl]-3-[(ethylamino)methyl]-	**79:** P78811r
4-[2-(2-chlorobenzoyl)-4-nitrophenyl]-3-[(ethylpropylamino)methyl]-	**79:** P78811r
3-(*p*-chlorobenzyl)-1-phenyl-	**63:** 11541d

TABLE 13 (*Continued*)

Compound	Reference
13.2. Alkyl- or Aryl-Disubstituted Δ^2(or Δ^3)-1,2,4-Triazolin-5-ones (*Continued*)	
4-[4-chloro-2-(2,6-difluorobenzoyl)phenyl]- 3-[(propylamino)methyl]-	**79:** P78811r
4-(2-chloroethyl)-3-(*p*-chlorophenyl)-	**68:** 78201q
3-(1-chloroethyl-2-phenyl-	**79:** 78693d
4-(chloroethyl)-3-phenyl-	**68:** 78201q
3-(chloromethyl)-1-phenyl-	**79:** 78693d
3-(chloromethyl)-4-phenyl-	**57:** P2229h
4-(*o*-chlorophenyl)-3-(*p*-chlorophenyl)-	**75:** 88541j
4-(4-chlorophenyl)-3-[(2,4-dichlorophenoxy)methyl]-	**80:** 59899e
4-(*o*-chlorophenyl)-3-(2,4-dichlorophenyl)-	**75:** 88541j
4-(*p*-chlorophenyl)-3-(2,4-dichlorophenyl)-	**75:** 88541j
4-(*p*-chlorophenyl)-1-[(diethylamino)methyl]-	**64:** 570f
3-(*p*-chlorophenyl)-4-ethyl-	**74:** 111968u
4-(*p*-chlorophenyl)-3-ethyl-	**55:** 10420b
4-(*p*-chlorophenyl)-1-[(hexahydro-1H-azepin- 1-yl)methyl]-	**64:** 570f
3-(*p*-chlorophenyl)-4-(2-hydroxyethyl)-	**68:** 78201q
3-(*p*-chlorophenyl)-4-(2-hydroxyethyl)-, acetate ester	**68:** 78201q
3-(*p*-chlorophenyl)-4-(2-iodoethyl)-	**68:** 78201q
3-(*o*-chlorophenyl)-4-(*p*-methoxyphenyl)-	**75:** 88541j
3-(*p*-chlorophenyl)-4-(*o*-methoxyphenyl)-	**75:** 88541j
3-(*p*-chlorophenyl)-4-(*p*-methoxyphenyl)-	**75:** 88541j
4-(*p*-chlorophenyl)-3-(*p*-methoxyphenyl)-	**58:** 4570a, **59:** 10053b
4-(*p*-chlorophenyl)-3-(*p*-methoxystyryl)-	**75:** 88541j
2-(3-chlorophenyl)-3-methyl-	**77:** P126637w
2-(4-chlorophenyl)-3-methyl-	**77:** P48632z, **77:** P126637w
3-(4-chlorophenyl)-1-methyl-	**77:** 87502b
4-(*o*-chlorophenyl)-3-methyl-	**61:** 10674a
4-(*m*-chlorophenyl)-3-methyl-	**55:** 8388a, **55:** 10420b
4-(*p*-chlorophenyl)-3-methyl-	**55:** 10420b, **61:** 10674a
4-(*o*-chlorophenyl)-3-[3,4-(methylenedioxy)phenyl]-	**75:** 88541j
4-(*p*-chlorophenyl)-3-[3,4-(methylenedioxy)phenyl]-	**75:** 88541j
2-(3-chlorophenyl)-3-(1-methylethyl)-	**77:** P48632z, **77:** P126637w
4-(*p*-chlorophenyl)-1-(morpholinomethyl)-	**64:** 570f
4-(*o*-chlorophenyl)-3-(*m*-nitrophenyl)-	**75:** 88541j
4-(*o*-chlorophenyl)-3-(*p*-nitrophenyl)-	**75:** 88541j
4-(*p*-chlorophenyl)-3-(*m*-nitrophenyl)-	**58:** 4570a
4-(*p*-chlorophenyl)-3-(*p*-nitrophenyl)-	**75:** 88541j
4-(4-chlorophenyl)-3-(phenoxymethyl)-	**80:** 59899e
1-(4-chlorophenyl)-3-phenyl-	**69:** 52073v, **69:** 52075x, **82:** 42707j, **85:** 46601e
2-(*p*-chlorophenyl)-3-phenyl-	**69:** 52073v, **69:** 52075x
3-(*o*-chlorophenyl)-1-phenyl-	**22:** 3660⁴
3-(*p*-chlorophenyl)-1-phenyl-	**72:** P132738t, **85:** 143036g
3-(*o*-chlorophenyl)-4-phenyl-	**75:** 88541j
3-(*p*-chlorophenyl)-4-phenyl-	**58:** 4569h, **59:** 10053a
4-(*o*-chlorophenyl)-3-phenyl-	**61:** 10674b
4-(*p*-chlorophenyl)-3-phenyl-	**57:** 16600h, **58:** 4570a, **59:** 10053b

TABLE 13 (*Continued*)

Compound	Reference

13.2. Alkyl- or Aryl-Disubstituted Δ^2(or Δ^3)-1,2,4-Triazolin-5-ones (*Continued*)

Compound	Reference
1-(3-[4-(3-chlorophenyl)-1-piperazinyl]propyl)-3-ethyl-	**83:** P179071n
4-(*p*-chlorophenyl)-1-(piperidinomethyl)-	**64:** 570f
4-(*p*-chlorophenyl)-3-(piperidinomethyl)-	**56:** 2440d, **57:** 809d
4-(*o*-chlorophenyl)-3-(3-pyridyl)-	**75:** 88541j
4-(*m*-chlorophenyl)-3-(3-pyridyl)-	**75:** 88541j
4-(*o*-chlorophenyl)-3-(2-thienyl)-	**75:** 88541j
4-(*m*-chlorophenyl)-3-(2-thienyl)-	**75:** 88541j
3-(*o*-chlorophenyl)-4-*p*-tolyl-	**58:** 4570a, **59:** 10053a
3-(*p*-chlorophenyl)-4-*p*-tolyl-	**75:** 88541j
4-(*p*-chlorophenyl)-3-*o*-tolyl-	**75:** 88541j
4-(*p*-chlorophenyl)-3-*p*-tolyl-	**75:** 88541j
4-(4-chlorophenyl)-3-[(2,4,5-trichlorophenoxy)methyl]-	**80:** 59899e
3-(5-chloro-2-thienyl)-4-phenyl-	**83:** 163180m
1-(2-cyanoethyl)-4-phenyl-	**64:** 570f, **75:** 20303c
4-cyclohexyl-3-cyclopropyl-	**57:** P2229h
4-cyclohexyl-3-ethyl-	**57:** P2229g
4-cyclohexyl-3-isopropyl-	**57:** P2229g
4-cyclohexyl-3-methyl-	**55:** 27284c, **57:** 2229g, **57:** 16599c
3-cyclohexyl-1-phenyl-	**73:** 77210h
3-cyclohexyl-4-phenyl-	**56:** 4748i, **57:** 16599h
4-cyclohexyl-3-phenyl-	**56:** 4748i, **57:** 16600b
4-cyclohexyl-3-propyl-	**57:** P2229g
3-cyclopropyl-1-propyl-	**80:** P82998v, **82:** P72995u
1,4-dibenzyl-	**76:** 4159a
1-(2,4-dibromophenyl)-3-phenyl-	**32:** 7457[1]
3-(2,2-dichloro-1,1-difluoroethyl)-4-(3,4-dichlorophenyl)-	**81:** 49647c
3-[2,4-dichlorophenoxy)methyl]-4-(1-naphthalenyl)-	**76:** 126869c
3-[(2,4-dichlorophenoxy)methyl]-2-phenyl-	**79:** 78693d
3-[(2,4-dichlorophenoxy)methyl]-4-phenyl-	**76:** 126869c
3-(2,4-dichlorophenyl)-4-(*p*-methoxyphenyl)-	**75:** 88541j
2-(2,4-dichlorophenyl)-3-methyl-	**77:** P126637w
2-(2,5-dichlorophenyl)-3-methyl-	**77:** P48632z, **77:** P126637w
4-(2,4-dichlorophenyl)-3-methyl-	**70:** P28924y
3-(2,4-dichlorophenyl)-4-phenyl-	**75:** 88541j
3-(2,4-dichlorophenyl)-4-*p*-tolyl-	**75:** 88541j
3,4-dicyclohexyl-	**57:** P2229g
1,4-diethyl-	**63:** 16340g
3,4-diethyl-	**61:** 10674c, **81:** P13553e, **83:** P179071n, **84:** 150568q
3-[1-(diethylamino)ethyl]-4-(4,5-dimethyl-1-[4-(1-methylethyl)benzoyl]phenyl)-	**79:** P78811r
1-[(diethylamino)methyl]-4-(*p*-ethoxyphenyl)-	**64:** 570f
3-[(diethylamino)methyl]-1-phenyl-	**57:** 4651f
3-[(diethylamino)methyl]-4-phenyl-	**57:** 809d
3-[3-(diethylamino)propyl]-4-phenyl-	**56:** 4748i, **57:** 16599i
3-(4,5-dihydro-5-oxo-3-pyrazolyl)-	**24:** 5751[4]
1,2-dimethyl-	**63:** 16340f, **70:** 87692u
1,3-dimethyl-	**63:** 16340a, **84:** 17237q

TABLE 13 (*Continued*)

Compound	Reference

13.2. Alkyl- or Aryl-Disubstituted Δ^2(or Δ^3)-1,2,4-Triazolin-5-ones (*Continued*)

Compound	Reference
1,4-dimethyl-	**63:** 16340f
2,3-dimethyl-	**84:** 17237q
3,4-dimethyl-	**82:** P72995u, **84:** 17237q
3-[(dimethylamino)methyl]-4-phenyl-	**57:** 809d
3-[*p*-(dimethylamino)phenyl]-4-phenyl-	**62:** 14661d
4-[4,5-dimethyl-2-[4-(1-methylethyl)benzoyl]phenyl]- 3-[1-(ethylamino)ethyl]-	**79:** P78811r
3-(2,4-dimethyl-3-quinolyl)-1-phenyl-	**32:** 9088[1]
1-(2,4-dinitrophenyl)-3-methyl-	**71:** 21770g

1,3-diphenyl	**21:** 743[3], **22:** 1337[5],
	  24: 365[8], **32:** 7456[9],
	  53: 8053c, **61:** 5640b,
	  61: 10674g, **63:** 11541d,
	  64: 15871h, **69:** 52073v,
	  69: 52075x, **72:** P132738t,
	  73: 77210h, **76:** 113185b,
	  85: 46601e, **85:** 143036g

1,4-diphenyl-	**61:** 10674g, **63:** 2976a

2,3-diphenyl-	**25:** 4614[5], **49:** 15916b,
	  51: 10506e, **64:** 15871h,
	  69: 52073v, **79:** 78693d,
	  80: 27175z, **82:** 42707j,
	  85: 46601e

3,4-diphenyl-	**51:** 5095b, **54:** 7695e,
	  55: 27284d, **56:** 4748i,
	  57: 12473b, **57:** 16599h,
	  57: 16600g, **58:** 4569h,
	  59: 10053a, **61:** 10674a,
	  62: 14661c,**67:** 43755a
	  71: 49864q

3-(diphenylmethyl)-1-phenyl-	**72:** P132738t
4-[2-(2,4-dipropylbenzoyl)-4-(ethylthio)phenyl]- 3-[(ethylpropylamino)methyl]-	**79:** P78811r
3,4-di-*p*-tolyl-	**58:** 4569h
-1-ethanol, 3-(5-nitro-2-furyl)-	**63:** P612b
3-(*o*-ethoxyphenyl)-4-ethyl-	**71:** 49853k
4-(*p*-ethoxyphenyl)-1-[(hexahydro-1H-azepin-1-yl)methyl]-	**64:** 570f
3-(4-ethyoxyphenyl)-1-methyl-	**85:** 143036g
4-(*p*-ethoxyphenyl)-3-methyl-	**57:** 16598i
3-(4-ethoxyphenyl)-1-phenyl-	**85:** 143036g
4-(*p*-ethoxyphenyl)-1-(pyrrol-1-ylmethyl)-	**64:** 570f
3-(α-ethylbenzyl)-4-methyl-	**57:** P2229h
4-(α-ethylbenzyl)-3-methyl-	**57:** P2229g
3-(α-ethylbenzyl)-1-phenyl-	**61:** 5640a, **63:** 2965b
4-ethyl-3-(α-ethylbenzyl)-	**57:** P2229h
4-ethyl-3-(*o*-hydroxyphenyl)-	**61:** 10674b
3-ethyl-2-(isopropyl)-	**77:** P34528h
4-ethyl-3-(*o*-methoxyphenyl)-	**61:** 10674b

TABLE 13 (*Continued*)

Compound	Reference
13.2. Alkyl- or Aryl-Disubstituted Δ^2(or Δ^3)-1,2,4-Triazolin-5-ones (*Continued*)	
2-ethyl-3-methyl-	**77:** P34528h
3-ethyl-1-methyl-	**63:** 16340b, **85:** 143036g
3-ethyl-4-methyl-	**84:** 150568q
4-ethyl-3-methyl-	**84:** 150568q
3-(1-ethyl-4-methylpentyl)-1-phenyl-	**61:** 5640b
1-ethyl-3-(5-nitro-2-furyl)-	**72:** 100607h
3-(1-ethylpentyl)-1-phenyl-	**61:** 5640b
2-ethyl-3-phenyl-	**77:** P34528h
3-ethyl-1-phenyl-	**61:** 10674f, **73:** 77210h, **85:** 143036g
3-ethyl-2-phenyl-	**77:** P48632z, **77:** P126637w
3-ethyl-4-phenyl-	**55:** 10420b, **61:** 10674a, **62:** 14437d
4-ethyl-3-phenyl-	**61:** 10674b
3-(4-ethylphenyl)-1-methyl-	**85:** 143036g
3-(4-ethylphenyl)-1-phenyl-	**85:** 143036g
2-ethyl-3-(2-propenyl)-	**85:** P46688p
4-ethyl-3-propyl-	**61:** 10674c
4-(1-ethylpropyl)-3-methyl-	**57:** P2229g
3-(1-ethylpropyl)-1-phenyl-	**61:** 5640b, **73:** 77210h
4-ethyl-3-(4-sulfamoylphenyl)-	**74:** 11968u
3-ethyl-4-*o*-tolyl-	**75:** 88541j
3-ethyl-4-*p*-tolyl-	**55:** 10420b, **75:** 88541j
3-(2-furanyl)-4-methyl-	**79:** 87327q
3-(2-furyl)-1-phenyl-	**69:** 52073v
3-(2-furyl)-2-phenyl-	**69:** 52073v
3-(2-furyl)-4-phenyl-	**75:** 88541j
5-(2-furyl)-1-phenyl-	**31:** 3917[4]
1-[(hexahydro-1H̲-azepin-1-yl)-methyl]-4-phenyl-	**64:** 570f
3-(4-hydroxy-*m*-anisyl)-2-phenyl-	**26:** 6018[5]
4-(2-hydroxyethyl)-3-(*p*-methoxyphenyl)-	**68:** 78201q
4-(2-hydroxyethyl)-3-(*p*-methoxyphenyl)-, acetate ester	**68:** 78201q
1-(2-hydroxyethyl)-3-(5-nitro-2-furyl)-	**72:** 100607h
1-(2-hydroxyethyl)-2-phenyl-	**76:** 126876c
1-(2-hydroxyethyl)-4-phenyl-	**76:** 126876c
4-(2-hydroxyethyl)-3-phenyl-	**68:** 78201q
4-(2-hydroxyethyl)-3-phenyl-, acetate ester	**68:** 78201q
3-(2-hydroxyphenyl)-4-methyl-	**84:** 105498f
3-(2-hydroxyphenyl)-2-phenyl-	**77:** 126508e, **83:** 97141w
1-(4-hydroxy-1,1,4-trimethyl-2-pentenyl)-3-phenyl-	**74:** 87894u
3-(1H̲-indol-3-yl)-1-phenyl-	**85:** 143036g
4-(2-iodoethyl)-3-(*p*-methoxyphenyl)-	**68:** 78201q
4-(2-iodoethyl)-3-methyl-	**68:** 78201q
4-(2-iodoethyl)-3-phenyl-	**68:** 78201q
3-(1-isobutyl-3-butenyl)-1-phenyl-	**61:** 5640c
3-isobutyl-1-phenyl-	**61:** 10674f, **73:** 77210h
3-isobutyl-4-phenyl-	**61:** 10674a
3-isopentyl-1-phenyl-	**73:** 77210h

TABLE 13 (*Continued*)

Compound	Reference
13.2. Alkyl- or Aryl-Disubstituted Δ^2(or Δ^3)-1,2,4-Triazolin-5-ones (*Continued*)	
3-(1-isopropyl-3-butenyl)-1-phenyl-	**61:** 5640c
2-isopropyl-3-methoxy-	**79:** P66363h
3-isopropyl-1-phenyl-	**73:** 77210h
3-isopropyl-4-phenyl-	**61:** 10674a
3-(4-isopropylphenyl)-1-methyl-	**85:** 143036g
3-(4-isopropylphenyl)-1-phenyl-	**85:** 143036g
-3-malonic acid, 1-phenyl-, diethyl ester	**30:** 4860[1]
3-[1-mercapto-2-(4-methoxyphenyl)ethenyl]-2-methyl-	**80:** 3422b
3-(mercaptomethyl)-2-methyl-	**76:** 3765h
3-(mercaptomethyl)-1-phenyl-	**76:** 3765h
3-(1-mercapto-2-phenylethenyl)-2-methyl-	**80:** 3422b
-1-methanol, 4-phenyl-	**64:** 570f
3-(2-methoxy-1-methylvinyl)-2-phenyl-	**52:** 11864b
4-(*o*-methoxyphenyl)-3-(*p*-methoxyphenyl)-	**75:** 88541j
1-(*p*-methoxyphenyl)-3-methyl-	**71:** 21770g
3-(4-methoxyphenyl)-1-methyl-	**85:** 143036g
4-(*p*-methoxyphenyl)-3-methyl-	**61:** 10674d
4-(*o*-methoxyphenyl)-3-(*m*-nitrophenyl)-	**75:** 88541j
4-(*p*-methoxyphenyl)-3-(*m*-nitrophenyl)-	**75:** 88541j
4-(*p*-methoxyphenyl)-3-(*p*-nitrophenyl)-	**75:** 88541j
1-(*p*-methoxyphenyl)-3-phenyl-	**71:** 21830b
3-(*o*-methoxyphenyl)-4-phenyl-	**75:** 88541j
3-(4-methoxyphenyl)-1-phenyl-	**85:** 143036g
3-(*p*-methoxyphenyl)-4-phenyl-	**54:** 7695e, **58:** 4569h, **59:** 10053a
4-(*o*-methoxyphenyl)-3-phenyl-	**75:** 88541j
4-(*p*-methoxyphenyl)-3-phenyl-	**61:** 10674b, **75:** 88541j
4-(*p*-methoxyphenyl)-3-(3-pyridyl)-	**75:** 88541j
4-(*p*-methoxyphenyl)-3-(2-thienyl)-	**75:** 88541j
3-(*p*-methoxyphenyl)-4-*p*-tolyl-	**54:** 7695e, **58:** 4570a, **59:** 10053a
4-(*p*-methoxyphenyl)-3-*p*-tolyl-	**75:** 88541j
3-(*p*-methoxystyryl)-4-phenyl-	**75:** 88541j
3-(*p*-methoxystyryl)-4-*m*-tolyl-	**75:** 88541j
3-(*p*-methoxystyryl)-4-*p*-tolyl-	**75:** 88541j
3-(α-methylbenzyl)-1-phenyl-	**61:** 5640b
3-[3,4-(methylenedioxy)phenyl]-4-phenyl-	**75:** 88541j
3-[3,4-(methylenedioxy)phenyl]-4-*p*-tolyl-	**75:** 88541j
3-methyl-2-(isopropyl)-	**77:** P34528h
1-methyl-3-(4-methylphenyl)-	**85:** 143036g
3-methyl-1-(1-naphthyl)-	**21:** 743[3]
3-methyl-4-(1-naphthyl)-	**61:** 10674a
4-methyl-3-(5-nitro-2-furanyl)-	**79:** 87327q, **81:** 163330u
1-methyl-3-(5-nitro-2-furyl)-	**63:** P611h, **72:** 100607h
1-methyl-3-[2-(5-nitro-2-furyl)vinyl]-	**65:** PC7188c, **68:** P39627p
3-methyl-1-(*p*-nitrophenyl)-	**25:** 3651[1], **71:** 21770g
3-methyl-2-(4-nitrophenyl)-	**77:** P126637w
3-[2-(2-methyl-10H-phenothiazin-10-yl)ethyl]-4-(2-methylphenyl)-	**77:** 34439e
3-[2-(2-methyl-10H-phenothiazin-10-yl)ethyl]-4-phenyl-	**77:** 34439e

TABLE 13 (*Continued*)

Compound	Reference

13.2. Alkyl- or Aryl-Disubstituted Δ^2(or Δ^3)-1,2,4-Triazolin-5-ones   (*Continued*)

Compound	Reference
3-[3-(4-methylphenoxy)propyl]-1-phenyl-	**79:** 78693d
1-methyl-2-phenyl-	**49:** 15916b
1-methyl-3-phenyl-	**49:** 11630h, **49:** 15916b,
	82: 42707j, **85:** 46601e,
	85: 143036g
1-methyl-4-phenyl-	**57:** 808i, **76:** 126876c
2-methyl-3-phenyl-	**79:** 78749b, **85:** 159988e
3-methyl-1-phenyl-	**25:** 3651², **30:** 4860²,
	49: 15916b, **49:** 11631a,
	53: 8053a, **61:** 10674f,
	71: 21770g, **73:** 77210h,
	73: P77253z, **81:** 37516w
3-methyl-2-phenyl-	**50:** 8651a, **77:** P126637w
3-methyl-4-phenyl-	**53:** 9197c, **55:** 10420b,
	55: 27284c, **56:** 2440d,
	56: 4748i, **57:** 809b,
	57: 16598g, **58:** 6827h,
	61: 10673h, **62:** 14437d,
	69: 67340d, **75:** 88541j
4-methyl-3-phenyl-	**61:** 10674b, **79:** 78749b,
	85: 159988e
1-(4-methylphenyl)-3-phenyl-	**85:** 46601e
3-(4-methylphenyl)-1-phenyl-	**81:** 24524x, **81:** 63263f,
	85: 143036g
4-methyl-3-propyl	**84:** 150568q
1-methyl-3-(1$\underline{H}$-pyrrol-2-yl)-	**85:** 143036g
1-methyl-3-(2-thienyl)-	**85:** 143036g
3-methyl-4-*o*-tolyl-	**61:** 10674a, **75:** 88541j
3-methyl-4-*m*-tolyl-	**55:** 10420b
3-methyl-4-*p*-tolyl-	**55:** 10420b, **61:** 10674a
1-methyl-3-(4$\underline{H}$-1,2,4-triazol-4-yl)-	**72:** 121448x
1-methyl-3-(trifluoromethyl)-	**80:** P82990m
4-methyl-3-(trifluoromethyl)-	**82:** P72995u, **80:** P82989t
3-methyl-2-[3-(trifluoromethyl)phenyl]-	**77:** P126637w
3-(morpholinomethyl)-1-phenyl-	**57:** 4651f, **57:** 4651g
3-(morpholinomethyl)-4-phenyl-	**57:** 809d
4-(1-naphthalenyl)-3-[(2-naphthalenyloxy)methyl]-	**80:** 59762e
3-[(2-naphthalenyloxy)methyl]-4-phenyl-	**80:** 59762e
4-(1-naphthalenyl)-3-(phenoxymethyl)-	**76:** 126869c
4-(1-naphthalenyl)-3-[(2,4,5-trichlorophenoxy)methyl]-	**76:** 126869c
3-(1-naphthalmethyl)-4-phenyl-	**74:** 141652d
4-(1-naphthyl)-3-(1-naphthylmethyl)-	**74:** 141652d
1-(1-naphthyl)-3-phenyl-	**21:** 743³
3-(1-naphtyl)-1-phenyl-	**72:** P132738t
4-(1-naphthyl)-3-phenyl-	**61:** 10674b
3-(5-nitro-2-furyl)-4-phenyl-	**75:** 88541j
1-(*p*-nitrophenyl)-3-phenyl-	**69:** 52073v, **69:** 52075x,
	71: 21830b, **81:** 24524x,
	81: 63263f, **82:** 42707j
2-(*p*-nitrophenyl)-3-phenyl-	**51:** 10506g, **69:** 52073v

TABLE 13 (*Continued*)

Compound	Reference
13.2. Alkyl- or Aryl-Disubstituted Δ²(or Δ³)-1,2,4-Triazolin-5-ones (*Continued*)	
3-(*o*-nitrophenyl)-4-phenyl-	**58:** 4569h
3-(*m*-nitrophenyl)-4-phenyl-	**54:** 7695e, **58:** 4569h, **59:** 10053a
3-(4-nitrophenyl)-1-phenyl-	**81:** 24524x, **81:** 63263f
3-(*p*-nitrophenyl)-4-phenyl-	**75:** 88541j
4-(*o*-nitrophenyl)-3-phenyl-	**75:** 88541j
3-(*o*-nitrophenyl)-4-*p*-tolyl-	**58:** 4570a, **59:** 10053a
3-(*m*-nitrophenyl)-4-*p*-tolyl-	**54:** 7695e, **58:** 4570a, **59:** 10053a
3-(*p*-nitrophenyl)-4-*p*-tolyl-	**75:** 88541j
3-pentyl-1-phenyl	**73:** 77210h
3-phenethyl-4-phenyl-	**61:** 10674b
3-(phenoxymethyl)-1-phenyl-	**57:** 4651e
3-(phenoxymethyl)-2-phenyl-	**79:** 78693d
3-(phenoxymethyl)-4-phenyl-	**76:** 126869c
1-phenyl-3-(1-phenyl-3-butenyl)-	**61:** 5640b
1-phenyl-3-[1-(1-phenyliminoethyl)acetonyl]-	**32:** 9088[1]
4-phenyl-1-(piperidinomethyl)-	**64:** 570f
4-phenyl-3-(piperidinomethyl)-	**56:** 2440d, **57:** 809c
1-phenyl-3-propenyl-	**61:** 5640b, **73:** 77210h
1-phenyl-3-propyl-	**61:** 10674f, **73:** 77210h
4-phenyl-3-propyl-	**55:** 10420b, **58:** 6827h, **61:** 10674a, **62:** 14437d, **75:** 88541j
4-phenyl-3-(2-pyridyl)-	**62:** 14661d
4-phenyl-3-(4-pyridyl)-	**57:** 8561d
4-phenyl-1-[2-(2-pyridyl)ethyl]-	**63:** 6996e
1-phenyl-3-(1H-pyrrol-2-yl)-	**85:** 143036g
1-phenyl-3-styryl-	**73:** 77210h
1-phenyl-3-(2-thienyl)-	**69:** 52073v, **85:** P143036g
2-phenyl-3-(2-thienyl)-	**69:** 52073v
4-phenyl-3-(2-thienyl)-	**75:** 88541j
1-phenyl-3-*p*-tolyl-	**21:** 743[4], **63:** 11541d
3-phenyl-1-*p*-tolyl-	**69:** 52073v, **69:** 52075x, **82:** 42707j
3-phenyl-2-*p*-tolyl-	**69:** 52073v, **69:** 52075x
3-phenyl-4-*o*-tolyl-	**61:** 10674b, **75:** 88541j
3-phenyl-4-*p*-tolyl-	**54:** 7695e, **57:** 16600g, **58:** 4569h, **59:** 10053a, **61:** 10674b
4-phenyl-3-*p*-tolyl-	**51:** 5095c, **75:** 88541j
4-phenyl-3-[(2,4,5-trichlorophenoxy)methyl]-	**76:** 126869c
-2-propanenitrile, 3-methyl-	**82:** P156321j
-1-propanimidamide, *N*-[(aminocarbonyl)oxy]-4-(2-chlorophenyl)-	**77:** 34430v
-1-propanimidamide, *N*-[(aminocarbonyl)oxy]-4-(4-methoxyphenyl)-	**77:** 34430v
-1-propanimidamide, *N*-[(aminocarbonyl)oxy]-4-(2-methylphenyl)-	**77:** 34430v

TABLE 13 (*Continued*)

Compound	Reference

13.2. Alkyl- or Aryl-Disubstituted Δ^2(or Δ^3)-1,2,4-Triazolin-5-ones (*Continued*)

Compound	Reference
-1-propanimidamide, N-[(aminocarbonyl)oxy]-4-(4-methylphenyl)-	**77:** 34430v
-1-propanimidamide, N-[(aminocarbonyl)oxy]-4-phenyl-	**77:** 34430v
-1-propanimidamide, 4-(2-chlorophenyl)-N-hydroxy-	**77:** 34430v
-1-propanimidamide, N-hydroxy-4-(4-methoxyphenyl)-	**77:** 34430v
-1-propanimidamide, N-hydroxy-4-(2-methylphenyl)-	**77:** 34430v
-1-propanimidamide, N-hydroxy-4-(4-methylphenyl)-	**77:** 34430v
-1-propanimidamide, N-hydroxy-4-phenyl-	**77:** 34430v
-1-propionitrile, 4-(o-chlorophenyl)-	**75:** 20303c
-1-propionitrile, 4-(m-chlorophenyl)-	**75:** 20303c
-1-propionitrile, 4-(p-methoxyphenyl)-	**75:** 20303c
-1-propionitrile, 4-p-tolyl-	**75:** 20303c
3-propyl-4-o-tolyl-	**75:** 88541j
3-propyl-4-p-tolyl-	**75:** 88541j
3-(3-pyridyl)-4-o-tolyl-	**75:** 88541j
3-(3-pyridyl)-4-m-tolyl-	**75:** 88541j
3-(4-pyridyl)-4-p-tolyl-	**57:** 8561d
3-(2-thienyl)-4-o-tolyl-	**75:** 88541j
3-(2-thienyl)-4-m-tolyl-	**75:** 88541j
3-(2-thienyl)-4-p-tolyl-	**75:** 88541j

13.3. Alkyl- or Aryl-Trisubstituted Δ^2(or Δ^1,Δ^3)-1,2,4-Triazolin-5-ones

Compound	Reference
-1-acetamide, α-benzoyl-N-(2-chloro-5-[(4-[2,4-bis(1,1-dimethylpropyl)phenoxy]-1-oxobutyl)amino]-phenyl)-3,4-dimethyl-	**83:** P155710u
-4-acetic acid, 3-(p-chlorophenyl)-1-phenyl-	**72:** P132738t
-4-acetic acid, 3-(p-chlorophenyl)-1-phenyl-, ethyl ester	**72:** P132738t
-4-acetic acid, 1,3-diphenyl-	**72:** P132738t
-4-acetic acid, 1,3-diphenyl-, ethyl ester	**72:** P132738t
-4-acetic acid, 3-(diphenylmethyl)-1-phenyl-	**72:** P132738t
-4-acetic acid, 3-(diphenylmethyl)-1-phenyl-, ethyl ester	**72:** P132738t
-1-acetic acid, 3-isopropyl-4-phenyl-, ethyl ester	**59:** P11507b, **62:** P1668e
-1-acetic acid, 4-methyl-3-(5-nitro-2-furanyl)-	**81:** 163330u
-1-acetic acid, 4-methyl-3-(5-nitro-2-furanyl)-, ethyl ester	**81:** 163330u
-4-acetic acid, 3-(1-naphthyl)-1-phenyl-	**72:** P132738t
-4-acetic acid, 3-(1-naphthyl)-1-phenyl-, ethyl ester	**72:** P132738t
3-(1-allyl-3-butenyl)-4-[2-(diethylamino)-ethyl]-1-phenyl-	**62:** 546f
4-allyl-3-(α-ethylbenzyl)-1-phenyl-	**62:** 546g
3-(o-aminophenyl)-1,4-diphenyl-	**23:** 3683[4]
1-(3-aminopropyl)-3,4-diethyl-	**83:** P179071n
3-(2-benzothiazolyl)-1-benzyl-4-phenyl-	**77:** 126518h
4-(2-benzoyl-4-chlorophenyl)-1-[2-(dimethylamino)ethyl]-3-[(dimethylamino)methyl]-	**79:** P92234m, **85:** 56539e

TABLE 13 (*Continued*)

Compound	Reference
13.3. Alkyl- or Aryl-Trisubstituted Δ^2(or Δ^1,Δ^3)-1,2,4-Triazolin-5-ones (*Continued*)	
4-(2-benzoyl-4-chlorophenyl)-3-[(dimethylamino)methyl]-1-methyl-	**79:** P78811r
4-(2-benzoyl-4-chlorophenyl)-1-methyl-3-[(methylamino)methyl]-	**79:** P78808v, **79:** P78811r, **85:** 56539e
4-benzyl-3-(α-ethylbenzyl)-1-phenyl-	**62:** 546g
1,2-bis(tert-butyl)-3-phenyl-	**81:** 13434s
1,2-(bis(1,1-dimethylethyl)-3-isopropyl-	**84:** P59486a
1,2-(bis(1,1-dimethylethyl)-3-methyl-	**84:** P59486a
1,2-(bis(1,1-dimethylethyl)-3-phenyl	**84:** P59486a
1-(3-[bis(2-hydroxyethyl)amino]propyl)-3,4-diethyl-	**81:** P13553e
1-(*p*-bromophenyl)-3-methyl-4-phenyl-	**75:** 110244p
4-(4-bromo-2-hydroxy-7,7-dimethyl-1-norbornyl)-3,3-dimethyl-	**70:** 106024w
3-tert-butyl-4-cyclohexyl-1-ethyl-	**59:** P11507b, **62:** P1668e
3-tert-butyl-4-cyclohexyl-1-methyl-	**59:** P11506h, **62:** P1668e
4-butyl-3-(α-ethylbenzyl)-1-phenyl-	**62:** 546g
4-sec-butyl-3-(α-ethylbenzyl)-1-phenyl-	**62:** 546g
3-butyl-4-methyl-1-phenyl-	**65:** 10579g
4-butyl-1-methyl-3-phenyl-	**55:** P4537i
3-(4-tert-butylphenyl)-1,2-dimethyl-	**85:** 143036g
4-[4-chloro-2-(2-chlorobenzoyl)phenyl]-3-[(dimethylamino)methyl]-1-(hydroxymethyl)-	**79:** P92234m, **85:** 56539e
1-(2-chloro-1-diethoxyphosphinylethenyl)-4-methyl-3-(trifluoromethyl)-	**80:** P82989t
1-(chloromethyl)-4-methyl-3-(trifluoromethyl)-	**82:** P72995u
4-(*o*-chlorophenyl)-1,3-dimethyl-	**59:** P11507a, **62:** P1668d
1-(4-chlorophenyl)-2,3-diphenyl-	**79:** 91736q
2-(4-chlorophenyl)-1,3-diphenyl-	**79:** 91736q
1-(*p*-chlorophenyl)-3-methyl-4-phenyl-	**75:** 110244p
1(3-[4-(3-chlorophenyl)-1-piperazinyl]propyl)-3,4-diethyl-,	**81:** P13553e
1-(3-chloropropyl)-3,4-diethyl-	**81:** P13553e
4-(4-chloro-*o*-tolyl)-1,3-dimethyl-	**59:** P11507a, **62:** P1668d
4-(4-chloro-*o*-tolyl)-3-ethyl-1-methyl-	**59:** P11507a, **62:** P1668d
4-(4-chloro-*o*-tolyl)-1-methyl-3-propyl-	**59:** P11507a, **62:** P1668d
4-cyclohexyl-1-ethyl-3-propyl-	**59:** P11507b, **62:** P1668e
4-cyclohexyl-3-isopropyl-1-methyl-	**59:** P11506h, **62:** P1668e
3-cyclohexyl-1-methyl-4-phenyl-	**59:** P11506h, **62:** P1668e
4-cyclohexyl-1-methyl-3-propyl-	**59:** P11506h, **62:** P1668e
4-decyl-3-(α-ethylbenzyl)-1-phenyl-	**62:** 546g
1-[2,2-dichloro-1-(diethoxyphosphinyloxy)ethyl]-4-methyl-3-(trifluoromethyl)-	**80:** P82989t
1-[2,2-dichloro-1-(dimethoxyphosphinyloxy)ethenyl]-4-methyl-3-(trifluoromethyl)-	**82:** P82989t
4-(2,4-dichlorophenyl-1,3-dimethyl-	**70:** P28924y
3,4-dicyclohexyl-1-[2-(dimethylamino)ethyl]-	**59:** P11507b, **62:** P1668e
3,4-dicyclohexyl-1-methyl-	**59:** P11506h, **62:** P1668e
1,4-dicyclohexyl-3-(5-nitro-2-furyl)-	**70:** 106436g

TABLE 13 (*Continued*)

Compound	Reference

13.3. Alkyl- or Aryl-Trisubstituted Δ^2(or Δ^1,Δ^3)-1,2,4-Triazolin-5-ones (*Continued*)

Compound	Reference
1-([[(diethoxyphosphinothioyl)oxy]methyl)-4-methyl-3-(trifluoromethyl)-	**82:** P72995u
1-[1-(diethoxyphosphinyl)ethenyl]-4-methyl-3-(trifluoromethyl)-	**80:** P82989t
4-[1-(diethoxyphosphinyl)ethenyl]-1-methyl-3-(trifluoromethyl)-	**80:** P82989t
4-[2-(diethylamino)ethyl]-1,3-diphenyl-	**62:** 546f
4-[2-(diethylamino)ethyl]-3-(α-ethylbenzyl)-1-phenyl-	**62:** 546e, **62:** 546f
4-[2-(diethylamino)ethyl]-3-(1-ethylpropyl)-1-phenyl-	**62:** 546f
3,4-diethyl-1-[3-(4-morpholinyl)propyl]-	**83:** P179071n
1,2-diethyl-3-phenyl-	**81:** 169457n
1-[2-(dimethylamino)ethyl]-3,4-diphenyl-	**59:** P11507b, **62:** P1668d
1-[2-(dimethylamino)ethyl]-3-isopropyl-4-phenyl-	**59:** P11507b, **62:** P1668d
1-[2-(dimethylamino)ethyl]-3-methyl-4-phenyl-	**59:** P11507b, **62:** P1668e
4-[2-(dimethylamino)ethyl]-1-phenyl-3-(1-phenylpropyl)-	**76:** P113218q
4-[2-(dimethylamino)-1-methylethyl]-3-(α-ethylbenzyl)-1-phenyl-	**62:** 546f
4-[3-(dimethylamino)propyl]-3-(α-ethylbenzyl)-1-phenyl-	**62:** 546f
4-[3-(dimethylamino)propyl]-3-methyl-1-phenyl-	**73:** 77210h
4-[3-(dimethylamino)propyl]-1-phenyl-2-(1-phenylpropyl)-	**76:** P113218q
4-[3-dimethylamino)propyl]-1-phenyl-3-(1-phenylpropyl)-	**76:** P113218q
1,2-dimethyl-3-[4-(1-methylethyl)phenyl]-	**85:** 143036g
1,2-dimethyl-3-(4-methylphenyl)-	**85:** 143036g
1,2-dimethyl-3-phenyl-	**79:** 78749b, **85:** 143036g, **85:** 159988e
1,3-dimethyl-4-phenyl-	**64:** 19614a
1,4-dimethyl-3-phenyl-	**59:** P11506h, **62:** P1668e, **79:** 78749b **85:** 159988e
3,3-dimethyl-4-phenyl-Δ^1-	**72:** 55339j
1,4-diphenyl-3-propyl-	**61:** 10674g
4-[2-(2,4-dipropylbenzoyl)-4-(ethylthio)phenyl]-1-methyl-3-[1-(propylamino)ethyl]-	**79:** P78811r
3-(4-ethoxyphenyl)-1,2-dimethyl-	**85:** 143036g
4-(*p*-ethoxyphenyl)-1,3-diphenyl-	**61:** 655h
3-(α-ethylbenzyl)-1,4-dimethyl-	**59:** P11506h, **62:** P1668e
3-(α-ethylbenzyl)-4-hexyl-1-phenyl-	**62:** 546g
3-(α-ethylbenzyl)-4-isobutyl-1-phenyl-	**62:** 546g
3-(α-ethylbenzyl)-4-isopropyl-1-phenyl-	**62:** 546g
3-(α-ethylbenzyl)-4-methyl-1-phenyl-	**62:** 546g
3-(α-ethylbenzyl)-4-[2-(4-methyl-1-piperazinyl)ethyl]-1-phenyl-	**65:** 11185a
3-(α-ethylbenzyl)-4-(2-morpholinoethyl)-1-phenyl-	**62:** 546e
3-(α-ethylbenzyl)-4-nonyl-1-phenyl-	**64:** P8199c
3-(α-ethylbenzyl)-4-pentyl-1-phenyl-	**62:** 546g
3-(α-ethylbenzyl)-1-phenyl-4-(2-piperidinoethyl)-	**62:** 546f
3-(α-ethylbenzyl)-1-phenyl-4-propyl-	**62:** 546g
3-(α-ethylbenzyl)-1-phenyl-4-[2-(1-pyrrolidinyl)ethyl]-	**62:** 546f
3-ethyl-1,4-diphenyl-	**61:** 10674g

TABLE 13 (*Continued*)

Compound	Reference

13.3. Alkyl- or Aryl-Trisubstituted Δ^2(or Δ^1,Δ^3)-1,2,4-Triazolin-5-ones (*Continued*)

Compound	Reference
4-ethyl-3-(α-ethylbenzyl)-1-phenyl-	**62:** 546g
1-ethyl-3-isopropyl-4-phenyl-	**59:** P11507b, **62:** P1668d
3-ethyl-4-(*o*-methoxyphenyl)-1-methyl-	**59:** P11507a, **62:** P1668d
4-ethyl-1-methyl-3-phenyl-	**59:** P11506h, **62:** P1668e
3-(4-ethylphenyl)-1,2-dimethyl-	**85:** 143036g
4-(*o*-ethylphenyl)-1,3-dimethyl-	**59:** P11506h, **62:** P1668d
3-(2-furyl-1,4-dimethyl-	**63:** 18071f
1-(hydroxymethyl)-4-methyl-3-(trifluoromethyl)-	**82:** P72995u
3-(2-hydroxyphenyl)-1,2-dimethyl-	**77:** 126508e, **83:** 97141w
3-isopropyl-1-methyl-4-phenyl-	**59:** P11506h, **62:** P1668e
4-(mercaptomethyl)-3-methyl-1-phenyl-, *S*-ester with *O,O*,-dimethyl phosphorodithioate	**64:** P3545h
4-(mercaptomethyl)-1-phenyl-3-propyl-, *S*-ester with *O,O*-diethyl phosphorodithioate	**64:** P3545h
3-(4-methoxyphenyl)-2,3-diphenyl-	**85:** 143036g
1-(4-methoxyphenyl)-2,3-diphenyl-	**79:** 91736q
1-methyl-3,4-diphenyl-	**59:** P11506h, **62:** P1668d, **69:** 67340d
3-methyl-1,4-diphenyl-	**61:** 10674g, **69:** 67340d, **73:** 77210h
3-(3,4-methylenedioxyphenyl)-1,4-dipenyl-	**23:** 3683[7]
3-methyl-3-(phenoxymethyl)-4-phenyl-Δ^1-	**72:** 78951s
1-(4-methylphenyl)-2,3-diphenyl-	**79:** 91736q
2-(4-methylphenyl)-1,3-diphenyl-	**79:** 91736q
4-(1-naphthyl)-1,3-diphenyl-	**61:** 655h
3-(*m*-nitrophenyl)-1,4-diphenyl-	**23:** 3683[4,5]
3-(*o*-nitrophenyl)-1,4-diphenyl-	**23:** 3683[4,5]
3-(*p*-nitrophenyl)-1,4-diphenyl-	**23:** 3683[4,5]
4-(*p*-nitrophenyl)-1,3-diphenyl-	**61:** 655g
-1-propanimidamide, *N*-[(aminocarbonyl)oxy]-4-(2-chlorophenyl)-3-methyl-	**77:** 34430v
-1-propanimidamide, *N*-[(aminocarbonyl)oxy]-4-(4-methoxyphenyl)-3-methyl-	**77:** 34430v
-1-propanimidamide, *N*-[(aminocarbonyl)oxy]-3-methyl-4-(2-methylphenyl)-	**77:** 34430v
-1-propanimidamide, *N*-[(aminocarbonyl)oxy]-3-methyl-4-(4-methylphenyl)-	**77:** 34430v
-1-propanimidamide, *N*-[(aminocarbonyl)oxy]-3-methyl-4-phenyl-	**77:** 34430v
-1-propanimidamide, 4-(2-chlorophenyl)-*N*-hydroxy-3-methyl-	**77:** 34430v
-1-propanimidamide, *N*-hydroxy-4-(4-methoxyphenyl)-3-methyl-	**77:** 34430v
-1-propanimidamide, *N*-hydroxy-3-methyl-4-(2-methylphenyl)-	**77:** 34430v
-1-propanimidamide, *N*-hydroxy-3-methyl-4-(4-methylphenyl)-	**77:** 34430v
-1-propanimidamide, *N*-hydroxy-3-methyl-4-phenyl-	**77:** 34430v

TABLE 13 (*Continued*)

Compound	Reference
13.3. Alkyl- or Aryl-Trisubstituted Δ^2(or Δ^1,Δ^3)-1,2,4-Triazolin-5-ones (*Continued*)	
-1-propionitrile, 4-(*o*-chlorophenyl)-3-ethyl-	**75:** 20303c
-1-propionitrile, 4-(*o*-chlorophenyl)-3-methyl-	**75:** 20303c
-1-propionitrile, 4-(*m*-chlorophenyl)-3-methyl-	**75:** 20303c
-1-propionitrile, 3-ethyl-4-(*p*-methoxyphenyl)-	**75:** 20303c
-1-propionitrile, 3-ethyl-4-phenyl-	**75:** 20303c
-1-propionitrile, 4-(*p*-methoxyphenyl)-3-methyl-	**75:** 20303c
-1-propionitrile, 3-methyl-4-phenyl-	**75:** 20303c
1,2,3-trimethyl-	**84:** 17237q
1,3,4-trimethyl-	**84:** 17237q
1,2,3-triphenyl-	**52:** 11823c, **64:** 15871h, **79:** 91736q
1,3,4-triphenyl-	**33:** 7287[7], **61:** 655f, **63:** 7000e

CHAPTER 14

Alkyl- or Aryl-1,2,4-Triazoline-5-thiones

14.1. Alkyl- or Aryl-Monosubstituted Δ^2 (or Δ^3)-1,2,4-Triazoline-5-thiones

s-*Triazoline-5-thione.* The condensation of hot formamide with thiosemicarbazide gave a 90% yield of s-triazoline-5-thione (**14.1-1**).[1,2] Also, the preparation of **14.1-1** was accomplished by the cyclization of thiosemicarbazide either with ethyl formate in the presence of sodium methoxide (65% yield)[3,4] or by heating with s-triazine (63% yield).[5] In addition, both the 1-formyl[1,6] and 1-(ethoxymethylene)[7] derivatives of thiosemicarbazide has been converted to **14.1-1** with aqueous sodium carbonate in excellent yields. Rearrangement of 2-amino-1,3,4-thiadiazole to **14.1-1** in 52% yield was effected in methanolic methylamine at 160°.[8]

14.1-1

Although few 1- or 2-substituted triazoline-5-thiones have been reported, good yields were obtained in the cyclization of both 2-methyl-[9] and 1-(4-nitrophenyl)-3-thiosemicarbazide[10] with formic acid to give the triazoline-5-thiones (**14.1-2** and **14.1-3**, respectively). In addition, the condensation of 1-benzyl-3-thiosemicarbazide (**14.1-5**) with ethyl formate in methanolic sodium methoxide gave an 83% yield of **14.1-4**.[11,12]

$$\begin{array}{c} \text{SC} \overset{\textstyle NH_2}{\underset{\textstyle N(Me)NH_2}{\Big\langle}} \quad \xrightarrow[\Delta]{HCO_2H} \quad S{=}\overset{\textstyle N}{\underset{\textstyle N{-}N}{\diagdown}}\overset{\textstyle }{\underset{R_1\ R_2}{}} \end{array}$$

$$\xrightarrow[\Delta]{HCO_2H}$$

14.1-2, $R_1 = Me$, $R_2 = H$
14.1-3, $R_1 = H$, $R_2 = 4\text{-}O_2NC_6H_4$
14.1-4, $R_1 = H$, $R_2 = PhCH_2$

$$\text{SC} \overset{\textstyle NH_2}{\underset{\textstyle NHNHC_6H_4\text{-}4\text{-}NO_2}{\Big\langle}}$$

$$\Big\uparrow \text{EtO}_2\text{CH/MeO}^-$$

$$\text{SC} \overset{\textstyle NH_2}{\underset{\textstyle NHNHCH_2Ph}{\Big\langle}}$$

14.1-5

In a basic medium a number of types of 1-acyl- and 1-aroyl-3-thiosemi-carbazides are dehydrated by the condensation of the 4-amino group with the carbonyl moiety to give good yields of triazoline-5-thiones. Although dehydration can occur in the presence of a strong mineral acid, the products are usually 1,3,4-thiadiazoles.[13,14] A simple explanation for the different modes of cyclization is provided by the consideration of the relative nucleophilicities of the atoms of the terminal thioamide function of the 1-substituted 3-thiosemicarbazide. In strong acid the 4-amino group is protonated and cannot enter into the condensation reaction, whereas in base the sulfur function is ionized, which increases the nucleophilicity of the 4-amino group and promotes triazoline-5-thione formation.

The majority of the monosubstituted triazoline-5-thiones are 3-alkyl and 3-aryl derivatives, which are readily obtained from thiosemicarbazide and its acylated products. For example, the methoxide catalyzed reaction of thiosemicarbazide with aliphatic carboxylic esters gave 3-alkyltriazoline-5-thiones (e.g., **14.1-6**) in 45 to 94% yield.[3] Similarly, aromatic carboxylic esters gave 3-aryltriazolin-5-thiones (e.g., **14.1-7**) in 54 to 86% yield.[3,15] In addition, the thermal condensation of the dithiocarboxylic acid ester (**14.1-15**) with thiosemicarbazide at 155° gave the corresponding 3-(3-pyrazolin-4-yl)triazoline-5-thione.[16]

The 1-acyl-3-thiosemicarbazides e.g., **14.1-11**) are converted in good yields with aqueous base to 3-alkyltriazoline-5-thiones (e.g., **14.1-8**).[17,18] Compound **14.1-8** was the precursor of 3-(aminomethyl)triazoline-5-thione.[17] Similarly, the cyclization of 1-benzoyl-3-thiosemicarbazide (**14.1-12**) to give **14.1-9**[19] was effected readily with aqueous sodium hydroxide and also with aqueous ammonia, trimethylamine,[20] and hydrazine.[21]

$$SC\begin{smallmatrix} NH_2 \\ \\ NHNH_2 \end{smallmatrix} + EtO_2CR_1 \xrightarrow{MeO^-} S=\begin{smallmatrix} H \\ N \\ | \\ N-N \\ H \end{smallmatrix}R_1$$

$$\xrightarrow{OH^-} SC\begin{smallmatrix} NH_2 \\ \\ NHNHCOR_1 \end{smallmatrix}$$

14.1-6, R_1 = Me
14.1-7, R_1 = 1-$C_{10}H_7$
14.1-8, R_1 = $PhCH_2O_2CNHCH_2$
14.1-9, R_1 = Ph
14.1-10, R_1 = 4-C_5H_4N

14.1-11, R_1 = $PhCH_2O_2CNHCH_2$
14.1-12, R_1 = Ph
14.1-13, R_1 = 4-C_5H_4N
14.1-14, R_1 = 3-C_5H_4N

H_2NCSNH_2

$$Ph\begin{smallmatrix} CS_2Et \\ \diagdown \diagup \\ N-N \\ Me\ Me \end{smallmatrix}=O$$

14.1-15

$$N\diagdown CONHNH_2$$

14.1-16

Also, this type of cyclization was successful for the preparation of triazoline-5-thiones substituted with a heteroaromatic ring in the 3-position.[22,23] In the fusion of hydrazides (e.g., **14.1-16**) with thiourea at 130°, the intermediate 1-aroyl-3-semicarbazide (e.g., **14.1-13**) was dehydrated to give the corresponding triazoline-5-thione (e.g., **14.1-10**).[24] In several instances only the 1-aroyl-3-semicarbazide (e.g., **14.1-14**) were isolated. In contrast to the base-catalyzed cyclization of 1-aroyl-3-thiosemicarbazides, the cyclization of the imino ether (**14.1-17**) in ethanolic ammonia at 150° gave a mixture of **14.1-18** and **14.1-19** in the ratio of 2:5.[25] Also, treatment of **14.1-20** with

$$SC\begin{smallmatrix} NH_2 \\ \\ NHN=C(OEt)C_6H_4-4-OMe \end{smallmatrix} \xrightarrow{NH_3} H_2N\begin{smallmatrix} S \\ \diagup \diagdown \\ N-N \end{smallmatrix}C_6H_4-4-OMe + S=\begin{smallmatrix} H \\ N \\ | \\ N-N \\ H \end{smallmatrix}C_6H_4-4-OMe$$

14.1-17   **14.1-18**   **14.1-19**

concentrated sulfuric acid gave a mixture of the 1,3,4-thiadiazole (**14.1-21**) and unexpectedly, the triazoline-5-thione (**14.1-22**).[26] The formation of **14.1-22** under these conditions was attributed to the electron-withdrawing 2-trifluoromethyl group, which increased the electrophilic nature of the carbonyl group. In the cyclization of **14.1-13** a higher yield of **14.1-10** was obtained by heating **14.1-13** in tetralin (see the preceding) than by treatment of **14.1-13** with ethoxide, which gave incomplete conversion.[27]

14.1-20   **14.1-21**   **14.1-22**

The 4-aroyl-3-thiosemicarbazides (e.g., **14.1-23**) must be intermediates in the condensation of aroyl isothiocyanates with excess hydrazine to give 3-aryl triazoline-5-thiones (e.g., **14.1-24**) in fair yields.[28,29]

4.1-23   **14.1-24**

In the reaction of 1,4-dibenzoyl-3-thiosemicarbazide (**14.1-25**) with aqueous sodium hydroxide, ^{14}C-labeling experiments indicated that the 4-benzoyl group was hydrolyzed and that the resulting 1-benzoyl-3-thiosemicarbazide intermediate (**14.1-12**) was converted to **14.1-9**.[30] This conclusion was supported by the cyclization of the isomeric acylated thiosemicarbazides, **14.1-26** and **14.1-27**, with ethoxide to give mainly **14.1-9** and **14.1-30**.[31] Both products are formed via the corresponding 1-substituted 3-thiosemicarbazide intermediates (**14.1-12** and **14.1-29**). However, in the cyclization of the isomeric 1,4-disubstituted 3-thiosemicarbazide derivatives (**14.1-28** and **14.1-31**), the major product in both reactions was 3-phenyl-triazoline-5-thione (**14.1-9**), indicating that these reactions were controlled by the relative stability of the amide functions toward hydrolysis. Apparently, **14.1-9** was formed from **14.1-28** via **14.1-12** and from **14.1-31** via **14.1-37**.[31] Also, in the conversion of the 4-benzoyl-3-semicarbazide derivatives (**14.1-32** to **14.1-34**) to **14.1-9**, the 1-substituent was removed during the reaction, although the cyclization of **14.1-34** to **14.1-9** was proposed to occur via the 2-ethoxycarbonyl derivative **14.1-38**.[32] In addition, the carbohydrazides derivatives (**14.1-35** and **14.1-36**) are converted under alkaline conditions to **14.1-9**.[33]

The formation of 4-methyltriazoline-5-thione (**14.1-40**) in good yields was effected both by dehydration of 1-formyl-4-methyl-3-thiosemicarbazide (**14.1-39**) at 190° and by cyclization of this thiosemicarbazide with aqueous sodium carbonate.[9,34] Similarly, the 4-aryltriazolin-5-thiones (e.g., **14.1-41**) are readily prepared by the condensation of 4-aryl-3-semicarbazides (e.g., **14.1-42**) with ethyl formate in the presence of ethoxide.[35,36]

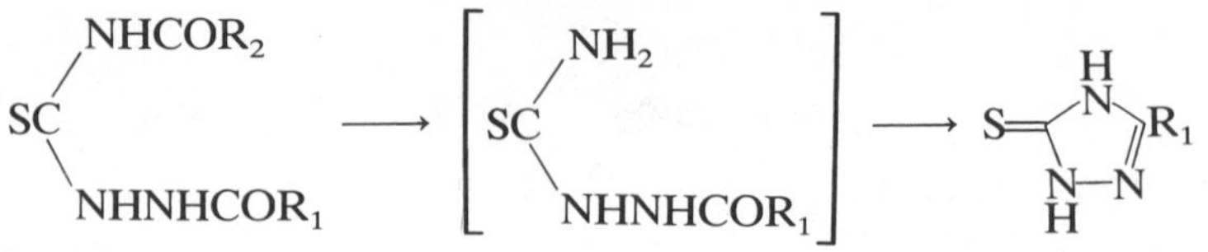

14.1-25, $R_1 = R_2 = Ph$
14.1-26, $R_1 = Ph$, $R_2 = CH_2Ph$
14.1-27, $R_1 = CH_2Ph$, $R_2 = Ph$
14.1-28, $R_1 = Ph$, $R_2 = Me$

14.1-12, $R_1 = Ph$
14.1-29, $R_1 = CH_2Ph$

14.1-9 $R_1 = Ph$
14.1-30, $R_1 = CH_2Ph$

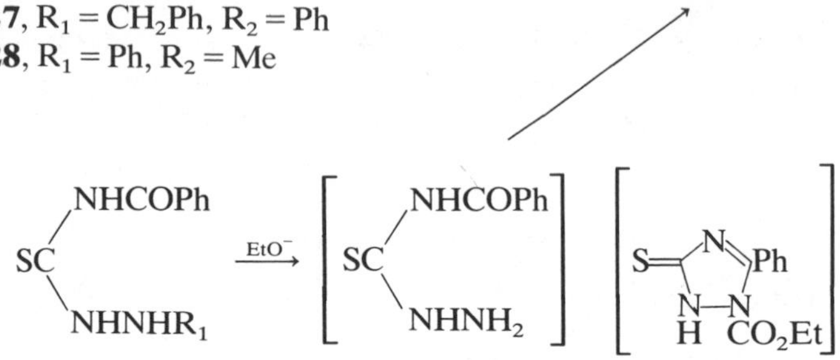

14.1-31, $R_1 = Ac$
14.1-32, $R_1 = CONH_2$
14.1-33, $R_1 = CSNH_2$
14.1-34, $R_1 = CO_2Et$
14.1-35, $R_1 = CONHNHPh$
14.1-36, $R_1 = CONHN=CHPh$

14.1-37

14.1-38

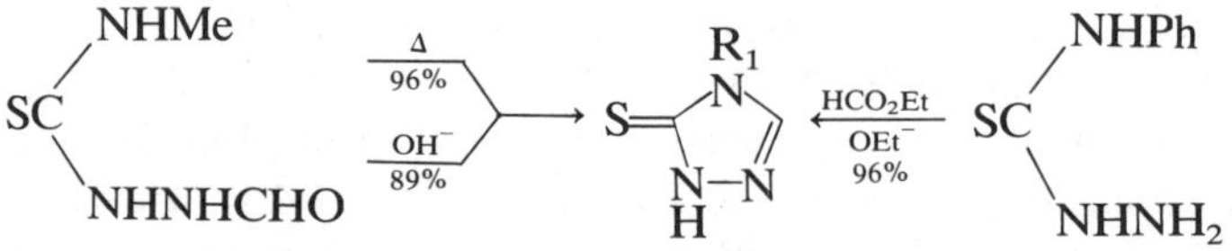

14.1-39

14.1-40, $R_1 = Me$
14.1-41, $R_1 = Ph$

14.1-42

Several monosubstituted triazoline-5-thiones have been prepared from amidrazones, as illustrated by the reaction of **14.1-43** with carbon disulfide in ethanolic potassium hydroxide to give an 83% yield of **14.1-44**.[37] In the cyclization of **14.1-45** with phenyl isothiocyanate to give **14.1-47**, aniline was eliminated from intermediate **14.1-46**.[38]

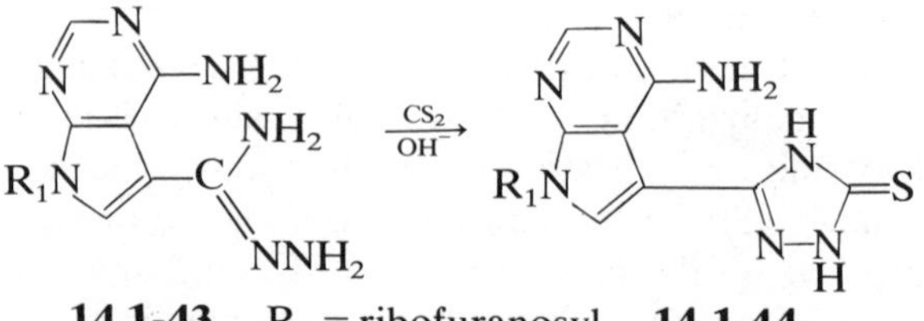

14.1-43 $R_1 = $ ribofuranosyl   **14.1-44**

14.1-45   **14.1-46**   **14.1-47**

The formation of monosubstituted triazoline-5-thiones from a variety of other ring systems has been observed. Treatment of the thiazolidine (**14.1-48**) with hydrazine hydrate at room temperature gave **14.1-49** (68%), which was heated with hydrazine to give **14.1-51** (50%).[39] Similarly, **14.1-51** was formed directly from **14.1-50** and refluxing aqueous hydrazine.[40] In the reaction of the benzoxazine (**14.1-52**) with 80% hydrazine

14.1-48   **14.1-49**

14.1-50   **14.1-51**

hydrate, a 72% yield of the 3-(2-hydroxyphenyl) triazoline-5-thione (**14.1-53**) was obtained.[41,42] The 1,3,4-thiadiazole (**14.1-54**) was rearranged in the presence of methylamine at 160° to give a good yield of **14.1-40**.[8] Apparently, this rearrangement proceeds to completion only at high temperatures since treatment of **14.1-55** with benzylamine in refluxing methanol gave a mixture of **14.1-56** and **14.1-57**, both in 30% yield.[43] The 4-methyl compound was also prepared via **14.1-59** by treatment of the hexahydro-1,2,4-triazine (**14.1-58**) with refluxing hydrochloric acid.[44] Of interest, the oxidative addition of sulfur to 4-phenyl-1,2,4-triazole at 200° gave a 54% yield of **14.1-41**.[45]

14.1-52   **14.1-53**

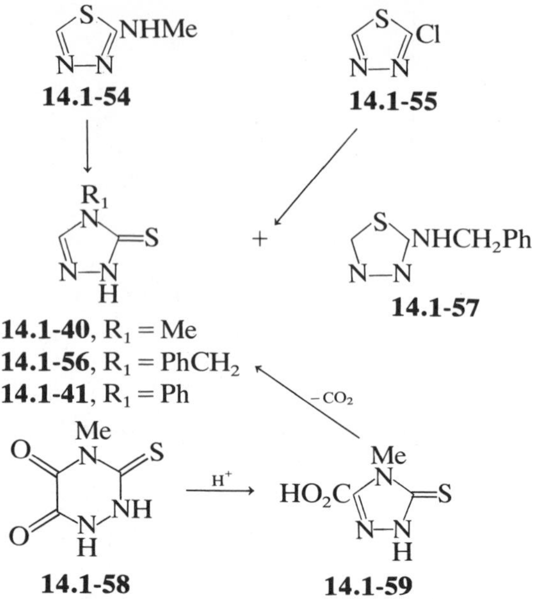

14.1-54

14.1-55

14.1-40, R_1 = Me
14.1-56, R_1 = PhCH$_2$
14.1-41, R_1 = Ph

14.1-57

$-CO_2$

14.1-58 $\xrightarrow{H^+}$ **14.1-59**

The preparation of monosubstituted triazoline-5-thiones from a pre-formed triazole system includes the deamination of **14.1-60** with nitrous acid to give **14.1-61**,[46] the reduction of the triazole-5-sulfonyl chloride

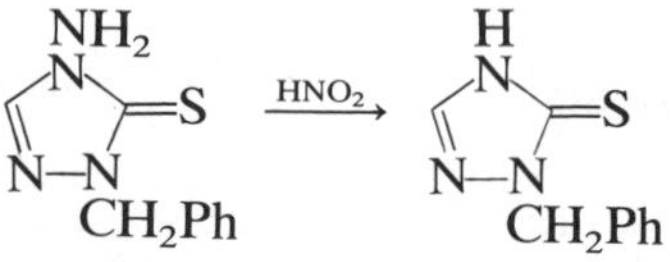

14.1-60   **14.1-61**

14.1-62 with hydrazine to give **14.1-63**,[47] and the removal of the alkene side chain of **14.1-64** with potassium permanganate to give **14.1-9**.[48] The sulfur moiety can also be removed, as demonstrated by the photolysis of triazoline-5-thione in methanol in a quartz container to give a 100% yield of 1,2,4-triazole.[49]

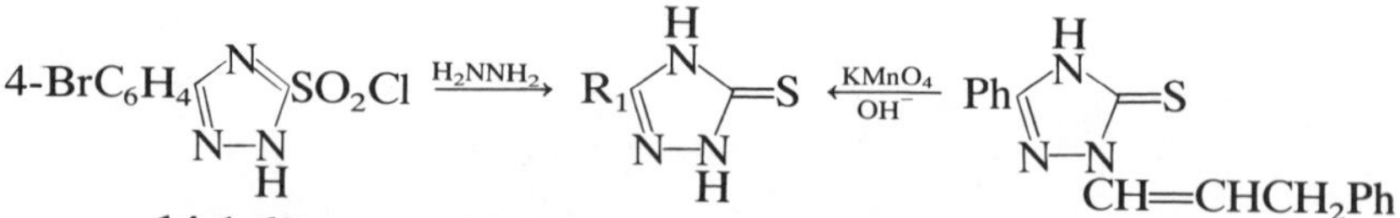

14.1-62   **14.1-63**, R_1 = 4-BrC$_6$H$_4$ **14.1-64**
                     14.1-9, R_1 = Ph

References

1. **57:** 804i
2. **68:** 49563d
3. **65:** 7169d
4. **67:** 64318a
5. **51:** 14751b
6. **55:** 4337i
7. **50:** 5647a
8. **51:** 12079d
9. **55:** 23508e
10. **65:** 704e
11. **67:** 32651n
12. **70:** 37729r
13. **79:** 115500m
14. **79:** 146466j
15. **85:** 159998h
16. **68:** 78208x
17. **66:** 2536b
18. **80:** 59762e
19. **64:** 5072h
20. **59:** 3930f
21. **58:** 13937f
22. **76:** 149140x
23. **83:** 114345k
24. **70:** 3966a
25. **51:** 17894e
26. **65:** 18576a
27. **52:** 1273g
28. **61:** 3064h
29. **70:** 77870m
30. **52:** 16363g
31. **53:** 8032i
32. **66:** 115652u
33. **75:** 129505w
34. **84:** 59327z
35. **61:** 4337h
36. **76:** 34182v
37. **83:** 22239a
38. **83:** 43245q
39. **75:** 35882q
40. **76:** 34220f
41. **77:** 126508e
42. **83:** 97141w
43. **56:** 7304h
44. **73:** 66517n
45. **65:** 16960f
46. **74:** 13070a
47. **78:** 43372a
48. **79:** 5298u
49. **72:** 121442r

14.2. Alkyl- or Aryl-Disubstituted Δ^2 (or Δ^3)-1,2,4-Triazoline-5-thiones

Many of the disubstituted triazoline-5-thiones are prepared from the appropriate 3-thiosemicarbazides by methods similar to those described for the preparation of monosubstituted triazoline-5-thiones in Section 14.1. Cyclization of 2,4-dimethyl-3-thiosemicarbazide (**14.2-1**) with formic acid gave an excellent yield of 1,4-dimethyltriazoline-5-thione (**14.2-2**),[1] and treatment of 4-phenyl-3-thiosemicarbazide (**13.2-3**) either with ethyl acetate in the presence of ethoxide[2–4] or with ethyl orthoacetate in refluxing xylene[5] gave good yields of 3-methyl-4-phenyltriazoline-5-thione (**14.2-4**). In the

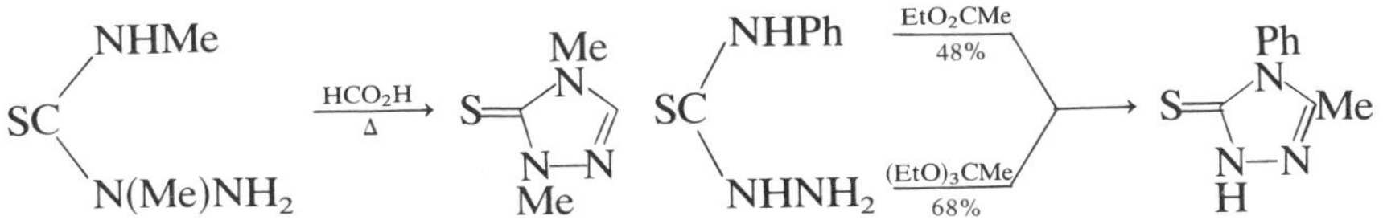

reactions of 4-phenyl-3-thiosemicarbazide (**14.2-3**) with the imino ether (**14.2-6**), the product obtained depended upon the nature of the solvent. In refluxing ethanol, a good yield of 1,3,4-thiadiazole (**14.2-5**) was isolated, whereas in refluxing pyridine a good yield of 3,4-diphenyltriazoline-5-thione (**14.2-7**) was isolated.[6]

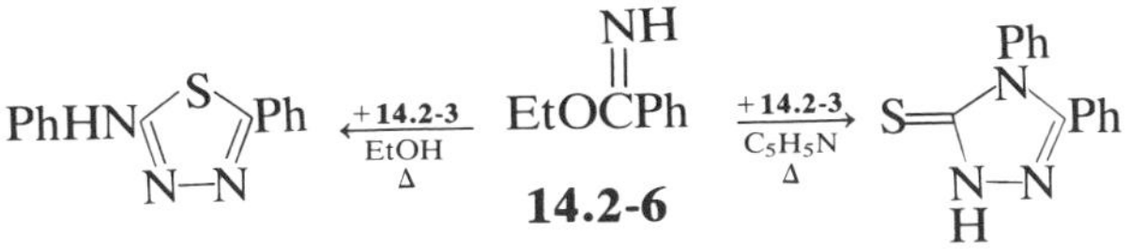

The dehydration of 1-acetyl-2-methyl-3-thiosemicarbazide (**14.2-8**) either in aqueous base or thermally at 200° gave 1,3-dimethyltriazoline-5-thione (**14.2-9**).[1] Also, the corresponding 3-(trifluoromethyl)-1,2,4-triazoline-5-thione was prepared from 1-(trifluoroacetyl)-2-methyl-3-thiosemicarbazide.[7]

14.2-8   **14.2-9**

The 1-acylthiosemicarbazide (**14.2-11**) was generated *in situ* in the condensation of the hydrazide (**14.2-10**) with thiourea to give **14.2-12**.[8] In

14.2-10   **14.2-11**   **14.2-12**

$R_1 = 2$-benzothiazolyl

other examples, the conversion of acyl- or aroyl-3-thiosemicarbazides to triazoline-5-thiones with aqueous hydroxide include the cyclization of 1-acetyl-4-methyl-3-thiosemicarbazide (**14.2-13**) to 3,4-dimethyltriazoline-5-thione (**14.2-16**) (55%),[9] and 1-(3,4,5-trimethoxybenzoyl)-4-propyl-3-thiosemicarbazide (**14.2-14**) to 3-(3,4,5-trimethoxyphenyl)-4-propyltriazoline-5-thione (**14.2-17**) (~80%).[10] Similar procedures were used for the preparation of 3-phenoxymethyl-,[11,12] 3-(benzoylaminomethyl)-,[13] 3-(2-ethoxyvinyl)-,[14] 3-(1-methylvinyl)-,[15] 3-(2,2-dichloro-1,1-difluoroethyl)-,[16] 3-(2-thiazolyl)-,[17] 3-(4-isoxazolyl)-,[18] 3-(4-pyridyl)-,[19] 3-(3-pyridazinyl)-,[20] 3-pyrazinyl-,[21] 3-(triazinyl)-,[22] and 3-(2-benzofuranyl)-[23] 4-substituted 1,2,4-triazoline-5-thiones. In contrast, treatment of 1-benzoyl-4-phenyl-3-thiosemicarbazide (**14.2-15**) with aniline gave a mixture of the triazoline-5-thione (**14.2-7**) and the 3-anilinotriazole (**14.2-18**). The latter probably

14.2-13, $R_1 = R_2 = $ Me   **14.2-16**, $R_1 = R_2 = $ Me
14.2-14, $R_1 = 3,4,5$-$(MeO)_3C_6H_2$ **14.2-17**, $R_1 = 3,4,5$-$(MeO)_3C_6H_2$,
       $R_2 = $ Pr $R_2 = $ Pr
14.2-15, $R_1 = R_2 = $ Ph   **14.2-7**, $R_1 = R_2 = $ Ph

resulted from an aminoguanidine intermediate formed by displacement of the sulfur moiety of **14.2-15** by aniline.[24] In addition, 1-acyl-3-thiosemi-carbazides are cyclized under acidic conditions to give 1,3,4-thiadia-zoles[23,25,26] and under oxidizing conditions to give 1,3,4-oxadiazoles.[23,26] An unusual reaction was observed in the base-catalyzed conversion of **14.2-19** to **14.2-21**, which probably involved the initial formation of the 1-acyl-3-thiosemicarbazide intermediate (**14.2-20**).[27]

The 2-alkyl-4-benzoyl-3-thiosemicarbazides (e.g., **14.2-23**) are the most probable intermediates in the treatment of alkyl hydrazines (e.g., **14.2-22**) with benzoyl isothiocyanate to give 1-alkyl-3-phenyltriazoline-5-thiones (e.g., **14.2-24**) in variable yields (20 to 61%).[28] In these reactions the yields

were lowered by a competing reaction, the benzoylation of the hydrazine by benzoyl isothiocyanate. Difficulties were also encountered in the reaction of equimolar amounts of benzoyl isothiocyanate and phenyl hydrazine, which gave a low yield of a mixture of 2,3-diphenyl- and 1,3-diphenyl-1,3,4-tria-zoline-5-thiones (**14.2-25** and **14.2-26**).[29] Based on the available informa-tion, higher yields of 1,2,4-triazoline-5-thiones are obtained in the cycliza-tion of 1-acyl-3-thiosemicarbazides than in the cyclization of 4-acyl-3-thio-semicarbazides.

PhNHNH$_2$ $\xrightarrow{\text{PhCONCS}}$ S=⟨structure⟩Ph + S=⟨structure⟩Ph

14.2-25   **14.2-26**

Oxidation of 4-aryl-1-benzylidene-3-thiosemicarbazones (e.g., **14.2-28** and **14.2-29**) with bromine in chloroform gave 2-(benzylidenehydrazino)-benzothiazoles (e.g., **14.2-27**) when the ortho positions of the aryl group were unsubstituted and a mixture of the benzothiazole and a 3,4-diaryltriazoline-5-thiones (e.g., **14.2-30**) when one of the ortho positions of the aryl

14.2-27   **14.2-28**, R$_1$ = H   **14.2-30**
14.2-29, R$_1$ = Cl

group was blocked.[30,31] In contrast, oxidation of 4-benzyl-3-thiosemicarbazones (e.g., **14.2-31**) with bromine in chloroform gave mixtures of 3-aryl-4-benzyl-1,2,4-triazoline-5-thiones (e.g., **14.2-32**) (50 to 80%) and 5-aryl-2-benzylamino-1,3,4-thiadiazoles (e.g., **14.2-33**) (10 to 30%).[32] Further investigation showed, however, that the major product was the triazoline-5-thiones when the oxidation was effected with bromine in acetic acid.[32] The double oxidation of **14.2-28** with ethanolic ferric chloride gave a tricyclic ring system.

14.2-31   **14.2-32**   **14.2-33**

Several other types of transformation of thiosemicarbazones to triazoline-5-thiones have been observed. In the addition of benzoyl isothiocyanate to 1-alkyl-2-isopropylidenehydrazines (e.g., **14.2-34**), the cyclic 1,3,4,6-oxa-triazepine-5-thiones (e.g., **14.2-35**) were isolated and converted on acid hydrolysis to triazoline-5-thiones (e.g., **14.2-36**).[33] Similarly, the reaction of phenyl hydrazones (e.g., **14.2-37**) with thiocyanic acid gave excellent yields of 1,2,4-triazolidine-5-thiones (e.g., **14.2-38**), which were readily oxidized with oxygen to 1,2,4-triazoline-5-thiones (e.g., **14.2-39**).[34] The intramolecu-

lar condensation of the thiosemicarbazone (**14.2-40**) with alkoxide gave the seven-membered heterocyclic, **14.2-41**, which on further reaction with base resulted in C—C bond cleavage and contraction of the ring to give 1-methyl-3-phenyltriazoline-5-thione (**14.2-36**).[35] A number of related compounds were prepared by this general method in lower yields.[36]

Several reports have appeared on the conversion of amidrazones to triazoline-5-thiones.[37,38] For example, the hydrochloride of amidrazone **14.2-42** was heated with ethoxycarbonylmethyl isothiocyanate to give a good yield of **14.2-43**.[39] In this reaction the 4—N of the triazole ring was provided by the isothiocyanate, indicating that the amino group of the amidrazone was displaced.

Disubstituted 1,2,4-triazoline-5-thiones have been generated from other ring systems by a number of interesting reactions. Treatment of the 2-thiazolin-5-one (**14.2-44**) with phenyl magnesium bromide in ether resulted in the formation of a low yield of the 3-(hydroxydiphenylmethyl)-1,2,4-triazoline-5-thione (**14.2-45**).[40] The conversion of 2-imino-1,3,4-oxadiazolines

14.2-44   **14.2-45**

(e.g., **14.2-46**) with ammonium sulfide to triazoline-5-thiones (e.g., **14.2-36**) presumably proceeded via 1-acyl-3-thiosemicarbazides (e.g., **14.2-47**).[41]

14.2-46   **14.2-47**   **14.2-36**

Although incomplete rearrangement of a 1,3,4-thiadiazole was observed when the nitramino group of **14.2-48** was displaced with benzylamine in refluxing xylene to give a mixture of **14.2-49** and **14.2-50**,[42] the base-

14.2-48   **14.2-49**
                                                14.2-50

catalyzed hydrolysis of the mesoionic 1,3,4-thiadiazole (**14.2-51**) gave an 80% yield of **14.2-52**.[43]

14.2-51   **14.2-52**

Although S- rather than N-alkylation of 1,2,4-triazoline-5-thiones occurs preferentially,[44,45] an exception to this generalization was demonstrated in the alkylation of 4-substituted 1,2,4-triazoline-5-thiones with several reagents. Reaction of 3-substituted 4-(ethoxycarbonyl)-1,2,4-triazoline-5-thiones (e.g., **14.2-53**) with ethyl bromoacetate,[46] followed by alkaline hydrolysis[47] of the 4-ethoxycarbonyl group of the product (e.g., **14.2-54**), provided a new route to 1,3-disbustituted 1,2,4-triazoline-5-thiones (e.g., **14.2-55**). Also, 4-phenyl-1,2,4-triazoline-5-thione (**14.2-57**) reacted with

14.2-53   **14.2-54**   **14.2-55**

acrylonitrile to give **14.2-56**[45,48] and with formaldehyde to give **14.2-58**.[48,49] In addition, cyanoethylation of 1-phenyl-1,2,4-triazoline-5-thione was reported to give the corresponding 4-(2-cyanoethyl)-1,2,4-triazoline-5-thione.[44]

14.2-56   **14.2-57**   **14.2-58**

References

1. **55:** 23508e	2. **53:** 21904h	3. **61:** 4337h	4. **72:** 100713q
5. **54:** 5629e	6. **59:** 6386c	7. **80:** 82990m	8. **77:** 126518h
9. **54:** 7695e	10. **57:** 16601e	11. **70:** 47369n	12. **76:** 126869c
13. **81:** 49624t	14. **52:** 11862e	15. **84:** 44710v	16. **81:** 49647c
17. **74:** 111955n	18. **70:** 11610v	19. **51:** 5055b	20. **68:** 105174s
21. **77:** 126568z	22. **83:** 114345k	23. **83:** 193185j	24. **51:** 5095b
25. **77:** 114312y	26. **81:** 3838n	27. **75:** 20371y	28. **66:** 37843r
29. **70:** 77870m	30. **60:** 14496d	31. **69:** 27312q	32. **71:** 49855n
33. **67:** 11476u	34. **79:** 92113w	35. **52:** 18379e	36. **55:** 1647b
37. **69:** 67340d	38. **83:** 43245q	39. **79:** 18625y	40. **76:** 59533y
41. **73:** 45427t	42. **56:** 7304h	43. **80:** 27175z	44. **74:** 141644c
45. **82:** 43276e	46. **79:** 18651d	47. **77:** 164587n	48. **76:** 34182v
49. **75:** 35891s			

14.3. Alkyl- or Aryl-Trisubstituted Δ^2 (or Δ^1, Δ^3)-1,2,4-Triazoline-5-thiones

Many methods have been used to prepare the relatively small number of trisubstituted triazoline-5-thiones reported. The reaction of 2,4-dimethyl-3-thiosemicarbazide (**14.3-2**) with refluxing acetic acid was claimed to give a high yield of 1,3,4-trimethyltriazoline-5-thione (**14.3-1**),[1] and the similarly substituted 1-methyl-3,4-diphenyltriazoline-5-thione (**14.3-4**) was obtained from 2-methyl-4-phenyl-3-thiosemicarbazide (**14.3-3**) and *N*-phenylbenzimidoyl chloride.[2]

14.3-1   **14.3-2**, $R_1 = Me$ **14.3-4**
                    14.3-3, $R_1 = Ph$

Cyclization of 1-benzoyl-1,2-dimethyl-3-thiosemicarbazide (**14.3-5**) with hydroxide gave a good yield of 1,2-dimethyl-3-phenyl-1,2,4-triazoline-5-thione (**14.3-6**),[3,4] and the interaction of 1,2-diphenylhydrazine with benzoyl

14.3-5   **14.3-6**

isothiocyanate in ether gave a 70% yield of 1,2,3-triphenyltriazoline-5-thione (**14.3-11**).[5] In the last reaction the intermediate, 4-benzoyl-3-thiosemicarbazide (**14.3-7**), has been isolated and was converted to **14.3-11** under both thermal and acidic conditions.[6] In addition, **14.3-7** rearranged in warm ethanol to the 1-benzoyl-3-thiosemicarbazide (**14.3-8**), which was also dehydrated either thermally or with hydroxide to give **14.3-11**.

14.3-7   **14.3-8**

14.3-9   **14.3-10** (90%)   **14.3-11**

Treatment of the thiosemicarbazone (**14.3-9**) with alkoxide resulted in C—C bond cleavage and cyclization to give an excellent yield of **14.3-10**.[3,7] This type of reaction was also observed in the base-catalyzed conversion of the indone (**14.3-12**) to **14.3-14**, which was probably formed via intermediate **14.3-13**.[8] The 4-phenylsemicarbazone of acetone (**14.3-15**) was oxidatively

14.3-12   **14.3-13**

14.3-14

cyclized in the presence of basic alumina to give a good yield of the Δ^1-1,2,4-triazoline-5-thione (**14.3-16**).[9,10] In contrast, treatment of **14.3-15** with maganese dioxide generated the isomeric 1,3,4-thiadiazoline. Furthermore, the formation of Δ^1-1,2,4-triazoline-5-thiones was hindered when the semicarbazone contained a 4-alkyl group or when the ketone moiety contained bulky substituents. Surprisingly, the addition of **14.3-16** to cyclopentadiene was unsuccessful, indicating that **14.3-16** was not a dienophile.

14.3-15   **14.3-16**

The generation of diphenylnitrilimine (**14.3-17**) from 2,5-diphenyltetrazole in the presence of phenyl isothiocyanate gave a 58% yield of the thiadiazole (**14.3-18**) and a 20% yield of 1,3,4-triphenyltriazoline-5-thione (**14.3-19**).[11] The thiadiazole resulted from bond formation between the sulfur of the isothiocyanate and the electrophilic carbon of the nitrilimine, whereas the triazoline-5-thione resulted from bond formation between the

$$[\text{Ph}\overset{+}{\text{C}}\!\!=\!\!\overset{-}{\text{N}}\text{NPh}] \xrightarrow[170°]{\text{PhNCS}} \text{PhN}\!\!=\!\!\overset{S}{\underset{\substack{N\!-\!N \\ Ph}}{\diamond}}\!\!\text{Ph} \; + \; S\!\!=\!\!\overset{\overset{Ph}{N}}{\underset{\substack{N\!-\!N \\ Ph}}{\diamond}}\!\!\text{Ph}$$

14.3-17   **14.3-18**   **14.3-19**

$$\text{PhNHN}\!\!=\!\!\text{C(Ph)NHPh} \xrightarrow[\Delta]{\underset{\text{DMSO}}{\text{PhNCS}}} \left[\text{SC} \overset{\text{NHPh}}{\underset{\text{N(Ph)N}=\text{C(Ph)NHPh}}{\diagdown}} \right]$$

14.3-20   **14.3-21**

nitrogen of the isothiocyanate and the electrophilic carbon of the nitrilimine. Also, the reaction of the amidrazone (**14.3-20**) with phenyl isothiocyanate gave **14.3-19**, presumably formed via **14.3-21**.[12,13] A similar reaction between **14.3-23** and ethoxycarbonylmethyl isothiocyanate (**14.3-22**) provided the 4-(ethoxycarbonylmethyl)-1,2,4-triazoline-5-thione (**14.3-24**).[14]

$$\underset{\textbf{14.3-22}}{\text{EtO}_2\text{CCH}_2\text{NCS}} + \underset{\textbf{14.3-23}}{\text{MeNHN}\!\!=\!\!\text{C(Ph)NH}_2\!\cdot\!\text{HCl}} \xrightarrow{\Delta} S\!\!=\!\!\overset{\overset{\text{CH}_2\text{CO}_2\text{Et}}{N}}{\underset{\substack{N\!-\!N \\ Me}}{\diamond}}\!\!\text{Ph}$$

14.3-24

The preparation of trisubstituted 1,2,4-triazoline-5-thiones from other ring systems includes the reaction of rhodanine (**14.3-25**) with methyl hydrazine to give **14.3-26**,[15] and the base-catalyzed ring cleavage of the

$$O\!\!=\!\!\overset{\overset{Et}{N}}{\underset{S}{\diamond}}\!\!=\!\!S \xrightarrow{\text{MeNHNH}_2} \text{HSCH}_2\!\!\overset{\overset{Et}{N}}{\underset{\substack{N\!-\!N \\ Me}}{\diamond}}\!\!=\!\!S$$

14.3-25   **14.3-26**

hexahydro-1,2,4-triazine (**14.3-27**) to give **14.3-28**.[16] As described previously, this reaction is related to the C—C bond cleavage observed in some semicarbazones to give triazoline-5-thiones.

$$\underset{\substack{\\ \text{Ph}}}{O}\!\!\diagdown\!\!\overset{\overset{Me}{N}}{\underset{\substack{N\!-\!NMe}}{\diamond}}\!\!S \xrightarrow{\text{MeO}^-} \text{Ph}\!\!\overset{\overset{Me}{N}}{\underset{\substack{N\!-\!N \\ Me}}{\diamond}}\!\!=\!\!S$$

14.3-27   **14.3-28**

Although the alkylation of **14.3-30** with 3-bromopropionic acid under alkaline conditions gave the S-substituted triazole (**14.3-29**), treatment of **14.3-30** with acrylonitrile gave the N-cyanoethylation product, **14.3-31**.[17,18]

14.3-29   **14.3-30**   **14.3-31**

Also, the addition of formaldehyde to **14.3-32** gave the 1-(hydroxymethyl)-1,2,4-triazoline-5-thione (**14.3-33**),[19,20] and to **14.3-34** the 4-(hydroxymethyl)-1,2,4-triazoline-5-thione (**14.3-35**).[21] The tetra-O-acetate of α-D-

14.3-32   **14.3-33**   **14.3-34**   **14.3-35**

glucopyranosyl bromide (**14.3-37**) reacted with **14.3-36** to give the (β-D-glucopyranosylthio)triazole (**14.3-38**), which rearranged in the presence of mercury (II) bromide to give the corresponding N-substituted triazoline-5-thione (**14.3-39**).[22,23] The latter was also formed directly from the bromide **14.3-37** and mercury salt of **14.3-36**.

14.3-36   **14.3-37**

14.3-38   **14.3-39**

References

1. **55:** 23508e   2. **62:** 2771d   3. **54:** 7695e   4. **79:** 78749b
5. **52:** 11822h   6. **79:** 91736q   7. **77:** 101468d   8. **75:** 20371y
9. **72:** 55339j   10. **72:** 78951s   11. **58:** 3415h   12. **63:** 7000e
13. **69:** 67340d   14. **79:** 18625y   15. **76:** 3765h   16. **77:** 61961q
17. **82:** 43276e   18. **83:** 164091v   19. **74:** 76428u   20. **83:** 147418q
21. **80:** 82998v   22. **80:** 146458d   23. **84:** 90504d

TABLE 14.   ALKYL- OR ARYL-1,2,4-TRIAZOLINE-5-THIONES

Compound	Reference

14.1. Alkyl- or Aryl-Monosubstituted Δ^2(or Δ^3)-1,2,4-Triazoline-5-thiones

Compound	Reference
Parent	**50:** 5647f, **51:** P3669b,
	51: 12079f, **51:** 14752c,
	55: 4337i, **57:** 805c,
	65: 7169e, **67:** 64318a,
	68: 49563d, **70:** P28922w,
	71: P81375s, **71:** P124443g
3-(3-acetamidophenyl)-	**79:** 91739t
3-(p-acetamidphenyl)-	**47:** 131g, **79:** 91739t
-3-acetic acid, ethyl ester	**49:** 13979c
3-(2-aminoethyl)-	**48:** 12739f
3-(aminomethyl)-	**66:** 2536b
3-(p-aminophenyl)-	**47:** 131g, **64:** 5072h,
	71: P22929c
3-(4-amino-7-β-D-ribofuranosyl-7H-pyrrolo[2,3-d]-pyrimidin-5-yl)-	**83:** 22239a
3-(1,2-benzisoxazol-3-ylmethyl)-	**85:** 46473q
3-(2-benzothiazolyl)-	**77:** 126518h
1-benzyl-	**74:** 13070a, **74:** 111161a
2-benzyl-	**74:** 111161a
3-benzyl-	**52:** 11822h, **52:** 11822f,
	53: 8033c, **53:** 10034d,
	72: 121442r
4-benzyl-	**56:** 7305b, **71:** 49855n,
	72: P100713q, **76:** 135847w,
	79: 146466j, **82:** P150485u
3-(5-bromo-2-butoxyphenyl)-	**85:** 159998h
3-(5-bromo-1-ethoxyphenyl)-	**85:** 159998h
3-(5-bromo-2-hydroxyphenyl)-	**68:** 59501w
3-(5-bromo-2-methoxyphenyl)-	**85:** 159998h
3-[5-bromo-2-(3-methylbutoxy)phenyl-	**85:** 159998h
3-(4-bromo-3-methyl-5-isothiazolyl)-	**63:** 10497c
3-[5-bromo-2-(2-methylpropoxy)phenyl]-	**85:** 159998h
3-[5-bromo-2-(pentyloxy)phenyl]-	**85:** 159998h
3-(2-bromophenyl)-	**78:** 43372a
3-(4-bromophenyl)-	**78:** 43372a
3-(5-bromo-2-propoxyphenyl)-	**85:** 159998h

TABLE 14   (*Continued*)

Compound	Reference

14.1. Alkyl- or Aryl-Monosubstituted Δ²(or Δ³)-1,2,4-Triazoline-5-thiones   (*Continued*)

Compound	Reference
3-(2-butoxyphenyl)-	**79:** 115500m
3-(*p*-butoxyphenyl)-	**52:** 12852b
3-butyl-	**65:** 7169e
4-butyl-	**72:** P100713q
3-tert-butyl-	**77:** 121989c, **84:** P135677f, **85:** P33028y
4-tert-butyl-	**72:** P100713q, **74:** 111161a
3-[4-[(2-carboxymethyl-4-octadecenoyl)amino]phenyl]-	**71:** P22929c
4-(2-chloro-6-methylphenyl)-	**84:** P44070m
3-(*p*-chlorophenyl)-	**44:** 2515d, **44:** 2517a, **45:** 1122b, **72:** 30148h, **85:** P177432y
4-(2-chlorophenyl)-	**82:** 16746b, **84:** P44070m
4-(4-chlorophenyl)-	**53:** 21905b, **76:** 135847w
1-cyclohexyl-	**74:** 111161a
2-cyclohexyl-	**74:** 111161a
3-cyclohexyl-	**72:** 121442r
4-cyclohexyl-	**72:** P100713q, **74:** 111161a
4-decyl-	**72:** P100713q
4-(2,4-dicarboxyphenyl)-	**73:** P36594t
3-(2,4-dichlorphenyl)-	**72:** 30148h
3-(3,4-dihydro-2,4-dimethyl-5(2H̲)-oxo-3-thioxo-1,2,4-triazin-6-yl)-	**83:** 114345x
3-(3,5-dimethyl-4-isoxazolyl)-	**55:** 7399g
3-(1,2dimethyl-5-oxo-3-phenyl-3-pyrazolin-4-yl)-	**68:** 78208x
3-(1,3dimethyl-5-oxo-2-phenyl-3-pyrazolin-4-yl)-	**68:** 78208x
3-(2,3dimethyl-5-oxo-1-phenyl-3-pyrazolin-4-yl)-	**68:** 78208x
3-[4-(1,3(2H̲)-dioxo-1H̲-isoindol-2-yl)butyl]-	**76:** P72514n
3-[(2(1H̲),4(3H̲)-dioxo-6-pyrimidinyl)methyl]-	**71:** 22101b
3-(2-ethoxyethyl)-	**49:** 13979h
3-(ethoxymethyl)-	**50:** 1785a
3-(2-ethoxyphenyl)-	**69:** 27340x, **71:** 49853k, **79:** 115500m
3-(*p*-ethoxyphenyl)-	**52:** 12852a
4-(2-ethoxyphenyl)-	**76:** 135847w
4-(4-ethoxyphenyl)-	**53:** 21905b, **63:** 4380a, **76:** 135847w
1-ethyl-	**74:** 111161a
2-ethyl-	**74:** 111161a
3-ethyl-	**47:** 3853d, **57:** 805e, **65:** 7169e, **72:** 121442r, **78:** P4286f
4-ethyl-	**72:** P100713q, **74:** 111968u, **79:** 146466j
3-(2-ethyl-4-pyridyl)-	**63:** 10497b
3-(2-ethyl-4-quinolinyl)-	**84:** 17287f
3-[*p*-(ethylsulfonyl)phenyl]-	**64:** 5072h
4-(2-fluorophenyl)-	**82:** 16746b, **84:** P44070m

TABLE 14 (*Continued*)

Compound	Reference
14.1. Alkyl- or Aryl-Monosubstituted Δ^2(or Δ^3)-1,2,4-Triazoline-5-thiones (*Continued*)	
3-(4-[(5-fluorosulfonyl-2-hexadecyloxyphenyl)ureylene]-phenyl)-	**71:** P22929c
3-(2-furyl-	**52:** 16341d, **76:** 149140x, **84:** 4868w
3-heptadecyl-	**65:** 7169e
3-heptyl-	**65:** 7169e, **81:** P129916e
3-(4-[(4-hexadecyloxy-3-fluorosulfonylbenzoyl)amino]-phenyl)-	**71:** P22929c
3-(4-[(4-hexadecyloxy-3-sulfobenzoyl)amino]phenyl)-	**71:** P22929c
4-hexyl-	**72:** P100713q
3-(2-hydroxyphenyl)-	**72:** 30148h, **77:** 126508e, **83:** 97141w
3-(*p*-isobutoxyphenyl)-	**52:** 12852b
4-isobutyl-	**69:** 77151g
3-[*p*-(isopentyloxy)phenyl]-	**52:** 12852c
3-(*p*-isopropoxyphenyl)-	**52:** 12852b
3-isopropyl-	**85:** P33028y
4-isopropyl	**77:** 74417c
3-(1-isoquinolyl)-	**71:** 89718b
3-(1-metcaptoethyl)-	**72:** P12733r, **74:** 125534m, **76:** 34220f
3-(mercaptomethyl)-	**71:** 70540r, **72:** P12733r, **74:** 125534m, **75:** 35882q, **76:** 34220f
3-(1-mercaptopropyl)-	**72:** P12733r, **74:** 125534m
3-(methoxymethyl)-	**50:** 1784i
3-(2-methoxyphenyl)-	**78:** 43372a, **79:** 115500m
3-(*p*-methoxyphenyl)-	**44:** 2515c, **44:** 2517a, **51:** 17894f, **52:** 12852a, **75:** 129505w
4-(3-methoxyphenyl)-	**76:** 135847w
4-(*p*-methoxyphenyl)-	**66:** 115650s, **76:** 135847w
1-methyl-	**55:** 23507f, **74:** 111161a
2-methyl-	**74:** 111161a
3-methyl-	**42:** 8799f, **50:** 1784d, **50:** 13886e, **51:** 12079f, **51:** 17895b, **55:** 23508g, **57:** 805e, **57:** 5903i, **58:** 4543f, **64:** 5072h, **65:** 7169e, **77:** 164587n, **80:** P89539e
4-methyl-	**51:** 12079e, **70:** 11686z, **72:** P100713q, **72:** P31806h, **73:** 66517n, **79:** 146466j, **80:** 52645k, **83:** P209412u, **84:** 59327z
3-[2-(3-methylbutoxy)phenyl]-	**79:** 115500m
4-[4-(3-methylbutoxy)phenyl]-	**76:** 135847w

TABLE 14 (*Continued*)

Compound	Reference

14.1. Alkyl- or Aryl-Monosubstituted Δ^2(or Δ^3)-1,2,4-Triazoline-5-thiones (*Continued*)

Compound	Reference
3-[2-(1-methylethoxy)phenyl]-	**79:** 115500m
3-(4-methyl-3-furazanyl)-	**79:** 105155p
3-(4-methyl-3-furazanyl)-, *N*-oxide	**79:** 105155p
3-(3-methyl-5-isothiazolyl)-	**63:** 10497c
3-(1-methyl-5-oxo-2-phenyl-3-pyrazolin-4-yl)-	**68:** 78208x
3-(2-methylphenyl)-	**81:** 136112t
4-(2-methylphenyl)-	**76:** 135847w
4-(4-methylphenyl)-	**76:** 135847w
3-(1-methyl-2-phthalimidoethyl)-	**49:** 13979a
3-[2-(2-methylpropoxy)phenyl]-	**79:** 115500m
3-(2-methyl-4-pyridyl)-	**52:** P1274a
3-(2-methyl-4-quinolinyl)-	**84:** 17287f
3-[(2-naphthalenyloxy)methyl]-	**80:** 59762e
3-(1-naphthyl)-	**65:** 7169f
3-(2-naphthyl)-	**65:** 7169f
3-(naphthylmethyl)-	**74:** 141652d
3-(5-nitro-2-furanyl)-	**84:** 4868w
2-(*p*-nitrophenyl)-	**65:** 704h, **74:** 111161a
3-(*o*-nitrophenyl)-	**65:** 704h
3-(*m*-nitrophenyl)-	**65:** 704g
3-(*p*-nitrophenyl)-	**58:** 13937g, **61:** 3065b, **64:** 5072h, **65:** 704g, **70:** 3966a
4-(*m*-nitrophenyl)-	**67:** P38283r
3-nonyl-	**65:** 7169e
3-(2-oxo-1-cyclopentylidene)-, thiosemicarbazone	**83:** 131519n
4-octyl-	**72:** P100713q
3-pentyl-	**57:** 7253i, **65:** 7169e
3-pentyl	**80:** P54523t, **83:** 50701p
4-pentyl-	**72:** P100713q
3-[2-(pentyloxy)phenyl]-	**79:** 115500m
3-(*p*-pentyloxyphenyl)-	**52:** 12852c
3-phenoxymethyl-	**50:** 1784i
1-phenyl-	**70:** 37729r, **74:** 141644c, **75:** 98505q
2-phenyl-	**70:** 37729r, **74:** 141644c
3-phenyl-	**44:** 2515a, **44:** 2517a, **44:** 6856a, **45:** 1122b, **47:** 8752c, **52:** 16364a, **52:** 11823e, **53:** 8033c, **53:** 10034b, **53:** 6894e, **58:** 13937g, **59:** 3930g, **61:** 3065b, **64:** 5072h, **64:** 9727g, **65:** 7169e, **66:** 115651t, **70:** 77870m, **72:** P100713q, **73:** 66517n, **75:** 129505w, **76:** 135847w, **77:** 164587n, **79:** 5298u, **85:** P177432y

TABLE 14 (*Continued*)

Compound	Reference

14.1. Alkyl- or Aryl-Monosubstituted Δ^2(or Δ^3)-1,2,4-Triazoline-5-thiones (*Continued*)

Compound	Reference
4-phenyl-	**53:** 21904i, **57:** 809a, **61:** 4337h, **64:** 8174b, **80:** 52645k, **82:** P150485u
3-(1-phthalimidoethyl)-	**49:** 13979a
3-(2-phthalimidoethyl)-	**48:** 12739e
3-phthalimidomethyl-	**49:** 13979a
3-[2-phthalimidopropyl]-	**49:** 13979a
3-[3-phthalimidopropyl]-	**49:** 13979a
2-piperonyl-	**67:** 32651n
3-piperonyl-	**67:** 32651n
4-(2-propenyl)-	**80:** P54523t
3-(2-propoxyphenyl)-	**79:** 115500m
3-*p*-propoxyphenyl-	**52:** 12852b
3-propyl-	**47:** 3853e, **65:** 7169e
4-propyl-	**72:** P100713q, **79:** 146466j
3-(pyrazinyl)-	**64:** 5072h
2-(4-pyridyl)-	**74:** 111161a
3-(2-pyridyl)-	**63:** 10497b, **64:** 5072h, **83:** 43245q
3-(3-pyridyl)-	**63:** 10497b, **64:** 5072h, **66:** 104948u, **72:** 30148h
3-(4-pyridyl)-	**49:** 3959f, **49:** 10937eg, **64:** 5072h, **67:** 54140x, **72:** 30148h, **83:** 97228e
4-(2-pyridyl)-	**57:** P9860e, **65:** 16960f, **74:** 111161a
4-(3-pyridyl)-	**74:** 111161a, **76:** 34182v
4-(4-pyridyl)-	**74:** 111161a
3-(4-quinolyl)-	**64:** 5072h
2-*p*-sulfamoylphenyl-	**49:** 1022f
4-(1,1,3,3-tetramethylbutyl)-	**72:** P100713q, **79:** 146466j
3-*p*-tolyl-	**58:** 13937g, **61:** 3065b
4-*o*-tolyl-	**61:** 4338a
4-*p*-tolyl-	**65:** 16960h
3-(α,α,α-trifluoro-*o*-tolyl)-	**65:** 18576a
3-(3,4,5-trimethoxyphenyl)-	**67:** P54140x, **72:** 30148h
3-undecyl-	**65:** 7169e
3-[*p*-(ureido)phenyl]	**64:** 5072h

14.2. Alkyl- or Aryl-Disubstituted Δ^2(or Δ^3)-1,2,4-Triazoline-5-thiones

Compound	Reference
-4-acetamide, 3-methyl-	**79:** 18625y
3-(4-acetamidophenyl)-4-ethyl-	**78:** P36245h, **80:** P76701b
3-(3-acetamidophenyl)-4-phenyl-	**79:** 91739t
3-(4-acetamidophenyl)-4-phenyl-	**79:** 91739t
-3-acetic acid, 4-benzyl-, [(2-hydroxyphenyl)methylene]-hydrazide	**82:** 140031d
-3-acetic acid, α-butyl-4-methyl-, hydrazide	**82:** 140031d

TABLE 14 (*Continued*)

Compound	Reference

14.2. Alkyl- or Aryl-Disubstituted Δ^2(or Δ^3)-1,2,4-Triazoline-5-thiones (*Continued*)

Compound	Reference
-3-acetic acid, 4-ethyl-, [(2-hydroxyphenyl)methylene]-hydrazide	**82:** 140031d
-3-acetic acid, α-ethyl-4-methyl-, hydrazide	**82:** 140031d
-1-acetic acid, 3-methyl-	**79:** 18651d
-4-acetic acid, 3-methyl-	**79:** 18625y
-4-acetic acid, 3-methyl-, ethyl ester	**79:** 18625y
-3-acetic acid, 4-methyl-, hydrazide	**82:** 140031d
-4-acetic acid, 3-methyl-, hydrazide	**79:** 18625y
-1-acetic acid, 3-methyl-α-phenyl-	**79:** 18651d
-1-acetic acid, 3-phenyl-	**79:** 5298u, **79:** 18651d
3-*o*-(acetylphenyl)-1-methyl-	**75:** 20371y
3-*o*-(acetylphenyl)-1-methyl-, hydrazone	**75:** 20371y
4-allyl-3-(5-bromo-2-hydroxyphenyl)-	**72:** 30148h
4-allyl-3-(*o*-chlorophenyl)-	**72:** 30148h
4-allyl-3-(*m*-chlorophenyl)-	**57:** 16601h
4-allyl-3-(*p*-chlorophenyl)-	**57:** 16601h
4-allyl-3-([(4-chloro-*o*-tolyl)oxy]methyl)-	**70:** 47369n
4-allyl-3-[(2,4-dichlorophenoxy)methyl]-	**70:** 47369n
4-allyl-3-(2,4-dichlorophenyl)-	**57:** 16601h
4-allyl-3-(1,3-dihydroxy-2-phenyl-)-	**73:** P50723r
4-allyl-3-(α-hydroxybenzyl)-	**70:** 47369n
4-allyl-3-(hydroxydiphenylmethyl)-	**70:** 47369n
4-allyl-3-(2-hydroxynaphthalen-3-yl)-	**73:** P50723r
4-allyl-3-(*m*-hydroxyphenyl)-	**72:** 30148h
4-allyl-3-(*p*-hydroxyphenyl)-	**72:** 30148h
4-allyl-3-(*o*-methoxyphenyl)-	**71:** 38865q
4-allyl-3-(*p*-methoxyphenyl)-	**57:** 16601h
4-allyl-3-methyl)-	**59:** P12340b
4-allyl-3-(1-naphthylmethyl)-	**74:** 141652d
4-allyl-3-(*m*-nitrophenyl)-	**68:** 59501w
4-allyl-5-[*p*-(pentyloxy)phenyl]-	**71:** 38865q
4-allyl-3-(phenoxymethyl)-	**70:** 47369n
4-allyl-3-(3-pyridyl)-	**57:** 16601h
4-allyl-3-(4-pyridyl)-	**57:** 16601h
4-allyl-3-(4-sulfamoylphenyl)-	**73:** 64790x
4-allyl-3-[(2,4,5-trichlorophenoxy)methyl]-	**70:** 47369n
4-allyl-3-(3,4,5-trimethoxyphenyl)-	**57:** 16601h
3-(*o*-aminophenyl)-4-(*p*-chlorophenyl)-	**63:** 13256h
3-(*p*-aminophenyl)-4-(*p*-ethoxyphenyl)-	**51:** 5055d
3-(4-aminophenyl)-4-ethyl-	**73:** 64790x
3-(*o*-aminophenyl)-4-phenyl-	**61:** 5645a, **63:** 13256h
3-(3-aminopyrazinyl)-4-ethyl-	**77:** 126568z
3-[(benzamido)methyl]-4-(4-chlorophenyl)-	**81:** 49624t
3-[(benzamido)methyl]-4-phenyl-	**81:** 49624t
3-[2-(1H-benzimidazol-2-yl)ethyl]-4-cyclohexyl-	**77:** 147663s
3-[2-(1H-benzimidazol-2-yl)ethyl]-4-(2-methoxyphenyl)-	**77:** 147663s
3-[2-(1H-benzimidazol-2-yl)ethyl]-4-(3-methylphenyl)-	**77:** 147663s
3-[2-(1H-benzimidazol-2-yl)ethyl]-4-phenyl-	**77:** 147663s

TABLE 14 (*Continued*)

Compound	Reference
14.2. Alkyl- or Aryl-Disubstituted Δ^2(or Δ^3)-1,2,4-Triazoline-5-thiones (*Continued*)	
3-[2-(1H-benzimidazol-2-yl)ethyl]-4-(2-propenyl)-	**77:** 147663s
3-(1,3-benzodioxol-5-yl)-5-(4-chlorophenyl)-	**85:** 186715v
3-(1,3-benzodioxol-5-yl)-4-(2,4-dimethylphenyl)-	**85:** 186715v
3-(1,3-benzodioxol-5-yl)-4-(2,6-dimethylphenyl)-	**85:** 186715v
3-(1,3-benzodioxol-5-yl)-4-(2-methoxyphenyl)-	**85:** 186715v
3-(1,3-benzodioxol-5-yl)-4-(4-methoxyphenyl)-	**85:** 186715v
3-(1,3-benzodioxol-5-yl)-4-(2-methylphenyl)-	**85:** 186715v
3-(1,3-benzodioxol-5-yl)-4-(4-methylphenyl)-	**85:** 186715v
3-(1,3-benzodioxol-5-yl)-4-phenyl-	**85:** 186715v
3-[4-(2-benzofuranyl)-2-thiazolyl]-4-phenyl-	**74:** 111955n
3-(2-benzothiazolyl)-1-benzyl-	**77:** 126518h
3-(2-benzothiazolyl)-4-benzyl-	**77:** 126518h
3-[2-benzothiazolyl]-4-butyl-	**84:** 4840f
3-(2-benzothiazolyl)-4-(4-chlorophenyl)-	**84:** 4840f
3-(2-benzothiazolyl)-1-(2,4-dichlorobenzyl)-	**77:** 126518h
3-(2-benzothiazolyl)-4-ethyl-	**77:** 126518h, **84:** 4840f
3-(2-benzothiazolyl)-1-(4-methoxybenzyl)-	**77:** 126518h
3-(2-benzothiazolyl)-4-(4-methoxyphenyl)-	**84:** 4840f
3-(2-benzothiazolyl)-4-methyl-	**77:** 126518h, **84:** 4840f
3-(2-benzothiazolyl)-4-phenyl-	**77:** 126518h, **84:** 4840f
3-(2-benzothiazolyl)-4-(2-propenyl)-	**77:** 126518h
3-(2-benzothiazolyl)-4-propyl-	**84:** 4840f
4-benzyl-3-(*o*-chlorophenyl)-	**71:** 49855n
4-benzyl-3-(*p*-chlorophenyl)-	**71:** 49855n
4-benzyl-3-cyanomethyl	**85:** 21243b
4-benzyl-3-(2,4-dichlorophenyl)-	**71:** 49855n
3-benzyl-4-ethyl-	**74:** 111968u
4-benzyl-3-ethyl-	**56:** 7305b
1,1'-[(benzylimino)dimethylene]bis[3-methyl-	**75:** P82399x
4-benzyl-3-isobutyl-	**56:** 7305b
4-benzyl-3-(*p*-methoxyphenyl)-	**71:** 49855n
1-benzyl-3-methyl-	**55:** 1647f
3-benzyl-4-methyl-	**53:** 21905b, **80:** 59899e
4-benzyl-3-[3,4-(methylenedioxy)phenyl]-	**71:** 49855n
4-benzyl-3-(*m*-nitrophenyl)-	**71:** 49855n
4-benzyl-3-(*p*-nitrophenyl)-	**71:** 49855n
1-benzyl-3-phenyl-	**66:** 37843r
3-benzyl-4-phenyl-	**53:** 21905a, **70:** 47369n, **71:** 49864q, **79:** 126404j
4-benzyl-3-phenyl-	**71:** 49855n
3-benzyl-4-(2-propenyl)-	**70:** 47369n
4-benzyl-3-propyl-	**56:** 7305b
3-benzyl-4-*p*-tolyl-	**70:** 47369n
4-benzyl-3-*p*-tolyl-	**71:** 49855n
3,4-bis(*p*-chlorophenyl-	**74:** 111968u
3,4-bis(4-fluorophenyl)-	**82:** 72886j
1,4-bis(hydroxymethyl)-	**75:** 35891s
1,4-bis(isopropyl)-	**78:** 15125d

TABLE 14   (*Continued*)

Compound	Reference
14.2. Alkyl- or Aryl-Disubstituted Δ^2(or Δ^3)-1,2,4-Triazoline-5-thiones (*Continued*)	
3-(5-bromo-2-benzofuranyl)-4-(4-bromophenyl)-	**83:** 193185j
3-(5-bromo-2-benzofuranyl)-4-(4-chlorophenyl)-	**83:** 193185j
3-(5-bromo-2-benzofuranyl)-4-cyclohexyl-	**83:** 193185j
3-(5-bromo-2-benzofuranyl)-4-(4-methoxyphenyl)-	**83:** 193185j
3-(5-bromo-2-benzofuranyl)-4-(2-methylphenyl)-	**83:** 193185j
3-(5-bromo-2-benzofuranyl)-4-(4-methylphenyl)-	**83:** 193185j
3-(5-bromo-2-benzofuranyl)-4-phenyl-	**83:** 193185j
3-(5-bromo-2-hydroxyphenyl)-4-(4-chlorophenyl)-	**68:** 59501w
3-(5-bromo-2-hydroxyphenyl)-4-cyclohexyl-	**68:** 59501w, **72:** 30148h
3-(5-bromo-2-hydroxyphenyl)-4-(3,4-dichlorophenyl)-	**68:** 59501w
3-(5-bromo-2-hydroxyphenyl)-4-ethyl-	**68:** 59501w, **72:** 30148h
3-(5-bromo-2-hydroxyphenyl)-4-heptyl-	**68:** 59501w
3-(5-bromo-2-hydroxyphenyl)-4-hexyl-	**68:** 59501w
3-(5-bromo-2-hydroxyphenyl)-4-isobutyl-	**72:** 30148h
3-(5-bromo-2-hydroxyphenyl)-4-isopentyl-	**68:** 59501w
3-(5-bromo-2-hydroxyphenyl)-4-isopropyl-	**72:** 30148h
3-(5-bromo-2-hydroxyphenyl)-4-pentyl-	**68:** 59501w
3-(5-bromo-2-hydroxyphenyl)-4-phenyl-	**72:** 30148h
3-(5-bromo-2-hydroxyphenyl)-4-propyl-	**68:** 59501w
3-[5-bromo-2-(methylthio)-4-pyrimidinyl]- 4-(4-bromophenyl)-	**85:** 94306j
3-[5-bromo-2-(methylthio)-4-pyrimidinyl]- 4-(4-ethoxyphenyl)-	**85:** 94306j
3-[5-bromo-2-(methylthio)-4-pyrimidinyl]- 4-(4-methoxyphenyl)-	**85:** 94306j
3-[5-bromo-2-(methylthio)-4-pyrimidinyl]- 4-(4-methylphenyl)-	**85:** 94306j
3-[5-bromo-2-(methylthio)-4-pyrimidinyl]- 4-phenyl-	**85:** 94306j
4-(*p*-bromophenyl)-3-[4-[(*p*-bromophenyl)- thioureylene]phenyl]-	**70:** 77880q
4-(4-bromophenyl)-3-[4-[(4-chlorophenyl)- sulfonyl]phenyl]-	**77:** 126522e
4-(4-bromophenyl)-3-[(2,4-dichlorophenoxy)methyl]-	**82:** 125327m
4-(4-bromophenyl)-3-(2-furanyl)-	**85:** 192627j
4-(4-bromophenyl)-3-(2-fluorophenyl)-	**82:** 72886j
4-(4-bromophenyl)-3-(3-fluorophenyl)-	**82:** 72886j
4-(4-bromophenyl)-3-(4-fluorophenyl)-	**82:** 72886j
4-(4-bromophenyl)-3-[[(4-fluorophenyl)sulfonyl]phenyl]-	**77:** 114312y
4-(*p*-bromophenyl)-5-(*o*-methoxyphenyl)-	**71:** 38865q
4-(4-bromophenyl)-3-[(2-methyl-1H̲-indol-3-yl)methyl]-	**82:** 118780a
4-(*p*-bromophenyl)-5-[*p*-(pentyloxy)phenyl]-	**71:** 38865q
3-[4-[(4-bromophenyl)sulfonyl]phenyl]- 4-(2-methylphenyl)-	**76:** 140655m
3-[4-[(4-bromophenyl)sulfonyl]phenyl]- 4-(1-naphthalenyl)-	**76:** 140655m
3-[4-(N-[(butylamino)carbonyl]sulfamoyl)phenyl]- 4-cyclohexyl-	**76:** 21123b

429

TABLE 14 (*Continued*)

Compound	Reference
14.2. Alkyl- or Aryl-Disubstituted Δ^2(or Δ^3)-1,2,4-Triazoline-5-thiones (*Continued*)	
3-[4-(N-[(butylamino)carbonyl]sulfamoyl)phenyl]-4-propyl-	**76:** 21123b
4-butyl-3-[4-(N-[(butylamino)carbonyl]sulfamoyl)-phenyl]-	**76:** 21123b
3-([4-(butylcarbamoyl)sulfamoyl]phenyl)-4-ethyl-	**76:** 21123b
4-butyl-3-(o-chlorophenyl)-	**72:** 30148h
4-butyl-3-(m-chlorophenyl)-	**57:** 16601h
4-butyl-3-(p-chlorophenyl)-	**57:** 16601h
4-butyl-1-(2-cyanoethyl)-	**72:** P100713q, **78:** P29774b
4-butyl-3-[4-(N-[(cyclohexylamino)carbonyl]-sulfamoyl)phenyl]-	**76:** 21123b
4-butyl-3-(2,4-dichlorophenyl)-	**57:** 16601h
4-butyl-3-(4,6-dimethyl-2-pyrimidinyl)-	**85:** 94310f
4-butyl-3-(4,6-diphenyl-2-pyrimidinyl)-	**85:** 94310f
4-butyl-5-(p-ethoxyphenyl)-	**71:** 38865q
3-butyl-1-ethyl-	**80:** P82998v, **82:** P72995u
4-butyl-3-(2-furyl)-	**72:** P100713q
4-butyl-3-(o-hydroxyphenyl)-	**72:** 30148h
4-butyl-3-(m-hydroxyphenyl)-	**72:** 30148h
4-butyl-3-(p-methoxyphenyl)-	**57:** 16601h
4-butyl-3-methyl-	**69:** 67340d, **72:** P100713q
4-tert-butyl-3-methyl-	**77:** 74417c
4-butyl-3-[(2-methyl-1H-indol-3-yl)methyl]-	**82:** 118780a
4-butyl-3-(m-nitrophenyl)-	**68:** 59501w
4-butyl-5-[p-(pentyloxy)phenyl]-	**71:** 38865q
4-butyl-3-phenyl-	**72:** 30148h
4-butyl-3-(3-pyridyl)-	**57:** 16601h
4-butyl-3-(4-pyridyl)-	**57:** 16601h
4-butyl-3-(4-sulfamoylphenyl)-	**72:** 78949x
4-butyl-3-p-tolyl-	**73:** 64790x
4-butyl-3-(3,4,5-trimethoxyphenyl)-	**57:** 16601h
3-(p-chlorobenzyl)-4-ethyl-	**74:** 111968u
3-(5-chloro-2-hydroxyphenyl)-4-phenyl-	**72:** 30148h
1-(chloromethyl)-4-methyl-	**72:** P31806h
1-(chloromethyl)-4-phenyl-	**63:** 4280c
4-(2-chloro-4-methylphenyl)-3-methyl-	**82:** 16746b, **84:** P44070m
4-(2-chloro-5-methylphenyl)-3-methyl-	**82:** 16746b, **84:** P44070m
4-(3-chloro-4-methylphenyl)-3-methyl-	**82:** 16746b
3-(2-chloro-4-nitrophenyl)-4-phenyl-	**54:** 8795b
3-[2-(2-chlorophenothiazin-10-yl)ethyl]-4-methyl-	**52:** 2021c
3-(4-chlorophenyl)-4-[5-(2-chlorophenyl)-1,3,4-oxadiazol-2-yl]-	**85:** 5568f
3-(4-chlorophenyl)-4-[5-(2-chlorophenyl)-1,3,4-oxadiazol-2-yl]-	**85:** 5568f
4-(p-chlorophenyl)-3-(4-[(p-chlorophenyl)-thioureylene]phenyl)-	**70:** 77880q
3-(1-chlorophenyl)-4-cyclohexyl-	**72:** 30148h
3-(o-chlorophenyl)-4-cyclohexyl-	**72:** 30148h

TABLE 14 (*Continued*)

Compound	Reference

14.2. Alkyl- or Aryl-Disubstituted Δ^2(or Δ^3)-1,2,4-Triazoline-5-thiones (*Continued*)

Compound	Reference
4-(4-chlorophenyl)-3-[(2,4-dichlorophenoxy)methyl]-	**82:** 125327m
4-(*p*-chlorophenyl)-1-[(diethylamino)methyl]-	**63:** 4280b
4-(*p*-chlorophenyl)-3-(diethylaminomethyl)-	**53:** 21905b
4-(4-chlorophenyl)-3-(2,4-dihydroxyphenyl)-	**82:** 125327m
4-(*p*-chlorophenyl)-3-(3,5-dimethyl-4-isoxazolyl)-	**55:** 7399f
3-(*o*-chlorophenyl)-4-ethyl-	**72:** 30148h
3-(*m*-chlorophenyl)-4-ethyl-	**72:** 30148h
3-(*p*-chlorophenyl)-4-ethyl-	**72:** 30148h
4-(*p*-chlorophenyl)-3-ethyl-	**74:** 111968u
4-(4-chlorophenyl)-3-(2-fluorophenyl)-	**82:** 72886j
4-(4-chlorophenyl)-3-(3-fluorophenyl)-	**82:** 72886j
4-(4-chlorophenyl)-3-(4-fluorophenyl)-	**82:** 72886j
4-(4-chlorophenyl)-3-(2-furanyl)-	**85:** 192627j
4-(2-chlorophenyl)-3-heptyl-	**82:** 16746b, **84:** P44070m
4-(4-chlorophenyl)-3-(4-hydroxyphenyl)-	**82:** 125327m
4-(*p*-chlorophenyl)-3-(*o*-hydroxyphenyl)-	**72:** 30148h
3-(4-chlorophenyl)-4-(4-iodophenyl)-	**78:** 159528k
3-(*m*-chlorophenyl)-4-isobutyl-	**72:** 30148h
3-(*p*-chlorophenyl)-4-isobutyl-	**72:** 30148h
3-(*o*-chlorophenyl)-4-isobutyl-	**72:** 30148h
3-(*o*-chlorophenyl)-4-isopropyl-	**72:** 30148h
3-(*m*-chlorophenyl)-4-isopropyl-	**57:** 16601h
3-(*p*-chlorophenyl)-4-isopropyl-	**57:** 16601h
3-(*p*-chlorophenyl)-4-(2-methoxyethyl)-	**73:** 64790x
3-(4-chlorophenyl)-1-methyl-	**73:** 45427t, **83:** 163373b
3-(4-chlorophenyl)-2-methyl-	**83:** 163373b
3-(*p*-chlorophenyl)-4-methyl-	**73:** 64790x, **83:** 163373b
4-(2-chlorophenyl)-3-methyl-	**82:** 16746b, **84:** P44070m
4-(*p*-chlorophenyl)-3-methyl-	**67:** P38283r
4-(4-chlorophenyl)-3-[(2-methyl-1H-indol-3-yl)methyl]-	**82:** 118780a
3-(4-chlorophenyl)-4-(4-methylphenyl)-	**78:** 159528k
4-(*p*-chlorophenyl)-3-morpholinomethyl)-	**57:** P9860g
4-(*p*-chlorophenyl)-3-(*p*-nitrophenyl)-	**68:** 59501w
4-[5-(4-chlorophenyl)-1,3,4-oxadiazol-2-yl]-3-(4-methylphenyl)-	**85:** 5568f
3-(*o*-chlorophenyl)-4-phenyl-	**72:** 30148h
3-(*m*-chlorophenyl)-4-phenyl-	**57:** 16601h
3-(*p*-chlorophenyl)-4-phenyl-	**54:** 8795a, **57:** 16601h
4-(*o*-chlorophenyl)-3-phenyl-	**60:** 14496f
4-(*p*-chlorophenyl)-1-(piperidinomethyl)-	**63:** 4280b
4-(*p*-chlorophenyl)-3-(piperidinomethyl)-	**57:** P9860h
3-(*o*-chlorophenyl)-4-propyl-	**72:** 30148h
3-(*m*-chlorophenyl)-4-propyl-	**57:** 16601h
3-(*p*-chlorophenyl)-4-propyl-	**57:** 16601h
4-(*p*-chlorophenyl)-3-(4-pyridyl)-	**51:** 5055d
4-(*p*-chlorophenyl)-1-[2-(2-pyridyl)ethyl]-	**63:** 6996e
4-(*p*-chlorophenyl)-1-(1-pyrrolidinylmethyl)-	**63:** 4280b
4-(*p*-chlorophenyl)-3-(*p*-sulfamoylphenyl)-	**74:** 111968u

TABLE 14 (*Continued*)

Compound	Reference
14.2. Alkyl- or Aryl-Disubstituted Δ^2(or Δ^3)-1,2,4-Triazoline-5-thiones (*Continued*)	
3-[4-[(4-chlorophenyl)sulfonyl]phenyl]-4-(4-iodophenyl)-	**77:** 126522e
3-(4-[(4-chlorophenyl)sulfonyl]phenyl)-4-(2-methylphenyl)-	**76:** 140655m
3-(4-[(4-chlorophenyl)sulfonyl]phenyl)-4-(1-naphthalenyl)-	**76:** 140655m
3-(4-[(4-chlorophenyl)sulfonyl]phenyl)-4-phenyl-	**77:** 126522e
3-([(4-chloro-*o*-tolyl)oxy]methyl)-4-phenyl-	**70:** 47369n
3-([(4-chloro-*o*-tolyl)oxy]methyl)-4-*p*-tolyl-	**70:** 47369n
4-[2-chloro-5-(trifluoromethyl)phenyl]-3-methyl-	**82:** 16746b, **84:** P44070m
1-(2-cyanoethyl)-4-phenyl-	**72:** P100713q
1-(2-cyanoethyl)-4-(3-pyridyl)-	**76:** 34182v
3-cyanomethyl-4-ethyl	**85:** 21243b
3-cyanomethyl-4-methyl-	**85:** 21243b
3-[4-(*N*-[(cyclohexylamino)carbonyl]sulfamoyl)phenyl]-4-ethyl-	**76:** 21123b
3-[4-(*N*-[(cyclohexylamino)carbonyl]sulfamoyl)phenyl]-4-propyl-	**76:** 21123b
4-cyclohexyl-3-[4-(*N*-[(cyclohexylamino)carbonyl]-sulfamoyl)phenyl]-	**76:** 21123b
4-cyclohexyl-3-(4,6-dimethyl-2-pyrimidinyl)-	**85:** 94310f
4-cyclohexyl-3-(4,6-diphenyl-2-pyrimidinyl)-	**85:** 94310f
4-cyclohexyl-5-(*p*-ethoxyphenyl)-	**71:** 38865q
4-cyclohexyl-3-(*o*-hydroxyphenyl)-	**72:** 30148h
4-cyclohexyl-3-(*m*-hydroxyphenyl)-	**72:** 30148h
4-cyclohexyl-3-(*p*-hydroxyphenyl)-	**72:** 30148h
4-cyclohexyl-5-(*o*-methoxyphenyl)-	**71:** 38865q
4-cyclohexyl-3-methyl-	**77:** 74417c
4-cyclohexyl-3-[(2-methyl-1H-indol-3-yl)methyl]-	**82:** 118780a
4-cyclohexyl-3-(*p*-nitrophenyl)-	**68:** 59501w
4-cyclohexyl-5-[*p*-(pentyloxy)phenyl]-	**71:** 38865q
4-cyclohexyl-3-phenyl-	**77:** 74417c
4-cyclohexyl-3-(3-pyridyl)-	**72:** 30148h
4-cyclohexyl-3-(4-pyridyl)-	**72:** 30148h
4-cyclohexyl-3-(4-sulfamoylphenyl)-	**72:** 78949x
4-cyclohexyl-3-*p*-tolyl-	**73:** 64790x
4-cyclohexyl-3-(3,4,5-trimethoxyphenyl)-	**72:** 30148h
4-cyclohexyl-3-[(2,4-dimethyl-3-quinolyl)methyl]-	**71:** 30411t
4-cyclohexyl-3-[(2-methyl-4-phenyl-3-quinolyl)methyl]-	**71:** 30411t
4-cyclohexyl-3-[(2-methyl-4-*p*-tolyl-3-quinolyl)methyl]-	**71:** 30411t
4-(3,5-dicarboxyphenyl)-3-methyl-	**82:** 49831d, **82:** 49832e, **82:** 9314f
3-(2,2-dichloro-1,1-difluoroethyl)-4-phenyl-	**81:** 49647c
3-[(2,4-dichlorophenoxy)methyl]-4-benzyl-	**70:** 47369n, **78:** 159528k
3-[(2,4-dichlorophenoxy)methyl]-4-(4-ethoxyphenyl)-	**82:** 125327m
3-[(2,4-dichlorophenoxy)methyl]-4-(4-iodophenyl)-	**82:** 125327m
3-[(2,4-dichlorophenoxy)methyl]-4-methyl-	**76:** 126869c
3-[(2,4-dichlorophenoxy)methyl]-4-(2-methylphenyl)-	**78:** 159528k, **82:** 125327m
3-[(2,4-dichlorophenoxy)methyl]-4-(3-methylphenyl)-	**78:** 159528k, **82:** 125327m

TABLE 14 (*Continued*)

Compound	Reference

14.2. Alkyl- or Aryl-Disubstituted Δ^2(or Δ^3)-1,2,4-Triazoline-5-thiones   (*Continued*)

Compound	Reference
3-[(2,4-dichlorophenoxy)methyl]-4-phenyl-	**70:** 47369n
3-(2,4-dichlorophenyl)-4-ethyl-	**73:** 64790x
4-(3,4-dichlorophenyl)-3-(4-fluorophenyl)-	**82:** 72886j
3-(2,4-dichlorophenyl)-4-isopropyl-	**57:** 16601h
4-(2,4-dichlorophenyl)-3-methyl-	**84:** P44070m
4-(2,4-dichlorophenyl)-5-methyl-	**82:** 16746b
4-(2,5-dichlorophenyl)-3-methyl-	**82:** 16746b, **84:** P44070m
4-(2,6-dichlorophenyl)-3-methyl-	**82:** 16746b, **84:** P44070m
3-(2,4-dichlorophenyl)-4-phenyl-	**57:** 16601h
3-(2,4-dichlorophenyl)-4-propyl-	**57:** 16601h
3,4-diethyl-	**74:** 111968u
3-[(diethylamino)methyl]-4-(*p*-ethoxyphenyl)-	**57:** P9860g
1-[(diethylamino)methyl]-3-phenyl-	**64:** PC208e, **75:** P82399x
3-(diethylaminomethyl)-4-phenyl-	**53:** 21905a
3-(3,4-dihydro-2,4-dimethyl-5(2H̲)-oxo-3-thioxo-1,2,4-triazin-6-yl)-4-methyl-	**83:** 114345k
3-(3,4-dihydro-2,4-dimethyl-5(2H̲)-oxo-3-thioxo-1,2,4-triazin-6-yl)-4-phenyl	**83:** 114345k
3-(3,4-dihydro-5(2H̲)-oxo-3-thioxo-1,2,4-triazin-6-yl)-4-methyl-	**83:** 114345k
3-(3,4-dihydro-5(2H̲)-oxo-3-thioxo-1,2,4-triazin-6-yl)-4-phenyl-	**83:** 114345k
4-(3,4-dihydro-2-phenyl-2H̲-1-benzopyran-4-yl)-3-(4-methoxyphenyl)-	**81:** 3838n
4-(3,4-dihydro-2-phenyl-2H̲-1-benzopyran-4-yl)-3-phenyl-	**81:** 3838n
3-(1,3-dihydroxyphenyl)-4-methyl-	**73:** P50723r
1-[(dimethoxyphosphinothioyl)thiomethyl]-4-(4-chlorophenyl)-	**72:** P31806h
1-[(dimethoxyphosphinothioyl)thiomethyl]-4-methyl-	**72:** P31806h
1,3-dimethyl-	**52:** 18379i, **55:** 23507h, **55:** 23508h, **78:** 124513z, **80:** P82998v, **82:** P72995u,
1,4-dimethyl-	**55:** 23508h, **70:** 68323h
2,3-dimethyl-	**78:** 124513z
3,4-dimethyl-	**54:** 7696a, **55:** 23508h, **77:** 74417c, **78:** 124513z
3-(3,5-dimethyl-4-isoxazolyl)-4-(*p*-ethoxyphenyl)-	**55:** 7399g
3-(3,5-dimethyl-4-isoxazolyl)-4-(*p*-methoxyphenyl)-	**55:** 7399g
3-(3,5-dimethyl-4-isoxazolyl)-4-(1-naphthyl)-	**55:** 7399g
3-[(2,4-dimethyl-3-quinolyl)methyl]-4-phenyl-	**71:** 30411t
3-[(2,4-dimethyl-3-quinolyl)methyl]-4-propyl-	**71:** 30411t
3-(2,4-dimethyl-5-thiazolyl)-4-phenyl-	**74:** 111955n
3-(5,6-dimethylpyrazinyl)-4-ethyl-	**77:** 126568z
3-(4,6-dimethyl-2-pyrimidinyl)-4-isopropyl-	**85:** 94310f
1,3-diphenyl-	**64:** 9727g, **70:** 77870m, **73:** 45427t, **77:** 164587n, **82:** 43276e
2,3-diphenyl-	**64:** 5072h, **70:** 77870m, **78:** 72015s

433

TABLE 14 (*Continued*)

Compound	Reference
14.2. Alkyl- or Aryl-Disubstituted Δ^2(or Δ^3)-1,2,4-Triazoline-5-thiones (*Continued*)	
3,4-diphenyl-	**44:** 2516f, **44:** 2517b, **51:** 5096b, **52:** 11822h, **53:** 21905a, **54:** 8795a, **57:** 12473c, **59:** 6386f, **71:** 49864q, **79:** 126404j, **81:** 91429h
3-(4,6-diphenyl-2-pyrimidinyl)-4-(1-methylethyl)-	**85:** 94310f
3-(4,6-diphenyl-2-pyrimidinyl)-4-(2-propenyl)-	**85:** 94310f
3-([(4,5-diphenyl-4$\underline{H}$-1,2,4-triazol-3-yl)thio]methyl)-4-phenyl-	**84:** 90086u
3-([(4,5-diphenyl-4$\underline{H}$-1,2,4-triazol-3-yl)thio]methyl)-(2-propenyl)-	**84:** 90086u
3-[(2(1$\underline{H}$),4(3$\underline{H}$)-dioxo-6-pyrimidinyl)methyl]-4-phenyl-	**71:** 22101b
3,4-di-*p*-tolyl-	**51:** 5095c
3-[4-(dodecanoylamino)phenyl]-4-ethyl-	**80:** P76701b
1-([ethoxy(ethyl)phosphinothioyl]thiomethyl)-4-methyl-	**72:** P31806h
3-[*o*-(ethoxymalonyl)phenyl]-1-methyl-	**75:** 20371y
3-(ethoxymethyl)-4-methyl-	**77:** P164704y, **85:** P21373u, **85:** P33015s
4-(*o*-ethoxyphenyl)-3-(4-[(*o*-ethoxyphenyl)-thioureylene]phenyl)-	**70:** 77880q
4-(*p*-ethoxyphenyl)-3-(4-[(*p*-ethoxyphenyl)-thioureylene]phenyl)-	**70:** 77880q
3-(*o*-ethoxyhenyl)-4-ethyl-	**71:** 49853k
4-(4-ethoxyphenyl)-3-(4-fluorophenyl)-	**82:** 72886j
4-(*p*-ethoxyphenyl)-1-(morpholinomethyl)-	**63:** 4280b
4-(*p*-ethoxyphenyl)-3-(morpholinomethyl)-	**57:** P9860g
4-(*p*-ethoxyphenyl)-3-(*p*-nitrophenyl)-	**68:** 59501w
3-(*o*-ethoxyphenyl)-4-phenyl-	**71:** 49853k
4-(*p*-ethoxyphenyl)-3-phenyl-	**51:** 5055d
4-(*p*-ethoxyphenyl)-1-piperidinomethyl-	**63:** 4280b
4-(*p*-ethoxyphenyl)-3-(piperidinomethyl)-	**57:** P9860g
4-(*p*-ethoxyphenyl)-3-[3-pyridyl]-	**51:** 5055d
4-(*p*-ethoxyphenyl)-3-[4-pyridyl]-	**51:** 5055d
4-(*p*-ethoxyphenyl)-1-[2-(2-pyridyl)ethyl]-	**63:** 6996e
5-(*p*-ethoxyphenyl)-4-*m*-tolyl-	**71:** 38865q
5-(*p*-ethoxyphenyl)-4-*o*-tolyl-	**71:** 38865q
3-(3-ethoxyvinyl)-2-phenyl-	**52:** 11862i
4-ethyl-3-(*p*-ethylphenyl)-	**73:** 64790x
4-ethyl-3-(*p*-fluorophenyl)-	**73:** 64790x
4-ethyl-3-[4-(hexamoylamino)phenyl]-	**80:** P76701b, **81:** P113665m
4-ethyl-3-(*o*-hydroxyphenyl)-	**72:** 30148h
4-ethyl-3-(*m*-hydroxyphenyl)-	**72:** 30148h
4-ethyl-3-(*p*-hydroxyphenyl)-	**72:** 30148h
3-(1-ethyl-1-hydroxypropyl)-1-phenyl-	**76:** 59533y
4-ethyl-3-(mercaptomethyl)-	**72:** P12733r, **74:** 125534m, **76:** 34220f
3-ethyl-1-methyl-	**55:** 23507i

TABLE 14 (*Continued*)

Compound	Reference
14.2. Alkyl- or Aryl-Disubstituted Δ^2(or Δ^3)-1,2,4-Triazoline-5-thiones (*Continued*)	

Compound	Reference
3-ethyl-4-methyl-	**55:** 23508i
4-ethyl-3-methyl-	**54:** 7696a
4-ethyl-3-(5-methylpyrazinyl)-	**77:** 126568z
4-ethyl-3-(*p*-nitrophenyl)-	**73:** 64790x
4-ethyl-3-pentyl-	**78:** P36245h, **80:** P76701b
1-ethyl-3-phenyl-	**73:** 45427t
3-ethyl-1-phenyl-	**79:** 92113w
3-ethyl-4-phenyl-	**54:** 5630e, **73:** P93613x, **75:** P114798v, **78:** P36245h, **80:** P76701b, **80:** P102274m
4-ethyl-3-phenyl-	**73:** P93613x
4-ethyl-3-propyl-	**80:** P54523t
4-ethyl-3-pyrazinyl-	**77:** 126568z
4-ethyl-3-(3-pyridyl)-	**72:** 30148h
4-ethyl-3-(4-pyridyl)-	**72:** 30148h
4-ethyl-3-(4-sulfamoylphenyl)-	**72:** 78949x
4-ethyl-3-*p*-tolyl-	**73:** 64790x
4-ethyl-3-(3,4,5-trimethoxyphenyl)-	**72:** 30148h
3-(2-fluorophenyl)-4-(4-fluorophenyl)-	**82:** 72886j
3-(3-fluorophenyl)-4-(4-fluorophenyl)-	**82:** 72886j
3-(2-fluorophenyl)-4-hexyl-	**82:** 72886j
3-(3-fluorophenyl)-4-hexyl-	**82:** 72886j
3-(4-fluorophenyl)-4-hexyl-	**82:** 72886j
3-(4-fluorophenyl)-4-(4-methylphenyl)-	**82:** 72886j
3-([(4-fluorophenyl)sulfonyl]phenyl)-4-(4-iodophenyl)--	**77:** 114312y
3-(4-[(4-fluorophenyl)sulfonyl]phenyl)-4-(2-methylphenyl)-	**76:** 140655m
3-(4-[(4-fluorophenyl)sulfonyl]phenyl)-4-(1-naphthalenyl)-	**76:** 140655m
3-([(4-fluorophenyl)sulfonyl]phenyl)-4-phenyl-	**77:** 114312y
3-(2-furanyl)-4-(4-methoxyphenyl)-	**85:** 192627j
3-(2-furanyl)-4-(4-methylphenyl)-	**85:** 192627j
3-(2-furanyl)-4-phenyl-	**85:** 192627j
3-(2-furanyl)-4-(2-propenyl)-	**85:** 192627j
3-(2-furyl)-4-methyl-	**72:** P100713q
3-(2-furyl)-1-phenyl-	**73:** 45427t
3-(2-furyl)-4-phenyl-	**72:** P100713q
3-[4-(2-furyl)-2-thiazolyl]-4-phenyl-	**74:** 111955n
3-[(4-hexadecyloxy-3-sulfo)phenyl]-	**71:** P22929c
4-hexadecyl-3-(3-sulfophenyl)-	**71:** P22929c
3-(α-hydroxybenzyl)-4-phenyl-	**70:** 47369n
3-(α-hydroxybenzyl)-4-*p*-tolyl-	**70:** 47369n
3-(hydroxydiphenylmethyl)-1-phenyl-	**76:** 59533y
3-(hydroxydiphenylmethyl)-4-phenyl-	**70:** 47369n
3-(hydroxydiphenylmethyl)-1-*p*-tolyl-	**76:** 59533y
4-(2-hydroxyethyl)-3-methyl-	**79:** 18625y
3-(2-hydroxy-4-mercaptophenyl)-4-phenyl-	**54:** 22468d
1-(hydroxymethyl)-3-methyl-	**64:** PC208e, **75:** P82399x

TABLE 14 (*Continued*)

Compound	Reference

14.2. Alkyl- or Aryl-Disubstituted Δ^2(or Δ^3)-1,2,4-Triazoline-5-thiones (*Continued*)

Compound	Reference
1-(hydroxymethyl)-4-methyl-	**72:** P31806h
1-(hydroxymethyl)-4-phenyl-	**75:** 35891s
1-(1-hydroxymethyl)-4-(3-pyridyl)-	**76:** 34182v
3-(*m*-hydroxyphenyl)-4-isobutyl-	**72:** 30148h
3-(*o*-hydroxyphenyl)-4-isobutyl-	**72:** 30148h
3-(*p*-hydroxyphenyl)-4-isobutyl-	**72:** 30148h
3-(*m*-hydroxyphenyl)-4-isopropyl-	**72:** 30148h
3-(*p*-hydroxyphenyl)-4-isopropyl-	**72:** 30148h
3-(*m*-hydroxyphenyl)-4-phenyl-	**72:** 30148h
3-(*p*-hydroxyphenyl)-4-phenyl-	**72:** 30148h
3-(*m*-hydroxyphenyl)-4-propyl-	**72:** 30148h
3-(*p*-hydroxyphenyl)-4-propyl-	**72:** 30148h
4-(*p*-iodophenyl)-3-(4-[(*p*-iodophenyl)thioureylene]- phenyl)-	**70:** 77880q
4-(4-iodophenyl)-3-[(2-methyl-1H-indol-3-yl)methyl]-	**82:** 118780a
4-(*p*-iodophenyl)-5-[*p*-(pentyloxy)phenyl]-	**71:** 38865q
4-isobutyl-3-phenyl-	**72:** 30148h
4-isobutyl-3-(3-pyridyl)-	**72:** 30148h
4-isobutyl-3-(4-pyridyl)-	**72:** 30148h
4-isobutyl-3-(4-sulfamoylphenyl)-	**73:** 64790x
4-isobutyl-3-(3,4,5-trimethoxyphenyl)-	**72:** 30148h
4-isopentyl-3-(*m*-nitrophenyl)-	**68:** 59501w
4-isopentyl-3-(*p*-nitrophenyl)-	**68:** 59501w
4-isopropyl-3-(*p*-methoxyphenyl)-	**57:** 16601h
4-isopropyl-3-(*m*-nitrophenyl)-	**68:** 59501w
3-isopropyl-1-phenyl-	**79:** 92113w
4-isopropyl-3-phenyl-	**72:** 30148h
4-isopropyl-3-(3-pyridyl)-	**57:** 16601h
4-isopropyl-3-(4-pyridyl)-	**57:** 16601h
4-isopropyl-3-(4-sulfamoylphenyl)-	**72:** 78949x
4-isopropyl-3-(3,4,5-trimethoxyphenyl)-	**57:** 16601h
3-(1-mercaptoethyl)-1-methyl-	**76:** 3765h
3-(mercaptomethyl)-1-methyl-	**76:** 3765h
3-(1-mercaptopropyl)-1-methyl-	**76:** 3765h
-1-methanol, 4-phenyl-	**63:** 4280c
-1-methanol, 4-(2-pyridyl)-	**65:** 16960h
4-(2-methoxyethyl)-3-(4-sulfamoylphenyl)-	**73:** 64790x
3-[*o*-(methoxymalonyl)phenyl]-1-phenyl	**75:** 20371y
3-(2-methoxy-1-methylvinyl)-2-phenyl-	**52:** 11863d
4-(*p*-methoxyphenyl)-3-(4-[(*p*-methoxyphenyl)- thioureylene]phenyl)-	**70:** 77880q
3-(4-methoxyphenyl)-1-methyl-	**83:** 163373b
3-(4-methoxyphenyl)-2-methyl-	**83:** 163373b
3-(*p*-methoxyphenyl)-4-methyl-	**44:** 2517b, **83:** 163373b
4-(2-methoxyphenyl)-3-[(2-methyl-1H-indol-3-yl)methyl]-	**82:** 118780a
4-(4-methoxyphenyl)-3-[(2-methyl-1H-indol-3-yl)methyl]-	**82:** 118780a
4-(*o*-methoxyphenyl)-3-(*p*-nitrophenyl)-	**68:** 59501w
4-(*p*-methoxyphenyl)-3-(*p*-nitrophenyl)-	**68:** 59501w

TABLE 14 (*Continued*)

Compound	Reference
14.2. Alkyl- or Aryl-Disubstituted Δ^2(or $\Delta^{3'}$)-1,2,4-Triazoline-5-thiones (*Continued*)	
4-[5-(4-methoxyphenyl)-1,3-4-oxadiazol-2-yl]-3-(4-methylphenyl)-	**85:** 5568f
4-[5-(4-methoxyphenyl)-1,3-4-oxadiazol-2-yl]-3-phenyl-	**85:** 5568f
3-(*p*-methoxyphenyl)-4-phenyl-	**57:** 16601h
4-(*o*-methoxyphenyl)-3-phenyl-	**69:** 27312q
3-(*p*-methoxyphenyl)-4-propyl-	**57:** 16601h
4-(*o*-methoxyphenyl)-3-(4-pyridyl)-	**51:** 5055d
3-(*o*-methoxyphenyl)-4-*m*-tolyl-	**71:** 38865q
3-(*o*-methoxyphenyl)-4-*o*-tolyl-	**71:** 38865q
3-(*o*-methoxyphenyl)-4-*p*-tolyl-	**71:** 38865q
3-(*p*-methoxyphenyl)-4-*o*-tolyl-	**60:** 14496f
3-[4-(3-methylbenzo[b]thien-2-yl)-2-thiazolyl]-4-phenyl-	**74:** 111955n
3-(1-methylethenyl)-4-phenyl-	**84:** 44710v
3-(1-methylethenyl)-4-(2-propenyl)-	**84:** 44710v
3-methyl-4-ethyl-	**74:** 111968u
3-[(2-methyl-1H-indol-3-yl)methyl]-4-(2-methylphenyl)-	**82:** 118780a
3-[(2-methyl-1H-indol-3-yl)methyl]-4-(3-methylphenyl)-	**82:** 118780a
3-[(2-methyl-1H-indol-3-yl)methyl]-4-(4-methylphenyl)-	**82:** 118780a
3-[(2-methyl-1H-indol-3-yl)methyl]-4-phenyl-	**82:** 118780a
3-[(2-methyl-1H-indol-3-yl)methyl]-4-(2-propenyl)-	**82:** 118780a
4-methyl-3-(1-methylethenyl)-	**84:** 44710v
1-methyl-3-(2-methyl-4-quinolinyl)-	**84:** 17287f
4-methyl-3-[(2-naphthalenyloxy)methyl]-	**80:** 59762e
4-methyl-3-(1-naphthylmethyl)-	**57:** P9860h, **80:** 59899e
1-methyl-3-(4-nitrophenyl)-	**83:** 163373b
2-methyl-3-(4-nitrophenyl)-	**83:** 163373b
4-methyl-3-(4-nitrophenyl)-	**83:** 163373b
4-methyl-3-(2-phenothiazin-10-yl-ethyl)-	**52:** 2020g
3-[2-(2-methyl-10H-phenothiazin-10-yl)ethyl]-4-phenyl-	**77:** 34439e
3-[2-(2-methyl-10H-phenothiazin-10-yl)ethyl]-4-(2-propenyl)-	**77:** 34439e
4-methyl-3-(phenoxymethyl)-	**76:** 126869c
1-methyl-3-phenyl-	**52:** 18380a, **66:** 37843r, **67:** 11476u, **73:** 45427t, **77:** 164587n
1-methyl-4-phenyl	**63:** 4280c
2-methyl-3-phenyl-	**66:** 37843r, **79:** 78749b, **80:** 27175z
3-methyl-1-phenyl	**52:** 18380a, **54:** 7695i, **80:** P89539e
3-methyl-4-phenyl-	**53:** 11356i, **53:** 21905a, **54:** 7696a, **61:** 4338a, **69:** 67340d
4-methyl-3-phenyl-	**44:** 2516i, **45:** 1122b, **74:** 110789z, **77:** 74417c, **84:** 90504d
3-(5-methyl-3-phenyl-4-isoxazolyl)-4-phenyl-	**70:** 11610v
4-(4-methylphenyl)-3-[(2-naphthalenyloxy)methyl]-	**80:** 59762e

TABLE 14 (*Continued*)

Compound	Reference
14.2. Alkyl- or Aryl-Disubstituted Δ^2(or Δ^3)-1,2,4-Triazoline-5-thiones (*Continued*)	
4-(2-methylphenyl)-3-[4-(phenylsulfonyl)phenyl]-	**76:** 140655m
3-methyl-1-(3-phenyl-1-propenyl)-	**79:** 5298u
3-[(2-methyl-4-phenyl-3-quinolyl)methyl]-4-phenyl-	**71:** 30411t
3-[(2-methyl-4-phenyl-3-quinolyl)methyl]-4-propyl-	**71:** 30411t
3-(4-methyl-2-phenyl-5-thiazolyl)-4-phenyl-	**74:** 111955n
3-methyl-1-(2-propenyl)-	**79:** 18651d
4-methyl-3-propyl-	**80:** P54523t
1-methyl-3-(4-pyridinyl)-	**83:** 163373b
4-methyl-3-(4-pyridinyl)-	**83:** 163373b
2-methyl-3-(2-pyridyl)-	**76:** 34180t
4-methyl-3-(2-pyridyl)-	**76:** 34180t
4-methyl-3-(3-pyridyl)-	**63:** 10497b
4-methyl-3-(4-pyridyl)-	**63:** 10497b
3-methyl-4-(4-sulfophenyl)-	**75:** P69541t, **75:** P69544w, **77:** P133193b
3-(4-methyl-2-thiazolyl)-4-phenyl-	**74:** 111955n
1-methyl-3-(2-thienyl)-	**73:** 45427t
4-methyl-3-[[(2-thiocarbamoyl)hydrazino]-carbonylmethyl]-	**82:** 140031d
4-methyl-1-(thiomorpholinomethyl)-	**64:** PC208e, **75:** P82399x
1-methyl-3-*p*-tolyl-	**55:** 1647f
3-[(2-methyl-4-*p*-tolyl-3-quinolyl)methyl]-4-phenyl-	**71:** 30411t
3-[(2-methyl-4-*p*-tolyl-3-quinolyl)methyl]-4-propyl-	**71:** 30411t
4-methyl-3-[(2,4,5-trichlorophenoxy)methyl]-	**76:** 126869c
1-methyl-3-(trifluoromethyl)-	**80:** P82990m
4-methyl-3-(trifluoromethyl)-	**80:** P82989t, **82:** P72995u
1-(morpholinomethyl)-4-phenyl-	**63:** 4280b, **75:** P82399x
3-(morpholinomethyl)-4-phenyl-	**57:** P9860g
1-(morpholinomethyl)-4-(2-pyridyl)-	**65:** 16960h
4-(1-naphthalenyl)-3-[4-(phenylsulfonyl)phenyl]-	**76:** 140655m
3-[(2-naphthalenyloxy)methyl]-4-phenyl-	**80:** 59762e
3-[(2-naphthalenyloxy)methyl]-4-(2-propenyl)-	**80:** 59762e
3-(1-naphthylmethyl)-4-phenyl-	**74:** 141652d
3-(1-naphthylmethyl)-4-*p*-tolyl-	**74:** 141652d
3-(*o*-[*p*-nitrobenzylidene)amino]phenyl)-4-phenyl-	**63:** 13256h
3-(*m*-nitrophenyl)-4-pentyl-	**68:** 59501w
3-(*p*-nitrophenyl)-4-pentyl-	**68:** 59501w
3-(*p*-nitrophenyl)-4-phenyl-	**54:** 8795b
4-(*m*-nitrophenyl)-3-(3-pyridyl)-, 3'-*N*-oxide	**52:** 6338c
5-[*p*-(pentyloxy)phenyl]-4-*m*-tolyl-	**71:** 38865q
5-[*p*-(pentyloxy)phenyl]-4-*o*-tolyl-	**71:** 38865q
3-[*p*-(pentyloxy)phenyl]-4-*p*-tolyl-	**71:** 38865q
3-pentyl-4-phenyl-	**73:** P93613x, **78:** P36245h, **80:** P76701b, **82:** P9971t
1-phenethyl-3-phenyl-	**66:** 37843r
3-(phenoxymethyl)-4-phenyl-	**70:** 47369n, **79:** 126404j
3-(phenoxymethyl)-4-*p*-tolyl-	**70:** 47369n
3-phenyl-1-benzyl-	**84:** 59312r

TABLE 14 (*Continued*)

Compound	Reference

14.2. Alkyl- or Aryl-Disubstituted Δ^2(or Δ^3)-1,2,4-Triazoline-5-thiones (*Continued*)

Compound	Reference
3-phenyl-4-(5-phenyl-1,2,4-oxadiazol-2-yl)-	**85:** 5568f
3-phenyl-1-(3-phenyl-1-propenyl)-	**79:** 5298u
4-phenyl-3-(4-phenyl-2-thiazolyl)-	**74:** 111955n
4-phenyl-3-[4-(phenylthioureylene)phenyl]-	**70:** 77880q
4-phenyl-1-([(4-phenyl-4H-1,2,4-triazol-2-yl)thio]methyl)-	**63:** 4280d
4-phenyl-1-(piperidinomethyl)-	**63:** 4280b, **75:** P82399x
4-phenyl-3-(piperidinomethyl)-	**57:** P9860g
3-phenyl-1-(2-propenyl)-	**79:** 18651d, **82:** 43276e
3-phenyl-4-(2-propenyl)-	**72:** 30148h, **81:** 91429h
1-phenyl-3-propyl-	**79:** 92113w
3-phenyl-4-propyl-	**72:** 30148h
4-phenyl-3-propyl-	**57:** P9860g
4-phenyl-3-(3-pyridazinyl)-	**68:** 105174s
4-phenyl-3-(2-pyridinyl)-	**83:** 43245q
4-phenyl-3-(3-pyridyl)-	**57:** 16601h
4-phenyl-3-(4-pyridyl)-	**57:** 8561d, **57:** 16601h
4-phenyl-1-[2-(2-pyridyl)ethyl]-	**63:** 6996e
4-phenyl-3-(4-sulfamoylphenyl)-	**72:** 78949x, **74:** 111968u
4-phenyl-3-[4-(2-thienyl)-2-thiazolyl]-	**74:** 111955n
3-phenyl-4-*p*-tolyl-	**51:** 5095c, **69:** 27312q
4-phenyl-3-[(2,4,5-trichlorophenoxy)methyl]-	**70:** 47369n
4-phenyl-3-tridecyl-	**85:** P134261z
4-phenyl-3-(3,4,5-trimethoxyphenyl)-	**57:** 16601h
4-phenyl-3-undecyl-	**78:** P22526p, **80:** P76701b
3-phenyl-1-[2-(2,6-xylyloxy)ethyl]-	**66:** 38743r, **67:** 11476u
1-(piperidinomethyl)-4-(2-pyridyl)-	**65:** 16960h
1-(1-piperidinylmethyl)-4-(3-pyridyl)-	**76:** 34182v
-1-propionic acid, 4-(*p*-chlorophenyl)-	**63:** 13242d
-1-propionic acid, 4-(*p*-ethoxyphenyl)-	**63:** 13242d
-4-propionic acid, 1-phenyl-	**74:** 141644c
-1-propionic acid, 4-phenyl-	**63:** 13242d
-1-propionitrile, 4-(*p*-chlorophenyl)-	**63:** 13242c
-1-propionitrile, 4-(*p*-ethoxyphenyl)-	**63:** 13242d
-4-propionitrile, 1-phenyl-	**74:** 141644c
-1-propionitrile, 4-phenyl-	**63:** 13242c
-1-propionitrile, 4-(2-pyridyl)-	**65:** 16961a
4-propyl-3-(3-pyridyl)-	**57:** 16601h
4-propyl-3-(4-pyridyl)-	**57:** 16601h
4-propyl-3-(4-sulfamoylphenyl)-	**72:** 78949x
4-propyl-3-*p*-tolyl-	**73:** 64790x
4-propyl-3-(3,4,5-trimethoxyphenyl)-	**57:** 16601h
4-(3-pyridinyl)-1-[2-(2-pyridinyl)ethyl]-	**76:** 34182v
4-(2-pyridyl)-1-[2-(2-pyridyl)ethyl]-	**65:** 16961a
3-(4-pyridyl)-4-*o*-tolyl-	**57:** 8561d, **63:** 2967c
3-(4-pyridyl)-4-*m*-tolyl-	**57:** 8561d
3-(4-pyridyl)-4-*p*-tolyl-	**57:** 8561d
3-(3-pyridyl)-4-*p*-tolyl-, 3'-*N*-oxide	**52:** 6338b
4-*o*-tolyl-3-*p*-tolyl-	**60:** 14496f

TABLE 14 (*Continued*)

Compound	Reference

14.2. Alkyl- or Aryl-Disubstituted Δ^2(or Δ^3)-1,2,4-Triazoline-5-thiones (*Continued*)

Compound	Reference
3-[4-[(*o*-tolyl)thioureylene]phenyl]-4-*o*-tolyl-	**70:** 77880q
3-[4-[(*m*-tolyl)thioureylene]phenyl]-4-*m*-tolyl-	**70:** 77880q
3-[4-[(*p*-tolyl)thioureylene]phenyl]-4-*p*-tolyl-	**70:** 77880q
4-(2,4-xylyl)-3-[4-[(2,4-xylyl)thioureylene]phenyl-	**70:** 77880q

14.3. Alkyl- or Aryl-Trisubstituted Δ^2(or Δ^1, Δ^3)-1,2,4-Triazoline-5-thiones

Compound	Reference
-4-acetamide, 1,3-diphenyl-	**79:** 18625y
-4-acetamide, 1-methyl-3-phenyl-	**79:** 18625y
-4-acetic acid, 1,3-diphenyl-	**79:** 18625y
-4-acetic acid, 1,3-diphenyl-, ethyl ester	**79:** 18625y
-4-acetic acid, 1,3-diphenyl-, hydrazide	**79:** 18625y
-4-acetic acid, 1-methyl-3-phenyl-	**79:** 18625y
-4-acetic acid, 1-methyl-3-phenyl-, ethyl ester	**79:** 18625y
-4-acetic acid, 1-methyl-3-phenyl-, hydrazide	**79:** 18625y
3-(*o*-acetylphenyl)-1,4-dimethyl-	**75:** 20371y
3-(2-benzothiazolyl)-1-benzyl-4-ethyl-	**77:** 126518h
3-(2-benzothiazolyl)-1-benzyl-4-methyl-	**77:** 126518h
3-(2-benzothiazolyl)-1-benzyl-4-phenyl-	**77:** 126518h
3-(2-benzothiazolyl)-1-benzyl-4-propyl-	**77:** 126518h
3-(2-benzothiazolyl)-1,4-dibenzyl-	**77:** 126518h
3-(2-benzothiazolyl)-1-(2,4-dichlorobenzyl)-4-ethyl-	**77:** 126518h
3-(2-benzothiazolyl)-1-(2,4-dichlorobenzyl)-4-phenyl-	**77:** 126518h
3-(2-benzothiazolyl)-4-ethyl-1-(4-methoxybenzyl)-	**77:** 126518h
3-(2-benzothiazolyl)-1-(4-methoxybenzyl)-4-phenyl-	**77:** 126518h
3-benzyl-1-([bis(2-hydroxyethyl)amino]methyl)-4-phenyl-	**83:** 147418q
3-benzyl-1-(4-morpholinylmethyl)-4-phenyl-	**83:** 147418q
3-benzyl-4-phenyl-1-(1-piperidinylmethyl)-	**83:** 147418q
1-([bis(2-hydroxyethyl)amino]methyl)-3,4-diphenyl-	**83:** 147418q
1-([bis(2-hydroxyethyl)amino]methyl)-3-phenyl-4-(2-propenyl)-	**83:** 147418q
4-(*o*-bromophenyl)-3,3-dimethyl-Δ^1-	**72:** 55339j
4-butyl-1-methyl-3-phenyl-	**69:** 67340d
4-butyl-3-methyl-1-phenyl-	**69:** 67340d
1-[2-chloro-1-(diethoxyphosphinyloxy)ethenyl]-4-methyl-3-trifluoromethyl-	**80:** P82989t
4-(chloromethyl)-1,3-dimethyl-	**80:** P82998v, **82:** P72995u
1-(chloromethyl)-3,5-dimethyl-	**69:** 75887c
4-(*p*-chlorophenyl)-3,3-dimethyl-Δ^1-	**72:** 55339j
4-([(diethoxyphosphinothioyl)oxo]methyl)-1,3-dimethyl-	**80:** P82998v, **82:** P72995u
4-([(diethoxyphosphinothioyl)thio]methyl)-1,3-dimethyl-	**80:** P82998v, **82:** P72995u
1-([(diethoxyphosphinothioyl)thio]methyl)-3,4-dimethyl-	**69:** 75887c
1-([(diethoxyphosphinothioyl)thio]methyl)-4-methyl-3-trifluoromethyl-	**74:** P76428u
1-[1-(diethoxyphosphinyloxy)ethenyl]-4-methyl-3-(trifluoromethyl)-	**80:** P82989t
1-[(diethoxyphosphinyloxy)methyl]-4-methyl-3-trifluoromethyl-	**74:** P76428u

440

TABLE 14 (*Continued*)

Compound	Reference
14.3. Alkyl- or Aryl-Trisubstituted Δ^2(or Δ^1, Δ^3)-1,2,4-Triazoline-5-thiones (*Continued*)	
1-[1-(diethoxyphosphinyloxy)-1-propenyl]-4-methyl-3-(trifluoromethyl)-	**80:** P82989t
4-[(diethoxyphosphinyloxy)-1-propenyl]-1-methyl-3-trifluoromethyl-	**80:** P82990m
1-[(diethoxyphosphinylthio)methyl]-4-methyl-3-trifluoromethyl-	**74:** P76428u
1-[(diethylamino)methyl]-3-phenyl-4-(2-propenyl)-	**83:** 147418q
1-[(dimethoxyphosphinothioyl)thiomethyl]-3,4-dimethyl-	**72:** P31806h
1-[1-(dimethoxyphosphinyloxy)ethenyl]-4-methyl-3-(trifluoromethyl)-	**80:** P82989t
4-[1-(dimethoxyphosphinyloxy)ethenyl]-1-methyl-3-trifluoromethyl-	**80:** P82990m
1-[1-(dimethoxyphosphinyloxy)ethenyl]-4-(2-propenyl)-3-trifluoromethyl-	**80:** P82989t
1-[(dimethylamino)methyl]-3,4-diphenyl-	**83:** 147418q
4-[p-(dimethylamino)phenyl]-3,3-dimethyl-Δ^1-	**72:** 55339j
3,3-dimethyl-4-(1-naphthyl)-Δ^1-	**72:** 55339j
1,2-dimethyl-3-phenyl-	**79:** 78749b, **85:** 159988e
1,3-dimethyl-2-phenyl	**54:** 7696h
1,3-dimethyl-4-phenyl-	**52:** 18380b, **72:** 78951s
1,4-dimethyl-3-phenyl-	**64:** 15716c, **75:** 20371y, **77:** 61961a, **77:** 164587n, **85:** 159988e
3,3-dimethyl-4-phenyl-Δ^1-	**72:** 55339j
3,4-dimethyl-1-phenyl-	**54:** 7696g
1,4-dimethyl-3-(2-pyridinyl)-	**78:** 72014r
1-([(diphenylphosphinothioyl)thio]methyl)-4-methyl-3-trifluoromethyl-	**74:** P76428u
3,4-diphenyl-1-(1-piperidinylmethyl)-	**83:** 147418q
1-([ethoxy(ethyl)phosphinothioyl]thiomethyl)-3,4-dimethyl-	**72:** P31806h
1-([ethoxy(ethyl)phosphinylthio]methyl)-4-methyl-3-trifluoromethyl-	**74:** P76428u
4-ethyl-3-(mercaptomethyl)-1-methyl-	**76:** 3765h
3-ethyl-1,4-dimethyl-	**55:** 23509g
3-ethyl-3-methyl-4-phenyl-Δ^1-	**72:** 55339j
1-β-D-glucopyranosyl-3-methyl-4-phenyl-	**80:** 146458d
1-β-D-glucopyranosyl-4-methyl-3-phenyl-	**84:** 90504d
4-(2-hydroxyethyl)-1,3-diphenyl-	**79:** 18625y
4-(2-hydroxyethyl)-1-methyl-3-phenyl-	**79:** 18625y
4-(hydroxymethyl)-1,3-dimethyl-	**80:** P82998v, **82:** P72995u
1-(hydroxymethyl)-3,4-diphenyl-	**83:** 147418q
1-(hydroxymethyl)-3-phenyl-4-(2-propenyl)-	**83:** 147418q
4-(p-hydroxyphenyl)-3,3, -dimethyl-Δ^1-	**72:** 55339j
1,1'-[(isopropylimino)dimethylene]bis[3,4-dimethyl-	**75:** P82399x
3-[o-(methoxymalonyl)phenyl]-1,4-dimethyl-	**75:** 20371y
3-[o-(methoxymalonyl)phenyl]-1-methyl-4-phenyl-	**75:** 20371y
1-methyl-3,4-diphenyl-	**62:** 2771d, **69:** 67340d, **71:** 2745m

TABLE 14 (*Continued*)

Compound	Reference
14.3. Alkyl- or Aryl-Trisubstituted Δ^2(or Δ^1, Δ^3)-1,2,4-Triazoline-5-thiones (*Continued*)	
3-methyl-1,4-diphenyl-	**52:** 18380b, **67:** 43755a, **69:** 67340d, **77:** 101468d
3-methyl-3,4-diphenyl-Δ^1-	**72:** 55339j
3-methyl-1-(morpholinomethyl)-4-phenyl-	**64:** PC208e, **75:** P82399x
3-methyl-1-(phenoxymethyl)-4-phenyl-	**72:** 78951s
3-methyl-3-(phenoxymethyl)-4-phenyl-Δ^1-	**72:** 78951s
1-[(methylphenylamino)methyl]-3,4-diphenyl-	**83:** 147418q
1-[(methylphenylamino)methyl]-3-phenyl-4-(2-propenyl)-	**83:** 147418q
3-methyl-4-phenyl-1-(2,3,4,6-tetra-O-acetyl-β-D-glucopyranosyl)-	**80:** 146458d
4-methyl-3-phenyl-1-(2,3,4,6-tetra-O-acetyl-β-D-glucopyranosyl)-	**84:** 90504d
3-methyl-4-phenyl-1-(2,3,4-tri-O-acetyl-β-D-xylopyranosyl)-	**80:** 146458d
4-methyl-3-phenyl-1-(2,3,4-tri-O-acetyl-β-D-xylopyranosyl)-	**84:** 90504d
4-methyl-3-phenyl-1-β-D-xylopyranosyl)-	**84:** 90504d
1-(4-morpholinylmethyl)-3-phenyl-4-(2-propenyl)-	**83:** 147418q
1-(4-morpholinylmethyl)-3,4-diphenyl-	**83:** 147418q
3-phenyl-1-(1-piperidinylmethyl)-4-(2-propenyl)-	**83:** 147418q
-1-propanamide, 3-benzyl-4-(2-propenyl)-	**83:** 164091v
-1-propanenitrile, 3-benzyl-4-phenyl	**83:** 164091v
-1-propanenitrile, 3-benzyl-4-phenyl	**83:** 164091v
-1-propanenitrile, 3-benzyl-4-(2-propenyl)-	**83:** 164091v
-1-propanenitrile, 3,4-diphenyl-	**82:** 43276e
-1-propanenitrile, 3-phenyl-4-(2-propenyl)-	**82:** 43276e
-1-propanoic acid, 3-benzyl-4-phenyl-	**83:** 164091v
-1-propanoic acid, 3,4-diphenyl-	**82:** 43276e
-1-propanoic acid, 3-phenyl-4-(2-propenyl)-	**82:** 43276e
1,2,3-trimethyl-	**78:** 124513z
1,3,4-trimethyl-	**55:** 23509f, **77:** 61961a **78:** 124513z, **83:** 178933h
3,3,4-trimethyl-Δ^1-	**72:** 55339j
1,2,3-triphenyl-	**52:** 11823c, **79:** 91736q
1,3,4-triphenyl-	**58:** 3417e, **63:** 7000e, **69:** 67340d

CHAPTER 15

1,2,4-Triazolin-5-ones Containing More Than One Representative Function

15.1. Acyl-Δ^2-1,2,4-Triazolin-5-ones

The direct synthesis of either *C*- or *N*-substituted acyl-1,2,4-triazolin-5-ones from a nontriazole reactant has been limited to several reports. Reaction of the thiazole (**15.1-2**) with hydrazine gave the hydrazone of 3-(phenylacetyl)triazolin-5-one (**15.1-1**), and treatment of **15.1-2** with methyl hydrazine gave the corresponding 3-(phenylthioacetyl)triazoline-5-one (**15.1-3**).[1] In both reactions the initial attack of the hydrazine occurred at

15.1-1   **15.1-2**   **15.1-3**

the 4-carbon of the thiazole ring. The dehydration of 1,4-dibenzoylsemicarbazide (**15.1-4**) with thionyl chloride was shown to give either 4-benzoyl- or 2-benzoyl-3-phenyl-1,2,4-triazolin-5-one (**15.1-5** or **15.1-6**).[2] Whatever the structure, treatment of this product with acid regenerated **15.1-4**.

15.1-4   **15.1-5**   **15.1-6**

The claim that treatment of the 1,2-dibenzoylsemicarbazide (**15.1-7**) with phosphorus oxychloride gave the 1-benzoyltriazolin-5-one (**15.1-8**) was later reported to be in error.[3] However, this compound and a series of 1-benzoyl-3,4-dialkyl-, 3-alkyl-4-aryl-, and 3,4-diaryltriazolin-5-ones were prepared by treatment of the preformed triazolinones (e.g., **15.1-9**) with benzoyl chloride in hot pyridine.[4] Because these compounds contain a 4-substituent, a nonionic compound can only be obtained by acylation at the 1-position.

Triazolin-5-ones substituted with a 1-acetyl group have been reported to result from the reaction of either 3- or 4-substituted triazolin-5-ones with refluxing acetic anhydride.[4,5] The structure of **15.1-11**, formed from **15.1-10** and acetic anhydride, was based on the procurement of **15.1-11** from **15.1-12** and acetic anhydride.[6] In addition, the products resulting from acetylation of 1,3-disubstituted triazolin-5-ones have been assigned as 4-acetyltriazolin-5-ones.[7,8]

References

1. **80:** 3422b   2. **68:** 39599f   3. **62:** 7747c   4. **61:** 10673g
5. **69:** 52075x   6. **74:** 141651c   7. **73:** 77210h   8. **75:** 110244p

15.2. Amino- and Diamino-Δ^2(or Δ^1)-1,2,4-Triazolin-5-ones

The N-amino-N'-carbamoylguanidine derivative (**15.2-1**) was cyclized in the presence of refluxing 2 N HCl to give a good yield of 3-aminotriazolin-5-one (**15.2-2**).[1] Also, N-(methoxycarbonylamino)guanidine (**15.2-3**) was readily cyclized with aqueous base to give a satisfactory yield of **15.2-2**.[2]

Similarly, the reaction of ethyl carbazate with both S-methyl thioureas and diarylcarbodiimides gave N-(carbonylamino)guanidines intermediates, which were converted with aqueous base to 3-substituted amino-4-aryltriazolin-5-ones.[3-6] A high yield of 3-benzamidotriazolin-5-one (**15.2-5**) was reported to result from the action of sulfuric acid on **15.2-4** at room

temperature.[7] The 1:2 adduct of ethyl carbazate and N,N'-diphenylcarbodiimide (**15.2-6**) was converted in refluxing ethanolic hydroxide to a mixture of **15.2-8** (~20%) and **15.2-9** (~62%). A higher yield of **15.2-9** (85%) was obtained by prior hydrolysis of the ester group of **15.2-6** with acid to give **15.2-7**, followed by cyclization of the latter under alkaline conditions.[8]

The intramolecular cyclization of N-(2-chlorobenzoylamino)-N'-ureido-guanidines (**15.2-10**) with 4 N sodium hydroxide gave a mixture of three products: the 3-[2-(2-chlorobenzoyl)hydrazino]- and 4-[(2-chlorobenzoyl)-amino]-3-aminotriazolin-5-ones (**15.2-11** and **15.2-12**) and the 3-hydrazino-triazole (**15.2-16**) formed via hydrolysis of **15.2-13** (see Section 6.2b).[9,10] In other benzoyl compounds related to **15.2-10**, the 3-hydrazinotriazoles (e.g., **15.2-16**) were consistently the major products. However, the yields of the acyl derivatives of the triazolin-5-ones were variable, being dependent on the position of the substituents in the N-(acylamino)-N'-ureidoguanidine (e.g., **15.2-10**). Hydrolysis of the benzoyl moiety of **15.2-11** with hydrochloric acid gave 3-hydrazinotriazolin-5-one (**15.2-15**), which was also

$$2\text{-ClC}_6\text{H}_4\text{CONHNHC} \begin{array}{c} \text{NH}_2 \\ \diagdown \\ \text{NNHCONH}_2 \end{array} \equiv \text{H}_2\text{NC} \begin{array}{c} \text{NHNHCOC}_6\text{H}_4\text{-2-Cl} \\ \diagdown \\ \text{NNHCONH}_2 \end{array} \equiv \text{H}_2\text{NCONHNHC} \begin{array}{c} \text{NH}_2 \\ \diagdown \\ \text{NNHCOC}_6\text{H}_4\text{-2-Cl} \end{array}$$

15.2-10

$\downarrow_{-\text{OH}}$

$$2\text{-ClC}_6\text{H}_4\text{CONHNH} \begin{array}{c} \text{H} \\ | \\ \text{N} \\ \diagdown \quad =O \\ \text{N}-\text{N} \\ | \\ \text{H} \end{array} + \text{H}_2\text{N} \begin{array}{c} \text{NHCOC}_6\text{H}_4\text{-2-Cl} \\ | \\ \text{N} \\ \diagdown \quad =O \\ \text{N}-\text{N} \\ | \\ \text{H} \end{array} + \text{H}_2\text{NCONHNH} \begin{array}{c} \text{N} \\ \diagdown \quad \text{C}_6\text{H}_4\text{-2-Cl} \\ \text{N}-\text{N} \\ | \\ \text{H} \end{array}$$

15.2-11   **15.2-12**   **15.2-13**

$\searrow_{\text{H}^+}$ $\downarrow$

$$\text{Me}_2\text{C}=\text{NNHC} \begin{array}{c} \text{NHCONHPh} \\ \diagdown \\ \text{NN}=\text{CMe}_2 \end{array} \xrightarrow{\text{H}^+} \text{H}_2\text{NNH} \begin{array}{c} \text{H} \\ | \\ \text{N} \\ \diagdown \quad =O \\ \text{N}-\text{N} \\ | \\ \text{H} \end{array} \qquad \text{H}_2\text{NNH} \begin{array}{c} \text{N} \\ \diagdown \quad \text{C}_6\text{H}_4\text{-2-Cl} \\ \text{N}-\text{N} \\ | \\ \text{H} \end{array}$$

15.2-14   **15.2-15**   **15.2-16**

formed directly in good yield by the cyclization of the N-(carbonyl)-N',N''-diaminoguanidine (**15.2-14**) in an acidic medium.[11] In the cyclization of the ethoxycarbonyl derivative (**15.2-17**), triazoles [**15.2-18** (25%) and **15.2-19** (40%)] were produced under neutral conditions and triazolin-5-ones [**15.2-20** (42%) and **15.2-21** (40%)] under alkaline conditions.[12,13] Simi-

$$\begin{array}{c} \nearrow \xrightarrow[\Delta]{\text{H}_2\text{O}} \quad \text{Ph} \begin{array}{c} \text{NHCO}_2\text{Et} \\ | \\ \text{N} \\ \diagdown \quad \text{NH}_2 \\ \text{N}-\text{N} \end{array} + \text{Ph} \begin{array}{c} \text{N} \\ \diagdown \quad \text{NHNHCO}_2\text{Et} \\ \text{N}-\text{N} \\ | \\ \text{H} \end{array} \\ \text{15.2-18} \qquad \qquad \text{15.2-19} \\ \text{PhCONHNHC(NH}_2)\text{NNHCO}_2\text{Et} \\ \textbf{15.2-17} \\ \searrow \xrightarrow{\text{OH}^-} \text{PhCOHNNH} \begin{array}{c} \text{H} \\ | \\ \text{N} \\ \diagdown \quad =O \\ \text{N}-\text{N} \\ | \\ \text{H} \end{array} + \text{H}_2\text{N} \begin{array}{c} \text{NHCOPh} \\ | \\ \text{N} \\ \diagdown \quad =O \\ \text{N}-\text{N} \\ | \\ \text{H} \end{array} \\ \text{15.2-20} \qquad \text{15.2-21} \end{array}$$

larly, the N-methyl derivative (**15.2-23**) gave **15.2-22** under neutral conditions and **15.2-24** under basic conditions.[14]

$$\text{Ph} \begin{array}{c} \text{N} \\ \diagdown \quad \text{NHNHCO}_2\text{Et} \\ \text{N}-\text{N} \\ | \\ \text{Me} \end{array} \xleftarrow{\text{H}_2\text{O}} \text{PhCONH(Me)NC} \begin{array}{c} \text{NH}_2 \\ \diagdown \\ \text{NNHCO}_2\text{Et} \end{array} \xrightarrow{\text{OH}^-}$$

15.2-22   **15.2-23**

$$\text{PhCONH(Me)N} \begin{array}{c} \text{H} \\ | \\ \text{N} \\ \diagdown \quad =O \\ \text{N}-\text{N} \\ | \\ \text{H} \end{array}$$

15.2-24

The rearrangement of 2-amino-5-arylamino-1,3,4-oxadiazoles (e.g., **15.2-25**) in refluxing aqueous potassium hydroxide proceeded via the semicarbazide intermediate (**15.2-26**) to give 4-aryl-3-aminotriazolin-5-ones (e.g., **15.2-27**) in good yields.[15,16] The method was also successful for the preparation of 3-amino-4-benzyl- and 3-anilino-4-phenyltriazolin-5-ones. Both 4-aryl- and 1,4-diaryl-3-aminotriazolin-5-ones (e.g., **15.2-27**) were obtained by treatment of the appropriate aryl or diarylsemicarbazide (e.g., **15.2-28**) with cyanogen bromide to give 1-cyanosemicarbazides (e.g., **15.2-29**), which were cyclized in a refluxing alkaline solution.[17,18] In a

modification of this route, cyanohydrazides have been treated with dialkyl amines to give 3-(dialkylamino)triazolin-5-ones.[19] Also, semicarbazides (e.g., **15.2-30**) were reacted with alkyl and aryl cyanates (e.g., **15.2-31**) to give 3-aminotriazolin-5-ones (e.g., **15.2-33**) formed via intermediate **15.2-32**.[20]

The addition of phosgene to the carbodiimide (**15.2-34**) gave **15.2-35**, which was treated with hydrazine hydrate to give the 3-anilinotriazolin-5-one (**15.2-36**).[21,22] In related reactions of this type, treatment of **15.2-37** with hydrazine gave **15.2-38**,[23] and **15.2-39** with phenyl hydrazine gave **15.2-40**.[24] N—N bond formation was observed in the reaction of the ureido-

15.2-37   **15.2-38**   **15.2-39**   **15.2-40**

1,2,4-oxadiazole (**15.2-41**) with alcoholic potassium hydroxide to give a high yield of **15.2-42**.[25] In addition, several 3-amino-1,2,4-triazolin-5-ones (e.g.,

15.2-41   **15.2-42**

15.2-2) have been prepared from 3-nitro-1,2,4-triazolin-5-ones (e.g., **15.2-43**) either by reduction with hydrazine[26] or by catalytic hydrogenation with platinum or palladium.[27,28]

15.2-43   **15.2-2**

Relative few methods have been developed for the preparation of 4-amino-1,2,4-triazolin-5-ones. The thermal self-condensation of **15.2-44** gave a 25% yield of 4-amino-1,2,4-triazolin-5-one (**15.2-45**).[29]

H$_2$NCONHNHCONH$_2$
15.2-44   **15.2-45**

Although the reaction of carbohydrazide (**15.2-46**) with refluxing formic acid also gave a low yield of **15.2-45**,[30] a higher yield was obtained by the condensation of **15.2-46** with ethyl orthoformate.[30,31] Similarly, the condensation of **15.2-46** with ethyl orthopropionate gave **15.2-47**, which was also obtained by alkaline ring closure of **15.2-48**.[30] In the cyclization of **15.2-49**

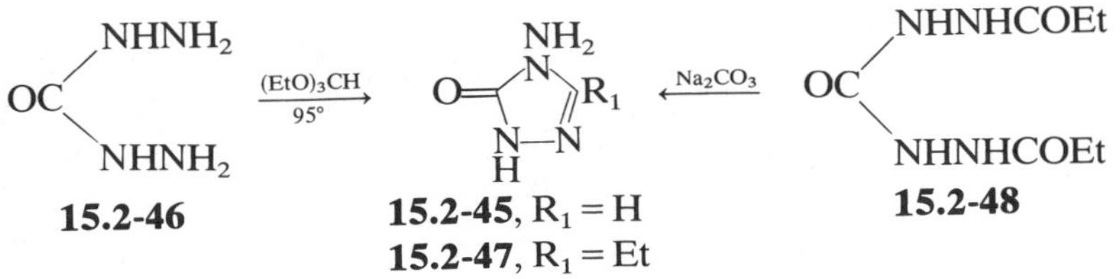

15.2-46 **15.2-45**, $R_1 = H$ **15.2-48**
15.2-47, $R_1 = Et$

with hydrazine, the sulfur moiety was displaced by hydrazine and hydrazine was eliminated from the carbonyl function to give **15.2-50**.[32]

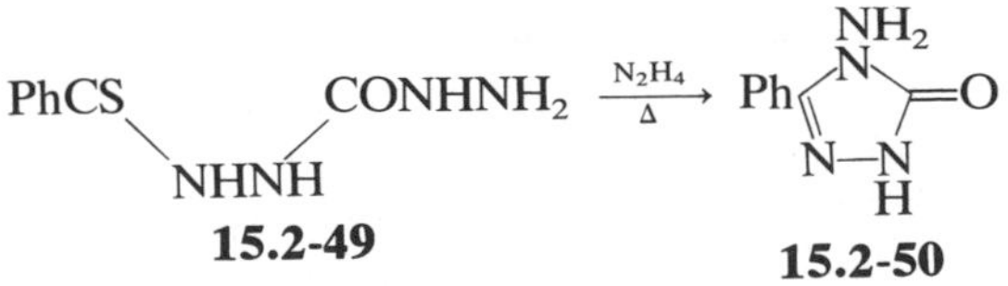

15.2-49 **15.2-50**

The base-catalyzed intramolecular cyclization of the chlorotriazole (**15.2-51**) gave **15.2-52**, which could also be prepared by the reaction of **15.2-54** with sodium azide.[33] Under acidic conditions, the oxadiazole ring of the bicyclic system was opened to give good yields of 4-benzoylamino-1,2,4-triazoline-5-ones (**15.2-53**).[33-35]

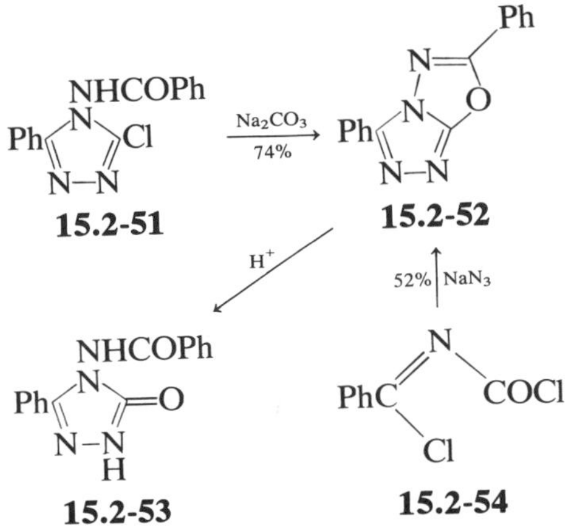

15.2-51 **15.2-52**

15.2-53 **15.2-54**

In the early literature the interaction of hydrazones of aldehydes with diethyl azodicarboxylate was reported to give derivatives of tetrazane, which underwent rearrangement and cyclization under acidic conditions to give tetrazinone compounds. A reinvestigation of this reaction showed that the reaction of benzaldehyde phenylhydrazone with diethyl azodicarboxylate gave the N-(acylamino)amidrazone (**15.2-55**), which was cyclized under acidic, basic, and thermal conditions to give good yields of the 4-(ethoxycarbonylamino)triazolin-5-one (**15.2-56**).[36,37] The carbamoyl linkage of **15.2-56** was cleaved with potassium hydroxide in refluxing ethylene glycol

PhCH + (N=N with CO₂Et groups) $\longrightarrow$ PhC(N(CO₂Et)NHCO₂Et / NNHPh) **15.2-55** $\xrightarrow{\Delta}$ (triazolinone NHCO₂Et) **15.2-56** (90%)

to give the corresponding 4-aminotriazolinone. Triazolinones were also obtained from phenylhydrazones derived from both aliphatic and heteroaromatic aldehydes but not from *N,N*-disubstituted hydrazones.[11,38] Some examples of the 4-amino-Δ^1-triazolinones ring system (e.g., **15.2-58**) have been prepared by the oxidation of carbohydrazide bishydrazones (e.g., **15.2-57**) with lead tetraacetate.[39] These compounds deteriorate in air or on exposure to light at 20° and are rapidly decomposed by acid.

OC(NHN=CMe₂ / NHN=CMe₂) **15.2-57** $\xrightarrow[0°]{Pb(OAc)_4}$ (structure) **15.2.58**

Two interconversions of preformed triazoles systems to 4-amino-1,2,4-triazolin-5-ones are of interest. The 1-benzyl-1,2,4-triazolium compound (**15.2-59**) in the presence of base underwent ring opening, presumably to **15.2-60**, followed by ring closure to give the 4-(benzylamino)-1,2,4-triazolin-5-one (**15.2-61**).[40] In the oxidation of **15.2-62** with lead tetraacetate in

15.2-59 $\xrightarrow{OH^-}$ [**15.2-60**] $\longrightarrow$ **15.2-61**

the presence of methyl acrylate, an 86% yield of the 4-(aziridinyl)triazolin-5-one (**15.2-63**) was produced.[41]

15.2-62 $\xrightarrow[Pb(OAc)_4]{CH_2=CHCO_2Me}$ **15.2-63**

In addition to the synthesis of 3,4-diamino-1,2,4-triazolin-5-ones de-scribed previously, treatment of **15.2-64** with hydrochloric acid resulted in partial decomposition (>50%), but some **15.2-64** was cyclized to give a low yield of **15.2-65**.[42] The latter was also obtained by hydrolysis of benzoyl group of **15.2-66** with hydrochloric acid.[9,10] The cyclization of the 1,5-disub-

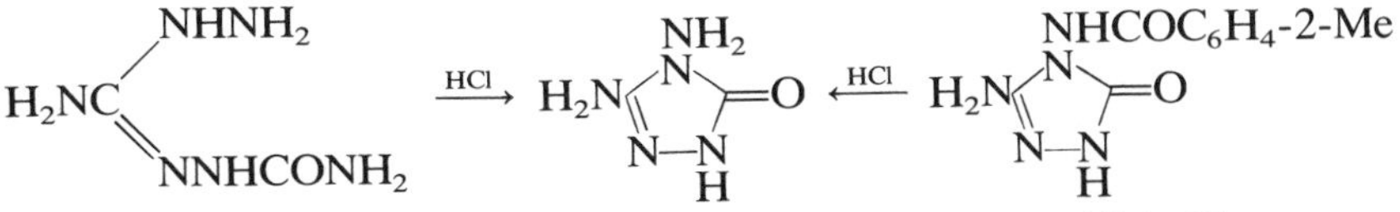

15.2-64   **15.2-65**   **15.2-66**

stituted carbohydrazide (**15.2-67**) under alkaline conditions gave a good yield of 3,4-dianilino-1,2,4-triazolin-5-one (**15.2-68**).[43] Also, a high yield of 4-amino-3-hydrazinotriazolin-5-one (**15.2-70**) was obtained by the con-

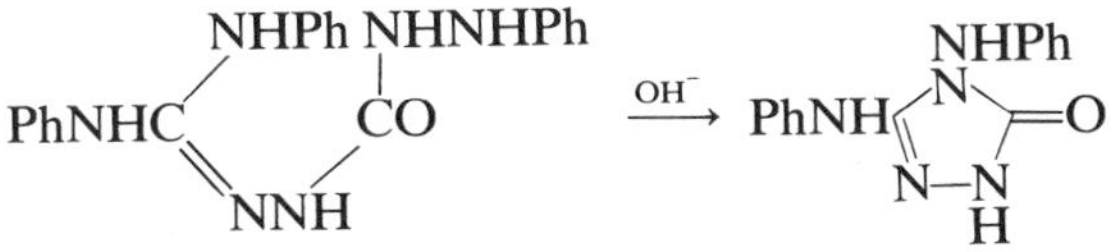

15.2-67   **15.2-68**

densation of N,N',N''-triaminoguanidine (**15.2-69**) with S,S'-dimethyl dithiocarbonate in refluxing dimethylformamide.[44]

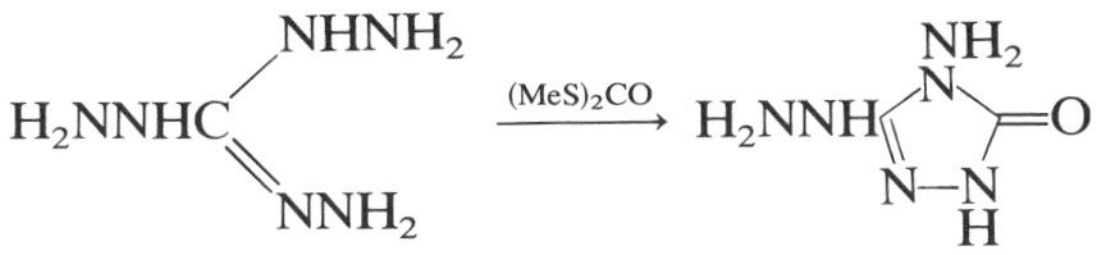

15.2-69   **15.2-70**

References

1. **55:** 2623g	2. **65:** 705a	3. **56:** 7306c	4. **59:** 603a
5. **65:** 13691f	6. **66:** 115651t	7. **53:** 10033i	8. **69:** 10398z
9. **60:** 1736d	10. **67:** 43756b	11. **64:** 8174b	12. **71:** 49947u
13. **72:** 78953u	14. **73:** 14772x	15. **63:** 5631g	16. **63:** 9936f
17. **54:** 11003f	18. **58:** 3415g	19. **85:** P46688p	20. **63:** 11549h
21. **59:** 2678d	22. **66:** 55070r	23. **80:** 82826n	24. **76:** 140832s
25. **73:** 66509m	26. **70:** 87692u	27. **70:** 115078u	28. **73:** 66847v
29. **70:** 11643h	30. **63:** 16339d	31. **65:** 12212h	32. **83:** 205712u
33. **81:** 3549u	34. **75:** 35892t	35. **76:** 85758v	36. **58:** 518f
37. **69:** 2623d	38. **61:** 11961g	39. **66:** 94958f	40. **76:** 85755s
41. **77:** 101467c	42. **66:** 104959y	43. **72:** 55347k	44. **63:** P4305f

15.3. S-Substituted Thio- and Sulfonyl-1,2,4-Triazolin-5-ones

The synthesis of S-substituted thio- and sulfonyl-1,2,4-triazolin-5-ones from noncyclic reactants has been limited to several reports. Fusion of the S-methyl thiosemicarbazide derivative (**15.3-1**) gave a good yield of the 3-(methylthio)triazolin-5-one (**15.3-2**).[1] Also, an ethoxycarbonyl intermediate

was involved in the conversion of **15.3-3** to **15.3-4** with ethyl chloroformate.[2] The 4-(4-toluenesulfonyl)triazolin-5-one (**15.3-6**) and several related compounds were prepared by the fusion of hydrazides (e.g., **15.3-5**) with 4-toluenesulfonyl isocyanate.[3]

The majority of the S-substituted thio-1,2,4-triazolin-5-ones are prepared by alkylation of preformed 1,2,4-triazolidin-3-one-5-thiones such as **15.3-7**, which with methyl iodide in an alkaline medium gave a 62% yield of

15.3-8.[4] The cephalosporin (5-thia-1-azabicyclo[4.2.0]oct-2-ene ring system) antibiotic (**15.3-11**), desired for structure-activity studies, was prepared by displacement of the acetoxy group of **15.3-9** with the thiol anion derived from **15.3-10**.[5] However, in the reaction of **15.3-12** with methyl iodide,

alkylation occurred on both the sulfur group and a ring nitrogen to give **15.3-13**.[6]

Oxidation of **15.3-10** with nitric acid in the presence of sodium nitrate gave the disulfide (**15.3-14**).[7] Similarly, treatment of **15.3-12** with iodine in ethanol gave **15.3-15**.[6]

References

1. **65:** 13691f   2. **64:** 8173g   3. **53:** 8053b   4. **73:** 35290d
5. **83:** 109119k   6. **53:** 18014f   7. **65:** 705d

15.4. Miscellaneous Δ^2(or Δ^1)-1,2,4-Triazolin-5-ones

The triazoline-5-ones not described in previous sections include carboxy-, acyl-, nitro-, alkoxy-, alkylthio-, and halo- derivatives and also derivatives of triazoline-3,5-dione.

The interaction of ethyl carbazate with the carbodiimide (**15.4-1**) proceeded via the monoadduct (**15.4-2**) and the diadduct (**15.4-5**) to give a mixture of the 3,5-dianilinotriazole (**15.4-4**) (<28% yield) and the triazolin-5-one (**15.4-3**) (<40%).[1] In this type of reaction the yields of the products depended upon the reaction conditions and the nature of the carbodiimide.

$$EtO_2CNHNH_2 + (4\text{-}BrC_6H_4N)_2C \xrightarrow[100°]{DMF} EtO_2CNHNHC\begin{matrix} NHC_6H_4\text{-}4\text{-}Br \\ \\ NC_6H_4\text{-}4\text{-}Br \end{matrix}$$

15.4-1   **15.4-2**

15.4-3   **15.4-4**   **15.4-5**

For example, a higher yield of **15.4-3** resulted from the condensation of **15.4-1** with the monoadduct (**15.4-2**) than with ethyl carbazate. The related triazoline-5-one (**15.4-7**) has been reported to result from the reaction of the imino chloride (**15.4-6**) with 4-phenylsemicarbazide.[2]

15.4-6   **15.4-7**

Another type of 1-carboxytriazolin-5-one was obtained from the quasi-1,3-dipole (**15.4-8**). Treatment of **15.4-8** with an equivalent amount of phenyl isocyanate gave an 80% yield of **15.4-9**.[3] The carbonyl chloride derivative (**15.4-11**), useful for the acylation of amines, was prepared by treatment of **15.4-10** with phosgene.[4]

15.4-8   **15.4-9**   **15.4-10**, R$_1$ = H
                                                           15.4-11, R$_1$ = COCl

The rearrangement of 2-amino-1,3,4-oxadiazole-5-carboxylic acid (**15.4-12**) and its ester derivatives occurred in base at room temperature to give the triazolinone (**15.4-13**);[5] the ethyl ester of **15.4-13** was generated in the amination of the 1,2,4-oxadiazoline (**15.4-14**) with alcoholic ammonia, a reaction in which a N—N bond is formed.[6] The methyl ester of **15.4-13** was

15.4-12   **15.4-13**   **15.4-14**

obtained in 79% yield by another type of rearrangement, one that involved ring opening of the 1-aminoimidazolidine-2,4,5-trione (**15.4-15**) with methanolic hydrochloric acid to give **15.4-16**, followed by cyclization of the latter to give **15.4-17**.[7] In the reaction of 2-ethoxy-4-phenylazo-2-thiazolin-

15.4-15   **15.4-16**   **15.4-17**

5-one (**15.4-18**) with aniline, amination and rearrangement occurred to give a 60% yield of the 3-carboxamide of the triazolin-5-one (**15.4-21**).[8,9] This reaction was postulated to proceed via the migration of the *O*-alkyl group of **15.4-19** to give the *S*-alkyl intermediate (**15.4-20**). In another method for the preparation of the 3-carboxy derivative of 1,2,4-triazole-5-one, nitrosation of **15.4-23** gave the 3-diazotriazole (**15.4-24**), which was converted by

15.4-18   **15.4-19**

15.4-20   **15.4-21**, R₁ = PhNH
15.4-22, R₁ = HO

$$\textbf{15.4-23} \qquad\qquad \textbf{15.4-24} \qquad\qquad \textbf{15.4-13}$$

acid hydrolysis to **15.4-13**.[10] The transformation of **15.4-25** to its carbox-amide derivatives (e.g., **15.4-26**) has been effected by treatment of **15.4-25** successively with phosphorus pentachloride and arylamines.[11] The hydrolysis of the amide linkage has been carried out in aqueous base, as illustrated by the conversion of **15.4-21** to **15.4-22**.[8,12]

$$\textbf{15.4-25} \qquad\qquad\qquad \textbf{15.4-26}$$

The *N*-acylation of aminotriazolinones occurred readily.[13,14] For example, brief treatment of the 3-aminotriazolin-5-ones (**15.4-27**) with hot acetic anhydride gave the 1-acetyl derivative (**15.4-28**).[13] When **15.4-27** was refluxed in acetic anhydride for several hours, the triacetyl derivative (**15.4-30**) was obtained, which was hydrolyzed with dilute base to give the 3-acetamidotriazolin-5-one (**15.4-29**).

$$\textbf{15.4-27} \qquad\qquad\qquad \textbf{15.4-28}$$

$$\textbf{15.4-29} \qquad\qquad\qquad \textbf{15.4-30}$$

The nitration of 1,2,4-triazolin-5-one (**15.4-31**) either with aqueous[15] or fuming[16] nitric acid gave a good yield of the 3-nitro derivative (**15.4-32**). Similarly, 1-methyl-, 4-methyl, and 1,4-dimethyl-1,2,4-triazolin-5-ones were transformed with fuming nitric acid to the corresponding 3-nitrotria-zolin-5-ones.[16] Although both 1-phenyl- and 4-phenyl-1,2,4-triazolin-5-ones gave the corresponding 3-nitrotriazolin-5-ones, these compounds also underwent nitration in the benzene ring. In contrast, the nitration of both

1,2-dimethyl- and 2-phenyl-1,2,4-triazolin-5-one was unsuccessful. However, both 2-phenyl- and 3-phenyl-1,2,4-triazolin-5-one were nitrated in the benzene ring. The methylation of **15.4-32** has been reported to give predominately either the 4-methyl or 1-methyl derivatives depending upon the conditions of the reaction. Treatment of **15.4-32** with methyl iodide in aqueous potassium hydroxide at reflux gave a 50% yield of the 4-methyl derivative (**15.4-33**), whereas treatment of **15.4-32** with dimethyl sulfate in aqueous potassium hydroxide at room temperature gave a 93% yield of the 1-methyl derivative (**15.4-35**).[17] The latter also resulted from the hydrolysis of the 5-nitro group of 1-methyl-3,5-dinitrotriazole (**15.4-34**) with triethylamine in aqueous dioxane.[18]

A simple method has been described for the synthesis of 3-methoxy-1,2,4-triazolin-5-ones. Treatment of ethyl N-[methoxy(methylthio)methylene]carbamate (**15.4-36**) with ethanolic hydrazine hydrate at reflux gave a 56% of **15.4-37**.[19] Similarly, reaction of **15.4-36** with phenylhydrazine gave a 49% yield of **15.4-38**. In a related reaction, a mixture of the chloride

(**15.4-39**) and 1,1-dimethylhydrazine was heated in toluene to give the 3-phenoxy-1,2,4-triazolin-5-one (**15.4-40**).[20] Alkoxy-1,2,4-triazolin-5-ones

$$(MeO)_4C \ + \ \underset{\substack{\text{H}_2\text{NHN}}}{\overset{\substack{\text{NHOH}}}{\diagup \text{CO}}} \ \xrightarrow[\Delta]{\text{EtOH}} \ MeO-\underset{\substack{\text{N—N}\\\text{H}}}{\overset{\substack{\text{N}}}{\diagdown}}=O$$

15.4-41   **15.4-42**

(e.g., **15.4-42**) are also formed in the condensation of 4-hydroxysemicarba-zide (**15.4-41**) with tetraalkyl orthocarbonates.[21] Several reports have de-scribed the alkylation of *N*-alkyl and aryl derivatives of triazolidin-3,5-dione. In the reaction of **15.4-43** with diazomethane, a low yield of the 3-methoxytriazolin-5-one (**15.4-38**) was obtained.[22] In a related reaction,

$$O=\underset{\substack{\text{N—N}\\\text{H} \quad \text{Ph}}}{\overset{\substack{\text{H}\\\text{N}}}{}}=O \ \xrightarrow[\text{Et}_2\text{O}]{\text{CH}_2\text{N}_2} \ MeO-\underset{\substack{\text{N—N}\\\text{Ph}}}{\overset{\substack{\text{H}\\\text{N}}}{}}=O$$

15.4-43   **15.4-38** (10%)

treatment of the blocked 1-(β-D-ribfuranosyl)-1,2,4-triazolidin-3,5-dione (**15.4-44**) with excess diazomethane was shown to give a 54% yield of the 3-methoxy-4-methyl derivative (**15.4-45**) and a trace amount of the isomeric 3,5-dimethoxy derivative.[23] Mixed results have also been reported

15.4-44   **15.4-45**

for the alkylation of 1,4-disubstituted 1,2,4-triazolidin-3,5-diones. Reaction of **15.4-46** with diethyl chlorothiophosphate was claimed to give the 3-*O*-phosphorus derivative (**15.4-47**),[24] whereas **15.4-48** with diazomethane gave a 1:3 mixture of **15.4-49** and **15.4-50**.[25] In the addition of benzonitrile oxide and the isomeric phenyl isocyanate to **15.4-51**, only the oxide gave an *O*-substituted product (**15.4-52**).[26] The isocyanate gave a ring *N*-substituted

$$O=\underset{\substack{\text{N—N}\\\text{H} \quad \text{Me}}}{\overset{\substack{\text{Me}\\\text{N}}}{}}=O \ \xrightarrow{\text{(EtO)}_2\text{P(S)Cl}} \ (EtO)_2P(S)O-\underset{\substack{\text{N—N}\\\text{Me}}}{\overset{\substack{\text{Me}\\\text{N}}}{}}=O$$

15.4-46   **15.4-47**

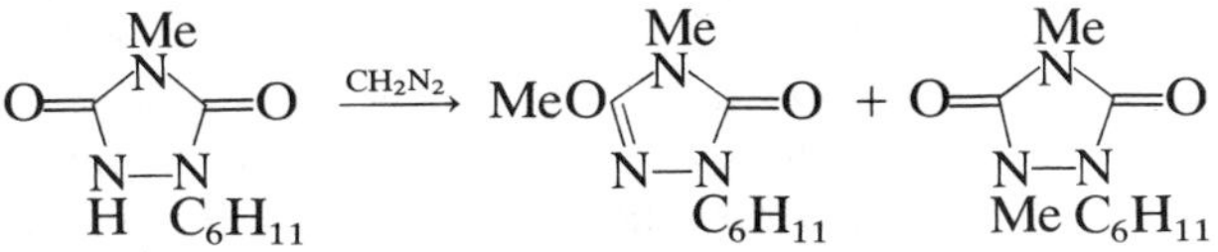

15.4-48   **15.4-49**   **15.4-50**

product, a result that was also observed in the alkylation of the 1,2-disubstituted triazolidin-3,5-dione (**15.4-53**) with diazomethane.[22] In contrast to the

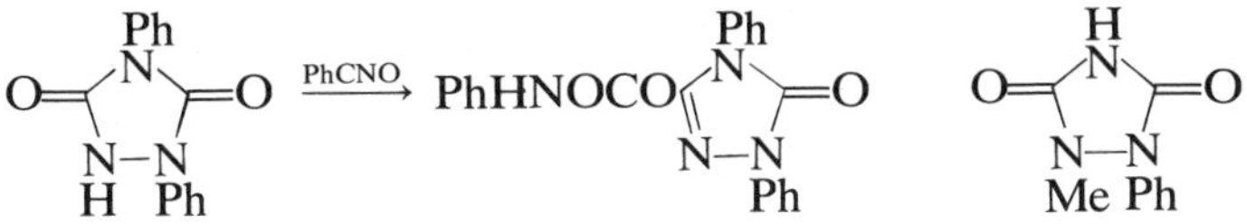

15.4-51   **15.4-52**   **15.4-53**

3-alkoxytriazolin-5-ones, little work has been carried out on the preparation of 3-alkylthiotriazolin-5-ones. Treatment of the product resulting from the methylation of the triazolidine-3,5-dithione (**15.4-54**) with base gave an 80% yield of the 3-(methylthio)triazolin-5-one (**15.4-55**).[27]

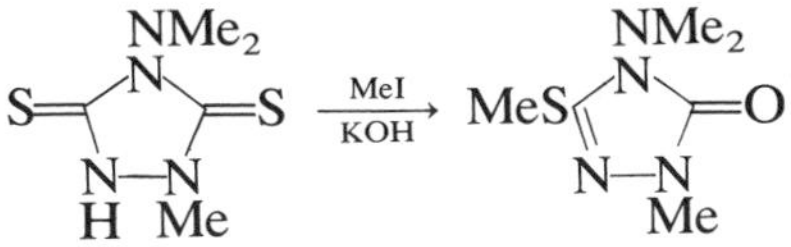

15.4-54   **15.4-55**

The synthesis of 3-halotriazolin-5-ones by a cyclization reaction has only been reported in two patents.[28,29] For example, reaction of the perchloro compound (**15.4-56**) with benzylhydrazine was reported to give the 3-chlorotriazolin-5-one (**15.4-57**).[28] The majority of the 3-halotriazolin-5-ones have been generated from a preformed triazole ring system. The direct bromination of 4-methyltriazolin-5-one (**15.4-58**) was found to give the corresponding 1,3-dibromo derivative (**15.4-59**).[30] Nucleophilic exchange of

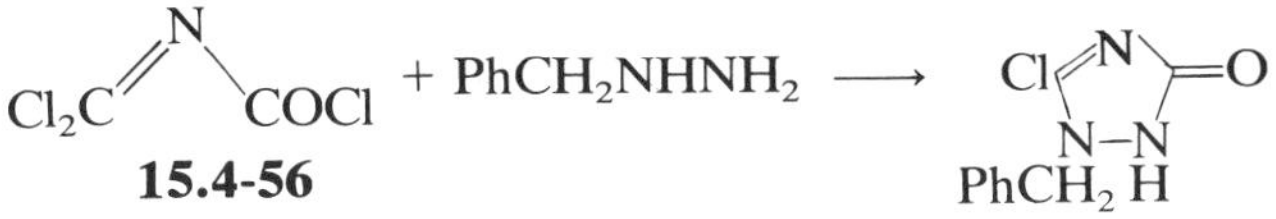

15.4-56   **15.4-57**

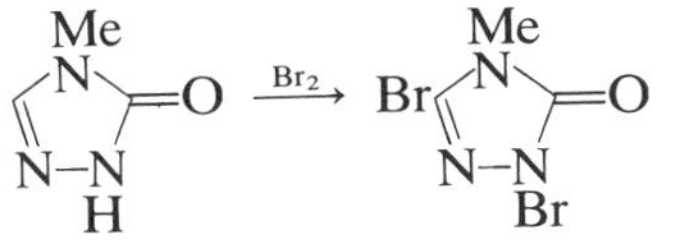

15.4-58   **15.4-59**

the nitro group of **15.4-32** was effected in high yield with hydrochloric acid to give the 3-chloro compound (**15.4-60**),[31] which was also prepared by the diazoation of 3-aminotriazolin-5-one (**15.4-63**) in concentrated hydrochloric acid in the presence of copper powder.[32] In addition, the 3-bromo derivative (**15.4-61**) was prepared by treatment of both **15.4-32** and **15.4-64** with hydrobromic acid.[31] Similarly, the 1-methyl derivative of **15.4-64** was converted to the 1-methyl derivative of **15.4-61**.[33] Recently, the preparation of the 3-fluorotriazolin-5-one (**15.4-62**) (21%) was accomplished by treatment of **15.4-32** at 100° with liquid hydrofluoric acid.[34]

Oxidation of the triazolidine-3,5-dione (**15.4-65**) gave 4-phenyl-1,2,4-triazoline-3,5-dione (**15.4-66**), a powerful dienophile that reacted instantly with dienes to give Diels–Alder adducts[35–38] and with monoolefins to give N-alkylated triazolidine-3,5-diones.[39] This oxidation has been effected with

lead tetraacetate,[38] tert-butyl hypochlorite,[40,41] N-bromosuccinimide,[42] and activated isocyanates.[43] In addition, a series of 4-alkyl-1,2,4-triazoline-3,5-diones have been prepared from the corresponding triazolidine-3,5-diones.[44–46] Recently, the parent ring system, 1,2,4-triazolin-3,5-dione (**15.4-67**), was prepared and reacted *in situ* with cyclopentadiene to give the corresponding Diels–Alder product.[47,48] In contrast to the triazoline-3,5-diones, which reacted spontaneously with dienes, the corresponding 3-iminotriazolin-5-ones (e.g., **15.4-68**) reacted with dienes only in the presence of boron trifluoride etherate.[49]

15.4-67   **15.4-68**

References

1. **69:** 10398z	2. **76:** 140832s	3. **71:** 61474v	4. **82:** 57676x
5. **65:** 7169g	6. **82:** 57623c	7. **83:** 193189p	8. **76:** 59533y
9. **78:** 136152u	10. **65:** 705c	11. **75:** 110244p	12. **75:** 98505q
13. **63:** 5631g	14. **65:** 12213a	15. **65:** 705a	16. **70:** 87692u
17. **70:** 115078u	18. **72:** 111380e	19. **80:** 82826n	20. **79:** P66363h
21. **70:** 68266s	22. **66:** 28303n	23. **73:** 66847v	24. **75:** 20410k
25. **74:** 54291s	26. **81:** 91438k	27. **71:** 21509d	28. **82:** 72994t
29. **84:** 105603m	30. **71:** 91399s	31. **67:** 82167e	32. **62:** 14437d
33. **73:** 45419s	34. **80:** 14898n	35. **58:** 2448f	36. **66:** 94960a
37. **67:** 108606z	38. **70:** 20009h	39. **67:** 116853t	40. **76:** 59534z
41. **82:** 87423e	42. **84:** 164686p	43. **82:** 16729y	44. **66:** 37837s
45. **67:** 116858y	46. **84:** 16586r	47. **79:** 115498s	48. **85:** 123874a
49. **80:** 108472w			

TABLE 15. 1,2,4-TRIAZOLIN-5-ONES CONTAINING MORE THAN ONE REPRE-SENTATIVE FUNCTION

Compound	Reference
15.1. Acyl-Δ²-1,2,4-Triazolin-5-ones	
1-acetyl-3-benzyl-4-(2-bromoethyl)-	**68:** 78201q
1-acetyl-3-benzyl-4-(2-iodoethyl)-	**68:** 78201q
1-acetyl-3-benzyl-4-phenyl-	**61:** 10674d
1-acetyl-4-(2-bromoethyl)-3-phenyl-	**68:** 78201q
1-acetyl-3,4-diethyl-	**61:** 10674d
1-acetyl-3,4-diphenyl-	**61:** 10674c
4-acetyl-1,3-diphenyl-	**75:** 110244p
1-acetyl-3-ethyl-4-phenyl-	**61:** 10674c
1-acetyl-4-ethyl-3-phenyl-	**61:** 10674d
1-acetyl-4-(2-iodoethyl)-3-phenyl-	**68:** 78201q
1-acetyl-3-methyl-4-phenyl-	**61:** 10674c
4-acetyl-3-methyl-1-phenyl-	**73:** 77210h
1-acetyl-3-methyl-4-p-tolyl-	**61:** 10674c
1-acetyl-4-phenyl-	**61:** 10674c
1-acetyl-3-phenyl-	**68:** 39599f, **69:** 52075x, **74:** 141651c
4-acetyl-3-phenyl-	**68:** 39599f
1-acetyl-4-phenyl-3-propyl-	**61:** 10674c
1-acetyl-3-phenyl-4-p-tolyl-	**61:** 10674c
1-benzoyl-3,4-diethyl-	**61:** 10674d
1-benzoyl-3,4-diphenyl-	**61:** 10674c

TABLE 15   (*Continued*)

Compound	Reference

15.1. Acyl-Δ^2-1,2,4-Triazolin-5-ones   (*Continued*)

Compound	Reference
1-benzoyl-3-ethyl-4-phenyl-	**61:** 10674c
1-benzoyl-3-methyl-4-phenyl-	**61:** 10674c
2(or 4)-benzoyl-3-phenyl-	**68:** 39599f
1-benzoyl-3-phenyl-4-*p*-tolyl-	**61:** 10674d, **62:** 7747c
2([2-carboxy-3,3-dimethyl-7-oxo-4-thia-1-aza-bicyclo[3.2.0]hept-6-yl)carbamoyl(phenyl)methyl]-carbamoyl)-3-ethyl-	**82:** P57676x
2-([2-carboxy-3,3-dimethyl-7-oxo-4-thia-1-azabicyclo[3.2.0]hept-6-yl)carbamoyl(phenyl)methyl]-carbamoyl)-3-methyl-	**82:** P57676x
2-([2-carboxy-3,3-dimethyl-7-oxo-4-thia-1-azabicyclo[3.2.0]hept-6-yl)carbamoyl(phenyl)methyl]-carbamoyl)-3-phenyl-	**82:** P57676x
1,2-diacetyl-	**47:** 6942c
3-[(2-methoxyphenyl)acetyl]-, benzylidene hydrazone	**80:** 3422b
3-[(2-methoxyphenyl)acetyl]-, hydrazone	**80:** 3422b
3-[(4-methoxyphenyl)thioacetyl]-2-methyl-	**80:** 3422b
3-(phenylacetyl)-, hydrazone	**80:** 3422b
3-(phenylthioacetyl)-2-methyl-	**80:** 3422b

15.2. Amino- and Diamino-Δ^2(or Δ^1)-1,2,4-Triazolin-5-ones

Compound	Reference
3-acetamido-	**63:** 5631g
3-acetamido-1-phenyl-	**73:** 66509m
3-acetamido-4-phenyl-	**63:** 5631g
3-amino-	**55:** 2623g, **62:** 14437d, **65:** 705b, **70:** 87692u
4-amino-	**44:** 2941i, **62:** 14437d, **63:** 16339e, **70:** 11643h
3-amino-4-benzamido-	**71:** P49947u, **72:** 78953u
3-amino-4-benzyl-	**63:** 5631g
3-amino-4-(benzylideneamino)-	**66:** 104959y
3-amino-4-(2-chlorobenzamido)-	**71:** P49947u
3-amino-4-(*o*-chlorophenyl)-	**58:** 3415g, **63:** 5631g
3-amino-4-(*m*-chlorophenyl)-	**58:** 3415g, **63:** 5631g
3-amino-4-(*p*-chlorophenyl)-	**58:** 3415g, **63:** 5631g
4-amino-1,3-dimethyl-	**63:** 16340a
3-amino-1,4-dimethyl-	**70:** 87692u
4-amino-1,3-diphenyl-	**58:** 518f
3-amino-1,4-diphenyl-	**54:** 11003f
3-amino-4-(*p*-ethoxyphenyl)-	**63:** 5631g
4-amino-3-ethyl-	**63:** 16339f
4-amino-3-ethyl-1-methyl-	**63:** 16340a
4-amino-3-ethyl-1-phenyl-	**64:** 8174g
4-amino-3-heptadecyl-	**71:** P22929c
4-amino-3-hydrazino-	**63:** P4305f
3-amino-4-(4-hydroxybenzamido)-	**72:** 78953u
3-amino-2-isopropyl-	**80:** P82651b

TABLE 15 (*Continued*)

Compound	Reference

15.2. Amino- and Diamino-Δ^2(or Δ^1)-1,2,4-Triazolin-5-ones (*Continued*)

Compound	Reference
3-amino-4-(*p*-methoxyphenyl)-	**63:** 5631g
3-amino-1-methyl-	**70:** 87692u, **70:** 115078u
3-amino-4-methyl-	**70:** 87692u, **70:** 115078u
4-amino-3-methyl-	**63:** 16339f
4-amino-1-methyl-	**63:** 16339h
3-amino-4-(3-methylbutyramido)-	**71:** P49947u, **72:** 78953u
3-amino-1-methyl-4-phenyl-	**63:** 11550f
3-amino-4-(1-naphthyl)-	**63:** 5631g
3-amino-1-phenyl-	**73:** 66509m
3-amino-4-phenyl-	**58:** 3415g, **63:** 5631g
4-amino-3-phenyl-	**83:** 205712u
4-amino-1-phenyl-3-(2-pyridyl)-	**61:** 11961h
4-amino-1-phenyl-3-(3-pyridyl)-	**61:** 11962a
4-amino-1-phenyl-3-(4-pyridyl)-	**61:** 11962b
4-amino-3-[2-(phenyl-2-pyridinylmethylene)hydrazino]-	**82:** P170960y
3-amino-1-β-D-ribofuranosyl-	**73:** 66847v
3-amino-4-(salicylamido)-	**71:** P49947u, **72:** 78953u
4-anilino-3-*p*-toluidino-	**72:** 55347k
3-amino-4-*o*-tolyl-	**63:** 5631g
3-amino-4-*p*-tolyl-	**58:** 3415g, **63:** 5631g
3-amino-4-[2-(*o*-tolyloxy)acetamido]-	**72:** 78953u
3-amino-4-(2,4-xylyl)-	**63:** 5631g
3-anilino-4-cyclohexyl-	**66:** P55070r
4-anilino-1,3-diphenyl-	**69:** 2623d
4-anilino-3-methyl-1-phenyl-	**69:** 2623d
3-anilino-4-phenyl-	**57:** 12473a, **63:** 9937b, **66:** 115651t, **80:** 108472w
4-anilino-3-phenyl-	**83:** 205712u
3-benzamido-	**53:** 10033i
4-benzamido-3-(4-bromophenyl)-	**75:** 35892t, **76:** 85758v
4-benzamido-3-(*p*-chlorophenyl)-	**75:** 35892t, **76:** 85758v
4-benzamido-3-(*p*-nitrophenyl)-	**75:** 35892t, **76:** 85758v
3-benzamido-1-phenyl-	**73:** 66509m
4-benzamido-3-phenyl-	**81:** 3549u
3-(2-benzoyl-1-ethylhydrazino)-	**73:** 14772x
3-(2-benzoyl-1-methylhydrazino)-	**73:** 14772x
4-benzylamino-	**76:** 85755s
4-benzylamino-3-phenyl	**83:** 205712u
3-benzylamino-1-(2,3,5-tri-*O*-benzoyl-β-D-ribofuranosyl)-	**83:** 97822u
4-benzyl-3-(benzylideneamino)-	**63:** 5631g
4-benzylideneamino-	**44:** 2941i, **44:** 2942a, **63:** 16339f
4-(benzylideneamino)-1,3-dimethyl-	**63:** 16340a
3-(benzylideneamino)-1,4-diphenyl-	**54:** 11003h
4-(benzylideneamino)-1,3-diphenyl-	**58:** 518f
4-(benzylideneamino)-3-ethyl-	**63:** 16339g
4-(benzylideneamino)-3-ethyl-1-methyl-	**63:** 16340a
4-(benzylideneamino)-1-methyl-	**63:** 16340a

TABLE 15   (*Continued*)

Compound	Reference

15.2. Amino- and Diamino-Δ^2(or Δ^1)-1,2,4-Triazolin-5-ones   (*Continued*)

Compound	Reference
4-(benzylideneamino)-3-methyl-	**63:** 16339f
3-(benzylideneamino)-4-phenyl-	**63:** 5631g
4-(benzylideneamino)-1-phenyl-3-(2-pyridyl)-	**61:** 11961h
4-(benzylideneamino)-1-phenyl-3-(3-pyridyl)-	**61:** 11962a
4-(benzylideneamino)-1-phenyl-3-(4-pyridyl)-	**61:** 11962b
3-(benzylideneamino)-4-*p*-tolyl-	**63:** 5631g
3-(benzylidene-1-methylhydrazino)-	**73:** 14772x
3-[benzyl(methyl)amino]-	**76:** P140832s
3-[benzyl(methyl)amino]-2-methyl-	**80:** P82651b
3-(*p*-bromoanilino)-4-(*p*-bromophenyl)-	**66:** 115651t
4-(4-bromobenzylamino)-	**76:** 85755s
4-(sec-butylideneamino)-3-ethyl-3-methyl-	**66:** 94958f
4-tert-butyl-3-(tert-butylamino)-	**71:** 38261q
-3-carbamic acid, ethyl ester	**28:** 1666[3]
-4-carbamic acid, 3-ethyl-1-phenyl-, ethyl ester	**64:** 8174g
-4-carbamic acid, 1,3-diphenyl-, benzyl ester	**58:** 518f
-4-carbamic acid, 1,3-diphenyl-, ethyl ester	**58:** 518f, **61:** 11961g
-4-carbamic acid, 3-(*o*-hydroxyphenyl)-1-phenyl-, ethyl ester	**69:** 2623d
-4-carbamic acid, 1-methyl-3-phenyl-, ethyl ester	**58:** 518f, **69:** 2623d
-4-carbamic acid, 3-methyl-1-phenyl-, ethyl ester	**58:** 518f
-4-carbamic acid, 3-(*o*-nitrophenyl)-1-phenyl-, ethyl ester	**69:** 2623d
-4-carbamic acid, 3-(*m*-nitrophenyl)-1-phenyl-, ethyl ester	**69:** 2623d
-4-carbamic acid, 3-(*p*-nitrophenyl)-1-phenyl-, ethyl ester	**69:** 2623d
-4-carbamic acid, 1-phenyl-3-(2-pyridyl)-, ethyl ester	**61:** 11961h
-4-carbamic acid, 1-phenyl-3-(3-pyridyl)-, ethyl ester	**61:** 11961h
-4-carbamic acid, 1-phenyl-3-(4-pyridyl)-, ethyl ester	**61:** 11962a
-4-carbamic acid, 3-phenyl-1-*m*-tolyl-, ethyl ester	**69:** 2623d
3-[2-(*p*-chlorobenzoyl)-1-methylhydrazino]-	**73:** 14772x
4-(4-chlorobenzylamino)-	**76:** 85755s
4-[(2-chlorobenzylidene)amino]-	**83:** P79251c
4-[(3-chlorobenzylidene)amino]-	**83:** P79251c
1-(chloromethyl)-4-[(5-nitrofurfurylidene)amino]-	**65:** P12213c
4-(*m*-chlorophenyl)-3-(diethylamino)-	**59:** 604a
4-(*p*-chlorophenyl)-3-(diethylamino)-	**56:** 7306f, **59:** 603g
1-(3-chloropropyl)-4-[(5-nitrofurfurylidene)amino]-	**65:** P12213c
4-cyclohexyl-3-(cyclohexylamino)-	**66:** 115651t, **80:** 108472w
4-cyclohexyl-3-(diethylamino)-	**56:** 7306f, **59:** 604a
4-cyclohexyl-3-morpholino-	**59:** 604a
4-cyclohexyl-3-piperidino-	**59:** 604a
2-cyclopentyl-3-(dimethylamino)-	**80:** P82651b
3,4-diamino-	**60:** 1736e, **66:** 104959y, **67:** 43756b, **71:** P49947u
3,4-dianilino-	**72:** 55347k

TABLE 15   (*Continued*)

Compound	Reference

15.2. Amino- and Diamino-Δ^2(or Δ^1)-1,2,4-Triazolin-5-ones   (*Continued*)

Compound	Reference
3-(dibutylamino)-4-phenyl-	**56:** 7306e, **59:** 603g
4-[(2,6-dichlorobenzylidene)amino]-	**83:** P79251c
4-[(3,4-dichlorobenzylidene)amino]-	**83:** P79251c
3-(diethylamino)-	**76:** P140832s
3-(diethylamino)-4-(*p*-ethoxyphenyl)-	**59:** 604a
3-(diethylamino)-2-methyl-	**76:** P140832s
3-(diethylamino)-1-phenyl-	**76:** P140832s
3-(diethylamino)-4-phenyl-	**56:** 7306e, **59:** 603c
3-(dimethylamino)-	**80:** 82826n
3-(dimethylamino)-4-(*p*-ethoxyphenyl)-	**59:** 604a
3-(dimethylamino)-2-ethyl-	**80:** P82651b
4-(dimethylamino)-4-(α-ethylbenzyl)-1-phenyl-	**64:** P8199b
3-(dimethylamino)-2-isopropyl-	**80:** P82651b, **82:** P156321j, **85:** P46688p
3-(dimethylamino)-2-methyl-	**80:** P82651b
3-(dimethylamino)-4-phenyl-	**56:** 7306e, **59:** 603c
3-(3,5-dimethylpyrazol-1-yl)-	**64:** 8174c
4-[(diphenylmethylene)amino]-3,3-diphenyl-	**66:** 94958f
3-(di-2-propenylamino)-2-methyl-	**76:** P140832s
3-(di-2-propenylamino)-2-isopropyl-	**80:** P82651b
-1-ethanol, 4-[(5-nitrofurfurylidene)amino]-	**65:** P12213c
4-(*p*-ethoxyphenyl)-3-morpholino-	**59:** 604a
4-(*p*-ethoxyphenyl)-3-piperidino-	**59:** 604a
3-(1-ethylhydrazino)-	**73:** 14772x
3-ethyl-1-(hydroxymethyl)-4-[(5-nitrofurfurylidene)-amino]-	**65:** 12213b
2-ethyl-3-(methylamino)-	**80:** P82651b
3-(1-ethyl-*p*-nitrobenzylidenehydrazino)-	**73:** 14772x
3-ethyl-4-[(5-nitrofurfurylidene)amino]-	**65:** P12213a, **69:** P59249r
3-ethyl-4-([3-(trifluoromethyl)benzylidene]amino)-	**83:** P79251c
4-[(2-fluorobenzylidene)amino]-	**83:** P79251c
3-hydrazino-	**60:** 1736e, **64:** 8174c, **67:** 43756b
3-(5-hydroxy-3-methylpyrazol-1-yl)-	**64:** 8174c
1-(3-hydroxypropyl)-4-[(5-nitrofurfurylidene)amino]-	**65:** P12213c
4-(isopropylideneamino)-3,3-dimethyl-	**66:** 94958f
4-isopropyl-3-(isopropylamino)-	**80:** 108472w
2-isopropyl-3-(methylamino)-	**80:** P82651b
2-isopropyl-3-(4-morpholinyl)-	**80:** P82651b
2-isopropyl-3-(1-pyrrolidinyl)-	**80:** P82651b
-1-methanol, 3-methyl-4-[(5-nitrofurfurylidene)amino]-	**65:** P12213c
-1-methanol, 4-[(5-nitrofurfurylidene)amino]-	**65:** P12213c
4-(2-methoxycarbonyl-1-aziridinyl)-3-methyl-1-phenyl-	**77:** 101467c
3-(methylamino)-	**73:** 14772x
3-(1-methylhydrazino)-	**73:** 14772x
3-methyl-4-(methylamino)-1-phenyl-	**82:** 12106v
3-methyl-4-[(α-methylbenzylidene)amino]-3-phenyl-	**66:** 94958f

TABLE 15 (*Continued*)

Compound	Reference

15.2. Amino- and Diamino-Δ^2(or Δ^1)-1,2,4-Triazolin-5-ones (*Continued*)

Compound	Reference
2-methyl-3-(methylphenylamino)-	**80:** P82651b
3-(1-methyl-*p*-nitrobenzylidenehydrazino)-	**73:** 14772x
3-[1-methyl-2-(*m*-nitrobenzoyl)hydrazino]-	**73:** 14772x
3-methyl-4-[(5-nitrofurfurylidene)amino]-	**65:** P12213a, **69:** P59249r
3-(methylphenylamino)-	**76:** P140832s
4-(4-methylphenyl)-3-[(4-methylphenyl)amino]-	**80:** 108472w
3-methyl-1-phenyl-4-(2-phenyl-1-aziridinyl)-	**77:** 101467c
3-[1-methyl-2-(phenylthiocarbamoyl)hydrazino]-	**73:** 14772x
3-methyl-4-[(3-trifluoromethyl)benzylidene]amino]-	**83:** P79251c
3-morpholino-4-phenyl-	**59:** 603e
3-(4-morpholinyl)-	**76:** P140832s
4-[(5-nitrofurfurylidene)amino]-	**65:** P12212h, **69:** P59249r
3-[(2-phenylethyl)amino]-1-(2,3,4,6-tetra-*O*-acetyl-β-D-glucopyranosyl)-	**83:** 97822u
4-phenyl-3-(2-phenylhydrazino)-	**23:** 1398³
4-phenyl-3-piperidino-	**59:** 603e
1-phenyl-3-(1-piperidinyl)-	**76:** P140832s
4-phenyl-3-(1-pyrrolidinyl)-	**59:** 604a
4-phenyl-3-sulfonamido-	**45:** 4671d, **45:** 4671i
1-phenyl-3-(4H-1,2,4-triazol-4-yl)-	**72:** 121448x
3-(1-pyrrolidinyl)-1-β-D-ribofuranosyl-	**83:** 97822u
3-(1-pyrrolidinyl)-1-(2,3,5-tri-*O*-benzoyl-β-D-ribofuranosyl)-	**83:** 97822u
3-*o*-toluidino-4-*o*-tolyl-	**59:** 2678g
3-*p*-toluidino-4-*p*-tolyl-	**69:** 10398z
3-(4H-1,2,4-triazol-4-yl)-	**72:** 121448x
4-([3-(trifluoromethyoxy)benzylidene]amino)-	**83:** P79251c
4-([3-(trifluoromethyl)benzylidene]amino)-	**83:** P79251c

15.3. S-Substituted Thio- and Sulfonyl-1,2,4-Triazolin-5-ones

Compound	Reference
3-[(7-amino-2-carboxy-8-oxo-5-thia-1-azabicyclo[4.2.0]oct-2-en-3-yl)methylthio]-	**80:** P120970q, **81:** P63648k
3-[(7-amino-2-carboxy-8-oxo-5-thia-1-azabicyclo[4.2.0]oct-2-en-3-yl)methylthio]-1-ethyl-	**81:** P63648k
3-[(7-amino-2-carboxy-8-oxo-5-thia-1-azabicyclo[4.2.0]oct-2-en-3-yl)methylthio]-4-ethyl-	**81:** P63648k
3-[(7-amino-2-carboxy-8-oxo-5-thia-1-azabicyclo[4.2.0]oct-2-en-3-yl)methylthio]-4-methyl-	**81:** P63648k
3-([7-([α-amino(4-hydroxyphenyl)acetyl]amino)-2-carboxy-8-oxo-5-thia-1-azabicyclo[4.2.0]oct-2-en-3-yl]methylthio)-	**81:** P63648k
3-[(7-[(α-aminophenylacetyl)amino]-2-carboxy-8-oxo-5-thia-1-azabicyclo[4.2.0]oct-2-en-3-yl)methylthio]-	**81:** P63648k
3-[(7-[(α-aminophenylacetyl)amino]-2-carboxy-8-oxo-5-thia-1-azabicyclo[4.2.0]oct-2-en-3-yl)methylthio]-4-ethyl-	**81:** P63648k

TABLE 15 (*Continued*)

Compound	Reference

15.3. *S*-Substituted Thio- and Sulfonyl-1,2,4-Triazolin-5-ones (*Continued*)

Compound	Reference
3-[(7-[(α-aminophenylacetyl)amino]-2-carboxy-8-oxo-5-thia-1-azabicyclo[4.2.0]oct-2-en-3-yl)methylthio]-4-methyl-	**81:** P63648k
3-(benzylthio)-2-isopropyl-	**85:** P46688p
3-(benzylthio)-4-phenyl-	**65:** 13691f, **66:** 115651t
3-(4-bromobenzylthio)-2-methyl-	**82:** P156321j
3-(2-butenylthio)-2-isopropyl-	**84:** P4965a
3-(2-butenylthio)-2-phenyl-	**84:** P74272w
3-(2-butenylthio)-2-methyl-	**81:** P91536r, **84:** P4965a, **84:** P74273x
2-butyl-3-(methylthio)-	**82:** P156321j
3-[(2-carboxyethyl)thio]-	**75:** P56785a
3-[(2-carboxyethyl)thio]-4-methyl-	**74:** P53817f
3-([2-carboxy-7-([α-hydroxy(4-hydroxyphenyl)acetyl]-amino)8-oxo-5-thia-1-azabicyclo[4.2.0]oct-2-en-3-yl]methylthio)-	**81:** P63648k
3-[(2-carboxy-7-[(α-hydroxyphenylacetyl)amino]-8-oxo-5-thia-1-azabicyclo[4.2.0]-oct-2-en-3-yl)-methylthio]-	**81:** P63648k
3-[(2-carboxy-7-[(α-hydroxyphenylacetyl)amino]-8-oxo-5-thia-1-azabicyclo[4.2.0]oct-2-en-3-yl)-methylthio]-4-methyl-	**81:** P63648k
3-([2-carboxy-7-([(methylsulfonyl)acetyl]amino)-8-oxo-5-thia-1-azabicyclo[4.2.0]oct-2-en-3-yl]-methylthio)-4-methyl-	**80:** P48019s, **83:** P10106s
3-[(2-carboxy-8-oxo-7-[([(trifluoromethyl)thio]acetyl)-amino]-5-thia-1-azabicyclo[4.2.0]oct-2-en-3-yl)-methylthio]-	**80:** P120970q, **83:** 109119k
3-[(4-chloro-2-butenyl)thio]-2-methyl-	**81:** P91536r
3-([4-chlorophenyl)methyl]thio)-2-isopropyl-	**85:** P46688p
3-([2-chlorophenyl)methyl]thio)-2-methyl-	**81:** P91540n
3-([(4-chlorophenyl)methyl]thio)-2-methyl-	**81:** P91540n
3-[(4-chlorophenylthio]-2-isopropyl-	**81:** P91540n
2-(2-cyanoethyl)-3-[(2-cyanoethyl)thio]-	**84:** P85647y
2-(2-cyanoethyl)-3-[(2-cyanomethyl)thio]-	**84:** P85647y
2-(2-cyanoethyl)-3-(ethylthio)-	**83:** P206281q
2-(2-cyanoethyl)-3-([2-(ethylthio)ethyl]thio)-	**84:** P85647y
2-(2-cyanoethyl)-3-(methylthio)-	**83:** P206281q, **84:** P85647y
2-(2-cyanoethyl)-3-(2-propenylthio)-	**84:** P85647y
3-[(2-cyanoethyl)thio]-2-phenyl-	**83:** P206285u
3-[(cyanomethyl)thio]-2-ethyl-	**81:** P120638h
3-[(cyanomethyl)thio]-2-methyl-	**82:** P156321j
3-[(cyanomethyl)thio]-2-phenyl	**83:** P206285u
1-cyclohexyl-4-methyl-3-(methylthio)-	**73:** 35290d
2-cyclopentyl-3-(methylthio)-	**85:** P46688p
2-cyclopentyl-3-(2-propenylthio)-	**81:** P91536r
3-(2,4-dichlorobenzylthio)-2-methyl-	**82:** P156321j
3-([2,6-dichlorophenyl)methyl]thio)-2-methyl-	**81:** P91540n

TABLE 15 (*Continued*)

Compound	Reference

15.3. *S*-Substituted Thio- and Sulfonyl-1,2,4-Triazolin-5-ones (*Continued*)

Compound	Reference
3-([3,4-dichlorophenyl)methyl]thio)-2-methyl-	**81:** P91540n
3-[diethoxyphosphinylthio)methylthio]-1,4-dimethyl-	**80:** P48002f
1,4-dimethyl-3-(methylthio)-	**73:** 35290d
3-([2,6-dinitro-4-(trifluoromethyl)phenyl]thio)-2-methyl-	**81:** P91540n
1,3-diphenyl-4-*p*-tolylsulfonyl-	**53:** 8053b
3-(dodecylthio)-2-methyl-	**82:** P156321j
2-ethyl-3-(ethylthio)-	**82:** P156321j, **85:** P46688p
3-ethyl-2-(methylsulfonyl)-	**82:** P156321j
2-ethyl-3-(methylthio)-	**81:** P13523v, **81:** P120638h, **84:** P4965a **85:** P46688p
2-ethyl-3-(2-propenylthio)-	**81:** P120638h, **81:** P91536r, **84:** P4965a
3-([2-(ethylthio)ethyl]thio)-2-isopropyl-	**84:** P85645w
3-([2-(ethylthio)ethyl]thio)-2-phenyl-	**84:** P85645w
3-(ethylthio)-1-isopropyl-	**82:** P156321j
3-(ethylthio)-4-methyl-	**81:** P13523v
3-([(ethylthio)methyl]thio)-2-isopropyl-	**84:** P85645w
3-([(ethylthio)methyl]thio)-2-methyl-	**84:** P85645w
3-([(ethylthio)methyl]thio)-2-phenyl-	**84:** P85645w
3-(ethylthio)-2-phenyl-	**82:** P156321j
3-([[(3-fluorophenyl)methyl]thio)-2-methyl-	**81:** P91540n
3-([[(4-fluorophenyl)methyl]thio)-2-methyl-	**81:** P91540n
3-(hexythio)-2-methyl-	**82:** P156321j
2-isopropyl-3-(isopropylthio)-	**79:** P66362g
1-isopropyl-4-methyl-3-(methylthio)-	**73:** 35290d
2-isopropyl-3-[(methylsulfonyl)thio]-	**82:** P156321j
1-isopropyl-3-(methylthio)-	**85:** P46688p
2-isopropyl-3-(methylthio)-	**79:** P66362g, **81:** P13523v, **81:** P120638h, **84:** P4965a
4-isopropyl-5-(methylthio)-	**81:** P13523v
2-isopropyl-3-(phenylthio)-	**79:** P66362g
2-(isopropyl)-3-(2-propenylthio)-	**81:** P91536r, **84:** P74273x
3-(isopropylthio)-	**79:** P66362g
3-(isopropylthio)-2-methyl-	**82:** P16845h
3-[(1-methoxycarbonyl-2-oxo-1-propyl)thio]-	**83:** P58834e
2-methyl-3-([[(2-methylphenyl)methyl]thio)-	**81:** P91540n
2-methyl-3-([[(3-methylphenyl)methyl]thio)-	**81:** P91540n
2-methyl-3-[(2-methylphenyl)thio]-	**82:** P156321j
2-methyl-3-[(2-methyl-2-propenyl)thio]-	**81:** P91536r
2-methyl-3-[(2-methylpropyl)thio]-	**82:** P156321j
1-methyl-3-(methylthio)-	**81:** P13523v
1-methyl-5-(methylthio)-	**84:** P4965a, **84:** P74273x
2-methyl-3-(methylthio)-	**81:** P13523v, **81:** P120638h, **85:** P46688p
4-methyl-3-(methylthio)-	**44:** 6415h, **81:** P13523v
1-methyl-5-([2-(methylthio)ethyl]thio)-	**84:** P175191z
2-methyl-3-([2-(methylthio)ethyl]thio)-	**82:** P156321j
1-methyl-3-(methylthio)-4-phenyl-	**21:** 2900[5], **44:** 6416g, **80:** P48002f

TABLE 15 (*Continued*)

Compound	Reference
15.3. *S*-Substituted Thio- and Sulfonyl-1,2,4-Triazolin-5-ones (*Continued*)	
1-methyl-3-(methylthio)-4-*p*-tolyl-	**53:** 18014f
1-methyl-3-(methylthio)-4-xylyl-	**21:** 2900[6]
2-methyl-3-[(2-naphthalenylmethyl)thio]-	**81:** P91540n
2-methyl-3-([(2-nitrophenyl)methyl]thio)-	**81:** P91540n
2-methyl-3-([(4-nitrophenyl)methyl]thio)-	**81:** P91540n
4-methyl-3-[(3-oxobutyl)thio]-	**74:** P53817f
2-methyl-3-[(2-oxopropyl)thio]-	**84:** P175191z
3-[(1-methyl-2-oxopropyl)thio]-	**83:** P164189h
2-methyl-3-(pentylthio)-	**82:** P16845h, **82:** P156321j
2-methyl-3-[(1-phenylethyl)thio]-	**81:** P91540n
3-methyl-1-phenyl-4-*p*-tolylsulfonyl-	**53:** 8053a
2-methyl-3-(2-propenylthio)-	**81:** P91536r, **84:** P74273x
3-[(2-methyl-2-propenyl)thio]-2-phenyl-	**84:** P74272w
2-(1-methylpropyl)-3-(methylthio)-	**85:** P46688p
2-methyl-3-(propylthio)-	**82:** P16845h, **82:** P156321j
2-methyl-3-(2-propynylthio)-	**81:** P91535q, **81:** P91536r
3-(methylsulfinyl)-2-phenyl-	**79:** P66362g
3-(methylsulfonyl)-2-phenyl-	**79:** P66362g
3-(methylthio)-	**44:** 6415h, **81:** P13523v
3-(methylthio)-1-[5-(methylthio)-4-phenyl-4H̲-1,2,4-triazol-3-yl]-4-phenyl-	**64:** 8173g
3-([1-(methylthio)-2-oxopropyl]thio)-	**83:** P164189h
3-(methylthio)-1-phenyl-	**79:** P53334z
3-(methylthio)-2-phenyl-	**79:** P66362g
3-(methylthio)-4-phenyl-	**44:** 6416g, **65:** 13691f, **66:** 115651t, **78:** 58320f
3-(methylthio)-2-propyl-	**82:** P156321j
3-(methylthio)-4-*p*-tolyl-	**53:** 18014f
3-(3-nitrobenzylthio)-2-methyl-	**82:** P156321j
3-[(3-oxobutyl)thio]-	**74:** P53817f
2-phenyl-3-(2-propenylthio)-	**84:** P74272w
1-phenyl-4-*p*-tolylsulfonyl-	**53:** 8052i
4-*p*-tolylsulfonyl-	**53:** 8053b
15.4. Miscellaneous Δ^2(or Δ^1)-1,2,4-Triazolin-5-ones	
1-acetyl-3-amino-4-benzyl-	**63:** 5631g
1-acetyl-3-amino-4-(*p*-chlorophenyl)-	**63:** 5631g
1-acetyl-3-amino-4-(1-naphthyl)-	**63:** 5631g
1-acetyl-3-amino-4-*o*-tolyl-	**63:** 5631g
1-acetyl-3-ethyl-4-[(5-nitrofurfurylidene)amino]-	**65:** P12213a
1-acetyl-4-methyl-3-(*m*-ethylthio)-	**44:** 6416c
1-acetyl-3-methyl-4-[(5-nitrofurfurylidene)amino]-	**65:** P12213a
1-acetyl-3-(methylthio)-	**44:** 6416f
1-acetyl-4-[(5-nitrofurfurylidene)amino]-	**65:** P12213a
4-allyl-3-[(diethoxyphosphinothioyl)oxy]-1-methyl-	**75:** P20410k
1-(*N*-allylthiocarbamoyl)-3-methyl-	**25:** 3651[2]
3-amino-1-benzoyl-4-phenyl-	**63:** 5632c
3-(6-amino-2-methoxy-*m*-tolylazo)-	**54:** P7162c

TABLE 15 (*Continued*)

Compound	Reference

15.4. Miscellaneous Δ^2(or Δ^1)-1,2,4-Triazolin-5-ones (*Continued*)

Compound	Reference
4-amino-3-(octylthio)-	**57:** P14623c
1-benzyl-3-bromo-	**70:** 87692u
4-benzyl-3-bromo-	**70:** 87692u
1-benzyl-3-chloro-	**70:** 87692u
2-benzyl-3-chloro-	**79:** P66364j, **82:** P72994t
4-benzyl-3-chloro-	**70:** 87692u
3-bromo-	**67:** 82167e
3-bromo-1,4-dimethyl-	**70:** 87692u
3-bromo-2-isopropyl-	**79:** P66364j
3-bromo-1-methyl-	**70:** 87692u, **73:** 45419s
3-bromo-4-methyl-	**70:** 87692u
3-bromo-1-phenyl-	**70:** 87692u
3-bromo-4-phenyl-	**70:** 87692u
2-butyl-3-chloro-	**79:** P66364j, **82:** P72994t
4-butyl-3-[(diethoxyphosphinothioyl)oxy]-1-isopropyl-	**75:** P20410k
4-butyl-3-[(diethoxyphosphinothioyl)oxy]-1-methyl-	**75:** P20410k
1-butyryl-4-[(5-nitrofurfurylidene)amino]-	**65:** P12213a
-3-carbonyl azide, 1-(*p*-nitrophenyl)-	**40:** 3444²
-2-carbonyl chloride, 3-ethyl-	**82:** P57676x
-2-carbonyl chloride, 3-methyl-	**82:** P57676x
-2-carbonyl chloride, 3-phenyl-	**82:** P57676x
-1-carbothioamide, 3-(4-morpholinyl)-	**76:** P140832s
-3-carboxamide, *N*-benzyl-1-(2-methylphenyl)-	**76:** 59533y, **78:** 136152u
-3-carboxamide, *N*-benzyl-1-(4-methylphenyl)-	**78:** 136152u
-3-carboxamide, *N*-benzyl-1-phenyl-	**76:** 59533y, **78:** 136152u
-4-carboxamide, 1-(1,1-dimethylethyl)-3-methyl-*N*-phenyl	**84:** P59485z
-3-carboxamide, *N*,*N*-diethyl-2-phenyl-	**77:** P140089a
-3-carboxamide, *N*,*N*-dimethyl-2-phenyl-	**82:** P156321j
-3-carboxamide, *N*-methyl-2-phenyl-	**77:** P140089a
-1-carboxamide, 3-(4-methylphenyl)-	**85:** 143036g
-3-carboxamide, 1-(4-methylphenyl)-*N*-(2-methylpropyl)-	**78:** 136152u
-3-carboxamide, 1-(2-methylphenyl)-*N*-phenyl-	**75:** 98505q, **78:** 136152u
-3-carboxamide, 1-(4-methylphenyl)-*N*-phenyl-	**78:** 136152u
-1-carboxamide, 3-(4-morpholinyl)-*N*-phenyl-	**76:** 140832s
-3-carboxamide, *N*-1-naphthyl-1-phenyl-	**75:** 110244p
-3-carboxamide, 2-phenyl-	**77:** P140089a
-1-carboxamide, 3-(1H-pyrrol-2-yl)-	**85:** 143036g
-1-carboxamidine, 3-anilino-*N*,*N'*,4-triphenyl-	**69:** 10398z
-1-carboxamidine, 3-(*p*-bromoanilino)-4-(*p*-bromophenyl)-*N*,*N'*-di-*p*-tolyl-	**69:** 10398z
-1-carboxamidine, 3-(*p*-bromoanilino)-*N*,*N'*,4-tris(*p*-bromophenyl)-	**69:** 10398z
-1-carboxamidine, 3-(*p*-toluidino)-*N*,*N'*,4-tri-*p*-tolyl-	**69:** 10398z
-1-carboxanilide, 3-ethoxy-4'-methylthio-	**24:** 1114⁵
-3-carboxanilide, 1-phenyl-	**74:** 140379h, **75:** 98505q, **75:** 110244p, **78:** 136152u
-1-carboxanilide, 4-*p*-tolylsulfonyl-	**53:** 8053b

TABLE 15 (*Continued*)

Compound	Reference

15.4. Miscellaneous Δ²(or Δ¹)-1,2,4-Triazolin-5-ones (*Continued*)

Compound	Reference
-3-carboxylic acid	**47:** 6941h, **65:** 705c
	65: 7169g, **67:** 64307w
	83: 193189p
-3-carboxylic acid, 1-(*p*-aminophenyl)-	**40:** 3444[1]
-3-carboxylic acid, 1-(*p*-arsonophenyl)-	**40:** 3444[1]
-3-carboxylic acid, 2-(3-chlorophenyl)-, ethyl ester	**77:** P140089a
-3-carboxylic acid, ethyl ester	**82:** 57623c
-3-carboxylic acid, 2-isopropyl-, ethyl ester	**77:** P140089a
-3-carboxylic acid, 3-methoxy-4-phenyl-, methyl ester	**71:** 61474v
-3-carboxylic acid, methyl ester	**83:** 193189p
-1-carboxylic acid, 4-[(5-nitrofurfurylidene)amino]-, ethyl ester	**65:** P12213b
-3-carboxylic acid, 1-(*p*-nitrophenyl)-	**33:** 2518[9], **40:** 3443[8]
-3-carboxylic acid, 1-phenyl-	**75:** 98505q, **78:** 136152u
-1-carboxylic acid, 3-phenyl-, ethyl ester	**56:** 8708h
-3-carboxylic acid, 2-phenyl-, ethyl ester	**77:** P140089a
-3-carboxylic acid, 2-phenyl-, methyl ester	**77:** P140089a
3-[*N*-(carboxymethyl)carbamoyl]-1-(2-methylphenyl)-	**76:** 59533y
3-[*N*-(carboxymethyl)carbamoyl]-1-(4-methylphenyl)-	**76:** 59533y
3-[*N*-(carboxymethyl)carbamoyl]-1-phenyl-	**76:** 59533y
-3-carboxy-*o*-toluidide, 1-phenyl-	**75:** 98505q
3-chloro-	**62:** 14437d, **67:** 82167e
3-chloro-2-cyclopentyl-	**79:** P66364j, **82:** P72994t
3-chloro-2-(2,5-dichlorophenyl)-	**79:** P66364j
3-chloro-2-(3,5-dichlorophenyl)-	**82:** P72994t
3-chloro-1,4-dimethyl-	**70:** 87692u
3-chloro-2-ethyl	**79:** P66364j, **82:** P72994t,
	84: P180232t
4-(2-chloroethyl)-3-[(diethoxyphosphinothioyl)oxy]-1-methyl-	**75:** P20410k
4-(3-chloro-2-hydroxypropyl)-1-methyl-3-nitro-	**84:** P150634h
3-chloro-2-isopropyl-	**79:** P66364j, **82:** P72994t,
	84: P105603m, **84:** P105604n
3-chloro-1-methyl-	**70:** 87692u
3-chloro-2-methyl-	**84:** 105604n
3-chloro-4-methyl-	**70:** 87692u
3-chloro-2-(1-methylpropyl)-	**79:** P66364j, **82:** P72994t,
	84: 105604n
3-chloro-2-(2-methylpropyl)-	**79:** P66364j, **82:** P72994t
5-chloro-1-neohexyl-	**82:** P156321j
3-chloro-2-sec-pentyl-	**79:** P66364j
3-chloro-2-phenyl-	**79:** P66364j, **82:** P72994t,
	84: P180232t
3-chloro-2-(1-phenylethyl)-	**79:** P66364j, **82:** P72994t
3-chloro-1-propyl-	**84:** P180232t
3-chloro-2-propyl-	**79:** P66364j, **82:** P72994t
3-chloro-2-(1,2,2-trimethylpropyl)-	**79:** P66364j, **82:** P72994t,
	84: P180232t

TABLE 15 (*Continued*)

Compound	Reference

15.4. Miscellaneous Δ^2(or Δ^1)-1,2,4-Triazolin-5-ones (*Continued*)

Compound	Reference
1-crotonoyl-4-[(5-nitrofurfurylidene)amino]-	**65:** P12213d
1-(2-cyclohexen-1-yl)-4-phenyl-3-[(phenylcarbamoyl)oxy]-	**81:** 91438k
4-cyclohexyl-3-[(diethoxyphosphinothioyl)oxy]-1-methyl-	**75:** P20410k
1-cyclohexyl-3-methoxy-4-methyl-	**74:** 54291s
1-cyclohexyl-4-phenyl-3-[(phenylcarbamoyl)oxy]-	**81:** 91438k
1,4-diacetyl-3-(methylthio)-	**44:** 6416f
3-[(diethoxyphosphinothioyl)oxy]-4-sec-butyl-1-methyl-	**75:** P20410k
3-[(diethoxyphosphinothioyl)oxy]-4-tert-butyl-1-methyl-	**75:** P20410k
3-[(diethoxyphosphinothioyl)oxy]-1,4-diisopropyl-	**75:** P20410k
3-[(diethoxyphosphinothioyl)oxy]-1,4-dimethyl-	**75:** P20410k
3-[(diethoxyphosphinothioyl)oxy]-4-(ethoxycarbonylmethyl)-1-methyl-	**75:** P20410k
3-[(diethoxyphosphinothioyl)oxy]-4-ethyl-1-isopropyl-	**75:** P20410k
3-[(diethoxyphosphinothioyl)oxy]-4-ethyl-1-methyl-	**75:** P20410k
3-[(diethoxyphosphinothioyl)oxy]-1-isopropyl-4-methyl-	**75:** P20410k
3-[(diethoxyphosphinothioyl)oxy]-4-isopropyl-1-methyl-	**75:** P20410k
3-[(diethoxyphosphinothioyl)oxy]-4-methyl-	**75:** P20410k
3-[(diethoxyphosphinothioyl)oxy]-1-methyl-4-phenyl-	**75:** P20410k
3-[(diethoxyphosphinothioyl)oxy]-1-methyl-4-propyl-	**75:** P20410k
3-[(diethoxyphosphinothioyl)oxy]-4-phenyl-	**75:** P20410k
1,4-diethyl-3-methoxy-	**79:** 53234s
3-([3,4-dihydro-2,5,7,8-tetramethyl-2-(4,8,12-trimethyltridecyl)-2H-1-benzopyran-6-yl]oxy)-1-phenyl-,	**83:** P10526d
3-[(dimethoxyphosphinothioyl)oxy]-1,4-dimethyl-	**75:** P20410k
3-[(dimethoxyphosphinothioyl)oxy]-1-methyl-4-propyl-	**75:** P20410k
4-(dimethylamino)-1-methyl-3-(methylthio)-	**71:** 21509d
3-(dimethylamino)-1-phenylsulfonyl)-	**76:** P140832s
1,4-dimethyl-3-nitro-	**70:** 87692u
1,4-dimethyl-3-[(phenylcarbamoyl)oxy]-	**81:** 91438k
1,4-diphenyl-3-[(phenylcarbamoyl)oxo]-	**81:** 91438k
3,3′-diselenobis-	**80:** P89539e
3,3′-dithiobis-	**65:** 705d
3,3′-dithiobis[4-(m-chlorophenyl)-	**53:** 18014d
3,3′-dithiobis[4-cyclohexyl-	**53:** 18014d
3,3′-dithiobis[4-methyl-	**44:** 6416b
3,3′-dithiobis[4-phenyl-	**21:** 2900[5]
3,3′-dithiobis[4-m-tolyl-	**53:** 18014d
3,3′-dithiobis[4-p-tolyl-	**21:** 2900[6], **53:** 18014d
3,3′-dithiobis[4-(2,3-xylyl)-	**53:** 18014d
1-(N,N′-di-p-tolylamidino)-3-ethoxy-	**24:** 1114[5]
3-ethoxy-	**24:** 1114[3], **82:** 86060x
3-ethoxy-1,4-diethyl-	**79:** 53234s
3-ethoxy-2-ethyl-	**85:** P46688p
3-ethoxy-4-hydroxy-	**70:** 68266s
3-ethoxy-2-methyl-	**81:** P120638h, **84:** P4965a
3-ethoxy-1-phenyl-	**24:** 1114[4], **66:** 28303n, **77:** P62007f, **79:** P66363h

TABLE 15 (*Continued*)

Compound	Reference

15.4. Miscellaneous Δ^2(or Δ^1)-1,2,4-Triazolin-5-ones (*Continued*)

Compound	Reference
4-ethyl-3-nitro-	**70:** 115078u
3-ethyl-4-[(5-nitrofurfurylidene)amino]-1-propionyl	**65:** P12213a
3-fluoro-	**80:** 14898n
3-fluoro-2-isopropyl-	**79:** P66364j
1-hexanoyl-4-[(5-nitrofurfurylidene)amino]-	**65:** P12213a
4-hydroxy-	**70:** 68266s
4-(2-hydroxyethyl)-1-methyl-3-nitro-	**84:** P150634h
4-hydroxy-3-methoxy-	**70:** 68266s
4-[2-hydroxy-3-(nitrooxy)propyl]-1-methyl-3-nitro-	**84:** P150634h
4-(2-hydroxypropyl)-1-methyl-3-nitro-	**84:** P150634h
2-isopropyl-3-[(2-methyl-2-propenyl)thio]-	**84:** P4965a
3-methoxy-	**80:** 82826n
3-methoxy-1,4-dimethyl-	**42:** 8191b
3-methoxy-2-methyl-	**81:** P120638h, **84:** P4965a
3-methoxy-4-methyl-	**42:** 8190i
3-methoxy-4-methyl-1-phenyl-	**42:** 8190g
3-methoxy-4-methyl-1-β-D-ribofuranosyl-, 2′,3′,5′-tribenzoate	**73:** 66847v
3-methoxy-1-phenyl-	**70:** 87692u, **70:** 115078u, **72:** 111380e, **73:** 45419s, **84:** P150634h
4-methyl-3-nitro-	**70:** 87692u, **70:** 115078u
3-methyl-4-[(nitrofurfurylidene)amino]-1-propionyl	**65:** P12213a
1-methyl-3-phenoxy-	**79:** P66363h
1-methyl-4-phenyl-3-[(phenylcarbamoyl)oxy]-	**81:** 91438k
4-methyl-1-phenyl-3-[(phenylcarbamoyl)oxy]-	**81:** 91438k
3-[N-(2-methylpropyl)carbamoyl]-1-phenyl-	**76:** 59533y, **78:** 136152u
3-[N-(2-methylpropyl)carbamoyl]-1-o-tolyl-	**76:** 59533y, **78:** 136152u
3-nitro-	**62:** 14437d, **65:** 705b, **70:** 87692u
4-[(5-nitrofurfurylidene)amino]-1-propionyl-	**65:** P12213a
4-[(5-nitrofurfurylidene)amino]-1-valeryl-	**65:** P12213a
3-nitro-1-(p-nitrobenzyl)-	**70:** 87692u
3-nitro-4-(p-nitrobenzyl)-	**70:** 87692u
3-nitro-1-(p-nitrophenyl)-	**70:** 87692u
3-nitro-4-(p-nitrophenyl)-	**70:** 87692u
-3-one	**70:** P28922w, **79:** 115498s, **85:** 123874a
-3-one, 4-benzyl-	**84:** 16586r, **84:** 164686p
-3-one, 4-(benzylideneamino)-	**84:** 16586r
-3-one, 4-(4-carboxyphenyl)-	**82:** P87423e
-3-one, 4-(3,4-dichlorophenyl)-	**84:** 164686p
-3-one, 4-(1,1-dimethylethyl)-	**84:** 16586r
-3-one, 4-(ethyl-	**79:** 115498s, **82:** 43365h, **84:** 16586r
-3-one, 4,4′-(1,6-hexanediyl)bis-	**78:** 72719z
-3-one, 4-(4-methoxyphenyl)-	**84:** 16586r
-3-one, 4-methyl-	**76:** 59534z, **84:** 16586r

TABLE 15 (*Continued*)

Compound	Reference

15.4. Miscellaneous Δ^2(or Δ^1)-1,2,4-Triazolin-5-ones (*Continued*)

Compound	Reference
-3-one, 4,4'-(methylenedi-4,1-phenylene)bis-	**78:** 72719z, **80:** 60251u
-3-one, 4-(1-naphthalenyl)-	**82:** P172343y
-3-one, 4-(4-nitrophenyl)-	**76:** 59534z, **84:** 16586r
-3-one, 4-phenyl-	**58:** 2448f, **66:** 94960a, **67:** 108606z, **70:** 20009h, **76:** 59534z, **82:** 16729y, **82:** 16751z
3-(1-piperidinecarbonyl)-1-phenyl-	**75:** 110244p, **76:** 59533y, **78:** 136152u
3-(1-piperidinecarbonyl)-1-*o*-tolyl-	**76:** 59533y, **78:** 136152u
3-(1-piperidinecarbonyl)-1-*p*-tolyl-	**76:** 59533y, **78:** 136152u
-3-selenonic acid	**77:** P48469b
-3-sulfonyl chloride, 4-phenyl-	**45:** 4671b
-1-thiocarbamide, 3-ethoxy-, *N*-phenyl-	**24:** 1114[5]
-1-thiocarbamide, 3-ethoxy-, *N*-*p*-tolyl-	**24:** 1114[5]
-1-thiocarbamide, 3-methyl-,	**25:** 3651[2]
-1-thiocarbamide, 3-methyl-, *N*-phenyl-	**25:** 3651[2]
-1-thiocarbamide, 3-methyl-, *N*-*p*-tolyl-	**25:** 3651[2]

CHAPTER 16

1,2,4-Triazoline-5-thiones Containing More Than One Representative Function

16.1. Amino- and Diamino-Δ^2(or Δ^3)-1,2,4-Triazoline-5-thiones

The successive treatment of *N*-aminoguanidine with acetone and phenyl isothiocyanate gave **16.1-1**, which in refluxing hydrochloric acid underwent hydrolysis and cyalization with elimination of aniline to give 3-amino-1,2,4-triazoline-5-thione (**16.1-2**) (75 to 80%).[1] The cyclization of the aminoguanidine derivative (**16.1-3**) in base was also reported to give a high yield of **16.1-2**.[2,3] In contrast, under alkaline conditions **16.1-1** underwent

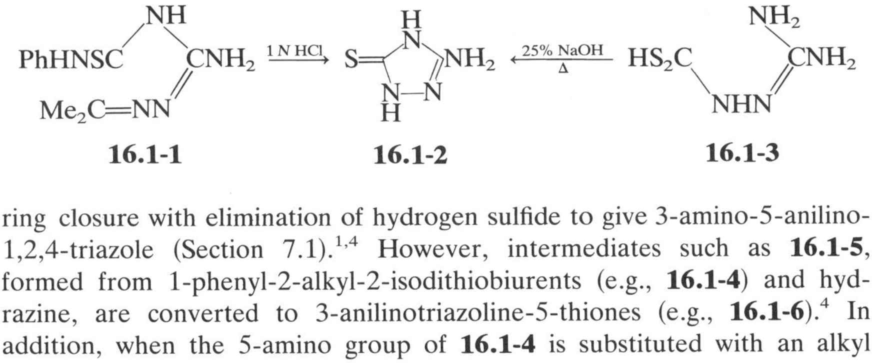

| **16.1-1** | **16.1-2** | **16.1-3** |

ring closure with elimination of hydrogen sulfide to give 3-amino-5-anilino-1,2,4-triazole (Section 7.1).[1,4] However, intermediates such as **16.1-5**, formed from 1-phenyl-2-alkyl-2-isodithiobiurents (e.g., **16.1-4**) and hydrazine, are converted to 3-anilinotriazoline-5-thiones (e.g., **16.1-6**).[4] In addition, when the 5-amino group of **16.1-4** is substituted with an alkyl group, 3-anilinotriazoline-5-thiones are formed in yields up to 90%, but when the 5-amino group of **16.1-4** is substituted with an aryl group, 3,5-dianilinotriazoles are formed as the major product.[5]

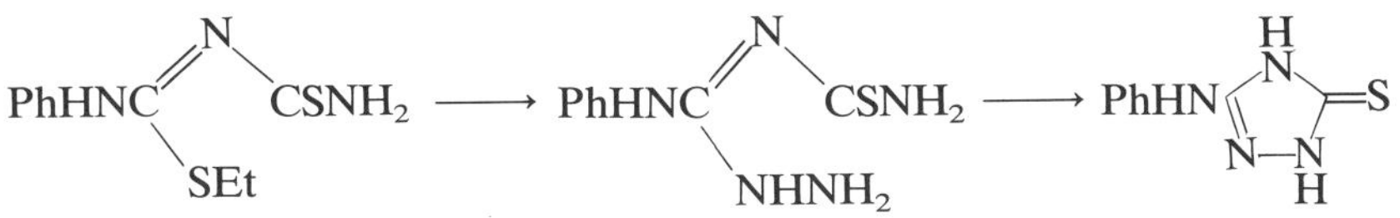

| **16.1-4** | **16.1-5** | **16.1-6** |

475

Reaction of N-aminoguanidine with both alkyl and aryl isothiocyanates in N,N-dimethylformamide gave 4-alkyl- and 4-aryl-1-amidino-3-thiosemicarbazides (e.g., **16.1-7** and **16.1-8**), which were readily isolated as toluenesulfonates. When heated in aqueous sodium hydroxide, these salts were converted with the loss of ammonia to give good yields of 4-substituted 3-aminotriazoline-5-thiones (e.g., **16.1-9** and **16.1-10**).[6,7] In an acidic medium these salts are converted to 1,3,4-thiadiazoles. Another method for the preparation of 3-aminotriazoline-5-thiones is illustrated by the cyclization of the N-(thiocarbamoyl)thiocarbohydrazide (**16.1-12**) in base to give the 3-(2-arylhydrazino)triazoline-5-thione (**16.1-13**). Steam distillation of the reaction mixture containing **16.1-13** resulted in cleavage of the 3-hydrazino group to give **16.1-11**.[8] The interaction of thiosemicarbazide with

16.1-7, $R_1 = CHMe_2$ **16.1-9**, $R_1 = Me_2CH$ (75%)
16.1-8, $R_1 = C_6H_4$-4-CF_3 **16.1-10**, $R_1 = 4$-$CF_3C_6H_4$ (70%)
                              16.1-11, $R_1 = 4$-MeC_6H_4 (78%)

16.1-12   **16.1-13**

diphenylcarbodiimide in hot N,N-dimethylformamide gave mainly the 1-(amidino)-3-thiosemicarbazide (**16.1-14**) (60 to 72%), and small amounts of the triazole (**16.1-16**) (5 to 7%) and the triazoline-5-thione (**16.1-6**) (6 to 9%).[9] The triazole (**16.1-16**) was postulated to form via the diaddition product, **16.1-15**, which cyclized with loss of aniline and the thiocarbamoyl group (Section 7.1). The cyclization of **16.1-14** *in situ* with the elimination of aniline gave the triazoline-5-thione (**16.1-6**), which was also obtained in 72% yield by pyrolysis of **16.1-14** at 200°. Of interest was the cyclization of **16.1-14** under acidic conditions to give the 1,3,4-thiadiazole (**16.1-18**) (85%) and under basic conditions to give a mixture of **16.1-19** (28%) and **16.1-17** (24%). In the reaction of 4-phenyl-3-thiosemicarbazide with diphenylcarbodiimide, the resulting intermediate corresponding to **16.1-14** underwent cyclization *in situ* to give **16.1-19** (52%) and **16.1-16** (15%).

$$\text{SC}\begin{array}{c}\text{NH}_2\\\\\text{NHNHC(NPh)NHPh}\end{array}\quad\longrightarrow\quad\left[\begin{array}{cc}\text{NPh}&\text{NPh}\\\|&\|\\\text{PhNHC}&\text{CNHPh}\\&\\\multicolumn{2}{c}{\text{N(CSNH}_2\text{)NH}}\end{array}\right]\quad\longrightarrow\quad \text{PhNH}\overset{\text{Ph}}{\underset{\text{N--N}}{\diagup\!\!\diagdown}}\text{NHR}_1$$

16.1-14   **16.1-15**

16.1-16, $R_1 = $ Ph
16.1-17, $R_1 = $ H

$-$PhNH$_2$ ↓

$-$PhNH$_2$ H$^+$

$$\text{S}=\overset{\text{H}}{\underset{\underset{\text{H}}{\text{N--N}}}{\diagup\!\!\diagdown}}\text{NHPh}$$

16.1-6

$$\text{H}_2\text{N}\overset{\text{S}}{\underset{\text{N--N}}{\diagup\!\!\diagdown}}\text{NHPh}$$

16.1-18

$$\text{S}=\overset{\text{Ph}}{\underset{\underset{\text{H}}{\text{N--N}}}{\diagup\!\!\diagdown}}\text{NHPh}$$

16.1-19

Also, **16.1-19** (46%) and a small amount of **16.1-21** (18%) were obtained from the reaction of N,N'-diamino-N''-phenylguanidine at 100° with phenyl isothiocyanate in N,N-dimethylformamide.[10] The formation of **16.1-19** resulted from the elimination of hydrazine from intermediate **16.1-20**, and **16.1-21** from the elimination of aniline. Compound **16.1-21** (11%) was also

$$(\text{H}_2\text{NNH})_2\text{C}{=}\text{NPh}\xrightarrow{\text{PhNCS}}\left[\text{SC}\begin{array}{c}\text{NHPh}\\\\\text{NHN}{=}\text{C(NHPh)NHNH}_2\end{array}\right]$$

16.1-20

$$\text{S}=\overset{\text{Ph}}{\underset{\underset{\text{H}}{\text{N--N}}}{\diagup\!\!\diagdown}}\text{NHPh}\quad+\quad \text{S}=\overset{\text{Ph}}{\underset{\underset{\text{H}}{\text{N--N}}}{\diagup\!\!\diagdown}}\text{NHNH}_2$$

16.1-19   **16.1-21**

formed by treatment of N,N'-diamino-N''-phenylguanidine with carbon disulfide, a reaction in which the product is formed by elimination of hydrogen sulfide in the cyclization of the intermediate adduct.[11] Higher yields (78 to 93%) of 4-aryl-3-anilinotriazoline-5-thiones (e.g., **16.1-23**) were obtained by the cyclization of 1-(N-arylthiocarbamoyl)-4-aryl-3-thiosemicarbazides (e.g., **16.1-22**) in refluxing aqueous sodium hydroxide.[12]

$$\text{SC}\begin{array}{c}\text{NHC}_6\text{H}_4\text{-4-Cl}\\\\\text{NHNHCSNHC}_6\text{H}_4\text{-4-Cl}\end{array}\quad\xrightarrow[\Delta]{2\text{ N NaOH}}\quad \text{S}=\overset{\text{4-ClC}_6\text{H}_4}{\underset{\underset{\text{H}}{\text{N--N}}}{\diagup\!\!\diagdown}}\text{NHC}_6\text{H}_4\text{-4-Cl}$$

16.1-22   **16.1-23**

When the aryl groups of **16.1-22** are different, two isomeric triazoline-5-thiones are obtained.[13]

Another approach was investigated for the preparation of 3-dialkylamino-4-phenyltriazoline-5-thiones. Treatment of *S*-methyl pseudothioureas with methyl dithiocarbazate and cyclization of the resulting intermediate (e.g., **16.1-24**) in refluxing base gave **16.1-25**.[14] An interesting rearrangement was

16.1-24   **16.1-25**

observed when the oxazole (**16.1-26**) was refluxed in ethanol to give a 95% yield of the 3-(phenacylamino)triazoline-5-thione (**16.1-27**).[15]

16.1-26   **16.1-27**

The condensation of thiosemicarbazides (e.g., **16.1-28**) with diethoxymethylenimine (**16.1-29**) under neutral conditions gave mainly 3-aminotriazoline-5-thiones (e.g., **16.1-30**).[16] In contrast, the hydrochloride of **16.1-28** condensed with **16.1-29** to give mainly 3-ethoxytriazoline-5-thione (**16.1-31**). With other thiosemicarbazide hydrochlorides such as the 4-methyl derivative, 1,3,4-thiadiazoles were formed.

16.1-28   **16.1-29**   **16.1-30**   **16.1-31**

Several methods have been investigated for the conversion of 1,3,4-thiadiazoles to triazoline-5-thiones. For example, 2-amino-5-anilino-1,3,4-thiadiazole (**16.1-32**) and related compounds of this type are rearranged readily in ethanolic potassium hydroxide to give the corresponding 3-amino-1,2,4-triazoline-5-thione (**16.1-30**).[17] However, if the amino groups of the 1,3,4-thiadiazole are substituted with different aryl groups, mixtures of isomeric triazoline-5-thiones are formed. In the reaction of the 1,3,4-thiadiazole (**16.1-33**) with aniline, a mixture of the 3-aminotriazoline-5-thione (**16.1-30**) and the 1,3,4-thiadiazole (**16.1-32**) are obtained.[18]

$$H_2N\text{—}\underset{N\text{—}N}{\overset{S}{\diagdown}}\text{—}NHPh \xrightarrow[\text{EtOH}]{5\% \text{ KOH}} H_2N\text{—}\underset{\underset{H}{N\text{—}N}}{\overset{\overset{Ph}{N}}{\diagdown}}\text{=}S \quad H_2N\text{—}\underset{\underset{H}{N\text{—}N}}{\overset{S}{\diagdown}}\text{=}S$$

16.1-32 **16.1-30** **16.1-33**

Some unusual reactants have also been used for the preparation of certain 3-amino-1,2,4-triazoline-5-thiones. These include the reaction of the triazolium iodide (**16.1-34**) with sulfur in pyidine to give **16.1-35**,[19] the

$$PhMeN\text{—}\underset{\underset{Ph}{N\text{—}N}}{\overset{\overset{Ph}{N}}{\diagdown}}\overset{+}{\diagdown}I^- \xrightarrow[C_5H_5N]{S} PhMeN\text{—}\underset{\underset{Ph}{N\text{—}N}}{\overset{\overset{Ph}{N}}{\diagdown}}\text{=}S$$

16.1-34 **16.1-35**

interaction of isoperthiocyanic acid (**16.1-36**) with 1,2-dimethylhydrazine to give **16.1-37**,[20] and the condensation of phenyl hydrazine with cyanogen

$$H_2N\text{—}\underset{S\text{—}S}{\overset{N}{\diagdown}}\text{=}S \xrightarrow{MeNHNHMe} H_2N\text{—}\underset{\underset{Me \ Me}{N\text{—}N}}{\overset{N}{\diagdown}}\text{=}S$$

16.1-36 **16.1-37**

isothiocyanate (**16.1-38**) to give **16.1-39**.[21] Also, the 3-chlorotriazoline-5-thione (**16.1-40**) has been reported to react with methylamine to give **16.1-41** in 64% yield.[22]

$$NCNCS \xrightarrow{PhNHNH_2} H_2N\text{—}\underset{\underset{Ph \ H}{N\text{—}N}}{\overset{N}{\diagdown}}\text{=}S \quad Cl\text{—}\underset{\underset{Ph}{N\text{—}N}}{\overset{\overset{Me}{N}}{\diagdown}}\text{=}S \xrightarrow{MeNH_2} MeNH\text{—}\underset{\underset{Ph}{N\text{—}N}}{\overset{\overset{Me}{N}}{\diagdown}}\text{=}S$$

16.1-38

16.1-39 **16.1-40** **16.1-41**

In work directed toward the preparation of 3-hydrazinotriazoline-5-thione (**16.1-46**), the cyclization of the N,N'-disubstituted diaminoguanidine (**16.1-42**) in refluxing water gave a 70% yield of **16.1-43** and a 10% yield of **16.1-44**.[23] Compound **16.1-43** resulted from the loss of ammonia from **16.1-42** and **16.1-44** from the loss of water. In an alkaline medium **16.1-42** gave a 51% yield of **16.1-44** and a 10% yield of **16.1-43**.[24] The benzoyl group of **16.1-43** was readily hydrolyzed in dilute hydrochloric acid to give **16.1-46**. Treatment of the N,N',N''-trisubstituted diaminoguanidine (**16.1-45**) with hydrochloric acid resulted in the loss of aniline to give a 65%

yield of **16.1-46** and in the loss of hydrogen sulfide to give a 10% yield of
16.1-47, both isolated by conversion to their isopropylidene derivatives.[25] A
higher yield of **16.1-47** (84%) was obtained on refluxing **16.1-45** in aqueous
sodium hydroxide (Section 7.1).

Direct treatment of N,N'-diaminoguanidine in N,N'-dimethylformamide
with an equimolar amount of phenyl isothiocyanate gave 3-hydrazino-4-
phenyltriazoline-5-thione (**16.1-21**) (35 to 40%), 3-amino-4-phenyl-
triazoline-5-thione (**16.1-30**) ($\sim$ 12%), and a small amount of **16.1-49**.[26]
Products **16.1-21** and **16.1-30** resulted from elimination of ammonia and
hydrazine from intermediate **16.1-48**, respectively, whereas **16.1-49** resulted
from reaction of **16.1-21** with phenyl isothiocyanate during the reaction.
Compound **16.1-49** was the major product when excess phenyl isothiocanate
was reacted with N,N'-diaminoguanidine.

The preparation of 4-aminotriazoline-5-thione (**16.1-51**) was accomplished by the cyclization of thiocarbohydrazide (**16.1-50**) with formic acid (79% yield), formamide (74% yield), and ethyl orthoformate (54% yield).[27] Also, reaction of 2-chloro-1,3,4-thiadiazole (**16.1-53**) with methanolic hydrazine hydrate resulted in displacement of the chloro group followed by rearrangement to give **16.1-51** in 78% yield.[28] In contrast, treatment of the

16.1-50   **16.1-51**, $R_1 = H$ **16.1-53**
                     16.1-52, $R_1 = Me$

monobenzylidene derivative of thiocarbohydrazide (**16.1-54**) with ethyl orthoformate gave only a 9% yield of **16.1-55** and a 47% yield of the 1,3,4-thiadiazole **16.1-57**.[29] The formation of **16.1-57** as the major product in this reaction was attributed to steric hindrance and the contribution of the conjugated ionic form of **16.1-54**. However, reaction of **16.1-54** with ethyl orthoacetate gave an 8% yield of the thiadiazole (**16.1-58**) and a 55% yield of **16.1-56**. The inconsistent results of the ethyl orthoformate and orthoacetate reactions are probably because the rearrangement of **16.1-58** to **16.1-56** occurred more readily during the reaction than the rearrangement of **16.1-57** to **16.1-55**.

16.1-54

16.1-55, $R_1 = H$ **16.1-57**, $R_1 = H$
16.1-56, $R_1 = Me$ **16.1-58**, $R_1 = Me$

The condensation of thiocarbohydrazide (**16.1-50**) with aliphatic carboxylic acids is also the method of choice for the preparation of 3-alkyl-4-aminotriazoline-5-thiones, as demonstrated by the reaction of **16.1-50** with acetic acid to give **16.1-52** in 75% yield.[27,30,31] In addition, reaction of

16.1-59 with formic and propionic acids gave **16.1-60** and **16.1-61**, respectively.[32]

16.1-59

16.1-60, $R_1 = H$
16.1-61, $R_1 = Et$

Hydrazinolysis of 2-aroyl- and 2-heteroaroyldithiocarbazates (e.g., **16.1-62**) is an important method for the preparation of 3-aryl-[30,33,34] and 3-heteroaryl-[35-37] 4-aminotriazoline-5-thiones (e.g., **16.1-64**) in good yields. In this reaction, the intermediate N-(benzoyl)thiocarbohydrazide (**16.1-63**) was dehydrated readily.[38] Treatment of the related intermediate (**16.1-65**)

16.1-62   **16.1-63**   **16.1-64**

with hydrazine was reported to give a 1,3,4-thiadiazole, but the product of this reaction was later identified as the 3-hydrazinotriazoline-5-thione (**16.1-21**).[39] Presumably the N-(thioisonicotinoyl) intermediate (**16.1-68**)

16.1-65   **16.1-66**   **16.1-21**

was formed in the reaction of **16.1-67** with thiocarbohydrazide (**16.1-50**) in aqueous base to give **16.1-70**.[40] However, fusion of **16.1-69** was reported to give mainly the 1,3,4-thiadiazoles (**16.1-71** and **16.1-72**) and a low yield of the triazoline-5-thione (**16.1-64**).[41]

Other methods for the preparation of 4-aminotriazoline-5-thiones include the interaction of the 1,3,4-oxadiazole (**16.1-73**) with hydrazine to give

16.1-64,[42] and the treatment of the triazolium chloride (**16.1-74**) with sulfur in the presence of an organic base to give **16.1-75**.[43,44] The oxidative

removal of the hydrazino group of **16.1-76** by oxygen in a basic medium gave a 78% yield of **16.1-51**.[45] Similarly, the condensed system (**16.1-78**) was converted to **16.1-77** in a hot alkaline medium.

Preformed 3-aminotriazoline-5-thiones have been acylated on the amino group,[46] alkylated at the sulfur group with alkyl halides,[47] and the sulfur group removed with bromine.[13] As expected, the 4-amino group of 4-aminotriazoline-5-thiones forms Schiff bases with aldehydes.[48] In the acetylation of **16.1-79** with acetic anhydride, a 46% yield of the diacetylamino derivative (**16.1-80**) was obtained.[31]

16.1-79   **16.1-80**

Although many of the reactions leading to 3,4-diaminotriazoline-5-thiones provided complex mixtures, the desired products can usually be isolated in moderate yields. Treatment of thiocarbohydrazide (**16.1-50**) with two molar proportions of diphenylcarbodiimide in N,N-dimethylformamide was postulated to proceed via addition of carbodiimide at the terminal 1- and 5-amino groups of **16.1-50** to give **16.1-81**, which was converted by loss of aniline to give the triazoline-5-thione (**16.1-82**).[11] When this reaction was carried out in methanol, the yield of **16.1-82** (20%) was lowered, and the yields of the by-products [**16.1-19** (12%) and **16.1-16** (15%)] were increased. A mechanism for the formation of these three products has been

16.1-50   **16.1-81**

16.1-19   **16.1-16**   **16.1-82**

described.[11] In a related reaction treatment of 1-phenylthiocarbohydrazide (**16.1-83**) with diphenylcarbodiimide in N,N-dimethylformamide lead to the isolation of the monoadduct (**16.1-84**).[49] This amidino derivative was cyclized in aqueous base to give the triazoline-5-thione (**16.1-85**). In contrast, treatment of **16.1-84** with ethanolic base resulted in dehydrogenation and cyclization to give the 1,3,4-thiadiazole (**16.1-86**).

The reaction of N,N'-bis(thiocarbamoyl)hydrazine (**16.1-87**) with hydrazine hydrate in refluxing water gave a 29% yield of 3,4-diaminotriazoline-5-thione (**16.1-89**).[50,51] In this reaction three additional products were isolated and identified as **16.1-46** (13%), **16.1-91** (9%), and **16.1-76** ($\sim$ 1%).[50] Presumably these products are formed from intermediates **16.1-90** and **16.1-88**, as shown in the following. The condensation of **16.1-89** with benzaldehyde gave a monobenzylidene derivative, presumably **16.1-91**,[50] and reaction of **16.1-89** with acetic anhydride gave the 3,4-bis(acetamido) derivative (**16.1-92**).[52]

Treatment of isobutyl isothiocyanate with methyl dithiocarbazate gave **16.1-93**, which reacted with ethanolic hydrazine to give **16.1-95**, presumably formed via **16.1-94**.[53] A compound (**16.1-96**) similar to this intermediate

was cyclized in hot pyridine to give the 1,3,4-thiadiazole (**16.1-100**) (15%), the triazolidine (**16.1-99**) (44%), and the two triazoline-5-thiones, **16.1-97** (7%) and **16.1-98** (32%).[54] Condensation of **16.1-98** with benzaldehyde gave the corresponding 4-benzylidene derivative, and oxidative deamination of **16.1-98** with nitrous acid gave the disulfide (**16.1-101**). The cyclization of **16.1-96** in aqueous base gave 4-phenyltriazolidine-3,5-dithione (see Section 20.2).

The interaction of thiocarbohydrazide (**16.1-50**) with hydrazine gave an 87% yield of 4-amino-3-hydrazinotriazoline-5-thione (**16.1-76**).[28] This compound was also obtained by hydrazinolysis of thiourea,[55] ethyl xanthate,[56] alkyl dithiocarbamates (e.g., **16.1-102**),[57] several pyrimidine-2-thiones (e.g., **16.1-103**),[58-61] and 1,3,4-thiadiazoles (e.g., **16.1-104**).[28] In addition, the fusion (160°) of **16.1-105**[29] and the treatment of $N,N''N'''$-triaminoguanidine with carbon disulfide in an alkaline medium[62] gave low yields of **16.1-76**. The condensation of **16.1-76** with benzaldehyde readily

16.1-103   **16.1-104**

MeS$_2$CNHNHCSNHNH$_2$
16.1-105

gave the bis(benzylidene) derivative (**16.1-106**).[28] Either of the monobenzylidene derivatives can be prepared, the 4-(benzylideneamino) compound (**16.1-91**) in hydrochloric acid, and the 3-(benzylidenehydrazino) compound (**16.1-107**) in aqueous sodium hydroxide. The latter is reversiblely cyclized to the condensed tetrazine (**16.1-108**), which in the presence of oxygen is oxidized to the purple-colored triazolotetrazine (**16.1-78**).[45] The formation of colored triazolotetrazines from **16.1-76** and both aliphatic and aryl aldehydes in an alkaline medium is the basis of a sensitive test for aldehydes.[55]

16.1-106, R$_1$ = R$_2$ = N═CHPh
16.1-91, R$_1$ = NH$_2$, R$_2$ = N═CHPh
16.1-107, R$_1$ = N═CHPh, R$_2$ = NH$_2$

16.1-108   **16.1-78**

References

1. **55:** 2623g	2. **50:** 13097d	3. **73:** P45513t	4. **68:** 12814t
5. **78:** 111227j	6. **56:** 12877h	7. **60:** 1754b	8. **67:** 90736e
9. **57:** 12472h	10. **66:** 115650s	11. **69:** 77170n	12. **59:** 13985f
13. **80:** 14879g	14. **59:** 603a	15. **64:** 19594g	16. **68:** 29639e
17. **70:** 3963x	18. **73:** 35283d	19. **74:** 53659f	20. **83:** 9914x
21. **83:** 28157u	22. **73:** 35290d	23. **66:** 104959y	24. **60:** 1736d
25. **63:** 11545c	26. **64:** 8174a	27. **57:** 804i	28. **58:** 4543d
29. **56:** 5948g	30. **66:** 2516c	31. **74:** 141645d	32. **55:** 23507d
33. **51:** 3579b	34. **55:** 7399c	35. **68:** 105174s	36. **70:** 11610v
37. **78:** 136235y	38. **71:** 22075w	39. **69:** 77151g	40. **64:** 5072h
41. **60:** 526a	42. **76:** 140741m	43. **76:** 85755s	44. **78:** 124510w
45. **83:** 97223z	46. **70:** 11640e	47. **84:** 30975y	48. **69:** P59249r
49. **72:** 55347k	50. **48:** 5181i	51. **61:** 10676a	52. **63:** 14850b

53. **65:** 5455b   54. **64:** 8173b   55. **81:** 49628x   56. **53:** 2126e
57. **51:** 3481e   58. **83:** 58739c   59. **84:** 17271w   60. **84:** 59374n
61. **84:** 180171x   62. **68:** 39547n

16.2. Miscellaneous Δ^2(or Δ^3)-1,2,4-Triazoline-5-thiones

The thermal dehydration of the oxamoyl derivative of 4-methylthiosemi-carbazide (**16.2-1**) gave a good yield of the carboxamide of the triazoline-5-thione (**16.2-4**), which was hydrolyzed in alcoholic potassium hydroxide to give **16.2-8**.[1] In the cyclization of **16.2-2** with sodium ethoxide, a mixture of **16.2-6** and **16.2-7** were produced. Treatment of the mixture with aqueous sodium hydroxide converted both components to **16.2-8**. In addition, the hydrazide (**16.2-3**) was cyclized to **16.2-5** under alkaline conditions.[2]

Carboxy derivatives of triazoline-5-thiones are also prepared by the rearrangement of 2-thiazolin-5-ones. For example, treatment of either **16.2-10** or **16.2-11** with piperidine gave **16.2-9**.[3-5] In these types of reactions, the structure of the product depends upon the nature of the alkoxy group in the thiazoline and the basicity of the amine as illustrated by the amination of the 2-ethoxy-2-thiazolin-5-one (**16.2-12**) with piperidine to give the triazolin-5-one (**16.2-13**).[4] Another method for the preparation of 3-carboxamide derivatives involved the displacement of chloride from **16.2-14** with sodium hydrosulfide to give **16.2-15**.[6]

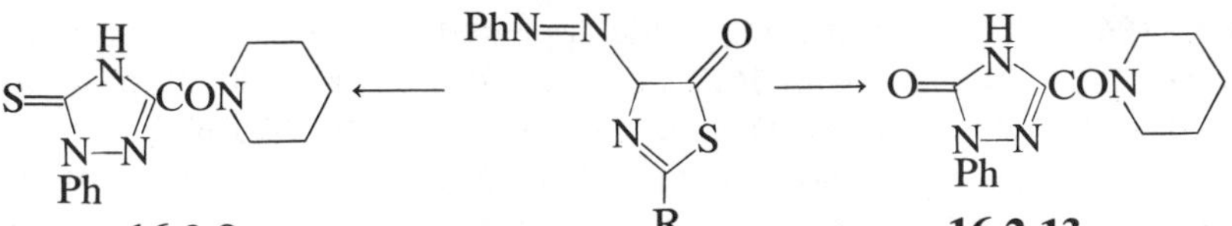

16.2-9

16.2-10, $R_1 = PhCH_2S$
16.2-11, $R_1 = PhCH_2O$
16.2-12, $R_1 = EtO$

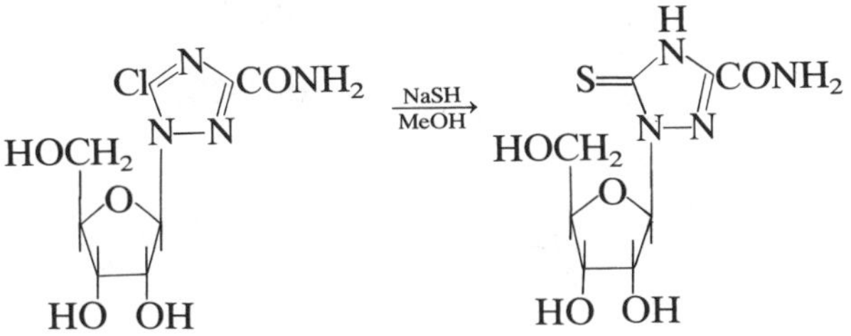

16.2-13

16.2-14   **16.2-15**

N-Carboxy derivatives of triazoline-5-thiones (e.g., **16.2-18**) are obtained directly by the condensation of amidrazones (e.g., **16.2-16**) with isothio-cyanates (e.g., **16.2-17**).[7] It has been demonstrated that *N*- rather than *S*-alkylation of **16.2-18** can be effect with certain reagents (e.g., ethyl bromoacetate) to give 1-substituted triazoline-5-thiones (e.g., **16.2-19**).[8,9]

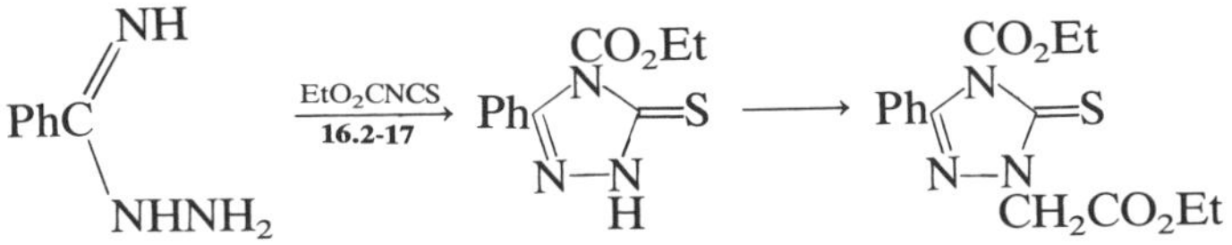

16.2-16   **16.2-18**   **16.2-19**

Also, the 3,4-dialkyl- and 3,4-diaryl-triazoline-5-thiones (e.g., **16.2-20**) are acylated with benzoyl chloride at reflux in benzene to give 1-benzoyltriazoline-5-thiones (e.g., **16.2-21**).[10-12] However, at room tempera-ture *S*-acyl derivatives are formed.[11] Acylation of the 4-aminotriazoline-5-

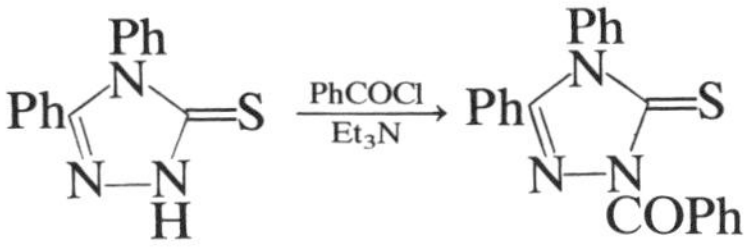

16.2-20   **16.2-21**

thione (**16.2-22**) with acetic anhydride appeared to give the triacetyl derivative (**16.2-23**), isomeric with **16.2-24** obtained from the reaction of thiocarbohydrazide with acetic anhydride.[13] Treatment of **16.2-23** with hot water
gave the 4-(diacetylamino) compound **16.2-25**, and hydrolysis of both
16.2-24 and **16.2-25** with hydrochloric acid gave **16.2-22**. Derivatives of 3-
formyltriazoline-5-thione have been prepared from the corresponding

16.2-22   **16.2-23**

16.2-24, R₁ = Ac
16.2-25, R₁ = H

acetal. Treatment of an acidic ethanolic solution of **16.2-26** with 2,4-
dinitrophenylhydrazine resulted in the formation of **16.2-27**.[14]

16.2-26   **16.2-27**

The direct preparation of 3-alkoxytriazoline-5-thiones was accomplished
by the base-catalyzed cyclization of **16.2-28** to give **16.2-29**.[15] The same
product was also obtained by the condensation of 4-methylthio-
semicarbazide (**16.2-30**) at 160° with tetraethyl orthocarbonate.[16] The 3-

16.2-28   **16.2-29**   **16.2-30**

(alkylthio)triazoline-5-thiones such as **16.2-32** have been prepared by treatment of triazolium iodides (e.g., **16.2-31**) with sulfur in refluxing pyridine.[17]

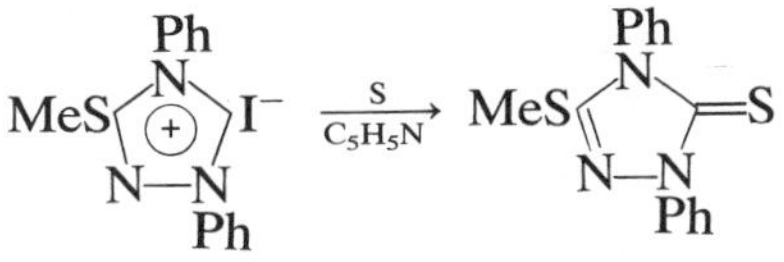

16.2-31   **16.2-32**

Also the monoaddition of acrylic acid to triazolidine-3,5-dithione (**16.2-33**) was reported to give a high yield of **16.2-34**.[18] The cycloaddition of

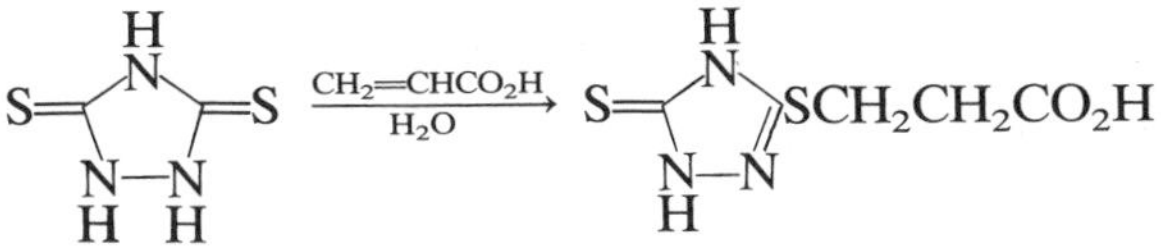

16.2-33   **16.2-34**

dimethylamino isothiocyanate (**16.2-35**) with phenyl isothiocyanate resulted in the formation of **16.2-37**, which was formed from intermediate **16.2-36** by an N to S migration of a methyl group.[19]

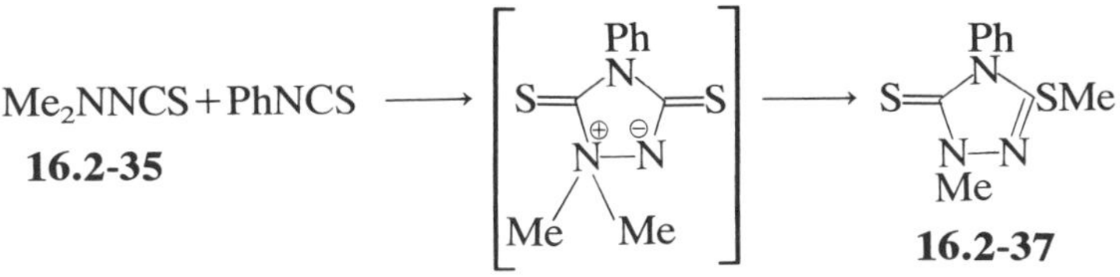

Me$_2$NNCS + PhNCS

16.2-35

16.2-37

16.2-36

The condensation of thiocarbohydrazide (**16.2-38**) with dimethyl dithioiminocarbonate in water at room temperature gave a 61% yield of 4-amino-3-(methylthio)-Δ^2-1,2,4-triazoline-5-thione (**16.2-39**).[20] The preparation of **16.2-39** by the methylation of the monopyridinium salt of **16.2-40** gave a sample that melted at a lower temperature, suggesting that this sample was contaminated with the dimethylated product.[21] Reaction of **16.2-39** with benzaldehyde gave the corresponding 4-benzylidene derivative (**16.2-41**), and deamination of **16.2-39** with nitrous acid gave a mixture of **16.2-42** (23%) and **16.2-43** (31%).[20] The pyrolysis of the carbazoylimidazole (**16.2-44**) gave dimethylamino isothiocyanate (**16.2-35**), which was converted to **16.2-46** via intermediate **16.2-45**.[22] Another type of compound in

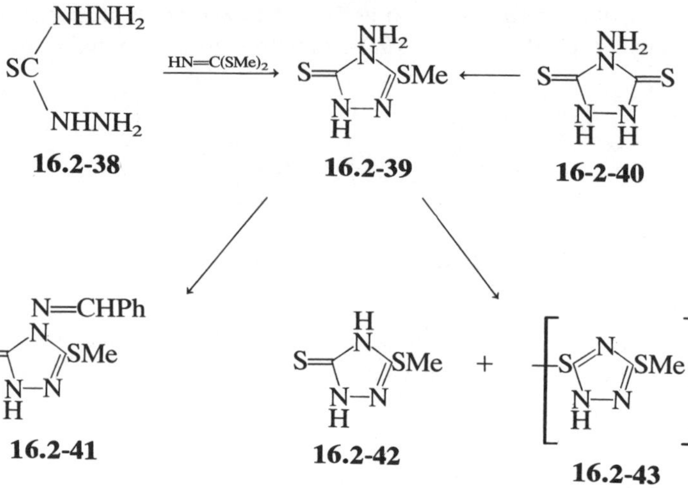

16.2-38 16.2-39 16-2-40

16.2-41 16.2-42 16.2-43

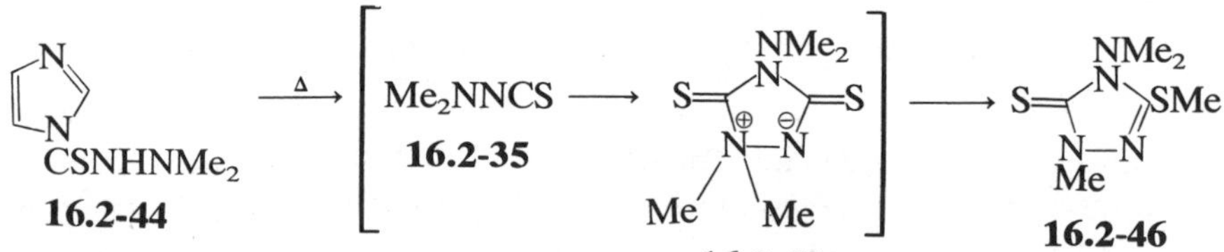

16.2-44 16.2-35 16.2-45 16.2-46

this group, the disulfide (**16.2-47**), was prepared by oxidation of **16.2-40** with ferric chloride.[23] The 3-chlorotriazoline-5-thione (**16.2-49**) was reported to result from the interaction 1-phenyl-4-methylthiosemicarbazide (**16.2-48**) with thiophosgene.[24]

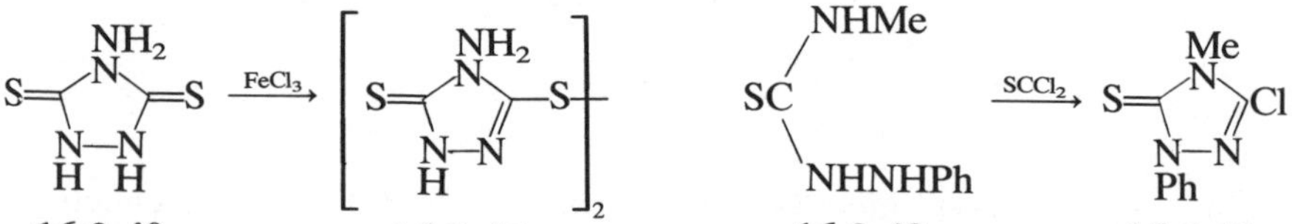

16.2-40 16.2-47 16.2-48 16.2-49

References

1. **73:** 66517n
2. **82:** 140031d
3. **75:** 98505q
4. **76:** 59533y
5. **78:** 136152u
6. **80:** 121258a
7. **77:** 164587n
8. **79:** 5298u
9. **79:** 18651d
10. **54:** 8794f
11. **81:** 91429h
12. **83:** 164091v
13. **57:** 804i
14. **74:** 125572x
15. **68:** 29639e
16. **82:** 72890f
17. **74:** 53659f
18. **74:** P53817f
19. **81:** 91435g
20. **61:** 10676a
21. **57:** 12471f
22. **69:** 51320m
23. **56:** 15513b
24. **73:** 35290d

Compound	Reference

16.1. Amino- and Diamino-Δ^2(or Δ^3)-1,2,4-Triazoline-5-thiones

Compound	Reference
3-acetamido-4-allyl-	**50:** 9913d
4-acetamido-3-amino-	**63:** 14850b
4-acetamido-3-ethyl-	**74:** 141645d
4-acetamido-3-heptyl-	**74:** P149248x, **80:** P126797k, **83:** P18957r
4-acetamido-3-methyl-	**74:** P8379d, **74:** P59399t, **80:** 126797k, **80:** P151126y, **83:** P18957r
3-acetamido-2-phenyl-	**81:** P179884p
4-acetamido-3-phenyl-	**51:** 3579d
-3-acetic acid, 4-amino-	**84:** 17271w
-3-acetic acid, 4-amino-, hydrazide	**84:** 17271w
-3-acetic acid, 4-(benzylideneamino)-	**84:** 17271w
4-allyl-3-amino-	**50:** P9913c
4-allyl-3-benzamido-	**50:** P9913d
3-amino-	**33:** 598⁵, **40:** 949¹, **55:** 2623g, **55:** 10452a, **68:** 12914t, **68:** 29639e, **73:** P45513t
4-amino-	**43:** 5396a, **57:** 805a, **58:** 4543d, **74:** 13070a, **83:** 97223z, **85:** P177501v
4-amino-3-anilino-	**64:** 8173d, **69:** 77170n
3-amino-4-benzyl-	**50:** P9913c
4-amino-2-benzyl-	**74:** 13070a, **78:** 124510w
4-amino-3-benzyl-	**66:** 2516c
3-amino-4-benzylideneamino-	**48:** 5182a, **58:** 4543g
4-amino-3-(benzylidenehydrazino)-	**83:** 97223z
3-amino-4-(p-bromophenyl)-	**60:** 1755a, **64:** 8174a, **67:** 90736e, **70:** 3963x
3-amino-4-butyl-	**56:** 12879d
4-amino-3-butyl-	**74:** 141645d, **84:** 30975y
4-amino-3-[[2-carboxy-8-oxo-7-[(2-thienylacetyl)-amino]-5-thia-1-azabicyclo[4.2.0]oct-2-en-3-yl]-methylenehydrazino]-	**83:** P58855n
4-amino-1-(2-chlorobenzyl)-	**80:** 82837s
4-amino-1-(3-chlorobenzyl)-	**78:** 124510w
4-amino-1-(4-chlorobenzyl)-	**78:** 124510w
3-amino-4-(p-chlorobenzylidene)amino-	**48:** 5182a
4-amino-3-[(o-chlorophenoxy)methyl]-	**64:** P2098a
3-amino-4-(p-chlorophenyl)-	**60:** 1755a, **70:** 3963x, **76:** 135847w
4-amino-3-(o-chlorophenyl)-	**66:** 2516c
3-amino-4-cyclohexyl-	**50:** P9913c, **60:** 1755a
4-amino-3-cyclopropyl-	**74:** 141645d, **85:** P177501v
4-amino-3-decyl-	**56:** P162a
4-amino-3-[(2,4-dichlorophenoxy)methyl]-	**64:** P2097h

493

TABLE 16 (*Continued*)

Compound	Reference

16.1. Amino- and Diamino-Δ²(or Δ³)-1,2,4-Triazoline-5-thiones (*Continued*)

Compound	Reference
3-amino-1,2-dimethyl-	**83:** 9914x
4-amino-1,3-dimethyl-	**55:** 23507g, **74:** 13070a
4-amino-3-(3,5-dimethyl-4-isoxazolyl)-	**55:** 7399e
4-amino-5-(2-ethoxyethyl)-	**85:** P177501v
4-amino-3-(ethoxymethyl)-	**85:** P177501v
3-amino-4-ethyl-	**50:** P9913c
4-amino-3-ethyl-	**57:** 805a, **58:** 4543e, **74:** 141645d, **81:** 136118z, **85:** P78172g, **85:** P177501v
4-amino-3-ethyl-1-methyl-	**55:** 23507i
4-amino-5-heptadecyl-	**55:** P26806i
4-amino-3-(heptafluoropropyl)-	**74:** 141645d, **79:** 92159r
4-amino-3-heptyl-	**83:** P50807c, **85:** P177501v
4-amino-3-hydrazino-	**48:** 4525a, **48:** 5181i, **49:** 10857g, **49:** 15869i, **51:** 3481f, **56:** 5948i, **58:** 4543e, **68:** 39547n, **80:** 82830j, **80:** 110110v, **81:** 49628x, **82:** 154565t, **83:** 58739c, **84:** 17271w, **84:** 180171x, **84:** 59374n
4-amino-3-[(2-hydrazinocarbonyl)phenyl-	**83:** 97223z
4-amino-3-(*o*-hydroxyphenyl)-	**71:** 22075w
4-amino-3-(*p*-hydroxyphenyl)-	**71:** 22075w
4-amino-3-(isobutylamino)-	**65:** 5455b
3-amino-4-isobutyl-	**60:** 1755a
4-amino-3-isobutyl-	**58:** 4543e, **74:** 141645d
3-amino-4-isopropyl-	**50:** P9913c, **60:** 1755a
3-amino-4-(*p*-methoxybenzylidene)amino-	**48:** 5182a
4-amino-3-(methoxymethyl)-	**74:** 141645d, **85:** P177501v
3-amino-4-(*p*-methoxyphenyl)-	**60:** 1775a, **70:** 3963x
4-amino-3-(*p*-methoxyphenyl)-	**48:** 4524i, **76:** 140741m
3-amino-4-methyl-	**56:** 12879a, **68:** 29639e
4-amino-1-methyl-	**55:** 23507e, **78:** 124510w
4-amino-3-methyl-	**48:** 4525a, **57:** 805a, **58:** 4543e, **74:** 141645d, **81:** 136118z, **84:** 17271w, **85:** P78173h, **85:** P177501v
4-amino-3-(3-methyl-5-isoxazolyl)-	**78:** 43419w
4-amino-3-(5-methyl-3-isoxazolyl)-	**78:** 43419w
4-amino-3-(5-methyl-3-phenyl-4-isoxazolyl)-	**70:** 11610v
4-amino-3-(5-methyl-1H-pyrazol-3-yl)-	**78:** 136235y
4-amino-3-[(2-naphthyloxy)methyl]-	**64:** P2098a
3-amino-4-[(5-nitrofurfurylidene)amino]-	**61:** 14663a
4-amino-3-[(*p*-nitrophenoxy)methyl]-	**64:** P2098a
4-amino-3-(4-nitrophenyl)-	**76:** 140741m
4-amino-1-octyl-	**78:** 124510w
4-amino-5-(pentafluoroethyl)-	**74:** 141645d, **79:** 92159r

TABLE 16 (*Continued*)

Compound	Reference

16.1. Amino- and Diamino-Δ^2(or Δ^3)-1,2,4-Triazoline-5-thiones (*Continued*)

Compound	Reference
4-amino-3-pentyl-	**74:** 141645d
4-amino-3-(phenoxymethyl)-	**64:** 2097h, **85:** P177501v
3-amino-1-phenyl-	**67:** P103933u
3-amino-2-phenyl-	**83:** 28157u
3-amino-4-phenyl-	**23:** 1639[9], **30:** 7574[7], **33:** 598[5], **56:** 12878b, **57:** 12473a, **64:** 8174a, **67:** 90736e, **68:** 29639e, **70:** 3963x, **76:** 135847w
4-amino-3-phenyl-	**48:** 4524i, **51:** 3579d, **61:** 14662h, **71:** 22075w, **74:** 141645d, **76:** 140741m, **78:** 43419w, **85:** P78173h, **85:** P177501v
4-amino-3-(5-phenyl-3-isoxazolyl)-	**78:** 43419w
4-amino-3-propyl-	**58:** 4543e, **74:** 141645d, **81:** 136118z, **85:** P177501v
4-amino-3-(3-pyridazinyl)-	**63:** 105174s
4-amino-3-(4-pyridyl)-	**49:** 10938e, **76:** 140741m, **83:** 97228e
4-amino-3-*p*-toluidino-	**69:** 77170n
3-amino-4-*o*-tolyl-	**30:** 7574[7], **33:** 598[6]
3-amino-4-*m*-tolyl-	**30:** 7574[7]
3-amino-4-*p*-tolyl-	**30:** 7574[7], **33:** 598[6], **56:** 12878g, **56:** 12879h, **67:** 90736e, **70:** 3963x
4-amino-3-*p*-tolyl-	**66:** 2516c
4-amino-3-[(2,4,5-trichlorophenoxy)methyl]-	**64:** P2097h
4-amino-3-tridecyl-	**85:** P177501v
4-amino-3-(trifluoromethyl)-	**72:** 3470a, **74:** 141645d, **79:** 92159r, **85:** P177501v
3-amino-4-(α,α,α-trifluoro-*m*-tolyl)-	**60:** 1755a
3-amino-4-(α,α,α-trifluoro-*p*-tolyl)-	**60:** 1755a
4-amino-3-(3,4,5-trimethoxyphenyl)-	**74:** 141645d
3-anilino	**57:** 12473a, **68:** 12914t
3-anilino-4-(benzylideneamino)-	**64:** 8173d, **69:** 77170n
3-anilino-4-(4-bromophenyl)-	**80:** 14879g
3-anilino-4-(*p*-chlorophenyl)-	**70:** 3963x
4-anilino-3-(4-chlorophenyl)-	**80:** 27172w
3-anilino-4-(2,3-diphenylguanidino)-	**69:** 77170n
3-anilino-4-(*p*-methoxyphenyl)-	**66:** 115650s, **70:** 3963x
3-anilino-4-phenyl-	**21:** 2900[5], **25:** 1504[2], **30:** 7574[7], **55:** 522a, **55:** 17626f, **57:** 12473a, **66:** 115650s, **69:** 77170n, **70:** 3963x, **79:** 42423b
4-anilino-3-phenyl-	**80:** 27172w
3-anilino-4-[1-(3-phenyl-2-thio)ureido]-	**64:** 8173d

TABLE 16 (*Continued*)

Compound	Reference

16.1. Amino- and Diamino-Δ²(or Δ³)-1,2,4-Triazoline-5-thiones (*Continued*)

4-anilino-3-*p*-toluidino-	**72:** 55347k
3-anilino-4-*p*-tolyl-	**70:** 3963x
3-*p*-anisidino-4-(*p*-methoxyphenyl)-	**70:** 3963x
3-*p*-anisidino-4-phenyl-	**70:** 3963x
3-*p*-anisidino-4-*p*-tolyl-	**70:** 3963x
3-benzamido-	**75:** P36046g
4-benzamido-	**83:** 97223z
4-benzamido-3-heptyl-	**74:** P149248x
4-benzamido-3-pentyl-	**74:** P149248x, **80:** P126797k, **83:** P18957r
3-benzamido-2-phenyl-	**81:** P179884p
4-benzamido-3-phenyl-	**51:** 3579e, **76:** 140741m
1-benzyl-4-(benzylideneamino)-	**74:** 13070a
1-benzyl-4-[(carbamoyl)amino]-	**76:** 85755s
4-(benzylideneamino)-	**56:** 5948h, **57:** 805b, **58:** 4543g
4-(benzylideneamino)-3-(benzylidenehydrazino)-	**83:** 97223z
3-(benzylideneamino)-1-(1-dimethoxyphosphinylethenyl)-3-(trifluoromethyl)-	**80:** P82989t
4-benzylideneamino-1,3-dimethyl-	**55:** 23507g
4-benzylideneamino-3-(3,5-dimethyl-4-isoxazolyl)-	**55:** 7399e
4-(benzylideneamino)-3-ethyl-	**57:** 805b, **58:** 4543g
4-(benzylideneamino)-3-ethyl-1-methyl-	**55:** 23507i
4-(benzylideneamino)-3-hydrazino-	**83:** 97223z
4-(benzylideneamino)-3-(*o*-hydroxyphenyl)-	**71:** 22075w
4-(benzylideneamino)-3-isobutyl-	**58:** 4543g
4-benzylideneamino-1-methyl-	**55:** 23507f
4-(benzylideneamino)-3-methyl-	**58:** 4543g
4-(benzylideneamino)-3-phenyl-	**71:** 22075w, **76:** 140741m
4-(benzylideneamino)-3-propyl-	**58:** 4543g
4-(benzylideneamino(-3-*p*-toluidino-	**69:** 77170n
4-benzyl-3-(methylamino)-	**73:** 35283d
4-benzyl-3-(phenylamino)-	**73:** 35283d
4-benzyl-3-(*p*-tolylamino)-	**73:** 35283d
3-(4-bromoanilino)-	**80:** 14879g
4-([α-(bromomethyl)benzylidene]amino)-3-ethyl-	**72:** 3470a, **81:** 136118z
4-(4-bromophenyl)-3-[(4-bromophenyl)amino]-	**68:** 59503y
4-(*p*-bromophenyl)-3-hydrazino-	**64:** 8174a
3-(*p*-butoxyanilino)-4-(*p*-butoxyphenyl)-	**59:** 13985g
3-(tert-butylamino)-4-phenyl-	**70:** 3963x
3-(tert-butylamino)-4-*p*-tolyl-	**70:** 3963x
4-[(butylcarbamoyl)amino]-3-ethyl-	**74:** 141645d
3-butyl-4-(2-phenylacetamido)-	**74:** P149248x
-3-carbamic acid, 4-methyl-, 2-chloro-1-(chloromethyl)ethyl ester	**75:** P36046g
3-(*o*-chloroanilino)-4-(*m*-chlorophenyl)-	**59:** 13985g
3-(*m*-chloroanilino)-4-(*o*-chlorophenyl)-	**59:** 13985g
3-(*p*-chloroanilino)-4-(*p*-chlorophenyl)-	**59:** 13985g, **70:** 3963x, **79:** 42423b

TABLE 16 (*Continued*)

Compound	Reference

16.1. Amino- and Diamino-Δ^2(or Δ^3)-1,2,4-Triazoline-5-thiones (*Continued*)

Compound	Reference
3-(*p*-chloroanilino)-4-phenyl-	**70:** 3963x
3-(*p*-chlorobenzamido)-4-methyl-	**75:** P36046g
1-(4-chlorobenzyl)-4-[(carbamoyl)amino]-	**76:** 85755s
4-[(2-chloro-5-nitrobenzylidene)amino]-3-ethyl-	**72:** 3470a, **74:** 141645d,
	75: 88589f
4-[(2-chloro-5-nitrobenzylidene)amino]-3-methyl-	**75:** 88589f
3-[(*o*-chlorophenoxy)methyl]-4-[(5-nitrofurfurylidene)-amino]-	**67:** P21917c
3-[(*p*-chlorophenoxy)methyl]-4-[(5-nitrofurfurylidene)-amino]-	**67:** P21917c
4-(4-chlorophenyl)-3-hydrazino-	**76:** 135847w
4-(*p*-chlorophenyl)-3-(2-phenylhydrazino)-	**60:** 526a
4-(4-chlorophenyl)-3-[2-(*N*-phenylthiocarbamoyl)-hydrazino]-	**76:** 135847w
4-(*p*-chlorophenyl)-3-(*p*-tolylamino)-	**73:** 35283d
3-[2-[(4-chloro-*o*-tolyl)oxy]acetamido]-4-phenyl-	**75:** P36046g
4-(cyclopentylideneamino)-3-ethyl-	**74:** 141645d
3,4-diacetamido-	**63:** 14850c
4-(diacetylamino)-3-ethyl-	**74:** 141645d
3,4-diamino-	**48:** 5181i, **58:** 4543f,
	61: 10676a, **83:** 193226y
3-(3,5-diamino-1H-1,2,4-triazol-1-yl)-	**77:** 164610q
3,4-dianilino-	**72:** 55347k
3-(2,2-dichloroacetamido)-4-methyl-	**75:** P36046g
3-[(2,4-dichlorophenoxy)methyl]-4-[(5-nitrofurfurylidene)amino]-	**67:** P21917c
3-(diethylamino)-4-phenyl-	**59:** 604c, **56:** 7306c
4-(1,4-dihydro-2-mercapto-4,4,6-trimethyl-1-pyrimidinyl)-3-(trifluoromethyl)-	**74:** 141645d
1-(1-dimethoxyphosphinylethenyl)-4-(dimethylamino)-3-(trifluoromethyl)-	**80:** P82989t
3-(dimethylamino)-4-phenyl-	**56:** 7306c
3-(3,5-dimethylpyrazol-1-yl)-	**63:** 11545c
4-(2,5-dimethylpyrrol-1-yl)-3-ethyl-	**74:** 141645d
3-(3,5-dimethylpyrazol-1-yl)-4-isobutyl-	**69:** 77151g
3-(3,5-dimethylpyrazol-1-yl)-4-phenyl-	**64:** 8174a
3-[(1,3-dioxo-2-isoindolinyl)acetamido]-	**70:** 11640e
3-(diphenylamino)-	**78:** 111227j
3-(*o*-ethoxyanilino)-4-(*o*-ethoxyphenyl)-	**59:** 13985g
3-(*p*-ethoxyanilino)-4-(*p*-ethoxyphenyl)-	**59:** 13985g
4-[(1-ethoxyethylidene)amino]-3-ethyl-	**72:** 3470a
4-[(ethoxymethylene)amino]-3-ethyl-	**74:** 141645d
4-(4-ethoxyphenyl)-3-[2-[*N*-(4-ethoxyphenyl)-thiocarbamoyl]hydrazino]-	**76:** 135847w
4-(*p*-ethoxyphenyl)-3-(*p*-ethylanilino)-	**59:** 13985g
4-(4-ethoxyphenyl)-3-hydrazino-	**76:** 135847w
4-(*p*-ethoxyphenyl)-3-(2-phenylhydrazino)-	**60:** 526a
4-(4-ethoxyphenyl)-3-[2-(*N*-phenylthiocarbamoyl)-hydrazino]-	**76:** 135847w

TABLE 16 (*Continued*)

Compound	Reference

16.1. Amino- and Diamino-Δ^2(or Δ^3)-1,2,4-Triazoline-5-thiones (*Continued*)

4-(*p*-ethoxyphenyl)-3-*p*-toluidino-	**67:** 90736e
3-(*N*-ethylanilino)-	**78:** 111227j
3-(*p*-ethylanilino)-4-(*p*-ethylphenyl)-	**59:** 13985g
3-ethyl-4-(1,4-dihydro-2-mercapto-4,4,6-trimethyl- 1-pyrimidinyl)-	**74:** 141645d
3-ethyl-4-(ethylideneamino)-	**74:** 141645d
3-ethyl-4-[(*p*-fluorobenzylidene)amino]-	**74:** 141645d
3-ethyl-4-(isopropylideneamino)-	**74:** 141645d
3-ethyl-4-[(*p*-methoxybenzylidene)amino]-	**74:** 141645d
3-ethyl-4-[(5-nitrofurfurylidene)amino]-	**69:** P59249r
3-ethyl-4-[(4-pyridylmethylene)amino]-	**74:** 141645d
3-ethyl-4-pyrrol-1-yl-	**74:** 141645d
3-ethyl-4-[(*p*-tolylsulfonylcarbamoyl)amino]-	**74:** 141645d
4-[(3-fluorosulfonylbenzoyl)amino]-3-heptadecyl-	**71:** P22929c
4-[(3-fluorosulfonylphenyl)ureylene]-3-heptadecyl-	**71:** P22929c
3-heptadecyl-4-[(3-sulfonbenzoyl)amino]-	**71:** P22929c
3-heptadecyl-4-[(3-sulfophenyl)ureylene]-	**71:** P22929c
3-(hexanoylamino)-2-phenyl-	**81:** P179884p
3-hydrazino-	**49:** 10858b, **60:** 1736e, **63:** 11545c, **66:** 104959y, **77:** 164610g
3-hydrazino-4-benzyl-	**76:** 135847w
3-hydrazino-4-isobutyl-	**69:** 77151g
3-hydrazino-4-(3-methoxyphenyl)-	**76:** 135847w
3-hydrazino-4-(*p*-methoxyphenyl)-	**64:** 8174a, **76:** 135847w
3-hydrazino-4-[4-(3-methylbutoxy)phenyl]-	**76:** 135847w
3-hydrazino-4-(2-methylphenyl)-	**76:** 135847w
3-hydrazino-4-(4-methylphenyl)-	**76:** 135847w
3-hydrazino-4-phenyl-	**64:** 8174a, **69:** 77151g, **76:** 135847w
3-hydrazino-4-phenyl-, 4-(dimethylamino)benzaldehyde- hydrazone	**76:** 135847w
3-hydrazino-4-*p*-tolyl-	**64:** 8174a
3-(*o*-iodoanilino)-4-(*o*-iodophenyl)-	**59:** 13985g
3-isopropyl-4-[(5-nitrofurfurylidene)amino]-	**69:** P59249r
3-(*o*-methoxyanilino)-4-(*o*-methoxyphenyl)-	**59:** 13985g
3-(*p*-methoxyanilino)-4-(*p*-methoxyphenyl)-	**59:** 13985g
4-(3-methoxyphenyl)-3-[2-(*N*-phenylthiocarbamoyl)- hydrazino]-	**76:** 135847w
4-(4-methoxyphenyl)-3-[2-(*N*-phenylthiocarbamoyl)- hydrazino]-	**76:** 135847w
4-(*p*-methoxyphenyl)-3-*p*-toluidino-	**70:** 3963x
3-(methylamino)-4-(1-methyl-2-propenyl)-	**82:** 43271z
3-(methylamino)-4-phenyl-	**70:** 3963x
3-(*N*-methylanilino)-	**78:** 111227j
3-(*N*-methylanilino)-1,4-diphenyl-	**74:** 53659f
4-[4-(3-methylbutoxy)phenyl]-3-[2-(*N*-phenylthiocar- bamoyl)hydrazino]-	**76:** 135847w
4-methyl-3-(methylamino)-	**70:** 3963x

TABLE 16 (*Continued*)

Compound	Reference

16.1. Amino- and Diamino-Δ^2(or Δ^3)-1,2,4-Triazoline-5-thiones (*Continued*)

Compound	Reference
4-methyl-3-(methylamino)-1-phenyl-	**73:** 35290d
4-methyl-3-[(1-methyl-2-propenyl)amino]-	**77:** 109606t, **82:** 43271z
3-methyl-4-[(5-nitrofurfurylidene)amino]-	**69:** P59249r
4-(4-methylphenyl)-3-[(4-methylphenyl)amino]-	**68:** 59503y, **70:** 3963x, **79:** 42423b
4-(2-methylphenyl)-3-[2-(*N*-phenylthiocarbamoyl)-hydrazino]-	**76:** 135847w
3-methyl-4-(4-sulfoanilino)-	**81:** 44054x
3-morpholino-4-phenyl-	**59:** 604c
4-*p*-nitrobenzamido-3-phenyl-	**51:** 3579e
4-(*p*-nitrobenzylideneamino)-	**55:** P19567b
4-[(5-nitrofurfurylidene)amino]-	**69:** P59249r
4-[(5-nitrofurfurylidene)amino]-3-(phenoxymethyl)-	**67:** P21917c
4-[(5-nitrofurfurylidene)amino]-3-phenyl-	**61:** 14663a
4-[(5-nitrofurfurylidene)amino]-3-[(2,4,5-trichlorophenoxy)methyl]-	**67:** P21917c
3-nonanamido-	**78:** P22526p
3-phenyl-4-anilino-	**84:** 4865t
3-(2-phenylhydrazino)-4-*p*-tolyl-	**60:** 526a
3-(2-phenylhydrazino)-4-(2,4-xylyl)-	**60:** 526a
4-phenyl-3-[α-(phenyl)phenacylamino]-	**64:** 19594g
1-phenyl-3-(β-phenylthiocarbamido)-	**23:** 2178[1]
4-phenyl-3-piperidino-	**59:** 604b
4-phenyl-3-*p*-toluidino-	**70:** 3963x
2-phenyl-3-(valerylamino)-	**81:** P179884p
4-(propionylamino)-3-heptyl-	**74:** P149248x
4-(propionylamino)-3-methyl-	**74:** P8379d, **74:** P149248x, **74:** P59399t
3-*o*-toluidino-	**30:** 7574[8]
3-*m*-toluidino-	**30:** 7574[8]
3-*p*-toluidino-	**30:** 7574[8]
3-*p*-toluidino-4-(2,3-di-*p*-tolylguanidino)-	**69:** 77170n
3-(*o*-toluidino)-4-*o*-tolyl-	**59:** 13985g, **70:** 3963x
3-(*p*-toluidino)-4-*p*-tolyl-	**21:** 2900[6], **59:** 13985g
3-xylidino-	**25:** 703[5], **30:** 7574[8]
3-(2,4-xylidino)-4-(2,4-xylyl)-	**59:** 13985g

16.2. Miscellaneous Δ^2(or Δ^3)-1,2,4-Triazoline-5-thiones

Compound	Reference
4-acetamido-1-acetyl-3-methyl-	**57:** 805c
-1-acetic acid, 4-(ethoxycarbonyl)-α,3-diphenyl-, ethyl ester	**79:** 18651d
-1-acetic acid, 4-(ethoxycarbonyl)-3-methyl-	**79:** 18651d
-1-acetic acid, 4-(ethoxycarbonyl)-3-methyl-, ethyl ester	**79:** 18651d
-1-acetic acid, 4-(ethoxycarbonyl)-3-methyl-α-phenyl-, ethyl ester	**79:** 18651d
-1-acetic acid, 4-(ethoxycarbonyl)-3-phenyl-	**79:** 18651d
-1-acetic acid, 4-(ethoxycarbonyl)-3-phenyl-, ethyl ester	**79:** 18651d

TABLE 16 (*Continued*)

Compound	Reference
16.2. Miscellaneous Δ^2(or Δ^3)-1,2,4-Triazoline-5-thiones (*Continued*)	
1-acetyl-4-allyl-3-(phenoxymethyl)-	**70:** 47369n
1-acetyl-4-allyl-3-[(2,4,5-trichlorophenoxy)methyl]-	**70:** 47369n
1-acetyl-3-(2-chloro-4-nitrophenyl)-4-phenyl-	**54:** 8795c
1-acetyl-3-(*p*-chlorophenyl)-4-phenyl-	**54:** 8795c
1-acetyl-3,4-diphenyl-	**54:** 8795b
1-acetyl-3-(*p*-hydroxyphenyl)-4-phenyl-, acetate	**54:** 8795c
1-acetyl-3-(*p*-nitrophenyl)-4-phenyl-	**54:** 8795c
1-acetyl-4-phenyl-3-[(2,4,5-trichlorophenoxy)methyl]-	**70:** 47369n
2-acetyl-3-*p*-toluidino-	**23:** 2974[9]
4-allyl-3-ethoxy-	**68:** 29639e
4-amino-3-(benzylthio)-	**57:** 12471i
4-amino-3-ethoxy-	**68:** 29639e
4-amino-3-(ethylthio)-	**80:** 82830j
4-amino-3-(methylthio)-	**57:** 12471i, **61:** 10676a, **79:** 31993m, **80:** 82830j
1-benzoyl-3-benzyl-4-phenyl-	**83:** 164091v
1-benzoyl-3-benzyl-4-(2-propenyl)-	**83:** 164091v
1-benzoyl-3,4-diphenyl-	**81:** 91429h
3-(benzoyloxy)-1,4-dimethyl-	**82:** 72890f
1-benzoyl-3-(phenoxymethyl)-4-phenyl-	**70:** 47369n
4-benzoyl-3-phenyl-	**58:** 13937g
1-benzoyl-3-phenyl-4-(2-propenyl)-	**81:** 91429h
3-(*N*-benzylcarbamoyl)-1-phenyl-	**76:** 59533y, **78:** 136152u
3-(*N*-benzylcarbamoyl)-1-*p*-tolyl-	**76:** 59533y, **78:** 136152u
4-(benzylideneamino)-3-(methylthio)-	**61:** 10676a, **79:** 31993m
3-benzyloxy-1,4-dimethyl-	**82:** 72890f
5-benzyl-1-(1-oxo-3-phenyl-2-propenyl)-4-phenyl-	**83:** 164091v
3-benzyl-1-(1-oxo-3-phenyl-2-propenyl)-4-(2-propenyl)-	**83:** 164091v
3-([1-(butoxycarbonyl)-2-ethyl]thio)-	**74:** P53817f
3-[(1-carbamoyl-2-ethyl)thio]-	**74:** P53817f
3-[(1-carbamoyl-1-methyl-2-ethyl)thio]-	**74:** P53817f
-3-carboxaldehyde, 3-(diethyl acetyl)	**74:** 125572x
-3-carboxaldehyde, 3-(dimethyl acetal)	**74:** 125572x
-3-carboxaldehyde, 3-[(2,4-dinitrophenyl)hydrazone]	**74:** 125572x
-3-carboxaldehyde, 3-(thiosemicarbazone)	**74:** 125572x
-3-carboxamide, *N,N*-diethyl-4-methyl-	**70:** P87820j, **73:** 66517n
-2-carboxamide, *N,N*-dimethyl-	**67:** P54140x
-3-carboxamide, *N*-methyl-1-phenyl-	**75:** 98505q, **78:** 136152u
-3-carboxamide, *N*-methyl-1-*o*-tolyl-	**75:** 98505q, **78:** 136152u
-3-carboxamide, 1-β-D-ribofuranosyl-	**80:** 121258a
-1-carboxanilide, 3-ethoxy-	**24:** 1114[5]
3-[(1-carboxy-2-ethyl)thio]-	**74:** P53817f
3-[(1-carboxy-2-ethyl)thio]-4-phenyl-	**74:** P53817f
3-(2-carboxyethylthio)-, thiosemicarbazide	**75:** P56785a
-3-carboxylic acid, 4-butyl-	**72:** P100713q
-3-carboxylic acid, 4-butyl-, ethyl ester	**72:** P100713q
-3-carboxylic acid, 1,4-dimethyl-	**70:** 68323h, **73:** 56069d
-4-carboxylic acid, 1,3-diphenyl-, ethyl ester	**77:** 164587n
-1-carboxylic acid, 3-methoxy-4-phenyl-, methyl ester	**71:** 61474v

TABLE 16 (*Continued*)

Compound	Reference
16.2. Miscellaneous Δ^2(or Δ^3)-1,2,4-Triazoline-5-thiones (*Continued*)	
-3-carboxylic acid, 4-methyl-	**70:** 11686z, **70:** 68323h, **72:** P100713q, **73:** 66517n
-3-carboxylic acid, 4-methyl-, ethyl ester	**70:** 68323h
-4-carboxylic acid, 3-methyl-, ethyl ester	**77:** 164587n
-3-carboxylic acid, 4-methyl-, [(2-hydroxyphenyl)-methylene]hydrazide	**82:** 140031d
-4-carboxylic acid, 1-methyl-3-phenyl-, ethyl ester	**77:** 164587n
-4-carboxylic acid, 3-methyl-1-(3-phenyl-1-propenyl)-, ethyl ester	**79:** 5298u
-4-carboxylic acid, 3-methyl-1-(2-propenyl), ethyl ester	**79:** 18651d
-3-carboxylic acid, 1-phenyl-	**75:** 98505q
-4-carboxylic acid, 2-phenyl-, ethyl ester	**77:** 164587n
-4-carboxylic acid, 3-phenyl-, ethyl ester	**77:** 164587n
-4-carboxylic acid, 3-phenyl-1-(3-phenyl-1-propenyl)-, ethyl ester	**79:** 5298u
-4-carboxylic acid, 3-phenyl-1-(3-phenyl-2-propenyl)-, ethyl ester	**79:** 5298u
-4-carboxylic acid, -3-phenyl-1-(2-propenyl)-, ethyl ester	**79:** 18651d
-3-carboxylic acid, 1-*o*-tolyl-	**75:** 98505q
3-[(2-carboxy-2-methylethyl)thio]-	**75:** P56785a
3-[(1-carboxy-2-propyl)thio]-	**74:** P53817f
3-chloro-4-methyl-1-phenyl	**73:** 35290d
3-[(chloromethyl)thio]-4-(dimethylamino)-1-methyl-	**80:** P48002f
3-(diethoxyphosphinylthio)-4-(dimethylamino)-1-methyl-	**80:** P48002f
3-[(diethoxyphosphinylthio)methylthio]-1,4-dimethyl-	**80:** P48002f
3-[(diethoxyphosphinylthio)methylthio]-4-(dimethylamino)-1-methyl-	**80:** P48002f
4-(dimethylamino)-3-([ethoxy(propylthio)phosphinylthio]-methylthio)-1-methyl-	**80:** P48002f
4-(dimethylamino)-3-(ethylthio)-1-methyl-	**80:** P48002f
4-(dimethylamino)-1-methyl-3-[(isopropyl)thio]-	**80:** P48002f
4-(dimethylamino)-1-methyl-3-(methylthio)-	**69:** 51320m
4-(dimethylamino)-1-methyl-d_3-3-(methyl-d_3-thio)-	**70:** 36934s
4-(dimethylamino)-1-methyl-d_3-3-(methylthio)-	**70:** 36934s
4-(dimethylamino)-1-methyl-3-(methyl-d_3-thio)-	**70:** 36934s
4-[di(methyl-d_3)amino]-1-methyl-d_3-3-(methyl-d_3-thio)-	**70:** 36934s
4-[di(methyl-d_3)amino]-1-methyl-d_3-3-(methylthio)-	**70:** 36934s
4-[di(methyl-d_3)amino]-1-methyl-3-(methylthio)-	**70:** 36934s
4-[di(methyl-d_3)amino]-1-methyl-3-(methyl-d_3-thio)-	**70:** 36934s
1,4-dimethyl-5-[(isopropyl)thio]-	**80:** P48002f
3-([1-(dimethylcarbamoyl)-2-ethyl]thio)-	**74:** P53817f
3-([1-(dimethylcarbamoyl)-2-ethyl]thio)-4-phenyl-	**74:** P53817f
3,3'-dithiobis[4-amino-	**56:** 15513b
3,3'-dithiobis[4-*o*-tolyl-	**33:** 599[3]
3,3'-dithiobis[4-*p*-tolyl-	**33:** 599[3]
3-[[1-(ethoxycarbonyl)-2-ethyl]thio]-	**74:** P53817f
3-[[2-(ethoxycarbonyl)-2-methylethyl]thio]-	**75:** P56785a
3-[[1-(ethoxycarbonyl)-2-propyl]thio]-	**74:** P53817f
3-ethoxy-4-(dimethylthiocarbamoyl)amino-	**68:** 29639e

TABLE 16 (*Continued*)

Compound	Reference

16.2. Miscellaneous Δ^2(or Δ^3)-1,2,4-Triazoline-5-thiones (*Continued*)

Compound	Reference
3-ethoxy-4-ethyl-	**68:** 29639e
3-ethoxy-4-methyl-	**68:** 29639e
3-ethoxy-4-phenyl-	**68:** 29639e
3-([2-(methoxycarbonyl)-2-methylethyl]thio)-	**75:** P56785a
1-methyl-3-[(isopropyl)thio]-4-phenyl-	**80:** P48002f
1-methyl-3-(methylthio)-4-phenyl-	**44:** 6416i, **80:** P48002f, **81:** 91435g
4-methyl-3-[1-(piperidino)carbonyl]-	**70:** 11686z, **70:** P87820j, **73:** 66517n
4-methyl-3-propoxy-	**68:** 29639e
4-methyl-3-[1-(pyrrolidino)carbonyl]-	**70:** P87820j, **73:** 66517n
3-(methylthio)-1,4-diphenyl-	**74:** 53659f
3-[(3-oxobutyl)thio]-	**74:** P53817f
3-[(3-oxobutyl)thio]-4-methyl	**74:** P53817f
3-[(3-oxobutyl)thio]-4-methyl-, 2-(thiosemicarbazone)	**74:** P53817f
3-[(3-oxobutyl)thio]-, oxime	**74:** P53817f
3-[(3-oxobutyl)thio]-4-phenyl-	**75:** P56785a
3-[(3-oxobutyl)thio]-, phenylhydrazone	**74:** P53817f
3-[(3-oxobutyl)thio]-, semicarbazone	**74:** P53817f
1-(1-oxo-3-phenyl-2-propenyl)-3,4-diphenyl-	**81:** 91429h
1-(1-oxo-3-phenyl-2-propenyl)-3-phenyl-4-(2-propenyl)-	**81:** 91429h
4-phenyl-3-[1-(piperidino)carbonyl]-	**70:** P87820j, **73:** 66517n
3-(1-piperidinecarbonyl)-1-phenyl-	**75:** 98505q, **78:** 136152u
3-(1-piperidinecarbonyl)-1-*o*-tolyl-	**75:** 98505q, **78:** 136152u
3-(1-piperidinecarbonyl)-1-*p*-tolyl-	**76:** 59533y, **78:** 136152u
-1-propanenitrile, 4-amino-3-[(2-cyanoethyl)thio]-	**79:** 31993m
-1-propanenitrile, 4-amino-3-(methylthio)-	**79:** 31993m
-1-propanenitrile, 4-(benzylideneamino)-3-(methylthio)-	**79:** 31993m
4-(propionyl)-3-methyl-	**74:** P8379d, **74:** P149248x, **74:** P59399t

502

Miscellaneous Δ^2(or Δ^1)-1,2,4-Triazolines Containing One or More Representative Functions

The condensation of amidrazones (e.g., **17.1-1**) with alkyl, aryl, and heteroaromatic aldehydes in ethanol gave good yields of Δ^2-1,2,4-triazolines (e.g., **17.1-2**).[1-3] The structure of **17.1-2** was confirmed by oxidation to the corresponding 1,2,4-triazole (**17.1-3**) with potassium permanganate.[1] Bis-(triazolines) were prepared either from diamidrazones (e.g., **17.1-4**) and aldehydes or from amidrazones and aryl dialdehydes (see Section 21.2).[4]

Triazolines (e.g., **17.1-7**) were also obtained in the acid-catalyzed condensation of **17.1-5** with ketones (e.g., **17.1-6**) in refluxing ethanol.[5,6] In the reaction of equilmolar amounts of **17.1-8** and ethyl orthoformate in refluxing 2-methoxyethanol, a 50% yield of the 5-ethoxytriazoline (**17.1-9**) was obtained.[7] Ethanol was eliminated on treatment of **17.1-9** with acid to give

the corresponding triazole. In addition, several 3-aryloxy-5-imino-1,2,4-triazolines (e.g., **17.1-13**) were prepared by the condensation of the methyl dithiocarbazate (**17.1-10**) with aryl cyanates (e.g., **17.1-11**) under mild conditions.[8] Apparently the ethers are formed via intermediates like **17.1-12** and **17.1-14**. The simple amidrazone (**17.1-16**) was postulated to be an

intermediate in the condensation of the aminomethylenemalonate (**17.1-15**) with hydrazine hydrate to give the 5-aminotriazoline (**17.1-17**).[9] In the

reaction of the amidrazone (**17.1-18**) with phenyl isothiocyanate in refluxing pyridine, a mixture of **17.1-19** and **17.1-20** was isolated with the latter being the major product.[10]

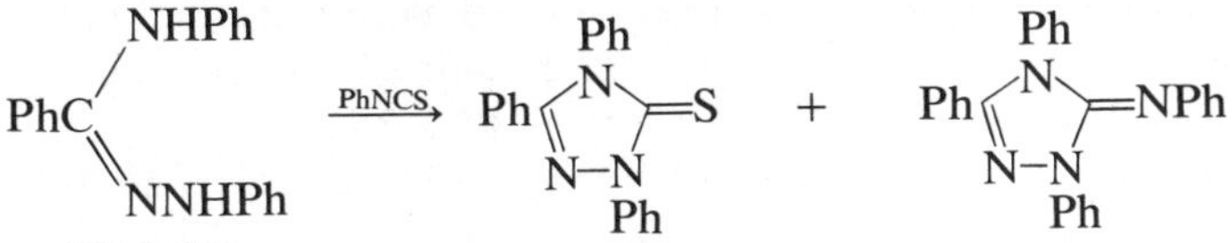

17.1-18   **17.1-19**   **17.1-20**

Imino-1,2,4-triazolines were also prepared by a variety of other methods. The condensation of N-amino-N,N'-dimethylguanidine (**17.1-21**) and N-amino-N,N',N''-trimethylguanidine (**17.1-22**) with the simple aliphatic carboxylic acid gave the 5-imino-1,2,4-triazolines (**17.1-23–25**) and the 5-methylimino-1,2,4-triazolines (**17.1-26–28**), respectively.[11] Also, compounds **17.1-23** and **17.1-26** were obtained by alkylation of 5-amino-1-methyl-1,2,4-triazole (**17.1-29**).[12] Treatment of **17.1-29** with dimethylsulfate resulted in preferential methylation of the 4-N position of the ring to give **17.1-23**, whereas treatment of a solution of **17.1-29** in ethanol with methyl iodide resulted in methylation of the 4-N position of the ring and the exocyclic imino group to give **17.1-26**. Preferential ring alkylation was also

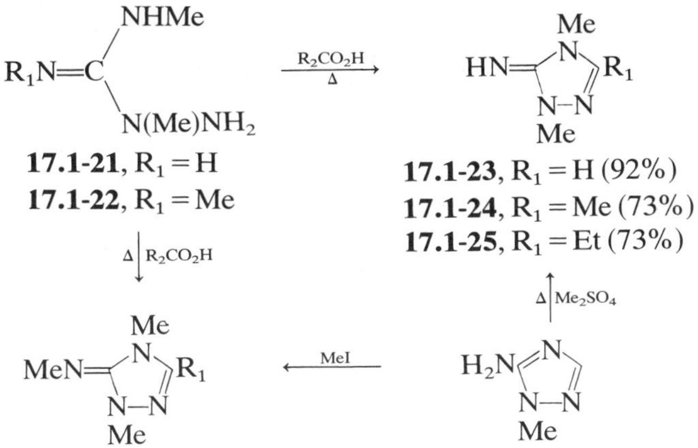

17.1-21, $R_1 = H$
17.1-22, $R_1 = Me$

17.1-23, $R_1 = H$ (92%)
17.1-24, $R_1 = Me$ (73%)
17.1-25, $R_1 = Et$ (73%)

17.1-26, $R_1 = H$ (85%)
17.1-27, $R_1 = Me$ (60%)
17.1-28, $R_1 = Et$ (28%)

17.1-29

shown in the reaction of either **17.1-30** or **17.1-32** with methyl iodide to give the same product (**17.1-31**).[13,14]

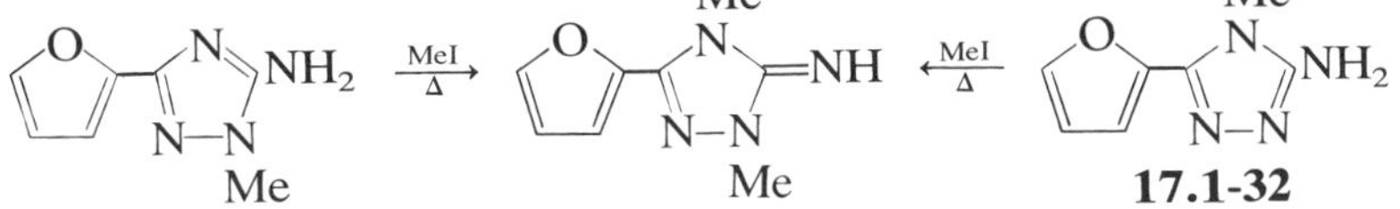

17.1-30   **17.1-31**   **17.1-32**

The rearrangement of 1,4-diphenyl-1,4-dihydrotetrazine (**17.1-33**) by N—N bond cleavage occurred readily in the presence of sodium ethoxide to give 5-imino-1,4-diphenyl-1,2,4-triazoline (**17.1-35**).[15]

The same product was obtained by decarboxylation of **17.1-36** in hot concentrated hydrochloric acid.[16] The related rearrangement of **17.1-34** to **17.1-23** was reported to occur via intermediates **17.1-37** and **17.1-38**.[17]

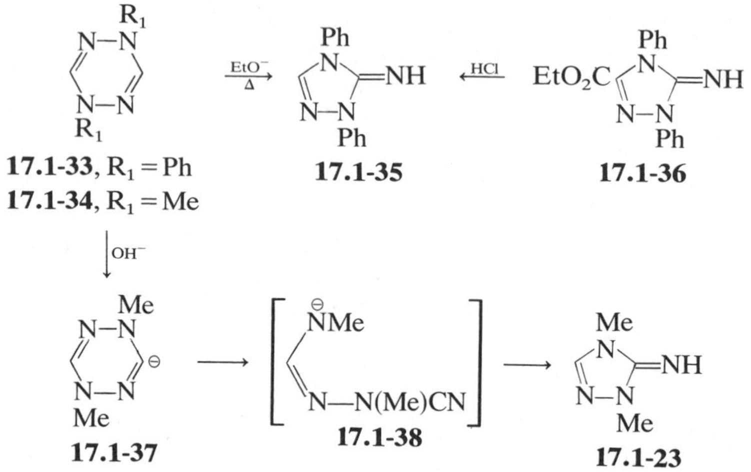

17.1-33, R$_1$ = Ph
17.1-34, R$_1$ = Me

17.1-35

17.1-36

17.1-37

17.1-38

17.1-23

The acylation of the 5-amino-1-methyl-1,2,4-triazole (**17.1-39**) with benzoyl chloride in pyridine gave the 4-benzoyl-5-imino-1,2,4-triazoline (**17.1-40**).[18,19] The same product was reported to result from the alkylation of 5-benzamido-3-phenyl-1,2,4-triazole (**17.1-41**) with dimethyl sulfate in an alkaline medium.[12] Reduction of the 4-benzoyl-1,2,4-triazoline (**17.1-40**) with lithium aluminum hydride gave the 4-benzyl compound (**17.1-42**), which was also prepared by alkylation of the 5-amino-1,2,4-triazole (**17.1-43**) with benzyl chloride at high temperatures in a sealed tube.[20]

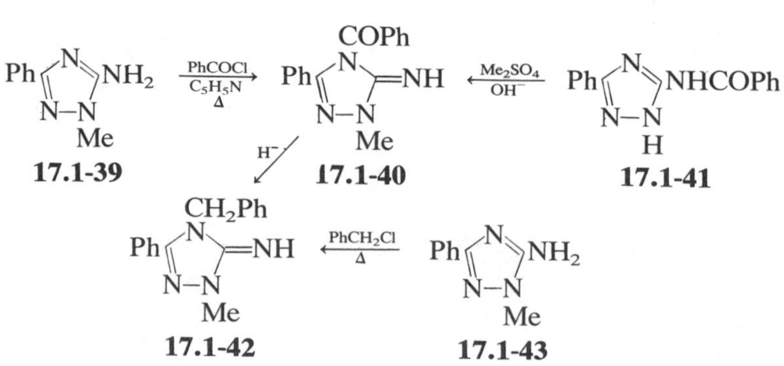

17.1-39

17.1-40

17.1-41

17.1-42

17.1-43

Derivatives of the imino function are readily formed as shown by the reaction of **17.1-23** with phenyl isothiocyanate to give **17.1-44** and with 4-nitrobenzoyl chloride to give **17.1-45**.[12]

17.1-44   **17.1-23**   **17.1-45**

The addition of N,N'-diphenylcarbodiimide to the hydrochloride of aminoguanidine (**17.1-46**) in N,N-dimethylformamide at 110° gave a mixture of **17.1-47** (38 to 45%) and the 5-phenylimino-1,2,4-triazoline **17.1-48** (30 to 35%).[21] The formation of the latter by the elimination of aniline from **17.1-49** was supported by the preparation of **17.1-48** in good yield from **17.1-47** and N,N'-diphenylcarbodiimide. The structure of **17.1-48** was based on its rapid conversion to 3,5-dianilino-4-phenyl-1,2,4-triazole in an

17.1-46   **17.1-47**

17.1-48   **17.1-49**

alkaline medium. Similarly, the interaction of **17.1-50** and diphenylcarbodiimide in N,N-dimethylformamide at 100° gave **17.1-51**, which under the conditions of the reaction was converted in part to **17.1-52** in yields up to 30%.[22,23] Treatment of **17.1-51** with ethanolic hydrogen chloride also gave **17.1-52** in 14 to 20% yield. As described above for the 5-aminotriazoles,

17.1-50   **17.1-51**   **17.1-52**

alkylation of 3,5-diamino-1,2,4-triazoles gave the ring-substituted product as demonstrated by the reaction of **17.1-53** with methyl iodide to give **17.1-54**.[24] The 4-amino-5-imino-1,2,4-triazolines (**17.1-56**, **17.1-57**, and

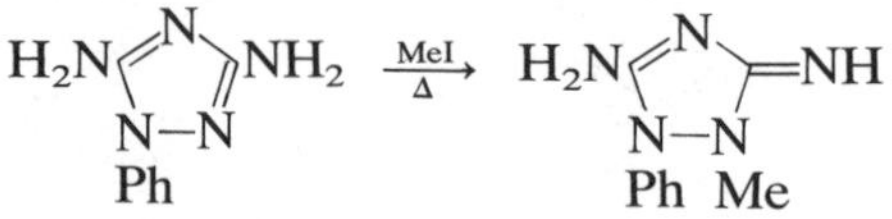

17.1-53   **17.1-54**

17.1-58) were obtained in good yields by the condensation of the hydrobromide of N,N'-diamino-N-methylguanidine (**17.1-55**) with the corresponding aliphatic carboxylic acids.[25] The unusual perfluoroamino-1,2,4-tria-

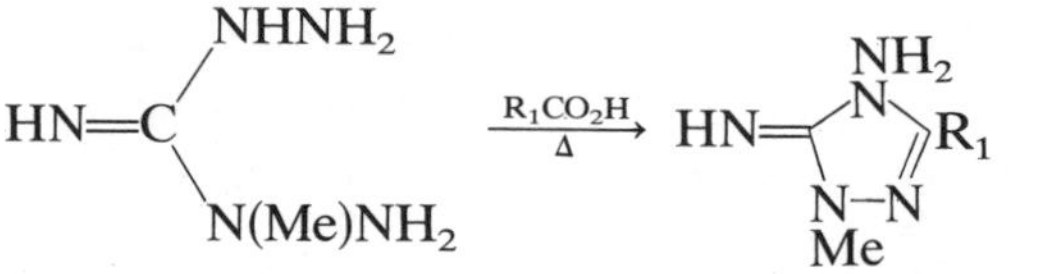

17.1-55

17.1-56, $R_1 = H$
17.1-57, $R_1 = Me$
17.1-58, $R_1 = Et$

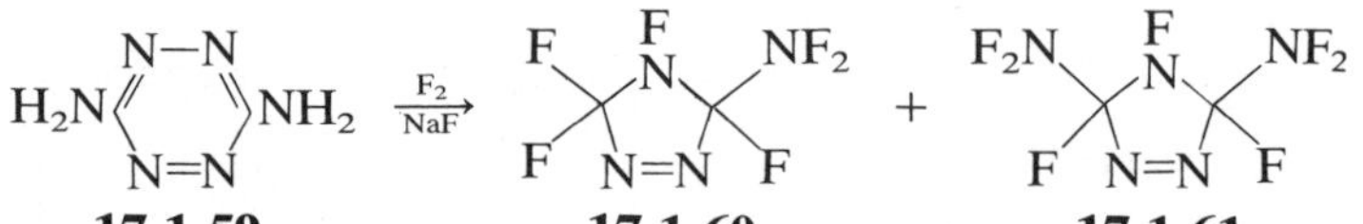

zolines (**17.1-60** and **17.1-61**) were components of a mixture resulting from the fluorination of 3,6-diamino-s-tetrazine (**17.1-59**).[26,27]

17.1-59   **17.1-60**   **17.1-61**

Good yields of Δ^2-1,2,4-triazolines are obtained by the cycloaddition of nitrilimines to azomethines.[28–31] For example, treatment of **17.1-62** in benzene containing triethylamine generated the nitrilimine (**17.1-63**), which reacted with benzylideneaniline to give 1,3,4,5-tetraphenyl-Δ^2-1,2,4-triazoline (**17.1-64**).[28] The same compound was obtained by the condensation of the amidrazone (**17.1-68**) with benzaldehyde in ethanol.[28] The reversibility of this reaction was shown by cleavage of the ring of **17.1-64** with 2,4-dinitrophenylhydrazine to give the corresponding hydrazone of benzaldehyde. Similarly, the addition of **17.1-63** to isopropylideneaniline gave the 5,5-dimethyl-Δ^2-1,2,4-triazoline (**17.1-67**).[10] Nitrilimine intermediates were also used to prepare the 3-acetyltriazoline (**17.1-65**)[32] and the triazoline-3-carboxylic acid ester (**17.1-66**).[33] The nitrilimine (**17.1-70**), generated by the thermolysis of the oxazaphosphole (**17.1-69**), was trapped with dimethyl azodicarboxylate to give the triazoline (**17.1-71**).[34] In a related procedure,

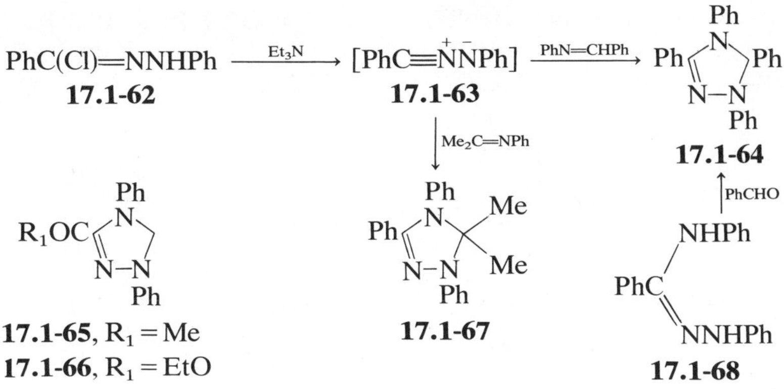

PhC(Cl)=NNHPh $\xrightarrow{\text{Et}_3\text{N}}$ [PhC$\overset{+}{\equiv}\overset{-}{\text{N}}$NPh] $\xrightarrow{\text{PhN=CHPh}}$

17.1-62   **17.1-63**

$\Big\downarrow \text{Me}_2\text{C=NPh}$

17.1-64

$\Big\uparrow \text{PhCHO}$

17.1-65, R_1 = Me
17.1-66, R_1 = EtO

17.1-67

17.1-68

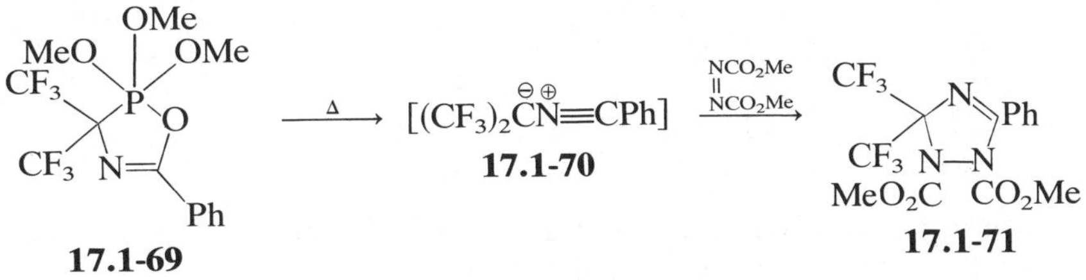

17.1-69 $\xrightarrow{\Delta}$ [(CF$_3$)$_2$$\overset{\ominus}{\text{C}}N\overset{\oplus}{\equiv}$CPh] $\xrightarrow[\text{NCO}_2\text{Me}]{\text{NCO}_2\text{Me}}$

17.1-70

17.1-71

irradiation of the azirine (**17.1-72**) gave **17.1-73**, which reacted with diethyl-azodicarboxylate to give a 70% yield of **17.1-74**.[3]

17.1-72 $\xrightarrow{h\nu}$ (Me)$_2$$\overset{\ominus}{\text{C}}$—$\overset{\oplus}{\text{N}}$$\equiv$CPh $\xrightarrow[\text{NCO}_2\text{Et}]{\text{NCO}_2\text{Et}}$

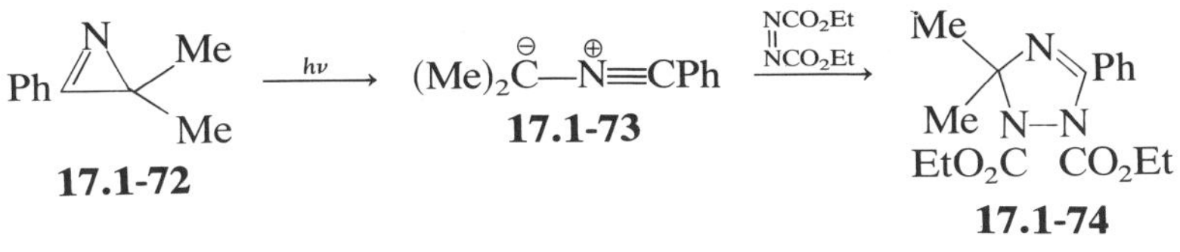

17.1-73

17.1-74

Recently several types of triazolines have been prepared by reduction of performed triazole systems. Treatment of triazolin-5-ones (e.g., **17.1-75**) with lithium aluminium hydride gave a triazolidin-5-one (e.g., **17.1-76**) (60%) when the ring double bond was conjugated with the exocyclic double bond of the oxo function. In contrast, triazolines (e.g., **17.1-78**) (50%) were obtained when the ring double bond was unconjugated as in **17.1-77**.[35] Also

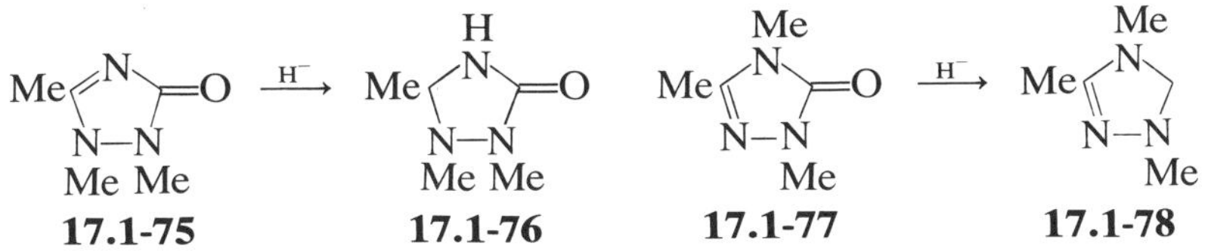

17.1-75 $\xrightarrow{\text{H}^-}$ **17.1-76**   **17.1-77** $\xrightarrow{\text{H}^-}$ **17.1-78**

hydride reduction of the triazolium compounds (**17.1-79** to **17.1-82**) gave the corresponding triazolines [**17.1-64** (97%),[36,37] **17.1-83** (55%),[38] **17.1-84**,[39] and **17.1-85** (92%),[40–42] respectively].

17.1-79   **17.1-80**   **17.1-81**   **17.1-82**

17.1-64   **17.1-83**   **17.1-84**   **17.1-85**

In contrast to this reduction process, the loss of a proton from triazolium salts containing an activated 5-methyl function (e.g., **17.1-86**) leads to 5-methylenetriazolines compounds (e.g., **17.1-87**).[43,44] The triazolium salt (**17.1-86**) was regenerated on treatment of **17.1-87** with perchloric acid.

17.1-86   **17.1-87**

Selenium derivatives of triazolines were prepared for use as accelerators for bleaching fixing baths and stabilizers for photographic silver halide emulsions.[45–47] Thus treatment of **17.1-88** with acetic anhydride gave **17.1-89**, which was cyclized to **17.1-90** in an alkaline medium. In the condensation of **17.1-88** with isoselenocyanates, selenodiazoles were obtained.[48]

17.1-88   **17.1-89**   **17.1-90**

References

1. **61:** 7007a
2. **74:** 125653z
3. **81:** 129811s
4. **76:** 85796f
5. **72:** 111882v
6. **79:** 66315u
7. **75:** 20356x
8. **64:** 5080f
9. **73:** 66512g
10. **63:** 7000c
11. **60:** 9264f
12. **59:** 12790e
13. **63:** 18071g
14. **79:** 87327q
15. **63:** 2975c
16. **63:** 7001b
17. **78:** 4224j
18. **58:** 10202f
19. **70:** 57860e
20. **59:** 12791g
21. **57:** 12472h
22. **69:** 10398z
23. **81:** 91439m
24. **53:** 3198e
25. **59:** 10050a
26. **67:** 64407d
27. **73:** 45550c
28. **61:** 654h
29. **62:** 13147e
30. **71:** 21771h
31. **85:** 192629m
32. **70:** 87679v
33. **70:** 96723g
34. **80:** 27179d
35. **76:** 59537c
36. **71:** 91389p
37. **78:** 159527j
38. **81:** 3839p
39. **84:** 105495c
40. **81:** 169245s
41. **82:** 169941m
42. **83:** 113599j
43. **72:** 121463y
44. **75:** 98508t
45. **75:** 76805k
46. **77:** 140086x
47. **77:** 146189e
48. **78:** 136178g

TABLE 17. MISCELLANEOUS Δ^2(OR Δ^1)-1,2,4-TRIAZOLINES CONTAINING ONE OR MORE REPRESENTATIVE FUNCTIONS

Compound	Reference
Parent	**44:** P9841e
4-(2-acetamidophenyl)-3-acetyl-1,5-diphenyl-	**70:** 87679v
1-acetoacetyl-3-amino-5-imino-2-(2-naphthyl)-	**45:** 613f
1-acetoacetyl-3-amino-5-imino-2-phenyl-	**45:** 613e, **47:** 12371g
1-acetoacetyl-3-amino-5-imino-2-p-tolyl-	**45:** 613f
1-acetonyl-5-hydroxy-3,5-dimethyl-4-phenyl-	**82:** 170807d
4-acetyl-5-(acetimido)-1-methyl-3-(5-nitro-2-furyl)-	**70:** P57860e
3-acetyl-4-isopropyl-1,5-diphenyl-	**70:** 87679v
3-acetyl-4-methyl-1,5-diphenyl-	**70:** 87679v
3-acetyl-1,4,5-triphenyl-	**70:** 96723g
3-amino-1-[(3-amino-5-imino-2-phenyl-Δ^3-1,2,4-triazolin-1-yl)crotonoyl]-5-imino-2-phenyl-	**45:** 613g
3-amino-1-[(3-amino-5-imino-2-p-tolyl-Δ^3-1,2,4-triazolin-1-yl)crotonoyl]-5-imino-2-p-tolyl-	**45:** 613h
3-amino-1-benzoylacetyl-5-imino-2-phenyl	**47:** 12371g
3-amino-1-benzoylacetyl-5-imino-2-p-tolyl-	**47:** 12371g
3-amino-1-benzyl-5-imino-2-phenyl-	**53:** 3198e
3-amino-1-[3-(3,7-dihydro-5-methyl-7-oxo-1-phenyl-s-triazolo[2,3-a]pyrimidin-2(1H)-ylideneamino)-crotonoyl]-5-imino-2-phenyl-	**46:** 4534f
3-amino-1-[N-(p-ethoxyphenyl)formimidoyl]-5-imino-2-phenyl-	**46:** 4535e
4-amino-3-ethyl-5-imino-1-methyl-	**59:** 10051h
3-amino-5-imino-1,2-dimethyl-	**55:** P15943i
4-amino-5-imino-1,3-dimethyl-	**59:** 10051g
4-amino-5-imino-1-methyl-	**59:** 10051f
3-amino-5-imino-1-phenyl-2-N-phenylformimidoyl-	**46:** 4535e
3-amino-1-methyl-5-imino-2-phenyl-	**53:** 3198e
3-anilino-4-(4-chlorophenyl)-1-methyl-5-phenyl-	**75:** 151734s, **81:** 3839p
3-anilino-1,4-diphenyl-5-(phenylimino)-	**64:** 5255e
3-anilino-1-methyl-4,5-diphenyl-	**81:** 3839p

TABLE 17 (*Continued*)

Compound	Reference
3-anilino-1-methyl-4-phenyl-5-(phenylimino)-	**81:** 91439m
4-benzoyl-5-imino-1-methyl-3-phenyl-	**58:** 10202f, **59:** 12791cg
5-benzoyl-3-(methylthio)-1,4-diphenyl-	**83:** 113599j
4-benzyl-5-(2-furyl)-1,3-diphenyl-	**61:** 655d
4-(benzylideneamino)-1,3-diphenyl-	**84:** 105495c
4-(benzylideneamino)-3-ethyl-5-imino-1-methyl-	**59:** 10051h
4-(benzylideneamino)-5-imino-1,3-dimethyl-	**59:** 10051h
4-(benzylideneamino)-5-imino-1-methyl-	**59:** 10050b
4-(benzylideneamino)-1-phenyl-	**84:** 105495c
4-benzyl-5-imino-1-methyl-3-phenyl-	**59:** 12791g
4-benzyl-1,3,5-triphenyl-	**61:** 655d
4,5-bis(*p*-nitrophenyl)-	**43:** 7022b
4,5-bis(*p*-nitrophenyl)-3,3-diphenyl-	**43:** 7022c
1-(4-bromophenacyl)-5-hydroxy-3,5-dimethyl-4-phenyl-	**79:** 115499t
5-(*p*-bromophenyl)-4-(*p*-nitrophenyl)-	**43:** 7022c
5-butyl-1,3-diphenyl-	**61:** 7007b
4-butyl-5-methyl-1,3-diphenyl-	**61:** 655d
5-*tert*-butyl-3-(methylthio)-1,4-diphenyl-	**82:** 169941m
-1-carbodithioic acid, 3-(*m*-chlorophenoxy)-5-imino-2-phenyl-, methyl ester	**64:** 5080f
-1-carbodithioic acid, 5-imino-3-phenoxy-2-phenyl-, methyl ester	**64:** 5080f
-1-carbodithioic acid, 5-imino-2-phenyl-3-(2,4-xylyloxy)-, methyl ester	**64:** 5080f
-1-carboxaldehyde, 3-amino-4-imino-2-phenyl	**46:** 4534h
-1-carboxamide, 3-amino-4-imino-2-phenyl-	**46:** 4535f
-1-carboxamidine, 3-anilino-4-phenyl-5-phenylimino-	**57:** 12472i
-1-carboxanilide, 3-amino-5-imino-2-phenyl-	**46:** 4535g
-5-carboxylic acid, 4-(*o*-acetamidophenyl)-3-acetyl-1-phenyl-, ethyl ester	**70:** 87679v
-1-carboxylic acid, 3-anilino-4-phenyl-5-phenylimino-, ethyl ester	**69:** 10398z
-3-carboxylic acid, 4-benzyl-1,5-diphenyl-, ethyl ester	**70:** 96723g
-3-carboxylic acid, 4-cyclohexyl-1,5-diphenyl-, ethyl ester	**70:** 96723g
-3-carboxylic acid, 5,5-dimethyl-1,4-diphenyl-, ethyl ester	**63:** 7001a
-1-carboxylic acid, 5,5-dimethyl-3-phenyl, ethyl ester	**81:** 129811s
-3-carboxylic acid, 5-imino-1,4-diphenyl-, ethyl ester	**63:** 7001c
-1-carboxylic acid, 5-methyl-3-phenyl-, ethyl ester	**81:** 129811s
-1-carboxylic acid, 3-phenyl-, ethyl ester	**81:** 129811s
-3-carboxylic acid, 1,4,5-triphenyl-, ethyl ester	**63:** 7001a, **70:** 96723g
5-(4-carboxyphenyl)-3-(methylthio)-1,4-diphenyl-	**81:** 169245s
3-(4-chloroanilino)-1-methyl-4,5-diphenyl-	**75:** 151734s, **81:** 3839p
4-(2-chlorobenzylideneamino)-1-phenyl-	**84:** 105495c
5-(*p*-chlorophenyl)-1,3-diphenyl-	**61:** 7007b, **62:** 13148b
1-(*o*-chlorophenyl)-3,4,5-triphenyl-	**61:** 655c
5-(3-chloropropyl)-3-(methylthio)-1,4-diphenyl-	**82:** 169941m
4-cyclohexyl-5-(cyclohexylimino)-3-[(phenylcarbamoyl)oxy]-	**81:** 91438k

TABLE 17 (*Continued*)

Compound	Reference
4-cyclohexyl-5-methyl-1,3-diphenyl-	**70:** 96723g
-1,2-dicarboxylic acid, 5-tert-butyl-3,3-bis(trifluoromethyl)-, dimethyl ester	**80:** 27179d
-1,2-dicarboxylic acid, 5-(4-chlorophenyl)-3,3-bis(trifluoromethyl)-, dimethyl ester	**80:** 27179d
-1,2-dicarboxylic acid, 3,3-dimethyl-5-phenyl-, diethyl ester	**81:** 129811s
-1,2-dicarboxylic acid, 3,5-diphenyl-, diethyl ester	**81:** 129811s
-1,2-dicarboxylic acid, 5-(4-methylphenyl)-3,3-bis(trifluoromethyl)-, dimethyl ester	**80:** 27179d
-1,2-dicarboxylic acid, 3-methyl-5-phenyl-, diethyl ester	**81:** 129811s
-1,2-dicarboxylic acid, 5-phenyl-3,3-bis(trifluoromethyl)-, dimethyl ester	**80:** 27179d
-1,2-dicarboxylic acid, 5-phenyl-, diethyl ester	**81:** 129811s
3-(difluoroamino)-3,4,5,5-tetrafluoro-	**67:** PC64407d, **73:** P45550c
5-[2,4-dihydro-5-(methylthio)-2,4-diphenyl-3$\underline{H}$-1,2,4-triazol-3-ylidene]-	
3-(methylthio)-1,4-diphenyl-	**83:** 113599j
4-(2,4-dimethoxyphenyl)-5-(*p*-nitrophenyl)-	**43:** 7022c
1,3-dimethyl-	**76:** 59537c
1,4-dimethyl-5-(methylimino)-	**59:** 12791d, **60:** 9265b
1,4-dimethyl-5-[(*p*-nitrobenzoyl)imino]-	**59:** 6227d
1,4-dimethyl-5-(nitroimino)-	**59:** 12791c
1,2-dimethyl-3-phenyl-	**74:** 99947b
1,4-dimethyl-5(*N*-phenylthiocarbamoyl)imino-	**59:** 12790e
5,5-dimethyl-1,3,4-triphenyl-	**63:** 7001a
5-(2,4-dinitrobenzylidene)-1,3,4-triphenyl-	**75:** 98508t
1,3-diphenyl-5-propyl-	**61:** 7007b
1,3-diphenyl-5-styryl-	**61:** 7007b
5-ethoxy-3-(5-nitro-2-furyl)-	**75:** 20356x
3-ethyl-1,4-dimethyl-5-(methylimino)-	**60:** 9265c
3-ethyl-1,4-dimethyl-5(*N*-phenylthiocarbamoyl)imino-	**59:** 12790e
5-ethyl-1,3-diphenyl-	**61:** 7007b
3-ethyl-5-imino-1,4-dimethyl-	**60:** 9265b
4-ethyl-5-imino-1-methyl-	**59:** 12790h
5-ethyl-3-(methylthio)-1,4-diphenyl-	**82:** 169941m
5-fluoren-9-ylidene-1,3,4-trimethyl-	**72:** 121463y
3-(2-furyl)-1,4-dimethyl-5-(methylimino)-	**63:** 18071f
5-(2-furyl)-4-methyl-1,3-diphenyl-	**61:** 655d
3-(2-furyl)-5-imino-1,4-dimethyl-	**63:** 18071e, **63:** 18071g
5-(2-furyl)-1,3,4-triphenyl-	**61:** 655d
-1-glyoxylic acid, 3-amino-5-imino-2-phenyl, ethyl ester	**46:** 4535c
5-hexyl-1,3-diphenyl-	**61:** 7007b
5-imino-1,2-bis(4,4,4-trichloro-3-hydroxy-1-oxobutyl)-	**76:** 140733k
5-imino-1,4-bis(4,4,4-trichloro-3-hydroxy-1-oxobutyl)-	**76:** 140733k
5-imino-1,4-dimethyl-	**59:** 12790h, **60:** 9265a, **78:** 4224j
3-imino-2,4-dimethyl-5-(5-nitro-2-furanyl)-	**79:** 87327q
5-imino-1,4-dimethyl-3-phenyl-	**59:** 12791a

TABLE 17 (*Continued*)

Compound	Reference
5-imino-1,4-diphenyl-	**63:** 2976b, **63:** 7001d
5-imino-1-methyl-4-(*p*-nitrobenzoyl)-3-phenyl-	**58:** 10202f
5-imino-1,3,4-trimethyl-	**60:** 9265b
4-isopropyl-5,5-dimethyl-1,3-diphenyl-	**61:** 655f
4-isopropyl-5-methyl-1,3-diphenyl-	**70:** 96723g
-1-methanol, 3-amino-5-imino-2-phenyl-α-(trichloromethyl)-	**46:** 4534i
4-(4-methoxybenzylideneamino)-3-(4-methoxyphenyl)-	**84:** 105495c
5-[4-(methoxycarbonyl)phenyl]-3-(methylthio)-1,4-diphenyl-	**81:** 169245s
5-(*p*-methoxyphenyl)-1,3-diphenyl-	**61:** 7007b, **62:** 13148b
5-(*p*-methoxyphenyl)-4-methyl-1,3-diphenyl-	**61:** 655d
5-methyl-3,5-bis(2-pyridinyl)-	**79:** 66315u
5-methyl-3-(4-methyl-2-pyridinyl)-5-phenyl-	**79:** 66315u
5-methyl-3-(4-methyl-2-pyridinyl)-5-(2-pyridinyl)-	**79:** 66315u
5-methyl-3-(methylthio)-1,4-diphenyl-	**81:** 169245s
4-methyl-5-(*p*-nitrobenzylidene)-1,3-diphenyl-	**75:** 98508t
3-methyl-1-(4-nitrophenyl)-4,5-diphenyl-	**85:** 192629m
5-methyl-3-(1,10-phenanthroline-2-yl)-5-phenyl-	**79:** 66315u
5-methyl-3-(1,10-phenanthrolin-2-yl)-5-(2-pyridinyl)-	**79:** 66315u
5-methyl-5-phenyl-3-(4-phenyl-2-pyridinyl)-	**79:** 66315u
5-methyl-5-phenyl-3-(2-pyridinyl)-	**72:** 111882v
5-methyl-3-(4-phenyl-2-pyridinyl)-5-(2-pyridinyl)	**79:** 66315u
5-methyl-5-(2-pyridinyl)-3-[2-(2-pyridinyl)-6-pyridinyl]-,	**79:** 66315u
3-(methylthio)-1,4-diphenyl-5-propyl-	**82:** 169941m
5-(methylthio)-4-phenyl-	**63:** 6996d
3-(methylthio)-1,4,5-triphenyl-	**81:** 169245s
1-methyl-4-(4-tolyl)-3-(4-toluidino)-5-(4-tolylimino)-	**81:** 91439m
4-methyl-1,3,5-triphenyl-	**61:** 655d
3-(*p*-nitrobenzyl)-5-(*p*-nitrobenzylidene)-1,4-diphenyl-	**75:** 98508t
4-(*m*-nitrophenyl)-5-(*p*-nitrophenyl)-	**43:** 7022b
4-(*p*-nitrophenyl)-5-(*m*-nitrophenyl)-	**43:** 7022b
3-(*p*-nitrophenylthio)-1,4-diphenyl-	**32:** 4545⁹
3-(*p*-nitrophenylthio)-1,4,5-triphenyl-	**32:** 4545⁹
1-(*p*-nitrophenyl)-3,4,5-triphenyl-	**61:** 655c
3-(1,10-phenanthroline-2-yl)-5-(2-pyridinyl)-	**79:** 66315u
5-phenyl-3-(2,2′-bipyridinyl-6-yl)-	**74:** 125653z
5-phenyl-3-pyrazinyl	**74:** 125653z
5-phenyl-3-(3-pyridazinyl)-	**74:** 125653z
5-phenyl-3-(2-pyridinyl)-	**72:** 111882v
5-phenyl-3-(4-pyrimidinyl)-	**74:** 125653z
-1-propionic acid, 3-amino-5-imino-β-oxo-2-phenyl-, ethyl ester	**46:** 4535a
3-(3-pyridazinyl)-5-(2-pyridyl)-	**74:** 125653z
4-(2-pyridinyl)-5-(2-pyridinylamino)-	**73:** 66512g
5-(2-pyridyl)-3-(2,2′-bipyridinyl-6-yl)-	**74:** 125653z
5-(2-pyridyl)-3-pyrazinyl-	**74:** 125653z
5-(2-pyridyl)-3-(4-pyrimidinyl)-	**74:** 125653z
-3-selone	**75:** P76805k, **77:** P146189e
-3-selone, 4-benzyl-5-methyl-	**75:** P76805k, **77:** P146189e

TABLE 17   (*Continued*)

Compound	Reference
-3-selone, 5-benzyl-	**77:** P146189e
-3-selone, 2,4-bis(isopropyl)-	**78:** 15125d
-3-selone, 4-tert-butyl-	**77:** 74417c
-3-selone, 4-cyclohexyl-	**77:** 74417c
-3-selone, 4-cyclohexyl-5-methyl	**77:** 74417c
-3-selone, 4-cyclohexyl-5-phenyl-	**77:** 74417c
-3-selone, 2,4-dimethyl-	**78:** 15125d
-3-selone, 2,5-dimethyl-	**77:** P146189e
-3-selone, 4,5-dimethyl-	**75:** P76805k, **77:** P146189e
-3-selone, 5-ethyl-	**77:** P146189e
-3-selone, 5-(2-furanyl)-	**75:** P76805k, **77:** P146189e
-3-selone, 4-isopropyl-	**77:** 74417c
-3-selone, 5-isopropyl-	**77:** P146189e
-3-selone, 4-methyl	**77:** 74417c
-3-selone, 5-methyl-	**75:** P76805k, **77:** 146189e
-3-selone, 4-methyl-5-phenyl-	**77:** 74417c
-5-selone, 3-methyl-1-phenyl-	**75:** P76805k, **77:** P146189e
-3-selone, 5-phenyl-	**75:** P76805k, **77:** P146189e, **78:** 136178g
-3-selone, 5-propyl-	**77:** P146189e
1,3,4,5-tetraphenyl-	**61:** 655b, **71:** 91389p
1,3,4-trimethyl-	**76:** 59537c
1,3,4-trimethyl-5-(methylimino)-	**60:** 9265c
1,3,5-trimethyl-4-phenyl-	**78:** 159527j
3,4,5-trimethyl-1-phenyl-	**78:** 159527j
1,3,4-trimethyl-5(*N*-phenylthiocarbamoyl)imino-	**59:** 12790e
1,3,5-triphenyl-	**61:** 7007b, **62:** 13148a
1,3,4-triphenyl-5-(phenylimino)-	**63:** 7000d
1,3,4-triphenyl-5-styryl-	**71:** 21771h

CHAPTER 18

1,2,4-Triazolidines

The sequential addition of aliphatic aldehydes and alkyl- or arylamines to a solution of 1,2-dimethylhydrazine in benzene in the presence of magnesium sulfate as dehydration agent and acetic acid as catalyst gave 1,2,4-triazolidines in 46 to 60% yields.[1] For example, treatment of **18.1-1** with valeraldehyde apparently gave the 1,3,4-oxadiazolidine intermediate (**18.1-2**), which was condensed with butylamine to give **18.1-3** in 55% yield. Higher yields of the 1,2,4-triazolidines were obtained by the reaction of the isolated 1,3,4-oxadiazolidines with primary amines.[2]

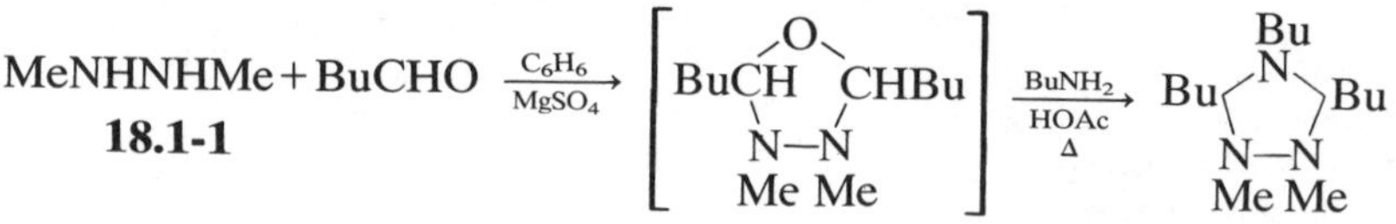

MeNHNHMe + BuCHO $\xrightarrow[\text{MgSO}_4]{\text{C}_6\text{H}_6}$

18.1-1

$$\left[\begin{array}{c} \text{BuCH} \quad \text{CHBu} \\ | \quad\quad | \\ \text{N—N} \\ \text{Me Me} \end{array}\right] \xrightarrow[\text{HOAc}]{\text{BuNH}_2} \begin{array}{c} \text{Bu} \\ \end{array}$$

18.1-2

18.1-3

The interaction of phenyl isocyanate with the azomethine imine (**18.1-4**) in carbon tetrachloride at room temperature gave the triazolidinone (**18.1-5**) in 86% yield.[3] In another type of ylide reaction, C—C bond scission

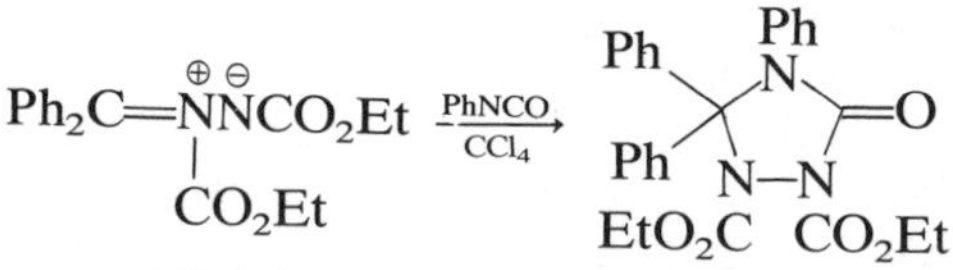

$$\text{Ph}_2\text{C}=\overset{\oplus}{\text{N}}\overset{\ominus}{\text{N}}\text{CO}_2\text{Et} \xrightarrow[\text{CCl}_4]{\text{PhNCO}}$$
$$\underset{\text{CO}_2\text{Et}}{|}$$

18.1-4

18.1-5

occurred on treatment of 1,2,4-triphenylaziridine (**18.1-6**) with diethyl azodicarboxylate in refluxing toluene to give via intermediate **18.1-7** an 82% yield of **18.1-8**.[4] Similarly, the aziridine (**18.1-9**) reacted with diethyl

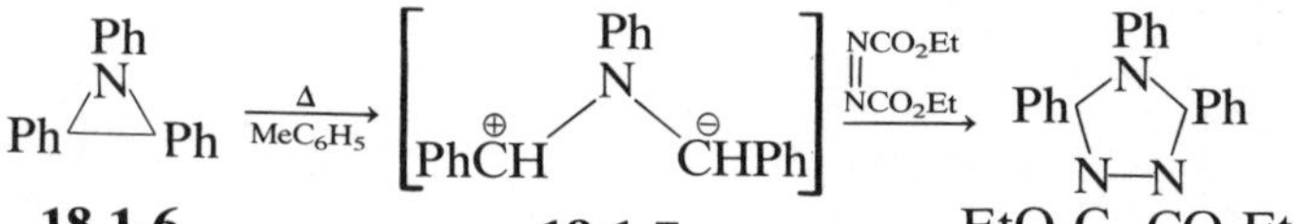

18.1-6

$$\left[\begin{array}{c} \text{Ph} \\ \overset{\oplus}{\text{PhCH}} \quad \overset{\ominus}{\text{CHPh}} \end{array}\right] \xrightarrow[\text{NCO}_2\text{Et}]{\text{NCO}_2\text{Et}}$$

18.1-7

18.1-8

516

azodicarboxylate in toluene at 100° to give a good yield of the triazolidine (**18.1-10**). In this reaction, the cis-dicarboxylate (**18.1-9**) gave mainly the trans-3,5-dicarboxylate (**18.1-10**), and trans-dicarboxylate (**18.1-9**) gave mainly the cis-3,5-dicarboxylate (**18.1-10**).[5-7] The ylide (**18.1-14**) was generated from **18.1-16**, which was formed from the interaction of tetracyano-ethylene oxide (**18.1-15**) with benzylidene aniline. Reaction of **18.1-14** with diethyl azodicarboxylate gave a 40% yield of **18.1-11**.[8] Similarly, azobenzene reacted with the lithium compound (**18.1-12**) under mild conditions to give the cycloadduct (**18.1-13**).[9] Diaziridines are also converted to triazolid-

ines as demonstrated by the addition of benzoyl isocyanate to **18.1-17** to give **18.1-19**, presumably formed via rearrangement of intermediate **18.1-18**.[10,11] In addition, the oxadiazolidine (**18.1-20**) underwent cleavage

of the ring N—O bond to give **18.1-21** followed by N—N bond formation to give **18.1-22** in 72% yield.[12]

18.1-20   **18.1-21**   **18.1-22**

Several 1,2,4-triazolidinones have been prepared from semicarbazones intermediates. In a patent report the condensation of formaldehyde with the semicarbazide (**18.1-23**) in methanolic potassium hydroxide to give **18.1-24** is described.[13,14] The same type of reaction has been reported for the

18.1-23   **18.1-24**

condensation of formaldehyde and 1,4-diphenylthiosemicarbazide.[15] The tautomeriaztion of semicarbazones (e.g., **18.1-25**) to triazolidinones (e.g., **18.1-26**) under acidic conditions has also been observed.[16,17] The facile

18.1-25   **18.1-26**

nature of this type of cyclization was demonstrated by the condensation of benzoyl isocyanate with the hydrazone (**18.1-27**) to give directly the 4-benzoyltriazolidinone (**18.1-29**).[18,19] Apparently, this reaction involves the initial formation of the semicarbazone (**18.1-28**). Similar reactions were

18.1-27   **18.1-28**   **18.1-29**

encountered in cycloaddition of thiocyanic acid with hydrazones (e.g., **18.1-30**) to give triazolidinethiones (e.g., **18.1-31**).[20-23] Thermolysis of **18.1-31** in 2-butanol gave the 2-phenylthiosemicarbazone (**18.1-32**).[22] In contrast, oxidation of **18.1-31** with either mercury (II) oxide or potassium permanganate resulted in ring opening between N-1 and C-5 to give **18.1-33**.[24]

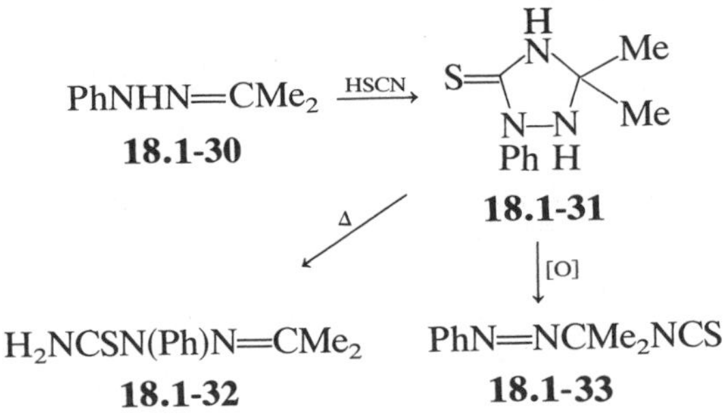

$$PhNHN{=}CMe_2 \xrightarrow{\text{HSCN}}$$

18.1-30

18.1-31

$$\downarrow \text{[O]}$$

$$H_2NCSN(Ph)N{=}CMe_2 \qquad PhN{=}NCMe_2NCS$$

18.1-32   **18.1-33**

Other methods for the preparation of triazolidinones include the reaction of the carbamoyl chloride (**18.1-34**) with 1,2-dimethylhydrazine (**18.1-1**) to give **18.1-35**,[25] and the condensation of the methyl carbazate (**18.1-36**) with

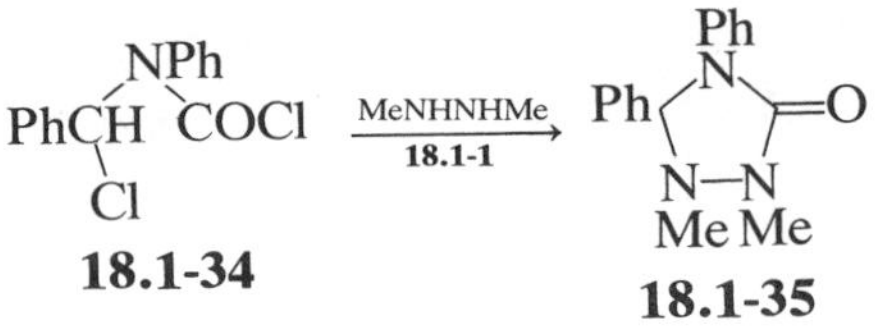

18.1-34   **18.1-35**

diphenylcarbodiimide to give **18.1-37**.[26] In the latter reaction, the triazoline (**18.1-38**) was formed as a by-product. Also, several triazolidinones (e.g.,

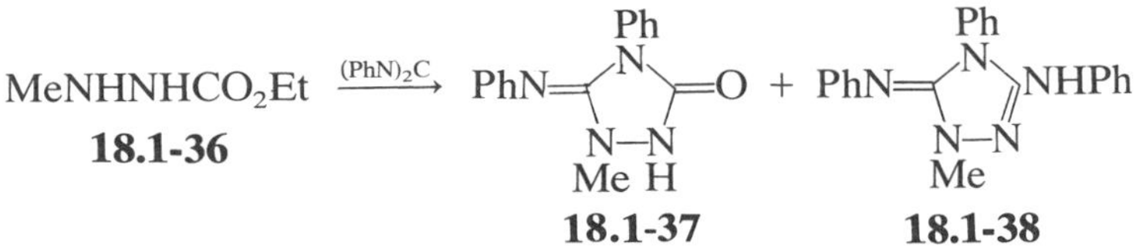

$$MeNHNHCO_2Et \xrightarrow{(PhN)_2C}$$

18.1-36

18.1-37   **18.1-38**

18.1-42) were prepared by the condensation of aryl cyanates (e.g., **18.1-40**) with 4-arylsemicarbazides (e.g., **18.1-39**).[27] Apparently, the triazolidinones were formed by the elimination of a phenol from the diaddition intermediate (**18.1-41**).

The triazolium compound (**18.1-43**)[28] and the corresponding sulfur analog (**18.1-44**)[29] are both reduced by lithium aluminum hydride in dioxane to give the triazolidines (**18.1-45** and **18.1-46**, respectively). In the reduction of unquaternary triazolin-5-ones, the position of the ring double bond determined the structure of the product. For example, the ring double bond of

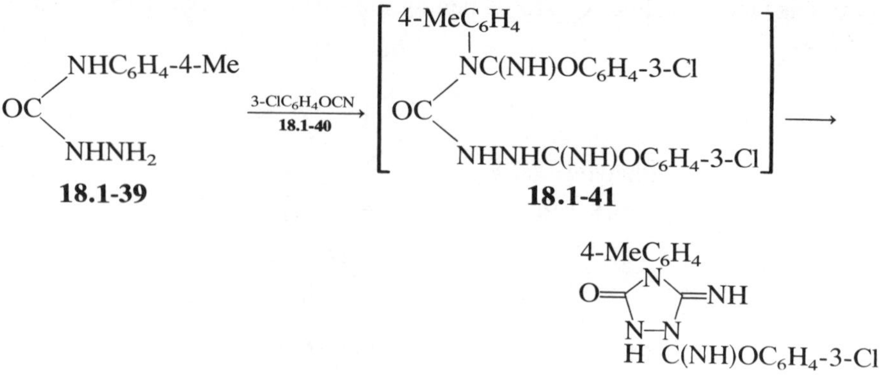

$$\underset{\textbf{18.1-39}}{\overset{\displaystyle NHC_6H_4\text{-}4\text{-}Me}{OC{\Big\langle}}{\underset{\displaystyle NHNH_2}{}}} \quad\xrightarrow[\textbf{18.1-40}]{3\text{-}ClC_6H_4OCN}\quad \left[\,\underset{\textbf{18.1-41}}{\overset{\displaystyle 4\text{-}MeC_6H_4}{\underset{\displaystyle NC(NH)OC_6H_4\text{-}3\text{-}Cl}{OC{\Big\langle}}}}\right]\longrightarrow$$

Reaction leading to **18.1-42**:

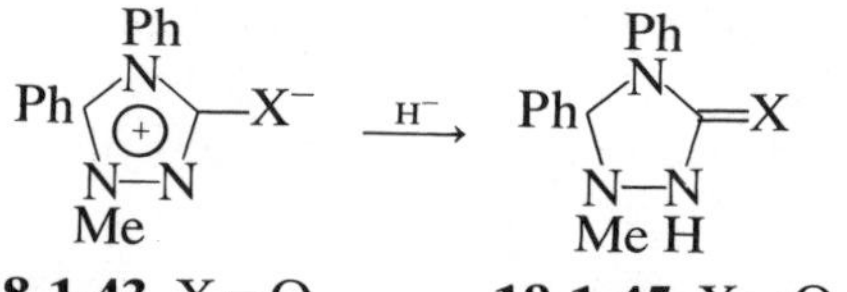

4-MeC$_6$H$_4$

18.1-42

Structures **18.1-43, X = O**; **18.1-44, X = S** converting via H⁻ to **18.1-45, X = O**; **18.1-46, X = S**:

Δ^3-triazolin-5-ones (e.g., **18.1-47**) was reduced with lithium aluminum hydride to give triazolidin-5-ones (e.g., **18.1-48**), whereas the carbonyl group of Δ^2-triazolin-5-ones (e.g., **18.1-49**) was reduced followed by elimination of water to give triazoles.[30] Although the intermediate 5-hydroxy-1,2,4-triazol-

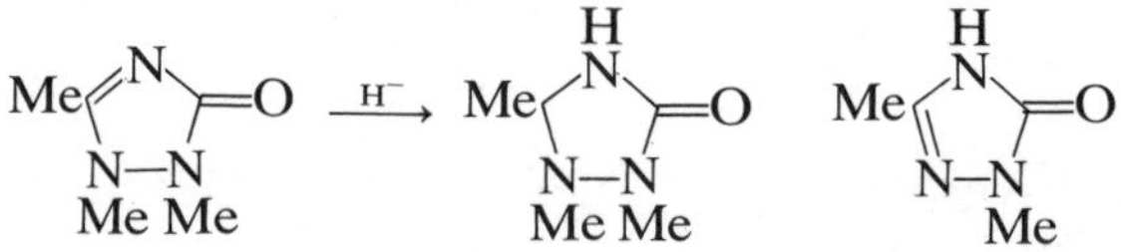

18.1-47   **18.1-48**   **18.1-49**

idine was unisolated in the latter reaction, the hydride reduction of the bicyclic heterocyclic (**18.1-50**) in dioxane was reported to give **18.1-51**.[31]

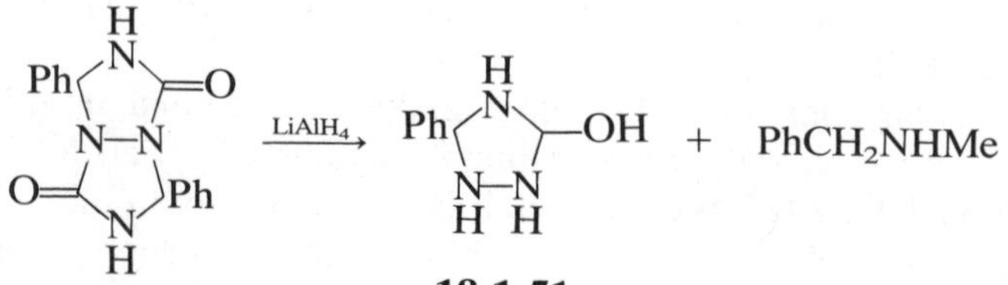

18.1-50   **18.1-51**

References

1. **61:** 1850d
5. **74:** 13072c
9. **74:** 53877a
13. **83:** 164191c
17. **76:** 98991r
21. **78:** 4193y
25. **78:** 124509c
29. **81:** 3847q

2. **62:** 14658d
6. **75:** 48993b
10. **79:** 137008j
14. **84:** 31088y
18. **82:** 42707j
22. **79:** 92113w
26. **81:** 91439m
30. **76:** 59537c

3. **63:** 11544d
7. **77:** 113608n
11. **81:** 169457n
15. **68:** 29649h
19. **85:** 46601e
23. **85:** 5647f
27. **63:** 11550g
31. **51:** 14700i

4. **66:** 28717a
8. **76:** 112984z
12. **77:** 152066d
16. **71:** 60385e
20. **78:** 147879n
24. **81:** 136069j
28. **81:** 3835j

TABLE 18. 1,2,4-TRIAZOLIDINES

Compound	Reference
-4-acetic acid, 1,2,3,5-tetramethyl-, ethyl ester	**61:** 1850e, **62:** 14658d
4-butyl-3,5-diethyl-1,2-dimethyl-	**61:** 1850e, **62:** 14658d
4-butyl-1,2-dimethyl-3,5-dipropyl-	**61:** 1850e
4-butyl-1,2,3,5-tetramethyl-	**61:** 1850e, **62:** 14658d
-1-carboxanilide, 3-oxo-4,5,5-triphenyl-	**63:** 11545a
-1-carboximidic acid, 5-imino-3-oxo-4-phenyl-, m-chlorophenyl ester	**63:** 11550g
-1-carboximidic acid, 5-imino-3-oxo-4-phenyl-, phenyl ester	**63:** 11550g
-1-carboximidic acid, 5-imino-3-oxo-4-p-tolyl-, m-chlorophenyl ester	**63:** 11550g
-1-carboximidic acid, 5-imino-3-oxo-4-p-tolyl-, 2,4-xylyl exter	**63:** 11550g
-1-carboxylic acid, 3-oxo-4,5,5-triphenyl-, ethyl ester	**63:** 11544f
1,4-diacetyl-3,5-bis(acetylimino)-2-phenyl-	**61:** P4541d
-1,2-dicarboxylic acid, 3-benzoyl-4-cyclohexyl-5-(3-nitrophenyl)-, diethyl ester, trans-	**77:** 113608n
-1,2-dicarboxylic acid, 3-benzoyl-4-cyclohexyl-5-(2-thienyl)-, dimethyl ester	**74:** 13072c
-1,2-dicarboxylic acid, 3-benzoyl-4-isopropyl-5-(3-nitrophenyl)-, diethyl ester, trans-	**77:** 113608n
-1,2-dicarboxylic acid, 3-benzoyl-4-isopropyl-5-(3-nitrophenyl)-, diphenyl ester, trans-	**77:** 113608n
-1,2-dicarboxylic acid, 4-cyclohexyl-3-(2-naphthoyl)-5-(2-thienyl)-, dimethyl ester	**74:** 13072c
-1,2-dicarboxylic acid, 3,3-dicyanodihydro-4,5-diphenyl-, diethyl ester	**76:** 112984z
-1,2-dicarboxylic acid, 4-(p-methoxyphenyl)-3,5-diphenyl-, dimethyl ester	**75:** 48993b
-3,5-dicarboxylic acid, 4-(p-methoxyphenyl)-1,2-diphenyl-, dimethyl ester	**75:** 48993b
-1,2-dicarboxylic acid, 5-oxo-3,3,4-triphenyl-, diethyl ester	**63:** 11544f
-1,2-dicarboxylic acid, 3,4,5-triphenyl-, diethyl ester	**66:** 28717a
3-hydroxy-5-phenyl-	**51:** 14700i
3,3-diphenyl-1,2-di(3-pyridinyl)-	**74:** 53877a

TABLE 18 (*Continued*)

Compound	Reference
3,3-diphenyl-1,2-di-*p*-tolyl-	**74:** 53877a
-3-one, 4-benzoyl-1,2-diethyl-5-methyl-	**81:** 169457n
-3-one, 4-benzoyl-5,5-dimethyl-2-(4-methylphenyl)-	**82:** 42707j, **85:** 46601e
-3-one, 4-benzoyl-5,5-dimethyl-2-(4-nitrophenyl)-	**85:** 46601e
-3-one, 4-benzoyl-5,5-dimethyl-2-phenyl-	**82:** 42707j, **85:** 46601e
-3-one, 4-benzoyl-1,2,5-triethyl-	**81:** 169457n
-3-one, 4-benzoyl-2,5,5-trimethyl-	**82:** 42707j, **85:** 46601e
-3-one, 4-(4-bromophenyl)-2-ethyl-	**83:** 164191c, **84:** P31088y, **84:** P55326a, **84:** P121839y
-3-one, 2-butyl-4-[5-(methylsulfonyl)-1,3,4-thiadiazol-2-yl]-	**84:** P31088y
-3-one, 4-(4-chloro-2-methylphenyl)-2-methyl-	**83:** 164191c, **84:** P31088y, **84:** P55326a, **84:** P121839y
-3-one, 4-(3-chlorophenyl)-2-methyl-	**83:** 164191c, **84:** P31088y, **84:** P55326a, **84:** P121839y
-3-one, 4-(5-cyclobutyl-1,3,4-thiadiazol-2-yl)-2-methyl-	**84:** P31088y
-5-one, 4-cyclohexyl-3-(cyclohexylimino)-2-methyl-	**81:** 91439m
-3-one, 1-cyclohexyl-5-methyl-4,5-diphenyl-	**79:** 137008j
-3-one, 4-(5-cyclohexyl-1,3,4-thiadiazol-2-yl)-2-methyl-	**84:** P31088y
-3-one, 4-(3,4-dichlorophenyl)-2-methyl-	**83:** 164191c, **84:** P31088y, **84:** P55326a
-3-one, 5-[4,7-dimethoxy-6-(1-methylethoxy)-5-benzofuranyl]-5-methyl-	**76:** 98991r
-5-one, 1,3-dimethyl-	**76:** 59537c
-3-one, 1,2-dimethyl-4,5-diphenyl-	**78:** 124509c
-3-one, 4-[5-(1,1-dimethylethyl)-1,3,4-thiadiazol-2-yl]-2-methyl	**84:** P31088y
-3-one, 1-(1,1-dimethylethyl)-2,4,5-triphenyl-	**77:** 152066d
-3-one, 5,5-dimethyl-2-(4-methylphenyl)-4-(phenyl-thioxomethyl)-	**82:** 42707j, **85:** 46601e
-3-one, 5,5-dimethyl-2-phenyl-	**71:** 60385e
-3-one, 5,5-dimethyl-2-phenyl-4-(phenylthioxomethyl)-	**82:** 42707j, **85:** 46601e
-3-one, 1,4-diphenyl-	**79:** P25673r
-3-one, 5-(6-ethoxy-4,7-dimethoxy-5-benzofuranyl)-5-methyl-	**76:** 98991r
-3-one, 5-ethylidene-4-(tetrahydro-2-oxo-2H̲-1,3-thiazin-4-yl)-1-methyl-2-phenyl-	**75:** 156464p
-3-one, 5-ethylidene-4-(tetrahydro-2-thioxo-2H̲-1,3-thiazin-4-yl)-1-methyl-2-phenyl-	**75:** 156464p
-3-one, 2-ethyl-4-(5-methyl-1,3,4-thiadiazol-2-yl)-	**84:** P31088y
-3-one, 5-(3-ethyl-4-oxo-2-thioxo-5-thiazolidinylidene)-1,2-dimethyl-4-phenyl-	**78:** P130586s
-3-one, 5-ethyl-2-phenyl-4-(phenylthioxomethyl)-	**85:** 46601e
-3-one, 2-hexyl-4-[5-(methylsulfinyl)-1,3,4-thiadiazol-2-yl]-	**84:** P31088y
-3-one, 5-hydroxy-5-(mercaptomethyl)-	**74:** 87899z
-3-one, 5-[7-hydroxy-4-methoxy-6-(1-methylethoxy)-5-benzofuranyl]-5-methyl-	**76:** 98991r
-5-one, 4-isopropyl-3-(isopropyl-3-(isopropylimino)-2-methyl-	**81:** 91439m

TABLE 18 (*Continued*)

Compound	Reference
-3-one, 5-(4-methoxyphenyl)-4-[3-(4-methoxyphenyl)-1-oxo-2-thioxopropyl]-1-phenyl-	**85:** 32926c
-3-one, 4-(2-methoxyphenyl-2-methyl-	**85:** 32926c **83:** 164191c, **84:** P31088y, **84:** P55326a, **84:** P121839y
-3-one, 4-[3-(4-methoxyphenyl)-1-oxo-2-thioxopropyl]-1,5-diphenyl-	**85:** 32926c
-3-one, 4-[3-(4-methoxyphenyl)-1-oxo-2-thioxopropyl]-1-phenyl-	**85:** 32926c
-3-one, 4-(5-methoxy-1,3,4-thiadiazol-2-yl)-2-propyl-	**84:** P31088y
-3-one, 1-methyl-4,5-diphenyl-	**75:** 151737v, **81:** 3835j
-3-one, 5-methyl-2,5-diphenyl-	**71:** 60385e
-3-one, 2-methyl-4-[4-(methylthio)phenyl]-	**83:** 164191c
-3-one, 3-methyl-4-[4-(methylthio)phenyl]-	**84:** P31088y, **84:** P55326a, **84:** P121839y
-3-one, 2-methyl-4-[5-(methylthio)-1,3,4-thiadiazol-2-yl]-	**84:** P31088v
-3-one, 2-methyl-4-(4-nitrophenyl)-	**83:** 164191c, **84:** P31088y, **84:** P55326a, **84:** P121839y
-5-one, 2-methyl-4-phenyl-3-(phenylimino)-	**81:** 91439m
-5-one, 2-methyl-4-(2-tolyl)-3-(2-tolylimino)-	**81:** 91439m
-3-one, 2-methyl-4-[4-(trifluoromethyl)phenyl]-	**83:** 164191c, **84:** P31088y, **84:** P55326a, **84:** P121839y
-3-one, 2-methyl-4-[5-(trifluoromethyl)-1,3,4-thiadiazol-2-yl]-	**84:** P31088y
-3-one, 4-methyl-1,2,5-triphenyl-	**78:** 124509c
-3-one, 4-(1-oxo-3-phenyl-2-thioxopropyl)-1,5-diphenyl-	**85:** 32926c
-3-one, 4-(1-oxo-3-phenyl-2-thioxopropyl)-1-phenyl-	**85:** 32926c
-3-one-5-selone	**75:** P76805k
-3-one, 1,2,4,5-tetraphenyl-	**77:** 152066d, **78:** 124509c
-3-one, 1,2,5-triethyl-	**81:** 169457n
-5-one, 1,2,3-trimethyl-	**76:** 59537c
-3-one, 1,2,4-trimethyl-5-phenyl-	**78:** 124509c
-3-one, 2,5,5-trimethyl-4-(phenylthioxomethyl)-	**82:** 42707j
-1,2,3,5-tetracarboxylic acid, 4-(*p*-methoxyphenyl)-, 1,2-diethyl dimethyl ester, *cis*-	**71:** 91194w, **75:** 48993b
-1,2,3,5-tetracarboxylic acid, 4-(*p*-methoxyphenyl)-, 1,2-diethyl dimethyl ester, *trans*-	**71:** 91194w, **75:** 48993b
-1,2,3,5-tetracarboxylic acid, 4-(*p*-methoxyphenyl)-, tetramethyl ester	**75:** 48993b
1,2,3,5-tetramethyl-4-phenyl-	**61:** 1850e, **62:** 14658d
1,2,3,3-tetraphenyl-	**74:** 53877a
1,2,3,5-tetraphenyl-	**74:** 53877a
-3-thione, 5-butyl-5-methyl-2-phenyl-	**78:** 4193y, **85:** P5647f
-3-thione, 5-tert-butyl-5-methyl-2-phenyl-	**78:** 4193y, **79:** 92113w, **85:** P5647f
-3-thione, 2-(3-chloro-4-methoxyphenyl)-5,5-dimethyl-	**85:** P5647f
-3-thione, 2-(3-chloro-4-methylphenyl)-5,5-dimethyl-	**85:** P5647f
-3-thione, 2-(4-chlorophenyl)-5,5-diethyl-	**85:** P5647f
-3-thione, 2-(2-chlorophenyl)-5,5-dimethyl-	**85:** P5647f
-3-thione, 2-(3-chlorophenyl)-5,5-dimethyl-	**85:** P5647f

TABLE 18 (*Continued*)

Compound	Reference
-3-thione, 2-(4-chlorophenyl)-5,5-dimethyl-	**85:** P5647f
-3-thione, 2-(4-chlorophenyl)-5-ethyl-5-methyl-	**85:** P5647f
-3-thione, 2-[(4-chlorophenyl)methyl]-5,5-dimethyl-	**85:** P5647f
-3-thione, 2-(3,4-dichlorophenyl)-5,5-dimethyl-	**85:** P5647f
-3-thione, 5,5-diethyl-	**78:** 147879n
-3-thione, 5,5-diethyl-2-(4-methylphenyl)-	**85:** P5647f
-3-thione, 5,5-diethyl-2-phenyl-	**78:** 4193y, **79:** 92113w, **85:** P5647f
-3-thione, 5,5-dimethyl-	**78:** 147879n
-3-thione, 5,5-dimethyl-2-(2-methylphenyl)-	**85:** P5647f
-3-thione, 5,5-dimethyl-2-(4-methylphenyl)-	**85:** P5647f
-3-thione, 5,5-dimethyl-2-(3-nitrophenyl)-	**85:** P5647f
-3-thione, 2,4-dimethyl-1-phenyl-	**72:** 37527q
-3-thione, 2,5-dimethyl-5-phenyl-	**85:** P5647f
-3-thione, 5,5-dimethyl-2-phenyl-	**78:** 4193y, **79:** 92113w, **81:** 136069j, **85:** P5647f
-3-thione, 5,5-dimethyl-2-(1-phenylethyl)-	**85:** P5647f
-3-thione, 1,4-diphenyl-	**68:** 29649h
-3-thione, 5,5'-(1,2-ethanediyl)bis[5-methyl-2-phenyl-	**78:** 4193y
-3-thione, 5-ethyl-	**78:** 147879n
-3-thione, 5-ethyl-2,5-dimethyl-	**85:** P5647f
-3-thione, 5-ethyl-2,5-diphenyl-	**78:** 4193y, **85:** P5647f
-3-thione, 5-ethyl-5-methyl-	**72:** 31701v, **73:** 25036m, **78:** 147879n
-3-thione, 5-ethyl-5-methyl-2-(4-methylphenyl)-	**85:** P5647f
-3-thione, 5-ethyl-5-methyl-2-phenyl-	**78:** 4193y, **79:** 92113w
-3-thione, 5-ethyl-2-phenyl-	**79:** 92113w, **85:** P5647f
-3-thione, 2-(4-fluorophenyl)-5,5-dimethyl-	**85:** P5647f
-5-thione, 3-hydroxy-3-mercapto-4-methyl-	**82:** 43271z
-3-thione, 5-isopropyl-5-methyl-2-phenyl-	**79:** 92113w
-3-thione, 5-isopropyl-2-phenyl-	**79:** 92113w
-3-thione, 5-mercapto-4-methyl-5-(methylamino)-	**79:** P137163f
-3-thione, 5-mercapto-4-methyl-5-[(1-methyl-2-propenyl)amino]-	**79:** P137163f, **82:** 43271z
-3-thione, 2-(4-methoxyphenyl)-5,5-dimethyl-	**85:** P5647f
-3-thione, 5-methyl-	**78:** 147879n
-3-thione, 1-methyl-4,5-diphenyl-	**81:** 3847q
-3-thione, 5-methyl-2,5-diphenyl-	**78:** 4193y, **85:** P5647f
-3-thione, 5-(1-methylethyl)-2-phenyl-	**85:** P5647f
-3-thione, 5-methyl-2-(4-methylphenyl)-5-propyl-	**85:** P5647f
-3-thione, 5-methyl-5-(2-methylpropyl)-2-phenyl-	**78:** 4193y, **79:** 92113w, **85:** P5647f
-3-thione, 5-methyl-5-pentyl-2-phenyl-	**78:** 4193y, **85:** P5647f
-3-thione, 5-methyl-2-phenyl-	**79:** 92113w, **85:** P5647f
-3-thione, 5-methyl-2-phenyl-5-(2-phenylethyl)-	**78:** 4193y, **85:** P5647f
-3-thione, 5-methyl-2-phenyl-5-propyl-	**78:** 4193y, **79:** 92113w, **85:** P5647f
-3-thione, 5-methyl-5-propyl-	**78:** 147879n
-3-thione, 2-phenyl-5,5-dipropyl-	**85:** P5647f
-5-thione, 1-phenyl-3-(1-phenyl-5-thioxo-1,2,4-triazolidin-3-yl)-	**78:** 4193y

TABLE 18 (*Continued*)

Compound	Reference
-3-thione, 2-phenyl-5-propyl-	**78:** 4193y, **79:** 92113w, **85:** P5647f
-3-thione, 5-propyl-	**78:** 147879n
-3-thione, 1,2,4-trimethyl-	**72:** 37527q
-3-thione, 2,5,5-trimethyl-	**85:** P5647f
3,4,5-tributyl-1,2-dimethyl-	**61:** 1850e
-1,2,3-tricarboxylic acid, 4-isopropyl-5-(3-nitrophenyl)-, trimethyl ester, *cis*-	**77:** 113608n
-1,2,3-tricarboxylic acid, 4-isopropyl-5-(3-nitrophenyl)-, trimethyl ester, *trans*-	**77:** 113608n
-1,3,5-tricarboxylic acid, 4-(*p*-methoxyphenyl)-2-phenyl-, trimethyl ester	**75:** 48993b

CHAPTER 19

1,2,4-Triazolidine-3,5-diones

19.1. Alkyl- or Aryl-1,2,4-Triazolidine-3,5-diones

The hydrazinolysis of a refluxing aqueous solution of methyl allophanate (**19.1-1**) with hydrazine hydrate gave 1,2,4-triazolidine-3,5-dione (**19.1-2**) in 78% yield.[1] This type of reaction was also successful in the condensation of substituted hydrazines with allophanates to give N-substituted 1,2,4-triazolidine-3,5-diones.[2,3] Also, hydrolysis of 3,5-dichloro-1,2,4-triazole with potassium hydroxide in ethylene glycol at 200° was reported to give a 80% yield of **19.1-2**.[4] Other methods for the preparation of **19.1-2** have been reviewed.[1]

Apparently, poor yields of 4-alkyl- and 4-aryl-1,2,4-triazolidine-3,5-diones (e.g., **19.1-3** and **19.1-4**) were obtained either by the fusion of 1-(carbamoyl)semicarbazide (**19.1-6**) with aliphatic or arylamines[5] or by the cyclization of 1-(N-substituted carbamoyl)semicarbazides (e.g., **19.1-7**) with aqueous sodium hydroxide.[6] However, a good yield of **19.1-5** resulted from the fusion of the N,N'-dibutyl derivative (**19.1-8**).[7] In some instances, good yields of 1,4-diaryl-1,2,4-triazolidine-3,5-diones (e.g., **19.1-9**) were obtained in the fusion of a mixture of an arylhydrazine (e.g., **19.1-10**), an aniline (e.g., **19.1-11**), and urea in the molar ration of $1:2:2$.[8–11] Presumably, these reactions involved the formation of a 1-(carbamoyl)semicarbazide intermediate such as **19.1-12**. In the reaction of urea with phenyl hydrazine, the product was found to depend upon the rate of heating. When urea and phenyl hydrazine were heated slowly to 160°, biuret was initially formed, which reacted with phenyl hydrazine to give **19.1-13**, followed by cyclization to give 1-phenyl-1,2,4-triazolidine-3,5-dione (**19.1-14**).[12,13] When the reactants were heated rapidly, 1-phenylsemicarbazide was initially formed, which after further reaction with phenyl hydrazine was converted to 4-anilino-1-phenyl-1,2,4-triazolidine-3,5-dione. Apparently, the 4-carbamoyl-semicarbazide (**19.1-15**) was an intermediate in the condensation of diethyl-carbamoyl isocyanate with 1,2-dimethyl hydrazine to give **19.1-16**.[14]

The 1-(ethoxycarbonyl)semicarbazides are readily converted either by fusion or in a basic medium to 1,2,4-triazolidine-3,5-diones. The precursor of 4-phenyl-1,2,4-triazoline-3,5-dione (see Section 15.4), 4-phenyl-1,2,4-triazolidine-3,5-dione (**19.1-28**), was prepared in yields up to 95% by the

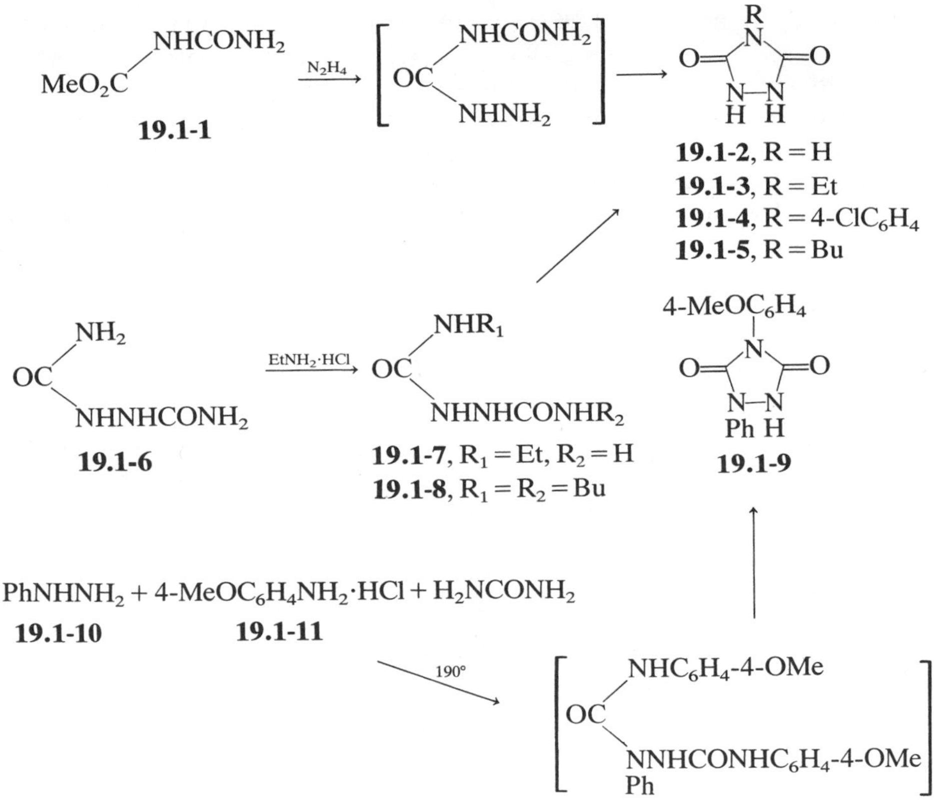

$$MeO_2C-NHCONH_2 \xrightarrow{N_2H_4} \left[OC\begin{array}{l} NHCONH_2 \\ NHNH_2 \end{array} \right] \longrightarrow$$

19.1-1

R-1,2,4-triazolidine-3,5-dione

19.1-2, R = H
19.1-3, R = Et
19.1-4, R = 4-ClC$_6$H$_4$
19.1-5, R = Bu

$$OC\begin{array}{l} NH_2 \\ NHNHCONH_2 \end{array} \xrightarrow{EtNH_2 \cdot HCl} OC\begin{array}{l} NHR_1 \\ NHNHCONHR_2 \end{array}$$

19.1-6

19.1-7, R$_1$ = Et, R$_2$ = H
19.1-8, R$_1$ = R$_2$ = Bu

4-MeOC$_6$H$_4$

19.1-9

PhNHNH$_2$ + 4-MeOC$_6$H$_4$NH$_2$·HCl + H$_2$NCONH$_2$

19.1-10   **19.1-11**

$$\xrightarrow{190°} \left[OC\begin{array}{l} NHC_6H_4\text{-}4\text{-}OMe \\ NNHCONHC_6H_4\text{-}4\text{-}OMe \\ Ph \end{array} \right]$$

19.1-12

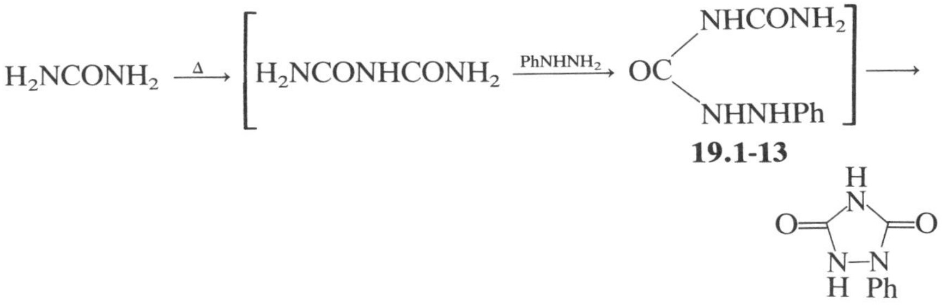

$$H_2NCONH_2 \xrightarrow{\Delta} \left[H_2NCONHCONH_2 \xrightarrow{PhNHNH_2} OC\begin{array}{l} NHCONH_2 \\ NHNHPh \end{array} \right] \longrightarrow$$

19.1-13

19.1-14

$$Et_2NCONCO + MeNHNHMe \longrightarrow \left[OC\begin{array}{l} NHCONEt_2 \\ NNHMe \\ Me \end{array} \right] \longrightarrow$$

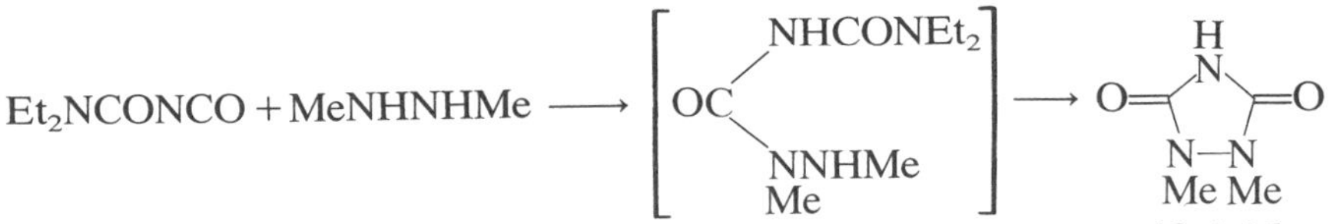

19.1-15   **19.1-16**

alkaline cyclization of **19.1-20**.[15] A similar procedure was used to prepare 4-alkyl-,[16] 1,2-dialkyl-,[17] 1,4-dialkyl-,[18,19] 1,2,4-trialkyl-,[17] 1-alkyl-4-aryl-,[20] 4-alkyl-1-aryl-,[21] 4-aryl-,[22] and 1,4-diaryl-1,2,4-triazolidine-3,5-diones[9,10,23] in good yields, as illustrated by the conversion of **19.1-21** to **19.1-29**.[17] In addition, a large number of 1,2,4-triazolidine-3,5-diones were obtained from 1-(ethoxycarbamoyl)semicarbazides prepared *in situ* from simpler reagents. The fusion of the ethyl carbazate (**19.1-17**) with cyclohexyl isocyanate at 200° provided **19.1-30** directly, presumably formed via **19.1-22**.[21]

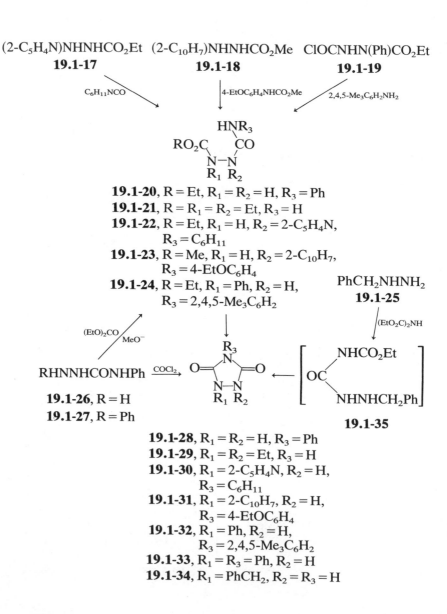

(2-C5H4N)NHNHCO2Et
19.1-17
(2-C10H7)NHNHCO2Me
19.1-18
ClOCNHN(Ph)CO2Et
19.1-19
C6H11NCO
4-EtOC6H4NHCO2Me
2,4,5-Me3C6H2NH2
HNR3
RO2C
CO
N—N
R1 R2
19.1-20, R = Et, R1 = R2 = H, R3 = Ph
19.1-21, R = R1 = R2 = Et, R3 = H
19.1-22, R = Et, R1 = H, R2 = 2-C5H4N, R3 = C6H11
19.1-23, R = Me, R1 = H, R2 = 2-C10H7, R3 = 4-EtOC6H4
19.1-24, R = Et, R1 = Ph, R2 = H, R3 = 2,4,5-Me3C6H2
PhCH2NHNH2
19.1-25
(EtO2C)2NH
(EtO)2CO
MeO−
RHNNHCONHPh
COCl2
R3
N
O
O
N—N
R1 R2
NHCO2Et
OC
NHNHCH2Ph
19.1-26, R = H
19.1-27, R = Ph
19.1-35

19.1-28, $R_1 = R_2 = H$, $R_3 = Ph$
19.1-29, $R_1 = R_2 = Et$, $R_3 = H$
19.1-30, $R_1 = 2\text{-}C_5H_4N$, $R_2 = H$, $R_3 = C_6H_{11}$
19.1-31, $R_1 = 2\text{-}C_{10}H_7$, $R_2 = H$, $R_3 = 4\text{-}EtOC_6H_4$
19.1-32, $R_1 = Ph$, $R_2 = H$, $R_3 = 2,4,5\text{-}Me_3C_6H_2$
19.1-33, $R_1 = R_3 = Ph$, $R_2 = H$
19.1-34, $R_1 = PhCH_2$, $R_2 = R_3 = H$

Similarly, fusion of the carbazate (**19.1-18**) with methyl 4-ethoxyphenyl-carbamate at 150 to 200° apparently gave intermediate **19.1-23**, from which **19.1-31** was obtained.[10,11,21] The reaction of the N-(chlorocarbonyl)carbazate (**19.1-19**) with 2,4,5-trimethylaniline in N,N-dimethylformamide was reported to give a solution of **19.1-24**, which was cyclized with aqueous alkali to give **19.1-32**.[10,11,21,24–26] In the base-catalyzed condensation of 4-phenylsemicarbazide (**19.1-26**) with diethyl carbonate, apparently intermediate **19.1-20** was first formed, followed by cyclization of the latter to give **19.1-28**.[21,27–30] In a related reaction, treatment of 1,4-diphenylsemi-carbazide (**19.1-27**) with the phosgene-pyridine complex in tetrahydrofuran gave **19.1-33** directly.[31,32] Also, triazolidine-3,5-diones (e.g., **19.1-34**) are obtained from 4-(ethoxycarbonyl)semicarbazide intermediates (e.g., **19.1-35**), which are formed *in situ* either by fusion of hydrazines (e.g., **19.1-25**) and diethyl iminodicarboxylate[11,33] or by heating a solution of these two reactants in xylene in the presence of base.[10,21,29,34–36]

The acid-catalyzed reaction of equimolar amounts of the oxadiazolin-5-one (**19.1-36**) with 4-ethoxyaniline at 200° resulted in ring-opening and reclosure to give a low yield of the 1,4-diaryl triazolidine-3,5-dione (**19.1-37**).[9,11,21,37] Similarly, the imino-1,3,4-oxadiazolidin-5-one (**19.1-38**)

rearranged to **19.1-39** under alkaline conditions.[38] In a related reaction, the exchange of the 4-NH of the 1,2,4-triazolidine-3,5-dione (**19.1-16**) for the

amino group of 4-methylaniline was reported to occur at high temperatures to give **19.1-40**.[9]

Dimethylamino isocyanate (**19.1-41**) was found to exist as a dimer, the mesoionic 4-aminotriazolidine-3,5-dione (**19.1-42**), which on equilibration with butyl isocyanate gave the meso-ionic triazolidine-3,5-dione (**19.1-44**).[39,40] This interesting compound underwent methyl group migration to give **19.1-45** on heating and underwent demethylation to give **19.1-43** in the presence of hydrochloric acid. The pyrolysis of the amino-2-benzoyl-

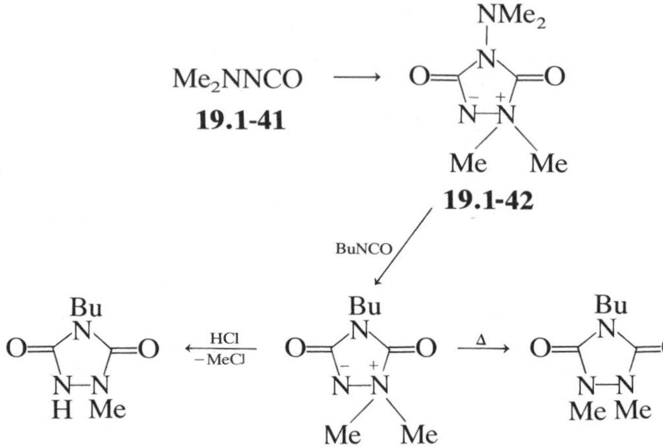

Me$_2$NNCO
19.1-41

19.1-42

BuNCO

19.1-43

$\xrightarrow[-\text{MeCl}]{\text{HCl}}$

19.1-44

$\xrightarrow{\Delta}$

19.1-45

imide (**19.1-46**) gave a number of products, including the triazolidine-3,5-dione (**19.1-48**).[30] The formation of the latter occurred by elimination of 1-hexene from the meso-ionic intermediate (**19.1-47**).

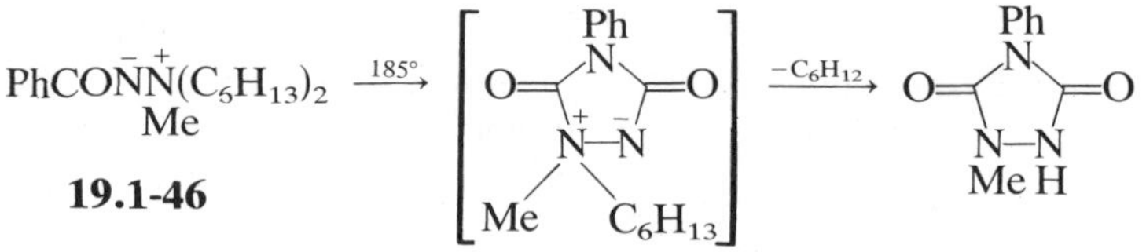

PhCONN(C$_6$H$_{13}$)$_2$
Me
19.1-46

$\xrightarrow{185°}$

19.1-47

$\xrightarrow{-\text{C}_6\text{H}_{12}}$

19.1-48 (20%)

Silylation of 1,2,4-triazolidine-3,5-dione (**19.1-49**) and its 1-methyl (**19.1-50**) and 4-methyl (**19.1-51**) derivatives gave trimethylsilyl compounds that reacted with 2,3,5-tri-O-benzoyl-D-ribofuranosyl bromide in acetonitrile to give the corresponding 1-(2,3,5-tri-O-benzoyl-β-D-ribofuranosyl)-1,2,4-triazolidine-3,5-diones (**19.1-52**, **19.1-53**, and **19.1-54**).[41] Although the oxidation of **19.1-55** gave the free radical **19.1-56**,[18] oxidation of **19.1-28** with such reagents as lead tetracetate generated the powerful cis-azo dienophile (**19.1-57**). The latter formed (4+2) cycloadducts with most dienes (Diels–Alder addition),[42] and reacted with a C=C double bond of vinyl-[43] and (4-phenylbutylidene)cyclopropanes,[44] 1-dimethylamino-2-methyl-1-propene,[45] tetramethoxyallene,[46] 4-aminopyrimidines,[47] 2,5-dimethyl-2,4-hexadiene,[48] sila- and germacyclopentenes,[49,50] 1-methylcyclo-

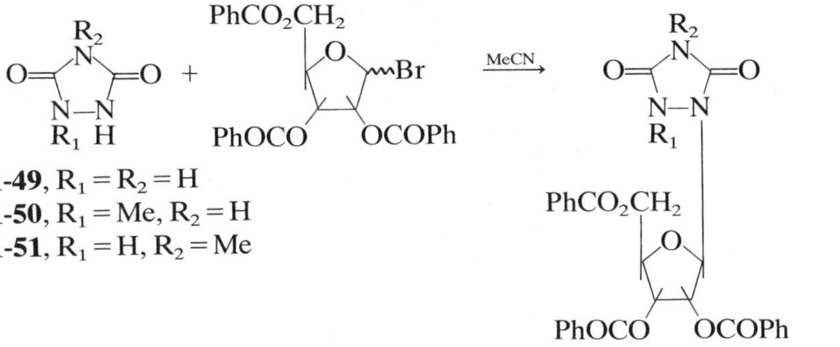

19.1-49, $R_1 = R_2 = H$
19.1-50, $R_1 = Me, R_2 = H$
19.1-51, $R_1 = H, R_2 = Me$

19.1-52, $R_1 = R_2 = H$
19.1-53, $R_1 = Me, R_2 = H$
19.1-54, $R_1 = H, R_2 = Me$

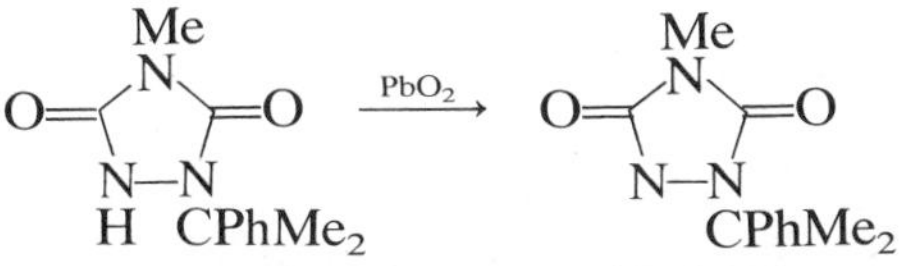

19.1-55   **19.1-56**

octene,[51] polycyclic hydrocarbons,[52] and steroids[49,53] to give 1-substituted-1,2,4-triazolidine-3,5-diones, as illustrated by its interaction with tropolone to give **19.1-59**.[54,55] This type of reaction is commonly called the "ene" reaction. In the reaction of **19.1-57** with indene the diazetidine (**19.1-58**) was obtained, which on catalytic hydrogenation over Raney nickel gave the corresponding 1-(2-indanyl)-1,2,4-triazolidine-3,5-dione.[56]

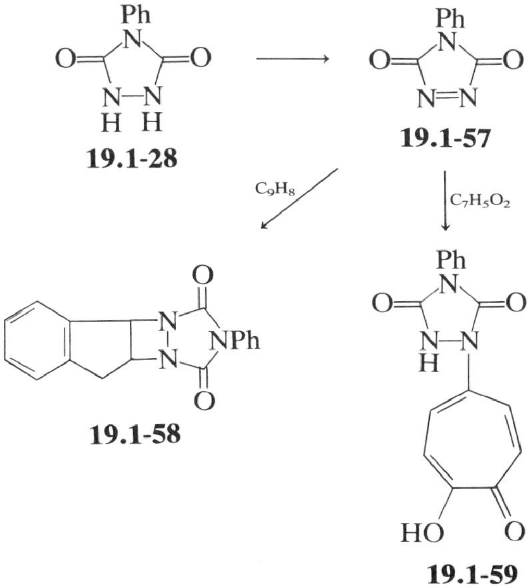

19.1-28   **19.1-57**

19.1-58

19.1-59

References

1. **52:** 945a	2. **76:** 34259a	3. **77:** 62007f	4. **71:** 81375s
5. **51:** 10587i	6. **51:** 12983d	7. **71:** 101796k	8. **58:** 1479g
9. **63:** 18105c	10. **64:** 2096g	11. **64:** 3555g	12. **78:** 29676w
13. **80:** 120848f	14. **78:** 135601c	15. **76:** 59534z	16. **66:** 37837s
17. **55:** 4537d	18. **81:** 25614p	19. **83:** 193683v	20. **76:** 14445e
21. **63:** 18106h	22. **55:** 22298h	23. **64:** 12685f	24. **57:** 11205a
25. **60:** 2949f	26. **64:** 3556f	27. **56:** 2440b	28. **57:** 808f
29. **64:** 11202c	30. **75:** 34816j	31. **51:** 14712a	32. **56:** 486e
33. **61:** 3094h	34. **56:** 10160f	35. **58:** 1479f	36. **60:** 8039g
37. **60:** 1764d	38. **83:** 178937n	39. **67:** 3039p	40. **81:** 91435g
41. **73:** 66847v	42. **84:** 16586r	43. **78:** 147403c	44. **77:** 114318e
45. **82:** 43283e	46. **77:** 126510z	47. **81:** 136110r	48. **70:** 20009h
49. **83:** 179393u	50. **84:** 74336v	51. **84:** 89679h	52. **82:** 156184s
53. **83:** 206479k	54. **72:** 66870b	55. **75:** 20306f	56. **72:** 121454w

19.2. Other 1,2,4-Triazolidine-3,5-diones

In the early literature 4-amino-1,2,4-triazolidine-3,5-dione (**19.2-3**) was denoted by both of the trivial names, 4-aminourazole and urazine. Unfortunately, the latter was also ascribed to compound **19.2-4**.[1] Many syntheses directed toward the preparation of **19.2-4** was later found to give **19.2-3**, which was easily obtained in 100% yield by heating an aqueous solution of phenyl carbazate (**19.2-1**) until the evaporation of phenol ceased.[2] In addition, a 73% yield of **19.2-3** was obtained by heating equimolar amounts of carbohydrazide (**19.2-2**) and concentrated hydrochloric acid.[3] The corresponding 1-phenyl derivative was prepared by the rapid heating of a mixture of urea and phenyl hydrazine.[4,5] Several interesting reactions have provided

routes to other N-amino-1,2,4-triazolidine-3,5-diones. A solution of methylhydrazine in dimethylformamide reacted with carbon monoxide in the presence of selenium to give the salt (**19.2-5**), which was further reacted with a mixture of methylhydrazine, carbon monoxide, and oxygen at 25° to give a 84% yield of **19.2-6**.[6] Also, treatment of oxamic acid hydrazide (**19.2-7**) with phosgene in a basic medium provided a high yield of the

$$MeNHNH_2 \xrightarrow[Se]{CO} [MeNHNH_3]^+[SeCONHNHMe]^- \xrightarrow[25°]{MeNHNH_2}{CO/O_2}$$

19.2-5

(structure **19.2-6**, bearing NHMe group)

$$H_2NCOCONHNH_2 \xrightarrow{COCl_2} [H_2NCOCONHNHCOCl] \longrightarrow$$

19.2-7

(structure **19.2-8**, bearing NHCOCONH₂ group)

4-amino derivative (**19.2-8**),[7] and heating of the meso-ionic compound **19.2-9** at 200° resulted in the migration of a methyl group to give **19.2-10**.[8]

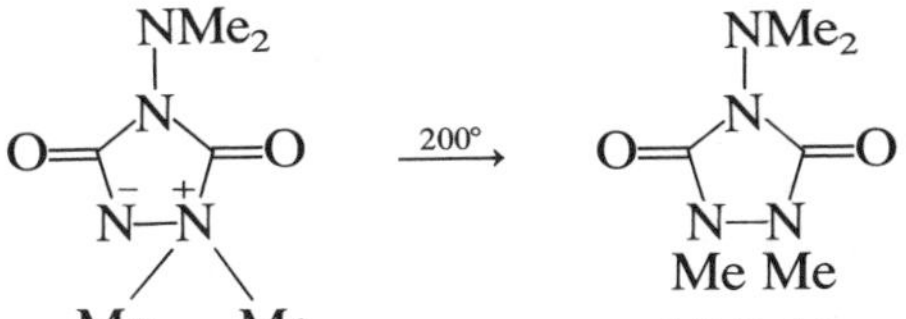

(structures **19.2-9** → **19.2-10**, 200°)

In addition, thermal decomposition of acyl azides can lead to either 1- or 4-amino derivatives. For example, pyrolysis of **19.2-11** gave a mixture of **19.2-12** and **19.2-13**,[9] whereas **19.2-14** gave the oxadiazolin-5-one

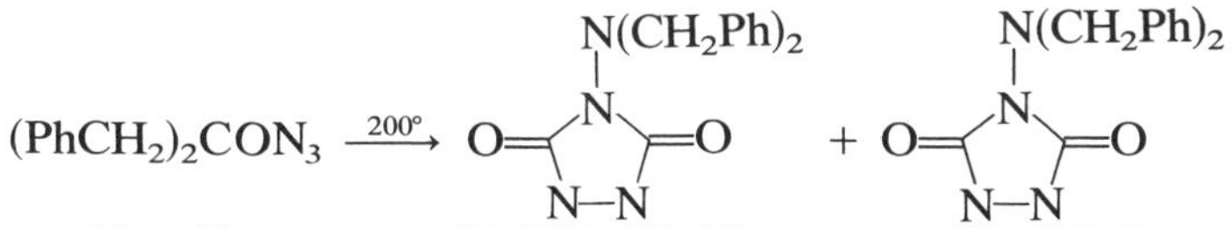

$$(PhCH_2)_2CON_3 \xrightarrow{200°}$$

19.2-11   (structure **19.2-12**) + (structure **19.2-13**)

(**19.2-15**) (see the following).[10] The latter was debenzylated and rearranged in concentrated hydrochloric acid to give **19.2-16**. Also, an unstable 1-amino-1,2,4-triazolidine-3,5-dione (**19.2-18**) was reported to result from the addition of piperidine to the triazoline-3,5-dione (**19.2-17**).[11,12] The formation of triazolidine-3,5-diones containing the N—S linkage was accomplished by the alkylation of **19.2-19** with trichloromethanesulfenyl chloride to give **19.2-20**[13,14] and by the oxidative removal of the hydrazino moiety of tosyl hydrazide with **19.2-17** to give **19.2-21**.[15]

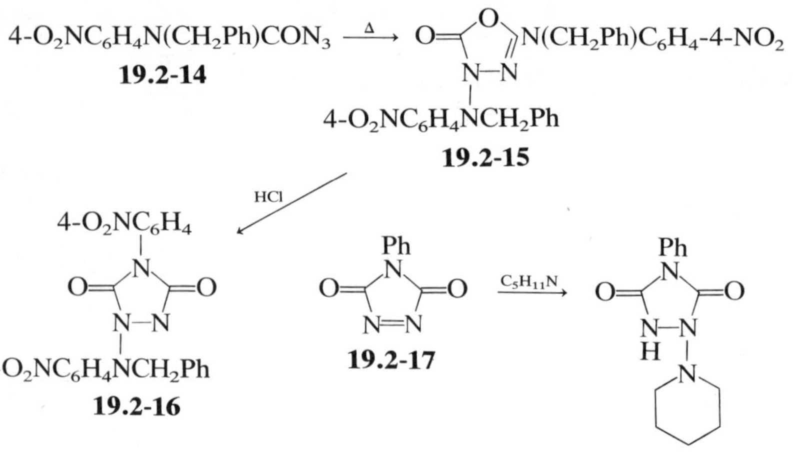

4-O$_2$NC$_6$H$_4$N(CH$_2$Ph)CON$_3$ $\xrightarrow{\Delta}$ O=C$\diagdown$C=N(CH$_2$Ph)C$_6$H$_4$-4-NO$_2$

19.2-14

19.2-15 structure with N—N, 4-O$_2$NC$_6$H$_4$NCH$_2$Ph

19.2-15

4-O$_2$NC$_6$H$_4$

$\xleftarrow{\text{HCl}}$

O=⟨N⟩=O (N—N, 4-O$_2$NC$_6$H$_4$NCH$_2$Ph)

19.2-16

O=⟨Ph-N⟩=O (N=N) $\xrightarrow{\text{C}_5\text{H}_{11}\text{N}}$ O=⟨Ph-N⟩=O (N—N, H, piperidine)

19.2-17   **19.2-18**

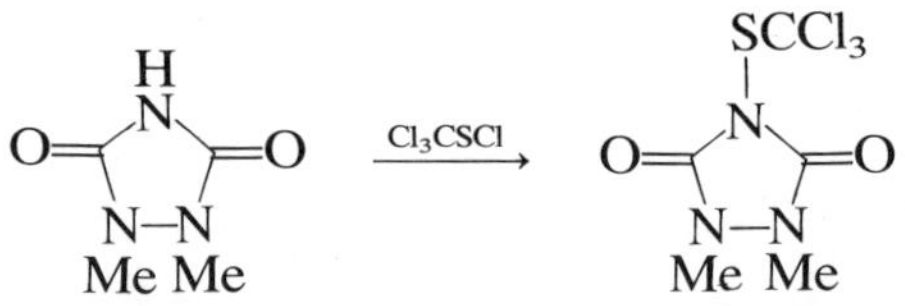

O=⟨H-N⟩=O (N—N, Me Me) $\xrightarrow{\text{Cl}_3\text{CSCl}}$ O=⟨SCCl$_3$-N⟩=O (N—N, Me Me)

19.2-19   **19.2-20**

O=⟨Ph-N⟩=O (N=N) $\xrightarrow{\text{Ts-NHNH}_2}$ O=⟨Ph-N⟩=O (N—N, Ts H)

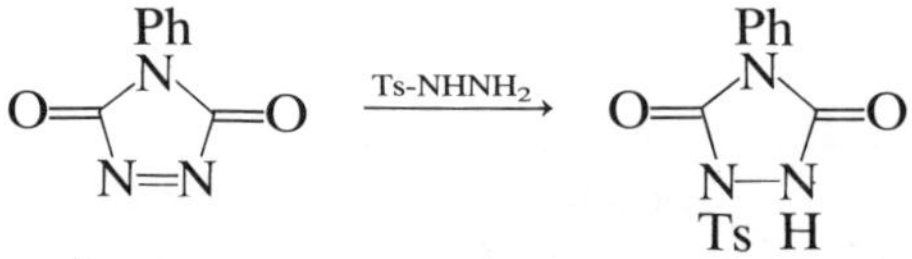

19.2-17   **19.2-21**

The direct formation of 4-benzoyl-1,2-di-*tert*-butyl-1,2,4-triazolidine-3,5-dione (**19.2-23**) (73%) was accomplished by treatment of *N,N*′-di-*tert*-butyl-diaziridinone (**19.2-22**) with benzoyl isocyanate.[16] This type of reaction was

Me$_3$C—N—N—CMe$_3$ diaziridinone with C=O $\xrightarrow{\text{PhCONCO}}$ O=⟨PhCO-N⟩=O (N—N, Me$_3$C CMe$_3$)

19.2-22   **19.2-23**

also successful with other isocyanates such as tosyl isocyanate.[17] A variety of conditions have been investigated for the formation of a N—CO linkage with a ring nitrogen of preformed 1,2,4-triazolidine-3,5-diones. The 1-phenyl derivative was acylated with acetic anhydride at 190° in a sealed tube

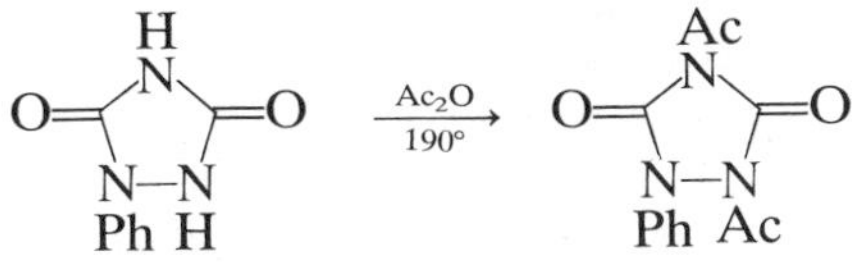

O=⟨N(H)⟩=O —Ac₂O/190°→ O=⟨N(Ac)⟩=O

N—N   N—N
Ph H   Ph Ac

19.2-24

to give a 60% yield of 1,4-diacetyl-2-phenyl-1,2,4-triazolidine-3,5-dione (**19.2-24**),[18] and the 4-phenyl derivative was converted in refluxing acetic anhydride to give a 96% yield of 1,2-diacetyl-4-phenyl-1,2,4-triazolidine-3,5-dione (**19.2-25**).[19] Also, N-acylation can be effected with acid

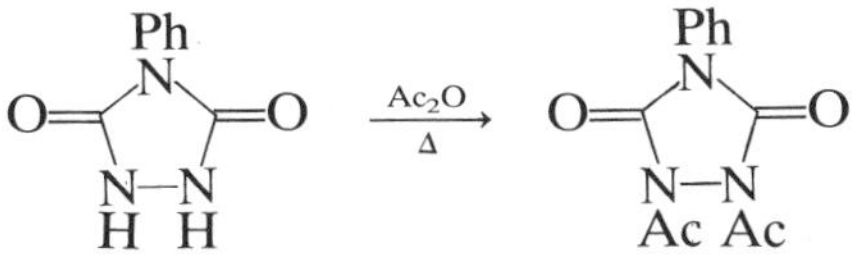

O=⟨N(Ph)⟩=O —Ac₂O/Δ→ O=⟨N(Ph)⟩=O

N—N   N—N
H H   Ac Ac

19.2-25

chlorides,[20–23] which is illustrated by the reaction of **19.2-26** with benzoyl chloride in pyridine to give a 73% yield of 4-benzoyl-1,2-di(2-propyl)-1,2,4-triazolidine-3,5-dione (**19.2-27**).[24] In the reaction of 4-phenyl-1,2,4-tria-

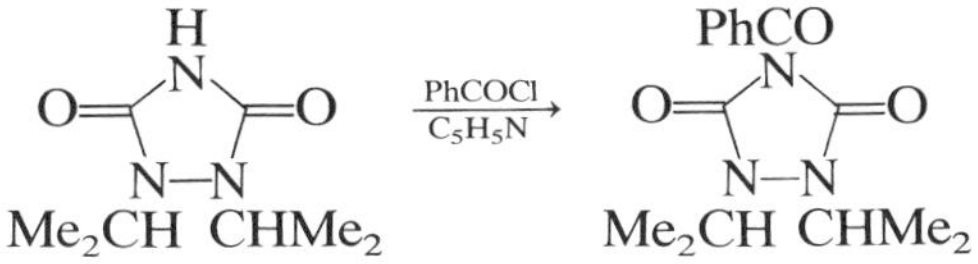

O=⟨N(H)⟩=O —PhCOCl/C₅H₅N→ O=⟨N(PhCO)⟩=O

N—N   N—N
Me₂CH CHMe₂   Me₂CH CHMe₂

19.2-26   **19.2-27**

zoline-3,5-dione (**19.2-17**) with vinyl acetate, the intermediate addition product (**19.2-28**) underwent intramolecular rearrangement to give the N-acetyl derivative (**19.2-31**).[25] The formation of the acyl derivative was subject to steric hindrance in the acyl moiety, as demonstrated by the reaction of **19.2-17** with vinyl pivalate to give both **19.2-29** (36%) and **19.2-30** (36%), in which the latter results from addition across the vinylene double bond. Furthermore, carboxylic acid derivatives have been observed to result from the decomposition of azidoformates in the presence of isocyanates. Although the photolysis of a mixture of benzoyl azide and ethyl isocyanate proceeded via a nitrene intermediate to give a 1,3,4-oxadiazolin-5-one, azidoformates were found to react predominately by addition of the azido group to the isocyanate. The proposed mechanism is illustrated by the reaction of **19.2-32** with ethyl isocyanate to give **19.2-33**, addition of a second molecule of isocyanate to the latter to give **19.2-34**, followed by N—N bond formation to give the carboxylic acid derivative (**19.2-35**) in 50% yield.[26] In the thermolysis of a mixture of **19.2-32** and ethyl isocyanate at 120°, a 68% yield of **19.2-35** was obtained.

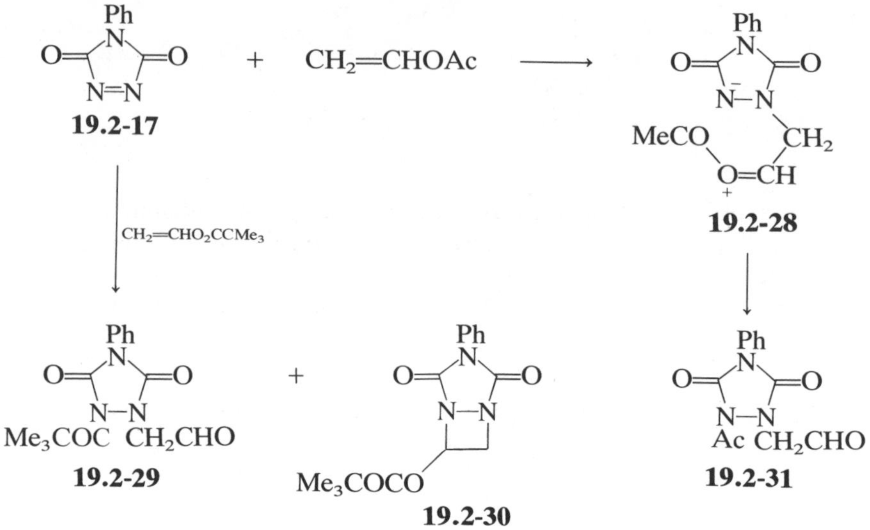

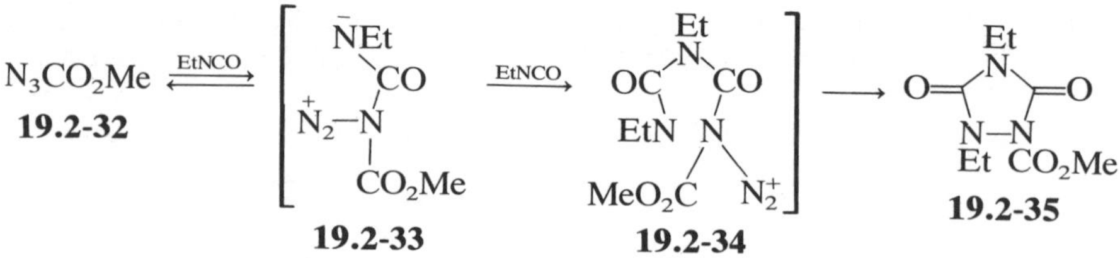

19.2-17 + CH$_2$=CHOAc ⟶ **19.2-28**

CH$_2$=CHO$_2$CCMe$_3$

19.2-29 Me$_3$COC CH$_2$CHO + **19.2-30** Me$_3$COCO

19.2-31 Ac CH$_2$CHO

N$_3$CO$_2$Me $\xleftrightarrow{\text{EtNCO}}$ **19.2-33** $\xrightarrow{\text{EtNCO}}$ **19.2-34** ⟶ **19.2-35**

19.2-32

Carbamoylation of 1,2,4-triazolidine-3,5-dione and its 1- and 4-phenyl derivatives with potassium isocyanate in the presence of hydrochloric acid gave both mono and dicarbamoyl derivatives, as shown by the conversion of **19.2-36** to **19.2-37** in 75% yield.[27] Although the addition of phenyl isocyan-

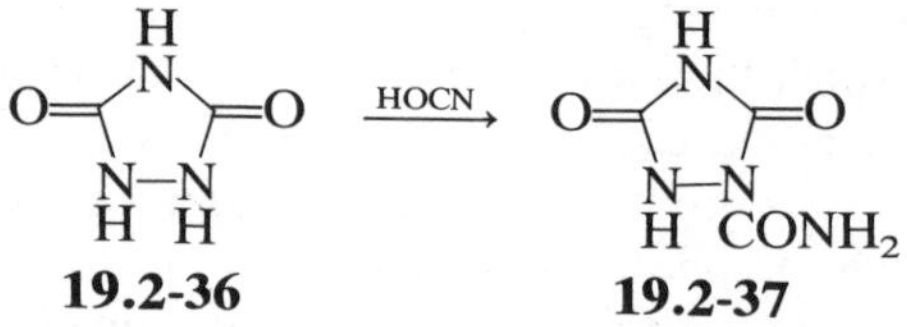

19.2-36 $\xrightarrow{\text{HOCN}}$ **19.2-37**

ate to **19.2-39** gave **19.2-38**, addition of phenyl cyanate to **19.2-39** gave the carbamic acid ester (**19.2-40**).[28] Treatment of **19.2-41** with phosgene was reported to give the carbonyl chloride (**19.2-42**), which was aminated with amines to give carboxamide derivatives.[29] Although **19.2-17** has been used to oxidize some alcohols (e.g., Me$_2$CHOH) to the corresponding carbonyl compounds, treatment of **19.2-17** with normal alcohols such as methanol

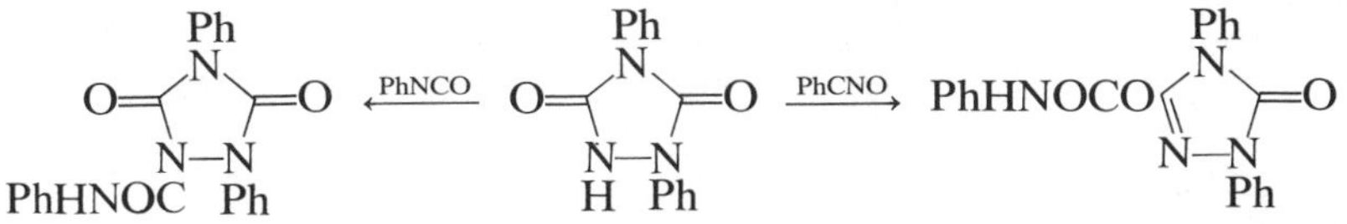

19.2-38 ←PhNCO **19.2-39** PhCNO→ **19.2-40**

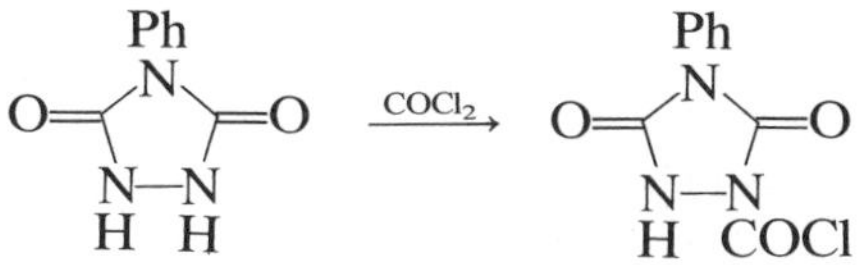

19.2-41 COCl₂→ **19.2-42**

resulted in the formation of triazolidine-1,2-dicarboxylic acids derivatives (e.g., **19.2-43**).[30]

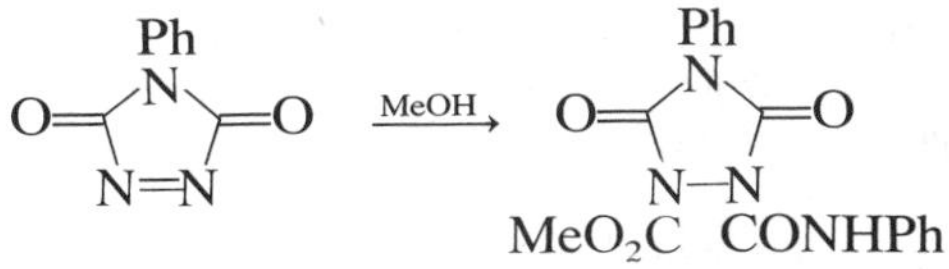

19.2-17 MeOH→ **19.2-43**

References

1. **61:** 10676a
2. **54:** 7625h
3. **50:** 6240b
4. **78:** 29676w
5. **80:** 120848f
6. **82:** 156185t
7. **53:** 11355a
8. **67:** 3039p
9. **63:** 6999b
10. **76:** 46146z
11. **76:** 140687y
12. **85:** 143053k
13. **72:** 78474p
14. **72:** 100656y
15. **78:** 29344m
16. **81:** 13434s
17. **85:** 94296f
18. **73:** 77210h
19. **65:** 3882d
20. **73:** 3842u
21. **73:** 119531w
22. **76:** 34259a
23. **77:** 62007f
24. **64:** 11202c
25. **77:** 75176k
26. **79:** 53234s
27. **61:** 8317c
28. **81:** 91438k
29. **82:** 57676x
30. **85:** 94283z

TABLE 19. 1,2,4-TRIAZOLIDINE-3,5-DIONES

Compound	Reference

19.1. Alkyl- or Aryl-1,2,4-Triazolidine-3,5-diones

Parent (1,2,4-triazolidine-3,5-dione)	**24:** 1114[4], **49:** 5452a, **52:** 945a, **61:** 3095b, **71:** P81375s, **71:** P124443g, **72:** 100656y, **78:** 29676w, **79:** P141516r, **80:** 102243a
Parent, 4-hydroxy-	**76:** 99573t
-1-acetaldehyde, α,α-dimethyl-4-phenyl-	**82:** 43283e
-4-acetamide, α-benzoyl-1,2-bis([4-(1,1-dimethylethyl)-phenyl]methyl)-N-phenyl-	**79:** P141516r
-4-acetamide, α-benzoyl-N-(5-[(4-[2,4-bis(1,1-dimethyl-propyl)phenoxy]-1-oxobutyl)amino]-2-chlorophenyl)-1-(2-chlorophenyl)-2-[(4-methylphenyl)methyl]-	**79:** P141516r
-4-acetamide, α-benzoyl-1-(2-chloro-4,6-dimethylphenyl)-N-phenyl-	**79:** P141516r
-4-acetamide, 1-benzyl-	**83:** 193683v
-4-acetamide, 2-benzyl-N-[(5-[(4-[2,4-bis(1,1-dimethyl-propyl)phenoxy]-1-oxobutyl)amino]-2-chlorophenyl)-α-(2,2-dimethyl-1-oxopropyl-1-phenyl)]-	**85:** 200518s
-4-acetamide, α,1-bis(benzyl)-	**83:** 193683v
-4-acetamide, N-(5-[(4-[2,4-bis(1,1-dimethylpropyl)-phenoxy]-1-oxobutyl)amino]-2-chlorophenyl)-1,2-bis(2-chlorophenyl)-α-(2,2-dimethyl-1-oxopropyl)-	**79:** P141516r
-4-acetamide, N-(5-[(4-[2,4-bis(1,1-dimethylpropyl)-phenoxy]-1-oxobutyl)amino]-2-chlorophenyl)-α-(2,2-dimethyl-1-oxopropyl)-1,2-diphenyl-	**79:** P141516r
-4-acetamide, N-(5-[(4-[2,4-bis(1,1-dimethylpropyl)-phenoxy]-1-oxobutyl)amino]-2-chlorophenyl)-α-(2,2-dimethyl-1-oxopropyl)-1-methyl-2-phenyl-	**79:** P141516r, **83:** P170844j
-4-acetamide, N-(2,5-dichlorophenyl)-α-(2,2-dimethyl-1-oxopropyl)-1,2-diphenyl-	**79:** P141516r
-4-acetamide, α-(2,2-dimethyl-1-oxopropyl)-1-methyl-N, 2-diphenyl-	**79:** P141516r
-4-acetamide, α-(2,2-dimethyl-1-oxopropyl)-1-methyl-N-phenyl-	**79:** P141516r
-4-acetamide, α-(2,2-dimethyl-1-oxopropyl)-N-phenyl-	**79:** P141516r
-4-acetamide, 1,2-diphenyl-	**53:** 7426i, **56:** P10161b
-4-acetamide, 1-methyl-	**83:** 193683v
-1-acetic acid, α-(dimethoxymethylene)-4-phenyl-, methyl ester	**77:** 126510z
-4-acetic acid, 1,2-diphenyl-	**53:** 7426i, **56:** P10161b
-1-acetic acid, α,4-diphenyl-α-[(phenylthioxomethyl)-amino]-	**85:** 32910t
-1-acetic acid, α-methyl-4-phenyl-α-[(phenylthioxo-methyl)amino]-	**85:** 32910t
-1-acetic acid, 2-phenyl-, ethyl ester	**53:** 7426i
1-acetonyl-4-phenyl-	**67:** 108606z
1-acetonyl-4-phenyl-, 1-[(2,4-dinitrophenyl)hydrazone]	**67:** 108606z

TABLE 19 *(Continued)*

Compound	Reference

19.1. Alkyl- or Aryl-1,2,4-Triazolidine-3,5-diones *(Continued)*

Compound	Reference
1-[(3α,5α,16α,17Z)-3-(acetyloxy)cholest-17(20)-en-16-yl]-4-phenyl-	**83:** 206479k
4-(*p*-acetylphenyl)-1-phenyl-	**64:** P3556f
4-(*m*-acetylphenyl)-1-(2-pyridyl)-	**60:** P2950f, **64:** PC12686b
4-allyl-	**67:** 116858y
4-allyl-1-methyl-	**75:** P20410k
4-[*p*-(allyloxy)phenyl]-1-(3-chloro-4-phenoxyphenyl)-	**64:** P2097d
4-[*p*-(allyloxy)phenyl]-1-phenyl-	**64:** P3556f
1-allyl-2-phenyl-	**54:** 4912h, **56:** P10160h
1-(4-amino-1,3-dimethyl-2(3H),6(H)-dioxopyrimidin-5-yl)-4-phenyl-	**81:** 136110r
1-(4-amino-2(3H),6(1H)-dioxopyrimidin-5-yl)-4-phenyl-	**81:** 136110r
1-(4-amino-6(1H)-oxo-2-phenylpyrimidin-5-yl)-4-phenyl-	**81:** 136110r
1-(*p*-aminophenyl)-	**61:** 3095e
4-(*m*-aminophenyl)-1-phenyl-	**64:** P3557a
1-(9-anthryl)-4-phenyl-	**67:** 108606z
1-(1,4-benzodioxan-6-yl)-4-(*p*-chlorophenyl)-	**64:** P2097e
1-(1,4-benzodioxan-6-yl)-4-cyclohexyl-	**63:** P18107h
1-(1,4-benzodioxan-6-yl)-4-(*p*-ethoxyphenyl)-	**64:** P2097e
1-(1,4-benzodioxan-6-yl)-4-[2-methoxy-4-(phenylthio)phenyl]-	**64:** P2097e
1-benzo[*a*]pyren-6-yl-4-phenyl-	**82:** 156184s
4-(2-benzoyl-4-chlorophenyl)-	**82:** 25638v
1-benzyl-	**61:** 3095b
4-benzyl-	**50:** 12032i, **59:** 8747h, **67:** 116858y, **84:** 16586r
4-(2-benzylcyclopentyl)-1-phenyl-	**63:** P18107h
1-benzyl-*N*-[2-(diethylamino)ethyl]-2-phenyl-	**58:** P11370e
1-benzyl-4-(2-hydroxyethyl)-2-phenyl-	**53:** 7426i, **76:** P34259a, **77:** P62007f
1-benzyl-2-methyl-	**61:** 3095e
1-benzyl-4-methyl-2-phenyl-	**53:** 7426i, **76:** P34259a, **77:** P62007f
4-[4-(benzyloxy)cyclohexyl]-1-phenyl-	**63:** P18107g
4-[4-(benzyloxy)-3-methoxyphenyl]-1-phenyl-	**63:** P18106b
1-[*p*-(benzyloxy)phenyl]-4-cyclohexyl-	**63:** P18107f
1-[*p*-(benzyloxy)phenyl]-4-(*p*-cyclohexylphenyl)-	**64:** P2097c
1-[*p*-(benzyloxy)phenyl]-4-(*p*-ethoxyphenyl)-	**64:** P2097b
1-[*p*-(benzyloxy)phenyl]-4-(*p*-hydroxyphenyl)-	**64:** P2097c
4-[*p*-(benzyloxy)phenyl]-1-(2-pyridyl)-	**60:** P2950f, **64:** P12686b
4-[*p*-(benzyloxy)phenyl]-1-(3,4-xylyl)-	**64:** P3556d
4-[6-(benzyloxy)-*m*-tolyl]-1-phenyl-	**60:** P2950a
1-benzyl-2-phenyl	**53:** 7426i
1-benzyl-2-phenyl-4-propyl-	**53:** 7426i, **76:** P34259a, **77:** P62007f
4-bicyclo[2.2.0]hex-5-en-2-yl-1-phenyl-	**63:** P18107f
1-(4-biphenylyl)-4-butyl-2-phenyl-	**56:** P486g

TABLE 19   (*Continued*)

Compound	Reference

19.1. Alkyl- or Aryl-1,2,4-Triazolidine-3,5-diones   (*Continued*)

Compound	Reference
1-(4-biphenylyl)-4-tert-butyl-2-phenyl-	**56:** P486h
1-(4-biphenylyl)-4-decyl-2-phenyl-	**56:** P486h
1-(4-biphenylyl)-4-[2-(diethylamino)ethyl]-2-phenyl-	**56:** P486g
1-(4-biphenylyl)-2,4-diphenyl-	**56:** P486h
1-(4-biphenylyl)-4-dodecyl-2-phenyl-	**56:** P486h
1-(4-biphenylyl)-4-ethyl-2-phenyl-	**56:** P486g
1-(4-biphenylyl)-4-heptyl-2-phenyl-	**56:** P486h
1-(4-biphenylyl)-4-hexadecyl-2-phenyl-	**56:** P486h
1-(4-biphenylyl)-4-hexyl-2-phenyl-	**56:** P486h
1-(4-biphenylyl)-4-isobutyl-2-phenyl-	**56:** P486g
1-(4-biphenylyl)-4-isopentyl-2-phenyl-	**56:** P486h
1-(4-biphenylyl)-4-(3-methoxypropyl)-2-phenyl-	**56:** P486h
1-(4-biphenylyl)-4-methyl-2-phenyl-	**56:** P486g
1-(4-biphenylyl)-4-[2-(methylthio)ethyl]-2-phenyl-	**56:** P486g
1-(4-biphenylyl)-4-nonyl-2-phenyl-	**56:** P486h
1-(4-biphenylyl)-4-octyl-2-phenyl-	**56:** P486h
1-(4-biphenylyl)-4-pentyl-2-phenyl-	**56:** P486h
1-(4-biphenylyl)-4-phenethyl-2-phenyl-	**56:** P486h
4-(4-biphenylyl)-1-phenyl-	**64:** P3556f
1-(4-biphenylyl)-2-phenyl-	**56:** P486g
1-(4-biphenylyl)-2-phenyl-4-propyl-	**56:** P486g
4-(4-biphenylyl)-1-(2-pyridyl)-	**60:** P2950f
4-[3,4-bis(benzyloxy)phenyl]-1-[*o*-(methylthio)phenyl]-	**64:** P2097c
4-[3,4-bis(benzyloxy)phenyl]-1-(2-napthyl)-	**63:** P18106b
4-[3,4-bis(benzyloxy)phenyl]-1-phenyl-	**63:** P18106b
1,2-bis(*p*-bromophenyl)-	**61:** 3095c
1,2-bis(tert-butyl)-4-(4-nitrophenyl)-	**85:** 94296f
1,4-bis(tert-butyl)-2-phenyl-	**83:** 178937n
1,2-bis(2-chlorophenyl)-	**79:** P141516r
1,4-bis(*p*-chlorophenyl)-	**57:** P11205c, **58:** P1479g
1,4-bis(*p*-cyclohexylphenyl)-	**64:** P2097d
1,4-bis(*p*-hydroxyphenyl)-	**64:** P2097c
1,4-bis(isopropyl)-2-phenyl-	**76:** P34259a, **77:** P62007f
1,2-bis(1-methylethyl)-	**76:** P34259a, **77:** 62007f
1,2-bis(1-methylethyl)-4-propyl-	**76:** P34259a, **77:** P62007f
1,2-bis(*p*-nitrophenyl)-	**61:** 3095d
1,2-bis(3-oxobutyl)-4-phenyl-	**65:** 718h
4-(2-bornyl)-1-phenyl-	**63:** P18107h
1-(3'-bromo-4-biphenylyl)-2-(*m*-bromophenyl)-4-propyl-	**56:** P486h
1-(4'-bromo-4-biphenylyl)-2-(*p*-bromophenyl)-4-propyl-	**56:** P486h
4-(2-bromoethyl)-1,2-diisopropyl-	**64:** 11202g
4-(2-bromoethyl)-1,2-diphenyl-	**64:** 11202g
4-(2-bromoethyl)-1-methyl-2-phenyl-	**58:** P11370d
4-[*p*-(2-bromoethyl)phenyl]-1-phenyl-	**64:** P3557c
1-(*p*-bromophenyl)-	**56:** P10160i, **61:** 3095b, **76:** P34259a, **77:** P62007f
4-(4-bromophenyl)-	**84:** P135700h

TABLE 19   (Continued)

Compound	Reference

19.1. Alkyl- or Aryl-1,2,4-Triazolidine-3,5-diones   (Continued)

Compound	Reference
4-(m-bromophenyl)-1-(p-cyclohexylphenyl)-	64: P2097d
1-(p-bromophenyl)-2-methyl-	54: 4912h, 56: P10161a, 61: 3095e, 76: P34259a, 77: P62007f
4-(m-bromophenyl)-1-(2-naphthyl)-	64: P3555g
1-(2-butenyl)-2-phenyl-	54: 4912h, 76: P34259a, 77: P62007f
1-(4-butoxy-5-chloro-m-tolyl)-4-(p-butoxyphenyl)-	64: P3555g
1-(4-butoxy-5-chloro-m-tolyl)-4-(p-ethoxyphenyl)-	64: P3555g
1-(4-butoxy-5-chloro-m-tolyl)-4-phenyl-	64: P3555g
4-(p-butoxyphenyl)-1-phenyl-	57: P11205c, 58: P1480a, 60: P1764fg
4-(4-butoxy-o-tolyl)-1-[m-(methylthio)phenyl]-	64: P2097d
4-(4-butoxy-o-tolyl)-1-phenyl-	60: P2949g
4-(4-butoxy-m-tolyl)-1-phenyl-	63: P18106a
4-butyl-	50: 12032i, 55: 22298i, 66: 37837s, 67: 116858y, 71: 101796k, 74: 54291s, 82: P172343y
4-tert-butyl-	76: 59534z
4-butyl-1,2-diisopropyl-	64: 11202f
4-butyl-1,2-dimethyl-	67: 3039p
4-butyl-1,2-diphenyl-	76: P34259a, 77: P62007f
1-tert-butyl-4-isopropyl-2-phenyl-	83: 178937n
4-(5-tert-butyl-4-methoxy-o-tolyl)-1-phenyl-	60: P2950a
4-butyl-1-methyl-	67: 3039p
1-[1-tert-butyl-1-methylgermacyclopent-4-en-3-yl]-4-phenyl-, cis-	84: 74336v
1-[1-tert-butyl-1-methylgermacyclopent-4-en-3-yl]-4-phenyl-, trans-	84: 74336v
4-(p-butylphenyl)-	50: 12032h
1-butyl-2-phenyl-	54: 4912h, 76: 34259a, 77: P62007f
4-butyl-1-phenyl-	55: P13758b, 76: P34259a, 77: P62007f
4-(p-butylphenyl)-1-phenyl-	64: P3556f
4-(p-tert-butylphenyl)-1-phenyl-	64: 3556f
4-(p-butylphenyl)-1-(2-pyridyl)-	60: P2950f, 64: PC12686b
4-[p-(butylthio)phenyl]-1-phenyl-	64: P3556f
-4-butyric acid, 1-phenyl-	66: 37845t
4-(carbamoylmethyl)-1,2-diphenyl-	76: P34259a, 77: P62007f
4-[1-(5-carboxy-2-chloroanilino)-4,4-dimethyl-1,3-dioxo-2-pentyl]-1-methyl-2-phenyl-	79: P141516r
4-(carboxymethyl)-1,2-diphenyl-	76: P34259a, 77: P62007f
4-carboxymethyl-1,2-diphenyl-, ethyl ester	76: P34259a, 77: P62007f
4-(4-carboxyphenyl)-	82: P87423e
1-(4'-chloro-4-biphenylyl)-2-(p-chlorophenyl)-4-isopropyl-	56: P486h

541

TABLE 19 (*Continued*)

Compound	Reference

19.1. Alkyl- or Aryl-1,2,4-Triazolidine-3,5-diones (*Continued*)

Compound	Reference
1-(2'-chloro-4-biphenylyl)-2-(*o*-chlorophenyl)-4-propyl-	**56:** P486h
1-(3'-chloro-4-biphenylyl)-2-(*m*-chlorophenyl)-4-propyl-	**56:** P486h
4-(2-chloro-*p*-cumenyl)-1-phenyl-	**60:** P2950a
4-(4-chloro-2,5-diethoxyphenyl)-1-phenyl-	**60:** P2950a
4-(4-chloro-2,5-dimethoxyphenyl)-1-(*p*-chlorophenyl)-	**60:** P2950c
1-(4-chloro-2,5-dimethoxyphenyl)-4-(4-chloro-*m*-tolyl)-	**63:** P18106b
1-(4-chloro-2,5-dimethoxyphenyl)-4-cyclohexyl-	**63:** P18107g
1-(4-chloro-2,5-dimethoxyphenyl)-4-(*p*-cyclohexylphenyl)-	**64:** P3556f
1-(4-chloro-2,5-dimethoxyphenyl)-4-(3,4-dimethoxy-phenyl)-	**63:** P18106b
1-(4-chloro-2,5-dimethoxyphenyl)-4-(*p*-ethoxyphenyl)-	**64:** P3555g
1-(4-chloro-2,5-dimethoxyphenyl)-4-phenyl-	**64:** P3555g
4-(4-chloro-2,5-dimethoxyphenyl)-1-phenyl-	**60:** P2949h
4-(4-chloro-2,5-dimethoxyphenyl)-1-(2-pyridyl)-	**60:** P2950f, **64:** PC12686a
1-(4-chloro-2,5-dimethoxyphenyl)-4-*p*-tolyl-	**64:** P3555g
4-(2-chloro-4-ethoxyphenyl)-1-phenyl-	**60:** P2950a
4-[*p*-(2-chloroethoxy)phenyl]-1-phenyl-	**64:** P3557g
4-(2-chloroethyl)-1,2-diphenyl-	**53:** 7426i, **76:** P34259a, **77:** P62007f
4-[2-chloro-4-(hexyloxy)phenyl]-1-(3-methyl-4-biphenylyl)-	**64:** P2097d
4-[2-chloro-4-(hexyloxy)phenyl]-1-phenyl-	**60:** P2950b
4-[2-chloro-4-(2-hydroxyethoxy)phenyl]-1-phenyl-	**60:** P2950b
4-(4-chloro-3-hydroxyphenyl)-1-phenyl-	**63:** P18105h
4-(5-chloro-2-hydroxy-*p*-tolyl)-1-phenyl-	**63:** P18105e
4-(2-chloro-4-methoxyphenyl)-1-phenyl-	**60:** P2950a
4-(4-chloro-6-methoxy-*m*-tolyl)-1-phenyl-	**60:** P2949h
4-(chloromethyl)-1,2-diisopropyl-	**64:** 11203c
1-(3-chloro-4-phenoxyphenyl)-4-cyclohexyl-	**63:** P18107f
1-(3-chloro-4-phenoxyphenyl)-4-(*p*-ethylphenyl)-	**64:** P2097d
1-(3-chloro-4-phenoxyphenyl)-4-(3-methoxy-2-dibenzofuranyl)-	**64:** P2097e
1-(2-chlorophenyl)-	**76:** P34259a, **77:** P62007f
1-(3-chlorophenyl)-	**76:** P34259a, **77:** P62007f
2-(4-chlorophenyl)-	**76:** P34259a, **77:** P62007f
4-(2-chlorophenyl)-	**82:** 30554u
4-(3-chlorophenyl)-	**80:** P146185n, **80:** P146210s, **83:** P58835f
4-(*p*-chlorophenyl)-	**51:** P10587i, **51:** P12983d, **56:** 2440d, **57:** 809e, **80:** P146185n, **80:** P146210s, **83:** P58835f
1-(2-chlorophenyl)-2-cyclohexyl-	**76:** P34259a, **77:** P62007f
1-(*p*-chlorophenyl)-4-cyclohexyl-	**63:** P18107f
1-(4-chlorophenyl)-4-(3,5-dichlorophenyl)-2-methyl-	**85:** P117969u
1-(*p*-chlorophenyl)-4-[*p*-(diethylamino)phenyl]-	**64:** P3556f
1-(*p*-chlorophenyl)-4-(3,4-dimethoxyphenyl)-	**63:** P18106b
1-(*p*-chlorophenyl)-4-(*p*-ethoxyphenyl)-	**57:** P11205c, **60:** P1764g

542

TABLE 19 (*Continued*)

Compound	Reference

19.2. Other 1,2,4-Triazolidine-3,5-diones (*Continued*)

Compound	Reference
1-(*p*-chlorophenyl)-4-(3-methoxy-2-dibenzofuranyl)-	**60:** P2950c
4-(*p*-chlorophenyl)-1-(4-methoxy-*o*-tolyl)-	**64:** P3555g
1-(*o*-chlorophenyl)-2-methyl-	**54:** 4912h, **76:** P34259a, **77:** P62007f
1-(*o*-chlorophenyl)-2-phenyl-	**54:** 4912h
4-(*p*-chlorophenyl)-1-phenyl-	**57:** P11205c, **58:** P1479g, **60:** P1764fg
4-(4-chloro-*m*-tolyl)-1-(*p*-cyclohexylphenyl)-	**64:** P2097d
4-(2-chloro-*p*-tolyl)-1-phenyl-	**60:** P2949g
4-(4-chloro-*m*-tolyl)-1-phenyl-	**63:** P18106a
4-(5-chloro-*m*-tolyl)-1-phenyl-	**63:** P18106a
4-(4-chloro-*m*-tolyl)-1-(5,6,7,8-tetrahydro-2-naphthyl)-	**64:** P2097d
4-[3-chloro-4-(trifluoromethyl)phenyl]-	**80:** P146156d, **80:** P146185n, **80:** P146210s, **83:** 58835f
4-[4-chloro-3-(trifluoromethyl)phenyl]-	**80:** 146185n, **80:** P146210s
1-(6-chloro-α,α,α-trifluoro-*m*-tolyl)-4-cyclohexyl-	**63:** P18107g
1-(6-chloro-α,α,α-trifluoro-*m*-tolyl)-4-(*p*-ethoxyphenyl)-	**64:** P2097e
1-(6-chloro-α,α,α-trifluoro-*m*-tolyl)-1-phenyl-	**60:** P2950b
4-(6-chloro-3,4-xylyl)-1-phenyl-	**60:** P2949g
4-*p*-cumenyl-1-phenyl-	**64:** P3556f
4-(2-cyanoethyl)-1-methyl-2-phenyl-	**58:** P11370f
4-cyclobutyl-1-phenyl-	**63:** P18107f
1-cycloheptyl-2-phenyl-	**76:** P34259a, **77:** P62007f
4-cycloheptyl-1-phenyl-	**63:** P18107h
1-(2-cyclohexen-1-yl)-4-phenyl-	**67:** 108606z, **85:** P21708g
4-(3-cyclohexen-1-yl)-1-phenyl-	**63:** P18107f
1-(cyclohexenyl)-4-methyl-	**67:** 116853t
1-(2-cyclohexen-1-yl)-4-phenyl-	**81:** 91438k
1-cyclohexyl-	**76:** P34259a, **77:** P62007f
4-cyclohexyl-	**61:** P8317e, **65:** 703g, **66:** 37837s
4-(4-cyclohexylcyclohexyl)-1-phenyl-	**63:** P18107g
1-cyclohexyl-2-(*p*-cyclohexylphenyl)-4-propyl-	**56:** P486h
4-cyclohexyl-1-(2,5-dimethoxy-3,4-xylyl)-	**63:** P18107g
1-cyclohexyl-2,4-dimethyl-	**74:** 54291s
4-cyclohexyl-1-(*p*-ethoxyphenyl)-	**63:** P18107f
1-cyclohexyl-2-(2-hydroxyethyl)-3-phenyl-	**53:** 7426i
1-cyclohexyl-4-(2-hydroxyethyl)-2-phenyl-	**76:** P34259a, **77:** P62007f
1-cyclohexyl-4-(hydroxymethyl)-2-phenyl-	**54:** 4912h, **76:** P34259a, **77:** P62007f
4-cyclohexyl-1-(*p*-hydroxyphenyl)-	**63:** P18107h
4-cyclohexyl-1-(2-methyl-4-biphenylyl)-	**63:** P18107f
1-cyclohexyl-2-(4-methylphenyl)-	**76:** P34259a, **77:** P62007f
1-cyclohexyl-4-methyl-2-phenyl-	**53:** 7426i, **76:** P34259a, **77:** P62007f
4-cyclohexyl-1-(2-naphthyl)-	**63:** P18107g
1-[2-(cyclohexyloxy)-2,3,3a,4,7,7a-hexahydro-4-benzofuranyl]-2-methyl-4-phenyl-	**79:** 136576f

543

TABLE 19 (*Continued*)

Compound	Reference

19.1. Alkyl- or Aryl-1,2,4-Triazolidine-3,5-diones (*Continued*)

Compound	Reference
1-[2-(cyclohexyloxy)-2,3,3a,4,7,7a-hexahydro-4-benzofuranyl]-4-phenyl-	**79:** 136576f
1-[2-(cyclohexyloxy)octahydro-5,6-dihydroxy-4-benzofuranyl]-2-methyl-4-phenyl-	**79:** 136576f
1-cyclohexyl-2-phenyl-	**53:** 7426i, **54:** 5933d, **76:** P34259a, **77:** P62007f
1-cyclohexyl-4-phenyl-	**85:** P21708g
4-cyclohexyl-1-phenyl-	**54:** 4912h, **63:** P18107c
1-(*p*-cyclohexylphenyl)-4-(2,5-diethoxy-4-phenoxyphenyl)-	**64:** P2097e
1-(*p*-cyclohexylphenyl)-4-(2′,3-dimethoxy-4-biphenylyl)-	**64:** P2097e
1-(*p*-cyclohexylphenyl)-4-(*p*-ethoxyphenyl)-	**64:** P2097a
4-(*p*-cyclohexylphenyl)-1-(*p*-hydroxyphenyl)-	**64:** P2097c
1-(*p*-cyclohexylphenyl)-4-(1-indanyl)-	**63:** P18107f
1-(*o*-cyclohexylphenyl)-4-phenyl-	**64:** P2097f
1-(*p*-cyclohexylphenyl)-4-phenyl-	**64:** P2097d
4-(*p*-cyclohexylphenyl)-1-phenyl)-	**64:** P3556f
1-cyclohexyl-2-phenyl-4-propyl-	**76:** P34259a, **77:** P62007f
1-cyclohexyl-3-phenyl-4-propyl-	**53:** 7426i
4-(*p*-cyclohexylphenyl)-1-(2-pyridyl)-	**60:** P2950f, **64:** P12686a
4-cyclohexyl-1-(5,6,7,8-tetrahydro-2-naphthyl)-	**63:** P18107g
1-cyclohexyl-2-*p*-tolyl-	**54:** 4912h
4-(4-cyclohexyl-*o*-tolyl)-1-phenyl-	**60:** P2949h
4-(4-cyclohexyl-2,6-xylyl)-1-phenyl-	**60:** P2949h
4-cyclooctyl-1-phenyl-	**63:** P18107h
1-(cyclopentenyl)-4-methyl-	**67:** 116853t
4-(2-cyclopenten-1-yl)-1-phenyl-	**63:** P18107f
4-cyclopentyl-1-[*m*-(methylthio)phenyl]-	**63:** P18107f
1-cyclopentyl-2-phenyl-	**54:** 4912h, **76:** P34259a, **77:** P62007f
4-cyclopentyl-1-phenyl-	**63:** P18107f
1-[6-cyclopropyl-5,6-dihydro-1,3(2H)-dioxo-2-phenyl-1H-[1,2,4]triazolo[1,2-*a*]cinnolin-6-yl]-4-phenyl	**78:** 147403c
4-decyl-	**50:** 12032i
1-decyl-2-phenyl-	**54:** 4912h, **56:** P10161b, **76:** P34259a, **77:** P62007f
1-(2-deoxy-β-D-erythro-pentofuranosyl)-2,4-dimethyl-, 3′,5′-di-*p*-toluate	**73:** 66847v
1-(2-deoxy-β-D-erythro-pentofuranosyl)-, 3′,5′-di-*p*-toluate	**73:** 66847v
4-(3,5-diacetylmesityl)-1-phenyl-	**60:** P2949h
1,2-dibenzyl-	**63:** 6999d
1,4-dibenzyl-2-phenyl-	**53:** 7426i, **56:** P10161c
4-(3,4-dibromocyclohexyl)-1-phenyl-	**63:** P18107h
1,2-di-sec-butyl-	**55:** P4537d
1,2-dibutyl-4-(3,5-dichlorophenyl)-	**85:** P117969u
4-(2,5-dichloro-*p*-cumenyl)-1-phenyl-	**60:** P2949h
4-[3,5-dichloro-4-(hexyloxy)phenyl]-1-(*p*-ethoxyphenyl)-	**64:** P3556f

TABLE 19   (*Continued*)

Compound	Reference

19.1. Alkyl- or Aryl-1,2,4-Triazolidine-3,5-diones   (*Continued*)

Compound	Reference
4-(3,5-dichloro-4-hydroxyphenyl)-1-phenyl-	**63:** P18105f
1-(2,4-dichlorophenyl)-	**56:** P10160g
1-(2,5-dichlorophenyl)-	**76:** P34259a, **77:** P62007f,
	80: 120848f
1-(3,4-dichlorophenyl)-	**80:** 120848f
4-(2,3-dichlorophenyl)-	**80:** 146185n, **80:** P146210s
4-(2,4-dichlorophenyl)-	**80:** 146156d, **80:** P146185n,
	80: P146210s, **83:** P58835f
4-(2,5-dichlorophenyl)-	**80:** 146185n, **80:** P146210s,
	83: P58835f
4-(3,4-dichlorophenyl)-	**80:** P146156d, **80:** P146185n,
	80: P146210s, **83:** P58835f
4-(3,5-dichlorophenyl)-	**80:** P146156d, **80:** P146185n,
	80: P146210s, **83:** P10088n,
	83: P58835f, **85:** P117969u
4-(3,5-dichlorophenyl)-1,2-diethyl-	**85:** P117969u
4-(3,5-dichlorophenyl)-1,2-dimethyl-	**85:** P117969u
4-(3,5-dichlorophenyl)-1,2-di-2-propenyl-	**85:** P117969u
4-(3,5-dichlorophenyl)-1,2-dipropyl-	**85:** P117969u
4-(3,5-dichlorophenyl)-1,2-di(2-propynyl)-	**85:** P117969u
4-(3,5-dichlorophenyl)-1-ethyl-2-methyl-	**85:** P117969u
4-(3,5-dichlorophenyl)-1-ethyl-2-phenyl-	**85:** P117969u
4-(3,5-dichlorophenyl)-1-isopropyl-	**85:** P117969u
4-(3,5-dichlorophenyl)-2-isopropyl-1-methyl-	**85:** P117969u
1-(2,5-dichlorophenyl)-2-methyl-	**54:** 4912h, **76:** P34259a,
	77: P62007f
4-(3,5-dichlorophenyl)-1-methyl-	**85:** P117969u
4-(3,5-dichlorophenyl)-1-methyl-2-(4-methylphenyl)-	**85:** P117969u
4-(3,5-dichlorophenyl)-1-methyl-2-(2-methylpropyl)-	**85:** P117969u
4-(3,5-dichlorophenyl)-1-methyl-2-phenyl-	**85:** P117969u
4-(3,5-dichlorophenyl)-1-methyl-2-propyl-	**85:** P117969u
4-(3,5-dichlorophenyl)-1-(2-methylpropyl)-	**85:** P117969u
4-(3,5-dichlorophenyl)-1-phenyl-	**85:** P117969u
4-(2,5-dichloro-*p*-tolyl)-1-phenyl-	**60:** P2949h
4-(4,6-dichloro-*m*-tolyl)-1-phenyl-	**60:** P2949h
1,2-dicyclohexyl-	**53:** 7426i, **56:** P10160g,
	76: P34259a
1,2-dicyclohexyl-4-(hydroxymethyl)-	**64:** 11203a
1,2-dicyclohexyl-4-[(methylmorpholinoamino)methyl]-	**65:** 3863e
1,2-dicyclohexyl-4-(piperidinomethyl)-	**65:** 3863d
1,2-dicyclohexyl-4-propyl-	**53:** 7426i, **76:** P34259a,
	77: P62007f
1,2-dicyclohexyl-4-(xanthen-9-yl)-	**64:** 11202h
1-[(5α)-7,8-didehydro-4,5-epoxy-3-methoxy-17-methyl-6-oxomorphinan-14-yl]-4-phenyl-	**80:** 83353t
4-(2,5-diethoxy-4-phenoxyphenyl)-1-phenyl-	**60:** P2950a
4-(2,5-diethoxyphenyl)-1-phenyl-	**60:** P2949h

TABLE 19 (*Continued*)

Compound	Reference

19.1. Alkyl- or Aryl-1,2,4-Triazolidine-3,5-diones (*Continued*)

Compound	Reference
1,2-diethyl-	**54:** 4912h, **55:** P4537d
1,4-diethyl-	**79:** 53234s
4-[2-(diethylamino)ethyl]-1,2-diisopropyl-	**64:** 11202g
4-[2-(diethylamino)ethyl]-1-isopropyl-2-phenyl-	**58:** P11370e
4-[2-(diethylamino)ethyl]-1-methyl-2-phenyl-	**58:** P11370d
4-[*p*-(diethylamino)phenyl]-1-phenyl-	**64:** P3556f
1,2-diethyl-4-methyl-	**55:** P4537d
1,4-diethyl-2-methyl-	**79:** 53234s
1,4-diethyl-2-phenyl-	**54:** 4912h, **76:** P34259a, **77:** P62007f
1-(5,6-dihydro-1,3(2H̲)-dioxo-1H̲-[1,2,4]triazolo-[1,2-*a*]cinnoline-6-yl)-4-phenyl-	**78:** 72719z, **80:** 60251u
1-(9,10-dihydro-9-methoxy-10-oxo-9-anthryl)-4-phenyl-	**66:** 94960a
1-(5,6-dihydro-2-methyl-1,3(2H̲)-dioxo-1H̲-[1,2,4]-triazolo[1,2-*a*]cinnolin-6-yl)-4-methyl-	**78:** 72719z
1-[1,4-dihydro-9-(1-methylethenyl)-1,4-methanonaphthalen-9̲-yl]-4-phenyl-	**79:** 66062j
1-[4,5-dihydro-4-(1-methylethyl)-5-oxo-2-phenyl-4-thiazolyl]-4-phenyl-	**85:** 32910t
1-(4,5-dihydro-5-oxo-2,4-diphenyl-4-thiazolyl)-4-phenyl-	**85:** 32910t
4-(3,4-dihydroxyphenyl)-1-(2-naphthyl)-	**63:** P18106c
4-(3,4-dihydroxyphenyl)-1-phenyl-	**63:** P18106c
4-(2,3-dihydroxypropyl)-1,2-diphenyl-	**53:** 7426i, **76:** P34259a, **77:** P62007f
1,2-diisobutyl-	**55:** P4537d
1,2-diisopropyl-	**53:** 7426i, **56:** P10161b
1,2-diisopropyl-4-[(methoxymethylamino)methyl]-	**65:** 3863b
1,2-diisopropyl-4-methyl-	**55:** P4537e, **64:** 11202e
1,2-diisopropyl-4-[(methylmorpholinoamino)methyl]-	**65:** 3863e
1,2-diisopropyl-4-[(4-methyl-1-piperazinyl)methyl]-	**65:** 3863b
1,2-diisopropyl-4-[(methylpiperidinoamino)methyl]-	**65:** 3863e
1,2-diisopropyl-4-(morpholinomethyl)-	**65:** 3863a
1,4-diisopropyl-2-phenyl-	**53:** 7426i
1,2-diisopropyl-4-(piperidinomethyl)-	**65:** 3863a
1,2-diisopropyl-4-propyl-	**53:** 7426i, **64:** 11202e
1,2-diisopropyl-4-(1-pyrrolidinylmethyl)-	**65:** 3863a
1,2-diisopropyl-4-[(trimethylhydrazino)methyl]-	**65:** 3863e
1,2-diisopropyl-4-(xanthen-9-yl)-	**64:** 11202h
4-(4,6-dimethoxy-*m*-cumenyl)-1-phenyl-	**60:** P2949h
4-(2,5-dimethoxy-4-phenoxyphenyl)-1-phenyl-	**60:** P2950a
1-(2,4-dimethoxyphenyl)-4-mesityl-	**60:** P2950c
4-(2,4-dimethoxyphenyl)-1-phenyl-	**60:** P2950b
4-(2,5-dimethoxyphenyl)-1-phenyl-	**60:** P2950c
4-(3,4-dimethoxyphenyl)-1-phenyl-	**63:** P18105e
1-(2,5-dimethoxy-3,4-xylyl)-4-(*p*-ethoxyphenyl)-	**64:** P3555g
4-(2,5-dimethoxy-3,4-xylyl)-1-phenyl-	**60:** P2950a
1,2-dimethyl-	**42:** 8190e, **78:** 135601c

TABLE 19 (*Continued*)

Compound	Reference

19.1. Alkyl- or Aryl-1,2,4-Triazolidine-3,5-diones (*Continued*)

Compound	Reference
1,4-dimethyl-	**42:** 8190e, **76:** 14445e
4-[2-(dimethylamino)ethyl]-1-isopropyl-2-phenyl-	**58:** P11370e
4-[2-(dimethylamino)ethyl]-1-methyl-2-phenyl-	**58:** P11370e
4-(*p*-dimethylaminophenyl)-1-phenyl-	**55:** P13758b
4-[3-(dimethylamino)propyl]-1-methyl-2-phenyl-	**58:** P11370g, **58:** P11370e
4,4′-dimethyl-2,2′-bis(1-methyl-1-phenylethyl)-1,1′-bis[-	**81:** 25614p
1-[1-(1,1-dimethylethyl)-1-methylgermacyclopent-4-en-3-yl]-4-phenyl-, *cis*-	**83:** 97441u
1-[1-(1,1-dimethylethyl)-1-methylgermacyclopent-4-en-3-yl]-4-phenyl-, *trans*-	**83:** 97441u
1-(1,1-dimethylgermacyclopent-4-en-3-yl)-4-phenyl-	**83:** 97441u, **84:** 74336v
1,2-dimethyl-4-[(methylmorpholinoamino)methyl]-	**65:** 3863e
4-(4,7-dimethyl-2-oxo-2H-1-benzopyran-6-yl)-1-phenyl	**60:** P2950c
2-[(7α)-4,4-dimethyl-3-oxocholesta-5,8(14)-dien-7-yl]-4-phenyl-	**83:** 179393u
1,2-dimethyl-4-phenyl-	**42:** 8190h, **67:** 3039p, **81:** 91435g
1,4-dimethyl-2-phenyl-	**42:** 8190g, **53:** 7426i, **66:** 28303n, **76:** P34259a, **77:** P62007f
1,4-dimethyl-2-β-D-ribofuranosyl-	**73:** 66847v
1-(1,1-dimethylsilacyclopent-4-en-3-yl)-4-phenyl-	**83:** 97441u, **84:** 74336v
1-(4,4-dimethyltricyclo[6.3.2$^{2.5}$]tridec-9-yl)-4-phenyl-	**84:** 89679h
1-*p*-dioxan-2-yl-4-phenyl-	**72:** 121454w
1-(2-dioxy-β-D-erythro-pentofuranosyl)-	**73:** 66847v
1-(2-dioxy-β-D-erythro-pentofuranosyl)-2-methyl-	**73:** 66847v
1-(2-dioxy-β-D-erythro-pentofuranosyl)-2-methyl-, 3′,5′-di-*p*-toluate	**73:** 66847v
4-(2,4-diphenoxyphenyl)-1-phenyl-	**60:** P2950a
1,2-diphenyl-	**53:** 7426i, **54:** 5933d, **61:** 3095c, **76:** P34259a, **77:** P62007f, **79:** P141516r
1,4-diphenyl-	**39:** 2068[9], **42:** 8190e, **44:** 6417a, **51:** 14712e, **76:** P34259a, **77:** P62007f
1-(1,1-diphenylgermacyclopent-4-en-3-yl)-4-phenyl-	**83:** 97441u, **84:** 74336v
1-(diphenylmethyl)-	**54:** 4912h, **56:** P10161b, **76:** P34259a, **77:** P62007f
1-diphenylmethyl-2-phenyl-	**54:** 4912h, **76:** P34259a, **77:** P62007f
1,2-diphenyl-4-(phenylmethyl)-	**76:** P34259a, **77:** P62007f
1,2-diphenyl-4-(piperidinomethyl)-	**65:** 3863d
1,2-diphenyl-4-propyl-	**53:** 7426i, **76:** P34259a, **77:** P62007f
1-(1,1-diphenylsilacyclopent-4-en-3-yl)-4-phenyl-	**83:** 97441u, **84:** 74336v
1,2-diphenyl-4-(xanthen-9-yl)-	**64:** 11202h
4-(α⁴,α⁴-diphenyl-2,4-xylyl)-1-phenyl-	**60:** P2949h

TABLE 19 (*Continued*)

Compound	Reference

19.1. Alkyl- or Aryl-1,2,4-Triazolidine-3,5-diones (*Continued*)

Compound	Reference
1,2-dipropyl-	**55:** P4537d
1-[3-(1,3-dithian-2-ylidene)2-oxopropyl]-4-phenyl-	**85:** 123830h
1,4-di-2,4-xylyl-	**60:** P2950c, **60:** P8040a
4-dodecyl-	**50:** 12032i, **51:** P10588a, **51:** P12983e
1-dodecyl-2-phenyl-	**54:** 5933d, **54:** 4912h
-4-ethanol, 1,2-diphenyl-	**53:** 7426i, **56:** P10161a
-4-ethanol, 1-isopropyl-2-phenyl-	**53:** 7426i
-4-ethanol, 1-methyl-2-phenyl-	**53:** 7426i
-4-ethanol, 1-phenyl-2-propyl-	**54:** 4912h
4-(4-ethoxycyclohexyl)-1-phenyl-	**63:** P18107g
1-(9-ethoxy-9,10-dihydro-10-oxo-9-anthryl)-4-phenyl-	**66:** 94960a
1-(ethoxydiphenylmethyl)-4-phenyl-	**78:** 29344m
4-[*p*-(2-ethoxyethoxy)phenyl]-1-phenyl-	**64:** P3556f
4-[(ethoxyethylamino)methyl]-1,2-diphenyl-	**65:** 3863e
4-(ethoxymethyl)-1,2-diisopropyl-	**64:** 11203c
4-[*p*-(*p*-ethoxyphenoxy)phenyl]-1-phenyl-	**64:** P3556f
4-[*p*-(*p*-ethoxyphenoxy)phenyl]-1-(2-pyridyl)-	**60:** P2950f, **64:** P12686a
1-(4-ethoxyphenyl)-	**76:** P34259a, **77:** P62007f
4-(*p*-ethoxyphenyl)-	**56:** 2440d, **57:** 809e, **61:** P8317e
4-(*p*-ethoxyphenyl)-1-(*p*-hydroxyphenyl)-	**64:** P2097b
4-(*p*-ethoxyphenyl)-1-(4-methoxy-*o*-tolyl)-	**64:** P3555g
1-(*p*-ethoxyphenyl)-2-methyl-	**54:** 4912h, **76:** P34259a, **77:** P62007f
4-(*p*-ethoxyphenyl)-1-(2-naphthyl)-	**64:** P3555h
4-(*p*-ethoxyphenyl)-1-phenyl-	**57:** P11205c, **58:** P1479gh, **60:** P1764fg
4-(*p*-ethoxyphenyl)-1-(2-pyridyl)-	**60:** P2950f, **64:** P12686a
4-(*p*-ethoxyphenyl)-1-*m*-tolyl-	**58:** P1479g
4-(*p*-ethoxyphenyl)-1-*p*-tolyl-	**57:** P11205c
1-(*p*-ethoxyphenyl)-4-(3,4,5-trimethoxyphenyl)-	**63:** P18106b
1-(*p*-ethoxyphenyl)-4-3,4-xylyl-	**63:** P18105f
1-ethyl-	**65:** P4905f
4-ethyl-	**50:** 12032i, **51:** P10587i, **51:** P12983de, **67:** 116858y, **83:** 42640c, **84:** 16586r
1-[4-(3-ethylbenzolthiazolium-2-yl)-1,3-butadienyl]-4-phenyl-, hydroxide, inner salt	**78:** P167001y
1-[2-(3-ethylbenzothiazolium-2-yl)ethenyl]-, hydroxide, inner salt	**78:** P167001y
1-[2-(3-ethylbenzothiazolium-2-yl)ethenyl]-4-phenyl-,	**78:** P167001y
4-(2-ethylcyclohexyl)-1-phenyl-	**63:** P18107g
4-ethyl-1,2-diisopropyl-	**64:** 11202e
1-ethyl-2,4-diphenyl-	**76:** P34259a, **77:** P62007f
4-ethyl-1,2-diphenyl-	**53:** 7426i, **77:** P62007f
1-ethyl-2-isopropyl-	**55:** P4537f

TABLE 19   (*Continued*)

Compound	Reference

19.1. Alkyl- or Aryl-1,2,4-Triazolidine-3,5-diones   (*Continued*)

Compound	Reference
1-ethyl-4-methyl-2-phenyl-	**54:** 4912h, **76:** P34259a, **77:** P62007f
1-(2-ethylphenyl)-	**76:** P34259a, **77:** P62007f
1-ethyl-2-phenyl-	**53:** 7426i, **54:** 5933d, **76:** P34259a,
1-ethyl-2-phenyl	**77:** P62007f
4-(*p*-ethylphenyl)-	**50:** 12032h, **51:** P10588a, **51:** P12983e
4-ethyl-1-phenyl-	**54:** 4912h, **76:** P34259a, **77:** P62007f
4-(*p*-ethylphenyl)-1-(2-naphthyl)-	**64:** P3556f
4-(*p*-ethylphenyl)-1-phenyl-	**64:** P3556f
4-ethyl-1-phenyl-2-propyl-	**54:** 4912h, **76:** P34259a, **77:** P62007f
4-(*o*-ethylphenyl)-1-(2-pyridyl)-	**60:** P2950f, **64:** P12686b
1-(1-ethylpropyl)-2-phenyl-	**54:** 4912h, **56:** P10160h, **76:** P34259a, **77:** P62007f
1-[2-(1-ethylquinolinium-2-yl)ethenyl]-4-phenyl-, hydroxide, inner salt	**78:** P167001y
4-[5-(ethylsulfonyl)-2-methoxyphenyl]-1-phenyl-	**60:** P2950b
4-[*p*-(ethylthio)phenyl]-1-phenyl-	**64:** P3556f
4-(*p*-fluorophenyl)-1-(2-pyridyl)-	**60:** P2950f, **64:** PC12686b
4-(*p*-fluorophenyl)-1-(5,6,7,8-tetrahydro-2-naphthyl)-	**64:** P2097d
1-heptyl-2-phenyl-	**54:** 4912h, **56:** P10161b, **76:** P34259a, **77:** P62007f
4-hexadecyl-	**50:** 12032i, **51:** P10588a
4-(α,α,α,α'α'α'-hexafluoro-3,5-xylyl)-1-phenyl-	**64:** P3556f
1-(2,4a,5,6,7,8-hexahydro-1,4a-dimethyl-2-naphthalenyl)-4-phenyl-, *trans-*	**84:** 4541j
1-(2,4a,5,6,7,8-hexahydro-2-naphthalenyl)-4-phenyl-	**78:** 158728p
4-hexyl-	**50:** 12032i, **51:** P10588a
4-(*p*-hexylphenyl)-	**50:** 12032i, **51:** P10588a
1-hexyl-2-phenyl-	**54:** 4912h
4-[*p*-(hexylthio)phenyl]-1-phenyl-	**64:** P3556f
4-[*p*-(hexylthio)phenyl]-1-(5,6,7,8-tetrahydro-2-naphthyl)-	**64:** P2097d
4-(2-hydroxycyclohexyl)-1-phenyl-	**63:** P18107g
4-(4-hydroxycyclohexyl)-1-phenyl-	**63:** P18107g
4-[*p*-(2-hydroxyethoxy)phenyl]-1-phenyl-	**64:** P3556f
4-(2-hydroxyethyl)-1,2-diphenyl-	**76:** P34259a, **77:** P62007f
4-(2-hydroxyethyl)-1-(1-methylethyl)-2-phenyl-	**76:** P34259a, **77:** P62007f
4-(2-hydroxyethyl)-1-methyl-2-phenyl-	**76:** P34259a, **77:** P62007f
4-(2-hydroxyethyl)-1-phenyl-2-propyl-	**76:** P34259a, **77:** P62007f
1-(1-hydroxy-2-indanyl)-4-phenyl-, *trans-*	**72:** 121454w
4-(4-hydroxy-3-methoxyphenyl)-1-phenyl-	**63:** P18106b
1-(4-hydroxy-5-oxo-1,3,6-cycloheptatrien-1-yl)-4-phenyl-	**72:** 66870b, **75:** 20306f
4-(*p*-hydroxyphenyl)-1-phenyl-	**57:** P11205c
4-(*p*-hydroxyphenyl)-1-(3,4-xylyl)-	**64:** P3555g

TABLE 19   (*Continued*)

Compound	Reference

19.1. Alkyl- or Aryl-1,2,4-Triazolidine-3,5-diones (*Continued*)

Compound	Reference
1-(4-hydroxy-1,1,4-trimethyl-2-pentenyl)-4-phenyl-, acetate ester	**70:** 20009h
1-(2-indanyl)-4-phenyl-	**72:** 121454w
4-(1-indanyl)-1-phenyl-	**63:** P18107f
4-(*p*-iodophenyl)-1-(3-methyl-4-biphenylyl)-	**64:** P2097d
4-(3-iodo-*p*-tolyl)-1-phenyl-	**63:** P18106a
4-(5-iodo-2,4-xylyl)-1-phenyl-	**60:** P2950a
4-isobutyl-	**67:** 116858y
1-isobutyl-2-phenyl-	**53:** 7426i, **54:** 4912h
4-isopentyl-1,2-diphenyl-	**53:** 7426i
4-[*p*-(isopentyloxy)phenyl]-1-(2-pyridyl)-	**60:** P2950h, **64:** P12686d
1-isopentyl-2-phenyl-	**53:** 7426i, **56:** P10161b
4-(*p*-isopentylphenyl)-1-phenyl-	**64:** P3556f
1-isopentyl-2-phenyl-4-propyl-	**53:** 7426i
4-(3-isopropoxymesityl)-1-phenyl-	**60:** P2950b
4-(4-isopropoxy-6-methoxy-*m*-tolyl)-1-phenyl-	**60:** P2950b
4-(4-isopropoxy-*o*-tolyl)-1-phenyl-	**60:** P2950b
4-(4-isopropoxy-2,5-xylyl)-1-phenyl-	**60:** P2950b
1-isopropyl-	**55:** P4537e, **76:** P34259a, **77:** P62007f
4-isopropyl-	**67:** 116858y
1-isopropyl-4-[(methylmorpholinoamino)methyl]-2-phenyl-	**65:** 3863e
1-isopropyl-4-methyl-2-phenyl-	**53:** 7426i
1-isopropyl-2-phenyl-	**53:** 7426i, **54:** 5933d
4-isopropyl-1-phenyl-	**83:** 178937n
1-isopropyl-2-phenyl-4-(piperidinomethyl)-	**65:** 3863d
1-isopropyl-2-phenyl-4-propyl-	**53:** 7426i, **76:** P34259a, **77:** P62007f
4-mesityl-1-(2-naphthyl)-	**60:** P2950c
4-[*p*-(mesityloxy)phenyl]-1-phenyl-	**60:** P2950b
4-mesityl-1-phenyl-	**60:** P2950c
4-mesityl-1-*o*-tolyl-	**60:** P2950c
4-mesityl-1-*m*-tolyl-	**60:** P2950c
-4-methanol, 1,2-diisopropyl-	**64:** 11203b
-4-methanol, 1,2-diisopropyl-, acetate	**64:** 11203b
-4-methanol, 1,2-diisopropyl-, carbaniliate	**64:** 11203b
-4-methanol, 1,2-diisopropyl-, chloroacetate	**64:** 11203b
-4-methanol, 1,2-dimethyl-	**64:** 11203b
-4-methanol, 1,2-diphenyl-	**64:** 11202h
-4-methanol, 1,2-diphenyl-, acetate	**64:** 11202h
-4-methanol, 1,2-diphenyl-, benzoate	**64:** 11203a
-4-methanol, 1,2-diphenyl-, chloroacetate	**64:** 11203a
-4-methanol, 1-isopropyl-2-phenyl-	**54:** 4912h, **64:** 11203b
-1-methanol, 4-phenyl-	**64:** 19597d
4-[4-[(methoxybenzyl)oxy]-*o*-tolyl]-1-phenyl-	**60:** P2950b
4-(3-methoxycyclohexyl)-1-(*m*-methoxyphenyl)-	**63:** P18107g
1-(4-methoxycyclohexyl)-2-phenyl-	**54:** 4912h, **76:** P34259a, **77:** P62007f

TABLE 19 (*Continued*)

Compound	Reference

19.1. Alkyl- or Aryl-1,2,4-Triazolidine-3,5-diones (*Continued*)

Compound	Reference
4-(3-methoxy-2-dibenzofuranyl)-1-phenyl-	**60:** P2950c
4-(3-methoxymesityl)-1-phenyl-	**60:** P2950b
4-[2-methoxy-4-(2-methoxyphenoxy)phenyl]-1-phenyl-	**60:** P2950a
4-[(methoxymethylamino)methyl]-1,2-diphenyl-	**65:** 3863e
4-(2-methoxy-4-phenoxyphenyl)-1-phenyl-	**60:** P2950b
4-(4-methoxy-2-phenoxyphenyl)-1-phenyl-	**60:** P2950a
4-(*p*-methoxyphenyl)-	**66:** 37837s, **84:** 16586r
4-(*p*-methoxyphenyl)-1-phenyl-	**58:** P1479h, **60:** P1764fg
4-(*o*-methoxyphenyl)-1-(2-pyridyl)-	**60:** P2950f, **64:** P12686b
4-(*m*-methoxyphenyl)-1-(2-pyridyl)-	**60:** P2950f, **64:** P12686b
4-(*p*-methoxyphenyl)-1-(2-pyridyl)-	**60:** P2950f, **64:** P12686b
4-[2-methoxy-5-(phenylsulfonyl)phenyl]-1-phenyl-	**60:** P2950a
4-[2-methoxy-4-(phenylthio)phenyl]-1-phenyl-	**60:** P2950a
4-[2-methoxy-4-(phenylthio)phenyl]-1-(2-pyridyl)-	**60:** P2950f, **64:** PC12686b
4-[2-methoxy-5-(1,1,3,3-tetramethylbutyl)phenyl]-1-phenyl-	**60:** P2950b
1-(4-methoxy-*o*-tolyl)-4-(*p*-phenoxyphenyl)-	**64:** P3556f
4-(4-methoxy-*o*-tolyl)-1-phenyl-	**60:** P2949h
4-(5-methoxy-*o*-tolyl)-1-phenyl-	**60:** P2950a
4-(6-methoxy-*m*-tolyl)-1-phenyl-	**60:** P2949h
1-(4-methoxy-*o*-tolyl)-4-(3,5-xylyl)-	**63:** P18105c
4-(6-methoxy-3,4-xylyl)-1-phenyl-	**60:** P2949h
1-methyl	**42:** 8190d, **53:** 7426i, **76:** 14445e, **80:** 120848f
4-methyl-	**42:** 8190d, **50:** 12032i, **53:** 7426i, **59:** 8747h, **66:** 37837s, **67:** 116858y
1-(2-methylallyl)-2-phenyl-	**54:** 4912h, **56:** P10160i
4-(1-methyl-5-benzimidazolyl)-1-phenyl-	**63:** P18105f
1-(α-methylbenzyl)-2-phenyl-	**54:** 4912h, **56:** P10160h
1-(3-methyl-4-biphenylyl)-4-(*p*-tert-pentylphenyl)-	**64:** P2097d
4-(3-methyl-4-biphenylyl)-1-(2-pyridyl)-	**60:** P2950g, **64:** P12686b
4-(3-methylbutyl)-1,2-bis(1-methylethyl)-	**76:** P34259a, **77:** P62007f
4-(3-methylbutyl)-1,2-diphenyl-	**76:** P34259a, **77:** P62007f
1-(1-methylbutyl)-2-phenyl-	**54:** 4912h, **56:** P10161b
1-(3-methylbutyl)-2-phenyl-	**76:** P34259a, **77:** P62007f
1-(3-methylbutyl)-2-phenyl-4-propyl-	**76:** P34259a, **77:** P62007f
1-[1-methylcyclohexyl]-2-phenyl-	**54:** 4912h
1-(2-methylcyclohexyl)-2-phenyl-	**76:** P34259a, **77:** P62007f
1-[4-methylcyclohexyl]-2-phenyl-	**54:** 4912h, **76:** P34259a, **77:** P62007f
4-(4-methylcyclohexyl)-1-phenyl-	**63:** P18107g
1-methyl-2,4-diphenyl-	**44:** 6417b, **76:** P34259a, **77:** P62007f
4-methyl-1,2-diphenyl-	**53:** 7426i, **76:** P34259a, **77:** P62007f
1-(2-methylenecyclooctyl)-4-phenyl-	**84:** 89679h
4-[3,4-(methylenedioxy)phenyl]-1-phenyl-	**60:** P2949g

TABLE 19 *(Continued)*

Compound	Reference

19.1. Alkyl- or Aryl-1,2,4-Triazolidine-3,5-diones *(Continued)*

Compound	Reference
1-(1-methylethyl)-2-phenyl-	**76:** P34259a, **77:** P62007f
4-(p-[(2-methylheptyl)oxy]phenyl)-1-(2-pyridyl)-	**64:** PC12686b
4-methyl-1-(1-methylallyl)-	**67:** 116853t
1-methyl-4-(3-methylbutyl)-2-phenyl-	**76:** P34259a, **77:** P62007f
4-methyl-1-(1-methylethyl)-2-phenyl-	**76:** P34259a, **77:** P62007f
1-methyl-2-(4-methylphenyl)-	**76:** P34259a, **77:** P62007f
4-methyl-2-(1-methyl-1-phenylethyl)-	**81:** 25614p
1-[1-methyl-1-(2-methylpropyl)germacyclopent-4-en-3-yl]-4-phenyl-, *cis-*	**83:** 97441u, **84:** 74336v
1-[1-methyl-1-(2-methylpropyl)germacyclopent-4-en-3-yl]-4-phenyl-, *trans-*	**83:** 97441u, **84:** 74336v
4-[(methylmorpholinoamino)methyl]-1,2-diphenyl-	**65:** 3863e
1-methyl-2-(p-nitrophenyl)-	**61:** 3095c
1-(4-methyl-2-nitrophenyl)-	**80:** 120848f
4-methyl-1-[2-oxo-2-phenyl-1-(triphenyl-phosphoranylidene)ethyl]-	**85:** 20278e
4-methyl-1-[2-oxo-1-(triphenylphosphoranylidene)propyl]-	**85:** 20278e
1-(α-methylphenethyl)-2-phenyl-	**54:** 4912h, **56:** P10161c
1-(2-methylphenyl)-	**76:** P34259a, **77:** P62007f
1-(4-methylphenyl)-	**76:** P34259a, **77:** P62007f
1-methyl-2-phenyl-	**42:** 8190g, **53:** 7426i, **61:** 3095c, **76:** P34259a, **77:** P62007f, **79:** P141516r
1-methyl-4-phenyl-	**75:** 34816j, **76:** 14445e, **81:** 91435g
4-methyl-1-phenyl-	**42:** 8190f, **53:** 7426i, **76:** P34259a, **77:** P62007f
1-methyl-2-phenyl-4-(2-piperidinoethyl)-	**58:** P11370e
1-methyl-2-phenyl-4-propyl-	**53:** 7426i, **76:** P34259a, **77:** P62007f
4-methyl-1-phenyl-2-propyl-	**76:** P34259a, **77:** P62007f
1-(1-methyl-1-phenylsilacyclopent-4-en-3-yl)-4-phenyl-, *cis-*	**83:** 97441u, **84:** 74336v
1-(1-methyl-1-phenylsilacyclopent-4-en-3-yl)-4-phenyl-, *trans-*	**83:** 97441u, **84:** 74336v
4-[(methylpiperidinoamino)methyl]-1,2-diphenyl-	**65:** 3683e
1-(2-methyl-2-propenyl)-2-phenyl-	**76:** P34259a, **77:** P62007f
1-(1-methylpropyl)-2-phenyl-	**76:** P34259a, **77:** P62007f
1-(2-methylpropyl)-2-phenyl-	**76:** P34259a, **77:** P62007f
1-methyl-2-β-D-robofuranosyl-	**73:** 66847v
1-methyl-2-β-D-ribofuranosyl-, 2′,3′,5′-tribenzoate	**73:** 66847v
4-methyl-1-β-D-ribofuranosyl-, 2′,3′,5′-tribenzoate	**73:** 66847v
4-[m-(methylthio)phenyl]-1-phenyl-	**64:** P3556f
4-[m-(methylthio)phenyl]-1-(2-pyridyl)-	**60:** P2950f, **64:** P12686b
1-[m-(methylthio)phenyl]-4-(3,4,5-trimethoxyphenyl)-	**64:** P2097d
1-methyl-2-p-tolyl-	**54:** 4912h

552

TABLE 19   (*Continued*)

Compound	Reference

19.1. Alkyl- or Aryl-1,2,4-Triazolidine-3,5-diones   (*Continued*)

4-methyl-1-(1,1,2-trimethylallyl)-	**67:** 116853t
1-naphtho[1,2,3,4-*def*]chrysen-8-yl-4-phenyl-	**82:** 156184s
1-(1-naphthyl)-	**65:** P4905f
4-naphthyl-	**67:** P69440u
1-(2-naphthyl)-4-(4-phenoxy-2,5-xylyl)-	**60:** P2950c
4-(2-naphthyl)-1-phenyl-	**63:** P18106a
4-(1-naphthyl)-1-(2-pyridyl)-	**60:** P2950g, **64:** P12686b
4-(2-naphthyl)-1-(2-pyridyl)-	**60:** P2950g, **64:** P12686b
1-(2-naphthyl)-4-*p*-tolyl-	**64:** P3555g
1-(*p*-nitrophenyl)-	**61:** 3095b, **76:** P34259a, **77:** P62007f
4-(*p*-nitrophenyl)-	**66:** 37837s, **76:** 59534z
1-(4-nitrophenyl)-2-benzyl-	**76:** 46146z
4-(*m*-nitrophenyl)-1-phenyl-	**64:** P3556f
4-(5-nitro-2,4-xylyl)-1-phenyl-	**60:** P2949g
1-nonyl-2-phenyl-	**54:** 4912h
4-octadecyl-	**50:** 12032i
4-octyl-	**50:** 12032i, **67:** 116858y
4-[*p*-(sec-octyloxy)phenyl]-1-phenyl-	**64:** P3556f
4-(*p*-octylphenyl)-	**50:** 12032i
1-octyl-2-phenyl-	**54:** 4912h
4-(*p*-octylphenyl)-1-phenyl-	**64:** P3556f
4-(*p*-octylphenyl)-1-(2-pyridyl)-	**60:** P2950f, **64:** P12686b
1-(3-oxobutyl)-2,4-diphenyl-	**65:** 718h
1-[2-oxo-2-phenyl-1-(triphenylphosphoranylidene)ethyl]-4-phenyl-	**85:** 20278e
4-pentyl-1,2-diphenyl-	**53:** 7426i, **76:** P34259a, **77:** P62007f
1-pentyl-2-phenyl-	**53:** 7426i, **56:** P10161b, **76:** P34259a, **77:** P62007f
4-(*p*-tert-pentylphenyl)-1-phenyl-	**64:** P3556f
1-phenethyl-2-phenyl-	**54:** 4912h
4-(*p*-phenethylphenyl)-1-phenyl-	**64:** P3556f
4-(*p*-phenethylphenyl)-1-(2-pyridyl)-	**60:** P2950f, **64:** PC12686b
4-[*p*-(2-phenoxyethoxy)phenyl]-1-phenyl-	**64:** P3556f
4-(*p*-phenoxyphenyl)-1-phenyl-	**64:** P3556f
4-(*p*-phenoxyphenyl)-1-(2-pyridyl)-	**60:** P2950f, **64:** PC12686b
4-(4-phenoxy-*o*-tolyl)-1-phenyl-	**60:** P2949h
4-(5-phenoxy-*o*-tolyl)-1-(2-pyridyl)-	**60:** P2950g
4-(4-phenoxy-2,5-xylyl)-1-phenyl-	**60:** P2950b
4-(5-phenoxy-2,4-xylyl)-1-phenyl-	**60:** P2950b
4-(4-phenoxy-2,5-xylyl)-1-(2-pyridyl)-	**64:** P12686b
1-phenyl-	**24:** 1114[4], **28:** 1665[9], **33:** 546[5], **42:** 8190de, **50:** 12033c, **53:** 7426i, **61:** 3095b, **73:** 77210h

TABLE 19   (*Continued*)

Compound	Reference

19.1. Alkyl- or Aryl-1,2,4-Triazolidine-3,5-diones   (*Continued*)

Compound	Reference
4-phenyl-	**41:** 948e, **42:** 8190de, **50:** 12033c, **55:** 22298i, **57:** 809e, **66:** 37837s, **76:** 59534z, **78:** 29344m, **78:** 29676w
1-phenyl-2,4-dibenzyl-	**76:** P34259a, **77:** P62007f
1-phenyl-2,4-dipropyl-	**53:** 7426i, **56:** P10161a, **76:** P34259a, **77:** P62007f
1-phenyl-4-[*p*-(phenylazo)phenyl]-	**64:** P3556f
4-phenyl-1-[2-(2-phenylcyclopropyl)-2-propenyl]-	**78:** 147403c
1-phenyl-2-(1-phenylethyl)-	**76:** P34259a
1-phenyl-2-(2-phenylethyl)-	**76:** P34259a, **77:** P62007f
1-phenyl-2-(phenylmethyl)-	**76:** P34259a, **76:** 46146z, **77:** P62007f
1-phenyl-4-(phenylmethyl)-	**76:** P34259a, **77:** P62007f
4-(phenyl-1-[1-(3-phenyl-1-propenyl)cyclopropyl]-	**77:** 114318e
1-phenyl-2-(3-phenylpropyl)-	**54:** 4912h
4-phenyl-1-(piperidinomethyl)-	**64:** 19597b
1-phenyl-2-(2-propenyl)-	**76:** P34259a, **77:** P62007f
1-phenyl-2-propyl-	**53:** 7426i, **54:** 5933d, **76:** P34259a, **77:** P62007f
1-phenyl-4-propyl-	**54:** 4912h, **76:** P34259a, **77:** P62007f
1-phenyl-2-(1-propylbutyl)-	**54:** 4912h, **76:** P34259a, **77:** P62007f
4-phenyl-1-(2-pyridyl)-	**60:** P2950e, **64:** P12686a
1-phenyl-4-[*p*-(2-pyridyloxy)phenyl]-	**64:** P3556f
4-phenyl-1-[(spiro[2.5]oct-4-en-4-yl)methyl]-	**80:** 107595b
1-phenyl-4-(*p*-styrylphenyl)-	**64:** P3556f
1-phenyl-4-(1,2,3,4-tetrahydro-2-naphthyl)-	**63:** P18107h
1-phenyl-4-(5,6,7,8-tetrahydro-2-naphthyl)-	**63:** P18106a
4-phenyl-1-(2,2,3,4-tetramethylsilacyclopent-4-en-3-yl)-	**83:** 97441u, **84:** 74336v
4-phenyl-1-[3-(triethylgermyl)-2-propenyl]-	**84:** 74336v
1-phenyl-4-(α,α,α-trifluoro-*m*-tolyl)-	**64:** P3556f
1-phenyl-4-(3,4,5-trimethoxyphenyl)-	**63:** P18106a
1-phenyl-4-(2,4,5-trimethylphenyl)-	**60:** P2949g
4-phenyl-1-[3-(trimethylsilyl)-2-propenyl]-	**84:** 74336v
1-phenyl-2-undecyl-	**54:** 4912h
1-phenyl-4-(2,4-xylyl)-	**60:** P2950c, **60:** P8040a
1-phenyl-4-(2,5-xylyl)-	**60:** P2950a
1-phenyl-4-(3,4-xylyl)-	**63:** P18105e, **63:** P18106c
1-phenyl-4-(3,5-xylyl)-	**63:** P18106a
4,4-(1,4-piperazinediylbis[(methylimino)methylene])-bis[1,2-diisopropyl-	**65:** 3863e
4,4-(1,4-piperazinediylbis[(methylimino)methylene])-bis[1,2-diphenyl-	**65:** 3863f
4,4-(1,4-piperazinediyldimethylene)bis[1,2-diisopropyl-	**65:** 3863b

TABLE 19   (*Continued*)

Compound	Reference

19.1. Alkyl- or Aryl-1,2,4-Triazolidine-3,5-diones   (*Continued*)

Compound	Reference
4-propyl-	**50:** 12032i
4-*p*-propylphenyl-	**50:** 12032h
1-(2-pyridyl)-	**64:** P12685f
1-(2-pyridyl)-4-*o*-tolyl-	**60:** P2950h, **64:** P12686d
1-(2-pyridyl)-4-*m*-tolyl-	**60:** P2950h, **64:** P12686d
1-(2-pyridyl)-4-*p*-tolyl-	**60:** P2950h, **64:** P12686d
1-(2-pyridyl)-4-(2,4-xylyl)-	**60:** P2950f, **64:** P12686a
1-β-D-ribofuranosyl-	**73:** 66847v
1-β-D-ribofuranosyl-, 2′,3′,5′-tribenzoate	**73:** 66847v
1-*p*-sulfamoylphenyl-	**43:** 601d
4-tetradecyl-	**50:** 12032i
4,4-tetramethylenebis[1,2-diisopropyl-	**64:** 11202f
4-*o*-tolyl-	**50:** 12032h
4-*p*-tolyl-	**50:** 12032h
1-*o*-tolyl-4-(2,4-xylyl)-	**60:** P2950c, **60:** P8040a
1-*p*-tolyl-4-(3,4-xylyl)-	**63:** P18106b
4-(2,4,5-trichlorophenyl)-	**80:** P146185n, **80:** P146210s
4-(3,4,5-trichlorophenyl)-	**80:** P146210s
1,2,4-triethyl-	**79:** 53234s
4-[3-(trifluoromethyl)phenyl]-	**80:** P146156d, **80:** P146185n, **80:** P146210s, **83:** P58835f
1,2,4-triisopropyl-	**53:** 7426i
1,2,4-trimethyl-	**42:** 8191a, **76:** P34259a, **77:** P62007f
1,2,4-triphenyl-	**53:** 7426i, **76:** P34259a, **84:** 30972v
-1-yl, 4-methyl-2-(1-methyl-1-phenylethyl)-	**81:** 25614p

19.2. Other 1,2,4-Triazolidine-3,5-diones

Compound	Reference
1-acetaldehyde, 2-acetyl-4-phenyl-	**77:** 75176k
1-acetaldehyde, 2-benzoyl-4-phenyl-	**77:** 75176k
-1-acetaldehyde, 2-(chloroacetyl)-4-phenyl-	**77:** 75176k
-1-acetaldehyde, 2-(2,2-dimethyl-1-oxopropyl)-4-phenyl-	**77:** 75176k
-1-acetaldehyde, 2-(2-methyl-1-oxopropyl)-4-phenyl-	**77:** 75176k
-4-acetamide, 1-acetyl-α-(2,2-dimethyl-1-oxopropyl)-N, 2-diphenyl-	**79:** P141516r
-4-acetamide, 1,2-diacetyl-α-(2,2-dimethyl-1-oxopropyl)-N-phenyl-	**79:** P141516r
4-acetyl-1-(4-biphenylyl)-2-phenyl-	**56:** P486g
1-acetyl-4-(3,5-dichlorophenyl)-	**85:** P117969u
1-acetyl-4-(3,5-dichlorophenyl)-2-methyl-	**85:** P117969u
1-acetyl-4-(3,5-dichlorophenyl)-2-(1-oxopropyl)-	**85:** P117969u
1-acetyl-2-(2-oxopropyl)-4-phenyl-	**77:** 75176k
1-acetyl-4-phenyl-	**66:** P120802x
4-acetyl-1-phenyl-	**74:** 140379h

TABLE 19 (*Continued*)

Compound	Reference

19.2. Other 1,2,4-Triazolidine-3,5-diones (*Continued*)

Compound	Reference
1-amino-	**47:** 8955e
4-amino-	**50:** 6240b, **54:** 7625i, **61:** 10676a, **78:** 29676w
4-amino-1,2-dibenzyl-	**63:** 6999d
4-anilino-1-phenyl-	**80:** 120848f
4-benzoyl-1,2-bis(tert-butyl)-	**81:** 13434s
1-benzoyl-4-butyl-2-methyl-	**67:** 3039p
4-benzoyl-1,2-diisopropyl-	**64:** 11202e
4-(benzylamino)-	**61:** 3098e
2-benzyl-1-[benzyl(4-chlorophenyl)amino]-4-(4-chlorophenyl)-	**76:** 46146z
2-benzyl-1-[(benzyl)phenylamino]-4-phenyl-	**76:** 46146z
1-benzyl-4-(dibenzylamino)-	**63:** 6999b
4-benzylideneamino-	**54:** 7625i, **63:** 16339g, **76:** 59534z
4-(benzylsulfonyl)-	**65:** PC16978g
1,2-bis(tert-butyl)-4-[(4-methylphenyl)sulfonyl]-	**85:** 94296f
1,2-bis([(4-chloro-*o*-tolyl)oxy]acetyl)-4-phenyl-	**73:** P119532x
1,2-bis[(2,4-dichlorophenoxy)acetyl]-4-phenyl-	**73:** P119532x
4-[(*p*-bromophenyl)sulfonyl]-	**65:** PC16978g
1-butyryl-4-phenyl-	**65:** 704b
-1-carbodithioic acid, 4-(3,5-dichlorophenyl)-2-methyl-, ethyl ester	**85:** P117969u
-1-carbonyl chloride, 4-ethyl-	**82:** P57676x
-1-carbonyl chloride, 4-methyl-	**82:** P57676x
-1-carbonyl chloride, 4-phenyl-	**82:** P57676x
-1-carbothioamide, 4-(3,5-dichlorophenyl)-*N*,2-dimethyl-	**85:** P117969u
-1-carboxaldehyde, 2-acetyl-4-phenyl-	**80:** 60251u
-1-carboxamide	**61:** P8317e
-1-carboxamide, 4-(3,5-dichlorophenyl)-*N*,2-dimethyl-	**85:** P117969u
-1-carboxamide, 4-(3,5-dichlorophenyl)-*N*-methyl-	**85:** P117969u
-1-carboxamide, *N*,4-dimethyl-	**83:** 42640c
-1-carboxamide, *N*,*N*-dimethyl-4-phenyl-	**61:** P8317e
-1-carboxamide, 4-(*p*-ethoxyphenyl)-*N*,*N*-dimethyl-	**61:** P8317e
-1-carboxamide, 4-phenyl-	**61:** P8317e
-1-carboxamide, *N*-(2-phenylbutyryl)-	**61:** 1781b
-1-carboxamide, *N*,2,4-triphenyl-	**81:** 91438k
-1-carboxamidine, *N*,*N*'-dicyclohexyl-4-phenyl-	**61:** P8317e
-1-carboxanilide, 4-phenyl-	**61:** P8317e
1-[[(2-carboxy-3,3-dimethyl-7-oxo-4-thia-1-azabicyclo-[3.2.0]hept-6-yl)carbamoyl(phenyl)methyl]carbamoyl]-4-ethyl-	**82:** P57676x
1-([2-carboxy-3,3-dimethyl-7-oxo-4-thia-1-(azabicyclo-[3.2.0]hept-6-yl)carbamoyl(phenyl)methyl]carbamoyl)-4-methyl-	**82:** P57676x
1-([[(2-carboxy-3,3-dimethyl-7-oxo-4-thia-1-azabicyclo-[3.2.0]hept-6-yl)carbamoyl(phenyl)methyl]-carbamoyl)-4-phenyl-	**82:** P57676x

TABLE 19   (*Continued*)

Compound	Reference
19.2. Other 1,2,4-Triazolidine-3,5-diones (*Continued*)	
1-carboxylic acid, 2-acetyl-4-(3,5-dichlorophenyl)-, ethyl ester	**85:** P117969u
1-carboxylic acid, 4-(3,5-dichlorophenyl)-, ethyl ester	**85:** P117969u
1-carboxylic acid, 4-(3,5-dichlorophenyl)-2-methyl-, ethyl ester	**85:** P117969u
1-carboxylic acid, 4-(3,5-dichlorophenyl)-2-([(1-methylethyl)amino]carbonyl)-, ethyl ester	**85:** P117969u
-1-carboxylic acid, 2,4-diethyl-, ethyl ester	**79:** 53234s
-1-carboxylic acid, 2,4-diethyl-, methyl ester	**79:** 53234s
-1-carboxylic acid, 2,4-dimethyl-, ethyl ester	**79:** 53234s
-4-carboxylic acid, 1,2-diphenyl-, ethyl ester	**53:** 7426i, **76:** P34259a, **77:** P62007f
-1-carboxylic acid, 4-phenyl-, methyl ester	**85:** 94283z
-1-carboxylic acid, 4-phenyl-2-[(phenylamino)carbonyl]-, benzyl ester	**85:** 94283z
-1-carboxylic acid, 4-phenyl-2-[(phenylamino)carbonyl]-, butyl ester	**85:** 94283z
-1-carboxylic acid, 4-phenyl-2-[(phenylamino)carbonyl]-, cyclopentyl ester	**85:** 94283z
-1-carboxylic acid, 4-phenyl-2-[(phenylamino)carbonyl]-, ethyl ester	**85:** 94283z
-1-carboxylic acid, 4-phenyl-2-[(phenylamino)carbonyl]-, isopropyl ester	**85:** 94283z
-1-carboxylic acid, 4-phenyl-2-[(phenylamino)carbonyl]-, methyl ester	**85:** 94283z
1-(4-chlorobutyryl)-4-phenyl-	**59:** 5152b
4-[(4-chloro-3-nitrophenyl)sulfonyl]-	**65:** PC16978g
4-(4-chlorophenyl)-1-[(4-chlorophenyl)amino]-	**76:** 46146z
4-[(*p*-chlorophenyl)sulfonyl]-	**65:** PC16978g
1-(*p*-chlorophenyl)-4-[(trichloromethyl)thio]-	**72:** 100656y
1-crotonoyl-4-methyl-	**73:** 3842u
1-crotonoyl-4-phenyl-	**73:** 3842u
1-cyclohexyl-4-phenyl-	**81:** 91438k
1,2-diacetyl-	**50:** 12033a
1,2-diacetyl-4-cyclohexyl-	**61:** P8317e
1,2-diacetyl-4-(3,5-dichlorophenyl)-	**85:** P117969u
1,2-diacetyl-4-phenyl-	**65:** 3882d
1,4-diacetyl-2-phenyl-	**73:** 77210h
1,2-dibenzyl-4-(benzylideneamino)-	**63:** 6999d
1,2-dibenzyl-4-(dibenzylamino)-	**63:** 6999b
-1,2-dicarboxamide, 4-cyclohexyl-*N*,*N*,*N'*,*N'*-tetramethyl-	**61:** P8317e
-1,2-dicarboxamide, 4-(3,5-dichlorophenyl)-*N*,*N'*-dimethyl-	**85:** P117969u
-1,2-dicarboxamide, 4-(*p*-ethoxyphenyl)-*N*,*N*,*N'*,*N'*-tetramethyl	**61:** P8317e
-1,2-dicarboxamide, 4,4-(4-methyl-*m*-phenylene)-bis[*N*,-*N*,*N'*,*N'*-tetramethyl]-	**61:** P8317f

TABLE 19   (Continued)

Compound	Reference

19.2. Other 1,2,4-Triazolidine-3,5-diones   (Continued)

Compound	Reference
-1,2-dicarboxamide, *N,N,N',N'*-tetramethyl-4-phenyl-	**61:** P8317e
-1,4-dicarboxamide, *N,N,N',N'*-tetramethyl-2-phenyl-	**61:** P8317f
-1,2-dicarboxylic acid, 4-(3,5-dichlorophenyl)-, diethyl ester	**85:** P117969u
-1,4-dicarboxylic acid, 2-phenyl-, diethyl ester	**53:** 7426i, **76:** P34259a, **77:** P62007f
4-(3,5-dichlorophenyl)-1,2-bis(1-oxopropyl)-	**85:** P117969u
4-([(2,6-dichlorophenyl)methylene]amino)-	**82:** P155791a
4-(3,5-dichlorophenyl)-1-methyl-2-(1-oxobutyl)-	**85:** P117969u
4-(3,5-dichlorophenyl)-1-methyl-2-(1-oxopropyl)-	**85:** P117969u
4-(3,5-dichlorophenyl)-1-(1-oxopropyl)-	**85:** P117969u
1,4-diethyl-2-(4-nitrobenzoyl)-	**79:** 53234s
1,2-diisopropyl-4-[(trichloromethyl)thio]-	**72:** P78474P, **72:** 100656y
4-(dimethylamino)-1,2-dimethyl-	**67:** 3039p
1,2-dimethyl-4-[(trichloromethyl)thio]-	**72:** P78474P, **72:** 100656y
-1,2-dithiocarboxamide, *N,N,N',N'*-tetramethyl-4-phenyl	**61:** P8317f
4-formamido-1-formyl	**53:** 11355f
-1-glyoxylamide, 4-oxamido-	**53:** 11355b
-1-glyoxylic acid, 4-(1-carboxyformamido)-	**53:** 11355d
-1-glyoxylic acid, 4-(1-carboxy-*N*-methylformamido)-2-methyl-, dimethyl ester	**53:** 11355e
1-hexanoyl-	**50:** 12033a
4-isopropylideneamino-	**54:** 7625i
4-[(*p*-methoxyphenyl)sulfonyl]-	**65:** PC16978g
1-methyl-4-(methylamino)-	**82:** 156185t
1-[(4-methylphenyl)sulfonyl]-	**78:** 29344m
4-[[(5-nitro-2-furanyl)methylene]amino]-	**78:** P16189q
4-[[3-(5-nitro-2-furanyl)-2-propenylidene]amino]-	**78:** P16189q
4-(4-nitrophenyl)-1-[(4-nitrophenyl)amino]-	**76:** 46146z
4-(4-nitrophenyl)-1-[(4-nitrophenyl)(benzyl)amino]-	**76:** 46146z
4-[(*o*-nitrophenyl)sulfonyl]-	**65:** PC16978g
4-[(*p*-nitrophenyl)sulfonyl]-	**65:** P16978g
1-[(3-oxo-5-indazolinyl)carbonyl]-	**73:** P125735g
1-(2-oxo-1-piperidinyl)-4-phenyl-	**85:** 143053k
1-(2-oxo-1(2H)-pyridinyl)-4-phenyl-	**85:** 143053k
4-phenyl-1-(1-piperidinyl)-	**76:** 140687y
4-(phenylsulfonyl)-	**65:** PC16978g
1-phenyl-4-[(trichloromethyl)thio]-	**72:** 100656y
-1-propionic acid, α,α-diethyl-β-oxo-4-phenyl-	**65:** 704c
-1-sulfonamide, 4-(3,5-dichlorophenyl)-*N,N*,2-trimethyl-	**85:** P117969u
4-(2-thienylsulfonyl)-	**65:** PC16978g
-1-thiocarboxanilide, 4-phenyl-	**61:** P8317e
4-(*p*-tolylsulfonyl)-	**65:** PC16978g
4-[(trichloromethyl)thio]-	**72:** 100656y
4-([4-(trifluoromethyl)benzylidene]amino)-	**83:** P79251c
4-[(α,α,α-trifluoro-*m*-tolyl)sulfonyl]-	**65:** PC16978g

CHAPTER 20

1,2,4-Triazolidin-3-one-5-thiones and 1,2,4-Triazolidine-3,5-dithiones

20.1. 1,2,4-Triazolidin-3-one-5-thiones

The triazolidin-3-one-5-thiones have been used as intermediates for the synthesis of insecticidal, acaricidal, and nematicidal agents.[1,2] The base-catalyzed intramolecular cyclization of 1-(ethoxycarbonyl)-3-thiosemicarbazide (**20.1-1**)[1,3–5] and the corresponding 1-(methylthiocarbonyl) compound (**20.1-2**)[6] gave the triazolidin-3-one-5-thione (**20.1-4**). In addition, alkaline cyclization of **20.1-3** gave mainly **20.1-4** by the elimination of aniline and a smaller amount of **20.1-5** by the elimination of ammonia.[6]

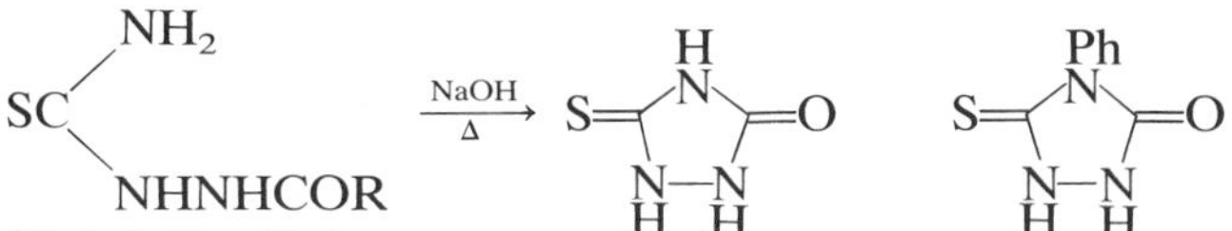

20.1-1, R = OEt
20.1-2, R = SMe
20.1-3, R = NHPh

The condensation of 4-aryl-3-thiosemicarbazides (e.g., **20.1-6**) with diethyl carbonate in the presence of methoxide gave 4-aryltriazolidines (e.g., **20.1-5**) formed via 4-aryl-1-(ethoxycarbonyl)-3-thiosemicarbazides intermediates (e.g., **20.1-7**).[7] Also, this type of intermediate was prepared by the reaction of ethyl carbazate (**20.1-8**) with phenylisothiocyanates and converted to 4-aryltriazolidin-3-one-5-thiones (e.g., **20.1-5**) with aqueous base.[8,9] Similar intermediates were used in the preparation of either 1-methyl- or 4-methyl-1,2,4-triazolidin-3-one-5-thiones.[10] Also, the cyclization of 4-(ethoxycarbonyl)thiosemicarbazides under acidic conditions to give 1-substituted 1,2,4-triazolidin-3-one-5-thiones has been reported in a patent.[11] In the reaction of thiosemicarbazides with phosgene in either chloroform or chlorobenzene, Δ^2-1,3,4-thiadiazolin-5-ones were isolated rather than 1,2,4-triazolidin-3-one-5-thiones.[12]

$$SC\begin{array}{c} NHPh \\ \\ NHNH_2 \end{array} \quad + \ (EtO)_2CO \ \xrightarrow{MeO^-} \ SC\begin{array}{c} NHPh \\ \\ NHNHCO_2Et \end{array}$$

20.1-6   **20.1-7**

$$PhNCS + H_2NNHCO_2Et$$
20.1-8

20.1-5

A variation of the methods described in the preceding was used to prepare 4-alkyltriazolidin-3-one-5-thiones (e.g., **20.1-11**) in good yields. Treatment of semicarbazide with alkyl isothiocyanates (e.g., **20.1-9**) gave 1-(*N*-substituted thiocarbamoyl)semicarbazides (e.g., **20.1-10**), which were readily cyclized in refluxing aqueous sodium hydroxide.[13] The trialkyl

$$BuNCS + H_2NNHCONH_2 \ \longrightarrow \ SC\begin{array}{c} NHBu \\ \\ NHNHCONH_2 \end{array} \ \xrightarrow[\Delta]{OH^-} \ \text{20.1-11}$$

20.1-9   **20.1-10**

triazolidine (**20.1-15**) was prepared by an unusual method. Pyrolysis of ethyl *N,N*-dimethylcarbazate (**20.1-12**) at 200° in the presence of phosphoric acid generated dimethylaminoisocyanate (**20.1-13**), which was trapped with butylisothiocyanate to give **20.1-14**, followed by methyl group migration to give **20.1-15**.[14] The dimethylaminoisocyanate (**20.1-13**) can

$$Me_2NNHCO_2Et \ \xrightarrow{\Delta} \ \left[Me_2NNCO \ \xrightarrow{BuNCS} \ \text{20.1-14} \right] \ \longrightarrow \ \text{20.1-15}$$

20.1-12   **20.1-13**   **20.1-14**   **20.1-15**

also be generated under mild conditions from **20.1-16**.[15] Treatment of an isolated sample of **20.1-17** with anhydrous hydrogen chloride gave the chloride (**20.1-18**), which on melting evolved methyl chloride to give **20.1-19**.[15] A methyl group was also removed from **20.1-17** when the latter was treated with benzyl chloride to give a mixture of **20.1-20** and **20.1-21**. Although these two isomeric products were not interconverted by heat, the

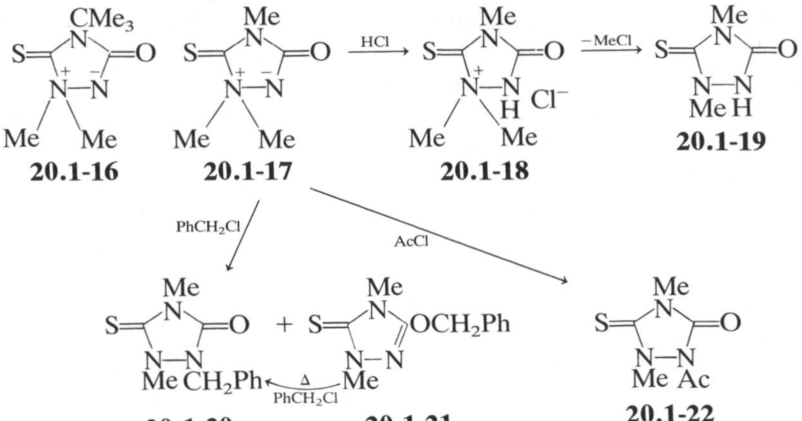

20.1-16   **20.1-17**   **20.1-18**   **20.1-19**

20.1-20   **20.1-21**   **20.1-22**

ether (**20.1-21**) was transformed to **20.1-20** in the presence of benzyl chloride. In the reaction of **20.1-17** with acetyl chloride, the 2-acetyl compound (**20.1-22**) was formed in 94% yield. Another unusual reaction was observed in the pyrolysis of the thiazolinedione (**20.1-23**) to give the triazolidine-3-one-5-thione (**20.1-24**).[16]

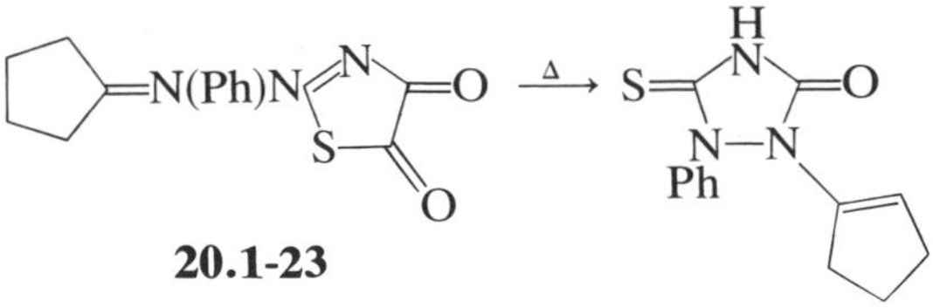

20.1-23

20.1-24

References

1. **84:** P4965a   2. **85:** 117969u   3. **17:** 3340   4. **83:** 58834e
5. **83:** 164189h   6. **16:** 2509⁵   7. **53:** 21904h   8. **53:** 18013i
9. **78:** 58320f   10. **70:** 115078u   11. **81:** 91535q   12. **73:** 35290d
13. **52:** 17244g   14. **67:** 3039p   15. **82:** 72890f   16. **82:** 43281c

20.2. 1,2,4-Triazolidine-3,5-dithiones

The cyclization of the N,N'-bis(thiocarbamoyl)hydrazine (**20.2-1**) in hot alkaline solution gave a 77% yield of 1,2,4-triazolidine-3,5-dithione (**20.2-9**).[1] Only a 47% yield of **20.2-9** was obtained on treatment of **20.2-1** with aqueous triethylamine. Similarly, the N-methylthiocarbamoyl (**20.2-2**) was converted with aqueous sodium hydroxide to the 4-methyltriazolidine

(**20.2-10**) in 80% yield. Apparently, methylamine was preferentially eliminated in the cyclization of **20.2-3** under basic conditions to give the 4-dodecyltriazolidine (**20.2-11**), which was isolated in 38% yield. In addition, the cyclization of **20.2-4** with poylphosphoric acid was reported to give **20.2-12** rather than a 1,3,4-thiadiazole.[2]

A large number of 4-aryl-1,2,4-triazolidine-3,5-dithiones were obtained by the general procedure described above. For example, treatment of the unsymmetrically substituted N,N'-bis(thiocarbamoyl)hydrazine (**20.2-5**) with aqueous base resulted in preferential elimination of ammonia to give **20.2-13** in 35% yield.[3] Similarly, the more complex **20.2-6** was converted to **20.2-14** with loss of 4-phenyl-3-thiosemicarbazide.[4] In the base-catalyzed cyclization of either **20.2-7** or **20.2-8**, aniline was reported to be eliminated to give **20.2-13** and **20.2-15**, respectively.[5] However, the melting points for some of

20.2-1, $R_1 = R_2 = H$
20.2-2, $R_1 = R_2 = Me$
20.2-3, $R_1 = C_{12}H_{15}$, $R_2 = Me$
20.2-4, $R_1 = R_2 = (CH_2)_2CO_2Me$
20.2-5, $R_1 = 4\text{-}MeC_6H_4$, $R_2 = H$
20.2-6, $R_1 = Ph$, $R_2 = NHCSNHPh$
20.2-7, $R_1 = 4\text{-}MeC_6H_4$, $R_2 = Ph$
20.2-8, $R_1 = C_6H_{11}$, $R_2 = Ph$

20.2-9, $R_1 = H$
20.2-10, $R_1 = Me$
20.2-11, $R_1 = C_{12}H_{15}$
20.2-12, $R_1 = (CH_2)_2CO_2Me$
20.2-13, $R_1 = 4\text{-}MeC_6H_4$
20.2-14, $R_1 = Ph$
20.2-15, $R_1 = C_6H_{11}$

the compounds produced by this procedure indicated that mixtures of triazolidines were obtained. Although the acetylation of **20.2-17** gave the N-acetyltriazolidine (**20.2-16**) (82%), methylation of **20.2-17** with dimethyl sulfate gave the 3,5-bis(methylthio)triazole (**20.2-18**) (94%).[6]

20.2-16   **20.2-17**   **20.2-18**

The reaction of thiocarbohydrazide (**20.2-19**) either with potassium ethyl xanthate or with carbon disulfide in a sealed tube was incorrectly reported to give dithio-p-urazine (**20.2-22**), an authentic sample of which was prepared from **20.2-19** and di(carboxymethyl)trithiocarbonate at room temperature.[7] The high-temperature reactions of **20.2-19** with ethyl xanthate and carbon disulfide, respectively, gave 4-aminotriazolidine-3,5-dithione (**20.2-20**,

50%)[8,9] and a mixture of **20.2-20** (50%) and 1,3,4-thiadiazolidine-2,5-dithione (**20.2-21**) (33 to 50%).[8,10] Also, **20.2-20** was prepared by the rearrangement of **20.2-22** in refluxing hydrochloric acid.[7] The structural assignment of **20.2-20** was confirmed by the preparation of the benzylideneamino derivative (**20.2-23**) and by deamination of the S-methylated derivative to give 3,5-bis(methylthio)triazole.[9] An excellent yield of the 4-

(dimethylamino)triazolidine-3,5-dithione (**20.2-25**) was obtained by the condensation of **20.2-24** with carbon disulfide in refluxing pyridine.[11]

References

1. **66:** 65474q   2. **53:** 384c   3. **58:** 2447e   4. **64:** 8173b
5. **63:** 6997e   6. **84:** 135551k   7. **56:** 14293h   8. **56:** 15513a
9. **61:** 10676a   10. **57:** 12471f   11. **69:** 51320m

TABLE 20.   1,2,4-TRIAZOLIDIN-3-ONE-5-THIONES AND 1,2,4-TRIAZOLIDINE-3,5-DITHIONES

Compound	Reference
20.1. 1,2,4-Triazolidin-3-one-5-thiones	
Parent	**16:** 2509[5], **44:** 6415g, **53:** 18013i, **81:** P13523v, **83:** P58834e, **83:** P164189h
2-acetyl-1,4-dimethyl-	**82:** 72890f
2-acetyl-4-methyl-	**44:** 6416d
2-benzoyl-1,4-dimethyl-	**82:** 72890f
4-benzyl-	**52:** 17245b, **53:** 18014d
2-benzyl-1,4-dimethyl-	**82:** 72890f
4-(p-bromophenyl)-	**53:** 18014d
1-butyl-	**81:** P91535q
1-tert-butyl-	**84:** P4965a, **84:** P59484y, **84:** P74273x, **84:** P85645w
4-butyl-	**52:** 17245b, **72:** P100713q
4-butyl-1,2-dimethyl-	**67:** 3039p
-1-carboxamide, N,N-dimethyl-4-phenyl-	**61:** P8317f
-2-carboxylic acid, 1,4-dimethyl-, benzyl ester	**82:** 72890f
4-(m-chlorophenyl)-	**53:** 18014d
4-(p-chlorophenyl)-	**53:** 18014d, **53:** 21905b
1-(2-cyanoethyl)-	**84:** 85647y
2-(1-cyclohepten-1-yl)-1-phenyl-	**82:** 43281c
2-(1-cyclohexen-1-yl)-1-phenyl-	**82:** 43281c
2-cyclohexyl-	**73:** 35290d
4-cyclohexyl-	**53:** 18014d
2-cyclohexyl-4-methyl-	**73:** 35290d
2-(1-cyclopenten-1-yl)-1-phenyl-	**82:** 43281c
2,4-diacetyl-	**44:** 6416e
4-(3,5-dichlorophenyl)-	**85:** P117969u
4-(3,5-dichlorophenyl)-1,2-dimethyl-	**85:** P117969u
1,4-dimethyl-	**82:** 72890f
2,4-dimethyl-	**73:** 35290d, **80:** P48002f
1,4-dimethyl-2-(2-methylbenzyl)-	**82:** 72890f
1-dodecyl-	**81:** P91535q
4-(p-ethoxyphenyl)-	**53:** 21905b
1-ethyl-	**81:** P13523v, **81:** P91535q, **81:** P120638h, **84:** P4965a, **84:** P74273x
2-ethyl-	**81:** P63648k
4-ethyl-	**52:** 17245b, **81:** P63648k
1-[2-(3-ethylbenzolthiazolium-2-yl)ethenyl]-4-phenyl-	**78:** P167001y
2-[2-(3-ethylbenzolthiazolium-2-yl)ethenyl]-4-phenyl-, hydroxide, inner salt	**78:** P167001y
4-heptyl-	**52:** 17245b
4-hexyl-	**52:** 17245b
1-isopropyl-	**81:** P13523v, **81:** P91535q, **81:** P120638h
2-isopropyl-	**73:** 35290d

TABLE 20   (*Continued*)

Compound	Reference

20.1. 1,2,4-Triazolidin-3-one-5-thiones   (*Continued*)

4-isopropyl-	**52:** 17245b, **81:** P13523v
2-isopropyl-4-methyl-	**73:** 35290d
4-mesityl-1-phenyl-	**60:** P2950d
4-(*p*-methoxyphenyl)-	**53:** 18014d
1-methyl-	**70:** 115078u, **81:** P91535q, **81:** P120638h, **84:** 4965a, **84:** P74273x, **84:** P85645w
4-methyl-	**44:** 6415h, **52:** 17245b, **70:** 115078u, **72:** P100713q, **81:** P13523v
2-methyl-4-phenyl-	**80:** P48002f
4-(3-methylphenyl)-	**78:** 58320f
4-(4-methylphenyl)-	**78:** 58320f
4-methyl-2-phenyl-	**73:** 35290d
4-pentyl-	**52:** 17245b
4-phenethyl-	**52:** 17245b
1-phenyl-	**81:** P91535q, **83:** P206285u, **84:** P74272w, **84:** P85645w
2-phenyl-	**53:** 21098e, **73:** 35290d
4-phenyl-	**44:** 6416f, **52:** 17245b, **53:** 21905b, **72:** P100713q, **78:** 58320f
1-propyl-	**81:** P91535q
4-propyl-	**52:** 17245b
4-(2-pyridyl)-	**53:** 21905b
4-*m*-tolyl-	**53:** 18014d
4-*p*-tolyl-	**53:** 18014d
1-(2,4-xylyl)-	**53:** 18014d

20.2. 1,2,4-Triazolidine-3,5-dithiones

Parent	**44:** 6416g, **66:** P65474q, **81:** P13523v, **83:** P58834e, **83:** P164189h,
1-acetyl-2-phenyl-	**84:** 135551k
4-allyl	**66:** P65474q
4-amino	**48:** 5181i, **56:** 14293i, **56:** 15513a, **61:** 10676b, **68:** 105175t, **83:** 27213x
4-benzyl	**63:** 6997f
4-(benzylideneamino)	**56:** 15513b, **61:** 10676a
4-(*p*-bromophenyl)	**58:** 2447g, **63:** 6997g
4-butyl	**63:** 6997g
4-(*p*-tert-butylphenyl)	**58:** 2447g
4-(4-carboxyphenyl)-	**79:** 38542x
1-(2-chlorophenyl)-	**81:** 86057f
1-(3-chlorophenyl)-	**81:** 86057f

TABLE 20   (*Continued*)

Compound	Reference

20.2. 1,2,4-Triazolidine-3,5-dithiones   (*Continued*)

Compound	Reference
4-(*o*-chlorophenyl)	**58:** 2447g
4-(3-chlorophenyl)-	**79:** 38542x, **79:** 61551h
4-(*p*-chlorophenyl)	**63:** 6997g, **79:** 38542x, **79:** 42423b
4-*p*-cumenyl	**58:** 2447g
4-cyclohexyl	**63:** 6997g
4-(3,5-dichlorophenyl)-1,2-dimethyl-	**85:** P117969u
4-(2,5-dimethoxyphenyl)	**63:** 6997g
1,4-dimethyl-	**80:** P48002f
4-(dimethylamino)-1-methyl-	**69:** 51320m
1,2-dimethyl-4-*o*-tolyl	**33:** 599[3]
1,2-dimethyl-4-*p*-tolyl	**33:** 599[3]
1,4-diphenyl	**59:** 10026c
4-dodecyl	**66:** P65474q
4-ethyl	**66:** P65474q
1-[2-(3-ethylbenzothiazolium-2-yl)ethenyl]-, hydroxide, inner salt	**78:** P167001y
1-[2-(3-ethylbenzothiazolium-2-yl)ethenyl]-4-phenyl-, hydroxide, inner salt	**78:** P167001y
4-hexyl	**63:** 6997g
4-(*p*-iodophenyl)	**63:** 6997g
4-isopropyl	**66:** P65474q, **66:** P110059w
4-(*o*-methoxyphenyl)	**58:** 2447g, **63:** 6997g
4-(*p*-methoxyphenyl)	**58:** 2447g, **63:** 6997g
4-methyl	**66:** P65474q, **72:** P95326j
1-(4-methylphenyl)-	**81:** 86057f
1-methyl-4-phenyl-	**80:** P48002f
4-(4-methylphenyl)-	**79:** 38542x, **79:** 42423b
4-(3-nitrophenyl)-	**79:** 38542x
1-(2-oxo-3-butenyl)-	**80:** P151113s
4-pentyl	**63:** 6997g
1-phenyl	**45:** P2350e, **81:** 86057f
4-phenyl	**29:** 7209[6], **33:** 599[7], **44:** 6416h, **63:** 6997f, **64:** 8173c, **66:** P65474q, **79:** 38542x, **79:** 42423b
-4-propionic acid, methyl ester	**53:** 385d
4-*o*-tolyl-	**33:** 599[2], **63:** 6997f
4-*m*-tolyl-	**63:** 6997f
4-*p*-tolyl-	**33:** 599[2], **58:** 2447g, **63:** 6997f
4-(2,4-xylyl)-	**58:** 2447g

Bi-, Di, Bis-, Tris-, and Poly(1,2,4-Triazoles, Triazolines, and Triazolidines) Linked Either Directly or with Chain Containing Carbon

21.1. Bi-, Di, Bis, and Tris(1,2,4-Triazoles)

Compounds containing two triazole rings joined directly or bridged with a chain containing carbon are usually prepared as model compounds in the synthesis of poly(1,2,4-triazoles). The synthesis of some bis(triazoles), those prepared from preformed triazoles (e.g., 3-amino-1,2,4-triazole), are straightforward, and these reactions are not described in the following.

Most of the routes to the bis(triazoles) use an amidrazone or diamidrazone as an intermediate. The formation of a bis(triazole) in which the triazole rings are joined directly was effected by the elimination of ammonia from **21.1-1** in refluxing acetic acid to give **21.1-2**.[1] Bis(triazoles)

21.1-1   **21.1-2**

linked by an aliphatic chain are readily obtained as demonstrated by thermal dehydration of **21.1-3** at 200° to give a 53% yield of **21.1-4**.[2] Similar methods were used for the preparation of **21.1-5**

21.1-3   **21.1-4**

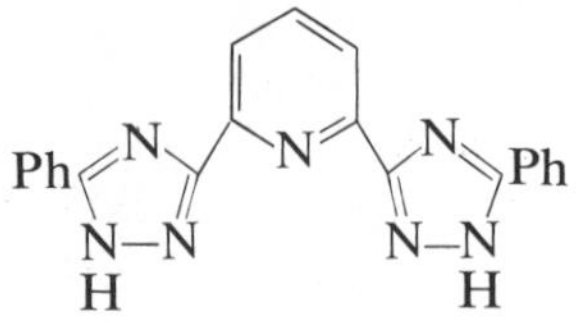

21.1-5 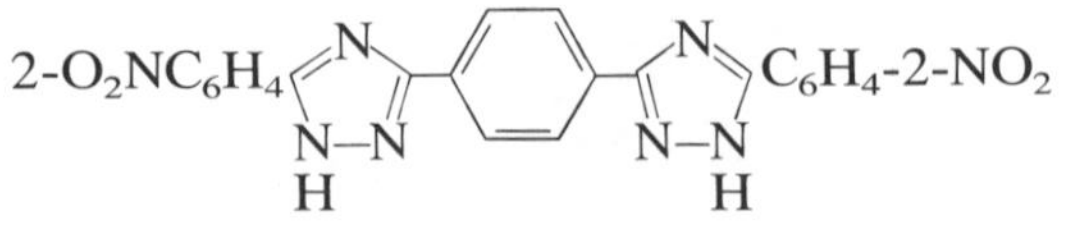   **21.1-6**

and **21.1-6**.[3–9] Compounds related to **21.1-6** were also prepared by the thermal condensation of terephathalic acid dihydrazide (**21.1-7**) with benzonitrile at 230°.[10] In addition, the condensation of the amidrazone (**21.1-8**) with aliphatic dicarboxylic acid chlorides gave the homologous series of bis(triazoles) (**21.1-9**) in 42 to 60% yields.[11] In the reaction of bis(alkylideneamino)oxamidines with acetic anhydride, the structure of the product depended upon the nature of the alkylidene moiety. When the

21.1-7   **21.1-8**   **21.1-9** (N = 0–8)

hydrocarbon side chain was saturated as in **21.1-10**, the product was the bis(1-vinyltriazole) (**21.1-11**).[12] In contrast, when the oxamidine side chain

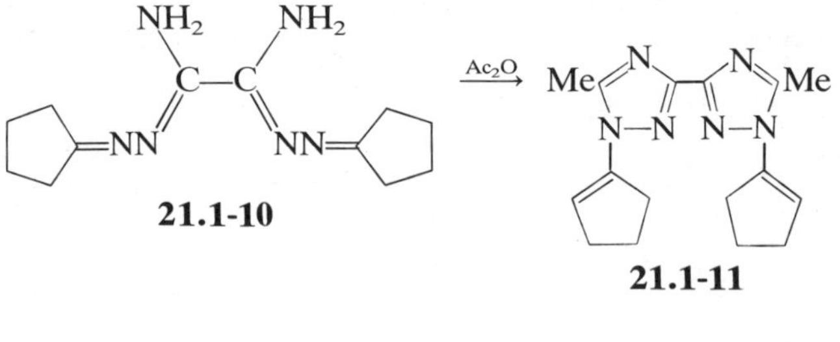

21.1-10

21.1-11

was unsaturated, the product was a bis(1,4-diacetyltriazoline) (see Section 21.2). The condensation of the hydrochloride of aminoguanidine with oxalic acid in boiling water gave **21.1-12**, which cyclized to 3,3'-bis(5-amino-1,2,4-triazole) (**21.1-13**) in an alkaline medium.[13] Similarly, other aliphatic dicarboxylic acids gave a series of bis(5-amino-1,2,4-triazoles) joined in the 3-position by a hydrocarbon chain. Although the cyclization of **21.1-14** to a 3,3'-perfluoroalkylbis(triazole) (**21.1-15**) was unsuccessful in a basic medium, this conversion was successful by thermal dehydration.[14,15]

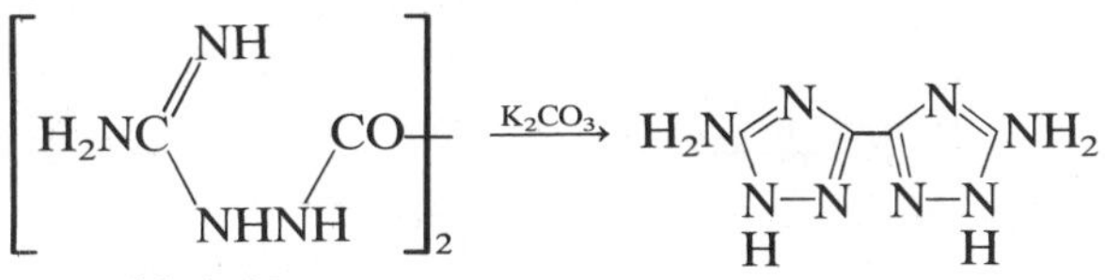

21.1-12 → **21.1-13**

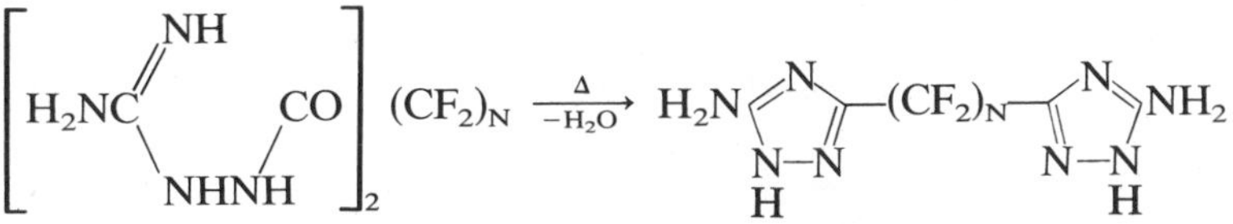

21.1-14   **21.1-15** (N = 3–6)

A solution of 1,4-di(5-tetrazolyl)benzene (**21.1-16**) in hot pyridine generated the bis(nitrilimine) (**21.1-18**), which underwent 1,3-dipolar addition with the imino chloride (**21.1-21**) to give 3,3′-p-phenylenebis[4,5-diphenyl-1,2,4-triazole] (**21.1-22**).[16] The preparation of the bis(nitrilimine) (**21.1-19**) was effected either by the pyrolysis of **21.1-17**[17,18] or by elimination of hydrogen chloride from the chloride (**21.1-20**) with triethylamine.[19] The 1,3-addition of **21.1-19** to benzonitrile gave the bis(triazoles) (**21.1-23**).

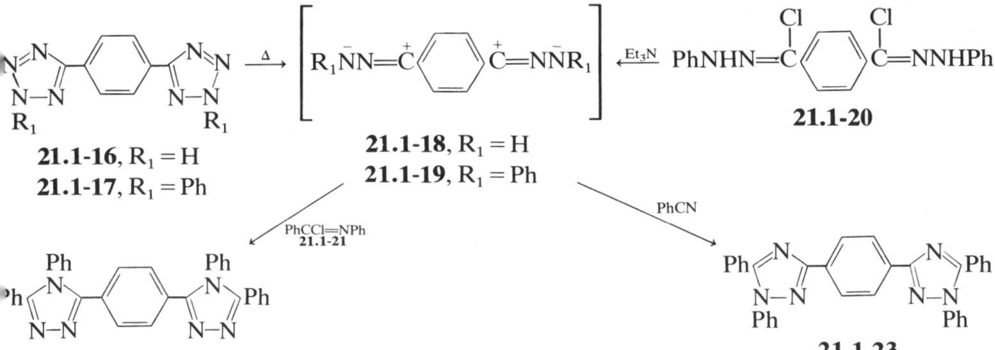

21.1-16, R₁ = H
21.1-17, R₁ = Ph

21.1-18, R₁ = H
21.1-19, R₁ = Ph

21.1-20

21.1-22

21.1-23

Similarly, benzoylphenylhydrazide chloride (**21.1-24**) reacted with perfluoroglutaronitrile to give **21.1-25**.

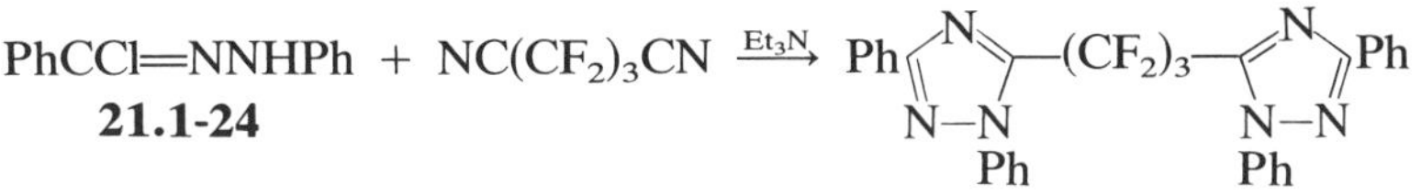

PhCCl=NNHPh + NC(CF₂)₃CN
21.1-24

21.1-25

The reaction of the 1,3,4-oxadiazole (**21.1-27**) with hexamethylenediamine at 210° gave a 24% yield of the bis(triazole) (**21.1-26**).[20] A similar reaction of **21.1-27** with 3-aminotriazole gave a 32% yield of **21.1-28**. In addition, bis (2-amino-1,3,4-oxadiazoles) (**21.1-29**) reacted with

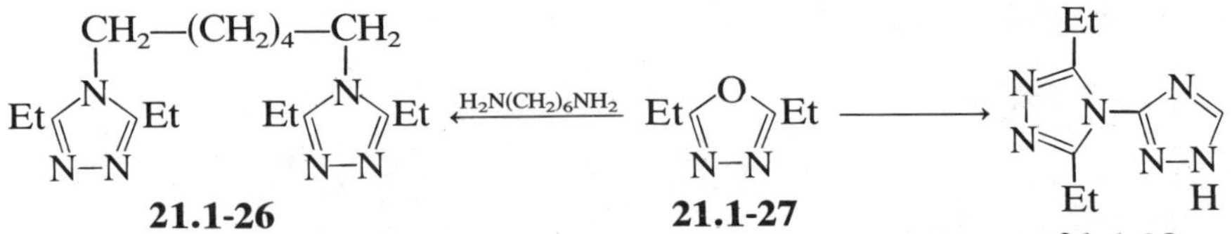

21.1-26   **21.1-27**   **21.1-28**

glycol in the presence of potassium hydroxide to give bis[5-(2-hydroxyethoxy)-1,2,4-triazoles] (**21.1-30**).[21] Also, a high yield of **21.1-32**

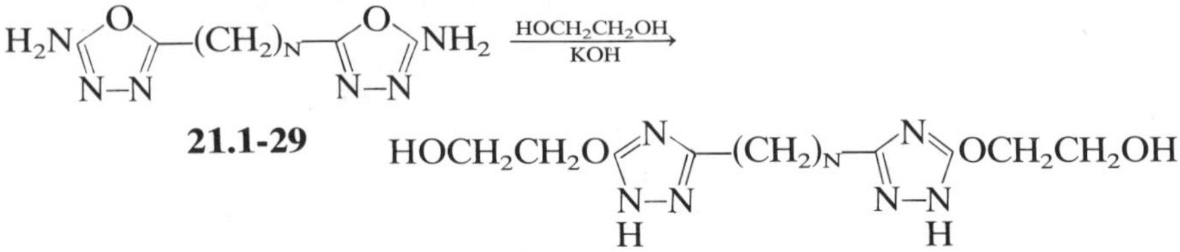

21.1-29

21.1-30 (N = 0–4, 8)

(79%) was reported to result from the condensation of **21.1-31** with aniline in presence of phosphorus trichloride at 198°.[22]

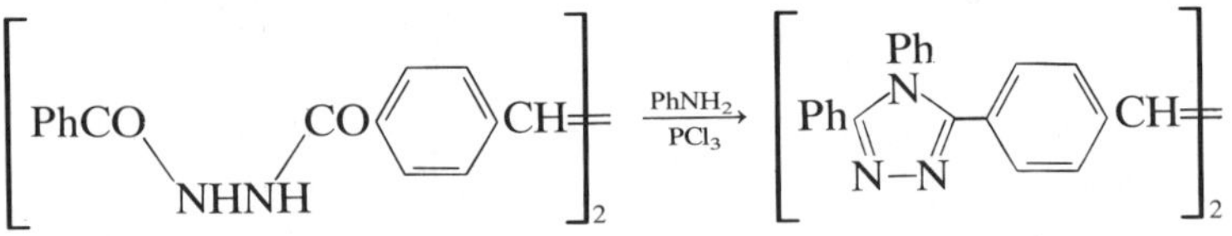

21.1-31   **21.1-32**

Bis(triazoles) (e.g., **21.1-34**), active as plant growth regulators, were prepared by the alkylation of 1,2,4-triazole (**21.1-33**) with dichlorodiphenylmethane in the presence of triethylamine and its hydrochloride.[23] The

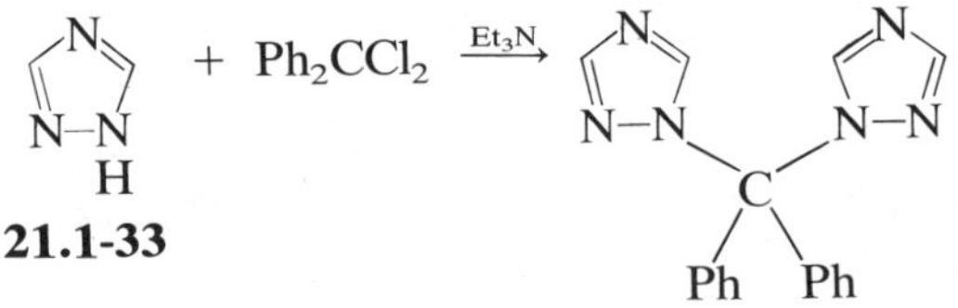

21.1-33   **21.1-34**

tris(triazoles) (**21.1-36**) resulted from the interaction of **21.1-35** with dichlorocarbene under neutral or basic conditions.[24] The bridged triazoles,

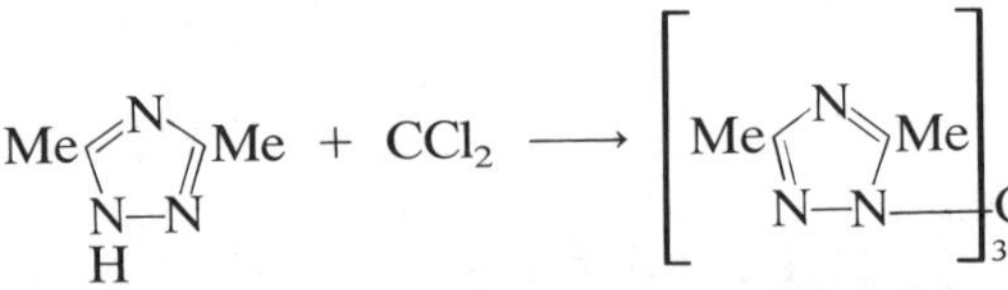

21.1-35   **21.1-36**

21.1-38[25] and **21.1-39**,[26] were prepared by the reaction of the triazoline **21.1-37** with triethyl orthoformate and **21.1-33** with cyanogen bromide in tetrahydrofuran, respectively.

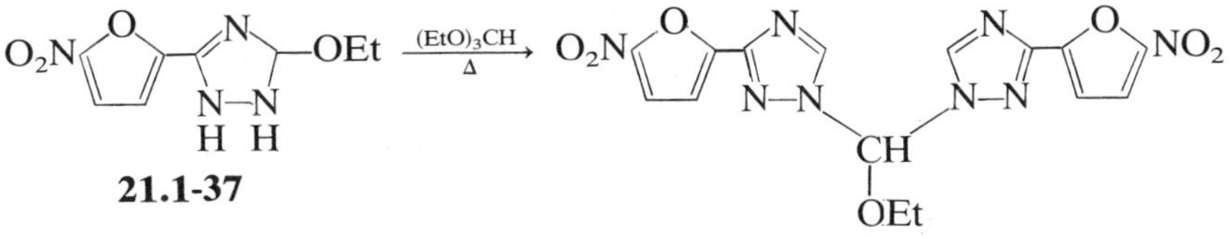

21.1-37

21.1-38

Treatment of either 1,2,4-triazole (**21.1-33**) in tetrahydrofuran or its trimethylsilyl derivative in benzene with phosgene gave a high yield of N,N'-carbonylbis(1,2,4-triazole), which has been described as the 4,4'-carbonyl compound but probably has the 1,1'-carbonyl structure (**21.1-40**).[27,28]

Similarly, acylation of 1,2,4-triazole (**21.1-33**) with either **21.1-41** or **21.1-42** in tetrahydrofuran was reported to give the bis compounds, **21.1-43** and **21.1-44**, respectively.[29]

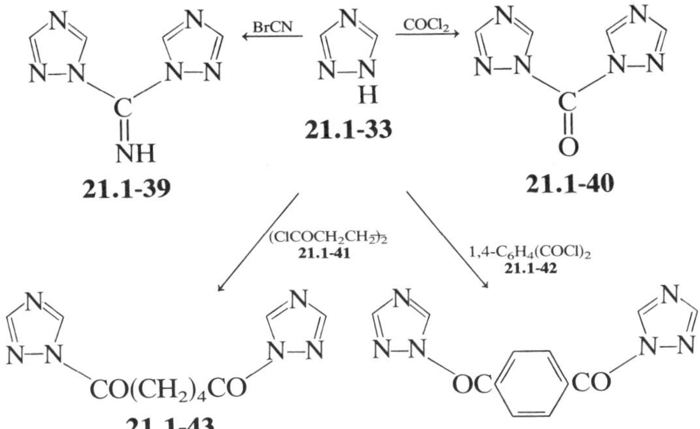

21.1-39

21.1-33

21.1-40

(ClCOCH₂CH₂)₂
21.1-41

1,4-C₆H₄(COCl)₂
21.1-42

21.1-43

21.1-44

References

1. **69:** 52078a
2. **65:** 15367g
3. **72:** 43572s
4. **72:** 90906p
5. **76:** 99614g
6. **80:** 15236g
7. **80:** 59912d
8. **80:** 96082d
9. **85:** 178022b
10. **59:** 6387c
11. **70:** 28874g
12. **83:** 193187m
13. **51:** 489h
14. **72:** 78954v
15. **81:** 136060z
16. **55:** 6473g
17. **68:** 50462b
18. **73:** 120949v
19. **69:** 67793d
20. **69:** 106627z
21. **68:** 78201q
22. **78:** 112683s
23. **77:** 5477k
24. **72:** 3454y
25. **75:** 20356x
26. **79:** 105150h
27. **52:** 7332e
28. **55:** 5484b
29. **54:** 12124c

21.2. Bi- and Bis(Triazolines and Triazolidines)

In contrast to the number of 3,3′-bis(triazoles), the 3,3′-bis(triazolines) are few in number and limited to those systems containing 5-alkyl, 5-aryl, 5-oxo, 5-thioxo, and 5-selenoxo substituents. Similarly, only a few bis(triazolidines) have been prepared. The condensation of the acetal of terephthaldicarboxaldehyde (**21.2-2**) with the amidrazone (**21.2-1**) under acidic conditions in refluxing benzene gave a 60% yield of the bis(triazoline)

(**21.2-3**).[1] Similarly, the pyridine amidrazone (**21.2-4**) reacted with **21.2-2** to give **21.2-5** in 74% yield.[2] The formation of bis(triazolines) was also effected

by the condensation of the pyridine diamidrazone (**21.2-7**) with either benzaldehyde to give **21.2-6** or with 2-pyridinecarboxaldehyde to give **21.2-8**.[2,3] When the oxamidine (**21.2-9**) was refluxed in acetic anhydride, the

bis(1,4-diacetyltriazoline) (**21.2-10**) was obtained.[4] In this type of reaction, oxamidines containing a saturated side chain gave bis(triazoles) (see Section 21.1).

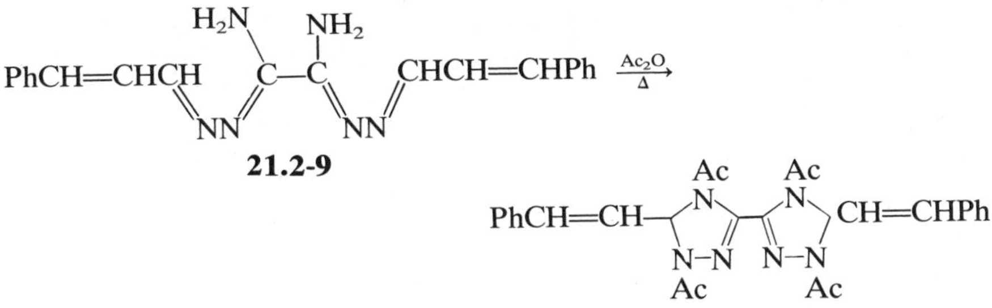

21.2-9

21.2-10

A variety of compounds containing two 1,2,4-triazoline-5-thiones moieties linked together in the 3-position by an aliphatic or aryl chain were prepared by the base-catalyzed condensation of thiosemicarbazide (**21.2-11**) with aliphatic or aryl dicarboxylic acids.[5,6] Examples are the reaction of thiosemicarbazide (**21.2-11**) with diethyl malonate, diethyl tartrate, and dimethyl terephthalate to give **21.2-12** (45%), **21.2-13** (38%), and **21.2-14** (65%), respectively. In the reaction of **21.2-11** with diethyl maleate or

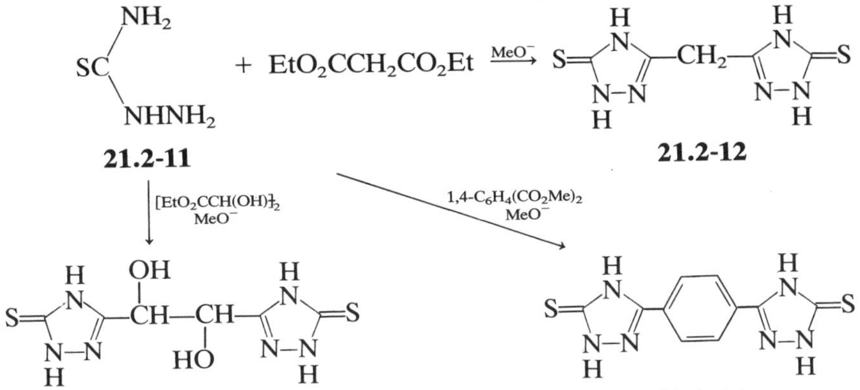

21.2-11   **21.2-12**

21.2-13   **21.2-14**

fumarate, the product was **21.2-15**, which resulted from the addition of methanol to the double bond of **21.2-16**. Bis(triazoline-5-thiones) (e.g.,

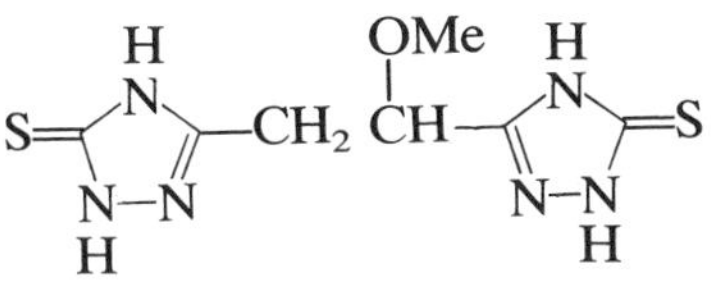

21.2-15 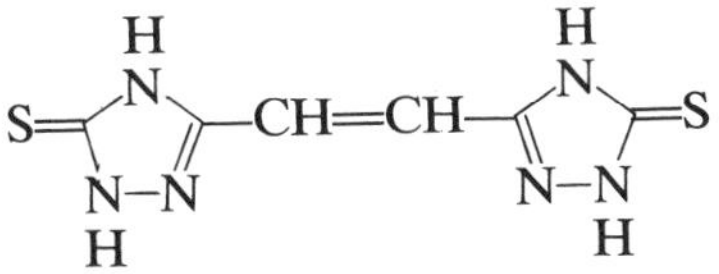   **21.2-16**

21.2-17   **21.2-18**

21.2-18) containing 4-alkyl and 4-aryl substituents are readily obtained by the base–catalyzed cyclization of preformed acylated thiosemicarbazides (e.g., **21.2-17**) derived from dicarboxylic acids.[7–11] The treatment of 4-methylthiosemicarbazide directly with diethyl oxalate gave a mixture of **21.2-18**, **21.2-19**, and **21.2-20**, the amount of each formed depending upon the concentration of ethoxide.[12,13] Although the cyclization of the 4-acylated

21.2-19   **21.2-20**

semicarbazide (**21.2-21**) with base was reported to give **21.2-22**, the thermal dehydration of **21.2-21** gave **21.2-24**.[14] The formation of the latter was attributed to the rearrangement of **21.2-21** to **21.2-23** before cyclization.

21.2-21   **21.2-22**

21.2-23   **21.2-24**

The bis(triazolines and triazolidines) connected by chains containing heteroatoms are prepared by methods described in previous sections and are not described here.

References

1. **70:** 96722f   2. **76:** 85796f   3. **72:** 111882v   4. **83:** 193187m
5. **65:** 7169f   6. **70:** 107521t   7. **70:** 11642g   8. **70:** 57742t
9. **70:** 67830j   10. **79:** 92120w   11. **83:** 192762h   12. **70:** 68323h
13. **73:** 66517n   14. **69:** 52073v

21.3. Poly(1,2,4-Triazoles, Triazolines, and Triazolidines)

Heterocyclic polymers containing 1,2,4-triazole units in the polymer chain were sought for their thermal stability and low sensitivity toward oxidative degradation.[1] Although these objectives were realized, the majority of the poly(triazoles) have low molecular weights. Apparently, the triazole polymers are used mainly as a component in a mixture with poly(α-olefins) to improve the dyeability of the latter. In addition, small amounts of either 3-chloro- or 3-bromo-1,2,4-1$\underline{H}$-triazole were reported to be good stabilizers for chlorinated polymers such as poly(vinylchloride).[1]

Poly(triazoles) are usually composed of the simple triazole nucleus or its 4-phenyl and 4-amino derivatives connected via the 3,5-positions with an aliphatic or aromatic bridge. An exception is the polymerization of a series of 3,5-dialkyl- and diaryl-1-vinyl-1,2,4-1$\underline{H}$-triazoles (e.g., **21.3-1**) in tetrahydrofuran at 50° with azobisisobutylnitrite to give poly(1-vinyl-1,2,4-1$\underline{H}$-triazoles).[2-4] Triazole polymers containing an amide linkage were prepared by the interaction of the 3,5-bis(3-aminophenyl)-1,2,4-4$\underline{H}$-triazole (**21.3-2**) with naphthalenedicarbonyl chloride.[5] The resulting polymer (**21.3-3**) was

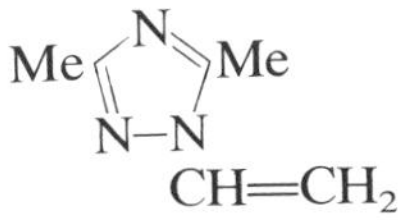

21.3-1

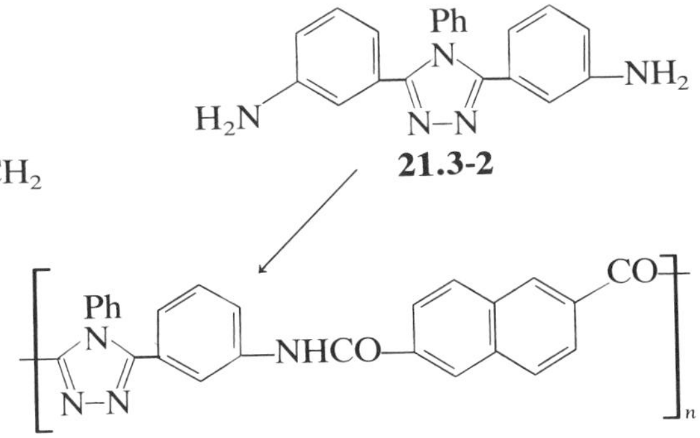

21.3-2

21.3-3

both film and fiber forming and was heat resistant. In contrast, triazole polymers containing both amide and ether linkages, obtained from **21.3-4** and terephthaloyl chloride, had low light resistance and poor thermal stability.[6]

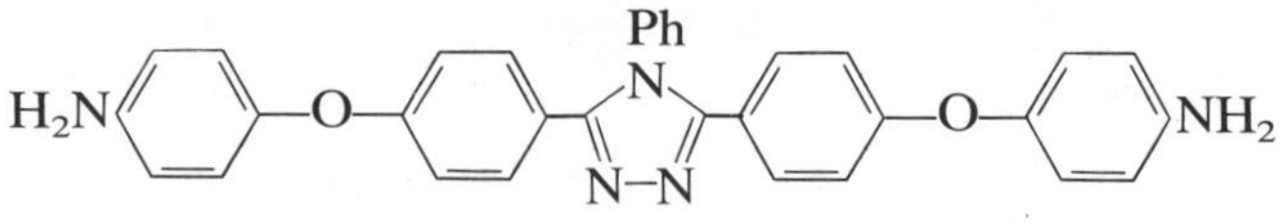

21.3-4

Many of the polymers containing a triazole backbone are formed from linear polymers by thermal dehydration. Treatment of poly(sebacic acid monohydrazide) (**21.3-5**) with ammonium hydroxide in an autoclave provided a polymer (**21.3-6**) in which the triazole units were connected by an

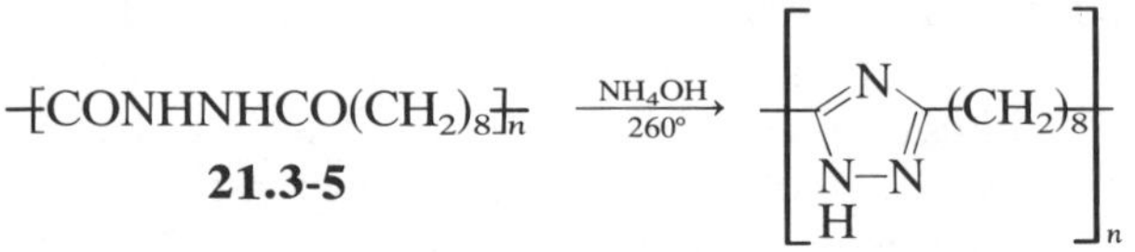

21.3-5 **21.3-6**

octamethylene bridge.[7] Also, aromatic amines such as aniline in the presence of polyphosphoric acid effected the cyclization of poly(hydrazides) to give polymers (e.g., **21.3-7**) containing the 4-phenyl-1,2,4-4 H-triazole moiety.[8,9] Although the latter had poor fiber-forming properties, the addition of **21.3-7** to poly(ethylene terephthalate) before melt spinning improved the

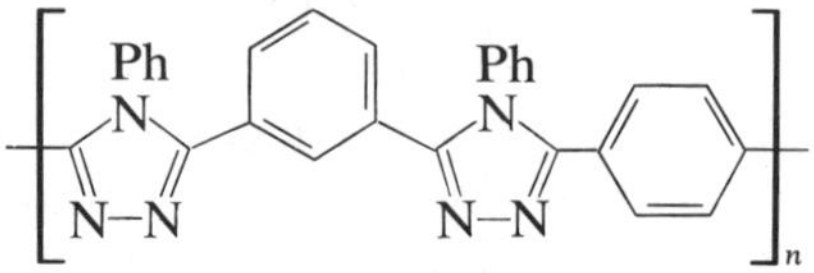

21.3-7

dyeability of the resulting fiber toward acid dyes.[9] In the thermal, intramolecular cyclization of linear polymers, **21.3-8** was converted to **21.3-9**[10] and **21.3-10** to **21.3-11**.[11] A polymer related to **21.3-9** has also

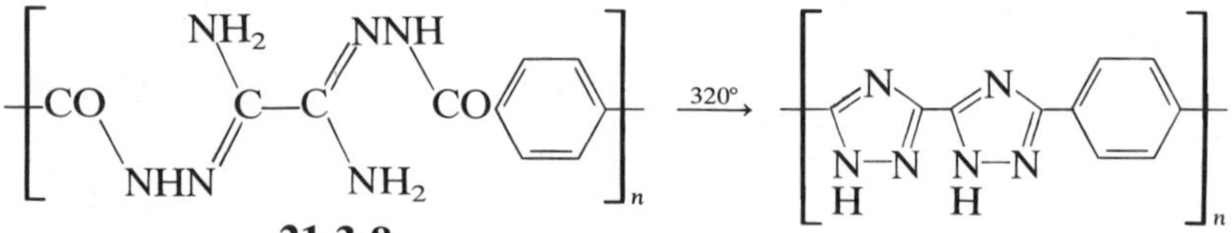

21.3-8 **21.3-9**

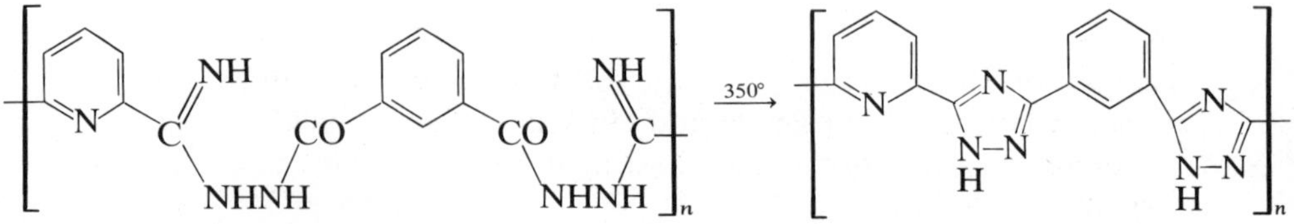

21.3-10 **21.3-11**

been prepared via a poly(amidrazone) intermediate.[12,13] Thermogravimetric analysis indicated that **21.3-11** underwent thermal decomposition between 450 and 520°, and other polymers of this type were also reported to have good thermal stability.[14-16] Intermediates related to **21.3-8** and **21.3-10** are formed in the conversion of poly(1,3,4-oxadiazoles) (e.g., **21.3-12**) with aniline in polyphosphoric acid to give 4-phenyltriazole polymers (**21.3-13**).[17-19] In addition polymers containing either a 4-(1-naphthyl)- or

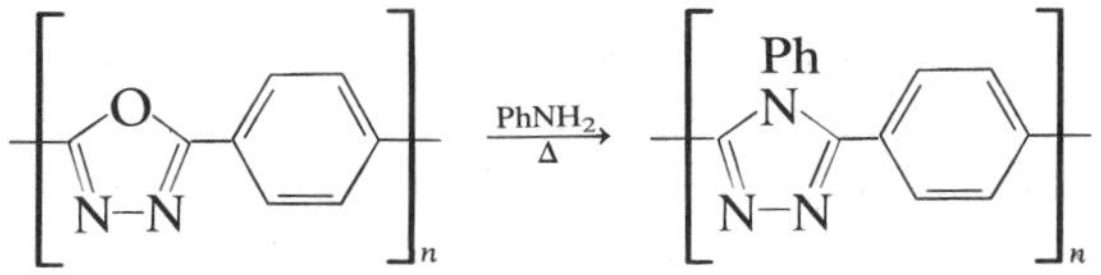

21.3-12   **21.3-13**

4-cyclohexyltriazole moiety and joined by a phenylene group were obtained from reaction of **21.3-12** with 1-naphthylamine and cyclohexylamine, respectively.[18,20]

Triazole polymers formed from monomers other than olefins include the 1,3-dipolar addition of phenylenebisnitrilimines with dinitriles. The nitrilimine intermediates (e.g., **21.3-15**) are generated either by pyrolysis of phenylenebis(2-phenyltetrazoles) (e.g., **21.3-14**)[21,22] or by elimination of hydrogen chloride from bis(acid hydrazide chlorides) (e.g., **21.3-16**) with triethylamine.[23] Typical examples are the condensation of **21.3-15** with

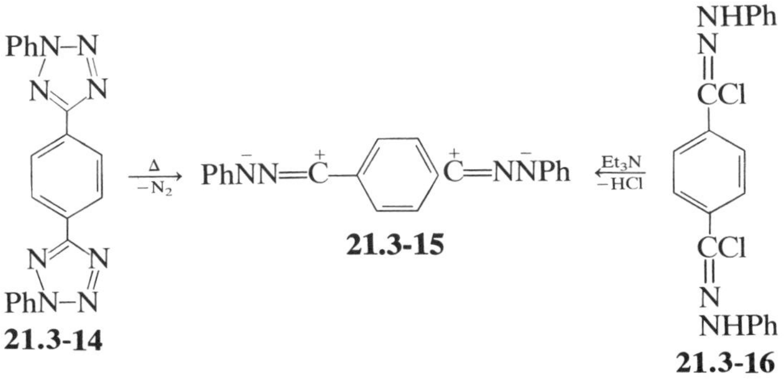

21.3-14   **21.3-15**   **21.3-16**

perfluoroglutaronitrile (**21.3-17**) to give **21.3-18** and with terephthalonitrile (**21.3-19**) to give **21.3-20** in good yields.[21,24] Although these polymers are thermally stable, they have low molecular weights.[25]

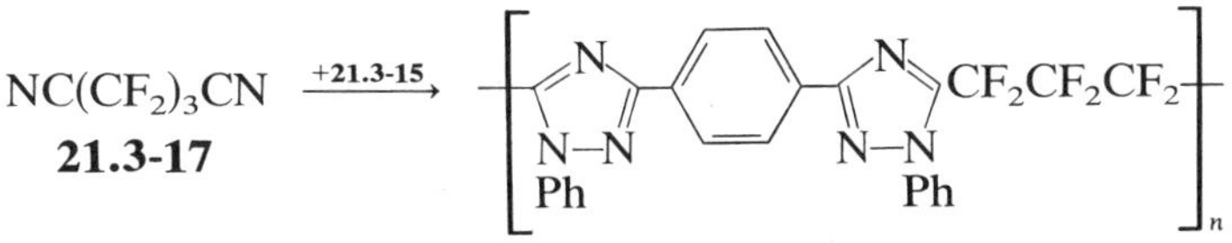

$NC(CF_2)_3CN$
21.3-17

21.3-18

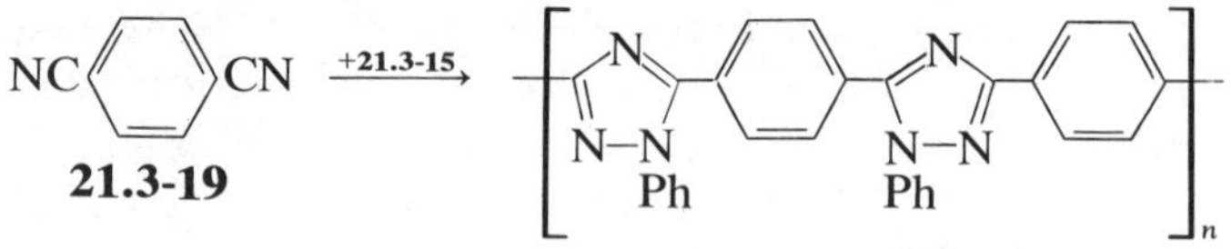

21.3-19

21.3-20

Poly(triazoles) with good thermal stability and dielectric properties and useful for the preparation of films and fibers are prepared by a two-stage condensation involving a 3,5-bis(3,4-diaminophenyl)-1,2,4-4H-triazole. Treatment of **21.3-21** with isophthaloyl chloride in N-methylpyrrolidinone gave **21.3-22**, which was thermally dehydrated to form the benzimidazole units and give **21.3-23**.[26] A similar procedure was used to convert 3-(2-aminophenyl)-1,2,4-triazoles (e.g., **21.3-24**) to a condensed triazole system, poly(triazoloquinazolines) (e.g., **21.3-25**).[27,28] The assignment of structure to this class of polymers is uncertain since cyclization can occur at both N-1 and N-4.[29]

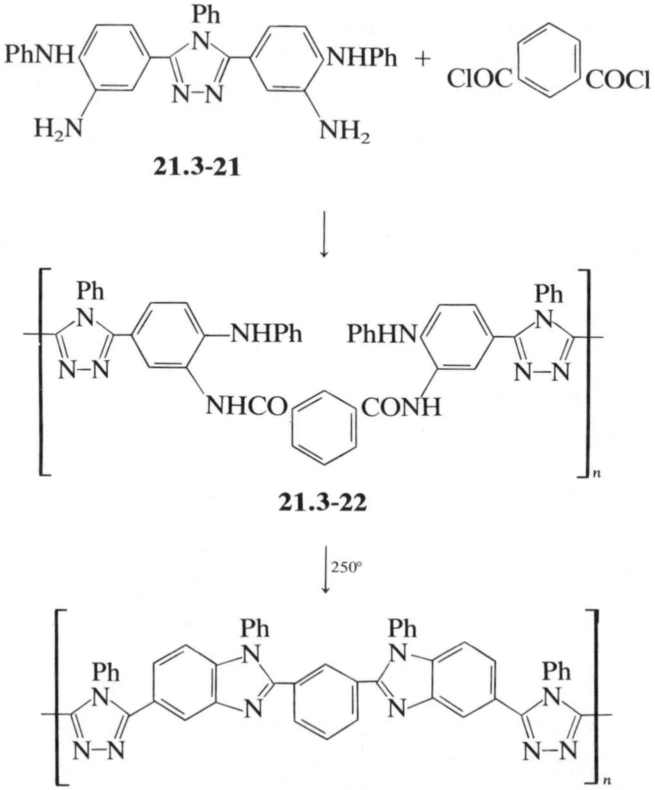

21.3-21

21.3-22

21.3-23

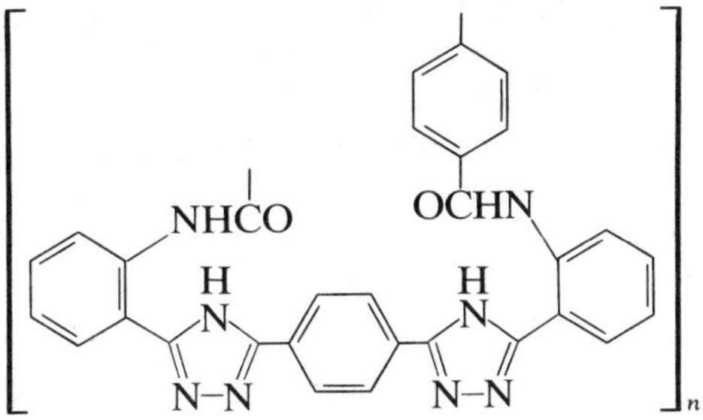

21.3-24

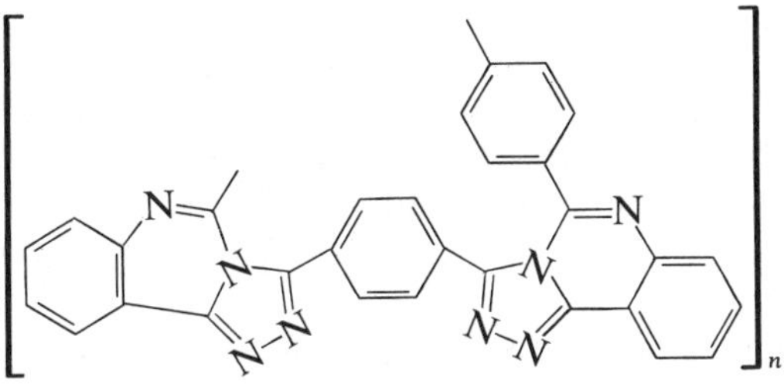

21.3-25

Two diverse types of compounds have been converted to poly(4-amino-1,2,4-4H-triazoles). In one procedure the condensation of a dibasic acid such as sebacic acid (**21.3-26**) with hydrazine gave the poly(4-aminotriazole) (**21.3-27**).[30,31] This transformation probably occurred via the intermediate

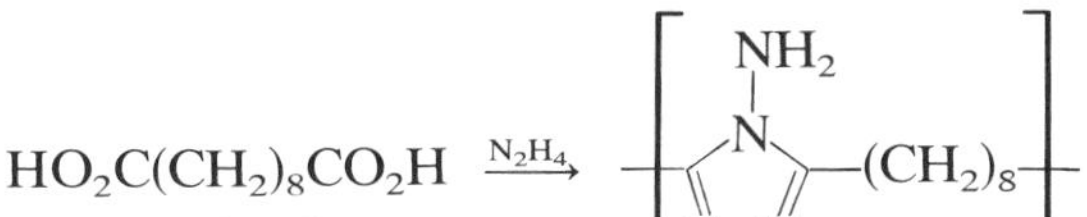

$$HO_2C(CH_2)_8CO_2H \xrightarrow{N_2H_4}$$

21.3-26

21.3-27

dihydrazide (**21.3-28**)[32-34] and poly(hydrazide) (**21.3-29**),[31] since both are known to undergo thermal dehydration in the presence of hydrazine to give **21.3-27**. Similar reactions to give polymers have been carried out with 5-oxosebacic[35] and 4,4'-oxydibutyric acid.[36] Also, mixed poly(4-aminotriazoles) in which the triazole units were linked by hydrocarbon chains of varying lengths were formed by the reaction of hydrazine with a mixture of adipic, azelaic, and sebacic acids.[37] Poly(4-aminotriazoles) connected by phenylene groups are obtained from aromatic diamidrazones and

$$(H_2NNHCO)_2(CH_2)_8$$ $$\{CO(CH_2)_8CONHNH\}_n$$
21.3-28   **21.3-29**

hydrazine at 300°.[38] Those polymers can be deaminated with nitrous acid to give poly(1,2,4-triazoles). Another method for the preparation of poly(4-aminotriazoles) (e.g., **21.3-31**) involved the rearrangement of poly-(dihydrotetrazines) (e.g., **21.3-30**) in refluxing hydrochloric acid.[39,40]

21.3-30   **21.3-31**

The blending of poly(4-aminotriazoles) with poly(α-olefins) has been reported to give fibers with antistatic properties[41] and with improved affinity for dyes and fastness to light, abrasion, and laundering.[42–46] In addition, the 4-aminotriazole polymers have been reported to be useful as toners in electrophotography[47] and as anticorrosion agents for metals in acidic solutions.[48]

Other triazole polymers were obtained by reaction of 3-amino-1,2,4-1$\underline{H}$-triazole with aromatic diisocyanates or diphenylcarbonate[49] and bi- or bistriazoles with aliphatic diisocyanates.[50] In **21.3-32** the introduction of the disiloxane group into the chain lowered the softening point and increased the solubility of the polymer in organic solvents.

21.3-32

A limited number reports have been published on the preparation of poly(triazolines) and poly(triazolidines). The condensation of pyridine-2,6-diamidrazole (**21.3-33**) with aromatic dialdehydes gave poly(triazolines),[51] and the cyclization of poly(1-acylthiosemicarbazides) with base gave poly(triazoline-5-thiones) (e.g., **21.3-34**).[52] The homopolymerization of 4-butyl-1,2,4-triazoline-3,5-dione (**21.3-35**) by visible irradiation gave the poly(triazolidine) **21.3-37**, which contains a nitrogen backbone.[53] This polymer was unstable in either carbon tetrachloride or base undergoing depolymerization to regenerate **21.3-35**. In addition, sublimation of the 4-phenyltriazoline (**21.3-36**) *in vacuo* gave a polymer.[54] Compound **21.3-36** has also been copolymerized with vinyl carbazole.

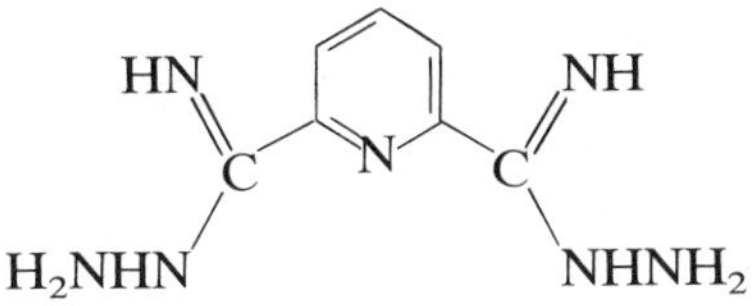

21.3-33

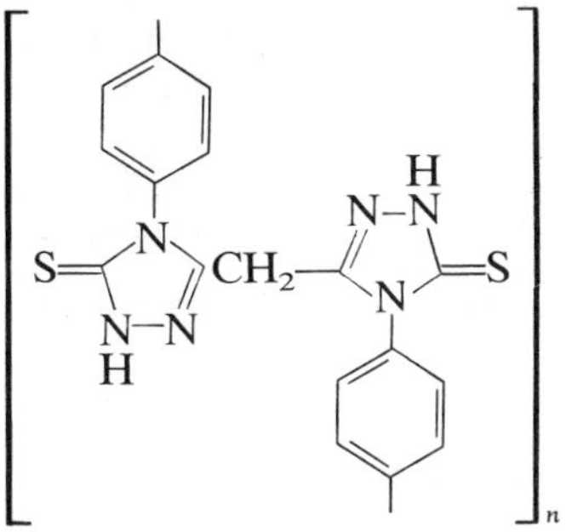

21.3-34

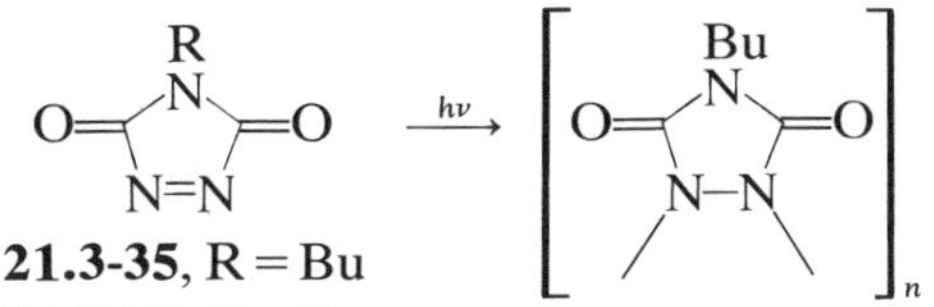

21.3-35, R = Bu
21.3-36, R = Ph

21.3-37

References

1. **65:** 7273e
2. **80:** 82823j
3. **81:** 106281e
4. **81:** 120637g
5. **70:** 20356u
6. **72:** 68047n
7. **60:** 10867f
8. **70:** 20885r
9. **76:** 4855z
10. **71:** 125885q
11. **72:** 90906p
12. **69:** 86973d
13. **69:** 107280t
14. **67:** 11793v
15. **73:** 25919b
16. **81:** 170229w
17. **68:** 13458j
18. **68:** 96231d
19. **75:** 152121b
20. **70:** 12143g
21. **71:** 125032j
22. **68:** 50462b
23. **69:** 67793d
24. **73:** 120949v
25. **55:** 20487f
26. **72:** 44284m
27. **76:** 14998n
28. **83:** 131973f
29. **80:** 15236g
30. **52:** 8625b
31. **54:** 1503e
32. **54:** 511b
33. **68:** 79571x
34. **68:** 79574a
35. **69:** 87631c
36. **71:** 125194p
37. **54:** 22603e
38. **64:** 9827e
39. **68:** 13448f
40. **71:** 113323s
41. **77:** 153755q
42. **77:** 36209d
43. **77:** 36213a
44. **80:** 60965e
45. **82:** 18517b
46. **82:** 113109d
47. **81:** 8417j
48. **80:** 111542z
49. **52:** 9656a
50. **83:** 79677w
51. **72:** 111882v
52. **70:** 57742t
53. **74:** 54291s
54. **80:** 60251u

TABLE 21. BI-, DI-, BIS-, TRIS-, AND POLY(1,2,4-TRIAZOLES, TRIAZOLINES, AND TRIAZOLIDINES) LINKED EITHER DIRECTLY OR WITH CHAIN CONTAINING CARBON

Compound	Reference

21.1. Bi-, Di-, and Tris(1,2,4-triazoles)

Compound	Reference
-5-acetonitrile, 3,3′-*p*-phenylenebis[1-phenyl-	**74:** P118381f, **75:** P93013t
1,1′-adipoylbis-	**54:** 12124c
3,3′-[*N*,*N*-anilinobis(2-ethylthio)]bis[-	**79:** P43665u
3,3′-[(1,3-benzenedicarboxamido)bis-2,1-phenylene]-bis[5-phenyl-	**80:** 15236g, **80:** 59912d
3,3′-[(1,4-benzenedicarboxamido)bis-2,1-phenylene]-bis[5-phenyl-	**80:** 15236g, **80:** 59912d
3,3′-bi[-	**70:** 114407p
3,3′-bi[1-acetyl-5-methyl-	**83:** 193187m
3,3′-bi[5-amino-	**65:** 12205b
3,3′-bi[5-azido-	**82:** 30784u
3,3′-[[6,6′-bi-1H-benz[*de*]isoquinoline]-1,1′3,3′(2H,2′H)-tetrone-2,2′-diyl]bis[-	**81:** P38960y
3,3-bi[5-[3-[3,5-bis(tert-butyl)-4-hydroxyphenyl]-propionylamino]-	**79:** P19662y
3,3′-bi[1,5-bis(2,4,6-trinitroanilino)-	**68:** 105111u
3,3′-bi[4,5-bis(2,4,6-trinitroanilino)-	**68:** 105111u
3,3′-bi[4-(2-carboxy-4-methylphenyl)-	**72:** 43550h
3,3′-bi[1-(1-cyclohepten-1-yl-5-methyl-	**83:** 193187m
3,3′-bi[1-(1-cyclopenten-1-yl)-5-methyl-	**83:** 193187m
3,3′-bi[1,5-diphenyl	**74:** P118381f, **75:** P93013t
3,3′-bi[5-(2-hydroxyethoxy)-	**68:** 78201q
4,4′-bi[3-methyl-	**78:** 123762t
3,3′-bi[5-methyl-1-(1-methyl-2-phenylethenyl-	**83:** 193187m
3,3′-bi[5-methyl-1-(2-phenylethenyl)-	**83:** 193187m
3,3′-bi[5-(6-methyl-2-pyridinyl)-	**69:** 52078a
3,3′-bi[5-(2-naphthyl)-	**69:** 52078a
3,3′-bi[5-phenyl-	**72:** P133573x
4,4′-bi[3-phenyl-	**78:** 123762t
3,3′-[1,1′-biphenyl]-4,4′-diylbis[4,5-diphenyl-	**78:** P112683s
1,1′-[[1,1′-biphenyl]-4,4′-diylbis[2,1-ethenediyl)-4,1-phenylene)]bis-	**84:** P91665u
3,3′-bi[5-(2-pyridinyl)-	**69:** 52078a
1,1′-[2,5-bis(1-aziridinyl)-1,4-dioxocyclohexa-2,5-diene-3,6-diyl]bis[-	**78:** 84321b
1,1′-[bis(*p*-chlorophenyl)methylene]bis-	**75:** P140860d, **77:** P5477k
1,1′-[2,5-bis(cyclohexylamino)-1,4-dioxocyclohexa-2,5-diene-3,6-diyl]bis[-	**78:** 84321b
3,3′-bis[5-diazo-	**73:** 45420k
3,3′-bis[4,5-diphenyl-	**70:** 28874g
1,1′-[bis(4-fluorophenyl)methylene]bis-	**75:** P140860D, **77:** P5477k
3,3′-bis[5-(2-hydroxybenzamido)-	**77:** P115460g
1,1′-[1,2-bis(methoxycarbonyl)ethylene]bis[-	**68:** 68937v
1,1′-[1,2-bis(methoxycarbonyl)ethylene]bis[3-methyl-	**68:** 68937v
3,3′-bis[5-methyl-	**80:** 96082d

TABLE 21 (*Continued*)

Compound	Reference

21.1. Bi-, Di-, Bis-, and Tris(1,2,4-triazoles) (*Continued*)

Compound	Reference
3,3'-bis[5-[2-(trimethylsilyl)ethyl]-	**80:** 96082d
3,3-bi[5-(2,4,6-trinitroanilino)-	**68:** 105111u
3,3'-bi[5-(4-tolyl)-	**69:** 52078a
1,1'-[3-bromophenyl)phenylmethylene]bis-	**84:** 169551a
5,5'-(1,4-butanediyl)bis[3-amino-	**78:** P147974q
3,3'-(1,4-butanediyl)bis[5-(2-hydroxybenzamido)-	**77:** P115460g, **78:** 147974q
3,3'-(1,4-butanediyl)bis[5-(2-hydroxy-5-octylbenzamido)-	**77:** P115460g
3,3'-(1,4-butanediyl)bis[5-methyl-	**80:** 96082d
5,5'-(1,4-butanediyl)bis[3-nitramino-	**79:** 78692c
3,3'-[3,3'-butylidenebis(6-hydroxybenzamido)]bis[-	**77:** P115460g
1,1'-carbonimidoylbis-	**79:** 105150h
3,3'-[3,3'-(carbonothioyldiimino)bis(6-hydroxy-benzamido)]bis[5-propyl-	**77:** P115460g
1,1'(or 4,4')-carbonylbis-	**52:** 7332f, **55:** 5484i
3,3'-([4,4'-carbonylbis(benzamido)]bis-2,1-phenylene)-bis[5-phenyl-	**80:** 15236g, **80:** 59912d
1,1'-(4-chloro-3-nitro-α-phenylbenzylidene)bis-	**74:** P125705t, **75:** P140860d, **77:** P5477k
1,1'-[(4-chloro-3-nitrophenyl)phenylmethylene]bis-	**84:** 169551a
1,1'-(o-chloro-α-phenylbenzylidene)bis-	**75:** P140860d, **77:** P5477k
1,1'-(p-chloro-α-phenylbenzylidene)bis-	**75:** P140860d, **77:** P5477k
1,1'-[(2-chlorophenyl)phenylmethylene]bis-	**84:** P169551a
1,1'-[(3-chlorophenyl)phenylmethylene]bis-	**75:** P140860d, **77:** P5477k
1,1'-[(4-chlorophenyl)phenylmethylene]bis-	**75:** P140860d, **77:** P5477k
1,1'-[(4-cyanophenyl)(4-nitrophenyl)methylene]bis-	**75:** P140860d, **77:** P5477k
1,1'-[(4-cyanophenyl)phenylmethylene]bis-	**75:** P140860d, **77:** P5477k
5,5'-(1,1,2,2,3,3,4,4,5,5-decafluoro-1,5-pentanediyl)-bis[3-amino-	**81:** 136060z
3,3'-decamethylenebis[5-(p-nitrophenyl)-	**65:** 15368d
1-dibutylstannyl bis[-	**60:** 6860f
1,1'-[(2,5-dichlorophenyl)phenylmethylene]bis-	**75:** P140860d, **77:** P5477k **84:** P169551a
1,1'-(2,5-diethoxy-1,4-dioxocyclohexa-2,5-diene-3,6-diyl)bis[-	**78:** 84321b
3-(3,5-diethyl-4H-1,2,4-triazol-4-yl)-	**69:** 106627z
3,3'-(1,2-dihydroxy-1,2-ethanediyl)bis[5-(2-hydroxy-benzamido)-	**77:** P115460g
1,1'-(1,2-dihydroxy-1,2-ethylene)bis[3-phenyl-	**74:** 125572x
1,1'-[2,5-di(isopropoxy)-1,4-dioxocyclohexa-2,5-diene-3,6-diyl]bis[-	**78:** 84321b
3,3'-[(1,2-dimethylethanediylidene)bis(nitrilomethylene)]-bis[5-(methylthio)-	**66:** 2536b
5,5'-[2,2'-(dimethylmalonyl)bishydrazino]bis[1-methyl-3-phenyl-	**82:** 125326k
3,3'-[(2,5-dimethyl-p-phenylene)bis[methylene-(ethylimino)-p-phenyleneazo]]bis[-	**69:** P28582q
3,3'-(4,6-dinitro-1,3-phenylene)bis[5-phenyl-	**85:** 123862v, **85:** 178022b
1-(1,4-dioxo-2,5-cyclohexadien-2,3,5,6-tetray)tetra[-	**77:** P164711y, **78:** 84321b

TABLE 21 (*Continued*)

Compound	Reference
21.1. Bi-, Di-, Bis-, and Tris(1,2,4-triazoles) (*Continued*)	
1,1′-(1,4-dioxocyclohexane-3,6-diyl)bis[-	**78:** 84321b
3,3′-(1,4-dioxo-2,5-cyclohexarlien-2,5-diyl)bis[5-amino-	**82:** P126605n
1,1′-[1,4-dioxo-2,5-di(1-piperidinyl)cyclohexa-2,5-diene-3,6-diyl]bis[-	**78:** 84321b
1,1′-(1,4-dioxo-2,5-dipropoxycyclohexa-2,5-diene-3,6-diyl)bis[-	**78:** 84321b
1,1′-[1,4-dioxonaphthalen-2,3-diylbis-	**77:** P164713a, **78:** 84321b
1,1′-(diphenylmethylene)bis-	**74:** P125705t, **75:** P140860d, **77:** P5477k, **84:** 169551a
1,5-diphenyl-5′-(4-pyridyl)-3,3′-bi[-	**70:** 114407p
1,1′-(2,2′-disulfostilben-4,4′-diyl)bis[-	**74:** P43528y
3,3′-[dithiobis(4-chloro-o-phenylene)]bis[5-methyl-	**69:** 52047q
3,3′-(dithiodi-o-phenylene)bis[1-acetyl-5-methyl-	**69:** 52047q
3,3′-(dithiodi-o-phenylene)bis[5-methyl-	**69:** 52047q
5,5′-(1,1,2,2,3,3,4,4,5,5,6,6-dodecafluoro-1,6-hexanediyl)bis[3-amino-	**81:** 136060z
3,3′-(1,2-ethanediyl)bis[5-amino-	**51:** P489i, **55:** P25991h, **56:** 10132e, **59:** 6386h, **65:** 12205b
1,1′-(1,2-ethanediyl)bis[3-[(4-amino-1-anthraquinonyl)-thio]-	**74:** P14185d
3,3′-(1,2-ethanediyl)bis[5-azido-	**82:** 30784u
3,3′-(1,2-ethanediyl)bis[4-(2-carboxyphenyl)-5-methyl-	**72:** 43550h
3,3′-(1,2-ethanediyl)bis[4,5-diphenyl-	**70:** 28874g
3,3′-(1,2-ethanediyl)bis[5-(2-hydroxybenzamido)-	**77:** P115460g
3,3′-(1,2-ethanediyl)bis[5-(2-hydroxyethoxy)-	**68:** 78201q
5,5′-(1,2-ethanediyl)bis[3-nitramino-	**79:** 78692c
3,3′-(1,2-ethanediyl)bis[5-nitro-	**73:** 87300b
3,3′-(1,2-ethanediyl)bis[4-(*p*-nitrophenyl)-5-phenyl-	**70:** 28874g
4,4′-(1,2-ethanediyl)di-*p*-phenylene)bis[3-(methylthio)-5-phenyl-	**70:** 57742t
3,3′-(1,2-ethanediyl)bis[5-(3-hydroxy-2-naphthalene-carboxamido)-	**77:** P115460g, **83:** P194350w
3,3′-[1,2-ethenediylbis(4,1-phenylene)]bis[4,5-di-(4-ethoxycarbonylphenyl)-	**78:** P112683s
3,3′-[1,2-ethenediylbis(4,1-phenylene)]bis[4-(4-ethoxycarbonylphenyl)-5-(4-methylphenyl)-	**78:** P112683s
3,3′-[1,2-ethenediylbis(4,1-phenylene)]bis[5-(4-ethoxycarbonylphenyl)-4-phenyl-	**78:** P112683s
3,3′-[1,2-ethenediylbis(4,1-phenylene)]bis[5-(4-methylphenyl)-4-(3-sulfophenyl)-	**78:** P112683s
3,3′-[1,2-ethenediylbis(4,1-phenylene)]bis[5-phenyl-4-(3-sulfophenyl)-	**78:** P112683s
3,3′-(1,2-ethenediyldi-4,1-phenylene)bis[5-[1,1′-biphenyl]-4-yl-4-phenyl-	**78:** P112683s
3,3′-(1,2-ethenediyldi-4,1-phenylene)bis[4,5-bis-(4-methylphenyl)-	**78:** P112683s
3,3′-(1,2-ethenediyldi-4,1-phenylene)bis[4-(4-chlorophenyl)-5-(4-methylphenyl)-	**78:** P112683s

TABLE 21 (*Continued*)

Compound	Reference

21.1. Bi-, Di-, Bis-, and Tris(1,2,4-triazoles) (*Continued*)

Compound	Reference
3,3'-(1,2-ethenediyldi-4,1-phenylene)bis[5-(4-chlorophenyl)-4-phenyl-	**78:** P112683s
3,3'-(1,2-ethenediyldi-4,1-phenylene)bis[4,5-diphenyl-	**78:** P112683s
3,3'-(1,2-ethenediyldi-4,1-phenylene)bis[5-(4-methylphenyl)-	**78:** P112683s
1,1'-(ethoxymethylene)bis[3-(5-nitro-2-furyl)-	**75:** 20356x
3,3'-ethylenebis[5-nitro-	**72:** 11383h
3,3'-(ethylenedithio)bis-	**71:** 124344a
1,1'-[(4-fluorophenyl)(4-nitrophenyl)methylene]bis-	**75:** P140860d, **77:** P5477k
1,1'-[(4-fluorophenyl)phenylmethylene]bis-	**75:** P140860d, **77:** P5477k
1,1'-fumaroylbis-	**54:** 12124d
3,3'-heptamethylenebis[5-amino-	**51:** P489i
3,3'-heptamethylenebis[4,5-diphenyl-	**70:** 28874g
3,3'-(hexafluorotrimethylene)bis[5-amino-	**72:** 78954v, **72:** P79061p
5,5'-(hexafluorotrimethylene)bis[1,3-diphenyl-	**69:** 67793d
3,3'-hexamethylenebis[5-amino-	**51:** P489i
4,4'-hexamethylenebis[3,5-diethyl-	**69:** 106627z
3,3'-hexamethylenebis[4,5-diphenyl-	**70:** 28874g
3,3'-(hexamethylenedioxy)bis-[5-ethoxy-	**44:** 6855d
3,3'-(hexanedioyldiamino)bis[-	**72:** P67723t
3,3'-[N,N'-(1,6-hexanediyl)bisureylene]bis[-	**82:** 156189x
4,4'-malonamidobis[-	**74:** 141669q
1,1'-(p-methoxy-α-phenylbenzylidene)bis-	**75:** P140860d
1,1'-[(4-methoxyphenyl)phenylmethylene]bis-	**77:** P5477k
3,3'-methylenebis[5-amino-	**51:** P489i, **65:** 12205b
3,3'-methylenebis[5-azido-	**82:** 30784u
3,3'-(methylene)bis[4-(2-carboxyphenyl)-5-methyl-	**72:** 43550h
3,3'-methylenebis–4,5-diphenyl-	**70:** 28874g
3,3'-[5,5'-methylenebis(2-hydroxybenzamido)]bis[-	**81:** 121645v
3,3'-[3,3'-methylenebis(6-hydroxybenzamido)]bis[5-ethyl-	**77:** P115460g
3,3'-[3,3'-methylenebis(6-hydroxybenzamido)]bis[5-methyl-	**77:** P115460g
3,3'-(methylene)bis[5-(2-hydroxyethoxy)-	**68:** 78201q
3,3'-methylenebis[5-methyl-	**80:** 96082d
3,3'-methylenebis[5-(methylthio)-4-phenyl-	**70:** 57742t
3,3'-methylenebis[5-(methylthio)-4-p-tolyl-	**70:** 57742t
5,5'-methylenebis[3-nitramino-	**79:** 78692c
3,3'-methylenebis[5-nitro-	**72:** 111383h
3,3'-methylenebis[5-(4-pyridinyl)-	**70:** 114407p
3,3'-methylenebis[5-(4-pyridyl)-	**70:** 114407p
3,3'-methylenebis[5-[2-(trimethylsilyl)ethyl]-	**80:** 96082d
3,3'-(methylenediimino)bis-	**72:** P67723t
3,3'-(methylenedithio)bis-	**71:** 124344a
3,3'-[2,2'-(methylenedithiobis(acetylamino)]bis[-	**72:** P67723t
1,1',1''-methylidynetris[3,5-dimethyl-	**72:** 3454y
1,1'-(4-methyl-3-nitro-α-phenylbenzylidene)bis-	**75:** P140860d, **77:** P5477k
3,3'-(N,N-(4-methyl-1,3-phenylene)bisureylene]bis[-	**82:** 156189x
3,3'-[(2,6-naphthalenedicarboxamido)bis-2,1-phenylene]-bis[5-phenyl-	**80:** 15236g, **80:** 59912d

TABLE 21 (*Continued*)

Compound	Reference
21.1. Bi-, Di-, Bis-, and Tris(1,2,4-triazoles) (*Continued*)	
1,1'-[2,6-naphthalenediylbis(2,1-ethenediyl-4,1-phenylene)]bis-	**84:** P91665u
3,3'-(octafluorotetramethylene)bis[5-amino-	**72:** 78954v, **72:** P79061p
3,3'-octamethylenebis[5-amino-	**51:** P489i
3,3'-octamethylenebis[4,5-diphenyl-	**70:** 28874g
3,3'-(octamethylene)bis[5-(4-hydroxybutoxy)-	**68:** 78201q
3,3'-(octamethylene)bis[5-(2-hydroxyethoxy)-	**68:** 78201q
3,3'-octamethylenebis[5-(*p*-nitrophenyl)-	**65:** 15368d
3,3'-(1,8-octanediyl)bis[5-[(2,4-dinitrophenyl)thio]-4-phenyl-	**81:** P97737d
5,5'-(1,8-octanediyl)bis[3-nitramino-	**79:** 78692c
3,3'-(*N*,*N*'-oxalyl)bis[3-amino-	**72:** P67723t
3,3'-oxamidobis[4-amino-5-benzyl-	**73:** 87862m
3,3'-oxamidobis[4-amino-5-isobutyl-	**73:** 87862m
3,3'-oxamidobis[4-amino-5-methyl-	**73:** 87862m
3,3'-oxamidobis[4-amino-5-phenyl-	**73:** 87862m
1,1'-(3-oxoglutaryl)bis[3,5-diamino-	**52:** 4607e
3,3'-([4,4'-oxybis(benzamido)]bis-2,1-phenylene)-bis[5-phenyl-	**80:** 15236g, **80:** 59912d
3,3'-pentamethylenebis[5-amino-	**51:** P489i
3,3'-pentamethylenebis[4,5-diphenyl-	**70:** 28874g
3,3'-[*N*,*N*'-(3,4,9,10-perylenetetracarboxy-3,4:9,10-diimido)]bis[5-methyl-	**73:** P36579s
3,3'-(3,4,9,10-perylenetetracarboxylic 3,4:9,10-diimido-bis[5-phenyl-	**73:** P36579s
3,3'-(1,3-phenylene)bis[5-(2-aminophenyl)-	**80:** 59912d, **81:** 169929t, **83:** 131973f
3,3'-(1,4-phenylene)bis[5-(2-aminophenyl)-	**80:** 15236g, **81:** 169929t, **83:** 131973f
3,3'-*p*-phenylenebis[5-anilino-	**64:** 9727e
3,3'-(1,3-phenylene)bis[5-(2-benzamidophenyl)-	**80:** 15236g, **80:** 59912d
3,3'-(1,4-phenylene)bis[5-(2-benzamidophenyl)-	**80:** 15236g, **80:** 59912d
3,3'-(1,3-phenylene)bis[5-[2-(1H-benzimidazol-2-yl)phenyl]-	**82:** 72892h
3,3'-(1,4-phenylene)bis[5-[2-(1H-benzimidazol-2-yl)-phenyl]-	**82:** 72892h
3,3'-(1,2-phenylenebis[2-(carbonylamino)phenyl])-bis[5-(2-pyridinyl)-	**76:** 14998n, **76:** 99614g
3,3'-*m*-phenylenebis[1,5-diphenyl-	**73:** 120949v
3,3'-*p*-phenylenebis[1,5-diphenyl-	**53:** 5253g, **55:** 6474c, **68:** 50462b, **73:** 120949v
3,3'-(1,4-phenylene)bis[5-[2-hydroxy-5-(oxtyloxy)-benzamido)]-	**77:** P115460g
4,4'-[*p*-phenylenebis(methylene nitrosimino)]bis-	**73:** 87860j
3,3'-(*p*-phenylene)bis[5,(6-methyl-2-pyridinyl)-	**69:** 52078a
4,4'-*p*-phenylenebis[3-(methylthio)-5-phenyl-	**70:** 57742t
3,3'-(1,3-phenylene)bis[5-(2-nitrophenyl)-	**80:** 15236g, **80:** 59912d, **81:** 169929t

TABLE 21   (*Continued*)

Compound	Reference

21.1. Bi-, Di-, Bis-, and Tris(1,2,4-triazoles)   (*Continued*)

Compound	Reference
3,3'-(1,4-phenylene)bis[5-(2-nitrophenyl)-	**80:** 15236g, **80:** 59912d, **81:** 169929t
3,3'-*p*-phenylenebis[5-phenyl-	**55:** P13450a, **59:** 6387c
3,3'-(*m*-phenylene)bis[5-(2-pyridinyl)-	**72:** 43572s, **72:** 90906p
3,3'-(*p*-phenylene)bis[5-(2-pyridinyl)-	**69:** 52078a
3,3'-*p*-phenylenebis[5-*p*-tolyl-	**69:** 52078a
1,1'-[*p*-phenylenebis(ureylene-*p*-phenylene)]bis[-	**72:** P100715s
4,4'-(1,4-piperazinediacetamide)bis[3,5-dimethyl-	**76:** 140662m
3,3'-(pyridine-2,6-diyl)bis[5-(2-aminophenyl)-	**76:** 14998n, **76:** 99614g, **80:** 15236g, **80:** 59912d, **81:** 169929t, **83:** 131973f
3,3'-(pyridine-2,6-diyl)bis[5-(2-benzamido)phenyl-	**76:** 99614g, **80:** 15236g, **80:** 59912d
3,3'-(pyridine-2,6-diyl)bis[5-(2-nitrophenyl)-	**76:** 14998n, **76:** 99614g, **80:** 15236g, **80:** 59912d, **81:** 169929t
3,3'-(pyridine-2,6-diyl)bis[5-phenyl-	**72:** 43572s, **72:** 90906p
3,3'-[[4,4'-sulfonylbis(benzamido)]bis-2,1-phenylene]-bis[5-phenyl-	**80:** 15236g, **80:** 59912d
3,3'-[3,3'-sulfonylbis(6-hydroxybenzamido)]bis[-	**77:** P115460g
3,3'-(*N,N'*-tartronoyl)bis[3-amino-	**72:** P67723t
1,1'-terephthaloylbis-	**54:** 12124c
4,4'-[terephthaloylbis(imino-*p*-phenylene)]bis[-	**72:** P100715s
3,3'-tetramethylenebis[5-amino-	**51:** P489i
3,3'-tetramethylenebis[4,5-diphenyl-	**70:** 28874g
3,3'-(tetramethylene)bis[5-(4-hydroxybutoxy)-	**68:** 78201q
3,3'-(tetramethylene)bis[5-(2-hydroxyethoxy)-	**68:** 78201q
3,3'-tetramethylenebis[5-phenyl-	**65:** 15368d
4,4'-(tetramethylenediimino)bis-	**70:** 87681q
3,3'-[2,2'-thiobis(acetylamino)]bis[-	**72:** P67723t
3,3'-[3,3'-thiobis(6-hydroxybenzamido)]bis[5-octyl-	**77:** P115460g
3,3'-[3,3-thiobis(propionylamino)]bis[-	**72:** P67723t
3,3'-(1,3-thioureylene)bis[-	**72:** P67723t
3,3'-(thioureylene)bis[5-(2-[3,5-bis(tert-butyl)-4-hydroxyphenyl)ethyl]-	**79:** P19662v
3,3'-trimethylenebis[5-amino-	**51:** P489i
3,3'-trimethylenebis[4,5-diphenyl-	**70:** 28874g
3,3'-(trimethylene)bis[5-(4-hydroxybutoxy)-	**68:** 78201q
3,3'-(trimethylene)bis[5-(2-hydroxyethoxy)-	**68:** 78201q
3,3'-(trimethylenedithio)bis-	**71:** 124344a
3,3'-vinylenebis[5-amino-	**51:** P489i
3,3'-(vinylenebis[(3-sulfo-*p*-phenylene)amino(6-amino-*s*-triazine-4,2-diyl)amino])di-	**75:** P65304v, **75:** P77857r
3,3'-(vinylenebis[(3-sulfo-*p*-phenylene)amino(6-anilino-*s*-triazine-4,2-diyl)amino])di-	**75:** P65304v
3,3'-(vinylenebis[(3-sulfo-*p*-phenylene)amino(6-[bis-(2-hydroxyethyl)amino]-*s*-triazine-4,2-diyl]-amino])di-	**75:** P65304v

TABLE 21 (*Continued*)

Compound	Reference

21.1. Bi-, Di-, Bis-, and Tris(1,2,4-triazoles) (*Continued*)

4,4′-(vinylenebis[(3-sulfo-*p*-phenylene)amino(6-[bis- (2-hydroxyethyl)amino]-*s*-triazine-4,2-diyl)amino])di-	**75:** P65304v
3,3′-(vinylenebis[(3-sulfo-*p*-phenylene)amino(6-[bis- (2-hydroxyethyl)amino]-*s*-triazine-4,2-diyl)- benzylamino])di-	**75:** P65304v
3,3′-(vinylenebis[(3-sulfo-*p*-phenylene)amino(6-(4- carboxyanilino)-*s*-triazine-4,2-diyl)amino])di-	**75:** P65304v
3,3′-(vinylenebis[(3-sulfo-*p*-phenylene)amino(6-[ethyl- (D-gluco-2,3,4,5,6-pentohydroxyhexyl)amino]-*s*- triazine-4,2-diyl)amino])di	**75:** P65306x
3,3′-(vinylenebis[(3-sulfo-*p*-phenylene)amino(6-[methyl- (D-gluco-2,3,4,5,6-pentahydroxy-hexyl)amino]-*s* triazine-4,2-diyl)amino])di-	**75:** P65306x
3,3′-(vinylenebis[(3-sulfo-*p*-phenylene)amino(6- morpholino-*s*-triazine-4,2-diyl)amino])di-	**75:** P65304v
3,3′-(vinylenebis[(3-sulfo-*p*-phenylene)amino[6- (*p*-sulfoanilino)-*s*-triazine-4,2-diyl]amino])di-	**75:** P65304v
4,4′-(vinylenebis[(3-sulfo-*p*-phenylene)amino[6- (*p*-sulfoanilino)-*s*-triazine-4,2-diyl]amino])di-	**75:** P65304v

21.2. Bi- and Bis(triazoline and triazolidines)

3,3′-bi[1,4-diacetyl-5-benzyl-5-phenyl-	**83:** 193187m
3,3′-bi[1,4-diacetyl-5-ethenyl-	**83:** 193187m
3,3′-bi[1,4-diacetyl-5-(2-phenylethenyl)-	**83:** 193187m
3,3′-(4,4′-biphenylylene)bis[5-methyl-5-(2-pyridinyl)-	**72:** 111882v
-3,5-dione, 4,4′-ethylenebis[1,2-diisopropyl-	**64:** 11202f
-3,5-dione, 4,4′-hexamethylenebis[1,2-diisopropyl-	**64:** 11202f
-3,5-dione, 4,4′-(1,6-hexanediyl)bis-	**78:** 72719z
-3,5-dione, 4,4′-methylenebis[1,2-diisopropyl-	**64:** 11202f
-3,5-dione, 4,4′-(methylene-di-*p*-phenylene)bis-	**75:** 89223u, **80:** 60251u, **82:** P172343y
-3,5-dione, 4,4′-(4-methyl-*m*-phenylene)bis-	**61:** P8317e
-3,5-dione, 4,4′-pentamethylenebis[1,2-disopropyl-	**64:** 11202f
-3,5-dione, 1,1-succinylbis[4-butyl-	**59:** 5152b
-3,5-dione, 4,4-trimethylenebis[1,2-diisopropyl-	**64:** 11202f
2,2′-malonylbis[3-amino-5-imino-1-phenyl-	**46:** 4535b
-5-one, 3,3′-[diselenobis(methylene)]bis-	**79:** 115502p
-5-one, 3,3′-[diselenobis(methylene)]bis[1-(4-nitrophenyl)-	**79:** 115502p
-5-one, 3,3′-[diselenobis(methylene)]bis[1-phenyl-	**79:** 115502p
-5-one, 3,3′-[dithiobis(methylene)]bis[-	**76:** 3765h
-5-one, 3,3′-[dithiobis(methylene)]bis[2-methyl-	**76:** 3765h
-5-one, 3,3′-[dithiobis(methylene)]bis[1-phenyl-	**76:** 3765h
-5-one, 3,3′-(*m*-phenylene)bis[1-phenyl-	**69:** 52073v
-5-one, 3,3′-(*m*-phenylene)bis[2-phenyl-	**69:** 52073v
-5-one, 3,3′-(*p*-phenylene)bis[1-phenyl-	**69:** 52073v
-5-one, 3,3′-(*p*-phenylene)bis[2-phenyl-	**69:** 52073v
2,2′-oxalylbis[3-amino-5-imino-1-phenyl-	**46:** 4535c

TABLE 21   (*Continued*)

Compound	Reference

21.2. Bi- and Bis(triazolines and triazolidines)   (*Continued*)

Compound	Reference
3,3′-[oxybis(*p*-phenylene)]bis[5-(2-pyridinyl)-	**72:** 111882v
3,3′-*m*-phenylenebis-	**65:** 7169f
3,3′-*p*-phenylenebis-	**65:** 7169f
3,3′-(1,4-phenylene)bis[5-(4-methyl-2-pyridinyl)-	**76:** 85796f
3,3′-(1,4-phenylene)bis[5-(4-phenyl-2-pyridinyl)-	**76:** 85796f
3,3′-(*m*-phenylene)bis[5-(2-pyridinyl)-	**72:** 111882v
3,3′-(*p*-phenylene)bis[5-(2-pyridinyl)-	**72:** 111882v, **76:** 85796f
5,5′-*p*-phenylenebis[1,3,4-triphenyl-	**70:** 96723g
5,5′-(pyridine-2,6-diyl)bis[5-methyl-3-(2-pyridyl)-	**72:** 111882v
3,3′-(pyridine-2,6-diyl)bis[5-(4-methyl-2-pyridinyl)-	**76:** 85796f
3,3′-(pyridine-2,4-diyl)bis[5-phenyl-	**76:** 85796f
3,3′-(pyridine-2,6-diyl)bis[5-(4-phenyl-2-pyridinyl)-	**76:** 85796f
3,3′-(pyridine-2,4-diyl)bis[5-(2-pyridinyl)-	**76:** 85796f
3,3′-(pyridine-2,6-diyl)bis[5-(2-pyridinyl)-	**76:** 85796f
-5-selone, 3,3′-bi-	**77:** P140086x, **77:** P146189e
-5-selone, 3,3′-bis[1-methyl-	**77:** P140086x
-5-selone, 3,3′-bis[4-methyl-	**77:** P140086x
-5-selone, 3,3′-bis[4-phenyl-	**77:** P140086x
-3-selone, 5,5′-(1,2-ethanediyl)bis[-	**77:** P140086x
-3-selone, 5,5′-(1,2-ethanediyl)bis[4-methyl-	**77:** P140086x
-3-selone, 5,5′-(1,2-ethanediyl)bis[4-phenyl-	**77:** P140086x
-3-selone, 5,5′-methylenebis[-	**77:** P140086x
-5-selone, 3,3′-methylenebis[1-methyl-	**77:** P140086x
-3-selone, 5,5′-methylenebis[4-methyl-	**77:** P140086x
-3-selone, 5,5′-(1,4-phenylene)bis[4-methyl-	**77:** P140086x
-3-selone, 5,5′-(thiodi-2,1-ethanediyl)bis[4-methyl-	**77:** P140086x
-5-thione, 3,3′-(acetamidomethylene)bis[4-methyl-	**82:** 140031d
-5-thione, 3,3′-[(4-amino-4$\underline{H}$-1,2,4-triazole-3,5-diyl)- bis(methylene)]bis[benzyl-	**85:** 21243b
-5-thione, 3,3′-[(4-amino-4$\underline{H}$-1,2,4-triazole-3,5-diyl)- bis(methylene)]bis[4-ethyl-	**85:** 21243b
-5-thione, 3,3′-[(4-amino-4$\underline{H}$-1,2,4-triazole-3,5-diyl)- bis(methylene)]bis[4-methyl-	**85:** 21243b
-5-thione, 3,3′-[(4-amino-4$\underline{H}$-1,2,4-triazole-3,5-diyl)- bis(methylene)]bis[4-propyl-	**85:** 21243b
-5-thione, 1,1′-[(benzylimino)dimethylene]bis[4-methyl-	**64:** PC208e
-5-thione, 3,3′-bi[-	**65:** 7169f, **70:** P107521t
-5-thione, 3,3′-bi[4-methyl-	**70:** 68323h, **73:** 66517n, **79:** 92120w
-5-thione, 3,3′-bi[4-phenyl-	**79:** 92120w
-5-thione, 3,3′-bi[4-(2-propenyl)-	**79:** 92120w

-5-thione, 3,3′-([4-([(4-chlorophenyl)methylene]amino)-
   4$\underline{H}$-1,2,4-triazole-3,5-diyl]bis(methylene))bis-
   [4-methyl-
	85: 21243b

-5-thione, 3,3′-(1,2-dihydroxy-1,2-ethanediyl)bis- [4-methyl-	**82:** 140031d
-5-thione, 3,3′-(1,2-dihydroxyethylene)di-	**65:** 7169f, **70:** P107521t
-5-thione, 3,3′-(ethoxyethylene)di-	**65:** 7169f, **70:** P107521t

TABLE 21 (*Continued*)

Compound	Reference

21.2. Bi- and Bis(triazolines and triazolidines) (*Continued*)

-5-thione, 3,3'-ethylenebis-	**70:** P107521t
-5-thione, 4,4'-(ethylenedi-*p*-phenylene)bis[3-phenyl-	**70:** 57742t
-5-thione, 3,3'-ethylidenedi-	**65:** 7169f, **70:** P107521t
-5-thione, 3,3'-[[4-([(2-hydroxyphenyl)methylene]amino)-4H-1,2,4-triazole-3,5-diyl]bis(methylene)]bis(4-methyl-	**85:** 21243b
-5-thione, 1,1'-[(isopropylimino)dimethylene]bis[3,4-dimethyl-	**64:** PC208e
-5-thione, 3,3'-(methoxyethylene)di-	**65:** 7169f, **70:** P107521t
-5-thione, 3,3'-methylenebis[4-phenyl-	**70:** 57742t
-5-thione, 3,3'-[methylenebis(thiomethylene)]bis-	**74:** P70245k
-5-thione, 3,3'-[methylenebis(thiomethylene)]bis[4-allyl-	**74:** P70245k
-5-thione, 3,3'-[methylenebis(thiomethylene)]bis[4-amino-	**74:** P18038f, **74:** P70245k
-5-thione, 3,3'-[methylenebis(thiomethylene)]bis[4-methyl-	**74:** P70245k
-5-thione, 3,3'-[methylenebis(thiomethylene)]bis[4-phenyl-	**72:** P138322z, **74:** P18038f
-5-thione, 3,3'-methylenebis[4-*p*-tolyl-	**70:** 57742t
-5-thione, 3,3'-methylenedi-	**65:** 7169f, **70:** P107521t
-5-thione, 3,3'-(4-methyl-2,5-thiazolediyl)bis[4-phenyl-	**74:** 111955n
-5-thione, 3,3'-(2,6-naphthylene)bis-	**65:** 7169f, **70:** P107521t
-5-thione, 3,3'-(nonylidene)bis[-4-methyl-	**82:** 140031d
-5-thione, 3,3'-oxtamethylenedi-	**65:** 7169f, **70:** P107251t
-5-thione, 3,3'-(oxydimethylene)bis[4-amino-	**72:** P138322z, **74:** P70245k
-5-thione, 3,3'-(pentylidene)bis[-4-methyl-	**82:** 140031d
-5-thione, 3,3'-phenethylidenebis[4-allyl-	**70:** 67830j
-5-thione, 3,3'-phenethylidenebis[4-methyl-	**70:** 67830j
-5-thione, 3,3'-*m*-phenylenebis-	**70:** P107521t
-5-thione, 3,3'-*p*-phenylenebis	**70:** P107521t
-5-thione, 3,3'-*m*-phenylenebis[4-allyl-	**70:** 11642g
-5-thione, 3,3'-*p*-phenylenebis[4-allyl-	**70:** 11642g
-5-thione, 3,3'-*p*-phenylenebis[1-methyl-	**70:** P107521t
-5-thione, 3,3'-*p*-phenylenebis[4-methyl-	**70:** P107521t
-5-thione, 3,3'-[1,2-phenylenebis(oxymethylene)]bis[4-methyl-	**83:** 192762h
-5-thione, 3,3'-[1,4-phenylenebis(oxymethylene)]bis[4-methyl-	**76:** 126869c
-5-thione, 3,3'-[1,2-phenylenebis(oxymethylene)]bis[4-(4-methylphenyl)-	**83:** 192762h
-5-thione, 3,3'-[1,2-phenylenebis(oxymethylene)]bis[4-phenyl-	**83:** 192762h
-5-thione, 3,3'-[1,4-phenylenebis(oxymethylene)]bis[4-phenyl-	**76:** 126869c
-5-thione, 3,3'-[1,2-phenylenebis(oxymethylene)]bis[4-(2-propenyl)-	**83:** 192762h
-5-thione, 3,3'-[1,4-phenylenebis(oxymethylene)]bis[4-(2-propenyl)-	**76:** 126869c
-5-thione, 3,3'-*m*-phenylenebis[4-phenyl-	**70:** 11642g
-5-thione, 3,3'-*p*-phenylenebis[4-phenyl-	**70:** 11642g
-5-thione, 4,4'-*p*-phenylenebis[3-phenyl-	**70:** 57742t

TABLE 21 (*Continued*)

Compound	Reference

21.2. Bi- and Bis(triazolines and triazolidines) (*Continued*)

Compound	Reference
-5-thione, 3,3'-*m*-phenylenebis[4-*p*-tolyl-	**70:** 11642g
-5-thione, 3,3'-propylidenedi-	**65:** 7169f, **70:** P107521t
-5-thione, 3,3'-propylidenebis[4-methyl-	**82:** 140031d
-5-thione, 3,3'-[thiobis(trimethylene)]bis-	**74:** P18038f
-5-thione, 3,3'-[thiobis(trimethylene)]bis[4-methyl-	**74:** P18038f, **74:** P70245k
-5-thione, 3,3'-[thiobis(trimethylene)]bis[4-phenyl-	**74:** 18038f, **74:** P70245k
-5-thione, 3,3'-(thiodiethylene)bis-	**72:** P138322z, **74:** P18038f
-5-thione, 3,3'-(thiodiethylene)bis[4-allyl-	**72:** P138322z
-5-thione, 3,3'-(thiodiethylene)bis[4-amino-	**72:** P138322z, **74:** P70245k
-5-thione, 3,3'-(thiodiethylene)bis[4-methyl-	**72:** P138322z, **74:** 18038f, **74:** P70245k
-5-thione, 3,3'-(thiodimethylene)bis-	**72:** P138322z
-5-thione, 3,3'-(thiodimethylene)bis[4-allyl-	**72:** P138322z, **74:** P70245k
-5-thione, 3,3'-(thiodimethylene)bis[4-amino-	**72:** P138322z, **74:** P70245k
-5-thione, 3,3'-(thiodimethylene)bis[4-methyl-	**72:** P138322z
-5-thione, 3,3'-tetramethylenedi-	**65:** 7169f, **70:** P107521t
-5-thione, 3,3'-trimethylenedi-	**65:** 7169f, **70:** P107521t

21.3. Poly(1,2,4-triazoles, triazolines, and triazolidines)

Compound	Reference
1,3-Benzenedicarbonitrile, polymer with 2,2'-[1,3-phenylenebis(1H-1,2,4-triazole-5,3-diyl)]bis[benzenamine]	**81:** 169929t
1,4-Benzenedicarbonitrile, polymer with 2,2'-[1,3-phenylenebis(1H-1,2,4-triazole-5,3-diyl)]bis[benzenamine]	**81:** 169929t
1,4-Benzenedicarbonitrile, polymer with 2,2'-[1,4-phenylenebis(1H-1,2,4-triazole-5,3-diyl)]bis[benzenamine]	**81:** 169929t
1,4-Benzenedicarbonitrile, polymer with 2,2'-[2,6-pyridinediylbis(1H-1,2,4-triazole-5,3-diyl)]bis[benzenamine]	**81:** 169929t
1,3-benzenedicarbonyl dichloride, polymer with 2,2'-[2,6-pyridinediylbis(4H-1,2,4-triazole-5,3-diyl)]bis[benzenamine]	**76:** 14998n, **80:** 15236g, **83:** 131973f
1,3-Benzenedicarboxylic acid, diphenyl ester, polymer with 2,2'-[2,6-pyridinediylbis(4H-1,2,4-triazole-3,5-diyl)]bis[benzenamine]	**76:** 14998n
Benzoyl chloride, 4,4'-carbonyldi-, polyamide with 3,5-bis(3-amino-4-anilinophenyl)-4-phenyl-4H-1,2,4-triazole	**71:** P62234d
Poly[(4-amino-4H-1,2,4-triazole-3,5-diyl)-1,4-butanediyl]	**80:** P60965e
Poly[(4-amino-4H-1,2,4-triazole-3,5-diyl)-1,4-butanediyliminocarbonylimino-1,8-octanediyl]	**82:** P18517b
Poly[(4-amino-4H-1,2,4-triazole-3,5-diyl)-decamethyleneoxydecamethylene]	**71:** P125194p

TABLE 21 (*Continued*)

Compound	Reference

21.3. Poly(1,2,4-triazoles, triazolines, and triazolidines) (*Continued*)

Compound	Reference
Poly[(4-amino-4H-1,2,4-triazole-3,5-diyl)-ethyleneoxy-*p*-phenyleneoxyethylene]	**71:** 113323s
Poly[(4-amino-4H-1,2,4-triazole-3,5-diyl)-heptamethylene]	**70:** P12144h
Poly[(4-amino-4H-1,2,4-triazole-3,5-diyl)-hexamethyleneoxyhexamethylene]	**71:** P125194p
Poly[(4-amino-4H-1,2,4-triazole-3,5-diyl)-methyleneoxy-*p*-phenyleneoxymethylene]	**71:** 113323s
Poly[(4-amino-4H-1,2,4-triazole-2,5-diyl)-methylene-*p*-phenylenemethylene]	**68:** 13448f
Poly[(4-amino-4H-1,2,4-triazole-3,5-diyl)-octamethylene]	**68:** P79571x, **68:** P79574a, **70:** P12144h, **70:** P48223x, **71:** P71822w, **77:** P36209d, **77:** P36213a, **77:** P153755q, **81:** P8417j, **82:** P113109d
Poly[(4-amino-4H-1,2,4-triazole-3,5-diyl)-octamethyleneoxyoctamethylene]	**71:** P125194p
Poly[(4-amino-4H-1,2,4-triazole-3,5-diyl)-(6-oxo-1,6-hexanediyl)imino-1,8-octanediyl]	**82:** P18517b
Poly[(4-amino-4H-1,2,4-triazole-3,5-diyl)-pentamethyleneoxypentamethylene]	**71:** P125194p
Poly[(4-amino-4H-1,2,4-triazole-2,5-diyl)-*p*-phenylene]	**68:** 13448f
Poly[(1-amino-*s*-triazole-3,5-diyl)trimethyleneoxy-tetramethylene]	**69:** P87631c
Poly[(4-amino-4H-1,2,4-triazole-3,5-diyl)-trimethyleneoxytrimethylene]	**70:** P38307p
Poly[(ar,ar′-diphenyl[3,3′-biphenyl-1,2,4-triazole]-ar, ar′-diyl)-1,4-phenyleneoxy(1,10-dioxo-1,10-decanediyl)oxy-1,4-phenylene]	**83:** 115119b
Poly([3,3′-bi-*s*-triazole]-5,5′-diyloctamethylene-	**69:** P107280t
Poly([3,3′-bi-1,2,4-triazole]-5,5′-diyl-*m*-phenylene)	**69:** P107280t
Poly([3,3′-bi-*s*-triazole]-5,5′-diyl-*p*-phenylene)	**71:** P125885q
Poly([3,3′-bi-*s*-triazole]-5,5′-diylvinylene)	**71:** P125885q
Poly(4-butyl-3,5-dioxo-1,2,4-triazolidine-1,2-diyl)	**74:** 54291s
Poly[(4-cyclohexyl-4H-1,2,4-triazole-3,5-diyl)-*p*-phenylene]	**68:** 96231d
Poly[(4,5-dihydro-4-phenyl-3H-1,2,4-triazole-3,5-diyl)-1,3-phenylene(4,5-dihydro-4-phenyl-3H-1,2,4-triazole-3,5-diyl)-1,4-phenylene]	**76:** P4855z
Poly[(5,7-dihydro-1,3,5,7-tetraoxobenzo[1,2-c:4,5-c′]-dipyrrole-2,6(1H,3H)-diyl)-*m*-phenylene-(4-phenyl-4H-1,2,4-triazole-3,5-diyl)-*m*-phenylene]	**70:** 20435u
Poly[(5,7-dihydro-1,3,5,7-tetraoxobenzo[1,2-c:4,5-c′]-dipyrrole-2,6(1H,3H)-diyl)-*s*-triazole-3,5-diyl]	**75:** 21112b
Poly[(5,7-dihydro-1,3,5,7-tetraoxodipyrrolo-[3,4-b:3′,4′-e]pyrazine-2,6(1H,3H)-diyl)-*s*-triazole-3,5-diyl]	**75:** 21112b

TABLE 21 (*Continued*)

Compound	Reference
21.3. Poly(1,2,4-triazoles, triazolines, and triazolidines) (*Continued*)	
Poly[(5,5'-dimethyl[3,3'-bi-4H-1,2,4-triazole]-4,4'-diyl)carbonylimino-1,6-hexanediyliminocarbonyl]	**83:** 79677w
Poly[(5,5'-dimethyl[3,3'-bi-4H-1,2,4-triazole]-4,4'-diyl)carbonylimino-1,3-propanediyl(1,1,3,3-tetramethyl-1,3-disiloxanediyl)-1,3-propanediyliminocarbonyl]	**83:** 79677w
Poly(1H-1,4-diyl-*p*-phenylene)	**68:** 87784j, **71:** 13464v
Poly[(5-mercapto-4H-1,2,4-triazole-4,3-diyl)-methylene(5-mercapto-4H-1,2,4-triazole-3,4-diyl)-*p*-phenylene]	**70:** 57742t
Poly[(5-mercapto-4H-1,2,4-triazole-4,3-diyl)-methylene(5-mercapto-4H-1,2,4-triazole-3,4-diyl)-*p*-phenyleneethylene-*p*-phenylene]	**70:** 57742t
Poly[[5-(methylthio)-4H-1,2,4-triazole-4,3-diyl]-methylene[5-(methylthio)-4H-1,2,4-triazole-3,4-diyl]-*p*-phenylene]	**70:** 57742t
Poly[[5-(methylthio)-4H-1,2,4-triazole-4,3-diyl]-methylene[5-(methylthio)-4H-1,2,4-triazole-3,4-diyl]-*p*-phenyleneethylene-*p*-phenylene]	**70:** 57742t
Poly[(5-methyl-4H-1,2,4-triazole-4,3-diyl)-1,4-butanediyl(5-methyl-4H-1,2,4-triazole-3,4-diyl)-carbonylimino-1,6-hexanediyliminocarbonyl]	**83:** 79677w
Poly[(5-methyl-4H-1,2,4-triazole-4,3-diyl)-1,4-butanediyl(5-methyl-4H-1,2,4-triazole-3,4-diyl)-carbonylimino(methyl-1,3-phenylene)iminocarbonyl]	**83:** 79677w
Poly[(5-methyl-4-1,2,4-triazole-4,3-diyl)-1,4-butanediyl(5-methyl-4H-1,2,4-triazole-3,4-diyl)-carbonylimino-1,3-propanediyl(1,1,3,3-tetramethyl-1,3-disiloxanediyl)-1,3-propanediyliminocarbonyl]	**83:** 79677w
Poly[(4-methyl-4H-1,2,4-triazole-3,5-diyl)-*m*-phenylene]	**73:** 25919b
Poly[(4-methyl-4H-1,2,4-triazole-3,5-diyl)-*p*-phenylene]	**73:** 25919b
Poly[(4-methyl-4H-1,2,4-triazole-3,5-diyl)-*m*-phenylene(4-methyl-4H-1,2,4-triazole-3,5-diyl)-*p*-phenylene]	**73:** 25919b
Poly[[4-(1-naphthyl)-4H-1,2,4-triazole-3,5-diyl]-*m*-phenylene]	**70:** P12143g
Poly[[4-(1-naphthyl)-4H-1,2,4-triazole-3,5-diyl]-*p*-phenylene]	**68:** 96231d
Poly[1,3,4-oxadiazole-2,5-diyl-1,8-oxtanediyl(4-amino-4H-1,2,4-triazole-3,5-diyl)-1,8-octanediyl]	**82:** P18517b
Poly[(5-oxo-4-phenyl-Δ^2-1,2,4-triazoline-1,3-diyl)-oxy(1-ethoxyethylene)]	**74:** 112476u
Poly[[1-*p*-phenoxyphenyl)-2,5-benzimidazolediyl][4-(*p*-phenoxyphenyl)-4H-1,2,4-triazole-3,5-diyl][1-(*p*-phenoxyphenyl)-5,2-benzimidazolediyl]-*m*-phenylene]	**72:** 44284m

TABLE 21 (*Continued*)

Compound	Reference
21.3. Poly(1,2,4-triazoles, triazolines, and triazolidines) (*Continued*)	
Poly[[4-(*p*-phenoxyphenyl)-4H̲-1,2,4-triazole-3,5-diyl]-[4-(*p*-phenoxyanilino)-*m*-phenylene]iminoisophthaloylimino[6-(*p*-phenoxyanilino)-*m*-phenylene]]	**72:** 44284m
Poly[(1-phenyl-2,5-benzimidazolediyl)(4-phenyl-4H̲-1,2,4-triazole-3,5-diyl)(1-phenyl-5,2-benzimidazolediyl)-*m*-phenylene]	**72:** 44284m
Poly[(1-phenyl-2,5-benzimidazolediyl)(4-phenyl-4H̲-1,2,4-triazole-3,5-diyl)(1-phenyl-5,2-benzimidazolediyl)-*p*-phenylene]	**72:** 44284m
Poly[(1-phenyl-2,5-benzimidazolediyl)(4-phenyl-4H̲-1,2,4-triazole-3,5-diyl)(1-phenyl-5,2-benzimidazolediyl)-*p*-phenyleneoxy-*p*-phenylenesulfonyl-*p*-phenyleneoxy-*p*-phenylene]	**72:** 44284m
Poly[(4-phenyl-4H-1,2,4-triazole-3,5-diyl)(4-anilino-*m*-phenylene)iminocarbonyl-*p*-phenyleneoxy-*p*-phenylenesulfonyl-*p*-phenyleneoxy-*p*-phenylenecarbonylimino(6-anilino-*m*-phenylene)]	**72:** 44284m
Poly[(4-phenyl-4H̲-1,2,4-triazole-3,5-diyl)(4-anilino-*m*-phenylene)iminoisophthaloylimino(6-anilino-*m*-phenylene)]	**72:** 44284m
Poly[(4-phenyl-4H̲-1,2,4-triazole-3,5-diyl)(4-anilino-*m*-phenylene)iminoterephthaloylimino(6-anilino-*m*-phenylene)]	**72:** 44284m
Poly[(1-phenyl-1H̲-1,2,4-triazole-3,5-diyl)(hexafluorotrimethylene)(1-phenyl-1H̲-1,2,4-triazole-5,3-diyl)-*p*-phenylene]	**68:** 50462b, **69:** 67793d, **71:** 125032j, **73:** 120949v
Poly[(4-phenyl-4H̲-1,2,4-triazole-3,5-diyl)oxy[1,1′-biphenyl]-4,4′-diyloxy]	**76:** 113771q
Poly[(4-phenyl-4H̲-1,2,4-triazole-3,5-diyl)-*m*-phenylene]	**68:** 13458j, **75:** 152121b
Poly[(4-phenyl-4H̲-1,2,4-triazole-3,5-diyl)-*p*-phenylene]	**68:** 13458j, **68:** 96231d, **75:** 152121b
Poly[(4-phenyl-4H̲-1,2,4-triazole-3,5-diyl)-*m*-phenyleneiminocarbonyl-2,6-naphthylenecarbonylimino-*m*-phenylene]	**70:** 20356u
Poly[(4-phenyl-4H̲-1,2,4-triazole-3,5-diyl)-*p*-phenyleneoxy-*p*-phenylene]	**68:** 87784j, **72:** 22220k
Poly[(4-phenyl-4H̲-1,2,4-triazole-3,5-diyl)-1,4-phenyleneoxy-1,4-phenyleneiminocarbonyl-1,4-phenylenecarbonylimino-1,4-phenylene]	**81:** 14549v
Poly[(4-phenyl-4H̲-1,2,4-triazole-3,5-diyl)-*m*-phenyleneoxy-*p*-phenyleneiminoisophthaloylimino-*p*-phenyleneoxy-*m*-phenylene]	**72:** 68047n
Poly[(4-phenyl-4H̲-1,2,4-triazole-3,5-diyl)-*p*-phenyleneoxy-*p*-phenyleneiminoisophthaloylimino-*p*-phenyleneoxy-*p*-phenylene]	**72:** 68047n
Poly[(4-phenyl-4H̲-1,2,4-triazole-3,5-diyl)-*p*-phenyleneoxy-*p*-phenyleneiminoterephthaloylimino-*p*-phenyleneoxy-*p*-phenylene]	**72:** 68047n

TABLE 21 (*Continued*)

Compound	Reference
21.3. Poly(1,2,4-triazoles, triazolines, and triazolidines) (*Continued*)	
Poly[(1-phenyl-1H̲-1,2,4-triazole-5,3-diyl)-*m*-phenylene(1-phenyl-1H̲-1,2,4-triazole-3,5-diyl)-4,4'-biphenylylene]	**71:** 125032j, **73:** 120949v
Poly[(1-phenyl-1H̲-1,2,4-triazole-5,3-diyl)-*p*-phenylene-(1-phenyl-1H̲-1,2,4-triazole-3,5-diyl)-4,4'-biphenylylene]	**71:** 125032j, **73:** 120949v
Poly[(1-phenyl-1H̲-1,2,4-triazole-5,3-diyl)-*m*-phenylene-(1-phenyl-1H̲-1,2,4-triazole-3,5-diyl)-(hexafluorotrimethylene)]	**73:** 120949v
Poly[(1-phenyl-1H̲-1,2,4-triazole-5,3-diyl)-*m*-phenylene-(1-phenyl-1H̲-1,2,4-triazole-3,5-diyl)-*p*-phenylene]	**71:** 125032j, **73:** 120949v
Poly[(1-phenyl-1H̲-1,2,4-triazole-3,5-diyl)-*p*-phenylene-(1-phenyl-1H̲-1,2,4-triazole-3,5-diyl)-*p*-phenylene]	**71:** 125032j, **73:** 120949v
Poly[(4-phenyl-4H̲-1,2,4-triazole-3,5-diyl)-*m*-phenylene(4-phenyl-4H̲-1,2,4-triazole-3,5-diyl)-*p*-phenylene]	**70:** 20885r
Poly[(1-phenyl-1H̲-1,2,4-triazole-5,3-diyl)-*m*-phenylene-(1-phenyl-1H̲-1,2,4-triazole-3,5-diyl)(2,3,5,6-tetrafluoro-*p*-phenylene)]	**71:** 125032j, **73:** 120949v
Poly(2,6-pyridinediylcarbonylimino-1,2-phenylene-1H̲-1,2,4-triazole-3,5-diyl-1,3-phenylene-1H̲-1,2,4-triazole-3,5-diyl-1,2-phenyleneiminocarbonyl)	**80:** 15236g, **81:** 169929t
Poly(2,6-pyridinediylcarbonylimino-1,2-phenylene-1H̲-1,2,4-triazole-3,5-diyl-1,4-phenylene-1H̲-1,2,4-triazole-3,5-diyl-1,2-phenyleneiminocarbonyl)	**80:** 15236g
Poly(2,6-pyridinediyl-1H̲-1,2,4-triazole-3,5-diyl-1,2-phenyleneiminocarbonyl[1,1'-biphenyl]-4,4'-diylcarbonylimino-1,2-phenylene-1H̲-1,2,4-triazole-3,5-diyl)	**80:** 15236g
Poly(2,6-pyridinediyl-1H̲-1,2,4-triazole-3,5-diyl-1,2-phenyleneiminocarbonyl-2,6-naphthalenediylcarbonylimino-1,2-phenylene-1H̲-1,2,4-triazole-3,5-diyl)	**80:** 15236g
Poly(2,6-pyridinediyl-1H̲-1,2,4-triazole-3,5-diyl-1,2-phenyleneiminocarbonyl-1,3-phenylenecarbonylimino-1,2-phenylene-1H̲-1,2,4-triazole-3,5-diyl)	**80:** 15236g
Poly(2,6-pyridinediyl-1H̲-1,2,4-triazole-3,5-diyl-1,2-phenyleneiminocarbonyl-1,4-phenylenecarbonylimino-1,2-phenylene-1H̲-1,2,4-triazole-3,5-diyl)	**80:** 15236g, **81:** 169929t
Poly(2,6-pyridinediyl-1H̲-1,2,4-triazole-3,5-diyl-1,2-phenyleneiminocarbonyl-1,4-phenylenecarbonyl-1,4-phenylenecarbonylimino-1,2-phenylene-1H̲-1,2,4-triazole-3,5-diyl)	**80:** 15236g
Poly(2,6-pyridinediyl-1H̲-1,2,4-triazole-3,5-diyl-1,2-phenyleneiminocarbonyl-1,4-phenyleneoxy-1,4-phenylenecarbonylimino-1,2-phenylene-1H̲-1,2,4-triazole-3,5-diyl)	**80:** 15236g
Poly(2,6-pyridinediyl-1H̲-1,2,4-triazole-3,5-diyl-1,2-phenyleneiminocarbonyl-1,4-phenylenesulfonyl-1,4-phenylenecarbonylimino-1,2-phenylene-1H̲-1,2,4-triazole-3,5-diyl)	**80:** 15236g

TABLE 21 (*Continued*)

Compound	Reference
21.3. Poly(1,2,4-triazoles, triazolines, and triazolidines) (*Continued*)	
Poly(2,6-pyridinediyl-1H̲-1,2,4-triazole-3,5-diyl-1,2-phenyleneiminocarbonyl-2,6-pyridinediylcarbonyl-imino-1,2-phenylene-1H̲-1,2,4-triazole-3,5-diyl)	**80:** 15236g
Poly(2,6-pyridinediyl-*s*-triazole-3,5-diyl-*m*-phenylene-*s*-triazole-3,5-diyl)	**72:** 90906p
Poly(2,6-pyridinediyl-Δ^2-1,2,4-triazoline-3,5-diyl-*p*-phenyleneoxy-*p*-phenylene-Δ^2-1,2,4-triazoline-5,3-diyl)	**72:** 111882v
Poly(2,6-pyridinediyl-Δ^2-1,2,4-triazoline-3,5-diyl-*m*-phenylene-Δ^2-1,2,4-triazoline-5,3-diyl)	**72:** 111882v
Poly(2,6-pyridinediyl)Δ^2-1,2,4-triazoline-3,5-diyl-*p*-phenylene-Δ^2-1,2,4-triazoline-5,3-diyl)	**72:** 111882v
Poly(1H̲-1,2,4-triazole-3,5-diyl)	**80:** 60356g
Poly[*s*-triazole-3,5-diyl(diazoamino)]	**70:** P96802g
Poly[1H̲-1,2,4-triazole-3,5-diyl-1,2-ethanediyl-(ethylimino)-1,2-ethanediyl]	**80:** P111542z
Poly(1H̲-1,2,4-triazole-3,5-diyl-1,3-phenylene)	**81:** P170229w
Poly(*s*-triazole-3,5-diyl-*p*-phenylene)	**69:** 86973d
Poly(1H̲-1,2,4-triazole-3,5-diyl-1,3-phenylene-1H̲-1,2,4-triazole-3,5-diyl-1,2-phenyleneimino-carbonyl[1,1′-biphenyl]-4,4′-diylcarbonylimino-1,2-phenylene)	**80:** 15236g, **81:** 169929t, **83:** 131973f
Poly(1H̲-1,2,4-triazole-3,5-diyl-1,4-phenylene-1H̲-1,2,4-triazole-3,5-diyl-1,2-phenyleneimino-carbonyl[1,1′-biphenyl]-4,4′-diylcarbonylimino-1,2-phenylene)	**80:** 15236g, **83:** 131973f
Poly(1H̲-1,2,4-triazole-3,5-diyl-1,3-phenylene-1H̲-1,2,4-triazole-3,5-diyl-1,2-phenyleneimino-carbonyl-2,6-naphthalenediylcarbonylimino-1,2-phenylene)	**80:** 15236g, **81:** 169929t, **83:** 131973f
Poly(1H̲-1,2,4-triazole-3,5-diyl-1,4-phenylene-1H̲-1,2,4-triazole-3,5-diyl-1,2-phenyleneimino-carbonyl-2,6-naphthalenediylcarbonylimino-1,2-phenylene	**80:** 15236g, **83:** 131973f
Poly(1H̲-1,2,4-triazole-3,5-diyl-1,3-phenylene-1H̲-1,2,4-triazole-3,5-diyl-1,2-phenyleneimino-carbonyl-1,3-phenylenecarbonylimino-1,2-phenylene)	**75:** 36790v, **80:** 15236g, **81:** 169929t, **83:** 131973f
Poly(1H̲-1,2,4-triazole-3,5-diyl-1,3-phenylene-1H̲-1,2,4-triazole-3,5-diyl-1,2-phenyleneimino-carbonyl-1,4-phenylenecarbonylimino-1,2-phenylene)	**75:** 36790v, **79:** 126839e, **80:** 15236g, **81:** 169929t, **83:** 131973f
Poly(1H̲-1,2,4-triazole-3,5-diyl-1,4-phenylene-1H̲-1,2,4-triazole-3,5-diyl-1,2-phenyleneimino-carbonyl-1,3-phenylenecarbonylimino-1,2-phenylene)	**80:** 15236g, **83:** 131973f
Poly(1H̲-1,2,4-triazole-3,5-diyl-1,4-phenylene-1H̲-1,2,4-triazole-3,5-diyl-1,2-phenyleneimino-carbonyl-1,4-phenylenecarbonylimino-1,2-phenylene)	**80:** 15236g, **81:** 169929t, **83:** 131973f
Poly(1H̲-1,2,4-triazole-3,5-diyl-1,4-phenylene-1H̲-1,2,4-triazole-3,5-diyl-1,2-phenyleneimino-carbonyl-1,4-phenylenecarbonyl-1,4-phenylenecarbonyl-imino-1,2-phenylene)	**80:** 15236g, **83:** 131973f

TABLE 21   (*Continued*)

Compound	Reference
21.3. Poly(1,2,4-triazoles, triazolines, and triazolidines) (*Continued*)	
Poly(1H-1,2,4-triazole-3,5-diyl-1,3-phenylene-1H-1,2,4-triazole-3,5-diyl-1,2-phenyleneimino-carbonyl-1,4-phenyleneoxy-1,4-phenylenecarbonyl-imino-1,2-phenylene)	**75:** 36790v, **80:** 15236g, **81:** 169929t, **83:** 131973f
Poly(1H-1,2,4-triazole-3,5-diyl-1,4-phenylene-1H-1,2,4-triazole-3,5-diyl-1,2-phenyleneimino-carbonyl-1,4-phenyleneoxy-1,4-phenylenecarbonyl-imino-1,2-phenylene)	**80:** 15236g, **83:** 131973f
Poly(1H-1,2,4-triazole-3,5-diyl-1,3-phenylene-1H-1,2,4-triazole-3,5-diyl-1,2-phenyleneimino-carbonyl-1,4-phenylenesulfonyl-1,4-phenylenecarbonyl-imino-1,2-phenylene)	**80:** 15236g, **81:** 169929t, **83:** 131973f
Poly(1H-1,2,4-triazole-3,5-diyl-1,4-phenylene-1H-1,2,4-triazole-3,5-diyl-1,2-phenyleneimino-carbonyl-1,4-phenylenesulfonyl-1,4-phenylenecarbonyl-imino-1,2-phenylene)	**80:** 15236g, **83:** 131973f
Poly(1H-1,2,4-triazole-3,5-diyl-1,3-phenylene-1H-1,2,4-triazole-3,5-diyl-1,2-phenyleneimino-carbonyl-2,6-pyridinediylcarbonylimino-1,2-phenylene)	**83:** 131973f
Poly(1H-1,2,4-triazole-3,5-diyl-1,4-phenylene-1H-1,2,4-triazole-3,5-diyl-1,2-phenyleneimino-carbonyl-2,6-pyridinediylcarbonylimino-1,2-phenylene)	**83:** 131973f
Poly(1H-1,2,4-triazole-3,5-diyl-2,6-pyridinediyl-1H-1,2,4-triazole-3,5-diyl-1,2-phenyleneimino-carbonyl[1,1'-biphenyl]-4,4'-diylcarbonylimino-1,2-phenylene)	**83:** 131973f
Poly(1H-1,2,4-triazole-3,5-diyl-2,6-pyridinediyl-1H-1,2,4-triazole-3,5-diyl-1,2-phenyleneimino-carbonyl-2,6-naphthalenediylcarbonylimino-1,2-phenylene)	**83:** 131973f
Poly(1H-1,2,4-triazole-3,5-diyl-2,6-pyridinediyl-1H-1,2,4-triazole-3,5-diyl-1,2-phenyleneimino-carbonyl-1,3-phenylenecarbonylimino-1,2-phenylene)	**83:** 131973f
Poly(1H-1,2,4-triazole-3,5-diyl-2,6-pyridinediyl-1H-1,2,4-triazole-3,5-diyl-1,2-phenyleneimino-carbonyl-1,4-phenylenecarbonylimino-1,2-phenylene)	**83:** 131973f
Poly(1H-1,2,4-triazole-3,5-diyl-2,6-pyridinediyl-1H-1,2,4-triäzole-3,5-diyl-1,2-phenyleneimino-carbonyl-1,4-phenylenecarbonyl-1,4-phenylenecarbonyl-imino-1,2-phenylene)	**83:** 131973f
Poly(1H-1,2,4-triazole-3,5-diyl-2,6-pyridinediyl-1H-1,2,4-triazole-3,5-diyl-1,2-phenyleneimino-carbonyl-1,4-phenyleneoxy-1,4-phenylenecarbonylimino-1,2-phenylene	**83:** 131973f
Poly(1H-1,2,4-triazole-3,5-diyl-2,6-pyridinediyl-1H-1,2,4-triazole-3,5-diyl-1,2-phenyleneimino-carbonyl-1,4-phenylenesulfonyl-1,4-phenylenecarbonyl-imino-1,2-phenylene)	**83:** 131973f

TABLE 21 (*Continued*)

Compound	Reference
21.3. Poly(1,2,4-triazoles, triazolines, and triazolidines) (*Continued*)	
Poly(1H-1,2,4-triazole-3,5-diyl-2,6-pyridinediyl-1H-1,2,4-triazole-3,5-diyl-1,2-phenyleneimino-carbonyl-2,6-pyridinediylcarbonylimino-1,2-phenylene)	**83:** 131973f
1H-1,2,4-triazol-3-amine, 1-ethenyl-, homopolymer	**80:** 82823j
1H-1,2,4-triazol-3-amine, polymer with formaldehyde	**80:** 16240r
1H-1,2,4-triazole-3,5-diamine, polymer with formaldehyde	**80:** 16240r
3H-1,2,4-triazole-3,5(4H)-dione, 4-phenyl-, homopolymer	**80:** 60251u
1H-1,2,4-Triazole, 1-ethenyl-3,5-diethyl-, homopolymer	**81:** P120637g
1H-1,2,4-Triazole, 1-ethenyl-3,5-diheptyl-, homopolymer	**81:** P120637g
1H-1,2,4-Triazole, 1-ethenyl-3,5-dimethyl-, homopolymer	**81:** P120637g
1H-1,2,4-Triazole, 1-ethenyl-3,5-diphenyl-, homopolymer	**81:** P120637g
1H-1,2,4-Triazole, 1-ethenyl-3,5-diundecyl-, homopolymer	**81:** P120637g

CHAPTER 22

1,2,4-Triazolium, Triazolinium, Triazolidinium, and Mesoionic Compounds

22.1. Alkyl- or Aryl-1,2,4-Triazolium Compounds

The triazolium salts (e.g., **22.1-1**) in this and the following sections are represented as a cationic center in which the positive charge is delocalized over the atoms of the ring.[1] Mesoionic (zwitterionic) compounds (e.g., **22.1-2**) in which the ionic centers are internally compensated are named by the imaginary addition of a molecule of water. In using this nomenclature a proton is added to the anionic center to give a positively charged triazole ring (e.g., **22.1-3**) to which the hydroxide anion serves as the counter ion.

Large molecules containing triazolium salts as substituents have been prepared and used both as dyes[2,3] and fluorescent whitening agents[4–9] for synthetic fibers. In general, alkylation of either 1- or 4-substituted 1,2,4-triazoles gave 1,4-disubstituted triazolium salts as illustrated in the following.

Quaternization of 4-substituted 3,5-dimethyl-1,2,4-triazoles (e.g., **22.1-4** and **22.1-5**) with methyl iodide in refluxing benzene occurred readily to give the corresponding 1-methyl-1,2,4-triazolium iodides (**22.1-6** and **22.1-7**, respectively).[10,11] In the quaternization of 3,5-dimethyl-1-phenyl-1,2,4-triazole (**22.1-8**), the product was shown to be the 4-methyl-1,2,4-triazolium iodide (**22.1-9**).[10] These compounds contain an activated 5-methyl group,

$$
\underset{\textbf{22.1-4, R = Me}}{\underset{\textbf{22.1-5, R = 4-ClC}_6\textbf{H}_4}{\underset{\text{N—N}}{\text{Me}\!\!\overset{\overset{\displaystyle \text{R}_1}{\text{N}}}{\diagdown}\!\!\text{Me}}}} \quad \xrightarrow{\text{MeI}} \quad \underset{\textbf{22.1-6, R = Me}}{\underset{\textbf{22.1-7, R = 4-ClC}_6\textbf{H}_4}{\underset{\underset{\text{Me}}{\text{N—N}}}{\text{Me}\!\!\overset{\overset{\displaystyle \text{R}}{\text{N}}}{\diagup\!\!\oplus\!\!\diagdown}\!\!\text{Me} \ \ \text{I}^-}}}
$$

$$
\underset{\textbf{22.1-8}}{\underset{\underset{\text{Ph}}{\text{N—N}}}{\text{Me}\!\!\overset{\overset{\displaystyle \text{N}}{}}{\diagdown}\!\!\text{Me}}} \quad \xrightarrow{\text{MeI}} \quad \underset{\textbf{22.1-9}}{\underset{\underset{\text{Ph}}{\text{N—N}}}{\text{Me}\!\!\overset{\overset{\displaystyle \text{Me}}{\text{N}}}{\diagup\!\!\oplus\!\!\diagdown}\!\!\text{Me} \ \ \text{I}^-}}
$$

which condense with derivatives of acetaldehyde substituted with an quaternized heterocyclic base in the presence of pyridine and triethylamine to give cyanine dyes. This approach has also been used for the synthesis of bicyclic systems.[12–14] For example, treatment of **22.1-10** with phenacyl bromide gave **22.1-11**, which underwent intramolecular cyclization in the presence of bicarbonate to give **22.1-12**.[15] In the quaternization of 3-methyl-4-phenyl-

$$
\underset{\textbf{22.1-10}}{\underset{\text{N—N}}{\text{Me}\!\!\overset{\overset{\displaystyle \text{Ph}}{\text{N}}}{\diagdown}\!\!\text{Me}}} \xrightarrow{\text{PhCOCH}_2\text{Br}} \underset{\textbf{22.1-11}}{\underset{\underset{\text{CH}_2\text{COPh}}{\text{N—N}}}{\text{Me}\!\!\overset{\overset{\displaystyle \text{Ph}}{\text{N}}}{\diagup\!\!\oplus\!\!\diagdown}\!\!\text{Me} \ \ \text{Br}^-}} \xrightarrow{\text{HCO}_3^-} \underset{\textbf{22.1-12}}{\underset{\text{N—N}}{\text{Me}\!\!\overset{\overset{\displaystyle \text{Ph}}{\text{N}}}{\diagdown}\!\!\diagup\!\!\diagdown\!\!\text{Ph}}}
$$

1,2,4-triazole (**22.1-13**) with either methyl iodide or phenacyl bromide, substitution occurred at both N-1 and N-2 to give mixtures of triazolium halides (e.g., **22.1-14** and **22.1-15**).[16] The formation of diquaternary salts

$$
\underset{\textbf{22.1-13}}{\underset{\text{N—N}}{\text{Me}\!\!\overset{\overset{\displaystyle \text{Ph}}{\text{N}}}{\diagdown}}} \xrightarrow{\text{PhCOCH}_2\text{Br}} \underset{\textbf{22.1.14}}{\underset{\underset{\text{CH}_2\text{COPh}}{\text{N—N}}}{\text{Me}\!\!\overset{\overset{\displaystyle \text{Ph}}{\text{N}}}{\diagup\!\!\oplus\!\!\diagdown}}} \quad + \quad \underset{\textbf{22.1-15}}{\underset{\underset{\text{PhCOCH}_2}{\text{N—N}}}{\text{Me}\!\!\overset{\overset{\displaystyle \text{Ph}}{\text{N}}}{\diagup\!\!\oplus\!\!\diagdown}}}
$$

has also been demonstrated. Treatment of 1-methyl-1,2,4-triazole (**22.1-16**) with triethyloxonium tetrafluoroborate gave **22.1-17**, which was fused with trimethyloxonium tetrafluoroborate to give **22.1-18**.[17] The structure of **22.1-18** was established by hydrolytic degradation in sulfuric acid to yield ethylamine and 1,2-dimethylhydrazine.

22.1-16   **22.1-17**   **22.1-18**

Several other approaches for the preparation of alkyl- and aryltriazolium salts have been developed. Cyclization of the acylamidrazone (**22.1-20**) in acetic acid gave the triazolium compound (**22.1-21**).[18] This procedure was also used to prepare related compounds, which, in addition, could be obtained directly from a 1,3,4-oxadiazolium salt and a primary amine in refluxing acetic acid.[19] In the reaction of **22.1-19** with aniline, it was shown that the acylamidrazone (**22.1-20**) was an intermediate indicating that C-2 of the oxadiazolium ring was the initial site of attack.[20] The oxadiazolium

22.1-19   **22.1-20**   **22.1-21**

salt (**22.1-22**) reacted with methylamine in ethanol to give a 77% yield of **22.1-23**.[21] This compound was deprotonated with triethylamine to give the triazoline (**22.1-24**), which in turn was protonated with perchloric acid to again give **22.1-23**. Several 3- and 5-(4-nitrobenzyl)triazolium salts (e.g.,

22.1-22   **22.1-23**   **22.1-24**

22.1-25 and **22.1-27**) were converted either to mesoionic compounds (e.g., **22.1-26**) or triazolines (e.g., **22.1-28**) with ethoxide in *N,N*-dimethylformamide.[19] Protonation of **22.1-26** and **22.1-28** gave the corresponding triazolium salts readily. A number of analogs of **22.1-26** and

22.1-25   **22.1-26**

22.1-28 were generated and found to undergo decomposition when purification was attempted.

22.1-27   **22.1-28**

Addition of the alkoxydiazenium tetrafluoroborate (**22.1-29**) to the Schiff base (**22.1-30**) gave the triazolium tetrafluoroborate (**22.1-21**), presumably formed via the elimination of ethanol and hydride from intermediate

22.1-29   **22.1-31**   **22.1-21**

22.1-31.[22–24] The facile oxidation of triazolines was demonstrated by treatment of **22.1-32** with dichromate in aqueous acetic acid to give a 80% yield of **22.1-33**.[25] A variety of triazolium salts were prepared by this procedure.

22.1-32   **22.1-33**

The sodium borohydride reduction of **22.1-34** and **22.1-35** was reported to give the unstable triazolines, **22.1-36** and **22.1-37**.[26] In addition, the

22.1-34, $R_1 = Ph$, $R_2 = Me$ **22.1-36**, $R_1 = Ph$, $R_2 = Me$
22.1-35, $R_1 = Me$, $R_2 = Ph$ **22.1-37**, $R_1 = Me$, $R_2 = Ph$

dethiation of **22.1-38** was effected with nitric acid to give a 100% yield of **22.1-39**.[27] A method that was directed toward the preparation of tetrazoles was found to yield triazolium salts as the major product: treatment of the nitrilium salt (**22.1-40**) with azide in toluene gave a 34% yield of the tetrazole (**22.1-42**) and a 65% yield of the triazolium tetrafluoroborate

22.1-38   **22.1-39**

(**22.1-44**).[28] Apparently, **22.1-44** is formed via intermediates **22.1-41** and **22.1-43**. On pyrolysis, the 4-ethyl rather than the 1-ethyl group was eliminated to give **22.1-45**.

22.1-40   **22.1-41**

22.1-42

22.1-43   **22.1-44**   **22.1-45**

References

1.	**51:** 7999c	2.	**78:** 148956x	3.	**82:** 126600g	4.	**71:** 14180m
5.	**72:** 22599r	6.	**72:** 91491m	7.	**73:** 100073b	8.	**74:** 14197j
9.	**77:** 90077x	10.	**54:** 7695e	11.	**72:** 43550h	12.	**78:** 58327p
13.	**80:** 133356u	14.	**81:** 13441s	15.	**79:** 115499t	16.	**82:** 170807d
17.	**77:** 75149d	18.	**69:** 77178w	19.	**75:** 98508t	20.	**74:** 111972r
21.	**72:** 121463y	22.	**67:** 90732a	23.	**71:** 91385j	24.	**71:** 91389p
25.	**70:** 96723g	26.	**78:** 159527j	27.	**74:** 53659f	28.	**76:** 85759w

22.2. Amino-1,2,4-Triazolium Compounds

Nitron, 3-anilino-1,4-diphenyltriazolium nitrate (**22.2-1**), has been known for years, and recently several groups have investigated the synthesis and preparation of analogs of this compound. The molecular structure of the conjugate acid of nitron was shown to be the 3-anilino tautomer, as represented by the cation portion of **22.2-1**.[1] This compound and its analogs have been used in analytical procedures for the spectrophotometric and gravimetric determination of a number of elements.[2–6]

Reaction of N-amino-N,N′-diphenylbenzamidine (**22.2-2**) with cyanogen bromide in refluxing methanol readily formed 3-amino-1,4,5-triphenyl-1,2,4-triazolium bromide (**22.2-3**), which underwent decomposition when

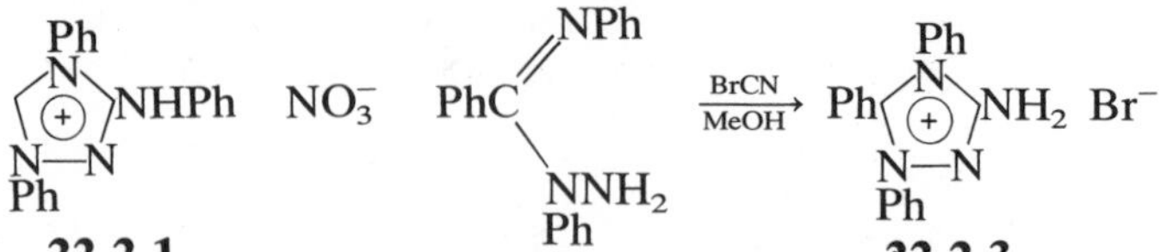

Ph
N
(+) NHPh   NO₃⁻
N—N
Ph
22.2-1

NPh
‖
PhC   BrCN/MeOH→
|
NNH₂
Ph
22.2-2

Ph
N
Ph (+) NH₂ Br⁻
N—N
Ph
22.2-3

the preparation of the corresponding mesoionic base was attempted.[7] In contrast, cyclization of the *N*-anilinoguanidine (**22.2-4**) with formic acid gave directly, after recrystallization from ethanol, the mesoionic compound, **22.2-5**, best known as nitron because of the insolubility of the corresponding 1,2,4-triazolium nitrate in water.[7,8] However, **22.2-5** was similar to **22.2-3** in that it was easily decomposed when treated with alcoholic potassium hydroxide.

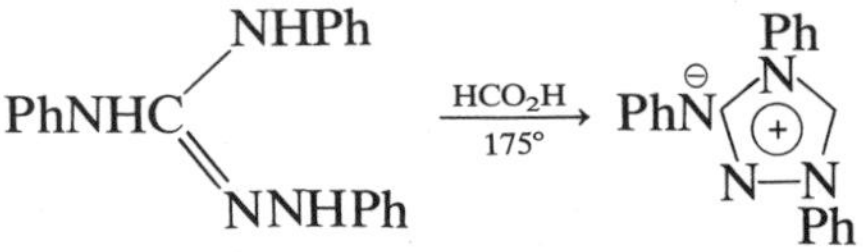

NHPh
/
PhNHC   HCO₂H/175°→
‖
NNHPh
22.2-4

Ph
N
PhN⁻ (+)
N—N
Ph
22.2-5

To circumvent the formation of a mixture when the amino groups of the amidine portion of an aminoguanidine contained different substituents, Ollis and coworkers developed an unambiguous route for the synthesis of aminotriazolium compounds. Treatment of **22.2-6** with the isocyanide dichloride (**22.2-7**), and **22.2-9** with **22.2-10** in refluxing toluene gave the isomeric mesoionic compounds, **22.2-8** (57%) and **22.2-11** (53%), respectively.[9,10] The isomers are readily distinguished by their mass spectra, which show a fragment ion containing the 4-substituent. Furthermore, **22.2-11** was quantitatively converted by ring opening and reclosure to **22.2-8** in refluxing ethanol, whereas **22.2-8** was recovered unchanged under the same conditions. In other compounds the position of equilibrium of isomeric pairs was dependent upon the nature of the substituent in the aryl ring. Methylation of

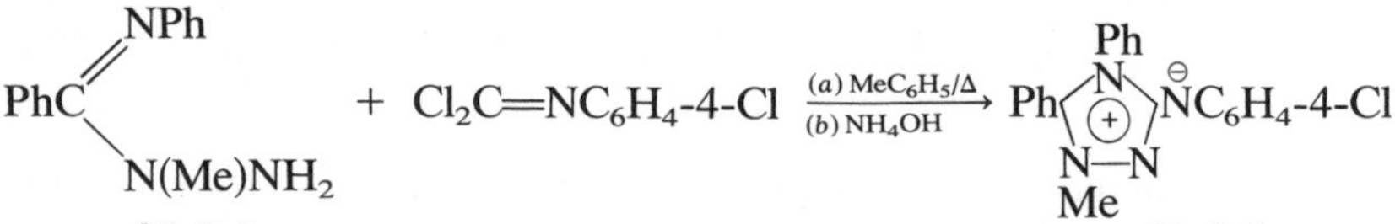

NPh
/
PhC
\
N(Me)NH₂
22.2-6

+ Cl₂C=NC₆H₄-4-Cl   (a) MeC₆H₅/Δ / (b) NH₄OH→

Ph
N
Ph (+) NC₆H₄-4-Cl
N—N
Me
22.2-8

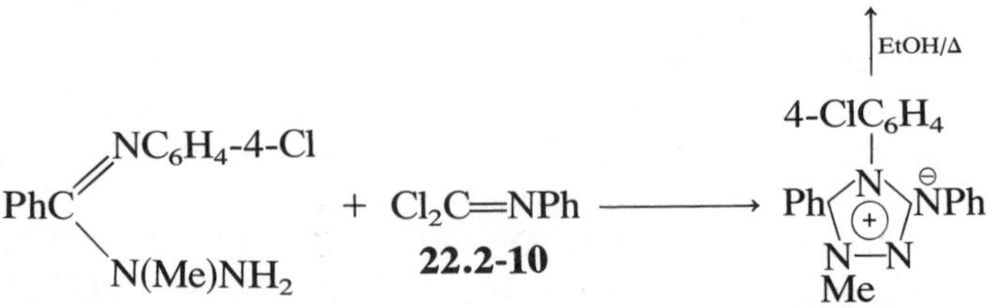

↑ EtOH/Δ

NC₆H₄-4-Cl
/
PhC
\
N(Me)NH₂
22.2-9

+ Cl₂C=NPh
22.2-10

→

4-ClC₆H₄
N
Ph (+) NPh
N—N
Me
22.2-11

these mesoionic compounds occurred on the exocyclic nitrogen, as demonstrated by the conversion of **22.2-12** with methyl iodide in refluxing benzene to **22.2-13**. In addition, **22.2-12** was reduced to the corresponding 3-

22.2-12   **22.2-13**

anilinotriazoline with lithium aluminum hydride in dioxane. A potentially useful route to 3,5-diaminotriazolium compounds (e.g., **22.2-15**) was provided by the reaction of the iminium salt (**22.2-14**) with 2-aminophenol.[11]

22.2-14   **22.2-15**

Although a variety of procedures have been investigated for the preparation of 4-aminotriazolium compounds, practically all the procedures are limited to some degree by the nature of the reactants. The condensation of 1,3,4-thiadiazolium salts (e.g., **22.2-16**) with phenylhydrazine gave a series of 4-anilino compounds (e.g., **22.2-18**) presumably formed via intermediate **22.2-17**.[12] With hydrazine and its alkyl derivatives, this reaction gave only

22.2-16   **22.2-17**   **22.2-18**

tetrazines. In contrast to **22.2-17**, the cyclization of the acyl N,N'-diaminobenzamidine (**22.2-19**) to **22.2-21** in acetic acid was unsuccessful.[13] However, when the phenyl group was substituted with a nitro group as in **22.2-20**, cyclization gave a 80% yield of **22.2-22**. Deprotonation of **22.2-22** with base gave the mesoionic compound, **22.2-23**, which was shown to undergo protonation, alkylation, and acylation on the exocyclic nitrogen to give the corresponding triazolium salts.[14] In the reaction of phenylhydrazine hydrochloride with ethyl orthoformate, a mixture of the tetrazine (**22.2-24**) and the 4-anilinotriazolium chloride (**22.2-25**) were obtained, the latter in 61% yield.[15] In their investigations of alkoxydiazenium salts, Mathur and

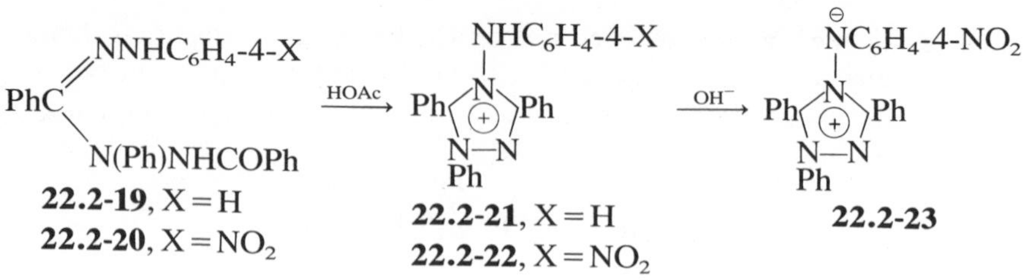

22.2-19, X = H
22.2-20, X = NO₂

22.2-21, X = H
22.2-22, X = NO₂

22.2-23

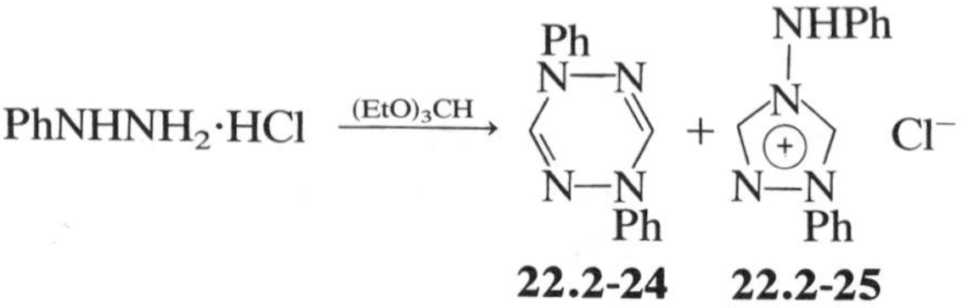

PhNHNH₂·HCl →(EtO)₃CH→

22.2-24   22.2-25

Suschitzky developed a straightforward method for the preparation of 4-(benzylideneamino)triazolium salts.[16] For example, reaction of **22.2-27** with the azine (**22.2-26**) generated the dipolar species (**22.2-28**), which underwent cycloaddition with the azine to give **22.2-30** and further transformation with the elimination of ethanol and hydride to give **22.2-29**.

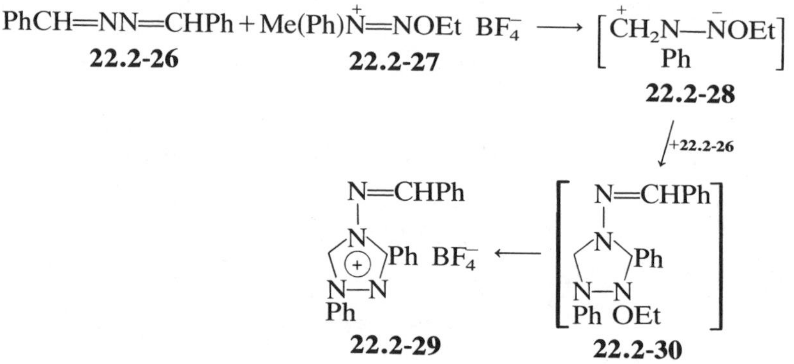

PhCH=NN=CHPh + Me(Ph)N⁺=NOEt BF₄⁻ ⟶ [C⁺H₂N—N⁻OEt]
                                              Ph

22.2-26   **22.2-27**   **22.2-28**

22.2-29   22.2-30

The preformed 4-aminotriazoles are versatile starting materials for the preparation of 4-aminotriazolium compounds. Although treatment of 4-aminotriazole (**22.2-31**) with nitric acid gave 1,2,4-triazole, reaction of **22.2-31** with ethyl nitrate in refluxing ethanolic sodium ethoxide gave the sodium salt (**22.2-32**).[17] Alkylation of **22.2-32** with benzyl chloride occurred at N-1 to give the mesoionic derivative (**22.2-36**). The latter was also prepared by nitration of **22.2-33** in a mixture of acetic acid and acetic anhydride.[18] Similarly, alkylation of 4-aminotriazoles and its 3,5-disubstituted derivatives with alkyl halides or tosylates produced the corresponding 4-aminotriazolium salts, as shown by the conversion of **22.2-31** to **22.2-34**.[19,20] The readiness in which 4-aminotriazoles alkylate was indicated by the treatment of the monocation (**22.2-38**) with methyl fluorosulfonate to give the dication (**22.2-39**).[21] Compound **22.2-34** was deaminated with nitrous acid to give 1-benzyltriazole and condensed with benzaldehyde to give the 4-(benzylideneamino)triazolium salt (**22.2-35**). Alkylation of 4-(benzylideneamino)-1,2,4-triazole also gave **22.2-35**.[22] The conversion of **22.2-34** with sulfur in a refluxing mixture of triethylamine and pyridine to the triazoline-5-thione (**22.2-37**)[23] in 92% yield is a reaction of preparative value.

Becker and coworkers have demonstrated that the 1-alkyl-4-(substituted amino)triazolium salts undergo several interesting rearrangement reactions. For example, treatment of **22.2-34** with dilute base gave a 68% yield of **22.2-42**.[20] Apparently, this product resulted from cleavage of the ring of **22.2-34** to give **22.2-40**, migration of the formyl group to give **22.2-41**, and recyclization. Crossing experiments indicated that the migration of the acyl moiety occurred intermolecularly. This route was supported by the conversion of the 4-acetylamino compound (**22.2-43**) to the 4-(benzylamino)triazole (**22.2-46**), in which the ring-cleavage product (**22.2-45**) corresponded to the rearranged intermediate (**22.2-41**) in the reaction described previously.[24] In addition, treatment of the 4-

(sulfonylamino)triazolium compound (**22.2-47**) with base gave an open-ring product (**22.2-48**) that underwent degradation rather than recyclization.[25] In

other work the 4-ureido derivative (**22.2-49**) gave the 4-aminotriazolin-5-one (**22.2-51**), formed via intermediate **22.2-50**.[26] In the reaction of the 4-

amidinotriazolium salt (**22.2-52**) with base, 3-phenyltriazole (**22.2-54**) was formed via intermediate **22.2-53**.[27] This reaction was also effected with acylating reagents.

N=C(NH₂)Ph

22.2-52

$\xrightarrow{OH^-}$

$\begin{bmatrix} & NHCHO \\ PhC & \diagdown \\ & NNHCH=NNHCH_2Ph \end{bmatrix}$

22.2-53

$\xrightarrow{90\%}$

Ph

22.2-54

References

1. **75:** 41667q
2. **72:** 74407w
3. **77:** 42777d
4. **78:** 92134h
5. **78:** 105462r
6. **80:** 103553p
7. **67:** 43755a
8. **51:** 7999c
9. **75:** 151734s
10. **81:** 3839p
11. **84:** 30981x
12. **80:** 146119u
13. **74:** 111972r
14. **75:** 98508t
15. **62:** 6476b
16. **84:** 105495c
17. **80:** 47913s
18. **76:** 14436c
19. **64:** 19596d
20. **70:** 87681q
21. **82:** 140030c
22. **76:** 25193x
23. **78:** 124510w
24. **72:** 66873e
25. **75:** 35893u
26. **76:** 85755s
27. **71:** 38867s

22.3. Azido-, Azo-, or Diazoamino(Triazeno)-1,2,4-Triazolium or Triazolinium Compounds Containing One or More Representative Functions

Two triazolium compounds, **22.3-1** and **22.3-2**, have been used for the spectrophotometric determination of antimony, bismuth, and copper.[1-3] However, most of the compounds in this section were prepared for use as dyes, which are reported to be suitable for the dyeing of synthetic and mordented cellulose fibers as well as silk, wool, and leather.

Quaternary azo dyes (e.g., **22.3-6**) are readily prepared by the diazotrization of 3-amino-1,2,4-triazole (**22.3-3**) and coupling of the resulting diazotriazole (**22.3-4**) with an aniline derivative (e.g., **22.3-5**) to give an azo compound (**22.3-7**), followed by methylation of the triazole ring of the latter with dimethylsulfate and isolation of the product as either the chloride or zinc chloride double salt.[4-11] A similar sequence of reactions was carried out by diazotization of one of the amino groups of 3,5-diamino-1-phenyl-1,2,4-triazole to give dyes containing an unsubstituted amino or imino moiety.[12-16] In addition to anilines, the coupler can be a mono-[17-19] or bicyclic heterocyclic system.[20-28] Organic oxides,[25,29-32] sulfides,[33] and acrylonitrile[34,35] have also been used as the alkylating agent. In the alkylation of **22.3-7** the two entering alkyl substituents might substitute to give a 1,2-dialkyl-3-azo-, 1,4-dialkyl-3-azo-, or 1,4-dialkyl-5-azotriazolium compound. In the absence of supporting evidence, the structure of most of the compounds given in Table 22.3 must be considered uncertain in regard to the location of the *N*-substituents.

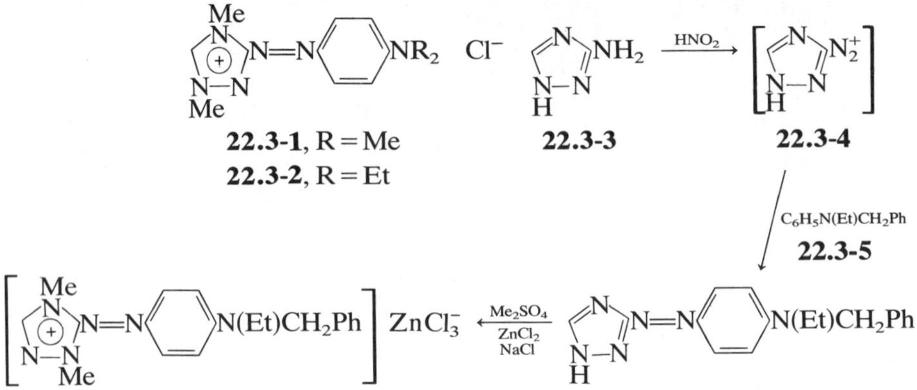

22.3-1, R = Me
22.3-2, R = Et

22.3-3

$\xrightarrow{\text{HNO}_2}$

22.3-4

$\diagdown$ C$_6$H$_5$N(Et)CH$_2$Ph

22.3-5

22.3-6

$\xleftarrow[\substack{\text{ZnCl}_2 \\ \text{NaCl}}]{\text{Me}_2\text{SO}_4}$

22.3-7

In a variation of the method using **22.3-4**, diazotized 3-amino-1,2,4-triazole-5-carboxylic acid (**22.3-8**) was coupled with an aromatic system to give an azo compound (e.g., **22.3-9**), followed by decarboxylation to give **22.3-11** and quaternization of the latter to give the dye (e.g., **22.3-10**).[36] Similar reactions were carried out with the couplers, 1,3-dimethoxy-benzene,[37] 2-phenylindole,[38] and 1-phenyl-3,5,5-trimethyl-2-pyrazoline.[39]

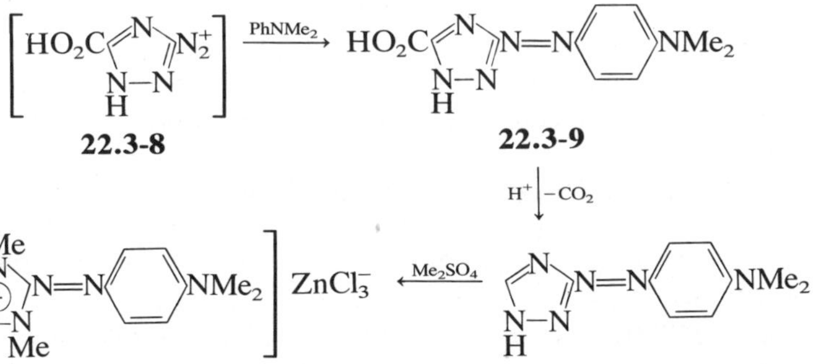

22.3-8

$\xrightarrow{\text{PhNMe}_2}$

22.3-9

$\text{H}^+ \Big| -\text{CO}_2$

22.3-10

$\xleftarrow{\text{Me}_2\text{SO}_4}$

22.3-11

Three methods are used to prepare dyes containing two azo linkages. Both azo functions are introduced simultaneously in the coupling of 3-diazo-1,2,4-triazole (**22.3-4**) with the N,N'-diphenyl-1,4-butanediamine (**22.3-12**) to give **22.3-13**, followed by quaternization of both triazole rings to give **22.3-14**.[40–43] Similarly, two azo groupings connected directly to the triazole ring are obtained by the use of tetrazolized guanazole, as demonstrated by the conversion of **22.3-15** to **22.3-16**.[44–50] The stepwise introduction of the azo groups is illustrated by the coupling of 3-diazo-1,2,4-triazole-5-carboxylic acid (**22.3-8**) with 2,5-dimethoxyaniline to give **22.3-17**, followed

$$\left[\begin{array}{c} \overset{N}{\underset{N-N}{\overset{|}{H}}}N_2^+ \end{array}\right] + \left[\begin{array}{c} \left\langle \right\rangle N(Et)CH_2CH_2 \end{array}\right]_2$$

22.3-4   **22.3-12**

$$\left[\begin{array}{c} \overset{N}{\underset{N-N}{\overset{|}{H}}}N=N\left\langle \right\rangle N(Et)CH_2CH_2 \end{array}\right]_2 \xrightarrow{Me_2SO_4} \left[\begin{array}{c} \overset{Me}{\underset{N-N}{\overset{|}{\underset{Me}{N}}}}\overset{\oplus}{N}=N\left\langle \right\rangle N(Et)CH_2CH_2 \end{array}\right]_2$$

22.3-13   **22.3-14**

$$\left[\begin{array}{c} {}^+N_2\overset{N}{\underset{N-N}{\overset{|}{H}}}N_2^+ \end{array}\right] \longrightarrow Et_2N\left\langle \right\rangle N=N\overset{Me}{\underset{N-N}{\overset{|}{\underset{Me}{\overset{\oplus}{N}}}}}N=N\left\langle \right\rangle NEt_2 \quad MeSO_4^-$$

22.3-15   **22.3-16**

by diazotization and coupling of this compound with N,N-diethyl-3-methylaniline to give **22.3-19**, which was decarboxylated and methylated to give **22.3-18**.[51,52]

$$\left[\begin{array}{c} HO_2C\overset{N}{\underset{N-N}{\overset{|}{H}}}N_2^+ \end{array}\right] \longrightarrow HO_2C\overset{N}{\underset{N-N}{\overset{|}{H}}}N=N\overset{MeO}{\underset{OMe}{\left\langle \right\rangle}}NH_2$$

22.3-8   **22.3-17**

$$\left[\begin{array}{c} \overset{MeO}{\underset{OMe}{\overset{Me}{\underset{N-N}{\overset{|}{\underset{Me}{\overset{\oplus}{N}}}}}N=N\left\langle \right\rangle N=N\left\langle \right\rangle NEt_2}} \end{array} ZnCl_3^- \right] \xleftarrow[Me_2SO_4]{-CO_2} HO_2C\overset{N}{\underset{N-N}{\overset{|}{H}}}N=N\overset{MeO}{\underset{OMe}{\left\langle \right\rangle}}N=N\overset{Me}{\left\langle \right\rangle}NEt_2$$

22.3-18   **22.3-19**

In the search for compounds with neuromuscular blocking activity, the 4,4′-azobis(triazolium salt) (**22.3-23**) was prepared by two routes.[53] Treatment of **22.3-20** with potassium bromate under acidic conditions gave **22.3-21** (27%), which was methylated with methyl fluorosulfonate to give **22.3-23** (70%). This compound was also obtained in 46% yield from the triazolium salt (**22.3-22**) and aqueous bromine.

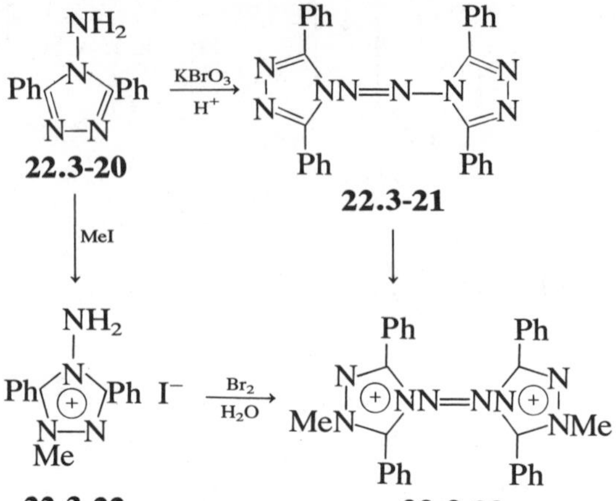

22.3-20 **22.3-21**

22.3-22 **22.3-23**

The triazenotriazolium salts are rare, with only two examples having been reported. The highly reactive chloro group of **22.3-24** was readily displaced in cold methanolic sodium azide to give the 5-azido-1,2,4-triazolium tetrafluoborate (**22.3-25**).[54] A mixture of **22.3-25** and sodium azide in a molar ratio of 2:1 in dimethylformamide evolved nitrogen and gave the triazeno compound (**22.3-27**), presumably formed from the interaction of intermediate **22.3-26** with **22.3-25**.[55] In addition, a triazeno dye (**22.3-30**) was

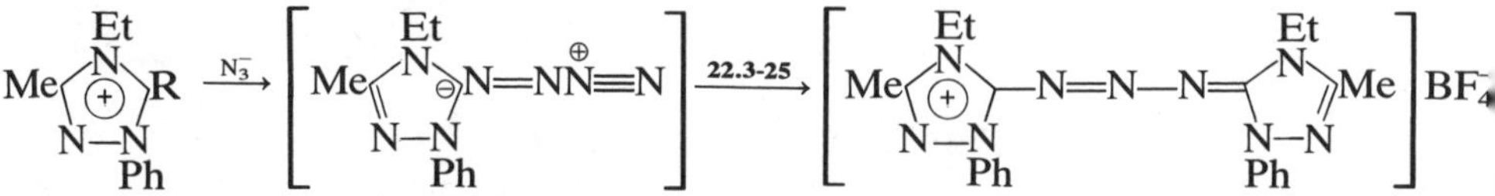

22.3-24, R = Cl   **22.3-26**   **22.3-27**
22.3-25, R = N$_3$

prepared by the coupling of 3-diazotriazole (**22.3-4**) with the 2-iminobenzothiazoline (**22.3-28**) to give **22.3-29**, followed by methylation of the latter.[56]

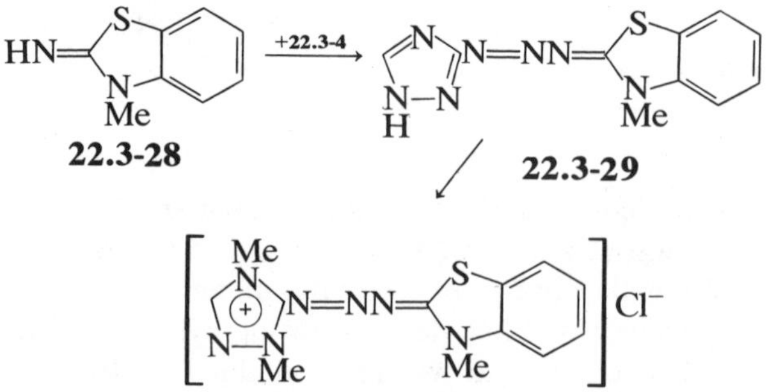

22.3-28   **22.3-29**

22.3-30

References

1. **73:** 62258f	2. **77:** 42829x	3. **80:** 90814x	4. **54:** 3970a
5. **61:** 5822g	6. **62:** 7910e	7. **84:** 6478t	8. **84:** 75683t
9. **85:** 34629g	10. **85:** 79672g	11. **85:** 95772v	12. **66:** 76923m
13. **75:** 130772n	14. **75:** 130778u	15. **78:** 17606k	16. **80:** 84685q
17. **72:** 122907h	18. **80:** 84684p	19. **83:** 99180g	20. **73:** 46632t
21. **75:** 119193d	22. **80:** 28454b	23. **80:** 28461b	24. **81:** 51113u
25. **81:** 79381d	26. **83:** 29853e	27. **83:** 81207m	28. **85:** 110086v
29. **79:** 93432m	30. **84:** 152199u	31. **85:** 22740y	32. **85:** 95798h
33. **81:** 93043b	34. **85:** 7255a	35. **85:** 22745d	36. **55:** P11862h
37. **57:** 13929c	38. **58:** 4673d	39. **62:** 5364f	40. **61:** 9614a
41. **65:** 7320e	42. **69:** 28582q	43. **70:** 58908p	44. **74:** 77396u
45. **75:** 50405e	46. **81:** 79380c	47. **81:** 93041z	48. **81:** 137569j
49. **81:** 171335q	50. **83:** 181082k	51. **59:** 12957c	52. **77:** 76669k
53. **82:** 4178e	54. **56:** 10133e	55. **59:** 11695e	56. **56:** 13123a

22.4. Hydroxy-1,2,4-Triazolium and Dioxo-1,2,4-Triazolidinium Compounds

Busch described the thermal cyclization of 1-formyl-1,4-diphenyl-semicarbazide (**22.4-1**) to give a product that was assigned the structure of the *endo*-triazoline **22.4-2**.* Subsequently, Baker and Ollis suggested that this compound could be represented as the mesoionic compound **22.4-3**,[1] which was supported by ^1H NMR and infrared spectral data.[2] This mesoionic compound has also been prepared by the elimination of carbon dioxide from **22.4-5**, which was a transient intermediate formed by the 1,3-dipolar addition of phenyl isocyanate with sydnone (**22.4-4**).[3]

---

* M. Busch and S. Schneider, *J. Prakt. Chem.*, **67**, 263 (1903).

Although the thermal cyclization of **22.4-6**, **22.4-7**, and **22.4-8** was unsuccessful, resulting in decomposition of these semicarbazides, this method was successful for the dehydration of **22.4-9** to give **22.4-10**.[4] In the reaction of compounds **22.4-6** to **22.4-8** with methoxide at room temperature, only **22.4-8** was converted to the triazolium compound (**22.4-11**).[4] Later the cyclization of **22.4-7** was effected in refluxing ethanolic ethoxide to give a 55% yield of **22.4-12**.[5,6] In addition, the intramolecular cyclization

of 1-benzoyl-1,4-diphenylsemicarbazide (**22.4-13**) in the presence of ethanolic sodium ethoxide gave a high yield of the inner salt of 3-hydroxy-1,4,5-triphenyl-1,2,4-triazolium hydroxide (**22.4-14**), which was also obtained in high yield by the reaction of the amidrazone **22.4-15** with phosgene in benzene.[4] Also, a 47% yield of **22.4-14** resulted from the amination of the 1,3,4-thiadiazole-2-thione (**22.4-17**) with aniline.[7] Compound **22.4-14** was decomposed in a refluxing mixture of methanol and

10% aqueous sodium hydroxide to give 1,4-diphenylsemicarbazide, but was stable in boiling 10% sodium bicarbonate. The phosgene procedure was used for the conversion of **22.4-16** to the chloride of the corresponding 1-methyl compound (**22.4-12**), which eliminated methyl chloride to give 3,4-diphenyl-1,2,4-triazolin-5-one on heating above its melting point. The reverse reaction, the alkylation of the triazolin-5-one (**22.4-18**), was reported to give a mixture of the triazolin-5-one (**22.4-19**) and the mesoionic compound **22.4-11**.[8] In contrast, the alkylation of **22.4-21** gave mainly the

22.4-18	**22.4-19**	**22.4-11**

triazolium hydroxide (**22.4-20**) with methyl iodide in an alkaline medium and mainly the 3-methoxytriazole (**22.4-22**) with diazomethane.[9] The stabil-

22.4-20	**22.4-21**	**22.4-22**

ity of the 5-hydroxy-1,2,4-triazolium compounds in base under mild conditions was illustrated by the cyclization of **22.4-23** in aqueous potassium hydroxide at room temperature to give **22.4-24**.[10]

22.4-23	**22.4-24**

The carboxylation of the phosphoramidate anion (**22.4-25**) resulted in the *in situ* formation of dimethylamino isocyanate (**22.4-26**), which underwent dimerization to give the dioxo-1,2,4-triazolidinium compound (**22.4-28**).[11] Although the monomer was not isolated, **22.4-26** was trapped with butyl isocyanate to give a 62% yield of **22.4-27**, also formed by heating **22.4-28** in butyl isocyanate. Two other methods have been discovered for the generation of **22.4-26**. Reaction of aryl isocyanates (e.g., phenylisocyanate) with the *N*-aminoformamidine (**22.4-29**) proceeded via **22.4-30** to give **22.4-32**, which underwent dissociation to **22.4-26** and **22.4-35**. The addition of **22.4-26** to a second molecule of phenyl isocyanate gave the triazolidinium compound (**22.4-34**) (81%).[12] In general, this reaction gave

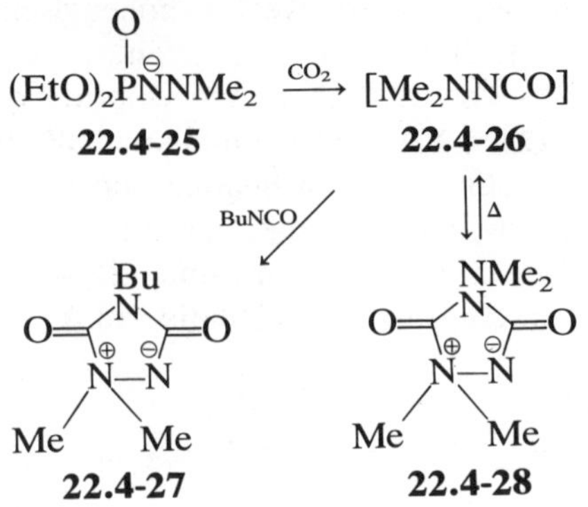

$$(EtO)_2\overset{\overset{\displaystyle O}{\|}}{P}\overset{\ominus}{N}NMe_2 \xrightarrow{CO_2} [Me_2NNCO]$$

22.4-25   **22.4-26**

BuNCO / $\qquad$ $\Updownarrow \Delta$

22.4-27   **22.4-28**

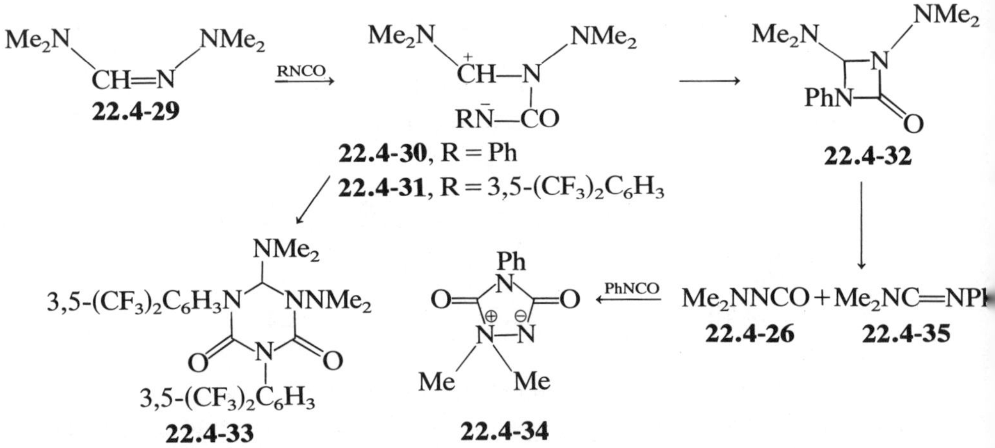

$$\underset{\textbf{22.4-29}}{\overset{Me_2N}{\underset{}{}}\diagdown_{CH=N}\diagup^{NMe_2}} \xrightarrow{RNCO} \underset{\textbf{22.4-30, R = Ph}}{\overset{Me_2N}{\underset{RN-CO}{}}\diagdown_{CH-N}^{+}\diagup^{NMe_2}} \longrightarrow \underset{\textbf{22.4-32}}{\overset{Me_2N}{\underset{PhN}{}}}$$

22.4-31, R = 3,5-$(CF_3)_2C_6H_3$

3,5-$(CF_3)_2C_6H_3$N$\qquad$NNMe$_2$

22.4-33

$O=\qquad=O \xleftarrow{PhNCO}$ Me$_2$NNCO + Me$_2$NC$=$NPh

22.4-34   **22.4-26**   **22.4-35**

good yields of products for those aryl isocyanates substituted with electron-donating groups. With some aryl isocyanates containing electron-withdrawing group (e.g., 3,5-bis[trifluoromethyl]phenyl isocyanate), inter-mediate **22.4-31** was trapped by a second molecule of the isocyanate to give a saturated s-triazine (e.g., **22.4-33**). A stabilized form of **22.4-26** resulted from the photolysis of the carbamoyl azide (**22.4-36**) in the presence of tert-butyl isocyanate to give **22.4-37**.[13] Dissociation of **22.4-37** under mild conditions offered a convenient source of **22.4-26**. The related triazolidinium compound (**22.4-39**) resulted from the interaction of 4-phenyl-1,2,4-triazoline-3,5-dione (**22.4-38**) with diphenyldiazomethane in ether.[14]

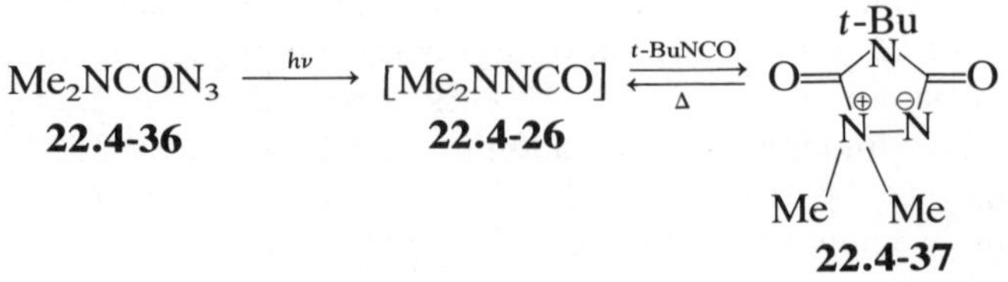

$$Me_2NCON_3 \xrightarrow{h\nu} [Me_2NNCO] \underset{\Delta}{\overset{t\text{-}BuNCO}{\rightleftarrows}}$$

22.4-36   **22.4-26**

22.4-37

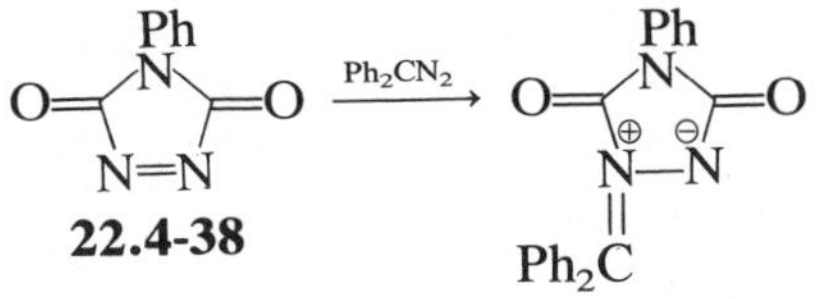

22.4-38 → (Ph₂CN₂) → **22.4-39**

The reactions of the 1,1-dialkyltriazolidinium compounds are illustrated by those conversions observed with **22.4-40**.[15] The $\overset{+}{N}$—$\bar{N}$ bond was cleaved on catalytic hydrogenation to give **22.4-41**, whereas an ethyl group was eliminated on pyrolysis to give ethylene and 1,4-diethyltriazolidine-3,5-dione (**22.4-43**). In contrast, in the presence of acid at 80° an ethyl group of **22.4-40** migrated to give a mixture of the triazoline (**22.4-44**) and triazolidine (**22.4-45**). In addition, treatment of analog **22.4-34** with methoxide resulted in the formation of the semicarbazide (**22.4-42**).[12]

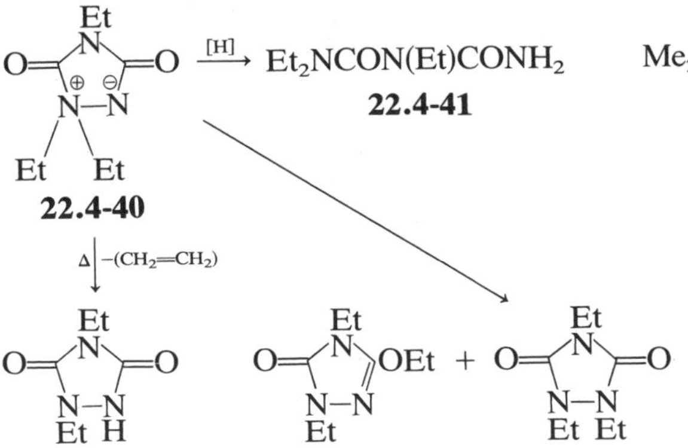

$Et_2NCON(Et)CONH_2$
22.4-41

$Me_2NNHCONHPh$
22.4-42

22.4-40

22.4-43   **22.4-44** + **22.4-45**

References

1. **51:** 7999c
2. **68:** 59504z
3. **68:** 114512c
4. **67:** 43755a
5. **75:** 151737v
6. **81:** 3835j
7. **67:** 100077g
8. **64:** 19614a
9. **85:** 159988e
10. **67:** 32651n
11. **67:** 3039p
12. **81:** 91435g
13. **82:** 3735x
14. **63:** 8346b
15. **75:** 20308h

22.5. Mercapto- or Alkylthio-1,2,4-Triazolium Compounds

The acyl-1,4-disubstituted thiosemicarbazides are readily dehydrated to give mercaptotriazolium compounds, as illustrated by the treatment of a methanolic solution of **22.5-1** with aqueous potassium carbonate to give **22.5-2**[1] (97%), and by the reaction of 4-methyl-1-phenylthiosemicarbazide

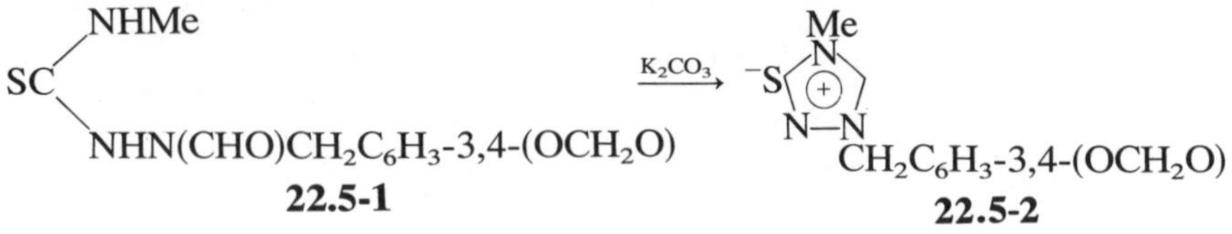

NHMe
SC
NHN(CHO)CH$_2$C$_6$H$_3$-3,4-(OCH$_2$O)
22.5-1

$\xrightarrow{\text{K}_2\text{CO}_3}$

Me
N
$^-$S(+)
N—N
CH$_2$C$_6$H$_3$-3,4-(OCH$_2$O)
22.5-2

(**22.5-3**) with acetyl chloride in benzene to give an acetyl intermediate, which under the conditions of the reaction was converted to the chloride (**22.5-4**).[2-4] Treatment of **22.5-4** with refluxing pyridine gave the corresponding mesoionic compound (**22.5-5**),[2] which was also obtained in good yield by the reaction of **22.5-3** with a mixture of acetic acid and acetic anhydride.[5] In addition, **22.5-5** was prepared by the alkylation of either **22.5-6** or **22.6-8** with methyl iodide to give the 3-(methylthio)triazolium iodide (**22.5-7**), followed by S-demethylation of **22.5-7** in refluxing pyridine.[5] Further support for the methylation of **22.5-8** (and **22.5-6**) at the

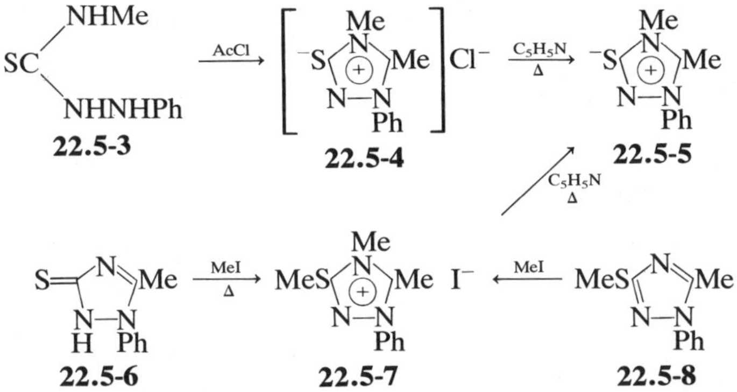

NHMe
SC
NHNHPh
22.5-3

$\xrightarrow{\text{AcCl}}$

$\left[\begin{array}{c} \text{Me} \\ \text{N} \\ ^-\text{S}(+)\text{Me} \\ \text{N—N} \\ \text{Ph} \end{array} \right]Cl^-$
22.5-4

$\xrightarrow{\text{C}_5\text{H}_5\text{N}}{\Delta}$

Me
N
$^-$S(+)Me
N—N
Ph
22.5-5

$\nearrow$ $\underset{\Delta}{\text{C}_5\text{H}_5\text{N}}$

S=⟨N⟩Me
N—N
H Ph
22.5-6

$\xrightarrow[\Delta]{\text{MeI}}$

Me
N
MeS(+)Me I$^-$
N—N
Ph
22.5-7

$\xleftarrow{\text{MeI}}$

MeS⟨N⟩Me
N—N
Ph
22.5-8

4-position, rather than the 2-position, was provided by the methylation of the 2-methyltriazoline-5-thione (**22.5-9**) to give the isomeric 3-(methylthio)triazolium iodide (**22.5-10**).[5] In addition, the 1-acyl-1,4-

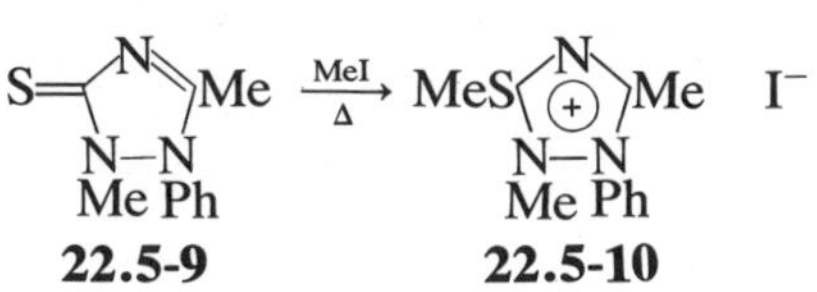

S=⟨N⟩Me
N—N
Me Ph
22.5-9

$\xrightarrow[\Delta]{\text{MeI}}$

MeS⟨N(+)⟩Me I$^-$
N—N
Me Ph
22.5-10

disubstituted thiosemicarbazides are converted directly to mesoionic compounds by heating above their melting points, as shown by the dehydration of **22.5-11** to give **22.5-12**.[2] In the methylation of **22.5-13**, substitution occurred on the ring N—N linkage at the nitrogen adjacent to the methyl group rather than the methylthio group to give **22.5-14**.[5]

The 1-(thioacyl)thiosemicarbazides are also converted to mesoionic mercaptotriazole compounds in high yields.[3,6] However, in the conversion of **22.5-15** to **22.5-17** (59%) and **22.5-16** to **22.5-18** (80%), minor amounts of the corresponding 1,3,4-thiadiazolium compounds (**22.5-19** and **22.5-20**) were formed by elimination of aniline.[3] The 4-alkylthiosemicarbazide

(**22.5-21**) also provided a mixture of **22.5-22** and **22.5-23**,[7] whereas the 4-benzoyl derivatives (e.g., **22.5-24**) invariably gave 2-benzamido-1,3,4-thiadiazolium compounds.[8,9]

In the early literature, structure **22.5-25** was assigned to the product resulting from the reaction of 1,4-diphenylthiosemicarbazide (**22.5-27**) with acetyl chloride.* Later this product was reassigned the structure of the isomeric triazolium compound (**22.5-28**),[2] but in more recent work Grashey

---

* M. Busch and S. Schneider, *J. Prakt. Chem.*, **67**, 246 (1903).

identified this product as the 1,3,4-thiadiazolium chloride (**22.5-26**).[6] A different mode of cyclization was observed either by treatment of **22.5-27** with acetic anhydride or by heating the 1-acetyl derivative of **22.5-27** above its melting point. Originally, the product of this reaction was thought to be **22.5-29**, formed via dehydration of a rearrangement product of acetylated **22.5-27**.[2] However, this reaction has been shown to give the normal cyclization product, **22.5-30**, identified by reductive cleavage with Raney nickel to give N,N'-diphenylacetamidine.[6] Also, the structure of **22.5-30** was supported by ^{1}H NMR data.[10]

In contrast to the reaction of **22.5-27** with acetyl chloride to give **22.5-26** as described previously, acylation of the S-methyl derivative (**22.5-31**) with carboxylic acid chlorides (e.g., **22.5-32**) in refluxing dioxane gave the 3-(methylthio)triazolium salts (e.g., **22.5-33**) in high yields.[11] These compounds are reduced with sodium borohydride in methanol to give the corresponding 3-(methylthio)triazoline compounds (e.g., **22.5-34**), which were hydrolyzed by aqueous acid to give aldehydes (e.g., **22.5-37**). Treatment of the triazolium salt (**22.5-35**) with sodium hydride in dimethylformamide generated the nucleophilic carbene (**22.5-36**), which reacted with butyl iodide, benzoyl chloride, and benzaldehyde to give **22.5-38**, **22.5-39**, and **22.5-40**.[12] In addition, the activated methyl group of **22.5-41** was methylated to give **22.5-42**.[13]

The preformed 2-arylamino-1,3,4-thiadiazolium compounds (e.g., **22.5-43**) are readily rearranged to the isomeric triazolium compounds (e.g., **22.5-17**) in refluxing ethanol.[13,14] In addition, 2-mercapto-1,3,4-thiadiazolium compounds (e.g., **22.5-44** and **22.5-45**) reacted thermally with aniline and its derivatives containing electron-donating groups to give satisfactory yields of mesoionic 3-mercaptotriazolium compounds (e.g., **22.5-17** and **22.5-18**).[7,15] However, with some anilines (especially those containing electron-withdrawing groups) and aliphatic amines, the conversion of thiadiazolium to triazolium compounds was unsuccessful.[16] In contrast, reaction of 2-(methylthio)thiadiazolium salts (e.g., **22.5-47**) with aliphatic amines at room temperature resulted in the elimination of methyl mercaptan and the formation of triazolium compounds (e.g., **22.5-46**).[16–18] The more basic anilines reacted with **22.5-47** under similar conditions to give the 3-(methylthio)triazolium salts (e.g., **22.5-48**)[15,16] and with an excess of amine to give the 3-mercaptotriazolium compounds (e.g., **22.5-18**).[18] The less basic anilines gave only open-ringed products of **22.5-47**.[16] Also, in the amination of the oxadiazolium compound **22.5-49** with cyclohexyl amine[19] and aniline,[16] only low yields of corresponding triazolium compounds (e.g., **22.5-18**) were obtained.

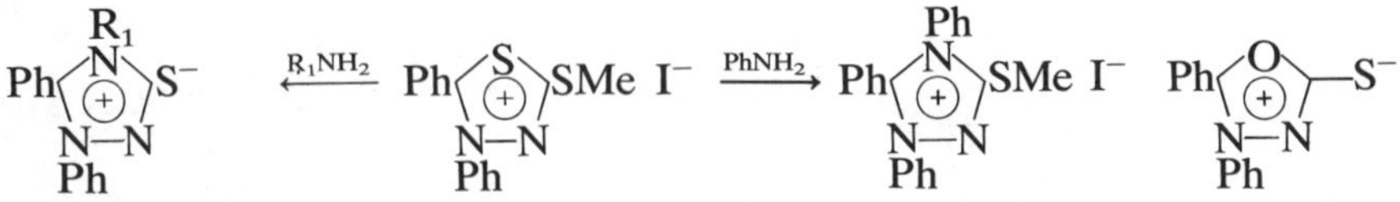

22.5-46, R₁ = Bu   **22.5-47**   **22.5-48**   **22.5-49**
22.5-18, R₁ = Ph

The amidrazone (**22.5-50**) reacted with thiophosgene in refluxing chloroform[2] and with carbon disulfide in the presence of dicyclohexyl carbodiimide[3] to give **22.5-18**. In the absence of dicyclohexyl carbodiimide, the carbon disulfide reaction gave only a 1,3,4-thiadiazolium compound. In contrast, the amidrazone (**22.5-52**) was thermally cyclized to give **22.5-51**.[8]

22.5-50   **22.5-18**, R = Ph   **22.5-52**
22.5-51, R = 4-O₂NC₆H₄

Other methods for the preparation of mesoionic triazoles include the methylation of **22.5-53** to give **22.5-54**,[20] and the condensation of the

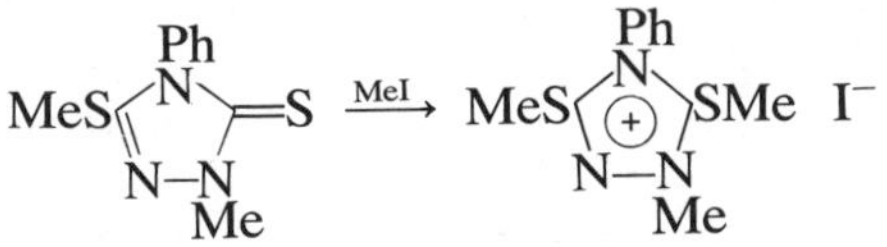

22.5-53   **22.5-54**

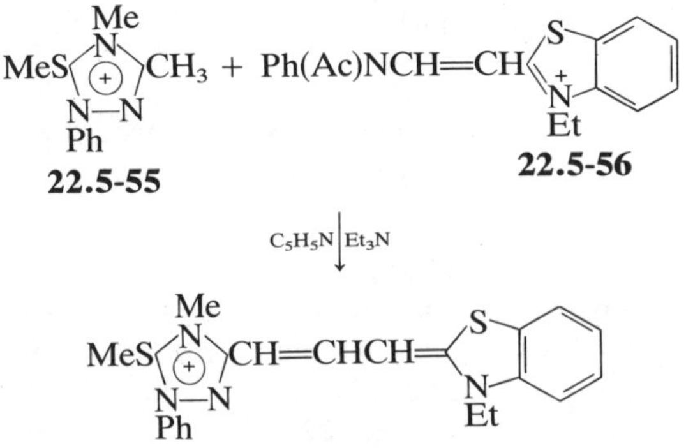

22.5-55   **22.5-56**

22.5-57

2-[(2-acetanilino)vinyl]heterocyclic quaternary salt (**22.5-56**) with the activated methyl group of **22.5-55** to give the trimethinecyanine dye (**22.5-57**).[5]

References

1. **67:** 32651n	2. **67:** 43755a	3. **77:** 101472a	4. **79:** 78701e
5. **54:** 7695e	6. **77:** 101468d	7. **74:** 42317s	8. **77:** 101473b
9. **78:** 84319g	10. **68:** 59504z	11. **81:** 169245s	12. **83:** 113599j
13. **82:** 169941m	14. **81:** 3847q	15. **76:** 153680r	16. **78:** 97565j
17. **67:** 100077g	18. **80:** 14340z	19. **77:** 139907j	20. **81:** 91435g

22.6. 4-Amino-5-Mercapto- or 5-(Alkylthio)-1,2,4-Triazolium Compounds

The 1-benzoylthiocarbohydrazide (**22.6-2**), prepared from the carboxymethylthio intermediate (**22.6-1**), was cyclodehydrated in the presence of hydrazine to give a good yield of the 4-amino-5-mercapto-1,2,4-triazolium compound (**22.6-3**).[1,2] The 4-amino group of **22.6-3** condensed with benzaldehyde to give mesoionic **22.6-4** and was removed with nitrous acid to give the triazoline-5-thione (**22.6-5**). The 4-(disubstituted amino) derivatives

(e.g., **22.6-8**) of **22.6-3** are prepared in good yields by the condensation of thiadiazolium iodides (e.g., **22.6-6**) with unsymmetrical disubstituted hydrazines (e.g., **22.6-7**).[3] This approach is limited in that monosubstituted hydrazines give tetrazines.

22.6-6   **22.6-8**   **22.6-9**

22.6-10   **22.6-11**

Alkylation of the exocyclic sulfur atom of **22.6-8** with methyl iodide gave the corresponding methylthio derivative (**22.6-9**).[3] The activated methylthio group of this type of compound is easily replaced, as demonstrated by the conversion of **22.6-10** in base to the polycyclic tetrazine (**22.6-11**).[4]

References

1. **78:** 72015s   2. **85:** 21235a   3. **80:** 27223p   4. **80:** 82837s

22.7. Miscellaneous 1,2,4-Triazolium, Triazolinium, and Triazolidinium Compounds Containing One or More Representative Functions

The pyrolysis in high vaccum of N,N-dimethylthiocarbazoylimidazole (**22.7-1**) generated dimethylaminoisothiocyanate (**22.7-2**), which dimerized apparently via **22.7-3** to give two products. One of the products was identified as **22.7-5** by independent synthesis and the other tentatively as **22.7-6**.[1] In support of this structure, one methyl group of **22.7-6** migrated readily to give **22.7-5**, which could be further alkylated with methyl iodide to give **22.7-4**. In addition, products related to mesoionic **22.7-6** were obtained from the dimerization of dialkylamino isocyanates. The carboxylation of the phosphoramidate anion (**22.7-7**) generated dimethylamino isocyanate (**22.7-8**), which was converted to **22.7-11**.[2] The formation of diethylamino isocyanate (**22.7-9**) was accomplished in high yields by photolysis of the carbamoyl azide (**22.7-10**). The transient **22.7-9** was rapidly dimerized to give **22.7-12**.[3]

A stabilized form of dimethylamino isocyanate (**22.7-8**) was prepared by its reaction with *tert*-butyl isocyanate to give **22.7-13**.[4] This compound

dissociated readily to form **22.7-8**, which reacted with diisopropyl-carbodiimide and methyl isothiocyanate to give **22.7-14** and **22.7-15**, respectively.[4,5] The addition of **22.7-8** to carbodiimides can occur by two pathways, as shown by the reaction of **22.7-8** with diethylcarbodiimide to give a mixture of **22.7-16** and **22.7-17**.[3]

Catalytic hydrogenation of **22.7-12** resulting in N—N cleavage to give a 70% yield of **22.7-18**, while pyrolysis of **22.7-12** formed ethylene and the triazolidine (**22.7-19**).[3]

22.7-14   **22.7-8**   **22.7-15**

22.7-16   **22.7-17**

22.7-18   **22.7-12**   **22.7-19**

Oxidation of the ethyl ester of the 1,2,4-triazolinecarboxylic acid (**22.7-20**) with potassium dichromate in acetic acid gave an 80% yield of the triazolium compound (**22.7-21**).[6] When **22.7-21** was heated in base, the ring N—N linkage was cleaved to form N-cyano-N,N'-diphenylbenzamidine (**22.7-22**). In contrast, aryl-substituted triazolium compounds undergo cleavage at N_1-C_5 to give N-amino-N'-acylamidines (see Section 22.1).

22.7-20   **22.7-21**   **22.7-22**

The triazolium compound (**22.7-23**) was O-alkylated with triethyloxonium fluoroborate to give a high yield of the corresponding ethoxytriazolium fluoroborate (**22.7-24**).[7] The mesoionic compound (**22.7-23**) was regenerated from **22.7-24** by heating and by passing the salt over a column of alumina.

22.7-23   **22.7-24**

In the alkylation of *s*-triazole with trimethyloxonium fluoroborate, a mixture of mono- and dimethylated products was obtained. In contrast, the alkylation of 1-acetyl-1,2,4-triazole (**22.7-25**) with this reagent was regiospecific, providing a high yield of the 4-methyltriazolium derivative (**22.7-26**), which was readily converted to 4-methyl-1,2,4-4*H*-triazole (**22.7-27**) by methanolysis.[8]

Treatment of the 5-chloro-1,2,4-triazole (**22.7-28**) with triethyloxonium fluoborate in 1,2-dichloroethylene gave an 82% yield of 5-chloro-4-ethyl-3-methyl-1-phenyl-1,2,4-triazolium fluoborate (**22.7-29**).[9] The activated chloro group of **22.7-29** was readily displaced with sodium azide to give **22.7-30**.

References

1. **69:** 51320m   2. **67:** 3039p   3. **75:** 20308h   4. **82:** 3735x
5. **82:** 72890f   6. **70:** 96723g   7. **73:** 77149v   8. **73:** 45425r
9. **56:** 10133e

TABLE 22.   1,2,4-TRIAZOLIUM,   TRIAZOLINIUM,   TRIAZOLIDINIUM   AND MESOIONIC COMPOUNDS

Compound	Reference

22.1. Alkyl- or Aryl-1,2,4-Triazolium Compounds

Compound	Reference
1-acetonyl-3,5-dimethyl-4-phenyl-, bromide	**82:** 170807d
1-[2-[(4-amino-3-chlorobenzoyl)amino]ethyl]-4-benzyl-, chloride	**78:** P148956x
5-(6-amino-1,3-dioxo-1H̲-benz[de]isoquinolin-2(3H̲)-yl)-1,2-dimethyl-, methyl sulfate	**84:** P75707d
1-[2-([(6-amino-5-[(4-nitrophenyl)azo]-2-naphthalenyl)-sulfonyl]amino)ethyl]-4-methyl-, methyl sulfate	**78:** P148956x
4-(p-aminophenyl)-1-methyl-, methyl sulfate	**72:** P100715s
1-[([(2-aminophenyl)sulfonyl]methylamino)methyl]-4-methyl, methyl sulfate	**78:** P148956x
1-[3-([4-amino-3-(trifluoromethyl)phenyl]amino)propyl]-4-methyl-, methyl sulfate	**78:** P148956x
5-benzyl-4-methyl-1,3-diphenyl-, perchlorate	**70:** 96723g
3-benzyl-1,4,5-triphenyl-, perchlorate	**74:** 111972r
4-benzyl-1,3-5-triphenyl-, perchlorate	**74:** 111972r
5-benzyl-1,3,4-triphenyl-, perchlorate	**74:** 111972r
4,4'-(4,4'-biphenylylene)bis[3,5-dimethyl-1-phenyl-, diperchlorate	**74:** 111972r
1,4-bis(p-chlorophenyl)-, tetrafluoroborate(1-)	**71:** 91385j
3,5-bis(p-nitrobenzyl)-1,4-diphenyl-, perchlorate	**75:** 98508t
1-[1-(4-bromobenzoyl)ethyl]-5-methyl-3,4-diphenyl-, bromide	**82:** 170807d
1-(4-bromophenacyl)-3,5-diethyl-4-phenyl-, perchlorate	**78:** 58327p
4-(4-bromophenacyl)-1,5-dimethyl-, bromide	**80:** 133356u
1-(4-bromophenacyl)-3,5-dimethyl-4-phenyl-, bromide	**78:** 58327p, **79:** 115499t
4-(4-bromophenacyl)-3,5-dimethyl-1-phenyl-, bromide	**81:** 13441s
1-(4-bromophenacyl)-3,5-dimethyl-4-phenyl-, hydroxide, inner salt	**79:** 115499t
1-(4-bromophenacyl)-3,4,5-trimethyl-, bromide	**79:** 115499t
4-(4-bromophenacyl-1,3,5-trimethyl-, bromide	**81:** 13441s
4-[2-(4-bromophenyl)-1-methyl-2-oxoethyl]-3,5-dimethyl-1-phenyl-, bromide	**81:** 13441s
1-[2-(4-bromophenyl)-2-oxoethyl]-3,5-dimethyl-4-benzyl-, bromide	**85:** 21220s
1-[2-(4-bromophenyl)-2-oxoethyl]-3,4,5-trimethyl-, salt with 4-methylbenzenesulfonic acid (1:1)	**85:** 21220s
1-[3-(p-butoxyphenyl)-2-oxo-2H̲-1-benzopyran-7-yl]-4-methyl-, methyl sulfate	**71:** P14180m
4-butyl-1,3,5-triphenyl-, perchlorate	**74:** 111972r, **79:** 31989q
1-[p-(carboxyamino)phenyl]-4-methyl-, methyl sulfate, diester with 2,2'-(p-phenylenedioxy)diethanol	**72:** P100715s
1-[p-(carboxyamino)phenyl]-4-(3-sulfopropyl)-, hydroxide, inner salt, 1-phenyl ester	**72:** P100715s
4-[1-carboxy-2-(3-carboxy-2-hydroxy-1-naphthyl)vinyl]-1-methyl, methyl sulfate, δ lactone, ethyl ester	**59:** P15420h

628

TABLE 22 *(Continued)*

Compound	Reference

22.1. Alkyl- or Aryl-1,2,4-Triazolium Compounds *(Continued)*

Compound	Reference
4-[1-carboxy-2-(2-hydroxy-1-naphthyl)vinyl]-1-methyl, methyl sulfate, δ lactone	**59:** P15420h
4-(o-carboxyphenyl)-3-(cyanomethyl)-1,5-dimethyl-, iodide methyl ester	**72:** 43550h
4-(m-carboxyphenyl)-3,5-dimethyl-1-phenyl-, perchlorate	**69:** 77178w, **74:** 111972r
4-(p-carboxyphenyl)-3,5-dimethyl-1-phenyl-, perchlorate	**74:** 111972r
4-(o-carboxyphenyl)-1,3,5-trimethyl-, iodide, methyl ester	**72:** 43550h
1-[2-[3-chloro-4-[[4-[(2-chloroethyl)ethylamino]-2-methylphenyl]azo]phenoxy]ethyl]-4-benzyl-, chloride	**78:** P148956x
1-[2-[3-chloro-4-(4-[(2-cyanoethyl)ethylamino]-2-methylphenylazo)benzoyl]aminoethyl]-4-benzyl-, chloride	**78:** P148956x
1-[2-[(3-chloro-4-[(4-[(2-cyanoethyl)ethylamino]phenyl)-azo]phenyl)sulfonyl]ethyl]-4-methyl-, methyl sulfate	**78:** P148956x
1-(2-[4-chloro-3-([4-(diethylamino)phenyl]azo)phenyl]-2-oxoethyl)-4-methyl-, methyl sulfate	**78:** P148956x
1-[7-(4-chloro-3-methylpyrazol-1-yl)-2-oxo-2H-1-benzo-pyran-3-yl]-4-methyl-, methyl sulfate	**74:** P127556u
1-(2-[(4-[(2-chloro-4-nitrophenyl)azo]-3-methylphenyl)-ethylamino]ethyl)-4-methyl-, methyl sulfate	**82:** P126600g
1-(2-[(4-[(2-chloro-4-nitrophenyl)azo]phenyl)ethylamino]-ethyl)-4-methyl-, chloride	**82:** P126600g
1-(2-[(4-[(2-chloro-4-nitrophenyl)azo]phenyl)ethylamino]-ethyl)-4-methyl-, -tetrachlorozincate(2-) (2:1)	**82:** P126600g
1-(4-chlorophenacyl)-3,5-dimethyl-4-phenyl-, bromide	**79:** 115499t
4-(o-chlorophenyl)-3,5-dimethyl-1-phenyl-, perchlorate	**74:** 111972r
1-(2-[(p-[3-(p-chlorophenyl)-2-pyrazolin-1-yl]phenyl)-sulfonyl]ethyl)-4-methyl-, methyl sulfate	**72:** P91491m
1-[7-(4-cyano-5-methyl-2H-1,2,3-triazol-2-yl)-2-oxo-2H-1-benzopyran-3-yl]-4-methyl-, methyl sulfate	**73:** P100073b
4-cyclohexyl-3,5-dimethyl-1-phenyl-, perchlorate	**74:** 111972r
4-cyclohexyl-1,5-diphenyl-, perchlorate	**70:** 96723g
4-cyclohexyl-5-methyl-1,3-diphenyl-, perchlorate	**70:** 96723g
4-cyclohexyl-1,3,5-triphenyl-, perchlorate	**74:** 111972r
1-[2-([(2,5-dichloro-4-[(4-[(2-cyanoethyl)ethylamino]-phenyl)azo]phenyl)sulfonyl]amino)ethyl]-4-methyl-, methyl sulfate	**78:** P148956x
4,4′-[(2,5-dichloro-p-phenylene)bis(ureylene-p-pheny-lene)]-bis[1methyl]-, bis(methyl sulfate)	**72:** P100715s
1-[2-([5-(diethylamino)-2-[(4-methylphenyl)azo]phenyl]-amino)-2-oxoethyl]-4-methyl-, methyl sulfate	**78:** P148956x
1-[2-([3-(diethylamino)phenyl]amino)-2-oxoethyl]-4-methyl-, methyl sulfate	**78:** P148956x
1-[2-([4-([4-(diethylamino)phenyl]azo)benzoyl]oxy)ethyl]-4-methyl-, methyl sulfate	**78:** P148956x
1-([([2-([4-(diethylamino)phenyl]azo)phenyl]sulfonyl)-methylamino]methyl)-4-methyl-, methyl sulfate	**78:** P148956x
1,4-diethyl-3,5-diphenyl-, tetrafluoroborate(1-)	**76:** 85759w

TABLE 22 (*Continued*)

Compound	Reference

22.1. Alkyl- or Aryl-1,2,4-Triazolium Compounds (*Continued*)

Compound	Reference
2,4-diethyl-1-methyl-, bis[tetrafluoroborate(1-)]	**77:** 75149d
1-[3-([4-[(4,5-dihydro-3-methyl-5-oxo-1-phenyl-1H-pyrazol-4-yl)azo]-3-(trifluoromethyl)phenyl]amino)-propyl]-4-methyl-, methyl sulfate	**78:** P148956x
1,4-dimethyl-, tetrafluoroborate(1-)	**78:** 4224j
1,5-dimethyl-3,4-diphenyl-, perchlorate	**69:** 77178w
3,5-dimethyl-1,4-diphenyl-, perchlorate	**69:** 77178w, **74:** 111972r
4,5-dimethyl-1,3-diphenyl-, perchlorate	**70:** 96723g
3,5-dimethyl-4-(*m*-nitrophenacyl)-1-phenyl-, bromide	**81:** 13441s
3,4-dimethyl-5-(3-methyl-2-benzothiazolinylidenemethyl)-1-phenyl, iodide	**54:** 7697b
3,5-dimethyl-1-(α-methylphenacyl)-4-phenyl-, bromide	**78:** 58327p
3,5-dimethyl-1-(3-nitrophenacyl)-4-phenyl-, bromide	**82:** 170807d
3,5-dimethyl-1-(4-nitrophenacyl)-4-phenyl-, bromide	**82:** 170807d
3,5-dimethyl-4-(*m*-nitrophenacyl)-1-phenyl-, bromide	**81:** 13441s
3,5-dimethyl-4-(*m*-nitrophenyl)-1-phenyl-, perchlorate	**69:** 77178w, **74:** 111972r
3,5-dimethyl-4-(*p*-nitrophenyl)-1-phenyl-, perchlorate	**74:** 111972r
3,5-dimethyl-1-phenacyl-4-phenyl-, bromide	**78:** 58327p, **79:** 115499t
3,5-dimethyl-4-phenacyl-1-phenyl-, bromide	**81:** 13441s
1,3-dimethyl-4-phenyl-, iodide	**82:** 170807d
1,5-dimethyl-4-phenyl-, iodide	**82:** 170807d
2,3(or 3,4)-dimethyl-1-phenyl, iodide	**54:** 7696d
3,5-dimethyl-1-phenyl-4-*o*-tolyl-, perchlorate	**69:** 77178w, **74:** 111972r
3,5-dimethyl-1-phenyl-4-*p*-tolyl-, perchlorate	**69:** 77178w, **74:** 111972r
3,4-dimethyl-1-phenyl-5-[3-(1,3,3-trimethyl-2-indolinylidene)propenyl], perchlorate	**54:** 7697i
3,5-dimethyl-1-phenyl-4-(2,6-xylyl)-, perchlorate	**69:** 77178w, **74:** 111972r
3-(2,4-dinitrobenzyl)-5-methyl-1,4-diphenyl-, hydroxide, inner salt	**75:** 98508t
3-(2,4-dinitrobenzyl)-5-methyl-1,4-diphenyl-, perchlorate	**75:** 98508t
3-(2,4-dinitrobenzyl)-1,4,5-triphenyl-, hydroxide, inner salt	**75:** 98508t
3-(2,4-dinitrobenzyl)-1,4,5-triphenyl-, perchlorate	**75:** 98508t
5-(2,4-dinitrobenzyl)-1,3,4-triphenyl-, perchlorate	**75:** 98508t
1,4-diphenyl-, perchlorate	**74:** 111972r
1,4-diphenyl-, tetrafluoroborate(1-)	**71:** 91385j
3-(diphenylmethyl)-1,4,5-triphenyl-, perchlorate	**72:** 121463y
4-(2-ethoxy-2-oxoethyl)-3,5-dimethyl-1-phenyl-, bromide	**81:** 13441s
1-[3-(*p*-ethoxyphenyl)-2-oxo-2H-1-benzopyran-7-yl]-4-methyl-, methyl sulfate	**71:** P14180m
5-[3-(3-ethyl-2-benzothiazolinylidene)propenyl]-1,3-dimethyl-4-phenyl, iodide	**54:** 7697i
5-[3-(3-ethyl-2-benzothiazolinylidene)propenyl]-3,4-dimethyl-1-phenyl, iodide	**54:** 7697h
5-[3-(3-ethyl-2-benzothiazolinylidene)propenyl]-4-(*p*-methoxyphenyl)-1,3-dimethyl, iodide	**54:** 7697i
5-[3-(3-ethyl-2-benzoxazolinylidene)propenyl]-1,3-dimethyl-4-phenyl, iodide	**54:** 7697h

TABLE 22 (*Continued*)

Compound	Reference

22.1. Alkyl- or Aryl-1,2,4-Triazolium Compounds (*Continued*)

Compound	Reference
5-[3-(3-ethyl-2-benzoxazolinylidene)propenyl]-3,4-dimethyl-1-phenyl, iodide	**54:** 7697h
5-[3-(3-ethyl-2-benzoxazolinylidene)propenyl]-4-(*p*-methoxyphenyl)-1,3-dimethyl, iodide	**54:** 7698a
5-[3-(3-ethyl-2-benzoxazolinylidene)propenyl]-1,3,4-trimethyl, perchlorate	**54:** 7697h
4-ethyl-1,2-dimethyl-, bis[tetrafluoroborate(1-)]	**77:** 75149d
1-ethyl-5-fluoren-9-yl-3,4-dimethyl-, tetrafluoroborate(1-)	**72:** 121463y
1-ethyl-5-fluoren-9-yl-3-methyl-4-phenyl-, tetrafluoroborate(1-)	**72:** 121463y
4-ethyl-1-methyl-, tetrafluoroborate(1-)	**77:** 75149d
1-ethyl-4-methyl-3,5-diphenyl-, tetrafluoroborate(1-)	**76:** 85013y
5-ethyl-4-methyl-1,3-diphenyl-, perchlorate	**70:** 96723g
1-[2-[ethyl(3-methylphenyl)amino]ethyl]-4-methyl-, methyl sulfate	**82:** P126600g
4-ethyl-1-[7-(3-methylpyrazol-1-yl)-2-oxo-2$\underline{H}$-1-benzopyran-3-yl]-, ethyl sulfate	**72:** P22599r
1-[7-(4-ethyl-5-methyl-2$\underline{H}$-1,2,3-triazol-2-yl)-2-oxo-2$\underline{H}$-1-benzopyran-3-yl]-4-(2-hydroxyethyl)-, acetate (salt)	**77:** P90077x
1-[7-(4-ethyl-5-methyl-2$\underline{H}$-1,2,3-triazol-2-yl)-2-oxo-2$\underline{H}$-1-benzopyran-3-yl]-4-methyl, methyl sulfate	**73:** P100073b, **74:** P14197j
4-ethyl-1-(2-oxo-7-pyrazol-1-yl-2$\underline{H}$-1-benzopyran-3-yl)-, ethyl sulfate	**72:** P22599r
1-[2-(ethylphenylamino)ethyl]-4-methyl-, methyl sulfate	**82:** P126600g
5-[3-(1-ethyl-2(1$\underline{H}$)-quinolylidene)propenyl]-3,4-dimethyl-1-phenyl, iodide	**54:** 7698a
5-fluoren-9-yl-1,3,4-trimethyl-, perchlorate	**72:** 121463y
1,1′-[hexamethylenebis(ureylene-*p*-phenylene)-bis[4-methyl-, bis(methyl sulfate)	**72:** P100715s
4,4′-[hexamethylenebis(ureylene-*p*-phenylene)]bis[1-methyl-, bis(methyl sulfate)	**72:** P100715s
1,1′-[hexamethylenebis(ureylene-*p*-phenylene)]bis[4-(3-sulfopropyl)-, bis(hydroxide, inner salt)	**72:** P100715s
4-(2-hydroxyethyl)-1-[7-(3-methyl-1$\underline{H}$-pyrazol-1-yl)-2-oxo-2$\underline{H}$-1-benzopyran-3-yl]-, formate	**77:** P90077x
2(or 4)-(2-hydroxyethyl)-1-methyl-3-[[1,2,3,4-tetrahydro-7-methyl-1-(phenylmethyl)-6-naphthalenyl]ethyl]-, tetrachlorozincate(2-) (2:1)	**85:** P125782t
4-(*p*-hydroxyphenyl)-5-methyl-1,3-diphenyl-, perchlorate	**69:** 77178w, **74:** 111972r
4-(*p*-hydroxyphenyl)-1,3,5-triphenyl-, perchlorate	**74:** 111972r
1,1′-[isophthaloylbis(imino-*p*-phenylene)]bis[4-ethyl-, dibromide	**72:** P100715s
4-isopropyl-5-methyl-1,3-diphenyl-, perchlorate	**70:** 96723g
4-(7-methoxy-2-oxo-2$\underline{H}$-1-benzopyran-3-yl)-1-methyl-, methyl sulfate	**59:** 15420g
1-(4-methoxyphenacyl)-3,5-dimethyl-4-phenyl-, bromide	**82:** 170807d

631

TABLE 22 (*Continued*)

Compound	Reference

22.1. Alkyl- or Aryl-1,2,4-Triazolium Compounds (*Continued*)

Compound	Reference
4-(4-methoxyphenacyl)-1,3,5-trimethyl-, bromide	**81:** 13441s
1-[3-(*p*-methoxyphenyl)-2-oxo-2H̲-1-benzopyran-7-yl]-4-methyl-, methyl sulfate	**71:** P14180m
1-[3-(*p*-methoxyphenyl)-2-oxo-2H̲-1-benzopyran-7-yl]-3,4,5-trimethyl-, methyl sulfate	**71:** P14180m
4-(*p*-methoxyphenyl)-1,3,5-trimethyl-, iodide	**54:** 7696e
4-methyl-1,3-diphenyl-, tetrafluoroborate(1-)	**69:** 96737u, **71:** 91385j
5-methyl-1,3-diphenyl-4-*p*-tolyl-, perchlorate	**69:** 77178w, **74:** 111972r
4,4'-[methylenebis(4',4-biphenylylene)]bis[1-methyl-, bis(methyl sulfate)	**72:** P100715s
5-methyl-1-(α-methylphenacyl)-3,4-diphenyl-, bromide	**78:** 58327p
4-methyl-1-[7-(3-methylpyrazol-1-yl)-2-oxo-2H̲-1-benzopyran-3-yl]-, methyl sulfate	**72:** P22599r
4-methyl-1-[7-(3-methylpyrazol-1-yl)-2-oxo-2H̲-1-benzopyran-3-yl]-, *p*-toluenesulfonate	**72:** P22599r
4-methyl-5-(*p*-nitrobenzyl)-1,3-diphenyl-, perchlorate	**75:** 98508t
5-methyl-3-(*p*-nitrobenzyl)-1,4-diphenyl-, perchlorate	**75:** 98508t
1-methyl-4-(*p*-nitrophenyl)-, methyl sulfate	**72:** P100715s
4-methyl-1-(2-oxo-3-phenyl-2H̲-1-benzopyran-7-yl)-, methyl sulfate	**71:** P14180m
4-methyl-1-[2-oxo-7-[4-(2-phenylethenyl)-2H̲-1,2,3-triazol-2-yl]-2H̲-1-benzopyran-3-yl]-, methyl sulfate	**85:** P194102h
4-methyl-1-(2-oxo-7-pyrazol-1-yl-2H̲-1-benzopyran-3-yl)-, methyl sulfate	**72:** P22599r
4-methyl-1-(2-oxo-3-*p*-tolyl-2H̲-1-benzopyran-7-yl)-, methyl sulfate	**71:** P14180m
3-methyl-1-phenacyl-4-phenyl-, bromide	**82:** 170807d
5-methyl-1-phenacyl-4-phenyl-, bromide	**82:** 170807d
4-(α-methylphenacyl)-1,3,5-trimethyl-, bromide	**81:** 13441s
3-methyl-1,4,5-triphenyl-, perchlorate	**74:** 111972r
4-methyl-1,3,5-triphenyl-, perchlorate	**74:** 111972r
5-methyl-1,3,4-triphenyl-, perchlorate	**69:** 77178w, **70:** 96723g, **74:** 111972r
1-methyl-1-vinyl, iodide	**62:** P13270h
4-(*p*-nitroanilino)-1,3,5-triphenyl-, perchlorate	**74:** 111972r, **75:** 98508t
3-(*p*-nitrobenzyl)-1,4,5-triphenyl-, hydroxide, inner salt	**75:** 98508t
3-(*p*-nitrobenzyl)-1,4,5-triphenyl-, perchlorate	**75:** 98508t
4,4'-*p*-phenylenebis[3,4-dimethyl-1-phenyl-, diperchlorate	**74:** 111972r
5,5'-*p*-phenylenebis[1,3,4-triphenyl-, diperchlorate	**70:** 96723g
1,1'-[*p*-phenylenebis(ureylene-*p*-phenylene)]bis[4-methyl-, bis(methyl sulfate)	**72:** P100715s
4,4'-[*p*-phenylenebis(ureylene-*p*-phenylene)]bis[1-methyl-, bis(methyl sulfate)	**72:** P100715s
1-phenyl-4-*p*-tolyl-, tetrafluoroborate(1-)	**71:** 91385j
1,1'-[terephthaloylbis(imino-*p*-phenylene)]bis[4-ethyl-, dibromide	**72:** P100715s
4,4'-[terephthaloylbis(imino-*p*-phenylene)]bis[1-methyl-, bis(methyl sulfate)	**72:** P100715s

TABLE 22 (Continued)

Compound	Reference

22.1. Alkyl- or Aryl-1,2,4-Triazolium Compounds (Continued)

Compound	Reference
1,3,4,5-tetramethyl, iodide	**49:** 14744i, **54:** 7696e
1,3,4,5-tetramethyl-, perchlorate	**82:** 4178e
1,3,4,5-tetramethyl-, (triiodide)	**82:** 4178e
1,3,4,5-tetraphenyl-, perchlorate	**69:** 77178w, **70:** 96723g, **74:** 111972r, **79:** 31989q
1,3,4,5-tetraphenyl-, tetrafluoroborate(1-)	**67:** 90732a, **71:** 91389p
1;2,4-trimethyl-, bis[tetrafluoroborate(1-)]	**77:** 75149d
1,3,5-trimethyl-, 2-(4-bromophenyl)-2-oxoethylide	**81:** 13441s
1,3,5-trimethyl-, 2-(3-nitrophenyl)-2-oxoethylide	**81:** 13441s
1,3,4-trimethyl-5-(3-methyl-2-benzothiazolinylidene-methyl), perchlorate	**54:** 7697f
1,3,5-trimethyl-4-(2-oxo-2-phenylethyl)-, bromide	**81:** 13441s
3,4,5-trimethyl-1-(2-oxo-3-p-tolyl-2H-1-benzopyran-7-yl)-, methyl sulfate	**71:** 14180m
1,3,5-trimethyl-4-[2-(3-nitrophenyl)-2-oxoethyl]-, bromide	**81:** 13441s
1,3,5-trimethyl-4-phenyl-, iodide	**54:** 7696e, **78:** 159527j
3,4,5-trimethyl-1-phenyl-, iodide	**54:** 7696d, **78:** 159527j
1,3,5-trimethyl-4-m-tolyl, iodide	**51:** P16163c
1,3,5-trimethyl-4-o-tolyl, iodide	**51:** P16163c
1,3,4-triphenyl-, tetrafluoroborate(1-)	**71:** 91385j
4,4'-(ureylene-di-p-phenylene)bis[1-methyl-, dichloride	**72:** P100715s

22.2. Amino-1,2,4-Triazolium Compounds

Compound	Reference
4-acetamido-1-benzyl-, chloride	**72:** 66873e
4-acetamido-1-benzyl-, nitrate	**76:** 14436c
4-acetamido-1-(p-bromophenacyl)-, hydroxide, inner salt	**72:** 66873e
4-acetamido-1-(p-bromophenacyl)-, bromide	**72:** 66873e
4-acetamido-1-butyl-, bromide	**72:** 66873e
4-acetamido-1-(p-chlorobenzyl)-, bromide	**72:** 66873e
4-acetamido-1-(1-naphthylmethyl)-, hydroxide, inner salt	**72:** 66873e
4-acetamido-1-(1-naphthylmethyl)-, bromide	**72:** 66873e
4-acetamido-1-(2-oxo-2-phenylethyl)-, hydroxide, inner salt	**72:** 66873e
4-acetamido-1-phenacyl-, bromide	**72:** 66873e
4-acetamido-1-propyl-, iodide	**72:** 66873e
1-allyl-4-(benzylideneamino), bromide	**64:** 19596g
3-[(4-amino-1-anthraquinonyl)amino]-4-methyl-, chloride	**68:** P106077z
4-amino-1-benzyl-, hydroxide, inner salt	**79:** 104400c
4-amino-1-benzyl-, chloride	**64:** 19596e
4-amino-1-benzyl-3,5-diheptyl-, chloride	**85:** P63073b
4-amino-1-benzyl-3,5-dimethyl, chloride	**64:** 19596e
4-amino-1-benzyl-3,5-diphenyl-, chloride	**85:** P63073b
4-amino-1-benzyl-3,5-diundecyl-, chloride	**85:** P63073b
4-[(α-aminobenzylidene)amino]-1-benzyl-, bromide	**71:** 38867s
4-[(α-aminobenzylidene)amino]-1-propyl-, bromide	**71:** 38867s
4-amino-1-benzyl-3-phenyl-, bromide	**77:** 139906h

633

TABLE 22 (*Continued*)

Compound	Reference

22.2. Amino-1,2,4-Triazolium Compounds (*Continued*)

Compound	Reference
4-amino-1-(*p*-bromophenacyl), bromide	**64:** 19596f
4-amino-1-[2-(4-bromophenyl)-1-methyl-2-oxoethyl)-3,5-diethyl-, bromide	**85:** 21220s
4-amino-1-[2-(4-bromophenyl)-1-methyl-2-oxoethyl]-3,5-dimethyl-, bromide	**85:** 21220s
4-amino-1-[2-(4-bromophenyl)-2-oxoethyl]-3,5-diethyl-, bromide	**85:** 21220s
4-amino-1-[2-(4-bromophenyl)-2-oxoethyl]-3,5-dimethyl-, bromide	**85:** 21220s
4-amino-1-(carboxymethyl), chloride	**64:** 19596f
4-amino-1-(carboxymethyl)-3,5-diundecyl-, chloride	**85:** P63073b
4-amino-1-(3-chlorobenzyl)-	**78:** 124510w
4-amino-1-(4-chlorobenzyl)-, hydroxide, inner salt	**79:** 104400c
4-amino-1-(4-chlorobenzyl)-, chloride	**78:** 124510w
4-amino-1-[(4-chlorophenyl)methyl]-3,5-diheptyl-, chloride	**85:** P63073b
4-amino-1-(2-cyanoethyl), bromide	**64:** 19596e
4-amino-1-(cyanomethyl), chloride	**64:** 19596e
4-amino-1-decyl-3,5-dimethyl-, chloride	**85:** P63073b
4-amino-3,5-diethyl-1-(2-oxo-2-phenylethyl)-, bromide	**85:** 21220s
4-amino-3,5-diethyl-1-(2-oxopropyl)-, bromide	**85:** 21220s
4-amino-3,5-diheptyl-1-(2-hydroxyethyl)-, chloride	**85:** P63073b
4-amino-3,5-diheptyl-1-methyl-, salt with 4-methyl-benzenesulfonic acid	**85:** P63073b
4-amino-3,5-diheptyl-1-(2-propenyl)-, chloride	**85:** P63073b
5-[(4-amino-9,10-dihydro-9,10-dioxo-1-anthracenyl)-amino]-4-methyl-, trichlorozincate(1-)	**84:** P6465m
5-[(4-amino-9,10-dihydro-3-methoxy-9,10-dioxo-1-anthracenyl)amino]-4-methyl-, trichlorozincate(1-)	**84:** 6465m
5-amino-1,4-dimethyl-, chloride	**70:** 91957b
4-amino-3,5-dimethyl-1-[2-(3-nitrophenyl)-2-oxoethyl]-, bromide	**85:** 21220s
4-amino-3,5-dimethyl-1-[2-(4-nitrophenyl)-2-oxoethyl]-, bromide	**85:** 21220s
4-amino-3,5-dimethyl-1-octadecyl-, chloride	**85:** P63073b
4-amino-3,5-dimethyl-1-octyl-, chloride	**85:** P63073b
4-amino-3,5-dimethyl-1-(2-oxo-2-phenylethyl)-, bromide	**85:** 21220s
4-amino-3,5-dimethyl-1-(2-oxopropyl)-, bromide	**85:** 21220s
4-amino-3,5-dimethyl-1-phenacyl, bromide	**64:** 19596e
4-amino-3,5-dimethyl-1-tetradecyl-, chloride	**85:** P63073b
4-amino-1-(2,4-dinitrophenyl)-3,5-dimethyl-, chloride	**85:** P63073b
4-amino-1-(2,4-dinitrophenyl)-3,5-diphenyl-, chloride	**85:** P63073b
3-amino-4,5-diphenyl-1-methyl, bromide	**67:** 43755a
4-amino-1-dodecyl-3,5-bis(isopropyl)-, chloride	**85:** P63073b
4-amino-1-dodecyl-, chloride	**85:** P63073b
4-amino-1-dodecyl-3,5-diethyl-, chloride	**85:** P63073b
4-amino-1-dodecyl-3,5-diheptyl-, chloride	**85:** P63073b
4-amino-1-dodecyl-3,5-dimethyl-, bromide	**85:** P63073b

TABLE 22 *(Continued)*

Compound	Reference
22.2. Amino-1,2,4-Triazolium Compounds *(Continued)*	
4-amino-1-dodecyl-3,5-dimethyl-, chloride	**85:** P63073b
4-amino-1-dodecyl-3,5-dimethyl-, iodide	**85:** P63073b
4-[(1-amino-3-ethoxy-3-oxopropylidene)amino]-1-benzyl-, chloride	**78:** 124510w
4-amino-4-[2-(*N*-ethylanilino)ethyl]-3,5-bis(hydroxy-methyl), chloride	**67:** P74426y
4-amino-1-hexadecyl-, chloride	**85:** P63073b
4-amino-1-hexadecyl-3,5-dimethyl-, chloride	**85:** P63073b
4-amino-1-(2-hydroxyethyl)-3,5-diundecyl-, chloride	**85:** P63073b
4-amino-1-[2-(4-methoxyphenyl)-2-oxoethyl]-3,5-di-methyl-, bromide	**85:** 21220s
4-amino-1-methyl-, chloride	**78:** 124510w
4-amino-1-methyl, toluenesulfonate	**64:** 19596e
4-amino-1-methyl-3,5-diphenyl-, bromide	**82:** 4178e
4-amino-1-methyl-3,5-diphenyl-, iodide	**82:** 4178e
4-amino-1-(*p*-nitrobenzyl), bromide	**64:** 19596e
4-amino-1-(4-nitrobenzyl)-, chloride	**78:** 124510w
4-(4-aminophenyl)-	**82:** 16751z
4-amino-1-(2-propenyl)-3,5-diundecyl-, chloride	**85:** P63073b
4-amino-1-propyl-, chloride	**78:** 124510w
4-amino-1,3,5-trimethyl-, salt with 2,4,6-trinitrophenol (1:1)	**82:** 4178e
3-amino-1,4,5-triphenyl, bromide	**67:** 43755a
4-amino-1-trityl, chloride	**64:** 19596e
3-anilino-4-(*p*-chlorophenyl)-1-methyl-5-phenyl-, hydroxide, inner salt	**75:** 151734s, **81:** 3839p
3-anilino-1,4-diphenyl, hydroxide, inner salt	**51:** 7999c, **68:** 59504z **81:** 2977v
3-anilino-1,4-diphenyl, dichloroiodate(1-)	**77:** 42777d
3-anilino-1,4-diphenyl, hexaiodorhodate(3-) (3:1)	**78:** 105462r
3-anilino-1,4-diphenyl, nitrate	**82:** 51075d
3-anilino-1,4-diphenyl, tetrabromoaurate(1-)	**80:** 103553p
3-anilino-1,4-diphenyl, tetrakis(isothiocyanato)-cobaltate(2-) (2:1)	**75:** 41667q
5-anilino-2,4-diphenyl-, hydroxide, inner salt	**83:** 141238j
3-anilino-4-(*p*-methoxyphenyl)-1-methyl-5-phenyl-, hydroxide, inner salt	**75:** 151734s, **81:** 3839p
4-anilino-3-(4-methoxyphenyl)-1-methyl-5-phenyl-, perchlorate	**80:** 146119u
4-anilino-5-(4-methoxyphenyl)-1-methyl-3-phenyl-, perchlorate	**80:** 146119u
3-anilino-1-methyl-4,5-diphenyl-, hydroxide, inner salt	**81:** 3839p
3-anilino-1-methyl-4,5-diphenyl-, nitrate	**81:** 3839p
3-anilino-5-methyl-1,4-diphenyl-, hydroxide, inner salt	**68:** 59504z
4-anilino-1-methyl-3,5-diphenyl-, chloride	**80:** 146119u, **83:** 188770s
4-anilino-1-methyl-3,5-diphenyl-, perchlorate	**80:** 146119u
4-anilino-1-methyl-5-phenyl-3-(2-thienyl)-, perchlorate	**80:** 146119u

TABLE 22 (*Continued*)

Compound	Reference

22.2. Amino-1,2,4-Triazolium Compounds (*Continued*)

Compound	Reference
3-anilino-1-methyl-5-phenyl-4-*p*-tolyl-, hydroxide inner salt	**75:** 151734s
4-anilino-1-phenyl, chloride	**62:** 6476c
3-anilino-1,4,5-triphenyl, hydroxide, inner salt	**39:** 2068[8], **81:** 3839p
4-benzamido-1-benzyl-, bromide	**72:** 66873e
4-benzamido-1-phenacyl-, hydroxide, inner salt	**72:** 66873e
4-benzamido-1-phenacyl-, bromide	**72:** 66873e
4-benzenesulfonamido-1-benzyl-, hydroxide, inner salt	**75:** 35893u
4-benzenesulfonamido-1-benzyl-, chloride	**75:** 35893u
4-benzenesulfonamido-1-(*p*-chlorobenzyl)-, hydroxide, inner salt	**75:** 35893u
4-benzenesulfonamido-1-(*p*-chlorobenzyl)-, bromide	**75:** 35893u
4-benzenesulfonamido-1-methyl-, hydroxide, inner salt	**75:** 35893u
4-benzenesulfonamido-1-methyl-, iodide	**75:** 35893u
4-(benzoylamino)-1-benzyl-, nitrate	**76:** 14436c
4-(benzylamino)-1-methyl-, iodide	**70:** 87681q
4-(benzylamino)-1-phenacyl-, bromide	**70:** 87681q
4-(benzylamino)-1,3,5-trimethyl-, iodide	**70:** 87681q
1-benzyl-4-(benzylamino)-, chloride	**70:** 87681q
1-benzyl-4-(benzylamino)-3,5-dimethyl-, chloride	**70:** 87681q
1-benzyl-4-(benzylideneamino), chloride	**64:** 19596f
1-benzyl-4-[(carbamoyl)amino]-, hydroxide, inner salt	**76:** 85755s
1-benzyl-4-[(carbamoyl)amino]-, bromide	**76:** 85755s
1-benzyl-4-formamido-, chloride	**72:** 66873e
1-benzyl-4-[(1-[(1-hydroxy-*N*-phenylformimidoyl)oxy]-ethylidene)amino]-, hydroxide, inner salt	**72:** 66873e
4-(benzylideneamino)-1-(*p*-bromophenacyl), bromide	**64:** 19596f
4-(benzylideneamino)-1-(2-cyanoethyl), bromide	**64:** 19596g
4-(benzylideneamino)-1-(cyanomethyl), chloride	**64:** 19596g
4-(benzylideneamino)-1-methyl, iodide	**64:** 19596f
4-(benzylideneamino)-1-methyl, methyl sulfate	**64:** 19596g
4-(benzylideneamino)-1-methyl, *p*-toluenesulfonate	**64:** 19596f
4-(benzylideneamino)-1-(*p*-nitrobenzyl), bromide	**64:** 19596f
4-(benzylideneamino)-1,3,5-trimethyl, iodide	**64:** 19596g
1-benzyl-4-isobutyramido-, chloride	**72:** 66873e
1-benzyl-4-([(4-methylphenyl)sulfonyl]amino)-, hydroxide, inner salt	**80:** 81776j
1-benzyl-4-(nitroamino)-, hydroxide, inner salt	**76:** 14436c
1-benzyl-4-(2-phenylacetamido)-, hydroxide, inner salt	**72:** 66873e
1-benzyl-4-(2-phenylacetamido)-, chloride	**72:** 66873e
1-benzyl-4-[(*N*-phenylcarbamoyl)amino]-, hydroxide, inner salt	**76:** 85755s
1-benzyl-4-[(*N*-phenylcarbamoyl)amino]-, bromide	**76:** 85755s
1-benzyl-4-[(*N*-phenylcarbamoyl)amino]-, chloride	**76:** 85755s
3,5-bis(dimethylamino)-4-(2-hydroxyphenyl)-1-methyl-, chloride	**84:** 30981x
4-[3,4-bis(ethoxycarbonyl)-2,5-dimethyl-1H̲-pyrrol-1-yl]-1-methyl-, perchlorate	**82:** 140030c

TABLE 22 (*Continued*)

Compound	Reference

22.2. Amino-1,2,4-Triazolium Compounds (*Continued*)

Compound	Reference
1-(4-bromobenzyl)-4-[(carbamoyl)amino]-, hydroxide, inner salt	**76:** 85755s
1-(4-bromobenzyl)-4-[(carbamoyl)amino]-, bromide	**76:** 85755s
4-(4-bromobenzylideneamino)-1-methyl-, iodide	**76:** 25193x
1-butyl-4-butyramido-, bromide	**72:** 66873e
1-butyl-4-[[1-[(1-hydroxy-*N*-phenylformimidoyl)oxy]-ethylidene]amino]-, hydroxide, inner salt	**72:** 66873e
4-[(carbamoyl)amino]-1-(4-chlorobenzyl)-, hydroxide, inner salt	**76:** 85755s
4-[(carbamoyl)amino]-1-(4-chlorobenzyl)-, bromide	**76:** 85755s
4-[(carbamoyl)amino]-1-propyl-, iodide	**76:** 85755s
3-(4-chloroanilino)-4-(4-chlorophenyl)-1-methyl-5-phenyl-, hydroxide, inner salt	**81:** 3839p
4-(*p*-chloroanilino)-1-(*p*-fluorophenyl)-, chloride	**72:** 74407w
3-(*p*-chloroanilino)-1-methyl-4,5-diphenyl-, hydroxide, inner salt	**75:** 151734s, **81:** 3839p
4-[(*p*-chlorobenzyl)amino]-1-methyl-, iodide	**70:** 87681q
4-(3-chlorobenzylideneamino)-1-methyl-, iodide	**76:** 25193x
4-(4-chlorobenzylideneamino)-1-methyl-, iodide	**76:** 25193x
1-(4-chlorobenzyl-4-[(*N*-methylcarbamoyl)amino]-, hydroxide, inner salt	**76:** 85755s
1-(4-chlorobenzyl)-4-[(*N*-methylcarbamoyl)amino]-, bromide	**76:** 85755s
1-(4-chlorobenzyl)-4-(nitroamino)-, hydroxide, inner salt	**76:** 14436c
1-(*p*-chlorobenzyl)-4-*p*-toluenesulfonamido-, hydroxide, inner salt	**75:** 35893u
1-(*p*-chlorobenzyl)-4-*p*-toluenesulfonamido-, bromide	**75:** 35893u
3-(*p*-chloro-*N*-methylanilino)-1-methyl-4,5-diphenyl-, iodide	**75:** 151734s, **81:** 3839p
4-[[(2-chlorophenyl)methylene]amino]-1-phenyl-, tetrafluoroborate(1-)	**84:** 105495c
4-(4-chlorophenyl)-1-methyl-3-(methylphenylamino)-5-phenyl-, iodide	**75:** 151734s, **81:** 3839p
3,3'-[1,10-decanediylbis(phenylimino)]bis[1,4-diphenyl-, 4-methylbenzenesulfonic acid (1:2)	**79:** P25673r
4-[(2,6-dichlorobenzylidene)amino]-1-ethyl, iodide	**67:** PC11327w
4-[(2,6-dichlorobenzylidene)amino]-1-methyl, iodide	**67:** PC11327w
1,4-diethyl-2-[2-(3-ethyl-2-benzothiazolinylidene)ethyli-deneamino]-5-[3-(3-ethyl-2-benzothiazolinylidene)pro-penyl], iodide	**44:** P6318f
4-[4-(dimethylamino)benzylideneamino]-1-methyl-, iodide	**76:** 25193x
3,5-dimethyl-4-(*p*-nitroanilino)-1-phenyl-, hydroxide, inner salt	**75:** 98508t
4-[(diphenylmethylene)amino]-1-phenyl-, tetrafluoro-borate(1-)	**84:** 105495c
1,3-diphenyl-4-[(phenylmethylene)amino-, chloride	**84:** 105495c
1,3-diphenyl-4-[(phenylmethylene)amino]-, tetrafluoro-borate(1-)	**84:** 105495c

TABLE 22 (*Continued*)

Compound	Reference

22.2. Amino-1,2,4-Triazolium Compounds (*Continued*)

Compound	Reference
4-(*N*-ethyl-*p*-nitroanilino)-3,5-dimethyl-1-phenyl-, tetrafluoroborate(1-)	**75:** 98508t
4-(*N*-ethyl-*p*-nitroanilino)-1,3,5-triphenyl-, tetrafluoroborate(1-)	**75:** 98508t
4-(*p*-fluoroanilino)-1-(*p*-fluorophenyl)-, chloride	**72:** 74407w
3-(2-furanyl)-4-[(2-furanylmethylene)amino]-1-phenyl-, tetrafluoroborate(1-)	**84:** 105495c
4-(2-hydroxybenzylideneamino)-1-methyl-, iodide	**76:** 25193x
4-(3-hydroxybenzylideneamino)-1-methyl-, iodide	**76:** 25193x
4-(4-hydroxybenzylideneamino)-1-methyl-, iodide	**76:** 25193x
3-[6-[(2-hydroxyethyl)amino]-1,3-dioxo-1H-benz[de]-isoquinolin-2(3H)-yl]-1,3-dimethyl-, methyl sulfate	**83:** P61703s
4-[4-(isopropyl)benzylideneamino]-1-methyl-, iodide	**76:** 25193x
4-(4-methoxybenzylideneamino)-1-methyl-, iodide	**76:** 25193x
3-(3-methoxyphenyl)-4-[[(3-methoxyphenyl)methylene]-amino]-1-phenyl-, tetrafluoroborate(1-)	**84:** 105495c
3-(4-methoxyphenyl)-4-[[(4-methoxyphenyl)methylene]-amino]-1-phenyl-, tetrafluoroborate(1-)	**84:** 105495c
4-([[(2-methoxyphenyl)methylene]amino)-1-phenyl-, tetrafluoroborate(1-)	**84:** 105495c
3-(*N*-methylanilino)-1,4-diphenyl-, perchlorate	**74:** 53659f
1-methyl-4,5-diphenyl-3-*p*-toluidino-, hydroxide, inner salt	**75:** 151734s, **81:** 3839p
1-methyl-4-(methylamino)-, iodide	**70:** 87681q
1-methyl-3-(*N*-methylanilino)-4,5-diphenyl-, iodide	**81:** 3839p
1-methyl-3-(4-methylanilino)-4-(4-methylphenyl)-5-phenyl-, hydroxide, inner salt	**81:** 3839p
1-methyl-4-(2-methyl-4,6-diphenyl-1-pyridinio)-, diperchlorate	**82:** 140030c
1-methyl-4-[(1-naphthylmethyl)amino]-, iodide	**70:** 87681q
1-methyl-4-(nitroamino)-, hydroxide, inner salt	**80:** 47913s
4-(*N*-methyl-*p*-nitroanilino)-1,3,5-triphenyl-, perchlorate	**75:** 98508t
1-methyl-4-(3-nitrobenzylideneamino)-, iodide	**76:** 25193x
1-methyl-4-(4-nitrobenzylideneamino)-, iodide	**76:** 25193x, **77:** 159501a
1-methyl-4-(2-phenylacetamido)-, iodide	**72:** 66873e
4-([[(4-methylphenyl)methylene]amino)-1-phenyl-, tetrafluoroborate(1-)	**84:** 105495c
3-(4-methylphenyl)-4-([[(4-methylphenyl)methylene]am-ino)-1-phenyl-, tetrafluoroborate(1-)	**84:** 105495c
1-(4-methylphenyl)-4-[(1-phenylethylidene)amino]-, tetrafluoroborate(1-)	**84:** 105495c
1-(4-methylphenyl)-4-[(phenylmethylene)amino]-, tetrafluoroborate(1-)	**84:** 105495c
1-(4-methylphenyl)-3-phenyl-4-[(phenylmethylene)am-ino]-, tetrafluoroborate(1-)	**84:** 105495c
1-methyl-4-(1H-pyrrol-1-yl)-, perchlorate	**82:** 140030c
1-methyl-4-(2,4,6-trimethyl-1-pyridinio)-, diperchlorate	**82:** 140030c
1-methyl-4-(2,4,6-triphenyl-1-pyridinio)-, diperchlorate	**82:** 140030c

TABLE 22 (*Continued*)

Compound	Reference

22.2. Amino-1,2,4-Triazolium Compounds (*Continued*)

Compound	Reference
3-(1-naphthalenyl)-4-[(1-naphthalenylmethylene)amino]-1-phenyl-, tetrafluoroborate(1-)	**84:** 105495c
4-(*p*-nitroanilino)-3,5-dimethyl-1-phenyl-, perchlorate	**74:** 111972r, **75:** 98508t
4-(*p*-nitroanilino)-1,3,5-triphenyl-, hydroxide, inner salt	**75:** 98508t
4-[*N*-(*p*-nitrophenyl)acetamido]-1,3,5-triphenyl-, perchlorate	**75:** 98508t
4-([(3-nitrophenyl)methylene]amino)-1-phenyl-, tetrafluoroborate(1-)	**84:** 105495c
4-[(*N*-phenylcarbamoyl)amino]-1-propyl-, hydroxide, inner salt	**76:** 85755s
4-[(*N*-phenylcarbamoyl)amino]-1-propyl-, iodide	**76:** 85755s
1-phenyl-4-[(1-phenylethylidene)amino]-, tetrafluoroborate(1-)	**84:** 105495c
1-phenyl-4-[(phenylmethylene)amino]-, tetrafluoroborate(1-)	**84:** 105495c
1-phenyl-3-(2-thienyl)-4-[(2-thienylmethylene)amino]-, chloride	**84:** 105495c
1-phenyl-3-(2-thienyl)-4-[(2-thienylmethylene)amino]-, tetrafluoroborate(1-)	**84:** 105495c
1,1'-tetramethylenebis[4-amino, bromide]	**64:** 19596e
1,1'-tetramethylenebis[4-(benzylideneamino), bromide]	**64:** 19596g
4-*o*-toluidino-1-*o*-tolyl, chloride	**62:** 6476c
4-*m*-toluidino-1-*m*-tolyl, chloride	**62:** 6476c, **72:** 74407w
4-*p*-toluidino-1-*p*-tolyl, chloride	**62:** 6476c, **72:** 74407w
1,1'-trimethylenebis[4-amino, bromide]	**64:** 19596e

22.3. Azido-, Azo-, or Diazoamino (Triazeno)-1,2,4-Triazolium or Triazolinium Compounds Containing One or More Representative Functions

Compound	Reference
5-([*p*-(2-acetyl-1-methylhydrazino)phenyl]azo)-1,4-dimethyl,	**62:** P7910e
3(or 5)-[(1-[2-(acetyloxy)ethyl]-1,2,3,4-tetrahydro-6-isoquinolinyl)azo]-1,4-dimethyl-, tetrachlorozincate(2-) (2:1)	**85:** P125782t
5-amino-3-[(*p*-[benzyl(2-cyanoethyl)amino]phenyl)azo]-2,4-diethyl-, tetrachlorozincate(2-)	**75:** P130778u
5-amino-3-[(*p*-[benzyl(ethyl)amino]phenyl)azo]-2,4-dimethyl-, chloride	**75:** P130778u
5-amino-3-[(*p*-[benzyl(ethyl)amino]phenyl)azo]-2,4-dimethyl-, tetrachlorozincate(2-) (2:1)	**75:** P130778u
5-([2-amino-6-(butylimino)-5-cyano-4-(ethylamino)-3-pyridinyl]azo)-1,4-dimethyl-, tetrachlorozincate(2-) (2:1)	**84:** P61174r
3-([2-amino-6-(butylimino)-5-cyano-1-ethyl-4-(ethylamino)-1,6-dihydro-3-pyridinyl]azo)-1,4-dimethyl-, tetrachlorozincate(2-) (2:1)	**84:** P61174r
3-[(4-[(2-[(4-amino-6-chloro-1,3,5-triazin-2-yl)amino]-ethyl)ethylamino]phenyl)azo]-1-benzyl-4-methyl-, methylsulfate	**77:** P103304w

TABLE 22   (*Continued*)

Compound	Reference

22.3. Azido-, Azo-, or Diazoamino (Triazeno)-1,2,4-Triazolium or Triazolinium Compounds
   Containing One or More Representative Functions (*Continued*)

5-amino-3-([*p*-(dibenzylamino)phenyl]azo)-2,4-diethyl-, tetrachlorozincate (2-) (2:1)	**75:** 130778u
5-amino-3-([*p*-(dimethylamino)phenyl]azo)-2,4-diethyl-, tetrachlorozincate (2-) (2:1)	**75:** P130778u
4-(2-aminoethyl)-1-benzyl-3-([4-(4-morpholinyl)phenyl]-azo)-5-phenyl-, chloride	**81:** P93043b
4-(2-aminoethyl)-1-benzyl-5-phenyl-3-[(2-phenyl-1H̲-indol-3-yl)azo]-, chloride	**81:** P79381d
2(or 4)-(3-amino-3-oxopropyl)-1,5-dimethyl-3-[(1-ethyl-1,2,3,4-tetrahydro-2,2,4-trimethyl-6-isoquinolinyl)-azo]-, tetrachlorozincate(2-) (2:1)	**85:** P110086v
2(or 4)-(3-amino-3-oxopropyl)-1,5-dimethyl-3-([1,2,3,4-tetrahydro-1-(phenylmethyl)-6-isoquinolinyl]-azo)-, tetrachlorozincate(2-) (2:1)	**85:** P125782t
1(2 or 4)-(3-amino-3-oxopropyl)-4(1 or 2)-ethyl-3-[(1-ethyl-2-phenyl-1H̲-indol-3-yl)azo]-, tetrachloro-zincate(2-) (2:1)	**85:** P7255a
1(2 or 4)-(3-amino-3-oxopropyl)-3-[(1-ethyl-2-phenyl-1H̲-indol-3-yl)azo]-4(1 or 2)-(2-hydroxy-3-methoxy-propyl)-, tetrachlorozincate(2-) (2:1)	**85:** P22740y
1(2 or 4)-(3-amino-3-oxopropyl)-3-[(1-ethyl-2-phenyl-1H̲-indol-3-yl)azo]-4(1 or 2)-(2-hydroxypropyl)-5-methyl-, tetrachlorozincate(2-) (2:1)	**85:** P144687g
1(2 or 4)-(3-amino-3-oxopropyl)-3-[[4-[ethyl(phenyl-methyl)amino]-2-methylphenyl]azo]-4(1 or 2)-(2-hydroxypropyl)-, tetrachlorozincate(2-) (2:1)	**85:** P79672g
1(2 or 4)-(3-amino-3-oxopropyl)-3-[(4-[ethyl(phenyl-methyl)amino]phenyl)azo]-4(1 or 2)-(2-hydroxycyclo-hexyl)-, tetrachlorozincate(2-) (2:1)	**85:** P22741z, **85:** P22745d
1(2 or 4)-(3-amino-3-oxopropyl)-3-[(4-[ethyl(phenyl]-methyl)amino]phenyl)azo]-4(1 or 2)-(phenylmethyl)-, tetrachlorozincate(2-) (2:1)	**85:** P22745d
2(or 4)-(3-amino-3-oxopropyl)-3-[(1-ethyl-1,2,3,4-tetrahydro-2-methyl-6-quinolinyl)azo]-1,5-dimethyl-, tetrachlorozincate(2-) (2:1)	**85:** P125783u
1(2 or 4)-(3-amino-3-oxopropyl)-4(1 or 2)-(2-hydroxy-3-phenoxypropyl)-3-[(1-methyl-2-phenyl-1H̲-indol-3-yl-azo]-, tetrachlorozincate(2-) (2:1)	**85:** P22740y
5-azido-4-ethyl-3-methyl-1-phenyl, tetrafluoroborate	**56:** 10134i
4,4′-azobis[1-methyl-3,5-diphenyl-, bis(fluorosulfate)	**82:** 4178e
4,4′-azobis[1-methyl-3,5-diphenyl-, diperchlorate	**82:** 4178e
4,4′-azobis[1,3,5-trimethyl-, bis(fluorosulfate)	**82:** 4178e
4,4′-azobis[1,3,5-trimethyl-, diperchlorate	**82:** 4178e
4,4′-azobis[1,3,5-trimethyl-, salt with 2,4,6-trinitro-phenol (1:2)	**82:** 4178e
3-([4-([2-(benzoyloxy)ethyl]ethylamino)phenyl]azo)-1,2(or 1,4)-bis(benzyl)-, trichlorozincate(1-)	**83:** P165797s

TABLE 22 *(Continued)*

Compound	Reference
22.3. Azido-, Azo-, or Diazoamino (Triazeno)-1,2,4-Triazolinium or Triazolium Compounds Containing One or More Representative Functions *(Continued)*	
3-([4-([2-(benzoyloxy)ethyl]ethylamino)phenyl]azo)-1,2(or-1,4)-dibutyl-, trichlorozincate(1-)	**83:** P165797s
3-([4-([2-(benzoyloxy)ethyl]ethylamino)phenyl]azo)-1,2(or 1,4)-diethyl-, trichlorozincate(1-)	**83:** P165797s
3-([4-([2-(benzoyloxy)ethyl]ethylamino)phenyl]azo)-1,2-(or 1,4)-dimethyl-, chloride	**83:** P165797s
1-benzyl-3-[(1-butyl-1,2,3,4-tetrahydro-3-hydroxy-6-quinolyl)azo]-4-methyl-5-phenyl-, methyl sulfate	**73:** P36546d
1-benzyl-4-(2-carbamoylethyl)-5-methyl-3-[[p-(N-methyl-anilino)phenyl]azo]-, chloride	**73:** P36546d
1-benzyl-4-(2-carbamoylethyl)-5-methyl-3-[(2-methyl-indol-3-yl)azo]-, chloride	**73:** P46632t
1-benzyl-4-(2-carboxyethyl)-5-[(1-methyl-2-phenyl-1H-indol-3-yl)azo]-, chloride	**80:** P28461b
1-benzyl-3-[[4-(diethylamino)phenyl]azo]-4-(2-hydroxybutyl)-, chloride	**81:** P93043b
1-benzyl-3-[[p-(diethylamino)phenyl]azo]-4-methyl-, methyl sulfate	**73:** P36546d
1-benzyl-3-[(2,3-dihydro-1,2,3,3-tetramethyl-1H-indol-5-yl)-azo]-4-(2-mercaptoethyl)-5-methyl-, chloride	**81:** P93043b
1-benzyl-4,5-dimethyl-3-[(2-methylindol-3-yl)azo]-, methyl sulfate	**73:** P46632t
5-[p-(benzylethylamino)phenylazo-1,4-dimethyl, p-toluenesulfonate or trichlorozincate	**54:** P3970a, **61:** P5822h
1-benzyl-4-ethyl-3-[(1-methyl-2-phenylindol-3-yl)azo]-, bromide	**73:** P46632t
1-benzyl-3-[(1-ethyl-2-phenyl-1H-indol-3-yl)azo]-4-(2-mercaptoethyl)-5-methyl-, chloride	**81:** P79381d
1-benzyl-4(or 2)-ethyl-3-[(1,2,4-triethyl-1,2,3,4-tetrahy-dro-2-methyl-6-quinolinyl)azo]-, bromide	**85:** P110086v
1-benzyl-4-(2-hydroxybutyl)-3-[(1-methyl-2-phenyl-1H-indol-3-yl)azo]-, chloride	**81:** P79381d
1-benzyl-4-methyl-3-[(1-methyl-2-phenylindol-3-yl)azo]-, methyl sulfate	**73:** P46632t
3-([4-([2-([1,1'-biphenyl]-4-yloxy)ethyl]ethylamino)phe-nyl]-azo)-1,2(or 1,4)-bis(2-hydroxyethyl)-, chloride	**83:** P165797s
3-[([1,1'-biphenyl]-4-yloxy)methyl]-5-([4-(diethylamino)-phenyl]azo)-1,2(or 1,4)-dimethyl, chloride	**81:** P65189k
1,4-bis(2-aminoethyl)-3,5-bis[(2-methyl-1H-indol-3-yl)-azo]-chloride	**81:** P79380c
1,4-bis(3-amino-3-oxopropyl)-3,5-bis([4-(diethylamino)-phenyl]azo)-, tetrachlorozincate(2-) (2:1)	**81:** P137569j
1,4-bis(3-amino-3-oxopropyl)-3,5-bis[(1-methyl-2-phenyl-1H-indol-3-yl)azo]-, chloride	**81:** P93041z
1,2(or 1,4)-bis(3-amino-3-oxopropyl)-3-[(4-[ethyl-(2-[(phenoxycarbonyl)oxy]ethyl)amino]phenyl)azo]-, chloride	**85:** P34629g

TABLE 22 (*Continued*)

Compound	Reference

22.3. Azido-, Azo-, or Diazoamino (Triazeno)-1,2,4-Triazolium or Triazolinium Compounds
Containing One or More Representative Functions (*Continued*)

Compound	Reference
1,4-bis(3-amino-3-oxopropyl)-3(or 5)-[(1-ethyl-1,2,3,4-tetra-hydro-2,7-dimethyl-6-quinolinyl)azo]-, tetrachloro-zincate(2-) (2:1)	**85:** P125783u
1,4-bis(3-amino-3-oxopropyl)-3(or 5)-[(1-ethyl-1,2,3,4-tetrahydro-2,2,4,7-tetramethyl-6-isoquinolinyl)azo]-, tetrachlorozincate(2-) (2:1)	**85:** P110086v
1,4-bis(3-amino-3-oxopropyl)-3-(3-hydroxypropyl)-5-[(2-methyl-1H-indol-3-yl)azo]-tetrachlorozincate(2-) (2:1)	**79:** P106129v
1,4-bis(3-amino-3-oxopropyl)-5-[[4-[methyl-(phenylmethyl)amino]phenyl]azo]-3-phenyl-, chloride	**81:** P65195j
1,4-bis(3-amino-3-oxopropyl)-3(or 5)-[[1,2,3,4-tetrahydro-7-methyl-1-(phenylmethyl)-6-isoquinolinyl]-azo]-, tetrachlorozincate(2-) (2:1)	**85:** P125782t
3-([2,6-bis(butylamino)-5-cyano-4-methyl-3-pyridinyl]azo)-1,4(or 2)-dimethyl-, chloride	**83:** P99180g
1,4-bis(2-carbamoylethyl)-3-[[4-[[2-([1,1'-biphenyl]-4-yloxy)ethyl]ethylamino]phenyl]azo]-, chloride	**83:** P165797s
3,5-bis[(5-chloro-1-ethyl-2-phenyl-1H-indol-3-yl)azo]-1,2(or 1,4)-dimethyl-, salt with 4-methylbenzene-sulfonic acid (1:1)	**83:** P181082k
3-[(4-[bis(2-cyanoethyl)amino]phenyl)azo]-1,4-dimethyl-, trichlorozincate(1-)	**85:** 95772v
5-[(*p*-[bis(2-cyanoethyl)amino]phenyl)azo]-1,4-dimethyl, trichlorozincate	**61:** P5823a
3,5-bis([4-(diethylamino)phenyl]azo)-1,4-bis(2-hydroxy-propyl)-, chloride	**81:** P171335q
3,5-bis([*p*-(diethylamino)phenyl]azo)1,4-dimethyl-, methyl sulfate	**75:** P50405e
3,5-bis[(2,3-dihydro-1,2,3,3-tetramethyl-1H-indol-5-yl)azo]-1,4-bis[3-(methylamino)-3-oxopropyl]-, chloride	**81:** P137569j
3,5-bis[(1,2-dimethyl-1H-indol-3-yl)azo]-1,2(or 1,4)-dimethyl-, methyl sulfate	**83:** P181082k
3,5-bis[(2-ethyl-1H-indol-3-yl)azo]-1,4-dimethyl-, benzenesulfonate	**83:** P181082k
3,5-bis[(1-ethyl-2-phenyl-1H-indol-3-yl)azo]-1,4-bis(2-mercaptoethyl)-, chloride	**81:** P79380c
3,5-bis[(1-ethyl-2-phenyl-1H-indol-3-yl)azo]-1,4-bis[3-(methylamino)-3-oxopropyl]-, chloride	**81:** P93041z
1,2(or 1,4)-bis(2-hydroxycyclohexyl)-3-[(1-methyl-2-phenyl-1H-indol-3-yl)azo]-, tetrachlorozincate(2-) (2:1)	**84:** P152200n
1,2(or 1,4)-bis(2-hydroxycyclohexyl)-3-[(2-phenyl-1H-indol-3-yl)azo]-, tetrachlorozincate(2-) (2:1)	**84:** P152200n
3-[(4-[bis(2-hydroxyethyl)amino]phenyl)azo]-1,4-dimethyl-, trichlorozincate(1-)	**85:** 95772v

642

TABLE 22 (*Continued*)

Compound	Reference

22.3. Azido-, Azo-, or Diazoamino (Triazeno)-1,2,4-Triazolium or Triazolinium Compounds Containing One or More Representative Functions (*Continued*)

Compound	Reference
3-(4-[(4-[bis(2-hydroxyethyl)amino]phenyl)azo]phenyl)-1,4-dimethyl-, methyl sulfate	**85:** 95772v
1,4-bis(2-hydroxy-3-methoxypropyl)-5-[[4-[methyl-(phenylmethyl)amino]phenyl]azo]-, chloride	**85:** P95798h
1,4-bis[3-[(hydroxymethyl)amino]-3-oxopropyl]-5-[[4-[methyl(phenylmethyl)amino]phenyl]azo]-3-phenyl-, chloride	**81:** P65195j
1,4-bis(2-hydroxy-3-phenoxypropyl)-5-[(1-methyl-2-phenyl-1H-indol-3-yl)azo]-, trichlorozincate(1-)	**79:** P93432m, **85:** P95798h
1,4-bis(2-hydroxypropyl)-3,5-bis[(1-methyl-2-phenyl-1H-indol-3-yl)azo]-, chloride	**81:** P79380c
[[4-[bis(phenylmethyl)amino]phenyl]azo]-(2-hydroxyethyl)methyl-, tetrachlorozincate(2-) (2:1)	**85:** P79672g
(3-butoxy-2-hydroxypropyl)ethyl-3-[(1-methyl-2-phenyl-1H-indol-3-yl)azo]-, tetrachlorozincate(2-) (2:1)	**85:** P22740y
3(or 5)-[(1-butyl-1,2,3,4-tetrahydro-2,4-dimethyl-6-quinolinyl)azo]-1,4-dimethyl-, tetrachlorozincate(2-) (2:1)	**85:** P125783u
3-[(1-butyl-1,2,3,4-tetrahydro-2-methyl-6-quinolinyl)-azo]-1,4(or 1,2)-dimethyl-, tetrachlorozincate(2-) (2:1)	**85:** P125783u
3(or 5)-[(1-butyl-1,2,3,4-tetrahydro-2,2,4-trimethyl-6-quinolinyl)azo]-1,4-dimethyl-, tetrachlorozincate(2-) (2:1)	**85:** P110086v
5-[[2-chloro-4-(diethylamino)phenyl]azo]-3-(3-hydroxypropyl)-1,4-bis(phenylmethyl)-, tetrachloro-zincate(2-) (2:1)	**79:** P106129v
5-[(4-[(2-chloroethyl)methylamino]phenyl)azo]-4-[2-hydroxy-3-(1-oxopropoxy)propyl]-1-benzyl-, chloride	**83:** P81203g
3-[(5-chloro-1-methyl-2-phenyl-1H-indol-3-yl)azo]-1,4-bis(2-hydroxycyclohexyl)-5-phenyl-, tetrachloro-zincate(2-) (2:1)	**84:** 152200n
3-[(5-chloro-1-methyl-2-phenyl-1H-indol-3-yl)azo]-1-(2-cyanoethyl)-4-(2-hydroxybutyl)-, tetrachloro-zincate(2-) (2:1)	**85:** P7255a
[(5-chloro-1-methyl-2-phenyl-1H-indol-3-yl)azo]-1-(2-hydroxyethyl)-4-methyl-, tetrachlorozincate(2-) (2:1)	**85:** P144687g
3-([4-([(4-chlorophenyl)methyl]methylamino)phenyl]-azo)-1(or 2), 4-dimethyl-, trichlorozincate(1-)	**85:** P7172w
1-(2-cyanoethyl)-3-([4-(diethylamino)-2-methylphenyl]-azo)-4-methyl-, tetrachlorozincate(2-) (2:1)	**85:** P22745d
1-(2-cyanoethyl)-3-([4-(diethylamino)phenyl]azo)-4-(2-hydroxypropyl)-, tetrachlorozincate(2-) (2:1)	**85:** P22745d
1-(2-cyanoethyl)-4-ethyl-3-[(4-[ethyl(2-phenoxyethyl)-amino]phenyl)azo]-, tetrachlorozincate(2-) (2:1)	**85:** P22745d

TABLE 22 (*Continued*)

Compound	Reference
22.3. Azido-, Azo-, or Diazoamino (Triazeno)-1,2,4-Triazolium or Triazolinium Compounds Containing One or More Representative Functions (*Continued*)	
1-(2-cyanoethyl)-4-ethyl-3-methyl-5-[(2-phenyl-1H-indol-3-yl)azo]-, tetrachlorozincate(2-) (2:1)	**85:** P7255a
1-(2-cyanoethyl)-3-[(1-ethyl-2-phenyl-1H-indol-3-yl)azo]-4-methyl-, tetrachlorozincate(2-) (2:1)	**85:** P7255a
1-(2-cyanoethyl)-3-[(4-[ethyl(phenylmethyl)amino]phenyl)azo]-4-(2-hydroxybutyl)-, tetrachlorozincate(2-) (2:1)	**85:** P22745d
1-(2-cyanoethyl)-3-[(4-[ethyl(phenylmethyl)amino]phenyl)azo]-4-methyl-, tetrachlorozincate(2-) (2:1)	**85:** P22745d
1-(2-cyanoethyl)-4-(2-hydroxy-3-methoxypropyl)-3-[(2-methyl-1H-indol-3-yl)azo]-5-phenyl-, tetrachlorozincate(2-) (2:1)	**85:** P7255a
1-(2-cyanoethyl)-3-[(4-[(2-methoxyethyl)methylamino]phenyl)azo]-4-methyl-, tetrachlorozincate(2-) (2:1)	**85:** P22745d
3-[(4-[(2-cyanoethyl)methylamino]phenyl)azo]-1,4-bis(2-hydroxycyclohexyl)-5-methyl-, tetrachlorozincate(2-) (2:1)	**84:** P152199u
5-([1-(2-cyanoethyl)-2-methyl-1H-indol-3-yl]azo)-1,4-dimethyl-, chloride	**81:** P51113u
1-(2-cyanoethyl)-4-methyl-3-[(1-methyl-2-phenyl-1H-indol-3-yl)azo]-, tetrachlorozincate(2-) (2:1)	**85:** P7255a
5-([4-[(2-cyanoethyl)(2-phenylethyl)amino]phenyl]azo)-1,4-dimethyl-, tetrachlorozincate(2-) (2:1)	**78:** P148964y
3(or 5)-([4-(cyclohexylmethylamino)phenyl]azo)-1,4-dimethyl-, trichlorozincate(1-)	**83:** P81186d
5-([4-(cyclohexylmethylamino)phenyl]azo)-1,4-dimethyl-, methyl sulfate	**84:** P6394n
5-([5-cyano-4-methyl-2,6-bis(methylamino)-3-pyridinyl]azo)-1,4-dimethyl-, chloride	**80:** P84684p
3(or 5)-[(2,6-diamino-5-cyano-4-methyl-3-pyridinyl)azo]-1,4-dimethyl-, chloride	**83:** P99180g
3(or 5)-[(1,3-diamino-4-isoquinolinyl)azo]-1,4-dimethyl-, tetrachlorozincate(2-) (2:1)	**83:** P81207m
5-([4-(dibutylamino)phenyl]azo)-1,4-bis(2-hydroxybutyl)-, chloride	**79:** P93432m, **85:** 95798h
1,4-dibutyl-3-[(4-[ethyl(2-[(phenoxycarbonyl)oxy]ethyl)amino]phenyl)azo]-, trichlorozincate(1-)	**85:** P34629g
3-[(4-[(2-[(4,6-dichloro-1,3,5-triazin-2-yl)amino]ethyl)ethylamino]phenyl)azo]-4-methyl-5-phenyl-1-benzyl-, methylsulfate	**77:** P103304w
5-([4-(diethylamino)-2,6-dimethylphenyl]azo)-1,2(or 1,4)-dimethyl-, trichlorozincate(1-)	**84:** P75683t
5-([4-(diethylamino)-2-(methoxycarbonyl)phenyl]azo)-1,4-dimethyl-, trichlorozincate(1-)	**77:** P90037j
3-([4-(diethylamino)-2-methylphenyl]azo)-1,4-bis(2-hydroxycyclohexyl)-, tetrachlorozincate(2-) (2:1)	**84:** P152199u
5-([4-(diethylamino)-2-methylphenyl]azo)-3-(3-hydroxypropyl)-1,4-dimethyl-, benzenesulfonate	**79:** P106129v

TABLE 22 (*Continued*)

Compound	Reference
22.3. Azido-, Azo-, or Diazoamino (Triazeno)-1,2,4-Triazolium or Triazolinium Compounds Containing One or More Representative Functions (*Continued*)	
3-([4-(diethylamino)phenyl]azo)-1,4-bis(2-hydroxy-cyclohexyl)-, tetrachlorozincate(2-) (2:1)	**84:** P152199u
3-([*p*-(diethylamino)phenyl]azo)-1,4-dimethyl-, chloride	**73:** 62258f, **77:** 42829x
3-([*p*-(diethylamino)phenyl]azo)-1,4-dimethyl-, hexachloroantimonate(1-)	**73:** 62258f
3-([4-(diethylamino)phenyl]azo)-1,4-dimethyl-, tetrafluoroborate(1-)	**82:** P113168x
3-([4-(diethylamino)phenyl]azo)-1,4-dimethyl-, thallium complex	**82:** 80069k
3-([4-(diethylamino)phenyl]azo)-1,4-dimethyl-, triiodocuprate(1-)	**80:** 90814x
5-([4-(diethylamino)phenyl]azo)-1,4(or 2,4)-dimethyl-, methyl sulfate	**54:** P3970a, **61:** P5823a, **77:** P166059r
3(or 5)-([4-(diethylamino)phenyl]azo)-1,4-dimethyl-, tetrafluoroborate(1-)	**84:** P166142g
5-([4-(diethylamino)phenyl]azo)-1,4-dimethyl-3-phenoxymethyl-, trichlorozincate(1-)	**81:** P65180a
5-([4-(diethylamino)phenyl]azo)-4-(2-hydroxybutyl)-1-methyl-, chloride	**83:** P81203g
5-([4-(diethylamino)phenyl]azo)-3-(3-hydroxypropyl)-1,4-dimethyl-, methyl sulfate	**79:** P106129v
3-([4-([4-(diethylamino)phenyl]azo)-5-methoxy-2-methylphenyl]azo)-4-methyl-1-benzyl-, methyl sulfate	**77:** P76669k
5-([4-([4-(diethylamino)-*o*-tolyl]azo)-2,5-dimethoxyphenyl]azo)-1,4-dimethyl, trichlorozincate	**59:** P12957d
1,4-diethyl-5-[(4-[benzyl(methyl)amino]phenyl)azo]-, salt with 4-methyl-1,3-benzenedisulfonic acid (1:1), sodium salt	**83:** P61660a
1,4-diethyl-3-[(1-benzyl-1,2,3,4-tetrahydro-6-quinolinyl)azo]-, bromide	**85:** P125782t
1,4-diethyl-3,5-bis[(1-ethyl-5-methoxy-2-phenyl-1H-indol-3-yl)azo]-, salt with 4-methylbenzene-sulfonic acid (1:1)	**83:** P181082k
1,4-diethyl-5-[[1-ethyl-2-(4-methylphenyl)-1H-indol-3-yl]azo]-, trichlorozincate(1⁻)	**80:** P28454b
1,4-diethyl-5-[(1-ethyl-2-phenyl-1H-indol-3-yl)azo]-, trichlorozincate(1-)	**80:** P28454b
5-[(2,4-diethyl-1,2,3,4-tetrahydro-1,2-dimethyl-6-quinolinyl)azo]-1,4-dimethyl-, tetrachlorozincate(2-) (2:1)	**85:** P110086v
5-[[2,5-dimethoxy-4-[(1-methyl-2-phenylindol-3-yl)azo]phenyl]azo]-1,4-dimethyl-, tetrachlorozincate	**59:** P12957d
5-[(2,4-dimethoxyphenyl)azo]-1,4-dimethyl, trichlorozincate	**57:** P13929h
5-(4-dimethylamino-2-methoxyphenylazo)-1,4-dimethyl, perchlorate	**54:** P3970a

TABLE 22. (*Continued*)

Compound	Reference
22.3. Azido-, Azo-, or Diazoamino (Triazeno)-1,2,4-Triazolium or Triazolinium Compounds Containing One or More Representative Functions (*Continued*)	

Compound	Reference
5-([*p*-(dimethylamino)phenyl]azo)-1,4-dimethyl-, methyl sulfate	**71:** P14170h
5-([*p*-(dimethylamino)phenyl]azo)-1,4-dimethyl-, perchlorate	**55:** P11863b
5-([4-(dimethylamino)phenyl]azo)-1,4-dimethyl-, trichlorozincate 1-)	**79:** 20277e
3-([4-(dimethylamino)phenyl]azo)-4(or 2)-methyl-1-(3-phenylpropyl)-, chloride	**84:** P6478t
5-([2-(dimethylamino)-4-phenyl-5-thiazolyl]azo)-1,4-dimethyl-, chloride	**72:** P122907h
1,2(or 1,4)-dimethyl-3[(4-[(benzyl)methylamino]phenyl)-azo]-, chloride	**76:** P60914e
1,4-dimethyl-5-[(4-[benzyl(methyl)amino]phenyl)azo]-, trichlorozincate(1-)	**77:** P166059r
1,4-dimethyl-3,5-bis[(2-methylindol-3-yl)azo]-, methyl sulfate	**74:** P77396u
1,2(or 1,4)-dimethyl-3,5-bis[[1-methyl-2-(4-methyl-phenyl)-1H-indol-3-yl]azo]-, methyl sulfate	**83:** P181082k
1,2(or 1,4)-dimethyl-3,5-bis[(1-methyl-2-phenyl-1H-indol-3-yl)azo]-, methyl sulfate	**74:** P77396u
1,4-dimethyl-3,5-bis[(*p*-morpholinophenyl)azo]-, methyl sulfate	**75:** P50405e
1,4-dimethyl-3,5-bis[(1,2,3,3-tetramethyl-5-indolinyl)azo]-, methyl sulfate	**75:** P50405e
1,2(or 1,4)-dimethyl-3,5-bis[(1,2,5-trimethyl-1H-indol-3-yl)azo]-, methyl sulfate	**83:** P181082k
1,4-dimethyl-5-[3-(3-methyl-2-benzothiazolinylidene)-1-triazeno], trichlorozincate	**55:** 17030b, **55:** 18125g, **56:** P13123e
1,2(or 1,4)-dimethyl-3[[4-(methylphenylamino)phenyl]-azo]-, chloride	**76:** P60914e
1,4-dimethyl-5-[(1-methyl-2-phenyl-1H-indol-3-yl)azo]-, chloride	**82:** 172419c, **84:** P67786p
1,4-dimethyl-5-[(1-methyl-2-phenyl-1H-indol-3-yl)azo]-, tetrachlorozincate-(2-) (2:1)	**83:** P29853e
1,4-dimethyl-5-[(4-[methyl(phenylmethyl)amino]phenyl)-azo]-, chloride	**81:** P14701p
1,4-dimethyl-5-[(4-[methyl(phenylmethyl)amino]phenyl)-azo]-, methyl sulfate	**84:** P19080g, **84:** P137129c, **84:** P137130w
1,4-dimethyl-3-[[4-[[2-(*N*-methyl-*N*-phenylsulfamoyl)-ethyl]amino]phenyl]azo]-, tetrachlorozincate(2) (2:1)	**82:** P100064k
5,5'-[(2,5-dimethyl-*p*-phenylene)bis[methylene-(ethylimino)-1-phenyleneazo]]bis[4-benzyl-1-methyl-	**69:** P28582q
1,4-dimethyl-5-[(6-phenylimidazo[2,1-*b*]thiazo-5-yl)azo]-	**84:** P91634h
1,4-dimethyl-5-[(2-phenylindol-3-yl)azo], chloride	**58:** P4673e
1,4-dimethyl-5-[(2-phenyl-1H-indol-3-yl)azo]-, methyl sulfate	**81:** P122623y
1,4-dimethyl-3(or 5)-phenyl-5(or 3)-[(1,2,3,4-tetrahydro-1,2,2,4-tetramethyl-6-quinolinyl)azo]-, tetrachlorozincate(2-) (2:1)	**85:** P110086v

TABLE 22.   (*Continued*)

Compound	Reference

22.3. Azido-, Azo-, or Diazoamino (Triazeno)-1,2,4-Triazolium or Triazolinium Compounds Containing One or More Representative Functions (*Continued*)

Compound	Reference
1,4-dimethyl-5-[(2-phenyl-1-[2-(vinylsulfonyl)ethyl]-indol-3-yl)azo]-, methyl sulfate	**75:** P89316b
1,4-dimethyl-3(or 5)-([1,2,3,4-tetrahydro-1-(1-methylethyl)-6-quinolinyl]azo)-, tetrachlorozincate(2-) (2:1)	**85:** P125782t
1,4-dimethyl-3(or 5)-[(1,2,3,4-tetrahydro-2-methyl-1-pentyl-6-quinolinyl)azo]-, tetrachlorozincate(2-) (2:1)	**85:** P125783u
1,2(or 1,4)-dimethyl-3-[[1,2,3,4-tetrahydro-7-methyl-1-(phenylmethyl)-6-isoquinolinyl]azo]-, tetrachloro-zincate(2-) (3:1)	**85:** P125782t
1,2(or 1,4)-dimethyl-3-[[1,2,3,4-tetrahydro-1-(phenylmethyl)-6-isoquinolinyl]azo]-, tetrachloro-zincate(2-) (2:1)	**85:** P125782t
1,2(or 1,4)-dimethyl-3-[[1,2,3,4-tetrahydro-1,2,2,4-tetramethyl-6-quinolinyl)azo]-, tetrachloro-zincate(2-) (2:1)	**85:** P110086v
1,4-dimethyl-5-[(4-thiomorpholino-*o*-toly)azo]-, methyl sulfate, *S,S*-dioxide	**75:** P89316b
1,4-dimethyl-5-[(1,2,4-triethyl-1,2,3,4-tetrahydro-2-methyl-6-quinolinyl)azo]-, tetrachlorozincate(2-) (2:1)	**85:** P110086v
1,4-dimethyl-5-([*p*-(3,5,5-trimethyl-2-pyrazolin-1-yl)-phenyl]azo), chloride	**60:** P16023a, **62:** P5364h
3-([4-([2-([1,1'-diphenyl]-4-yloxy)ethyl]ethylamino)-phenyl]azo)-1,4-bis(2-hydroxybutyl)-, chloride	**83:** P165797s
3-([4-([2-(ethenylsulfonyl)ethyl]amino)-3-ethylphenyl]-azo)-4-methyl-5-phenyl-1-benzoyl, methyl sulfate	**79:** P6749d
3-([4-([2-(ethenylsulfonyl)ethyl]amino)-3-methylphenyl]-azo)-4-methyl-1-(phenylmethyl)-, methyl sulfate	**79:** P6749d
5,5'-(ethylenebis[(ethylimino)-*p*-phenyleneazo])bis-[1,4-dimethyl, sulfate	**65:** P7321a
4-ethyl-5-[3-(3-ethyl-2-benzothiazolinylidene)-1-triazeno]-3-methyl-1-phenyl, tetrafluoroborate	**59:** 11696e
4-ethyl-5-[3-(4-ethyl-3-methyl-1-phenyl-Δ^2-1,2,4-triazolin-5-ylidene)-1-triazeno]-3-methyl-1-phenyl, tetrafluoroborate	**59:** 11695g
1-ethyl-5-[(4-[ethyl(phenylmethyl)amino]phenyl)azo]-4-(2-hydroxy-3-methoxypropyl)-, chloride	**83:** P81203g
5-[[*p*-[ethyl(2-hydroxyethyl)amino]phenyl]azo]-1,4-dimethyl-, iodide, 4,4'-methylenedicarbanilate (2:1) (ester)	**70:** P58908p
5-([4-(ethyl[3-methoxy-1-(methoxycarbonyl)-3-oxopropyl]amino)-2-methylphenyl]azo)-1,4-dimethyl-, tetrachlorozincate(2-) (2:1)	**83:** P181099w
5-[(4-[ethyl(3-methoxy-3-oxopropyl)amino]phenyl)-azo]-1,4-dimethyl-, tetrachlorozincate(2-) (2:1)	**80:** P84685q
4-ethyl-1-methyl-3-[(*p*-morpholinophenyl)azo]-, ethyl sulfate	**73:** P36546d
5-([4-(ethyl[2-(2-naphthalenyloxy)ethyl]amino)-phenyl]azo)-1,2(or 1,4)-dimethyl-, trichlorozincate	**81:** P122759x

TABLE 22 (*Continued*)

Compound	Reference
22.3. Azido-, Azo-, or Diazoamino (Triazeno)-1,2,4-Triazolium or Triazolinium Compounds Containing One or More Representative Functions (*Continued*)	
3-[(4-[ethyl(2-[(phenoxycarbonyl)oxy]ethyl)amino]-phenyl)azo]-1,2(or 1,4)-bis(2-hydroxybutyl)-, chloride	**85:** P34629g
3-[(4-[ethyl(2-[(phenoxycarbonyl)oxy]ethyl)amino]-phenyl)azo]-1,2(or 1,4)-dimethyl-, chloride	**85:** P34629g
5-[(4-[ethyl(2-phenoxyethyl)amino]phenyl)azo]-1,2(or 1,4)-dimethyl-, trichlorozincate	**81:** P122759x
3(or 5)-([4-(ethyl[2-(phenoxysulfonyl)ethyl]amino)-phenyl]azo)-1,4-dimethyl-, tetrachloro-zincate(2-) (2:1)	**83:** P81198j
3-[(1-ethyl-2-phenyl-1H̲-indol-3-yl)azo]-1,4-bis(2-hydroxycyclohexyl)-5-methyl-, tetrachloro-zincate(2-) (2:1)	**84:** P152200n
3-[(1-ethyl-2-phenyl-1H̲-indol-3-yl)azo]-1-(2-hydroxybutyl)methyl-, tetrachlorozincate(2-) (2:1)	**85:** P22740y
3-[(4-[ethyl(phenylmethyl)amino]phenyl)azo]-1,4-bis(2-hydroxycyclohexyl)-, tetrachlorozincate(2-) (2:1)	**84:** P152199u
5-[(4-[ethyl(phenylmethyl)amino]phenyl)azo]-3-(3-hydroxypropyl)-1,4-dimethyl-, 4-methyl-benzene-sulfonate (1:1)	**79:** P106129v
3(or 5)-[(1-ethyl-1,2,3,4-tetraethyl-2-methyl-6-quinolinyl)azo]-1,4-dimethyl-, tetrachloro-zincate(2-) (2:1)	**85:** P125783u
3-[(1-ethyl-1,2,3,4-tetrahydro-2-methyl-6-quinolinyl)-azo]-1-(2-hydroxyethyl)-4-methyl-, tetrachloro-zincate(2-) (2:1)	**85:** P125783u
3-[(1-ethyl-1,2,3,4-tetrahydro-2-methyl-6-quinolinyl)-azo]-1-methyl-4-(phenylmethyl)-, tetrachloro-zincate(2-) (2:1)	**85:** P125783u
3(or 5)-[(1-ethyl-1,2,3,4-tetrahydro-2,2,4-trimethyl-6-quinolinyl)azo]-1,4-dimethyl-, tetrachloro-zincate(2-) (2:1)	**85:** P110086v
1-(2-hydroxyethyl)methyl-3(or 5)-[(1-methyl-2-phenyl-1H̲-indol-3-yl)azo]-5(or 3)-phenyl-, chloride	**85:** P144687g
2(or 4)-(2-hydroxyethyl)-1-methyl-3-[(1,2,4-triethyl-1,2,3,4-tetrahydro-2-methyl-6-isoquinolinyl)azo]-, tetrachlorozincate(2-) (2:1)	**85:** 110086v
3-(3-hydroxypropyl)-1,4-dimethyl-5-[(4-[methyl-(phenylmethyl)amino]phenyl)azo]-, methyl sulfate	**79:** P106129v
(2-hydroxypropyl)methyl-3-[(1-methyl-2-phenyl-1H̲-indol-3-yl)azo]-, tetrachlorozincate(2-) (2:1)	**85:** P22740y
5-(imidazo[2,1-*b*]thiazol-5-ylazo)-1,4-dimethyl-, methyl sulfate	**75:** P119193d
-inium, 3-([2-chloro-4-(diethylamino)phenyl]azo)-5-imino-2,4-dimethyl-1-phenyl-, chloride	**75:** P130772n
-inium, 3-([2-chloro-4-(diethylamino)phenyl]azo)-5-imino-2,4-dimethyl-1-phenyl—methylsulfate	**66:** P76923m
-inium, 3-([*p*-(dimethylamino)phenyl]azo)-2,4-diethyl-5-imino-1-phenyl-, chloride	**75:** P130772n

TABLE 22 (*Continued*)

Compound	Reference

22.3. Azido-, Azo-, or Diazoamino (Triazeno)-1,2,4-Triazolium or Triazolinium Compounds Containing One or More Representative Functions (*Continued*)

-inium, 3-([*p*-(dimethylamino)phenyl]azo)-2,4-diethyl-5-imino-1-phenyl—ethyl sulfate	**66:** P76923m
-inium, 3-([*p*-(dimethylamino)phenyl]azo)-5-imino-2,4-dimethyl-1-phenyl-, chloride	**75:** 130772n
-inium, 3-([*p*-(dimethylamino)phenyl]azo)-5-imino-2,4-dimethyl-1-phenyl—methyl sulfate	**66:** P76923m
-inium, 3-([4-[ethyl(3-methoxy-3-oxopropyl)amino]-phenyl]azo)-5-imino-2,4-dimethyl-1-phenyl-, chloride	**80:** P84685q
-inium, 5-imino-1,4-dimethyl-3-[(2-methyl-1H-indol-3-yl)azo]-2-phenyl-, methyl sulfate	**78:** 17606k
-inium, 5-imino-2,4-dimethyl-3-[(1-methyl-2-phenyl-1H-indol-3-yl)azo]-1-phenyl-, methyl sulfate	**78:** P17606k
4-[(2-methoxy-4-piperidinophenyl)azo]-1,4-dimethyl, trichlorozincate	**57:** P13929h
3(or 5)-[(1-methyl-2-phenyl-1H-indol-3-yl)azo]-1,4-bis[2-(2-pyridinyl)ethyl]-, trichlorozincate	**81:** P65203k
2(or 4)-methyl-1-(phenylmethyl)-3-([1,2,3,4-tetrahydro-1-(phenylmethyl)-6-isoquinolinyl]azo)-, tetrachlorozincate(2-) (2:1)	**85:** P125782t
2(or 4)-methyl-1-(phenylmethyl)-3-[(1,2,4-triethyl-1,2,3,4-tetrahydro-2-methyl-6-quinolinyl)azo]-, tetrachlorozincate(2-) (2:1)	**85:** P110086v
5,5'-[oxybis[ethylene(ethylimino)-*p*-phenyleneazo]]-bis[1,4-dimethyl-, sulfate]	**65:** P7320h
5,5'-[oxybis[ethylimino)-*p*-phenyleneazo]]bis[1,4-dimethyl-, sulfate]	**61:** P9614e

22.4. Hydroxy-1,2,4-Triazolium and Dioxo-1,2,4-Triazolidinium Compounds

3-hydroxy-1,4-dimethyl-5-phenyl, hydroxide, inner salt	**67:** 43755a, **85:** 159988e
3-hydroxy-1,5-dimethyl-4-phenyl, hydroxide, inner salt	**63:** 2967a, **64:** 19614a, **67:** 43755a
3-hydroxy-4,5-dimethyl-1-phenyl, hydroxide, inner salt	**63:** 2967a, **78:** 96707v
3-hydroxy-1,4-diphenyl-, hydroxide, inner salt	**63:** 2967a, **68:** 59504z, **68:** 114512c, **73:** P14857d, **74:** 53662b, **82:** 4164x, **85:** 32927d
3-hydroxy-1-methyl-4,5-diphenyl-, hydroxide, inner salt	**75:** 151737v, **81:** 3835j
3-hydroxy-1-methyl-4,5-diphenyl, chloride	**67:** 43755a
3-hydroxy-4-methyl-1,5-diphenyl, hydroxide	**63:** 2967a
3-hydroxy-5-methyl-1,4-diphenyl, hydroxide, inner salt	**39:** 2068[8], **85:** 32927d
5-hydroxy-2-methyl-4-(4-methylphenyl)-3-phenyl-, hydroxide, inner salt	**81:** 3835j

TABLE 22 (*Continued*)

Compound	Reference
22.4. Hydroxy-1,2,4-Triazolium and Dioxo-1,2,4-Triazolidinium Compounds (*Continued*)	
3-hydroxy-4-methyl-1-phenyl, hydroxide, inner salt	**63:** 2967a, **78:** 96707v
3-hydroxy-4-(4-methylphenyl)-1,5-diphenyl-, hydroxide, inner salt	**85:** 32927d
3-hydroxy-1-(4-methylphenyl)-4-phenyl-, hydroxide, inner salt	**74:** 53662b, **85:** 32927d
3-hydroxy-4-(4-methylphenyl)-1-phenyl-, hydroxide, inner salt	**74:** 53662b, **85:** 32927d
3-hydroxy-4-methyl-1-piperonyl, hydroxide, inner salt	**67:** 32651n
3-hydroxy-1,4,5-triphenyl, hydroxide, inner salt	**63:** 2966h, **67:** 43755a, **67:** 100077g
-idinium, 4-(4-bromophenyl)-1,1-dimethyl-3,5-dioxo-, hydroxide, inner salt	**81:** 91435g
-idinium, 4-tert-butyl-1,1-diethyl-3,5-dioxo-, hydroxide, inner salt	**82:** 3735x, **82:** 72890f
-idinium, 4-butyl-1,1-dimethyl-3,5-dioxo-, hydroxide, inner salt	**67:** 3039p
-idinium, 4-tert-butyl-1,1-dimethyl-3,5-dioxo-, hydroxide, inner salt	**82:** 3735x, **82:** 72890f
-idinium, 4-(3-chloro-4-methoxyphenyl)-1,1-dimethyl-3,5-dioxo-, hydroxide, inner salt	**81:** 91435g
-idinium, 4-(3-chlorophenyl)-1,1-dimethyl-3,5-dioxo-, hydroxide, inner salt	**81:** 91435g
-idinium, 4-(4-chlorophenyl)-1,1-dimethyl-3,5-dioxo-, hydroxide, inner salt	**81:** 91435g
-idinium, 4-(3,4-dichlorophenyl)-1,1-dimethyl-3,5-dioxo-, hydroxide, inner salt	**81:** 91435g
-idinium, 1,1-diethyl-3,5-dioxo-4-phenyl-, hydroxide, inner salt	**82:** 3735x
-idinium, 1,1-diethyl-4-methyl-3,5-dioxo-, hydroxide, inner salt	**75:** 20308h
-idinium, 4-(3,5-dimethoxyphenyl)-1,1-dimethyl-3,5-dioxo-, hydroxide, inner salt	**81:** 91435g
-idinium, 4-(dimethylamino)-1,1-dimethyl-3,5-dioxo-, hydroxide, inner salt	**67:** 3039p
-idinium, 1,1-dimethyl-3,5-dioxo-4-phenyl-, chloride	**81:** 91435g
-idinium, 1,1-dimethyl-3,5-dioxo-4-phenyl-, hydroxide, inner salt	**81:** 91435g
-idinium, 1,1-dimethyl-3,5-dioxo-4-(1,1,3,3-tetra-methylbutyl)-, hydroxide, inner salt	**67:** 3039p
-idinium, 1,1-dimethyl-3,5-dioxo-4-[3-(trifluoromethyl)-phenyl]-, hydroxide, inner salt	**81:** 91435g
-idinium, 1,1-dimethyl-4-(4-methylphenyl)-3,5-dioxo-, hydroxide, inner salt	**81:** 91435g
-idinium, 1,1-dimethyl-4-[3-(methylthio)phenyl]-3,5-dioxo-, hydroxide, inner salt	**81:** 91435g
-idinium, 4-(3,5-dimethylphenyl)-1,1-dimethyl-3,5-dioxo-, hydroxide, inner salt	**81:** 91435g

TABLE 22 (*Continued*)

Compound	Reference

22.4. Hydroxy-1,2,4-Triazolium and Dioxo-1,2,4-Triazolidinium Compounds (*Continued*)

-idinium, 3,5-dioxo-4-phenyl-, diphenylmethylide	**78:** 29344m
-idinium, 1-(diphenylmethylene)-3,5-dioxo-4-phenyl-	**63:** 8346b
-idinium, 4-(4-fluorophenyl)-1,1-dimethyl-3,5-dioxo-, hydroxide, inner salt	**81:** 91435g
-idinium, 4-[3-(methoxycarbonyl)phenyl]-1,1-dimethyl- 3,5-dioxo-, hydroxide, inner salt	**81:** 91435g
-idinium, 4-(2-methoxyphenyl)-1,1-dimethyl-3,5-dioxo-, hydroxide, inner salt	**81:** 91435g
-idinium, 4-(3-methoxyphenyl)-1,1-dimethyl-3,5-dioxo-, hydroxide, inner salt	**81:** 91435g
-idinium, 4-(4-methoxyphenyl)-1,1-dimethyl-3,5-dioxo-, hydroxide, inner salt	**81:** 91435g
-idinium, 1,1,4-triethyl-3,5-dioxo-, hydroxide, inner salt	**75:** 20308h
-idinium, 1,1,4-trimethyl-3,5-dioxo-, hydroxide, inner salt	**81:** 91435g

22.5. Mercapto- or Alkylthio-1,2,4-Triazolium Compounds

1-benzyl-4-butyl-3-mercapto-, hydroxide, inner salt	**79:** 126378d
1-benzyl-3-mercapto-4,5-diphenyl-, hydroxide, inner salt	**77:** 101472a
4-benzyl-3-mercapto-1,5-diphenyl-, hydroxide, inner salt	**77:** 101472a
4-benzyl-3-mercapto-1-methyl-5-phenyl-, hydroxide, inner salt	**77:** 101472a
1-benzyl-3-mercapto-4-phenyl-, hydroxide, inner salt	**79:** 126378d
5-benzyl-3-(methylthio)-1,4-diphenyl-, iodide	**83:** 113599j
1,4-bis(benzyl)-3-mercapto-5-phenyl-, hydroxide, inner salt	**77:** 101472a
4-(4-bromophenyl)-3-mercapto-1,5-diphenyl-, hydroxide, inner salt	**79:** 78701e
5-(4-bromophenyl)-3-mercapto-1,4-diphenyl-, hydroxide, inner salt	**79:** 78701e
4-butyl-3-mercapto-1,5-diphenyl-, hydroxide, inner salt	**78:** 97565j, **79:** 126378d, **80:** 14340z
4-butyl-3-mercapto-1-methyl-, hydroxide, inner salt	**79:** 126378d
5-butyl-3-(methylthio)-1,4-diphenyl-, iodide	**83:** 113599j
5-tert-butyl-3-mercapto-1-phenyl-4-(2-propenyl)-, hydroxide, inner salt	**78:** 84319g
5-tert-butyl-3-(methylthio)-1,4-diphenyl-, iodide	**82:** 169941m
5-(4-carboxyphenyl)-3-(methylthio)-1,4-diphenyl-, iodide	**81:** 169245s
4-(2-chlorophenyl)-3-mercapto-1,5-diphenyl-, hydroxide, inner salt	**78:** 97565j
4-(2-chlorophenyl)-5-mercapto-2,3-diphenyl-, hydroxide, inner salt	**76:** 153680r
4-(3-chlorophenyl)-3-mercapto-1,5-diphenyl-, hydroxide, inner salt	**78:** 97565j

TABLE 22 (*Continued*)

Compound	Reference
22.5. Mercapto- or Alkylthio-1,2,4-Triazolium Compounds (*Continued*)	
4-(3-chlorophenyl)-5-mercapto-2,3-diphenyl-, hydroxide, inner salt	**76:** 153680r
4-(4-chlorophenyl)-3-mercapto-1,5-diphenyl-, hydroxide, inner salt	**78:** 97565j
4-(4-chlorophenyl)-5-mercapto-2,3-diphenyl-, hydroxide, inner salt	**76:** 153680r
5-(3-chlorophenyl)-3-mercapto-1,4-diphenyl-, hydroxide, inner salt	**79:** 78701e
4-(*p*-chlorophenyl)-3-mercapto-1-methyl-5-phenyl-, hydroxide, inner salt	**75:** 151735t, **81:** 3847q
5-(3-chloropropyl)-3-(methylthio)-1,4-diphenyl-, iodide	**82:** 169941m
5-(4-cyanophenyl)-3-mercapto-1,4-diphenyl-, hydroxide, inner salt	**79:** 78701e
4-cyclohexyl-3-mercapto-1,5-diphenyl-, hydroxide, inner salt	**77:** 139907j
4-cyclohexyl-3-mercapto-1-methyl-5-phenyl-, hydroxide, inner salt	**77:** 101472a
5-cyclopropyl-3-(methylthio)-1,4-diphenyl-, iodide	**83:** 113599j
3,3′-[1,10-decanediylbis(thio)]bis[1,4-diphenyl-, 4-methylbenzenesulfonic acid (1:2)	**79:** P25673r
4,5,-dihydro-5,5-dimethyl-3-(methylthio)-2-phenyl-, iodide	**78:** 4193y
1,4-dimethyl-5-(3-methyl-2-benzothiazolinylidenemethyl)-3-(methylthio)-, perchlorate	**54:** 7697g
2,5-dimethyl-3-(methylthio)-1-phenyl, iodide	**54:** 7696i
3,4-dimethyl-5-(methylthio)-1-phenyl, iodide	**54:** 7696d
4,5-dimethyl-3-(methylthio)-1-phenyl, iodide	**54:** 7696d
1,4-diphenyl-3-mercapto-, hydroxide, inner salt	**68:** 29649h, **81:** 2977v
5-(diphenylmethyl)-3-(methylthio)-1,4-diphenyl-, iodide	**82:** 169941m
5-(3-ethoxyphenyl)-3-mercapto-1,4-diphenyl-, hydroxide, inner salt	**79:** 78701e
5-[3-(3-ethyl-2-benzothiazolinylidene)propenyl]-1,4-dimethyl-3-(methylthio), perchlorate	**54:** 7697i
5-[3-(3-ethyl-2-benzothiazolinylidene)propenyl]-3-(ethylthio)-1-	**54:** 7697i
5-[3-(3-ethyl-2-benzothiazolinylidene)propenyl]-3-(ethylthio)-1-methyl-4-phenyl, iodide	**54:** 7697i
5-[3-(3-ethyl-2-benzothiazolinylidene)propenyl]-3-(ethylthio)-4-methyl-1-phenyl, iodide	**54:** 7697i
5-[3-(3-ethyl-2-benzothiazolinylidene)propenyl]-1-methyl-3-(methylthio)-4-phenyl, iodide	**54:** 7697i
5-[3-(3-ethyl-2-benzothiazolinylidene)propenyl]-4-methyl-3-(methylthio)-1-phenyl, iodide	**54:** 7697i
5-[3-(3-ethyl-2-benzoxazolinylidene)propenyl]-1,4-dimethyl-3-(methylthio), iodide	**54:** 7697h
5-[3-(3-ethyl-2-benzoxazolinylidene)propenyl]-3-(ethylthio)-1,4-dimethyl, perchlorate	**54:** 7698a
5-[3-(3-ethyl-2-benzoxazolinylidene)propenyl]-3-(ethylthio)-4-methyl-1-phenyl, iodide	**54:** 7698a

TABLE 22 *(Continued)*

Compound	Reference
22.5. Mercapto- or Alkylthio-1,2,4-Triazolium Compounds *(Continued)*	
5-[3-(3-ethyl-2-benzoxazolinylidene)propenyl]- 1-methyl-3-(methylthio)-4-phenyl, iodide	**54:** 7697h
5-[3-(3-ethyl-2-benzoxazolinylidene)propenyl]- 4-methyl-3-(methylthio)-1-phenyl, iodide	**54:** 7697h
4-ethyl-5-[3-(3-ethyl-2-benzothiazolinylidene)- propenyl]-1-methyl-3-(methylthio), perchlorate	**54:** 7697i
4-ethyl-5-[3-(3-ethyl-2-benzoxazolinylidene)- propenyl]-1-methyl-3-(methylthio), perchlorate	**54:** 7698a
5-ethyl-3-mercapto-1,4-diphenyl, chloride	**67:** 43755a
5-ethyl-3-mercapto-4-methyl-1-phenyl, chloride	**67:** 43755a
4-ethyl-3-(methylthio)-1,5-diphenyl, iodide	**67:** 100077g
5-ethyl-3-(methylthio)-1,4-diphenyl-, iodide	**82:** 169941m, **83:** 113599j
3-(ethylthio)-4-methyl-1-phenyl-5-[3-(1,3,3- trimethyl-2-indolinylidene)propenyl], perchlorate	**54:** 7698a
3-(ethylthio)-1,4,5-triphenyl-, tetrafluoroborate(1-)	**77:** 139907j
5-(α-hydroxybenzyl)-3-(methylthio)-1,4-diphenyl-, iodide	**83:** 113599j
5-(1-hydroxyethyl)-3-(methylthio)-1,4-diphenyl-, iodide	**83:** 113599j
5-(4-iodophenyl)-3-mercapto-1,4-diphenyl-, hydroxide, inner salt	**79:** 78701e
5-isopropyl-3-(methylthio)-1,4-diphenyl-, iodide	**82:** 169941m
3-mercapto-1,4-bis(*p*-nitrophenyl), hydroxide, inner salt	**67:** 43755a
3-mercapto-1,4-dimethyl-, hydroxide, inner salt	**79:** 126378d
3-mercapto-1,4-dimethyl-5-(morpholinocarbonyl)-, hydroxide, inner salt	**70:** 11686z
3-mercapto-4,5-dimethyl-1-(phenoxymethyl)-, hydroxide, inner salt	**72:** 78951s
3-mercapto-1,4-dimethyl-5-phenyl, hydroxide, inner salt	**67:** 43755a, **74:** 42317s, **75:** 151735t
3-mercapto-1,5-dimethyl-4-phenyl, hydroxide, inner salt	**63:** 2967a, **78:** 96707v
3-mercapto-4,5-dimethyl-1-phenyl, chloride	**67:** 43755a
3-mercapto-4,5-dimethyl-1-phenyl, hydroxide, inner salt	**54:** 7696e, **63:** 2967a, **67:** 43755a
3-mercapto-1,4-diphenyl, hydroxide, inner salt	**63:** 2967a, **68:** 59504z
3-mercapto-1,5-diphenyl-4-propyl-, hydroxide, inner salt	**78:** 97565j
3-mercapto-4-(2-methoxyphenyl)-1,5-diphenyl-, hydroxide, inner salt	**78:** 97565j
3-mercapto-4-(4-methoxyphenyl)-1,5-diphenyl-, hydroxide, inner salt	**78:** 97565j
3-mercapto-5-(4-methoxyphenyl)-1,4-diphenyl-, hydroxide, inner salt	**79:** 78701e
5-mercapto-4-(2-methoxyphenyl)-2,3-diphenyl-, hydroxide, inner salt	**76:** 153680r
5-mercapto-4-(4-methoxyphenyl)-2,3-diphenyl-, hydroxide, inner salt	**76:** 153680r, **80:** 14340z
3-mercapto-4-(4-methoxyphenyl)-1-methyl-5-phenyl-, hydroxide, inner salt	**77:** 101472a

TABLE 22 (*Continued*)

Compound	Reference

22.5. Mercapto- or Alkylthio-1,2,4-Triazolium Compounds (*Continued*)

Compound	Reference
3-mercapto-1-methyl-4,5-diphenyl, hydroxide, inner salt	**67:** 43755a, **74:** 42317s, **75:** 151735t, **77:** 101472a, **81:** 3844m, **81:** 3847q
3-mercapto-4-methyl-1,5-diphenyl, hydroxide, inner salt	**57:** P1791i, **67:** 100077g, **77:** 101472a, **78:** 97565j, **79:** 126378d
3-mercapto-5-methyl-1,4-diphenyl, hydroxide, inner salt	**39:** 2068[8], **63:** 2967a, **68:** 59504z, **77:** 101468d
3-mercapto-5-methyl-1,4-diphenyl, chloride	**63:** 2967b, **67:** 43755a
3-mercapto-5-methyl-1,4-diphenyl, picrate	**67:** 43755a
3-mercapto-1-methyl-4-(4-methylphenyl)-5-phenyl-, hydroxide, inner salt	**75:** 151735t, **81:** 3847q
3-mercapto-1-methyl-4-(4-nitrophenyl)-5-phenyl-, hydroxide, inner salt	**77:** 101473b
3-mercapto-5-methyl-1-(phenoxymethyl)-4-phenyl-, hydroxide, inner salt	**72:** 78951s
3-mercapto-4-methyl-1-phenyl, hydroxide, inner salt	**54:** 7696f, **63:** 2967a
3-mercapto-4-(2-methylphenyl)-1,5-diphenyl-, hydroxide, inner salt	**78:** 97565j
3-mercapto-4-(4-methylphenyl)-1,5-diphenyl-, hydroxide, inner salt	**78:** 97565j
3-mercapto-5-(4-methylphenyl)-1,4-diphenyl-, hydroxide, inner salt	**79:** 78701e
5-mercapto-4-(2-methylphenyl)-2,3-diphenyl-, hydroxide, inner salt	**76:** 153680r
5-mercapto-4-(4-methylphenyl)-2,3-diphenyl-, hydroxide, inner salt	**76:** 153680r, **80:** 14340z
3-mercapto-1-methyl-4-phenyl-, hydroxide, inner salt	**79:** 126378d
3-mercapto-4-methyl-1-(phenylmethyl)-, hydroxide, inner salt	**79:** 126378d
3-mercapto-4-methyl-1-piperonyl, hydroxide, inner salt	**67:** 32651n
3-mercapto-4-(4-nitrophenyl)-1,5-diphenyl-, hydroxide, inner salt	**77:** 101472a
3-mercapto-5-(4-nitrophenyl)-1,4-diphenyl-, hydroxide, inner salt	**79:** 78701e
3-mercapto-5-(3-nitrophenyl)-1-phenyl-4-(2-propenyl)-, hydroxide, inner salt	**78:** 84319g
3-mercapto-1,4,5-trimethyl, hydroxide, inner salt	**54:** 7696f, **63:** 2967a, **67:** 43755a, **83:** 178933h
3-mercapto-1,4,5-triphenyl-, hydroxide, inner salt	**63:** 2966h, **67:** 43755a, **67:** 100077g, **76:** 153680r, **77:** 101472a, **78:** 97565j, **79:** 126378d, **80:** 14340z, **81:** 3844m
3-mercapto-1,4,5-tris(*p*-nitrophenyl), hydroxide, inner salt	**67:** 43755a
5-[2-(methoxycarbonyl)ethyl]-3-(methylthio)-1,4-diphenyl-, iodide	**82:** 169941m, **83:** 113599j

TABLE 22 (*Continued*)

Compound	Reference

22.5. Mercapto- or Alkylthio-1,2,4-Triazolium Compounds (*Continued*)

Compound	Reference
5-[4-(methoxycarbonyl(phenyl]-3-(methylthio)-1,4-diphenyl-, iodide	**81:** 169245s
4-(4-methoxyphenyl)-3-(methylthio)-1,5-diphenyl-, iodide	**76:** 153680r, **78:** 97565j
1-methyl-3,5-bis(methylthio)-4-phenyl-, iodide	**81:** 91435g
3-methyl-2,4-diphenyl-5-mercapto-, hydroxide, inner salt	**79:** 126378d
4-methyl-5-(3-methyl-2-benzothiazolinylidenemethyl)-3-(methylthio)-1-phenyl, iodide	**54:** 7697f
1-methyl-3-(methylthio)-4,5-diphenyl-, iodide	**81:** 3847q
3-methyl-5-(methylthio)-1,4-diphenyl, chloride	**67:** 43755a
3-methyl-5-(methylthio)-1,4-diphenyl-, iodide	**77:** 101468d
4-methyl-3-(methylthio)-1,5-diphenyl, iodide	**67:** 100077g
5-methyl-3-(methylthio)-1,4-diphenyl-, chloride	**81:** 169245s
5-methyl-3-(methylthio)-1,4-diphenyl-, iodide	**77:** 101468d, **82:** 169941m, **83:** 113599j
4-methyl-3-(methylthio)-1-phenyl, iodide	**54:** 7696d
4-methyl-3-(methylthio)-1-phenyl-5-[3-(1,3,3-trimethyl-2-indolinylidene)propenyl], perchlorate	**54:** 7697i
4-(4-methylphenyl)-3-(methylthio)-1,5-diphenyl-, iodide	**78:** 97565j
3-(methylthio)-1,4-diphenyl-, hydroxide, inner salt	**83:** 113599j
3-(methylthio)-1,4-diphenyl, iodide	**67:** 43755a, **83:** 113599j
3-(methylthio)-1,4-diphenyl-5-(2-phenylethenyl)-, iodide	**81:** 169245s
3-(methylthio)-1,4-diphenyl-5-propyl-, iodide	**82:** 169941m
3-(methylthio)-5-(3-nitrophenyl)-1-phenyl-4-(2-propenyl)-, iodide	**78:** 84319g
3-(methylthio)-1,4,5-triphenyl-, chloride	**81:** 169245s
3-(methylthio)-1,4,5-triphenyl-, iodide	**76:** 153680r, **78:** 97565j, **80:** 14340z
1,4,5-trimethyl-3-(methylthio), iodide	**63:** 2967a, **67:** 43755a

22.6. 4-Amino-5-Mercapto- or 5-(Alkylthio)-1,2,4-Triazolium Compounds

Compound	Reference
4-amino-1-benzyl-5-[(cyanomethyl)thio]-, chloride	**80:** 82837s
4-amino-1-benzyl-5-(methylthio)-, iodide	**80:** 82837s
4-amino-1-(4-chlorobenzyl)-5-(methylthio)-, iodide	**80:** 82837s
4-amino-5-(4-chlorophenyl)-1-phenyl-3-mercapto-, hydroxide, inner salt	**85:** 21235a
4-amino-5-[(cyanomethyl)thio]-1-methyl-, chloride	**80:** 82837s
4-amino-3-mercapto-1-benzyl-, hydroxide, inner salt	**85:** 21235a
4-amino-3-mercapto-1,5-diphenyl-, hydroxide, inner salt	**78:** 72015s, **82:** 43270y, **85:** 21235a
4-amino-3-mercapto-1-methyl-, hydroxide, inner salt	**85:** 21235a
4-amino-1-methyl-5-(methylthio)-, iodide	**80:** 82837s
4-amino-5-(methylthio)-1-octyl-, iodide	**80:** 82837s
4-benzamido-3-mercapto-1-methyl-5-phenyl-, hydroxide, inner salt	**80:** 27223p
1-benzyl-4-(benzylideneamino)-3-mercapto-, hydroxide, inner salt	**85:** 21235a
4-benzylideneamino-3-mercapto-1,5-diphenyl-, hydroxide, inner salt	**78:** 72015s

TABLE 22 (*Continued*)

Compound	Reference

22.6. 4-Amino-5-Mercapto- or 5-(Alkylthio)-1,2,4-Triazolium Compounds (*Continued*)

Compound	Reference
4-(benzylideneamino)-1-methyl-3-mercapto-, hydroxide, inner salt	**85:** 21235a
4-(benzylideneamino)-3-(methylthio)-1,5-diphenyl-, iodide	**85:** 21235a
4-(tert-butoxycarbonylamino)-3-mercapto-1-methyl-5-phenyl-, hydroxide, inner salt	**80:** 27223p
4-(dimethylamino)-3-mercapto-1,5-diphenyl-, hydroxide, inner salt	**80:** 27223p
4-(dimethylamino)-3-mercapto-5-(4-methoxyphenyl)-1-methyl-, hydroxide, inner salt	**80:** 27223p
4-(dimethylamino)-3-mercapto-1-methyl-5-phenyl-, hydroxide, inner salt	**80:** 27223p
4-(dimethylamino)-3-mercapto-5-methyl-1-phenyl-, hydroxide, inner salt	**80:** 27223p
4-(dimethylamino)-5-(4-methoxyphenyl)-1-methyl-3-(methylthio)-, iodide	**80:** 27223p
4-(dimethylamino)-1-methyl-3,5-bis(methylthio)-, iodide	**71:** 21509d
4-(dimethylamino)-1-methyl-3-(methylthio)-5-phenyl-, iodide	**80:** 27223p
4-(dimethylamino)-3-(methylthio)-1,5-diphenyl-, iodide	**80:** 27223p
4-[([4-(dimethylamino)phenyl]methylene)amino]-3-mercapto-1,5-diphenyl-, hydroxide, inner salt	**85:** 21235a
4-[(2-furanylmethylene)amino]-3-mercapto-1,5-diphenyl-, hydroxide, inner salt	**85:** 21235a
4-([(4-hydroxy-3-methoxyphenyl)methylene]amino)-3-mercapto-1,5-diphenyl-, hydroxide, inner salt	**85:** 21235a
4-([(2-hydroxyphenyl)methylene]amino)-3-mercapto-1,5-diphenyl-, hydroxide, inner salt	**85:** 21235a
3-mercapto-1-methyl-4-(methylphenylamino)-5-phenyl-, hydroxide, inner salt	**80:** 27223p

22.7. Miscellaneous 1,2,4-Triazolium, Triazolinium, and Triazolidinium Compounds Containing One or More Representative Functions

Compound	Reference
3-acetyl-4-hydroxy-1,5-diphenyl-, hydroxide, inner salt	**78:** 15224k
3-acetyl-4-isopropyl-1,5-diphenyl-, chloride	**70:** 87679v
1-acetyl-4-methyl-, tetrafluoroborate(1-)	**73:** 45425r
3-acetyl-4-methyl-1,5-diphenyl-, chloride	**70:** 87679v
3-acetyl-4-methyl-1,5-diphenyl-, tetraphenylborate(1-)	**70:** 87679v
3-acetyl-1,4,5-triphenyl-, perchlorate	**70:** 96723g
4-amino-1-[(3-chlorophenyl)methyl]-5-(dithiocarboxy)-, hydroxide, inner salt	**80:** 82837s
3-amino-1,4-dimethyl-5-[(1-methyl-2-phenyl-1H-indol-3-yl)azo]-, tetrachlorozincate(2-) (2:1)	**79:** P147415d
4-amino-5-(dithiocarboxy)-1-methyl-, hydroxide, inner salt	**80:** 82837s
3-benzoyl-4-hydroxy-1,5-diphenyl-, hydroxide, inner salt	**78:** 15224k

TABLE 22   (*Continued*)

Compound	Reference
22.7. Miscellaneous 1,2,4-Triazolium, Triazolinium, and Triazolidinium Compounds Containing One or More Representative Functions (*Continued*)	
4-benzoyl-3-mercapto-1-methyl-5-phenyl-, hydroxide, inner salt	**74:** 42317s
4-benzyl-3-carboxy-1,5-diphenyl-, perchlorate, ethyl ester	**70:** 96723g
1-benzyl-4-phenyl-3-selenyl-, hydroxide, inner salt	**85:** 108584n
3,5-bis[(5-chloro-1,2-dimethyl-1H̱-indol-3-yl)azo]-1,4-diethyl-, ethyl sulfate	**83:** P181082k
3,5-bis[[4-(diethylamino)-*o*-tolyl]azo]-1,4-dimethyl-, methyl sulfate	**75:** P50405e
3,5-bis[(2-phenyl-1H̱-indol-3-yl)-azo]-1,4-dibenzyl-, chloride	**83:** P181082k
1-bromo-3,4-dimethyl-, bromide	**83:** 114298x
1-bromo-4-methyl-, bromide	**83:** 114298x
1-bromo-3-methyl-4-phenyl-, bromide	**83:** 114298x
1-bromo-4-phenyl-, bromide	**83:** 114298x
3-carboxy-4-cyclohexyl-1,5-diphenyl-, perchlorate, ethyl ester	**70:** 96723g
3-carboxy-4-cyclohexyl-5-methyl-1-phenyl-, perchlorate, ethyl ester	**70:** 96723g
3-carboxy-1,4-diethyl-5-[(1-ethyl-2-phenyl-1H̱-indol-3-yl)azo]-, trichlorozincate(1⁻)	**80:** P28454b
3-carboxy-1,4,5-triphenyl-, perchlorate, ethyl ester	**70:** 96723g
5-chloro-4-ethyl-3-methyl-1-phenyl-, tetrafluoroborate	**56:** 10134d
4-(dimethylamino)-1,1-dimethyl-3,5-dioxo-, hydroxide, inner salt	**82:** 3735x
3-ethoxy-4-methyl-1-phenyl-, tetrafluoroborate(1-)	**73:** 77149v
4-ethyl-1-(thiobenzoyl)-, tetrafluoroborate(1-)	**79:** 78696g
-idinium, 4-benzyl-1,1-diethyl-3-hydroxy-5-thioxo-hydroxide, inner salt	**82:** 72890f
-idinium, 4-benzyl-1,1-dimethyl-3-hydroxy-5-thioxo-, hydroxide, inner salt	**82:** 72890f
-idinium, 1-(dibenzylhydrazono)-3,5-dioxo-4-phenyl-, hydroxide, inner salt	**73:** 109240m, **77:** 87997s
-idinium, 4-(diethylamino)-1,1-diethyl-3,5-dioxo-, hydroxide, inner salt	**75:** 20308h, **83:** 146782s
-idinium, 1-(diethylhydrazono)-3,5-dioxo-4-phenyl-, hydroxide, inner salt	**73:** 109240m
-idinium, 1,1-diethyl-3-hydroxy-4-methyl-3-oxo-5-thioxo-hydroxide, inner salt	**82:** 72890f
-idinium, 4-(dimethylamino)-3-mercapto-1,1-dimethyl-5-thioxo, hydroxide, inner salt	**69:** 51320m
-idinium, 1-(dimethylhydrazono)-3,5-dioxo-4-phenyl-, hydroxide, inner salt	**73:** 109240m
-idinium, 1-(diphenylhydrazono)-3,5-dioxo-4-phenyl-, hydroxide, inner salt	**73:** 109240m
-idinium, 4-ethyl-5-(ethylimino)-1,1-dimethyl-3-oxo-, hydroxide, inner salt	**75:** 20308h

TABLE 22 (*Continued*)

Compound	Reference
22.7. Miscellaneous 1,2,4-Triazolium, Triazolinium, and Triazolidinium Compounds Containing One or More Representative Functions (*Continued*)	
-idinium, 3-hydroxy-1,1,4-trimethyl-5-thioxo-chloride	**82:** 72890f
-idinium, 3-hydroxy-1,1,4-trimethyl-5-thioxo-hydroxide, inner salt	**82:** 3735x, **82:** 72890f
-idinium, 4-isopropyl-5-(isopropylimino)-1,1-dimethyl-3-oxo-, hydroxide, inner salt	**75:** 20308h, **82:** 3735x
-idinium, 1-(pentamethylenehydrazono)-3,5-dioxo-4-phenyl-, hydroxide, inner salt	**73:** 109240m
-idinium, 1,1,4-triethyl-5-(ethylimino)-3-oxo-, hydroxide, inner salt	**75:** 20308h
3-mercapto-1,4-dimethyl-5-(morpholinocarbonyl)-, hydroxide, inner	**70:** 11686z
1-methyl-4-phenyl-3-selenyl-, hydroxide, inner salt	**85:** 108584n
1,4,5-triphenyl-3-selenyl-, hydroxide, inner salt	**85:** 108584n

CHAPTER 23

Metal-triazole Complexes

The compounds in this section are composed of 1,2,4-triazoles attached to metal atoms by both pi and sigma bonds. In addition to their preparation, a number of applications that have been claimed for this class of compounds are described.

The reaction of 4-butyl-1,2,4-4H-triazole (**23.1-1**) with zinc chloride in the presence of the dithiocarbonyl derivative of ethylenediamine, **23.1-2**, gave the zinc complex (**23.1-3**).[1] Although both **23.1-1** and its zinc complex were effective fungicides against leaf rust in wheat seedings, the complex, in contrast to **23.1-1**, caused no injury to the seedings.

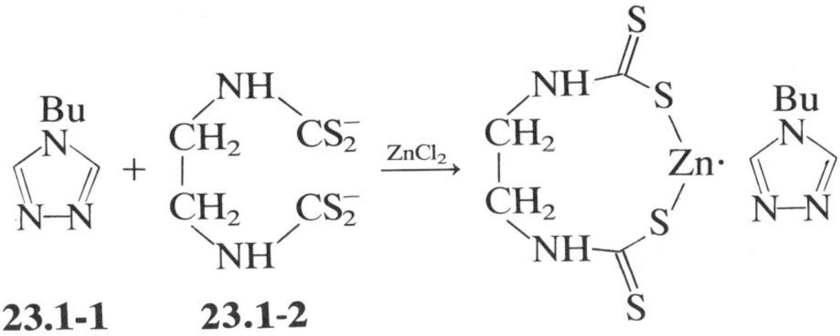

23.1-1   23.1-2   **23.1-3**

A convenient route to the tin (II) derivative (**23.1-6**) involved the protolysis of the metal carbon bonds of dicyclopentadienyltin (**23.1-4**) with s-triazole (**23.1-5**) in benzene or tetrahydrofuran at room temperature.[2] The product was characterized by ^{119m}Sn Mössbauer spectra, which confirmed that the metal had retained its lower oxidation state. Similarly, the addition

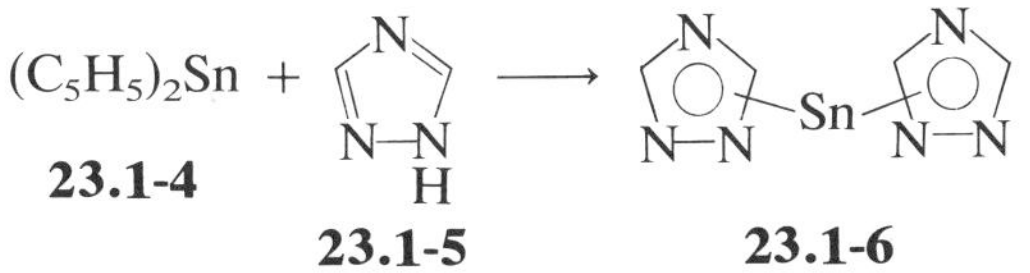

$(C_5H_5)_2Sn$ +

23.1-4

23.1-5   **23.1-6**

of bis(η^5-2,4-cyclopentadien-1-yl)dihydromolybdenum (**23.1-7**) to 4-phenyl-1,2,4-triazoline-3,5-dione (**23.1-8**) resulted in the formation of the metalated triazoline (**23.1-9**).[3]

$$(C_5H_5)_2MoH_2 + \underset{\textbf{23.1-8}}{O=\overset{\overset{\displaystyle Ph}{N}}{\underset{N=N}{\bigcirc}}=O} \longrightarrow HO\overset{\overset{\displaystyle Ph}{N}}{\underset{N-N}{\bigcirc}}=O \quad O=\overset{\overset{\displaystyle Ph}{N}}{\underset{N-N}{\bigcirc}}-OH$$

23.1-7 **23.1-8**

23.1-9

Treatment of the dimethylsulfide complex of gold (I) chloride (**23.1-10**) with s-triazole (**23.1-5**) in tetrahydrofuran in the presence of triethylamine gave the gold (I) derivative (**23.1-11**), useful for the decoration of glass.[4] In

$$\underset{\substack{N-N\\H\\ \textbf{23.1-5}}}{\overset{N}{\bigcirc}} + \underset{\textbf{23.1-10}}{Me_2S \cdot AuCl} \longrightarrow \underset{\substack{N-N\\ \textbf{23.1-11}}}{\overset{N}{\bigcirc}}-Au$$

the reaction of 1,2,4-triazoline-5-thione (**23.1-12**) with H_2AuCl_4, gold (II) was reduced to gold (I) and the thione was oxidized either to the sulfonic acid with excess gold (II) or to the disulfide with excess ligand.[5] The gold (I) formed *in situ* reacted with the thione (**23.1-12**) to give the colorless complex, **23.1-13**.

$$\underset{\substack{N-N\\H\\ \textbf{23.1-12}}}{\overset{\overset{\displaystyle H}{N}}{\bigcirc}}=S + [Au^+] \longrightarrow \left[\underset{\substack{N-N\\H\\ \textbf{23.1-13}}}{\overset{\overset{\displaystyle H}{N}}{\bigcirc}}=S\right]_2 Au^+$$

An aqueous solution of nickel (II) nitrate and s-triazole (**23.1-5**) in the molar ratio of 1:2 deposited purple-blue plates of the nickel complex (**23.1-14**).[6] X-ray diffraction techniques showed that the complex contained

$$Ni(NO_3)_2 + \underset{\substack{N-N\\H\\ \textbf{23.1-5}}}{\overset{N}{\bigcirc}} \longrightarrow \left[(H_2O)_3\left[\underset{\substack{N-N\\H}}{\overset{N}{\bigcirc}}\right]_3 Ni\right]_2 Ni(NO_3)_6(H_2O)_2$$

23.1-14

a trinuclear cation in which the nickel atoms are octahedrally coordinated and colinear. Treatment of a solution of the nickel (II) compound (**23.1-15**) in aqueous pyridine with an alcoholic solution of 3-(4-pyridinyl)-1,2,4-tria-zoline-5-thione (**23.1-16**) at reflux gave a product that was assigned the structure of the octahedral complex (**23.1-17**), in which the metal ion and the plane of the ligand molecules are in one plane and the pyridine

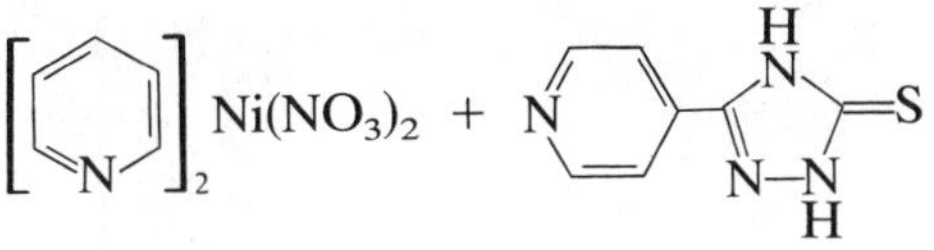

23.1-15 **23.1-16**

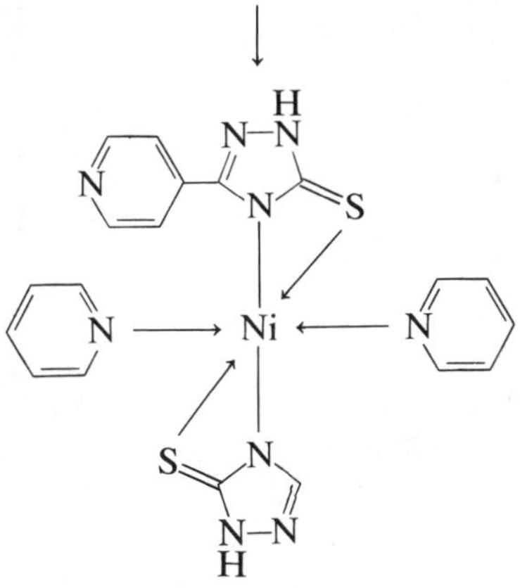

23.1-17

molecules are coordinated perpendicularly above and below this plane.[7] This thione (**23.1-16**) also formed octahedral complexes with cobalt (II), palladium (II), and platinum (II),[7,8] whereas zinc (II), cadmium (II), and mercury (II) formed tetrahedral complexes.[9] The interaction of rhodium (III) chloride hydrate with the thiones (**23.1-16**) in hot alcohol gave an orange-colored precipitate, which was assigned the structure of the chloro-bridged complex, **23.1-18**, based on solubility and diamagnetic properties.[8]

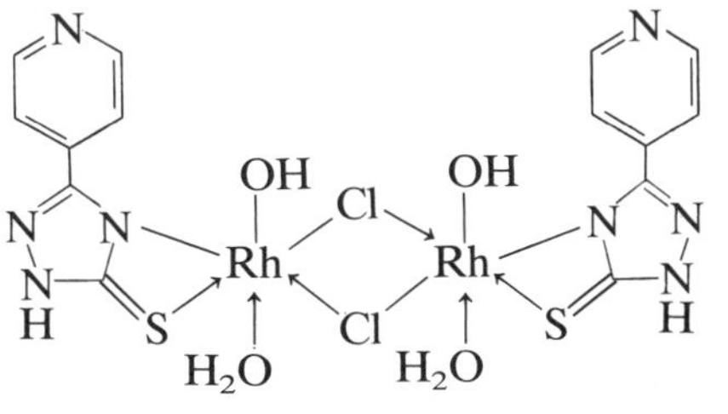

23.1-18

In contrast to the results with platinum (II) described previously, treatment of platinum (IV) with thione **23.1-12**, not containing a 3-(4-pyridinyl) group, resulted in reduction of the metal to platinum (II).[10] With a deficiency of thione the generated platinum (II) formed an 1:2 metal-thione complex. However, in the presence of excess thione a 1:4 complex was obtained, which was represented as structure **23.1-19**.

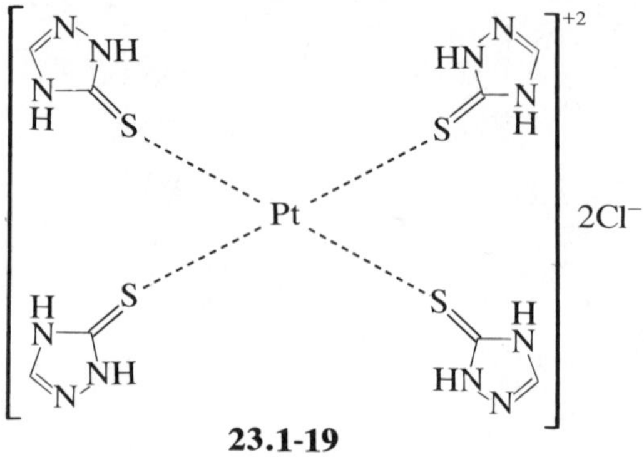

23.1-19

Several organolead compounds containing a triazole moiety were prepared for use as antiwear additives in lubricating oils and greases. For
example, the tributylplumbyl compound, **23.1-20**, in ether was treated with
3-amino-1,2,4-triazole to give **23.1-22**.[11]

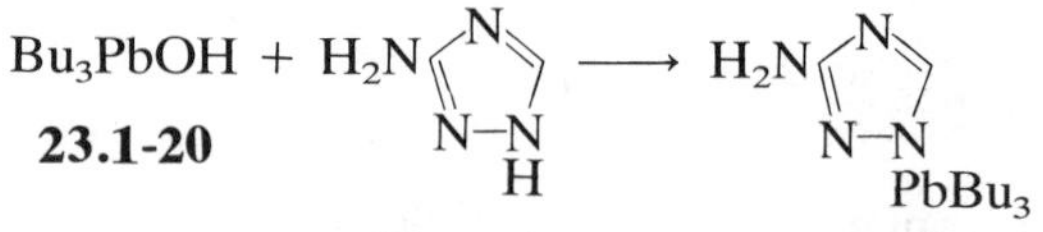

23.1-20   **23.1-21**   **23.1-22**

The adsorption of s-triazoline-5-thiones on silver halide were found to act
as photographic stabilizers, and this prompted an investigation into the
nature of the products of this interaction.[12] Treatment of 3-methyl-4-
phenyl-1,2,4-triazoline-5-thione (**23.1-23**) with excess silver nitrate in water
at 85° gave the salt (**23.1-24**), which was complexed with about an equimolar amount of silver nitrate. Other s-triazoline-5-thiones formed silver salts
complexed with varying amounts of silver nitrate. The infrared spectra of
these products indicated that two types of complexes were obtained: one
type contained nitrate groups readily replaceable by chloride, and the second
contained nitrato groups not readily replaceable by chloride.

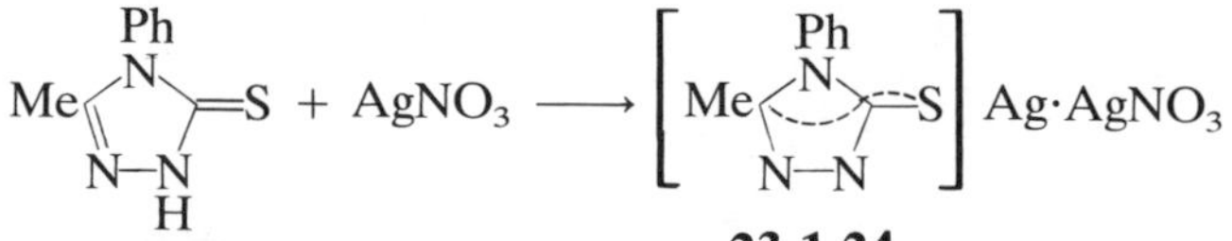

23.1-23   **23.1-24**

Bisazotriazoles are metalated to give dyes that are used to dye various
fibers in fast shades.[13] For example, reaction of **23.1-25** with chromium (III)
formate in dimethylformamide at 110° followed by treatment of the resulting
complex with ammonia was reported to give the reddish blue dye, **23.1-26**.

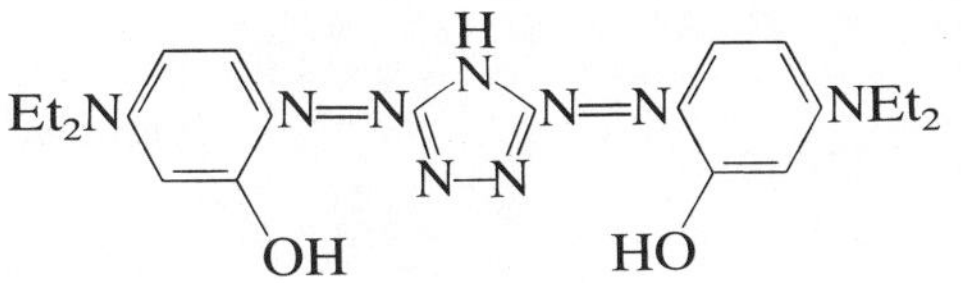

23.1-25

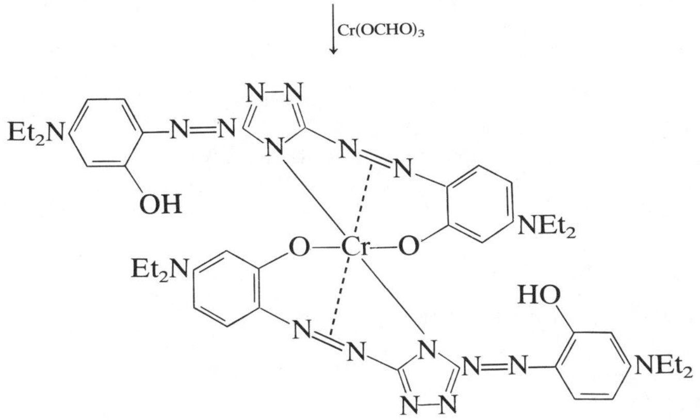

Cr(OCHO)₃

23.1-26

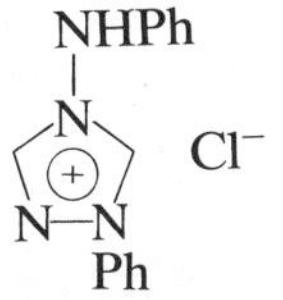

23.1-27

Several *s*-triazole derivatives have been used in analytical procedures for the determination of metals. The triazolium chloride (**23.1-27**) was proposed for the spectrophotometric determination of cobalt,[14] and the thione (**23.1-12**) for the simultaneous spectrophotometric determination of rhodium and platinum.[15]

References

1. **76:** 113224p
2. **77:** 55814p
3. **81:** 25762k
4. **79:** 5460r
5. **80:** 74824v
6. **68:** 46313m
7. **77:** 158352r
8. **81:** 130305m
9. **76:** 8560x
10. **75:** 26054u
11. **71:** 30588f
12. **79:** 98822y
13. **81:** 65182c
14. **69:** 92632h
15. **78:** 66590j

TABLE 23. METAL-TRIAZOLE COMPLEXES

Compound	Reference
Arsonium, triphenyl[2-(1H1-,2,4-triazol-1-yl)ethyl]-, tetraphenylborate(1-)	**83:** 179253y
Cadmium, bis[2,4-dihydro-5-(4-pyridinyl)-3H-1,2,4-triazole-3-thionato-N⁴, S]-, (T-4)	**76:** 8560x
Cadmium, (4-butyl-4H-1,2,4-triazole)[(1,2-ethanediylbis-[carbamodithioato])(2-)]-	**76:** P113224p
Chromate(1-), bis[(2,2′-[4H-1,2,4-triazole-3,5-diylbis-(azo)]bis[5-(diethylamino)phenolato])(2-)], hydrogen	**81:** P65182c
Chromium, pentacarbonyl(2,4-dihydro-2,4-dimethyl-3H-1,2,4-triazol-3-ylidene)-	**82:** 85553y
Cobalt, bis(4-anilino-1-phenyl-1H-1,2,4-triazole)tetrakis-(isothiocyanato)-	**69:** 92632h
Cobalt, bis[hydrotris(1H-1,2,4-triazolato-N¹)borato(1-)-N², N²*, N²**]-	**77:** 157129m
Cobalt, bis(pyridine)bis[5-(4-pyridinyl)-3H-1,2,4-triazole-3-thione]-	**81:** 161627x
Cobalt, diaquabis[2,4-dihydro-5-(4-pyridinyl)-3H-1,2,4-triazole-3-thionato]-	**77:** 158352r
Cobalt, diaquabis[2,4-dihydro-5-(4-pyridinyl)-3_H-1,2,4-triazole-3-thionato-N⁴, S³]-	**79:** 25270g
Cobalt, dichlorobis(1-ethenyl-1H1-,2,4-triazol-3-amine)-	**79:** 66253x
Cobalt, dichlorobis(4-ethenyl-4H-1,2,4-triazol-3-amine)-	**79:** 66253x
Cobalt, dichlorobis(1-ethenyl-1H-1,2,4-triazol-3-amine-N⁴)-	**80:** 82823j
Cobaltate(1-), bis[(1,1′-[4H-1,2,4-triazole-3,5-diylbis-(azo)]bis[2-naphthalenolato])(2-)]-, hydrogen	**81:** P65182c
Copper, aquobis[1,2-dihydro-5-(4-pyridinyl)-3H-1,2,4-triazole-3-thionato](pyridine)-, homopolymer	**77:** 158352r
Copper, bis(4-butyl-4H-1,2,4-triazole-N¹)dichloro-	**76:** P113224p
Copper, bis[5-4-[(4-chlorophenyl)sulfonyl]phenyl]-2,4-dihydro-4-(4-idophenyl)-3H-1,2,4-triazole-3-thionato-N²,S³-	**80:** 77774q
Copper, (μ-[N,N′-bis(2,3-dihydro-3-imino-1H-isoindol-1-ylidene)-1H-1,2,4-triazole-3,5-diaminato(4-)])di-	**79:** 93420f
Copper, bis(1H-1,2,4-triazolato-N¹)-, homopolymer	**82:** 92444s
Copper, (4-butyl-4H-1,2,4-triazole)([1,2-ethanediylbis-	**76:** P113224p
Copper, dibromo(4H-1,2,4-triazol-4-amine-N¹)-, homopolymer	**77:** 119903b
Copper, dibromo(4H-1,2,4-triazole-N¹)-, homopolymer	**77:** 119903b
Copper, dichloro(4H-1,2,4-triazol-4-amine-N¹)-, homopolymer	**77:** 119903b
Copper, dichloro(s-triazole)-, polymers	**69:** 55854m
Cuprate(2-), tetrakis(1H-1,2,4-triazolato-N¹)-, disodium	**82:** 92444s
Ferrate(1-), chloro[dihydrogen 3,7,12,17-tetramethyl-8,13-divinyl-2,18-porphinedipropionato(2-)]-(s-triazolato)-, diethyl ester	**69:** 16270y
Gold(1⁺), bis(2,4-dihydro-3H-1,2,4-triazole-3-thione)-	**80:** 74824v
Gold, [3-cyclohexyl-4,5-dihydro-1H-1,2,4-triazolato(2-)]-	**79:** P5460r
Gold, [4,5-dihydro-3-methyl-1H-1,2,4-triazolato(2-)]-	**79:** P5460r

TABLE 23 *(Continued)*

Compound	Reference
Gold, [2,4-dihydro-5-(4-pyridinyl)-3H-1,2,4-triazole-3-thionato-N⁴]-, homopolymer	**79:** 25270g
Gold, [4,5-dihydro-1H-1,2,4-triazolato(2-)]-	**79:** P5460r
Gold, [4,5-dihydro-3-undecyl-1H-1,2,4-triazolato(2-)]-	**79:** P5460r
Iron, bis(4-butyl-4H-1,2,4-triazole-N)dichloro-	**76:** P113224p
Iron, (4-butyl-4H-1,2,4-triazole)-[(1,2-ethanediylbis[carbamodithioato])(2-)]-	**76:** P113224p
Iron, [dihydrogen 3,7,12,17-tetramethyl-8,13-divinyl-2,18-porphinedipropionato-(2-)](2,4-lutidine)-(s-triazolato)-	**68:** 100283s
Iron, [dihydrogen 3,7,12,17-tetramethyl-8,13-divinyl-2,18-porphinedipropionato(2-)](2,6-lutidine)-(s-triazolato)-	**68:** 100283s
Iron(1+), [dihydrogen 3,7,12,17-tetramethyl-8,13-divinyl-2,18-porphinedipropionato(2-)](2,4-lutidine)-(s-triazole)-, ion	**68:** 100283s
Iron(1+), [dihydrogen 3,7,12,17-tetramethyl-8,13-divinyl-2,18-porphinedipropionato(2-)](2,6-lutidine)-(s-triazole)-, ion	**68:** 100283s
Iron, [dihydrogen 3,7,12,17-tetramethyl-8,13-divinyl-2,18-porphinedipropionato(2-)](2-picoline)-(s-triazolato)-	**68:** 100283s
Iron(1+), [dihydrogen 3,7,12,17-tetramethyl-8,13-divinyl-2,-8-porphinedipropionato(2-)](2-picoline)-(s-triazole)-, ion	**68:** 100283s
Iron(1+), [dihydrogen 3,7,12,17-tetramethyl-8,13-divinyl-2,18-porphinedipropionato(2-)](s-triazole)-, ion, diethyl ester	**68:** 46313m
Iron, [dihydrogen 3,7,12,17-tetramethyl-8,13-divinyl-2,18-prophinedipropionato(2-)](pyridine)-(s-triazolato)-	**68:** 100283s
Iron(1+), [dihydrogen 3,7,12,17-tetramethyl-8,13-divinyl-12,18-porphinedipropionato(2-)](pyridine)-(s-triazole)-, ion	**68:** 100283s
Iron, tetracarbonyl(2,4-dihydro-2,4-dimethyl-3H-1,2,4-triazole-3-ylidene)-	**82:** 85553y
Manganese, (4-butyl-4H-1,2,4-triazole)([1,2-ethane-diylbis[carbamodithioato])(2-)]-	**76:** P113224p
Manganese, dichlorobis(1-ethenyl-1H-1,2,4-triazol-3-amine-N⁴)-	**80:** 82823j
Mercury, bis[2,4-dihydro-5-(4-pyridinyl)-3H-1,2,4-triazole-3-thionato-N⁴,S]-, (T-4)-	**76:** 8560x
Mercury, (4-butyl-4H-1,2,4-triazole-N¹)dichloro-	**76:** P113224p
Mercury, chloro(2,4-dihydro-5-methyl-4-phenyl-3H-1,2,4-triazole-3-thionato-S)-	**80:** 146458d
Molybdenum, bis(∗⁵-2,4-cyclopentadien-1-yl)bis(4-phenyl-1,2,4-triazolidine-3,5-dionato-N¹)-	**81:** 25762k
Nickel, bis(4-butyl-4H-1,2,4-triazole-N¹)bis(nitrato-O)-	**76:** P113224p
Nickel, bis(4-butyl-4H-1,2,4-triazole-N¹)dichloro-	**76:** P113224p

TABLE 23 *(Continued)*

Compound	Reference
Nickel, bis[2,4-dihydro-5-(4-pyridinyl)-3H̲-1,2,4-triazole-3-thionato]bis(pyridine)-	**77:** 158352r
Nickel, bis(3-[(1-[(dimethylamino)carbonyl]1H̲-1,2,4-triazol-3-yl)amino]-2,4,5-cycloheptatrien-1-ylidene)]amino]-*N*,*N*-dimethyl-1H̲-1,2,4-triazole-3-carboxamidato)-	**77:** 113628u
Nickel, bis[4-(3-methylphenyl)-5-thioxo-1,2,4-triazolidin-3-onato]-	**78:** 58320f
Nickel, bis[4-(4-methylphenyl)-5-thioxo-1,2,4-triazolidin-3-onato]-	**78:** 58320f
Nickel, bis(4-phenyl-5-thioxo-1,2,4-thioxo1,2,4-triazolidin-	**78:** 58320f
Nickel, bis(pyridine)bis[5-(4-pyridinyl)-3H̲-1,2,4-triazole-3-thione]-	**81:** 161627x
Nickel, (4-butyl-4H̲-1,2,4-triazole)([1,2-ethanediylbis-carbamodithioato)](2-))-	**76:** P113224p
Nickel(6 +), hexaaquahexakis(μ-1H̲-1,2,4-triazole)tri-, hexanitrate, dihydrate	**68:** 108827t, **75:** 41620u
Palladium, bis[2,4-dihydro-5-(4-pyridinyl)-3H̲-1,2,4-triazole-3-thionato-N^4, S^3]-	**81:** 130305m
Palladium(2 +), tetrakis(2,4-dihydro-3H̲-1,2,4-triazole-3-thione)-	**80:** 74824v
Platinum, bis[2,4-dihydro-5-(4-pyridinyl)-3H̲-1,2,4-triazole-3-thionato-N^4, S^3]-	**81:** 130305m
Platinum, dichlorobis[2,4-dihydro-5-(4-pyridinyl)-3H̲-1,2,4-triazole-3-thionato]-, homopolymer	**81:** 130305m
Platinum(2 +), tetrakis(Δ^2-1,2,4-triazoline-5-thione)-, dichloride	**75:** 26054u
Plumbane, (3-amino-5-phenyl-1H̲-1,2,4-triazol-1-yl)-tributyl-	**75:** P142574n
Plumbane, (3-amino-1H̲-1,2,4-triazol-1-yl)tributyl-	**71:** P30588f, **71:** 102015s, **75:** P142574n
Plumbane, (3-amino-1H̲-1,2,4-triazol-1-yl)triethyl-	**71:** P30588f, **71:** P102015s
Plumbane, (3-anilino-5-isopropyl-1H̲-1,2,4-triazol-1-yl)-tributyl-	**75:** P142574n
Plumbane, triethyl-1H̲-1,2,4-triazol-1-yl-	**71:** P30588f, **71:** P102015s
Rodium, aquadi-μ-chlorobis[2,4-dihydro-5-(4-pyridinyl)-3H̲-1,2,4-triazole-3-thione-S]di-	**81:** 130305m
Rhodium, diaquadi-μ-chlorobis[2,4-dihydro-5-(4-pyridinyl)-3H-1,2,4-triazole-3-thionato-N^4,S^3]-dihydroxydi-	**81:** 130305m
Rohodium(3 +), hexakis(2,4-dihydro-3H̲-1,2,4-triazole-3-thione S)-	**78:** 66590j
Silver, bis(2,4-dihydro-4-phenyl-3H̲-1,2,4-triazole-3-thionato)(nitrato)tri-	**79:** 98822y
Silver, (4-butyl-4H̲-1,2,4-triazole-N^1)(nitrato-O)-	**76:** P113224p
Silver, dichlorotris(2,4-dihydro-4-phenyl-3H̲-1,2,4-triazole-3-thionato)penta-	**79:** 98822y
Silver, dichlorotris(2,4-dihydro-4-phenyl-3H̲-1,2,4-triazole-3-thionato)undeca	**79:** 98822y

TABLE 23 (*Continued*)

Compound	Reference
Silver, (2,4-dihydro-5-methyl-4-phenyl-3$\underline{H}$-1,2,4-triazole-3-thionato)(nitrato)di-	**79:** 98822y
Silver, pentachlorohexakis(2,4-dihydro-4-methyl-3$\underline{H}$-1,2,4-triazole-3-thionato)undeca-	**79:** 98822y
Silver, pentachlorotetrakis(2,4-dihydro-5-methyl-4-phenyl-3$\underline{H}$-1,2,4-triazole-3-thionato)nona-	**79:** 98822y
Silver, pentakis(2,4-dihydro-4-methyl-3$\underline{H}$-1,2,4-triazole-3-thionato)tetrakis(nitrato)nona-	**79:** 98822y
Silver, pentakis (4-ethyl-2,4-dihydro-5-methyl-3$\underline{H}$-1,2,4-triazole-3-thionato)hexakis(nitrato)undeca-	**79:** 98822y
Silver, [μ-(*s*-triazole-3,5-dicarboxylato)]di-	**73:** P120638t
Silver, trichlorobis(4-ethyl-2,4-dihydro-5-methyl-3$\underline{H}$-1,2,4-triazole-3-thionato)penta-	**79:** 98822y
Thallium, [2,4-dihydro-5-(4-pyridinyl-3$\underline{H}$-1,2,4-triazole-3-thionato-N^4]-, homopolymer	**79:** 25270g
Tin, bis(1$\underline{H}$-1,2,4-triazolato-N^1)-	**77:** 55814p
Tin, (4-butyl-4$\underline{H}$-1,2,4-triazole)[(1,2-ethanediylbis ×[carbamodithioato])(2-)]-	**76:** P113224p
Zinc, bis(acetato-O)bis(4-butyl-4$\underline{H}$-1,2,4-triazole-N^1), (T-4)-	**76:** P113224p
Zinc, bis(4-butyl-4$\underline{H}$-1,2,4-triazole-N^1)bis(nitrato-O)-, (T-4)-	**76:** P113224p
Zinc, bis(4-butyl-4$\underline{H}$-1,2,4-triazole-N^1)dichloro-, (T-4)-	**76:** P113224p
Zinc, bis(4-butyl-4$\underline{H}$-1,2,4-triazole-N^1)[sulfato(2-)-O,O^1]-, (T-4)-	**76:** P113224p
Zinc, bis[2,4-dihdyro-5-(4-pyridinyl)-3$\underline{H}$-1,2,4-triazole-3-thionato-N^4,S]-, (T-4)-	**76:** 8560x
Zinc, bis(1$\underline{H}$-1,2,4-triazolato-N^1)-, homopolymer	**82:** 924445
Zinc, (4-butyl-4$\underline{H}$-1,2,4-triazole)[(2-ethanediylbis carbamodithioato])(2-)]-	**76:** P11322p
Zinc, dichlorobis(1-ethenyl-1$\underline{H}$-1,2,4-triazol-3-amine)-	**79:** 66253x
Zinc, dichlorobis(4-ethenyl-4$\underline{H}$-1,2,4-triazol-3-amine)-	**79:** 66253x
Zinc, dichlorobis(1-ethenyl-1$\underline{H}$-1,2,4-triazol-3-amine-N^4), (T-4)-	**80:** 82823j
Zinc, dichlorobis[4-(3-methylbutyl)-3$\underline{H}$-1,2,4-triazole-N^1]-, (T-4)-	**76:** P113224p
Zinc, dichlorobis(4-octyl-4$\underline{H}$-1,2,4-triazole-N^1)-, (T-4)-	**76:** P113224p
Zinc, dichlorobis[4-(phenylmethyl)-4$\underline{H}$-1,2,4-triazole-N^1]-, (T-4)-	**76:** P113224p
Zinc, dichlorobis(4-phenyl-4$\underline{H}$-1,2,4-triazole-N^1), (T-4)-	**76:** P113224p
Zinc, dichlorobis[4-(2-pyridinyl)-4$\underline{H}$-1,2,4-triazole-N^1]-, (T-4)-	**76:** P113224p

Chemical Abstract
and Journal References
for 1,2,4-Triazoles

9: 210   K. Brunner, *Berichte*, **47**, 2671 (1914).

9: 3058   K. Brunner, *Monatsh. Chem.* **36**, 509 (1915).

11: 1412   H. Wolchowe, *Monatsh. Chem.* **37**, 473 (1916).

16: 2509[5]   F. Arndt, E. Milde, and F. Tschenscher, *Berichte*, **55B**, 341 (1922).

17: 3340   E. Fromm and E. Nehring, *Berichte*, **56B**, 1370 (1923).

18: 51   M. Busch, H. Muller, and E. Schwarz, *Berichte*, **56**, 1600 (1923).

21: 742[7]   C. Gastaldi and E. Princivalle, *Gazz. Chim. Ital.*, **56**, 557 (1926).

21: 2900[4]   P. C. Guha and P. C. Sen, *J. Indian Chem. Soc.*, **4**, 43 (1927).

21: 3200[4]   K. Brunner and J. Medweth, *Monatsh. Chem.*, **47**, 741 (1927).

21: 3200[7]   F. Hernler and F. Matthes, *Monatsh. Chem.*, **47**, 791 (1927).

21: 3200[9]   R. Grüner, Z. Benes, E. Schubert, and M. Arman, *Monatsh. Chem.*, **48**, 37 (1927).

21: 3620[4]   F. Hernler, *Monatsh. Chem.*, **48**, 391 (1927).

22: 1337[4]   G. T. Whyburn and J. R. Bailey, *J. Amer. Chem. Soc.*, **50**, 905 (1928).

22: 3660[3]   R. Stollé and N. Merkle, *J. Prakt. Chem.*, **119**, 275 (1928).

22: 4123[4]   S. C. De and S. K. Roy-Choudhury, *J. Indian Chem. Soc.*, **5**, 269 (1928).

22: 4123[9]   S. C. De, *J. Indian Chem. Soc.*, **5**, 373 (1928).

23: 113[7]   C. Naegeli and G. Stefanovitsch, *Helv. Chim. Acta*, **11**, 609 (1928).

23: 835[9]   G. Heller, W. Köhler, S. Gottfried, H. Arnold, and H. Herrmann, *J. Prakt. Chem.*, **120**, 49 (1929).

23: 836[4]   S. N. Chakrabarty and S. Dutt, *J. Indian Chem. Soc.*, **5**, 555 (1928).

23: 1397[9]   P. C. Guha and S. K. Roy-Choudhury, *J. Indian Chem. Soc.*, **5**, 163 (1928).

23: 1639[8]   P. Fantl and H. Silbermann, *Justus Liebigs Ann. Chem.*, **467**, 274 (1928).

23: 2177[1]   E. Fromm, R. Kapeller-Alder, W. Friedenthal, L. Stangel, J. Edlitz, E. Braumann, and J. Nussbaum, *Justus Liebigs Ann. Chem.*, **467**, 240 (1928).

23: 2974[6]   P. C. Guha and T. K. Chakraborty, *J. Indian Chem. Soc.*, **6**, 99 (1929).

23: 3226[6]   F. Hernler, *Monatsh. Chem.*, **51**, 267 (1929).

23: 3470[9]   J. Reilly and D. Madden, *J. Chem. Soc.*, **1929**, 815.

23: 3682[8]   G. Minunni, S. D'Urso, S. Guglielmino, P. Salanitro, D. Torrisi, and M. Vasta, *Gazz. Chim. Ital.*, **58**, 820 (1928).

24: 363[3]   G. Ponzio and M. Torres, *Gazz. Chim. Ital.*, **59**, 461 (1929).

24: 368[3]   F. Hernler, *Monatsh. Chem.*, **53/54**, 668 (1929).

24: 1113[9]   P. C. Guha, S. Rao, and A. Saletore, *J. Indian Chem. Soc.*, **6**, 565 (1929).

24: 2130[5]   F. Hernler, *Monatsh. Chem.*, **55,** 3 (1930).

24: 2747[9]   G. Heller and R. Mecke, *J. Prakt. Chem.*, **126,** 76 (1930).

24: 3216[1]   A. Benckiser, *J. Prakt. Chem.*, **125,** 236 (1930).

24: 3216[5]   W. Wirbatz, *J. Prakt. Chem.*, **125,** 267 (1930).

24: 4031[2]   H. Aspelund, *Chem. Zentralbl.*, **1929,** 100 (I), 2414.

24: 5751[2]   S. C. De and D. N. Dutt, *J. Indian Chem. Soc.*, **7,** 473 (1930).

25: 295[2]   K. L. Bhagat and J. N. Rây, *J. Chem. Soc.*, **1930,** 2357.

25: 703[5]   I. K. Matzurevich, *Bull. Soc. Chim.* **47,** 1160 (1930).

25: 1503[8]   S. M. Mistry and P. C. Guha, *J. Indian Chem. Soc.*, **7,** 793 (1930).

25: 2119[7]   S. C. De and T. K. Chakravorty, *J. Indian Chem. Soc.*, **7,** 875 (1930).

25: 2442[3]   G. Scheuing and B. Walach, U.S. Patent 1,796,403.

25: 3650[9]   T. N. Ghosh and M. V. Betrabet, *J. Indian Chem. Soc.*, **7,** 899 (1930).

25: 4557[1]   A. Boehringer, G. Scheuning, and B. Walach, British Patent 340, 237 (1929).

25: 4614[5]   C. Pietra, *Arch. Int. pharmacodyn. Ther.*, **36,** 236 (1929).

26: 258[4]   G. Scheuing and B. Walach, U.S. Patent 1,825,549.

26: 2469[4]   G. Scheuing and B. Walach, German Patent 541,700 (1928).

26: 3263[9]   Ger. Patent 543,026 (1928).

26: 3516[5]   G. Scheuing and B. Walach, German Patent 544,892 (1930).

26: 6018[5]   M. L. Duca, *Arch. Int. pharmacodyn. Ther.* **40,** 408 (1931).

27: 4541[2]   R. Meyer, German Patent 574,944 (1933).

28: 1665[8]   J. A. Murray and F. B. Dains, *J. Amer. Chem. Soc.*, **56,** 144 (1934).

28: 2679[5]   W. Oberhummer, *Monatsh. Chem.*, **63,** 285 (1933).

28: 2714[1]   R. Stollé and W. Dietrich, *J. Prakt. Chem.*, **139,** 193 (1934).

28: 2714[8]   R. Stollé and W. Dietrich, *J. Prakt. Chem.*, **139,** 193 (1934).

28: 4418[8]   J. Vŏrišek, *Collect. Czech. Chem. Commun.* **6,** 69 (1934).

28: 5457[7]   H. Aspelund, *Chem. Zentralbl.* **1933,** I, 780.

29: 1817[6]   H. Goldstein and F. Chastellian, *Helv. Chim. Acta*, **17,** 1481 (1934).

29: 5088[7]   H. Aspelund and A.-M. Augustson, *Acta Acad. Aboensis Math. Phys.*, **7,** (No. 10), 1 (1933).

29: 7209[5]   J. V. Dubsky and J. Trtilek, *Chem. Listy*, **29,** 33 (1935).

30: 4859[9]   T. N. Ghosh, *J. Indian Chem. Soc.*, **13,** 86 (1936).

30: 7574[6]   I. K. Matzurevich, *Univ. État Kiev, Bull. Sci., Receuil Chim.*, **1,** (No. 4), 9 (1935).

31: 687[9]   P. W. Neber and H. Wörner, *Justus Liebigs Ann. Chem.* **526,** 173 (1936).

31: 3053[8]   H. Wuyts and A. Lacourt, *Bull. Soc. Chim. Belges*, **45,** 685 (1936).

31: 3917[3]   A. Sanna, *Atti V Congr. Nazl. Chim. Pura Applicata, Rome* **1935,** Pt. I, 528 (1936).

31: 7849[3]   L. E. Hinkel, G. O. Richards, and O. Thomas, *J. Chem. Soc.*, **1937,** 1432.

32: 3399[1]   H. G. Underwood and F. B. Dains, *Univ. Kansas Sci. Bull.*, **24,** 5 (1936).

32: 4545[7]   M. Busch and K. Schulz, *J. Prakt. Chem.*, **150,** 173 (1938).

32: 7456[5]   R. Fusco and C. Musante, *Gazz. Chim. Ital.*, **68,** 147 (1938).

32: 9087[8]   T. N. Ghosh, *J. Indian Chem. Soc.*, **15,** 240 (1938).

33: 546[2]   E. Jolles and G. Ragni, *Gazz. Chim. Ital.*, **68,** 516 (1938).

33: 598[3]   P. C. Guha and D. R. Mehta, *J. Indian Inst. Sci.*, **21A,** 41 (1938).

33: 599[2]   P. C. Guha and D. R. Mehta, *J. Indian Inst. Sci.*, **21A,** 56 (1938).

33: 599[6]   P. C. Guha and S. L. Janniah, *J. Indian Inst. Sci.*, **21A,** 60 (1938).

33: 2518[1]   R. Fusco and C. Musante, *Gazz. Chim. Ital.*, **68,** 665 (1938).

33: 7287[4]   P. Grammaticakis, *C. R. Acad. Sci., Paris*, **208,** 1910 (1939).

35: 3598[6]   R. Kuhn and O. Westphal, *Berichte*, **73B,** 1109 (1940).

36: 427[9]   G. W. Raiziss, L. W. Clemence, and M. Freifelder, *J. Amer. Chem. Soc.*, **63,** 2739 (1941).

36: 5793[2]   K. A. Jensen, *Dansk Tids. Farm.*, **15,** 299 (1941).

36: 6511[3]	S. Rajagopalan, *Curr. Sci.*, **11**, 146 (1942).
37: 1130[5]	G. F. D'Alelio and J. W. Underwood, U.S. Patent 2,295,563 (1942).
37: 1402[1]	G. W. Anderson, H. E. Faith, H. W. Marson, P. S. Winnek, and R. O. Roblin, Jr, *J. Amer. Chem. Soc.*, **64**, 2902 (1942).
37: 2737[9]	M. Girard, *C. R. Acad. Sci., Paris*, **212**, 547 (1941).
37: 3091[8]	M. Girard, *Ann. Chim.*, **16**, 326 (1941).
37: 3092[4]	E. Ghigi and T. Pozzo-Balbi, *Ann. Chim. Farm.*, **1940**, 141.
38: 730[2]	S. Rajagopalan, *Proc. Indian Acad. Sci.*, **18A**, 100 (1943).
38: 970[3]	L. Birkofer, *Berichte*, **76B**, 769 (1943).
38: 2938[1]	H. J. Backer and J. de Jonge, *Rec. Trav. Chim. Pays-Bas*, **62**, 158 (1943).
38: 5828[3]	C. F. H. Allen and A. Bell, *Org. Syntheses*, **24**, 12 (1944).
38: 6321[7]	K. A. Jensen and K. Schmith, *Z. Immun.*, **105**, 40 (1944).
39: 1254[9]	G. F. D'Alelio, U.S. Patent 2,352,944 (1944).
39: 2068[5]	K. A. Jensen and A. Friediger, *Kgl. Danske Videnskab. Selskab, Math.-fys. Medd.*, **20**, 1 (1943).
39: P2382[4]	H. W. Marson, U.S. Patent 2,367,037 (1945).
39: 2741[4]	M. J. S. Dewar and F. E. King, *J. Chem. Soc.*, **1945**, 114.
39: 3536[4]	G. F. D'Alelio, U.S. Patent 2,369,949 (1945).
39: 5552[9]	G. F. D'Alelio, U.S. Patent 2,369,948 (1945).
40: 368[9]	D. W. Kaiser, U.S. Patent 2,382,156 (1945).
40: 948[8]	D. A. McGinty and W. G. Bywater, *J. Pharmacol.*, **84**, 342 (1945).
40: 1285[4]	G. F. D'Alelio, U.S. Patent 2,374,335 (1945).
40: 3348[9]	A. Bavley, U.S. Patent 2,395,776 (1946).
40: 3443[7]	D. B. Sharp and C. S. Hamilton, *J. Amer. Chem. Soc.*, **68**, 588 (1946).
40: 4227[8]	D. W. Kaiser, U.S. Patent 2,399,598 (1946).
41: 42h	A. Bavley, U.S. Patent 2,406,654 (1946).
41: 755g	C. F. H. Allen and A. Bell, *Org. Syntheses*, **26**, 11 (1946).
41: 948c	A. T. d'Arcangelo, *Rev. Fac. Cienc. Quim.* (Univ. Nacl. La Plata), **18**, 81 (1943).
41: 2607a	British Patent 583,367 (1946).
42: 1232d	R. Fusco and R. Romani, *Gazz. Chim. Ital.*, **76**, 439 (1946).
42: 6335i	H. W. Grimmel and J. F. Morgan, *J. Amer. Chem. Soc.*, **70**, 1750 (1948).
42: 8190d	F. Arndt, L. Loewe, and A. Tarlan-Akön, *Rev. Fac. Sci. Univ. Istanbul*, **13A**, 127 (1948).
42: 8799e	M. Girard, *C. R. Acad. Sci., Paris*, **225**, 458 (1947).
43: 52h	N. Heimbach and W. Kelly, Jr., U.S. Patent 2,449,225 (1948).
43: 601a	M. Amorosa, *Farm. Sci. Tec.* (*Pavia*), **3**, 389 (1948).
43: 3843c	J. K. Simons, U.S. Patent 2,456,090 (1948).
43: 5396a	H. Feichtinger, *Z. Naturforsch.*, **36**, 377 (1948).
43: 7018d	G. B. Bachman and L. V. Heisey, *J. Amer. Chem. Soc.*, **71**: 1985 (1949).
43: 7022a	A. Mustafa, *J. Chem. Soc.*, **1949**, 234.
43: 7730b	Swiss Patent 244,579 (1947).
43: 8365b	R. A. Henry and W. M. Dehn, *J. Amer. Chem. Soc.*, **71**, 2297 (1949).
44: 66b	N. Heimbach and W. Kelly, Jr., U.S. Patent 2,476,549 (1949).
44: 2514i	E. Hoggarth, *J. Chem. Soc.*, **1949**, 1160.
44: 2515i	E. Hoggarth, *J. Chem. Soc.*, **1949**, 1163.
44: 2517c	E. Hoggarth, *J. Chem. Soc.*, **1949**, 1918.
44: 2941i	G. D. Buckley and N. H. Ray, *J. Chem. Soc.*, **1949**, 1156.
44: P6316e	A. W. Anish and C. A. Clark, U.S. Patent 2,476,525 (1949).
44: 6415g	L. Loewe and M. Türgen, *Rev. Fac. Sci. Univ. Istanbul*, **14A**, 227 (1949).
44: 6854e	H. Gehlen, *Justus Liebigs Ann. Chem.*, **563**, 185 (1949).
44: 6855e	E. Hoggarth, *J. Chem. Soc.*, **1950**, 612.

44: 6856a   E. Hoggarth, *J. Chem. Soc.*, **1950,** 614.
44: 7849d   F. H. S. Curd, D. G. Davey, D. N. Richardson, and R. de B. Ashworth, *J. Chem. Soc.*, **1949,** 1739.
44: 8904a   D. Jerchel and R. Kuhn, *Justus Liebigs Ann. Chem.* **568,** 185 (1950).
44: 9287d   J. D. Kendall and G. F. Duffin, British Patent 634,952 (1950).
44: 9841d   R. E. Stauffer, U.S. Patent 2,497,917 (1950).
45: 613c   P. Papini and S. Checchi, *Gazz. Chim. Ital.*, **80,** 100 (1950).
45: 629c   J. W. Cook, R. P. Gentles, and S. H. Tucker, *Rec. Trav. Chim.*, **69,** 343 (1950).
45: 1121h   E. Hoggarth, *J. Chem. Soc.*, **1950,** 1579.
45: 2350d   D. J. T. Howe, U.S. Patent 2,534,599 (1950).
45: 4670h   R. O. Roblin, Jr. and J. W. Clapp, *J. Amer. Chem. Soc.* **72,** 4890 (1950).
45: 4671f   W. H. Miller, A. M. Dessert, and R. O. Roblin, Jr., *J. Amer. Chem. Soc.*, **72,** 4893 (1950).
45: 6211e   W. Baker, W. D. Ollis, and V. D. Poole, *J. Chem. Soc.*, **1950,** 3389.
45: 8560i   G. D. Buckley, N. H. Ray, British Patent 649,445 (1951).
46: 4534d   P. Papini and S. Checchi, *Gazz. Chim. Ital.*, **80,** 850 (1950).
46: 4534g   P. Papini and M. Manganelli, *Gazz. Chim. Ital.*, **80,** 855 (1950).
46: 5581g   H. Beyer and E. Kreutzberger-Reese, *Chem. Ber.*, **84,** 478 (1951).
46: 6088e   R. A. Henry, *J. Amer. Chem. Soc.*, **72,** 5343 (1950).
46: 8633g   D. Jerchel and H. Fischer, *Justus Liebigs Ann. Chem.* **574,** 85 (1951).
46: 11147a   Q. E. Thompson, *J. Amer. Chem. Soc.*, **73,** 5914 (1951).
47: 131d   R. Duschinsky and H. Gainer, *J. Amer. Chem. Soc.*, **73,** 4464 (1951).
47: 496h   H. M. Curry and J. P. Mason, *J. Amer. Chem. Soc.*, **73,** 5043 (1951).
47: 1625c   W. Ried and F. Müller, *Chem. Ber.*, **85,** 470 (1952).
47: 3819i   K. A. Jensen and S. A. K. Christensen, *Acta Chem. Scand.*, **6,** 166 (1952).
47: 3853c   H. Wojahn, *Arch. Pharm.*, **285,** 122 (1952).
47: 3929a   H. A. Offe, W. Siefken, and G. Domagk, *Z. Naturforsch.*, **7b,** 446 (1952).
47: 4335a   M. R. Atkinson and J. B. Polya, *J. Chem. Soc.*, **1952,** 3418.
47: 6941h   H. Gehlen, *Justus Liebigs Ann. Chem.*, **577,** 237 (1952).
47: 8035f   K. Miyatake, *J. Pharm. Soc. Japan*, **72,** 1486 (1952).
47: 8751b   D. A. Peak and F. Stansfield, *J. Chem. Soc.*, **1952,** 4067.
47: 8955e   H. W. Gausman, C. L. Rhykerd, H. R. Hinderliter, E. S. Scott, and L. F. Audrieth, *Botan. Gaz.*, **114,** 292 (1953).
47: 9862g   W. R. McBride, R. A. Henry, and S. Skolnik, *Anal. Chem.*, **25,** 1042 (1953).
47: 12370i   P. Papini and S. Checchi, *Gazz. Chim. Ital.*, **82,** 735 (1952).
48: 187b   British Patent 681,376 (1952).
48: 598e   F. L. Scott, D. A. O'Sullivan, and J. Reilly, *J. Appl. Chem.* (London), **2,** 184 (1952).
48: 675c   J. P. Thurston and J. Walker, *J. Chem. Soc.*, **1952,** 4542.
48: 1275i   D. W. Kaiser and G. A. Peters, *J. Org. Chem.*, **18,** 196 (1953).
48: 1343d   E. Lieber, S. Schiff, R. A. Henry, and W. G. Finnegan, *J. Org. Chem.*, **18,** 218 (1953).
48: 2050b   R. A. Henry, S. Skolnik, and G. B. L. Smith, *J. Amer. Chem. Soc.*, **75,** 955 (1953).
48: 2719h   G. H. Hitchings, A. Maggiolo, P. B. Russell, H. V. Werff, and I. M. Rollo, *J. Amer. Chem. Soc.*, **74,** 3200 (1952).
48: 4524f   E. Hoggarth, *J. Chem. Soc.*, **1952,** 4811.
48: 5181h   E. Hoggarth, *J. Chem. Soc.*, **1952,** 4817.
48: 5184a   V. M. Rodionov and V. K. Zvorykina, *Izv. Akad. Nauk. SSSR, Otdel. Khim. Nauk*, 70 (1953).
48: 7006c   R. M. Herbst and J. A. Garrison, *J. Org. Chem.*, **18,** 872 (1953).

48: 8268i   D. W. Kaiser and J. J. Roemer, U.S. Patent 2,648,670 (1953).
48: 10468a   H. Furuseki and S. Ohba, Japanese Patent 4723 (1953).
48: 12092c   R. H. Wiley and A. J. Hart, *J. Org. Chem.*, **18**, 1368 (1953).
48: 12739c   C. Ainsworth and R. G. Jones, *J. Amer. Chem. Soc.*, **75**, 4915 (1953).
48: 13687a   D. W. Kaiser, G. A. Peters, and V. P. Wystrach, *J. Org. Chem.*, **18**, 1610 (1953).
49: 1022e   M. Amorosa, *Boll. Sci. Fac. Chim. Ind. Bologna*, **10**, 159 (1952).
49: 1710g   B. R. Brown, D. Ll. Hammick, and S. G. Heritage, *J. Chem. Soc.*, **1953**, 3820.
49: 1710i   M. R. Atkinson and J. B. Polya, *J. Chem. Soc.*, **1954**, 141.
49: 2426i   R. Oda and S. Tanimoto, *J. Chem. Soc. Japan*, Ind. Chem. Sect. **55**, 595 (1952).
49: 2446f   F. L. Scott, *Chem. Ind.* (London), **1954**, 158.
49: 3949c   L. W. Hartzel and F. R. Benson, *J. Amer. Chem. Soc.*, **76**, 667 (1954).
49: 3950c   M. R. Atkinson and J. B. Polya, *J. Amer. Chem. Soc.*, **75**, 1471 (1953).
49: 3959e   H. C. Beyerman, J. S. Bontekoe, W. J. van der Burg, and W. L. C. Veer, *Rev. Trav. Chim.*, **73**, 109 (1954).
49: 5452a   P. G. Gordon, *Dissertation Abstr.*, **14**, 926 (1954).
49: 5532f   Swiss Patent 281,950 (1952).
49: 6239f   J. Goerdeler, K. Wember, and G. Worsch, *Chem. Ber.*, **87**, 57 (1954).
49: 6928c   M. R. Atkinson and J. B. Polya, *Chem. Ind.* (London), **1954**, 462.
49: 10857e   L. F. Audrieth, E. S. Scott and P. S. Kippur, *J. Org. Chem.*, **19**, 733 (1954).
49: 10857h   E. S. Scott and L. F. Audrieth, *J. Org. Chem.*, **19**, 742 (1954).
49: 10937d   S. Yoshida and M. Asai, *J. Pharm. Soc. Japan*, **74**, 946 (1954).
49: 10937f   S. Yoshida and M. Asai, *J. Pharm. Soc. Japan*, **74**, 948 (1954).
49: 10938b   S. Yoshida and M. Asai, *J. Pharm. Soc. Japan*, **74**, 951 (1954).
49: 11630g   M. R. Atkinson and J. B. Polya, *J. Chem. Soc.*, **1954**, 3319.
49: 12451a   R. A. Henry and Wm. G. Finnegan, *J. Amer. Chem. Soc.*, **76**, 290 (1954).
49: 13227a   K. T. Potts, *J. Chem. Soc.*, **1954**, 3461.
49: 13978h   C. Ainsworth and R. G. Jones, *J. Amer. Chem. Soc.*, **76**, 5651 (1954).
49: 14744g   R. Giuliano and G. Leonardi, *Il. Farmaco* (Pavia), *Ed. sci.*, **9**, 529 (1954).
49: 14744i   G. F. Duffin, J. D. Kendall, and H. R. J. Waddington, *Chem. Ind.* (London), **1954**, 1548.
49: 15869h   H. Beyer, W. Lässig, and U. Schultz, *Chem. Ber.*, **87**, 1401 (1954).
49: 15915i   M. R. Atkinson, E. A. Parkes, and J. B. Polya, *J. Chem. Soc.*, **1954**, 4256.
49: 15916f   M. R. Atkinson, A. A. Komzak, E. A. Parkes, and J. B. Polya, *J. Chem. Soc.*, **1954**, 4508.
50: 966i   P. Papini, S. Checchi, and M. Ridi, *Gazz. Chim. Ital.*, **84**, 769 (1954).
50: 971i   T. Sasaki, *Pharm. Bull.* (Japan), **2**, 123 (1954).
50: 1784b   R. G. Jones and C. Ainsworth, *J. Amer. Chem. Soc.*, **77**, 1538 (1955).
50: 1785c   C. Ainsworth and R. G. Jones, *J. Amer. Chem. Soc.*, **77**, 621 (1955).
50: 3417i   R. H. Wiley, N. R. Smith, D. M. Johnson, and J. Moffat, *J. Amer. Chem. Soc.*, **77**, 2572 (1955).
50: 5647a   M. Kanaoka, *J. Pharm. Soc. Japan*, **75**, 1149 (1955).
50: 6189c   P. Grammaticakis, *C. R. Acad. Sci., Paris*, **241**, 1049 (1955).
50: 6240b   L. F. Audrieth and E. B. Mohr, *Inorganic Syntheses*, **4**, 29 (1953).
50: 6437h   H. G. Mautner and W. D. Kumler, *J. Amer. Chem. Soc.*, **77**, 4076 (1955).
50: 8649g   C. W. Whitehead and J. J. Traverso, *J. Amer. Chem. Soc.*, **77**, 5872 (1955).
50: 9396d   H. Gehlen and H. Elchlepp, *Justus Liebigs Ann. Chem.*, **594**, 14 (1955).
50: 9913b   D. J. Fry and A. J. Lambie, British Patent 741,228 (1955).

50: 12032d   T. Tsuji, *Pharm. Bull.* (Japan), **2**, 403 (1954).
50: P12115i   C. Ainsworth and R. G. Jones, U.S. Patent 2,719,849 (1955).
50: 12116d   J. S. Webb and A. S. Tomcufcik, U.S. Patent 2,714,110 (1955).
50: 13097d   D. J. Fry and A. J. Lambie, British Patent 736,568 (1955).
50: 13886b   C. Ainsworth, *J. Amer. Chem. Soc.*, **78**, 1973 (1956).
50: 16789b   J. Klosa, *Arch. Pharm.*, **288**, 452 (1955).
51: 489h   R. N. Shreve and R. K. Charlesworth, U.S. Patent 2,744,116 (1956).
51: 899a   J. D. Kendall, G. F. Duffin, and H. R. J. Waddington, British Patent 743,133 (1956).
51: 1957e   D. D. Libman and R. Slack, *J. Chem. Soc.*, 2253 (1956).
51: 2874h   R. W. Young, U.S. Patent 2,744,907 (1956).
51: 3481e   M. Kulka, *Can. J. Chem.*, **34**, 1093 (1956).
51: 3579b   M. Kanaoka, *J. Pharm. Soc. Japan*, **76**, 1133 (1956).
51: 3669b   C. J. Grundmann and A. Kreutzberger, U.S. Patent 2,763,661 (1956).
51: 4440d   C. W. Huffman, U.S. Patent 2,762,816 (1956).
51: 5055b   I. Ya. Postovskiĭ and N. N. Vereshchagina, *Zh. Obshch. Khim.*, **26**, 2583 (1956).
51: 5095b   W. Dymek, *Ann. Univ. Mariae Curie-Sklodowska*, Lublin-Polonia Sect AA, **9**, 61 (1954).
51: 6607f   G. Leandri and P. Rebora, *Ann. Chim.* (Rome), **46**, 953 (1956).
51: 7999c   W. Baker and W. D. Ollis, *Quart. Rev.* (London), **11**, 15 (1957).
51: 8085e   H. Beyer and T. Pyl, *Chem. Ber.*, **89**, 2556 (1956).
51: 8732g   H. A. Staab, *Chem. Ber.*, **89**, 1927 (1956).
51: 10504i   H. A. Staab, *Chem. Ber.*, **89**, 2088 (1956).
51: 10506b   C. L. Arcus and B. S. Prydal, *J. Chem. Soc.*, **1957**, 1091.
51: 10506f   F. Mužík, *Chem. Listy*, **51**, 515 (1957).
51: 10587i   S. Kato, Japanese Patent 3428 (1956).
51: 12079a   G. W. Sawdey, *J. Amer. Chem. Soc.*, **79**, 1955 (1957).
51: 12079d   J. Goerdeler and J. Galinke, *Chem. Ber.*, **90**, 202 (1957).
51: 12150c   British Patent 774,867 (1957).
51: 12983d   S. Kato, Japanese Patent 3121 (1956).
51: 13934h   R. Zweidler and E. Keller, U.S. Patent 2,784,184 (1957).
51: 14698f   M. Häring and T. Wagner-Jauregg, *Helv. Chim. Acta*, **40**, 852 (1957).
51: 14712a   C. Scholtissek, *Chem. Ber.*, **89**, 2562 (1956).
51: 14751b   C. Grundmann and A. Kreutzberger, *J. Amer. Chem. Soc.*, **79**, 2839 (1957).
51: 15699h   S. Akiya, *Japan J. Exp. Med.*, **26**, 91 (1956).
51: 16159f   Y. Kuwabara and K. Aoki, *Yakugaku Zasshi*, **77**, 906 (1957).
51: 16162i   J. D. Kendall, G. F. Duffin, and H. R. J. Waddington, British Patent 766,380 (1957).
51: 17453c   E. A. Steck and F. C. Nachod, *J. Amer. Chem. Soc.*, **79**, 4411 (1957).
51: 17894e   C. G. Raison, *J. Chem. Soc.*, **1957**, 2858.
51: 17923e   A. Mustafa and A. E. A. A. Hassan, *J. Amer. Chem. Soc.*, **79**, 3846 (1957).
51: 18009a   C. J. Grundmann and R. F. W. Ratz, U.S. Patent 2,800,486 (1957).
51: 18613d   K. Harada and K. Murata, *Hiroshima Daigaku Kôgakubu Kenkyû Hôkoku*, **6**, 69 (1957).
52: 364c   C. Grundmann and R. Rätz, *J. Org. Chem.*, **21**, 1037 (1956).
52: 364e   E. Lieber, T.-S. Chao and C. N. R. Rao, *J. Org. Chem.*, **22**, 654 (1957).
52: 945a   P. G. Gordon and L. F. Audrieth, *Inorganic Syntheses* **5**, 52 (1957).
52: 1273g   W. Wilde, British Patent 776,118 (1957).
52: 2020d   E. F. Godefroi and E. L. Wittle, *J. Org. Chem.*, **21**, 1163 (1956).
52: 4297g   H. A. Staab, W. Otting, and A. Ueberle, *Z. Elektrochem.*, **61**, 1000 (1957).

52: 4606e	P. Papini, S. Checchi, and M. Ridi, *Gazz. Chim. Ital.*, **87**, 931 (1957).
52: 6337h	M. Eckstein, M. Gorczyca, A. Kocwa, and A. Zejc, *Diss. Pharm.*, **9**, 197 (1957).
52: 6345f	V. G. Yashunskiĭ, L. N. Pavlov, V. G. Ermolaelva, and M. N. Shchukina, *Khim. Nauka Prom.*, **2**, 658 (1957).
52: 6803e	British Patent 784,665 (1957).
52: 7332e	H. A. Staab, *Justus Liebigs Ann. Chem.* **609**, 75 (1957).
52: 8625b	T. Lieser and H. Gehlen, German Patent 854,576 (1952).
52: 9656a	E. B. Towne, J. W. Wellman, and J. B. Dickey, U.S. Patent 2,824,086 (1958).
52: 11822d	S. Takagi and A. Sugii, *Yakugaku Zasshi*, **78**, 280 (1958).
52: 11822h	A. Sugii, *Yakugaku Zasshi*, **78**, 283 (1958).
52: 11823c	A. Sugii, *Yakugaku Zasshi*, **78**, 306 (1958).
52: 11862e	G. Shaw and R. N. Warrener, *J. Chem. Soc.*, **1958**, 153
52: 11863e	G. Shaw and R. N. Warrener, *J. Chem. Soc.*, **1958**, 157.
52: 12851i	A. L. Mndzhoyan, V. G. Afrikyan, and A. A. Dokhikyan, *Izv. Akad. Nauk Armyan SSR, Ser Khim. Nauk*, **10**, 357 (1957).
52: 16341b	A. L. Mndzhoyan, V. G. Afrikyan, and V. E. Badalyan, *Izv. Akad. Nauk Armyan SSR, Ser. Khim. Nauk*, **10**, 421 (1957).
52: 16341e	H. A. Staab, *Justus Liebigs Ann. Chem.* **609**, 83 (1957).
52: 16363g	R. Nakai, M. Sugii, and H. Nakao, *Pharm. Bull.* (Tokyo), **5**, 576 (1957).
52: 17244f	M. Sekiya and S. Ishikawa, *Yakugaku Zasshi*, **78**, 549 (1958).
52: 17245b	V. Chmátal, J. Poskočil, and Z. J. Allan, *Chem. Listy*, **52**, 1156 (1958).
52: 18379e	G. Losse, W. Hessler, and A. Barth, *Chem. Ber.*, **91**, 150 (1958).
52: 20151g	V. G. Yashunskiĭ, L. N. Pavlov, V. G. Ermolaelva, and M. N. Shchukina, *Med. Prom.*, **11** (No. 12), 38 (1957).
53: 384c	A. F. McKay, S. Gelblum, E. J. Tarlton, P. R. Steyermark, and M. A. Mosley, *J. Amer. Chem. Soc.*, **80**, 3335 (1958).
53: 1734b	British Patent 791,932 (1958).
53: 1747h	K. Saftien and E. Anton, German Patent 923,028 (1955).
53: 2126e	R. E. Strube and C. Lewis, *J. Amer. Pharm. Ass.*, **48**, 73 (1959).
53: 2264c	J. F. Tinker and J. Sagal, Jr., U.S. Patent 2,835,581 (1958).
53: 3197g	E. A. Steck, R. P. Brundage, and L. T. Fletcher, *J. Amer. Chem. Soc.*, **80**, 3929 (1958).
53: 3198f	A. Dornow and K. Rombusch, *Chem. Ber.*, **91**, 1841 (1958).
53: 5253d	W. Kirmse and L. Horner, *Ann. Chem.*, **614**, 1 (1958).
53: 5253e	R. Huisgen, J. Sauer, and M. Seidel, *Chem. Ind.* (London), **1958**, 1114.
53: 5267i	G. Pala, *Farmaco, Ed. Sci.*, **13**, 461 (1958).
53: 6215i	M. Pesson and G. Polmanss, *C. R. Acad. Sci., Paris*, **247**, 787 (1958).
53: 6894e	A. Sugii and Y. Yamazaki, *Nippon Daigaku Yakugaku Kenkyu Hôkoku*, **2**, 6 (1958).
53: 7426g	C. L. Mitchell, H. H. Keasling, and E. G. Gross, *J. Amer. Pharm. Ass., Sci. Ed.* **48**, 122 (1959).
53: 8032i	A. Sugii, *Nippon Daigaku Yakugaku Kenkyu Hôkoku*, **2**, 10 (1958).
53: 8052i	W. Logemann, *Chem. Ber.*, **91**, 2578 (1958).
53: 9197b	H. Gehlen and E. Benatzky, *Justus Liebigs Ann. Chem.*, **615**, 60 (1958).
53: 9197g	A. N. Kost and F. Gents, *Zh. Obshch. Khim.*, **28**, 2773 (1958).
53: 9200b	E. Klingsberg, *J. Amer. Chem. Soc.*, **80**, 5786 (1958).
53: 10033i	A. Sugii, *Yakugaku Zasshi*, **79**, 100 (1959).
53: 11354d	K. Biemann and H. Bretschneider, *Monatsh. Chem.*, **89**, 603 (1958).
53: 11355a	R. Rätz and H. Schroeder, *J. Org. Chem.*, **23**, 2017 (1958).
53: 11355h	R. M. Herbst and J. E. Klingbeil, *J. Org. Chem.*, **23**, 1912 (1958).
53: 11398h	R. Rätz and H. Schroeder, *J. Org. Chem.*, **23**, 1931 (1958).
53: 14125d	W. Schäfer, R. Wegler, and L. Eue, U.S. Patent 2,870,144 (1959).

53: 16121a   E. Klingsberg, *J. Org. Chem.*, **23,** 1086 (1958).

53: 17020e   H. Stetter, R. Engl, and H. Rauhut, *Chem. Ber.*, **92,** 1184 (1959).

53: 17151e   M. Niese and D. Delfs, U.S. Patent 2,875,209 (1959).

53: 18013i   M. Tišler, *Arch. Pharm.* **292,** 90 (1959).

53: 19169d   T. Zsolnai, *Zentralbl. Bakteriol. Parasitenk.* **175,** 269 (1959).

53: 21098e   A. Brändström, *Ark. Kemi*, **11,** 567 (1957).

53: 21326c   I. V. Yanitskiĭ and V. I. Zelionkaĭte, *Zh. Neorg. Khim.*, **3,** 1755 (1958).

53: 21901h   H. A. Staab and G. Seel, *Chem. Ber.*, **92,** 1302 (1959).

53: 21904h   M. Pesson, G. Polmanss, and S. Dupin, *C. R. Acad. Sci., Paris*, **248,** 1677 (1959).

54: 510e   N. N. Vereshchagina and I. Ya. Postovskiĭ, *Nauch. Dokl. Vyssheĭ Shkoly, Khim. Khim. Tekhnol.*, **2,** 341 (1959).

54: 511b   V. V. Korshak, G. N. Chelnokova, and M. A. Shkolina, *Izv. Akad. Nauk SSSR, Otdel. Khim. Nauk*, **1959,** 925 (1959).

54: 515a   R. Huisgen, M. Seidel, J. Sauer, J. W. McFarland, and G. Wallbillich, *J. Org. Chem.*, **24,** 892 (1959).

54: 1503e   V. V. Korshak, G. N. Chelnokova, and M. A. Shkolina, *Izv. Akad. Nauk SSSR, Otdel. Khim. Nauk*, **1959,** 929 (1959).

54: 1534i   C. F. H. Allen, H. R. Beilfuss, D. M. Burness, G. A. Reynolds, J. F. Tinker, and J. A. Van Allan *J. Org. Chem.*, **24,** 787 (1959).

54: 1535i   C. F. H. Allen, H. R. Beilfuss, D. M. Burness, G. A. Reynolds, J. F. Tinker, and J. A. VanAllan, *J. Org. Chem.*, **24,** 793 (1959).

54: 1536h   C. F. H. Allen, H. R. Beilfuss, D. M. Burness, G. A. Reynolds, J. F. Tinker, and J. A. VanAllan, *J. Org. Chem.*, **24,** 796 (1959).

54: 1552e   British Patent 816,531 (1959).

54: 2332i   S. Checchi and M. Ridi, *Gazz. Chim. Ital.*, **87,** 597 (1957).

54: 3970a   W. Bossard, J. Voltz and F. Favre, U.S. Patent 2,883,373 (1959).

54: 4623b   W. B. Hardy and J. F. Hosler, U.S. Patent 2,914,536 (1959).

54: 4912g   C. L. Mitchell, H. H. Keasling, and C. W. Hirschler, *J. Amer. Pharm. Ass., Sci. Ed.*, **48,** 671 (1959).

54: 5629e   G. A. Reynolds and J. A. VanAllan, *J. Org. Chem.*, **24,** 1478 (1959).

54: 5933c   C. L. Mitchell and H. H. Keasling, *J. Pharmacol. Exp. Ther.*, **128,** 79 (1960).

54: 6376d   S. P. Popeck and J. R. Sottysiak, U.S. Patent 2,915,395 (1959).

54: 7162b   British Patent 822,858 (1959).

54: 7625h   F. Eloy and C. Moussebois, *Bull. Soc. Chim. Belges*, **68,** 423 (1959).

54: 7695d   G. Ramachander and V. R. Srinivasan, *Curr. Sci. (India)*, **28,** 368 (1959).

54: 7695e   G. F. Duffin, J. D. Kendall, and H. R. J. Waddington, *J. Chem. Soc.*, **1959,** 3799.

54: 7959d   L. I. Korolev, V. A. Voĭtekhova, and L. D. Stonov, *Materialy Mezhrespublik. Soveshchaniya Koordinatsii Nauch-Issledovatel. Rabot po Khlopkovodstvu, Akad. Nauk Uzbek. SSR, Tashkent*, **1957,** 215.

54: 8794f   A. Silberg and N. Cosma, *Acad. Rep. Pop. Rom. Filiala Cluj, Stud. Cercetări Chim.*, **10,** 151 (1959).

54: 9898c   I. Y. Postovskiĭ and N. N. Vereshchagina, *Zh. Obshch. Khim.*, **29,** 2139 (1959).

54: 9937a   V. C. Chambers, *J. Amer. Chem. Soc.*, **82,** 605 (1960).

54: 11003f   H. Gehlen and G. Blankenstein, *Justus Liebigs Ann. Chem.*, **627,** 162 (1959).

54: 12118h   J. Goerdeler, H. Haubrich, and J. Galinke, *Chem. Ber.*, **93,** 397 (1960).

54: 12124b   H. A. Staab, *Chem. Ber.*, **90,** 1326 (1957).

54: 14272g   H. A. Staab, German Patent 1,033,210 (1958).

54: 15371d   J. Goerdeler and H. Horstmann, *Chem. Ber.*, **93,** 663 (1960).

54: 18484g   M. Adámek, *Collect. Czech. Chem. Commun.*, **25,** 1694 (1960).

54: 19641e   R. Rips, C. Derappe, and N. P. Buu-Hoï, *J. Org. Chem.*, **25,** 390 (1960).

54: 19693i   C. F. H. Allen, G. A. Reynolds, J. F. Tinker, and L. A. Williams, *J. Org. Chem.*, **25,** 361 (1960).

54: 20217e   M. L. Seidenfaden, German Patent 1,028,714 (1958).

54: 21063g   R. Fusco and S. Rossi, *Ann. Chim.* (Rome), **50,** 277 (1960).

54: 21498h   E. Jeney and T. Zsolnai, *Zentralbl. Bakteriol. Parasit.*, **177,** 220 (1960).

54: 22468c   A. Silberg and I. Simiti, *Acad. Rep. Pop. Rom. Filiala Cluj, Stud. Cercetări Chim.*, **10,** 319 (1959).

54: 22602f   C. Runti, V. D'Osualdo, and F. Ulian, *Ann. Chim.* (Rome), **49,** 1668 (1959).

54: 22603e   V. V. Korshak, G. N. Chelnokova, and M. A. Shkolina, *Vysokomolekulyarnye Soedineniya*, **1,** 1772 (1959).

54: 24680e   F. Maggio, G. Werber, and G. Lombardo, *Ann. Chim.* (Rome). **50,** 491 (1960).

54: 24790h   F. Dallacker, *Monatsh. Chem.*, **91,** 294 (1960).

55: 522a   H. Foster, *J. Amer. Chem. Soc.*, **82,** 3780 (1960).

55: 1004e   H. Illy and R. Rüegg, Swiss Patent 341,247 (1959).

55: 1645g   E. Schmitz and R. Ohme, *Justus Liebigs Ann. Chem.*, **635,** 82 (1960).

55: 1647b   G. Losse and W. Farr, *J. Prakt. Chem.*, (4) **8,** 298 (1959).

55: 2623g   L. E. A. Godfrey and F. Kurzer, *J. Chem. Soc.*, **1960,** 3437.

55: 2625b   E. J. Browne and J. B. Polya, *Chem. Ind.* (London), **1960,** 1085.

55: 2625d   E. J. Browne and J. B. Polya, *Chem. Ind.* (London), **1960,** 1086.

55: 2990d   W. B. Hardy and J. F. Hosler, U.S. Patent 2,953,491 (1960).

55: 3565g   B. G. van den Bos, *Rec. Trav. Chim.*, **79,** 836 (1960).

55: 4337i   C. Ainsworth, "Org. Syntheses," Vol. 40., Wiley, New York, 1960 p. 99.

55: P4537b   W. J. Close and D. A. Dunnigan, U.S. Patent 2,944,060 (1960).

55: 4537f   H. Baumann and F. Arnemann, German Patent 1,049,381 (1959).

55: 5484b   L. Birkofer, P. Richter and A. Ritter, *Chem. Ber.*, **93,** 2804 (1960).

55: 5855h   W. W. Allen, German Patent 972,405 (1959).

55: 6473g   R. Huisgen, J. Sauer, and M. Seidel, *Chem. Ber.*, **93,** 2885 (1960).

55: 6474c   H. C. Brown and D. Pilipovich, *J. Amer. Chem. Soc.*, **82,** 4700 (1960).

55: 6495c   M. Lipp, F. Dallacker, and J. Thoma, *Monatsh. Chem.*, **91,** 595 (1960).

55: 7399c   S. V. Sokolov and I. Y. Postovskiĭ, *Zh. Obshch. Khim.*, **30,** 1781 (1960).

55: 8387d   H. Gehlen and G. Blankenstein, *Justus Liebigs Ann. Chem.*, **638,** 136 (1960).

55: 8387f   H. Gehlen, J. Dost, and J. Cermak, *Justus Liebigs Ann. Chem.* **638,** 141 (1960).

55: 8393d   B. G. van den Bos, *Rec. Trav. Chim.*, **79,** 1129 (1960).

55: 8394a   M. Ruccia, *Ann. Chim.* (Rome), **50,** 1363 (1960).

55: 10420a   H. Gehlen, J. Dost, and J. Cermak, *Justus Liebigs Ann. Chem.* **639,** 100 (1961).

55: 10450e   K. Shirakawa, *Yakugaku Zasshi*, **80,** 1542 (1960).

55: 10452h   K. Shirakawa, *Yakugaku Zasshi*, **80,** 1550 (1960).

55: 11397a   P. Grünanger, P. V. Finzi, and E. Fabbri, *Gazz. Chim. Ital.*, **90,** 413 (1960).

55: 11862h   British Patent 837,471 (1960).

55: 12392f   K. T. Potts, *Chem. Rev.*, **61,** 87 (1961).

55: 13449i   H. Weidinger and J. Kranz, German Patent 1,076,136 (1960).

55: 13758a   A. Müller and G. Pampus, German Patent 1,081,713 (1960).

55: 14440f   D. R. Liljegren and K. T. Potts, *J. Chem. Soc.*, **1961,** 518.

55: 14478a   H. Gold, *Angew. Chem.*, **72,** 956 (1960).

55: 15943g   J. Eibl and M. Niese, German Patent 1,080,243 (1960).

55: 16522h   B. G. van den Bos, M. J. Koopmans, and H. O. Huisman, *Rec. Trav. Chim.*, **79,** 807 (1960).

55: 17009h   S. Hünig and K.-H. Oette, *Justus Liebigs Ann. Chem.*, **641,** 94 (1961).

55: 17030a   W. Bossard and J. Voltz, U.S. Patent 2,978,290 (1961).
55: 17626d   A. Silberg, I. Simiti, N. Cosma, and I. Proinov, *Acad. Rep. Pop. Rom. Filiala Cluj, Stud. Cercetari Chim.*, **8**, 315 (1957).
55: 17626g   C. Pedersen, *Acta Chem. Scand.*, **13**, 888 (1959).
55: 18125e   J. Voltz and W. Bossard, German Patent 1,069,563 (1959).
55: 18778e   M. Adámek, Czechoslovakian Patent 94,768 (1960).
55: 19566i   E. Günther and W. Lässig, German Patent 1,064,343 (1959).
55: 20487f   C. I. Abshire and C. S. Marvel, *Makromol. Chem.*, **44**, 388 (1961).
55: 22298h   G. Zinner and W. Deucker, *Arch. Pharm.*, **294**, 370 (1961).
55: 22300h   H. Hellman and W. Schwiersch, *Chem. Ber.*, **94**, 1868 (1961).
55: 22335b   K. Futaki and S. Tosa, *Chem. Pharm. Bull.* (Tokoyo), **8**, 908 (1960).
55: 23504e   H. Gehlen and J. Dost, *Justus Liebigs Ann. Chem.*, **643**, 118 (1961).
55: 23507a   V. Grïnšteĭn and G. I. Chipen, *Zh. Obshch. Khim.*, **31**, 886 (1961).
55: 23507d   C.-F. Kröger, E. Tenor, and H. Beyer, *Justus Liebigs Ann. Chem.* **643**, 121 (1961).
55: 23508e   C.-F. Kröger, W. Sattler, and H. Beyer, *Justus Liebigs Ann. Chem.*, **643**, 128 (1961).
55: 23509g   J. C. Howard and H. A. Burch, *J. Org. Chem.*, **26**, 1651 (1961).
55: 24728c   H. Gehlen, J. Dost, and J. Cermak, *Justus Liebigs Ann. Chem.* **643**, 116 (1961).
55: 24729e   W. Asker and Z. E. Elagroudi, *J. Org. Chem.*, **26**, 1440 (1961).
55: 25991f   H. Weidinger and J. Kranz, German Patent 1,073,499 (1960).
55: 26806g   W. Lässig and E. Günther, German Patent 1,058,844 (1959).
55: 27047c   G. Berkelhammer, S. DuBreuil, and R. W. Young, *J. Org. Chem.*, **26**, 2281 (1961).
55: 27282f   S. S. Novikov, V. A. Rudenko, and V. M. Brusnikina, *Izv. Akad. Nauk SSSR, Otdel. Khim. Nauk*, **1961**, 1148.
55: 27284a   M. Pesson, S. Dupin, and M. Antoine, *C. R. Acad. Sci.*, Paris, **253**, 285 (1961).
56: 472b   K. Nagakubo, *Nippon Kagaku Zasshi*, **81**, 305 (1960).
56: 486e   J. Thesing and G. Mohr, German Patent 1,103,342 (1958).
56: 1388a   R. Huisgen, J. Sauer, and M. Seidel, *Chem. Ber.*, **94**, 2503 (1961).
56: 1388h   R. Huisgen and M. Seidel, *Chem. Ber.*, **94**, 2509 (1961).
56: 2440b   M. Pesson and S. Dupin, *C. R. Acad. Sci.*, Paris, **252**, 3830 (1961).
56: 2440e   A. A. Ponomarev and M. D. Lipanova, *Uchenye Zapiski Saratov. Univ.*, **71**, 151 (1959).
56: 4748g   M. Pesson, S. Dupin, and M. Antoine, *C. R. Acad. Sci.*, Paris, **253**, 992 (1961).
56: 4776e   D. V. Sickman, U.S. Patent 2,987,520 (1961).
56: 5948g   J. Sandström, *Acta Chem. Scand.*, **14**, 1037 (1960).
56: 7304h   H. Saikachi and M. Kanaoka, *Yakugaku Zasshi*, **81**, 1333 (1961).
56: 7306c   M. Pesson and S. Dupin, *C. R.*, Acad. Sci., Paris, **253**, 1974 (1961).
56: 7307a   K. Matsuda and L. T. Morin, *J. Org. Chem.*, **26**, 3783 (1961).
56: 8708e   T. Bacchetti, *Gazz. Chim. Ital.*, **91**, 866 (1961).
56: 9421c   E. H. Browne and J. B. Polya, *Anal. Chem.*, **34**, 298 (1962).
56: 10131h   J. Kranz and H. Weidinger, *Festschrift Carl Wurster zum 60. Geburtstag*, **1960**, 119.
56: 10133e   H. Balli and F. Kersting, *Justus Liebigs Ann. Chem.*, **647**, 1 (1961).
56: 10160f   C. R. Jacobson, A. D'Adamo, and C. E. Cosgrove, German Patent 1,104,965 (1959).
56: 11111h   British Patent 836,148 (1960).
56: 12877h   L. E. A. Godfrey and F. Kurzer, *J. Chem. Soc.*, **1961**, 5137.
56: 12907g   J. Norris, British Patent 883,732 (1961).
56: 13506e   K. Medne, V. Grinsteins, and G. Cipens, *Latv. PSR Zinatnu Akad. Vestis*, **1961**, 85.

56: 13123a	British Patent 884,582 (1961).
56: 14193c	F. F. Ebetino and G. Gever, *J. Org. Chem.*, **27**, 188 (1962).
56: 14293h	J. Sandström, *Acta Chem. Scand.*, **15**, 1575 (1961).
56: 15513a	N. Petri, *Z. Naturforsch.*, **16b**, 767 (1961).
56: 15949e	J. F. Fredrick, *Physiol. Plantarum*, **14**, 734 (1961).
57: 804i	H. Beyer, C.-F. Kroger, and G. Busse, *Justus Liebigs Ann. Chem.*, **637**, 135 (1960).
57: 805h	E. J. Browne and J. B. Polya, *J. Chem. Soc.*, **1962**, 575.
57: 808f	M. Pesson and S. Dupin, *Bull. Soc. Chim. France*, **1962**, 250.
57: 1791c	Belgian Patent 610,096 (1961).
57: 2229d	K. Hauptmann and K. Zeile, German Patent 1,126,882 (1962).
57: 3423i	H. Gehlen and G. Blakenstein, *Justus Liebigs Ann. Chem.* **651**, 128 (1962).
57: 3425c	H. Gehlen and G. Blankenstein, *Justus Liebigs Ann. Chem.*, **651**, 137 (1962).
57: 3436i	D. R. Liljegren and K. T. Potts, *Chem. Ind.* (London), **1961**, 2049.
57: 3730d	J. F. Fredrick, *Phyton* (Buenos Aires), **17**, 147 (1961).
57: 3982a	C. Ainsworth, N. R. Easton, M. Livezey, D. E. Morrison, and W. R. Gibson, *J. Med. Pharm. Chem.*, **5**, 383 (1962).
57: 4651d	S. Bellioni, *Ann. Chim.* (Rome), **52**, 187 (1962).
57: 5897b	D. G. Farnum and P. Yates, *J. Org. Chem.*, **27**, 2209 (1962).
57: 5903d	F. Schneider and W. Schaeg, *Z. Physiol. Chem.*, **327**, 74 (1962).
57: 5907i	H. A. Staab, M. Lueking, and F. H. Duerr, *Berichte*, **95**, 1275 (1962).
57: 5909c	R. Huisgen, J. Sauer, and M. Seidel, *Justus Liebigs Ann. Chem.*, **654**, 146 (1962).
57: 5911g	A. J. Boulton and A. R. Katritzky, *J. Chem. Soc.*, **1962**, 2083.
57: 6072h	H. Baumann and S. Huenig, German Patent 1,117,233 (1961).
57: 7253h	B. G. van den Bos, C. J. Schoot, M. J. Koopmans, and J. Meltzer, *Rec. Trav. Chim.*, **80**, 1040 (1961).
57: 8561c	W. Dymek and M. Dziewonska, *Dissert. Pharm.*, **13**, 313 (1961).
57: 8585b	H. Hellmann and W. Schwiersch, German Patent 1,123,331 (1962).
57: 9860b	H. Heller, E. Keller, H. Gysling, and F. Mindermann, U.S. Patent 3,004,896 (1961).
57: 11191c	E. J. Browne and H. B. Polya, *Tetrahedron*, **18**, 539 (1962).
57: 11198d	R. Huisgen, R. Grashey, M. Seidel, G. Wallbillich, H. Knupfer, and R. Schmidt, *Justus Liebigs Ann. Chem.*, **653**, 105 (1962).
57: 11205a	H. Ruschig, K. Schmitt, L. Ther, and W. Pfaff, German Patent 1,129,498 (1962).
57: 12470i	H. C. Brown and M. T. Cheng, *J. Org. Chem.*, **27**, 3240 (1962).
57: 12471f	J. Sandström, *Acta Chem. Scand.*, **15**, 1295 (1961).
57: 12472e	K. T. Potts and T. H. Crawford, *J. Org. Chem.*, **27**, 2631 (1962).
57: 12472h	L. E. A. Godfrey and F. Kurzer, *J. Chem. Soc.*, **1962**, 3561.
57: 12473b	S. Naqui and V. R. Srinivasan, *J. Sci. Ind. Res.* (India), **21B** (No. 4), 195 (1962).
57: 13927c	K. H. Schuendehuette, F. Suckfuell, H. Nickel, and K. H. Schmidt, German Patent 1,130,098 (1962).
57: 13929c	H. Baumann and J. Dehnert, British Patent 891,515 (1962).
57: 14623b	E. Guenther and W. Laessig, German Patent 1,128, 296 (1962).
57: 15096c	H. Hellmann and W. Elser, *Berichte*, **95**, 1955 (1962).
57: 16598e	M. Pesson, S. Dupin, and M. Antoine, *Bull Soc. Chim. France*, **1962**, 1364.
57: 16600d	G. Ramachander and V. R. Srinivasan, *J. Sci. Ind. Res.* (India), **21C** (No. 2), 44 (1962).
57: 16601e	M. H. Shah, V. M. Patki, and M. Y. Mhasalkar, *J. Sci. Ind. Res.* (India), **21C** (No. 3), 76 (1962).

57: 16607a	Y. Makisumi, *Chem. Pharm. Bull.* (Tokyo), **9,** 808 (1961).
57: 16613g	L. A. Williams, *J. Chem. Soc.,* **1962,** 3854.
58: 518f	B. T. Gillis and F. A. Daniher, *J. Org. Chem.,* **27,** 4001 (1962).
58: 518g	G. Cipens, V. Grinsteins, and R. Preimans, *Zh. Obshch. Khim.,* **32,** 454 (1962).
58: 519b	G. Cipens and V. Grinsteins, *Zh. Obshch. Khim.,* **32,** 460 (1962).
58: 1469g	Belgian Patent 610,039 (1962).
58: 1479f	K. Schmitt and G. Driesen, German Patent 1,135,479 (1962).
58: 1479g	K. Schmitt, German Patent 1,135,478 (1962).
58: 2447e	R. G. Dubenko and P. S. Pel'kis, *Zh. Obshch. Khim.,* **32,** 939 (1962).
58: 2447h	E. J. Browne and J. B. Polya, *J. Chem. Soc.,* **1962,** 5149.
58: 2448a	S. Naqui and V. R. Srinivasan, *J. Sci. Ind. Res.* (India), **21B,** 456 (1962).
58: 2448f	R. C. Cookson, S. S. H. Gilani, and I. D. R. Stevens, *Tetrahedron Lett.,* **1962,** 615.
58: 2449a	S. Kubota and M. Ohtsuka, *Tokushima Daigaku Yakugaku Kenkyu Nempo,* **9,** 15 (1960).
58: 2451h	J. K. Stille and T. Anyos, *J. Org. Chem.,* **27,** 3352 (1962).
58: 3415g	H. Gehlen and W. Schade, *Naturwissenschaften,* **46,** 667 (1959).
58: 3415h	R. Huisgen, R. Grashey, M. Seidel, H. Knupfer, and R. Schmidt, *Justus Liebigs Ann. Chem.,* **658,** 169 (1962).
58: 4543d	H. Saikachi and M. Kanaoka, *Yakugaku Zasshi,* **82,** 683 (1962).
58: 4569g	V. R. Srinivasan, G. Ramachander, and S. Naqui, *Arch. Pharm.,* **295,** 405 (1962).
58: 4673d	H. Pfitzner and H. Baumann, Belgian Patent 611,906 (1962).
58: 5568c	J. Kotler-Brajtburg, A. Swirska, and A. Raczka, *Roczniki Chem.,* **36,** 971 (1962).
58: 5661c	H. Behringer and H. J. Fischer, *Berichte,* **95,** 2546 (1962).
58: 5703f	F. A. Grunwald and W. B. Lacefield, U.S. Patent 3,056,780 (1962).
58: 6827g	H. Gehlen and W. Schade, *Naturwissenschaften,* **46,** 667 (1959).
58: 7924d	K. T. Potts, *J. Org. Chem.,* **28,** 543 (1963).
58: 8071d	F. Wiloth, German Patent 1,135,920 (1962).
58: 9051f	H. Gehlen and G. Röbisch, *Justus Liebigs Ann. Chem.,* **660,** 148 (1962).
58: 9071e	A. A. Ponomarev and M. D. Lipanova, *Zh. Obshch. Khim.,* **32,** 2974 (1962).
58: 10202c	G. Cipens and V. Grinsteins, *Zh. Obshch. Khim.,* **32,** 3811 (1962).
58: 10220g	H. P. Burchfield and D. K. Gullstrom, U. S. Patent 3,054,800 (1962).
58: 11370b	H. Ruschig, K. Schmitt, L. Ther, and G. Vogel, German Patent 1,139,846 (1962).
58: 13937f	Y.-T. Ch'en and T.-I. Chang, *K'o Hsueh T'ung Pao,* **1962,** 37.
58: 13965f	K. Shirakawa, Y. Usui, and T. Tsujikawa, Japanese Patent 6834 (1962).
59: 161b	W. Laessig and E. Guenther, German Patent 1,146,367 (1963).
59: 603a	S. Dupin and M. Pesson, *Bull. Soc. Chim. France,* **1963,** 144.
59: 1623a	F. Baumbach, H. G. Henning, and G. Hilgetag, *Z. Chem.,* **2,** 369 (1962).
59: 1624b	M. Taniyama, T. Nagaoka, T. Takata, and T. Okauchi, *Kogyo Kagaku Zasshi,* **65,** 1694 (1962).
59: 2678d	H. Ulrich and A. A. R. Sayigh, *J. Org. Chem.,* **28,** 1427 (1963).
59: 2803g	S. Naqui and V. R. Srinivasan, *Tetrahedron Lett.* **1962,** 1193.
59: 2815h	A. Mustafa, S. A. Khattab, and W. Asker, *Can. J. Chem.,* **41,** 1813 (1963).
59: 2820h	H. Gehlen and G. Roebisch, *Justus Liebigs Ann. Chem.,* **663,** 119 (1963).
59: 2821h	F. Kurzer and P. M. Sanderson, *J. Chem. Soc.,* **1963,** 3333.
59: 3918f	Y. Makisumi and H. Kanō, *Chem. Pharm. Bull.* (Tokyo), **11,** 67 (1963).
59: 3927a	E. R. Bissell, *J. Org. Chem.,* **28,** 1717 (1963).
59: 3927g	K. R. Huffman and F. C. Schaefer, *J. Org. Chem.,* **28,** 1816 (1963).

59: 3930f	M. Ohta and H. Ueda, *Nippon Kagaku Zasshi*, **82,** 1525 (1961).
59: 5151h	G. Zinner and W. Deucker, *Arch. Pharm.*, **296,** 13 (1963).
59: 6227c	G. Cipens and V. Grinsteins, *Latv. PSR Zinat. Akad. Vestis, Kim. Ser.*, **1962,** 401.
59: 6227d	G. Cipens and V. Grinsteins, *Latv. PSR Zinat. Akad. Vestis, Kim. Ser.*, **1962,** 411.
59: 6386c	H. Weidinger and J. Kranz, *Berichte*, **96,** 1059 (1963).
59: 6386f	H. Weidniger and J. Kranz, *Berichte*, **96,** 1064 (1963).
59: 6729e	P. Massini, *Acta Botan. Neerl.*, **12,** 64 (1963).
59: 7530e	B. Tornetta, *Ann. Chim.* (Rome), **53,** 244 (1963).
59: 7531a	K. R. Huffman and F. C. Schaefer, *J. Org. Chem.*, **28,** 1812 (1963).
59: 8589e	M. B. Frankel, E. A. Burns, J. C. Butler, and E. R. Wilson, *J. Org. Chem.*, **28,** 2428 (1963).
59: 8747g	E. Nachbaur and W. Gottardi, *Monatsh. Chem.*, **94,** 584 (1963).
59: 8759b	British Patent 919,458 (1963).
59: 8880c	H. Hopff and M. Lippay, *Makromol. Chem.*, **66,** 157 (1963).
59: 10025g	F. Runge, Z. El-Hewehi, and E. Taeger, *J. Prakt. Chem.*, **18,** 262 (1962).
59: 10048c	Y. Makisumi, *Chem. Pharm. Bull.* (Tokyo), **11,** 851 (1963).
59: 10048d	Y. Makisumi, *Chem. Pharm. Bull.* (Tokyo), **11,** 859 (1963).
59: 10048f	C.-F. Kroeger, G. Etzold, and H. Beyer, *Justus Liebigs Ann. Chem.*, **664,** 146 (1963).
59: 10050a	C.-F. Kroeger, G. Etzold, H. Beyer, and G. Busse, *Justus Liebigs Ann. Chem.*, **664,** 156 (1963).
59: 10052c	L. E. A. Godfrey and F. Kurzer, *J. Chem. Soc.*, **1963,** 4558.
59: 10052e	H. J. Shine, L.-T. Fang, H. E. Mallory, N. F. Chamberlain, and F. Stehling, *J. Org. Chem.*, **28,** 2326 (1963).
59: 10052g	V. R. Srinivasan and G. Ramachander, *Indian J. Chem.*, **1,** 234 (1963).
59: 10053c	B. M. Lynch, *Can. J. Chem.*, **41,** 2380 (1963).
59: 10065h	M. Taniyama, T. Nagaoka, T. Okauchi, and T. Takada, Japanese Patent 9221 (1962).
59: 11392a	I. P. Tsukervanik and S. A. Israilova, *Dokl. Akad. Nauk Uz. SSR*, **20,** 25 (1963).
59: 11473d	H. Gehlen and J. Dost, *Justus Liebigs Ann. Chem.* **665,** 144 (1963).
59: 11493d	H. Weidinger and J. Kranz, *Berichte*, **96,** 2070 (1963).
59: 11494e	P. Yates and D. G. Farnum, *J. Amer. Chem. Soc.*, **85,** 2967 (1963).
59: 11506g	Belgian Patent 621,842 (1963).
59: 11507b	O. W. Webster, U.S. Patent 3,093,653 (1963).
59: 11695e	H. Balli and F. Kersting, *Justus Liebigs Ann. Chem.* **663,** 96 (1963).
59: 11696b	H. Balli and F. Kersting, *Justus Liebigs Ann. Chem.* **663,** 103 (1963).
59: 12340b	E. Weyde, H. V. Rintelen, and A. V. Koenig, German Patent 1,153,247 (1963).
59: 12787f	E. R. Ward and D. D. Heard, *J. Chem. Soc.*, **1963,** 4794.
59: 12787h	T. Okamoto, M. Hirobe, and Y. Tamai, *Chem. Pharm. Bull.* (Tokyo), **11,** 1089 (1963).
59: 12788b	H. Gehlen and G. Röbisch, *Justus Liebigs Ann. Chem.*, **665,** 132 (1963).
59: 12789f	G. Cipens and V. Grinsteins, *Latv. PSR Zinatnu Akad. Vestis, Kim. Ser.*, **1962,** 255.
59: 12790e	G. Cipens and V. Grinsteins, *Latv. PSR Zinatnu Akad. Vestis, Kim. Ser.*, **1962,** 263.
59: 12791d	G. Cipens, V. Grinsteins, and A. Grinvalde, *Latv. PSR Zinatnu Akad. Vestis, Kim. Ser.*, **1962,** 495.
59: 12791g	G. Cipens and V. Grinsteins, *Latv. PSR Zinatnu Akad. Vestis, Kim. Ser.*, **1962,** 503.
59: 12826f	C. Holstead, Belgian Patent 619,423 (1962).

59: 12957c   H. Baumann and J. Dehnert, British Patent 932,022 (1963).
59: 13277e   M. H. Woolford, Jr., *J. Ass. Offic. Agr. Chem.* **46,** 463 (1963).
59: 13985f   R. G. Dubenko and P. S. Pel'kis, *Zh. Obshch. Khim.,* **33,** 2220 (1963).
59: 15276h   A. A. Ponomarev and M. D. Lipanova, *Uch. Zap. Saratovsk. Gos. Univ.,* **75, 35** (1962).
59: 15278h   V. M. Brusnikina, S. S. Novikov, and V. A. Rudenko, *Izv. Akad. Nauk SSSR, Ser. Khim.,* **1963,** 1681.
59: 15420d   R. Raue and H. Gold, Belgian Patent 621,482 (1962).
59: 15420h   A. F. Strobel and S. C. Catino, French Patent 1,316,471 (1963).
60: 520h   R. Gompper, H. E. Noppel, and H. Schaefer, *Angew. Chem.,* **75,** 918 (1963).
60: 526a   J. Sandström, *Acta Chem. Scand.,* **17,** 1595 (1963).
60: 1736d   H. Gehlen and F. Lemme, *Naturwissenschaften,* **50,** 645 (1963).
60: 1752h   W. Logemann, D. Artini, L. Canavesi, and G. Tosolini, *Berichte,* **96,** 2909 (1963).
60: 1753f   W. Logemann, D. Artini, and L. Canavesi, *Berichte,* **96,** 2914 (1963).
60: 1754b   F. Kurzer and J. Canelle, *Tetrahedron,* **19,** 1603 (1963).
60: 1764f   K. Schmidt and F. Soldam, German Patent 1,151,257 (1963).
60: 2949f   H. Ruschig, K. Schmitt, G. Driesen, L. Ther, and W. Pfaff, German Patent 1,153,759 (1963).
60: 2950d   French Patent 1,334,586 (1963).
60: 2951c   R. H. Wiley and N. R. Smith, U.S. Patent 3,111,524 (1963).
60: 2951h   British Patent 888,686 (1962).
60: 4131c   A. Spassov, E. Golovinsky, and G. Russev, *Berichte,* **96,** 2996 (1963).
60: 4133a   M. Regitz and B. Eistert, *Berichte,* **96,** 3120 (1963).
60: 4141e   G. W. Miller and F. L. Rose, *J. Chem. Soc.,* **1963,** 5642.
60: 4155f   E. K. Gladding and D. C. Remy, U.S. Patent 3,102,889 (1963).
60: 5484g   Y. Makisumi, S. Notsumoto, and H. Kano, *Shionogi Kenkyusho Nempo,* **13,** 37 (1963).
60: 5512g   D. C. Remy, U.S. Patent 3,115,498 (1963).
60: 5536c   L. Birkofer, A. Ritter, and P. Richter, *Berichte,* **96,** 2750 (1963).
60: 6852c   T. R. Hopkins and J. W. Pullen, U.S. Patent 3,116,322 (1963).
60: 6860c   J. G. A. Luijten and G. J. M. van der Kerk, *Rec. Trav. Chim.,* **82,** 1181 (1963).
60: 6963a   C. Taube, E. Messmer, and K. H. Freytag, British Patent 936,895 (1963).
60: 8039g   K. Schmitt and G. Driesen, German Patent 1,159,460 (1963).
60: 9263a   K. J. Farrington, *Aust. J. Chem.,* **17,** 230 (1964).
60: 9264f   C.-F. Kroeger, G. Schoknecht, and H. Beyer, *Berichte,* **97,** 396 (1964).
60: 9383e   H. R. Hensel, *Berichte,* **97,** 96 (1964).
60: 10674d   W. P. Norris and R. A. Henry, *J. Org. Chem.,* **29,** 650 (1964).
60: 10867f   Y. Kobayashi, J. Kitaoka, H. Tsujimoto, K. Akamatsu, T. Taneda, and N. Fukuma, Japanese Patent 26,767 (1963).
60: 12006c   F. Baumbach, H. G. Henning, and G. Hilgetag, *Z. Chem.,* **4,** 67 (1964).
60: 13236g   J. Klosa, *J. Prakt. Chem.,* **23,** (4) 34 (1964).
60: 13238g   H. Kano and E. Yamazaki, *Tetrahedron,* **20,** 159 (1964).
60: 13242g   Ya. A. Levin, A. P. Fedotova, and V. A. Kukhtin, *Zh. Obshch. Khim.,* **34,** 499 (1964).
60: 13243a   Ya. A. Levin and V. A. Kukhtin, *Zh. Obshch. Khim.,* **34,** 502 (1964).
60: 14496d   V. R. Rao and V. R. Srinivasan, *Experientia,* **20,** 200 (1964).
60: 14508e   P. Grammaticakis, *C. R. Acad. Sci., Paris,* **258,** 1262 (1964).
60: 14513h   British Patent 947,485 (1964).
60: 14667a   J. Dehnert, W. Grosch, G. Luetzel, and W. Rohland, Belgian Patent 631,585 (1963).
60: P14668d   J. Dehnert, W. Grosch, G. Luetzel, and W. Rohland, Belgian Patent 631,586 (1963).

60: 15861e   M. Hauser and O. Logush, *J. Org. Chem.*, **29**, 972 (1964).

60: 16022g   D. Leuchs, H. R. Hensel, and H. Baumann, Belgian Patent 630,494 (1963).

61: 654h   R. Huisgen, R. Grashey, H. Knupfer, R. Kunz, and M. Seidel, *Berichte*, **97**, 1085 (1964).

61: 655h   P. Westermann, H. Paul, and G. Hilgetag, *Berichte*, **97**, 528 (1964).

61: 750b   J. Dehnert and G. Hansen, Belgian Patent 631,121 (1963).

61: 1781a   Ph. Gold-Aubert and L. Toribio, *Arch. Sci.* (Geneva), **16**, 405 (1963).

61: 1786a   E. Jucker, A. Lindenmann, E. Schenker, E. Flueckiger, and M. Taeschler, *Arzneimittel-Forsch.* **13**, (4), 269 (1963).

61: 1850d   J. Strating, W. E. Weening, and B. Zwanenburg, *Rec. Trav. Chim.*, **83**, 387 (1964).

61: 1873f   R. Mohr and M. Zimmermann, German Patent 1,168,437 (1964).

61: 1982a   J. Dehnert, W. Grosch, G. Luetzel, and W. Rohland, Belgian Patent 631,511 (1963).

61: 2637e   British Patent 951,106 (1964).

61: 3064h   Y.-T. Chen and T.-I. Chang, *Hua Hsueh Hsueh Pao*, **30**, 10 (1964).

61: 3094h   J. Bourdais, F. Cugniet, J.-C. Prin, and P. Chabrier, *Bull. Soc. Chim. France*, **1964**, 500.

61: 3098e   D. Kaminsky, B. Dubnick, and F. E. Anderson, *J. Med. Chem.*, **7**, 367 (1964).

61: 3105c   E. Tenor and C.-F. Kroeger, *Berichte*, **97**, 1373 (1964).

61: 4337h   V. Zotta and A. Gasmet, *Farmacia* (Bucharest), **11**, 731 (1963).

61: 4338e   J. Kinugawa, M. Ochiai, C. Matsumura, and H. Yamamoto, *Chem. Pharm. Bull.* (Tokyo), **12**, 433 (1964).

61: 4364d   T. R. Hopkins and J. W. Pullen, U.S. Patent 3,132,150 (1964).

61: 4515d   I. Yamase, N. Kuroki, and K. Konishi, *Kôgyô Kagaku Zasshi*, **67**, 102 (1964).

61: 4541b   Belgian Patent 634,526 (1963).

61: 5624b   R. Kuhn and H. Trischmann, *Monatsh. Chem.*, **95**, 457 (1964).

61: 5626f   R. Huisgen and H. Blaschke, *Tetrahedron Lett.*, **1964**, 1409.

61: 5640a   Ph. Gold-Aubert, D. Melkonian, and L. Toribio, *Helv. Chim. Acta*, **47**, 1188 (1964).

61: 5645a   N. N. Vereshchagina, I. Ya. Postovskii, and S. L. Mertsalov, *Zh. Obshch. Khim.*, **34**, 1689 (1964).

61: 5822g   W. Kuster, Swiss Patent 373,840 (1964).

61: 7007a   B.-G. Baccar and F. Mathis, *C. R. Acad. Sci.*, **258**, 6470 (1964).

61: 7403g   V. Grinsteins, K. Medne, and G. Cipens, *Latv. PSR Zinat. Akad. Vestis, Khim. Ser.*, **1963**, 593.

61: 8298c   A. L. Mndzhoyan, A. A. Aroyan, M. A. Kaldrikyan, T. R. Ovsepyan, and R. Sh. Arshakyan, *Izv. Akad. Nauk Arm. SSR. Khim. Nauki*, **17**, 204 (1964).

61: 8317c   E. W. Bousquet, U.S. Patent 3,141,023 (1964).

61: 8784a   F. Zajdela, R. Royer, E. Bisagni, and A. Ennuyer, *Acta Unio Int. Contra Cancrum*, **20**, 233 (1964).

61: 9490f   I. Saikawa, *Yakugaku Zasshi*, **84**, 566 (1964).

61: 9614a   Belgian Patent 634,376 (1963).

61: 10673g   H. Gehlen and W. Schade, *Justus Liebigs Ann. Chem.* **675**, 180 (1964).

61: 10676a   A. W. Lutz, *J. Org. Chem.*, **29**, 1174 (1964).

61: 11961g   L. Pentimalli and P. Bruni, *Ann. Chim.* (Rome), **54**, 180 (1964).

61: 13232d   B.-G. Baccar and J. Barrans, *C. R. Acad. Sci., Paris.*, **259**, 1340 (1964).

61: 13232f   H. G. O. Becker and H. J. Timpe, *Z. Chem.*, **4**, 304 (1964).

61: 14662g   Y. Kato and I. Hirao, *Kyushu Kogyo Daigaku Kenkyu Hokoku*, No. 14, 35 (1964).

61: 14675c   W. Ried and A. Czack, *Justus Liebigs Ann. Chem.* **676**, 121 (1964).

61: 14690g   G. H. Harris, P. S. Ichioka, and B. C. Fischback, U.S. Patent 3,152,136 (1964).

61: 14829a   H. Raab, Belgian Patent 635,490 (1964).

61: 16717b   F. H. Cisneros, I. Jimenez, R. Leon, and A. Caffarena, *Agronomia* (Lima), **29,** 175 (1962).

62: 546e   Ph. Gold-Aubert, D. Melkonian, and L. Toribio, *Helv. Chim. Acta*, **47,** 2068 (1964).

62: 567d   British Patent 970,480 (1964).

62: 1647a   P. Westermann, H. Paul, and G. Hilgetag, *Berichte*, **97,** 3065 (1964).

62: 1668c   K. H. Hauptmann and K. Zeile, British Patent 971,606 (1964).

62: 2771c   H. Najer, J. Menin, and J.-F. Giudicelli, *C. R. Acad. Sci., Paris,* **259,** 2868 (1964).

62: 4024c   M. Hauser, *J. Org. Chem.*, **29,** 3449 (1964).

62: 5221h   E. Grigat and R. Puetter, *Berichte*, **97,** 3027 (1964).

62: 5360h   A. Spiliadis, E. Badic, V. Cornea, M. Hilsenrath, D. Bretcan, and R. Neagoe, *Rev. Chim.* (Bucharest), **15,** 23 (1964).

62: 5364f   D. Leuchs, H. R. Hensel, and H. Baumann, British Patent 972,956 (1964).

62: 6392a   T. Kauffmann, S. Spaude, and D. Wolf, *Berichte*, **97,** 3436 (1964).

62: 6476b   C. Runti and C. Nisi, *J. Med. Chem.*, **7,** 814 (1964).

62: 7747b   K. T. Potts and H. R. Burton, *Proc. Chem. Soc.*, **1964,** 420.

62: 7747c   K. Moeckel, *Z. Chem.*, **4,** 428 (1964).

62: 7770d   N. R. Smith and R. H. Wiley, U.S. Patent 3,165,753 (1965).

62: 7910e   British Patent 974,523 (1964).

62: 9122a   J. F. Geldard and F. Lions, *J. Org. Chem.*, **30,** 318 (1965).

62: 9124b   F. Kurzer and K. Douraghi-Zadeh, *J. Chem. Soc.*, **1965,** 932.

62: 10857b   A. K. Williams, S. T. Cox, and R. G. Eagon, *Biochem. Biophys. Res. Commun.*, **18,** 250 (1965).

62: 11803c   H. Gehlen and V. Winzer, *Justus Liebigs Ann. Chem.*, **681,** 100 (1965).

62: 11804b   M. Ochiai and T. Kamikado, *Chem. Pharm. Bull.* (Tokyo), **12,** 1515 (1964).

62: 11804g   M. Dziewonska and M. Polanska, *Dissert. Pharm.*, **16,** 507 (1964).

62: 11805b   H. H. Takimoto, G. C. Denault, and S. Hotta, *J. Org. Chem.*, **30,** 711 (1965).

62: 11829c   French Patent M2723 (1964).

62: 13147e   R. Huisgen, R. Grashey, E. Aufderhaar, and R. Kunz, *Chem. Ber.*, **98,** 642 (1965).

62: 13148g   M. Ruccia and S. Cusmano, *Atti Accad. Sci., Lett. Arti Palermo. Pt. I,* **23,** 149 (1964).

62: 13270g   French Patent 1,389,363 (1965).

62: 14437c   H. Gehlen and J. Schmidt, *Justus Liebigs Ann. Chem.*, **682,** 123 (1965).

62: 14658d   B. Zwanenburg, W. E. Weenig, and J. Strating, *Rec. Trav. Chim.*, **84,** 408 (1965).

62: 14661b   A. Spassov, E. Golovinsky, and G. Demirov, *Chem. Ber.*, **98,** 932 (1965).

62: 14661e   A. Spassov and S. Robev, *Chem. Ber.*, **98,** 928 (1965).

62: 14688a   French Patent 1,385,856 (1965).

62: 16232d   H. Eilingsfeld, *Chem. Ber.*, **98:** 1308 (1965).

62: 16259b   A. V. Koenig, German Patent 1,189,380 (1965).

62: 16259e   British Patent 974,713 (1964).

62: 16278b   French Patent 1,379,479 (1964).

63: 611e   L. E. Benjamin, French Patent 1,381,647 (1964).

63: 1783b   R. G. Child, *J. Heterocyclic Chem.*, **2,** 98 (1965).

63: 1783c   D. S. Deorha and S. P. Sareen, *J. Indian Chem. Soc.*, **41,** 793 (1964).

63: 1786f   E. C. Taylor and R. W. Hendess, *J. Amer. Chem. Soc.*, **87,** 1980 (1965).

63: 1887g   A. P. Grekov, S. A. Sukhorukova, and K. A. Kornev, *Vysokomolekul. Soedin.*, **7**, 255 (1965).

63: 2965b   Ph. Gold-Aubert, *Arch. Sci.* (Geneva), **17**, 323 (1964).

63: 2966h   K. T. Potts, S. K. Roy, and D. P. Jones, *J. Heterocyclic Chem.*, **2**, 105 (1965).

63: 2967c   W. Dymek, M. Dziewonska, and M. Polanska, *Dissert. Pharm.*, **16**, 495 (1964).

63: 2975c   R. Huisgen, E. Aufderhaar, and G. Wallbillich, *Chem. Ber.*, **98**, 1476 (1965).

63: 4280a   I. L. Shegal and I. Ya. Postovskii, *Khim. Geterotsikl. Soedin., Akad. Nauk Latv. SSR*, **1965**, 133.

63: 4281g   C.-F. Kroeger and H. Frank, *Angew. Chem.*, **77**, 429 (1965).

63: 4305f   R. J. Harder, U.S. Patent 3,184,471 (1965).

63: 4306e   French Addition 84,814 (1965).

63: 5631c   B. M. Lynch, and Y.-Y. Hung, *J. Heterocyclic Chem.*, **2**, 218 (1965).

63: 5631g   H. Gehlen and W. Schade, *Justus Liebigs Ann. Chem.*, **683**, 149 (1965).

63: 5632c   H. Seeboth, D. Baerwolff, and B. Becker, *Justus Liebigs Ann. Chem.*, **683**, 85 (1965).

63: 6817e   G. Cipens, J. Eiduss, Ya. S. Bobovich, and V. Grinsteins, *Zh. Strukt. Khim.*, **6**, 53 (1965).

63: 6996d   I. Ya. Postovskii and V. L. Nirenburg, *Khim. Geterotsikl. Soedin., Akad. Nauk Latv. SSR*, **1965**, 309.

63: 6997e   M. S. Abbasi and J. P. Trivedi, *J. Indian Chem. Soc.*, **42**, 333 (1965).

63: 6999b   G. Palazzo and G. Corsi, *Ann. Chim.* (Rome), **55**, 126 (1965).

63: 7000c   R. Huisgen, R. Grashey, R. Kunz, G. Wallbillich, and E. Aufderhaar, *Chem. Ber.*, **98**, 2174 (1965).

63: 7001b   R. Huisgen and E. Aufderhaar, *Chem. Ber.*, **98**, 2185 (1965).

63: 7012g   F. Kurzer and K. Douraghi-Zadeh, *J. Chem. Soc.*, **1965**, 3912.

63: 7119b   M. Lippay-Faller, *Promotionsarb.* (Zurich), **3324**, 70 pp (1963).

63: 8345h   G. F. Bettinetti and L. Capretti, *Gazz. Chim. Ital.*, **95**, 33 (1965).

63: 9936f   H. Gehlen and K. Moeckel, *Justus Liebigs Ann. Chem.*, **685**, 176 (1965).

63: 10497b   D. H. Jones, R. Slack, S. Squires, and K. R. H. Wooldridge, *J. Med. Chem.*, **8**, 676 (1965).

63: 11541a   B.-G. Baccar and F. Mathis, *C. R. Acad. Sci., Paris*, **261**, 174 (1965).

63: 11544d   G. F. Bettinetti and P. Gruenanger, *Tetrahedron Lett.* **1965**, 2553.

63: 11545b   F. Kruzer and K. Douraghi-Zadeh, *J. Chem. Soc.*, **1965**, 4448.

63: 11549h   E. Grigat and R. Puetter, *Chem. Ber.*, **98**, 2619 (1965).

63: 13027f   E. Akerblom and M. Sandberg, *Acta Chem. Scand.*, **19**, 1191 (1965).

63: 13048b   J. Schmidt and H. Gehlen, *Z. Chem.*, **5**, 304 (1965).

63: 13242b   I. Ya. Postovskii and I. L. Shegal, *Khim. Geterotsikl. Soedin., Akad. Nauk Latv. SSR*, **1965**, 443.

63: 13242f   I. L. Shegal and I. Ya. Postovskii, *Khim. Geterotsikl. Soedin., Akad. Nauk Latv. SSR*, **1965**, 449.

63: 13243b   V. L. Nirenburg, I. L. Shegal, and I. Ya. Postovskii, *Khim. Geterotsikl. Soedin., Akad. Nauk Latv. SSR*, **1965**, 470.

63: 13243f   G. E. Cipens and V. Grinsteins, *Latv. PSR Zinatnu Akad. Vestis, Kim. Ser.*, **1965**, 204.

63: 13256f   N. N. Vereshchagina, I. Ya. Postovskii, and S. L. Mertsalov, *Zh. Organ. Khim.*, **1**, 1154 (1965).

63: 13275d   H. Gehlen and N. Pollok, East German Patent 36,433 (1965).

63: 14850b   R. G. Child and A. S. Tomcufcik, *J. Heterocyclic Chem.*, **2**, 302 (1965).

63: 15025g   H. Leister and E. H. Rohe, Belgian Patent 647,655 (1964).

63: 16335g   G. Palazzo and L. Baiocchi, *Ann. Chim.* (Rome), **55**, 935 (1965).

63: 16336b   W. Ried and M. Schoen, *Chem. Ber.*, **98**, 3142 (1965).

63: 16339d   C.-F. Kroeger, L. Hummel, M. Mutscher, and H. Beyer, *Chem. Ber.*, **98,** 3025 (1965).

63: 16340e   C.-F. Kroeger, P. Seditz, and M. Mutscher, *Chem. Ber.*, **98,** 3034 (1965).

63: 18071a   E. Akerblom, *Acta Chem. Scand.*, **19,** 1135 (1965).

63: 18071g   E. Akerblom, *Acta Chem. Scand.*, **19,** 1142 (1965).

63: 18084h   C. Temple, Jr., C. L. Kussner, and J. A. Montgomery, *J. Org. Chem.*, **30,** 3601 (1965).

63: 18105c   H. Ruschig, K. Schmitt, G. Driesen, L. Ther, and W. Pfaff, German Patent 1,200,824 (1965).

63: 18106h   K. Schmitt, G. Driesen, L. Ther, and W. Pfaff, German Patent 1,200, 825 (1965).

64: 208e   Belgian Patent 659,366 (1965).

64: 570f   I. L. Shegal, V. L. Nirenburg, V. F. Degtyarev, and I. Ya. Postovskii, *Khim. Geterotsikl. Soedin., Akad. Nauk Latv. SSR,* **1965,** 580.

64: 646b   G. P. Sharnim and B. I. Buzykin, USSR Patent 172,735 (1965).

64: 687f   U. P. Basu and S. Dutta, *J. Org. Chem.*, **30,** 3562 (1965).

64: 2081c   H. C. Brown and C. R. Wetzel, *J. Org. Chem.*, **30,** 3729 (1965).

64: 2096g   H. Ruschig, K. Schmitt, G. Driesen, L. Ther, and W. Pfaff, German Patent 1,203,272 (1965).

64: 2097h   M. Kuranari and H. Takeuchi, Japanese Patent 21, 420 (1965).

64: 3545g   H. Sakamoto, M. Nakagawa, T. Mizutani, H. Tsuchiya, T. Fujimoto, and Y. Okuno, Japanese Patent 18,747 (1965).

64: 3555g   H. Ruschig, K. Schmitt, G. Driesen, L. Ther, and W. Pfaff, German Patent 1,200,313 (1965).

64: 3556e   H. Ruschig, K. Schmitt, G. Driesen, L. Ther, and W. Pfaff, German Patent 1,200,826 (1965).

64: 3557c   H. Gehlen and F. Lemme, (East) German Patent 30, 868 (1965).

64: 4189h   M. C. Carter, *Physiol. Plantarum*, **18,** 1054 (1965).

64: 4919f   C. F. Kroeger and W. Freiberg, *Z. Chem.*, **5,** 381 (1965).

64: 4935f   G. J. Durant, G. M. Smith, R. G. W. Spickett, and S. H. B. Wright, *J. Med. Chem.*, **9,** 22 (1966).

64: 5072b   K. T. Potts and H. R. Burton, *J. Org. Chem.*, **31,** 251 (1966).

64: 5072d   K. T. Potts, H. R. Burton, and J. Bhattacharyya, *J. Org. Chem.*, **31,** 260 (1966).

64: 5072e   K. T. Potts, H. R. Burton, and S. K. Roy, *J. Org. Chem.*, **31,** 265 (1966).

64: 5072h   H. L. Yale and J. J. Piala, *J. Med. Chem.*, **9,** 42 (1966).

64: 5074f   G. Cipens and V. Grinsteins, *Khim. Geterotsikl. Soedin., Akad. Nauk Latv. SSR* **1965,** 624.

64: 5079h   E. Grigat, R. Puetter, and E. Muehlbauer, *Chem. Ber.*, **98,** 3777 (1965).

64: 5255e   R. H. Hansen, T. De Beuedictis, and W. M. Martin, *Polymer Eng. Sci.*, **5,** 223 (1965).

64: 5437c   A. Aaronson, H. Baker, B. Bensky, O. Frank, and A. C. Zahalsky, *Develop. Ind. Microbiol.*, **6,** 48 (1964).

64: 6644a   M. S. Solanki and J. P. Trivedi, *J. Indian Chem. Soc.*, **42,** 817 (1965).

64: 8170g   D. Martin and A. Weise, *Chem. Ber.*, **99,** 317 (1966).

64: 8173b   A. Dornow and H. Paucksch, *Chem. Ber.*, **99,** 81 (1966).

64: 8173f   A. Dornow and H. Paucksch, *Chem. Ber.*, **99,** 85 (1966).

64: 8174a   F. Kurzer and K. Douraghi-Zadeh, *J. Chem. Soc.*, C, **1966,** 1.

64: 8174b   F. Kurzer and K. Douraghi-Zadeh, *J. Chem. Soc.*, C, **1966,** 6.

64: 8174e   L. Pentimalli and S. Bozzini, *Ann. Chim.* (Rome), **55,** 441 (1965).

64: 8175c   J. Preston, *J. Heterocyclic Chem.*, **2,** 441 (1965).

64: 8196c   Netherlands Appl. 301,383 (1965).

64: 8198g   P. Gold-Aubert, D. Melkonian, and R. Y. Mauvernay, Netherlands Appl. 6,504,121 (1965).

64: 9727d R. Sgarbi, *Chim. Ind.* (Milan), **48,** 18 (1966).

64: 9827e J. K. Stille and F. E. Arnold, *J. Polymer Sci.*, Pt.A, **3,** 4284 (1965).

64: 11119b R. Huisgen, H. Blaschke, and E. Brunn, *Tetrahedron Lett.* **1966,** 405.

64: 11132e J. Yates and E. Haddock, British Patent 1,019,120 (1966).

64: 11202c G. Zinner and H. Boehlke, *Arch. Pharm.*, **299,** 81 (1966).

64: 12533h W. Schulze, G. Letsch, and H. Fritzsche, *J. Prakt. Chem.*, **30,** (4) 302 (1965).

64: 12662e H. C. Brown, H. J. Gisler, Jr., and M. T. Cheng, *J. Org. Chem.* **31,** 781 (1966).

64: 12662f W. Carpenter, A. Haymaker, and D. W. Moore, *J. Org. Chem.*, **31,** 789 (1966).

64: 12685d L. I. Bagal, M. S. Pevzner, V. A. Lopyrev and E. A. Yurchak, USSR Patent 176,913 (1965).

64: 12685e British Patent 1,020,070 (1966).

64: 12685f K. Schmitt, G. Driesen, and W. Pfaff, U.S. Patent 3,236,853 (1966).

64: 12686e C. E. Pawloski, U.S. Patent 3,231,579 (1966).

64: 14100f N. K. Bliznyuk, A. F. Kolomiets, and R. N. Golubeva, USSR Patent 175,054 (1965).

64: 15716c J. Sandström and I. Wennerbeck, *Acta Chem. Scand.*, **20,** 57 (1966).

64: 15870g J. Goerdeler and H. Schenk, *Chem. Ber.*, **99,** 782 (1966).

64: 15886g E. Grigat and R. Puetter, *Chem. Ber.*, **99,** 958 (1966).

64: 16886c M. Tomanek and F. Uhlig, U.S. Patent 3,230,081 (1966).

64: 17631f F. Rijkens, M. J. Janssen, and G. J. M. van der Kerk, *Rec. Trav. Chim.*, **84,** 1597 (1965).

64: 19458a P. Westermann, H. Paul, and G. Hilgetag, *Chem. Ber.*, **99,** 1111 (1966).

64: 19594g H. Beyer, E. Bulka, and K. Dittrich, *J. Prakt. Chem.*, **30** (4), 280 (1965).

64: 19596c H. A. Burch and W. O. Smith, *J. Med. Chem.*, **9,** 405 (1966).

64: 19596d H. G. O. Becker, H. Boettcher, T. Roethling, and H. J. Timpe, *Wiss. Z. Tech. Hochsch. Chem. Leuna-Merseburg*, **8,** 22 (1966).

64: 19596h G. R. Harvey, *J. Org. Chem.*, **31,** 1587 (1966).

64: 19597a G. Zinner, B. Boehlke, R. O. Weber, R. Moll, and W. Deucker, *Arch. Pharm.*, **299,** 222 (1966).

64: 19613f G. Palazzo and L. Baiocchi, *Ann. Chim.* (Rome), **56,** 190 (1966).

65: 703g G. Zinner and B. Boehlke, *Arch. Pharm.*, **299,** 43 (1966).

65: 704d H. H. Takimoto, G. C. Denault, and S. Hotta, *J. Heterocyclic Chem.*, **3,** 119 (1966).

65: 704e G. Cipens, D. Duka, and V. Grinsteins, *Khim. Geterotsikl. Soedin., Akad. Nauk Latv. SSR,* **1966,** 117.

65: 705a G. Cipens, R. Bokalders, and V. Grinsteins, *Khim. Geterotsikl. Soedin., Akad. Nauk Latv. SSR,* **1966,** 110.

65: 718g G. Zinner, *Arch. Pharm.*, **299,** 312 (1966).

65: 725f Netherlands Appl. 6,510,386 (1966).

65: 2274g E. Muehlbauer and W. Theuer, U.S. Patent 3,248,398 (1966).

65: 3862h G. Zinner and H. Boehlke, *Arch. Pharm.*, **299,** 411 (1966).

65: 3881g G. Zinner, R. Moll, and B. Boehlke, *Arch. Pharm.*, **299,** 441 (1966).

65: P4905e A. E. Anderson and W. W. Rees, French Patent 1, 417,782 (1965).

65: 5404e M. Asscher, D. Vofsi, and A. Katchalsky, Belgian Patent 654,544 (1965).

65: 5455b R. S. McElhinney, *J. Chem. Soc.*, C, **1966,** 1256.

65: 7027a B. G. van den Bos, A. Schipperheyn, and F. W. van Deursen, *Rec. Trav. Chim.*, **85,** 429 (1966).

65: 7169c J. F. Willems and A. Vandenberghe, *Bull. Soc. Chim. Belges*, **75,** 358 (1966).

65: 7169g G. Werber, F. Maggio, and F. Buccheri, *Ric. Sci.*, **36,** 332 (1966).

65: 7188c Netherlands Appl. 6,513,451 (1966).

65: 7273e	J. Sambeth and F. Grundschober, *Inform. Chim.*, **64,** 71 (1966).
65: 7320e	W. Yamatani, M. Honda, and S. Fujino, Japanese Patent 1793 (1966).
65: 7387e	H. Hartel, German Patent 1,217,062 (1966).
65: 8897e	T. Okamoto, M. Hirobe, and E. Yabe, *Chem. Pharm. Bull.* (Tokyo), **14,** 523 (1966).
65: 8926d	E. Grigat and R. Puetter, German Patent 1,219,031 (1966).
65: 10579g	G. Palazzo and G. Picconi, *Boll. Chim. Farm.*, **105,** 217 (1966).
65: 11185a	J. Vacher, Ph. Gold-Aubert, and P. Duchene-Marullaz, *Int. Symp. Nonsteroidal Anti-Imflammatory Drugs, Milan,* **1964,** 299.
65: 12204h	L. I. Bagal, M. S. Pevzner, and V. A. Lopyrev, *Khim. Geterotsikl. Soedin.*, *Akad. Nauk Latv. SSR,* **1966,** 440.
65: 12212h	Netherlands Appl. 6,511,837 (1966).
65: 13691f	F. Kurzer and D. R. Hanks, *Chem. Ind.* (London), **1966,** 1143.
65: 15367g	T. Kauffmann and L. Bán, *Chem. Ber.*, **99,** 2600 (1966).
65: 16960f	I. Ya. Postovskii and I. L. Shegal, *Khim. Geterotsikl. Soedin.*, *Akad. Nauk Latv. SSR,* **1966,** 443
65: 16978e	M. Wolf, U.S. Patent 3,267,114 (1966).
65: 18448g	J. F. K. Wilshire, *Aust. J. Chem.*, **19,** 1935 (1966).
65: 18575h	I. Lalezari and N. Sharghi, *J. Heterocyclic Chem.*, **3,** 336 (1966).
65: 18576b	P. K. Kadaba, *Tetrahedron*, **22,** 2453 (1966).
65: 18576h	M. Regitz and A. Liedhegener, *Chem. Ber.*, **99,** 2918 (1966).
65: 18595e	L. E. Benjamin, U.S. Patent 3,272,840 (1966).
65: 20138h	British Patent 1,040,551 (1966).
66: 2516c	K. T. Potts and R. M. Huseby, *J. Org. Chem.*, **31,** 3528 (1966).
66: 2536b	S. E. Mallett and F. L. Rose, *J. Chem. Soc., C,* **1966,** 2038.
66: 10878f	W. Ried and H. Lohwasser, *Angew. Chem.*, **78,** 904 (1966).
66: 10907m	H.-G. Schmelzer, E. Degener, and H. Holtschmidt, *Angew. Chem. Int. Ed. Engl.*, **5,** 960 (1966).
66: 28303n	A. A. Gordon, A. R. Katritzky, and F. D. Popp, *Tetrahedron Suppl.* No. 7, 213 (1966).
66: 28717a	H. W. Heine, R. Peavy, and A. J. Durbetaki, *J. Org. Chem.*, **31,** 3924 (1966).
66: 37837s	J. C. Stickler and W. H. Pirkle, *J. Org. Chem.*, **31,** 3444 (1966).
66: 37843r	G. J. Durant, *J. Chem. Soc., C,* **1967,** 92.
66: 37845t	G. Palazzo and L. Baiocchi, *Ann. Chim.* (Rome), **56,** 199 (1966).
66: 37902j	H. H. Takimoto and G. C. Denault, *Tetrahedron Lett.*, **1966,** 5369.
66: 46372p	B. C. Ennis, G. Holan, and Mrs. E. L. Samuel, *J. Chem. Soc., C,* **1967,** 33.
66: 46376t	A. Spassov, E. Golovinsky, and G. Demirov, *Chem. Ber.*, **99,** 3728 (1966).
66: 46412b	A. Spassov, E. Golovinsky, and G. Demirov, *Chem. Ber.*, **99,** 3734 (1966).
66: 46427k	K. Thomae, British Patent 1,053,085 (1966).
66: 46705z	J. Preston and W. B. Black, *Amer. Chem. Soc., Div. Polymer Chem., Preprints,* **6,** 757 (1965).
66: 55070r	British Patent 1,049,111 (1966).
66: 55438s	W. Ried and H. Lohwasser, *Justus Liebigs Ann. Chem.*, **699,** 88 (1966).
66: 55489j	M. Wolf, U.S. Patent 3,293,259 (1966).
66: 65474q	B. R. D. Whitear and D. J. Fry, British Patent 1,049,053 (1966).
66: 73426r	D. J. Cummings, V. A. Chapman, S. S. DeLong, and L. Mondale, *J. Virol.*, **1,** 193 (1967).
66: 75958q	A. Mugnaini, L. P. Friz, E. Provinciali, P. Rugarli, A. Olivi, E. Zelfelippo, L. Almirante, and W. Murmann, *Boll. Chim. Farm.*, **105,** 596 (1966).

66: 76923m   Netherlands Appl. 6,600,940 (1966).

66: 85757n   R. G. W. Spickett and S. H. B. Wright, *J. Chem. Soc., C,* **1967,** 503.

66: 85767r   E. Steininger, *Monatsh. Chem.,* **97,** 1195 (1966).

66: 94958f   J. Warkentin and P. R. West, *Tetrahedron Lett.,* **1966,** 5815.

66: 94960a   J. Sauer and B. Schroeder, *Chem. Ber.,* **100,** 678 (1967).

66: 94962c   R. Fusco, F. D'Alo, and A. Masserini, *Gazz. Chim. Ital.,* **96,** 1084 (1966).

66: 94965f   K. T. Potts and C. A. Hirsch, *Chem. Ind.* (London), **1966,** 2168.

66: 104601a   I. B. Mazheika, G. Cipens, and S. Hillers, *Khim. Geterotsikl. Soedin.,* **1966,** 776.

66: 104948u   A. J. Blackman, E. J. Browne, and J. B. Polya, *J. Chem. Soc., C,* **1967,** 661.

66: 104959y   H. Gehlen and F. Lemme, *Justus Liebigs Ann. Chem.,* **702,** 101 (1967).

66: 110059w   D. H. Cole and D. J. Fry, British Patent 1,049,054 (1966).

66: 115650s   F. Kurzer and K. Douraghi-Zadeh, *J. Chem. Soc., C,* **1967,** 742.

66: 115652u   G. Palazza and L. Baiocchi, *Gazz. Chim. Ital.,* **96,** 1020 (1966).

66: 115651t   F. Kurzer and D. R. Hanks, *J. Chem. Soc., C,* **1967,** 746.

66: 115714r   I. Saikawa, T. Wada, Y. Matsubara, S. Nakayama, Y. Suzuki, and A. Takai, Japanese Patent 3096 (1967).

66: P120802x   K. M. Milton, U.S. Patent 3,295,981 (1967).

67: 3039p   W. S. Wadsworth and W. D. Emmons, *J. Org. Chem.,* **32,** 1279 (1967).

67: 11327w   French Patent 1,455,835 (1966).

67: 11476u   G. J. Durant, *J. Chem. Soc., C,* **1967,** 952.

67: 11793v   G. Caraculacu and I. Zugravescu, *Rev. Roum. Chim.,* **12,** 291 (1967).

67: 16500g   G. B. Barlin and T. J. Batterham, *J. Chem. Soc., B,* **1967,** 516.

67: 21858j   A. E. Siegrist, *Helv. Chim. Acta,* **50,** 906 (1967).

67: 21871h   H. C. Brown and R. J. Kassal, *J. Org. Chem.,* **32,** 1871 (1967).

67: 21917c   M. Kuranari and H. Takeuchi, Japanese Patent 6343 (1967).

67: 32651n   S. G. Boots and C. C. Cheng, *J. Heterocyclic Chem.,* **4,** 272 (1967).

67: 37867x   M. Dziewońska, *Spectrochim. Acta, Part A,* **23,** 1195 (1967).

67: 38283r   British Patent 1,066,215 (1967).

67: 43097n   R. Huisgen, H. Knupfer, R. Sustmann, G. Wallbillich, and V. Weberndörfer, *Chem. Ber.,* **100,** 1580 (1967).

67: 43155e   C. Temple, Jr., C. L. Kussner, and J. A. Montgomery, *J. Org. Chem.,* **32,** 2241 (1967).

67: 43755a   K. T. Potts, S. K. Roy, and D. P. Jones, *J. Org. Chem.,* **32,** 2245 (1967).

67: 43756b   H. Gehlen and F. Lemme, *Justus Liebigs Ann. Chem.,* **703,** 116 (1967).

67: 44809h   French Patent 1,464,255 (1966).

67: 44841n   E. Schick, H. W. Modrow, and K. Lippold, (East) German Patent 53,267 (1967).

67: 54140x   B. C. McKusick, U.S. Patent 3,308,131 (1967).

67: 64306v   G. Cipens, *Metody Poluch. Khim. Reaktivov Prep.,* **14,** 9 (1966).

67: 64307w   G. Cipens, *Metody Poluch. Khim. Reaktivov Prep.* **14,** 119 (1966).

67: 64314w   R. Miethchen and C.-F. Kroeger, *Z. Chem.,* **7,** 184 (1967).

67: 64315x   W. Freiberg, C.-F. Kroeger, and R. Radeglia, *Tetrahedron Lett.,* **1967,** 2109.

67: 64318a   I. L. Shegal and I. Ya. Postovskii, *Metody Poluch. Khim. Reaktivov Prep.* **14,** 116 (1966).

67: 64407d   H. A. Brown, U.S. Patent 3,326,889 (1967).

67: 65353v   E. Sommer and F. Wiloth, German Patent 1,238,212 (1967).

67: 69440u   W. W. Rees and W. H. Russell, U.S. Patent 3,295,980 (1967).

67: 72992f   G. B. Barlin, *J. Chem. Soc., B,* **1967,** 641.

67: 73570d   Th. Kauffmann, J. Albrecht, D. Berger, and J. Legler, *Angew. Chem. Int. Ed. Engl.,* **6,** 633 (1967).

67: P74426y   J. Carbonell and W. Wehrli, French Patent 1,466,796 (1967).

67: 82167e   C.-F. Kroeger and R. Miethchen, *Chem. Ber.*, **100,** 2250 (1967).

67: 90732a   Th. Eicher, S. Huenig, and P. Nikolaus, *Angew. Chem. Int. Ed. Engl.*, **6,** 699 (1967).

67: 90736e   R. G. Dubenko, I. M. Bazavova, and P. S. Pel'kis, *Ukr. Khim. Zh.*, **33,** 638 (1967).

67: 90738g   R. K. Bartlett and I. R. Humphrey, *J. Chem. Soc., C*, **1967,** 1664.

67: 90815e   G. Palazzo, German Patent 1,235,930 (1967).

67: 90816f   E. Grigat and R. Puetter, German Patent 1,241,835 (1967).

67: 97910r   L. K. Kulikova and I. G. Shuidenko, *Zh. Mikrobiol. Epidemiol. Immunobiol.*, **44,** 145 (1967).

67: 99764v   U. Anthoni, C. Larsen, and P. H. Nielsen, *Acta Chem. Scand.*, **21,** 1201 (1967).

67: 100068e   M. Ruccia and N. Vivona, *Ann. Chim.* (Rome), **57,** 671 (1967).

67: 100076f   Ya. A. Levin and M. S. Skorobogatova, *Khim. Geterotsikl. Soedin.*, **3,** 339 (1967).

67: 100077g   M. Ohta, H. Kato, and T. Kaneko, *Bull. Chem. Soc. Jap.*, **40,** 579 (1967).

67: 100157h   R. G. W. Spickett and S. H. B. Wright, British Patent 1,070,243 (1967).

67: 103933u   French Patent 1,467,384 (1967).

67: 108606z   R. C. Cookson, S. S. H. Gilani, and I. D. R. Stevens, *J. Chem. Soc., C*, **1967,** 1905.

67: 116333y   R. Huisgen, K. Adelsberger, E. Aufderhaar, H. Knupfer, and G. Wallbillich, *Monatsh. Chem.*, **98,** 1618 (1967).

67: 116853t   W. H. Pirkle and J. C. Stickler, *Chem. Commun.*, **1967,** 760.

67: 116858y   M. Furdik, S. Mikulasek, M. Livar, and S. Priehradny, *Chem. Zvesti*, **21,** 427 (1967).

68: P2902y   V. A. Lopyrev, L. M. Kogan, K. L. Krupin, and M. S. Pevzner, USSR Patent 189,867 (1966).

68: 12914t   J. S. Davidson, *J. Chem. Soc., C*, 2471 (1967).

68: 13448f   L. Stoicescu-Crivat, E. Mantaluta, G. Neamtu, and I. Zugravescu, *Rev. Roumaine Chim.*, **11,** 1127 (1966).

68: 13458j   A. V. D'yachenko, V. V. Korshak, and E. S. Krongauz, *Vysokomol. Soedin., Ser. A*, **9,** 2231 (1967).

68: 21884k   K. Kottke, F. Friedrich, and R. Pohloudek-Fabini, *Arch. Pharm.*, **300,** 583 (1967).

68: P21922w   Netherlands Patent 6,607,822 (1966).

68: 21961h   Netherlands Patent 6,615,211 (1967).

68: 29639e   R. Clarkson and J. K. Landquist, *J. Chem. Soc., C*, **1967,** 2700.

68: 29645d   R. Jacquier, M. L. Roumestant, and P. Viallefont, *Bull. Soc. Chim. Fr.*, 2630 (1967).

68: 29646e   R. Jacquier, M. L. Roumestant, and P. Viallefont, *Bull. Soc. Chim. Fr.*, 2634 (1967).

68: 29649h   G. W. Evans and B. Milligan, *Aust. J. Chem.*, **20,** 1783 (1967).

68: 39547n   K. T. Potts and C. Hirsch, *J. Org. Chem.*, **33,** 143 (1968).

68: 39599f   T. Sasaki and K. Minamoto, *J. Heterocyclic Chem.*, **4,** 571 (1967).

68: 39627p   L. E. Benjamin, U.S. Patent 3,314,947 (1967).

68: P39628q   E. R. Lynch and W. Cummings, British Patent 1,096,600 (1967).

68: 46313m   P. Mohr, W. Scheler, and K. Fraenk, *Naturwissenschaften*, **54,** 227 (1967).

68: 49563d   A. A. Aroyan and A. S. Azaryan, *Sin. Geterotsikl. Soedin., Akad. Nauk Arm. SSR, Inst. Tonkoi Org. Khim.*, **7,** 43 (1966).

68: 50462b   J. K. Stille and L. D. Gotter, *J. Polym. Sci., Part B*, **6,** 11 (1968).

68: 58906v   H. G. O. Becker, H. Huebner, H. J. Timpe, and M. Wahren, *Tetrahedron*, **24,** 1031 (1968).

68: 59500v   S. P. Dutta, B. P. Das, B. K. Paul, A. K. Acharyya, and U. P. Basu, *J. Org. Chem.*, **33**, 858 (1968).

68: 59501w   A. K. Bhat, R. P. Bhamaria, R. A. Bellare, and C. V. Deliwala, *Indian J. Chem.*, **5**, 397 (1967).

68: 59503y   I. Simiti and I. Proinov, *Farmacia*, **15**, 415 (1967).

68: 59504z   G. W. Evans and B. Milligan, *Aust. J. Chem.*, **20**, 1779 (1967).

68: 59506b   S. Hauptmann, H. Wilde, and K. Moser, *Tetrahedron Lett.*, 3295 (1967).

68: 68937v   R. M. Acheson and M. W. Foxton, *J. Chem. Soc., C*, **1968**, 389.

68: 68940r   H. C. Van der Plas and H. Jongejan, *Tetrahedron Lett.*, 4385 (1967).

68: P68975f   H. E. Faith, U.S. Patent 3,337,551 (1967).

68: 69071v   E. Ettenhuber and K. Ruehlmann, *Chem. Ber.*, **101**, 743 (1968).

68: 78201q   H. Gehlen and J. Stein, *J. Prakt. Chem.*, **37**, 168 (1968).

68: 78203s   H. Gehlen and R. Neumann, *J. Prakt. Chem.*, **37**, 182 (1968).

68: 78208x   V. W. Dymek, B. Janik, A. Cygankiewicz, and H. Gawron, *Acta Pol. Pharm.*, **24**, 97 (1967).

68: 78612z   E. D. Nicolaides, H. DeWald, R. Westland, M. Lipnik, and J. Posler, *J. Med. Chem.*, **11**, 74 (1967).

68: P79571x   British Patent 1,082,584 (1967).

68: P79574a   French Patent 1,469,140 (1967).

68: 87784j   V. V. Korshak, E. S. Krongauz, A. P. Travnikova, A. V. D'yachenko, A. A. Askadskii, and V. P. Sidorova, *Dokl. Akad. Nauk USSR*, **178**, 607 (1968).

68: 95761q   H. G. O. Becker and V. Eisenschmidt, *Z. Chem.*, **8**, 105 (1968).

68: 95763s   V. Zota and A. Tataru-Gamset, *Farmacia*, **15**, 479 (1967).

68: 96231d   E. S. Krongauz, V. V. Korshak, and A. V. D'yachenko, *Vysokomol. Soedin., Ser. B*, **10**, 108 (1968).

68: 100283s   P. Mohr, W. Scheler, J. Gallasch, and Ch. Garleb, *Acta Biol. Med. Ger.*, **18**, 655 (1967).

68: 104067x   G. Matolesy and E. A. A. Gomaa, *Weed Res.*, **8**, 1 (1968).

68: 105101r   H. Gehlen and G. Zeiger, *J. Prakt. Chem.*, **37**, 269 (1968).

68: 105106w   M. H. Shah, M. Y. Mhasalkar, N. A. Varaya, R. A. Bellare, and C. V. Deliwala, *Indian J. Chem.*, **5**, 391 (1967).

68: 105111u   M. D. Coburn and T. E. Jackson, *J. Heterocyclic Chem.*, **5**, 199 (1968).

68: 105162m   E. J. Browne and J. B. Polya, *J. Chem. Soc., C*, **1968**, 824.

68: 105174s   G. P. Sokolov and S. Hillers, *Khim. Geterotsikl. Soedin.*, **3**, 556 (1967).

68: 105175t   Ch. Larsen and E. Binderup, *Acta Chem. Scand.*, **21**, 1984 (1967).

68: 106077z   H. Sanielevici, F. Brodman, S. Rosenberg, and C. Stuleanu, British Patent 1,104,053 (1968).

68: P106539b   P. D. Oja and E. T. Niles, U.S. Patent 3,375,230 (1968).

68: 108827t   C. W. Reimann and M. Zocchi, *Chem. Commun.*, 272 (1968).

68: 114391n   U. P. Basu and S. P. Dutta, *Ind. Chim. Belge*, **32**, 1224 (1967).

68: 114512c   H. Kato, S. Sato, and M. Ohta, *Tetrahedron Lett.*, 4261 (1967).

68: P114607n   E. Grigat and R. Puetter, German Patent 1,242,234 (1967).

68: P115683w   A. Spiliadis, M. Hilsenrath, M. Panfil, and G. Molau, Romanian Patent 48,706 (1967).

69: 2623d   E. Fahr and H. D. Rupp, *Justus Liebigs Ann. Chem.*, **712**, 100 (1968).

69: 2896v   L. M. Rice, K. R. Scott, and C. H. Grogan, *J. Med. Chem.*, **11**, 183 (1968).

69: 2912x   F. Russo and M. Ghelardoni, *Boll. Chim. Farm.*, **105**, 911 (1966).

69: 10398z   F. Kurzer and D. R. Hanks, *J. Chem. Soc., C*, 1375 (1968).

69: P10473v   A. J. Lindenmann and C. Brueschweiler, Swiss Patent 436288 (1967).

69: P11257q   J. J. D'Amico, U. S. Patent 3,379,700 (1968).

69: P11393f   K. Ueda, S. Ando, and T. Kinoshita, British Patent 1,112,949 (1968).

69: P11964t   A. C. B. McPhail, F. R. Heather, and G. Ellis, British Patent 1,111,680 (1968).

69: 16270y   P. Mohr, W. Scheler, and K. Fraenk, *Acta Biol. Med. Ger.*, **20**, 263 (1968).

69: 16431b   G. W. Chang and E. E. Snell, *Biochemistry*, **7**, 2005 (1968).

69: 18568a   H. Paul, G. Hilgetag, and G. Jaehnchen, *Chem. Ber.*, **101**, 2033 (1968).

69: 19082z   R. Kraft, H. Paul, and G. Hilgetag, *Chem. Ber.*, **101**, 2028 (1968).

69: 27312q   V. R. Rao and V. R. Srinivasan, *Symp. Syn. Heterocycl. Compounds Physiol. Interest, Hyderabad, India*, 137 (1964).

69: 27340x   F. Russo and M. Ghelardoni, *Boll. Chim. Farm.*, **106**, 826 (1967).

69: 27342z   W. Ried and J. Valentin, *Chem. Ber.*, **101**, 2117 (1968).

69: 27360d   H. Becker and H. Boettcher, *Wiss. Z. Tech. Hochsch. Chem. Leuna-Merseburg*, **8**, 122 (1966).

69: 27374m   W. Ried and J. Valentin, *Chem. Ber.*, **101**, 2106 (1968).

69: P27451j   G. H. Harris, P. I. Traylor, and B. C. Fishback, U.S. Patent 3,360,516 (1967).

69: P28582q   W. Yamatani and K. Bando, Japanese Patent 68 03,506 (1968).

69: 35596a   A. M. Roe, R. A. Burton, G. L. Willey, M. W. Baines, and A. C. Rasmussen, *J. Med. Chem.*, **11**, 814 (1968).

69: 36039h   J. Sauer and K. K. Mayer, *Tetrahedron Lett.*, 325 (1968).

69: P37104n   H. Sanielevici, F. Brodman, C. Stuleanu, and S. Rosenberg, Romanian Patent 48,578 (1967).

69: 43249e   S. Hoffmann and E. Muchle, *Z. Chem.*, **8**, 222 (1968).

69: 43772v   J. Gardent and G. Hazebroucq, *Bull. Soc. Chim. Fr.*, 600 (1968).

69: 51320m   U. Anthoni, C. Larsen, and P. H. Nielsen, *Acta Chem. Scand.*, **22**, 309 (1968).

69: 51807a   M. Regitz and D. Stadler, *Chem. Ber.*, **101**, 2351 (1968).

69: 52047q   H. Boeshagen and W. Geiger, *Chem. Ber.*, **101**, 2472 (1968).

69: 52073v   O. Tsuge, T. Itoh, and S. Kanemasa, *Nippon Kagaku Zasshi*, **89**, 69 (1968).

69: 52075x   O. Tsuge, S. Kanemasa, and M. Tashiro, *Tetrahedron*, **24**, 5205 (1968).

69: 52078a   W. Ried and P. Schomann, *Justus Liebigs Ann. Chem.*, **714**, 122 (1968).

69: P52142s   K. Sugino and R. Kitawaki, Japanese Patent 67 25,895 (1967).

69: 55854m   M. J. Campbell, R. Grzeskowiak, and M. Goldstein, *Spectrochim. Acta, Part A*, **24**, 1149 (1968).

69: 59162g   J. B. Polya and B. M. Lynch, *Can. J. Chem.*, **46**, 2629 (1968).

69: P59245m   L. I. Bagal and M. S. Pevzner, USSR Patent 210,175 (1968).

69: P59249r   J. R. Green, G. A. Howarth, and W. Hoyle, British Patent 1,110,469 (1968).

69: P59251k   V. Grinsteins and A. Veveris, USSR Patent 215,225 (1968).

69: P60044b   H. Vollmann and H. Leister, British Patent 1,116,176 (1968).

69: P67151t   W. F. Bruce, U.S. Patent 3,377,353 (1968).

69: 67293r   H. Gehlen and B. Simon, *J. Prakt. Chem.*, **38**, 107 (1968).

69: 67337h   J. Weinstock, H. Graboyes, G. Jaffe, I. J. Pachter, K. Snader, C. B. Karash, and R. Y. Dunoff, *J. Med. Chem.*, **11**, 560 (1968).

69: 67340d   T. Bany, *Rocz. Chem.*, **42**, 247 (1968).

69: P67388a   G. Cipens, R. Bokaldere, and V. Grinsteins, USSR Patent 213,887 (1968).

69: 67793d   J. K. Stille and F. W. Harris, *J. Polym. Sci., Part A-1*, **6**, 2317 (1968).

69: 75887c   M. Pianka, *J. Sci. Food Agr.*, **19**, 475 (1968).

69: P77107x   A. A. Ponomarev and M. D. Lipanova, USSR Patent 203,692 (1967).

69: 77151g   F. Kurzer and M. Wilkinson, *J. Chem. Soc., C*, 2108 (1968).

69: 77170n   F. Kurzer and M. Wilkinson, *J. Chem. Soc., C*, 2099 (1968).

69: 77177v   P. M. Theus, W. Weuffen, and H. Tiedt, *Arch. Pharm.*, **301**, 401 (1968).

69: 77178w   G. V. Boyd and A. J. H. Summers, *Chem. Commun.*, 549, (1968).

69: P77265x   M. Wolf, U.S. Patent 3,394,143 (1968).

69: 83184m   M. Adamek and A. Cee, *Chem. Prum.*, **18**, 332 (1968).

69: 86504b   G. A. Russell, R. Konaka, E. Strom, W. C. Danen, K. Chang, and G. Kaupp, *J. Amer. Chem. Soc.*, **90**, 4646 (1968).

69: 86973d   T. Mukaiyama and S. Ono, *Tetrahedron Lett.*, 3569 (1968).

69: P87631c   K. Saotome, Japanese Patent 68 13,061 (1968).

69: 92632h   C. Calzolari and L. Favretto, *Analyst* (London), **93**, 494 (1968).

69: 96593u   H. Gehlen and H. Segeletz, *J. Prakt. Chem.*, **38**, 150 (1968).

69: 96601v   M. Fujimori, E. Haruki, and E. Imoto, *Bull. Chem. Soc. Jap.*, **41**, 1372 (1968).

69: P96734r   H. Becker, K. Wehner, and V. Eisenschmidt, British Patent 1,123,947 (1968).

69: 96737u   S. Huenig and T. Eicher, French Patent 1,481,761 (1967).

69: 106627z   M. S. Skorobogatova, N. P. Zolotareva, and Y. A. Levin, *Khim. Geterotsikl. Soedin.*, **4**, 372 (1968).

69: 107280t   T. Shono, M. Izumi, S. Matsummura, and N. Asano, Japanese Patent 68 14,479 (1968).

69: 110617p   G. C. Denault, P. C. Marx, and H. H. Takimoto, *J. Chem. Eng. Data*, **13**, 514 (1968).

69: P111675t   P. Sohar, *Magy. Kem. Foly.*, **74**, 298 (1968).

69: P112206c   A. Von Koenig, W. Berthold, A. Voigt, and W. Acharnke, Belgian Patent 695,282 (1967).

70: 3963x   Y. R. Rao, *Indian J. Chem.*, **6**, 287 (1968).

70: 3966a   S. P. Dutta, A. K. Acharyya, and U. P. Basu, *J. Indian Chem. Soc.*, **45**, 338 (1968).

70: 11610v   G. K. Khisamutdinov, I. T. Strukov, G. V. Golubkova, and V. E. Chistyakov, *Khim.-Farm. Zh.*, **2**, 35 (1968).

70: 11640e   E. J. Browne and J. B. Polya, *J. Chem. Soc.*, C, 2904 (1968).

70: 11642g   I. Matei and E. Comanita, *Bul. Inst. Politeh. Iasi*, **13**, 269 (1967).

70: 11643h   I. B. Romanova, K. V. Anan'eva, and L. K. Freidlin, *Uzb. Khim. Zh.*, **12**, 35 (1968).

70: 11686z   M. Pesson and M. Antoine, *C. R. Acad. Sci., Paris Ser. C*, **267**, 904 (1968).

70: 11687a   G. Barnikow and G. Strickmann, *Z. Chem.*, **8**, 385 (1968).

70: P11703c   A. Strazdina and V. Grinsteins, USSR Patent 203,693 (1967).

70: P12143g   V. V. Korshak, E. S. Kröngauz, and A. V. D'yachenko, USSR Patent 221,277 (1968).

70: 12144h   I. Aijima, A. Tanaka, A. Kitaoka, and Y. Katayama, Japanese Patent 68 13,062 (1968).

70: 20009h   B. T. Gillis and J. D. Hagarty, *J. Org. Chem.*, **32**, 330 (1967).

70: 20356u   L. Mortillaro, *Mater. Plast. Elastomeri*, **34**, 389 (1968).

70: 20435u   L. Mortillaro, *Mater. Plast. Elastomeri*, **34**, 305 (1968).

70: 20885r   R. G. Spain and L. G. Picklesimer, *Text. Res. J.*, **36**, 619 (1966).

70: 28872e   M. Fujimori, E. Haruki, and E. Imoto, *Nippon Kagaku Zasshi*, **89**, 900 (1968).

70: 28874g   A. Spassov and G. Demirov, *Chem. Ber.*, **101**, 4238 (1968).

70: P28922w   H. G. O. Becker, V. Eisenschmidt, and K. Wehner, (East) German Patent 59, 288 (1967).

70: P28924y   C. H. S. Boehringer, French Patent 1,505,095 (1967).

70: 36934s   J. Moeller, U. Anthoni, C. Larsen, and P. H. Nielsen, *Acta Chem. Scand.*, **22**, 2493 (1968).

70: 37729r   E. G. Kovalev and I. Y. Postovskii, *Khim. Geterotsikl. Soedin.*, **4**, 740 (1968).

70: 38307p K. Saotome and T. Yamazaki, Japanese Patent 68 15,634 (1968).

70: 43660c A. V. Radushev, A. N. Chechneva, and E. G. Kovalev, *Zh. Anal. Khim.*, **23,** 1410 (1968).

70: 47368m B. K. Paul, S. N. Ghosh, A. N. Bose, and U. P. Basu, *Indian J. Chem.*, **6,** 618 (1968).

70: 47369n I. Matei and E. Comanita, *Bul. Inst. Politeh. Iasi*, **12,** 175 (1966).

70: 47370f G. Chipens, R. Bokaldere, and V. Grinsteins, *Khim. Geterotsikl. Soedin.*, **4,** 743 (1968).

70: 47462n H. G. O. Becker, V. Eisenschmidt, and K. Wehner, German Patent 60,762 (1968).

70: 48223x British Patent 1,138,261 (1968).

70: 57739x V. A. Lopyrev, N. K. Beresneva, and V. L. Akimova, *Tr. Sev.-Zap. Zaoch. Politekh. Inst.*, **2,** 135 (1967).

70: 57742t C. H. Budeanu, *An. Stiint. Univ. "Al. I. Cuza" Iasi, Sect. Ic*, **13,** 179 (1967).

70: P57860e British Patent 1, 135, 498 (1968).

70: P58908p J. G. Fisher and C. A. Coates, Jr., U.S. Patent 3,415,809 (1968).

70: 67830j H. Braeuniger and K. Breuer, *Pharm. Zentralh.*, **107,** 795 (1968).

70: 68266s G. Zinner, *Arch. Pharm.* (Weinheim), **301,** 827 (1968).

70: 68268u H. Gehlen and K. H. Uteg, *Arch. Pharm.* (Weinheim), **301,** 911 (1968).

70: 68272r J. S. Davidson, *J. Chem. Soc., C*, 194 (1969).

70: 68301z H. Guglielmi, *Hoppe-Seyler's Z. Physiol. Chem.*, **349,** 1733 (1968).

70: 68323h M. Pesson and M. Antoine, *C. R. Acad. Sci., Paris, Ser. C*, **267,** 1726 (1968).

70: P68372y W. Kummer, H. Staehle, H. Koeppe, and K. Zeile, South African Patent 68 00,417 (1968).

70: P68377d H. Gehlen and K. H. Uteg, (East) German Patent 63,507 (1968).

70: 77870m R. Pohloudek-Fabini and E. Schroepl, *Pharm. Zentralh.*, **107,** 736 (1968).

70: 77876t L. I. Bagal, M. S. Pevzner, and V. A. Lopyrev, *Khim. Geterotsikl. Soedin., Sb.* 1: *Azotsoderzhashchie Geterotsikly*, 180 (1967).

70: 77877u L. I. Bagal, M. S. Pevzner, V. A. Lopyrev, and E. A. Yurchak, *Khim. Geterotsikl. Soedin., Sb.* 1: *Azotsoderzhashchie Geterotsikly*, 184 (1967).

70: 77879w J. P. Critchley and J. S. Pippett, *Gt. Brit., Roy. Aircraft Estab., Tech. Rep.*, TR-68026 (1968).

70: 77880q C. H. Budeanu and C. Ciugureanu, *An. Stiint. Univ. "Al. I. Cuza" Iasi, Sect 1c*, **14,** 81 (1968).

70: 77881r G. Cipens, R. Bokalders, and V. Grinsteins, *Khim. Geterotsikl. Soedin.*, **4,** 1105 (1968).

70: P87037j H. A. Barch, U.S. Patent 3,427,387 (1969).

70: 87679v H. Daniel, *Chem. Ber.*, **102,** 1028 (1969).

70: 87681q H. G. O. Becker and H. J. Trimpe, *J. Prakt. Chem.*, **311,** 9 (1969).

70: 87692u C. F. Kroeger, R. Miethchen, H. Frank, M. Siemer, and S. Pilz, *Chem. Ber.*, **102,** 755 (1969).

70: P87820j M. Pesson, French Patent 1,512,421 (1968).

70: 91957b E. A. Lipatova, I. N. Shokhor, V. A. Lopyrev, and M. S. Pevzner, *Khim. Geterotsikl. Soedin., Sb.* 1: *Azotsoderzhashchie Geterotsikly*, 186 (1967).

70: P92277s E. Lind and W. Wiedemann, South African Patent 68 02, 334 (1968).

70: 96723q R. Fusco and P. Dalla Croce, *Gazz. Chim. Ital.*, **99,** 69 (1969).

70: 96755u K. L. Shepard, J. W. Mason, O. W. Woltersdorf, Jr., J. H. Jones, H. James, and E. J. Cragoe, Jr., *J. Med. Chem.*, **12,** 280 (1969).

70: P96798k British Patent 1,143,279 (1969).

70: P96800e L. I. Bagal, M. S. Pevzner, V. Ya Samarenko, and A. P. Egorov, USSR Patent 227,328 (1968).

70: P96802g M. Hauser, U.S. Patent 3,431,251 (1969).

70: P97943r   H. Noack, G. Puhlmann, and F. Wolf, (East) German Patent 62,804 (1968).

70: 103962q   V. Grinsteins, K. Medne, A. Sausins, G. Cipens, and R. Bokalders, *Latv. PSR Zinat. Akad. Vestis, Kim. Ser.*, 691 (1968).

70: 106024w   H. O. Larson, T. Ooi, W. K. H. Luke, and K. Ebisu, *J. Org. Chem.*, **34**, 525 (1969).

70: 106436g   T. Sasaki and T. Yoshioka, *Bull. Chem. Soc. Jap.*, **42**, 258 (1969).

70: P107521t   J. F. Willems and A. L. Vandenberghe, British Patent 1,138,587 (1969).

70: 114407p   E. J. Browne and J. B. Polya, *J. Chem. Soc.*, *C*, 1056 (1969).

70: 115073p   H. Gehlen and R. Neumann, *J. Prakt. Chem.*, **311**, 213 (1969).

70: 115076s   S. S. Novikov, V. M. Brusnikina, and V. A. Rudenko, *Khim. Geterotsikl. Soedin.*, **5**, 157 (1969).

70: 115078u   G. Cipens and R. R. Bokalders, *Khim. Geterotsikl. Soedin.*, **5**, 159 (1969).

70: 115080p   H. Gehlen and H. Effert, *J. Prakt. Chem.*, **311**, 231 (1969).

71: 2032b   S. Petersen, W. Gauss, H. Kiehne, and L. Juehling, *Z. Krebsforsch.*, **72**, 162 (1969).

71: 2745m   J. F. Giudicelli, J. Menin, And H. Najer, *Bull. Soc. Chim. Fr.*, 870 (1969).

71: 13464v   V. V. Rode, E. M. Bondarenko, A. V. D'yachenko, E. S. Krongauz, and V. V. Korshak, *Vysokomol. Soedin., Ser. A*, **11**, 828 (1969).

71: 13549b   M. I. Shtil'man, O. Ya. Fedotova, G. S. Kolesnikov, and M. S. Ustinova, USSR Patent 230,417 (1968).

71: 14170h   W. W. Robbins, U.S. Patent 3,438,963 (1969).

71: P14180m   C. W. Schellhammer, British Patent 1,145,966 (1969).

71: 21509d   U. Anthoni, C. Larsen, and P. H. Nielsen, *Acta Chem. Scand.*, **23**, 537 (1969).

71: 21770g   H. Neunhoeffer, M. Neunhoeffer, and W. Litzius, *Justus Liebigs Ann. Chem.*, **722**, 29 (1969).

71: 21771h   N. Singh, S. Mohan, and J. S. Sandhu, *J. Chem. Soc., D*, 387 (1969).

71: 21830b   H. Nenuhoeffer, *Justus Liebigs Ann. Chem.*, **722**, 38 (1969).

71: 22075w   F. Kurzer and M. Wilkinson, *J. Chem. Soc., C*, 1218 (1969).

71: 22101b   T. Sasaki and M. Ando, *Yuki Gosei Kagaku Kyokai Shi*, **27**, 169 (1969).

71: P22929c   J. F. Willems, R. J. Thiers, and F. C. Heugebaert, Belgian Patent 710,798 (1968).

71: 30411t   D. Kreysig, H. H. Stroh, and G. Kempter, *Z. Chem.*, **9**, 187 (1969).

71: 30588f   W. L. Perilstein and H. A. Beatty, French Patent 1,525,268 (1968).

71: P35023c   A. G. Kalle, French Patent 1,527,253 (1968).

71: 38261q   R. S. Neale and N. L. Marcus, *J. Org. Chem.*, **34**, 1808 (1969).

71: 38865q   J. S. Shukla, H. H. Singh, and S. S. Parmar, *J. Prakt. Chem.*, **311**, 523 (1969).

71: 38867s   G. O. H. Becker, L. Krahnert, G. Rasch, W. Riediger, and J. Witthauer, *J. Prakt. Chem.*, **311**, 477 (1969).

71: 38870n   H. Schildknecht and G. Hatzmann, *Angew. Chem., Int. Ed. Engl.*, **8**, 456 (1969).

71: 38872q   M. Adamek and J. Klicnar, *Sb. Ved. Pr., Vys. Sk. Chemickotechnol., Pardubice*, **16**, 17 (1967).

71: P38973y   A. A. Hyatt, U.S. Patent 3,444,162 (1969).

71: P49571s   J. E. Roberston, J. K. Harrington, and A. Mendel, South African Patent 68 04,125 (1968).

71: 49853k   F. Russo and M. Ghelardoni, *Boll. Chim. Farm.*, **108**, 128 (1969).

71: 49855n   T. R. Vakula, V. R. Rao, and V. R. Srinivasan, *Indian J. Chem.*, **7**, 577 (1969).

71: 49864q   I. Matei and E. Comanita, *Bul. Inst. Politeh. Iasi*, **14**, 255 (1968).

71: P49947u   H. Gehlen, K. H. Uteg, and F. Lemme, (East) German Patent 64,970 (1968).

71: 50108c   M. Regitz and W. Anschuetz, *Chem. Ber.*, **102**, 2216 (1969).

71: 56483t   A. Cee and M. Adamek, *Sb. Ved. Pr., Vys. Sk. Chemickotechnol., Pardubice,* **17,** 5 (1968).

71: 60385e   H. Schildknecht and G. Hatzmann, *Justus Liebigs Ann. Chem.,* **724,** 226 (1969).

71: 61474v   E. Brunn and R. Huisgen, *Angew. Chem., Int. Ed. Engl.,* **8,** 513 (1969).

71: P62234d   French Patent 1,552,942 (1969).

71: P66072x   D. J. Beavers and W. F. Smith, *Offic. Gaz.,* **864,** 1407 (1969).

71: 70540r   O. P. Shvaika, S. N. Baranov, and V. N. Artemov, *Dokl. Akad. Nauk SSSR* **186,** 1102 (1969).

71: 70546x   M. Ruccia and N. Vivona, *Ann. Chim.* (Rome), **59,** 434 (1969).

71: P70607t   S. I. Burmistrov and L. I. Limarenko, USSR Patent 228,692 (1969).

71: P71822w   K. Tanabe and K. Matsubayashi, Japanese Patent 69 00,914 (1969).

71: 81268j   A. Spasov and G. Demirov, *Chem. Ber.,* **102,** 2530 (1969).

71: P81374r   French Patent 1,536;979 (1968).

71: P81375s   H. Becker and K. Wehner, British Patent 1,157,256 (1969).

71: P82422s   J. J. D'Amico, U.S. Patent 3,454,572 (1969).

71: P82637r   J. Gmaj, H. Scibsiz, and L. Wojciechowski, Polish Patent 55,858 (1968).

71: 89718b   K. C. Agrawal and A. C. Sartorelli, *J. Med. Chem.,* **12,** 771 (1969).

71: P90233w   P. C. Hamm, U.S. Patent 3,458,634 (1969).

71: P91056c   H. Martin, A. Lukaszczyk, D. Duerr, G. Pissiotas, O. Rohr, A. Hubele, and S. Janiak, German Patent 1,802,739 (1969).

71: 91194w   R. Huisgen, W. Scheer, H. Maeder, and E. Brunn, *Angew. Chem., Int. Ed. Engl.,* **8,** 604 (1969).

71: 91375f   G. Westphal and P. Henklein, *Z. Chem.,* **9,** 305 (1969).

71: 91382f   H. Gehlen and R. Drohla, *J. Prakt. Chem.,* **311,** 539 (1969).

71: 91385j   T. Eicher, S. Huenig, H. Hansen, and P. Nikolaus, *Chem. Ber.,* **102,** 3159 (1969).

71: 91386k   H. G. O. Becker, J. Witthauer, N. Sauder, and G. West, *J. Prakt. Chem.,* **311,** 646 (1969).

71: 91389p   T. Eicher, S. Huenig, and P. Nikolaus, *Chem. Ber.,* **102,** 3176 (1969).

71: 91395n   I. Simiti and I. Proinov, *J. Prakt. Chem.,* **311,** 687 (1969).

71: 91398r   V. J. Bauer, G. E. Wiegand, W. J. Fanshawe, and S. R. Safir, *J. Med. Chem.,* **12,** 944 (1969).

71: 91399s   R. Miethchen, H. U. Seipi, and C. F. Kroeger, *Z. Chem.,* **9,** 300 (1969).

71: 91400k   K. H. Uteg and H. Gehlen, *J. Prakt. Chem.,* **311,** 701 (1969).

71: P91505y   H. Bickel, R. Bosshardt, B. Fechtig, E. Menard, J. Mueller, and H. Peter, German Patent 1,816,824 (1969).

71: 101441r   T. Eicher, S. Huenig, and H. Hansen, *Chem. Ber.,* **102,** 2889 (1969).

71: 101781b   H. Gehlen, K. H. Uteg, and J. Vieweg, *Arch. Pharm.* (*Weinheim*), **302,** 105 (1969).

71: 101796k   M. Furdik, S. Mikulasek, and S. Priehradny, *Acta Fac. Rerum Natur. Univ. Comenianae, Chim.,* 239 (1968).

71: P101861c   O. Scherer and H. Mildenberger, South African Patent 68 03,471 (1968).

71: 102015s   L. C. Willemens, French Patent 1,547,920 (1968).

71: 112866j   E. J. Browne, *Aust. J. Chem.,* **22,** 2251 (1969).

71: 113155p   G. Defaye and M. Fetizon, *Bull. Soc. Chim. Fr.,* 2835 (1969).

71: 113323s   L. Stoicescu-Crivat, E. Mantaluta, G. Neamtu, and I. Zugravescu, *J. Polym. Sci., Part C,* **22** (2), 761 (1967).

71: 124310n   V. Zotta and A. Tataru-Gasmet, *Farmacia* (Bucharest), **17,** 461 (1969).

71: 124342y   S. Nakagawa, H. Saito, A. Kato, K. Kuno, and S. Sekimoto, *Nichidai Igaku Zasshi,* **27,** 256 (1968).

71: 124344a   W. Rudnicka and J. Sawlewicz, *Gdansk. Tow. Nauk., Rozpr. Wydz. 3,* **4,** 327 (1967).

71: 124346c   W. Rudnicka and J. Sawlewicz, *Gdansk. Tow. Nauk., Rozpr. Wydz. 3,* **4,** 335 (1967).

71: P124441e   H. G. O. Becker, W. Riediger, L. Krahnert, and K. Wehner, (East) German Patent 67,130 (1969).

71: P124443g   French Patent 1,523,729 (1968).

71: P124444h   British Patent 1,166,690 (1969).

71: P124458r   T. Takano, South African Patent 68 02,695 (1968).

71: 125032j   J. K. Stille and L. D. Gotter, *J. Polym. Sci., Part A*-1, **7,** 2493 (1969).

71: P125194p   K. Saotome and T. Yamazaki, Japanese Patent 69 07,956 (1969).

71: P125885q   A.-G. Glanzstoff, French Patent 1,567,327 (1969).

72: 3301w   L. Weinberger, P. Unger, and P. Cherin, *J. Heterocyclic Chem.,* **6,** 761 (1969).

72: 3437v   H. G. O. Becker, *Z. Chem.,* **9,** 325 (1969).

72: 3440r   H. Gehlen and K. H. Uteg, *Z. Chem.,* **9,** 338 (1969).

72: 3454y   R. Jones and C. W. Rees, *J. Chem. Soc., C,* 2251 (1969).

72: 3470a   T. George, R. Tahilramani, and D. A. Dabholkar, *Indian J. Chem.,* **7,** 959 (1969).

72: P3491h   A. P. Fedotova, A. V. Kazymov, and E. P. Shchelkina, USSR Patent, 245,554 (1969).

72: 12648s   C. F. Kroeger and R. Miethchen, *Z. Chem.,* **9,** 378 (1969).

72: 12671u   K. T. Potts and C. Lovelette, *J. Org. Chem.,* **34,** 3221 (1969).

72: P12733r   V. N. Artemov, S. N. Baranov, and O. P. Shvaika, USSR Patent 245,120 (1969).

72: 21667z   A. L. Mndzhoyan, A. S. Azaryan, and A. A. Aroyan, *Arm. Khim. Zh.,* **22,** 488 (1969).

72: 22220k   S. Rafikov, R. Rode, V. V. Zhuravleva, E. M. Bodarenko, and P. N. Gribkova, *Vysokomol. Soedin., Ser. A,* **11,** 2043 (1969).

72: P22599r   C. W. Schellhammer, A. Dorlars, and W. D. Wirth, British Patent 1,163,876 (1969).

72: 30148h   M. H. Shah, M. Y. Mhasalkar, V. M. Patki, C. V. Deliwala, and U. K. Sheth, *J. Pharm. Sci.,* **58,** 1398 (1969).

72: 31701v   I. Arai, Y. Satoh, I. Muramatsu, and A. Hagitani, *Bull. Chem. Soc. Jap.,* **42,** 2739 (1969).

72: 31710x   N. S. Zefirov and N. K. Chapovskaya, *Vestn. Mosk. Univ., Khim.,* **24,** 113 (1969).

72: 31711y   W. Rudnicka and J. Sawlewicz, *Gdansk. Tow. Nauk., Rozpr. Wydz. 3,* **5,** 257 (1968).

72: 31712z   W. Rudnicka, A. Czarnocka-Janowicz, P. Zoga, and J. Sawlewicz, *Gdansk. Tow. Nauk., Rozpr. Wydz. 3,* **5,** 265 (1968).

72: P31806h   H. Timmler, I. Hammann, and R. Wegler, British Patent 1,161,381 (1969).

72: 37527q   J. Sandstrom, *Spectrochim. Acta, Part A,* **25,** 1865 (1969).

72: P43185t   Y. Yokoi, S. Wakabayashi, M. Tsunoda, T. Suzuki, and M. Tsuda, Japanese Patent 69 28,484 (1969).

72: 43550h   W. Ried and B. Peters, *Justus Liebigs Ann. Chem.,* **729,** 124 (1969).

72: 43572s   P. M. Hergenrother, *J. Heterocyclic Chem.,* **6,** 965 (1969).

72: 43579z   G. Westphal and P. Henklein, *Z. Chem.,* **9,** 425 (1969).

72: P43682c   G. Hegar, German Patent 1,925,654 (1969).

72: P43692f   F. Reisser, German Patent 1,918,795 (1969).

72: 44284m   G. Lorenz, M. Gallus, W. Giessler, F. Bodesheim, H. Wieden, and G. E. Nischk, *Makromol. Chem.,* **130,** 65 (1969).

72: P45020j   G. Hegar, German Patent 1,909,107 (1969).

72: 55339j   J. K. Landquist, *J. Chem. Soc., C,* 63 (1970).

72: 55340c   M. A. Khan and J. B. Polya, *J. Chem. Soc., C,* 85 (1970).

72: 55345h   F. Kurzer and M. Wilkinson, *J. Chem. Soc., C,* 19 (1970).

72: 55347k   F. Kurzer and M. Wilkinson, *J. Chem. Soc., C,* 26 (1970).

72: P55458x   M. D. Coburn, U.S. Patent 3,483,211 (1969).

72: P56717t   A. Froehlich and H. Loeffel, German Patent 1,901,422 (1969).

72: P56720p   P. Marx, U. Hess, R. Otto, W. Pusechel, and W. Pelz, Belgian Patent 713,448 (1968).

72: 66185g   A. Heesing, G. Iinsieke, G. Maleck, R. Peppmoeller, and H. Schulze, *Chem. Ber.*, **103**, 539 (1970).

72: 66870b   T. Sasaki, K. Kanematsu, and K. Hayakawa, *J. Chem. Soc., D*, 82 (1970).

72: 66873e   H. G. O. Becker, N. Sauder, and H. J. Timpe, *J. Prakt. Chem.*, **311**, 897 (1969).

72: P67723t   M. Minagawa, K. Nakagawa, and M. Goto, German Patent 1,926,547 (1970).

72: P67724u   M. Minagawa and K. Nakagawa, German Patent 1,927,447 (1970).

72: 68047n   H. E. Kuenzel, G. D. Wolf, F. Bentz, G. Blankenstein, and G. E. Nischk, *Makromol. Chem.*, **130**, 103 (1969).

72: 74407w   C. Calzolari, L. Favretto, G. L. Favretto, and M. G. Pertoldi, *Rass Chim.*, **21**, 135 (1969).

72: 74965b   D. Aures, R. Hakanson, and C. Wiseman, *Eur. J. Pharmacol.*, **8**, 232 (1969).

72: P78474p   M. W. Moon, U.S. Patent 3,484,451 (1969).

72: 78949x   B. K. Paul and U. P. Basu, *J. Indian Chem. Soc.*, **46**, 1121 (1969).

72: 78951s   J. K. Landquist, *J. Chem. Soc., C*, 323 (1970).

72: 78953u   H. Gehlen and K. H. Uteg, *Arch. Pharm.* (Weinheim), **302**, 877 (1969).

72: 78954v   V. A. Lopyrev, L. P. Sidorova, O. A. Netsetskaya, and M. P. Grinblat, *Zh. Obshch. Khim.*, **39**, 2525 (1969).

72: P79052m   K. Sugino and K. Shirai, Japanese Patent 69 30,506 (1969).

72: P79056r   J. C. Kauer, U.S. Patent 3,489,761 (1970).

72: P79061p   V. A. Lopyrev, M. P. Grinblat, O. A. Netsetskaya, M. M. Fomicheva, L. P. Sidorova, and A. M. Lundstrem, USSR Patent 245,118 (1969).

72: P79854z   French Patent 1,575,720 (1969).

72: 90378t   J. H. Boyer and P. J. A. Frints, *J. Heterocyclic Chem.*, **7**, 71 (1970).

72: 90379u   J. H. Boyer and P. J. A. Frints, *J. Heterocyclic Chem.*, **7**, 59 (1970).

72: P90516m   E. J. Cragce, Jr., German Patent 1,808,677 (1970).

72: 90906p   P. M. Hergenrother, *Macromolecules*, **3**, 10 (1970).

72: P91491m   French Patent 2,001,069 (1969).

72: P95326j   A. E. Harris and I. Blundell, British Patent 1,177,287 (1970).

72: 100607h   L. E. Benjamin, A. Homer, and R. Dobson, *J. Med. Chem.*, **13**, 291 (1970).

72: 100656y   G. Matolcsy and B. Bordas, *Acta Phytopathol.*, **4**, 197 (1969).

72: P100713q   S. A. Greenfield, M. C. Seidel, and W. C. Von Meyer, German Patent 1,943,915 (1970).

72: P100715s   S. Petersen, H. Striegler, H. Gold, and A. Haberkorn, South African Patent 69 01,244 (1969).

72: 100864q   S. S. Washburne and W. R. Peterson, Jr., *J. Organometal. Chem.*, **21**, 427 (1970).

72: 111375g   T. N. Vereshchagina, V. A. Lopyrev, M. S. Pevzner, and L. M. Kogan, *Khim. Geterotsikl. Soedin.*, **5**, 913 (1969).

72: 111380e   L. I. Bagal, M. S. Pevzner, and V. Y. Samarenko, *Khim. Geterotsikl. Soedin.*, **6**, 269 (1970).

72: 111383h   L. I. Bagal, M. S. Pevzner, A. N. Frolov, and N. I. Sheludyakova, *Khim. Geterotsikl. Soedin.*, **6**, 259 (1970).

72: 111384j   L. I. Bagal, M. S. Pevzner, N. I. Sheludyakova, and V. M. Kerusov, *Khim. Geterotsikl. Soedin.*, **6**, 265 (1970).

72: 111882v   P. M. Hergenrother and L. A. Carlson, *J. Polym. Sci., Part A*-1, **8**, 1003 (1970).

72: P112771v   K. Ueda, S. Ando, and T. Kinoshita, British Amended Patent 1,112,949 (1968).

72: P121183g   J. K. Harrington, D. C. Kvam, A. Mendel, and J. E. Robertson, French Patent 1,579,473 (1969).

72: 121442r   A. J. Blackman, *Aust. J. Chem.*, **23**, 631 (1970).

72: 121448x   N. K. Beresneva, V. A. Lopyrev, and K. L. Krupin, *Khim. Geterotsikl. Soedin.*, **5**, 118 (1969).

72: 121454w   E. Koerner von Gustorf, D. V. White, B. Kim, D. Hess, and J. Leitich, *J. Org. Chem.*, **35**, 1155 (1970).

72: 121456y   V. Grinsteins and A. Strazdina, *Khim. Geterotsikl. Soedin.*, **5**, 1114 (1969).

72: 121463y   G. V. Boyd and M. D. Harms, *J. Chem. Soc.*, *C*, 807 (1970).

72: 121541x   H. Illy, German Patent 1,943,800 (1970).

72: P122907h   J. G. Fisher and C. A. Coates, Jr., German Patent 1,812,982 (1969).

72: 131663c   N. P. Bednyagina, G. N. Lipunova, A. P. Novikova, A. P. Zeif, and L. N. Shchegoleva, *Zh. Org. Khim.*, **6**, 619 (1970).

72: 132636h   E. J. Browne, *Tetrahedron Lett.*, 943 (1970).

72: P132735q   H. Timmler, I. Hammann, and R. Wegler, British Patent 1, 149,159 (1969).

72: P132738t   A. L. Langis, U.S. Patent 3,499,000 (1970).

72: P133573x   K. Arayoshi, T. Shono, M. Izumi, S. Matsumura, and N. Asano, Japanese Patent 70 03,395 (1970).

72: P138322z   French Patent 2,004,995 (1969).

73: 3842u   G. Zinner and D. Boese, *Arch. Pharm.* (Weinheim), **303**, 222 (1970).

73: P3923w   British Patent 1,187,323 (1970).

73: 14233x   M. Yokoi, O. Wakabayashi, M. Tsunoda, Y. Suzuki, and M. Tsunda, Japanese Patent 70 07,521 (1970).

73: 14605v   R. L. N. Harris, *Aust. J. Chem.*, **23**, 1199 (1970).

73: 14770v   H. Gehlen, P. Demin, and K. H. Uteg, *Arch. Pharm.* (Weinheim), **303**, 310 (1970).

73: 14771w   R. N. Butler, T. Lambe, and F. L. Scott, *Chem. Ind.* (London), 628 (1970).

73: 14772x   H. Gehlen and P. Demin, *Arch. Pharm.* (Weinheim), **303**, 263 (1970).

73: 14857d   M. Ota, Japanese Patent 70 07,741 (1970).

73: P16264g   H. Burkhard, R. Entschel, and W. Steinemann, German Patent 1,901,711 (1970).

73: 25036m   I. Arai and A. Hagitani, *Nippon Kogaku Zasshi*, **91**, 262 (1970).

73: 25364s   F. P. Woerner and H. Reimlinger, *Chem. Ber.*, **103**, 1908 (1970).

73: P25462x   F. Reisser, Swiss Patent 484,920 (1969).

73: P25480b   G. Asato, G. Berkelhammer, W. H. Gastrock, I. Starer, C. C. Papaioamiou, J. D. Albright, and R. G. Shepherd, German Patent 1,951,259 (1970).

73: P25482d   M. Dukes, German Patent 1,946,315 (1970).

73: 25919b   J. Sundquist, *Makromol. Chem.*, **134**, 287 (1970).

73: 26583z   D. J. Gale and J. F. K. Wilshire, *Aust. J. Chem.*, **23**, 1063 (1970).

73: 31239w   P. P. Kish and Y. K. Onishchenko, *Zh. Anal. Khim.*, **25**, 500 (1970).

73: 35283d   M. V. Konher, *Indian J. Chem.*, **8**, 391 (1970).

73: 35286g   L. A. Vlasova, V. I. Minkin, and I. Ya. Postovskii, *Zh. Obshch. Khim.*, **40**, 372 (1970).

73: 35288j   L. Florvall, *Acta Pharm. Suecica*, **7**, 87 (1970).

73: 35290d   K. Sasse, *Justus Liebigs Ann. Chem.*, **735**, 158 (1970).

73: 35293g   F. L. Scott and T. A. F. O'Mahony, *Tetrahedron Lett.*, 1841 (1970).

73: P36546d   M. Iizuka, N. Igari, and S. Maeda, German Patent 1,925,475 (1970).

73: P36579s   G. A. Klein, German Patent 1,955,455 (1970).

73: P36594t   O. Riester, Belgium Patent 719,338 (1969).

73: 45419s   L. I. Bagal, M. S. Pevzner, V. Y. Samarenko, and A. P. Egorov, *Khim. Geterotsikl. Soedin.*, **6**, 702 (1970).

73: 45420k   A. N. Frolov, M. S. Pevzner, I. N. Shokhor, A. G. Gal'kovskaya, and L. I. Bagal, *Khim. Geterotsikl. Soedin.*, **6**, 705 (1970).

73: 45425r   R. A. Olofson and R. V. Kendall, *J. Org. Chem.*, **35**, 2246 (1970).

73: 45427t   H. Gehlen and P. Demin, *Z. Chem.*, **10**, 189 (1970).

73: P45513t   L. G. Merrifield and N. J. Iungius, German Patent 1,960,981 (1970).

73: P45550c   H. A. Brown, J. G. Erickson, D. R. Husted, and C. D. Wright, U.S. Patent 3,515,603 (1970).

73: P46632t   M. Iizuka, N. Igari, and S. Maeda, German Patent 1,925,491 (1969).

73: P50723r   French Patent 2,005,870 (1969).

73: 56038t   E. J. Browne, E. E. Nunn, and J. B. Polya, *J. Chem. Soc., C*, 1515 (1970).

73: 56050r   H. Gehlen and B. Simon, *Arch. Pharm.* (Weinheim), **303**, 511 (1970).

73: 56069d   M. Pesson and M. Antoine, *Bull. Soc. Chim. Fr.*, 1599 (1970).

73: 56073a   P. Scheiner and J. F. Dina, Jr., *Tetrahedron*, **26**, 2619 (1970).

73: P57084s   F. Favre and J. Voltz, Swiss Patent 487,231 (1970).

73: P57178a   French Patent 2,006,084 (1969).

73: 62258t   P. P. Kish and Y. K. Onishchenko, *Zh. Anal. Khim.*, **25**, 112 (1970).

73: 64790x   M. Y. Mhasalkar, M. H. Shah, S. T. Nikam, K. G. Anantanarayanan, and C. V. Deliwala, *J. Med. Chem.*, **13**, 672 (1970).

73: 66509m   M. Ruccia and N. Vivona, *J. Chem. Soc., D*, 866 (1970).

73: 66511f   P. Grammaticakis, *C. R. Acad. Sci., Paris, Ser. C*, **271**, 75 (1970).

73: 66512g   C. M. Gupta, A. P. Bhaduri, and N. M. Khanna, *Tetrahedron*, **26**, 3069 (1970).

73: 66513h   R. Miethchen, H. Albrecht, and E. Rachow, *Z. Chem.*, **10**, 220 (1970).

73: 66517n   M. Pesson and M. Antoine, *Bull. Soc. Chim. Fr.*, 1590 (1970).

73: 66847v   J. T. Witkowski and R. K. Robins, *J. Org. Chem.*, **35**, 2635 (1970).

73: P67700d   G. Mingasson and A. Domergue, French Patent 1,541,050 (1968).

73: 75210j   E. Schroeder, R. Albrecht, and C. Rufer, *Arzneim.-Forsch.*, **20**, 737 (1970).

73: 77149v   K. T. Potts, E. Houghton, and S. Husain, *J. Chem. Soc., D*, 1025 (1970).

73: 77155u   V. Evdokimoff, *Boll. Chim. Farm.*, **109**, 240 (1970).

73: 77206m   J. Kobe, B. Stanovnik, and M. Tisler, *Tetrahedron*, **26**, 3357 (1970).

73: 77210h   T. Kametani, K. Sota, and M. Shio, *J. Heterocyclic Chem.*, **7**, 821 (1970).

73: 77251x   S. Okano and K. Yasujaga, Japanese Patent 70 17,191 (1970).

73: P77253z   K. Soda, K. Nishide, and M. Shio, Japanese Patent 70 17,190 (1970).

73: 87300b   L. I. Bagal and M. S. Pevzner, *Khim. Geterotsikl. Soedin.*, **6**, 588 (1970).

73: 87717t   L. F. Miller and R. E. Bambury, *J. Med. Chem.*, **13**, 1022 (1970).

73: 87860j   H. J. Anacker, C. Gericke, and H. J. Timpe, *Wiss. Z. Tech. Hochsch. Chem. "Carl Schorlemmer" Leuna-Merseburg*, **12**, 72 (1970).

73: 87861k   H. C. Van der Plas and H. Jongejan, *Rec. Trav. Chim. Pays-Bas*, **89**, 680 (1970).

73: 87862m   H. Gehlen and R. Drohla, *Arch. Pharm.* (Weinheim), **303**, 650 (1970).

73: 87864p   K. H. Uteg and H. Gehlen, *Arch. Pharm.* (Weinheim), **303**, 634 (1970).

73: 87897b   R. Bokaldere and V. Grinsteins, *Khim. Geterotsikl. Soedin*, **6**, 563 (1970).

73: P93613x   R. Ohi, H. Iwano, T. Shishido, and I. Shimamura, German Patent 1,935,310 (1970).

73: 98876v   H. Reimlinger, W. R. F. Lingier, J. J. M. Vandewalle, and E. Goes, *Synthesis* 2, 433 (1970).

73: P100007k   H. Burkhard, German Patent 1,946,431 (1970).

73: P100073b   French Patent 2,004,074 (1969).

73: 109240m   K. H. Koch and E. Fahr, *Angew. Chem., Int. Ed. Engl.*, **9**, 634 (1970).

73: 109744d   H. Reimlinger, *Chem. Ind.* (London), 1082 (1970).

73: P109799a   V. C. Lintzenich, German Patent 2,008,843 (1970).

73: 115028q   J. F. Willems, G. C. Heugebaert, and R. J. Pollet, German Patent 1,908,217 (1969).

73: P119532x   A. Zschocke and A. Fischer, South African Patent 69 05,463 (1970).

73: 120582v   K. T. Potts and E. G. Brugel, *J. Org. Chem.*, **35**, 3448 (1970).

73: P120632m   H. G. O. Becker, T. Roethling, H. Lehmann, and R. Kleiner, German Patent 1,943,285 (1970).

73: P120633n   S. S. Adams, B. J. Armitage, B. V. Heathcote, and N. W. Bristow, German Patent 2,014,847 (1970).

73: P120638t   N. B. Vinogradova, I. A. Kuleshova, V. L. Gol'dfarb, and N. V. Khromov-Borisov, USSR Patent 253,807 (1970).

73: P120647v   J. G. Lombardino, German Patent 1,943,265 (1970).

73: 120949v   L. D. Gotter and J. K. Stille, *Polym. Prepr., Amer. Chem. Soc., Div. Polym. Chem.*, **10**, 168 (1969).

73: 125735g   R. J. Noe, J. F. Willems, A. L. Poot, and K. E. Verhille, French Patent 2,010,309 (1970).

73: 130941a   J. Lange and H. Tondys, *Diss. Pharm. Pharmacol.*, **22**, 217 (1970).

73: 130946f   H. Gehlen and R. Drohla, *Arch. Pharm.* (Weinheim), **303**, 709 (1970).

73: 130967p   H. Reimlinger and M. A. Peiren, *Chem. Ber.*, **103**, 3266 (1970).

74: 3557a   J. B. Polya and A. J. Blackman, *J. Chem. Soc., C*, 2403 (1970).

74: 3559c   M. Gattuso and G. Lo Vecchio, *Atti Soc. Peloritana Sci. Fis. Mat. Natur.*, **14**, 371 (1968).

74: P3636a   M. Murakami, K. Takahashi, A. Matsumoto, and K. Tamazawa, Japanese Patent 70 28,374 (1970).

74: P8379d   British Patent 1,204,931 (1970).

74: 13070a   J. Sandstrom, *Z. Chem.*, **10**, 406 (1970).

74: 13072c   J. W. Lown and K. Matsumoto, *Can. J. Chem.*, **48**, 3399 (1970).

74: 13074e   P. Grammaticakis, *C. R. Acad. Sci., Ser. C*, **271**, 940 (1970).

74: 13103p   F. H. Case, *J. Heterocyclic Chem.*, **7**, 1001 (1970).

74: P14185d   M. A. Weaver and R. R. Giles, German Patent 2, 008, 881 (1970).

74: P14197j   A. Dorlars, German Patent 1,906,662 (1970).

74: P18038f   W. Berthold, A. Von Koenig, and H. Timmler, German Patent 1,919,045 (1970).

74: 22770f   M. D. Coburn, E. D. Loughran, and L. C. Smith, *J. Heterocyclic Chem.*, **7**, 1149 (1970).

74: 39709r   N. L. Couse, D. J. Cummings, V. A. Chapman, and S. S. DeLong, *Virology*, **42**, 590 (1970).

74: 42317s   A. Ya. Lazaris and A. N. Egorochkin, *Zh. Org. Khim.*, **6**, 2342 (1970).

74: 42320n   K. T. Potts and S. Husain, *J. Org. Chem.*, **36**, 10 (1971).

74: P43528y   A. Horn, E. Schinzel, and G. Roesch, German Patent 1,919, 209 (1970).

74: 53659f   R. Walentowski and H. W. Wanzlick, *Z. Naturforsch. B*, **25**, 1421 (1970).

74: 53660z   I. Mir, M. T. Siddiqui, and A. Comrie, *Tetrahedron*, **26**, 5235 (1970).

74: 53661a   H. Gehlen and K. H. Uteg, *Arch. Pharm.* (Weinheim), **303**, 945 (1970).

74: 53662b   H. Kato, T. Shiba, H. Yoshida, and S. Fujimori, *J. Chem. Soc., D*, 1591 (1970).

74: 53664d   M. A. Khan and B. M. Lynch, *J. Heterocyclic Chem.*, **7**, 1237 (1970).

74: 53671d   H. Reimlinger, W. R. F. Lingier, and R. Merenyl, *Chem. Ber.*, **103**, 3817 (1970).

74: 53817f   J. W. Gates, A. W. Wise, D. A. J. Beavers, and P. E. Miller, German Patent 2,016,082 (1970).

74: 53877a   T. Kauffmann, H. Berg, E. Ludorff, and A. Woltermann, *Agnew. Chem., Int. Ed. Engl.*, **9**, 960 (1970).

74: 54291s   W. H. Pirkle and J. C. Stickler, *J. Amer. Chem. Soc.*, **92**, 7497 (1970).

74: 59399t   Y. Ohyama and S. Miyazawa, German Patent 2,015,344 (1970).

74: P65592s   French Patent 2,014,274 (1970).

74: 65595v   W. D. Wirth, C. W. Schellhammer, and A. Dorlars, German Patent 1,919,181 (1970).

74: 70245k   A. Von Koenig, W. Liebe, and H. Timmler, German Patent 1,930,339 (1970).

74: 76376a   L. I. Bagal, M. S. Pevzner, A. P. Egorov, and V. Ya. Samarenko, *Khim. Geterotsikl. Soedin.*, **6,** 997 (1970).

74: 76428u   T. Cebalo, German Patent 2,029,375 (1970).

74: P77385q   A. Roueche, German Patent 2,021,326 (1970).

74: P77396u   M. Ozutsumi and S. Maeda, German Patent 2,022, 625 (1970).

74: P81774h   French Patent 2,015, 184 (1970).

74: 87894u   B. T. Gillis and J. G. Dain, *J. Org. Chem.*, **36,** 518 (1971).

74: 87898y   H. Reimlinger, W. R. F. Lingier, and J. J. M. Vandewalle, *Chem. Ber.* **104,** 639 (1971).

74: 87899z   O. P. Shvaika, V. N. Artemov, and S. N. Baranov, *Khim. Geterotsikl. Soedin.*, **6,** 991 (1970).

74: 87901u   M. Tisler, A. Andolsek, B. Stanovnik, M. Kikar, and P. Schauer, *J. Med. Chem.*, **14,** 53 (1971).

74: P88677n   F. Fleck and H. R. Schmid, German Patent 2,025,792 (1970).

74: 99946a   D. J. McCaustland, W. H. Burton, and C. C. Cheng, *J. Heterocyclic Chem.*, **8,** 89 (1971).

74: 99947b   H. Boehme and K. H. Ahrens, *Tetrahedron Lett.*, 149 (1971).

74: 99948c   L. I. Bagal, M. S. Pevzner, V. Ya. Samarenko, and A. P. Egorov, *Khim. Geterotsikl. Soedin.*, **6,** 1701 (1970).

74: 99951y   T. N. Vereshchagina and V. A. Lopyrev, *Khim. Geteotsikl. Soedin.*, **6,** 1695 (1970).

74: 100062t   French Patent 2,016,526 (1970).

74: 100065w   E. Grigat and R. Puetter, German Patent 1,940,366 (1971).

74: 100066x   H. G. O. Becker, G. West, and K. Wehner, (East) German Patent 74,545 (1970).

74: 100068z   K. T. J. Skagius and E. B. Akerblom, U.S. Patent 3,557,137 (1971).

74: 110789z   R. D. Speirs and J. H. Lang, *U.S. Dept. Agric. Marketing Res. Rep.*, No. 885 (1970).

74: 111161a   J. B. Polya and A. J. Blackman, *J. Chem. Soc.*, C, 1016 (1971).

74: 111182h   H. G. O. Becker, D. Beyer, and H. J. Timpe, *J. Prakt. Chem.*, **312,** 869 (1971).

74: 111955n   G. Kempter, H. Schafer, and G. Sarodnick, *Z. Chem.*, **10,** 460 (1970).

74: 111967t   H. G. O. Becker and H. J. Timpe, *J. Prakt. Chem.*, **312,** 586 (1970).

74: 111968u   C. V. Deliwala, M. Y. Mhasalkar, M. H. Shah, S. T. Nikam, and K. G. Anantanarayanan, *J. Med. Chem.*, **14,** 260 (1971).

74: 111972r   G. V. Boyd, A. J. H. Summers, *J. Chem. Soc.*, C, 409 (1971).

74: 112048f   G. Jaeger and K. H. Buechel, German Patent 1,940,626 (1971).

74: 112476u   G. B. Butler, L. J. Guilbault, and S. R. Turner, *J. Polym. Sci., Part B*, **9,** 115 (1971).

74: 118381f   D. A. Brooks, French Patent 2,020,524 (1970).

74: 124598y   C.-H. Chang, R. F. Porter, and S. H. Bauer, *J. Mol. Struct.*, **7,** 89 (1971).

74: 125500x   J. F. Magalhaes, Q. Mingoia, and M. G. Piros, *Rev. Farm. Bioquim. Univ. Sao Paulo*, **8,** 125 (1970).

74: 125534m   O. P. Shvaika, V. N. Artemov, and S. M. Baranov, *Ukr. Khim. Zh.*, **37,** 172 (1971).

74: 125572x   E. J. Browne, *Aust. J. Chem.*, **24,** 393 (1971).

74: 125574z   E. Haruki, *Yuki Gosei Kagaku Kyokai Shi*, **28,** 1072 (1970).

74: 125575a   A. G. M. Willems, A. Tempel, D. Hamminga, and B. Stork, *Rec. Trav. Chim. Pays-Bas*, **90,** 97 (1971).

74: 125578d   M. Ruccia, N. Vivona, and G. Cusmano, *J. Heterocyclic Chem.*, **8,** 137 (1971).

74: 125579e   M. E. Derieg, R. I. Fryer, and S. S. Hillery, *J. Heterocyclic Chem.*, **8,** 181 (1971).

74: 125651x   H. G. O. Becker, D. Beyer, G. Israel, R. Mueller, W. Riediger, and H. J. Timpe, *J. Prakt. Chem.*, **312**, 669 (1970).

74: 125653z   F. H. Case, *J. Heterocyclic Chem.*, **8**, 173 (1971).

74: 125697s   W. Draber and K. H. Buechel, German Patent 1,940,627 (1971).

74: 125698t   K. H. Buechel and W. Draber, German Patent 1,940,628 (1971).

74: 125705t   E. Regel, K. H. Buechel, R. Schmidt, L. Eue, German Patent 1,949,012 (1971).

74: 127556u   C. W. Schellhammer, German Patent 1,942,926 (1971).

74: 140379h   T. Kametani, S. Hirata, S. Shibuya, and M. Shio, *Org. Mass Spectrom.*, **5**, 117 (1971).

74: 141216q   F. Fujikawa, K. Hirai, T. Hirayama, T. Matsunashi, Y. Nakanishi, K. Kumoto, T. Shimizu, C. Sakaki, and Y. Hamuro, *Yakugaku Zasshi*, **91**, 159 (1971).

74: 141644c   E. G. Kovalev and I. Ya. Postovskii, *Khim. Geterotsikl. Soedin.*, **6**, 1138 (1970).

74: 141645d   T. George, D. V. Mehta, R. Tahilramani, J. David, and P. K. Talwalker, *J. Med. Chem.*, **14**, 335 (1971).

74: 141651c   H. G. O. Becker, G. Goermar, and H. J. Timpe, *J. Prakt. Chem.*, **312**, 610 (1970).

74: 141652d   E. Comanita and M. Tutoveanu, *Bul. Inst. Politeh. Iasi*, **15**, 89 (1969).

74: 141669q   H. G. O. Becker, H. Boettcher, R. Ebisch, and G. Schmoz, *J. Prakt. Chem.*, **312**, 780 (1971).

74: P141811e   V. L. Gol'dfarb, N. B. Vinogradova, and N. V. Khromov-Borisov, USSR Patent 289,090 (1970).

74: 141814h   M. W. Moon, U.S. Patent 3,555,152 (1971).

74: 141817m   A. Dobbelstein, H. D. Hille, and H. Holfort, German Patent 1,937,310 (1971).

74: 142323j   B. Tanacas, I. Mezo, L. Bursics, and I. Teplan, *Nov. Metody Poluch. Radioaktiv. Prep., Sb. Dokl. Simp.*, 290 (1969).

74: 149220g   French Patent 2,018,611 (1970).

74: 149248x   N. Itoh, S. Tosa, H. Tsukahara, and N. Kobayashi, German Patent 2,040,801 (1971).

75: 5441c   R. S. Shadbolt, *J. Chem. Soc., C*, 1667 (1971).

75: 13530w   G. F. Mitchell, German Patent 2,028,861 (1970).

75: 14233p   Y. P. Kitaev, I. M. Skrebkova, and L. I. Maslova, *Izv. Akad. Nauk SSSR, Ser. Khim.*, 2194 (1970).

75: 19520q   S. Beveridge and R. L. N. Harris, *Aust. J. Chem.*, **24**, 1229 (1971).

75: 20254n   E. Funke, R. Huisgen, and F. C. Schaefer, *Chem. Ber.*, **104**, 1550 (1971).

75: 20303c   S. Somasekhara, N. V. Upadhyaya, and S. L. Mukherjee, *J. Indian Chem. Soc.*, **48**, 289 (1971).

75: 20306f   T. Sasaki, K. Kanematsu, and K. Hayakawa, *J. Chem. Soc., C*, 2142 (1971).

75: 20308h   W. Lwowski, R. A. DeMauriac, R. A. Murray, and L. Lunow, *Tetrahedron Lett.*, 425 (1971).

75: 20356x   I. Ilirao, Y. Kato, T. Hayakawa, and H. Tateishi, *Bull. Chem. Soc. Jap.*, **44**, 780 (1971).

75: 20371y   W. A. Mosher and R. B. Toothill, *J. Heterocyclic Chem.*, **8**, 209 (1971).

75: P20399p   J. W. Black, G. J. Durant, J. C. Emmett, and C. R. Ganellin, German Patent 2,052,692 (1971).

75: P20410k   W. Dietsch, H. Adolphi, P. Beutel, and K. H. Koenig, German Patent 1,949,490 (1971).

75: P20438a   E. J. Cragoe, Jr., U.S. Patent 3,573,306 (1971).

75: 21112b   G. B. Vaughan, J. C. Rose, and G. P. Brown, *J. Polym. Sci., Part A-1*, **9**, 1117 (1971).

75: 26054u   A. V. Radushev, A. N. Chechneva, and I. I. Mudretsova, *Zh. Anal. Khim.*, **26,** 796 (1971).

75: 29276d   Y. P. Kitaev, I. M. Skrebkova, V. V. Zverev, and L. I. Maslova, *Izu. Akad. Nauk SSSR, Ser. Khim.*, 28 (1971).

75: 34816j   S. Wawzonek and E. E. Paschke, *J. Org. Chem.*, **36,** 1474 (1971).

75: 35361u   R. Pohloudek-Fabini, K. Kottke, and F. Friedrich, *Pharmazie*, **26,** 283 (1971).

75: 35845e   E. Brunn, E. Funke, H. Gotthardt, and R. Huisgen, *Chem. Ber.*, **104,** 1562 (1971).

75: 35882q   O. P. Shvaika, V. N. Artemov, and S. N. Baranov, *Khim. Geterotsikl. Soedin.*, **7,** 39 (1971).

75: 35891s   L. A. Vlasova, E. M. Shamaeva, L. A. Chechulina, G. B. Atanas'eva, and I. Y. Postovskii, *Puti Sin. Izyskaniya Protivoopukholevykh Prep.*, 3*th*, 295 (1968).

75: 35892t   F. L. Scott, T. M. Lambe, and R. N. Butler, *Tetrahedron Lett.*, 1729 (1971).

75: 35893u   H. G. O. Becker and H. J. Timpe, *J. Prakt. Chem.*, **312,** 1112 (1970).

75: 35894v   G. Henning and F. Wolf, *Z. Chem.*, **11,** 153 (1971).

75: 35895w   F. L. Scott and J. K. O'Halloran, *J. Chem. Soc., D*, 426 (1971).

75: P36046g   E. Grigat and R. Puetter, German Patent 1,956,508 (1971).

75: 36790v   V. V. Korshak, A. L. Rusanov, E. L. Baranov, Ts. G. Iremashvili, and T. I. Bezhuashvili, *Dokl. Akad. Nauk*, **196,** 1357 (1971).

75: 41620u   C. W. Reimann and M. Zocchi, *Acta Crystallogr., Sect. B*, **27,** 682 (1971).

75: 41667q   L. Zambonelli, S. Cerrini, M. Colapietro, and R. Spagna, *J. Chem. Soc., A*, 1375 (1971).

75: 48993b   E. Brunn and R. Huisgen, *Tetrahedron Lett.*, 473 (1971).

75: P49083c   E. J. Cragoe, Jr., U.S. Patent 3,585,199 (1971).

75: P50405e   M. Ozutsumi, T. Nakajyo, and S. Maeda, German Patent 2,022,624 (1971).

75: P56785a   A. W. Wise, J. W. Gates, Jr., and D. A. J. Beavers, German Patent 2,016,083 (1970).

75: P65304v   H. Moeller, H. Bloching, and C. Werner, German Patent 1,955,431 (1971).

75: P65306x   H. Moeller, M. Dohr, H. Bloching, and C. Werner, German Patent 1,955,430 (1971).

75: P69541t   French Patent 2,037,403 (1970).

75: P69544w   G. Bach, K. W. Junge, H. Engelmann, and G. Fischer, German Patent 1,962,744 (1971).

75: 75910d   L. A. Neiman, M. M. Shemyakin, S. V. Zhukova, Y. S. Nekrasov, T. Pehk, and E. Lippmaa, *Tetrahedron*, **27,** 2811 (1971).

75: 76355p   Y. Kubota and T. Tatsuno, *Chem. Phar. Bull.*, **19,** 1226 (1971).

75: 76665q   S. Huenig, G. Kiesslich, and H. Quast, *Justus Liebigs Ann. Chem.*, **748,** 201 (1971).

75: 76692w   I. Simiti and D. Ghiran, *Farmacia* (Bucharest), **19,** 199 (1971).

75: 76693x   F. L. Scott, T. M. Lambe, and R. N. Butler, *Tetrahedron Lett.*, 2669 (1971).

75: P76791c   H. Timmler, K. H. Buechel, R. Schmidt, and L. Eue, German Patent 1,964,996 (1971).

75: P76798k   W. Draber, R. Schmidt, E. Regel, K. H. Buechel, and L. Eue, German Patent 1,964,995 (1971).

75: P76805k   J. C. Brown, German Patent 2,063,575 (1971).

75: 77857r   H. Moeller, K. Rehnelt, and M. Voss, German Patent 1,961,590 (1971).

75: P82399x   F. Nishio, R. Ohi, and T. Tajima, U.S. Patent 3,577,240 (1971).

75: 88541j   S. Husain, V. R. Srinivasan, and T. G. Surrendra, *Indian J. Chem.*, **9,** 642 (1971).

75: 88589f   T. George and R. Tahilramani, *J. Org. Chem.*, **36**, 2190 (1971).

75: P88615m   B. Boehner and K. Gubler, German Patent 2,057,170 (1971).

75: 89223u   B. Saville, *J. Chem. Soc., D*, 635 (1971).

75: 89316b   J. M. Straley and J. G. Fischer, U.S. Patent 3,585,182 (1971).

75: 93013t   D. A. Brooks, U.S. Patent 3,592,656 (1971).

75: 96997c   P. F. Periti, *Pharmacol. Res. Commun.*, **2**, 309 (1970).

75: 97926r   F. L. Scott, D. A. Cronin, and J. K. O'Halloran, *J. Chem. Soc., C*, 2769 (1971).

75: 98505q   A. Mustafa, A. H. Harhash, M. H. Elnagdi, and F. Abd El-All, *Justus Liebigs, Ann. Chem.*, **748**, 79 (1971).

75: 98508t   G. V. Boyd and A. J. H. Summers, *J. Chem. Soc., B*, 1648 (1971).

75: P98576p   N. A. Shvink, Y. A. Levin, A. V. Kazymov, and N. M. Abzalova, USSR Patent 306,128 (1971).

75: P103658y   R. J. Pollet, H. A. Philippaerts, J. F. Williams, F. Jozef, and F. H. Claes, German Patent 2,053,023 (1971).

75: 110242m   H. J. Timpe, M. DeBois, B. Fuerstenberg, and W. Herrmann, *Z. Chem.*, **11**, 258 (1971).

75: 110244p   T. Kametani, S. Shibuya, and M. Shio, *J. Heterocyclic Chem.*, **8**, 541 (1971).

75: P110320k   H. Timmler, K. H. Buechel, and M. Plempel, German Patent 2,004,697 (1971).

75: 110997m   J. Schroeder and C. W. Schellhammer, German Patent 1,955,068 (1971).

75: P114798v   H. Iwano, T. Hatano, and I. Schimamura, German Patent 2,105,397 (1971).

75: 118255p   H. Moeller, *Justus Liebigs Ann. Chem.*, **749**, 1 (1971).

75: 118258s   K. C. Joshi and V. Singh, *J. Prakt. Chem.*, **313**, 169 (1971).

75: 118283w   S. Yanagida, M. Yokoe, M. Ohoka, and S. Komori, *Bull. Chem. Soc. Japan*, **44**, 2182 (1971).

75: P118317k   J. S. Black, G. Durant, J. C. Emmett, and C. R. Ganellin, German Patent 2,053,175 (1971).

75: P119193d   J. G. Fisher and J. M. Straley, British Patent 1,236,656 (1971).

75: 129025q   A. N. Frolov, M. S. Pevzner, and L. I. Bagal, *Zh. Org. Khim*, **7**, 1519 (1971).

75: 129505w   F. Kurzer, *J. Chem. Soc., C*, 2927 (1971).

75: 129725t   V. Mozolis and S. Jokubaityte, *Liet. TSR Mokslu Akad. Darb., Ser. B*, 79 (1971).

75: P129817z   M. Matter and K. Michel, German Patent 2,047,146 (1971).

75: P130772n   R. Mohr and J. Ostermeier, U.S. Patent 3,597,412 (1971).

75: P130778u   M. Ozutsumi, K. Arakawa, M. Yamamoto, and O. Narukawa, German Patent 2,104,624 (1971).

75: 140767d   N. S. Zefirov, N. K. Chapovskaya, and V. V. Kolesnikov, *J. Chem. Soc., D*, 1001 (1971).

75: 140782e   H. Reimlinger R. Jacquier, and J. Daunis, *Chem. Ber.*, **104**, 2702 (1971).

75: 140792h   E. J. Browne, *Aust. J. Chem.*, **24**, 2389 (1971).

75: P140860d   E. Regel, K. H. Buechel, R. R. Schmidt, and L. Eue, South African Patent 70 06,389 (1971).

75: P142574n   W. L. Pevilstem and H. A. Beatty, German Patent 2,008,067 (1971).

75: 151734s   W. D. Ollis and C. A. Ramsden, *J. Chem. Soc., D*, 1224 (1971).

75: 151735t   W. D. Ollis and C. A. Ramsden, *J. Chem. Soc., D*, 1222 (1971).

75: 151737v   W. D. Ollis and C. A. Ramsden, *J. Chem. Soc., D*, 1223 (1971).

75: 152121b   E. S. Krongauz, A. V. D'yanchenko, A. P. Travnikova, and Z. O. Virpsha, *Kinet. Mech. Polyreactions, Int. Symp. Macromol Chem. Prepr.*, **5**, 37 (1969).

75: 156464p   N. M. Turkevich, L. G. Kolosova, V. N. Kulitskii, and O. I. Khoma, *Zh. Prikl. Spektrosk.*, **15**, 557 (1971).

75: 157035m   D. J. Beavers and W. F. Smith, French Patent 2,042,748 (1971).

76: 3765h   O. P. Shvaika, V. N. Artemov, and S. N. Baranov, *Zh. Org. Khim.*, **7,** 1968 (1971).

76: 3766j   T. N. Vereshchagina, K. L. Krupin, and V. A. Lopyrev, *Tr. Sev.-Zap. Zaoch. Politekh. Inst.*, **6,** 49 (1969).

76: P3866s   K. Odo and E. Ichikawa, Japanese Patent 71 35,055 (1971).

76: P3867t   K. Mizushima, M. Saheki, H. Shilina, and A. Hamada, Japanese Patent 71 35,056 (1971).

76: 4159a   A. Novacek and F. Pavel, *Collect. Czech. Chem. Commun.*, **36,** 3507 (1971).

76: P4855z   T. Shima, A. Kawase, and M. Ohshima, Japanese Patent 71 05,226 (1971).

76: 7941k   K. T. Potts, R. Armbruster, and E. Houghton, *J. Heterocyclic Chem.*, **8,** 773 (1971).

76: 8560x   B. Singh and R. Singh, *Indian J. Chem.*, **9,** 1013 (1971).

76: 10155n   P. W. Aschbacher, V. J. Feil, and R. G. Zaylskie, *J. Animal Sci.*, **33,** 638 (1971).

76: 11489m   A. Wiater, D. Hulanicka, and T. Klopotowski, *Acta Biochim. Pol.*, **18,** 289 (1971).

76: 14220c   V. S. Egorova, V. N. Ivanova, and N. I. Putokhin, in "*Khimiya*," A. G. Sarkisov (Ed), Kuibyshev, USSR, 1969, p. 140.

76: 14436c   H. J. Timpe, *Z. Chem.*, **11,** 340 (1971).

76: 14437d   H. Gehlen and H. Segeletz, *J. Prakt. Chem.*, **313,** 294 (1971).

76: 14444d   G. V. Boyd and S. R. Dando, *J. Chem. Soc. C*, 3873 (1971).

76: 14445e   G. Zinner and U. Gebhardt, *Arch. Pharm.* (Weinheim), **304,** 706 (1971).

76: 14500u   L. Capuano and H. J. Schrepfer, *Chem. Ber.*, **104,** 3039 (1971).

76: 14998n   P. M. Hergenrother, *J. Polym. Sci., Part A*-1, **9,** 2377 (1971).

76: 21123b   C. V. Deliwala, M. Y. Mhasalkar, M. H. Shah, P. D. Pilankar, S. T. Nikam, and K. G. Anantanarayanan, *J. Med. Chem.*, **14,** 1000 (1971).

76: P24217w   P. Ten Haken and B. P. Armitage, German Patent 2,117,807 (1971).

76: 25193x   Y. P. Kitaev, V. I. Savin, V. V. Zverev, and G. V. Popova, *Khim Geterotsikl. Soedin.*, **7,** 559 (1971).

76: P25296h   M. Iizuka, S. Maeda, and I. Shimmura, Japanese Patent 71 39, 341 (1971).

76: 30510c   L. A. Vlasova, E. M. Shamaeva, G. B. Afanas'eva, and I. Ya. Postovskii, *Khim.-Farm. Zh.*, **5,** 25 (1971).

76: 34177x   H. Wamhoff and P. Sohar, *Chem. Ber.*, **104,** 3510 (1971).

76: 34180t   S. Kubota, M. Uda, and M. Ohtsuka, *Chem. Pharm. Bull.*, **19,** 2331 (1971).

76: 34182v   I. L. Shegal, *Vop. Kuznechno-Pres. Liteinogo Proizvod.*, 155 (1970).

76: 34220f   V. N. Artemov, N. I. Geonya, S. N. Baranov, L. R. Kolomoitsev, and O. P. Shvaika, *Khim.-Farm. Zh.*, **5,** 6 (1971).

76: P34258z   M. Nakanishi and S. Sakuragi, Japanese Patent 71 39,861 (1971).

76: P34259a   C. R. Jacobson, A. D'Adamo, and C. E. Cosgrove, U.S. Patent 3,621,099 (1971).

76: 42580p   M. A. DeWaard, *Meded. Fac. Landbouwwetensch, Rujksuniv. Gent.*, **36,** 113 (1971).

76: 46146z   T. Kametani, S. Shibuya, and M. Shio, *J. Heterocyclic Chem.*, **8,** 889 (1971).

76: 46484h   J. Gante and G. Mohr, *Angew. Chem. Int. Ed. Engl.*, **10,** 807 (1971).

76: 58494z   S. Nicholson, G. J. Stacey, and P. J. Taylor, *J. Chem. Soc., Perkin Trans.* 2, 4 (1972).

76: 59533y   A. H. Harhash, M. H. Elnagdi, and E. A. A. Hafez, *J. Prakt. Chem.*, **313,** 706 (1971).

76: 59534z   R. C. Cookson, S. S. Gupte, I. D. R. Stevens, and C. T. Watts, *Org. Syn.*, **51,** 121 (1971).

76: 59537c   J. Daunis, Y. Guindo, R. Jacquier, and P. Viallefont, *Bull. Soc. Chim. Fr.*, 3296 (1971).

76: 59542a   O. P. Shvaika and V. I. Fomenko, *Dokl. Akad. Nauk SSSR*, **200,** 134 (1971).

76: 59576q   J. Lee and W. W. Paudler, *J. Chem. Soc., D*, 1636 (1971).

76: P59632e   W. Draber and E. Regel, German Patent 2,009,020 (1971).

76: P59636j   K. Meiser, K. H. Buechel, and M. Plemple, German Patent 2,022,206 (1971).

76: P60914e   P. Caveng, German Patent 2,111,013 (1971).

76: P60936p   French Patent 2,042,445 (1971).

76: P66326m   R. J. Pollet, H. A. Philippaerts, J. F. Willems, and F. H. Claes, Belgian Patent 758,103 (1971).

76: 67703u   D. N. Pachadzhanov and M. Yu. Yusupov, *Proc. Conf. Appl. Phys. Chem.*, *2nd*, **1,** 345 (1971).

76: 71575w   P. Bassinet, J. Pinson, and J. Armand, *C. R. Acad. Sci., Paris, Ser. C*, **274,** 189 (1972).

76: 72456v   H. Reimlinger, W. R. F. Lingier, J. J. M. Vandewalle, and R. Merenyi, *Chem. Ber.*, **104,** 3965 (1971).

76: 72459y   H. Reimlinger, W. R. F. Lingier, and J. J. M. Vandewalle, *Chem. Ber.*, **104,** 3976 (1971).

76: P72514n   G. J. Durant, J. C. Emmett, C. R. Ganelin, and G. R. White, German Patent 2,131,625 (1971).

76: P72528v   R. F. Brookes, D. H. Godson, D. Greenwood, S. B. Wakerley, and M. Tulley, German Patent 2,132,618 (1972).

76: 80560q   M. Yu. Yusupov, D. N. Pachadzhanov, and M. A. Israilov, *Dokl. Akad. Nauk Tadzh., SSR*, **14,** 31 (1971).

76: 81641d   R. J. Adamski, S. Numajiri, P. Satoru, R. Poe, L. J. Spears, and F. C. Bach, *J. Med. Chem.*, **14,** 1244 (1971).

76: 85013y   L. A. Lee and J. W. Wheeler, *J. Org. Chem.*, **37,** 348 (1972).

76: 85193g   H. Reimlinger, J. M. Gilles, G. Anthoine, J. J. M. Vandewalle, W. R. F. Lingier, E. DeRuiter, R. Merenyi, and A. Hubert, *Chem. Ber.*, **104,** 3925 (1971).

76: 85755s   H. G. O. Becker, K. Heimburger, and H. J. Timpe, *J. Prakt. Chem.*, **313,** 795 (1971).

76: 85756t   S. Hauptmann, H. Wilde, and K. Moser, *J. Prakt. Chem.*, **313,** 882 (1971).

76: 85758v   R. N. Butler, T. M. Lambe, and F. L. Scott, *J. Chem. Soc., Perkin Trans. 1*, 269 (1972).

76: 85759w   L. A. Lee, R. Evans, and J. W. Wheeler, *J. Org. Chem.*, **37,** 343 (1972).

76: 85792b   T. Hirata, L.-M. Twanmoh, H. B. Wood, Jr., A. Goldin, and J. Driscoll, *J. Heterocyclic Chem.*, **9,** 99 (1972).

76: 85796f   F. H. Case, *J. Heterocyclic Chem.*, **8,** 1043 (1971).

76: 98991r   M. Anteunis, F. Borremans, W. Tadros, A. H. A. Zaher, and S. S. Gliobrial, *J. Chem. Soc., Perkin Trans. 1*, 616 (1972).

76: 99568v   K. H. Uteg, *Pharmazie*, **26,** 735 (1971).

76: 99569w   C. L. Habraken and P. Cohen-Fernandes, *J. Chem. Soc., Chem. Commun.*, 37 (1972).

76: 99572s   A. Colautti, R. J. Ferlauto, V. Maurich, M. DeNardo, C. Nisi, F. Rubessa, and C. Runti, *Chim. Ther.*, **6,** 367 (1971).

76: 99573t   K.-Y. Zee-Cheng and C. C. Cheng, *Experientia*, **28,** 10 (1972).

76: 99614g   P. M. Hergenrother, *J. Heterocyclic Chem.*, **9,** 131 (1972).

76: P99678f   M. L. Hoefle and A. Holmes, German Patent 2,133,038 (1972).

76: P99686g   K. Heusler, H. Bickel, B. Fechtig, H. Peter, and R. Scartazzini, German Patent 2,127,287 (1971).

76: 101078f   R. L. N. Harris, J. L. Huppatz, R. M. Hoskinson, and I. M. Russell, *J. Text. Inst.*, **62,** 700 (1971).

76: 112984z   J. W. Lown and K. Matsumoto, *Can. J. Chem.*, **50,** 534 (1972).

76: 113138p   H. G. O. Becker, B. Fielitz, M. Harwart, M. Kodeih, and N. Saunder, *J. Prakt. Chem.*, **313,** 768 (1971).

76: 113183z   K. Lempert and K. Zauer, *Acta Chim.* (Budapest), **71,** 371 (1972).

76: 113185b   F. A. Neugebauer, W. Otting, H. O. Smith, and H. Trischmann, *Chem. Ber.*, **105,** 549 (1972).

76: P113218q   P. Gold-Aubert and D. Melkonian, French Patent 2,051,506 (1971).

76: P113224p   H. O. Bayer, R. S. Cook, and W. C. Von Meyer, South African Patent 70 04,373 (1971).

76: 113771q   L. Holliday and W. A. Holmes-Walker, *J. Appl. Polym. Sci.*, **16,** 139 (1972).

76: 126869c   M. Tutoveanu, E. Comanita, and A. Andrei, *Bul. Inst. Politeh. Iasi*, **17,** 89 (1971).

76: 126876c   L. A. Vlasova and I. Y. Postovskii, *Khim. Geterotsikl. Soedin.*, **7,** 700 (1971).

76: 126882b   I. Hirao, Y. Kato, and H. Tateishi, *Bull. Chem. Soc. Jap.*, **45,** 208 (1972).

76: 126949d   I. Y. Postovskii and V. A. Ershov, *Khim. Geterotsikl. Soedin*, **7,** 708 (1971).

76: 126953a   M. M. Kochhar, *J. Heterocyclic Chem.*, **9,** 153 (1972).

76: 135847w   S. Tsutsumi, Y. Sakamoto, K. Nakamura, K. Nakamura, S. Hisai, and E. Kashima, *Repura*, **39,** 33 (1970).

76: 140587r   H. H. Ong and E. L. May, *J. Org. Chem.*, **37,** 712 (1972).

76: 140655m   C. Demetrescu and I. Saramet, *Rev. Roum. Chim.*, **17,** 115 (1972).

76: 140662m   M. R. Chaurasia, *Curr. Sci.*, **41,** 142 (1972).

76: 140687y   P. DeMayo, J. B. Stothers, and M. C. Usselman, *Can. J. Chem.*, **50,** 612 (1972).

76: 140711b   B. Simon and K. H. Uteg, *Z. Chem.*, **12,** 20 (1972).

76: 140733k   F. I. Luknitskii and F. P. Shvartsman, *Khim. Geterotsikl. Soedin.*, **8,** 260 (1972).

76: 140741m   V. N. Artemov and O. P. Shvaika, *Khim. Geterotsikl. Soedin.*, **7,** 905 (1971).

76: P140832s   E. Grigat, German Patent 2,042,660 (1972).

76: 149140x   M. M. Kochhar and B. B. Williams, *J. Med. Chem.*, **15,** 332 (1972).

76: 149945p   J. E. Oliver, S. C. Chang, R. T. Brown, J. B. Stokes, and A. B. Borkovec, *J. Med. Chem.*, **15,** 315 (1972).

76: 153661k   Z. Cojocaru and C. Bicleseanu, *Rev. Med.* (Tirgu-Mures), **17,** 415 (1971).

76: 153680r   P. B. Talukdar, S. K. Sengupta, and A. K. Datta, *Indian J. Chem.*, **9,** 1417 (1971).

76: P153796h   T. Oshima, S. Hata, H. Takamatsu, M. Okuda, and M. Yokota, Japanese Patent 72 08,820 (1972).

77: 341b   R. B. Moffett, A. Rober, and L. L. Skaletzky, *J. Med. Chem.*, **14,** 963 (1971).

77: 5410h   K. Byalkovskii-Krupin and V. A. Lopyrev, *Metody Poluch. Khim. Reaktiv. Prep.*, **22,** 92 (1970).

77: P5477k   E. Regel, K. H. Buechel, R. R. Schmidt, and L. Eue, British Patent 1,269,619 (1972).

77: P5480f   J. J. Baldwin and F. Novello, German Patent 2,147,794 (1972).

77: P5482h   M. Nakanishi, S. Sakuragi, and S. Inamasu, Japanese Patent 72 07,377 (1972).

77: P5487p   J. J. Baldwin and F. C. Novello, German Patent 2,147,882 (1972).

77: P6570x   G. Minakawa and M. Akutsu, Japanese Patent 71 23,896 (1971).

77: 19615r   M. F. G. Stevens, J. Chem. Soc., *Perkin Trans.* 1, 1221 (1972).

77: 28922r   M. A. Hahn, *J. Natl. Cancer Inst.*, **48**, 783 (1972).

77: 30239s   S. Motoyama, M. Watanabe, A. Aoki, and Y. Ueda, *Sankyo Kenkyusho Nempo*, **23**, 233 (1971).

77: 34426y   V. S. Garkusha-Bozhko, S. M. Baranov, and O. P. Shvaika, *Ukr. Khim. Zh.*, **38**, 171 (1972).

77: 34430v   S. Somasekhara and N. V. Upadhyaya, *J. Indian Chem. Soc.*, **49**, 201 (1972).

77: 34431w   J. Lang and H. Tondys, *Diss. Pharm. Pharmacol.*, **24**, 59 (1972).

77: 34439e   B. Albrecht and H. Braeuniger, *Pharmazie*, **27**, 282 (1972).

77: 34469q   O. Riobe, *C. R. Acad. Sci., Paris, Ser. C*, **274**, 1462 (1972).

77: P34528h   D. Dawes and B. Boehner, German Patent 2,150,098 (1972).

77: P34529j   Y. Hasegawa and M. Dotani, Japanese Patent 72 09,580 (1972).

77: 34570r   E. J. Cragoe, Jr. and K. L. Shepard, French Patent 2,034,542 (1971).

77: P36209d   K. Chimura, K. Ito, S. Takajima, and M. Kawajima, Japanese Patent 71 27,582 (1971).

77: P36213a   K. Chimura, K. Ito, S. Takajima, and M. Kawajima, Japanese Patent 71 27,579 (1971).

77: 42777d   N. Ganchev, V. Kanazirska, and D. Atanasova, *Talanta*, **19**, 692 (1972).

77: 42829x   A. I. Busev, N. L. Shestidesyatnaya, and P. P. Kish, *Zh. Anal. Khim.*, **27**, 298 (1972).

77: 48344g   J. S. Davidson, *Chem. Ind.* (London), **11**, 464 (1972).

77: 48469b   E. Piccotti, German Patent 2,148,633 (1972).

77: P48632z   D. Dawes and B. Boehner, German Patent 2,150,169 (1972).

77: 55814p   P. G. Harrison, *J. Chem. Soc., Chem. Commun.*, 554 (1972).

77: 56420u   B. Blank, D. M. Nichols, and P. D. Vaidya, *J. Med. Chem.*, **15**, 694 (1972).

77: 61916q   G. Pifferi and P. Consonni, *J. Heterocyclic Chem.*, **9**, 581 (1972).

77: 61961a   J. Daunis, Y. Guindo, R. Jacquier, and P. Viallefont, *Bull. Soc. Chim. Fr.*, 1511 (1972).

77: P62007f   C. R. Jacobson, A. D'Adamo, and C. E. Cosgrove, U.S. Patent 3,663,564 (1972).

77: 67602f   Y. Kato and I. Hirao, *Bull. Chem. Soc. Jap.*, **45**, 1876 (1972).

77: 70181y   H. Ozawa and T. Katsuragi, *Yakugaku Zasshi*, **92**, 471 (1972).

77: 74417c   U. Svanholm, *Acta Chem. Scand.*, **26**, 459 (1972).

77: 74519n   M. S. Pevzner, V. Ya. Samarenko, and L. I. Bagal, *Khim. Geterotsikl. Soedin.*, **8**, 568 (1972).

77: 75149d   T. J. Curphey and K. S. Prasad, *J. Org. Chem.*, **37**, 2259 (1972).

77: 75176k   K. B. Wagener, S. R. Turner, and G. B. Butler, *J. Org. Chem.*, **37**, 1454 (1972).

77: 75194q   W. Skorianetz and E. Sz. Kovats, *Helv. Chim. Acta*, **55**, 1404 (1972).

77: 76506e   G. N. Antipina, L. N. Smirnova, T. I. Tikhonova, T. A. Miroshkina, Yu. B. Zimin, Ya. A. Gurvich, P. I. Levin, S. I. Burmistrov, and L. I. Limarenko, *Proizvod. Sin. Volokon*, 132 (1971).

77: P76669k   M. Ozutsumi, S. Maeda, Y. Kawada, and T. Kurahashi, Japanese Patent 72 08,468 (1972).

77: 87362f   T. A. F. O'Mahony, R. N. Butler, and F. L. Scott, *J. Chem. Soc., Perkin Trans.* 2, 1319 (1972).

77: 87502b   F. L. Scott, T. M. Lambe, and R. N. Butler, *J. Chem. Soc., Perkin Trans.* 1, 1918 (1972).

77: 87997s   R. Ahmed and J. P. Anselme, *Can. J. Chem.*, **50**, 1778 (1972).

77: 88381y   T. S. Gloria, *Rev. Farm. Bioquim. Univ. Sao Paulo*, **9**, 193 (1971).

77: 88421m   R. A. Bowie and D. A. Thomason, *J. Chem. Soc., Perkin Trans.* 1, 1842 (1972).

77: P88557k   K. Meguro, U. Kuwada, and H. Tawada, German Patent 2,153,519 (1972).

77: 90026e   E. Balta, *Lucr. Conf. Nat. Chim. Anal.,* 3rd, **4,** 145 (1971).

77: P90037j   P. Moser and V. Ramanathan, German Patent 2,147,809 (1972).

77: P90077x   G. Boehmke and H. Theidel, German Patent 2,050,726 (1972).

77: 101459b   D. Matthies, *Synthesis,* 380 (1972).

77: 101467c   K. K. Mayer, F. Schroeppel, and J. Sauer, *Tetrahedron Lett.,* 2899 (1972).

77: 101468d   R. Grashey and M. Baumann, *Tetrahedron Lett.,* 2947 (1972).

77: 101472a   R. Grashey, M. Baumann, and R. Hamprecht, *Tetrahedron Lett.,* 2939 (1972).

77: 101473b   R. Grashey, R. Hamprecht, N. Keramaris, and M. Baumann, *Tetrahedron Lett.,* 2943 (1972).

77: 101474c   M. Look and L. R. White, *J. Agr. Food Chem.,* **20,** 824 (1972).

77: 101475d   F. L. Scott, J. K. O'Halloran, J. O'Driscoll, and A. F. Hegarty, *J. Chem. Soc., Perkin Trans.* 1, 2224 (1972).

77: 101478g   J. K. O'Halloran and F. L. Scott, *J. Chem. Soc., Perkin Trans.* 1, 2219 (1972).

77: P101690v   A. Gagneux, R. Heckendorn, and R. Meier, German Patent 2,159,527 (1972).

77: P101899v   K. H. Buechel, B. Hamburger, H. Klesper, W. Paulus, and O. Pauli, German Patent 2,056,652 (1972).

77: P103304w   M. Kotei, S. Maeda, and T. Nakajo, Japanese Patent 72 15,475 (1972).

77: 109080s   M. Yu. Yusupov, M. A. Israilov, D. N. Pachadzhanov, and A. V. Vakhobov, *Izv. Akad. Nauk Tadzh. SSR, Otd. Fiz.-Mat. Geol.-Khim. Nauk,* 60 (3), (1971).

77: 109606t   P. W. Aschbacher, V. J. Feil, and R. G. Zaylskie, *J. Animal Sci.,* **35,** 97 (1972).

77: 113608n   J. W. Lown and M. H. Akhtar, *Can. J. Chem.,* **50,** 2236 (1972).

77: 113628u   D. R. Eaton, R. E. Benson, C. G. Bottomley, and A. D. Josey, *J. Amer. Chem. Soc.,* **94,** 5996 (1972).

77: 114143u   M. D. Lipanova, A. I. Gontar, and V. V. Baksova, *Issled. Obl. Geterotsikl. Soedin.,* 22 (1971).

77: 114312y   C. Demetrescu, *Rev. Roum. Chim.,* **17,** 1013 (1972).

77: 114314a   R. Kraft and H. Paul, *Monatsh. Chem.,* **103,** 156 (1972).

77: 114318e   D. J. Pasto and A. Fu-Tai Chen, *Tetrahedron Lett.,* 2995 (1972).

77: 114397e   K. Heusler, H. Bickel, B. Fechtig, H. Peter, and R. Scartazzini, South African Patent 71 02,523 (1971).

77: P115048d   A. Dobbelstein, H. D. Hille, and H. Holfort, German Patent 1,937,310 (1972).

77: P115460g   M. Minagawa, M. Akutsu, and K. Nakagawa, German Patent 2,164,234 (1972).

77: 119903b   S. Emori, M. Inoue, and M. Kubo, *Bull. Chem. Soc. Jap.,* **45,** 2259 (1972).

77: 121989c   G. Niebch and F. Schneider, *Z. Naturforsch. B,* **27,** 675 (1972).

77: 126492v   S. Kano, *Yakugaku Zasshi,* **92,** 935 (1972).

77: 126508e   G. Wagner and S. Leistner, *Z. Chem.,* **12,** 175 (1972)

77: 126510z   R. W. Hoffmann and W. Schaefer, *Chem. Ber.,* **105,** 2437 (1972).

77: 126515e   J. P. Critchley and J. S. Pippet, *J. Fluorine Chem.,* **2,** 137 (1972).

77: 126518h   F. Russo, M. Santagati, and G. Pappalardo, *Ann. Chim.* (Rome), **62,** 351 (1972).

77: 126522e   C. Demetrescu and V. Manu, *Pharmazie,* **27,** 439 (1972).

77: 126523f   A. K. Pan'kov, M. S. Pevzner, and L. I. Bagal, *Khim. Geterotsikl. Soedin.,* **8,** 713 (1972).

77: 126568z   V. Ambrogi, K. Bloch, S. Daturi, W. Logemann, and M. A. Parenti, *J. Pharm. Sci.*, **61**, 1483 (1972).

77: 126637w   D. Dawes and B. Boehner, German Patent 2,200,436 (1972).

77: 126641t   H. G. O. Becker and T. Roethling, (East) German Patent 89,617 (1972).

77: P126707u   J. B. Hester, Jr., U.S. Patent 3,681,343 (1972).

77: P126708v   J. B. Hester, Jr., German Patent 2,203,782 (1972).

77: 133193b   G. Bach, H. Engelmann, G. Fischer, and K. W. Junge, British Patent 1,275,701 (1972).

77: 135025x   S. S. Yang, F. M. Herrera, R. G. Smith, M. Reitz, G. Lancini, R. C. Ting, and R. C. Gallo, *J. Natl. Cancer Inst.*, **49**, 7 (1972).

77: 139906h   H. G. O. Becker, G. Goermar, H. Haufe, and H. J. Timpe, *J. Prakt. Chem.*, **314**, 101 (1972).

77: 139907j   A. Ya. Lazaris and A. N. Egorochkin, *Zh. Org. Khim.*, **8**, 1538 (1972).

77: 139909m   H. J. Timpe and H. G. O. Becker, *J. Prakt. Chem.*, **314**, 325 (1972).

77: 139992h   M. Zupan, B. Stanovnik, and M. Tisler, *J. Org. Chem.*, **37**, 2960 (1972).

77: P140086x   J. C. Brown and P. J. Keogh, German Patent 2,162,855 (1972).

77: P140089a   D. Dawes and B. Boehner, German Patent 2,200,467 (1972).

77: 140109g   G. L. Dunn and J. R. E. Hoover, German Patent 2,162,575 (1972).

77: 146070j   B. L. Davydov, V. F. Zolin, L. G. Koreneva, and M. A. Somakhina, *Zh. Prikl. Spektrosk.*, **17**, 413 (1972).

77: P146189e   J. C. Brown, D. J. Fry, and P. J. Keogh, German Patent 2,162,856 (1972).

77: P146201c   J. Matsuyama, Y. Nakazawa, Y. Nakamura, R. Ohi, and T. Kondo, German Patent 2,160,972 (1972).

77: 147663s   S. S. Parmar, A. K. Gupta, H. H. Singh, and T. K. Gupta, *J. Med. Chem.*, **15**, 999 (1972).

77: 147920y   C. Gurgo, R. Ray, and M. Green, *J. Natl. Cancer Inst.*, **49**, 61 (1972).

77: 152042t   L. M. Shegal and I. L. Shegal, *Khim. Geterotsikl. Soedin*, **8**, 1062 (1972).

77: 152066d   M. Komatsu, Y. Ohshiro, and T. Agawa, *J. Org. Chen.*, **37**, 3192 (1972).

77: 152067e   K. Bast, M. Christl, R. Huisgen, and W. Mack, *Chem. Ber.*, **105**, 2825 (1972).

77: 152070a   H. Blaschke, E. Brunn, R. Huisgen, and W. Mack, *Chem. Ber.*, **105**, 2841 (1972).

77: 152071b   T. Kauffmann, J. Legler, E. Ludorff, and H. Fischer, *Angew. Chem. Int. Ed. Engl.*, **11**, 846 (1972).

77: 152103p   M. H. Nosseir, N. N. Messiha, and G. G. Gabra, *U.A.R.J. Chem.*, **13**, 379 (1970).

77: 152130v   R. A. Bowie, M. D. Gardner, D. G. Neilson, K. M. Watson, S. Mahmood, and V. Ridd., *J. Chem. Soc., Perkin Trans. 1*, 2395 (1972).

77: 152186t   W. Buck and R. Sehring, German Patent 2,107,146 (1972).

77: P152192s   B. Boehner and K. Gubler, U.S. Patent 3,689,500 (1972).

77: P153755q   S. Matsuda and H. Ohtake, Japanese Patent 72 19,599 (1972).

77: 157129m   T. W. McGaughy and B. M. Fung, *Inorg. Chem.*, **11**, 2728 (1972).

77: 158352r   B. Singh and R. Singh, *J. Inorg. Nucl. Chem.*, **34**, 3449 (1972).

77: 159501a   A. A. Vafina, A. V. Il'yasov, I. D. Morozova, I. M. Skrebkova, and Y. P. Kitaev, *Izu. Akad. Nauk. SSR, Ser. Khim.*, 1731 (1972).

77: 164587n   T. Bany and M. Dobosz, *Rocz. Chem.*, **46**, 1123 (1972).

77: 164601n   M. I. Ali, A. A. El-Sayed, and A.-E. M. Abd-Elfattah, *J. Org. Chem.*, **37**, 3209 (1972).

77: 164610q   K. H. Uteg, *Z. Chem.*, **12**, 291 (1972).

77: P164704y   G. J. Durant, J. C. Emmett, and C. R. Ganellin, German Patent 2,211,454 (1972).

77: P164709d   H. Moeller, German Patent 2,113,731 (1972).

77: P164710x   M. L. Hoefle and A. Holmes, French Patent 2,100,863 (1972).

77: P164711y   W. Gauss and S. Petersen, German Patent 2,106,845 (1972).

77: P164713a   W. Gauss and S. Petersen, German Patent 2,106,845 (1972).
77: P166059q   J. J. Leucker and H. Mollet, German Patent 2,161,177 (1972).
77: P166993r   V. C. Lintzenich, German Patent 2,206,933 (1972).
77: P171214s   W. Gauss, W. Mueller-Bardoff, A. Von Koening, F. Moll, and W. Saleck, German Patent 2,042,533 (1972).
78: P4136g   K. Takamura, Japanese Patent 72 34,712 (1972).
78: 4193y   H. Schildknecht and G. Renner, *Justus Liebigs Ann. Chem.*, **761,** 189 (1972).
78: 4224j   H. Kohn and R. A. Olofson, *J. Org. Chem.*, **37,** 3504 (1972).
78: P4261u   B. Boehner and D. Dawes, German Patent 2,202,471 (1972).
78: P4286f   P. G. Hughes and J. P. Verge, German Patent 2,213,558 (1972).
78: P4289j   A. Gagneux, R. Heckendorn, and R. Meier, German Patent 2,215,943 (1972).
78: 12030h   J. T. Witkowski, R. K. Robins, R. W. Sidwell, and L. N. Simon, *J. Med. Chem.*, **15,** 1150 (1972).
78: 15125d   U. Svanholm, *Ann. N.Y. Acad. Sci.*, **192,** 124 (1972).
78: 15224k   A. Chinone and M. Ohta, *Chem. Lett.*, 969 (1972).
78: 16143v   M. G. Luchina, V. A. Lopyrev, and K. L. Byalkovskii-Krupin, *Metody Poluch. Khim. Reaktiv. Prep.*, **23,** 75 (1971).
78: P16189q   H. Uoda and T. Kosugi, Japanese Patent 72 37,435 (1972).
78: P17606k   R. Mohr and J. Ostermeier, U.S. Patent 3,635,942 (1972).
78: P22526p   M. J. Bevis, H. J. Bello, C. F. Holtz, and W. S. Lane, German Patent 2,163,665 (1972).
78: P28517h   H. H. Scott, U.S. Patent 3,701,645 (1972).
78: 28599m   K. H. Uteg, *Z. Chem.*, **12,** 328 (1972).
78: 29344m   R. Ahmed and J. P. Anselme, *Tetrahedron*, **28,** 4939 (1972).
78: 29676w   J. A. Lenoir, L. D. Colebrook, and D. F. Williams, *Can. J. Chem.*, **50,** 2661 (1972).
78: 29701a   A. Bezeg, B. Stanovnik, B. Sket, and M. Tisler, *J. Heterocyclic Chem.*, **9,** 1171 (1972).
78: P29774b   S. A. Greenfield, M. C. Seidel, and W. C. Von Meyer, German Patent 1,966,473 (1972).
78: P29776d   A. Kotone, M. Hoda, T. Hori, and Y. Nakane, Japanese Patent 72 31,982 (1972).
78: 29956n   S. Kozima, T. Itano, N. Mihara, K. Sisido, and T. Isida, *J. Organometal. Chem.*, **44,** 117 (1972).
78: P30495t   H. E. Kuenzel, F. Bentz, G. Lorenz, and G. Nischk, British Patent 1,238,511 (1971).
78: P36245h   British Patent 1,297,473 (1972).
78: P36272q   H. Hayashi, R. Ohi, and T. Shishido, German Patent 1,148,667 (1972).
78: 43372a   A. A. Tsurkan, V. I. Efremenko, and V. V. Groshev, *Farm. Zh.* (Kiev), **27,** 72 (1972).
78: 43384f   H. J. Timpe, *Z. Chem.*, **12,** 333 (1972).
78: 43419w   E. Ajello, O. Migliara, and V. Spiro, *J. Heterocyclic Chem.*, **9,** 1169 (1972).
78: 43610b   N. I. Shvetsov-Shilovskii, N. P. Ignatova, R. G. Bobkova, V. Y. Man-yukhina, and N. N. Mel'nikov, *Zh. Obshch. Khim.*, **42,** 1939 (1972).
78: P57044g   H. Matarai and H. Iezaka, Japanese Patent 72 04,963 (1972).
78: 58320f   C. H. Budeanu and A. Gavriliuc, *An. Stiint. Univ. "Al. I. Cuza" Iasi*, Sect. 1c, **18,** 183 (1972).
78: 58327p   V. A. Kovtunenko and F. S. Babichev, *Ukr. Khim. Zh.*, **38,** 1142 (1972).
78: 66590j   A. V. Radushev and E. N. Prokhorenko, *Zh. Anal. Khim.*, **27,** 2209 (1972).
78: 71045q   E. M. Burgess and J. P. Sanchez, *J. Org. Chem.*, **38,** 176 (1973).

78: 72011n   A. Kreutzberger and B. Meyer, *Chem. Ber.*, **105,** 3974 (1972).

78: 72014r   S. Kubota and M. Uda, *Chem. Pharm. Bull.*, **20,** 2096 (1972).

78: 72015s   A. Ya. Lazaris, S. M. Shmuilovich, and A. N. Egorochkin, *Zh. Org. Khim.*, **8,** 2621 (1972).

78: 72019w   M. A. Khan and J. B. Polya, *Rev. Latinoamer. Quim.*, **3,** 119 (1972).

78: 72075m   S. Somasekhara, R. K. Thakkar, and G. F. Shah, *J. Indian Chem. Soc.*, **49,** 1057 (1972).

78: P72154m   G. Henning, H. Weinelt, and F. Wolf, (East) German Patent 84,388 (1971).

78: P72161m   D. Guenther, German Patent 2,126,839 (1972).

78: P72232k   J. B. Hester, German Patent 2,221,790 (1972).

78: 72719z   K. B. Wagener, S. R. Turner, and G. B. Butler, *J. Polym. Sci., Polym. Lett. Ed.*, **10,** 805 (1972).

78: P73641c   H. Noack, W. Schroth, W. Guddat, and M. Zeiler, (East) German Patent 87,108 (1972).

78: 77189m   A. J. Blackman and J. H. Bowie, *Org. Mass Spectrom.*, **7,** 57 (1973).

78: 84319g   P. Thieme, M. Patsch, and H. Konig, *Justus Liebigs Ann. Chem.*, **764,** 94 (1972).

78: 84321b   W. Gauss, H. Heitzer, and S. Petersen, *Justus Liebigs Ann. Chem.*, **764,** 131 (1972).

78: 84327h   H. G. O. Becker and R. Ebisch, *J. Prakt. Chem.*, **314,** 923 (1972).

78: 84378a   D. G. Neilson, S. Mahmood, and K. M. Watson, *J. Chem. Soc., Perkin Trans. 1*, 335 (1973).

78: P84766a   J. T. Witkowski and R. Robins, German Patent 2,220,246 (1972).

78: 92134h   N. P. Ganchev and D. Atanasova, *Dokl. Bolg. Akad. Nauk*, **25,** 1669 (1972).

78: 92411w   A. Dauvarte, A. Zidermane, R. Kalnbergs, M. Lidaks, K. Venters, N. M. Sukhova, R. Verpele, and S. Hillers, in *Protivoopukholevye Soedin.*, 5-Nitrofuranovogo Ryada, S. A. Giller (ed), Zinatne, Riga, USSR, 1972, p 17.

78: 92412x   A. Dauvarte, A. Zidermane, R. Kalnbergs, M. Lidaks, B. Kurgane, N. M. Sukhova, and S. Hillers, *Protivoopukholevye Soedin. 5-Nitrofuranovoga Ryada*, S. A. Giller (ed), Zinatne, Riga, USSR, 1972, p 33.

78: 96707v   K. T. Potts, R. Armbruster, E. Houghton, and J. Kane, *Org. Mass. Spectrom.*, **7,** 203 (1973).

78: 97563g   Z. Janousek and H. G. Viehe, *Angew. Chem.*, **85,** 90 (1973).

78: 97565j   P. B. Talukdar, S. K. Sengupta, and A. K. Datta, *Indian J. Chem.*, **10,** 1070 (1972).

78: 97680t   A. W. J. Chow, G. L. Dunn, J. R. E. Hoover, and J. A. Weisbach, German Patent 2,230,316 (1973).

78: P98971a   E. N. Abrahart and H. R. Bolliger, German Patent 2,204,725 (1973).

78: 105462r   L. A. Mineeva and A. S. Babenko, *Izv. Vyssh. Ucheb. Zaved., Khim. Khim. Tekhnol,* **16,** 13 (1973).

78: 111227j   J. S. Davidson, *J. Prakt. Chem.*, **314,** 663 (1972).

78: P112683s   A. Horn, G. Roesch, and G. Schnabel, German Patent 2,133,159 (1973).

78: 120168t   R. L. Fye, C. W. Woods, A. B. Borkovec, and P. H. Terry, *J. Econ. Entomol.*, **66,** 38 (1973).

78: 123762t   I. B. Lundina, L. E. Deev, E. G. Kovalev, and I. Ya. Postovskii, *Khim. Geterotsikl. Soedin.*, **9,** 285 (1973).

78: 124509c   K. Koyano and C. R. McArthur, *Can. J. Chem.*, **51,** 333 (1973).

78: 124510w   H. G. O. Becker, D. Nagel, and H. J. Timpe, *J. Prakt. Chem.*, **315,** 97 (1973).

78: 124513z   J. L. Barascut, J. Daunis, and R. Jacquier, *Bull. Soc. Chim. Fr.*, 323 (1973).

78: P124642r A. Gagneux, R. Heckendorn, and R. Meier, German Patent 2,234,620 (1973).

78: P124646v A. Gagneux, R. Heckendorn, and R. Meier, German Patent 2,234,652 (1973).

78: P130578r A. Baldassarri and L. Cellone, German Patent 2,212,550 (1972).

78: 130586s A. Von Koenig, H. Kampfer, E. M. Brinckman, and F. C. Heugebaert, German Patent 2,140,462 (1973).

78: 131702p D. N. Pachadzhanov, M. A. Israilov, M. Yu. Yusupov, and A. Tabarov, *Izv. Akad. Nauk Tadzh. SSR, Otd. Fiz.-Mat. Geol.-Khim. Nauk,* 42 (1972).

78: 135601c G. Zinner and G. Isensee, *Chem.-Ztg.,* **97,** 73 (1973).

78: P135625p M. Okada and Y. Asami, Japanese Patent 73 08,706 (1971).

78: 136152u A. H. Harhash, M. H. Elnagdi, and Ch. A. S. Elsannib, *J. Prakt. Chem.,* **315,** 211 (1973).

78: P136177f J. E. Oliver and P. E. Sonnet, *J. Org. Chem.,* **38,** 1437 (1973).

78: 136178g E. Bulka and D. Ehlers, *J. Prakt. Chem.,* **315,** 155 (1973).

78: 136179h A. Pollak, S. Polanc, B. Stanovnik, and M. Tisler, *Monatsh. Chem.,* **103,** 1591 (1972).

78: 136235y E. Ajello and C. Arnone, *J. Heterocyclic Chem.,* **10,** 103 (1973).

78: 136237a R. Bokaldere and A. Liepina, *Khim. Geterotsikl. Soedin.,* **9,** 276 (1973).

78: P136290n J. W. Black, G. J. Durant, J. C. Emmett, and C. R. Ganellin, British Patent 1,305,549 (1973).

78: P136302t M. Gall, German Patent 2,240,043 (1973).

78: P136305w A. Kotone and M. Hoda, Japanese Patent 73 19,576 (1973).

78: P136354m J. B. Hester, Jr., M. Gall, German Patent 2,240,198 (1973).

78: 147403c D. J. Pasto and A. Fu-Tai Chen., *Tetrahedron Lett.,* 713 (1973).

78: 147712c K. I. Vakhreeva, A. E. Lipkin, T. B. Ryskina, and N. I. Skachkova, *Khim.-Farm. Zh.,* **7,** 24 (1973).

78: 147879n I. Arai, S. Abe, and A. Hagitani, *Bull. Chem. Soc. Jap.,* **46,** 677 (1973).

78: 147898t R. Bokaldere and A. Liepina, *Khim. Geterotsikl. Soedin,* **9,** 423 (1973).

78: 147921r R. Bokaldere, A. Liepina, I. Mazeika, I. S. Yankovskaya, and E. Liepins, *Khim. Geterotsikl. Soedin,* **9,** 419 (1973).

78: P147974q M. Minagawa and M. Akutsu, Japanese Patent 73 08,779 (1973).

78: 148008q K. Dickore, R. Wegler, and G. Hermann, U.S. Patent 3,686,301 (1972).

78: P148956x M. Wiesel and G. Wolfrum, German Patent 2,141,987 (1973).

78: P148964y K. Mix, E. Mundlos, and R. Mohr, German Patent 2,142,565 (1973).

78: 158728p B. M. Jacobson, *J. Amer. Chem. Soc.,* **95,** 2579 (1973).

78: 159355b F. A. Trofimov, N. G. Tsyshkova, and A. N. Grinev, *Khim. Geterotsikl. Soedin.,* **9,** 308 (1973).

78: 159527j T. Isida, T. Akiyama, N. Mihara, S. Kozima, and K. Sisido, *Bull. Chem. Soc. Jap.,* **46,** 1250 (1973).

78: 159528k V. J. Ram and H. N. Pandey, *Agr. Biol. Chem.,* **37,** 575 (1973).

78: 159619r A. Omodei-Sale, G. Pifferi, P. Consonni, A. Diena, and B. Rosselli del Turco, German Patent 2,235,544 (1973).

78: P159694m J. B. Hester, Jr., German Patent 2,242,918 (1973).

78: P159868w K. H. Buechel and I. Hammann, German Patent 2,143,252 (1973).

78: P167001y P. W. Jenkins, U.S. Patent 3,718,477 (1973).

79: 5280g D. Mathies and R. Wolff, *Pharm. Acta Helv.,* **48,** 44 (1973).

79: 5298u T. Bany and M. Dobosz, *Ann. Univ. Mariae Curie-Sklodowska, Sect. AA,* **26/27,** 359 (1971).

79: 5307w T. Okabe, E. Taniguchi, and K. Maekawa, *Agr. Biol. Chem.,* **37,** 441 (1973).

79: 5460r W. J. Chambers, German Patent 2,245,447 (1973).

79: 5560y G. Bram, *Tetrahedron Lett.,* 469 (1973).

79: P6749d   M. Ozutsumi, S. Maeda, and T. Kurahashi, Japanese Patent 72 14,350 (1972).

79: 18620t   T. Bany and M. Dobosz, *Ann. Univ. Mariae Curie-Sklodowska Sect. AA,* **26/27,** 17 (1971).

79: 18625y   T. Bany and M. Dobosz, *Ann. Univ. Mariae Curie-Sklodowska, Sect. AA,* **26/27,** 23 (1971).

79: 18648h   G. Rembarz, E. Fischer, and F. Tittelbach, *Wiss. Z. Univ. Rostock, Math.-Naturwiss. Reihe,* **21,** 119 (1972).

79: 18651d   T. Bany and M. Dobosz, *Ann. Univ. Mariae Curie-Sklodowska, Sect. AA,* **26/27,** 289 (1971).

79: 18723d   D. Dawes and B. Boehner, German Patent 2,251,074 (1973).

79: P19662v   M. Minagawa and M. Akutsu, Japanese Patent 73 08,667 (1973).

79: 20277e   E. Balta, *Chim. Anal.* (Bucharest), **2,** 262 (1972).

79: P20314q   S. Maeda, M. Kozutsumi, T. Kurahashi, and I. Niimura, Japanese Patent 73 15,881 (1973).

79: 25270g   B. Singh, R. Singh, R. V. Chaudhary, and K. P. Thakur, *Indian J. Chem.,* **11,** 174 (1973).

79: P25673r   G. M. Dappen, French Patent 2,106,053 (1972).

79: P28414t   D. Dawes and B. Boehner, Swiss Patent 533,949 (1973).

79: P28415u   D. Dawes and B. Boehner, Swiss Patent 533,948 (1973).

79: 31989q   J. Elguero and A. J. H. Summers, *An. Quim.,* **69,** 411 (1973).

79: 31993m   A. D. Sinegibskaya, E. G. Kovalev, and I. Ya. Postovskii, *Khim. Geterotsikl. Soedin.,* **9,** 562 (1973).

79: P32461y   A. Kotone, T. Hori, and M. Hoda, Japanese Patent 72 35,099 (1972).

79: 38542x   P. N. Kulyabko, R. G. Dubenko, and V. D. Konysheva, *Fiziol. Aktiv. Veshchestva,* **4,** 87 (1972).

79: 38876j   D. G. Streeter, J. T. Witkowski, G. P. Khare, R. B. Sidwell, R. J. Bauer, R. K. Robins, and L. N. Simon, *Proc. Nat. Acad. Sci. U.S.,* **70,** 1174 (1973).

79: 42421z   T. Hirata, H. B. Wood, Jr., and J. S. Driscoll, *J. Chem. Soc., Perkin Trans. 1,* 1209 (1973).

79: 42423b   M. V. Konher, *Indian J. Chem.,* **11,** 321 (1973).

79: P42513f   D. Dawes and B. Boehner, German Patent 2,251,075 (1973).

79: P42515h   D. Dawes and B. Boehner, German Patent 2,251,096 (1973).

79: P43665u   M. A. Weaver and C. A. Coates, Jr., German Patent 2,216,391 (1973).

79: P47838a   R. Moraw and G. Schaedlich, German Patent 2,154,313 (1973).

79: 53234s   S. M. A. Hai and W. Lwowski, *J. Org. Chem.,* **38,** 2442 (1973).

79: P53334z   B. Boehner, D. Dawes, and W. Meyer, German Patent 2,259,983 (1973).

79: P53343b   J. R. E. Hoover, German Patent 2,259,776 (1973).

79: 54855g   J. B. Polya and M. Woodruff, *Aust. J. Chem.,* **26,** 1585 (1973).

79: P59979k   W. Gauss, T. H. Gys, A. Von Koenig, F. Moll, W. Mueller-Bardorff, H. Philipaerts, R. Pollet, and W. Saleck, German Patent 2,161,045 (1973).

79: 61551h   E. E. Chebotarev, P. N. Kulyabko, V. A. Kuz'menko, and V. I. Fedorchenko, *Radiobiologiya,* **13,** 197 (1973).

79: 61983a   M. Kurita, T. Teraji, Y. Saito, H. Harada, K. Hattori, T. Kamiya, M. Nishida, and T. Takano, *Advan. Antimicrob. Antineoplastic Chemother., Proc. Int. Congr. Chemother., 7th,* **1,** 1055 (1971).

79: 66062j   P. L. Watson and R. N. Warrener, *Aust. J. Chem.,* **26,** 1725 (1973).

79: 66253x   L. P. Makhno, E. S. Domnina, and G. G. Skvortsova, *Dokl. Vses. Konf. Khim. Atsetilena, 4th,* **1,** 493 (1972).

79: 66255z   G. Just and G. Reader, *Tetrahedron Lett.,* 1525 (1973).

79: 66260x   C. Derycke, V. Jaeger, J. P. Putzeys, M. Van Meerssche, and H. G. Viehe, *Angew. Chem.,* **85,** 447 (1973).

79: 66285j   F. A. Neugebauer and H. Fischer, *Chem. Ber.,* **106,** 1589 (1973).

79: 66307t   E. J. Browne, *Aust. J. Chem.*, **26,** 1809 (1973).

79: 66315u   F. H. Case, *J. Heterocyclic Chem.*, **10,** 353 (1973).

79: P66362g   B. Boehner, D. Dawes, and W. Meyer, German Patent 2,259,960 (1973).

79: P66363h   B. Boehner, D. Dawes, and W. Meyer, German Patent 2,259,974 (1973).

79: P66364j   D. Dawes and B. Boehner, German Patent 2,260,015 (1973).

79: 67690z   E. Warno and M. Nowacki, *Proc. Inst. Przem. Org.*, **3,** 87 (1971).

79: P74954s   F. Suzuki, Y. Mera, F. Motohashi, M. Hayashi, and Y. Iwabuchi, Japanese Patent 73 26,935 (1973).

79: P78436x   E. Gigat and R. Puetter, German Patent 1,568,627 (1973).

79: 78692c   Zh. N. Fidler, M. G. Luchina, V. A. Lopyrev, and A. A. Stotskii, *Zh. Org. Khim.*, **9,** 1205 (1973).

79: 78693d   K. A. Nuridzhanyan and G. V. Kuznetsova, *Zh. Org. Khim.*, **9,** 1171 (1973).

79: 78696g   W. Walter and M. Radke, *Justus Liebigs Ann. Chem.*, 636 (1973).

79: 78698j   R. N. Butler, T. M. Lambe, J. C. Tobin, and F. L. Scott, *J. Chem. Soc., Perkin Trans. I*, 1357 (1973).

79: 78701e   P. B. Talukdar, S. K. Sengupta, and A. K. Datta, *J. Indian Chem. Soc.*, **50,** 154 (1973).

79: 78749b   S. Kubota and M. Oda, *Chem. Pharm. Bull.*, **21,** 1342 (1973).

79: P78808v   J. B. Hester, Jr.. German Patent 2,257,627 (1973).

79: P78811r   J. B. Hester, Jr., U.S. Patent 3,749,733 (1973).

79: P78824x   H. Bickel, R. Bosshardt, and J. Mueller, Swiss Patent 535,261 (1973).

79: 78965u   K. H. Buchel, B. Hamburger, H. Klesper, W. Paulus, and O. Pauli, British Patent 1,319,889 (1973).

79: 80297w   B. Klimek, M. Nowacki, and L. Szuster, *Pr. Inst. Przem. Org.*, **3,** 75 (1971).

79: 87327q   E. B. Akerblom and D. E. S. Campbell, *J. Med. Chem.*, **16,** 312 (1973).

79: 87336s   J. G. Lombardino, E. H. Wiseman, and J. Chiaini, *J. Med. Chem.*, **16,** 493 (1973).

79: 91273m   L. A. Aleksandrova, A. A. Kraevskii, A. A. Arutyunyan, V. A. Spivak, and B. P. Gottikh, *Izv. Akad. Nauk SSSR, Ser. Khim.*, 1321 (1973).

79: 91736q   O. Tsuge and S. Kanemasa, *J. Org. Chem.*, **38,** 2972 (1973).

79: 91739t   M. Tutoveanu, E. Comanita, and S. Steinberg, *Bul. Inst. Politeh. Iasi*, **18,** 113 (1972).

79: 92107x   N. N. Vereshchagina, G. S. Melkozerova, N. N. Frolova, A. V. Bedrin, and I. Y. Postovskii, *Khim.-Farm. Zh.*, **7,** 18 (1973).

79: 92113w   I. Arai, *Bull. Chem. Soc. Jap.*, **46,** 2215 (1973).

79: 92120w   M. Tutoveanu, *Bul. Inst. Politeh. Iasi*, **18,** 103 (1972).

79: 92159r   H. Golgolab, I. Lalezari, and L. Hosseini-Gohari, *J. Heterocyclic Chem.*, **10,** 387 (1973).

79: 92176u   K. Meguro, H. Tawada, and Y. Kuwada, *Chem. Pharm. Bull.*, **21,** 1619 (1973)

79: P92234m   M. Gall, U.S. Patent 3,748,339 (1973).

79: P92235n   German Patent 2,264,159 (1973).

79: 93420f   V. A. Gnedina, R. P. Smirnov, and G. A. Matyushin, *Izu. Vyssh. Ucheb. Zaved., Khim. Khim. Tekhnol.*, **16,** 589 (1973).

79: P93432m   K. Ohkawa, Y. Takeda, and K. Hirabayashi, Japanese Patent 73 18,316 (1973).

79: 98822y   A. Parentich, L. H. Little, and R. H. Ottewill, *J. Inorg. Nucl. Chem.*, **35,** 2271 (1973).

79: 104394d   R. N. Butler, *Can. J. Chem.*, **51,** 2315 (1973).

79: 104400c   H. J. Timpe, *J. Prakt. Chem.*, **315,** 775 (1973).

79: 105150h   H. G. O. Becker and V. Eisenschmidt, *J. Prakt. Chem.*, **315,** 640 (1973).

79: 105154n   J. L. Barascut, R. M. Claramunt, and J. Elguero, *Bull. Soc. Chim. Fr.*, 1849 (1973).

79: 105155p   A. Gasco, V. Mortarini, and E. Reynaud, *Farmaco, Ed. Sci.*, **28,** 624 (1973).

79: P105257y   W. Meiser, K. H. Buechel, W. Kraemer, and F. Grewe, German Patent 2,201,063 (1973).

79: P106129v   S. Maeda, M. Ozutsumi, T. Kurahashi, and I. Niimura, Japanese Patent 73 28,529 (1973).

79: 115418r   E. Sato, T. Chiba, and Y. Takada, *Hokkaido Daigaku Kogakubu Kenkyu Hokoku*, 87 (1973).

79: 115498s   J. E. Herweh and R. M. Fantazier, *Tetrahedron Lett.*, 2101 (1973).

79: 115499t   H. G. O. Becker, H. D. Steinleitner, and H. J. Timpe, *Synthesis*, 414 (1973).

79: 115500m   A. A. Aroyan and N. S. Iradyan, *Arm. Khim. Zh.*, **26,** 499 (1973).

79: 115502p   O. P. Shvaika, V. N. Artemov, V. E. Kononenko, and S. N. Baranov, *Khim. Geterotsikl. Soedin.*, **9,** 930 (1973)

79: 126378d   A. Ya. Lazaris, A. N. Egorochkin, S. M. Shmuilovich, and A. I. Bunov, *Khim. Geterotsikl. Soedin.*, **9,** 1136 (1973).

79: 126404j   Z. Cojocaru, C. Nistor, E. Chindris, and D. Ghiran, *Farmaco, Ed. Sci.*, **28,** 691 (1973).

79: 126446z   L. M. Werbel, E. F. Elslager, and V. P. Chu, *J. Heterocyclic Chem.*, **10,** 631 (1973).

79: P126535c   A. Gagneux, R. Heckendorn, and R. Meier, German Patent 2,304,307 (1973).

79: 126839e   N. V. Karyakin, A. N. Mochalov, V. N. Sapozhnikov, and I. B. Rabinovich, *Tr. Khim. Khim. Tekhnol.*, 134 (1972).

79: 136576f   E. J. Corey and B. B. Snider, *Tetrahedron Lett.*, 3091 (1973).

79: 137008j   A. Nabeya, Y. Tamura, I. Kodama, and Y. Iwakura, *J. Org. Chem.*, **38,** 3758 (1973).

79: 137155e   J. W. Black, G. J. Durant, J. C. Emmett, and C. R. Ganelin, U.S. Patent 3,759,944 (1973).

79: P137163f   U. Olthoff and K. Matthey, (East) German Patent 96,948 (1973).

79: P137937m   H. W. Cressman, German Patent 2,247,496 (1973).

79: P141516r   T. Kojima, H. Imamura, M. Fujiwhara, W. Fujimatsu, and T. Endo, German Patent 2,261,361 (1973).

79: 146129b   F. A. Neugebauer, H. Fischer, and W. Otting, *Chem. Ber.*, **106,** 2530 (1973).

79: 146466j   R. A. Coburn, B. Bhooshan, and R. A. Glennon, *J. Org. Chem.*, **38,** 3947 (1973).

79: P147415d   M. Kozutsumi, T. Arakawa, M. Yamamoto, and N. Koyano, Japanese Patent 72 50,208 (1972).

79: 152590m   D. N. Pachadzhanov, M. A. Israilov, and M. Yu. Yusupov, *Izv. Akad. Nauk Tadzh. SSR, Otd. Fiz.-Mat. Geol.-Khim. Nauk*, 61 (1973).

80: 65b   J. P. Henichart, R. Houssin, C. Lespagnol, J. C. Cazin, M. Cazin, and F. Rousseau, *Chim. Ther.*, **8,** 358 (1973).

80: 3422b   V. N. Artemov, S. N. Baranov, N. A. Kovach, and O. P. Shvaika, *Dokl. Akad. Nauk, SSSR*, **211,** 1369 (1973).

80: 3472t   Z. Csuros, R. Soos, A. Antus-Ercsenyi, I. Bitter, and J. Tamas, *Acta Chim.* (Budapest), **78,** 419 (1973).

80: 10254b   J. T. Witkowski, R. K. Robins, G. P. Khare, and R. W. Sidwell, *J. Med. Chem.*, **16,** 935 (1973).

80: 10273g   D. W. Henry, V. H. Brown, M. Cory, J. G. Johansson, and E. Bueding, *J. Med. Chem.*, **16,** 1287 (1973).

80: 14221m   F. Capitan, F. Salinas, and E. J. Alonso, *Bol. Soc. Quim. Peru*, **39,** 1 (1973).

80: 14340z   P. B. Talukdar, S. K. Sengupta, and A. K. Datta, *Indian J. Chem.*, **11,** 753 (1973).

80: 14879g   I. Simiti and A. Marie, *Arch. Pharm.* (Weinheim), **306,** 659 (1973).

80: 14898n   S. R. Naik, J. T. Witkowski, and R. K. Robins, *J. Org. Chem.*, **38,** 4353 (1973).

80: P14934w   J. B. Hester, Jr., German Patent 2,233,682 (1973).

80: 15236g   V. V. Korshak, A. L. Rusanov, Ts. G. Iremashvili, I. V. Zhuravleva, and E. L. Baranov, *Macromolecules*, **6,** 483 (1973).

80: 16240r   F. W. Jones, J. D. Leeder, and A. M. Wemyss, *J. Text. Inst.*, **64,** 587 (1973).

80: 26672x   V. V. Mel'nikov, V. V. Stolpakova, S. A. Zacheslavskii, M. S. Pevzner, and B. V. Sidaspov, *Khim. Geterotsikl. Soedin*, **9,** 1423 (1973).

80: 27172w   R. Grashey and C. Knorn, *Chem.-Zig.*, **97,** 566 (1973).

80: 27175z   A. Ya. Lazaris, S. M. Shmuilovich, and A. N. Egorochkin, *Khim. Geterotsikl. Soedin.*, **9,** 1345 (1973).

80: 27179d   K. Burger and K. Einhelig, *Chem. Ber.*, **106,** 3421 (1973).

80: 27223p   R. Grashey, C. Knorn, and M. Weidner, *Chem.-Ztg.*, **97,** 565 (1973).

80: P27263b   R. D. Haugwitz and V. L. Narayanan, U.S. Patent 3,767,654 (1973).

80: P28454b   M. Ohkawa and K. Hirabayashi, Japanese Patent 73 48,527 (1973).

80: 28461b   M. Kotsutsumi, S. Maeda, and T. Nakajo, Japanese Patent 73 00,851 (1973).

80: 36682q   V. Mozolis and S. Jokubaityte, *Liet. TSR Mokslu Akad. Darb.*, *Ser. B*, 75 (1973).

80: P37116v   M. Gall, German Patent 2,322,045 (1973).

80: P37118x   S. Kano, O. Nomura, and T. Taniguchi, Japanese Patent 73 68,589 (1973).

80: 47060m   B. Adler and H. J. Timpe, *J. Prakt. Chem.*, **315,** 949 (1973).

80: 47869g   K. Yamauchi and M. Kinoshita, *J. Chem. Soc., Perkin Trans.* 1, 2506 (1973).

80: 47911q   V. Mozolis, A. Ceika, and L. Rastenyte, *Liet. TLR Mokslu Akad. Darb.*, *Ser. B*, 83 (1973).

80: 47913s   A. R. Katritzky and J. W. Mitchell, *J. Chem. Soc. Perkin Trans.* 1, 2624 (1973).

80: P48002f   J. Zielinski, U.S. Patent 3,767,666 (1973).

80: P48019s   R. M. DeMarinis and J. R. E. Hoover, German Patent 2,322,127 (1973).

80: P49275j   A. Domergue and R. F. M. Sureau, German Patent 2,319,828 (1973).

80: 52645k   A. Parentich, L. H. Little, and R. H. Ottewill, *Kolloid-Z. Z. Polym.*, **251,** 494 (1973).

80: P54523t   S. Ebato and N. Itoh, German Patent 2,326,333 (1973).

80: 59602c   J. Perronnet and P. Girault, *Bull. Soc. Chim. Fr.*, 2843 (1973).

80: P59693h   W. F. Bruce, U.S. Patent 3,775,405 (1973).

80: 59762e   M. Tutoveanu, A. Kasper, and S. Steinberg, *Bul. Inst. Politeh. Iasi*, **18,** 125 (1972).

80: 59899e   M. Tutoveanu, C. Constantinescu, and V. Maerean, *Rev. Chim.* (*Bucharest*), **24,** 155 (1973).

80: 59912d   V. V. Korshak, A. L. Rusanov, Ts. G. Iremashvili, I. V. Zhuravleva, S. S. Gitis, E. L. Vulakh, and V. M. Ivanova, *Khim. Geterotsikl. Soedin.*, **9,** 1574 (1973).

80: P59944r   J. B. Hester, Jr., U.S. Patent 3,772,317 (1973).

80: 60251u   L. J. Guilbault, *Nuova Chim.*, **49,** 98 (1973).

80: 60356g   I. V. Stankevich and O. B. Tomilin, *Izv. Akad. Nauk SSSR, Ser. Khim*, 2515 (1973).

80: P60965e   K. Matsui, S. Matsuda, K. Akamatsu, and T. Taneda, Japanese 72 48,312 (1972).

80: 61056w   S. Yamada and K. Tsunemitsu, Japanese Patent 73 09,094 (1973).

80: 62968g   A. Kotone and T. Hori, Japanese Patent 73 62,634 (1973).

80: 67324d   L. W. Smith and F. Y. Chang, *Weed Res.*, **13,** 339 (1973).

80: 70754z   L. Al Yusofi, V. N. Bubnovskaya, and F. S. Babichev, *Ukr. Khim. Zh.* (*Russ. Ed.*), **39,** 1289 (1973).

80: P70813t   Spanish Patent 382,492 (1972).

80: 74824v   L. P. Romanenko and A. V. Radushev, *Zh. Anal. Khim.*, **28,** 1908 (1973).

80: P76701b   T. Masuda, K. Ohkubo, and T. Shishido, German Patent 2,321,218 (1973).

80: 77774q   N. Teodorescu, C. Demetrescu, and C. Chirita, *Rev. Chim.* (*Bucharest*), **24,** 560 (1973).

80: 78028m   P. P. Kish and I. I. Pogoida, *Zh. Anal. Khim.*, **28,** 1923 (1973).

80: 81776j   R. A. Abramovitch, T. D. Bailey, T. Takaya, and V. Uma, *J. Org. Chem.*, **39,** 340 (1973).

80: 82270b   R. Grashey and M. Weidner, *Chem.-Ztg.*, **97,** 623 (1973).

80: 82651b   D. Dawes, W. Meyer, and B. Boehner, German Patent 2,330,089 (1974).

80: 82818m   S. Conde, C. Corral, and R. Madronero, *Synthesis*, 28 (1974).

80: 82823j   G. G. Skvortsova, E. S. Dommina, L. P. Makhno, V. K. Voronov, O. D. Taryashinova, and N. N. Chipanina, *Khim. Geterotsikl. Soedin*, **9,** 1566 (1973).

80: 82826n   P. R. Atkins, S. E. J. Glue, and I. T. Kay, *J. Chem. Soc., Perkin Trans. 1*, 2644 (1973).

80: 82830j   A. D. Sinegibskaya, E. G. Kovalev, and I. Ya. Postovskii, *Khim. Geterotsikl. Soedin.*, **9,** 1708 (1973).

80: 82837s   H. G. O. Becker, D. Nagel, and H. J. Timpe, *J. Prakt. Chem.*, **315,** 1131 (1973).

80: 82840n   P. Dubus, B. Decroix, J. Morel, and C. Paulmier, *C. R. Acad. Sci., Ser. C*, **278,** 61 (1974).

80: P82954c   R. L. Rosati, German Patent 2,316,170 (1973).

80: P82989t   T. Cebalo and J. L. Miesel, U.S. Patent 3,780,052 (1973).

80: P82990m   J. L. Miesel, U.S. Patent 3,780,054 (1973).

80: P82992p   H. Kristinsson and K. Ruefenacht, German Patent 2,332,721 (1974).

80: P82998v   J. L. Miesel, U.S. Patent 3,780,053 (1973).

80: P83012f   P. Crooij, and A. Colinet, German Patent 2,331,599 (1974).

80: 83019p   R. Scartazzini and H. Bickel, German Patent 2,331,133 (1974).

80: P83040p   D. E. O'Brien, T. Novinson, and R. H. Springer, German Patent 2,327,133 (1974).

80: P83083e   J. B. Hester, Jr., U.S. Patent 3,781,289 (1973).

80: 83353t   R. Giger, R. Rubinstein, and D. Ginsburg, *Tetrahedron*, **29,** 2387 (1973).

80: 84684p   E. Fleckenstein, R. Mohr, and E. Heinrich, German Patent 2,222,099 (1973).

80: 84685q   R. Mohr and K. Hohmann, German Patent 2,222,042 (1973).

80: P89539e   J. M. Dyke, German Patent 2,331,220 (1974).

80: 90814x   P. P. Kish, A. I. Busev, and I. I. Pogoida, *Zavod. Lab.*, **39,** 1302 (1973).

80: P95962d   H. Kristinsson and K. Ruefenacht, German Patent 2,330,606 (1974).

80: 96082d   G. S. Gol'din, M. V. Maksakova, V. G. Poddubnyi, A. N. Kol'tsova, A. V. Kisin, V. N. Torocheshnikov, and A. A. Simonova, *Zh. Obshch. Khim.*, **44,** 115 (1974).

80: 102243a   R. M. Fantazier and J. E. Herweh, *J. Amer. Chem. Soc.*, **96,** 1187 (1974).

80: P102274m   H. Iwano, I. Shimamura, Y. Iijima, and K. Hayashi, German Patent 2,321,401 (1973).

80: 103553p   A. Dimitrova and V. Bogdanova, *Dokl. Bolg. Akad. Nauk*, **26,** 1367 (1973).

80: 107585y   A. F. Hegarty, T. A. F. O'Mahony, P. Quain, and F. L. Scott, *J. Chem. Soc., Perkin Trans.* 2, 2047 (1973).

80: 107595b   S. Sarel, A. Felzenstein, and J. Yovell, *J. Chem. Soc., Chem. Commun.*, 859 (1973).

80: 107696k   B. V. Ioffe and V. A. Chernyshev, *Dokl. Akad. Nauk SSSR*, **214,** 336 (1974).

80: 108456u   G. Westphal, C. Milbradt, P. Henklein, and R. Kraft, *Z. Chem.*, **13,** 468 (1973).

80: 108472w   R. Sunderdiek and G. Zinner, *Chem. Ber.*, **107,** 454 (1974).

80: 110110v   J. M. Harkin, J. R. Obst, and W. F. Lehmann, *Forest Prod. J.* **24,** 27 (1974).

80: P111542z   A. Koton, T. Hori, M. Hoda, and Y. Nakane, Japanese Patent 73 89,141 (1973).

80: 120848f   J. A. Lenoir and B. L. Johnson, *Tetrahedron Lett.*, 5123 (1973).

80: P120970q   R. M. DeMarinis and J. R. E. Hoover, German Patent 2,336,345 (1974).

80: 121258a   S. R. Naik, J. T. Witkowski, and R. K. Robins, *J. Heterocyclic Chem.*, **11,** 57 (1974).

80: 126797k   S. Suzuki, N. Kobayashi, and K. Futaki, U.S. Patent 3,776,728 (1973).

80: 133347s   J. A. Maddison, P. W. Seale, E. P. Tiley, and W. K. Warburton, *J. Chem. Soc., Perkin Trans.* 1, 81 (1974).

80: 133356u   F. S. Babichev, V. A. Kovtunenko, and L. N. Didenko, *Ukr. Khim. Zh.*, **40,** 245 (1974).

80: 133389g   E. M. Burgess and J. P. Sanchez, *J. Org. Chem.*, **39,** 940 (1974).

80: P133443v   T. Tahara, H. Matsuki, K. Araki, and M. Shiroki, German Patent 2,340,217 (1974).

80: 144999g   R. E. Ardrey, J. Emsley, A. J. B. Robertson, and J. K. Williams, *J. Chem. Soc., Dalton Trans.*, 2641 (1973).

80: 146119u   O. P. Shavaika and V. I. Fomenko, *Zh. Org. Khim.*, **10,** 377 (1974).

80: P146156d   B. Von Bredow and H. U. Brechbuehler, German Patent 2,343,547 (1974).

80: P146169k   W. Meiser, W. Kraemer, K. H. Buechel, and M. Plempel, German Patent 2,247,186 (1974).

80: P146175j   H. G. Viehe and Z. Janousek, German Patent 2,340,378 (1974).

80: P146185n   B. Von Bredow and H. U. Brechbuehler, German Patent 2,343,526 (1974).

80: P146210s   B. Von Bredow and H. E. Brechbuehler, German Patent 2,343,347 (1974).

80: 146456b   P. Dea, M. P. Schweizer, and G. P. Kreishman, *Biochemistry*, **13,** 1862 (1974).

80: 146458d   G. Wagner, G. Valz, B. Dietzsch, and G. Fischer, *Pharmazie*, **29,** 90 (1974).

80: P146980t   A. E. Siegrist, P. Liechti, E. Maeder, and H. R. Meyer, Swiss Patent 542,266 (1973).

80: P146981u   A. E. Siegrist, P. Liechti, E. Maeder, L. Guglielmetti, H. R. Meyer, and K. Weber, Swiss Patent 542,212 (1973).

80: P151113s   C. A. Goffe, R. D. Lindholm, and P. M. Stacy, Def. Publ., U.S. Pat. Off. T 917,003 (1973).

80: P151126y   T. Noda and K. Futaki, Japanese Patent 73 102,621 (1973).

81: 2977v   W. D. Ollis and C. A. Ramsden, *J. Chem. Soc., Perkin Trans.* 1, 645 (1974).

81: 3549u   O. Tsuge, M. Yoshida, and S. Kanemasa, *J. Org. Chem.*, **39,** 1226 (1974).

81: 3835j   W. D. Ollis and C. A. Ramsden, *J. Chem. Soc., Perkins Trans.* 1, 642 (1974).

81: 3838n   R. Bognar, C. Jaszberenyi, and A. L. Tokes, *Magy. Kem. Foly.*, **80,** 114 (1974).

81: 3839p   W. D. Ollis and C. A. Ramsden, *J. Chem. Soc., Perkin Trans.* 1, 638 (1974).

81: 3844m   A. R. McCarthy, W. D. Ollis, and C. A. Ramsden, *J. Chem. Soc., Perkin Trans.* 1, 627 (1974).

81: 3847q   W. D. Ollis and C. A. Ramsden, *J. Chem. Soc., Perkin Trans.* 1, 633 (1974).

81: 3897f   V. Mozolis and S. Jokubaityte, *Liet. TSR Mokslu Akad. Darb., Ser. B*, 95 (1973).

81: 4188n   G. R. Revankar, R. K. Robins, and R. L. Tolman, *J. Org. Chem.*, **39,** 1256 (1974).

81: P8417j   S. Nagashima, Y. Tsuchiya, and T. Kyoda, Japanese Patent 73 37,380 (1973).

81: P8424j   M. K. Reichel and M. Tomanek, German Patent 1,477,008 (1974).

81: P8428p   W. Gauss, S. Petersen, E. Ranz, W. Himmelmann, and H. Von Rintelen, German Patent 2,228,110 (1973).

81: 13434s   Y. Ohshiro, M. Komatsu, Y. Yamamoto, K. Takaki, and T. Agawa, *Chem. Lett.*, 383 (1974).

81: 13441s   V. A. Kovtunenko, L. N. Didenko, and F. S. Babichev, *Ukr. Khim. Zh.*, **40,** 402 (1974).

81: 13442t   J. Stein, *Z. Chem.*, **14,** 95 (1974).

81: P13520s   B. Boehner, W. Meyer, and D. Dawes, German Patent 2,352,140 (1974).

81: P13521t   B. Boehner, W. Meyer, and D. Dawes, German Patent 2,352,141 (1974).

81: P13523v   H. Hoffmann and I. Hammann, German Patent 2,250,572 (1974).

81: P13524w   N. A. Dahle and W. C. Doyle, Jr., U.S. Patent 3,808,223 (1974).

81: P13553e   G. Palazzo, German Patent 2,351,739 (1974).

81: P14321q   M. Minagawa and M. Akutsu, Japanese Patent 73 34,370 (1973).

81: P14549v   D. O. Hummel, H. Siesler, E. Zoschke, I. Vierling, U. Morlock, and T. Stadtlaender, *Melliand Textilber Int.*, **54,** 1340 (1973).

81: P14701p   R. Haeberli and H. Mollet, German Patent 2,338,505 (1974).

81: 24524x   Y. Hayashi, *Mem. Fac. Ind. Arts, Kyoto Tech*, **22,** 30 (1973).

81: 25598m   P. Bouchet, C. Coquelet, G. Joncheray, and J. Elguero, *Syn. Commun.*, **4,** 57 (1974).

81: 25609r   I. L. Shegal, K. V. Stanovkina, N. G. Kovalenko, and L. M. Shegal, *Khim. Geterotsikl. Soedin.*, **10,** 422 (1974).

81: 25611k   A. Kreutzberger and B. Meyer, *Arch. Pharm.* (Weinheim), **307,** 281 (1974).

81: 25613n   B. T. Heitke and C. G. MacCarty, *J. Org. Chem.*, **39,** 1522 (1974).

81: 25614p   P. L. Gravel and W. H. Pirkle, *J. Amer. Chem. Soc.*, **96,** 3335 (1974).

81: P25678n   A. Kotone, M. Hoda, T. Hori, and Y. Nakane, German Patent 2,248,257 (1974).

81: 25762k   A. Nakamura, M. Aotake, and S. Otsuka, *J. Amer. Chem. Soc.*, **96,** 3456 (1974).

81: P27329y   S. H. Roth, U.S. Patent 3,783,017 (1974).

81: P27330s   S. H. Roth, U.S. Patent 3,783,018 (1974).

81: P37390a   A. Lauton, R. Putzar, H. U. Berendt, and G. Scherer, German Patent 2,353,691 (1974).

81: 37516w   G. S. Gol'din, V. G. Poddubnyi, and E. S. Smirnova, *Khim. Geterotsikl. Soedin.*, **10,** 562 (1974).

81: 37521u   S. Kimoto, M. Okamoto, T. Kawabata, and S. Ohta, *Yakugaku Zasshi*, **94,** 55 (1974).

81: 37556j   K. Heusler, H. Bickel, B. Fechtig, H. Peter, and R. Scartazzini, British Patent 1,353,326 (1974).

81: P38960y   French Patent 2,172,402 (1973).

81: 44054x   H. D. Meissner, *Photogr. Process., Proc. Symp.*, 183 (1971).
81: 44127y   I. Endo and T. Yamanouchi, U.S. Patent 3,765,883 (1973).
81: 49309u   J. Beger, H. Fuellbier, and W. Gaube, *J. Prakt. Chem.*, **316,** 173 (1974).
81: P49560u   P. Stoss, M. Herrmann, and G. Satzinger, German Patent 2,258,036 (1974).
81: 49624t   K. Meguro and Y. Kuwada, *Heterocycles*, **2,** 335 (1974).
81: 49628x   R. G. Dickinson and N. W. Jacobsen, *Org. Prep. Proced. Int.*, **6,** 156 (1974).
81: 49629y   G. M. Golubushina, G. N. Poshtaruk, and V. A. Chuiguk, *Khim. Geterotsikl. Soedin.*, **10,** 565 (1974).
81: 49647c   H. Roechling and G. Hoerlein, *Justus Liebigs Ann. Chem.*, 504 (1974).
81: P49689t   M. Gall and J. B. Hester, Jr., U.S. Patent 3,813,412 (1974).
81: P51113u   S. Fujino, H. Honda, and T. Hattori, Japanese Patent 74 09,532 (1974).
81: 62940n   P. Bouchet, C. Coquelet, and J. Elguero, *J. Chem. Soc., Perkin Trans. 2*, 499 (1974).
81: 63263f   Y. Hayashi, Y. Iwagami, A. Kadoi, T. Shono, and D. Swern, *Tetrahedron Lett.*, 1071 (1974).
81: P63478e   L. H. Sarett and W. V. Ruyle, South African Patent 72 01,129 (1973).
81: P63636e   J. T. A. Boyle and J. C. Saunders, U.S. Patent 3,813,400 (1974).
81: P63640b   B. Boehner, W. Meyer, and D. Dawes, German Patent 2,352,142 (1974).
81: P63648k   D. A. Berges, German Patent 2,356,704 (1974).
81: P65180a   B. Henzi, German Patent 2,338,682 (1974).
81: P65182c   T. Nakajo, S. Maeda, and T. Kurahashi, Japanese Patent 73 89,932 (1973).
81: P65189k   B. Henzi, German Patent 2,338,730 (1974).
81: P65195j   H. Kenmochi, M. Yamamoto, T. Ikeda, Y. Korenaga, Y. Takeda, and T. Ohkawa, Japanese Patent 73 37,967 (1973).
81: P65203k   H. L. Dorsch and R. Raue, German Patent 2,248,738 (1974).
81: P68556p   J. R. Guarini, U.S. Patent 3,796,801 (1974).
81: P68558r   G. Laber and E. Schuetze, German Patent 2,340,709 (1974).
81: P72295v   M. Yu. Yusupov, D. N. Pachadzhanov, M. Israilov, and T. F. Novosel'tseva, U.S.S.R. Patent 415,236 (1974).
81: P73392m   I. Chiyomaru, E. Yoshinaga, and G. Dohke, Japanese Patent 73 39,460 (1973).
81: 77848n   R. L. N. Harris and L. T. Oswald, *Aust. J. Chem.*, **27,** 1531 (1974).
81: P77930h   T. P. Kofman, M. S. Pevzner, and V. I. Manuilova, U.S.S.R. Patent 432,147 (1974).
81: P79380c   T. Kurahashi, M. Ozutsumi, and S. Maeda, Japanese Patent 74 24,228 (1974).
81: P79381d   S. Maeda, M. Ozutsumi, and T. Kurahasi, Japanese Patent 74 24,229 (1974).
81: P79384g   K. Anderton and R. Porter, German Patent 2,358,080 (1974).
81: 86057f   P. N. Kulyabko, V. A. Kuz'menko, V. I. Fedorchenko, R. G. Dubenko, P. S. Pel'kis, I. M. Bazavova, and V. M. Andrianov, *Fiziol. Aktiv Veshchestva*, **5,** 114 (1973).
81: 91429h   M. M. Tsitsika, S. M. Khripak, and I. V. Smolanka, *Khim. Geterotsikl. Soedin.*, **10,** 851 (1974).
81: 91435g   K. Seckinger, *Helv. Chim. Acta*, **56,** 2061 (1973).
81: 91438k   R. Sunderdiek and G. Zinner, *Arch. Pharm.* (Weinheim), **307,** 504 (1974).
81: 91439m   R. Sunderdiek and G. Zinner, *Arch. Pharm.* (Weinheim), **307,** 509 (1974).
81: 91461n   L. Godefroy, A. Decormeille, G. Queguiner, and P. Pastour, *C. R. Acad. Sci., Ser. C*, **278,** 1421 (1974).

81: P91535q   B. Boehner, D. Dawes, and W. Meyer, German Patent 2,360,623 (1974).

81: P91536r   B. Boehner, D. Dawes, and W. Meyer, German Patent 2,360,631 (1974).

81: P91539u   U. Petersen, S. Petersen, H. Scheinpflug, and B. Hamburger, German Patent 2,260,704 (1974).

81: P91540n   B. Boehner, D. Dawes, and W. Meyer, German Patent 2,360,687 (1974).

81: P91733c   K. H. Buechel and I. Hammann, German Patent 2,261,455 (1974).

81: 91849v   N. Katagiri, K. Itakura, and S. A. Narang, *J. Chem. Soc., Chem. Commun.*, **9**, 325 (1974).

81: P93041z   S. Maeda, M. Ozutsumi, and T. Kurahashi, Japanese Patent 74 24,225 (1974).

81: P93043b   S. Maeda, M. Ozutsumi, and I. Niimura, Japanese Patent 74 24,227 (1974).

81: P93049h   J. Pchmeze and R. Sureau, German Patent 2,355,687 (1974).

81: P97737d   M. Hinata, K. Shiba, R. Ohi, and T. Shishido, German Patent 2,363,308 (1974).

81: 104454q   G. Scherowsky and H. Franke, *Tetrahedron Lett.*, 1673 (1974).

81: 105370w   D. Clerin and J. P. Fleury, *Bull. Soc. Chim. Fr.*, 211 (1974).

81: 105402h   A. A. Aroyan and M. A. Iradyan, *Arm. Khim. Zh.*, **27**, 525 (1974).

81: 105408q   A. M. Abd-Elfattah, S. M. Hussain, and M. I. Ali, *Tetrahedron*, **30**, 987 (1974).

81: 105467h   A. Etienne and B. Bonte, *Bull. Soc. Chim. Fr.*, **7–8**, 1497 (1974).

81: P105517z   A. Kotone, T. Fujita, and J. Hoda, Japanese Patent 74 20,175 (1974).

81: P105518a   A. Omodei-Sale, German Patent 2,358,011 (1974).

81: P106281e   Y. Minoura, A. Kotone, T. Fujita, and M. Hoda, Japanese Patent 74 34,990 (1974).

81: P113665m   T. Masuda and K. Okubo, Japanese Patent 74 13,224 (1974).

81: 114436z   M. Fuertes, J. T. Witkowski, D. G. Streeter, and R. K. Robins, *J. Med. Chem.*, **17**, 642 (1974).

81: 120541w   T. P. Kofman, G. A. Zykova, V. I. Manuilova, T. N. Timofeeva, and M. S. Pevzner, *Khim. Geterotsikl. Soedin*, **10**, 997 (1974).

81: P120637g   A. Kotone, T. Fujita, and M. Hoda, Japanese Patent 74 35,384 (1974).

81: P120638h   H. Hoffmann, I. Hammann, W. Behrenz, B. Homeyer, and W. Stendel, German Patent 2,301,400 (1974).

81: P120640c   A. Kotone, M. Hoda, and T. Hori, Japanese Patent 74 30,363 (1974).

81: 121645v   M. Minagawa, M. Akutsu, and N. Kubota, *Soc. Plast. Eng., Tech. Pap.*, **20**, 328 (1974).

81: P122623y   E. Daubach and G. Meyer, German Patent 2,257,945 (1974).

81: P122759x   B. Henzi, German Patent 2,338,681 (1974).

81: P122761s   K. Konishi, A. Kotone, Y. Nakane, T. Hori, M. Hoda, T. Kitao, and I. Yamase, German Patent 2,347,849 (1974).

81: 129811s   P. Gilgen, H. Heimgartner, and H. Schmid, *Helv. Chim. Acta*, **57**, 1382 (1974).

81: P129840a   A. Von Koenig, H. Reuss, and A. Mueller, German Patent 2,304,322 (1974).

81: P129916e   F. P. Serrien, B. H. Tavernier, R. J. Pollet, and F. J. Sel, German Patent 2,349,527 (1974).

81: 130305m   B. Singh, R. Singh, K. P. Thakur, and M. Chandra, *Indian J. Chem.*, **12**, 631 (1974).

81: P135150s   M. Nakanishi, T. Aiya, S. Sakuragi, A. Khirose, M. Takahashi, and A. Chiba, Japanese Patent 74 13,266 (1974).

81: 136060z   V. A. Lopyrev. L. P. Sidorova, E. D. Maiorova, and M. P. Grinblat, *Zh. Org. Khim.*, **10**, 1648 (1974).

81: 136062b   R. Schmidt, G. Westphal, and B. Froelich, *Z. Chem.*, **14**, 270 (1974).

81: 136067g   S. Leistner, G. Wagner, and H. Richter, *Z. Chem.*, **14**, 267 (1974).

81: 136069j   J. Schantl, *Monatsh. Chem.*, **105,** 427 (1974).

81: 136110r   F. Yoneda, S. Matsumoto, and M. Higuchi, *J. Chem. Soc., Chem. Commun.*, 551 (1974).

81: 136112t   K. S. Dhaka, J. Mohan, V. K. Chadha, and H. K. Pujari, *Indian J. Chem.*, **12,** 485 (1974).

81: 136118z   K. S. Dhaka, J. Mohan, V. K. Chadha, and H. K. Pujari, *Indian J. Chem.*, **12,** 287 (1974).

81: 137569j   S. Maeda, M. Ozutsumi, and I. Niimura, Japanese Patent 74 24,224 (1974).

81: 137590j   M. Takahashi and M. Fumishi, German Patent 2,362,649 (1974).

81: P144229b   K. Lohmer, A. Von Koenig, and A. Meueller, German Patent 2,304,321 (1974).

81: 147547q   P. Mildner and B. Mihanovic, *Croat. Chem. Acta*, **46,** 79 (1974).

81: P152238u   M. Shiroki, T. Tahara, K. Araki, and H. Matsuki, Japanese Patent 74 69,667 (1974).

81: P152287j   M. Gall and J. B. Hester, Jr., German Patent 2,403,808 (1974).

81: P152289m   M. Gall and J. B. Hester, Jr., German Patent 2,402,363 (1974).

81: P154525j   S. Fujimo, H. Honda, and T. Hattori, Japanese Patent 74 23,823 (1974).

81: 161627x   B. Singh and K. P. Thakur, *J. Inorg. Nucl. Chem.*, **36,** 1735 (1974).

81: 163330u   E. B. Akerblom, *J. Med. Chem.*, **17,** 756 (1974).

81: P164763z   T. Takematsu, Y. Hasegawa, M. Doya, and M. Furushima, Japanese Patent 74 08,051 (1974).

81: 169245s   G. Doleschall, *Tetrahedron Lett.*, 2649 (1974).

81: 169457n   M. Komatsu, N. Nishikaze, M. Sakamoto, Y. Ohshiro, and T. Agawa, *J. Org. Chem.*, **39,** 3198 (1974).

81: 169484u   T. L. Gilchrist, C. J. Harris, C. J. Moody, and C. W. Rees, *J. Chem. Soc., Chem. Commun.*, 486 (1974).

81: P169567y   Y. Kuwada, H. Tawada, and K. Meguro, German Patent 2,407,731 (1974).

81: 169748h   G.-F. Huang, T. Okamoto, M. Maeda, and Y. Kawazoe, *Chem. Pharm. Bull.*, **22,** 1938 (1974).

81: 169929t   V. V. Korshak, A. L. Rusanov, Ts. G. Iremanshvili, I. V. Zhuravleva, and E. L. Baranov, *Vysokomol. Soedin., Ser. A*, **16,** 1516 (1974).

81: P170229w   M. Kurabayashi, Japanese Patent 74 01,959 (1974).

81: 171335q   I. Niimura, M. Ozutsumi, and S. Maeda, Japanese Patent 74 26,329 (1974).

81: P179884p   H. Stark, K. Kueffner, F. Mueller, H. Glockner, and F. Sommer, German Patent 2,308,530 (1974).

82: 3735x   W. J. S. Lockley, V. T. Ramakrishnan, and W. Lwowski, *Tetrahedron Lett.*, 2621 (1974).

82: 4164x   S. Nakazawa, T. Kiyosawa, K. Hirakawa, and H. Kato, *J. Chem. Soc., Chem. Commun.*, 621 (1974).

82: 4178e   E. E. Glover and K. T. Rowbottom, *J. Chem. Soc., Perkin Trans. 1*, 1792 (1974).

82: 4183c   M. Woodruff and J. B. Polya, *Aust. J. Chem.*, **27,** 2447 (1974).

82: P4260a   E. Winkelmann and W. Raether, German Patent 2,262,554 (1974).

82: P4534t   J. T. Witkowski and R. K. Robins, German Patent 2,411,823 (1974).

82: 9314f   D. Rahner, G. Fischer, and K. W. Junge, *J. Signalaufzeich-nungsmaterialien*, **2,** 201 (1973).

82: P9971t   K. Shiba, T. Hirose, T. Aono, R. Ohi, and T. Shishido, German Patent 2,417,914 (1974).

82: 11679x   J. M. Essery, U. Corbin, V. Sprancmanis, L. B. Crast, Jr., R. G. Graham, P. F. Misco, Jr., D. Willner, D. N. McGregor, and L. C. Cheney, *J. Antibiot.*, **27,** 573 (1974).

82: 12106v   A. Trebst and E. Harth, *Z. Naturforsch., Teil C*, **29,** 232 (1974).

82: P12285c   M. M. Fawzi and B. Quebedeaux, Jr., U.S. Patent 3,836,350 (1974).

82: 16729y   J. A. Moore, R. Muth, and R. Sorace, *J. Org. Chem.*, **39,** 3799 (1974).

82: 16746b   J. H. Wikel and C. J. Paget, *J. Org. Chem.*, **39,** 3506 (1974).

82: 16751z   S. W. Moje and P. Beak, *J. Org. Chem.*, **39,** 2951 (1974).

82: 16754c   A. M. Khalil, I. I. Abd El-Gawad, and H. M. Hassan, *Aust. J. Chem.*, **27,** 2509 (1974).

82: P16845h   B. Boehner, D. Dawes, and W. Meyer, German Patent 2,412,564 (1974).

82: P18517b   K. Matsui, T. Nakanishi, and I. Aijima, Japanese Patent 74 07,054 (1974).

82: 18628p   H. Harnisch, German Patent 2,300,488 (1974).

82: 25627r   D. E. Duggan, J. J. Baldwin, B. H. Arison, and R. E. Rhodes, *J. Pharmacol. Exp. Ther.*, **190,** 563 (1974).

82: 25638v   Y. Kanai, *Xenobiotica*, **4,** 441 (1974).

82: 30522g   V. V. Mel'nikov, L. F. Baeva, V. V. Stolpakova, M. S. Pevzner, M. N. Martynova, and B. V. Gidaspov, *Khim. Geterotsikl. Soedin.*, **10,** 1283 (1974).

82: 30784u   M. S. Pevzner, M. N. Martynova, and T. N. Timofeeva, *Khim. Geterotsikl. Soedin,* **10,** 1288 (1974).

82: 31304z   S. Polanc, B. Vercek, B. Sek, B. Stanovnik, and M. Tisler, *J. Org. Chem.*, **39,** 2143 (1974).

82: P31324f   C. C. Beard, J. A. Edwards, and J. H. Fried, German Patent 2,411,295 (1974).

82: P32330e   Austrian Patent 315,169 (1974).

82: P31344n   H. Bickel and J. Mueller, Swiss Patent 553,821 (1974).

82: 31364u   Y. Kuwada, K. Meguro, and H. Tawada, Japanese Patent 74 85,095 (1974).

82: 42707j   O. Tsuge and S. Kanemasa, *Bull. Chem. Soc. Jpn.*, **47,** 2676 (1974).

82: 43269e   A. A. Stotskii and N. P. Tkacheva, *Zh. Org. Khim.*, **10,** 2232 (1974).

82: 43270y   A. Y. Lazaris, S. M. Shmuilovich, and A. N. Egorochkin, *Zh. Org. Khim.*, **10,** 2236 (1974).

82: 43271z   U. Olthoff and K. Matthey, *Pharmazie*, **29,** 20 (1974).

82: 43275d   B. I. Buzykin, L. P. Sysoeva, and Yu. P. Kitaev, *Zh. Org. Khim.*, **10,** 2200 (1974).

82: 43276e   M. M. Tsitsika, S. M. Khripak, and I. V. Smolanka, *Khim. Geterotsikl. Soedin.*, **10,** 1425 (1974).

82: 43281c   R. Neidlein and H. G. Hege, *Chem.-Ztg.*, **98,** 512 (1974).

82: 43283e   R. H. Rynbrandt, *J. Heterocyclic Chem.*, **11,** 787 (1974).

82: 43365h   E. E. Knaus, F. M. Pasutto, and C. S. Giam, *J. Heterocyclic Chem.*, **11,** 843 (1974).

82: P43428f   A. A. Stotskii and N. P. Tkacheva, U.S.S.R. Patent 446,507 (1974).

82: P43429g   K. Dickore, R. Wegler, and G. Hermann, German Patent 1,770,920 (1974).

82: P43443g   T. Ozasa, M. Murakami, and I. Isaka, Japanese Patent 74 48,690 (1974).

82: P43479y   A. Gagneux, R. Heckendorn, and R. Meier, Swiss Patent 551, 993 (1974).

82: 43697t   A. A. Kraevskii, N. B. Tausova, T. L. Tsilevich, and B. P. Gottikh, *Izu. Akad. Nauk SSSR, Ser. Khim.*, 1856 (1974).

82: 49831d   D. Rahner, G. Fischer, and K. W. Junge, *J. Signalaufzeichnungsmaterialien*, **2,** 39 (1974).

82: 49832e   D. Rahner, G. Fischer, and K. W. Junge, *J. Signalaufzeichnungsmaterialien*, **1,** 367 (1973).

82: 51075d   A. Tateda and H. Murakami, *Bull. Chem. Soc. Jpn.*, **47,** 2885 (1974).

82: 57055u   A. F. Hegarty, P. Quain, T. A. O'Mahony, and F. L. Scott, *J. Chem. Soc.*, *Perkin Trans.* 2, 997 (1974).

82: 57623c   R. Un, *Chim. Acta Turc.*, **2,** 1 (1974).

82: 57642h   E. Lippmann and V. Becker, *Z. Chem.*, **14,** 405 (1974).

85: 57659u   R. M. Claramunt, J. M. Fabrega, and J. Elguero, *J. Heterocyclic Chem.*, **11,** 751 (1974).

82: P57676x   W. Schroeck, K. G. Metzger, and B. Hans, German Patent 2,320,039 (1974).

82: P57711e   S. Seki and T. Ishimaru, German Patent 2,418,088 (1974).

82: P57724m   F. G. Kathawala, U.S. Patent 3,847,918 (1974).

82: P58077q   M. Sunagawa, M. Yamamoto, S. Morooka, M. Koshiba, S. Inaba, and H. Yamamoto, Japanese Patent 74 80,070 (1974).

82: P59896t   A. Kotone, Y. Nakane, T. Hori, M. Hoda, E. Kita, K. Nakamura, and T. Kuroki, Japanese Patent 74 69,765 (1974).

82: 67667j   H. Cinatlova, L. Sucha, and M. Suchanek, *Sb. Vys. Sk. Chem. Technol. Praze, Anal. Chem.*, **H10,** 75 (1974).

82: 72886j   K. C. Joshi and D. S. Mehta, *J. Indian Chem. Soc.*, **51,** 613 (1974).

82: 72890f   V. T. Ramakrishnan and W. Lwowski, *Tetrahedron Lett.*, 3249 (1974).

82: 72892h   V. V. Korshak, A. L. Rusanov, S. N. Leont'eva, and T. K. Dzhashiashvili, *Khim. Geterotsikl. Soedin.*, **10,** 1569 (1974).

82: 72909u   J. J. Barr, R. C. Storr, and J. A. Rimmer, *J. Chem. Soc., Chem. Commun.*, 657 (1974).

82: P72994t   B. Boehner, D. Dawes, H. Kny, W. Meyer, and J. Perchais, German Patent 2,417,970 (1974).

82: P72995u   J. L. Misel, French Patent 2,200,276 (1974).

82: P72997w   W. Kraemer, K. H. Buechel, W. Meiser, and M. Plempel, German Patent 2,324,424 (1974).

82: P72999y   E. Regel, L. Eue, and R. R. Schmidt, German Patent 2,321,330 (1974).

82: P73409t   B. Shimizu and A. Saito, Japanese Patent 74 86,372 (1974).

82: P73413q   M. Sunagawa, M. Yamamoto, S. Morooka, M. Koshiba, S. Inaba, and H. Yamamoto, Japanese Patent 74 80,071 (1974).

82: 80069k   P. P. Kish, S. G. Kremeneva, and E. E. Monich, *Zh. Anal. Khim.*, **29,** 1741 (1974).

82: 85553y   J. Mueller, K. Oefele, and G. Krebs, *J. Organomet. Chem.*, **82,** 383 (1974).

82: 86060x   G. Zinner, I. Holdt, and G. Nebel, *Arch. Pharm.* (Weinheim), **307,** 889 (1974).

82: P87423e   G. A. Pope, P. J. Smewin, and G. B. Stokes, German Patent 2,342,929 (1974).

82: 92444s   F. Sell and J. Rodrian, *Justus Liebigs Ann. Chem.*, 1784 (1975).

82: 97303e   E. A. Zvezdina, M. P. Zhdanova, V. A. Bren, and G. N. Dorofeenko, *Khim. Geterotsikl. Soedin.*, **10,** 1461 (1974).

82: 97518d   V. V. Makarskii, M. G. Voronkov, V. P. Feshin, V. A. Lopyrev, N. I. Berestennikov, E. F. Shibanova, and L. I. Volkova, *Dokl. Akad. Nauk SSSR* **220,** 101 (1975).

82: 98308y   W. J. Gensler, S. Chan, and D. B. Ball, *J. Amer. Chem. Soc.*, **97,** 436 (1975).

82: 98356m   J. Gante and G. Mohr, *Chem. Ber.*, **108,** 174 (1975).

82: P100064k   E. Fleckenstein and P. Mischke, German Patent 2,322,404 (1974).

82: 107475d   B. Boehner, D. Dawes, and W. Meyer, *Kem.-Kemi*, **1,** 585 (1974).

82: 111099b   J. Rebeck, S. F. Wolf, and A. B. Mossman, *J. Chem. Soc., Chem. Commun.*, 711 (1974).

82: 111992n   M. Woodruff and J. B. Polya, *Aust. J. Chem.*, **28,** 421 (1975).

82: P113109d   S. Matsuda and K. Takagi, Japanese Patent 74 24,190 (1974).

82: 113168x   E. Fleckenstein, German Patent 2,308,263 (1974).

82: 118780a   V. K. Rastogi, V. K. Agarwal, J. N. Sinha, A. Chaudhari, and S. S. Parmar, *Can. J. Pharm. Sci.*, **9,** 107 (1974).

82: P124906n   J. Beger and H. Fuellbier, (East) German Patent 107,893 (1974).

82: 125326k   M. Woodruff and J. B. Polya, *Aust. J. Chem.*, **28**, 133 (1975).

82: 125327m   V. J. Ram and H. N. Pandey, *Chem. Pharm. Bull.*, **22**, 2778 (1974).

82: 125368a   A. D. Sinegibskaya, G. A. Vshivkova, and I. Ya. Postovskii, *Khim. Khim. Tekhnol., Obl. Nauchno-Tekh. Konf.*, **2**, 41 (1973).

82: P126600g   S. C. Rennison, British Patent 1,374,801 (1974).

82: P126605n   P. Gangneux, German Patent 2,305,550 (1973).

82: 132249r   P. Bassinet and C. Bois, *C. R. Acad. Sci., Paris, Ser. C*, **279**, 627 (1974).

82: 139570x   R. N. Butler and W. B. King, *J. Chem. Soc., Perkin Trans.* 1, 61 (1975).

82: 139995q   F. Russo, M. Santagati, and M. Alberghina, *Farmaco, Ed. Sci.*, **30**, 70 (1975).

82: 140028h   V. A. Kovtunenko, V. N. Bubnovskaya, and F. S. Babichev, *Khim Geterotsikl. Soedin.*, **11**, 138 (1975).

82: 140030c   A. R. Katritzky and J. W. Suwinski, *Tetrahedron Lett.*, 4123 (1974).

82: 140031d   C. N. O'Callaghan, *Proc. R. Ir. Acad., Sect. B*, **74**, 455 (1974).

82: 140033f   E. Fluck and E. Beuerle, *Z. Anorg. Allg. Chem.*, **411**, 125 (1975).

82: P140121h   M. Nakanishi, Y. Naka, and R. Kobayashi, Japanese Patent 74 101,373 (1974).

82: 142281j   R. Rogers, *Thermochim. Acta* **11**, 131 (1975).

82: P150485u   S. A. Greenfield, M. C. Seidel, and W. C. Von Meyer, German Patent 1,966,806 (1974).

82: 154565t   M. Sulzbadher, *Manuf. Chem. Aerosol News*, **46**, 61, 63, 65 (1975).

82: P155791a   W. F. Bruce, U.S. Patent 3,850,915 (1974).

82: 156003g   T. L. Gilchrist, C. W. Rees, and C. Thomas, *J. Chem. Soc., Perkin Trans.* 1, 12 (1975).

82: 156162h   O. Dann, H. Fick, B. Pietzner, E. Walkenhorst, R. Fernbach, and D. Zeh, *Justus Liebigs Ann. Chem.*, 160 (1975).

82: 156184s   M. Zander, *Chem.-Ztg.*, **99**, 92 (1975).

82: 156185t   K. Kondo, N. Sonoda, and H. Sakurai, *Synth. Commun.*, **5**, 131 (1975).

82: 156189x   E. Comanita, M. Tutoveanu, and A. Kasper, *Bul. Inst. Politeh. Iasi, Sect.* 2, **19**, 123 (1973).

82: 156237m   V. Nair, *J. Heterocyclic Chem.*, **12**, 183 (1975).

82: P156320h   G. Tanaka, Japanese Patent 74 95,973 (1974).

82: P156321j   B. Boehner, D. Dawes, W. Meyer, H. Kristinsson, and K. Ruefenacht, German Patent 2,428,204 (1975).

82: P156325p   W. Kraemer, K. H. Buechel, W. Meiser, H. Kaspers, and P. E. Frohberger, German Patent 2,324,010 (1975).

82: P156401k   J. B. Hester, Jr., German Patent 2,439,209 (1975).

82: P157819q   A. Kotone, Y. Nakane, T. Hori, and M. Hoda, Japanese Patent 74 116,129 (1974).

82: 169941m   G. Doleschall, *Tetrahedron Lett.*, 681 (1975).

82: 170243s   M. G. Voronkov, T. I. Yushmanova, E. N. Medvedeva, I. D. Kalikhman, V. V. Keiko, V. V. Makarskii, and V. A. Lopyrev, *Zh. Org. Khim.*, **11**, 902 (1975).

82: 170800w   B. I. Buzykin, L. P. Sysoeva, and Y. P. Kitaev, *Zh. Org. Khim.*, **11**, 173 (1975).

82: 170807d   F. S. Babichev and V. A. Hevtunenko, *Ukr. Khim. Zh.*, **41**, 252 (1975).

82: P170960y   H. E. Eriksson and G. L. Florvall, German Patent 2,427,208 (1975).

82: P170962a   A. Kotone, M. Hoda, and T. Fujita, Japanese Patent 74 126,679 (1974).

82: P170966e   K. H. Buechel, W. Kraemer, and P. E. Frohberger, German Patent 2,335,020 (1975).

82: P170967f   W. Kraemer, K. H. Buechel, W. Brandes, and P. E. Frogberger, German Patent 2,334,352 (1975).

82: P170991j   G. R. Fosker and G. Burton, German Patent 2,428,139 (1975).

82: P171068a   T. Kishimoto, H. Kochi, and Y. Kaneda, German Patent 2,434,310 (1975).
82: P171103h   F. G. Kathawala, U.S. Patent 3,862,137 (1975).
82: P171367x   R. Schmidt and K. Klemm, German Patent 2,426,279 (1975).
82: P172343y   E. Cutts and G. T. Knight, German Patent 2,434,426 (1975).
82: 172419c   E. J. Siepmann, *Text. Chem. Color.*, **7,** 46 (1974).
82: P172596h   K. Konishi, A. Kotone, Y. Nakane, T. Hori, and M. Hoda, Japanese Patent 74 116,127 (1974).
82: P172621n   A. Dorlars, O. Neuner, and U. Claussen, German Patent 2,340,237 (1975).
83: 9910t   J. Vilarrasa and R. Granados, *J. Hererocyclic Chem.*, **11,** 867 (1974).
83: 9914x   O. M. Astakhova, G. S. Supin, O. S. Martynova, N. P. Lyalyakina, and N. I. Shvetsov-Shilovskii, *Zh. Obshch. Khim.*, **45,** 485 (1975).
83: 9917a   L. P. Sysoeva, B. I. Buzykin, and Yu. P. Kitaev, *Zh. Org. Khim.*, **11,** 348 (1975).
83: 9953j   D. E. Klinge, H. C. Van der Plas, G. Geurtsen, and A. Koudijs, *Rec. Trav. Chim. Pays-Bas*, **93,** 236 (1974).
83: 9958q   M. Kovacic, S. Polac, B. Stanovnik, and M. Tisler, *J. Heterocyclic Chem.*, **11,** 949 (1974).
83: 10006j   J. Kobe, D. E. O'Brien, R. K. Bobins, and T. Novinson, *J. Heterocyclic Chem.*, **11,** 991 (1974).
83: 10007k   A. H. Harhash, M. H. Elnagdi, and A. A. A. Elbanani, *Tetrahedron*, **31,** 25 (1975).
83: P10088n   B. Von Bredow and E. Sturm, German Patent 2,429,042 (1975).
83: P10106s   R. M. DeMarinis and J. R. E. Hoover, U.S. Patent 3,865,819 (1975).
83: P10526d   H. Scholz, German Patent 2,331,300 (1975).
83: P18957r   British Patent 1,383,693 (1975).
83: 22239a   K. H. Schram and L. B. Townsend, *J. Carbohydr., Nucleosides, Nucleotides*, **1,** 39 (1974).
83: 27047w   V. N. Sheinker, L. G. Tischenko, A. D. Granovskii, and O. A. Osipov, *Khim. Geterotsikl. Soedin.*, **11,** 571 (1975).
83: 27213x   I. Ya. Postovskii, A. D. Sinegibskaya, and E. G. Kovalev, *Khim Geterotsikl. Soedin.*, **11,** 566 (1975).
83: 27827a   J. C. Jagt and A. M. van Leusen, *Rec. Tran. Chim. Pays-Bas.*, **94,** 12 (1975).
83: 28157u   R. Neidlein and H. G. Reuter, *Arch. Pharm.* (Weinheim), **308,** 189 (1975).
83: 28165v   S. F. Vasilevskii, A. N. Sinyakov, M. S. Shvartsberg, and I. L. Kotlyarevskii, *Izv. Akad. Nauk SSSR, Ser. Khim.*, 690 (1975).
83: P28252w   W. J. Gottstein, M. A. Kaplan, and A. P. Granatek, German Patent 2,422,702 (1975).
83: P29853e   Y. Saito, M. Nakad, T. Sakamoto, S. Kawamura, and K. Okada, Japanese Patent, 75 03,122 (1975).
83: P29887u   A. E. Siegrist, P. Liechti, E. Maeder, L. Guglielmetti, H. R. Meyer, and K. Weber, Swiss Patent 557,917 (1975).
83: P35680n   W. Wiedemann, German Patent 2,336,094 (1975).
83: 42640c   R. F. Smith, S. B. Kaldor, E. D. Laganis, and R. F. Oot, *J. Org. Chem.*, **40,** 1854 (1975).
83: 43117z   S. Yoshina and K. Yamamoto, *Yakugaku Zasshi*, **95,** 219 (1975).
83: 43245q   M. Takahashi and N. Sugawara, *Nippon Kagaku Kaishi*, 334 (1975).
83: P43336v   W. Kraemer and K. H. Buechel, German Patent 2,348,663 (1975).
83: P43337w   K. H. Buechel, M. Plempel, and W. Kraemer, German Patent 2,347,057 (1975).
83: P43338x   K. H. Buechel, W. Kraemer, and M. Plempel, German Patent 2,350,121 (1975).

83: P43339y   K. Meguro, H. Tawada, and Y. Kywada, British Patent 1,374,476 (1974).

83: P43387n   S. C. Lang, Jr., E. Cohen, and A. E. Sloboda, U.S. Patent 3,863,010 (1975).

83: 43657a   G. Just and M. Ramjeesingh, *Tetrahedron Lett.*, 985 (1975).

83: 50701p   K. Shiba, T. Aono, S. Kubodera, T. Hirose, R. Ohi, and T. Shishido, German Patent 2,437,405 (1975).

83: P50807c   British Patent 1,373,415 (1974).

83: 52878a   M. Israilov, D. N. Pachadzhanov, and M. Yu. Yusupov, *Khimiya v Tadzhikistane*, 197 (1974).

83: 53854b   H. Ozawa and H. Oyama, *Yakugaku Zasshi*, **95,** 373 (1975).

83: 58722s   B. I. Buzykin, G. D. Lezhnina, and Yu. P. Kitaev, *Zh. Org. Khim*, **11,** 1078 (1975).

83: 58739c   R. G. Dickinson, N. W. Jacobsen, and R. G. Gillis, *Aust. J. Chem.*, **28,** 859 (1975).

83: P58834e   H. Hoffmann, I. Hammann, W. Behrenz, and W. Stendel, German Patent 2,350,631 (1975).

83: P58835f   B. Von Bredow and H. U. Brechbuehler, German Patent 2,446,900 (1975).

83: P58837h   K. H. Buechel, W. Kraemer, H. Kaspers, and W. Brandes, German Patent 2,350,122 (1975).

83: 58855n   M. Yoshioka, Y. Sendo, M. Murakami, S. Miyazaki, and K. Ishikura, German Patent 2,450,618 (1975).

83: P58895a   K. Meguro, H. Tawada, and Y. Kuwada, U.S. Patent 3,876,634 (1971).

83: P61660a   M. Ohkawa and G. Suzuki, Japanese Patent 75 10,319 (1975).

83: P61703s   K. Loehe and T. Papenfuchs, German Patent 2,341,289 (1975).

83: P79251c   K. H. G. Pilgram, U.S. Patent 3,884,910 (1975).

83: P79262g   M. Numata, M. Yamaoka, Y. Imashiro, and I. Minamida, German Patent 2,446,901 (1975).

83: P79295v   Netherlands Appl. 73 10,900 (1974).

83: P79302v   J. B. Hester, Jr., U.S. Patent 3,880,878 (1975).

83: 79677w   G. S. Gol'din, V. G. Poddubnyi, and A. N. Kol'tsova, *Vysokomol. Soedin., Ser. B*, **17,** 322 (1975).

83: P80456e   A. Kotone, M. Hoda, and T. Hori, Japanese Patent 75 38,743 (1975).

83: P81186d   M. Ohkawa, M. Hanai, S. Abeta, and S. Kawabata, Japanese Patent 74 109,682 (1974).

83: P81197h   A. Kotone, M. Hoda, T. Fujita, and T. Hori, Japanese Patent 75 22,026 (1973).

83: P81198j   E. Fleckenstein and P. Mischke, German Patent 2,338,135 (1975).

83: P81203g   K. Tsunemitsu, M. Ban, and E. Ozaki, Japanese Patent 75 28,525 (1975).

83: P81207m   E. Fleckenstein and R. Mohr, German Patent 2,347,756 (1975).

83: 97141w   G. Wagner and S. Leistner, *Pharmazie*, **30,** 141 (1975).

83: 97150y   A. Leonardi, D. Nardi, and M. Veronese, *Boll. Chim. Farm.*, **114,** 70 (1975).

83: 97223z   R. G. Dickinson and N. W. Jacobsen, *J. Chem. Soc., Perkin Trans. 1*, 975 (1975).

83: 97228e   L. Giammanco, *Atti Accad. Sci., Lett. Arti Palermo, Parte* 1, **33,** 235 (1973).

83: P97302z   A. Kotone, Y. Nakane, M. Hoda, and T. Hori, Japanese Patent 75 04,075 (1975).

83: 97441u   M. Lesbre, A. Laporterie, J. Dubac, and G. Manuel, *C. R. Acad. Sci., Ser. C*, **280,** 787 (1975).

83: 97822u   H. Vorbrueggen, K. Krolikiewicz, and U. Niedballa, *Justus Liebigs Ann. Chem.*, 988 (1975).

83: P99180g   J. Dehnert and G. Lamm, German Patent 2,352,831 (1975).

83: 105984r   B. Steyer and F. P. Schaefer, *Appl. Phys.*, **7,** 113 (1975).

83: 109119k　　R. M. DeMarinis, J. R. E. Hoover, G. L. Dunn, P. Actor, J. V. Uri, and J. A. Weisbach, *J. Antibiot.*, **28,** 463 (1975).

83: 113229p　　K. Bolton, R. D. Brown, F. R. Burden, and A. Mishra, *J. Mol. Struct.*, **27,** 261 (1975).

83: 113599j　　G. Doleschall, *Tetrahedron Lett.*, 1889 (1975).

83: 114298x　　A. Bernardini, P. Viallefont, J. Daunis, M. L. Roumestant, and A. B. Soulami, *Bull. Soc. Chim. Fr.*, 647 (1975).

83: 114345k　　J. Daunis and M. Follet, *Bull. Soc. Chim. Fr.*, 864 (1975).

83: P114412e　　E. Winkelmann and W. Raether, German Patent 2,362,171 (1975).

83: P114414g　　A. Kotone, M. Hoda, and T. Fujita, Japanese Patent 75 04,074 (1975).

83: 114416j　　L. P. Makhno, T. G. Ermakova, E. S. Damnina, L. A. Tatarova, G. G. Skvortsova, and V. A. Lopyrev, U.S.S.R. Patent 464,584 (1975).

83: P114420f　　D. Dawes, B. Boehner, and W. Meyer, German Patent 2,457,146 (1975).

83: 114441p　　K. Heusler, H. Bickel, B. Fechtig, H. Peter, and R. Scartazzini, Swiss Patent 562,250 (1975).

83: 114784c　　J. T. Witkowski, M. Fuertes, P. Cook, and R. K. Robins, *J. Carbohydr., Nucleosides, Nucleotides*, **2,** 1 (1975).

83: 115119b　　P. M. Hergenrother, *J. Polym. Sci., Polym. Chem. Ed.*, **12,** 2857 (1974).

83: P116930w　　A. Kotone, M. Hoda, T. Fujita, and T. Hori, Japan. Patent 75 34,623 (1975).

83: P116968q　　T. Papenfuhs, E. Spietschka, and H. Troester, German Patent 2,360,705 (1975).

83: 126026y　　J. Zulalian, D. A. Champagne, R. S. Wayne, and R. C. Blinn, *J. Agric. Food Chem.*, **23,** 724 (1975).

83: 126268a　　L. L. Grechishkin and L. K. Gavrovskaya, *Byull. Eksp. Biol. Med.*, **79,** 71 (1975).

83: P126969e　　Y. Hamada, T. Matsuno, T. Ishii, K. Imai, and M. Mano, Japanese Patent 75 63,119 (1975).

83: 131395u　　A. R. Katritzky and J. W. Suwinski, *Tetrahedron*, **31,** 1549 (1975).

83: 131519n　　T. Takeshima, N. Fukada, E. Okabe, F. Mineshima, and M. Muraoka, *J. Chem. Soc., Perkin Trans. 1*, 1277 (1975).

83: 131874z　　G. F. Huang, M. Maeda, T. Okamoto, and Y. Kawazoe, *Tetrahedron*, **31,** 1363 (1975).

83: 131973f　　V. V. Korshak, A. L. Rusanov, T. G. Iremashvili, L. Kh. Plieva, and T. V. Lekae, *Makromol. Chem.*, **176,** 1233 (1975).

83: 141238j　　J. Janickis, J. Valanciunas, V. Zelionkaite, V. Janickis, and S. Grevys, *Liet. TSR Mokslu Akad. Darb., Ser. B*, 63 (1975).

83: 146700p　　A. Bernardini, P. Viallefont, and J. Daunis, *J. Heterocyclic Chem.*, **12,** 655 (1975).

83: 146782s　　W. Lwowski, R. A. De Mauriac, M. Thompson, and R. E. Wilde, *J. Org. Chem.*, **40,** 2608 (1975).

83: 146905j　　M. Woodruff and J. B. Polya, *Aust. J. Chem.*, **28,** 1583 (1975).

83: 147414k　　L. Kramberger, P. Lorencak, S. Polac, B. Vercek, B. Stanovnik, M. Tisler, and F. Povazanec, *J. Heterocyclic Chem.*, **12,** 337 (1975).

83: 147418q　　S. M. Khripak, M. M. Tsitsika, I. V. Smolanka, *Khim. Geterotsikl. Soedin.*, **11,** 1000 (1975).

83: P155710u　　W. Fujimatsu, S. Sato, T. Kojima, S. Itoh, and T. Endo, German Patent 2,433,812 (1975).

83: 163180m　　R. Kallury, M. R. Krishna, T. G. S. Nath, and V. R. Srinivasan, *Aust. J. Chem.*, **28,** 2089 (1975).

83: 163373b　　S. Kubota and M. Uda, *Chem. Pharm. Bull.*, **23,** 955 (1975).

83: 164088z　　J. Heindl, E. Schroeder, and H. W. Kelm, *Eur. J. Med. Chem.-Chim. Ther.*, **10,** 121 (1975).

83: 164091v　　S. M. Khripak, M. M. Tsitsika, and I. V. Smolanka, *Khim. Geterotsikl. Soedin.*, **11,** 844 (1975).

83: 164137q   A. B. Katunina, E. G. Kovalev, and G. A. Kitaev, *Khim. Geterotsikl. Soedin.*, **11**, 847 (1975).

83: P164183b   A. Kotone, M. Hoda, and T. Fujita, Japanese Patent 74 117,470 (1974).

83: P164184c   J. J. Baldwin and F. C. Novello, German Patent 2.424,404 (1974).

83: P164189h   H. Hoffmann, I. Hammann, W. Behrenz, and W. Stendel, German Patent 2,361,451 (1975).

83: 164191c   J. Krenzer, U.S. Patent 3,890,342 (1975).

83: 164195g   R. Aries, French Patent 2,240,219 (1975).

83: 164206m   R. M. De Marinis and J. R. E. Hoover, German Patent 2,426,970 (1974).

83: P165712k   J. J. Leucker and H. Mollet, Swiss Patent 564,068 (1975).

83: P165797s   H. P. Kuehlthau and H. Beecken, German Patent 2,355,076 (1975).

83: P170844j   M. Fujiwara, R. Sato, T. Masukawa, and T. Endo, Japanese Patent 75 36,125 (1975).

83: 172416v   J. H. Sellstedt, C. J. Guinosso, A. J. Begany, S. C. Bell, and M. Rosenthale, *J. Med. Chem.*, **18**, 926 (1975).

83: 173146u   A. Rastelli, P. G. De Benedetti, G. Gavioli Battistuzzi, and A. Albasini, *J. Med. Chem.*, **18**, 963 (1975).

83: 178256q   R. Un, *Chim. Acta Turc.*, **2**, 115 (1974).

83: P178845f   J. R. Beck and R. G. Suhr, U.S. Patent 3,897,440 (1975).

83: 178919h   J. VIlarrasa, C. Galvez, and M. Calafell, *An. Quim.*, **71**, 631 (1975).

83: 178931f   E. Melendez and J. Vilarrasa, *An. Quim.*, **70**, 966 (1974).

83: 178933h   M. Gannon, J. E. McCormick, and R. S. McElhinney, *Proc. R. Ir. Acad., Sect. B*, **74**, 331 (1974).

83: 178937n   C. J. Wilkerson and F. D. Greene, *J. Org. Chem.*, **40**, 3112 (1975).

83: 178946q   R. M. Claramunt, R. Granados, J. Elguero, and R. Lazaro, *An. Quim.*, **70**, 986 (1974).

83: 178980w   L. Lorente, R. Madronero, and S. Vega, *An. Quim.*, **70**, 974 (1974).

83: P179068s   A. Kotone, Japanese Patent 75 35,165 (1975).

83: P179070m   W. Kraemer, C. Stoelzer, K. H. Buechel, and W. Meiser, German Patent 2,401,715 (1975).

83: P179071n   Belgium Patent 821,084 (1975).

83: P179072p   C. Fauran, C. Douzon, G. Raynaud, and B. Pourrias, French Patent 2,244,462 (1975).

83: P179133j   R. Aries, French Patent 2,248,274 (1975).

83: 179253y   R. Manske and J. Gosselck, *Tetrahedron*, **31**, 2121 (1975).

83: 179393u   J. Brynjolffssen, A. Emke, D. Hands, J. M. Midgley, and W. B. Whalley, *J. Chem. Soc., Chem. Commun.*, 633 (1975).

83: P181082k   M. Ozutsumi, S. Maeda, and Y. Dawada, U.S. Patent 3,822,247 (1974).

83: 181099w   K. Hohmann, German Patent 2,357,448 (1975).

83: 186250y   Anon., *Res. Discl.*, **136**, 39 (1975).

83: 188199n   J. J. Baldwin, P. A. Kasinger, F. C. Novello, J. M. Sprague, and D. E. Duggan, *J. Med. Chem.*, **18**, 895 (1975).

83: 188770s   V. I. Fomenko, N. I. Geonya, V. F. Lipnitskii, O. P. Shvaika, G. P. Kondratenko, and S. N. Baranov, *Fiziol. Akt. Veshchestva*, **7**, 64 (1975).

83: P189336s   K. H. Buechel, W. Kraemer, W. Meiser, and K. Luerssen, German Patent 2,407,143 (1975).

83: 192762h   M. Tutoveanu and E. Comanita, *Dokl. Bolg. Akad. Nauk*, **28**, 763 (1975).

83: 193182f   M. O. Lozinskii, T. N. Kudrya, and S. V. Bonadyk, *Zh. Org. Khim.*, **11**, 1573 (1975).

83: 193185j   G. L. Maheshwari, R. P. Mahesh, and P. Singh, *Curr. Sci.*, **44**, 549 (1975).

83: 193187m   M. J. Cooper, R. Hull, and M. Wardleworth, *J. Chem. Soc., Perkin Trans. 1*, 1433 (1975).

83: 193188n   A. Walser, T. Flynn, and R. I. Fryer, *J. Heterocyclic Chem.*, **12,** 717 (1975).

83: 193189p   T. J. Schwan and R. L. White, Jr., *J. Heterocyclic Chem.*, **12,** 771 (1975).

83: 193193k   K. Matsumoto, M. Suzuki, M. Tomie, N. Yoneda, and M. Miyoshi, *Synthesis*, 609 (1975).

83: 193226y   G. N. Tyurenkova, N. V. Serebryakova, and I. I. Mudretsova, *Zh. Org. Khim.*, **11,** 1669 (1975).

83: 193255g   E. M. Essassi, J. P. Lavergne, Ph. Viallefont, and J. Daunis, *J. Heterocyclic Chem.*, **12,** 661 (1975).

83: 193337k   A. Hubele, German Patent 2,506,598 (1975).

83: 193683v   A. S. Dutta and J. S. Morley, *J. Chem. Soc., Perkin Trans.* 1, 1712 (1975)

83: P194350w   P. Strop, O. Mikes, and J. Coupek, German Patent 2,504,031 (1975).

83: P195211p   A. Kotone, M. Hoda, T. Fujita, and T. Hori, Japanese Patent 75 26,829 (1975).

83: P195215t   A. Kotone, M. Hoda, T. Fujita, and T. Hori, Japanese Patent 75 34,029 (1975).

83: 205712u   R. Esmail and F. Kurzer, *J. Chem. Soc., Perkin Trans.* 1, 1781 (1975).

83: 206077c   S. Yoshina, A. Tanaka, C.-H. Wu, and H. Kuo, *Yakugaku Zasshi*, **95,** 883 (1975).

83: 206172e   J. Lange and H. Tondys, *Pol. J. Pharmacol. Pharm.*, **27,** 203 (1975).

83: P206281q   D. Dawes, B. Boehner, and W. Meyer, German Patent 2,457,147 (1975).

83: P206285u   H. Hoffmann, I. Hammann, and W. Stendel, German Patent 2,403,711 (1975).

83: P206286v   German Patent 2,424,670 (1974).

83: P206289y   C. Stoelzer, W. Kraemer, K. H. Buechel, and W. Meiser, German Patent 2,406,665 (1975).

83: P206322d   M. Yamamoto, S. Morooka, M. Koshiba, S. Inaba, and H. Yamamoto, German Patent 2,508,333 (1975).

83: P206340h   Y. Kuwada, K. Meguro, Y. Sato, H. Tawada, T. Sohda, and H. Natsugari, German Patent 2,508,329 (1975).

83: 206479k   B. B. Snider, R. J. Corcoran, and R. Breslow, *J. Amer. Chem. Soc.*, **97,** 6580 (1975).

83: 207574z   H. Scheuermann, W. Mach, and D. Augart, German Patent 2,406,220 (1975).

83: P209412u   J. R. Guarini, U.S. Patent 3,903,278 (1975).

84: 4425z   R. Banks, R. F. Brookes, and D. H. Godson, *J. Chem. Soc., Perkin Trans.* 1, 1836 (1975).

84: 4541j   I. Sasson and J. Labovitz, *J. Org. Chem.*, **40,** 3670 (1975).

84: 4840f   S. N. Sawhney, J. Singh, and O. P. Bansal, *Indian J. Chem.*, **13,** 804 (1975).

84: 4861p   R. Un and A. Ikizler, *Chim. Acta Turc.*, **3,** 1 (1975).

84: 4865t   R. Esmail and F. Kurzer, *J. Chem. Soc., Perkin Trans.* 1, 1787 (1975).

84: 4866u   H. Behringer and R. Ramert, *Justus Liebigs Ann. Chem.*, 1272 (1975).

84: 4868w   M. D. Lipanova, E. V. Burov, and L. K. Kulikova, *Issled. V Obl. Sinteza Katal. Organ. Soedin.*, 29 (1975).

84: 4965a   H. Hoffmann, I. Hammann, B. Homeyer, and W. Stendel, German Patent 2,407,304 (1975).

84: 4966b   J. T. Witkowski and R. K. Robins, German Patent 2,511,829 (1975).

84: P6394n   S. Abeta and T. Okaniwa, Japanese Patent 75 77,675 (1975).

84: P6465m   F. Urseanu, C. Tarabasanu-Mihaila, L. Negoita, and V. Zdrentu, *Rev. Roum. Chim.*, **20,** 791 (1975).

84: P6478t   B. Henzi, German Patent 2,509,095 (1975).

84: 16586r   M. E. Burrage, R. C. Cookson, S. S. Gupte, and I. D. R. Stevens, *J. Chem. Soc., Perkin Trans.* 2, 1325 (1975).

84: 17043y   K. Yamamoto and S. Yoshina, *Yakugaku Zasshi*, **95,** 1218 (1975).

84: 17236p   J. Lange and H. Tondys, *Pol. J. Pharmacol. Pharm.*, **27**, 211 (1975).

84: 17237q   A. Bernardini, P. Viallefont, J. Daunis, and M. L. Roumestant, *Bull. Soc. Chim. Fr.*, 1191 (1975).

84: 17238r   D. S. Deshpande, T. G. S. Nath, and V. R. Srinivasan, *Indian J. Chem.*, **13**, 851 (1975).

84: 17241m   J. L. Barascut, P. Viallefont, and J. Daunis, *Bull. Soc. Chim. Fr.*, 1649 (1975).

84: 17271w   R. G. Dickinson and N. W. Jacobsen, *Aust. J. Chem.*, **28**, 2435 (1975).

84: 17287f   G. Doleschall and K. Lempert, *Tetrahedron*, **30**, 3997 (1974).

84: 17652w   N. Katagiri, K. Itakura, and S. A. Narang, *J. Amer. Chem. Soc.*, **97**, 7332 (1975).

84: P19080g   T. Okaniwa and S. Abeta, Japanese Patent 75 107,278 (1975).

84: 30967x   T. P. Kofman, V. I. Manuilova, M. S. Pevzner, and T. N. Timofeeva, *Khim. Geterotsikl. Soedin.*, **11**, 705 (1975).

84: 30972v   J. Drapier, A. J. Hubert, and P. Teyssie, *Synthesis*, 649 (1975).

84: 30975y   A. S. Azaryan, N. S. Iradyan, and A. A. Aroyan, *Arm. Khim. Zh.*, **28**, 709 (1975).

84: 30981x   G. Germain, J. P. Declercq, M. VanMeerssche, F. Hervens, H. G. Viehe, and A. Michel, *Bull. Soc. Chim. Belg.*, **84**, 1005 (1975).

84: 31078v   R. Ludwig and E. Tenor, (East) German Patent 111,074 (1975).

84: P31084u   M. Gall and J. B. Hester, Jr., U.S. Patent 3,910,943 (1975).

84: 31088y   J. Krenzer, German Patent 2,510,573 (1975).

84: P31146r   J. B. Hester, Jr., U.S. Patent 3,903,103 (1975).

84: P31151p   Y. Kuwada, H. Natsugari, K. Meguro, and H. Tawada, German Patent 2,516,674 (1975).

84: P44070m   C. J. Paget, Jr. and J. H. Wikel, German Patent 2,509,843 (1975).

84: P44197h   Y. Kuwada, K. Meguro, Y. Sato, H. Tawada, T. Sohda, and H. Natsugari, German Patent 2,514,663 (1975).

84: P44206k   A. Gagneux, R. Heckendorn, and R. Meier, Swiss Patent 569,018 (1975).

84: P44603n   J. T. Witkowski and R. K. Robins, German Patent 2,511,828 (1975).

84: 44710v   Cr. Simionescu, E. Comanita, and M. Vata, *Acta Chim. Acad. Sci. Hung.*, **86**, 459 (1975).

84: P55326a   J. Krenzer, U.S. Patent 3,925,054 (1975).

84: 59312r   H. Behringer and R. Ramert, *Justus Liebigs Ann. Chem.*, 1264 (1975).

84: 59317w   E. J. Browne, *Aust. J. Chem.*, **28**, 2543 (1975).

84: 59321t   A. Kreutzberger and B. Meyer, *Arch. Pharm.* (Weinheim), **308**, 868 (1975).

84: 59327z   V. I. Efremenko, A. A. Tsurkan, and L. I. Kozhevnikova, *Sb. Nauchn. Tr., Ryazan. Med. Inst.*, **50**, 68 (1975).

84: 59334z   P. Wolkoff, S. T. Nemeth, and M. S. Gibson, *Can. J. Chem.*, **53**, 3211 (1975).

84: 59374n   R. G. Dickinson and N. W. Jacobsen, *Aust. J. Chem.*, **28**, 2741 (1975).

84: 59484y   H. Hoffmann and I. Hammann, German Patent 2,420,068 (1975).

84: 59485z   T. Agawa, Y. Ohshiro, K. Takagi, and M. Komatsu, Japanese Patent 75 117,775 (1975).

84: 59486a   T. Agawa, Y. Ohshiro, K. Takagi, and M. Komatsu, Japanese Patent 75 117,774 (1975).

84: P61174r   J. Dehnert and W. Juenemann, German Patent 2,418,081 (1975).

84: P67786p   K. Kinjo, H. Matsuno, and T. Hasegawa, German Patent 2,448,824 (1975).

84: 74076k   W. L. Mock and J. H. McCausland, *J. Org. Chem.*, **41**, 242 (1976).

84: P74272w   H. Hoffmann, I. Hammann, B. Homeyer, and W. Stendel, German Patent 2,419,623 (1975).

84: P74273x   H. Hoffmann, I. Hammann, and W. Stendel, German Patent 2,423,765 (1975).

84: P74275z   J. J. Baldwin and F. C. Novello, U.S. Patent 3,928,361 (1975).

84: 74336v   A. Laporterie, J. Dubac, and M. Lesbre, *J. Organomet. Chem.*, **101,** 187 (1975).

84: P75683t   V. Kaeppeli, German Patent 2,518,345 (1975).

84: P75707d   K. Loehe and T. Papenfuhs, German Patent 2,423,547 (1975).

84: P85617p   K. H. Buechel, W. Kraemer, and H. Kaspers, German Patent 2,423,987 (1975).

84: P85645w   H. Hoffmann, I. Hammann, and B. Homeyer, German Patent 2,422,548 (1975).

84: P85647y   H. Hoffmann, I. Hammann, B. Homeyer, and W. Stendel, German Patent 2,423,683 (1975).

84: 89679h   A. J. Bridges and G. H. Whitham, *J. Chem. Soc., Perkin Trans.* 1, 2264 (1975).

84: 90078t   A. M. Van Leusen, B. E. Hoogenboom, and H. A. Houwing, *J. Org. Chem.*, **41,** 711 (1976).

84: 90081p   J. K. Fraser, D. G. Neilson, L. R. Newlands, K. M. Watson, and M. I. Butt, *J. Chem. Soc., Perkin Trans* 1, 2280 (1975).

84: 90086u   M. M. Tsitsika, S. M. Khripak, and I. V. Smolanka, *Khim. Geterotsikl. Soedin.*, **11,** 1564 (1975).

84: 90089x   J. L. Closset, A. Copin, Ph. Dreze, F. Alderweireldt, F. Durant, G. Evrard, and A. Michel, *Bull. Soc. Chim. Belg.*, **84,** 1023 (1975).

84: 90123d   J. T. A. Boyle, M. F. Grundon, and M. D. Scott, *J. Chem. Soc., Perkin Trans.* 1, 207 (1976).

84: 90156s   E. P. Krysin, S. N. Levchenko, L. G. Andronova, and M. M. Nabenina, U.S.S.R. Patent 491,636 (1975).

84: 90504d   G. Wagner, B. Dietzsch, and U. Krake, *Pharmazie*, **30,** 694 (1975).

84: 91634h   J. G. Fisher and J. M. Straley, U.S. Patent 3,928,311 (1975).

84: P91665u   E. Schinzel, H. Frischkorn, and G. Roesch, German Patent 2,524,677 (1976).

84: 92352b   P. E. Rouse, Jr., *J. Chem. Eng. Data*, **21,** 16 (1976).

84: P100878v   K. H. Buechel and W. Stendel, German Patent 2,424,891 (1975).

84: 105494b   A. Novacek and J. Gut, *Collect. Czech. Chem. Commun.*, **40,** 3512 (1975).

84: 105495c   S. S. Mathur and H. Suschitzky, *J. Chem. Soc., Perkin Trans.* 1, 2474 (1975).

84: 105498f   G. Wagner, J. Kamecki, and S. Leistner, *Pharmazie*, **30,** 803 (1975).

84: 105501b   A. A. Stotskii and N. P. Tkacheva, *Zh. Org. Khim.*, **12,** 235 (1976).

84: 105503d   S. A. Lang, Jr., B. D. Johnson, and E. Cohen, *J. Heterocyclic Chem.*, **12,** 1143 (1975).

84: P105603m   B. Boehner, D. Dawes, W. Meyer, J. Perchais, and H. Fischer, German Patent 2,525,851 (1976).

84: P105604n   B. Boehner, D. Dawes, W. Meyer, J. Perchais, and H. Fischer, German Patent 2,525,852 (1976).

84: P105667k   A. Gagneux, R. Heckendorn, and R. Meier, Swiss Patent 569,739 (1975).

84: 105990k   Yu. A. Berlin, V. A. Efimov, M. N. Kolosov, V. G. Korobko, O. G. Chakhmakhcheva, and L. N. Shingarova, *Bioorg. Khim.*, **1,** 1121 (1975).

84: P121506n   K. Sasse and L. Eue, German Patent 2,423,536 (1975).

84: 121737p   M. Hayashi, K. Yamauchi, and M. Kinoshita, *Bull. Chem. Soc. Jap.*, **49,** 283 (1976).

84: P121839y   J. Krenzer, U.S. Patent 3,922,162 (1975).

84: P121844w   R. Heckendorn and R. Meier, German Patent 2,525,691 (1976).

84: P121845x   M. Gall, German Patent 2,524,119 (1975).

84: 121930w   G. S. Gol'din, V. G. Poddubnyi, and A. A. Simonova, *Zh. Obschch. Khim.*, **45**, 2665 (1975).

84: 130811k   T. Novinson, T. Okabe, R. K. Robins, and T. R. Matthews, *J. Med. Chem.*, **19**, 517 (1976).

84: 134733j   A. Maquestiau, Y. Van Haverbeke, R. Flammang, R. Huisgen, and Y. B. Chae, *Bull. Soc. Chim. Belg.*, **84**, 1179 (1975).

84: 135551k   N. P. Lyalyakina, G. S. Supin, N. I. Shvetsov-Shilovskii, O. M. Astakhova, and O. S. Martynova, *Zh. Obschch. Khim.*, **46**, 176 (1976).

84: 135600a   G. Tennant and R. J. S. Vevers, *J. Chem. Soc., Perkin Trans.* 1, 421 (1976).

84: 135602c   H. Neunhoeffer and V. Boehnisch, *Justus Liebigs Ann. Chem.*, 153 (1976).

84: 135677f   T. I. Watkins and D. M. Weighton, German Patent 2,530,287 (1976).

84: 135700h   O. Wakabayashi, K. Matsuya, H. Ohta, T. Jikihara, and S. Suzuki, German Patent 2,526,358 (1976).

84: P135730t   Y. Kuwada, K. Meguro, H. Tawada, T. Sohda, H. Natsugari, and Y. Sato, German Patent 2,521,326 (1975).

84: 135765h   D. Quane and S. D. Roberts, *J. Organomet. Chem.*, **108**, 27 (1976).

84: 136515p   R. L. Ellis, U.S. Patent 3,932,444 (1976).

84: P137129c   T. Okaniwa and S. Obeta, Japanese Patent 75 145,681 (1975).

84: P137130w   T. Okaniwa and S. Abeta, Japanese Patent 75 145,680 (1975).

84: 145644g   W. Grunow, H. J. Altmann, and C. Boehme, *Arch. Toxicol.*, **34**, 315 (1975).

84: 150568q   R. Un and A. Ikizler, *Chim. Acta Turc.*, **3**, 113 (1975).

84: P150633g   A. Kotone, M. Hoda, T. Hori, and Y. Nakane, Japanese Patent 75 82,066 (1975).

84: P150634h   T. P. Kofman, M. S. Pevzner, and I. V. Vasil'eva, U.S.S.R. Patent 497,298 (1975).

84: P150636k   H. Timmler, W. Kraemer, K. H. Buechel, H. Kaspers, and W. Brandes, German Patent 2,431,407 (1976).

84: P150671t   Y. Kuwada, H. Natsugari, and K. Meguro, German Patent 2,520,143 (1975).

84: P152199u   K. Arakawa, M. Yamamoto, and N. Suzuki, Japanese Patent 75 156,539 (1975).

84: 152200n   K. Arakawa, M. Yamamoto, and T. Kurahashi, Japanese Patent 75 156,540 (1975).

84: 164588h   I. A. Mazur, R. I. Katkevich, L. I. Borodin, A. A. Martynovskii, and E. G. Knish, *Farm. Zh.* (Kiev), **31**, 63 (1976).

84: 164664e   J. Heindl, E. Schroeder, and H. W. Kelm, *Eur. J. Med. Chem.–Chim. Ther.*, **10**, 591 (1975).

84: 164686p   H. Wamhoff and K. Wald, *Org. Prep. Proced. Int.*, **7**, 251 (1975).

84: 164789z   C. Fuaran, C. Douzon, G. Raynaud, and B. Pourrias, French Patent 2,269,938 (1975).

84: 164874y   A. Gagneux, R. Heckendorn, and R. Meier, Swiss Patent 572,057 (1976).

84: P166142g   M. Haehnke and E. Fleckenstein, German Patent 2,433,233 (1976).

84: P169551a   K. H. Buechel and M. Plempel, German Patent 2,430,039 (1976).

84: P174980u   R. F. Brookes and L. G. Copping, *Pest. Sci.*, **6**, 665 (1975).

84: 175174w   Austrian Patent 326,410 (1975).

84: P175191z   B. Boehner, W. Meyer, and D. Dawes, Swiss Patent 571,817 (1976).

84: 180140m   T. I. Yushmanova, E. N. Medvedeva, L. I. Volkova, V. V. Makarskii, V. A. Lopyrev, and M. G. Voronkov, *Khim. Geterotsikl. Soedin.*, **12**, 421 (1976).

84: 180171x   R. G. Dickinsor and N. W. Jacobsen, *Aust. J. Chem.*, **29,** 459 (1976).

84: 180232t   B. Boehner, W. Meyer, and D. Dawes, Swiss Patent 573,206 (1976).

84: 180518r   E. J. Prisbe, J. Smejkal, J. P. H. Verheyden, and J. G. Moffatt, *J. Org. Chem.*, **41,** 1836 (1976).

84: 184927w   J. J. Baldwin and F. C. Novello, U.S. Patent 3,947,577 (1976).

85: 191t   M. V. Pickering, J. T. Witkowski, and R. K. Robins, *J. Med. Chem.*, **19,** 841 (1976).

85: 5552w   D. W. Wiley, O. W. Webster, and E. P. Blanchard, *J. Org. Chem.*, **41,** 1889 (1976).

85: 5568f   H. Singh and L. D. S. Yadav, *Agric. Biol. Chem.*, **40,** 759 (1976).

85: P5644c   G. Asato, G. Berkelhammer, and W. H. Gastrock, U.S. Patent 3,940,411 (1976).

85: P5647f   W. Lunkenheimer, G. Jaeger, and F. Hoffmeister, German Patent 2,440,378 (1976).

85: P5701u   A. Gagneux, R. Heckendorn, and R. Meier, Swiss Patent 573,931 (1976).

85: P7172w   M. Hanai, G. Suzuki, M. Ohkawa, Y. Matsuo, and S. Kawabata, Japanese Patent 75 138,188 (1975).

85: P7255a   K. Arakawa, M. Yamamoto, and T. Kurahashi, Japanese Patent 76 23,517 (1976).

85: P10088d   J. Haase, Q. Bowes, and R. F. Wurster, German Patent 2,539,979 (1976).

85: P12288t   S. Kloetzer, E. Moisar, and H. Von Rintelen, German Patent 2,416,814 (1975).

85: 20278e   A. Hassner, D. Tang, and J. Keogh, *J. Org. Chem.*, **41,** 2102 (1976).

85: 21220s   F. S. Babichev, V. A. Kovtunenko, and E. Ya. Shapiro, *Ukr. Khim. Zh.*, **41,** 1053 (1975).

85: 21230v   H. E. Eriksson and L. G. Florvall, *Acta Pharm. Suec.*, **13,** 79 (1976).

85: 21235a   A. Yu. Lazaris, S. M. Shmuilovich, and A. N. Egorochkin, *Khim. Geterotsikl. Soedin.*, **12,** 424 (1976).

85: 21243b   C. N. O'Callaghan, *Proc. R. Ir. Acad., Sect. B*, **76,** 37 (1976).

85: 21300t   B. S. Drach, V. A. Kovalev, and A. V. Kirsanov, *Zh. Org. Khim.*, **12,** 673 (1976).

85: P21373u   G. J. Durant, J. C. Emmett, and C. R. Ganellin, U.S. Patent 3,950,353 (1976).

85: P21507r   A. Gagneux, R. Heckendorn, and R. Meier, Swiss Patent 574,426 (1976).

85: P21708g   D. J. Aberhart and A. C. T. Hsu, *J. Org. Chem.*, **41,** 2098 (1976).

85: 21797k   H. R. Kricheldorf, M. Fehrle, and J. Kaschig, *Angew. Chem.*, **88,** 337 (1976).

85: P22740y   K. Arakawa, M. Yamamoto, and T. Kurahashi, Japanese Patent 76 18,739 (1976).

85: P22741z   K. Arakawa, M. Yamamoto, and N. Suzuki, Japanese Patent 76 18,738 (1976).

85: P22745d   K. Arakawa, M. Yamamoto, and N. Suzuki, Japanese Patent 76 23,516 (1976).

85: 29406c   L. G. Copping and R. F. Brookes, *Proc. Br. Weed Control Conf.*, **2,** 809 (1974).

85: 32910t   G. C. Barrett and R. Walker, *Tetrahedron*, **32,** 571 (1976).

85: 32926c   N. A. Kassab, S. O. Abdalla, and H. A. R. Ead, *Z. Naturforsch.: Anorg. Chem., Org. Chem.*, **31B,** 380 (1976).

85: 32927d   H. Kato, T. Shiba, E. Kitajima, T. Kiyosawa, F. Yamada, and T. Nishiyama, *J. Chem. Soc., Perkin Trans. 1*, 863 (1976).

85: P33015s   G. J. Durant, J. C. Emmett, and C. R. Ganellin, U.S. Patent 3,950,333 (1976).

85: P33028y   W. C. Doyle, Jr. and J. L. Kirkpatrick, German Patent 2,545,245 (1976).

85: P34629g   H. P. Kuehlthau, German Patent 2,437,549 (1976).

85: 42100t   M. Mazza, L. Montanari, and F. Payanetto, *Farmaco, Ed. Sci.*, **31**, 334, (1976).

85: 45813p   R. Faure, E. J. Vincent, R. M. Claramunt, J. M. Fabrega, and J. Elguero, *Tetrahedron*, **32**, 341 (1976).

85: 46328w   S. Yoshina, A. Tanaka, C.-H. Wu, and H.-S. Kuo, *Yakugaku Zasshi*, **95**, 883 (1975).

85: 46473q   H. Uno, M. Kurokawa, K. Natsuka, Y. Yamato, and Nishimura, *Chem. Pharm. Bull.*, **24**, 632 (1976).

85: 46523f   B. V. Ioffe and V. A. Chernyshev, *Khim. Geterotsikl. Soedin.*, **12**, 563 (1976).

85: 46578c   B. Vercek, B. Stanovnik, and M. Tisler, *Heterocycles*, **4**, 943 (1976).

85: 46601e   O. Tsuge and S. Kanemasa, *Asahi Garasu Kogyo Gijutsu Shoreikai Kenkyu Hokoku*, **26**, 101 (1975).

85: 46603g   C. P. Joshua and V. P. Rajan, *Aust. J. Chem.*, **29**, 1051 (1976).

85: P46688p   B. Boehner, D. Dawes, and W. Meyer, German Patent 2,537,973 (1976).

85: P46692k   H. Allgeier, German Patent 2,547,942 (1976).

85: 56539e   M. Gall, J. B. Hester, Jr., A. D. Rudzik, and R. A. Lahti, *J. Med. Chem.*, **19**, 1057 (1976).

85: P58083a   H. Timmler, W. Draber, K. H. Buechel, and K. Luerssen, German Patent 2,448,060 (1976).

85: P58084b   H. Timmler, W. Draber, K. H. Buechel, and K. Luerssen, German Patent 2,448,062 (1976).

85: P63073b   A. Kotone and M. Hoda, Japanese Patent 76 16,671 (1976).

85: 77387u   M. G. Voronkov, T. V. Kashik, V. V. Makarskii, V. A. Lopyrev, S. M. Ponomareva, and E. F. Shibanova, *Dokl. Akad. Nauk SSSR*, **227**, 1116 (1976).

85: 77541q   H. Neunhoeffer, H. J. Degen, and J. J. Koehler, *Justus Liebigs Ann. Chem.*, 1120 (1975).

85: 78052t   A. A. Stotskii and N. P. Tkacheva, *Zh. Org. Khim.*, **12**, 655 (1976).

85: P78172g   W. L. Albrecht and F. W. Sweet, U.S. Patent 3,954,983 (1976).

85: P78173h   W. L. Albrecht, U.S. Patent 3,954,984 (1976).

85: 78185p   S. Fischer, L. K. Peterson, and J. F. Nixon, *Can. J. Chem.*, **52**, 3981 (1974).

85: 79672g   K. Arakawa, M. Yamamoto, and N. Suzuki, Japanese Patent 76 56,828 (1976).

85: 94283z   L. H. Dao and D. Mackay, *J. Chem. Soc., Chem. Commun.*, 326 (1976).

85: 94296f   C. A. Renner and F. D. Greene, *J. Org. Chem.*, **41**, 2813 (1976).

85: 94306j   S. C. Bennur, V. B. Jigajinni, and V. V. Badiger, *Rev. Roum. Chim.*, **21**, 757 (1976).

85: 94307k   A. N. Kost, R. S. Sagitullin, and G. G. Danagulyan, *Khim. Geterotsikl. Soedin.*, **12**, 706 (1976).

85: 94310f   P. Vainilavicius, G. Mekuskiene, and L. Jasinskas, *Zh. Vses. Khim. Obshchest.*, **21**, 352 (1976).

85: P94368f   G. Van Reet, J. Heeres, and L. Wals, German Patent 2,551,560 (1976).

85: P94369g   G. Winters, G. Odasso, G. Galliani, and L. J. Lerner, German Patent 2,551,879 (1976).

85: 94637t   T. E. England and T. Neilson, *Can. J. Chem.*, **54**, 1714 (1976).

85: 95772v   G. Alberti, A. Cerniani, and G. Seu, *Ann. Chim.* (Rome), **65**, 305 (1975).

85: P95798h   K. Ohkawa, Y. Takeda, and K. Hirabayashi, Japanese Patent 76 07,173 (1976).

85: P102316e   A. Quaglia, German Patent 2,528,638 (1976).

85: P104199t   J. W. Black and M. E. Parsons, U.S. Patent 3,954,982 (1976).

85: 108584n   A. Ya. Lazaris and A. N. Egorochkin, *Izv. Akad. Nauk SSSR, Ser. Khim.*, 1191 (1976).

85: P110086v   D. M. Fawkes, German Patent 2,555,334 (1976).

85: P117969u   T. Shigematsu, M. Tomida, T. Shibahara, M. Nakazawa, and T. Munakata, German Patent 2,554,866 (1976).

85: 123823h   O. P. Shvaika, N. I. Korotkikh, G. F. Tereshchenko, and N. A. Kovach, *Khim. Geterotsikl. Soedin.*, **12,** 853 (1976).

85: 123830h   S. Danishefsky, R. McKee, and R. K. Singh, *J. Org. Chem.*, **41,** 2934 (1976).

85: 123862v   V. V. Korshak, L. A. Rusanov, M. K. Kereselidze, and T. K. Dzhashiash-vili, *Izu. Akad. Nauk Gruz, SSR, Ser. Khim.*, **2,** 39 (1976).

85: 123874a   E. E. Knaus, F. M. Pasutto, C. S. Giam, and E. A. Swinyard, *J. Heterocyclic Chem.*, **13,** 481 (1976).

85: 123878e   V. I. Fomenko and O. P. Shvaika, *Khim. Geterotsikl. Soedin.*, **12,** 629 (1976).

85: P123928w   T. Misawa, Y. Shimizu, and Y. Nakacho, Japanese Patent 76 36,455 (1976).

85: P123930r   R. Aries, French Patent 2,275,456 (1976).

85: P125782t   D. M. Fawkes, German Patent 2,555,332 (1976).

85: P125783u   D. M. Fawkes, German Patent 2,555,333 (1976).

85: P134261z   S. Sakanoue, M. Tsubota, Y. Fuseya, K. Adachi, and T. Scishido, German Patent 2,528,616 (1976).

85: 138489c   G. Guenther, Ch. Bergstedt, and B. Kelner, *Mekh. Deistviya Gerbits.*, *Dokl. Simp.*, 163 (1971).

85: 138543r   G. Holan, J. J. Evans, J. M. Haslam, and W. J. Roulston, *Environ. Qual. Saf.*, *Suppl.*, **3,** 450 (1975).

85: 143036g   B. George and E. P. Papadopoulos, *J. Org. Chem.*, **41,** 3233 (1976).

85: 143053k   V. V. Kane, H. Werblood, and S. D. Levine, *J. Heterocyclic Chem.*, **13,** 673 (1976).

85: P144687g   K. Arakawa, M. Yamamoto, and T. Kurahashi, Japanese Patent 76 56,829 (1976).

85: P147920p   A. Kotone, T. Hori, and M. Yasuda, Japanese Patent 76 10,187 (1976).

85: P151754e   I. Boie, K. Kueffner, and D. Lowski, German Patent 2,442,703 (1976).

85: P155072d   German Patent 2,600,655 (1976).

85: 159988e   S. Kubota and M. Uda, *Chem. Pharm. Bull.*, **24,** 1336 (1976).

85: 159991a   L. P. Sysoeva, B. I. Buzykin, and Yu. P. Kitaev, *Zh. Org. Khim.*, **12,** 1557 (1976).

85: 159994d   T. P. Kofman, T. L. Uspenskaya, N. Yu. Medvedeva, and M. S. Pevzner, *Khim. Geterotsikl. Soedin.*, **12,** 991 (1976).

85: 159997g   M. M. Tsitsika, S. M. Khripak, and I. V. Smolanka, *Ukr. Khim. Zh.*, **42,** 841 (1976).

85: 159998h   A. A. Aroyan, N. S. Iradyan, R. V. Agababyan, and M. A. Iradyan, *Arm. Khim. Zh.*, **29,** 545 (1976).

85: P160095t   E. Regel, W. Draber, K. H. Buechel, and M. Plempel, German Patent 2,461,406 (1976).

85: 177322n   A. H. Harhash, N. A. L. Kassab, A. A. A. Elbanani, and A. F. M. Abu Dayyah, *Indian J. Chem.*, *Sect. B*, **14B,** 268 (1976).

85: 177325r   A. M. Khalil, I. I. Abd El-Gawad, and H. M. Hassan, *Aust. J. Chem.*, **29,** 1627 (1976).

85: 177373e   B. Decroix, P. Dubus, J. Morel, and P. Pastour, *Bull. Soc. Chim. Fr.*, 621 (1976).

85: P177430w   F. C. Novello and J. J. Baldwin, U.S. Patent 3,963,731 (1976).

85: P177432y   J. T. A. Boyle and J. C. Saunders, U.S. Patent 3,962,237 (1976).

85: P177435b   J. J. Baldwin and F. C. Novello, U.S. Patent 3,978,054 (1976).

85: P177501v   W. L. Albrecht and W. D. Jones, U.S. Patent 3,954,981 (1976).

85: 178022b   V. V. Korshak, A. L. Rusanov, L. Kh. Plieva, M. K. Kereselidze, and T. V. Lekae, *Macromolecules*, **9,** 626 (1976).

95: 178509r   A. Kotone, M. Hoda, and T. Hori, Japanese Patent 76 93,999 (1976).

85: 186070f   N. N. Chipanina, L. P. Makhno, D. D. Taryashinova, E. S. Domnina, and G. G. Skvortsova, *Tezisy Dikl.-Vses. Chugaevskoe Soveshch. Khim. Kompleksn. Soedin., 12th,* **2,** 279 (1975).

85: 186427c   I. Bornschein, H. H. Borchert, A. Wierer, and S. Pfeifer, *Pharmazie*, **31,** 578 (1976).

85: 186715v   S. S. Parmar, S. K. Chaudhary, M. Chaudhary, A. Chaudhari, and T. K. Auyong, *Proc. West. Pharmacol. Soc.*, **19,** 183 (1976).

85: 192627j   V. B. Jigajinni, S. C. Bennur, R. S. Bennur, and V. V. Badiger, *J. Karnatak Univ., Sci.*, **20,** 1 (1975).

85: 192629m   A. S. Shawali and H. M. Hassaneen, *Indian J. Chem., Sect. B,* **14B,** 425 (1976).

85: 192674x   E. J. Gray and M. F. G. Stevens, *J. Chem. Soc., Perkin Trans.* 1, 1492 (1972).

85: P192740r   A. Kotone, M. Hoda, T. Hori, and Y. Nakane, Japanese Patent 76 16,670 (1976).

85: 193010q   M. Fuertes, R. K. Robins, and J. T. Witkowski, *J. Carbohydr., Nucleosides, Nucleotides,* **3,** 169 (1976).

85: P194102h   R. Zweidler, G. Kabas, H. Schlaepfer, and I. J. Fletcher, Swiss Patent 580,192 (1976).

85: 200518s   M. Sato, H. Ishikawa, M. Ishihara, and K. Nakazato, German Patent 2,526,481 (1976).

Author Index

Numbers in parenthesis are reference numbers cited at each section within the chapter; numbers in brackets are "Chemical Abstract and Journal References" numbers listed separately on pages 669-739 of the book.

Kinugawa, J., [61:4338e], 257 (51)
Kirsanov, A. V., [85:21300t], 63 (10)
Kish, P. P., [73:62258t], 609 (1);
[77:42829x], 609 (2); [80:90814x],
609 (3)
Kisin, A. V., [80:96082d], 6 (5), 44 (24),
568 (8)
Kitaev, G. A., [83:164137q], 214 (19)
Kitaev, Y. P., [76:25193x], 607, 609 (22);
[82:170800w], 158 (37)
Kitaev, Yu. P., [83:9917a], 310 (13);
[85:159991a], 42 (22), 159 (39), 158
(38)
Kitaoka, J., [60:10867f], 573, 574, 576 (7)
Klingbeil, J. E., [53:11355h], 255 (42)
Klinge, D. E., [83:9953j], 8 (26)
Klingsberg, E., [53:16121a], 69 (3); [53:-
9200b], 69, 70, 71 (1)
Knaus, E. E., [85:123874a], 460 (48)
Knorn, C., [80:27223p], 623, 624 (3)
Knupfer, H., [57:11198d], 65, 66 (26),
96 (2); [58:3415h], 419 (11); [61:-
654h], 383 (9), 508 (28)
Kny, H., [82:72994], 459 (28)
Kobayashi, Y., [60:10867f], 573, 574,
576 (7)
Kochlar, M. M., [76:126953a], 211 (4);
[76:149140x], 406 (22)
Kodama, I., [79:137008j], 517 (10)
Kodeih, M., [76:113138p], 162 (63)
Koehler, J. J., [85:77541q], 159 (40)
Koenig, K. H., [75:20410k], 458 (24)
Koerner von Gustorf, E., [72:121454w],
531 (56)
Kofman, T. P., [81:120541w], 319 (12);
[81:P77930h], 229 (29); [84:30967x],
229 (31)
Kogan, L. M., [72:111375g], 103 (49), 306
(20)
Kohn, H., [78:4224j], 506 (17)
Kolomoitsev, L. R., [76:34220f], 369
(41), 409 (40)
Kol'tsova, A. N., [80:96082d], 6 (5), 44
(24), 568 (8); [83:79677w], 580 (50)
Komatsu, M., [77:152066d], 517 (12);
[81:13434s], 385 (15), 534 (16);
[81:169457n], 381 (2), 517 (11)
Komori, S., [75:118283ew], 369 (50)
Kondo, K., [82:156185t], 532 (6)
Konher, M. V., [73:35283d], 478 (18)
Kononenko, V. E., [79:115502p], 369
(44)
Koopmans, M. J., [57:7253h], 288 (3)

Korotkikh, N. I., [85:123823h], 160 (47),
228 (18)
Korshak, V. V., [54:1503e]; [54:511b];
[54:22603e], 579 (37); [68:13458j],
577 (17); [[68:96231d], 577 (18);
[70:12143g], 577 (20); [80:15236g],
44 (32), 568 (6), 578 (29); [80:59912d],
568 (7); [83:131973f], 578 (28); [85:-
178022b], 568 (9)
Kost, A. N., [85:94307k], 134 (32)
Kotlyarevskaya, I. L., [83:28165v], 31 (9),
34 (9), 69 (42)
Koton, A., [80:111542z], 580 (48)
Kotone, A., [78:29776d], 156 (19);
[81:105517z], 68 (40); [81:120640c];
[85:P192740r], 572, 575 (3), 156 (20),
156 (21); [82:170962a], 68 (39); [83:-
114414g], 5 (12, 16), 68 (41)
Kotsutsumi, M., [80:28461b], 609 (23)
Kottke, K., [68:21884k], 257 (53)
Koudijs, A., [83:9953j], 8 (26)
Kovach, N. A., [80:3422b], 369 (42), 443
(1); [85:123823h], 160 (47), 228 (18)
Kovalev, E. G., [70:37729r], 149 (26), 290,
291 (14), 370 (53), 404 (12); [74:-
141644c], 253 (32), 416, 417 (44);
[83:164137q], 214 (19)
Kovalev, V. A., [85:21300t], 63 (10)
Kovtunenko, V. A., [78:58327p], 600 (12);
[80:133356u], 600 (13); [81:13441s],
600 (14); [82:140028h], 251 (12)
Kozhevnikova, L. I., [84:59327z], 407 (34)
Kozima, S., [78:159527j], 510 (37), 602
(26)
Kraft, R., [69:19082z], 70 (15)
Krahnert, L., [71:38867s], 6 (9), 102 (38),
161, 162 (62), 608 (27)
Krake, U., [84:90504d], 421 (23)
Kranz, J., [55:25991f], 132 (20); [59:-
11493d], 48 (55), 65 (24); [59:-
6386c]; [59:6386f], 40 (10), 132 (21),
367 (30), 411 (6), 568 (10)
Krenzer, J., [83:164191c], 518 (13);
[84:31088y], 518 (14)
Kreutzberger, A., [79:2839], 2 (12); [51:
14751b], 130 (4), 155 (6), 365 (4),
Kristinsson, H., [80:P82992p]; [80:-
P95962d], 315 (27); [80:P95962d],
300 (14), 305 (18), 315 (28); [82:-
156321j], 241 (22), 303 (4), 318 (7)
Kroeger, C.-F., [59:10048f]; [59:10050a],
153 (17), 213 (10, 13), 508 (25);
[60:9264f], 145 (7), 152 (8), 505 (11);

Subject Index

THE CHEMISTRY OF HETEROCYCLIC COMPOUNDS

A SERIES OF MONOGRAPHS

ARNOLD WEISSBERGER and EDWARD C. TAYLOR

Editors

TRIAZOLES

1,2,4

This is the Thirty-Seventh Volume in the Series

THE CHEMISTRY OF HETEROCYCLIC COMPOUNDS